Acidic Deposition and Aquatic Ecosystems

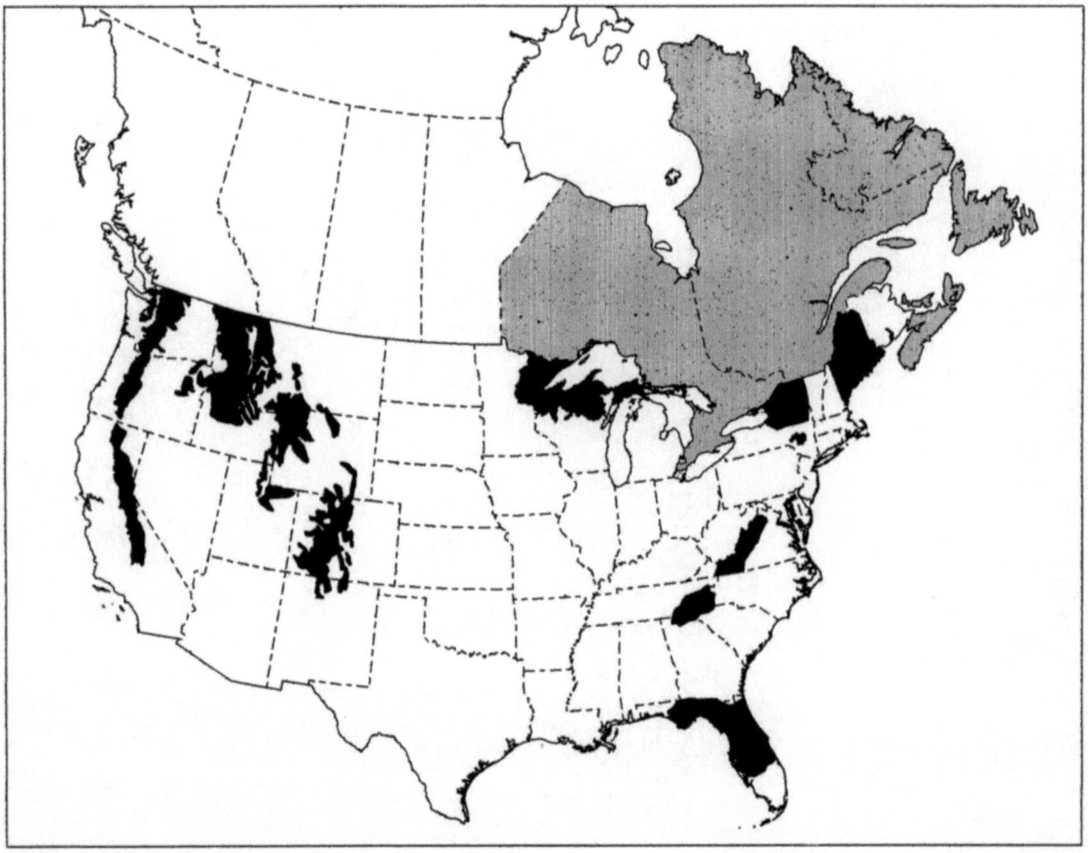

Donald F. Charles
Editor

Susan Christie
Technical Editor

Acidic Deposition and Aquatic Ecosystems

Regional Case Studies

With 373 Illustrations

Springer-Verlag
New York Berlin Heidelberg London
Paris Tokyo Hong Kong Barcelona

Donald F. Charles
Department of Biology
Indiana University
Bloomington, Indiana 47405
USA

currently working under cooperative agreement at:
U.S. EPA Environmental Research Laboratory
200 SW 35th Street
Corvallis, Oregon 97333
USA

Cover illustration: Oblique aerial photograph of seepage lakes and wetlands in Burnett County, northwestern Wisconsin, USA. Photograph taken by Jim Omernik, U.S. EPA Environmental Research Laboratory, Corvallis, Oregon, USA.

Frontispiece: Map of case study areas prepared by Tony Selle, Mantech Environmental Technologies, Inc., U.S. EPA Environmental Research Laboratory, Corvallis, Oregon, USA.

Library of Congress Cataloging-in-Publication Data
Acidic deposition and aquatic ecosystems: regional case studies/
D.F. Charles editor.
p. cm.
Includes bibliographical references and index.
1. Acid deposition—Environmental aspects—United States. 2. Acid pollution of rivers, lakes, etc.—Environmental aspects—United States. 3. Aquatic ecology—United States. 4. Aquatic animals—United States—Effect of water pollution on. I. Charles, Donald F. (Donald Franklin), 1949-
QH545.A17A238 1990
574.5'2632'0973—dc20 90-9847

Printed on acid-free paper.

Typeset by Publishers Service of Montana, Bozeman, Montana.

9 8 7 6 5 4 3 2 1

ISBN-13: 978-1-4613-9040-4 e-ISBN-13: 978-1-4613-9038-1
DOI: 10.1007/978-1-4613-9038-1

Acknowledgments

In preparing this book, we have summarized and synthesized a wide range of information and tried to present it in a consistent format so that readers can easily compare results from one chapter to the next. I especially thank the authors for their commitment and cooperation in preparing an integrated book rather than a set of loosely related chapters. Many people have participated in this project and devoted considerable time and energy. They have communicated effectively and cooperated closely. I thank all involved.

The idea for a Regional Case Studies book originated within the U.S. EPA Aquatic Effects Research Program (AERP), particularly with Dixon Landers, Rick Linthurst, and Courtney Riordan. They were also responsible for providing funding for the project.

Susan Christie, NSI Senior Technical Editor at the EPA Environmental Research Laboratory in Corvallis, Oregon, has made a significant contribution to the overall editorial quality of this book. She has had responsibility for many of the duties normally associated with a managing editor. She has kept track of the various versions and components of chapters, communicated frequently with authors and reviewers, edited manuscripts for grammar, consistency, format, style, and readability, and made many suggestions for improving the design and layout of the book. Susan has devoted considerable creative energy to this book. I have enjoyed working with her very much.

Several NSI personnel at the EPA Environmental Research Laboratory in Corvallis, Oregon, have provided significant contributions and support. Tony Selle used geographic information system (GIS) software to prepare most of the maps in this book. Jeremy Smith obtained and reformatted several of the computer databases used by chapter authors and carried out many of the standard data analyses presented in the book, including the enrichment factor analyses. He, Jim Blick, Mark DeHaan, Danny Kugler, and Joe Bernert performed many statistical analyses and prepared many computer graphics. Kathy Hurley provided the considerable and complicated administrative support necessary to make this project happen. Roze Royce typed and revised versions of some of the chapters and many letters, memos, and other project related communications. Sharon Ziminski helped organize and coordinate the four meetings of project participants and arranged for the drafting of final versions of many figures. Sue Brenard assisted with compilation and assembly of the final draft manuscript. I thank Kathy Fox for preparing the index.

In addition to being coauthors on several chapters, Joan Baker and Sigurd Christensen provided material on biota, especially fish, for several of the case study chapters. Rudy Husar, also coauthor of a chapter and an appendix, provided critical information on precipitation chemistry and loading for several types of analyses and maps used in case study chapters, overview sections, and integration chapters.

Scientific peer review was a very important part of the writing process. We are indebted to Dr. David Schindler and Dr. Hans Martin Seip for reviewing nearly all the chapters and making many useful comments about the science and about how the chapters could be revised to present a more integrated and cohesive assessment. Dr. Robert Cook, one of the authors, thoroughly reviewed most of the chapters and provided many helpful suggestions. We also owe considerable thanks to all the people who reviewed individual chapters and appendices:

David Armstrong, University of Wisconsin
Gerald Aubertin, Southern Illinois University
Charles Barnes, U.S. Geological Survey
David Brakke, Western Washington University
Mark Brenner, Florida State Museum, University of Florida
Edward Deevey, Florida State Museum, University of Florida
Peter Dillon, Ontario Ministry of the Environment
Charles A. Donkers, Northern States Power Company
James Drever, University of Wyoming
Clinton Duncan, Central Washington University
Keith Eshleman, University of Virginia
Thomas Frost, University of Wisconsin
David Helvey, U.S. Forest Service, Northeastern Experiment Station
Charles Hendry, Environmental Science & Engineering, Inc.
Walter Kretser, Adirondack Lake Survey Corp., New York
James LaBaugh, U.S. Geological Survey
William McFee, Purdue University
Tilden Meyers, National Oceanographic and Atmospheric Administration
Stephen C. Nodvin, National Park Service
Stephen A. Norton, University of Maine
Anthony Olsen, Battelle Northwest National Laboratory
Steve Paulsen, University of Nevada
Karen Roy, Adirondack Park Agency, New York
David W. Schindler, Canada Department of Fisheries and Oceans
Carl Schofield, Cornell University
Hans Martin Seip, Center for Industrial Research, Norway
John P. Smol, Queen's University at Kingston
Robert Stauffer, Private Consultant, Lexington, Kentucky
Kent Thornton, FTN Associates
Kathy Tonnesson, California Air Resources Board
Richard Vong, U.S. EPA Environmental Research Laboratory, Corvallis, Oregon
Jack Waide, U.S. Forest Service, Southeastern Experiment Station
Donna Wales, Ministry of Natural Resources, Canada
Eugene Welch, University of Washington
Denis White, NSI, U.S. EPA Environmental Research Laboratory, Corvallis, Oregon

The research described in the chapters of this book has been funded wholly or in part by the U.S. Environmental Protection Agency. The book has been subjected to the Agency's peer and administrative review and has been approved for publication.

Contents

Part III. Synthesis and Integration

Appendices

Contributors

Linda C. Bacon Maine Department of Environmental Protection, Augusta, Maine 04333, USA
Joan P. Baker Western Aquatics, Inc., Durham, North Carolina 27713, USA
Lawrence A. Baker Water Resources Research Center, University of Minnesota, St. Paul, Minnesota 55108, USA; *currently working under cooperative agreement at:* U.S. EPA Environmental Research Laboratory, Corvallis, Oregon 97333, USA
Glenn F. Cada Environmental Sciences Division, Oak Ridge National Laboratory, Oak Ridge, Tennessee 37831-6036, USA
Daniel E. Canfield, Jr. Department of Fisheries and Aquaculture, University of Florida, Gainesville, Florida 32611, USA
Donald F. Charles Department of Biology, Indiana University, Bloomington, Indiana 47405; *currently working under cooperative agreement at:* U.S. EPA Environmental Research Laboratory, Corvallis, Oregon 97333, USA
Sigurd W. Christensen Environmental Sciences Division, Oak Ridge National Laboratory, Oak Ridge, Tennessee 37831-6036, USA
Robert B. Cook Environmental Sciences Division, Oak Ridge National Laboratory, Oak Ridge, Tennessee 37831-6036, USA
B. Jack Cosby Department of Environmental Sciences, Clark Hall, University of Virginia, Charlottesville, Virginia 22901, USA; *current address:* School of Forestry and Environmental Sciences, Duke University, Durham, North Carolina 27706, USA
Christopher S. Cronan Department of Botany and Plant Pathology, University of Maine, Orono, Maine 04469, USA
Charles T. Driscoll Department of Civil and Environmental Engineering, Syracuse University, Syracuse, New York 13244-1190, USA
Joseph M. Eilers E&S Environmental Chemistry, Inc., Corvallis, Oregon 97330, USA
Jerry W. Elwood Environmental Sciences Division, Oak Ridge National Laboratory, Oak Ridge, Tennessee 37831-6036; *current address:* U.S. Geological Survey, Reston, Virginia 22092, USA
Ivan J. Fernandez Department of Plant and Soil Sciences, University of Maine, Orono, Maine 04469, USA
James N. Galloway Department of Environmental Sciences, Clark Hall, University of Virginia, Charlottesville, Virginia 22901, USA
Steven A. Gherini Tetra Tech, Inc., Lafayette, California 94549, USA
Chad P. Gubala Department of Civil and Environmental Engineering, Syracuse University, Syracuse, New York 13244-1190, USA

Terry A. Haines Department of Zoology, University of Maine, Orono, Maine 04469, USA

Alan T. Herlihy Utah Water Research Laboratory, Utah State University, Logan, Utah 84321; *currently working under cooperative agreement at:* U.S. EPA Environmental Research Laboratory, Corvallis, Oregon 97333, USA

George M. Hornberger Department of Environmental Sciences, Clark Hall, University of Virginia, Charlottesville, Virginia 22901, USA

Rudolf B. Husar CAPITA, Washington University, St. Louis, Missouri 63130, USA

Henriette I. Jager Environmental Sciences Division, Oak Ridge National Laboratory, Oak Ridge, Tennessee 37831-6036, USA

Dean S. Jeffries National Water Research Institute, Department of Environment, Burlington, Ontario L7R 4A6, Canada

Jeffrey S. Kahl Department of Geological Sciences, Sawyer Environmental Research Center, University of Maine, Orono, Maine 04469, USA

Philip R. Kaufmann Utah Water Research Laboratory, Utah State University, Logan, Utah 84321; *currently working under cooperative agreement at:* U.S. EPA Environmental Research Laboratory, Corvallis, Oregon 97333, USA

John M. Melack Department of Biological Sciences and Marine Science Institute, University of California, Santa Barbara, California 93106, USA

Ronald K. Munson Tetra Tech, Inc., Hadley, Massachusetts 01035, USA

Peter S. Murdoch U.S. Geological Survey, Water Resources Division, Albany, New York 12201, USA

Peter O. Nelson Environmental and Water Resources Engineering Program, Department of Civil Engineering, Oregon State University, Corvallis, Oregon 97331, USA

Robert M. Newton Department of Geology, Smith College, Northampton, Massachusetts 01063, USA

Stephen A. Norton Department of Geological Sciences, University of Maine, Orono, Maine 04469, USA

Curtis D. Pollman KBN Engineering and Applied Sciences, Inc., Gainesville, Florida 32605, USA

Patrick F. Ryan Science Applications International Corp., Oak Ridge National Laboratory, Oak Ridge, Tennessee 37831-6036, USA

Michael J. Sale Environmental Sciences Division, Oak Ridge National Laboratory, Oak Ridge, Tennessee 37831-6036, USA

Anthony R. Selle NSI Technology Services, Inc., U.S. EPA Environmental Research Laboratory, Corvallis, Oregon 97333, USA

Norman E. Spahr U.S. Geological Survey, Water Resources Division, Denver Federal Center, Lakewood, Colorado 80225, USA

John L. Stoddard New York City Department of Environmental Protection, Valhalla, New York 10595; *current address:* NSI Technology Services, U.S. EPA Environmental Research Laboratory, Corvallis, Oregon 97333, USA

Timothy J. Sullivan E&S Environmental Chemistry, Inc., Corvallis, Oregon 97330, USA

John T. Turk U.S. Geological Survey, Water Resources Division, Denver Federal Center, Lakewood, Colorado 80225, USA

James R. Webb Department of Environmental Sciences, Clark Hall, University of Virginia, Charlottesville, Virginia 22901, USA

Part I
Background

Introduction—Acidic Deposition and Aquatic Ecosystems: Regional Case Studies

Donald F. Charles

Acidic deposition and its effects on aquatic ecosystems have been major scientific and public policy issues in the United States since the early 1970s. The high level of concern arises primarily from the potential for wide-scale and long-term changes in water chemistry and the resulting loss of aquatic biota, especially fish. Considerable research has been performed and hundreds of scientific papers and several major books and reports have been written on the subject. We now have a good understanding of the nature and extent of at least wet acidic deposition, and we know that current low pH values for wetfall are caused primarily by sulfur and nitrogen oxide emissions from combustion of fossil fuels. From numerous surveys and studies, including the U.S. Environmental Protection Agency's (EPA) National Surface Water Survey (NSWS), we have learned much about the current status of low-alkalinity surface waters in many regions of the country, and the extent to which water chemistry has been affected by acidic deposition. However, the many and diverse studies completed by state and federal agencies, universities, and other organizations and individuals have not been thoroughly analyzed and integrated on a regional or national scale.

This book is the first comprehensive, integrated synthesis of available information on the current and potential effects of acidic deposition on lakes and streams in geographic regions of the United States having significant numbers of low-alkalinity surface waters. It presents and evaluates data for entire regions and is national in scope. We have brought together in one volume, and with similar organization and format, the results of diverse studies of aquatic ecosystems. We stress the current status of water chemistry and the processes important in controlling water chemistry. Authors of case study chapters have characterized these processes on a regional basis by using, assessing, and comparing high-quality data sets. A major conclusion demonstrated by these regional comparisons is that there is substantial diversity among regions with respect to the nature of surface waters and the processes affecting them. The most important contributions of this book are the descriptions of intra- and interregional variations in surface water chemistry and its controlling factors, and the assessment of the significance of these variations to our understanding of the acidic deposition phenomenon and its implications for public policy decisions.

Organization of the Book

Chapters in the book are organized into three major sections: (1) background, (2) case studies, and (3) synthesis and integration. The first five introductory chapters provide background and basic information on lake/watershed processes (Chapter 1) and how to assess their importance (Chapter 2), methods for characterizing sulfur deposition trends and assessing long-term trends in lake chemistry (Chapter 3), a brief review of the effects of surface water acidification on aquatic biota and assessment methods (Chapter 4), and a geographic overview and general description of the nature of surface waters in the case study regions (Chapter 5).

Chapters 6 to 16 present the case studies grouped by major geographic region. A brief overview of each major region (Northeast, Southeast, Florida,

Upper Midwest, and West) is provided immediately before each group of case study chapters. These overview sections put each case study into perspective relative to the larger region, point out major differences and similarities among case study regions within the group, and offer additional information on the spatial patterns of surface water alkalinity and atmospheric deposition within the larger regions.

The last two chapters provide a synthesis and integration of the information on current water chemistry and important lake/watershed processes (Chapter 17), and a discussion of the historical trends and future considerations (Chapter 18) presented in the case study chapters. These last chapters make comparisons among the regions, highlighting regional differences, similarities, and diversity, and attempt to synthesize our current state of knowledge regarding the regional effects of acidic deposition on aquatic ecosystems.

A glossary and list of abbreviations are included to help readers not familiar with the terminology. All references appear at the end of the book. The appendices describe procedures used to (1) interpolate atmospheric wet deposition to lakes, (2) estimate dry deposition, and (3) perform the enrichment factor calculations presented in each case study chapter.

Audience

This book has been written primarily for scientists in academia, government, and industry who are interested in the effects of acidic deposition specifically, and the nature and functioning of aquatic ecosystems generally. It is also directed toward knowledgeable policy analysts and decision makers, although the objectives of the book are based on scientific issues, not policy concerns. We have emphasized material that is relevant to assessing the effects of acidic deposition and the potential response of systems to future changes in atmospheric deposition loading.

Objectives of the Book

The main objectives of this book are to:

- Describe current limnological conditions within the case study areas and characterize the important factors that control surface water chemistry (especially pH, alkalinity, and related parameters) under present and past (pre-industrial) conditions.
- Assess the past and current status of biological components, particularly fish, of low-alkalinity lakes and streams within the regions.
- Compare conditions among regions and emphasize key regional similarities and differences with respect to aquatic ecosystem response to acidic precipitation.
- Assess briefly and qualitatively the past and the possible future effects of changes in acidic atmospheric deposition on aquatic ecosystems, based on existing data and current modeling efforts.

We used the following questions as a basis for selecting the material to be presented and the data analyses required to meet our objectives:

- What are the current limnological conditions relevant to acidic deposition phenomena within the case study regions?
- What are the key ecosystem components and processes controlling water chemistry? How does the importance of these factors vary among regions?
- For individual regions, what are the relationships between observed surface water physico- chemical characteristics and variables such as land use, hydrology, and atmospheric deposition?
- What were the physical, chemical, and biological characteristics of surface waters under pre-industrial conditions (prior to the onset of acidic deposition)?
- What is the relative importance of natural versus cultural factors in controlling the acid-base status of surface waters and contributing to their acidification?
- What has been the effect of atmospheric deposition of strong acids on lake and stream water chemistry and biota? What do historical data (chemistry and fisheries), current studies of species distributions, paleoecological data, and computer model simulations indicate about recent trends in lake and stream chemistry and biota?
- In general, how might lake and stream chemistry and biota within the regions respond to continued or changing levels of atmospheric deposition of strong acids? What are the key factors that might affect the recovery of recently acidified systems?

Selection of Regions and Data Sets

The selection of case study regions was based on the availability of (1) high-quality survey data sets that were adequate to characterize water chemistry and acidic deposition effects on a regional basis, (2) site-specific studies detailed enough to assess temporal variability and the importance of processes affecting water chemistry, and (3) scientists with the necessary expertise and familiarity with regional data to write the chapters. The chapter on southeastern Canada is shorter and less detailed than those on areas in the United States, but it is included because it provides basic information for a useful comparison between eastern Canada and regions in the United States. A treatment of Canadian aquatic resources at the level of detail presented for the United States would require a separate volume.

Although authors refer to most major water chemistry data sets available for their respective regions, they analyzed and compared only a few data sets in detail. We used the following general criteria to select these data sets:

- The data should be of high quality. Methods and results should be fully documented and verified.
- Data sets should contain information on at least 30 lakes or streams, or cover geographical areas not well represented by other data sets.
- Data sets should be complete and contain all information necessary to complete the basic analyses planned for all regions.
- Data should generally not require considerable manipulation before use.

National Surface Water Survey

The primary water chemistry data base used for analyses presented in this book was the U.S. EPA's National Surface Water Survey. The NSWS is the most important regional source of information on acidic lakes and streams in the United States. At least four main features set it apart from other studies and explain why it is central to the book. First, it included hundreds of lakes and streams in most regions of the country that have low ANC (acid neutralizing capacity) surface waters susceptible to acidification. Similar sampling procedures and analytical methods were used, so that data are comparable among regions. Second, the characterization of water chemistry is fairly complete and includes variables of biological relevance. Third, quality assurance and quality control were a major component of the study, so that data quality is high. Fourth, and perhaps most important, the lakes and streams sampled were a probability sample from an explicitly defined population, allowing statistical extrapolation to the target population of surface waters in the regions surveyed.

The NSWS is the only multi-region data base that can be used to characterize regional surface water chemistry statistically. However, though the NSWS results are statistically representative, they do not necessarily characterize certain subsets of lakes or streams as well as some other smaller scale regional surveys. Therefore, one important objective of the case studies authors was to compare results of NSWS and other studies and point out and discuss similarities and differences.

The NSWS has three major components, the Eastern Lake Survey (ELS), the Western Lake Survey (WLS), and the National Stream Survey (NSS). Data from each are used in this book. Chapter 5, the Geographic Overview, contains a map and a general description of the NSWS regions. The ELS Phase I included 1,612 lakes sampled in fall 1984 (Landers et al. 1988b, Linthurst et al. 1986a, Overton et al. 1986, Kanciruk et al. 1986, Blick et al. 1987, Hillman et al. 1987, Drousé et al. 1987, Best et al. 1986). In the WLS, 720 lakes were sampled in fall 1985 (Landers et al. 1987, Eilers et al. 1987a, Kerfoot and Faber 1987). The NSS survey of 504 stream reaches was conducted in spring 1986 (Kaufmann et al. 1988, Sale et al. 1988, Hagley et al. 1988, Cougan et al. 1988). It was preceded by a pilot survey of 54 streams in the Southern Blue Ridge, conducted during the spring and summer of 1985 (Messer et al. 1986, 1988, Drousé 1987, Knapp et al. 1987). Some regions of the United States have significant numbers of low ANC systems that were not included in the NSWS, for example, the South Atlantic and Gulf Coastal Plains; in other areas, lakes were sampled, but not streams, as in the Northeast, Upper Midwest, and West.

The purpose of the NSWS was to quantify the current surface water chemistry in regions of the United States potentially susceptible to the effects of acidic deposition. The NSWS was designed to determine the percentage (of total number and lake surface area or stream length) and location of surface waters with pH and ANC below certain values, particularly those considered acidic. Acidic surface waters, in the NSWS and this book, are defined as waters having ANC $\leq$ 0 µeq L^{-1}. Stratified sampling designs were used for the surveys to ensure that adequate numbers of low ANC surface waters were included in the survey. There were three levels of stratification: first, region (e.g., Upper Midwest), second, subregion (e.g., Upper Peninsula of Michigan), and third, ANC category (< 100, 100 to 200, > 200 µeq L^{-1}). A minimum of 50 lakes was initially selected in each ANC category.

In both the ELS and WLS, only one sample was taken per lake, so that a maximum number of lakes could be sampled. Sampling was conducted after fall turnover, when within-lake variability was assumed to be relatively small for most lakes, although mixing had not occurred in some lakes at the time of sampling. Fall was selected rather than spring because it was unlikely that all lakes could have been sampled within the shorter period of lake mixing in the spring. The fall sample is intended to be an "index" of lake chemistry. Comparisons with samples taken at other times of the year show good correlation with the fall index sample (Newell 1987).

Lakes in the ELS were selected from medium-scale (1:250,000) topographic maps. Because lakes with a surface area < 4 ha are not well represented on these maps, these small lakes are not well represented in the data set (Johnson et al., 1989). There are relatively large numbers of smaller lakes in some of the regions and they often have lower ANC than the larger lakes. Thus the number and percentage of acidic and low ANC lakes in these regions is underestimated in ELS results. This issue of "small lakes" is discussed in detail in the case study chapters [e.g., Adirondacks, Chapter 6 (Driscoll et al.); Maine, Chapter 7 (Kahl et al.)]. A different map scale (1:100,000) was used in selecting lakes for the WLS, so that lakes as small as 1 ha in surface area are represented.

The NSS focused on the stream resource consisting of small- to mid-size streams in the size range managed for sport fisheries. This group of streams was identified as those draining land areas less than 155 km², but large enough to be represented as blue lines on 1:250,000-scale U.S. Geological Survey topographic maps. This size range was judged to include streams large enough to be perennial and important for fish habitat, yet still small enough to represent the principal resource at risk from acidic deposition.

For streams, the NSS index water chemistry measurements were made during one, two, or three sampling visits, depending on the subregion, at both the upstream and downstream ends of each sample reach. The samples were collected during a spring baseflow indexing period between snowmelt and leafout; sampling was excluded during high flow events. Results from the NSS pilot survey indicated that chemical variability during this indexing period was small enough to allow calculation of robust population estimates (Messer et al. 1986, 1988).

Not all NSWS results were available to RCS authors at the time of writing. The National Stream Survey results (Kaufmann et al. 1988) and database (Sale et al. 1988) had not yet been released, although results had been published for the NSS Pilot Survey (Messer et al. 1986, 1988). For this reason, the case studies describing streams used synoptic data sets available to the authors at the time of writing. We were unable to make use of the NSS results to their fullest potential in the treatment of individual regions, although the data are used quite heavily in the Synthesis and Integration chapter (Chapter 17, Baker et al., this volume). Also, the Northern Appalachian Plateau, a region with many acidic streams, and the Mid-Atlantic Coastal Plain were not treated in this book because NSS data were not yet available.

Integration and Intercomparison

One of our major goals was to make it easy for readers to compare similar data and results among chapters. To this end, we standardized certain components of all case study chapters (except the southeastern Canada overview chapter).

- Major topic headings: Introduction, Regional Characteristics, Regional Surface Water Chemistry, Processes Influencing Surface Water Chemistry, Evidence of Historical Trends, Future Considerations, Biological Status, Conclusions.
- Maps showing distribution of lakes and streams divided into three categories each for pH, ANC, and SO_4/SBC, where SBC = sum of base cations.

- Figures showing (1) frequency histograms of nine chemical and morphometric characteristics, (2) spline plots of concentrations of major cations and anions versus ANC, (3) plots of ANC versus SO_4/C_B, and (4) enrichment factor diagrams showing differences between precipitation input ion concentrations and lake water chemistry caused by the influence of in-lake and watershed processes.

The NSWS results were used for all chapters where applicable, in order to provide a basis for meaningful comparisons among regions. In addition, we have tried to be consistent in use of terms and units. A glossary and list of abbreviations are included in the back for quick reference.

The last two chapters (17 and 18) are the most important contribution to synthesis and integration. They summarize and compare results from the case study chapters, and assess the major similarities and differences among the regions in current chemistry, important processes, historical change, and outlook for the future.

Despite our best attempts, we were unable to achieve the degree of integration, consistency, and comparability that we would have liked. Part of this is due to differences among regions in nature and number of studies and data sets, relative importance of the various in-lake watershed processes, and differences in atmospheric deposition.

Maps

Many maps are used throughout the book to illustrate the primary spatial relationships among factors important in assessing surface water acidification. Most of the maps were prepared at the U.S. EPA Environmental Research Laboratory in Corvallis by Tony Selle with GIS ARC/INFO software. All maps showing all of North America or any part of Canada, are Lambert conic conformal projected, except for Figures 9.2 and 9.9 in Chapter 9, which are Mercator projection. All other maps are in Albers conic equal-area projection. Map scales vary with geographic region; however, most maps for any one region are at the same scale. Most data presented on the maps are from the NSWS.

Role of EPA

Preparation of this book was initiated and funded by the U.S. EPA Aquatic Effects Research Program (AERP) through the U.S. EPA Environmental Research Laboratory in Corvallis, Oregon. This book and associated contributions provide the EPA and the National Acid Precipitation Assessment Program (NAPAP) with a more thorough understanding of the effects of acidic deposition on several geographic regions of national concern. This book also provides further evaluation and interpretation of data collected as part of the EPA's NSWS project. Coordination, standard data analysis done for all case study areas, and preparation of most of the maps and many of the figures occurred at the U.S. EPA Environmental Research Laboratory in Corvallis.

Writing and Review Process

Our intent was to make this book as integrated as possible, to facilitate comparison among regions. All authors met as a group four times to plan the book's contents and organization. Subgroups also met to discuss particular book components. Drafts of chapters were exchanged at early stages. Each chapter was reviewed twice by two different sets of authors of other chapters to promote author interaction and overall quality and internal consistency of the book. In addition, "topic" coordinators were assigned to review treatment of major subjects (e.g., fish, organic acidity) to help promote consistency and integration of material.

External scientific review occurred at two levels. Each chapter was reviewed by two experts familiar with the relevant case study area or subject (see Acknowledgments for list of names). In addition, Drs. David Schindler and Hans Martin Seip reviewed the outline for the book and nearly all chapters, and made many helpful comments.

1
Processes Influencing the Acid-Base Chemistry of Surface Waters

Ronald K. Munson and Steven A. Gherini

ABSTRACT. A wide variety of processes influence the acid-base characteristics of precipitation as it flows through a watershed and into streams and lakes. These include deposition, hydrologic and chemical processes occurring in terrestrial systems, and biological and chemical interactions that take place within streams and lakes.

Atmospheric deposition includes both wet and dry components. Dry deposition is further subdivided into particulate and gaseous components. Dry deposition is an important component of ion budget calculations, which is often ignored in surface water acidification studies because it is difficult to quantify.

Processes occurring in terrestrial systems that influence solution chemistry include routing of precipitation, vegetation processes, and chemical and biological processes that take place in soils and wetlands. The routing of precipitation is determined by depth of permeable media, hydraulic conductivity, evapotranspiration, land slope, and other factors. Important vegetation processes include the enhanced collection of dry deposition, foliar exudation, nutrient uptake, and root respiration. Soil processes that influence surface water quality include cation exchange, anion adsorption, litter decay, organic acid production, nitrification, redox processes, mineral weathering, and aluminum precipitation, dissolution, and exchange. Important wetland processes include the production and decay of organic matter and redox processes. Changes in land use patterns can also influence the acid-base characteristics of surface waters.

Hydrologic, chemical, and biological processes occurring in streams and lakes influence water quality. Important hydrologic properties include mixing, stratification, evaporation, ice formation and melting, and the distributions of inflows and outflows. Important chemical processes include nitrification, gas transfer, and solution-phase proton transfer and speciation. Biologically influenced reactions include uptake of nitrogen species by plants and algae, decay of organic matter, and microbially mediated reduction of sulfate and nitrate.

The interactions of these diverse processes determine the ultimate acid-base characteristics of surface waters. Surface waters bear the history of the processes that have influenced solute concentrations at various points along the flowpaths. These interactions are described and illustrated with examples.

Introduction

Many processes influence the response of surface waters to acidic deposition. These include the deposition processes themselves (wet and dry), hydrologic processes, which determine where water flows (see Figure 1.1), and biogeochemical processes, which influence the acid-base chemistry of water flowing through a basin. The interaction of all of these processes determines water quality.

This chapter describes various processes that can influence the acid-base characteristics of surface waters. In addition, it provides background information on acid-base chemistry and on the concept of acid neutralizing capacity (ANC). The processes are also discussed in each of the case study chapters, inasmuch as they influence surface water characteristics in a particular geographic area. The relative importance of each process across all case study regions is discussed in Chapter 17 (Baker et al. this volume).

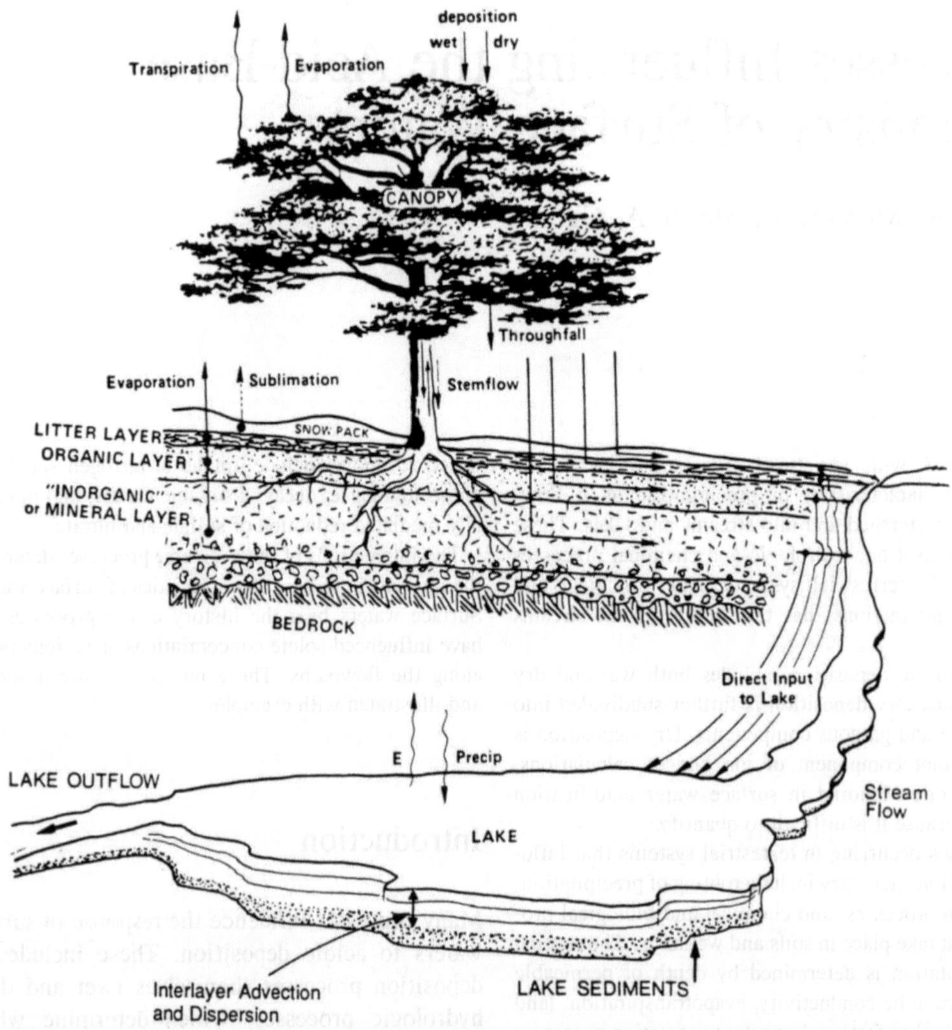

FIGURE 1.1. Lake-watershed system components and selected hydrologic processes that influence the acid-base characteristics of surface waters (Goldstein et al. 1987).

Acid-Base Chemistry

A substance is an acid if it donates hydrogen ions (protons) to solution (Bronsted 1923). Likewise, a base is a substance that accepts protons from solution. The ease with which an acid donates protons to an aqueous solution determines the strength of that particular acid in water. For example, hydrochloric acid (HCl) readily donates protons to solution, even when protons are already abundant in solution, and is thus considered a strong acid. The chlorine atom bonded to hydrogen has very little affinity for the hydrogen nucleus (proton). This is reflected in the large acid dissociation constant for hydrochloric acid, which is calculated as the ratio of the product of the dissociated hydrogen and chloride ion activities (concentrations in dilute solutions) divided by the activity of the undissociated form:

$$\frac{\{H^+\}\{Cl^-\}}{\{HCl\}} = K_A = 1000 \qquad (1\text{-}1)$$

where: { } = constituent activities

Acetic acid, on the other hand, is generally considered a weak acid. Acetate ion has a relatively strong affinity for hydrogen ion; thus its dissociation constant is relatively small: ($K_A = 10^{-4.75}$).

The coexistence of strong and weak acids makes mass balance calculations of free hydrogen ion concentration, or pH, very difficult. When two dilute sodium chloride solutions are mixed in equal volumes, the sodium concentration of the resulting solution is simply the average of the two original solution concentrations. This is rarely the case for free hydrogen ion concentration for two major reasons:

1. The degree of dissociation of acids, and in particular weak acids, is dependent upon the existing hydrogen ion concentration in solution.
2. The analytical procedure for determining hydrogen ion concentration (the pH meter and the glass electrode) measures only those hydrogen ions free in solution. The hydrogen ions associated or complexed with other species (e.g., weak acid anions such as HCO_3^-) are not measured. This is not the case in the analysis of most other solutes. For example, the analytical determination of magnesium by atomic adsorption analysis quantifies not only the hydrated Mg^{2+} ion but also its complexes (e.g., $Mg(OH)^+$, $Mg(OH)_2^0$, $MgCl^+$, and $Mg \cdot R_2$, where R represents an organic ligand).

The computational difficulty associated with the nonconservative behavior of free hydrogen ion can be overcome by using the concept of ANC, or alkalinity. Acid neutralizing capacity can be measured or calculated and can be used directly in mass balance calculations (such as those discussed earlier for sodium chloride). Solution pH can then be calculated based on ANC and the analytical concentrations of the buffer system components (weak acids and bases). In dilute waters, these include total dissolved inorganic carbon (C_T), total organic acid anion (R_T), and total monomeric aluminum (Al_T). The concept of ANC and the determination of solution phase pH are discussed in more detail in the next section.

Acid Neutralizing Capacity (ANC)

Acid neutralizing capacity, or alkalinity, is a measurable parameter that indicates the net strong base in solution or net strong acid if its value is negative. It is routinely measured by titration with strong acid or base to an equivalence point (inflection point). At the equivalence point, the ANC is zero and all ions are in the zero proton reference level (see section on Proton Reference Level for additional explanation). Several analytical methods, such as the Gran procedure (Gran 1952) have been developed to help identify this point. The higher the ANC of a water body, the more strong acid it takes to reduce the pH to the equivalence point. This ability to resist decreases in pH (buffering) derives from aqueous species that combine with the added hydrogen ions and thereby offset the increase in free hydrogen ion concentration. For example, bicarbonate ion, HCO_3^-, the major acid neutralizing species in most surface waters, combines with hydrogen ion to form aqueous carbon dioxide and water:

$$HCO_3^- + H^+ \rightarrow CO_2(aq) + H_2O \quad (1\text{-}2)$$

The ANC of a water body can be defined as the concentration of all such hydrogen ion acceptors minus the concentration of hydrogen ion donors (Stumm and Morgan 1981) as follows:

$$\begin{aligned} ANC = {} & [HCO_3^-] + 2[CO_3^{2-}] + [OH^-] \\ & + [\text{other } H^+ \text{ ion acceptors}] - [H+ \text{ ion donors}] \end{aligned} \quad (1\text{-}3)$$

where: HCO_3^- = bicarbonate ion
CO_3^{2-} = carbonate ion
OH^- = hydroxide ion
[ ] = molar concentrations

For most fresh waters, all the terms on the right hand side of Equation 1-3 except the first ($[HCO_3^-]$) are negligible and can be set equal to zero. In low ANC waters, however, the concentration of the other hydrogen ion acceptors often becomes large relative to the total concentration of bicarbonate, carbonate, and hydroxide. These other hydrogen ion acceptors include organic substances with carboxyl (–COOH) and enolic (–C=C–OH) or phenolic (e.g., C_6H_5OH) hydroxyl groups, for example:

$$R{-}COO^- + H^+ \rightarrow R{-}COOH \quad (1\text{-}4)$$

and the monomeric aluminum species and their complexes, for example:

$$Al(OH)_2^+ + 2H^+ \rightarrow Al^{3+} + 2H_2O \quad (1\text{-}5)$$

$$Al \cdot R_3 + 3H^+ \rightarrow Al^{3+} + 3HR \quad (1\text{-}6)$$

Acid Neutralizing Capacity (ANC) Definition

Titration

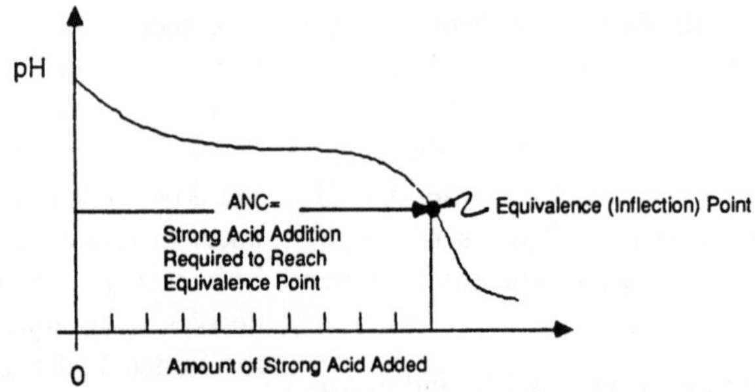

Calculation

$$ANC = [HCO_3^-] + 2[CO_3^{2-}] + [OH^-] + [\text{other proton acceptors}] - \left[\begin{matrix}\text{proton}\\ \text{donors}\end{matrix}\right] \quad \text{(a)}$$

$$ANC = \Sigma C_B - \Sigma C_A \quad \text{(b)}$$

$$\Delta ANC = \Delta\Sigma C_B - \Delta\Sigma C_A \quad \text{(c)}$$

where

ΣC_B = the sum of the base cation molarities times charge at equivalence point* $\equiv 2[Ca_T] + 2[Mg_T] + [Na_T] + [K_T] + [NH_{4T}] + 2[Al_T] + 2[Mn_T] + 2[Fe_T]$ (d)

ΣC_A = the sum of the strong acid anion molarities times charge at equivalence point* $\equiv 2[SO_{4T}] + [NO_{3T}] + [Cl_T] + [F_T]$ (e)

*Proton Reference Level Includes CO_2 (aq), $Al(OH)^{2+}$, HR (fully protonated organic acid)

FIGURE 1.2. Acid neutralizing capacity (ANC) definition. ANC can be measured by titration or determined by calculation as shown in Equation b. Addition of base cations, C_B, and removal of strong acid anions, C_A, both increase the ANC of a solution. Concentrations shown above are in mol L^{-1}.

An alternative representation of solution-phase ANC that is mathematically equivalent to Equation 1-3 is derived as follows (Gherini et al. 1985). The derivation is based upon the ANC definition and the solution electroneutrality condition. The derivation is for a metal that hydrolyzes, $M(OH)^{n-}_{(z+n)}$, and a strong acid, HA. This approach can be readily extended to yield a general form for the ANC equation. The solution charge balance gives

$$[H^+] + z[M^{z+}] + (z-1)[M(OH)^{(z-1)}] + \ldots + [M(OH)^+_{(z-1)}] = [OH^-] + n[M(OH)^{n-}_{(z+n)}] + \ldots + [A^-] \quad (1\text{-}7)$$

where: z = valence state of the metal M
n = species charge

From the ANC definition (ANC = concentration of the hydrogen ion acceptors minus the concentration of hydrogen ion donors; proton reference level = H_2O and M^{z+}), we have

$$ANC = [OH^-] + [M(OH)^{(z-1)+}] + \ldots + (z-1)[M(OH)^+_{(z-1)}] + z[M(OH)_z] + \ldots + (z+n)[M(OH)^{n-}_{(z+n)}] + \ldots - [HA] - [H^+] \quad (1\text{-}8)$$

Solving the charge balance (Equation 1-7) for $[OH^-] - [H^+]$ and substituting into the ANC definition (Equation 1-8) yields

$$\begin{aligned} \text{ANC} = & - [\text{HA}] - [\text{A}^-] \\ & + z\,[\text{M}^{z+}] + (z-1)[\text{M(OH)}^{(z-1)+}] \\ & + [\text{M(OH)}^{(z-1)+}] + \ldots + [\text{M(OH)}^{+}{}_{(z-1)}] \\ & + (z-1)[\text{M(OH)}^{+}{}_{(z-1)}] + z[\text{M(OH)}_z] \\ & + \ldots + (z+n)[\text{M(OH)}^{n-}{}_{(z+n)}] \\ & - n[\text{M(OH)}^{n-}{}_{(z+n)}] \end{aligned} \tag{1-9}$$

Letting

$$\begin{aligned} \text{M}_T = & [\text{M}^{z+}] + [\text{M(OH)}^{(z-1)+}] + \ldots \\ & + [\text{M(OH)}^{+}{}_{(n-1)}] + [\text{M(OH)}_z] + \ldots \\ & + [\text{M(OH)}^{n-}{}_{(z+n)}] \end{aligned} \tag{1-10}$$

and

$$\text{A}_T = [\text{HA}] + [\text{A}^-] \tag{1-11}$$

yields

$$\text{ANC} = z \cdot \text{M}_T - \text{A}_T \tag{1-12}$$

Expansion to multiple acid and base systems yields

$$\text{ANC} = \sum_{i=1}^{n} |z_i| \text{M}_{T_i} - \sum_{j=1}^{m} |z_j| \text{A}_{T_j} \equiv \Sigma C_B - \Sigma C_A \tag{1-13}$$

This equation relates ANC to the difference between the sum of the base cation concentrations times their predominant charge at the equivalence point minus the sum of the strong acid anion concentrations times their predominant charge at the equivalence point. In most natural waters, these sums can be represented as

$$\begin{aligned} \sum_{i=1}^{n} |z_i| \text{M}_{T_i} = \; & \Sigma C_B = 2[\text{Ca}_T] + 2[\text{Mg}_T] \\ & + [\text{Na}_T] + [\text{K}_T] + [\text{NH}_{4_T}] + 2[\text{Al}_T] \\ & + 2[\text{Mn}_T] + 2[\text{Fe}_T] \end{aligned} \tag{1-14}$$

and

$$\begin{aligned} \sum_{j=1}^{m} |z_j| \text{A}_{T_j} = \; & \Sigma C_A = 2[\text{SO}_{4_T}] + [\text{NO}_{3_T}] \\ & + [\text{Cl}_T] + [\text{F}_T] \end{aligned} \tag{1-15}$$

where the subscript T stands for the analytical totals, which are readily and commonly measured. The terms ΣC_B and ΣC_A (as opposed to $\Sigma|z_i|\text{M}_{T_i}$ and $\Sigma|z_j|\text{A}_{T_j}$) are most commonly used. Figure 1.2 summarizes the definition of ANC. As shown in Equation (c) in the figure, changes in ANC result from changes in ΣC_B and/or changes in ΣC_A.

TABLE 1.1. Acid neutralizing capacity.

a. *Proton surplus/deficit table*

+1	0 (Proton reference level)	−1	−2	−3	−4
H^+	H_2O	OH^-			
	$H_2CO_3^*$	HCO_3^-	CO_3^{2-}		
	HR	R^-			
Al^{3+}	$Al(OH)^{2+}$	$Al(OH)_2^+$	$Al(OH)_3^0$	$Al(OH)_4^-$	
		AlR^{2+}	AlR_2^+	AlR_3^0	

b. *ANC expressions*

$\text{ANC}_W = [\text{OH}^-] - [\text{H}^+]$

$\text{ANC}_C = [\text{HCO}_3^-] + 2[\text{CO}_3^{2-}]$

$\text{ANC}_R = [\text{R}^-]$

$\text{ANC}_{\text{Al}} = [\text{Al(OH)}_2^+] + 2[\text{Al(OH)}_3^0] + 3[\text{Al(OH)}_4^-] - [\text{Al}^{3+}]$

$\text{ANC}_{\text{Al-O}} = [\text{AlR}^{2+}] + 2[\text{AlR}_2^+] + 3[\text{AlR}_3^0]$

$\text{ANC}_T = \text{ANC}_W + \text{ANC}_C + \text{ANC}_R = \text{ANC}_{\text{Al}} + \text{ANC}_{\text{Al-O}}$

Proton Reference Level

In Equation 1-14, the molar concentration of the total monomeric aluminum species is multiplied by 2 because the ANC proton reference level was set to include $Al(OH)^{2+}$. The proton reference level is defined as a group of species that, upon addition to solution either individually or together, can influence pH but, by definition, do not change ANC (Stumm and Morgan 1981).

In strong acid-strong base-aqueous CO_2 systems (i.e., the carbonate system), the proton reference level for ANC is taken to include water and aqueous CO_2 (H_2CO_3*). In more complex systems, which contain proton acceptors other than HCO_3^-, CO_3^{2-}, and OH^-, the proton reference level must be clearly stated. The proton reference level that coincides with the ANC definition presented earlier consists of water, aqueous CO_2 (CO_2(aq) + H_2CO_3), HR (fully protonated organic acid), and $Al(OH)^{2+}$ (see Table 1.1). ANC calculated in this manner correlates well with the Gran titrated ANC as shown in Figure 1.3a. In the past, a proton reference level that included Al^{3+} was commonly used, but as Figure 1.3b indicates, a proton reference level that includes $Al(OH)^{2+}$ provides a better fit between $\Sigma C_B - \Sigma C_A$ and Gran titrated ANC (Sullivan et al. 1989).

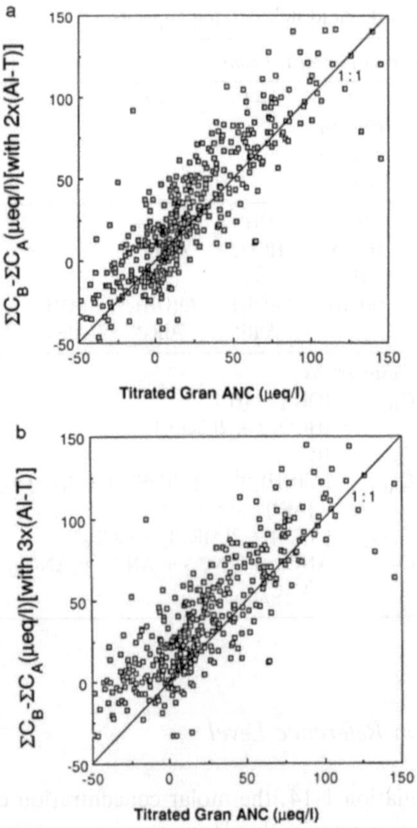

FIGURE 1.3. Comparison of $\Sigma C_B - \Sigma C_A$ and Gran titrated ANC. ΣC_B includes (a) $2[Al_T]$, and (b) $3[Al_T]$. The former ($2[Al_T]$) yields results that agree more closely with Gran titrated ANC (Gherini et al. 1989, Sullivan et al. 1989).

Another alternative proton reference level that is sometimes used includes $Al(OH)_3^0$. A convenience associated with this reference level is that one need not keep track of aluminum concentration when calculating ANC. However, in low ANC waters with appreciable aluminum concentrations, ANC calculated in this manner does not match the Gran ANC as closely as either method described above (Figure 1.4).

In summary, ANC is a conservative parameter, independent of temperature, pressure, and the partial pressure of CO_2 (for a solution not reacting with the solid phase). Although these factors change the relative sizes of the terms in Equation 1-3, they do not change the sum. For example, increasing the partial pressure of CO_2 increases the hydrogen ion concentration but also increases the concentration of bicarbonate and carbonate in a directly offsetting manner, and thereby yields no net change in ANC.

$$CO_2 + H_2O \rightleftharpoons H^+ + HCO_3^- \rightleftharpoons 2H^+ + CO_3^{2-} \tag{1-16}$$

Transport and mass balance equations can be readily written for ANC, which can then be used in the determination of the H^+ ion concentration and pH.

Buffer Systems

Water with only strong acid and strong base can change from a high to a low pH with only a small addition of strong acid (e.g., adding 40 μeq L^{-1} of strong acid to such a system with an initial ANC = 20 μeq L^{-1} [all as OH^-], will take it from pH = 9.3 down to 4.7). If small amounts of weak acids such as CO_2, organic acids, or aluminum hydroxide species are added to such a solution, they will serve as buffers and resist such drastic changes in pH. For example, in Figure 1.5, titration curves are shown for a pure strong acid-strong base system (H_2O, H^+, OH^-), a strong acid-strong base system kept in equilibrium with atmospheric CO_2 (and thus including $H_2CO_3^*$, HCO_3^- and CO_3^{2-}), and a strong acid-strong base-atmospheric CO_2 system that has organic acid in solution (HR plus R^-, as indicated by 3 mg L^{-1} DOC; total charge density = 6.5 μeq mg^{-1} DOC; pK_A = 4.5). Note the much more gradual change in pH for the latter systems upon the addition of strong acid or strong base. This results from the amphoteric behavior of the weak acid species. In other words, they can act both as acids and bases and thereby minimize changes in pH upon addition of strong acid or strong base. For example, upon addition of strong acid, HCO_3^- undergoes the following reaction:

$$HCO_3^- + H^+ \rightarrow CO_2 + H_2O \tag{1-17}$$

Upon addition of strong base, HCO_3^- reacts as follows:

$$HCO_3^- + OH^- \rightarrow CO_3^{2-} + H_2O \tag{1-18}$$

As Figure 1.5 indicates, the addition of organic acids supplies additional buffering beyond that provided by the carbonate system. In addition to buffering changes in pH upon addition of strong acid or base, the organic acids depress the pH of

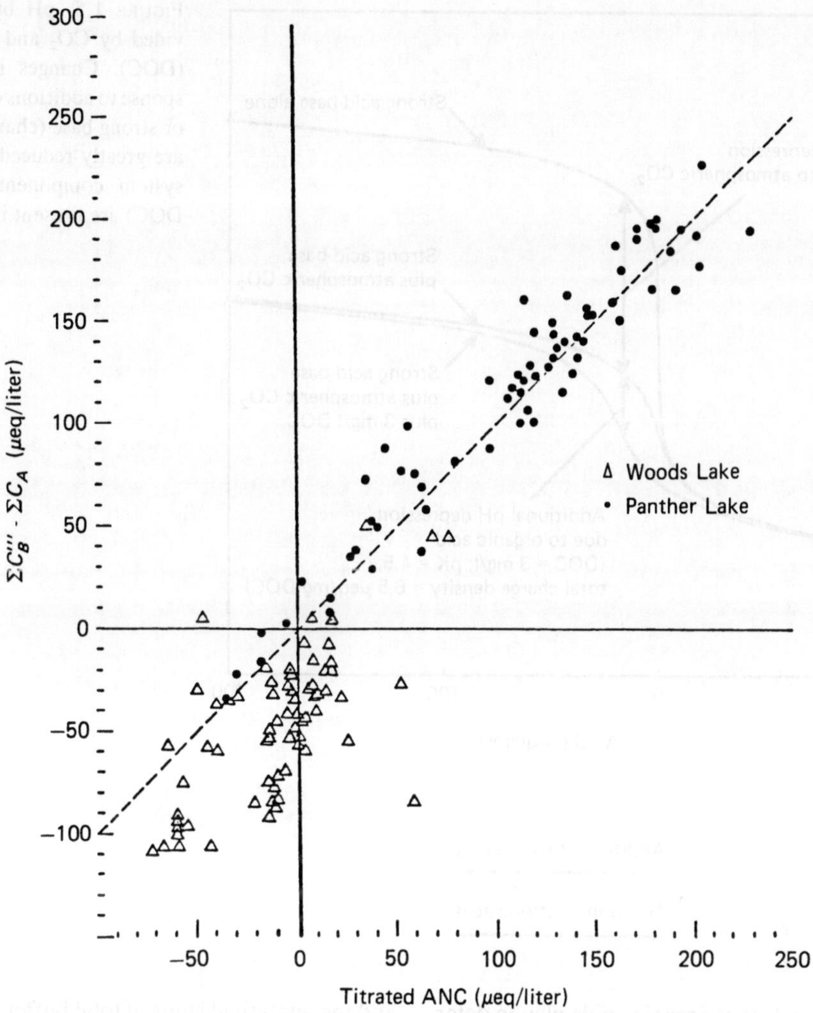

FIGURE 1.4. Comparison of $\Sigma C_B - \Sigma C_A$ with titrated ANC for the inlets and outlets of Panther and Woods Lakes. ΣC_B here does not include Al_T (i.e., the reference level is $Al(OH)_3$). The dashed line represents 1:1 equivalence. Note the divergence of calculated and measured values in low ANC waters with high aluminum concentrations.

the solution (i.e., they act as weak acids). Even the addition of the small quantity indicated in Figure 1.5 (3 mg L^{-1} DOC) can lead to a depression of over one pH unit.

In addition to serving as acid-base buffers in solution, organic acids also form complexes with metal ions, including aluminum, in solution. The toxicity of monomeric aluminum species to fish is mitigated by organic complexation (Baker 1984).

One difficulty associated with the presence of organics in solution is that they tend to obscure the inflection point in the ANC titration. This results in the underestimation of the Gran ANC relative to the ANC calculated from $\Sigma C_B - \Sigma C_A$ (Chen et al. 1988, Linthurst et al. 1986, Oliver et al. 1983, Sullivan et al. 1989). There are two hypotheses about the reason for this observation: first, the possibility that the organics produce curvature in the F1 Gran function—a fitting function to help identify the inflection point (Stumm and Morgan 1981, Gran 1952)—and thus make identification of the inflection point difficult; second, some organic acids may be strong enough to act as strong acid anions and should thus be included in ΣC_A (Davis et al. 1987).

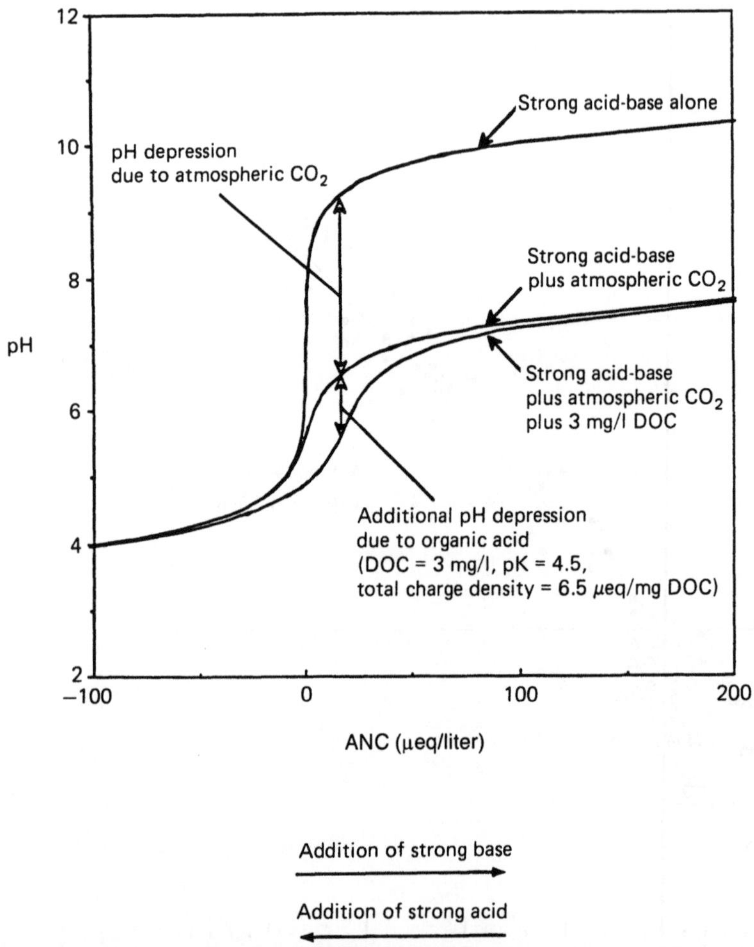

FIGURE 1.5. pH buffering provided by CO_2 and organic acid (DOC). Changes in pH in response to additions of strong acid or strong base (changes in ANC) are greatly reduced when buffer system components (CO_2 and DOC) are present in solution.

The precise role that organic acids play in determining lake and stream acid-base characteristics is an area of active research (e.g., Kramer and Davies 1988). Several empirical approaches have been used to quantify the effect of organic acids on solution pH. These include the Oliver model (Oliver et al. 1983), in which an equivalent acid dissociation constant for organic acid (K_A) is represented as a function of the solution hydrogen ion concentration. Several organic acid analogue approaches have been developed using monoprotic representations (e.g., Schecher and Driscoll 1987; pK_A = 4.5) and polyprotic acids with several functional groups (e.g., Gherini et al. 1985; pK_{A_1} = 4.0, pK_{A_2} = 6.1, pK_{A_3} = 9.25).

Determining Solution-phase pH

Free hydrogen ion concentration can be directly measured or determined by calculation from ANC and the analytical sums of total buffer system components: dissolved inorganic carbon, C_T, total monomeric aluminum, Al_T, and total organic acid, R_T, that is,

$$H^+ = f(ANC, C_T, Al_T, R_T) \qquad (1\text{-}19)$$

where: C_T = $[CO_2(aq)] + [HCO_3^-] + [CO_3^{2-}]$

Al_T = the total analytical concentration of monomeric aluminum species in solution, including the complexes with OH^-, F^-, SO_4^{2-}, and organic acid ligand

R_T = the total organic acid anion in solution, including complexes and protonated forms

An expression for ANC can be derived in which the individual ANC components are all represented by their analytical totals (e.g., C_T) and ionization and complexation constants. For example, the concentration of uncomplexed HCO_3^- is represented by

$$[HCO_3^-] = C_T\left(\frac{[H^+]}{K_1} + 1 + \frac{K_2}{[H^+]}\right)^{-1} \quad (1\text{-}20)$$

where K_1 and K_2 are the first and second acid dissociation constants for the carbonate system. When all such representations are substituted into the ANC expression (final equation in Table 1.1), the only unknown in the resulting equation is $[H^+]$. This equation can be solved by trial and error or by the use of numerical techniques (e.g., Newton-Raphson, Secant Method). Equations similar to this for the other species are presented elsewhere (e.g., Stumm and Morgan 1981) and have been included in solution equilibrium models such as PHCALC and ALCHEMI, which are discussed in the following chapter.

Acidic Deposition

The atmospheric acids and bases reaching lake and stream watershed systems enter in both wet and dry forms. Wet deposition is taken to be that directly entering the terrestrial system and the surface water (i.e., basin) with precipitation. This flux is relatively easy to quantify. It is calculated by multiplying measured precipitation amount by observed precipitation solute concentrations. There are many stations at which precipitation amount is measured, but relatively few where solution composition is also determined. Fortunately, however, studies indicate that variation in deposition across a small region (e.g., the Adirondacks) or across a subregion of a larger region, such as the Upper Midwest, results mainly from variation in precipitation amount (i.e., solution concentrations are relatively constant throughout a region; Johannes et al. 1986). Thus reasonable estimates of deposition can be made for a particular basin even if precipitation concentration values are not available from directly within or adjacent to the basin.

Dry deposition occurs as gases, particles, or a combination of the two (aerosols). Gases such as SO_2 and NO_x are absorbed by vegetation and are also converted to SO_4 and NO_3 on leaf surfaces. Particles made up of a variety of cations and anions [e.g., empirically $H_3NH_4(SO_4)_2$] also enter basins, are dissolved by precipitation, and contribute to the acid-base chemistry of the surface and ground waters. In addition, the deposition of nitric acid vapor can constitute significant acid input (Lindberg et al. 1986).

Measurement of dry deposition is an area of active research. Dry deposition is currently being quantified by a variety of methods, including collection in buckets that are covered during precipitation events (dry bucket collection), the use of Hi-vol samplers, and various micrometeorological techniques. Although dry bucket sampling has been criticized (e.g., Sievering and Ton 1985), one particular method that has been used successfully is to relate dry bucket measurements to solute concentrations in throughfall (Chen et al. 1983, Lindberg et al. 1986).

The quantification of dry deposition is an important aspect of any lake-watershed acidification study. If dry deposition is not quantified or is underestimated, the rate of export of ions from the watershed can be overestimated, leading to errors in predictions regarding future acidification. Unfortunately, this important component, because of its complexity, has often been ignored.

Hydrologic Processes

Hydrologic processes play an important role in determining the acid-base characteristics of surface waters. These processes dictate where water flows and thereby determine which biogeochemical processes influence solution chemistry. The hydrologic processes occurring in soils, streams, and lakes are depicted in Figure 1.6, and their consequences are described in the following sections.

Soil Hydrologic Processes

It has been demonstrated that in drainage basins, the routing of precipitation, or the path along which water flows (flow path), is a major determinant of surface water quality (Goldstein et al. 1984, Bricker and Rice 1989). Field observations indicate that soil solution pH increases with increasing depth in the soil profile, as indicated in Figure 1.7. Drainage basins with thick permeable media tend to have alkaline surface waters, whereas those with thin soils tend to have lower, or even negative, ANC values (Newton et al. 1987). This is because the thicker soils in general provide

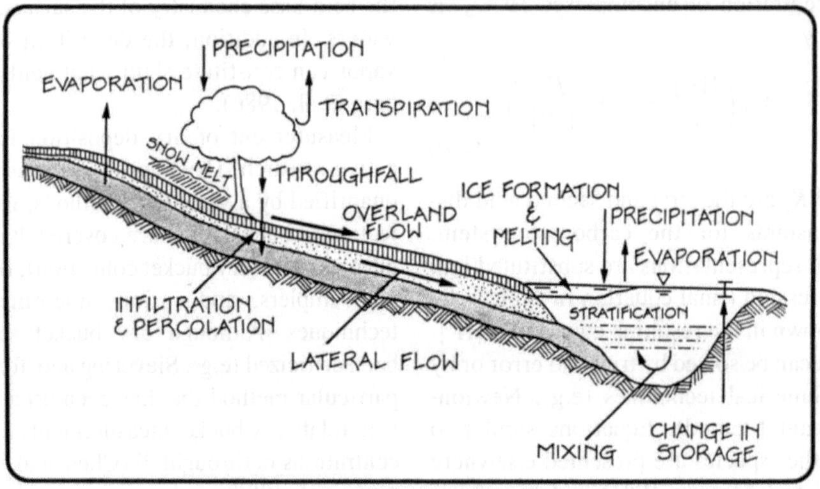

FIGURE 1.6. Hydrologic processes that influence surface water acid-base characteristics.

for deeper flow routing through alkaline soil horizons. Flow paths are determined by depth of the permeable media, hydraulic conductivity, land slope, precipitation-evapotranspiration, and other factors. The importance of these factors in the determination of flow paths through the soil are described in the following sections.

Flow Routing in Soils

The lateral flow carrying capacity of a soil horizon can be estimated using Darcy's Law (1856):

$$Q = -KAS \quad (1\text{-}21)$$

where: Q = volumetric flow
K = hydraulic conductivity
A = soil horizon cross sectional area perpendicular to flow
S = hydraulic energy gradient (change in head per unit length)

Hydraulic conductivity is a function of soil physical characteristics (e.g., grain size) and varies over at least 14 orders of magnitude (Hough 1957). The energy gradient in most forested basins can be estimated by the land slope along the direction of flow. The soil horizon cross-sectional area depends on soil depth and width of the catchment.

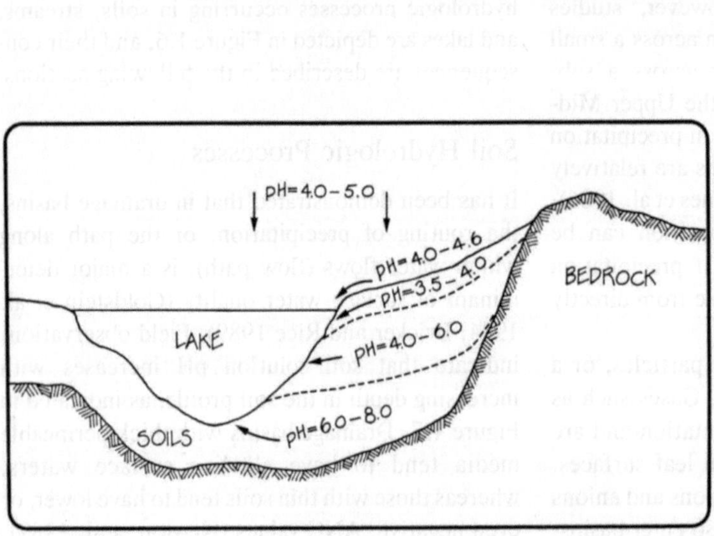

FIGURE 1.7. Changes in solution pH with vertical distance through the soil profile for an Adirondack lake-watershed system. (After Gherini et al. 1985.)

If the lateral flow carrying capacity of a particular soil layer is exceeded, infiltrating water will back up and begin to flow laterally through more shallow horizons. The soil solution in shallow horizons is typically more acidic than at depth, and thus water is not neutralized to the extent it would be if it had reached a lower horizon before flowing laterally. This phenomenon can occur during intense rainstorms and during snowmelt, when several months of precipitation (snowpack) is applied to the soil over a very short time period (e.g., 1 to 2 weeks). In the limit, the water can back up though the entire soil column (i.e., the entire soil column reaches saturated conditions) and create overland flow (Chen et al. 1984). Increased lateral flow through upper horizons also results when limitations to vertical percolation exist (e.g., hardpans and frozen soil).

Evapotranspiration

Another hydrologic process that has a direct effect on surface water quality is evapotranspiration, which increases the concentration of solutes. Evapotranspiration includes both direct evaporation and the movement of water through plants and into the atmosphere. One important point to remember is that evapotranspiration alone cannot lead to a change from acid to alkaline waters or vice versa. If an acidic water becomes alkaline, this indicates that processes other than evapotranspiration are acting to produce the change in ANC. Evapotranspiration alone can only make acidic water more acidic or akaline water more alkaline. A process that has the same effect on acid-base chemistry as evapotranspiration is sublimation of snow (direct change from solid to vapor phase). Here again, this process tends to concentrate the solutes in the snowpack while reducing the total volume of water applied to the soil during melt.

Stream Hydrologic Processes

Most streams where acidic deposition effects are of interest are located in steep catchments and thus have high energy gradients and fast flow. The high velocities induce turbulence that keeps the waters well mixed both vertically and across the width of the stream. Concentration gradients are observed, however, from the upstream headwaters to the downgradient reaches (Driscoll et al. 1987, Kaufmann et al. 1988). In general, the upgradient waters are more acidic because they drain thinner soils, and the downstream reaches are more alkaline because they are fed by thicker deposits. Because of the high velocities in these systems, the detention times in streams can be on the order of hours or minutes as opposed to groundwater detention times of months to years.

Lake Hydrologic Processes

Many processes that impact water quality occur within lakes. The importance of the processes varies depending upon lake type, trophic status, and hydraulic detention time. An overview of the various lake types is presented in the following section, along with descriptions of the hydrologic processes that influence lake acid-base characteristics.

Lake Types

Three major classes of lakes have been delineated based on the source of water to the lake. The three classes are drainage lakes, groundwater flow-through seepage lakes, and groundwater recharge lakes (Figure 1.8).

Drainage lakes have observable outlets. The precipitation that falls within the topographic basin of a drainage lake either exits through the outlet or is evapotranspired.

Seepage lakes have no observable inlets or outlets. Groundwater flow-through lakes receive a significant amount of their total inflow from groundwater. Unlike drainage lakes, the groundwater source area of these lakes often has little relation to the topographically defined watershed, as is indicated for East Eightmile Lake, Wisconsin (Figure 1.9). These lakes often have moderate values of ANC (e.g., > 100 μeq L^{-1}) because they receive an appreciable amount of high ANC groundwater. Groundwater flow-through lakes also often have relatively high concentrations of terrestrially derived solutes such as sodium and silica. These solutes indicate the importance of weathering in determining the water quality of these lakes (Wentz et al. 1987).

Groundwater recharge lakes, or mounded seepage lakes, also lack surface inlets or outlets. But unlike the flow-through systems, recharge

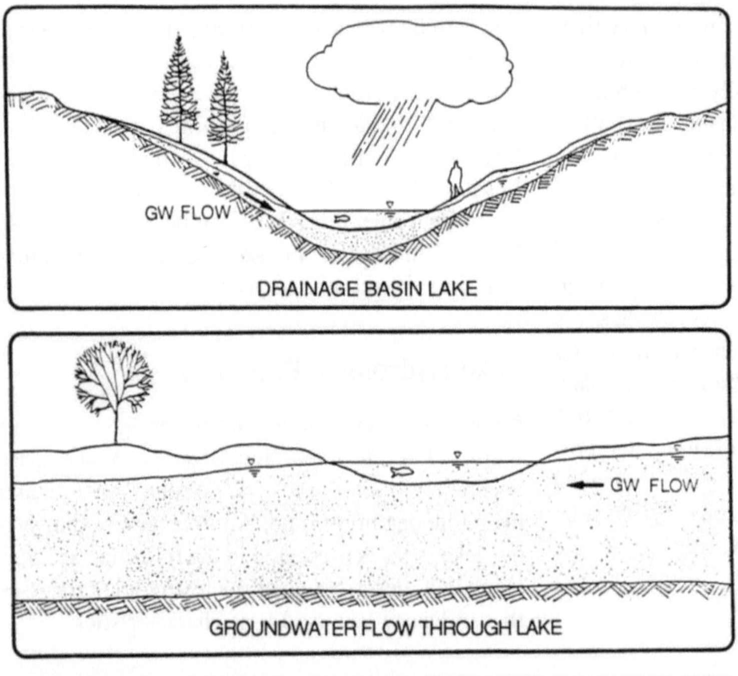

FIGURE 1.8. Lake types. (After Goldstein et al. 1987.)

lakes receive very little groundwater inflow. In these lakes, nearly all the water comes directly from precipitation. The production of ANC from in-lake processes is very important in recharge systems (Kelly et al. 1987). Often these lakes have long hydraulic detention times (e.g., > 3 years) that allow slow processes such as in-lake sulfate reduction to produce significant amounts of ANC (Kelly et al. 1987). Recent work has also shown that in a lake that receives only a small percentage of its inflow from groundwater (Crystal Lake, Wisconsin), the ANC associated with that groundwater contributes significantly (30 to 50%) to the observed lake ANC value (Greb et al. 1988).

Flow Routing in Lakes

Hydrologic processes occurring in lakes influence the distribution of the water in the lake. Because of these processes, the water quality at one depth is often markedly different from that at another. Stratification and mixing, two processes that are important in determining the movement of water through lakes, are described here.

Stratification

Lakes of interest in terms of acidic deposition effects generally have fairly small surface areas. The small surface areas and the low rate of horizontal

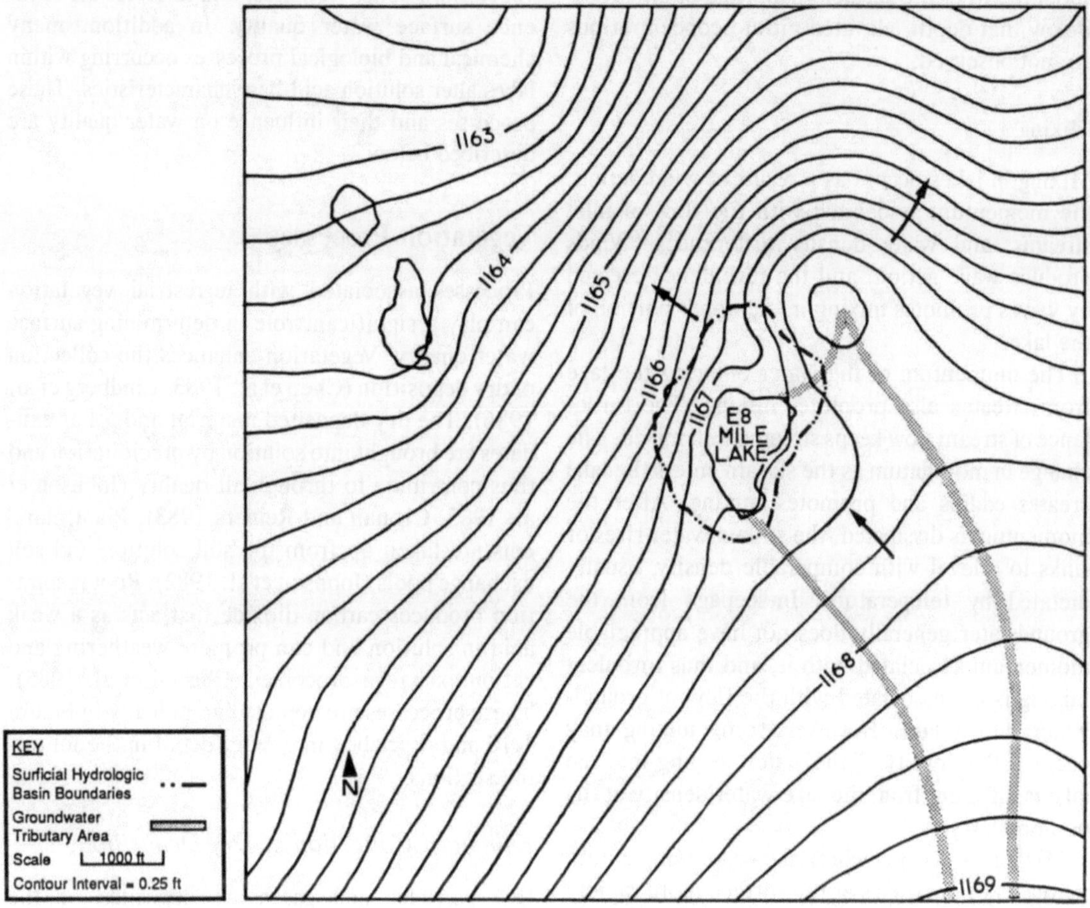

FIGURE 1.9. Topographically defined watershed versus groundwater contributing area for East Eightmile Lake, Wisconsin. The contours shown are groundwater elevation. As indicated, there is very little relationship between the topographically defined watershed and the groundwater tributary area for this flow-through seepage lake. (From Goldstein et al. 1987.)

turbulent diffusion lead to the observation that across-lake differences in water quality are typically quite small (Quay et al. 1980). However, vertical concentration gradients do exist as a result of thermal stratification. Water has its highest density at approximately 4°C. Since heat does not transfer instantaneously throughout a lake, there are times during the year when the temperature at one depth can be significantly different than at another. The water with the temperature nearest 4°C is always found at the deepest point within the lake, because that water is the most dense.

This can result in interesting chemical phenomena. High ANC waters may build up in the hypolimnion of lakes during summer and winter stratification because the thermal structure does not allow high ANC groundwater inputs and reduction products (e.g., Fe^{2+}, NH_4) to mix throughout the lake (Cook et al. 1986). Stratification also influences the routing of water associated with spring melt, as seen at Panther Lake, New York. Due to decay processes occurring under the snowpack and accumulation in the snowpack, the meltwater contains high concentrations of nitrate (Rascher et al. 1987). The temperature of the meltwater that enters the lake is near 0°C. At this time of year, the lake is reverse stratified with water near 4°C at the bottom of the lake and near 0°C just under the ice. The melt water enters the lake and flows directly to the outlet, staying within about 0.5 m of the surface (Schofield et al. 1985). Below that depth, elevated nitrate concentrations are not observed.

Mixing

Mixing in lakes occurs as a result of wind action, the momentum associated with the flow of inlet streams, and water density differences. Winds produce wave action, and the turbulence created by waves promotes mixing in the upper portion of the lake.

The momentum of the water entering the lake from streams also promotes mixing. The turbulence of stream flow keeps streams well mixed. The change in momentum as the stream enters the lake creates eddies and promotes mixing. After the momentum is dissipated, the stream water rises or sinks to a level with comparable density, usually dictated by temperature. In-seepage from the groundwater generally does not have appreciable momentum associated with it, and thus turbulent mixing is not associated with the flow of groundwater into a lake. However, some mixing may occur if the density of the water seeping into the lake is different from the lake water density at the point of entry.

Mixing also occurs when lake water temperature increases or decreases at the surface more rapidly than at depth. This happens in many lakes in temperate climates two times during the year (spring and fall), and lakes that have this type of mixing pattern are called dimictic. In the fall when the ambient air temperature begins to drop, the water at the lake surface cools. But at depth, the insulation of the upper layers of water slows the rate of temperature change. Eventually the upper water cools to such an extent that its density is greater than the deep water, and it sinks, displacing the less dense deep water to the surface. In the spring, the opposite pattern is often observed. This is commonly referred to as lake turnover and provides considerable mixing (Wetzel 1975).

Biogeochemical Processes

Several biogeochemical processes occur in hydrologic basins that alter the acid-base chemistry of water. In terrestrial systems, processes involving vegetation, soils, wetlands, and land use all influence surface water quality. In addition, many chemical and biological processes occurring within lakes alter solution acid-base characteristics. These processes and their influence on water quality are described below.

Vegetation Processes

Processes associated with terrestrial vegetation can play a significant role in determining surface water quality. Vegetation enhances the collection of dry deposition (Chen et al. 1983, Lindberg et al. 1986). The dry deposited material and foliar exudates are brought into solution by precipitation and thus contribute to throughfall quality (Johnson et al. 1985, Cronan and Reiners 1983). Plant nutrients are taken up from the soil solution and soil exchange pools (Johnson et al. 1982). Root respiration produces carbon dioxide that acts as a weak acid in solution and can promote weathering and cation exchange processes (Gherini et al. 1985). These processes are depicted graphically in Figure 1.10 and described in greater detail in the following sections.

Enhanced Collection of Dry Deposition

Dry particles and gases are deposited on the vegetative canopy by gravitational settling and inertial impaction among other processes. Gases can be absorbed directly through leaf stomata. Vegetation provides additional surface area beyond the land surface for the absorption, settling, and impaction to take place. The extent of the enhancement provided by forest canopy is a function of the additional surface area available, which is indicated by the leaf area index (LAI). The LAI is simply the ratio of the total surface area of leaves over the vertically projected area of the tree on the ground surface. For deciduous trees, typical LAI values range from about 1 $m^2\ m^{-2}$ during the winter to 6 $m^2\ m^{-2}$ during the summer. For coniferous trees, LAIs range from about 12 $m^2\ m^{-2}$ to 15 $m^2\ m^{-2}$ (Cronan 1985). In part, because coniferous trees serve as such effective collectors of dry deposition, the throughfall below coniferous trees is often more acidic than the incident precipitation in areas receiving significant quantities of acidic dry deposition, such as the Adirondacks. In the same

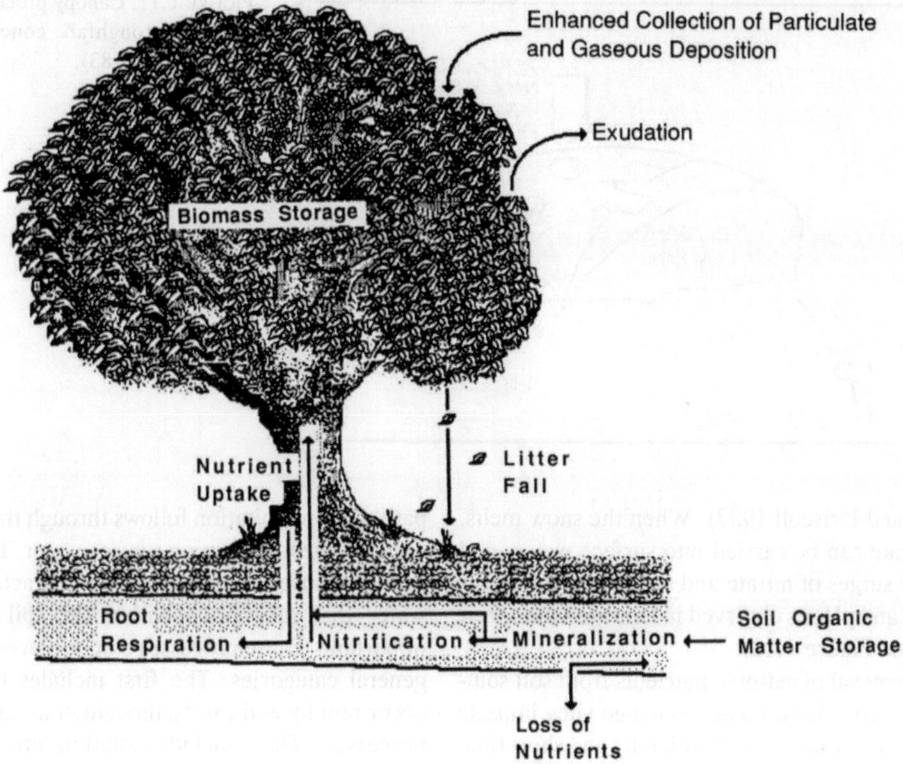

FIGURE 1.10. Vegetation processes influencing surface water acid-base characteristics.

region, throughfall from deciduous trees is generally more alkaline than incident precipitation (Johannes et al. 1985).

Foliar Exudation and Leaching

The alkalizing effect observed below deciduous canopy is due to foliar exudation and leaching. The products of these processes are generally alkaline in both coniferous and deciduous trees. When mixed with precipitation and acidic dry deposited materials, the products contribute to the observed solute concentrations in throughfall. In deciduous stands, the products compensate for the acidic dry deposited materials, and the throughfall produced can be more alkaline than the incident precipitation (Johannes et al. 1985). In Adirondack coniferous stands, the additional dry deposited materials can overwhelm this alkalinity source, causing throughfall to be more acidic than precipitation (Johannes et al. 1985). In areas receiving smaller quantities of acidic dry deposition (e.g., Wisconsin), the throughfall below both tree types is more alkaline than the precipitation (Knauer et al. 1989). The canopy processes are depicted in Figure 1.11.

Nutrient Uptake

Plants remove nutrients from soil solution and the soil exchange complex to support growth and create biomass. Removal of these nutrients from the solution and solid phase pools impacts the quality of soil and surface waters.

Nitrogen is an important plant nutrient. During the growing season, nearly all of the nitrogen that enters a watershed or is produced internally by decay processes is taken up to support plant growth (Johnson 1985). For this reason, the concentrations of nitrogen compounds in forested catchment waters are typically quite low during periods of growth. During the winter dormant season, however, nitrogen concentrations (especially NO_3^-) can build up in the upper soil horizons and snowpack

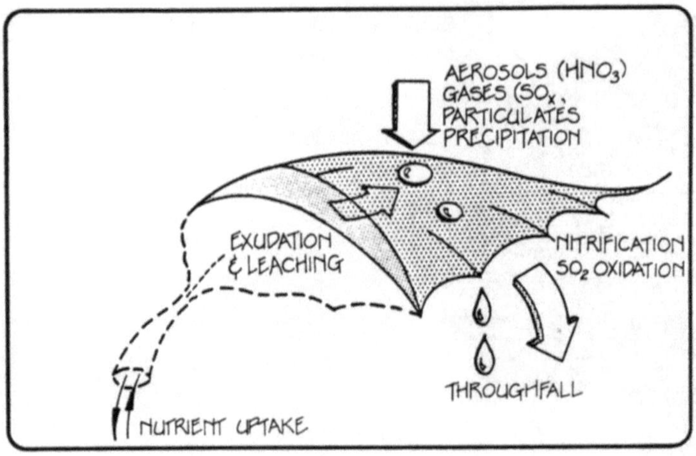

FIGURE 1.11. Canopy processes influencing throughfall concentrations (Chen et al. 1983).

(Peters and Driscoll 1987). When the snow melts, this nitrate can be carried into surface waters and produce surges of nitrate and significant decreases in ANC and pH, as observed in Panther Lake, New York (See Figure 1.12).

The removal of cationic nutrients from soil solution and from solid phase exchange sites impacts surface water quality over both long and short time frames. In the short term, if base cations are removed from solution in excess of strong acid anions, the ANC of the soil solution is reduced. In the long term, if nutrients are removed from the cation exchange complex and converted into biomass, those cations are no longer available to neutralize acid inputs via cation exchange reactions. This can lead to soil acidification and, eventually, to decreases in surface water ANC.

Root Respiration

One of the important consequences of root respiration is that it supplies CO_2 to the soil gas and solution phases. Aqueous CO_2 is a weak acid, and the hydrogen ions it provides to solution can help drive mineral weathering and cation exchange reactions. Root respiration along with microbial activity produces significant quantities of CO_2, such that partial pressures of carbon dioxide in ground waters are often 3 to 300 times atmospheric P_{CO2} levels.

Soil Processes

As indicated previously, a major determinant of surface water quality in drainage basins is the flowpath that precipitation follows through the soil system enroute to becoming surface water. Therefore, processes affecting the acid-base characteristics of water that take place within the soil are very important. These processes can be placed into two general categories. The first includes those that occur rapidly and can be thought of as equilibrium processes. They include solution phase proton transfer, cation exchange, and anion adsorption. The other general category includes slower processes that can be thought of as rate limited. These include litter decay, nitrification, and mineral weathering. Aluminum dissolution and precipitation can be conceptualized as either an equilibrium or rate limited process. Any or all of these processes may be important in an individual basin, and determining which are the most influential is essential in predicting how a basin will respond to acidic deposition over time. These soil processes are depicted schematically in Figure 1.13 and are described in the following sections.

Cation Exchange

Cation exchange is a fast process that attempts to maintain constant cation concentrations in solution (Figure 1.14). A soil has cation exchange sites that include functional groups on immobile organic matter, the surfaces and edges of fine particles, etc. The exchange sites are occupied by base cations (including aluminum) and hydrogen ions. In order to minimize changes in solution-phase hydrogen ion concentration, when acids enter a soil, hydrogen ions in the solution are exchanged for base

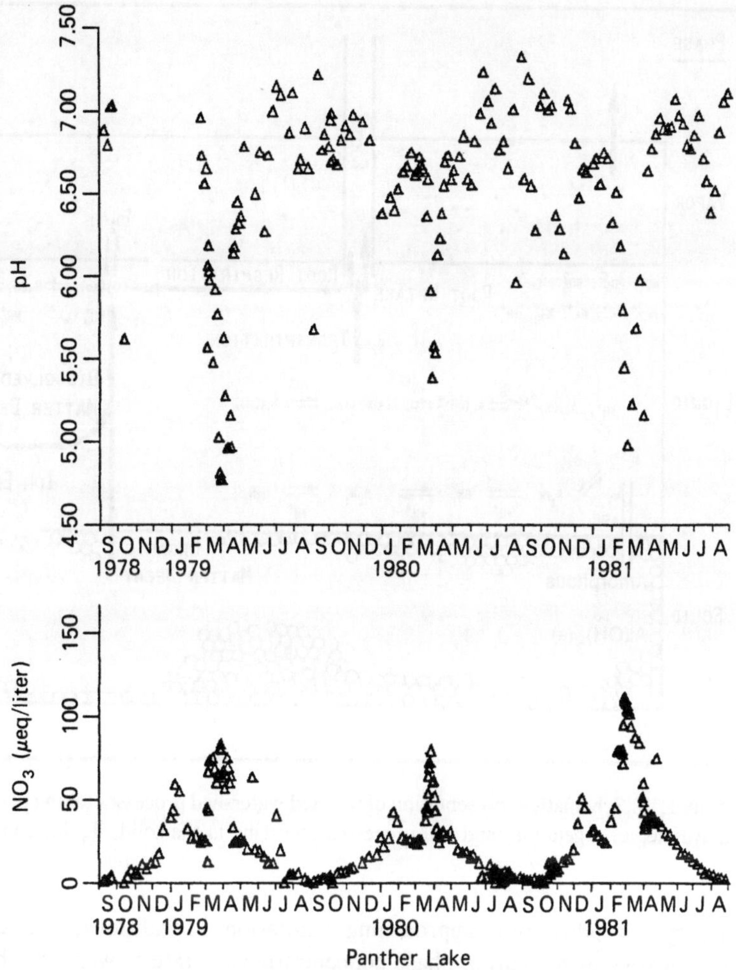

FIGURE 1.12. pH and nitrate concentration at Panther Lake. pH depressions coincide with nitrate peaks (Galloway et al. in ILWAS data base, ed. by Valentini and Gherini 1986).

cations sorbed to the immobile phases. This produces ANC (by increasing ΣC_B). If for some reason an excess of base cations were added to solution (e.g., by liming the soil), hydrogen ions would desorb from the exchange complex and enter solution. In this manner, cation exchange serves as an effective buffering process that minimizes changes in soil solution hydrogen ion concentrations.

The cation exchange complex of a soil has a finite capacity. Studies have indicated that soils in the northeastern United States have adequate sorbed base cations to continue to buffer acid inputs for anywhere from 50 to 200 years (Schnoor and Stumm 1984) to as many as several hundred years (Gherini et al. 1985). If exchangeable base cations become depleted, the supply of ANC by this process would no longer be effective and lake or stream ANC could decrease.

Anion Adsorption

Adsorption of heterovalent anions is an important process, particularly in old, highly weathered soils containing high concentrations of iron and aluminum oxides and hydroxides. Anion adsorption effectively maintains constant solution anion concentrations, at least over short periods. For example, if sulfate ion is added to a sorbing soil, the added ions are adsorbed, thereby producing ANC (decrease in ΣC_A). Conversely, if a solution with a very low sulfate ion concentration were added to a sorbing soil, anions would be desorbed in an attempt to maintain constant solution sulfate concentration. The anion adsorption process is depicted schematically in Figure 1.15.

Anion adsorption behavior typically follows a Langmuir isotherm (see Figure 1.15), with the

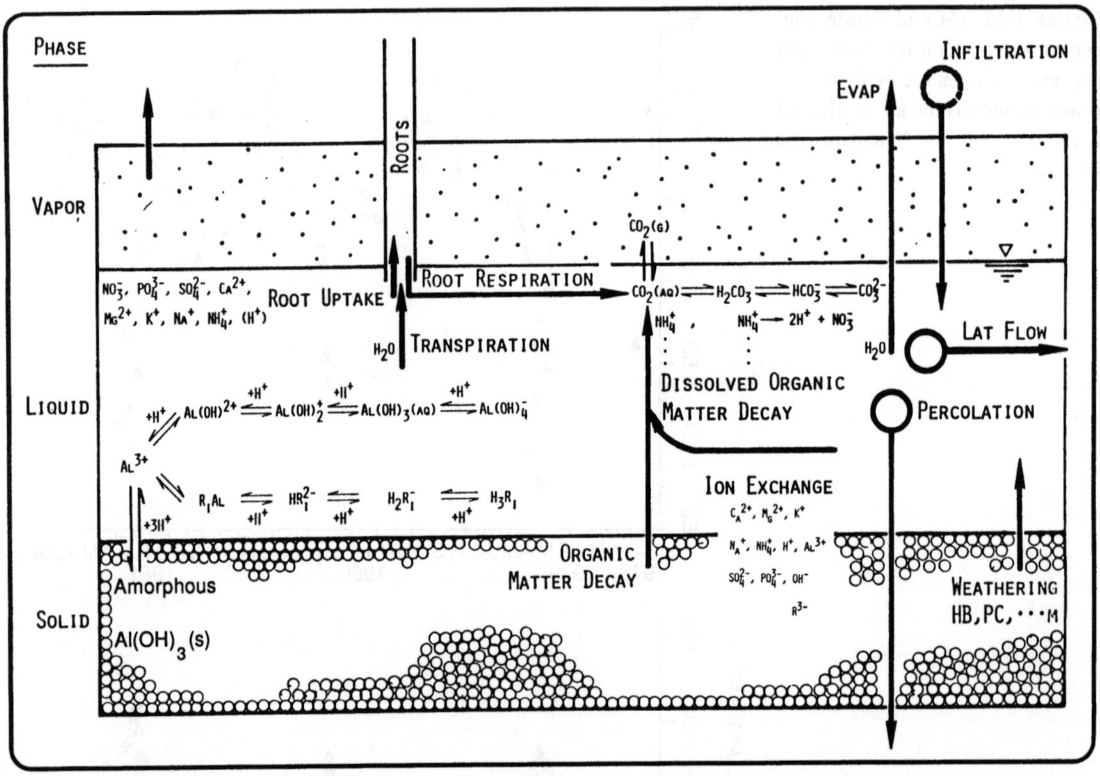

FIGURE 1.13. Schematic representation of selected watershed processes taking place within a single soil layer. Included are hydrologic, vegetation, and soil processes occurring in the solid, liquid, and vapor phases (Gherini et al. 1985).

capacity for adsorption approaching saturation levels at very high solution phase concentrations. If these saturation levels were reached, anion adsorption would not be effective in supplying ANC to the throughflowing waters. A steady-state condition with respect to sulfate inputs would most likely be reached before the soil adsorption sites became saturated. Under these conditions, sulfate would be desorbed at the same rate it was adsorbed, resulting in no net change in soil solution ANC. If sulfate steady-state conditions were reached in a system with previously net sorbing soils, lake or stream ANC might decrease depending on the response of other compensation processes (e.g., weathering and cation exchange).

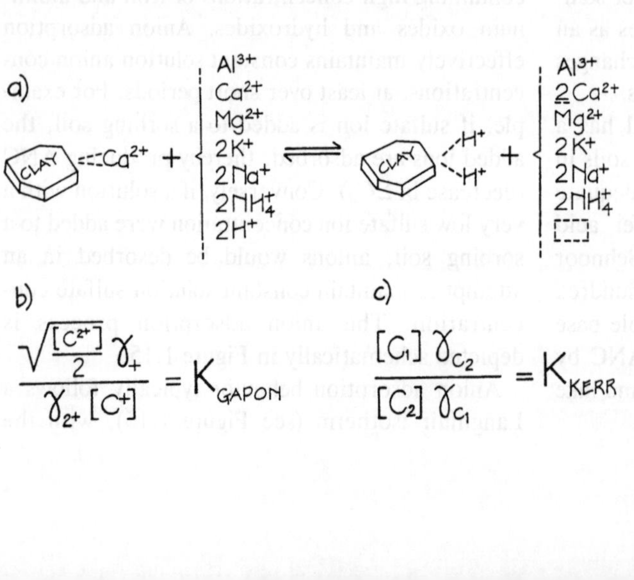

FIGURE 1.14. Competitive cation exchange: (a) Idealized representation, (b) Heterovalent exchange equation, (c) Homovalent exchange equation. [C] represents solution-phase ion concentration and γ represents solid-phase ion concentration. Alternate representations (e.g., Gaines-Thomas) can also be used (Reuss and Johnson 1986).

Aluminum Behavior

Aluminum hydroxide is an effective chemical buffer that resists changes in pH at lower pH values (pH < ≈4.5). Aluminum solubility increases rapidly with decreasing pH (< 5.5) as indicated in Figure 1.16 (solubility is also temperature dependent). At higher pH values, it is soluble but polymerizes. Although aluminum is an effective chemical buffer, its monomeric inorganic species (e.g., Al^{3+}, $Al(OH)^{2+}$, $Al(OH)_2{}^{+}$) are toxic to fish (Schofield 1976). Aluminum dissolution in mineral

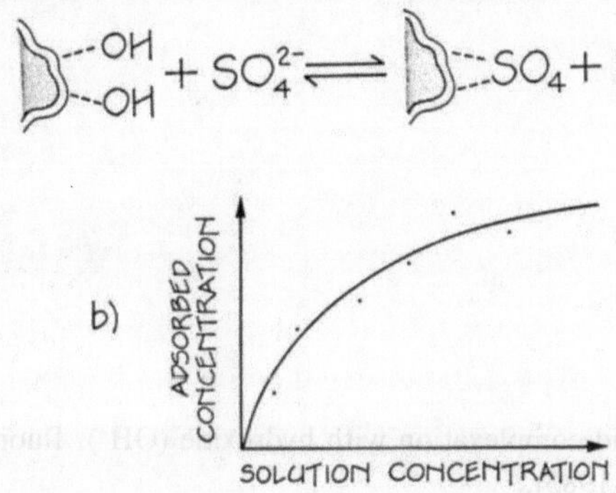

FIGURE 1.15. Anion adsorption: (a) Idealized representation, (b) Adsorption isotherm.

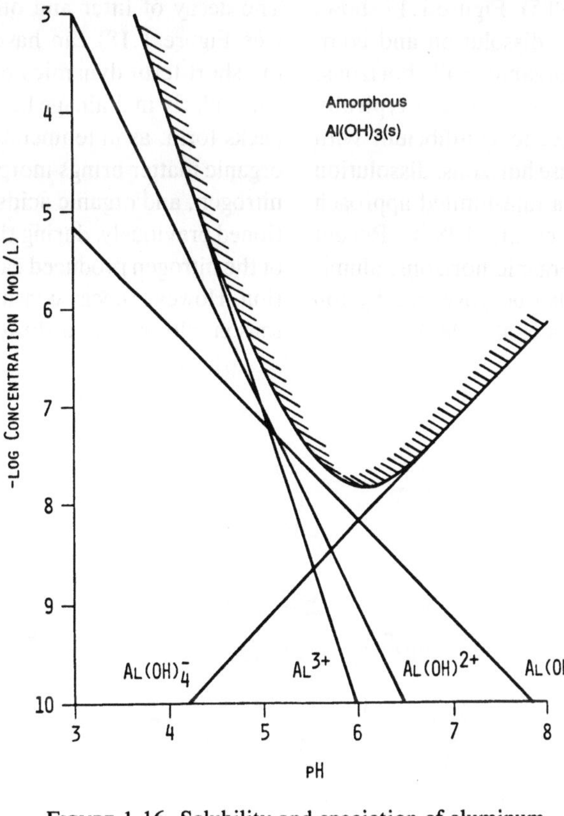

FIGURE 1.16. Solubility and speciation of aluminum.

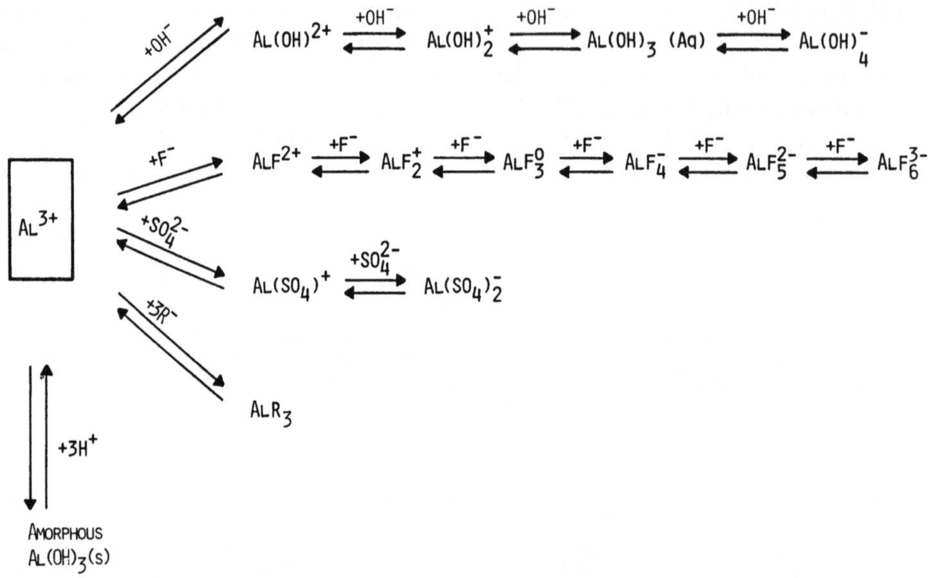

FIGURE 1.17. Aluminum dissolution and complexation with hydroxide (OH^-), fluoride (F^-), sulfate (SO_4^{2-}), and organic acid ligand (R^-) (Gherini et al. 1985).

soil horizons can be adequately described using equilibrium constants associated with amorphous $Al(OH)_3$ (Schofield et al. 1985). Figure 1.17 shows several common aluminum dissolution and complexation reactions. In organic soil horizons, observations indicate that solutions are typically undersaturated with respect to equilibrium with amorphous $Al(OH)_3$. In these horizons, dissolution has been calculated using a rate-limited approach to equilibrium (Gherini et al. 1985). Recent research indicates that in organic horizons, aluminum concentration may also be governed by ion exchange reactions (Cronan et al. 1987).

Decay Processes

The decay of litter and other soil organic matter (see Figure 1.18) can have important impacts on the short-term dynamics of surface water quality, particularly in drainage basin systems where snowpacks form, as in temperate regions. The decay of organic matter brings inorganic species, especially nitrogen, and organic acids into solution. As mentioned previously, during the growing season much of the nitrogen produced can be taken up by vegetation. However decay continues during the dormant season, although at a slower rate, in spite of lower

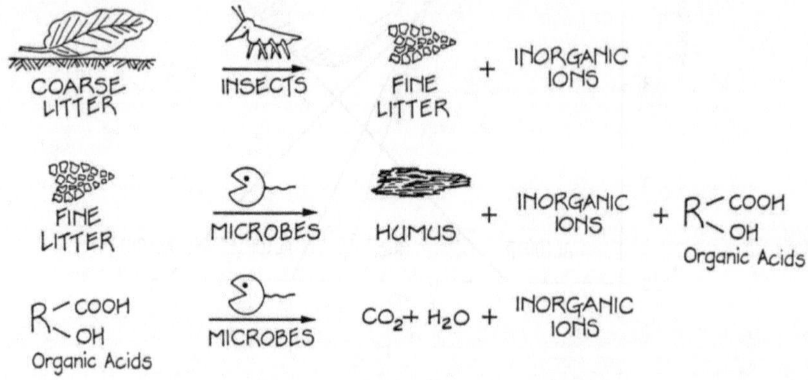

FIGURE 1.18. Sequential decay of litter and organic matter.

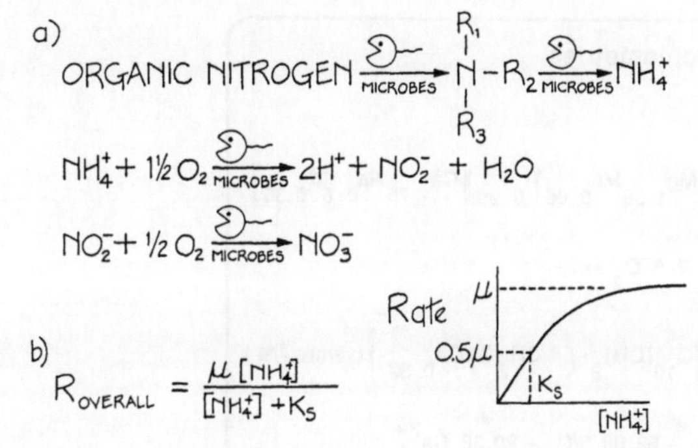

FIGURE 1.19. Mineralization and oxidation of organic nitrogen: (a) General stoichiometry, (b) Rate expression, with NH_4^+ oxidation limiting reaction.

air and soil temperatures (Strayer et al. 1981). With minimal infiltration (precipitation is in the form of snow) and the lack of nutrient uptake, nitrate can build up in the upper soil horizons under the snowpack (Rascher et al. 1987). When the snow melts, the nitrate is flushed from the system and contributes to the spikes shown previously in Figure 1.12 (Schofield et al. 1985).

Decay processes also produce organic acids. The temporal behavior of these organic acids can be similar in some respects to that of nitrate. The highest surface water concentrations are associated with spring snowmelt, again in temperate region drainage systems. This is the period when the largest volume of water flows laterally through the upper soil horizons where the organic acids are produced and accumulate under the snowpack. One of the major differences between the organic acid behavior and nitrate behavior is that surface water nitrate concentrations are generally negligible during the growing season, whereas significant organic acid concentrations, as indicated by dissolved organic carbon concentrations on the order of ≥ 2 mg L^{-1}, can be observed throughout the year.

Nitrification

Ammonium enters soil solution via deposition and organic matter decay processes (see Figure 1.19) and is oxidized to nitrate and hydrogen ion according to the following overall stoichoimetry:

$$NH_4^+ + 2O_2 \rightarrow NO_3^- + 2H^+ + H_2O \qquad (1\text{-}27)$$

This reaction has been observed to proceed, even in soils with low solution phase pH (e.g., pH < 3.6; Strayer et al. 1981). The reaction results in the loss of a base cation (ΣC_B decreases) and the addition of a strong acid anion (ΣC_A increases) to solution and thus a decrease in ANC. As stated above, most of the nitrate in solution is taken up to support plant growth, but significant pulses can be observed associated with spring melt and large rain storms (Driscoll et al. 1987, Schofield et al. 1985).

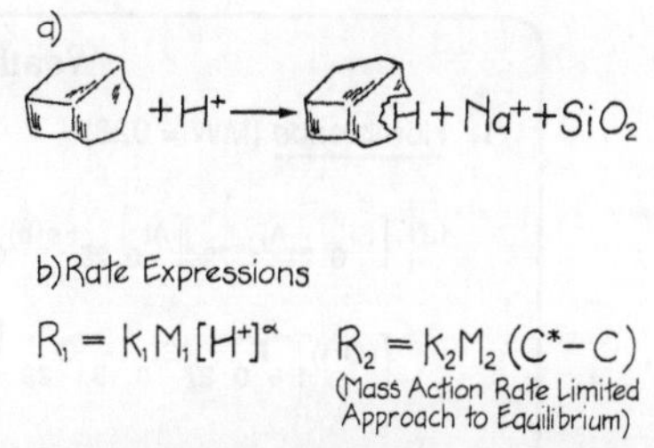

FIGURE 1.20. Mineral weathering: (a) Idealized representation, (b) Rate expressions. C* = equilibrium concentration of reaction product.

Mineral Weathering

Mineral weathering is a slow process that adds base cations (and thus ANC; Figure 1.20) in specific stoichiometric ratios (Figure 1.21) to solution (April and Newton 1984). Since the total content of minerals in most systems is very large, the potential supply of base cations from mineral weathering is nearly infinite. However, the rate of supply can be quite slow. The rate of weathering can be represented mathematically as follows:

$$R = kM[H^+]^{\alpha} \qquad (1\text{-}28)$$

where: R = mineral weathering rate
k = specific reaction rate coefficient
M = the mass of mineral present
$[H^+]$ = the solution phase hydrogen ion concentration
α = the functional dependence of weathering on solution hydrogen ion concentration

The value of α is reported to range from 0.0 to 0.8 (Bloom 1988, Schnoor and Stummn 1984, Drever and Hurcomb 1986). Thus the weathering rate is weakly dependent on hydrogen ion concentration and the response to any change in solution pH would be less than linear. In addition, since cation exchange responds rapidly to limit any

Weathering Stoichiometries

1. Hornblende (MW = 928)

$$12\left\{[Si_{6.21}Al_{1.79}][Al_{0.25}Fe(II)_{2.05}Fe(III)_{0.88}Mg_{1.36}Mn_{0.05}Ti_{0.26}][Ca_{1.75}Na_{0.6}K_{0.35}]\right.$$
$$\left.[(OH)_{1.6}F_{0.27}Cl_{0.13}]O_{22}\right\} + 79.96\ H^+ + 9.6\ O_2 \rightarrow$$
$$2\left\{[Fe(II)_{2.36}Fe(III)_{0.48}Al_{0.16}][Al_{1.28}Si_{2.72}]O_{10}(OH)_2 \cdot (4.3H_2O)\,Ca_{0.32}\right\}\text{(vermiculite)}$$
$$+ 14.74\ Fe_2O_3 + 21.6\ Al(OH)_3 + 3.12\ TiO_2 + 69.08\ SiO_2 + 20.36\ Ca^{2+}$$
$$+ 16.32\ Mg^{2+} + 4.2\ K^+ + 7.2\ Na^+ + 3.24F^- + 1.56\ Cl^-$$
$$+ 0.6\ MnO_2 + 6.58\ H_2O$$

2. Plagioclase (MW = 266)

$$\left\{Fe(II)_{0.01}K_{0.01}Na_{0.84}Ca_{0.17}Al_{1.17}Si_{2.82}O_8\right\} + 1.19\ H^+ + 0.575\ H_2O$$
$$+ 0.0025\ O_2 \rightarrow 0.005\ Fe_2O_3 + 0.01\ K^+ + 0.84\ Na^+ + 0.17\ Ca^{2+}$$
$$+ 0.585\ \underset{\text{(kaolinite)}}{H_4Al_2Si_2O_9} + 1.65\ SiO_2$$

3. K-Feldspar (MW = 278)

$$2KAlSi_3O_8 + 2H^+ + H_2O \rightarrow \underset{\text{(kaolinite)}}{H_4Al_2Si_2O_9} + 2K^+ + 4SiO_2$$

FIGURE 1.21. Weathering stoichiometries for a hornblende, plagioclase and K-feldspar. These formulas have been derived by Dr. Robert Newton of Smith College and Dr. Richard April of Colgate University. (From Chen et al. 1983.)

change in solution hydrogen ion concentration, short-term changes in weathering rate in response to changes in deposition acidity would be expected to be small.

Wetland Processes

Important wetland processes include those associated with inundation and with the production and decay of organic matter. Inundation, by limiting oxygen transfer, provides an excellent environment for nitrate and sulfate reduction reactions. These reactions remove strong acid anions from solution and thus produce ANC (ΣC_A decreases). However, if the system dries and is exposed to air, the reduced sulfur compounds can oxidize and thereby produce surges of strong acid in solution (ΣC_A increases) (Bayley et al. 1986). Wetlands also provide adequate substrate for microbial activity. This, along with the limitation of oxygen transfer and the absence of flow through permeable media (which provides surface for the sorption of dissolved organics), leads to high concentrations of DOC in solution. The organic acids associated with this DOC can depress solution pH. They also readily complex with the naked aquo aluminum ion (Al^{3+}), which is in desperate search of a partner. This complexation mitigates the toxicity of aquo aluminum

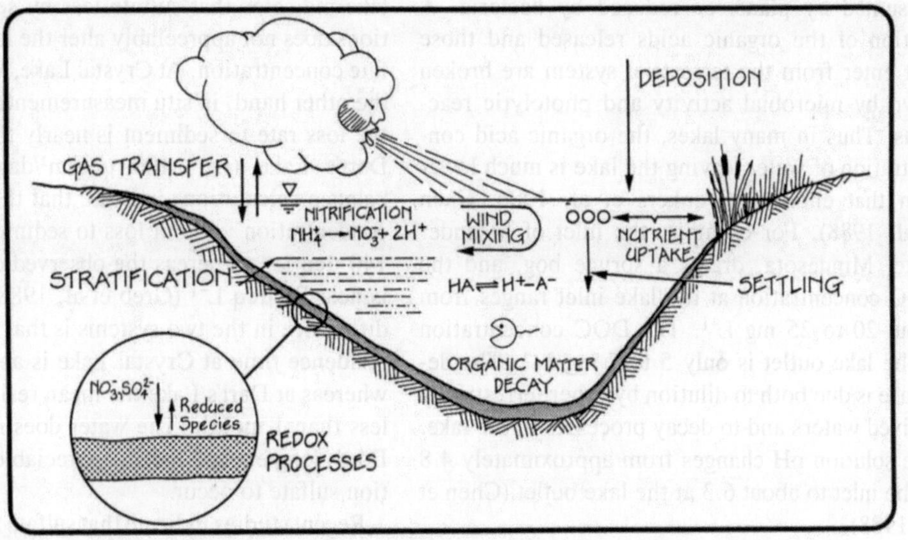

FIGURE 1.22. Lake biogeochemical processes influencing solution acid-base characteristics.

to fish (Schofield et al. 1985). Organic acids also serve as an aqueous phase chemical buffer.

Land Use Changes

Soil and surface water acidification can occur through a variety of processes other than atmospheric deposition. Changes in land use can have a dramatic impact on the acid-base characteristics of natural waters. Application of nitrogen fertilizers to soils can increase the nitrate concentrations in soil solution. Hydrogen ions associated with the nitrate can promote cation exchange and mineral weathering reactions. If the nitrate is not taken up by plants, it has an acidifying effect (ΣC_A increases), whereas the base cations that enter the soil solution via exchange and weathering have an alkalizing effect.

Clear cutting can also impact surface water quality. Through a series of interactions stimulated by the removal of vegetation, the surface water ANC and pH initially fall, and then, as vegetation begins to be re-established, they eventually rise (Lawrence et al. 1988). However, over the long term, Krug and Frink (1983) have hypothesized that revegetation may have an acidifying effect.

Lake Biogeochemical Processes

Many chemical and biological processes occurring in lakes influence acid-base characteristics of the solution phase (Figure 1.22). These include plant growth, decay processes, gas transfers, and anion reduction. Some of these processes play an important role in determining the acid-base characteristics of almost all lakes, whereas others exert observable influences mostly in longer detention time systems. All of these processes are described in the following sections.

Plant Growth

Nitrogen species serve as nutrients to aquatic plants, including algae. Uptake of nitrogen species has an alkalizing effect to the extent that nitrate is taken up in excess of ammonium ($\Delta ANC = \Delta C_B - \Delta C_A$). Most low ANC lakes have fairly low levels of productivity, but nitrogen uptake by plants can still be an important process in determining lake acid-base characteristics.

Decay Processes

Decay processes occurring in lakes include the breakdown of organic acids and plant matter. The decay of plant matter in lakes results in the release of small quantities of inorganic species (e.g., nitrogen compounds) and dissolved organic matter (e.g., organic acids). Although nitrification processes within the lake convert some of the ammonium released to nitrate, producing a short-term decrease in ANC, most of the nitrogen released is consumed by plants or reduced by bacteria. A portion of the organic acids released and those that enter from the terrestrial system are broken down by microbial activity and photolytic reactions. Thus in many lakes, the organic acid concentration of water leaving the lake is much lower than that entering (Donkers et al. 1986, Chen et al. 1988). For example, the inlet of Meander Lake, Minnesota, drains a spruce bog, and the DOC concentration at the lake inlet ranges from about 20 to 25 mg L^{-1}. The DOC concentration at the lake outlet is only 5 to 7 mg L^{-1}. The decrease is due both to dilution by other terrestrially derived waters and to decay processes in the lake. The solution pH changes from approximately 4.8 at the inlet to about 6.3 at the lake outlet (Chen et al. 1988).

Gas Transfers

As indicated above, CO_2 partial pressures in soil solution can build up to significant levels (e.g., from 3 to 300 times atmospheric CO_2 levels). One reason for the buildup is that CO_2 is produced by microbial activity and root respiration, and the diffusion path for the CO_2 out of the soil is occluded by solids and soil moisture. Thus high concentrations are observed. When groundwater enters surface waters, turbulent transport leads to the release of CO_2 from solution. The effect of this is that the pH rises relative to the groundwater as the weak acid CO_2 ($H_2CO_3^*$) is lost from solution. The observed CO_2 level in the epilimnion of most lakes is generally within 2 to 3 times atmospheric levels. During the winter, CO_2 can build up under ice cover due to respiration in the lake leading to depressions of pH (Kratz et al. 1987).

Anion Reduction

Anion reduction reactions take place in the sediments of most lakes. However, they are relatively slow processes (sulfate reduction in particular) and have their largest impact on anion concentrations in lakes with longer hydraulic residence times (Kelly et al. 1987). For example, in Dart's Lake, New York, in situ measurements indicate that sulfate reduction is occurring at a rate of about 130 μeq/m²/day (Kelly et al. 1987). However, applying a mass balance for sulfate in the lake indicates that sulfate loss by sediment reactions does not appreciably alter the lake water sulfate concentration. At Crystal Lake, Wisconsin, on the other hand, in situ measurements indicate that the loss rate to sediment is nearly the same as in Dart's Lake (i.e., 100 μeq/m²/day), but mass balance calculations indicate that the lake sulfate concentration without loss to sediments would be 140 μeq L^{-1}, whereas the observed concentration is near 75 μeq L^{-1} (Greb et al. 1988). The major difference in the two systems is that the hydraulic residence time at Crystal Lake is about 10 years, whereas at Dart's Lake the mean residence time is less than 1 month. The water does not remain in Dart's long enough for an appreciable loss of solution sulfate to occur.

Recent studies indicate that sulfate reduction can be represented mathematically as a rate-limited reaction that is first order with respect to the sulfate concentration. Table 1.2 shows a sulfate loss rate expression as well as the references upon which the formulation was based. The reaction also appears to be temperature dependent. Arrhenius activation energy analysis indicates that at low temperatures ($\leq$ 20°C), the process is chemical reaction (i.e., energy) limited, but at higher temperatures ($>$ 20°C), the reaction is diffusion limited (Munson and Gherini 1986).

Interaction of Processes

Although all the processes described in the previous sections can have important acid-base consequences, the interaction of the processes determines the ultimate water quality of lakes and streams receiving acidic deposition. These processes are largely coupled through the solution phase, and the action of one process can dictate the degree to which other processes influence solution acid-base characteristics. As indicated in the section on soil processes, cation exchange responds quickly to minimize the changes in soil solution hydrogen ion concentration that may result from changes in acidic deposition. This, in essence, has the effect of suppressing any short-term change in mineral weathering rate in response to the change in deposition. A similar phenomenon occurs in soils where sulfate is adsorbed. If sulfate enters a

TABLE 1.2. Sulfate loss to sediments.[a]

Formulation:

$$-V\frac{d[SO_4^{2-}]}{dt} = K\ A\ [SO_4^{2-}]^{1}$$

(Boxed reference numbers in the formulation: 5,7 and 9 point to K and A; 1,2,8 and 10 point to the exponent 1.)

Possible concerns:
a. Reaction order with respect to sulfate – 1,2,8
b. Limiting reactants (e.g. reductants, CH_2O) – 6,2
c. Reaction reversability; products formed; stability of latter; outward diffusion; (Fe_xS_y; HS^- + H_2S; C-S organics, alkyl sulfate esters) – 3,4
d. Toxicity effects – 2,8
e. Others

[a] Numbers in solid circles refer to below listed references. Complete citations (including journal name and volume) are included in the reference list.

1. Cook, R.B. and D.W. Schindler, 1983. The Biogeochemistry of Sulfur in an Experimentally Acidified Lake.
2. Kelly, C.A. and J.W.M. Rudd, 1984. Epilimnetic Sulfate Reduction and Its Relationship to Lake Acidification.
3. David, M.B. and Myron J. Mitchell, 1985. Sulfur Constituents and Cycling in Waters, Seston, and Sediments of an Oligotrophic Lake.
4. Nriagu, J.O. and Y.K. Soon, 1985. Distribution and Isotopic Composition of Sulfur in Lake Sediments of Northern Ontario.
5. Cook, R.B., C.A. Kelly, D.W. Schindler, and M.A. Turner, 1984. Mechanisms of Hydrogen Ion Neutralization in an Experimentally Acidified Lake.
6. Rudd, J.W.M., C.A. Kelly, V. St. Louis, R.H. Hesslein, A. Furutani, and M. Holoka, 1985. Microbial Consumption of Nitric and Sulfuric Acids in Sediments of Acidified Lakes in Four Regions of the World.
7. Rudd, J.W.M. and C.A. Kelly, 1987. Factors Controlling Accumulation of Organic and Inorganic Sulfur in the Epilimnetic Sediments of Four Adirondack Lakes.
8. Baker, L.A., T.E. Perry, and P.L. Brezonik, 1985. Neutralization of Acid Precipitation in Softwater Lakes.
9. Kelly, C.A., J.W.M. Rudd, R.H. Hesslein, D.W. Schindler, P.J. Dillon, C.T. Driscoll, S.A. Gherini, R.E. Hecky, 1987. Prediction of Biological Acid Neutralization in Lakes.
10. Knauer, D.R., J.G. Bockheim, A.J. Prey, J. Flickinger, J.E. Leide, J.A. Morton, L.C. Sheng, E.A. Jepsen, J.M. Esser, D.A. Wentz, W.A. Rose, J.T. Krohelski, P.J. Garrison, S.A. Gherini, and C. Chen, 1985. Wisconsin RILWAS.

sorbing soil associated with hydrogen ions, the sulfate is removed from solution and the hydrogen ions can be considered to combine with the desorbed hydroxyl ions to form water (see Figure 1.15). This removes hydrogen ions from solution that may have driven cation exchange or mineral weathering reactions.

The interaction of processes can also be seen by examining the effects of manipulations in watersheds. One such manipulation is a clear-cut of all of the trees in a system. Once the trees have been removed, the following series of events may take place:

- Total deposition may be reduced because enhanced collection of dry deposition by the forest canopy no longer occurs.
- Incoming precipitation is not exposed to the alkalizing effects of foliar exudation and leaching.
- The ground surface is warmed because the shading from the trees is no longer present. The warmer soil temperatures lead to accelerated microbial activity that results in increased production of nitrate and thus lower soil solution pH and ANC levels.
- The increased hydrogen ion concentration in soil solution leads to cation exchange or weathering reactions that produce elevated alkali cation concentrations in solution.
- To the extent that hydrogen ion is not consumed by exchange or weathering, it may lead to dissolution of aluminum hydroxide and elevated monomeric aluminum concentrations in solution.

TABLE 1.3. Dependencies of selected ANC producing reactions.

Process	Type of reaction	Hydrogen ion dependence [H]$^{\alpha}$	Other dependencies
Cation exchange	Equilibrium	Heterovalent = 2.0 Homovalent = 1.0	$[Ca^{2+}]$, $[Mg^{2+}]$, $[K^{+}]$, $[Na^{+}]$, $[NH_4^{+}]$, $([Al^{n+}])$
Weathering	Rate limited	0.2–0.8	
SO_4 absorption	Equilibrium	~0	$\frac{k[SO_4^{2-}]}{k+[SO_4^{2-}]}$
SO_4 reduction	Rate limited	~0	$[SO_4^{2-}]$
Algal uptake	Rate limited	~0	$[NO_3^{-}]$
Tree growth	Rate limited	(Through CEC)	$[NO_3^{-}]$, $[NH_4^{+}]$ $[Ca^{2+}]$, $[Mg^{2+}]$ $[K^{+}]$, $[Na^{+}]$ $[H_nPO_4^{n-3}]$

The net effect is that, in the short term, surface water concentrations of nitrate and to a lesser extent alkali cations and aluminum, are elevated. This results in lower ANC and pH. As vegetation is re-established within the watershed, the nitrate is removed from soil solution to support growth and ANC and pH rise (Lawrence et al. 1988).

These examples illustrate that the processes occurring in natural systems are interlinked. For this reason, it is important to consider all of the processes described in this chapter and their various functional dependencies (see examples in Table 1.3) when making an assessment of the acid-base status of surface waters and when predicting any future response to changes in acidic deposition. The following chapter describes several methods of making such assessments. It also discusses complex biogeochemical models that have been developed to account for the interaction of processes by considering their observed functional interdependencies.

2
Hydrochemical Assessment Methods for Analyzing the Effects of Acidic Deposition on Surface Waters

Ronald K. Munson and Steven A. Gherini

ABSTRACT. In analyzing the effects of acidic deposition on surface waters, a variety of assessment methods can be used. They include methods that illustrate the distribution of physical and chemical characteristics of surface waters across a geographic region and those that lend themselves to the analysis of individual basins. Within these two general areas, a variety of specific analyses can be performed.

The distribution of characteristics across a geographic region can be analyzed in many ways. One approach is to separate the information into the components making up the strong acid-base system and those comprising the weak acid-base system. The data can be presented in the form of histograms, spline diagrams, and frequency distributions.

The analysis of individual systems can provide a great deal of information regarding the processes influencing the acid-base characteristics of surface waters and groundwaters and the mechanisms by which those processes operate. Several different methods can be applied to extract this type of information, including the analysis of evapotranspiration and mass balance (ET/MB) within a system. This approach identifies various constituents that are produced or consumed within the basin, factoring in the concentrating effects of evapotranspiration. In conjunction with an ET/MB approach, an analysis of the dynamics of individual base cations and strong acid anions (C_B/C_A analysis) can help point out interactions among ions, which provides a means of process identification. Buffer system analysis can help to quantify the contributions of weak acids and bases to a system's observed water quality. A rigorous assessment that combines all these analyses involves the application of dynamic models that take into account the interactions of the major processes supplying acid and base to the water flowing through a basin. All these assessment methods are discussed in detail and specific examples of their application are presented.

Introduction

Many different approaches can be used to assess the influence of acidic deposition on surfacc water systems. The particular approach used in any single instance can be a function of the type of data available and the questions being addressed. For example, survey data lends itself to the analysis of the distribution of characteristics, and can be used to develop a general understanding of the status of surface waters throughout a particular region (Linthurst et al. 1986). Data from intensive studies on a few lakes or streams are useful in the analysis of individual systems and associated processes (e.g., Gherini et al. 1985, Munson et al. 1987, Rascher et al. 1987, Bricker and Rice 1989). The knowledge gained from these studies can then be used in the analysis of other systems.

This chapter describes hydrochemical assessment methods that can be applied to multiple or individual systems. In general, the methods are presented in order of complexity, starting with simple analyses and moving toward methods that require the use of advanced analysis tools, such as dynamic models. The first methods presented involve analyzing the distributions of components of the strong and weak acid systems. These are followed by descriptions of individual system analyses, looking at evapotranspiration and mass balance (ET/MB), analysis of strong acid-base system components (C_B/C_A analysis), ion ratio analyses, and analysis of the influence of buffer systems on aqueous phase pH (buffer system analysis), using graphical methods and solution equilibrium models. Examples are presented depicting the use of these methods on an individual basis,

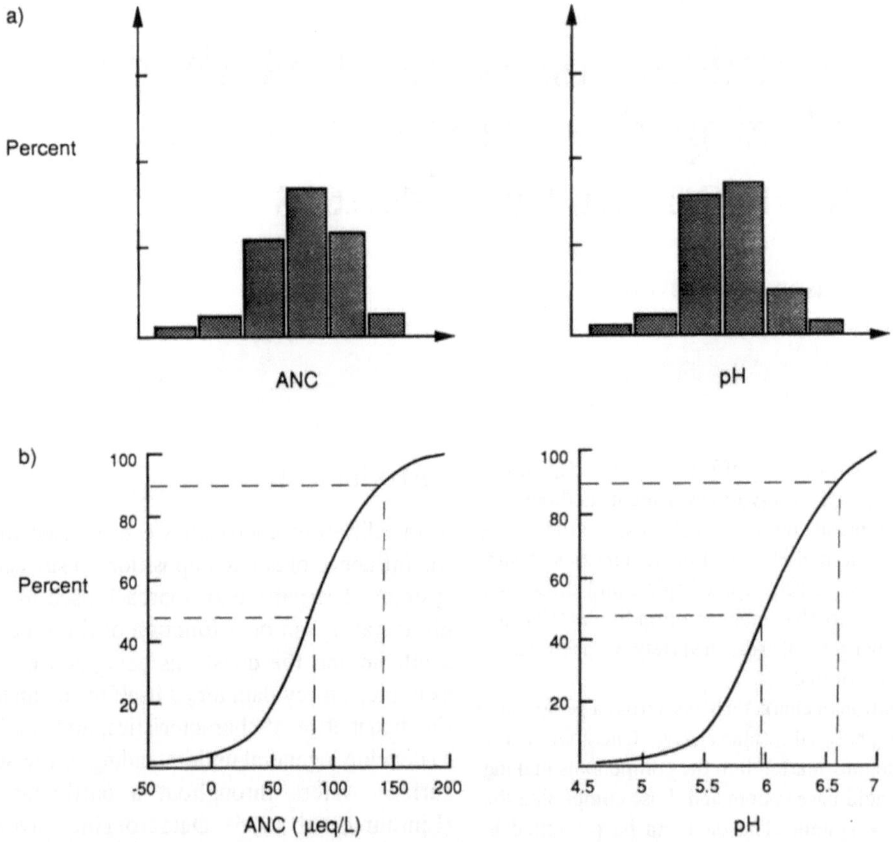

FIGURE 2.1. This figure presents (a) histograms and (b) cumulative frequency distributions of ANC and pH for lakes from a hypothetical region. The dashed lines on the cumulative frequency diagrams indicate that in this hypothetical region, 50% of the lakes have ANC < 75 µeq L^{-1} and pH < 6.0 and 90% of the lakes have ANC < 150 µeq L^{-1} and pH < 6.5.

and an example showing how these methods can be combined and integrated is also given. Finally, a brief description of three commonly used dynamic watershed models (ILWAS, MAGIC, and ETD) is presented. Many of the assessment methods described in this chapter are used extensively in the individual case study chapters, whereas others are used less frequently. However, all the methods described can provide important information for the evaluation of the acid-base status of surface waters.

Distributions of Characteristics

The analysis of data from a large number of lakes or streams can provide information regarding the distributions of various acid-base characteristics of surface waters throughout a region. The analysis can be further refined by analyzing the solutes making up both the strong and weak acid-base systems.

Strong Acid-Base System

A regional analysis of the components of the strong acid-base system can be performed using histograms and cumulative frequency distributions, as was done extensively in the U.S. EPA National Surface Water Survey reports (e.g., Linthurst et al. 1986, Landers et al. 1987, Kaufmann et al. 1988). These graphical tools provide a simple means of displaying the distribution of water quality parameters. They give a visual indication of the number of survey lakes or streams in a region that fall into a given pH or ANC range relative to the total resources surveyed. For example, Figure 2.1 shows the distributions of pH and ANC for lakes surveyed in a hypothetical region. Data for other constituents and basin characteristics can also be depicted in this manner. In the case study chapters in this book, a common set of histograms is presented showing distributions of lake area, lake depth, watershed area to lake area ratio, pH, ANC, SBC, SO_4^{2-}, DOC, and SiO_2. In areas dominated by streams, the first three parameters are replaced by watershed area, elevation, and gradient.

Spline diagrams provide an effective means of evaluating the relative importance of various constituents that contribute to ANC. In the example diagrams shown in Figure 2.2a, the total base cation concentrations increase with increasing ANC. This is a result of the increasing concentration of Ca^{2+} and, to a lesser extent, increases in Mg^{2+} and Na^+ concentrations. The figure indicates that as ANC increases, the contribution of monomeric Al decreases, due to its lower solubility at higher pH, which increases with increasing ANC. Figure 2.2b is equivalent to a plot of the uppermost and lowermost lines of Figure 2.2a plotted on the same graph. This shows how the sum of base cations (SBC) and strong acid anions vary in relation to one another and cause the observed differences in ANC. For example, lake ANC values increase due mostly to increasing levels of SBC. This same type of figure can be used with other constituents, such as hydrogen ion, to show their importance in a particular region.

Finally, Figure 2.3 shows variations in the concentrations of acid-base constituents thought to be particularly important in determining the toxicity of low pH waters to fish (see Chapter 4, Baker and Christensen, this volume). Both inorganic monomeric Al and free H^+ are thought to contribute to fish toxicity. Calcium is thought to have a mitigating effect.

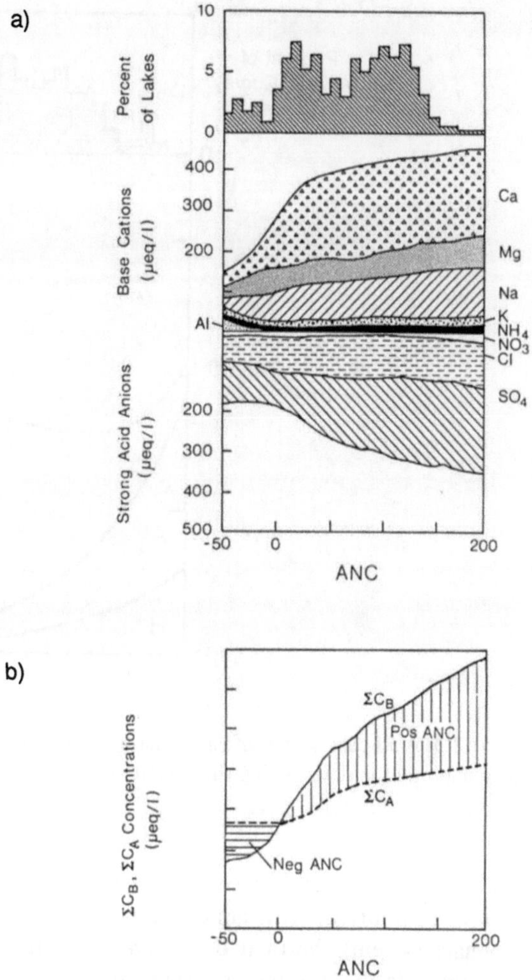

FIGURE 2.2. Strong acid-base system analysis: (a) strong acid-base system components and (b) determinants of ANC. ANC increases are due to increases in ΣC_B and/or decreases in ΣC_A. Decreases in ANC are due to decreases in ΣC_B and/or increases in ΣC_A (graphical format similar to that used by Brakke et al. 1988).

Weak Acid-Base System

As with the strong acid-base system, the analysis of the weak acid-base system throughout a region can involve the use of histograms and other frequency distribution diagrams. In addition, the comparison of regional field data to theoretically expected values based on observed ANC can provide estimates of the extent to which weak acids and bases contribute to observed water quality. This in turn provides an indication of the processes operating in a region that influence surface water pH.

Figure 2.4a shows field pH versus ANC for 23 lakes in the Adirondacks that were sampled approximately monthly over a 2-year period as part of the RILWAS project (Gherini et al. 1989). The solid line on the plot is the theoretical curve for pH

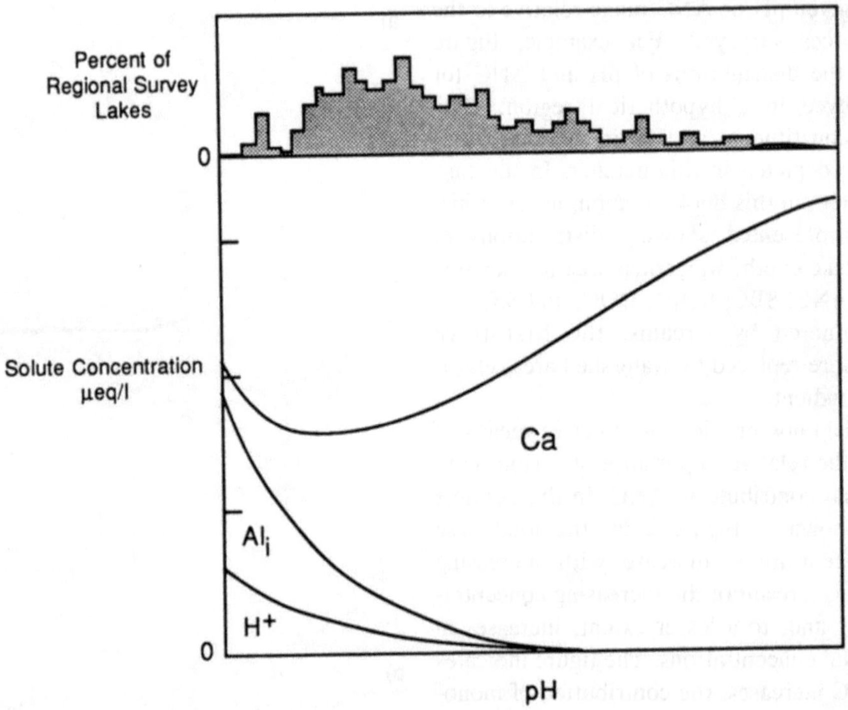

FIGURE 2.3. Acid-base solutes of particular biological interest. Distribution of acid-base solutes thought to be particularly important in determining the toxicity of low pH waters to fish.

versus ANC for a water body with strong acids and bases in equilibrium with atmospheric CO_2 (the carbonate system). Vertical deviations from the curve indicate the presence of weak acids in the system. The weak acids include aqueous CO_2, organic substances, and aluminum species, among others. In Figure 2.4b, the effect of these weak acids on pH (field pH minus theoretical air equilibrated pH) is displayed as a function of ANC. Naturally occurring weak acids have a major effect on pH at ANC values near zero, because at these low values the solutions are poorly buffered. As ANC becomes negative, the solution is dominated by strong acids, and free H^+ concentrations are too high for the weak acids to donate a proton. Thus, their effect on pH is not as great in negative ANC solutions.

The weak acid effect can be further defined by delineating the effect that results from volatile versus nonvolatile acids. Volatile weak acids consist largely of CO_2. Their effect on pH can be calculated as the measured field pH minus measured air-equilibrated pH. Nonvolatile weak acids include organic substances and Al species. Their effect can be calculated as measured air-equilibrated pH minus theoretical air-equilibrated pH. The effects of volatile and nonvolatile weak acids on the RILWAS lakes and streams are depicted in Figures 2.5a and 2.5b, respectively.

Frequency distributions such as those shown in Figures 2.6a, 2.6b, and 2.6c can be used to show concentration distributions for the three major weak acid systems. These types of plots are convenient for showing comparisons among regions, as indicated for two hypothetical regions depicted in Figure 2.6. Figure 2.6a indicates the degree of CO_2 saturation. Values for the ratio $P_{CO_2}/P_{CO_2} \cdot$ Atm greater than 1.0 indicate waters that are supersaturated relative to atmospheric P_{CO_2} levels; values less than 1.0 indicate undersaturation. As shown in Figure 2.6a, hypothetical Region I has waters with P_{CO_2} levels generally greater than those for hypothetical Region II. Figure 2.6b shows that hypothetical Region I has higher DOC levels and Figure 2.6c

FIGURE 2.4. Weak acid-base system analysis: (a) field pH versus ANC; vertical deviations from curve indicate depressions or elevations in pH due to weak acids and bases respectively; curve is for strong acid-base system in equilibrium with atmospheric CO_2; (b) change in pH due to weak acids (measured field pH minus theoretical air equilibrated pH). Data from the RILWAS investigation (Gherini et al. 1989).

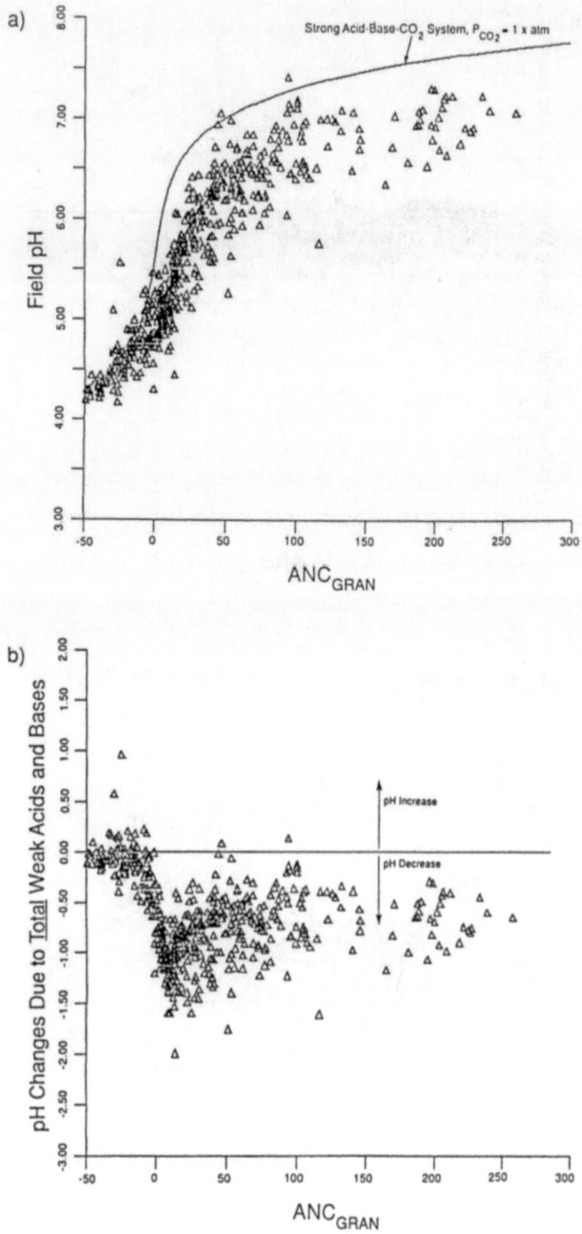

indicates that the total monomeric Al concentrations for Region I are lower than those for Region II.

Individual Systems

A great deal of information regarding the processes influencing the acid-base characteristics of surface waters and groundwaters and the mechanisms by which these processes occur can be gained through the analysis of individual systems. This is not to say that survey data cannot be used in this type of analysis. The survey data must simply be analyzed on a lake-by-lake basis, and then distributions of the results of individual system analyses can be studied.

Several different analysis methods can be applied on an individual basis. They include evapotranspiration calculations and solute mass balances (ET/MB)

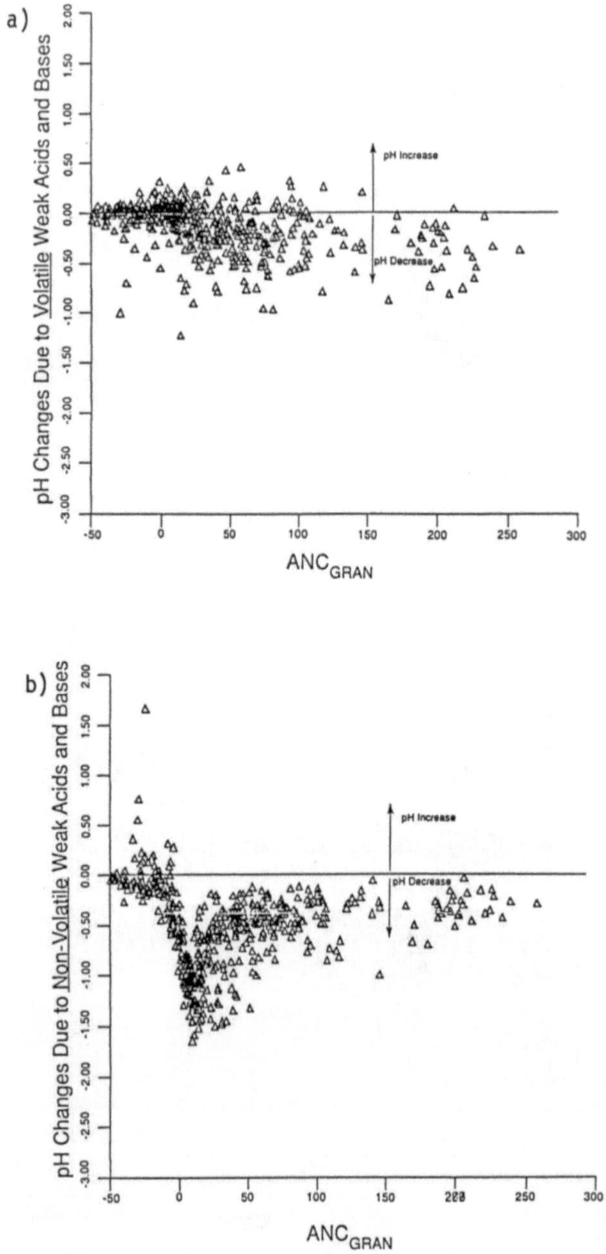

FIGURE 2.5. This figure presents (a) change in pH due to volatile weak acids (mostly CO_2; measured field pH minus measured air-equilibrated pH), and (b) change in pH due to nonvolatile weak acids (mostly organics and aluminum; measured air-equilibrated pH minus theoretical air-equilibrated pH). Note that at ANC values between 0 and 100 $\mu eq\ L^{-1}$, pH depressions due to weak acids can be as large as 2 pH units. Data from the RILWAS investigation (Gherini et al. 1989).

that together can be used to identify the production or consumption of solutes within hydrologic basins. In conjunction with an ET/MB approach, an analysis of the dynamics of individual base cations and strong acid anions (C_B/C_A analysis) can help elucidate interactions among ions, and provides a means of process identification. Buffer system analysis can help to quantify the contributions of strong and weak acids and bases to a water body's observed pH. The application of dynamic models combines all the above analyses. The dynamic models also take into account the interactions of the major processes supplying acid and base to the water flowing through a basin. These assessment methods are described in greater detail in the following sections.

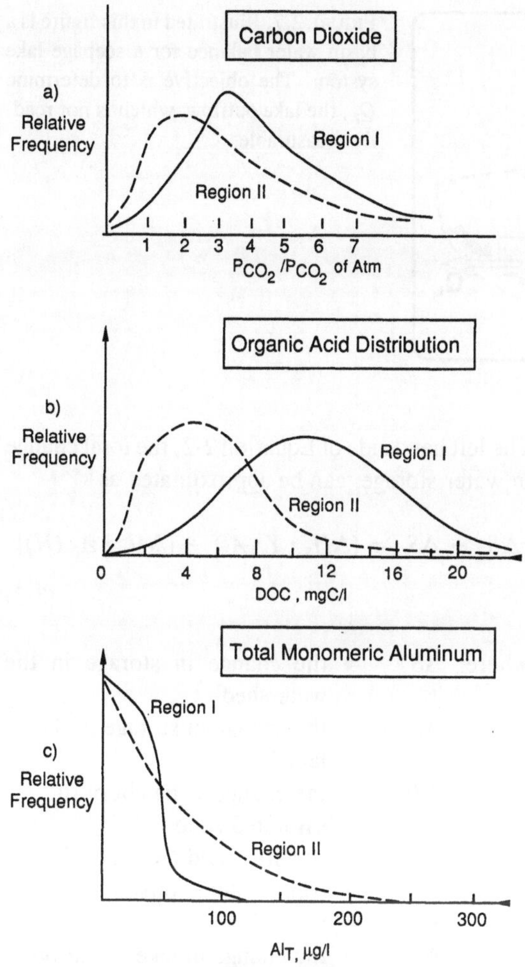

FIGURE 2.6. Hypothetical distribution of weak acid system components: (a) carbon dioxide distribution, as indicated by partial pressure of CO_2 in solution relative to that in the atmosphere, (b) organic acid distribution, as indicated by dissolved organic carbon (DOC, mg L^{-1}), and (c) total monomeric aluminum distribution (µg L^{-1}).

Evapotranspiration/Mass Balance Approach

This section describes the basic theory of the ET/MB approach and provides an example of how it can be applied. Essentially the same type of analysis is used in each case study chapter to evaluate the importance of processes for single lakes or groups of lakes. The calculations are referred to as "ion enrichment analysis" in other parts of the book and are briefly described in the following paragraphs. The specific procedures and equations used are described in Appendix A.

The first step in applying the ET/MB approach is to estimate evapotranspiration. This can be done by making hydrologic budget calculations or by applying regional estimates to the specific system of interest.

Estimating Evapotranspiration

Calculated hydrologic budgets are best suited for lake and stream drainage basin systems. The following water balance equation is used:

$$\Delta S = P - Q - E \tag{2-1}$$

where: ΔS = change in storage over time
P = precipitation amount
Q = basin outflow
E = evapotranspiration

The change in storage term is particularly important when calculating short-term (e.g., monthly) budgets. By using a longer period of record (e.g., several hydrologic years) and by carefully selecting the beginning and ending dates for the hydrologic year, the change in storage term can be minimized relative to the other terms in the expression. This allows its value to be approximated as zero, and evapotranspiration can be calculated as the difference between precipitation and basin outflow.

For example, the evapotranspiration for Woods Lake basin, a small drainage basin lake located in the Adirondack Mountains of New York, would be calculated using the following information (Peters and Murdoch 1985):

- Precipitation summed over two years = 246 cm (October 1, 1979–September 30, 1981)
- Runoff (basin outflow) summed over two years = 152 cm
- $E = \dfrac{P - Q}{2\ yrs}$
- E = 47 cm/yr (38% of precipitation)

This value coincides well with regional estimates.

Water Budget Calculations

Water budget calculations, together with solute mass balance calculations, can be used to identify potential sources and sinks for various acid-base constituents within a basin. The first step in making a

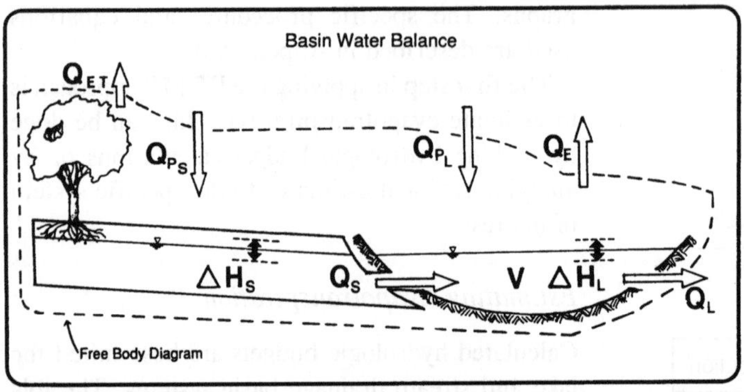

FIGURE 2.7. Illustrated in this figure is a basin water balance for a seepage lake system. The objective is to determine Q_L, the lake outflow, which is not readily measurable.

water budget calculation involves establishing the system boundary (free body diagram). This boundary ideally should be selected so that only one unknown flow or flux crosses it at a time. This unknown can then be determined using the known flows or fluxes and budget or solute balance equations. Changes in concentration that cannot be accounted for by considering inputs, outputs, and changes in storage are attributed to sources and sinks (chemical reactions), and the mechanisms underlying these sources and sinks can then be investigated further.

The method used in establishing boundaries for budget calculations also depends on one's objectives and on data availability. For example, if the objective is to make an overall water balance, the system boundary might be drawn to encompass the entire basin as indicated in Figure 2.7. The overall water balance equation is

$$\Delta S = I - O \tag{2-2}$$

where: ΔS = change in water volume
I = inputs
O = outputs

The left hand side of Equation 2-2, the total change in water storage, can be approximated as

$$\Delta S_S + \Delta S_L = (\Delta H_S \cdot Y \cdot A_S) + [\Delta H_L \cdot A_L (H)] \tag{2-3}$$

where: ΔS_S = the change in storage in the watershed
ΔS_L = the change in storage in the lake
ΔH_S = the change in height of the saturated zone
Y = specific yield (volume of water/volume of soil)
A_S = watershed area
ΔH_L = the change in lake elevation
$A_L(H)$ = lake area (which is a function of lake stage)

The right-hand side, inputs minus outputs, becomes

$$I - O = Q_{P_S} + Q_{P_L} - Q_{ET} - Q_E - Q_L \tag{2-4}$$

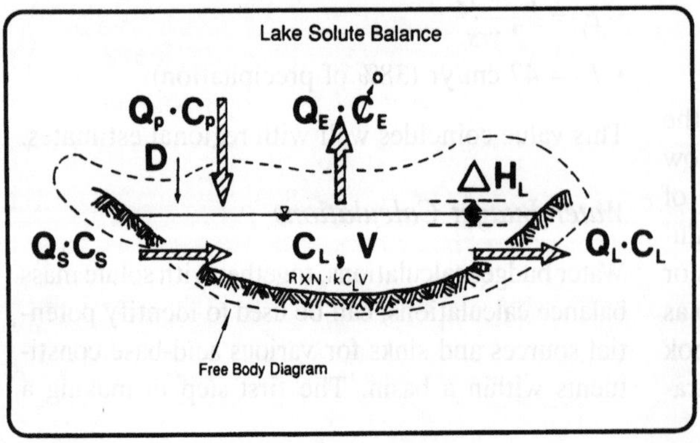

FIGURE 2.8. Illustrated in this figure is a solute balance for a seepage lake. The unknowns are inflow from groundwater, Q_S, and the lake outflow, Q_L. Both these flows are very difficult to measure for a seepage system.

where: Q_{P_S} = precipitation onto the watershed (volume/time)
Q_{P_L} = precipitation onto the lake surface (volume/time)
Q_{ET} = volume of water evapotranspired from the watershed
Q_E = volume of water evaporated from the lake
Q_L = lake outflow volume

Substitution of Equations 2-3 and 2-4 into Equation 2-2 yields the basin water budget equation:

$$\Delta S_S + \Delta S_L = Q_{P_S} + Q_{P_L} - Q_{ET} - Q_E - Q_L \qquad (2\text{-}5)$$

Solute Balance Calculations

If, on the other hand, the objective was to calculate the groundwater inflow to and outflow from a seepage lake, a dynamic solute balance could be used to provide the necessary information. The system boundary would be drawn around the lake itself, as shown in Figure 2.8. The water budget equation is again Equation 2-5, but the change in storage is simply the change in lake volume (ΔV) which is

$$\Delta V = \Delta H_L \cdot A_L(H) = Q_S + Q_{P_L} - Q_E - Q_L \qquad (2\text{-}6)$$

where Q_S is the volume of water entering the lake as groundwater and all other terms are the same as indicated previously. The solute balance can be defined as

$$\frac{d(C_L V)}{dt} = Q_S C_S + Q_{P_L} C_P + D - Q_E C_E - Q_L C_L - kC_L V \qquad (2\text{-}7)$$

where: $C_L V$ = total mass of solute in the lake
C_S = the solute concentration in water entering the lake as groundwater
C_P = solute concentration in precipitation
D = dry deposition input to the lake
C_E = solute concentration in the water evaporating from the lake ($C_E = 0$)
C_L = solute concentration in the lake
k = first order reaction rate constant for removal of solute from lake water
V = lake volume

TABLE 2.1. Physical and chemical characteristics of Crystal Lake region.

Lake characteristics[a]			
Surface area	= 36 ha	ANC	= 15–25 μeq L^{-1}
Volume	= 3.7 × 10^6 m^3	pH	= 5.9
Average depth	= 10.5 m	Ca^{2+}	= 55 μeq L^{-1}
Annual precipitation	= 95 cm	NH_4^+	= 2 μeq L^{-1}
Hydraulic detention time	= 10.5 years	NO_3^-	= 0.5 μeq L^{-1}
Chlorophyll α	= 1 mg L^{-1}	SO_4^{2-}	= 75 μeq L^{-1}

Input characteristics

Solute	Precipitation concentrations[b] (μeq L^{-1})	Ground water concentrations[a] (μeq L^{-1})	Dry deposition[c] (eq ha^{-1} yr^{-1})
Ca^{2+}	13	210	43
Mg^{2+}	4	140	16
Na^+	2.5	90	15
K^+	0.8	25	8
NH_4^+	24	1.2	31
SO_4^{2-}	50	110	69
NO_3^-	20	1.5	37
Cl^-	4	20	5

[a] From Greb et al. 1988.
[b] From NADP, Trout Lake Station.
[c] From Knauer et al. 1985.

In the absence of reaction, Equation 2-7 can be approximated as

$$C_L \Delta V = C_L \cdot \Delta H_L \cdot A_L(H) = Q_S C_S + Q_P C_{P_L} + D - Q_L C_L \qquad (2\text{-}8)$$

An example of using solute balances and water budgets together to determine groundwater inflow and outflow has been done for Crystal Lake, an oligotrophic seepage lake located in northcentral Wisconsin. Table 2.1 summarizes important characteristics for Crystal Lake, including typical ion concentrations in precipitation and groundwater as well as typical values for dry deposition to the lake. If the change in volume for the lake is assumed to be negligible, the water budget (Equation 2-5) can be approximated as follows:

$$Q_S + Q_{P_L} = Q_E + Q_L \qquad (2\text{-}9)$$

The solute balance can be written as shown here, assuming no change in storage and no reaction:

$$Q_S C_S + Q_{P_L} C_P + D = Q_L C_L \qquad (2\text{-}10)$$

Crystal Lake Water and Solute Budgets

Given that

$$Q_E = 0.6\ Q_{P_L} \qquad \text{(Greb } \underline{\text{et al}}\text{., 1988),}$$

Equation 9 is solved for Q_L and substituted into Equation 10 which is then solved for Q_s as follows,

$$Q_S = \frac{Q_{P_L}(0.4\ C_L - C_P) - D}{C_S - C_L}$$

Using calcium as a conservative ion in the lake, and taking the following values from Table 1,

$$C_L = 55\ \mu eq / l$$

$$C_P = 13\ \mu eq / l$$

$$C_S = 210\ \mu eq / l$$

$$Q_{P_L} = 95\ cm (36\ ha)\left(\frac{10{,}000\ m^2}{ha}\right)\left(\frac{m}{100\ cm}\right)\left(\frac{1000 l}{m^3}\right) = 342 \bullet 10^6\ l / yr$$

$$D = 43 \frac{eq}{ha} (36\ ha)\left(\frac{10^6\ \mu eq}{eq}\right) = 1548 \bullet 10^6\ \mu eq$$

Then,

$$Q_S = 10 \bullet 10^6\ l / yr, \text{ or}$$

$$Q_S = 3 \text{ percent of total inflow}$$

FIGURE 2.9. Calculating outflow from seepage lake using water budget and solute balance.

Here we have two equations but three unknowns: Q_S, Q_L and Q_E. However, the value of Q_E can be estimated from regional data. In the vicinity of Crystal Lake, lake evaporation is approximately 60% of annual precipitation (i.e., $Q_E = 0.6\ Q_P$; Greb et al. 1988). Given this information, Equation 2-9 can be solved for Q_L and the result substituted into Equation 2-10, which is then solved for Q_S. A sample calculation is shown in Figure 2.9 for Crystal Lake. This calculation can be repeated using the concentrations of several other constituents (Mg^{2+}, K^+, Na^+, Cl^-) thought to be conserved in the lake. The resulting calculated values for groundwater inflow as a fraction of total inflow cluster around 3%. If the value of 3% groundwater inflow is used to calculate solute budgets, the expected concentrations (in the absence of in-lake sources and sinks) are those listed in Table 2.2.

TABLE 2.2. Predicted and observed solute concentrations for Crystal Lake.

Solute	Lake solute concentrations: Mass balance concentration w/o in-lake rxn (μeq L^{-1})	Observed concentration (μeq L^{-1})
Ca^{2+}	56	55
Mg^{2+}	23	23
K^{+}	6	8
Na^{+}	16	14
NH_4^{+}	63	2
SO_4^{2-}	140	75
NO_3^{-}	36	0.5
Cl^{-}	12	12

As indicated in the table, the expected concentrations are similar to those observed for Ca^{2+}, Mg^{2+}, K^+, Na^+, and Cl^-. The mass balance based concentrations of NH_4^+, NO_3^-, and SO_4^{2-} diverge widely from the measured values. This indicates that in-lake processes play a major role in determining the concentrations of these ions. Ammonium and NO_3^- are both consumed by plants and algae in the lake. Sulfate and NO_3^- concentrations are decreased by redox processes occurring in lake sediments.

Ion Enrichment Analysis

The results of using a simple ET/MB approach, referred to as ion enrichment analysis in this book, can be depicted graphically as indicated in Figure 2.10. This figure represents the contributions of various acid/base constituents to observed ANC, taking into account the concentrating effects of evapotranspiration. Estimates of evapotranspiration are based solely on Cl^- concentrations in lakes and atmospheric deposition in this analysis (this method of estimating evapotranspiration is discussed further in the following section). Acidic deposition provides an input of negative ANC. To the extent that base cation concentrations are higher in the surface water than in deposition, again accounting for the effects of evapotranspiration, they constitute a positive contribution to ANC. If the SO_4^{2-} concentration in the lake differs from the evapotranspiration corrected deposition concentration, this contribution is also noted in the figure. For example, if SO_4^{2-} is being adsorbed or reduced in the basin, its surface water concentration will be lower than expected and will result in a positive contribution to ANC. If there is a SO_4^{2-} source within the basin (e.g., pyritic deposits), the lake concentration will be higher than expected and will lead to a negative contribution to ANC in the figure. The removal of nitrogen species results in a positive contribution to ANC if NO_3^- is removed in excess of NH_4^+. Finally, the observed lake ANC is compared to the ANC that would result from the calculated contributions of the individual constituents (i.e., $C_B - C_A$). Ion enrichment analysis has been used in this book to analyze acid-base reactions for various groups of lakes throughout the case study regions. In graphical presentations, illustrated by Figure 2.10, regional medians are represented by bars and quartile ranges are shown with shading. Details of ion enrichment calculations as used in this book, including subsetting of lakes and data inputs, are presented in Appendix A.

C_B/C_A Analysis

An analysis of the behavior of the individual base cations and strong acid anions can provide a great deal of information regarding the acid-base processes occurring in a hydrologic basin. This type of analysis yields much more information than an analysis of ANC alone. The processes that influence ANC do so by changing the concentration of one or more of its components (i.e., one or more of the base cations or strong acid anions). A description of the behavior of each of the major strong acid anions and base cations is presented in the following sections, along with examples of the use of C_B/C_A analysis.

Strong Acid Anions

The strong acid anions that influence ANC to the greatest extent are SO_4^{2-}, NO_3^-, and Cl^-. An analysis of each individual ion can provide insight into the processes occurring within a basin.

Chloride can be released from watershed soils via the weathering of minerals (e.g., some amphiboles), but the major source in low ANC basins is usually atmospheric deposition. Aside from weathering reactions, very few processes have a significant impact on Cl^- concentration except evapotranspiration. Thus, in systems that do not contain significant amounts of chloride-bearing

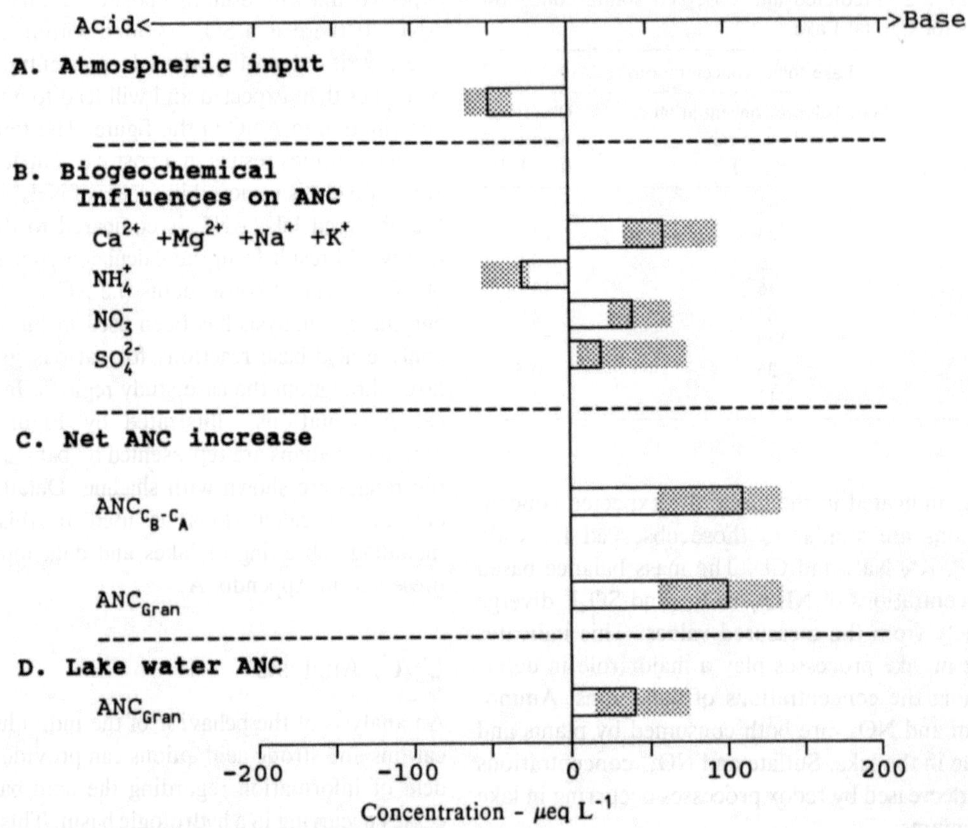

FIGURE 2.10. Modification of atmospheric deposition by watershed and lake biogeochemical processes (basis 1 liter of lake water). A. Precipitation [H⁺] adjusted for estimated dry deposition and evapotranspiration (ET); equivalent to expected lake water $ANC_{C_B\text{-}C_A}$ in the absence of any process other than ET. B. Reactions influencing ANC; bars to the right indicate ANC productions; those to the left show ANC consumption. C. Net ANC production; $ANC_{C_B\text{-}C_A}$ is the sum of ANC production terms for measured anions and cations including those shown in B; ANC_{Gran} is determined by difference between input ANC (A.) and lake water ANC (D.). D. Lake water ANC represents the net reaction of atmospheric inputs with the lake watershed system; shaded bars represent 25% and 75% quartiles.

minerals and are not impacted by road salting, comparing observed lake or stream Cl^- concentrations with Cl^- in deposition provides an estimate of evapotranspiration. This is how evapotranspiration is calculated in the ion enrichment analyses. A point to consider, however, is that the accuracy of the analytical determination of Cl^- is less than desirable; a relatively high degree of scatter occurs between split samples. The inaccuracy is due in part to low precision in ion chromatograph determinations.

Many processes influence surface water NO_3^- concentrations (see Chapter 1, Munson and Gherini, this volume). These include organic matter decay, nitrification, and redox processes. Nitrate also serves as a nutrient for plants and algae. Thus in many forested lake-watershed systems, the NO_3^- from deposition and that produced by nitrification is largely removed from solution. However, organic matter decay and nitrification processes continue during the winter dormant season, and NO_3^- concentrations can build up under the snowpack. During snowmelt, the NO_3^- built up under the snowpack and that which accumulates in the snowpack as a result of atmospheric deposition can be flushed from the soil into surface waters, where it contributes to the spring pulses of acidity observed in some watersheds. This spring pulse of NO_3^- and acidity was illustrated using data from Panther Lake, New York, in Chapter 1, Figure 1.12 (Munson and Gherini, this volume).

Like NO_3^-, SO_4^{2-} is a product of the decay of organic matter, but the majority of the SO_4^{2-} in surface waters and groundwaters in the systems of interest comes from atmospheric deposition. One process that removes SO_4^{2-} from solution is sorption on watershed soils. This is generally important in watersheds with old (not recently glaciated), highly weathered soils. These soils typically have high concentrations of iron and aluminum oxides and hydroxides that serve as adsorption sites for the SO_4^{2-}. An important in-lake process that can influence SO_4^{2-} concentration is reduction in lake sediments. This process is relatively slow and is usually important only in longer residence time systems (i.e., hydraulic residence times longer than about 3 years). However, under certain circumstances it can also impact short residence time systems, as is illustrated by data from Pancake-Hall Creek, New York, in a later section.

Base Cations

The base cations generally most important in surface water acidification studies include Ca^{2+}, Mg^{2+}, Na^+, K^+, NH_4^+, and Al. The first three (Ca^{2+}, Mg^{2+}, and Na^+) are all influenced by similar processes, but subtle differences can help elucidate the relative importance of these processes. Potassium behavior is highly influenced by its role as a plant nutrient. Ammonium behavior is closely linked with organic matter decay, and Al is normally important only in low pH waters.

Calcium, Mg^{2+}, Na^+, and K^+ all undergo cation exchange reactions and are produced by mineral weathering. Both processes can supply base cations to solution, but determining which is the major source in a particular basin can have important implications regarding future acidification. As indicated in Chapter 1, the potential supply from mineral weathering is very large, but the process is relatively slow and unresponsive to changes in H^+ concentration. Cation exchange is a fast process that responds rapidly to offset changes in H^+ concentration. However, the supply of base cations from the exchange complex is of shorter duration (e.g., hundreds of years) than that from weathering, and thus a system that receives most of its base from cation exchange rather than weathering might be more susceptible to long-term acidification. One method of estimating the relative supply of base from weathering versus cation exchange is to examine SiO_2 and nonsea salt Na^+ concentrations. Both are products of mineral weathering. Sodium exchange is sluggish compared to the other exchangeable ions, and SiO_2 does not exchange at all. Thus, high Na^+ and SiO_2 concentrations in surface waters suggest that much of the base cation supply is coming from weathering. If the concentrations of these two constituents are low, the majority of base cations supplied may be coming from exchange.

One complicating factor in using this approach is that SiO_2 is taken up in lakes by diatoms and chrysophytes. Thus, the SiO_2 observed in some lakes may not represent the amount actually produced by weathering reactions. This becomes particularly important in lakes with low SiO_2 concentrations, where the amount of uptake may represent a significant fraction of the total available. In lakes with higher SiO_2, the uptake does not make as significant an impact on the total observed concentration. Therefore, in almost all systems SiO_2 provides an indication of the lower limit of primary mineral weathering taking place within the basin.

Much of the NH_4^+ that enters a basin as deposition or is released upon the decay of organic matter is converted to NO_3^- or taken up directly by plants or algae. As with NO_3^-, in many basins, elevated concentrations of NH_4^+ are observed in surface waters only during spring runoff.

Appreciable Al concentrations are found only in low pH waters that have come into contact with soils. Barnes Lake, New York, for example, has low pH (pH = 4.7), but it has very little aluminum (Al_T concentration = 10 to 50 $\mu g\ L^{-1}$). This is because Barnes is a mounded seepage lake and receives limited groundwater inflow. Woods Lake, New York, on the other hand, also has low pH (pH = 4.7), but it has much higher Al concentrations (Al_T concentration = 200 to 1,000 $\mu g\ L^{-1}$). The Al in Woods Lake comes from dissolution of amorphous aluminum hydroxide and aluminum silicates in the watershed soils. Even in lakes that typically have high pH, elevated Al concentrations can occur

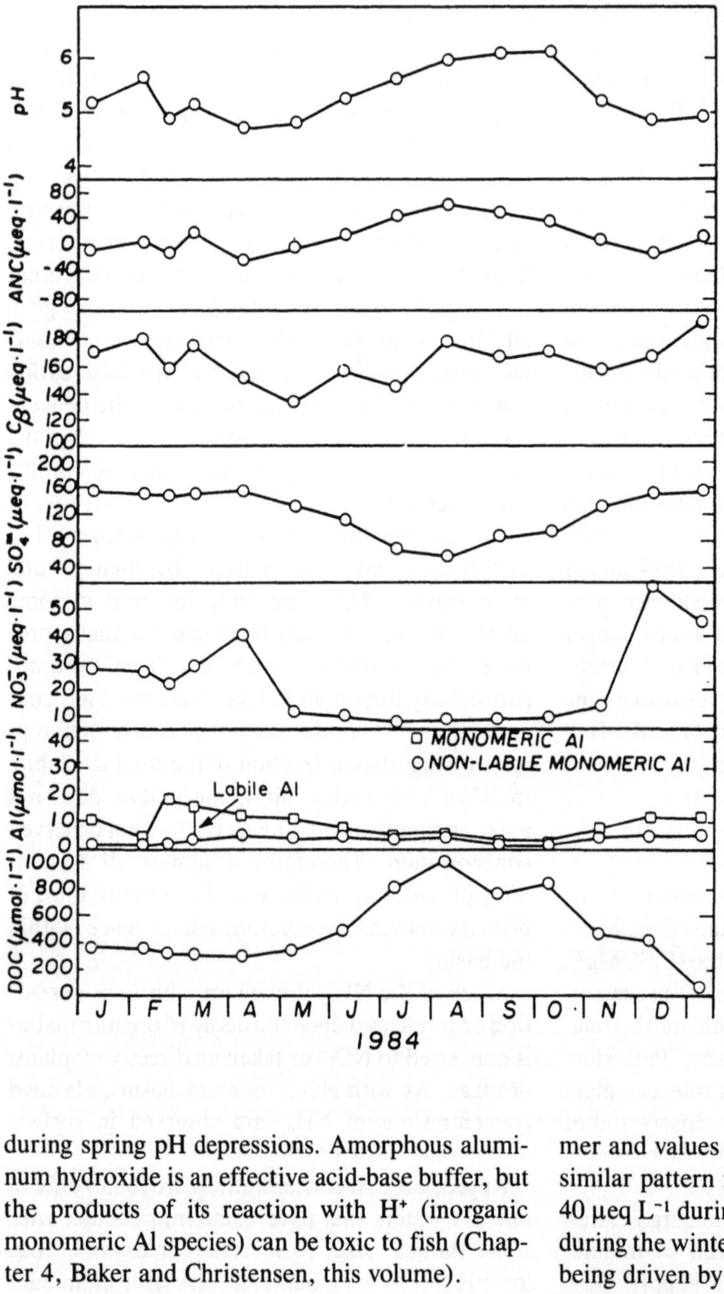

FIGURE 2.11. Concentration versus time traces for Pancake-Hall Creek. Note that low summer sulfate concentration produces increased ANC and increased pH (Driscoll et al. 1987).

during spring pH depressions. Amorphous aluminum hydroxide is an effective acid-base buffer, but the products of its reaction with H^+ (inorganic monomeric Al species) can be toxic to fish (Chapter 4, Baker and Christensen, this volume).

Example of C_B/C_A Analysis

C_B/C_A analysis can be used to evaluate the behavior of the Pancake-Hall Creek system in the Adirondacks. Pancake-Hall Creek is a small stream located in a steep catchment. The acid-base characteristics of the stream water are strongly influenced by a beaver dam that has formed a pond in the system. As indicated in Figure 2.11, the pH shifts seasonally, with values near 6.0 in the summer and values less than 5.0 during the winter. A similar pattern is seen for ANC, with values near 40 μeq L^{-1} during the summer and negative values during the winter. The changes in pH are obviously being driven by changes in ANC, and the changes in ANC can be related to the concentrations of its components.

As indicated in Figure 2.11, the concentration of C_β, which includes Ca^{2+}, Mg^{2+}, Na^+, and K^+, shows no discernible pattern over the year. However, SO_4^{2-} and NO_3^- show definite seasonal patterns with high winter concentrations (150 μeq L^{-1} and 30 μeq L^{-1}, respectively) and low summer concentrations (60 μeq L^{-1} and 5 μeq L^{-1}, respectively). As indicated in Chapter 1, low summer NO_3^- concentrations are expected, due to plant uptake.

FIGURE 2.12. Example ion ratio analysis depicting ANC versus the ratio of sulfate to the sum of the base cation concentrations for a hypothetical data set.

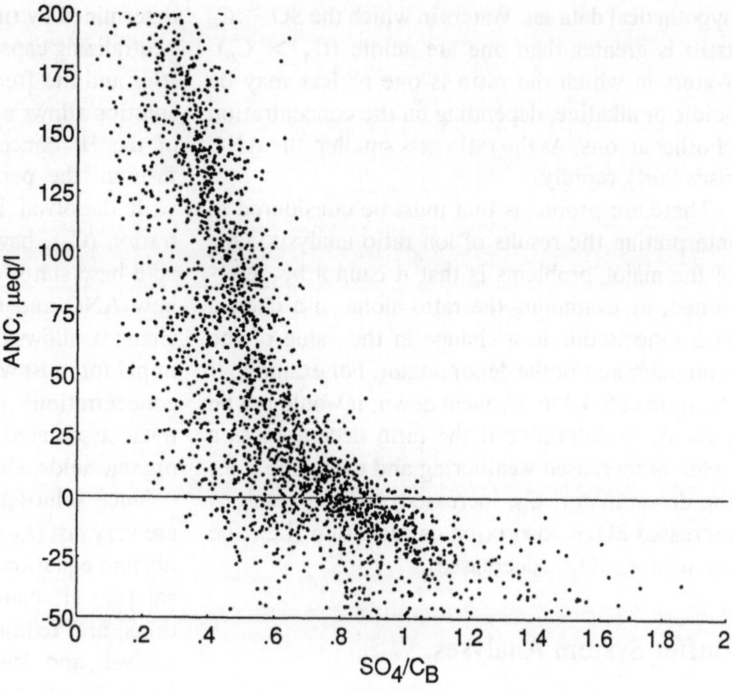

However, the SO_4^{2-} dynamics are not typically observed in this region (Schofield et al. 1985), and SO_4^{2-} appears to be the major factor driving the ANC dynamics in the stream.

Thus, a process that influences SO_4^{2-} concentration plays a major role in determining the acid-base characteristics of Pancake-Hall Creek. The Adirondacks lie within a zone that was glaciated within the past 10,000 to 12,000 years, and thus the soils are not highly weathered, which suggests that SO_4^{2-} adsorption is not a major factor in controlling stream SO_4^{2-} concentration. This indicates that SO_4^{2-} is being lost to reductive processes, most likely within the beaver pond.

Sulfate reduction is generally most important in long residence time systems. The residence time within the beaver pond at Pancake-Hall Creek is short (< 1 year), but reduction still plays a major role in determining summer SO_4^{2-} concentrations, probably because the pond waters are shallow and warm, with a large surface area in relation to depth. In addition, the beaver activity in the area helps to make the waters high in organic content. All these factors combine to produce ideal conditions for reductive processes to occur in the summer months.

As indicated in the bottom plot of Figure 2.11, the DOC concentration rises in the summer, most likely due to the effects of beaver activity. This rise in DOC, and associated organic acids, results in a smaller rise in pH (in response to the change in ANC) than would be expected if only ANC had changed. This DOC also forms complexes with Al that mitigate the Al toxicity to fish.

Ion Ratio Analysis

Another commonly used analysis technique involves examining the ratios of individual ions or sums of ions in a system, a technique that allows the changes in mass through the system to be tracked. For example, if Cl^- is assumed to be conservative (as cited earlier, this assumption may not always be accurate), the changes in the ratio of individual ions to Cl^- may provide an indication of which processes are active within the basin. For instance, if the ratio of SO_4^{2-} to Cl^- were 10:1 in deposition, but 9:1 in lake outlet waters, it would indicate that SO_4^{2-} was being removed from solution, which might result from SO_4^{2-} adsorption onto watershed soils or reduction in lake sediments.

Examination of the distributions of ion ratios across a region is another commonly used analysis technique. One particular ratio that is used in all the case study chapters is ANC versus the ratio of sulfate to base cations ($SO_4^{2-}:C_B$). A plot demonstrating this analysis is shown in Figure 2.12 for a hypothetical data set. Waters in which the $SO_4^{2-}:C_B$ ratio is greater than one are acidic ($C_A > C_B$). Waters in which the ratio is one or less may be acidic or alkaline, depending on the concentration of other anions. As the ratio gets smaller, the ANC rises fairly rapidly.

There are problems that must be considered in interpreting the results of ion ratio analysis. One of the major problems is that it cannot be determined, by examining the ratio alone, if a change in a ratio is due to a change in the value of the numerator and/or the denominator. For example, if the ratio of SO_4^{2-} to C_B went down, it would not be possible to determine if the ratio decreased as a result of increased weathering and exchange (i.e., the denominator, C_B, increased) or as a result of increased SO_4^{2-} adsorption or reduction (i.e., the numerator, SO_4^{2-}, decreased).

Buffer System Analyses

In general, the ANC of precipitation entering a hydrologic basin increases as it flows through the watershed and into streams and lakes. The processes involved and the dynamics of these systems are complex and are best analyzed using simulation models that take into account the concurrent interaction of the processes. In certain cases, natural waters can be analyzed using simpler tools. For example, the acid-base chemistry of water isolated from the solid phase can be analyzed using graphical techniques. A slightly more complex analysis can be used for mounded seepage lakes, which receive essentially all of their water directly from precipitation. In these types of systems, the major processes influencing ANC and pH are anion loss and solute concentration by evaporation. These types of approaches are addressed in the following sections.

Aqueous Phase Buffer Systems

The principal objective of aqueous phase acid-base analysis is to determine H^+ concentration. As mentioned in Chapter 1, the H^+ concentration that results when two equal volumes of water with different H^+ concentrations are mixed is rarely the arithmetic mean of their individual concentrations. To predict H^+ concentration, a measure of its total quantity is necessary. This is commonly determined by titration and is termed base or acid neutralizing capacity. Knowing one of these capacities and the free H^+ concentration (pH) for each solution allows a unique determination to be made of the H^+ concentration of the mixture. In this fashion, the parameters pH and ANC, together with dissolved inorganic carbon (DIC) concentration (C_T), have come to be used to define the acid-base status of most natural waters. Knowing how ANC and C_T change, given other solution changes, allows determination of H^+ concentration or pH for most waters. In very dilute solutions, the concentrations of proton acceptors other than those associated with the carbonate system (e.g., organic acids, aluminum) must also be considered.

Since solution phase proton transfer reactions are very fast ($t_{1/2} \approx 10^{-10}$ sec) (Laidler 1965), equilibrium equations are commonly used to calculate solution H^+ concentrations. Since these calculations are tedious, graphical solution methods evolved, and later the actual equations were solved for each application using computer models. Both methods are extremely useful and are summarized in the following sections.

Graphical Methods

The early work of Deffeyes (1965) and others has produced ANC-C_T-pH diagrams of the type shown in Figure 2.13. Given such graphs, we can determine how the pH of a water body will change upon addition or removal of strong acid or strong base, CO_2, and bicarbonate and carbonate salts.

Given initial ANC and C_T for a water body, or ANC and pH, or C_T and pH, a point is plotted on the diagram. The effects of the chemical perturbations described in the previous section are then determined by simply making vectoral displacements from that point, as shown at the bottom of Figure 2.13.

The original diagram prepared by Deffeyes was designed for waters with a much higher range of ANC than those of concern in lake-watershed acidification investigations. The diagram has been revised for waters with ANC of −50 to 250 μeq L^{-1}. Also, the original diagram was prepared for waters that are buffered by the carbonate system. A modification of the Deffeyes diagram is presented in Figure 2.14 for waters that contain organic acids as indicated here by 4 mg L^{-1} (333

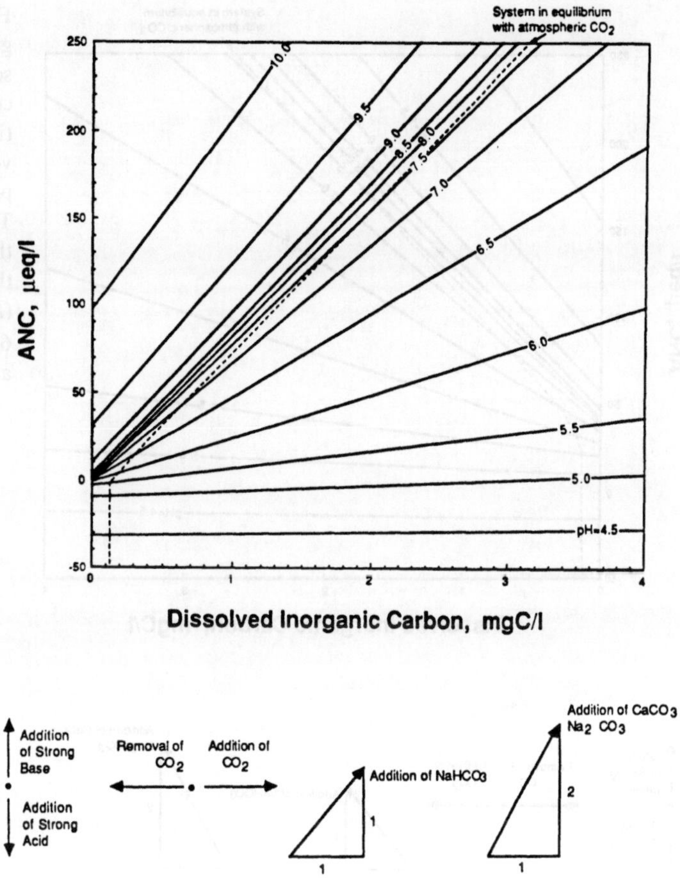

FIGURE 2.13. ANC versus dissolved inorganic carbon (DIC), with lines of constant pH shown. A point defining a solution composition moves as shown by the vectors for the indicated solution alterations. pH lines have been plotted using PHCALC. The dashed line has been added to determine the response of air equilibrating a water body that is over- or undersaturated with CO_2. (After Deffeyes 1965, Stumm and Morgan 1981.)

µmol L^{-1}) of DOC. The calculations necessary to produce the enhanced figures were made using a solution equilibrium model (PHCALC), which also can be used directly to analyze waters that contain both organic acids and Al.

Solution Equilibrium Models

Solution equilibrium models have been developed specifically for performing solution phase acid-base calculations (e.g., PHCALC, Gherini et al. 1985; ALCHEMI, Schecher and Driscoll 1987). Because of their very specific objectives, they are unlike many geochemical models, which can be difficult to use (Kincaid et al. 1984).

The code used to generate the enhanced Deffeyes plots and other figures in this chapter (PHCALC) has four major functions. It calculates pH, ANC, C_T (dissolved inorganic carbon), and Al_T. In addition to making these calculations, it evaluates complexation of Al with organic ligand, F^-, and SO_4^{2-} (important for fish toxicity analysis). By examining the short output table (Figure 2.15), we can quantify the proton acceptors and donors for a given water body. The difference between the carbonate system acceptors and the ANC will give the sum of the organic ligand and Al hydrolysis species proton acceptors.

Perhaps the major use for solution equilibrium models is to check the consistency of acid-base water chemistry data and perform sensitivity analysis such as that shown in Figure 2.16. Here the effects of various manipulations on a water body are demonstrated. The encircled point at ANC = 100 µeq L^{-1} represents a low ANC water body in the northeastern United States with a chemical composition as noted in the figure. Air equilibration of the water (setting the aqueous CO_2 system to be in equilibrium with atmospheric CO_2 levels) results in no change in ANC (proton refer-

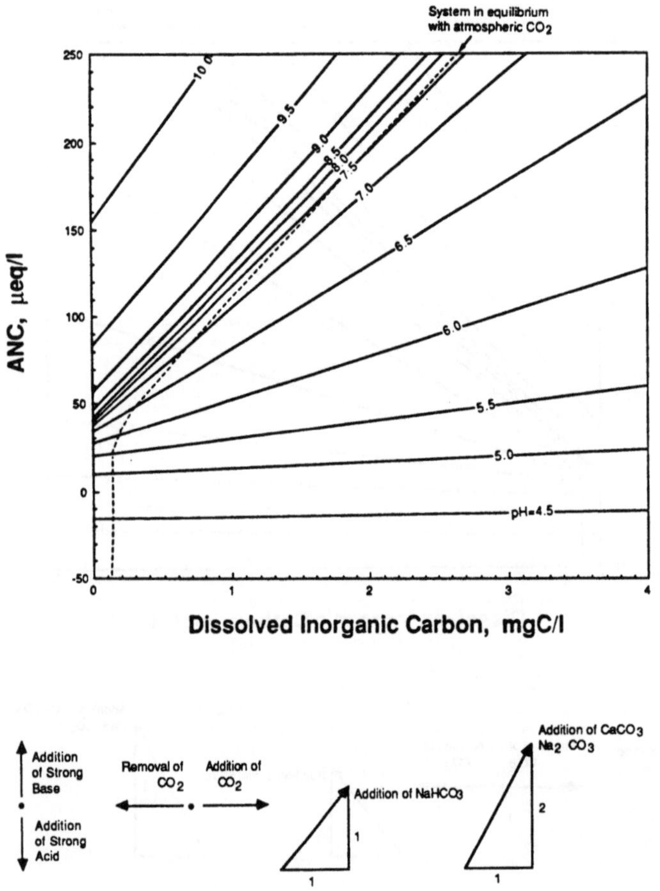

FIGURE 2.14. ANC versus dissolved inorganic carbon (DIC) with 4 mg L^{-1} of dissolved organic carbon (DOC), with lines of constant pH shown. A point defining a solution composition moves as shown by the vectors for the indicated solution alterations. pH lines have been plotted using PHCALC. The dashed line has been added to determine the response of air equilibrating a water body that is over- or undersaturated with CO_2. (R_T = 15 μeq mg^{-1} DOC; $pK_{A_{1,2,3}}$ = 4.0, 6.1, 9.25). (After Deffeyes 1965, Stumm and Morgan 1981.)

```
 **** PHCALC.CONV ****
NEW EQUILIBRIUM CONDITIONS AT  25.0 DEGREES C ARE:
      PH=     6.32
     ALK=    50.00(UEQ/L)
     TIC=    750.(UG/L)    PCO2= 9.482E-04(ATM)    PCO2/ATM. CO2=, 3.000
          HCO3=   2.96168E 01 (UEQ/L)        CO3=  5.76958E-03 (UEQ/L)
     ALT=     5.000(UEQ/L)
   OACD1=    30.000(UEQ/L)
   OACD2=     0.000(UEQ/L)
       F=     3.000(UEQ/L)
     SO4=   120.000(UEQ/L)
    AL-O=     1.598(UEQ/L)
    AL-F=     0.160(UEQ/L)
    AL-S=     0.000(UEQ/L)
```

FIGURE 2.15. Output table for solution equilibrium model PHCALC.

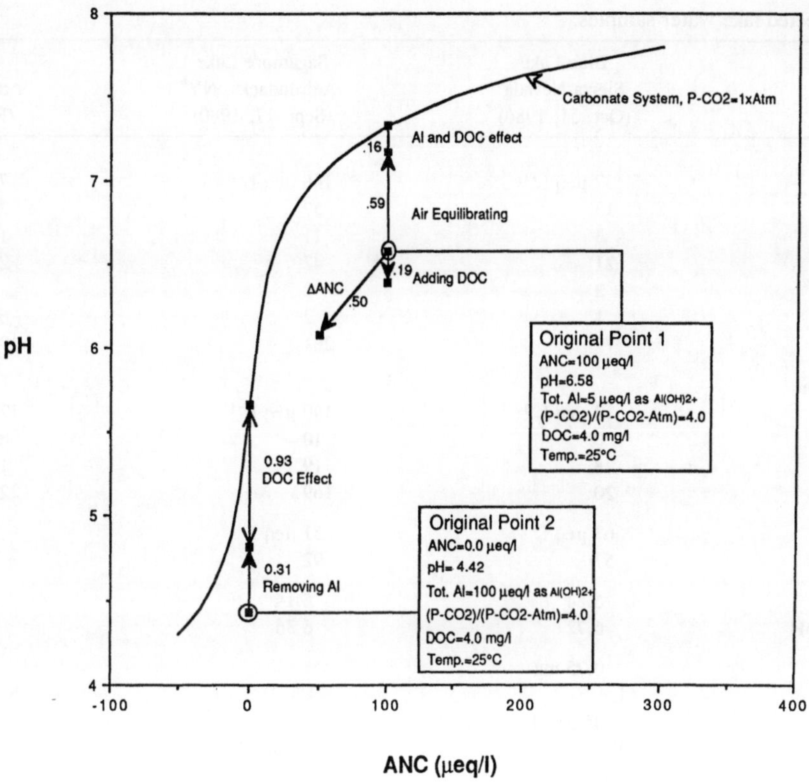

FIGURE 2.16. Acid-base system sensitivity analysis using a solution equilibrium model (PHCALC).

ence level includes aqueous CO_2), but a 0.59 unit pH increase. The remaining difference in pH from the theoretical curve for a carbonate system in equilibrium with the atmosphere (0.16 units) is due to the nonvolatile acids (organic acid, DOC = 4 mg L^{-1}, and Al_T = 45 μg L^{-1}). If 50 μeq L^{-1} of strong acid were added to this solution, the alkalinity would drop by 50 μeq L^{-1} and the new calculated pH would be 0.50 units lower. Adding additional organic acid (an additional 4 mg L^{-1} of DOC) to the original water drops its pH by 0.19 units. For a water body with ANC = 0.0 and the chemical composition noted under "Original Point 2" in Figure 2.16, removal of Al as $Al(OH)^{2+}$ in solution leads to a 0.31 unit increase in pH. The pH of this solution is 0.93 units below the theoretical expected value due to the presence of DOC. Numerous other changes and combinations of variables can be rapidly considered. These types of calculations can be extremely helpful in deducing the relative importance of strong and weak acid-base systems in determining the H^+ activity of natural waters.

Examples of Aqueous Buffer System Analysis

Solution equilibrium models can also be used to analyze the buffer systems operating within a given water body. Waters from three different lakes have been selected to illustrate this.

Blue Lake is a low productivity lake (chlorophyl-*a* < 1 μg L^{-1}) located in the Sierra Nevada at an elevation of 5,900 feet. Both Sagamore Lake and Woods Lake are located in the Adirondack Mountains of New York state. Although Sagamore and Woods are within 20 km of each other and receive similar deposition, their water quality characteristics are quite different. These three drainage basin lakes all have low ANC. Table 2.3 presents the water quality of the lakes on specific dates.

As discussed in Chapter 1, HCO_3^- is the major H^+ acceptor in most surface waters. However, for waters with moderate to low ANC, other buffer systems can be quantitatively important. Given the water quality parameters in Table 2.3, PHCALC

TABLE 2.3. Selected lake water samples.

	Blue Lake Sierra Nevada[a] (Oct. 31, 1986)	Sagamore Lake Adirondacks, NY[b] (Sept. 17, 1980)	Woods Lake Adirondacks, NY[b] (Feb. 23, 1981)
Base cations			
Ca^{2+}	37 μeq L^{-1}	146 μeq L^{-1}	75 μeq L^{-1}
Mg^{2+}	13	57	20
K^+	3	11	10
Na^+	21	42	25
NH_4^+	2	2	3
Monomeric Al	1	2	62
ΣC_B	77	261	195
Strong acid anions			
SO_4^{2-}	12 μeq L^{-1}	140 μeq L^{-1}	127 μeq L^{-1}
NO_3^-	0	10	81
Cl^-	8	19	13
ΣC_A	20	169	221
Measured ANC	63 μeq L^{-1}	51 μeq L^{-1}	−24 μeq L^{-1}
$C_B - C_A$ ANC	57	92[c]	−26
pH	6.84	6.14	4.42
Air-equilibrated pH	6.72	6.36	4.45
DIC	0.75 mg L^{-1}	NA[d]	NA[d]
DOC	1045 μmol L^{-1}	617 μmol L^{-1}	125 μmol L^{-1}
Silica	147	134	42

[a] Data from Gilbert and Sagraves, PG&E Co. (Gilbert et al. 1988).
[b] Data from Galloway, Schofield, and Hendry; ILWAS investigation (Valentini and Gherini 1986).
[c] Difference between measured ANC and $C_B - C_A$ results from relatively high levels of DOC in the water (DOC = 7.4 mgC L^{-1}; 617 μmol L^{-1}, Sullivan et al. 1989).
[d] NA = not available.

has been used to help elucidate the buffer systems operating in each of the three lakes (Table 2.4).

Although the water of Blue Lake has a relatively low alkalinity, the predominant proton acceptor there is HCO_3^- (80% of total), similar to most surface waters in the United States. However, because of the low ANC, a small amount of organic ligand (DOC = 1.25 mg L^{-1}; 104 μmol L^{-1}) is enough to provide most of the remaining 20%.

At Sagamore Lake, the water passes through extensive wetlands before entering the lake and has high organic carbon content (DOC = 4 to 10 mg L^{-1}; 333 to 833 μmol L^{-1}). The acid-base functional groups (e.g., carboxyl, phenolic hydroxyl) of this organic matter are an important component of the aqueous buffer system, and represent over half of the total proton acceptors. Bicarbonate ion makes up most of the difference with a small component supplied by Al species [e.g., $Al(OH)_2^+$].

The water of Woods Lake has more proton donors than acceptors and thus has a negative ANC (net positive mineral acidity). The few proton acceptors in the water are associated mostly with the high Al concentration and occur as such species as $Al(OH)_2^+$ and Al · R.

This analysis shows that the classical carbonate approach to ANC, which works well for most waters, is not sufficient for many low ANC waters, where other acid-base buffering systems become important.

Combined Analysis – Crum Lake

The analyses described earlier, in the sections on evapotranspiration/mass balance, C_B/C_A, and buffer systems, when applied individually, can assist in the characterization of a lake-watershed system. For some hydrologically simple systems, such as

TABLE 2.4. Aqueous phase buffer system analysis.[a]

	Blue Lake (ANC = 56 μeq L^{-1})	Sagamore Lake (ANC = 92 μeq L^{-1})	Woods Lake (ANC = 26 μeq L^{-1})
Proton acceptors			
HCO_3^-	44	35	0.3
CO_3^{2-}	0.03	0.004	0.0000004
OH^-	0.1	0.01	.0001
Al species	<3	<10	7–12
Organic ligands	9–12	47–57	<5
Proton donors			
H^+	0.14	0.72	38

[a] See Table 2.3 for associated water quality data. Units are all μeq L^{-1}.

mounded seepage lakes, these analyses can be combined, and predictions regarding the response of such systems to changes in acidic deposition loading can be made. This type of analysis has been applied to Crum Lake, a seepage lake in northcentral Minnesota. Crum Lake has ANC of 45 μeq L^{-1}, pH of 5.8, and DOC and SO_4^{2-} concentrations of 5 to 6 mg L^{-1} (417 to 500 μmol L^{-1}) and 18 μeq L^{-1}, respectively. The analysis procedure and results are presented in the following sections.

Water Balance

Installation and monitoring of piezometers at Crum Lake show that the groundwater table elevation around the lake is below the elevation of the surface of the lake (Donkers et al. 1986). Concentrations of terrestrially derived solutes in the lake were also found to be very low (e.g., $SiO_2 < 1$ mg L^{-1}; 83 μmol L^{-1}). This indicates that Crum Lake is hydrologically mounded (Chapter 1, Munson and Gherini, this volume) and essentially receives zero groundwater inflow.

Groundwater outflow from the lake can be estimated using Equation 2-6:

$$\Delta V = Q_S + Q_{P_L} - A_E - Q_L \qquad (2\text{-}11)$$

With no groundwater inflow, Q_S equals zero. As discussed previously, the time interval for which the water balance is made can be selected so that the change in lake volume, ΔV, is essentially zero. Solving for the groundwater outflow, Q_L, gives

$$Q_L = Q_{P_L} - Q_E \qquad (2\text{-}12)$$

The precipitation onto the lake surface (Q_{P_L}), and Q_E, the lake evaporation, can be estimated from local hydrologic monitoring data. Also, the lake evaporation can be cross-checked using lake and deposition Cl^- levels as discussed previously. The hydrologic data used in evaluating Crum Lake are summarized in Table 2.5. The results of the foregoing calculation are shown in Figure 2.17. With the

TABLE 2.5. Hydrologic and chemical data for Crum Lake.[a]

Hydrologic data		
Mean lake area	7.49 ha	(1)
Mean lake volume	1.73×10^8 L	(1)
Average annual precipitation	72.9 cm	(1)
Average annual lake evaporation	51.3 cm	(2)
Groundwater outflow rate	1.62×10^7 L yr^{-1}	(3)
Groundwater inflow rate	0 L yr^{-1}	(4)
Chemical data		
Average annual wet sulfate loading	10.8 kg SO_4^{2-} ha^{-1} yr^{-1}	(1)
Average annual dry sulfate loading	2.3 kg SO_4^{2-} ha^{-1} yr^{-1}	(1)
Average precipitation sulfate concentration	31 μeq L^{-1}	(1)
Average groundwater sulfate concentration	21 μeq L^{-1}	(5)
Current lake sulfate concentration	18.1 μeq L^{-1}	(5)

References: (1) Donkers et al. 1986. (2) Calculated from Cl^- ratios and ΣC_β ratios. (3) Calculated from water budget on lake. (4) Estimate, based upon piezometric heads and lake water chemistry. (5) C. Driscoll, Syracuse University, personal communication.

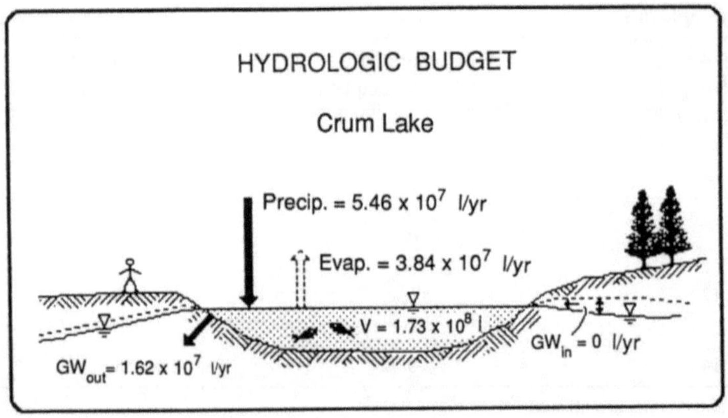

FIGURE 2.17. Hydrologic budget for Crum Lake.

measured precipitation, calculated groundwater outflow, and known lake volume, the hydraulic residence time for the lake is calculated ($t_d = V/Q_{P_L}$) to be 3.2 years. The chemical residence time ($t_c = V/Q_L$) is 10.7 years.

Solute Balances

In a similar fashion, balances can be made for solutes found in the lake using Equation 2-10, as discussed previously. An example of such a calculation for SO_4^{2-} is shown in Figure 2.18. Without considering chemical reaction in the lake, a concentration for SO_4^{2-} of 127 μeq L^{-1} would be predicted. The actual concentration is about 18 μeq L^{-1}.

As discussed in Chapter 1, SO_4^{2-} is lost from the water column to the lake sediments. The mass balance calculations and water quality data indicate that this is occurring at Crum Lake. The

Mass Balance Calculations for Crum Lake

Example for Sulfate:

Inputs:

Wet Deposition $= 5.46 \times 10^7 \frac{l}{yr} \times 30.9 \frac{\mu eq}{l} = 1.7 \times 10^9 \frac{\mu eq}{yr}$

Dry Deposition $= 4.80 \times 10^7 \frac{\mu eq}{ha \bullet yr} \times 7.49\,ha = 0.36 \times 10^9 \frac{\mu eq}{yr}$

Ground Water $\cong 0 \times 21 \frac{\mu eq}{l} = 0$

Water Budget:

Outflow $= (5.46 + 0 - 3.84) \times 10^7 \frac{l}{yr} = 1.62 \times 10^7 \frac{l}{yr}$

Expected* Lake Concentrlation

$$C = \frac{2.06 \times 10^9 \frac{\mu eq}{yr}}{1.62 \times 10^7 \frac{l}{yr}} = 127\ \mu eq/l, \quad \text{Actual} \cong 18\ \mu eq/l$$

*Expected concentration: Concentration in lake without any in-lake sinks or sources.

FIGURE 2.18. Solute balance for Crum Lake.

SOLUTE BALANCE FOR SULFATE CONSIDERING LOSS TO SEDIMENTS

Governing Equation (Equation 6)

$$\frac{d(C_L V)}{dt} = Q_s C_s + Q_{P_L} C_P + D - Q_E C_E - Q_L C_L - kC_L V$$

where: Sediment Sink = $FA_B = kC_L V$

F = net flux of SO_4^{2-} to the sediments

A_B = bottom area of lake and all other terms as defined previously

$$V\frac{dC_L}{dt} = \overbrace{Q_S C_S + Q_{P_L} C_P + D}^{=R} - (Q_L + kV)C_L$$

$$\frac{dC_L}{dt} = \frac{R}{V} - \frac{(Q_L + kV)}{V} C_L$$

let $\beta = \frac{Q_L + kV}{V}$

Then

$$\frac{dC_L}{dt} = \frac{R}{V} - \beta C_L$$

Separation of variables yields

$$\int_{C_i}^{C} \frac{-\beta dC_L}{\frac{R}{V} - \beta C_L} = \int_{0}^{t} -\beta t$$

and then

$$C_L = \frac{R}{V\beta}(1 - e^{-\beta t}) + C_i e^{-\beta t}$$

FIGURE 2.19. Predicting sulfate behavior in lakes.

effects of such reactions can be quantified using Equation 2-7, which takes into account change in concentration due to chemical reaction. This is illustrated further in Figure 2.19, where an integrated form of the mass balance equation is presented (Donkers et al. 1986).

Similar equations have been formulated for other acid-base solutes that undergo reaction in the lake. These include the coupled system of NH_4^+ and NO_3^- (Donkers et al. 1986). These mass balance equations and their integrated forms are shown in Figure 2.20.

Given these equations and those shown for SO_4^{2-} in Figure 2.19, calculations can be made to determine how the lake solute concentrations would respond to changes in the amount of solutes entering the lake as deposition. This approach takes into account the compensating effects of in-lake reactions. By calculating the changes in concentration of the base cations and strong acid anions, change in ANC can also be determined (Chapter 1, Munson and Gherini, this volume). The change in pH resulting from the calculated change in ANC can then be determined, using solution equilibrium models.

The response of Crum Lake to a 50% increase in the total deposition of all the C_B/C_A solutes has been predicted in this manner. Table 2.6 shows the change in lake ANC versus time as well as the

COUPLED NITROGEN BEHAVIOR IN MOUNDED SEEPAGE LAKES

$$V\frac{dC}{dt} = \overbrace{Q_pC_p + D_CA}^{a} - Q_oC - k_1CV - k_2CV$$

$$V\frac{dB}{dt} = \overbrace{Q_pB_p + D_BA}^{b} - Q_oB + k_1CV - k_3BV$$

$$C(t) = C_ie^{-\left(k_1+k_2+\frac{Q_o}{V}\right)t} + \frac{a}{\left(k_1+k_2+\frac{Q_o}{V}\right)}\left\{1 - e^{-\left(k_1+k_2+\frac{Q_o}{V}\right)t}\right\}$$

$$B(t) = B_ie^{-\left(k_3+\frac{Q_o}{V}\right)t} + \frac{b}{\left(k_3+\frac{Q_o}{V}\right)}\left(1 - e^{-\left(k_3+\frac{Q_o}{V}\right)t}\right) - \frac{k_1C_i}{(k_1+k_2-k_3)}\left\{e^{-\left(k_1+k_2+\frac{Q_o}{V}\right)t} - e^{-\left(k_3+\frac{Q_o}{V}\right)t}\right\}$$

$$+ \frac{ak_1}{\left(k_1+k_2+\frac{Q_o}{V}\right)\left(k_3+\frac{Q_o}{V}\right)}\left\{1 - e^{-\left(k_3+\frac{Q_o}{V}\right)t}\right\} + \frac{ak_1}{\left(k_1+k_2+\frac{Q_o}{V}\right)(k_1+k_2-k_3)}\left\{e^{-\left(k_1+k_2+\frac{Q_o}{V}\right)t} - e^{-\left(k_3+\frac{Q_o}{V}\right)t}\right\}$$

Where: $C = [NH_4^+]$

$B = [NO_3^-]$

Q_P = precipitation volume

D = dry deposition

V = lake volume

k_1 = rate constant for NH_4^+ oxidation

k_2 = rate constant for algal uptake of NH_4^+

k_3 = rate constant for algal uptake of NO_3^-

t = time

FIGURE 2.20. Nitrogen simulation equations.

associated changes in C_B/C_A constituents. The changes in the concentrations of Cl^- and the base cations, which do not typically undergo reaction in the lake (C'_B = Ca^{2+}, Mg^{2+}, K^+, and Na^+), are fairly slow because their concentrations are only influenced by evaporation and washout (groundwater outflow). The concentrations of constituents that do undergo reaction in the lake (NH_4^+, NO_3^-, and SO_4^{2-}) reach new steady state values rapidly.

TABLE 2.6. Predicted changes in ANC components with a 50% increase in atmospheric load of sulfur and a proportional increase in all other ions (μeq L^{-1}).

t =	Δ $(C'_B - Cl)$	Δ NH_4^+	Δ SO_4^{2-}	Δ NO_3^-	Δ ANC[a]
0 yrs.	0	0	0	0	0
5	14	1.5	8.7	0.2	6.6
10	22	1.5	9.0	0.2	14.3
50	37	1.5	9.1	0.2	29.2
100	38	1.5	9.1	0.2	30.2

[a] Δ ANC = Δ $(C'_B - Cl^-)$ + Δ NH_4^+ − Δ SO_4^{2-} − Δ NO_3^-.

As indicated in Table 2.6, the NH_4^+ and NO_3^- concentrations change very little in response to the changes in deposition. The predicted changes in SO_4^{2-}, C'_B, and Cl^-, on the other hand, are significant. The deposition of the additional base cations and Cl^- leads to an increase in $(C'_B - Cl^-)$ of 38 μeq L^{-1}. The influence of the increased sulfur deposition is mitigated by an increase in the SO_4^{2-} reduction rate, which is simulated as being first order with respect to SO_4^{2-} concentration (see Table 1.2, Chapter 1, Munson and Gherini, this volume). Thus, as SO_4^{2-} concentration increases in response to increased deposition, the rate of loss increases as well. The net result is that the final increase in SO_4^{2-} concentration is less than would be the case in the absence of reaction. Taking the changes in the concentrations of the individual constituents together, the change in ANC in response to the change in deposition is an increase of 30 μeq L^{-1}. Figure 2.21, which shows the results of applying PHCALC, indicates that this would lead to a gradual *increase* in pH from 5.83 to 6.29, or about 0.46 units over 100 years.

FIGURE 2.21. Predicted ANC and pH with a 50% increase in atmospheric sulfur load and a proportional increase in the deposition of all other ions at Crum Lake.

CRUM LAKE - pH RESPONSE TO CHANGE IN DEPOSITION

Time	ΔANC	ANC*	pH**
0 yr	0 µeq/l	45 µeq/l	5.83
5	6.6	51.6	5.97
10	14.3	59.3	6.09
50	29.2	74.2	6.28
100	30.2	75.2	6.29

** pH calculated with PHCALC

PHCALC Input Parameters

ANC = as above

TIC : 3.0 x Atm P_{CO_2}

Al_T = 0.8 µM

Total Organic Acid : = 75 µeq/l

Temp = 10°C

The results of this simplified approach compare well with the results of applying the dynamic process model, ILWAS (see next section), to the same lake. For the same deposition scenario, the ILWAS model predicted a change of 0.37 pH units in 20 years compared to 0.35 units predicted using this simplified approach for the same time period. For a 50% increase in the deposition of sulfur alone, the ILWAS model predicted a pH decrease of 0.12 units in 20 years compared to a 0.23 unit decrease predicted by this simplified approach.

Dynamic Modeling

Dynamic lake-watershed acidification models have been developed to predict responses of surface water quality to changes in deposition. They combine both hydrologic and chemical analyses and typically simulate multiple variables, including volumetric water flows, and water quality parameters, including pH and ANC.

The three models most commonly applied in the United States include ILWAS (Integrated Lake Watershed Acidification Study Model; Chen et al. 1983), MAGIC (Model of Acidification of Groundwater in Catchments; Cosby et al. 1985b), and Trickle Down (Schnoor et al. 1984). The model characteristics have been summarized in Table 2.7 by Malanchuk and Turner (1987).

One of the chief uses of these models has been the prediction of how a lake or stream will respond to changes in deposition. Figure 2.22 illustrates the predicted response of Woods and Panther Lakes to a 50% decrease in total sulfur deposition (ILWAS model predictions; Gherini et al. 1985). As indicated, the predicted change in pH at Woods Lake is about 0.7 units, and that at Panther Lake is less than 0.05 pH units. This result occurs in part because Woods Lake is a low ANC system (Table 2.3) and a given change in ANC produces a large change in pH. Panther Lake ANC (about 150 µeq L^{-1}) is in the range in which large changes in ANC are required to produce significant changes in pH.

Another use of the dynamic simulation models is to identify the processes supplying acid or base to the water flowing through a basin. This has been

TABLE 2.7. Comparison of processes and resolution in the ILWAS, MAGIC, and Trickle Down Models.[a]

	Model[b]		
Processes	ILWAS	MAGIC	Trickle Down
Atmospheric input	*E, A*	*A, LT*	*A, LT*
Hydrology	*E, A*	*A, LT*	*A, LT*
Weathering	*A, LT*	*LT*	*LT*
Anion retention			
SO_4^{2-}	+	+	−
Nitrification	+	−(+)	−
Denitrification	+	−(+)	−
Base cation buffering			
Percent base saturation	+	+	+
Aluminum kinetics	+	+	+
Biological			
Uptake	+	−	−
Excretion, decomposition	+	−	−
Transformation	+	−	−
Respiration	+	−	−
SO_4^{2-}	+	−	−(+)
Spatial resolution	*V, H*	*Pt*	*V*
Temporal resolution	*D*	*M*	*D*

[a] Malanchuk and Turner, 1987.
[b] +, process is included in model structure; −, process is not included in model structure; *E*, episodic time scale; *A*, annual time scale; *LT*, long-term time scale (i.e., > 10 years); *Pt*, point; *V*, vertical; *H*, horizontal; *M*, Month; *D*, Day.

done for the six systems indicated in Figure 2.23, using the ILWAS model. As indicated, Woods and Panther Lakes receive heavy deposition loads, and most of their bases are produced by cation exchange and mineral weathering reactions. Panther Lake has thicker soils than Woods Lake and thus more base is supplied at Panther, offsetting the acidic deposition and producing the observed ANC. Woods receives more acid than base and thus the observed ANC is negative. Coweeta is a stream system in western North Carolina that has old, highly weathered soils. Sulfate adsorption is the dominant ANC producing process in this system. Crystal and Round Lakes are seepage systems in northern Wisconsin. Crystal is a mounded lake and thus anion reduction is an important process in this system. Round Lake is a flow-through seepage system and receives a significant amount of base from weathering as well as from anion reduction. Meander Lake is a drainage lake in northcentral Minnesota. It has very thin till in much of the watershed but still has positive ANC due to base supplied by weathering and cation exchange. In addition, the acid input from deposition there is very small (Munson et al. 1987).

All three models have been described by Jenne et al. (1989) and are summarized briefly in the following sections. Each has different characteristics and levels of detail. Finally, a short comparison of the models is provided.

ILWAS (Integrated Lake Watershed Acidification Study Model)

The ILWAS model, the most process intensive of the three models, focuses on hydrologic and water quality dynamics. In order to capture the dynamic behavior of hydrologic and chemical parameters, the model takes into account the spatial heterogeneity of lake watershed systems. This is done by dividing the hydrologic basin into subcatchments, stream segments (if streams are present), and a lake. In each subcatchment, there are compartments to represent the canopy, snowpack, and soil layers. The lake is divided into horizontally mixed layers to allow for calculation of temperature and water quality profiles.

The model routes precipitation through the forest canopy, up to five soil horizons, streams, and lakes using mass balance concepts and equations that relate flow to hydraulic gradients. The physical and chemical processes that change the acid-base characteristics of the water are simulated by rate (kinetic) and equilibrium expressions and include mass transfers between gas, liquid, and solid phases. The aqueous constituents simulated include pH, ANC, the major cations (Ca^{2+}, Mg^{2+}, K^+, Na^+, and NH_4^+) and anions (SO_4^{2-}, NO_3^-, Cl^-), aqueous monomeric Al and its organic complexes, an organic acid analogue, and dissolved inorganic carbon. The concentration of free H^+ is derived from the solution ANC and the total concentrations of inorganic carbon, organic acid, and monomeric Al.

ILWAS has been used to evaluate the response of lakes and streams throughout the United States and abroad to changes in deposition. It has been applied successfully in both drainage (e.g., Gherini et al. 1985, Davis et al. 1987) and seepage lake systems (e.g., Knauer et al. 1989, Greb et al. 1987). It has also been used to help guide research projects and to test scientific hypotheses.

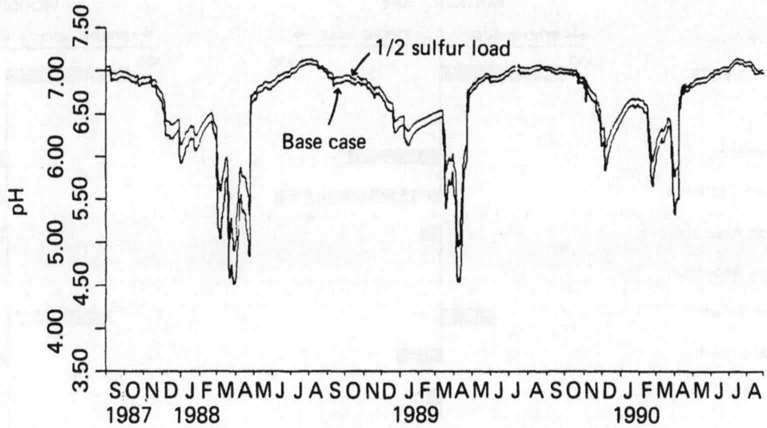

(a) Panther Lake Outlet pH

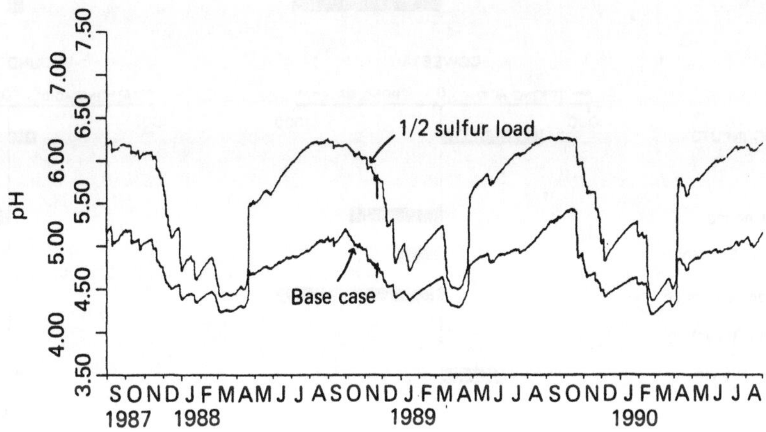

(b) Woods Lake Outlet pH

FIGURE 2.22. Predicted response (ILWAS) of Woods and Panther Lakes to a halving in the total atmospheric sulfur deposition (wet and dry). Input data for 1978, 1979, and 1980 were run back to back 4 times to create 12 years of simulation. The 3 years shown above correspond to years 10, 11, and 12 of the simulation (Gherini et al. 1985).

MAGIC (Model of Acidification of Groundwater in Catchments)

MAGIC is a relatively simple model designed to predict the average watershed response to long-term acidic deposition. Typical simulations are done using 140 years of estimated background deposition data. The model is run under these conditions to establish a steady state for the watershed in terms of sorbed and solution-phase cation concentrations. A historical trend in SO_2 emissions is then used to estimate the increase in deposition over time, and the model is run forward from 140 years ago to the present. If present-day simulated and observed concentrations are consistent at that point, the simulation is continued to estimate any future changes in surface water characteristics.

MAGIC uses lumped parameters based on basin physical and chemical characteristics. These parameters are aggregated horizontally over the entire watershed and are broken out vertically into two soil layers and a completely mixed lake. Canopy processes are not considered. Flow through the basin is calculated using the companion model called TOPMODEL, which is based upon the concept of a variable source area for input to the lake or stream. Geochemical processes are simulated using both rate-limited and equilibrium expressions. MAGIC has been applied to lakes and streams in the eastern United States and in Scandanavia.

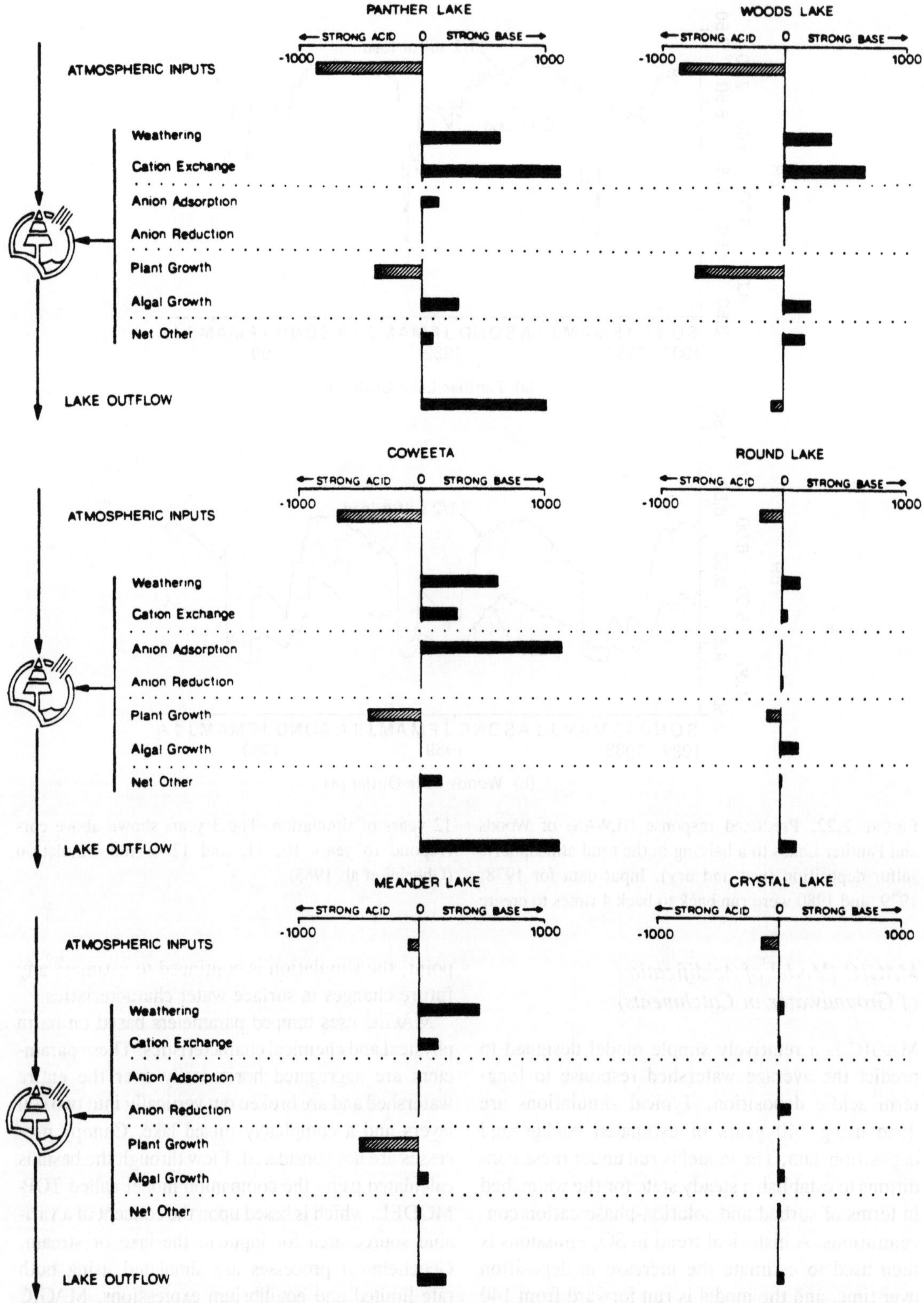

FIGURE 2.23. Contributions of various processes to the observed ANC in different basins (based on ILWAS model simulations). Units are eq ha^{-1} yr^{-1} (Munson et al. 1987).

Enhanced Trickle Down (ETD)

The Enhanced Trickle Down model is a lumped parameter model based on the concept of ANC mass balance. All the processes included in the model are formulated in terms of their direct effects on ANC. The focus of this model is to predict seasonal fluctuations in lake acid-base characteristics based on a single master variable.

As with MAGIC, parameters are lumped horizontally over the watershed, but in the Enhanced Trickle Down model, they are broken out vertically into three soil layers and a completely mixed lake. All of the chemical processes in this model are represented as being rate-limited, except for SO_4^{2-} adsorption and CO_2-carbonic acid equilibrium. Mass balances are included only for ANC, SO_4^{2-} (as a check on SO_4^{2-} reduction rate), and Cl^- (as a check on hydrologic calibration). The other constituents are not explicitly tracked by the model. The Enhanced Trickle Down model has been applied to seepage lakes in the midwest (Schnoor et al. 1986), and to drainage lakes in the eastern United States (Nikolaidis 1987).

Model Comparisons

The major differences in the three models involve the degree of spatial heterogeneity and level of process representation. For example, the number of soil layers represented varies from up to five in ILWAS down to two in MAGIC. In addition, ILWAS explicitly simulates several processes that MAGIC and ETD do not, including vegetative processes (enhanced collection of dry deposition, foliar exudation, root respiration, nutrient uptake, litter fall and decay), microbial processes (nitrification, uptake of nitrogen compounds in the lake), and lake physical processes (stratification and mixing).

The level of detail used in the simulation of some processes common to all three models varies as well. For example, weathering in the ILWAS model is constrained such that base cations (and SiO_2) are released only in the stoichiometric ratios observed in the field. In MAGIC, no such constraint exists and weathering rates are established on the basis of individual ions. Weathering in Enhanced Trickle Down simply produces ANC and thus there is no individual ion constraint. This points out another significant difference, in that Enhanced Trickle Down tracks only ANC, SO_4^{2-}, and Cl^-, whereas both of the other models track multiple species.

Another difference is that the models include different Al species in their proton reference levels. ILWAS includes Al^{3+} in its proton reference level, whereas MAGIC includes $Al(OH)_3^0$. The only time this difference is significant is when there are appreciable Al concentrations in solution (i.e., at low pH). Since ETD does not track Al, its proton reference level does not explicitly include any Al species.

Although the three models do have differences, many of those differences result from the intended use of the models and the focus of the developers. These factors must be evaluated when the use of any of these models is being considered for a particular investigation.

3
Historical Trends in Atmospheric Sulfur Deposition and Methods for Assessing Long-Term Trends in Surface Water Chemistry

Rudolf B. Husar, Timothy J. Sullivan, and Donald F. Charles

ABSTRACT. This chapter presents historical sulfur emission and deposition trends for regions in the United States and describes methods for assessing changes in water chemistry based on current spatial patterns, ion ratios and empirical models, and paleolimnological approaches. Reconstruction of sulfur deposition trends shows that current deposition to case study regions ranges from a factor of about 1 (Upper Midwest) to a factor of 10 (Catskills) above natural background. Deposition in the Northeast was high during the 1920s, 1940s, and 1960s, and has declined significantly since 1970. Sulfur deposition in the Southeast was low before the 1950s, but has increased significantly since then. Change in surface water chemistry can be assessed using simple empirical models, ion ratios, and analysis of current spatial patterns of chemistry. Many assumptions are implicit in these methods, so results should be interpreted carefully. Paleolimnological reconstructions of chemistry and biota from lake sediment records provide more direct evidence of past change than other approaches. Quantitative analyses of diatom and chrysophyte assemblages can be used to reconstruct past lakewater pH with a mean standard error of about ± 0.25 pH units.

Introduction

Three questions are central to the ongoing assessment of acidic deposition effects on aquatic ecosystems:

1. How has acidic deposition changed from preindustrial times to the present in the regions of concern?
2. How has the acid-base chemistry of surface waters changed in response to changes in atmospheric loading?
3. How have the changes in aquatic chemistry affected biota?

Despite the importance of these basic questions, few data exist with which to make quantitative regional assessments (National Research Council 1986).

The objectives of this chapter are to synthesize historical deposition data and describe methods that have been used to assess historical trends in water chemistry. The material presented here provides more background and in-depth description than is possible within the individual case study chapters. The chapter is divided into three sections: first, historical sulfur deposition patterns, derived from emissions data and simple atmospheric transport and conversion models (Husar); second, methods commonly used to estimate historical changes in surface water acid-base chemistry, based on simple empirical models and ion ratios (Sullivan); third, paleolimnological reconstruction techniques, using diatom and chrysophyte sediment assemblages (Charles). Some methods used to evaluate historical change in some case study regions are not described here, for example, procedures for comparing historical and modern water chemistry and for using computer models to hindcast changes. These methods are described in the case study chapters where results are presented, and some are described in Chapter 18.

Evidence for past chemical and biological changes that occurred within the case study regions in response to acidic deposition are presented within each of the regional chapters. Historical changes in biology and chemistry are summarized and evaluated by Baker and Christensen (Chapter 4, this volume) and Sullivan (Chapter 18, this volume), respectively.

Historical Trends in Atmospheric Sulfur Deposition

This section presents estimates of historical sulfur deposition trends for the case study regions discussed in this book, and describes the rationale and procedures by which they were derived. The estimates are based on emissions of SO_x, and include wet deposition only for the period 1860–1984.

Atmospheric deposition of sulfur provides the most significant contribution of strong acid anions to aquatic ecosystems in the case study regions on an annual basis (Chapter 5, Eilers and Selle, this volume). It is therefore important to know about spatial patterns and temporal trends in sulfur deposition in order to understand and assess causes and magnitudes of changes in water chemistry. Reconstructing historical trends is difficult, however, partly because emissions have varied significantly in source type and amount over the past century. As a consequence, both deposition rates and trends are significantly different from one region to another. Emissions have been highest in states east of the Mississippi River, and for this reason and because data coverage is better, analyses in this chapter focus on the eastern United States. Estimates of wet deposition are probably a meaningful surrogate for the pattern of total sulfur deposition, because differences among regions in wet deposition probably exceed differences in the ratio of wet to dry deposition.

Nitrogen compounds are also significant contributors to acidity of precipitation, but emissions are more difficult to reconstruct than S trends. Important, but not considered here, are potential changes in deposition of cations such as Ca^{2+}, Mg^{2+}, and NH_4^+.

The following sections describe the methodology for estimating wet deposition trends, including the emission estimates, atmospheric transmission between sources and receptors, and resulting deposition trends.

Methodology for Estimating Long-Term Deposition

Reconstructing the chemical history or the "chemical climate" for a given location requires several steps:

1. **Gathering of historical data on chemical concentrations in precipitation**. Such data are limited to certain time periods and geographical locations, but provide significant anchor points for reconstructions. Data should not be accepted without scrutiny, but neither should they be discarded without appropriate justification.
2. **Reconstruction of emission trends**. In general, the region for which emission trends should be reconstructed cannot be determined a priori. This is because the region of influence of each source depends strongly on the form of sulfur being considered. For example, the airborne concentration of SO_2 is influenced mostly by local sources located within 100 km of a receptor. Sulfate aerosol and sulfur wet deposition are determined by sources within 500 to 1000 km.
3. **Calculation of regional background concentrations and deposition**. The contribution of sources to the regional background can be calculated using a long range transport model, along with suitable historical emission inventories.
4. **Calculation of local concentration**. Sulfur dioxide concentration and dry deposition are strongly influenced by sources on a local scale (100 km), requiring a local emission inventory and a local scale dispersion model. Alternatively, the existing precipitation monitoring data may be used to develop the local concentration field.
5. **Reconciliation of precipitation monitoring data with model estimates**. This is the feedback stage at which the atmospheric model parameters are modified so that calculated deposition more closely matches the observations.

The art and science of chemical climate reconstruction is in its infancy. This is a modest effort toward such a reconstruction. Earlier attempts included the work of Husar and Patterson (1984a,b) and Fay et al. (1986).

Emission Trends

Sulfur Oxides

Sulfur emissions data have been compiled for states on an annual basis by Gschwandtner et al. (1985) and Husar (1986) for the 20th century (through 1980) and on a monthly basis by Knudson (1986) for the period 1975 through 1984. The sulfur emission trend data presented in this chapter have been combined from these three data sources. The procedures for deriving this integrated seasonal sulfur emission data set are described elsewhere (Husar 1988a,b). Sulfur emissions data for eastern Canada were taken from the U.S.-Canada Memorandum of Intent (MOI, 1982) and extrapolated for pre-1955 years using smelter production data (Husar 1986).

Sulfur emission trends for the United States covering the period 1860–1984 (Figure 3.1) show an initial period of steady increase followed by periods of increasingly larger peaks in the 1920s, 1940s, and early 1970s. During the most recent period (1970–84), there was a significant reduction to levels comparable to the low emission periods of the 1950s, 1930s, and pre-1910. Figure 3.1 shows the range of uncertainty, depicted as plots of the high and low estimates of sulfur emissions. During the past century, the fraction of sulfur emitted east of the Mississippi River has remained roughly constant at about 70% of total United States emissions.

There are substantial differences in emission trends among eastern subregions. The emission trend for eastern Canada (Figure 3.1b) shows an increase since the turn of the century leading to a peak at about 3.6 million tons in the mid-1960s, followed by a decline. Sulfur emissions for the northeastern states, including Pennsylvania, (Figure 3.2) increased strongly beginning in the 1890s, leading to a peak between 1910 and 1930, followed by a depression in the 1930s. Another increase occurred in the 1940s, and the emission levels have fluctuated around that level ever since. Current SO_x emissions are comparable to those in the 1930s and 1910s, but substantially below the 1920s, 1940s, and 1960s. Emission trends for the eastern states south of the Ohio River (Figure 3.2) have doubled since the turn of the century, with peaks occurring in the 1920s, 1940s, and 1980s and depressions in the 1930s and 1950s. Most significant is the increase in the 1960s.

The sulfur emission trend for the industrialized Midwest, including Illinois, Indiana, Michigan, Missouri, and Ohio (Figure 3.3) increased strongly before the turn of the century, with peaks in the 1920s, 1940s, and 1970s, and depressions in the 1930s and 1950s and during the past decade. The sulfur emissions for the Upper Midwest (Iowa, Minnesota, and Wisconsin) were low in the early part of this century, and have remained at low levels into the 1980s.

More detailed examination reveals additional significant qualitative changes in the trends: (1) a shift in a seasonal emission pattern from a strong winter peak to a winter and summer peak and (2) a drift of the geographic center of emissions from the populated and industrial Northeast toward the Southeast. The January and July emission trends for the Northeast (Illinois, Indiana, Ohio, Pennsylvania, and New York), show that over the Northeast, emissions have varied within less than a factor of 2 for the past 80 years. Also, until the late 1970s, the winter emissions were significantly higher. In the 1980s, the summer-winter difference have become less pronounced.

Comparable data for the Southeast (Tennessee, North Carolina, South Carolina, Georgia, Alabama, and Louisiana) (Figure 3.2) show a significant increase in emissions since the 1950s and a slight decline since the late 1970s. They also illustrates a shift from a strong winter peak to a winter and summer peak. In the Northeast, much of the winter sulfur deposition accumulates in the snowpack. The strong winter emissions peak implies higher winter deposition prior to the 1970s. Thus, it is likely that the sulfur content of the snowpack and the resulting meltwater were historically higher than the current values.

Nitrogen Oxides

Historical emissions of NO_x were reconstructed by Gschwandtner et al. (1985) and Husar (1986). The two emission estimates use different methodologies but are generally similar. The following figures are based on Husar's estimates. Historical NO_x emission trends have significantly higher uncertainty (about a factor of 2) compared to the sulfur emission trends, which are believed to be within $\pm$ 30%.

Nitrogen oxide emissions in the northeastern states (Figure 3.4) have increased monotonically

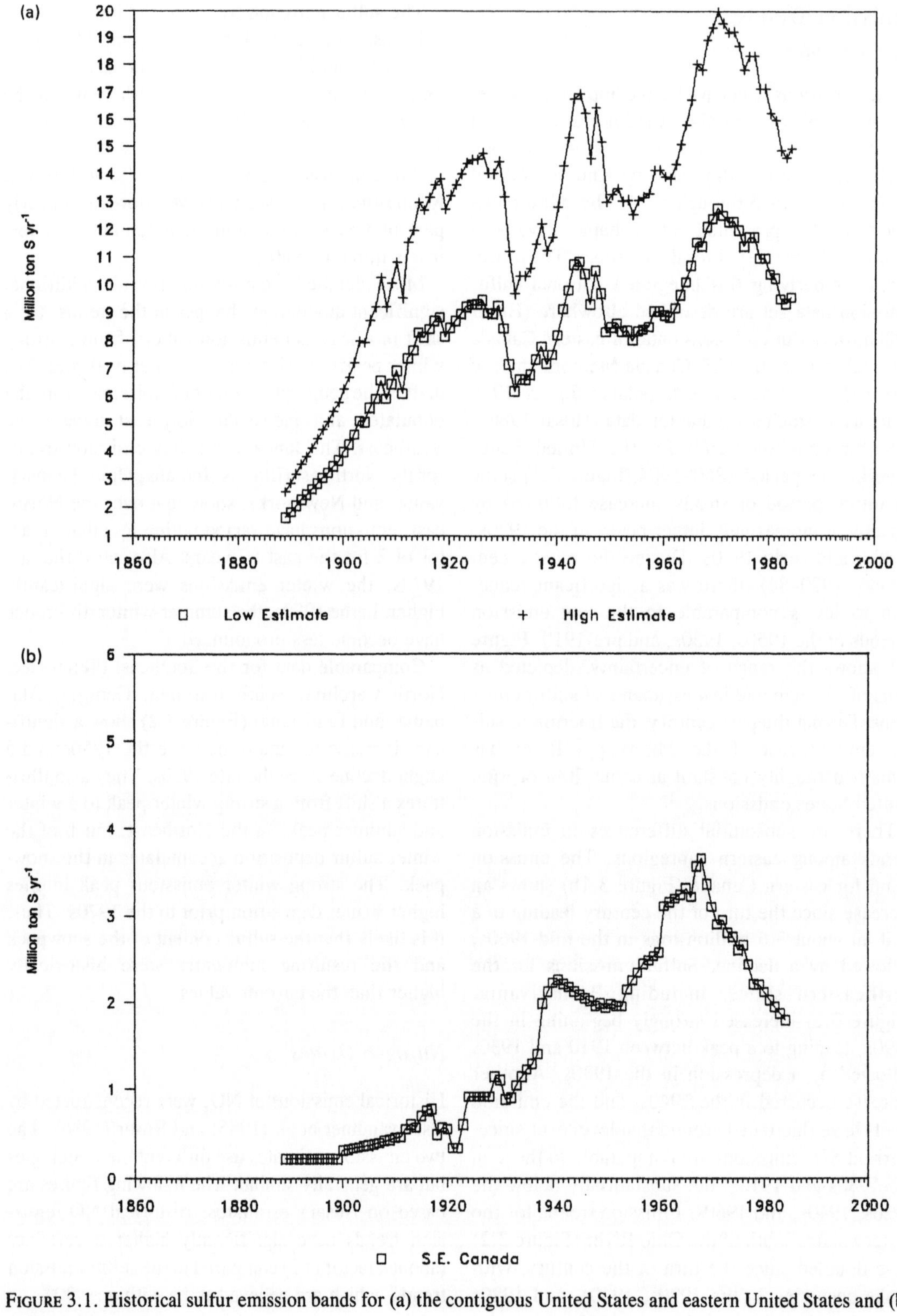

FIGURE 3.1. Historical sulfur emission bands for (a) the contiguous United States and eastern United States and (b) southeast Canada.

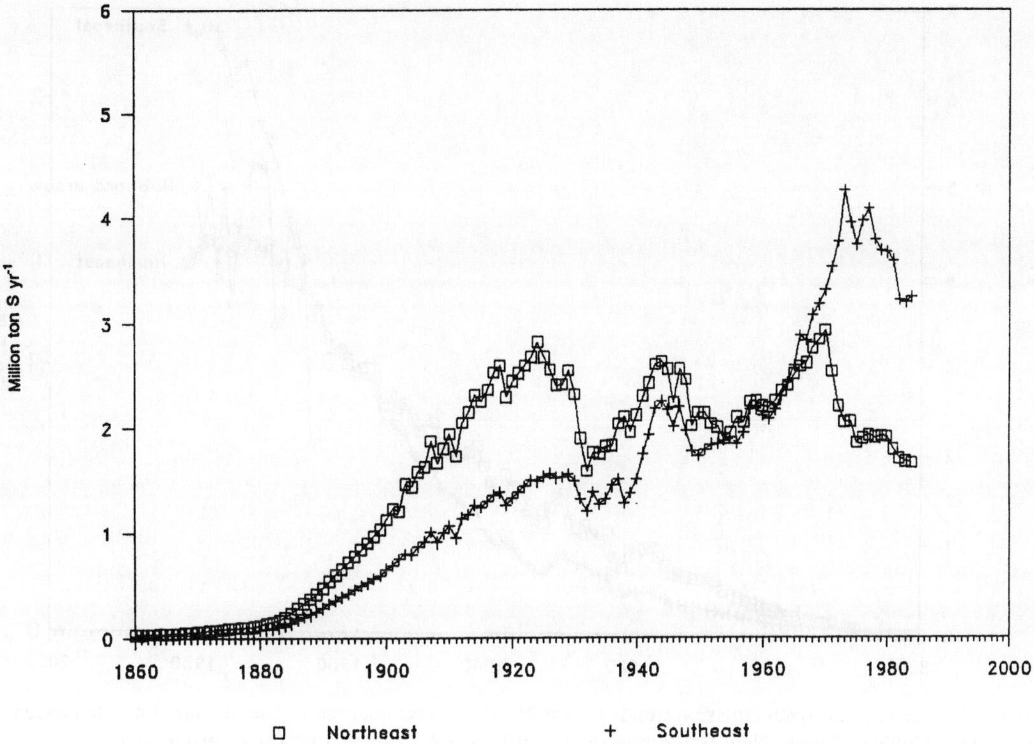

FIGURE 3.2. Historical sulfur emission trends for the northeastern and southeastern United States.

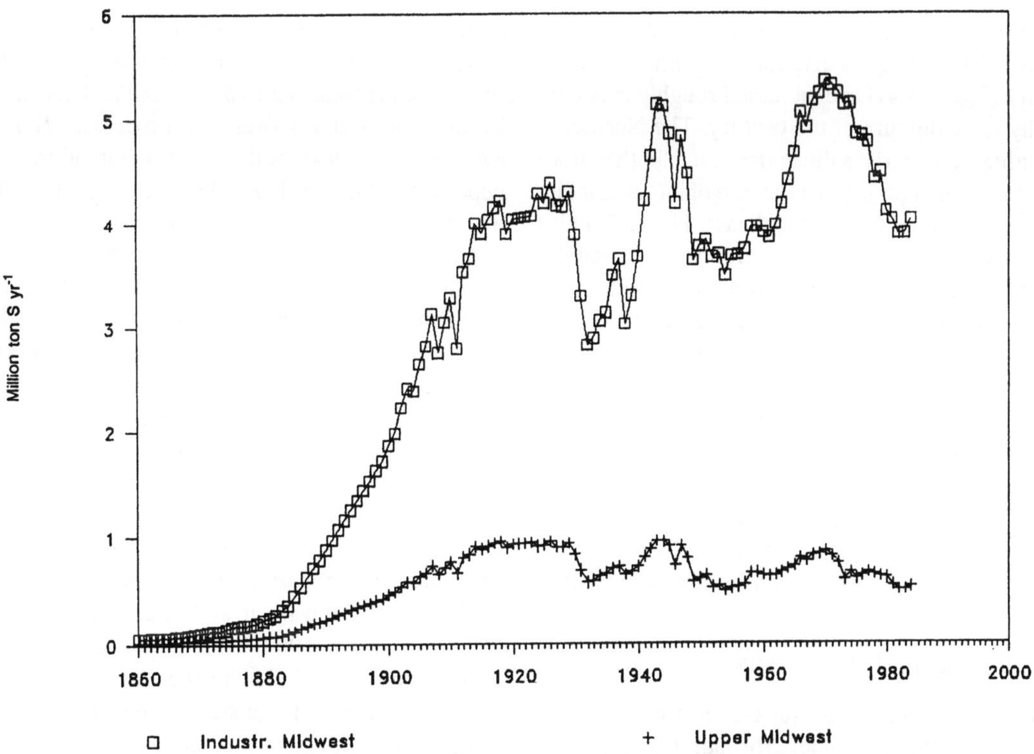

FIGURE 3.3. Historical sulfur emission trends for the industrialized Midwest and Upper Midwest.

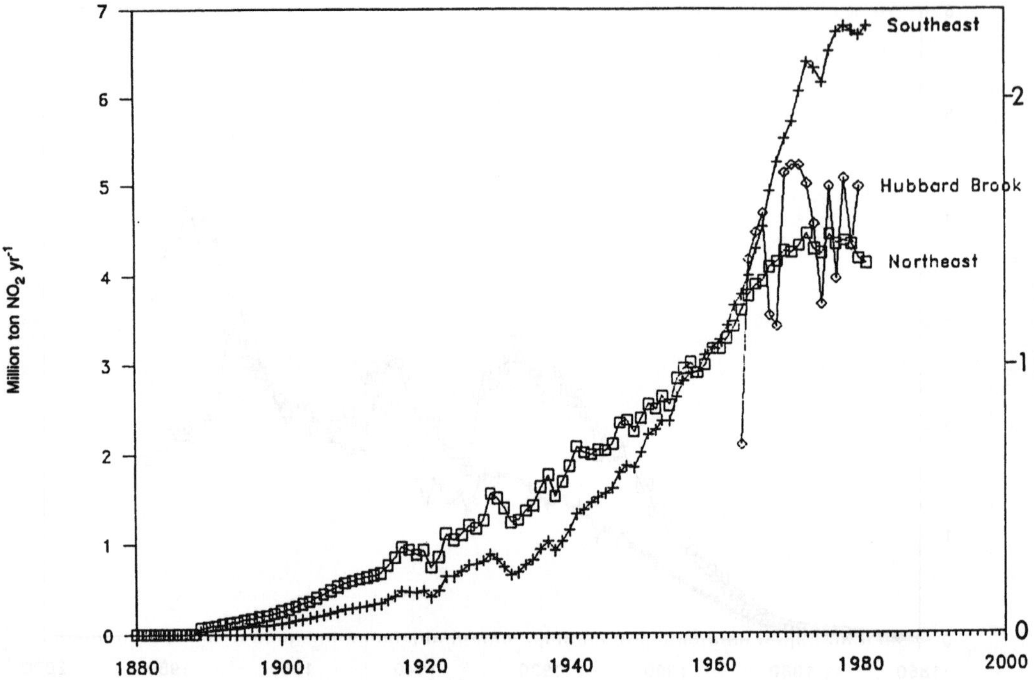

FIGURE 3.4. Historical nitrogen emission trends for the Northeast and Southeast. The measured nitrate concentration trend for Hubbard Brook, New Hampshire (National Research Council 1983) is superimposed.

since the mid-1800s, and have leveled off since the mid-1970s. Emissions in the states south of the Ohio River (Figure 3.4) have increased roughly exponentially since the turn of the century. The Northeast-Southeast comparison illustrates that in both regions the trends are upward, but the rate of increase is significantly higher for the Southeast. Figure 3.4 also contains NO_3^- concentration trends for precipitation chemistry measured at Hubbard Brook, New Hampshire (NRC 1983). The qualitative trend comparison of NO_x emissions and precipitation concentrations shows that the two are not inconsistent. However, the year-to-year variability of precipitation nitrate concentration precludes a firm conclusion. The NO_x emission trends for the industrialized Midwest and Upper Midwest have patterns qualitatively similar to patterns in the Northeast, but the magnitude of the emissions is significantly less.

Atmospheric Transmission

Sulfur emissions from various sources are distributed by the atmosphere through transport processes that include winds and clouds. The range of atmospheric transport for sulfur compounds is 500 to 1,000 km. The corresponding atmospheric sulfur residence time is 3 to 5 days and is determined by dry and wet removal rate processes. Hence, every source has an oval-shaped region of greatest deposition (500 to 1,000 km), elongated in the direction of the prevailing winds.

Atmospheric transmission and deposition may be calculated using regional scale models that incorporate measured meteorological windfield, precipitation amounts and rates, chemical and physical removal rate constants, and a sulfur emission field. With these inputs, a regional model can be used to estimate the deposition pattern for any location and time within its domain.

The CAPITA regional Monte Carlo model (Patterson et al. 1981) was used for estimating historical source receptor relationships presented in this book. The model was run at 3-hour increments for the entire year in 1980, using the wind, precipitation, and emission fields for that year. This procedure yielded a unit transfer matrix, that is, the quantity of sulfur deposited at any given receptor per unit emission at the source.

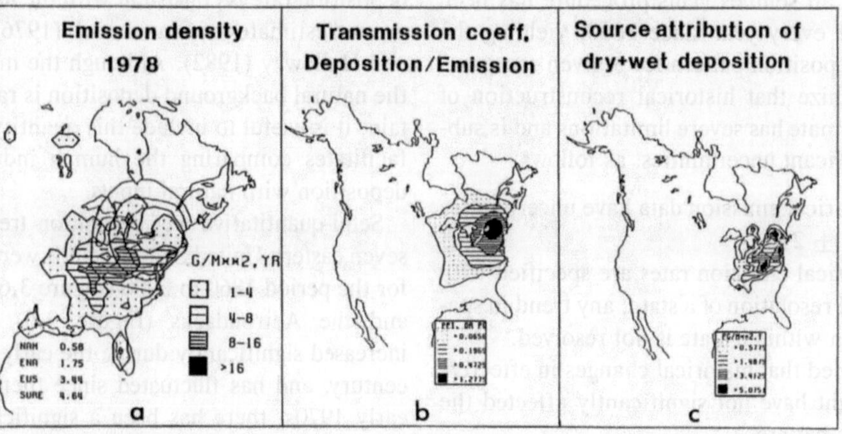

FIGURE 3.5. Source attribution of deposition in the Adirondacks, showing (a) emission density for 1978, (b) unit transfer matrix, and (c) source attribution map.

The attribution of total sulfur deposition, $D(i,j)$ at receptor site j to the responsible sources $S(i)$ can be calculated as follows:

$$D(i,j) = t(i,j) * S(i)$$

Thus, the deposition $D(i,j)$ (g S m^{-2} yr^{-1}) is proportional to the source strength, $S(i)$ (g S m^{-2} yr^{-1}), where the proportionality constant is the element of a transfer or source-receptor matrix, $t(i,j)$. The physical interpretation of the transfer matrix is that it states the fraction of the matter emitted from a source at location (grid) i and deposited at receptor j. A transfer matrix element may also be envisioned as the probability that a molecule emitted at i will be deposited at j. The deposition at j from all i sources may be calculated as:

$$D(j) + \Sigma D(i,j) = \Sigma t(i,j) * S(i)$$

The total deposition $D(j)$ (g S m^{-2} yr^{-1}) is the sum of all the source strengths $S(i)$, each weighed by its transfer matrix element $t(i,j)$. This linearized formulation of source-receptor relationship assumes that deposition from one source occurs independently of other emissions. This assumption has been criticized in the past (Rodhe et al. 1981), on the grounds that the chemical compounds from one source influence the fate of emissions from other sources. However, both the participants of the 1982 Stockholm Conference on the Acidification of the Environment (Acidification 1982) and the NAS report, Acid Deposition: Atmospheric Processes in Eastern North America (NRC 1983), concluded that the total deposition over an industrialized region (1,000 km) will respond proportionally to emission changes.

The procedure of regional source attribution is illustrated graphically in the three maps shown in Figure 3.5. The emission density for SO_x shows highest values (> 8 g S m^{-2} yr^{-1}) over the Ohio River Valley. The center map shows the contours of constant transfer matrix elements for dry plus wet deposition for a receptor located in the Adirondack Mountains. The third map, known as the source attribution matrix, is the product of the SO_2 emission density matrix times the transfer matrix. It reveals the location of source areas that contribute most heavily to the sulfur deposition in the Adirondacks. The blackened areas account for 25% of the deposition to the Adirondacks, whereas the lowest source attribution contour includes 90% of the sources of deposition. According to these calculations, sources south of the Ohio River and west of the Mississippi River contribute less than 10% of the sulfur deposition to the Adirondacks.

For historical reconstruction it is assumed that the transfer matrix is invariant in time. This seems reasonable, given that the climatological average transport wind field and the precipitation field probably have not changed dramatically over the past century. For such a case, the changes in deposition field are due exclusively to changes in the emission pattern. The historical deposition at any receptor is then obtained by summing the product of the emission rate multiplied by the unit transfer matrix over all sources. This procedure has been repeated for every year since 1860, yielding the historical deposition pattern for a given receptor.

We recognize that historical reconstruction of chemical climate has severe limitations and is subject to significant uncertainties, as follows:

- State historical emission data have uncertainties of at least ± 25%.
- The historical emission rates are specified with the spatial resolution of a state; any trend in spatial pattern within a state is not resolved.
- It is assumed that historical changes in effective stack height have not significantly affected the deposition pattern.
- Meteorology for the year 1980 was assumed to be representative of atmospheric transmission during the period 1860–1984.
- Possible nonlinear chemical interactions between sulfur and nonsulfur compounds in the atmosphere are ignored.
- There have been significant shifts in the seasonality of the emissions, whereas the deposition calculations discussed in the empirical methods section of this chapter use yearly average emissions and transfer matrix coefficients.

Due to these limitations and uncertainties of emission trends, it is recommended that the reconstructed historical deposition pattern be used only in semi-quantitative analyses.

Deposition Trends

Wet deposition trends were estimated for seven eastern United States regions covered in this book: northern New England, the Adirondacks, the Catskill Mountains, W. Virginia, the Southern Blue Ridge, northern Florida, and the Upper Midwest. Deposition estimates were calculated for a receptor site chosen to represent the geographic center of each case study region (see Chapter 5, Eilers and Selle, this volume, for map of regions). Because of uncertainties in atmospheric transmission, reconstructions were restricted to sulfur oxides. It was assumed that deposition at each location is the sum of natural background deposition and human induced deposition. The background estimate of pre-industrial wet deposition was taken to be invariant in time and location and set at 0.10 g S m^{-2} yr^{-1} or 3 kg SO_4^{2-} ha^{-1} yr^{-1} (31 eq ha^{-1} yr^{-1}). This deposition rate is consistent with the natural background estimates of Granat et al. (1976) and Husar and Holloway (1982). Although the magnitude of the natural background deposition is rather uncertain, it is useful to include this quantity because it facilitates comparing the human induced sulfur deposition with natural inputs.

Semi-quantitative wet deposition trends for the seven eastern United States regions were calculated for the period 1860 to 2000 (Figure 3.6). In Maine and the Adirondacks (Figure 3.6), deposition increased significantly during the early part of this century, and has fluctuated since then. Since the early 1970s, there has been a significant decline, which corresponds to the reduction in sulfur emissions in the northeastern United States and adjacent Canada. Recent deposition ranging between 15 and 20 kg SO_4^{2-} ha^{-1} yr^{-1} (160 to 210 eq ha^{-1} yr^{-1}) is five to eight times the estimated natural deposition. Sulfur deposition at Hubbard Brook, New Hampshire (National Research Council 1983), indicates a decline over the past 20 years. This is consistent with the estimated historical trends for the Adirondacks and New England.

Of the case study regions discussed in this book, the Catskills (Figure 3.6) shows the highest historical and current deposition rates because it is near industrialized regions of Pennsylvania and New York. The anthropogenic sulfur deposition there is estimated to be about a factor of 10 higher than the natural deposition. The shape of the trend graph closely resembles the northeastern emission trends (Figure 3.2). The deposition pattern for W. Virginia (Figure 3.6) exhibits a deposition pattern that is a combination of the Northeast and the Southeast. There was significant deposition prior to the 1960s, but also significant increase since the 1960s. Evidently, this region is influenced by emissions from the industrialized Northeast as well as from the recently developing Southeast.

The deposition estimates for the Southern Blue Ridge and Florida (Figure 3.6) show a qualitatively different pattern. Prior to the 1950s, deposition was not significantly different from the natural background. However, beginning with the 1960s, there was a rapid rise through the late 1970s. Hence, in the Southeast, rapid deposition increase is of more recent origin and until 1984 there was no evidence of significant decline.

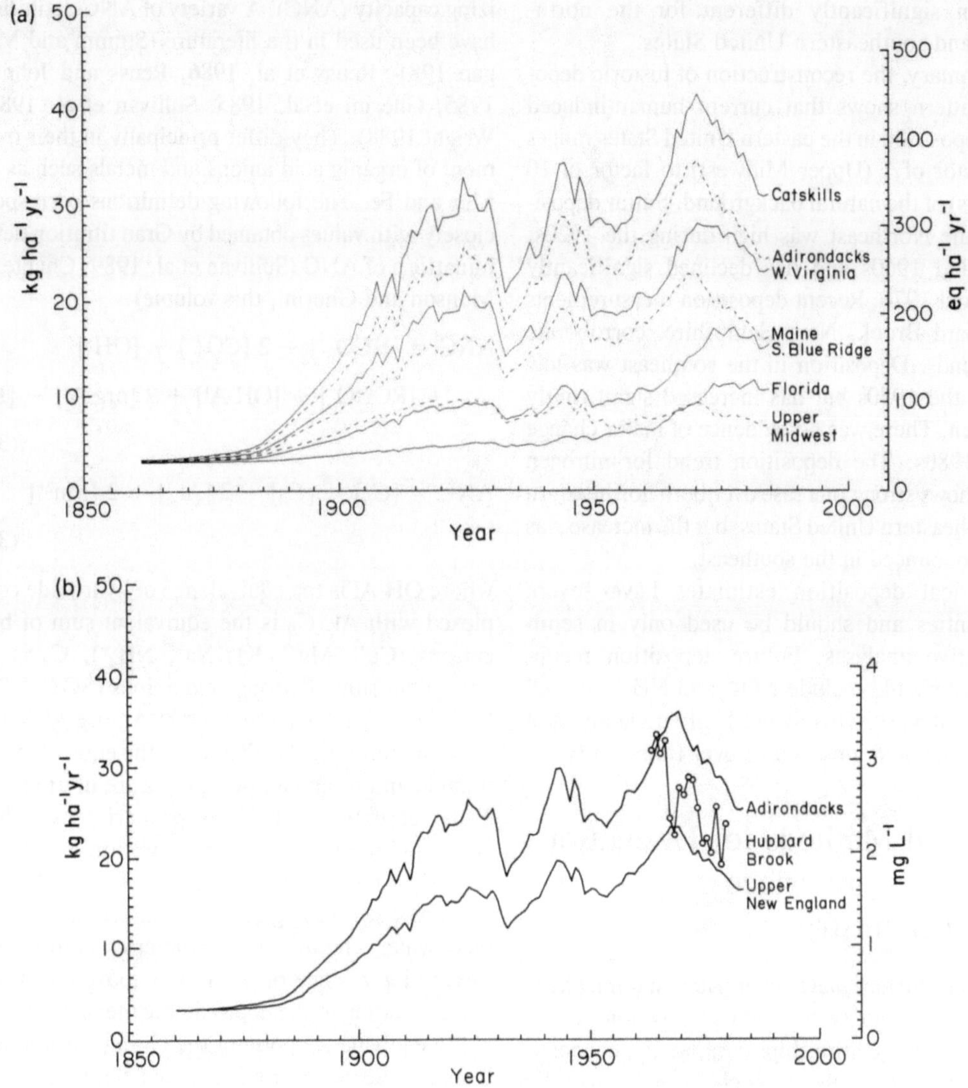

FIGURE 3.6. Historical wet sulfur deposition trends for seven regions in the eastern United States: Catskills; Adirondacks; W. Virginia; Maine; S. Blue Ridge; Florida; Upper Midwest. See text for explanation of calculations. (b) Historical wet deposition for the Adirondacks, New England, and Hubbard Brook (HB data from National Research Council 1983)

The deposition pattern for the lake region of the Upper Midwest (Figure 3.6) shows an increase during the early part of the century, reaching values three to four times the natural background in the 1920s, followed by a slow fluctuating decline. Currently, the anthropogenic wet sulfur deposition in the western parts of the Upper Midwest is comparable to the natural background. As noted in Chapter 13 (Cook and Jager, this volume), there is a significant east-west gradient of sulfur deposition, with highest deposition in eastern Michigan.

The comparison of wet deposition trends for the Northeast and the Southeast shows that over the Northeast, sulfur deposition was high during most of this past century, whereas in the Southeast, much of the increase over the natural background has occurred since the 1960s. Undoubtedly, historical sulfur-induced stress on aquatic ecosystems has been significantly different for the northeastern and southeastern United States.

In summary, the reconstruction of historic deposition pattern shows that current human-induced sulfur deposition in the eastern United States ranges from factor of 1 (Upper Midwest) to factor of 10 (Catskills) of the natural background. Sulfur deposition in the Northeast was high during the 1920s, 1940s, and 1960s and has declined significantly since about 1970. Recent deposition measurements at Hubbard Brook, New Hampshire, corroborate these trends. Deposition in the southeast was low prior to the 1950s but has increased significantly since then. There was no evidence of major change in the 1980s. The deposition trend for nitrogen oxides shows strong increase over both northeastern and southeastern United States, but the increase was more pronounced in the southeast.

Historical deposition estimates have severe uncertainties and should be used only in semi-quantitative analyses. Future deposition reconstructions should include NO_3^- and NH_4^+, as well as Ca^{2+} and Mg^{2+}. This would facilitate analysis of the full chemical ion balance over time.

Empirical Methods for Evaluation of Changes in Surface Water Chemistry

Changes in surface water chemistry that may have occurred in response to acidic deposition can be inferred from relationships among ionic constituents. Analyses of historical change are generally presented in the form of simple empirical models (Henriksen 1979, 1980, Wright 1983, Eshleman and Kaufmann 1988) or ion ratios (Schindler 1988b, Kaufmann et al. 1988). Several assumptions are implicit in the evaluation of empirical relationships, however, and results must therefore be interpreted with caution (Kramer and Tessier 1982). This section describes both the uses and limitations of empirical models and ratios. Together they constitute a valuable assessment tool, but they can also easily be overinterpreted.

Acid Neutralizing Capacity: Definition and Measurement

Empirical evaluations are simply a logical extension of a charge balance definition of acid neutralizing capacity (ANC). A variety of ANC definitions have been used in the literature (Stumm and Morgan 1981, Reuss et al. 1986, Reuss and Johnson 1985, Gherini et al. 1985, Sullivan et al. 1988a, Wright 1988). They differ principally in their treatment of organic acid anions and metals such as Al, Mn, and Fe. The following definitions correspond closely with values obtained by Gran titration determinations of ANC (Sullivan et al. 1989; Chapter 2, Munson and Gherini, this volume):

$$\mathrm{ANC} = [\mathrm{HCO_3^-}] + 2\,[\mathrm{CO_3^{2-}}] + [\mathrm{OH^-}] + [\mathrm{RCOO^-}] + [\mathrm{OH\text{-}Al}] + 2\,[\mathrm{org\text{-}Al}] - [\mathrm{H^+}] \quad (3\text{-}1)$$

$$\mathrm{ANC} = [C_B] - [C_A] + 2\,[\mathrm{Al}_m] + 2\,[\mathrm{Mn^{2+}}] \quad (3\text{-}2)$$

where OH-Al is the equivalence of hydroxide complexed with Al, C_B is the equivalent sum of base cations (Ca^{2+}, Mg^{2+}, K^+, Na^+, NH_4^+), C_A is the equivalent sum of strong acid anions (SO_4^{2-}, NO_3^-, Cl^-, F^-), Al_m is total monomeric Al, org-Al is non-labile monomeric Al, $RCOO^-$ is the equivalence of organic anions and all other units are in µmol L^{-1}. Thus, cationic Al behaves primarily as a base cation with respect to Gran titration ANC, and ΔAl must be evaluated separately from ΔANC in order to assess biologically relevant changes in surface water chemistry. Monomeric aluminum is assigned a valence of +2 in the above equations, corresponding to the approximate mean Al valence at the equivalence point of the Gran titration (Sullivan et al. 1989). Fe is ignored because, in the waters of interest, most Fe is likely to be bound to organic matter (e.g., Perdue et al. 1976). The contribution of organo-Al and organo-Fe complexes to Gran ANC are probably small, but uncertainties regarding organo-metal binding strength and electrostatic interactions preclude more rigorous treatment of these complexes (Barnard and Bisogni 1985, Driscoll et al. 1984). Where organic anions are present in significant concentrations, Gran titration values will underestimate the definition of ANC using Equation 3-2 because of curvature in the Gran function used to estimate the equivalence point of the titration (Sullivan et al. 1989) and low pK values for some organic functional groups.

The underestimation of Gran ANC, as compared to a defined ANC using Equation 3-2, will vary with DOC and the endpoint used in the titration.

Titrating to low pH (e.g., pH = 3) helps to minimize the curvature in the Gran function, and thus reduces the organic effects on Gran ANC (Sullivan et al. 1989). It also reduces the concentration of organic anions that are not protonated in the titration. An organic term ($RCOO_G^-$) can be included in C_A in Equation 3-2 to represent the influence of organic acids on Gran ANC. The value of $RCOO_G^-$ depends on organic interference in the Gran titration and the strength of organic acid functional groups relative to the endpoint used in the Gran titration. $RCOO_G$ was estimated for northeastern United States lake waters, using data from Phase II of the Eastern Lake Survey, as the difference between Gran ANC and ANC as defined by Equation 3-2. The resulting estimate varied with DOC (Figure 3.7):

$$RCOO_G^- \ (\mu eq\ L^{-1}) = 7.5 * DOC\ (mg\ L^{-1}) \quad (3\text{-}3)$$

and agrees with earlier estimates (Eshleman and Hemond 1985). It should be noted that $RCOO_G^-$ is not equivalent to the sum of total organic anions. Oliver et al. (1983) determined the carboxyl contents of humic materials from surface and groundwater samples collected throughout the United States and Canada. The carboxyl content ranged from 5 to 13 μeq mg^{-1} organic carbon, with a mean and standard deviation of 10 and 1.7, respectively. The concentration of $RCOO_G^-$ in Equation 3-1 will be somewhat lower than the total organic anion concentration, principally because some organic acids will remain unprotonated at the titration endpoint. Because organic acid anions affect measurement of ANC using both the definitions presented in Equations 3-1 and 3-2, they have been divided into "weak" and "strong" organic acid components (e.g., Eshleman and Hemond 1985). Such a distinction is subjective, however, because it depends on both the concentration and acid characterics (pK distribution) of the organic materials and also on the conditions under which the titration was performed.

Interpretation of the exact cause of the discrepancy between Gran ANC and ANC defined according to Equation 3-2 (Figure 3.7) is not important if we are interested in quantifying the influence of organic acids on Gran ANC. Regardless of the relative importance of curvature in the Gran function (Sullivan et al. 1989), protonation of adjacent functional groups of polyprotic organic acids as solution hydrogen ion concentration increases (Oliver et al. 1983, Stevenson 1982), or strength of carboxylic acidity (Eshleman and Hemond 1985), the end result is the same. Measured Gran ANC is somewhat lower than defined ANC because of the presence of organic acids. Whether the data points are shifted to the left (e.g., curvature) or shifted up (e.g., low pKa) in the data presented in Figure 3.7, the bias is largely attributable to the presence of organic acids and can be approximated for northeastern United States lakes by Equation 3-3.

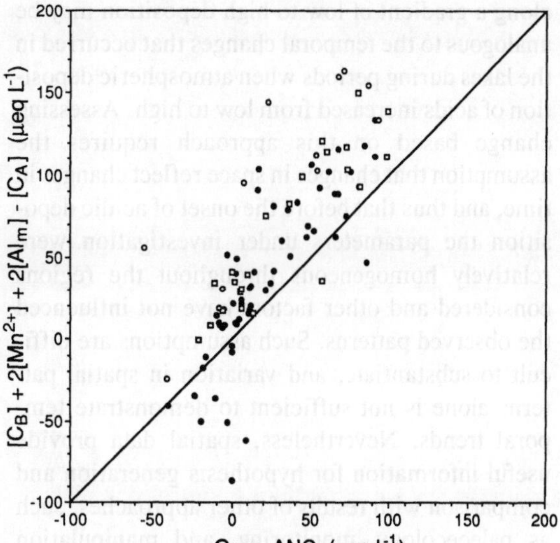

FIGURE 3.7. Acid neutralizing capacity defined as $[C_B] + 2[Mn^{2+}] + 2[Al_m] - [C_A]$ versus ANC determined by Gran titration. C_B is the equivalent sum of base cations ($Ca^{2+} + Mg^{2+} + Na^+ + K^+ + NH_4^+$), C_A is the equivalent sum of acid anions ($SO_4^{2-} + NO_3^- + Cl^- + F^-$), Al_m is total monomeric Al. The multiplier used for Al_m, +2, corresponds to the average valence of monomeric Al near the equivalence point of Gran titrations for dilute waters (pH 4.8 to 5.2) (Sullivan et al. 1989). Data points are coded by dissolved organic carbon: > 6 mg L^{-1} (500 μmol L^{-1}), ○ ; 4 to 6 L^{-1}, □; ≤ 4 mg L^{-1} (333 μmol L^{-1}), ●. The vertical distance from each data point to the 1:1 line is an estimate of organic acid influence on Gran titration estimation of ANC.

Spatial Patterns

Current spatial patterns in water chemistry parameters across a gradient of atmospheric deposition provide useful information for evaluating historical change. Differences in surface water chemistry along a gradient of low to high deposition may be analogous to the temporal changes that occurred in the lakes during periods when atmospheric deposition of acids increased from low to high. Assessing change based on this approach requires the assumption that changes in space reflect changes in time, and thus that before the onset of acidic deposition the parameters under investigation were relatively homogeneous throughout the regions considered and other factors have not influenced the observed patterns. Such assumptions are difficult to substantiate, and variation in spatial patterns alone is not sufficient to demonstrate temporal trends. Nevertheless, spatial data provide useful information for hypothesis generation and comparison with results of other approaches, such as paleoecology, monitoring, and manipulation experiments.

Spatial analyses are most easily interpreted when performed separately on data from groups of lakes and streams that had relatively homogeneous chemical characteristics before the onset of atmospheric deposition. It is difficult to define these groups, however, because surface waters and watersheds vary in their current chemistry and response to acidic inputs (see Chapter 1, Munson and Gherini, this volume). In areas suspected of having little historical change, the sensitivity of spatial analyses can be optimized by focusing on designated subsets of homogeneous surface waters and deleting systems that are unlikely to have changed or that have particularly large uncertainties. The results of such spatial analyses would then apply only to the subset under investigation, but much information could be extracted for the systems of greatest biological concern.

Aquatic systems within a relatively homogeneous region can be subset in a variety of ways, such as hydrologic type (drainage, groundwater recharge, groundwater flowthrough), DOC, ANC, and base cation concentrations. Particular care must be taken, however, to ensure that patterns are not introduced into spatial investigations by the way in which subsets are chosen.

An example of analyzing spatial patterns to learn about historical change is the analysis by Sullivan et al. (1988) of spatial variations in ANC and SO_4^{2-}, based on $Ca^{2+} + Mg^{2+}$ stratification, using NSWS data. This analysis showed that measured SO_4^{2-} and ANC were inversely related, in agreement with Equation 3-2, provided base cation concentrations were generally constant. Regression relationships were not presented, however, because it was unclear to what extent the observed patterns were attributable to increased mobilization of base cations by SO_4^{2-}. Neary and Dillon (1988) presented data for low conductance lakes ($\leq$ 50 $\mu S\ cm^{-1}$) across a large depositional gradient in Ontario, showing a spatial pattern of decreasing ANC and pH with increasing SO_4^{2-}. However, this relationship may have been due in part to the subset conductance criteria. For example, an increase in lakewater $[SO_4^{2-}]$ by up to 100 $\mu eq\ L^{-1}$ was simulated by Sullivan (unpublished) in low-deposition subregions of the National Lake Survey (northeastern Minnesota and western subregions). The added SO_4^{2-} was charge-balanced on an equivalent basis in the simulation with $[Ca^{2+} + Mg^{2+}]$. The spatial data suggested that median ANC had decreased by 5 to 30 $\mu eq\ L^{-1}$ in all subregions after deletion of lakes calculated to have exceeded the conductance limit of 50 $\mu S\ cm^{-1}$, because increased $[SO_4^{2-}]$, $[Ca^{2+}]$, and $[Mg^{2+}]$ caused some of the higher ANC lakes to exceed the conductance criterion, resulting in lower median ANC for the remaining subset that had received large SO_4^{2-} and base cation additions.

Empirical Models

Empirical models for estimating historical changes in surface water chemistry can be derived by assuming that certain of the parameters included in Equation 3-2 may have changed appreciably in response to changing concentrations of SO_4^{2-} and/or NO_3^-, whereas other parameters are likely to have been relatively unaffected. For example, Henriksen (1979, 1980) initially assumed that only the ANC change was appreciable in response to increased $[SO_4^{2-}]$. (Henriksen's definition of ANC treats Al^{n+} as an acidic cation, similar to H^+ in Equation 3-1, and change in labile Al concentration from preindustrial times to the present, ΔAl^{n+}, is therefore not included explicitly as a model parameter). Subsequently, Henriksen (1982) and Wright (1983) presented evidence for Norwegian and North American lakes, respectively, suggesting that increased base cation release accounted for up to 40% of added SO_4^*, whereas the additional 60% to 100% of SO_4^* input replaced bicarbonate alkalinity. The proportional change in base cations relative to $[SO_4^*]$ is referred to as the *F*-factor.

$$F = \Delta[C_B^*]/\Delta[SO_4^*] \quad (3\text{-}4)$$

It is generally assumed that most or all the base cation change is attributable to changes in $[Ca^{2+}]$ and $[Mg^{2+}]$. Incorporation of the *F*-factor concept into the ANC definition presented in Equation 3-2 yields an empirical model for estimating historical change in ANC as

$$\begin{aligned} \Delta ANC &= \Delta[C_B^*] + \Delta[Al_i] - \Delta[SO_4^*] \\ &= (F \times \Delta[SO_4^*]) + \Delta[Al_i] - \Delta[SO_4^*] \end{aligned} \quad (3\text{-}5)$$

This estimator of ΔANC (Equation 3-5) will generally yield results similar to that proposed by Henriksen (1979), ΔANC = ([Ca*] + [Mg*] − ANC), if *F* is assumed equal to 0, but requires less restrictive assumptions (Eshleman and Kaufmann 1988) and includes the Al change in a manner consistent with Gran ANC. It assumes that Cl^-, organic acid anions, and nitrogen species have not changed, that marine sources of SO_4^{2-} and base cations can be subtracted where appropriate using measured or estimated marine Cl^- and the ionic composition of sea water, that pre-industrial concentrations of labile monomeric Al (Al_i) in surface waters were negligible (Chapter 18, Sullivan, this volume), and that pre-acidification background SO_4^* can be estimated for the waters of interest (Chapter 18, Sullivan, this volume).

The major uncertainties associated with estimating loss of ANC using Equation 3-5 are (1) estimation of an appropriate *F*-factor to account for mineral acid neutralization via base cation release, (2) estimation of an appropriate regional background $[SO_4^*]$ from which to estimate $\Delta[SO_4^*]$, and (3) the assumption of temporally constant organic acid anion concentrations. Although increased deposition of nitrogen species has occurred (Gschwandtner et al. 1985), these compounds are generally rapidly assimilated by vegetation, and were present in low concentrations in most ELS and WLS lakes during autumn sampling (Landers et al. 1987, 1988, Eilers et al. 1988c). Where $[NO_3^-]$ is elevated, as, for example, in Catskill streams (Chapter 8, Stoddard and Murdock, this volume), $[NO_3^-]$ should be included in an empirical evaluation. It is likely that base cation neutralization varies as a function of initial base cation concentration, and *F* will be low in watersheds where carbonic acid weathering yields low surface water base cation concentrations. In contrast, watersheds in which carbonic acid weathering yields relatively high surface water $[C_B]$ are more likely to exhibit greater neutralization of SO_4^{2-} acidity by increasing base cation release. Thus, *F* should approach 1.0 at higher initial C_B concentrations. Unfortunately, there is little basis at present for describing the distribution of *F*-factors for a population of lakes, and there is no justification for the choice of a single *F*-factor to describe population-level change in base cation release in North American surface waters. In some cases, useful information can be obtained by assuming that $F = 0$ in order to assess the maximum possible change in ANC attributable to SO_4^{2-} input (Eshleman and Kaufmann 1988). In other cases, a maximum "reasonable" *F* can be calculated for a given lake or stream by assuming a lower reasonable level for initial pre-acidification C_B. Maximum *F* is then estimated as

$$F_{max} = ([C_{B_1}^*] - [C_{B_0}^*])/\Delta SO_4^* \quad (3\text{-}6)$$

where subscripts 1 and 0 refer to current concentration and a reasonable lower limit for initial concentration, respectively.

The concentration of SO_4^{2-} in precipitation has been estimated as approximately 5 to 7 $\mu eq\ L^{-1}$ in remote areas of the world (Galloway et al. 1984, 1987). Brakke et al. (1989) used these data and an assumed 50% evapotranspiration rate to estimate an upper bound of 10 to 15 $\mu eq\ L^{-1}$ for background SO_4^{2-} in eastern United States lakes having low concentrations of base cations. This assumes that watershed sources of SO_4^{2-} are inconsequential for low base cation systems, which is supported by analyses of lakewater $[SO_4^{2-}]$/precipitation $[SO_4^{2-}]$ in low ANC northeastern lakes (Sullivan et al. 1988; Chapter 18, Sullivan, this volume). Watershed sources of $[SO_4^{2-}]$ are probably of greater significance in lakes having higher ANC and base cation concentrations (Wright 1983, Sullivan et al. 1988).

The issue of potential change in organic acid anion concentrations in response to acidic deposition has largely been ignored in the acid deposition literature. A loss of dissolved organic carbon (DOC) and $RCOO^-$ has been hypothesized (Almer et al. 1974, Krug and Frink 1983) and the topic was recently reviewed by Marmorek et al. (1988), who concluded that there are inconsistencies in the available data, but most evidence suggests that organic acids have been lost from surface waters in response to increased acidic deposition. Proposed mechanisms include decreased organic acid mobilization and dissociation, and increased loss of DOC to sediments through chelation with metals (e.g., Al, Fe) (Marmorek et al. 1988). The best evidence available to date that bears on this issue is a paleoecological study of diatom inferred ΔDOC for two lakes in Norway by Davis et al. (1985), suggesting a historic decrease of 3 to 6 mg L^{-1} DOC. Diatom transfer functions are not yet available for the United States, but regional DOC distributions across a sulfur deposition gradient suggest that lakewater organic concentrations may have decreased in the northeastern United States in response to sulfur deposition (Chapter 18, Sullivan, this volume).

Although inorganic Al change can be estimated on the basis of current concentrations, assessment of the proportionality constants for base cations and organic acid anions requires use of an additional tool, such as paleolimnology, for quantification. In the absence of regional quantification that employs an independent tool such as paleolimnology or long-term monitoring, empirical models are of limited value for quantification of historical change in water chemistry.

Ion Ratios

Whereas empirical models are used to quantify change in water chemistry under a particular suite of assumptions, ion ratios offer a more qualitative assessment of chemical change. Perhaps the most commonly used are two ratios that reflect the interrelationships between SO_4^{2-}, base cations, and ANC:

$$ANC/[C_B] \quad \text{and} \quad [SO_4^{2-}]/[C_B]$$

Interpretation of both ratios is often based on the assumptions that pristine low DOC surface waters typically exhibit a near 1:1 ratio of base cations (corrected for marine contributions) to ANC (Henriksen 1979) and that the principal determinants of ANC are base cations and SO_4^{2-}. The C_B term in these ratios is generally limited to [Ca* + Mg*]. If Na^+ and/or K^+ are associated with appreciable alkalinity sources in a particular region, then these cations should also be included (Kramer and Tessier 1982). For example, ANC is approximately equal to $[Ca^{2+} + Mg^{2+}]$ in low ANC drainage lakes in the Pacific Northwest subregion of the Western Lake Survey (Landers et al. 1987), whereas $[Ca^{2+} + Mg^{2+}]$ values are lower than ANC in the California subregion (Figure 3.8). These data and the very low K^+ concentrations of California lakes (Landers et al. 1987) suggest that Na^+ is associated with alkalinity production in California lakes, for example, via carbonic acid weathering of albite ($NaAlSi_3O_8$) (Stumm and Morgan 1970; Chapter 15, Melack and Stoddard, this volume). The ANC/$[C_B]$ ratio has been misinterpreted in the acid deposition literature. A ratio $<<$ 1 does not necessarily indicate that acidification–loss of ANC, as defined by Galloway et al. (1984)–has occurred. Two assumptions are implicit in the use of this ratio. The ratio approximates 1.0 only where surface water organic acid anion concentrations ($RCOO^-$) are low and watershed sources of SO_4^{2-} are minimal. Organic acid anions tend to lower Gran ANC relative to base cation concentrations, as appears to be the case in northeastern Minnesota (Figure 3.8). Watershed sources of SO_4^{2-} are derived from weathering reactions yielding base cations charge balanced by SO_4^{2-}, rather than HCO_3^-. This also lowers the ANC/$[C_B]$ ratio. Even in low DOC waters lacking watershed sources of SO_4^{2-}, a ratio $<<$ 1 implies only that surface water chemistry has changed. The change could be due to increased $[C_B]$, decreased ANC, or a combination of both. Although one could argue that there is a finite limit to increased base cation release, the ratio alone does not demonstrate acidification.

The ratio $[SO_4^{2-}]/[C_B]$ quantifies the SO_4^{2-} concentration relative to surface water susceptibility to acidification. The most important factor among those determining whether or not adverse effects will occur from acidic deposition is the inherent susceptibility of the watershed, as reflected in surface water base cation concentrations (Wright 1988; Chapter 2, Munson and Gherini, this volume). High $[SO_4^{2-}]$ generally is biologically significant only where $[C_B]$ is low. Where $[SO_4^{2-}]/[C_B] > 1$, water is acidic (ignoring Al) because of high $[SO_4^{2-}]$, irrespective of organic acid anion concentrations. Similarly, if $[RCOO_G^-]/[C_B] > 1$, where $[RCOO_G^-]$

is defined as in Equation 3-3, water is acidic because of high organic acid concentrations.

In summary, a variety of empirical approaches has been used to assess historical changes in surface water chemistry. These are generally presented in the form of ratios between ionic constituents and simple empirical models, both of which are based either implicitly or explicitly on a charge-balance definition of ANC. Ion ratios constitute a very useful qualitative tool for historical assessment. Empirical models are more quantitative, but are of limited value in the absence of independent data with which to specify parameter values. The most significant limitation is the scarcity of information regarding changes in base cation release in response to changes in mineral acid inputs.

Paleoecological Methods for Inferring Recent Acidification Trends

The use of paleoecological techniques to assess acidification trends in aquatic ecosystems is increasing rapidly (Charles et al. 1990). Analysis of the chronostratigraphic record contained in lake sediments is currently the best approach for determining change in lake acidity status from pre-industrial times to the present. Of the various sediment components, the remains of diatoms and scaled chrysophytes provide the best data for quantitatively reconstructing past lakewater pH and alkalinity (Battarbee 1984, Charles and Norton 1986, Smol et al. 1986, Davis 1987). Diatoms have been used to infer acidification trends in lakes in several countries, including the United States, Canada, Norway, Sweden, Finland, West Germany, Netherlands, England, Scotland, and Wales (Charles and Norton 1986, Battarbee and Charles 1987, Smol et al. 1986, Charles et al. 1990). In the United States and Canada, sediment cores from over 100 lakes have been analyzed stratigraphically to reconstruct historical lake pH trends (Charles et al. 1990). A large proportion of the lakes, including those mentioned in the case study chapters of this book, were studied as part of the PIRLA project (Paleoecological Investigation of Recent Lake Acidification) (Charles and Whitehead 1986). Several sediment characteristics

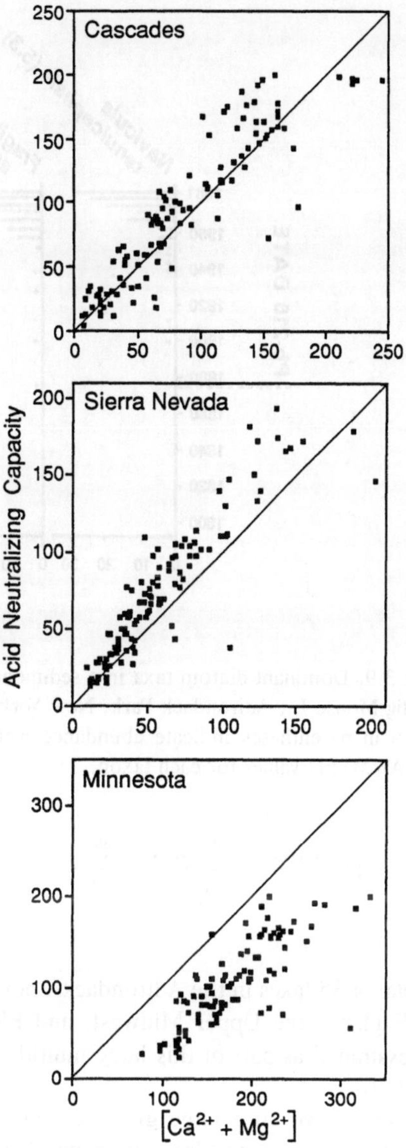

FIGURE 3.8. Acid neutralizing capacity (µeq L^{-1}) versus [Ca^{2+} + Mg^{2+}] (µeq L^{-1}) for lakewater in three case study regions that receive low levels of acidic deposition. The slight deviation above the 1:1 line for Sierra Nevada waters suggests another base cation (e.g., Na^+) is associated with alkalinity production in Sierra Nevada watersheds. The deviation below the 1:1 line for lakewater in Minnesota illustrates organic anion influence on ANC. Median DOC in lakewater in Minnesota was 9.1 mg L^{-1}, compared with 0.8 and 1.0 mg L^{-1} in the Cascades and Sierra Nevada, respectively.

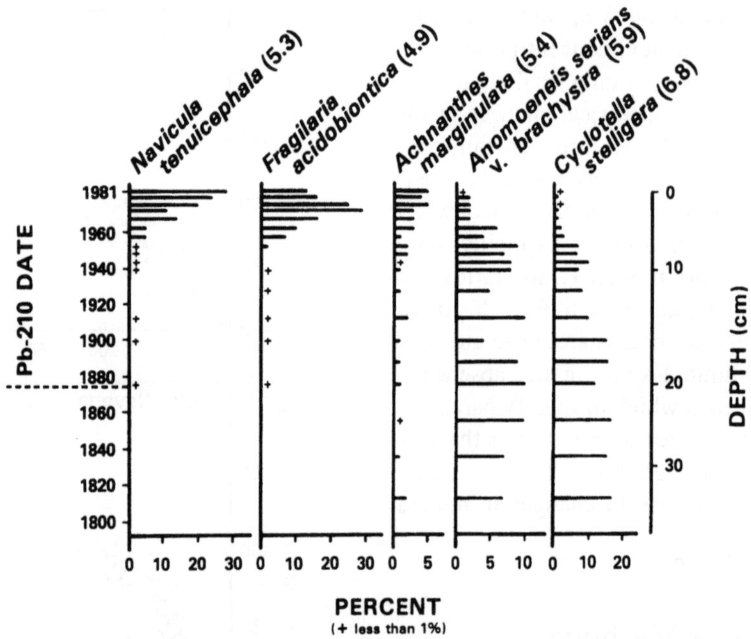

FIGURE 3.9. Dominant diatom taxa in a sediment core from Big Moose L., Adirondack Park, New York. The numbers in parentheses indicate abundance weighted mean (AWM) pH values for each taxon,

$$\text{pH AWM} = \sum_{i=1}^{56} P_i(X_i) / \sum_{i=1}^{56} P_i$$

P_i = the percentage occurrence of the taxon in the sediment assemblage from lake i; X_i = the mean air-equilibrated surface sample pH of lake i (determined from mean of H^+ concentrations), calculated from distribution of the taxa in 56 Adirondack lakes. ^{210}Pb dates determined by Stephen Norton, University of Maine, Orono. Dates below dashed line are extrapolations of ^{210}Pb dates.

of a total of 35 lakes in the Adirondacks, northern New England, the Upper Midwest, and Florida were examined as part of this large multidisciplinary study.

Diatoms make up a large group of single-cell freshwater and marine algae (division Bacillariophyta). They have silicious cell walls and are formed of two halves or valves. Scaled chrysophytes (Chrysophyceae; Mallomonadaceae) are primarily freshwater plankton. They have flagella and an external covering of overlapping siliceous scales and bristles or spines.

Diatom assemblages in sediment are good indicators of past lakewater pH because (1) diatoms are common in nearly all freshwater habitats, (2) distributions of diatom taxa are strongly correlated with lakewater pH, (3) diatom remains are preserved well in sediment and can be identified to the lowest taxonomic level, (4) their remains are usually abundant in sediment (10^4 to 10^8 valves/cm^3 of sediment) so that rigorous statistical analyses are possible, and (5) many taxa are usually represented in sediment assemblages (20 to 100 taxa per count of 500 valves is typical), so that inferences are based on the ecological characteristics of many taxa. Some disadvantages in using diatoms as pH indicators are that (1) diatom identification requires considerable taxonomic expertise, (2) occasionally diatoms are not well preserved because of dissolution—for example, in some peaty and some calcareous sediments, (3) sometimes the number of taxa is low, as in some bog lakes, (4) calibration data sets (used to quantify the relationships between water chemistry and surface sediment diatom assemblages) are not always available for the lake region studied, and (5) good ecological data are not always available for all dominant taxa. For regional studies of lake acidity change, the advantages greatly outweigh the disadvantages.

In general, the use of chrysophyte scales for pH reconstructions involves the same advantages and disadvantages as diatoms (Smol et al. 1986), except that the number of chrysophyte taxa in a sediment assemblage is typically about one-tenth the number of taxa of diatoms. All scaled chrysophyte taxa are euplanktonic (normally suspended in the water). Chrysophytes are sometimes more abundant than diatoms in acidic lakes.

Techniques for Determining pH Trends

Several techniques have been developed that use diatom assemblages to assess trends in acidification and to derive equations for inferring lakewater pH. The simplest and most straightforward approach is to count sediment-core diatom and chrysophyte assemblages and prepare depth profiles of percentages of the dominant taxa. Changes in the profiles are then interpreted in light of the ecological data available on the taxa. An example is the diatom stratigraphy of Big Moose Lake in the Adirondack Mountains, New York (Figure 3.9).

Hustedt (1939) made one of the first significant steps toward establishing a more quantitative approach for using diatoms as pH indicators, by defining the following pH occurrence categories:

- Acidobiontic (ACB)—optimum distribution at pH less than 5.5
- Acidophilic (ACP)—widest distribution at pH less than 7.0
- Circumneutral/indifferent (IND)—distributed equally above and below pH 7.0
- Alkaliphilic (ALK)—widest distribution at pH greater than 7.0
- Alkalibiontic (ALKB)—occurs only at pH greater than 7.0

Assignments of diatom taxa to categories can be based on literature references and on the distribution of taxa within waters of particular geographic regions. Changes in the percentages of diatom valves in each pH category in a sediment core can be used to estimate trends in lakewater pH.

Several types of equations have been developed for calculating lakewater pH from diatom data using these categories. Predictive equations are calibrated for a study region using measured lakewater pH and surface sediment diatom assemblage data for at least 20 to 30 lakes within the

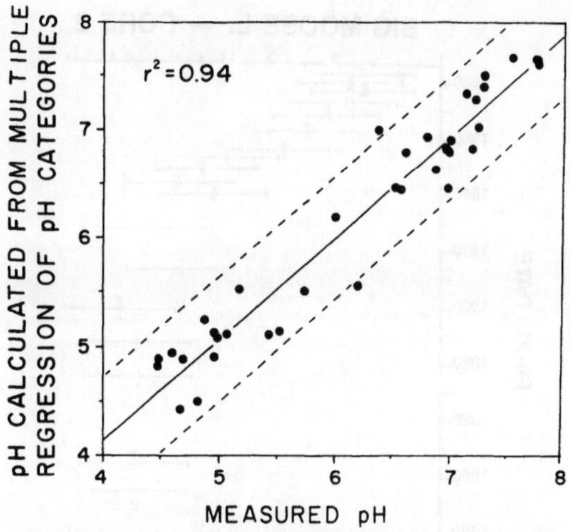

FIGURE 3.10. Diatom inferred lakewater pH versus measured pH for 37 Adirondack lakes, and 95% confidence intervals for an individual prediction of pH from diatom data; pH = 8.14 − 0.041 ACB − 0.034 ACF − 0.0098 IND − 0.0034 ALKF; r^2 = 0.94, SE = ± 0.28 pH units.

region (Figure 3.10). In the United States and Canada, at least 15 calibration data sets have been developed, involving more than 700 lakes (Charles et al. 1990). Until recently, the most successful approach was the use of multiple linear regression analysis of measured lakewater pH with percentage of diatoms in each Hustedt pH category (Charles 1985, Charles et al. 1990). The r^2 values for the best sets of equations range from 0.80 to 0.94; standard errors of estimates from ± 0.4 to ± 0.25 pH units (Battarbee and Charles 1987). The multiple regression approach is generally better than using the earlier indices calculated as ratios of the percentages of diatom valves in Hustedt pH categories [indices α, ω, and ε of Nygaard, and Renberg and Helberg's index B; Battarbee et al. (1986) discuss the historical development of indices].

Several new techniques have been developed that provide inference equations based on abundance of individual taxa instead of categories. They can be used to infer a variety of water chemistry characteristics. The most successful are based on weighted averaging and canonical correspondence analysis (CCA) (Ter Braak 1986, Oksanen et al. 1988, Stevenson et al. 1989). Other new techniques have been developed that use chrysophytes

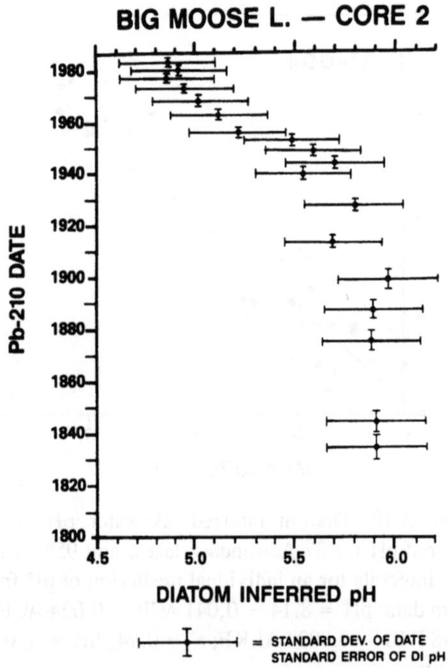

FIGURE 3.11. Diatom inferred pH for Big Moose Lake, Adirondack Mountains, New York, using the predictive relationship described in Figure 3.10. Horizontal bars represent standard error estimate of pH; vertical bars are standard deviation of ^{210}Pb dates (Michael Binford, Harvard School of Design, pers. comm.).

both alone and with diatoms for inferring pH (Charles and Smol 1988).

Once developed, predictive equations can be applied to diatom assemblage data from lake sediment cores to infer past pH conditions (Figure 3.11). Trends within cores can be analyzed statistically using, for example, change-point analysis to determine if they are significant. Inferred pH data can be compared with stratigraphies of other lake sediment characteristics such as pollen, charcoal, coal and oil soot, polycyclic aromatic hydrocarbons, Pb, Zn, Cu, V, Ca, Mg, Ti, Al, Si, S, and others that provide a record of watershed disturbance and atmospheric inputs of materials associated with the combustion of fossil fuels (Charles and Norton 1986). With these data, in addition to knowledge of watershed events and some historical information on regional atmospheric emissions of sulfur and nitrogen, it is often possible to assess with reasonable certainty whether lakes have been affected by acidic deposition, and to what extent.

Dates associated with sediment core depths are nearly always determined from measurements of ^{210}Pb activity (half-life 23 years). This method is used widely in paleolimnological studies to date recent sediments. Errors associated with the method can be estimated. Whenever possible, ^{210}Pb dates are compared with independent chronostratigraphic markers to evaluate the accuracy of the dates. Comparisons that have been made between ^{210}Pb dates and dates derived from pollen profiles (land use changes), charcoal (fires), liming events, soil disturbance, or varved laminations, have nearly always shown good agreement.

Comparisons between results of paleoecological studies and output from lake and watershed computer models can be particularly informative. Past water chemistry inferred from the sediment record can be compared with that from model hindcasts to help validate the models. The more complete water chemistry data that can be derived from the models can be used to help explain trends inferred from the sediments.

Paleoecological assessments of lake pH trends are summarized in this volume for several regions: Adirondacks (Chapter 6, Driscoll et al.), Maine (Chapter 7, Kahl et al.), Florida (Chapter 12, Pollman and Canfield), the Upper Midwest (Chapter 13, Cook and Jager), the Rocky Mountains (Chapter 14, Turk and Spahr), and the Sierra Nevada (Chapter 15, Melack and Stoddard).

Paleolimnological studies involving diatom and chrysophyte analysis will no doubt continue to be a primary and increasingly important tool in lake acidification studies.

4
Effects of Acidification on Biological Communities in Aquatic Ecosystems

Joan P. Baker and Sigurd W. Christensen

ABSTRACT. Extensive information is available on the effects of acidification on biological communities in aquatic ecosystems. Whole-system experiments, mesocosm experiments, and field surveys have demonstrated major shifts in species composition and decreases in species richness with increasing acidity. Sensitive species may be lost even at moderate levels of acidity. For example, effects on important zooplankton predators, such as *Mysis relicta* and *Epischura lacustris*, occur at pH 5.6 to 5.9; acid sensitive mayfly and stonefly species (e.g., *Baetis lapponicus*) are affected at pH levels near 6.0; and sensitive fish species, such as the fathead minnow (*Pimephales promelas*), experience recruitment failure and extinction at pH 5.6 to 5.9. At the same time, however, acid tolerant species may appear or increase in abundance, resulting in little or no overall decrease in standing crop. Attached filamentous algae, especially *Mougeotia* spp., often become visibly more abundant in acidified lakes and streams. Many ecosystem level processes (primary productivity, rates of decomposition, and nutrient concentrations) also appear relatively robust and unaffected at least at moderate levels of acidification (pH 5.0 to 5.6). Specific mechanisms for effects of acidification on aquatic biota have been studied particularly for fish communities. The toxicity of acidic waters to fish is determined primarily by pH, calcium, and inorganic monomeric aluminum. Fish population declines associated with acidification may result most commonly from recruitment failure. Relatively few quantitative models are available for predicting biological responses to acidification; most deal with effects on fish populations. All these models require further development and testing prior to their use for regional assessments of effects from acidic deposition.[1]

Introduction

Changes in surface water acid-base chemistry can alter the structure and function of aquatic communities. The objectives of this chapter are to briefly summarize the types of biological effects expected to occur as a result of surface water acidification, the approximate magnitude and significance of these effects, and the levels of acidity (and associated parameters) at which specific changes occur. Methods that can be used to make regional assessments of effects on aquatic biota are also reviewed. This information provides a framework for further discussions of biological effects and status in each regional case study chapter.

Biological responses of interest include measures of both *pattern*, such as species composition, and *process*. Important processes include (1) system production, for example, primary production, (2) nutrient cycling, which in turn influences system production, and (3) decomposition (the rate of decomposition is one factor limiting rates of nutrient cycling). The components of the biological community considered are plankton (zooplankton and phytoplankton), periphyton, macrophytes, benthic macroinvertebrates, and fish. Other organisms, for example, amphibians and waterfowl, also may be affected by surface water acidification, but are beyond the scope of this brief review.

The case study analyses presented elsewhere in this book cover a broad range of aquatic resources across North America. Our understanding of biological responses to acidification is probably best, however, for lakes in northern climates, based on studies in the northern United States, eastern Canada, and Scandinavia. For streams, and also for ecosystems in the southeastern and western United States, the available data are more limited and our understanding less complete. Thus, to a degree, the following summary of biological effects

TABLE 4.1. Summary of results from the experimental acidification of Lake 223, Northwestern Ontario.[a,b]

Aquatic biota	1977 6.13	1978 5.93	1979 5.64	1980 5.59	1981 5.02	1982 5.09	1983 5.13
Phytoplankton	Some shift in relative abundance of species; increase in green algae.	Changes in species composition; no change or slight increase in production.	Changes in species composition more pronounced; decline in chrysophyceae.	Increase in total biomass and abundance of dinoflagellates and blue-green algae.	Drastic decline in chrysophyceae; replaced by dinoflagellates and blue-green algae; productivity remains above normal.	Species composition stabilizes; total biomass and productivity near "normal."	Species composition and biomass stable.
Zooplankton		Loss of copepod species: *Diaptomus sicilis*.	Sharp decline in abundance of copepod *Epischura lacustris*.	Loss of *E. lacustris*; first appearance of *Daphnia catawba*; increase in total biomass.	Marked increase in abundance of cladocera; balanced by loss of other species.	Species composition stabilizes; total biomass near "normal."	Species composition and biomass stable.
Benthic algae			Thick mats of filamentous algae (*Mougeotia*) appear in littoral areas.			Mats partially cover trout spawning beds; shift in spawning location observed.	Periphyton production similar to reference despite abundance of filamentous algae.
Invertebrates	Increase in chironomid emergence.	Sharp decline in abundance of *Mysis relicta*; further increase in chironomids.	Exoskeletons of crayfish harden slowly after molting.	Crayfish recruitment failure; crayfish diseased; chironomid abundance at all-time high.	Decline in crayfish abundance; recruitment failure.	Crayfish nearly extinct; chironomid emergence decreases to pre-acidification levels.	Crayfish absent; mayfly *Hexagenia* sp. absent.
Forage fish		Fathead minnow experience recruitment failure.	Fathead minnow near extinction; slimy sculpin decline in abundance.	Increase in abundance of pearl dace; suckers very abundant.	White sucker recruitment failure; no effect on adult growth and survival.	Recruitment failure for all species.	Recruitment failure for all species.
Lake trout	Increase in condition; lake trout "fatter."		Increased abundance of young-of-the-year.	Lake trout recruitment failure; condition similar to pre-acidification.	Recruitment failure; no effect on adult growth and survival.	Lake trout condition poor; recruitment failure; reduced adult survival.	Lake trout condition very poor; recruitment failure; reduced adult survival.

[a] Shown by year and pH units.
[b] Data from Schindler et al. 1985a, Mills et al. 1987; adapted from Malanchuk and Turner 1987.

TABLE 4.2. Preliminary results from the experimental acidification of Little Rock Lake, Wisconsin.[a]

pH (year)	Observed change in the acidified basin relative to the reference basin at pH 6.1
5.6 (1985 and 1986)	Increased prevalence of benthic algae (principally *Mougeotia*), covering 12 to 15% of the littoral zone.
	Increased lake transparency; spectral shift towards shorter wavelengths.
	No overall change in epilimnetic chlorophyll *a*, ^{14}C measures of primary productivity, or nutrient concentrations.
	No clear change in phytoplankton species composition in 1985; second year of acidification accompanied by an increase in proportion of cyanophytes and pyrrhophytes, and a decline in chrysophytes.
	Change in zooplankton species composition: (1) increased abundance of *Keratella taurocephala*, (2) decrease in *Keratella cochlearis*, (3) substantial decline in *Epischura lacustris*, and (4) distinct increase in *Daphnia schoedleri*.
	No change in abundance or species composition of benthic invertebrate community.
	Reduced emergence of chironomid subfamily Tanytarsini, one of four important chironomid subfamilies in the lake.
	No change in macrophyte species or distribution.
	Decreased fungal growth on decomposing leaves, but reduced rates of leaf decomposition observed for only one of four species of leaves examined.
	No change in the reproductive activity of yellow perch, largemouth bass, black crappie, and rock bass.
5.1–5.2 (1987)	Further increase in extent of coverage by benthic algae, covering almost all of the littoral zone.
	Loss of *Leptodora kindtii*; no change in abundance of *Chaoborus*.

[a] Data from Brezonik et al. 1986, Detenbeck 1987, Sierszen 1988, Frost and Montz 1988, Watras and Frost 1989, Frost et al. 1988.

from acidification reflects this bias toward northern lakes. Where differences in response for other types of systems or regions are recognized, they are noted. Otherwise, more detailed region-specific information is provided in the individual chapters for each region.

Biological responses to acidification are complex. Acidification may influence biological communities and processes both directly (through direct physiological stress and toxicity) and indirectly (via changes in food availability or predation). Community interactions are extremely important. As a result, whole-system acidification experiments, which incorporate these community interactions, are an essential component of studies on the biological effects of acidification (Ravera 1986, Schindler 1987). Bioassays and microcosm/mesocosm experiments may be used to define physiological limits or to test specific mechanisms of response. Field surveys provide information on biological responses in natural settings, useful for evaluating the validity of experimental results across a broad range of environmental conditions. Together these approaches form a complementary body of information (Watras and Frost 1989).

Unfortunately, relatively few whole-system acidification experiments have been conducted. One of the best examples involved Lake 223 in the Experimental Lakes Area (ELA) of Ontario. Over a period of eight years, Lake 223 was gradually acidified, with sulfuric acid, from pH 6.8 to 5.0; observed biological responses are summarized in Table 4.1 and in Schindler et al. (1985a). A similar whole-lake experiment is underway at Little Rock Lake, Wisconsin (Brezonik et al. 1986, Watras and Frost 1989). The lake was divided in half in late summer 1984, and the pH in the treatment basin was lowered with sulfuric acid from pH 6.1 to pH 5.6 in spring 1985, maintained at pH 5.6 through 1986, and then lowered to pH 5.1–5.2 in spring 1987. Preliminary results from this experiment are summarized in Table 4.2. Preliminary results are also available for Lake 302 at the Experimental Lakes Area, Ontario (Table 4.3). Divided in half in 1982, the south basin of the lake (302S) was acidified with sulfuric acid from pH 6.6 to 6.9 in 1982 to pH 5.6 in 1984; the north basin (302N) was acidified with nitric acid over the same time period to pH 6.2 in 1984 (Turner et al. 1987). Finally, Hall et al. (1980) presented results from the experimental acidification of Norris Brook, New Hampshire (Table 4.4). The pH of the stream was lowered abruptly from pH 5.7–6.4 to pH 4.0, and held at pH 4.0 for six months (April–September 1977).

TABLE 4.3. Preliminary results from the experimental acidification of Lake 302, South Basin.[a]

pH (year)	Observed change in the basin acidified with sulfuric acid (natural pH 6.5 to 6.7)
6.2 (1982)	No consistent trends.
5.9 (1983)	Increased proliferation of attached filamentous algae (principally *Mougeotia*) in the upper littoral zone.
	Increased lake transparency; increase in photosynthetically active radiation (PAR) at 4.0 m from 10.9 to 12.7 (pre-acidification) to 15.7 (as % of surface irradiance).
5.6 (1984)	Continued high lake transparency; PAR 14.5 (% of surface irradiance).
	Epilimnetic phytoplankton productivity rates at or above values for reference lakes.
	Mean euphotic zone production similar to that in reference lakes; deeper euphotic zone.
	Phytoplankton biomass similar to pre-acidification levels.
	Continued proliferation of benthic algae in upper littoral zone.
	Net photosynthesis decline in epilithic periphyton in the middle littoral zone.
	Inhibition of nitrification during winter.

[a] Data from Shearer and DeBruyn 1986, Turner et al. 1987, Rudd et al. 1988.

TABLE 4.4. Biological responses to the experimental acidification of Norris Brook, New Hampshire, from pH 5.7–6.4 to pH 4.0.[a]

Decreased emergence (by 37% overall) of adult mayflies (Ephemeroptera), some stoneflies (Plecoptera), and some true flies (Diptera).
Increased invertebrate drift (by up to 13-fold) during the first week after acid addition; following the first week, total numbers of organisms drifting in the acidified reach were similar to values for the reference area.
Decreased invertebrate density (by 75%) in the acidified area.
Increased periphyton biomass in the acidified area.
Decreased density of hyphomycete fungus; increased density of basidiomycete fungus.

[a] Data from Hall et al. 1980.

Results from whole-system manipulation experiments provide the primary basis for the following discussions of the biological effects of acidification, supplemented with data from smaller scale laboratory and field experiments and from field surveys. Our review provides an overview of important concepts, major findings, and remaining uncertainties, but it is not intended to be a comprehensive listing of all available research results.

Ecosystem-Level Processes

Many ecosystem processes appear relatively robust and unaffected by acidification, at least at moderate levels of acidity (pH 5.0 to 5.6) (Schindler 1987). In Lake 223, acidification to pH 5.0 caused no measurable decrease in primary productivity, rates of decomposition, or nutrient concentrations (Kelly et al. 1984, Schindler et al. 1985a, Shearer et al. 1987). In Little Rock Lake, acidification to pH 5.6 resulted in no clear trends in levels of chlorophyll *a*, primary productivity, or nutrients in the acidified basin (Detenbeck 1987, Frost et al. 1988 unpublished manuscript). Likewise, acidification of Lake 302S to pH 5.6 caused no decline in epilimnetic primary productivity (Shearer and DeBruyn 1986). Field survey data also indicate no consistent relationship between lake pH and algal biomass or primary productivity (National Research Council Canada 1981, Altshuller and Linthurst 1984).

Some exceptions to this general conclusion deserve note, however. Rudd et al. (1988) provided evidence that the experimental acidification of Lakes 223 and 302S altered the in-lake nitrogen cycle. At pH 5.4 to 5.7 and below, increased overwinter accumulations of NH_4^+ and the absence of any increase in NO_3^- suggest an inhibition of nitrification, that is, the oxidation of NH_4^+ to NO_3^- by nitrifying bacteria. The long-term effects of disruption of the nitrogen cycle on ecosystem productivity are unknown. Both NH_4^+ and NO_3^- may be used directly by algae as a nutrient source.

Although the experimental acidification of Lake 223 indicated no system-level changes in decomposition, results from other studies, including field surveys, in situ measurements, and smaller scale field and laboratory experiments, have been somewhat contradictory. Lower rates of decomposition or microbial production in acidic waters (pH 4.5 to 5.2) have been reported by some investigators (Traaen 1980, Baker et al. 1983, Francis et al. 1984, Palumbo et al. 1987); whereas others have reported no pH-related effects or correlations (Andersson et al. 1978, McKinley and Vestal 1982, Boylen et al. 1983). Accumulations of detrital materials have been observed in acidic lakes in

Scandinavia (Grahn et al. 1974, Leivestad et al. 1976); lake liming and increased pH reduced the quantity of detritus (Andersson et al. 1974, Traaen 1980), suggesting an adverse impact of low pH on detrital processing and decomposition.

These reported differences in the effects of acid stress on decomposition may result in part from different study methods, but may also reflect true variations in response with varying environmental conditions (Jensen et al. 1988, Elwood and Mulholland 1989). In Lake 223, as acidification progressed, no change occurred in the whole-lake decomposition rate during winter or the decomposition rate in the hypolimnion during summer, as measured by the accumulation of CO_2 in the water column (Kelly et al. 1984). At the same time, however, sediment pH levels remained near neutral, protecting sediment microbial communities from acid stress. In laboratory experiments, low pH waters caused decreased rates of organic matter decomposition only when sediments were actively suspended in the water column (Jensen et al. 1988). In addition, the effect of low pH on decomposition varied with substrate type; pH effects were greatest for leaf litter and recently deposited organic matter (Kelly et al. 1984, Jensen et al. 1988). Thus, acidification may cause measurable declines in the decomposition of leaf litter and other recently deposited organic material in surface sediments, lake littoral zones, and streams with active sediment-water interaction. However, acidification may have no effect on rates of decomposition in deeper sediments or in the lake hypolimnion as measured in Lake 223.

Acidic deposition and acidification may also have subtle effects on nutrient supplies, which may be difficult to detect in field experiments or field surveys but could result in a long-term gradual decline in system productivity. For example, Broberg and Persson (1984) hypothesized that increased mobilization of aluminum and changes in soil chemistry resulting from acidic deposition would decrease phosphorus leaching from watersheds. Nalewajko and O'Mahony (1988) examined the influence of short-term changes in pH on water column concentrations of phosphate. Water samples collected from two Ontario lakes over a 15-month period (pH 4.8 to 6.9) were acidified with sulfuric acid to pH 4.0–5.8. Minimum phosphate concentrations occurred at pH 5.2 to 5.8, maximum concentrations at pH 4.0. Experimental additions of iron and aluminum suggested that iron-phosphate and aluminum-phosphate co-precipitation may account for the decrease in phosphate levels observed at pH 5.2 to 5.8. The increase in phosphate levels at lower pH resulted largely from the release of phosphate from particles > 0.45 m in size, and perhaps to some degree from a decrease in phosphate uptake by phytoplankton and bacteria at low pH (< 5.0). It is not known whether these processes cause biologically significant changes in the phosphorus cycle or phosphate availability with acidification of surface waters in the field.

Increases in water clarity and transparency with acidification have been reported consistently in whole-lake manipulations and field surveys. Grahn et al. (1974) originally proposed that this increased water transparency resulted from a decreased number of phytoplankton (and lower primary productivity) in acidic waters. Studies such as the Lake 223 and Little Rock Lake experiments have shown, however, that these changes in light attenuation are not associated with changes in phytoplankton abundance (Schindler et al. 1985a, Shearer and DeBruyn 1986, Shearer et al. 1987). Instead, increases in water transparency with acidification appear to result either from a change in levels of dissolved organic carbon (DOC) (Effler et al. 1985) or from a change in the chemical nature and light absorption capacity of dissolved organics in the water (Schindler et al. 1985a).

Independent of the cause, an increase in transparency and light penetration may indirectly influence biological responses to acidification; for example, it may enhance primary productivity at lower water depths (Shearer and DeBruyn 1986) and alter other surface water characteristics, such as thermal stratification (Yan 1983, Effler and Owens 1985, Ruggaber 1988). Changes in the lake thermal regime would have substantial implications for a number of physical, chemical, and biological parameters. An increase in lake transparency would be expected to cause an increase in the depth of the epilimnion, increased hypolimnetic heating, and decreased thermal stability. This decreased thermal stability, in turn, may decrease the suitability of the lake for cold-water fisheries (salmonids, for example), increase the vertical mixing of dissolved constituents (potentially increasing the availability of nutrients in the epilimnion and thus increasing primary production), and increase oxygen inputs to the hypolimnion.

Although further research is needed, the information that is available suggests that biological processes, per se, may be poor indicators of acidification stress (Mills and Schindler 1986, Schindler 1987). Changes that do occur may be subtle, long term, and difficult to measure.

Plankton

Major changes in plankton composition and species richness (numbers of species per lake) with acidification have been consistently demonstrated in whole-lake experiments, mesocosm experiments, and field surveys (Altshuller and Linthurst 1984, Schindler et al. 1985a, Mierle et al. 1986, Stokes 1986; see also in this volume: Chapter 12, Pollman and Canfield; Chapter 13, Cook and Jager; and Chapter 15, Melack and Stoddard). Decreases in species richness by as much as 30% to 70% between pH 6.0 and 5.0 are common (Figure 4.1; also Chapter 12, Pollman and Canfield, this volume).

Zooplankton species for which experimental data exist to confirm their sensitivity to acidic conditions include the following (Marmorek and Bernard 1987):

- *Mysis relicta* (approximate threshold for population decline and loss, pH 5.6 to 5.9; Nero and Schindler 1983)
- *Diaptomus sicilis* (pH 5.8; Malley et al. 1982)
- *Epischura lacustris* (pH 5.6; Sprules 1975, Malley et al. 1982, Confer et al. 1983, Frost et al. 1988)
- *Tropocyclops prasinus mexicanus* (pH 5.3 to 5.6; Malley et al. 1982, Keller and Pitblado 1984, Price and Swift 1985, Havens and DeCosta 1986)
- *Daphnia galeata mendotae* (pH 5.2 to 5.4; Hobaek and Raddum 1980, Keller and Pitblado 1984, Price and Swift 1985)
- *Daphnia pulex* (pH 5.0 to 5.4; Keller and Pitblado 1984, Price and Swift 1985)
- *Leptodora kindtii* (pH 5.1 to 5.2; Sprules 1975, Hobaek and Raddum 1980, Keller and Pitblado 1984, Frost et al. 1988).

These species are often important components of zooplankton communities in circumneutral lakes with low acid neutralizing capacity (ANC), potentially susceptible to acidification from acidic deposition. Field surveys have identified additional zooplankton species that may be sensitive to acidic conditions.

While species sensitive to acidity decrease in abundance, tolerant zooplankton species often increase in abundance with increasing acidity. As a result, total biomass may be relatively unchanged, even when species richness is reduced (Dillon et al. 1984, Yan 1986). Species that frequently occur at higher densities in acidic waters (pH $\leq$ 5.0) include *Diaptomus minutus*, *Bosmina longirostris*, *Holopedium gibberum*, *Diaphanasoma* sp., and *Keratella taurocephala* (Malley et al. 1982, Bleiwas et al. 1984, Keller and Pitblado 1984, Siegfried et al. 1984, Havens and DeCosta 1986, MacIsaac et al. 1987). During the experimental acidification of both Lake 223 and Little Rock Lake, important invertebrate predators, such as *Mysis relicta* and *Epischura lacustris*, declined sharply or were lost at relatively high pH levels of 5.5 to 6.0. Other zooplankton species, including *Bosmina longirostris*, *Keratella taurocephala*, and selected species of *Daphnia* (*D. catawba, D. schoedleri*), increased in abundance (Tables 4.1 and 4.2).

Similar patterns of decreasing species richness and a marked change in species composition have been observed for phytoplankton. In Lake 223, particularly at pH < 5.6, the phytoplankton community shifted from a largely chrysophycean community to one dominated by peridinean (dinoflagellates) and cyanophycean (blue-greens) species (Table 4.1; Findlay and Kasian 1986). In numerous surveys of low ANC and acidic lakes in the eastern United States, eastern Canada, and Scandinavia, altered species composition and reduced species richness in the phytoplankton community were consistently correlated with low pH levels, with major changes occurring in the pH interval 6.0 to 5.0 (Almer et al. 1974, Leivestad et al. 1976, Kwiatkowski and Roff 1976, Yan 1979, Crisman et al. 1980).

In general, species are lost from all classes of algae as pH declines. However, proportionally larger losses occur within some groups than in others. As a result, the dominant algae in acidic lakes are often different from those characteristic of circumneutral lakes. The types of phytoplankton that dominate at low pH vary among systems,

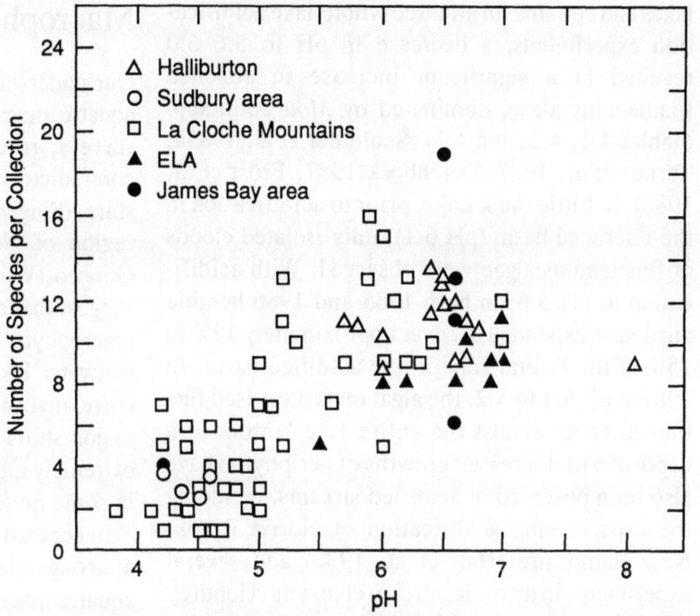

FIGURE 4.1. The relationship between the average number of species of planktonic crustaceans identified in each collection and pH of various lakes located on the Canadian Shield. (Data from National Research Council Canada 1981.)

although dinoflagellates, for example, *Peridinium inconspicuum* and *P. limbatum*, are frequently reported (Stokes 1986).

Some types of algae, in particular diatoms and chrysophytes, are well preserved in lake sediments. As a result, sediment core analyses provide a record of changes in these algal assemblages over time. Strong correlations have been observed between the distribution of diatom and chrosphyte species (based on specimens preserved in surface sediments) and lake pH (Meriläinen 1967, Battarbee 1984, Smol et al. 1984b, Davis and Anderson 1985, Charles 1985, Charles and Norton 1986, Hartmann and Steinberg 1987). Using this information on the influence of environmental conditions on community structure, predictive indices and equations have been developed to infer historical trends in lake pH from data on sediment diatom and chrysophyte assemblages. Methods for paleoecological analyses are described further in Husar et al. (Chapter 3, this volume). Results from such analyses are presented in many of the regional case study chapters in this volume (Chapter 6, Driscoll et al.; Chapter 7, Kahl et al.; Chapter 12, Pollman and Canfield; Chapter 13, Cook and Jager; Chapter 14, Turk and Spahr; Chapter 15, Melack and Stoddard).

It is not yet clear to what degree observed changes in the structure of plankton communities result from direct toxic effects of acidity as opposed to indirect factors such as shifts in grazing or predation, increasing water clarity, or changes in nutrient availability that may be caused by acidification (Stokes 1986, Brezonik et al. 1986). Direct toxic pH effects have been cited as the reason for the disappearance of *Mysis relicta* (Nero and Schindler 1983) and *Epischura lacustris* (Malley et al. 1982) in acidified waters. On the other hand, Henrikson et al. (1980) observed that many changes in the plankton community associated with acidification parallel changes that occur in nonacidic waters with the loss or decline of planktivorous fish populations. Yan and Strus (1980) hypothesized that changes in phytoplankton community structure may adversely affect energy transfer to the zooplankton community. Sierszen (1988), however, measured no detectable effects of acidification to pH 5.2 on zooplankton community grazing rates in Little Rock Lake. Some of the zooplankton species lost with acidification are important food items for fish, yet the degree to which changes in food quality and quantity measurably affect fish populations in acidified lakes remains uncertain (see discussion on fish).

Periphyton

One of the most visible responses to acidification is the development of unusually high densities of filamentous periphyton or benthic algae in acidic lakes and streams. In all three whole-lake acidification experiments, a decrease in pH to 5.6–6.0 resulted in a significant increase in attached filamentous algae, dominated by *Mougeotia* spp. (Tables 4.1, 4.2, and 4.3) (Schindler et al. 1985a, Turner et al. 1987, Detenbeck 1987, Frost et al. 1988). In Little Rock Lake, prior to acidification in the reference basin (pH 6.1), only isolated clouds of filamentous algae were observed. With acidification to pH 5.6, in both 1985 and 1986 benthic algal mats expanded to cover approximately 12% to 15% of the littoral zone in the acidified basin. In 1987 at pH 5.1 to 5.2, the algal mats increased further to cover almost the entire lake bottom to a depth of 6 m. Increased growths of periphyton have also been observed in acidified streams, including the experimental acidification of Norris Brook, New Hampshire (Hall et al. 1980) and several experiments in artificial stream channels (Hendrey 1976, Parent et al. 1986). Field surveys have also indicated generally higher levels of periphyton biomass in streams with pH $\leq$ 5.0 for extended periods of time (Mulholland et al. 1986, Elwood and Mulholland 1989).

Mechanisms responsible for the increased abundance of periphyton in acidic waters are still uncertain (Stokes 1986). Laboratory bioassays indicate that growth rates of at least some isolates of *Mougeotia* are enhanced at lower pH (5.0 to 5.5) (Jackson 1985, Detenbeck 1987). Turner et al. (1987) proposed that *Mougeotia* adapt to the low levels of inorganic carbon in acidic waters, and thus are able to outcompete other algal species. Studies by Detenbeck (1987) suggest that neither increased light penetration nor decreased grazing pressure was responsible for the increased accumulation of periphyton in Little Rock Lake and associated mesocosm experiments. In streams, on the other hand, the abundance of periphyton may be more directly linked to changes in grazing pressure (Elwood and Mulholland 1989). Experimental reductions in the density of grazing invertebrates in circumneutral streams resulted in significant increases in periphyton biomass (Mulholland et al. 1983, Lamberti and Resh 1983). In addition, field surveys indicate that acidic streams with high periphyton biomass also tend to have lower numbers of grazing invertebrates (Hendrey 1976, Mulholland et al. 1986; see discussion on benthic macroinvertebrates).

Macrophytes

Our understanding of the effects of acidification on aquatic macrophytes is based principally on field surveys; results from the available studies appear contradictory. Surveys conducted in New Hampshire (Singer and Boylen 1984), the Adirondack region of New York (Roberts et al. 1985), and Ontario (Wile and Miller 1983, Yan et al. 1985a) suggest that acidification has little, if any, effect on macrophyte species composition and biomass in softwater lakes. Lake acidity was not significantly correlated with macrophyte species richness or major shifts in species composition. In addition, extremely high levels of biomass have been reported in some quite acidic lakes (Wile et al. 1985).

In contrast, Jackson and Charles (1988) observed a strong relationship between the composition of aquatic macrophyte communities and lake pH in a survey of 31 Adirondack lakes. Detrended correspondence analysis identified pH, and associated chemical variables (Ca^{2+}, Mg^{2+}, alkalinity, and conductivity) as the dominant environmental gradient determining macrophyte composition. Other lake characteristics such as elevation, lake morphometry, substrate, and trophic status had only a secondary effect or no detectable effect. Of the 45 submersed and floating-leaved species collected in the 31 lakes, 27 (60%) were restricted to lakes with pH > 6.0. No information on macrophyte biomass was collected for these lakes.

In many Scandinavian lakes, increases in lake acidity have been accompanied by the development of extensive beds of submersed *Sphagnum* and the consequent reduction of populations of other macrophyte species (Grahn 1977). Although *Sphagnum* spp. occur in some acidic lakes in North America (Hendrey and Vertucci 1980, Catling et al. 1986), invasions of *Sphagnum* similar to those in Scandinavia have rarely been reported.

Benthic Macroinvertebrates

Several important taxa of benthic macroinvertebrates are adversely impacted by acidification and are consistently absent or occur only in reduced numbers in acidic lakes and streams (Altshuller and Linthurst 1984, Okland and Okland 1986, Mierle et al. 1986). France (1987) documented the decline of the crayfish *Orconectes virilis* during the experimental acidification of Lake 223 (Table 4.1). Population recruitment failure occurred at pH 5.6, principally as a result of increased egg loss during oviposition. Similar effects of low pH (5.0 to 5.6) on crayfish reproductive success have been demonstrated in laboratory experiments with *Astacus astacus*, a crayfish common in Scandinavian waters (Appelberg 1987). Populations of *A. astacus* rarely occur in the field at pH < 6.0, and have never been collected at pH < 5.5 (Okland and Okland 1986). In contrast, other crayfish species appear relatively tolerant of acidic conditions. Reproducing populations of *Cambarus bartoni* have been observed in Ontario lakes with pH 4.6 to 4.9 (Collins et al. 1981). In field and laboratory exposures to low pH and elevated Al, stage III juveniles of *C. robustus* molted and survived at pH levels near 4.0, whereas juvenile stages of *O. rusticus* and *O. propinquus* experienced increased mortality at pH 5.4 to 6.1 (Berrill et al. 1985).

Most mollusc species are also sensitive to low pH. In field surveys of lakes in Norway, acidic lakes consistently had fewer species and smaller numbers of molluscs; no snails occurred at pH < 5.2, no mussels at pH < 4.7 (summer surface pH) (Okland and Okland 1986). Rooke and Mackie (1984) observed reproductive failure of the snail *Amnicola limosa* in Heeney Lake, Ontario, associated with a spring pH depression to 4.7. Mackie (1987) proposed that the most significant effects of acidification on populations of molluscs are changes in shell composition and morphology, such as production of thin shells, and decreased reproduction and growth, rather than direct mortality. *A. limosa* grew more slowly in lakes with pH 5.8 to 5.9 in Ontario than in waters with pH > 6.0, and exhibited impaired egg development in laboratory experiments at pH 5.0 and delayed development at pH 5.5 (Servos et al. 1985).

Effects of acidity (pH ≤ 6.0) on amphipods have been demonstrated in field studies and laboratory experiments (Okland and Okland 1986). The amphipod *Hyalella azteca* is widely distributed among lakes on the Canadian shield, but is generally absent from waters with pH < 5.6 (Stephenson and Mackie 1986). France and Stokes (1987) found that pH levels < 5.5 were toxic to *H. azteca* juveniles in laboratory experiments. In Europe, common species of amphipods (*Gammarus pulex*, *G. lacustris*, and *G. fossarum*) rarely occur at pH levels below 5.7 to 6.0; laboratory experiments have confirmed the sensitivity of these organisms to low pH (Okland and Okland 1986).

Insect taxa vary markedly in their sensitivity to acidity. Mayflies (Ephemeroptera) are, in general, quite sensitive; several species, for example, *Baetis lapponicus*, are affected at pH levels near 6.0 (Sutcliffe and Carrick 1973, Engblom and Lingdell 1984, Okland and Okland 1986). Stoneflies (Plecoptera) and caddisflies (Trichoptera) exhibit a wide range of response; some species are affected at moderate pH levels (5.0 to 5.5), whereas others are relatively acid tolerant (occurring at pH levels as low as 4.5) (Sutcliffe and Carrick 1973, Leivestad et al. 1976, MacKay and Kersey 1985, Ormerod and Edwards 1987). On the other hand, midges (Chironomidae) and other families of true flies (Diptera), for example, black flies (Simulidae), as well as dragonflies and damselflies (Odonata) are generally more acid tolerant and frequently occur in waters with pH 4.5 to 5.0 and below (Havas and Hutchinson 1982, Berrill et al. 1987). In addition, although not strictly members of the benthic community, several taxa of the true bugs (Hemiptera), such as water striders, water boatmen, and backswimmers, and beetles (Coleoptera), for example, whirligig beetles and predacious diving beetles, commonly occur in waters with pH < 5.0 (Nilssen 1980, Althsuller and Linthurst 1984, Okland and Okland 1986). Several investigators have noted that shifts in the relative abundance of these major taxa provide a useful index of a change in the acid-base chemistry of surface waters (Raddum and Fjellheim 1984, Sharpe et al. 1987, Singer and Smith 1987).

Effects of acidification on the composition of the benthic insect community have been demonstrated experimentally and observed during time trend studies. Hall and Ide (1987) documented the loss (over the period 1937–42 to 1984–85) of acid sensitive mayfly and stonefly species (acid sensitivity defined from prior studies) from two streams in central Ontario with spring pH depressions to pH 4.9–5.6. At other sites in the area, where current pH fluctuations are small (pH 6.1 to 6.4), no major changes in the composition of the benthic community occurred over the same time period. Raddum and Fjellheim (1987) documented the loss of the mayfly *B. rhodani* from Vikedahl River, Norway, associated with minimum river pH values of 5.0 to 5.3. Experiments in artificial stream channels with *B. rhodani* confirmed that pH levels < 5.5 caused increased drift and mortality. Simulations of acidic episodes to pH 5.2 caused increased drift and mortality and a subsequent decline in the density of *Baetis* sp. in experimental stream channels operated near Emerald Lake, California (Chapter 15, Melack and Stoddard, this volume). Hall et al. (1980) observed an increase in the drift and decreased emergence of mayflies, some stoneflies, and some true flies during the experimental acidification of Norris Brook, New Hampshire, to pH 4.0 (Table 4.4).

Although changes in species composition and reduced species richness are consistently associated with acidic conditions, the effects of acidification on benthic invertebrate abundance and productivity are less clear. Acid tolerant species and taxa such as dragonflies and aquatic beetles may be unaffected or even increase in abundance in acidic waters, perhaps as a result of indirect effects associated with shifts in community interactions—decreased predation, for example. Results from acidification of Lake 223 (to pH 5.0) and Little Rock Lake (to pH 5.6) indicate no significant decline in the total abundance of benthic invertebrates (Schindler et al. 1985a, Frost et al. 1988). Likewise, field surveys for lakes suggest that in most cases no significant correlation exists between acidity and benthic abundance or biomass (Crisman et al. 1980, Collins et al. 1981, Dermott 1985, Harvey and McArdle 1986).

In streams, the pattern is less consistent. Harriman and Morrison (1982) and Smith et al. (1990) found no relationship between stream pH and total invertebrate density in 12 streams in Scotland (pH < 4.4 to 5.8) and 8 streams in the Adirondacks of New York (pH 4.5 to 7.0), respectively. Kimmel et al. (1985) and Rosemond (1987), on the other hand, reported a lower density of macroinvertebrates in streams with lower pH in the Laurel Hill region of Pennsylvania (pH 4.6 to 7.4) and in the Great Smoky Mountain National Park, Tennessee (pH 4.5 to 6.8). Hall et al. (1980) (Table 4.4) and Allard and Moreau (1986) documented a decrease in benthic invertebrate density following experimental acidification to pH 4.0. However, Allard and Moreau (1986) noted no change in invertebrate biomass in the same experiment. Acidified stream channels were populated with a large acid tolerant genus of Chironomidae (*Microtendipes*). Acidification to pH 4.0 caused increased insect drift, decreased emergence, and/or decreased stream colonization in the two experiments. Zischke et al. (1983) acidified a moderately hardwater system (pH 8.0) to pH 6.0 and pH 5.0 for a period of 17 weeks in experimental stream channels. During most of the acidification period, the density of macroinvertebrates in the acidified channels was lower than in the reference channel. By the end of the study, however, populations of acid tolerant taxa had increased in abundance in the acidified channels and the total invertebrate densities were actually higher in the treated channels at pH 5.0 and 6.0 than in the reference channel at pH 8.0.

Benthic invertebrates play a significant role in the physical breakdown of organic matter and nutrient regeneration in aquatic ecosystems and also provide an important food source for fish. Elwood and Mulholland (1989) concluded, however, that reductions in the decomposition of leaves reported in acidic streams result principally from reduced microbial activity rather than reduced macroinvertebrate shredding. MacKay and Kersey (1985) observed lower mass loss rates of leaves incubated in acidic streams relative to circumneutral streams, independent of the influence of macroinvertebrates. van Frankenhuyzen et al. (1985) found that the caddisfly *Clistoronia magnifica* actually grew faster when raised on leaves conditioned at low pH (4.1). Incubation of the leaves at pH 4.1 resulted in a doubling of the fungal biomass, which apparently improved the quality of the food for invertebrate growth. Additional studies are needed, however, to fully evaluate the influence of acidification on macroinvertebrate energetics and the indirect effects of acid induced changes in benthic communities on aquatic processes of decomposition and nutrient cycling.

Fish

Numerous laboratory and field experiments have demonstrated that acidic conditions are toxic to fish and that surface water acidification can result in the decline and loss of fish populations (Altshuller and Linthurst 1984, Schindler et al. 1985a). Here we briefly review potential mechanisms of the effects on fish populations, key chemical characteristics that influence fish population response, and approximate pH levels at which adverse effects on fish populations may occur.

Mechanisms of Fish Population Response

Acidification may impact fish populations directly, by direct toxicity, or indirectly, through effects on the food chain. Direct effects of acidification on fish populations may be classified as involving either recruitment failure or reduced adult survival. Recruitment failure may result from (1) increased mortality of early life stages, (2) impaired reproductive physiology and ovarian maturation, and/or (3) inhibition of spawning behavior. Explanations reported for reduced adult survival have included (1) fish kills associated with acidic episodes, (2) high mortality of sensitive adult stages, such as immediately following spawning, and (3) out-migration of adults from acidic systems, which is a behavioral avoidance reaction.

One of the best studies to date of fish population response to acidification involved the experimental acidification of Lake 223 (Table 4.1). With gradual acidification of the lake from pH 6.5 in 1976 to pH 5.1 in 1983, the following changes and effects were observed (Mills et al. 1987):

- All fish species in the lake (lake trout [*Salvelinus namaycush*], white sucker [*Catostomus commersoni*], fathead minnow [*Pimephales promelas*], pearl dace [*Semotilus margarita*], and slimy sculpin [*Cottus cognatus*]) declined in abundance; one species (fathead minnow) was lost.
- The primary mechanism for population decline and extinction was recruitment failure. All fish species in the lake had stopped reproducing by 1981–82 at pH 5.0 to 5.1. With time, the lack of reproduction presumably would have resulted in extinction of these populations from the lake.
- Recruitment failure apparently resulted principally from increased mortality of early life stages. There was no evidence of increased egg resorption or adverse effects on reproductive physiology. Lake trout successfully spawned each year.
- Growth, condition, and survival of lake trout and white sucker remained constant or increased with acidification through 1982, despite the loss of several potentially important fish prey organisms (e.g., *Mysis relicta*) (Table 4.1).
- In 1982, two years after the onset of reproductive failure (in 1980), lake trout growth decreased slightly. By 1983, following a collapse of the lake trout food base, adult trout were severely emaciated and adult survival declined sharply. These indirect effects of acidification, via the food chain, may have accelerated the decline of the lake trout population, but were not considered the primary mechanism of effect.

These experimental results from Lake 223 are consistent with patterns observed in field surveys. Many investigators have reported that fish populations in acidic waters consist of only large, older fish, indicative of recent successive recruitment failures (Altshuller and Linthurst 1984, Beggs et al. 1985). With improvements in water quality – as a result of liming, for example – successful recruitment can recur, as evidenced by the reappearance of very young fish (Andersson et al. 1984, Gunn and Keller 1984). Observations of reduced fecundity or spawning inhibition are relatively rare (e.g., Beamish and Harvey 1972), suggesting that increased mortality of early life stages may be the primary mechanism of effect in these situations. In addition, adult fish in acidic waters generally have growth rates and condition factors comparable to or exceeding those of fish in circumneutral waters (Almer et al. 1978, Rosseland et al. 1980, Schofield et al. 1981, Trippel and Harvey 1987, Rask and Raitaniemi 1988), suggesting that food shortages do not play a dominant role in fish population declines with acidification.

A contrasting pattern, with missing older age classes and populations dominated by young fish, has been observed in some acidic systems in Norway (Rosseland et al. 1980, Andersen et al. 1984) and Ontario (Trippel and Harvey 1987). Researchers speculated that this shift in age class structure, termed juvenilization, resulted from high mortality of mature fish immediately following spawning. Direct evidence in support of this hypothesis is lacking, however (Muniz et al. 1985). Reasons for the absence of older fish in certain stressed populations are uncertain, as is the regional significance of this response as a mechanism for the effects of acidification on fish populations.

Adult fish kills associated with acidic episodes have also been observed, but principally for Atlantic salmon (*Salmo salar*) and brown trout (*Salmo trutta*) in Norwegian rivers (e.g., Skogheim et al. 1984, Hesthagen 1985). Similar effects have been reported for native fish populations in the wild at only one location in North America: fish kills of pumpkinseed sunfish (*Lepomis gibbosus*) in Plastic Lake, Ontario (Harvey and Lee 1982). The ability of adult fish to detect and avoid lethal levels of acidity may mitigate, in part, the extent and severity of fish kills caused by acidic episodes (Marmorek et al. 1986). Movement of fish into refuge areas with more suitable water chemistry has been observed in several systems (Muniz and Leivestad 1980, Hall et al. 1980). The availability and distribution of such refuge areas during acidic episodes, and the long-term implications of such shifts in spatial distribution, have not been determined, however.

Available evidence supports the hypothesis that recruitment failure occurs frequently and is perhaps the most common cause of fish population decline with acidification. In addition, most studies have emphasized increased mortality of early life stages as the primary cause for recruitment failure. Clearly, however, responses vary in different lakes and streams for different fish species.

Key Chemical Characteristics that Influence Fish Population Response

The chemical parameters identified as primary determinants of fish survival in acidic waters are pH, inorganic Al, and Ca^{2+}. Field and laboratory evidence in support of this conclusion was summarized in Altshuller and Linthurst (1984). The toxicity of low pH levels, < 5.0 to 6.0, depending on the fish species, results in most cases from impaired body salt regulation (McDonald 1983). In some regions, acidic surface waters with pH < 5.2 to 5.5 generally have elevated levels of inorganic (labile monomeric) Al (> 50 to 100 $\mu g\ L^{-1}$) (Wright et al. 1980, Driscoll et al. 1980), reflecting the influence of pH on Al solubility. Numerous studies have documented the toxicity of Al in acidic waters (Baker and Schofield 1982, Brown 1983a, Neville 1985, Ingersoll 1986, Mount 1987). Calcium, perhaps via its influence on membrane permeability, mitigates pH and Al toxicity. Fish tolerate lower pH levels and higher Al concentrations in waters with higher Ca^{2+} concentrations (Brown 1983b, Ingersoll 1986).

Aluminum toxicity is highly dependent on Al speciation, pH, and fish life history stage. Over the range of conditions anticipated with acidic deposition, Al may have no effect, a severe detrimental effect, or even a beneficial effect on fish survival (Baker and Schofield 1982). Several studies have confirmed that complexation of Al with organic ligands markedly reduces or eliminates Al toxicity (Driscoll et al. 1980, Karlsson-Norrgren et al. 1986). Thus, Al toxicity is generally attributed to the inorganic species: the free ion (Al^{3+}) or complexes with fluoride or hydroxide ligands. Variations in inorganic Al toxicity with pH probably reflect both variations in the concentration of inorganic Al species and interactions between H^+ and Al at biological surfaces (Campbell and Stokes 1985).

Other metals may also play an important role in fish toxicity under certain conditions, particularly in waters with pH > 5.2 and, thus, without significant mobilization of Al. For example, Hutchinson and Sprague (1986) identified Al, Zn, and Cu as the primary factors controlling fish toxicity in soft waters in Ontario at pH 5.8. Potential contributing effects from metals other than Al, at concentrations anticipated in acidic surface waters without point source inputs, have not been adequately quantified; further research is needed.

Levels of pH, Al, Ca^{2+}, and other chemical and physical parameters vary markedly over time and space within any given lake or stream. In addition, fish life stages vary in their sensitivity to acidity and associated parameters. It is possible, therefore, that the effects of acidification on fish populations are controlled not by annual average conditions but instead by chemical conditions occurring at specific times of the year, and perhaps in specific locales, that coincide with the occurrence of particularly sensitive life stages (Gunn 1986). Annual average chemistry, or other indices of the integrated chemical exposure, may be relevant to fish population response only to the extent that these average concentrations or indices are correlated with chemical conditions during critical seasons or periods or with the frequency, magnitude, and timing of acidic episodes.

Unfortunately, detailed field studies have not been conducted to confirm these hypotheses or to identify specific critical times and locales for fish population survival. These critical times and locations undoubtedly vary among fish species, as a result of different life history strategies, and also, to a degree, among lakes and streams with different patterns of temporal and spatial chemical variability.

Approximate pH Thresholds for Effects on Fish Populations

The large body of data from laboratory and field bioassays, field experiments, and field surveys can aid in identifying acidity levels or ranges likely to cause adverse effects on fish populations. Several problems arise, however, that make quantitative integration of these data difficult:

- The large number of variables and factors that influence fish response to acidity and fish population response to acidification, including biological interactions and the inherent productivity or resiliency of the system or population.
- The lack of consistency among studies, in terms of response variables measured, experimental or survey design, and the types and ranges of chemical parameters and environmental conditions examined.
- The uncertainties associated with extrapolating from laboratory bioassays, or field bioassays, to predict population responses in the field. Key concerns or issues include (1) cumulative effects through the life cycle, (2) potential interactions and additive effects between sublethal responses to acidity and other environmental stresses, (3) compensatory processes (density-dependent survival, growth, and fecundity), and (4) temporal and spatial heterogeneity in water chemistry.

To take full advantage of the available information, we used a fairly simple, qualitative approach to identify approximate pH levels at which effects on fish populations may occur. Altshuller and Linthurst (1984) used a similar procedure; the results presented here are an update of that report.

We reviewed 88 papers and reports that we considered relevant for determining pH levels associated with adverse effects on fish populations from acidic deposition. For each study, the approximate threshold pH for significant adverse effects was identified (Table 4.5). As noted earlier, fish response variables and procedures for data analysis varied greatly among studies. As a result, no single criterion could be used for all studies to define "significant adverse effects." For bioassays, thresholds for effects were generally defined either by statistically significant increases in mortality or by survival rates less than 50% of survival rates in control waters. For field surveys, values reported represent pH levels consistently associated with population absence or loss.

Aluminum and Ca^{2+} were not explicitly treated in selection of the threshold value for effects. Because of the complex interactions among pH, Al, and Ca^{2+}, we chose not to develop multivariate threshold values. We included, however, only those laboratory tests with Al concentrations similar to those measured in low pH lakes in the northeastern United States (Driscoll et al. 1980). Thus, these estimates of pH thresholds may be less accurate for waters with distinctly higher or lower levels of Al, such as seepage lakes in Florida or the Upper Midwest.

Based on the pH thresholds from individual studies, an overall threshold for population effects was estimated for 25 fish species (Figure 4.2). Particular difficulties arose in combining data from bioassays and field surveys. Chemistry values (and pH thresholds) for field surveys frequently represent one or a few measurements taken during the summer or fall season. Thresholds for effects in laboratory and field bioassays, on the other hand, reflect the sensitivity of specific life stages. In many cases, pH thresholds for specific life stages were somewhat lower than pH thresholds from field surveys (Table 4.5). Our intent was to determine a pH threshold for effects on fish populations for use with regional measures of surface water chemistry collected during field surveys. Thus, in selecting the reported critical pH range for effects, we assumed that fish early life stages may occur at times when pH levels are substantially lower than values measured during field surveys. In combining results from surveys and bioassays, we therefore considered an offset of up to 0.5 pH units between apparent thresholds for effects from bioassays and field surveys to be reasonable. These reported pH ranges should be considered only as guidelines for estimating the potential nature and severity of effects of acidification on fish populations.

TABLE 4.5. Summary of field experiments, field bioassays, and laboratory bioassays relating critical pH values to fish response.*

	Brook trout	Lake trout	Rainbow trout	Atlantic salmon	White sucker	Creek chub	Blacknose dace
Field survey							
Population absence	5.1[a] 4.9[b] 5.8[c] 4.6–4.7[d] 5.1–5.3[e] 5.6[g] 5.0[j] 5.0[l] 4.6–4.7[k] <5.2[o] 4.6[f] 5.0[n]	4.9[b] 4.4[c] 5.1[f]	4.9–5.0[c] 4.7–5.7[e] 5.4[f]	5.3[d]	4.8–4.9[a] 4.9[b] 4.2–4.3[c] 4.6–4.7[d] 5.4[f] 4.9[h] 5.4–5.5[k] 5.2[m] 4.6[f] 5.5[n]	5.2[a] 5.1[b] 6.0[c] 5.9[g] 5.6[h] 5.0[i] 4.6[f] 5.7[n]	6.5[a] 5.6[f]
Population loss	<5.1[a] 5.0[j] 4.3–5.0[r] 4.7–5.0[s] 5.1–5.6[v] 4.6–5.6[w] 4.6–5.1[m]	5.4[a] 5.2[j] 5.3–5.5[r] 5.2–5.3[u] 5.2–5.6[v] 5.0–5.2[x] 5.2–5.8[y]		5.0–5.1[p] 4.7–5.0[q]	<5.1[a] 4.5–4.8[r] 4.9–5.1[s] 5.4–5.7[t] 4.9–5.2[u] 4.7–5.2[y]	5.4[a] 4.7–5.0[s]	6.2[a] 5.7–6.1[t]
Recruitment failure	4.9–5.4[s]	6.9[c] 5.5[j] 5.5–5.6[u] 5.2–5.5[z]			4.8–4.9[a] 4.9–5.0[c] 4.9–5.5[s] 4.7–5.2[z]		
Stocking failure	4.8–5.0[j] <4.0[aa] 4.5–4.8[cc] 4.8–5.0[dd]	5.0–5.2[u] 5.0–5.6[bb]	4.4[aa] 5.0–5.5[cc]				
Fish kill				5.2–5.5[ee] 3.9–4.2[ff]			
Field experiment							
Population loss	4.8–5.0[gg]						
Recruitment failure		5.6[hh]			5.1[hh]		
Field bioassay							
Embryo mortality	4.3–4.6[ii] 5.2–5.3[jj] 4.4–5.0[kk] 4.5–4.6[nn]	5.0–5.3[ii] 5.2–5.6[jj] 4.4–5.0[ll]	>5.4[jj]	5.0–5.5[ff] 4.7[kk] <5.1[mm] 4.4–4.9[oo]		5.3–5.7[ii]	5.3–5.9[ii]
Fry mortality	5.9[dd] 5.0[ii] 5.0–5.4[hh] 4.8[ss]	5.0[ii] 5.0–5.4[ll] <4.5–5.0[rr]	5.4–5.5[pp]	5.0–5.2[qq]		5.3–5.9[ii]	
Fingerling/young-of-the-year mortality	5.0–5.1[gg] 4.4–4.7[ii]	<5.0[ii]	5.9[e]	5.1–5.3[mm] 4.7–5.0[tt]		4.5–4.8[ii]	5.4–5.5[ii]
Yearling/adult mortality	4.7–5.1[e] 4.4–4.8[ii] 4.7–5.1[nn] 4.8[uu] 4.6–4.8[vv]	5.5–5.7[ii] 5.0–5.6[vv]	5.5–5.9[uu]			4.4–4.7[ii]	5.0–5.3[ii]

	Brook trout	Lake trout	Rainbow trout	Atlantic salmon	White sucker	Creek chub	Blacknose dace
Laboratory bioassay							
Embryo mortality	5.1–5.2[ww] <4.2[xx] 4.2[aaa] 4.5[eee] 4.5[fff]	4.2–4.8[xx]	<4.5[yy] 5.0–5.5[bbb] 5.0–5.6[ddd]	5.5[zz]	5.2[aaa] 4.8–5.1[ccc]		
Fry mortality	4.4–5.9[dd] 4.2–4.4[ww] 4.2–4.5[xx] 4.4–5.9[aaa] 4.5[eee] 5.5[fff]	4.4–4.8[xx]	4.5–5.5[yy]		5.02–5.2[aaa] 5.3[ggg] 5.1[ccc]		
Juvenile/yearling mortality	<4.6[hhh] 4.4–4.9[jjj]		4.5–5.0[iii]	4.7–5.3[hhh]			
Reproductive inhibition			5.5[bbb]				

	Brown trout	Arctic char	Smallmouth bass	Walleye	Yellow perch	Fathead minnow	Mud-minnow
Field survey							
Population absence	4.4–4.8[oo] 4.6[f]		5.6[a] 4.9[b] 4.4[c] 5.2[h] 5.5–5.7[k] 5.6[f] 7.0[n]	5.2[b] 5.2[c] 5.5[g]	4.8–4.9[a] 4.6[b] 4.2–4.3[c] 4.4[d] 5.5–5.7[k] 4.4[x] 4.9[m] 4.5[f] 4.5[n]	5.1[b] 5.5[i] 6.3[f]	4.8–4.9[c] 4.0[h] 4.2[f] 4.5[n]
Population loss	5.4–5.6[t] 4.8–6.0[kkk] 4.6–5.3[lll] 5.0–5.1[mmm]		5.1–5.4[a] 5.5[u] 5.2–5.4[x] >5.5[y]	5.5[u] 5.2–5.9[x] 5.2–5.8[y]	<5.1[a] <4.7[y]		
Recruitment failure	4.9–5.3[ff] 5.0[nnn]	5.2[nnn]	4.9–5.0[c] 5.5–6.0[z]	5.9[c] 5.5–6.0[z]	4.4[c] 4.5–4.7[z] 4.4–5.0[ooo]		
Stocking failure	4.8–5.0[cc] 4.7–4.8[ppp]						
Fish kill	5.0[p] 4.6[ppp]						
Field experiment							
Population loss		5.0[qqq]				5.6[hh]	
Recruitment failure		5.0[qqq]				5.9[hh]	
Field bioassay							
Embryo mortality	4.5[ff] 5.1[rrr]			5.4[pp] 5.5[sss]	5.0[a]		
Fry mortality			5.15[ss]				
Yearling/adult mortality	5.0–5.2[vv]						<4.6[a]
Laboratory bioassay							
Embryo mortality			4.4–5.0[ddd]	4.9[ccc]		6.0[ttt]	
Fry mortality	4.5–5.1[uuu]		5.1[vvv] 4.8–5.1[ccc] 5.0–5.6[ddd]	5.1[ccc]	4.5[www]		
Juvenile/yearling mortality	4.7–5.1[xxx]			4.7[ccc]			
Reproduction inhibition							6.0[ttt]

continued

TABLE 4.5. (*Continued*).

	Largemouth bass	Rock bass	Pumpkinseed sunfish	Brown bullhead	Golden shiner	Common shiner
Field survey						
Population absence	4.9[b]	5.2[b]	4.8–4.9[a]	4.8–4.9[a]	4.8–4.9[a]	5.6[a]
	4.4[c]	4.2–4.3[c]	4.6[b]	4.8[b]	5.1[b]	5.1[b]
	4.6[h]	5.2[h]	4.2–4.3[c]	4.6–4.7[c]	4.8–4.9[c]	5.7[c]
	<4.1–4.5[yyy]	4.6[f]	5.4[g]	4.8[d]	4.8[d]	5.3[d]
	4.7[f]		4.9[h]	4.5[h]	5.5[g]	6.0[g]
	5.0[n]		5.5–5.7[k]	4.6–4.7[k]	5.2[h]	6.2[h]
			4.6[f]	5.4[zzz]	5.0–5.1[k]	5.4[i]
			4.9[n]	4.5[f]	4.7[i]	4.9[f]
				4.7[n]	5.9[m]	6.1[n]
					4.5[f]	
					5.1[n]	
Population loss	4.9–5.4[ss]	4.7–5.2[y]	4.8–4.9[r]	<5.1[a]	<5.1[a]	
	<3.7[aaaa]		4.7–5.2[y]	4.2–4.5[r]	4.6–4.8[r]	
				4.7–5.1[s]	4.7–5.0[s]	
				4.7–5.2[y]		
Recruitment failure	5.2[c]	4.9–5.0[c]		4.9[c]	5.2[c]	
	<4.1–4.5[yyy]	4.7–5.2[z]		4.7–5.2[z]		
		4.8–5.0[bbbb]				
Laboratory bioassay						
Embryo mortality						6.0[ccc]
Fry mortality	4.5[www]	5.0[www]				5.1[rrr]
Juvenile/yearling mortality	<4.2[ccc]					

	Northern pike	Redbelly dace	Bluntnose minnow	Blacknose shiner	Slimy sculpin	Killifish
Field survey						
Population absence	5.1[b]	6.5[a]	5.6[b]	5.9–6.0[c]	5.8[i]	4.9[a]
	4.2–4.3[c]	5.1[b]	5.7[c]	6.5[h]		5.4[f]
	5.1[h]	4.9–5.0[c]	6.2[h]	5.8[i]		
	4.0[zzz]	5.3[h]	5.6[i]	6.1[m]		
	5.2[m]	5.0[i]	6.6[f]			
	5.6[f]	6.1[m]	5.7[n]			
	5.9[n]	4.7[f]				
Population loss	4.7–5.2[y]	6.2[a]			5.4[a]	<5.1[a]
	4.6[cccc]					
Recruitment failure	4.9–5.0[c]		6.0[c]	6.0[c]		
	4.4–4.9[nnn]					
	5.0[cccc]					
Field experiment						
Recruitment failure					5.6–5.9[hh]	

*All reported pH values take into account anticipated calcium and aluminum levels in acid softwater systems potentially sensitive to acidic deposition.

[a] Schofield and Driscoll 1987
[b] Green et al. 1987
[c] Harvey 1979
[d] Smith et al. 1986
[e] Elwood et al. 1985
[f] Kretser et al. 1989
[g] Haines et al. 1986
[h] Wiener and Eilers 1987
[i] Matuszek et al. 1988
[j] Beggs and Gunn 1986
[k] Langdon 1984
[l] Singer and Boylen 1984
[m] Frenette et al. 1986
[n] Cusimano et al. 1988
[o] Langdon 1985
[p] Leivestad et al. 1976
[q] Watt et al. 1983
[r] Baker and Harvey 1984
[s] Schofield 1976a,c
[t] Halliwell 1989
[u] Beggs et al. 1985
[v] Reckhow et al. 1987
[w] Sharpe et al. 1984
[x] Harvey and Lee 1982
[y] Beamish et al. 1975
[z] Beamish 1976
[aa] Kretser and Colquhoun 1984
[bb] Gunn et al. 1987
[cc] Grande et al. 1978
[dd] Schofield and Trojnar 1980
[ee] Skogheim et al. 1984
[ff] Jensen and Snekvik 1972
[gg] Schofield et al. 1986
[hh] Mills et al. 1987
[ii] Johnson et al. 1987
[jj] Gunn and Keller 1984
[kk] Lacroix 1985
[ll] Hutchinson et al. 1985
[mm] Henriksen et al. 1984
[nn] Schofield 1965
[oo] Harriman and Morrison 1982
[pp] Hulsman et al. 1983
[qq] Farmer et al. 1980
[rr] Beggs et al. 1987a
[ss] Beggs et al. 1987b
[tt] Lacroix and Townsend 1987
[uu] Sharpe et al. 1983
[vv] Colquhoun et al. 1983
[ww] Ingersoll 1986
[xx] Hutchinson et al. 1989
[yy] Holtze 1984
[zz] Peterson et al. 1980
[aaa] Baker and Schofield 1982
[bbb] Wiener et al. 1986
[ccc] Holtze and Hutchinson 1989
[ddd] Schweinforth et al. 1989
[eee] Cleveland et al. 1986
[fff] Hunn et al. 1987
[ggg] Trojnar 1977
[hhh] van Coillie et al. 1983
[iii] Neville 1985
[jjj] Siddens et al. 1986
[kkk] Andersson and Andersson 1984
[lll] Muniz et al. 1984
[mmm] Sevaldrud and Skogheim 1986
[nnn] Almer et al. 1978
[ooo] Ryan and Harvey 1980
[ppp] Overrein et al. 1980
[qqq] Lindstrom et al. 1984
[rrr] Muniz and Leivestad 1980
[sss] Wales and Liimatainen 1987
[ttt] McCormick et al. 1989
[uuu] Brown 1983b
[vvv] Kane and Rabeni 1987
[www] Eaton, EPA unpublished data
[xxx] Sadler and Lynam 1986
[yyy] Keller 1984
[zzz] Leuven et al. 1987
[aaaa] Canfield et al. 1985
[bbbb] Ryan and Harvey 1977
[cccc] Grahn et al. 1974

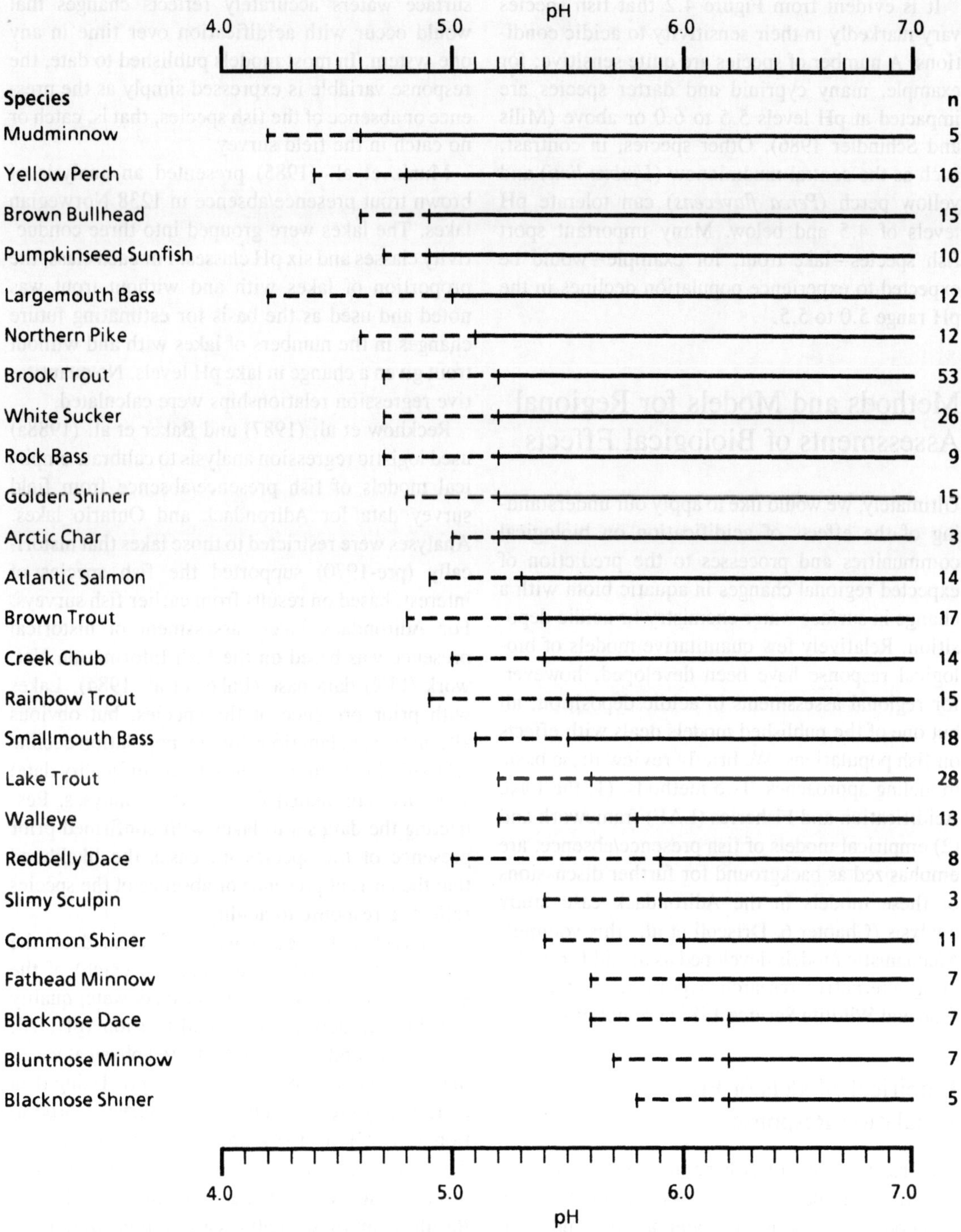

FIGURE 4.2. Estimated "critical" pH values for effects on fish populations, based on qualitative literature review. Dashed lines reflect the approximate range of uncertainty in the estimated "critical" pH. Number of observations (*n*) used to derive the estimated "critical" pH noted.

It is evident from Figure 4.2 that fish species vary markedly in their sensitivity to acidic conditions. A number of species are quite sensitive; for example, many cyprinid and darter species are impacted at pH levels 5.5 to 6.0 or above (Mills and Schindler 1986). Other species, in contrast, such as the central mudminnow (*Umbra limi*) and yellow perch (*Perca flavecens*) can tolerate pH levels of 4.5 and below. Many important sport fish species–lake trout, for example–would be expected to experience population declines in the pH range 5.0 to 5.5.

Methods and Models for Regional Assessments of Biological Effects

Ultimately, we would like to apply our understanding of the effects of acidification on biological communities and processes to the prediction of expected regional changes in aquatic biota with a change in surface water chemistry or acidic deposition. Relatively few quantitative models of biological response have been developed, however, for regional assessments of acidic deposition; all but one of the published models deals with effects on fish populations. We briefly review these basic modeling approaches. Two methods, (1) the Lake Acidification and Fisheries (LAF) framework and (2) empirical models of fish presence/absence, are emphasized as background for further discussions of these models in the Adirondack case study analysis (Chapter 6, Driscoll et al., this volume). Mechanistic models developed as an aid for evaluating alternative research hypotheses (e.g., U.S. Fish and Wildlife Service 1982) are not discussed.

Empirical Models of Fish Population Response

Empirical models of fish population response to acidification, based on the association between fish population status and water acidity (and associated parameters) measured in synoptic surveys, have been proposed by several investigators. As for any empirical approach to predictive modeling, it is assumed that (1) the systems surveyed are at steady state and (2) the observed association between fish population status and acidity among surface waters accurately reflects changes that would occur with acidification over time in any one system. In most models published to date, the response variable is expressed simply as the presence or absence of the fish species, that is, catch or no catch in the field survey.

Muniz et al. (1985) presented an analysis of brown trout presence/absence in 1238 Norwegian lakes. The lakes were grouped into three conductivity classes and six pH classes. For each class, the proportion of lakes with and without trout was noted and used as the basis for estimating future changes in the numbers of lakes with and without trout given a change in lake pH levels. No quantitative regression relationships were calculated.

Reckhow et al. (1987) and Baker et al. (1988a) used logistic regression analysis to calibrate empirical models of fish presence/absence from field survey data for Adirondack and Ontario lakes. Analyses were restricted to those lakes that historically (pre-1970) supported the fish species of interest, based on results from earlier fish surveys. For Adirondack lakes, assessment of historical presence was based on the Fish Information Network (FIN) data base (Baker et al. 1984). Lakes with prior presence of the species, but obvious alternative explanations for any population decline (classified without reference to water quality data) were also eliminated from further analyses. Restricting the data set to lakes with confirmed prior presence of the species increases the likelihood that the current presence or absence of the species reflects a response to acidity.

For each fish species with sufficient data, the current presence/absence (catch/no catch) of the species was regressed as a function of water quality variables associated with acidification (pH, Al, Ca^{2+}, DOC, and ANC) and selected physical variables (lake area, depth, and elevation). Using data collected by the Adirondack Lake Survey Corporation (ALSC) in 1984–85, Baker et al. (1988a) developed empirical models for three fish species: brook trout, white sucker, and brown bullhead. Results from these analyses are summarized in the Adirondack region case study chapter (Chapter 6, Driscoll et al., this volume). Model results for lake trout, derived from surveys of 192 Ontario lakes that had historically supported lake trout (Beggs et al. 1985), are presented in Figure 4.3. Lake trout presence/absence was closely associated

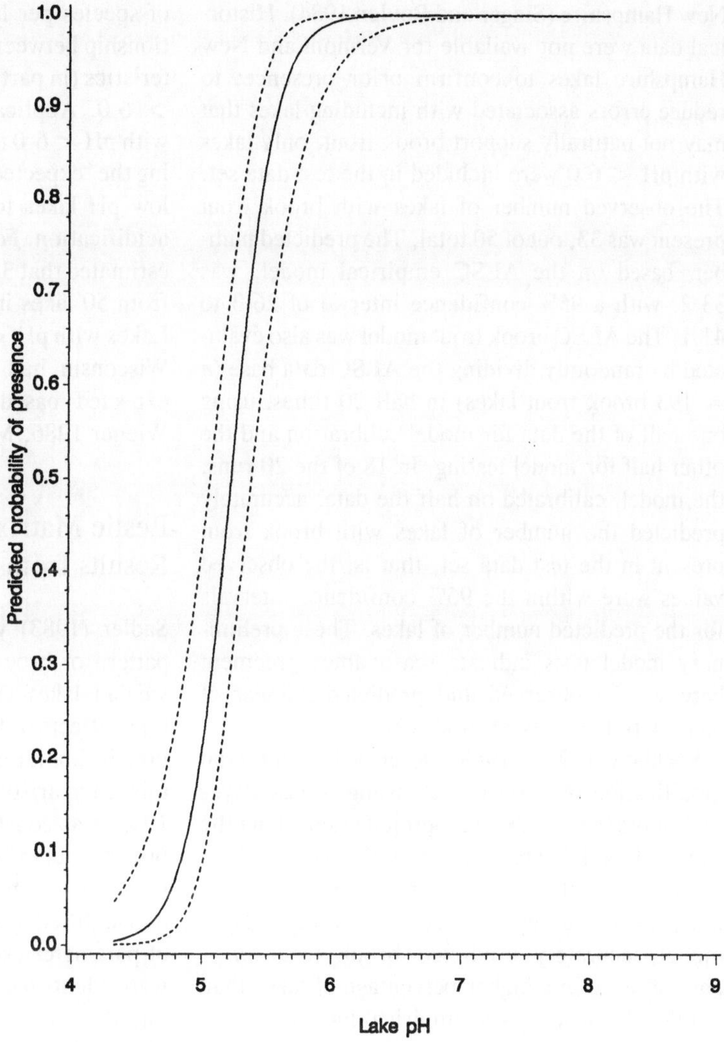

FIGURE 4.3. Ontario lake trout predicted probability of presence (with 95% confidence interval) as a function of lake pH.

($p \leq 0.001$) with lake pH; no other variables were significantly associated (all p values > 0.05) with lake trout status after adjusting for effects related to lake pH. Thus, the final lake trout model includes only pH:

$$P_{LT} = \frac{1}{1 + \exp\,[32.2 - 6.18(\text{pH})]}$$

where P_{LT} is the predicted probability of lake trout presence, plotted in Figure 4.3 with 95% confidence intervals. The absence of a significant effect for Ca^{2+} or Al, after adjusting for pH effects, may reflect the high degree of correlation ($p < 0.0001$) between pH and both Ca^{2+} and Al in the field survey data set.

The lake trout and brook trout empirical models (Chapter 6, Driscoll et al., this volume) have been tested using several independent data sets (Baker et al. 1988a). For lake trout, the empirical model derived from the Ontario data set was assessed relative to the 20 ALSC lakes with confirmed presence of lake trout in the past and no obvious alternative explanations for the loss of lake trout. The model predicted that 19.0 of these lakes would have lake trout, with a 95% confidence interval of 17.6 to 20.4. Seventeen of the 20 lakes actually had lake trout present.

For brook trout, the test data set includes 23 lakes in Ontario (Beggs and Gunn 1986), 18 lakes in Vermont (Langdon 1983, 1984), and 9 lakes in

New Hampshire (Singer and Boylen 1984). Historical data were not available for Vermont and New Hampshire lakes to confirm prior presence; to reduce errors associated with including lakes that may not naturally support brook trout, only lakes with pH < 6.0 were included in the test data set. The observed number of lakes with brook trout present was 33, out of 50 total. The predicted number, based on the ALSC empirical model, was 33.2, with a 95% confidence interval of 26.3 to 41.1. The ALSC brook trout model was also evaluated by randomly dividing the ALSC data base (n = 193 brook trout lakes) in half 20 times, using one-half of the data for model calibration and the other half for model testing. In 18 of the 20 runs, the model, calibrated on half the data, accurately predicted the number of lakes with brook trout present in the test data set, that is, the observed values were within the 95% confidence intervals for the predicted number of lakes. These preliminary model tests indicate reasonable agreement between the observed and predicted number of lakes with fish present or absent.

Reckhow (1987, 1988) described a potential modification of these models using robust Bayes estimators that combine expert judgment about the expected model parameters with the basic logistic model. In a preliminary test of this approach on an independent survey data set, the robust Bayes model accurately predicted brook trout presence/absence in a higher percentage of lakes than did the classical logistic empirical model.

Empirical Models of Fish Species Richness

As an alternative to predicting the presence or absence of an individual fish species, empirical regression based models may also be used to estimate species richness, that is, the total number of fish species occurring in a given lake or stream. To date, this approach has been used largely to evaluate existing effects, such as the number of species populations lost to date in a region.

Harvey and Lee (1982), Rago and Wiener (1986), and Matuszek and Beggs (1988) applied fairly similar approaches to evaluating the effects of acidification on species richness in Ontario and northern Wisconsin lakes. The "expected" number of species per lake was calculated from the relationship between species richness and lake characteristics (in particular, lake area) for lakes with pH > 6.0. Application of this relationship to lakes with pH < 6.0 then provided the basis for comparing the "expected" and observed species richness in low pH lakes to determine possible effects from acidification. For example, Harvey and Lee (1982) estimated that 388 populations of fish had been lost from 50 lakes in the LaCloche region of Ontario. Lakes with pH < 6.0 in both Ontario and northern Wisconsin had significantly fewer species than expected based on their lake area (Rago and Wiener 1986, Matuszek and Beggs 1988).

Leslie Matrix Models Incorporating Results from Laboratory Bioassays

Sadler (1983) wished to determine whether the pattern of brown trout presence/absence in Norwegian lakes could be explained by the known toxic effects of low pH at low Ca^{2+} levels as measured in laboratory bioassays. As a starting point for this comparison, a Leslie matrix model (Leslie 1945) was constructed using existing literature on brown trout in Norwegian lakes to define expected ranges for fish fecundity and annual survival in nonacidified systems. Twenty-two alternative sets of parameter estimates for fecundity and survival were identified in the literature; it was assumed that these 22 sets of parameters reflected the true distribution of population characteristics among brown trout populations in Norwegian lakes.

Results from laboratory bioassays with brown trout eggs were used to define the expected additional fish mortality caused by low pH conditions. Increasing the mortality of any life stage in the Leslie matrix may result in a population that is not self-sustaining, that is, the toxic effects of low pH are predicted to cause extinction of the fish population over time. The sensitivity of the population to added stress and mortality is a direct function of the original parameter estimates selected for fecundity and natural survival rates. Sadler (1983) assumed that the proportion of the 22 alternative Leslie population matrices (with different fecundity/survival estimates) projected to become extinct at a given pH and calcium level (and associated expected egg mortality) could be interpreted as the proportion of

Norwegian lakes at that pH and Ca^{2+} level expected to lose brown trout.

Comparison of the model predictions with the observed pattern of brown trout presence or absence in Norwegian lakes indicated relatively poor agreement. In general, the model underestimated the number of lakes with brown trout absent. While there are a number of reasonable explanations for the apparent bias in model predictions, two in particular deserve note. First, chemical conditions during brown trout spawning and egg development in Norwegian lakes may be much more severe than the chemical conditions measured during synoptic surveys that were used to estimate egg mortality. Second, the available parameter estimates for fecundity and natural survival rates may poorly represent true conditions in Norwegian lakes susceptible to effects from acidic deposition. Both of these limitations may be countered, at least in part, by tying the Leslie matrix more directly to observed patterns of fish distribution in the field. Such a modification has been incorporated into the LAF framework, described below.

Lake Acidification and Fisheries Framework

The Electric Power Research Institute (EPRI) funded the University of Wyoming and Oak Ridge National Laboratory to develop models of fish population response to acidification for several species of fish as part of the Lake Acidification and Fisheries (LAF) project. These activities are ongoing, and only a preliminary version of the model for brook trout, termed the LAF framework, is available at this time. The framework is currently considered a research tool only and has not been completely tested.[1]

The LAF framework has three major components:

1. The PHALCA Model – Integrates laboratory bioassay data to predict conditional rates of mortality, reduction in growth, or reduction in fecundity as a function of chemical stress, that is, for a given pH, Ca^{2+}, and inorganic Al combination, for a given life stage.
2. The FISHEGGS Model – Integrates predicted impacts on individual life stages from the PHALCA model over the entire life cycle using a Leslie matrix and predicts "relative recruitment potential" [nee "relative reproductive potential" (Christensen et al. 1988)] for a given population of fish.
3. Logistic Regression Analysis – "Calibrates" the overall model for predicting population response based on correlations between observed fish presence/absence in field surveys and the predicted "relative recruitment potential" for a given fish population from PHALCA and FISHEGGS.

The major components and linkages in the LAF framework are presented in Figure 4.4; further details on the framework are provided in Van Winkle et al. (1986) and Christensen et al. (1988).

Relative recruitment potential (RRP), the output from the FISHEGGS model, is defined as the mean number of first-generation yearling females produced per present generation yearling female. In a population at equilibrium, RRP equals one, that is, each female replaces herself. RRP will be less than one for populations declining and greater than one for an increasing population. The FISHEGGS model is calibrated such that RRP equals one with no additional mortality related to acidity, that is, without the PHALCA model.

Calibration or application of the framework requires a set of lakes for which pH, Al, and Ca^{2+} have been measured or can be estimated. The chemical conditions are used as input to PHALCA. For acidic lakes, mortality of fry and adults resulting from the chemical conditions are predicted. These predictions, in turn, are used as input to FISHEGGS. The RRPs calculated in FISHEGGS for acidic lakes will be below one.

Populations with RRPs below one, if they were initially at equilibrium, would be expected to decline and eventually disappear. Biological compensation is not explicitly incorporated in FISHEGGS because of insufficient information. In reality, however, compensatory processes will tend to offset reduced RRPs (Goodyear 1980, Saila et al. 1987). For lakes where the added stress (mortality) did not exceed the compensatory capacity of the population, these processes would restore the RRPs to one, in the equilibrium case. Populations in these lakes would persist, although probably at a somewhat reduced level. For lakes where the added stress did exceed the compensatory capacity, the RRPs would

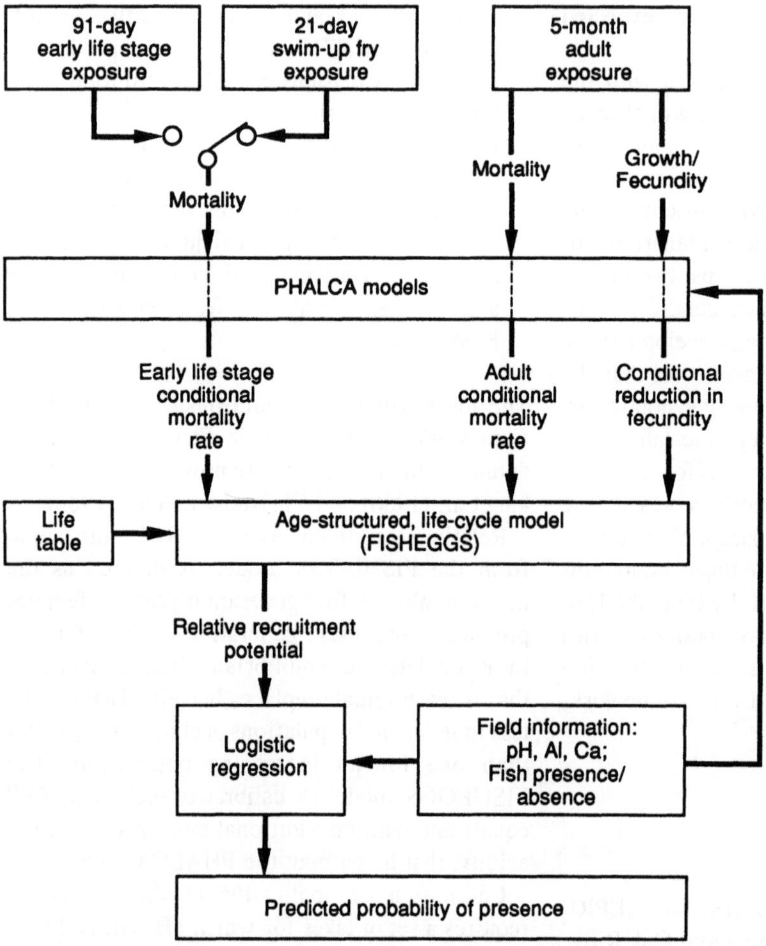

FIGURE 4.4. Lake Acidification and Fisheries (LAF) framework for modeling fish population response to acidification.

remain below one; the populations would decline and then disappear. The level of stress that these biological processes can compensate for will vary from lake to lake, depending on the overall quality of the lake as habitat for the species. It is, therefore, most useful to think of RRP not as a direct predictor of presence/absence per se, but rather as an integrated measure of population level stress due to chemical conditions.

Although biological compensation is not explicitly incorporated into the FISHEGGS model, its operation is accounted for in the final calibration step of the framework. In this step, logistic regression is used to establish a relationship between fish presence and the predicted RRPs using the same set of ALSC brook trout lakes that was used for empirical modeling described earlier. In the fitted model, lakes with lower values of RRP are associated with a lower predicted probability of brook trout presence. This reflects the increasing likelihood that, as acidic stress increases (indicated by lower RRPs), the compensatory capacity of the population will be exceeded. An interim calibrated relationship between RRP and the predicted probability of brook trout presence for the ALSC lakes is presented in the chapter on the Adirondacks (Chapter 6, Driscoll et al., this volume).

Predictions of the probability of brook trout presence from the LAF framework were evaluated by randomly dividing the ALSC data ($n = 193$) in half 20 times, using one-half of the data for model calibration and the remaining half for model testing. The observed number of lakes with brook trout present in the test data set was correctly predicted, that is, within the 95% confidence interval for the predicted number of lakes with brook trout, in 19 of 20 model runs. Tests with independent data sets have not yet been conducted, due largely to the lack of values for inorganic Al in field survey data sets.

Models of Fishery Yield

Minns et al. (1986) described two models developed by the Canadian Department of Fisheries and Oceans to make regional assessments of effects from acidic deposition in eastern Canada. Both include a fish subcomponent model estimating fisheries yield. While the procedures for calculating fish biomass and yield are fairly complex, adverse effects from acidic conditions are handled quite directly and simply. In both cases, species specific pH (or pH-Al) threshold values are defined from the existing literature on fish sensitivity to acidification (e.g., pH 5.5 for lake trout). Predicted pH levels (from the chemistry submodels) below these set thresholds result in a loss of the population in the first model. In the second model, complete recruitment failure occurs. Continued recruitment failure, that is, continued pH levels below the threshold, eventually results in population extinction; periodic recruitment failures cause only a decrease in fishery yield for the species. The detailed procedures for tracking changes in fishery yield are not described here since, for the most part, these calculations occur independently of any acidification effects.

Benthic Invertebrate Community

Ormerod et al. (1988) linked a set of biological response models to the hydrochemical Model of Acidification of Groundwaters in Catchments (MAGIC) to predict past and future changes in benthic invertebrate and fish communities in Welsh streams given alternative deposition scenarios. The invertebrate model (Weatherley and Ormerod 1987) was empirical, derived from the present-day association between invertebrate community composition and stream chemical and physical characteristics for 18 streams in the region. Three classes of invertebrate communities were defined based on species composition and relative species abundance. The relationship between invertebrate community type and stream characteristics was then determined using multiple discriminant analysis. The final model predicts the likely invertebrate assemblage type from the stream Al concentration, total hardness, and catchment area. In a similar manner, an empirical model of trout density was calibrated from a stepwise multiple regression analysis of data for 88 streams in the area, predicting trout density as a function of the stream Al concentration, total hardness, and average daily flow. No information was provided on model testing or sensitivity analyses.

All the models and modeling approaches described require further development and testing. Quantitative regional assessments of effects from acidic deposition on aquatic biota have not yet been completed for any of the regions discussed in the subsequent case study chapters.[1]

Summary and Conclusions

1. Knowledge of effects of acidification on aquatic biota relies on whole-system manipulation experiments; laboratory, mesocosm, and in-situ experiments; and field surveys.
2. Key ecosystem level attributes such as nutrient levels, decomposition rates, and primary production rates seem relatively insensitive to acidic conditions.
3. Biological species composition does, however, shift in response to acidification. Acid tolerant periphyton increase markedly. Acid sensitive phytoplankton, zooplankton, and benthic macroinvertebrate species disappear, and acid tolerant species appear or increase; the result is a decrease in species richness. Overall standing crops at various trophic levels do not, however, seem to be reduced.
4. Survival of fish in acidic waters is governed primarily by levels of pH, inorganic Al, and Ca^{2+}, which mitigates toxicity. Other potentially important considerations include indirect (food web) effects of acidification; temporal and spatial variation in chemistry in relation to occurrence of sensitive life stages; and behavioral avoidance. Acidification is expected to lead to reduced fish species richness.
5. Estimated thresholds, with ranges of uncertainty, for effects of pH on 25 fish species are presented, based on a qualitative evaluation of relevant literature.

6. Several modeling approaches for predicting regional effects of acidification have been developed. These require further development and testing.
7. Quantitative assessments of the effects of acidic deposition on aquatic biota have not yet been completed for any of the regions discussed in the subsequent case study chapters.[1]

Acknowledgments. Joan P. Baker's contribution to this chapter was sponsored by the U.S. Environmental Protection Agency (USEPA) under Contract No. 68-03-3439 with Kilkelly Environmental Associates. Sigurd W. Christensen's contribution was also sponsored by the USEPA, under Interagency Agreement DW89932112-01-7 with the U.S. Department of Energy under Contract No. DE-AC05-84OR21400 with Martin Marietta Energy Systems, Inc. This chapter does not necessarily reflect the opinions of the U.S. EPA, and no official endorsement should be inferred. Publication No. 3528, Environmental Sciences Division, Oak Ridge National Laboratory.

[1]Since submission of this manuscript in 1988, fish response models have been developed, tested, and applied regionally as part of the National Acid Precipitation Assessment Program Integrated Assessment.

5
Geographic Overview of the Regional Case Study Areas

Joseph M. Eilers and Anthony R. Selle

ABSTRACT. Evaluation of low ANC surface waters can be aided by defining study areas that minimize intraregional differences in physiography and water chemistry. Major factors, including geology, soils, land use, vegetation, landforms, climate, and hydrology, were selected to best represent important processes distinguishing surface water chemistry among regions. Four of the case study regions currently receive high levels of acidic deposition (Adirondack Mountains, Catskill Mountains, western Virginia, and the Southern Blue Ridge), but they differ markedly in the extent of acidic waters, climate, geology, soils, land use, and vegetation. The regional case study areas receiving moderate levels of acidic deposition (Maine, Upper Midwest, and Florida) also vary widely in the major factors affecting surface water chemistry. Only among the three low deposition case study regions in the West (Sierra Nevada, Cascade Mountains, Rocky Mountains) is there a high degree of similarity in physical characteristics and lake chemistry. The diversity of physiography in the United States underscores the need for analysis of aquatic resources on a regional basis.

Introduction

Limnology and Regional Classification

One of the challenges in evaluating the influence of acidic deposition on lakes and streams has been to distinguish between surface water chemistry occurring under natural conditions and patterns induced by acidic deposition. The task is made more difficult by the need to separate naturally acidic systems (Patrick et al. 1981) from those that are acidic because of atmospheric deposition or other anthropogenic disturbances, such as acid mine drainage. Another issue in assessing the effects of acidic deposition is how to best summarize the massive amount of information available on acidification of surface waters. By organizing the information within a systematic geographic framework, we can achieve a reasonable solution to both problems. To some degree, the seminal work, *Limnology in North America* (Frey 1966), used a similar strategy, whereby the status of limnological research was synthesized in the context of political divisions that bore strong relationships to physiography. The value of analyzing natural and and cultural factors on the basis of regions was developed in Europe during the 19th century (Hartshorne 1939; Holt-Jensen 1980). The concept of regionalism and its application to ecology was later recognized for its power in integrating regions, not just in studying the regions themselves (Odum 1951). Regionalization is rapidly gaining acceptance as a useful tool for natural resource management in the United States and has been reviewed by Gallant et al. (1989). In this chapter, we address the need to synthesize research on the aquatic impacts of acidic deposition that has been organized explicitly to reflect regional differences in limnology.

An underlying assumption in any analysis of natural systems is that the groups being compared have some commonality that makes the comparison meaningful. As an extreme example, most limnologists would agree that lakes in the Sierra Nevada are radically different from those in Florida, and a comparison of these two sets of lakes without an appreciation of these differences would be misleading. Classification of surface waters into groups with similar properties minimizes within-group

TABLE 5.1. Mean percentage land use in watersheds associated with lakes from the EPA's National Lake Survey for the case study regions, excluding the Catskill Mountains and western Virginia.[a]

	Watershed land use (%)						
Case study region	Forest[b]	Range	Barren[c]	Wetland	Water	Agriculture	Urban
Adirondack Mountains	90	<1	<1	4	3	2	1
Maine	88	<1	<1	5	2	3	1
Upper Midwest[d]	79	<1	<1	12	3	6	1
Rocky Mountains	45	14	51	<1	1	<1	<1
Cascade Mountains[e]	84	<1	12	<1	<1	<1	<1
Sierra Nevada[e]	37	2	60	<1	<1	0	<1
Florida	32	<1	<1	6	1	45	8
Southern Blue Ridge Province	64	<1	<1	<1	1	22	11

[a] Based on unpublished data from the National Surface Water Survey, U.S. Environmental Protection Agency, Corvallis, Oregon. Land use information derived from U.S. Geological Survey topographic maps (1:24,000; 1:62,500 scale).
[b] Combines deciduous and coniferous types.
[c] Includes tundra classification for western sites.
[d] Based on lakes in the revised boundaries for the Upper Midwest shown in Cook and Jager (Chapter 13, this volume).
[e] Based on lakes in the boundaries for these mountain ranges as represented in Chapters 15 and 16, this volume.

variance, which aids in distinguishing natural variation from various perturbations. In addition, the use of more homogeneous groups of lakes and streams presumably results in identifying systems that are more similar with respect to key biogeochemical processes, thus making it easier to understand the factors controlling surface water chemistry. In an attempt to compare ecologically similar systems, we have classified surface waters into groups that exhibit homogeneity with respect to various physical properties. In most cases these groups strongly reflect major physiographic provinces.

The process of defining regions for the purpose of assessing surface water chemistry requires specification of important limnological characteristics. These characteristics are selected on the basis of research that indicates the relative importance of various ecological processes for the issue under consideration. Therefore, a system designed to classify plant communities (e.g., Bailey 1976) would necessarily use processes to distinguish among terrestrial ecosystems that may be quite different from the processes used here to evaluate aquatic ecosystems. Even a classification system developed in part for grouping surface waters (e.g., Omernik 1987) may yield patterns entirely different from a classification system developed specifically to assess lake and stream chemistry related to the effects of acidic deposition. A definition for a region, based on key factors, constitutes a regional hypothesis describing the interrelationship between characteristics that distinguish the aquatic systems of the region. Consequently, regional boundaries are amenable to testing and refinement based on new information.

Purpose of Chapter

This chapter describes the geographic basis for delineating the regions used in this book and provides an assessment of the key factors associated with these regions. We describe the factors considered important on a regional scale, based to some degree on the processes described in Chapter 2 (Munson and Gherini, this volume). Factors of greater importance for distinguishing among systems *within a region* are discussed in the individual case study chapters. We then summarize the characteristics of the regional case study areas. Detailed treatment of issues such as uncertainty of regional boundaries, effects of scale, and selection and integration of factors in defining regions is beyond the scope of this presentation. In keeping with the focus of the book, the geographic scope of this chapter is restricted principally to the United States. Resource classification has been pursued in Canada, however, and is treated in other publications (Hunt 1974, Shilts 1981, Thie and Ironside 1976, Wiken 1986).

TABLE 5.2. Summary of selected climatic characteristics and precipitation chemistry of regional case study areas in the United States.

	Climate[a]				Precipitation chemistry[d] (average annual, 1985)		
Case study region	Annual precipitation[b] range (cm)	Annual evaporation[c] (cm)	Average temperature (°C) January	July	pH	$[SO_4^{2-}]$	$[NO_3^-]$
Northeast							
Adirondack Mountains	101–123	60	−12	18	4.37	39.8	23.9
Maine	92–109	55	−6	20	4.50	30.6	15.0
Catskill Mountains	107	70	−5	20	4.28[e]	48.5	23.9
Southeast							
Western Virginia	101–142	70	0	25	4.49[e]	34.1	16.8
Southern Blue Ridge	142–162	80	3	24	4.46	37.9	14.7
Florida	131–152	105	12	28	4.72[e]	22.1	10.5
Upper Midwest	61–81	55	−10	17	5.10	21.4	13.6
					4.55[e]	35.8	24.4
West							
Rocky Mountains	60–120	40	−7	17	4.99	19.0	15.3
					5.52	6.7	5.3
Sierra Nevada	93–153	40	−2	16	5.35	6.9	7.7
Cascade Mountains	80–250	40	−4	13	5.31	5.0	1.9

[a] Data from NOAA, National Climatic Summaries.
[b] 1931 through 1960 annual averages. NOAA; Department of Commerce.
[c] Pan evaporation (Baldwin 1973) × 0.7 (see Linsley et al. 1975, p. 168). Estimates of evapotranspiration based on differences in precipitation and surface runoff may yield substantially different values than evaporation from open water surfaces.
[d] Source: NADP/NTN (1987) precipitation weighted means for selected stations. Units for SO_4^{2-} and NO_3^- are in μeq L^{-1}. Low and high deposition sites have been selected for the Upper Midwest and Rocky Mountains because of the wide range in deposition chemistry for these regions. The stations selected were: Huntington Wildlife (NY), Greenville Station (ME), Biscuit Brook (NY), Marcell Experimental Forest (MN), Raco (MI), Sequoia National Park (CA), H.J. Andrews Experimental Forest (OR), Rocky Mountain National Park-Beaver Meadows (CO), Yellowstone National Park (WY), Bradford Forest (FL), Horton's Station (VA), Great Smoky Mountains National Park (TN).
[e] Data from 1987 (NADP/NTN 1988).

Major Factors in Defining Regions for Regional Case Studies

The factors used to define and describe the case study regions were chosen because they strongly influence the relative importance of major processes influencing surface water chemistry (see Chapter 2, Munson and Gherini, this volume). Some processes, such as base cation mineralization from the watersheds, have been represented using crude surrogates; other factors, such as hydrologic flow path, operate at scales that cannot be generalized on a regional basis.

We selected the following factors to describe the case study regions: bedrock geology, surficial geology, soils, topography, climate, vegetation, and land use. Together, these factors determine the hydrologic properties (and other characteristics) of lakes and streams, which in turn represent other key determinants affecting surface water chemistry. These characteristics are summarized for the case study regions in Tables 5.1 through 5.3 and are discussed in greater detail in the following paragraphs.

Bedrock Geology

Bedrock geology has long been considered a key terrigenous factor for identifying areas likely to contain low ANC surface waters (Norton et al. 1982), because low ANC lakes and streams are commonly associated with bedrock highly resistant to chemical weathering. This is largely the case for many watersheds in the western United

TABLE 5.3. Summary of landscape characteristics of regional case study areas in the United States.

Case study region	Topography	Bedrock geology	Surficial geology/soils	Vegetation land use
Northeast				
Adirondack Mountains	Low mountains and high hills	Granite, mixed metamorphic anorthosite center (Canadian Shield)	Glaciated, Spodosols	Forest/low population density
Maine	Plains with high hills and low mountains	Heterogenous metasedimentary, granite intrusive	Glaciated (deep till), Spodosols, Inceptisols	Forest/low population density
Catskill Mountains	Low mountains	Sandstone, mudstone	Glaciated, Incepticols	Forest with some cropland and pasture/low population density
Southeast				
Western Virginia	Low mountains	Granites, metabasalts along crests; sedimentary along valley and ridge	Unglaciated, Ultisols	Forest, pasture, and cropland/low population density
Southern Blue Ridge	Low mountains	Precambrian metamorphic and granitic in east; sandstones and limestones in west	Unglaciated, Inceptisols and Ultisols	Forest, pasture and cropland/low population density
Florida	Plains	Limestone/Dolomite, (Floridan aquifer)	Unglaciated, marine sands (Entisols)	Forest, cropland, and pasture/low to moderate population density
Upper Midwest	Irregular plains	Mafic, intermediate and felsic igneous (Canadian Shield)	Glaciated (deep till), Spodosol	Forest, wetland and cropland/low population density
West				
Rocky Mountains	High mountains	Mixed metamorphic and igneous	Glaciated, Alfisols, Entisols	Forest (montane-alpine)
Sierra Nevada	High mountains	Granitic batholith	Glaciated, Ultisols	Forest (montane-alpine)
Cascade Mountains	High volcanic peaks, moderately high mountains	Andesite, basalt in south; mixed metamorphic in north	Glaciated, Entisols	Forest (montane-subalpine)

States and Norway that typically have little overburden or vegetative cover to modify the weathering products. Where overburden is extensive and deep, bedrock composition becomes less important as a classification factor. For example, in Florida, the bedrock is predominantly limestone and dolomite, yet many lakes have low ANC because they are situated on highly weathered quartzite sand (Chapter 12, Pollman and Canfield, this volume). Major bedrock types for the United States are illustrated in Figure 5.1. Although the bedrock types are grouped on the basis of age of the rocks, the map indicates that most low ANC waters are found in areas with noncalcareous bedrock.

Surficial Geology

Overlying till can substantially modify expected regional surface water ANC patterns inferred from bedrock geology alone. For example, in northeastern Wisconsin, many lakes have high ANC, despite the underlying noncalcareous rocks of the

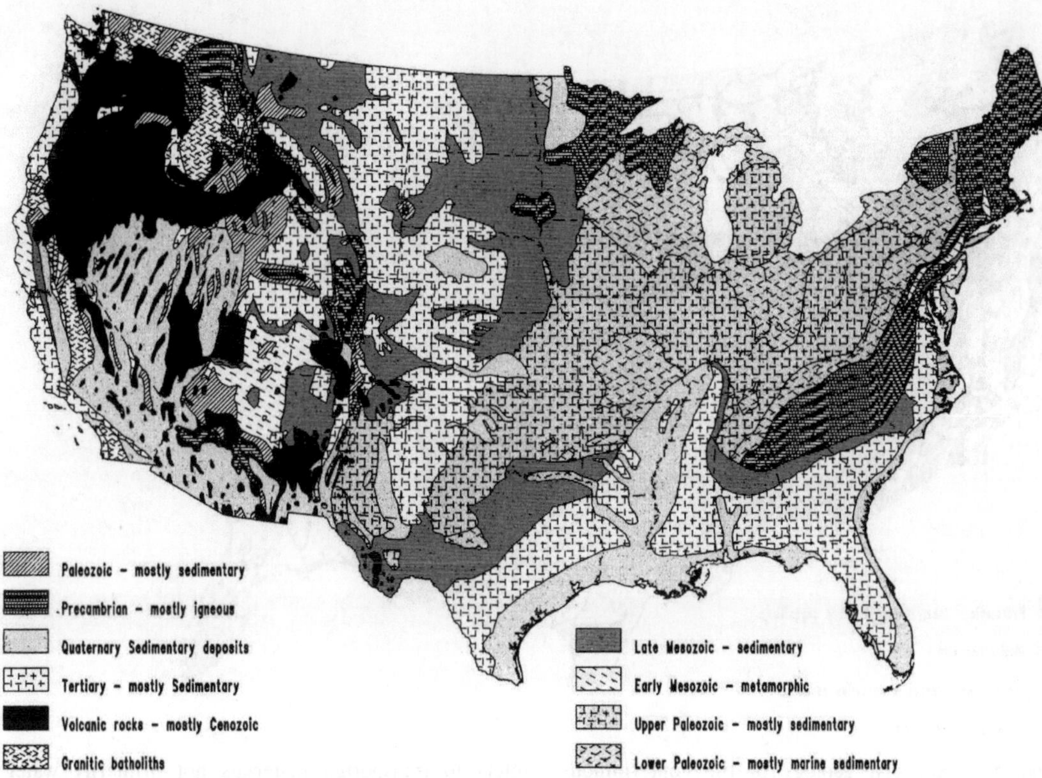

FIGURE 5.1. Bedrock geology of the conterminous United States derived from Hunt (1967). All classification categories shown are as developed by Hunt, but legend explanations have been condensed to major epoch and/or rock type. More detailed explanations for each class can be found in Hunt.

Canadian Shield, because of the presence of till containing calcite (Simpkins et al. 1978). The depth of till can also be important in influencing the chemistry of surface waters. Lakes in the Adirondack Mountains draining watersheds containing thick till are more likely to have high ANC values than comparable lakes in watersheds with shallow till present (Chen et al. 1984, Driscoll and Newton 1985, Booty et al. 1988). Most of the regional case study areas have been glaciated (Figure 5.2); only the Southeast and Florida were unaffected by recent glaciation. In these areas of the Southeast, surface water chemistry is strongly affected by soil characteristics.

Soils

Soil types have also been used to infer broad-scale sensitivity to acidification in the United States (McFee 1980, Kaplan et al. 1981). Low ANC surface waters are generally associated with soils low in cation exchange capacity (CEC) and base saturation. Where soils have high CEC, production of base cations exceeds deposition of strong mineral acids, and surface waters are able to maintain high ANC and pH. The case study regions display a wide variety of soils, but most of the soils have low CEC (Figure 5.3). Most northern soils associated with low ANC waters are young Spodosols and Histosols, whereas the southern soils are primarily much older Inceptisols, Alfisols, and Ultisols. The Spodosols and Histosols are often associated with moist, cool climates and have the capacity to generate significant quantities of organic acidity (Krug 1989). The Ultisols, and to a lesser extent the Inceptisols and Alfisols, which are common in the Southern Blue Ridge Province, show substantial retention of sulfur (Rochelle et al. 1987). In contrast, Spodosols in the Northeast showed little retention of sulfur in intensively studied water-

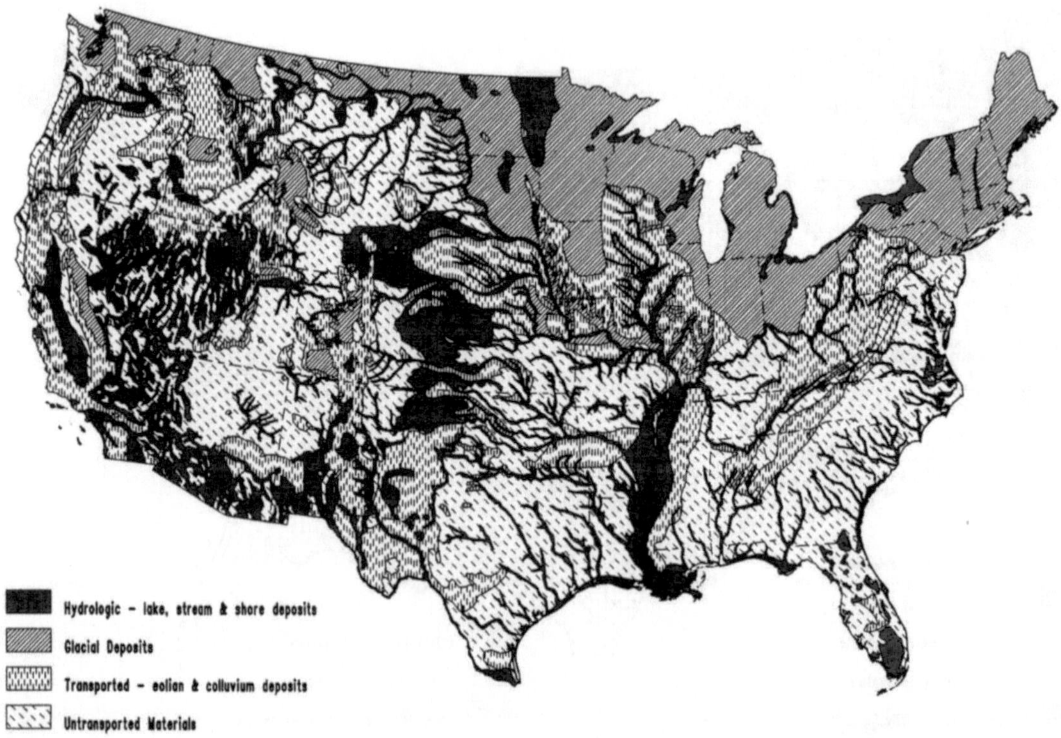

FIGURE 5.2. Surficial geology of the conterminous United States derived from Hunt (1986). Classes have been aggregated from Hunt's divisions. Hydrologic refers to water induced transport or deposition and includes lake, stream, and shore deposits. Glacial refers to all glacial deposits as depicted by Hunt. Transported refers to transported materials not primarily water induced and includes eolian, and colluvium. Untransported refers to all other categories identified by Hunt, including exposed bedrock, organic deposits, and saprolitic residuum. Depth to bedrock is not indicated on this map.

sheds (Rochelle et al. 1987). Sulfur retention has been offered as a hypothesis to explain regional-scale differences in surface water acidification from atmospheric deposition (Gallowày et al. 1983). The Florida sites have highly unreactive Entisols and soils in mountainous areas of the West, where present, are generally frigid, cryic subgroups of Ultisols, Inceptisols, and Entisols.

Land Use and Vegetation

Land use within the regional case study areas is predominantly forest, although important variations in land use activities occur in some case study regions (Table 5.1). Agriculture is common in the Florida watersheds sampled in Phase I of the U.S. EPA Eastern Lake Survey (ELS-I) and, to a lesser extent, in the Southern Blue Ridge. The positive association between agricultural land use and lake ANC is strong (see Chapter 12, Pollman and Canfield, this volume), but it is not clear in all areas whether this is a causal relationship or simply reflects the association between agricultural activities and favorable soil characteristics. Wetlands are an important feature in the Upper Midwest, the Northeast, and Florida, contributing to high concentrations of DOC in many lakes in these regions. Widespread intensive urban land use is uncommon in most watersheds with low ANC lakes sampled in the U.S. EPA National Surface Water Survey (NSWS), although it is comparatively high in the Southern Blue Ridge and Florida.

Landform and Elevation

Topography influences water chemistry at both local and regional scales. High local relief increases the rate at which water moves through the

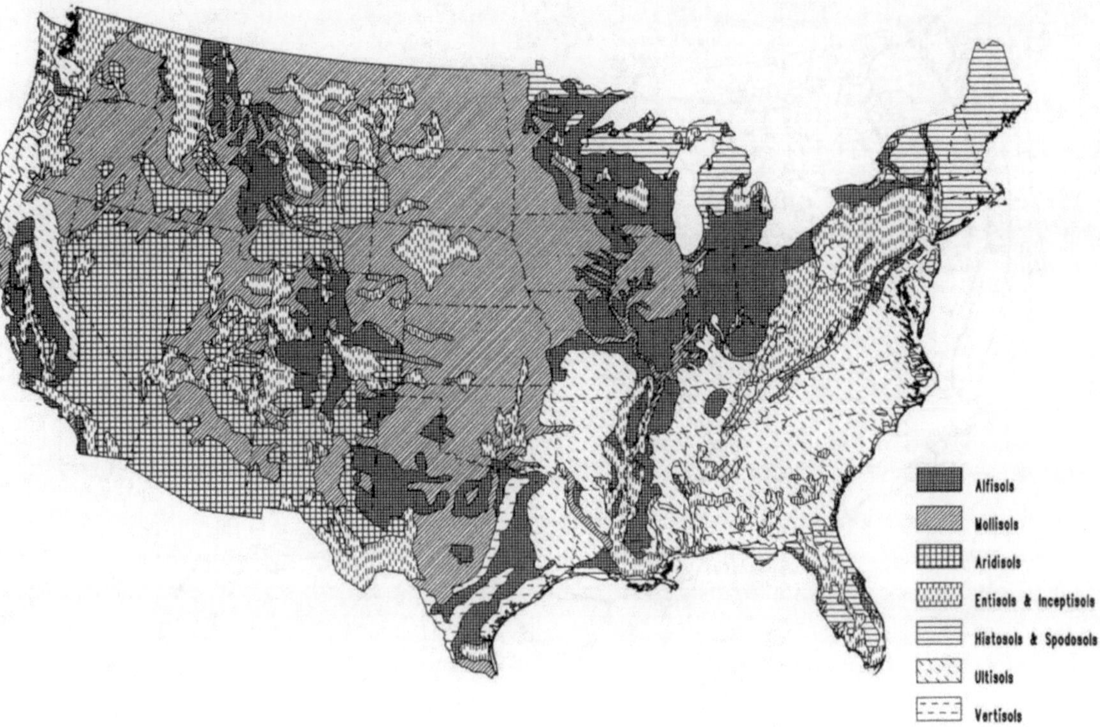

FIGURE 5.3. Major soil orders in the conterminous United States derived from a map by the USDA, Soil Conservation Service (USGS 1970). Classes were arranged generally from high to low cation exchange capacity (CEC), with Entisols and Inceptisols combined and Histosols and Spodosols combined. Soil temperature and moisture suborder classes are not shown. Cool or frigid soils would generally have a lower CEC than the same soils in a warmer climate.

watershed. High-elevation sites are generally associated with high local relief, although this is not always the case. For example, in the southern Oregon Cascades, some lakes are located on low relief glacial plateaus despite elevations approaching 2,000 m. An indication of the regional topographic relief is provided in the landforms diagram of Raisz (1957; Figure 5.4). A measure of local relief for lakes sampled in the NSWS is provided in Table 5.4 on an ordinal scale from 1 to 5. Local relief is lowest in Florida and the Upper Midwest, intermediate in the Northeast, and greatest in the West. Elevation is often associated with other factors that have a major influence on surface water chemistry, including geology, soils, climate, vegetation, and hydrology. For example, within the Adirondack Mountains, the older uplifted areas generally have the least weatherable exposed bedrock at higher elevations. The elevation extremes are greatest in the West, and highly developed soils and forests are not present at elevations greater than 2,000 to 3,500 m depending on the latitude. Orographic effects have a pronounced influence on local precipitation patterns, particularly in the West where the western slopes of the mountains often receive substantially greater precipitation than the eastern slopes.

Climate and Deposition Chemistry

Total ionic inputs to a given site are determined by precipitation volume and chemistry, and by dry deposition. These inputs are further concentrated by evapotranspiration. With the exception of dry deposition, which is poorly documented, we can estimate these inputs quite well on a regional basis. Total precipitation volume is highest in the Northwest and the Southeast (Figure 5.5). Wet deposition (the product of precipitation volume and ionic concentration) of H^+, SO_4^{2-}, NO_3^-, and Ca^{2+},

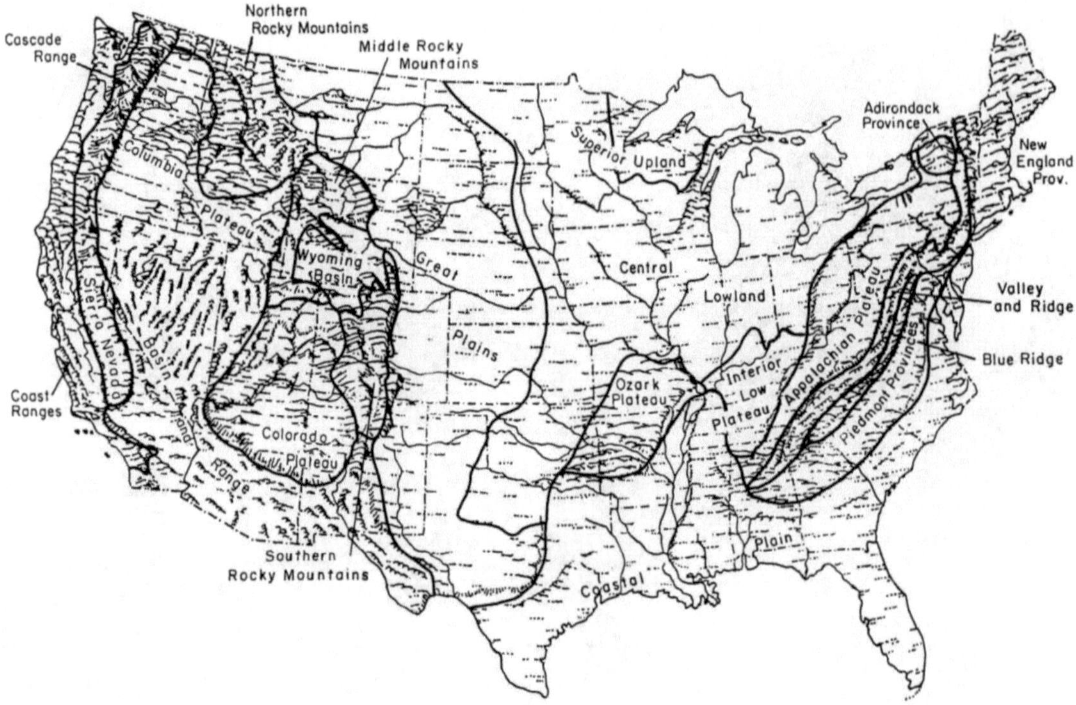

FIGURE 5.4. Landforms of the conterminous United States after Raisz (1957) with physiographic provinces from Fenneman (1946) superimposed. Province names primarily from Raisz, supplemented by Fenneman where needed.

however, is consistently higher in the Northeast (Figure 5.6). Evapoconcentration increases the ionic concentrations of the deposition inputs to a given watershed and rates of evaporation are important in assessing the effects of acidic deposition, particularly in the Southeast (Figure 5.7).

Hydrology

The most important hydrologic distinction among the case study regions is the relative abundance of lakes compared to streams. Three of the case study regions, the Catskill Mountains, western Virginia, and the Southern Blue Ridge Province, contain few natural lakes and, like much of the Appalachian system, are dominated by streams. Lakes are uncommon in the Appalachians, because they have not recently been subject to glaciation, which forms lakes through a variety of processes including scouring, damming of valleys by moraines, and melting of ice blocks in outwash and till (Hutchinson 1957). Streams are also abundant in the glaciated areas of the Northeast, Upper Midwest, and West, but synoptic data on stream chemistry are uncommon. Consequently, most of the information presented on surface waters outside of the Southeast highlands focuses on lakes.

Several aspects of hydrology are relevant to the response of lakes to acidic deposition, including hydrologic flow paths (the routing of water through the watershed), hydraulic residence time (τ_w; the average length of time water resides in a lake), runoff response (especially to snowmelt events), and the degree of coupling of the lake to the watershed. Collectively, these hydrologic properties determine the degree to which the surface water chemistry is affected by terrestrial processes versus in-lake processes.

In-lake processes can substantially modify lake chemistry (Kelly and Rudd 1984, Baker et al. 1986, Schiff and Anderson 1987), but these processes are difficult to represent on a regional scale. Hydraulic residence time is one key determinant that affects the influence of in-lake processes (Baker et al. 1986). Many low ANC seepage lakes have hydraulic residence times of 5 to 10 years (Lin

TABLE 5.4. Summary of selected lake and watershed characteristics of regional case study areas in the United States.[a]

Case study region	Area (km²)	Number of lakes[b]	Lake density (#/100 km²)	Median lake area (ha)[c]	Median lake depth (m)	Median watershed area (ha)	Median lake elevation (m)	Watershed relief[d]
Northeast								
Adirondack Mountains[e]	50,600	1672	3.3	21	5.8	270	490	2.4
Maine	86,000	2967	3.4	25	4.9	440	170	2.4
Catskill Mountains	3,100	29	0.9	13	3.6	190	300	NA[f]
Southeast								
Western Virginia	33,300	18	0.05	–	–	–	–	–
Southern Blue Ridge	28,800	97	0.3	37	13.1	4913	470	NA[f]
Florida	102,400	7959	7.8	18	3.0	–	20	~1
Upper Midwest	121,800	10,273	8.4	17	5.5	180	410	1.4
West								
Rocky Mountains	353,300	3558	1.0	8	11.0	190	2720	4.4
Sierra Nevada	44,700	1187	2.7	9	13.7	270	3060	4.4
Cascade Mountains	80,600	758	0.9	10	16.2	190	1430	4.4

[a] Values derived from EPA's National Lake Survey (Kanciruk et al. 1986; Eilers et al. 1987). Values for the Upper Midwest and Southern Blue Ridge are modified to reflect the boundaries shown in Cook and Jager (Chapter 13, this volume).
[b] Estimated number of lakes with surface area > 4 ha. The numbers of lakes include all lakes within the statistical sampling frame. The actual target populations of lakes are less than these values because of exclusion criteria applied before and during lake sampling.
[c] For lakes with surface area ≥ 4 ha.
[d] Watershed relief expressed on an ordinal scale within 2.5 miles of the lake center where: 1 = plain (< 100 ft); 2 = rolling hills (100–500 ft); 3 = low hills (500–1,000 ft); 4 = high hills/low mountains (1,000–2,000 ft); 5 = mountains (> 2,000 ft).
[e] Statistics were computed for the area defined as subregion 1A from ELS-I (Linthurst et al. 1986). The area for the Adirondack Park is 23,300 km² and lake density for the park is about twice that listed for subregion 1A; other statistics for lakes in the Adirondack Park differ little from those for subregion 1A.
[f] NA = not available.

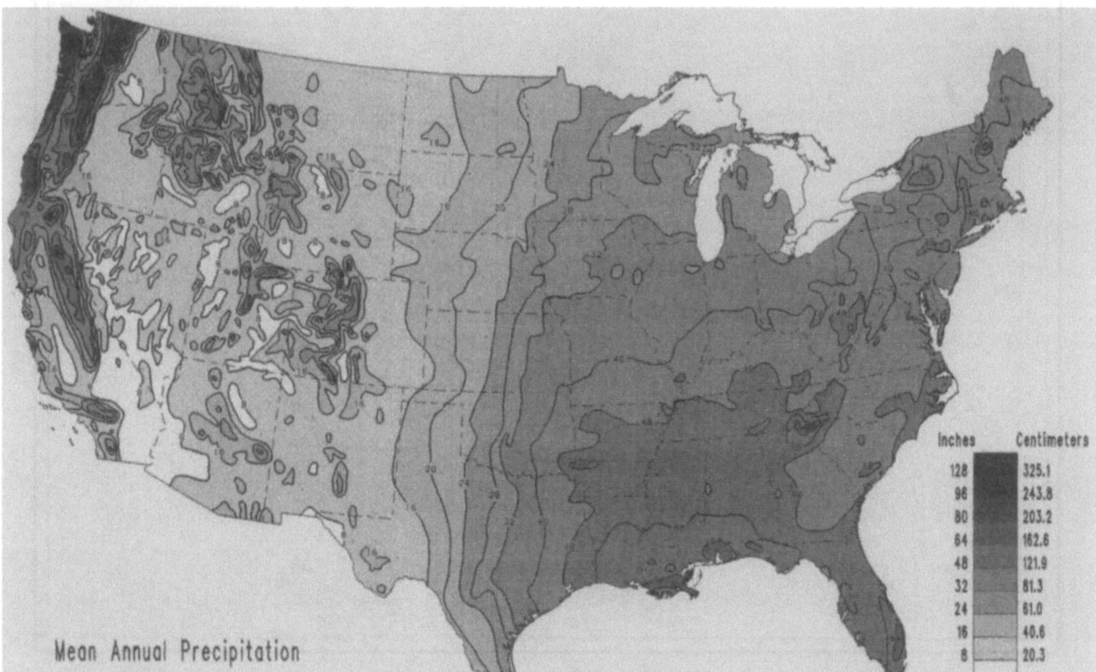

FIGURE 5.5. Mean annual precipitation for the conterminous United States for the period 1931–1960 (after USGS 1970). Note the effect of topography on precipitation patterns, particularly in the western United States.

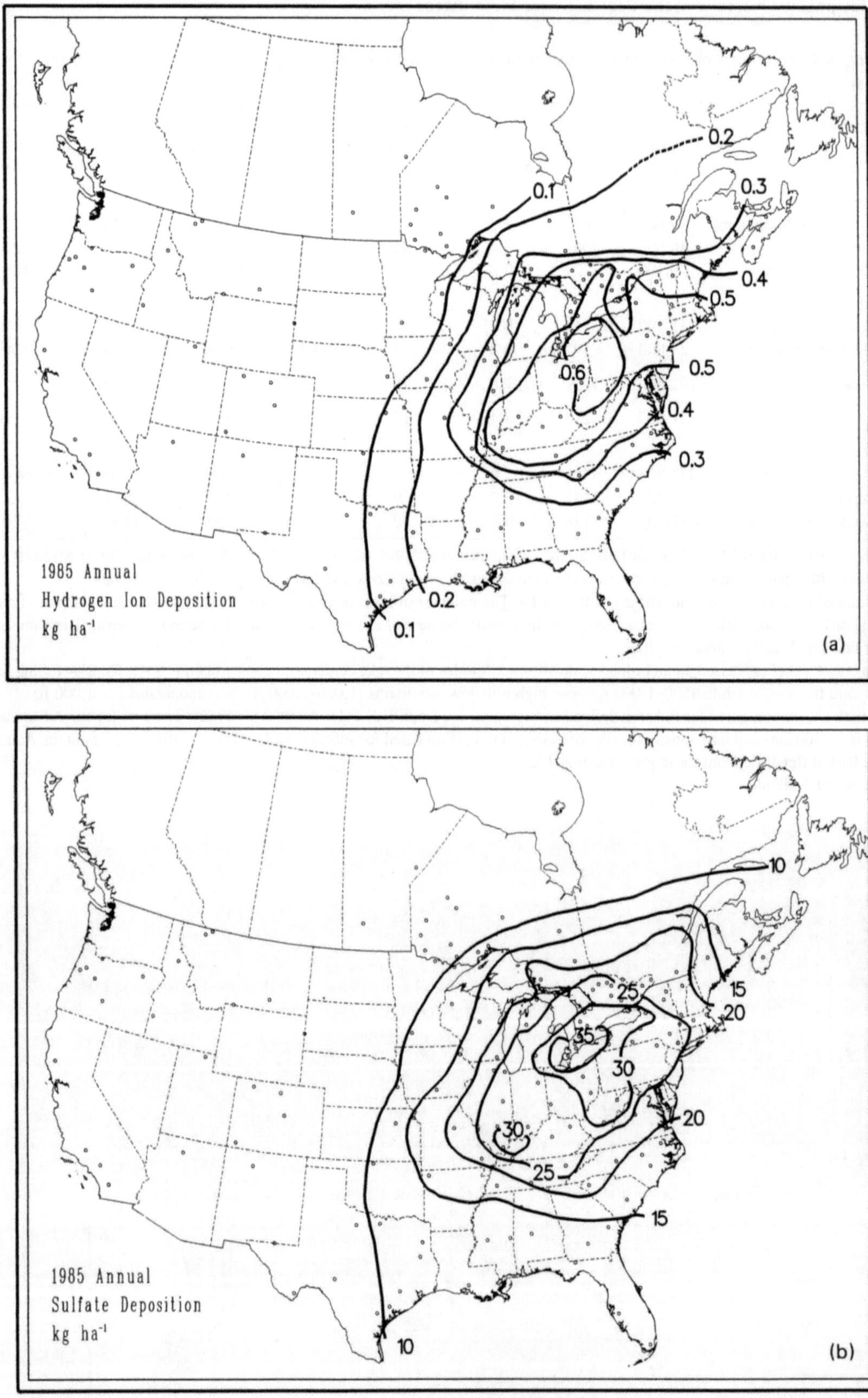

FIGURE 5.6. Wet deposition of: (a) H^+, (b) SO_4^{2-}, (c) NO_3^-, and (d) Ca^{2+}, in kg ha^{-1} for the calendar year 1985.

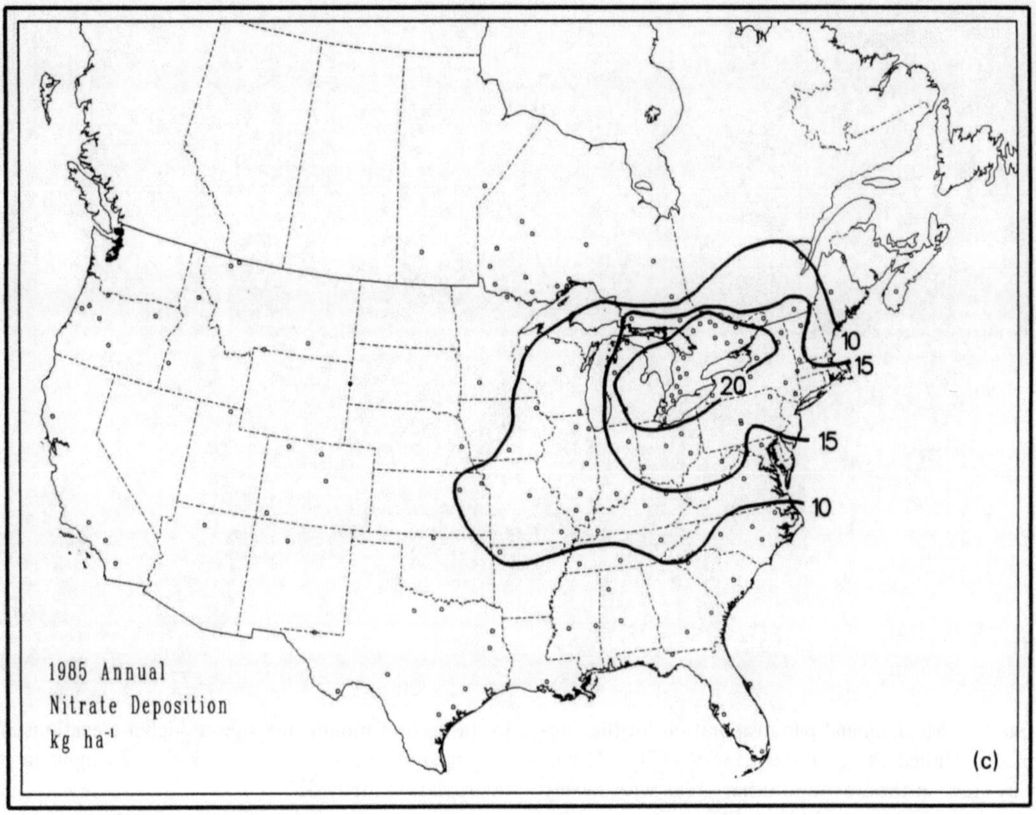

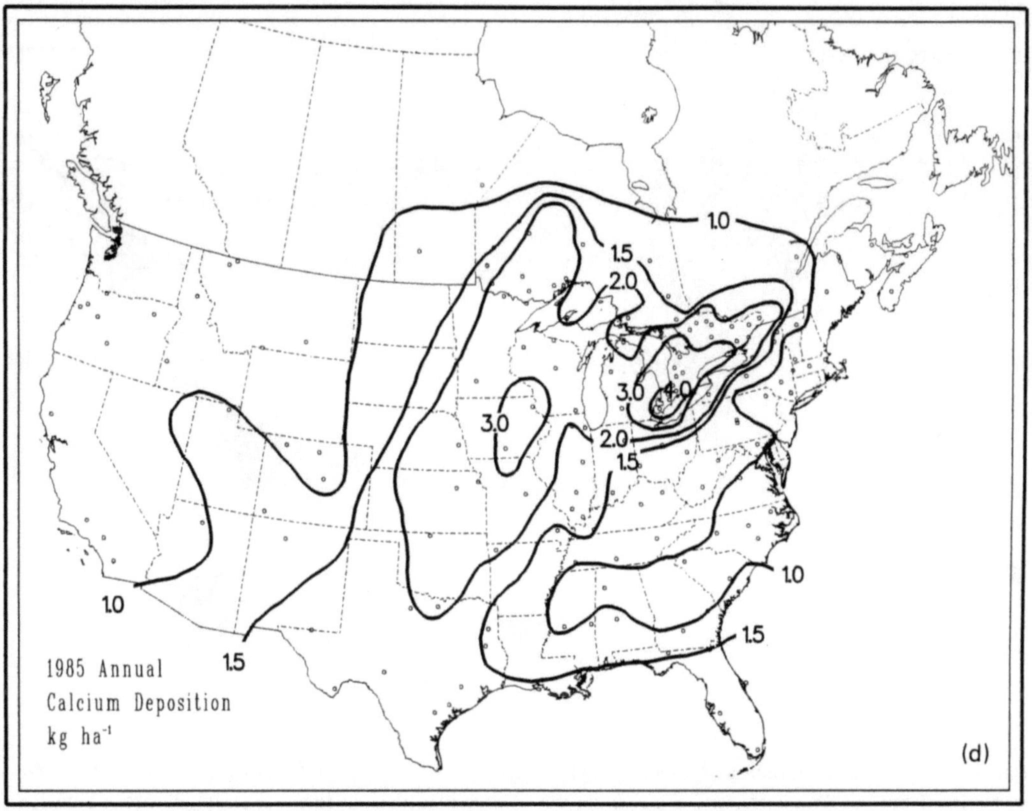

FIGURE 5.6. *Cont.* Points represent NADP monitoring sites used in calculating isopleths. Data from NADP (1987).

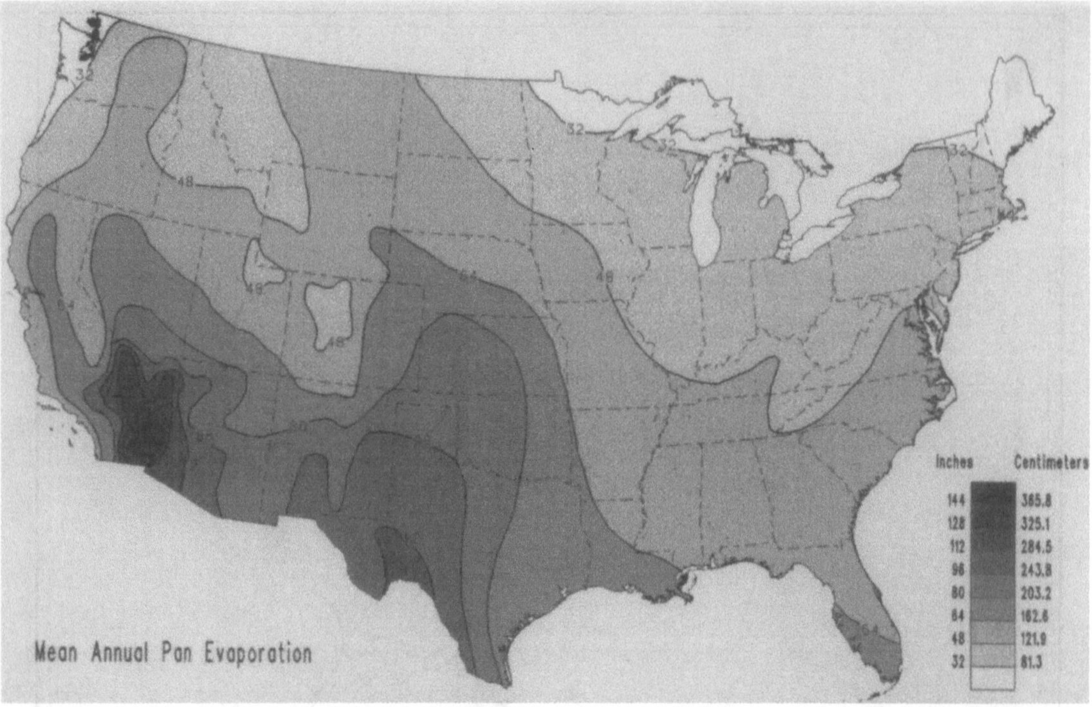

FIGURE 5.7. Mean annual pan evaporation for the conterminous United States (after USGS 1970). Mountainous areas in the west and other areas with locally high topographic relief are likely to be overestimated due to: (a) lack of monitoring sites at higher elevations and (b) apparent omission of elevation as an input in the interpolation method.

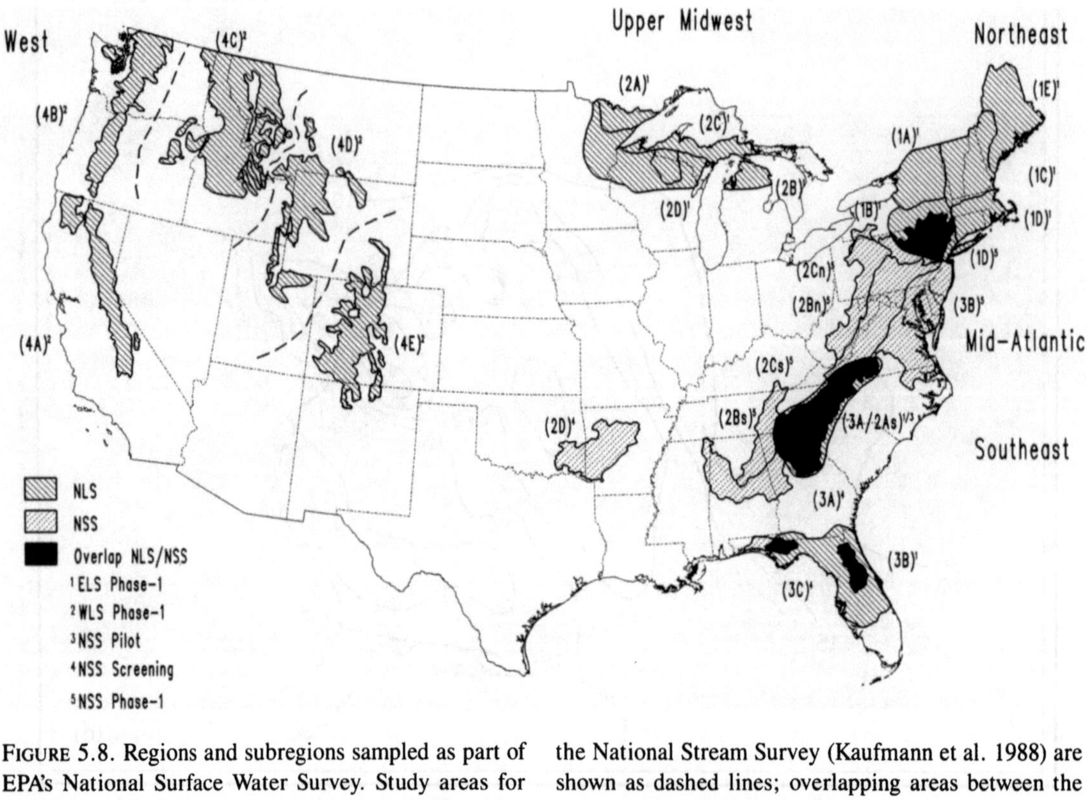

FIGURE 5.8. Regions and subregions sampled as part of EPA's National Surface Water Survey. Study areas for the National Lake Survey (Linthurst et al. 1986, Landers et al. 1987) are shown as solid diagonals; areas for the National Stream Survey (Kaufmann et al. 1988) are shown as dashed lines; overlapping areas between the two projects are shown in black.

and Schnoor 1986, Kenoyer 1986, Wentz and Rose 1989) and in-lake processes are critical to predicting the future acid-base status of these lakes. The dominance of seepage lake types and long residence time drainage lakes in the Upper Midwest and Florida suggest that in-lake processes are especially important in these areas. The degree to which in-lake processes are important in regulating the acid-base status of drainage lakes in other regions of the United States and Canada is unresolved (Schafran and Driscoll 1987, Shaffer et al. 1988, Stoddard 1987, 1988, Schindler 1988, Kelly 1988). Although in-lake processes have not been explicitly identified as a factor in establishing hydrologic classification, physiographic characteristics often reflect the dominant hydrologic properties in these study regions.

Integration of Factors for Defining Regions

Previous Attempts at Regionalization

Single factors such as soils (McFee 1980) and bedrock geology (Norton et al. 1982) have been used to delineate regions in the United States sensitive to acidic deposition. Kaplan et al. (1981) combined county-level information on soils and geology to illustrate their relationship with water quality in the northeastern United States, whereas Shilts (1981) classified regions of southern Canada on the basis of surficial geology and sensitivity to acid loading. A different approach was used to delineate areas of low ANC surface waters for the NSWS, whereby existing lake and stream ANC data ($n \simeq 2{,}500$), in part, were used to circumscribe areas likely to contain low ANC waters (defined as ANC $\leq$ 400 $\mu eq\ L^{-1}$). Areas with little available ANC data were assigned to ANC classes on the basis of land use, vegetation, soils, and geology, and the perceived associations of these factors with low ANC waters (Omernik and Powers 1983).

The original map of low ANC areas developed by Omernik and Powers (1983) was key to developing the hierarchical stratification used in the design of the NSWS. The region and subregion boundaries for the NSWS are shown in Figure 5.8. Not shown are the individual ANC strata that served as sampling units within the subregions (Linthurst et al. 1986, Landers et al. 1987, Kaufmann et al. 1988). Boundaries within some of the NSWS regions were found to be unsatisfactory for analysis of acidic deposition impacts. In particular, Eilers et al. (1988) noted that the original four subregion boundaries for the Upper Midwest were inappropriate because two of these subregions crossed important physiographic units and had an orientation parallel with the deposition gradient. We have since redrawn the subregion boundaries for the Upper Midwest and made minor adjustments for the Florida study area as well. The revised maps of the NSWS subregions are shown in the individual case study chapters.

The results of the NSWS and other surveys have been used to refine the delineation of low ANC areas in the United States (Figure 5.9). The evolution of ANC maps from Omernik and Powers (1983) to the map shown in Figure 5.9, which was prepared with about 39,000 observations, illustrates that as we obtain additional data on surface water chemistry, the perceived patterns in ANC become far more heterogeneous. Large-scale maps shown in the individual regional case studies further demonstrate the heterogeneous patterns in lake and stream ANC.

An alternative approach for displaying low ANC areas was selected for Canada. Here, knowledge of the soils and geology was integrated to produce a map representing areas with surface waters susceptible to acidification (Environment Canada 1988). The Canadian map is conceptually attractive in that it explicitly identifies variables used in the classification, although, like the ANC maps used in the NSWS, it does not clarify the relative importance of input variables in a given area. Dasymetric maps such as these examples for the United States and Canada are important for focusing research efforts, but have limited applications for quantitative assessments.

Relationship of Regional Case Study Areas to Low ANC Areas in the United States

The regional case studies, as noted in the Introduction (Charles, this volume), were selected on the basis of a variety of considerations. As a consequence, the spatial coverage of low ANC waters in the United States is not complete and the level of detail varies, in large part, because of the wide

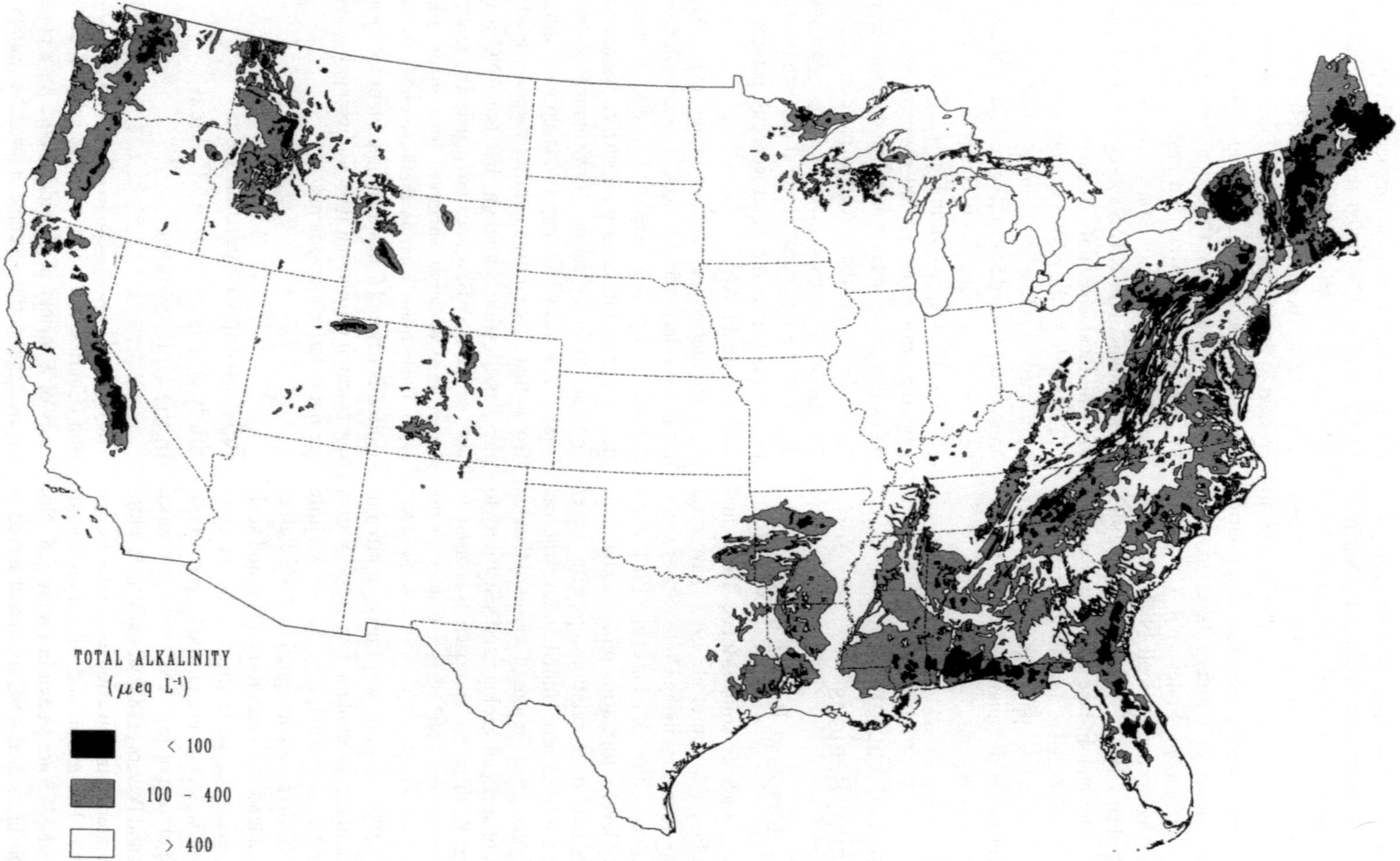

FIGURE 5.9. Total alkalinity (in μeq L^{-1}) of the conterminous United States. This map represents a more detailed update of an earlier map by Omernik and Powers (1983). The present map is adapted from one prepared in 1988 by J. M. Omernik, G. E. Griffith, J. T. Irish, and C. B. Johnson (U.S. EPA Environmental Research Laboratory, Corvallis, Oregon), which used data from the National Surface Water Survey and an additional 36,000 observations.

range of geographic scales of the individual chapters. In describing the resources of the Rocky Mountains (Chapter 14, Turk and Spahr, this volume), which covers more than 380,000 km^2, the level of treatment obviously needs to be different from that used for the Catskill Mountains (Chapter 8, Stoddard and Murdoch, this volume), which represent less than 1% of the area of the Rocky Mountains. Although each chapter is presented as a regional case study, the topical area shown in Figure 5.9 may represent (using the terminology of Fenneman, 1946) a division (e.g., Rocky Mountains), a province (e.g., Adirondack Mountains), or a section (e.g., Catskill Mountains). Maine, although represented by the state political boundary, is contained within the New England province. Its presentation by Kahl et al. (Chapter 7, this volume), is convenient for organizing this chapter without compromising the regional orientation of this book.

A major low ANC physiographic group that is shown in Figure 5.9 but not treated in this book is the Atlantic Coastal Plain. It is apparent that additional research is required to characterize the acid-base status of aquatic systems in the Atlantic Coastal Plain, as well as the processes that govern the response of the systems to changes in atmospheric deposition.

Physiographic Characteristics of Case Study Regions

Selected climatic, landscape, and morphometric features of the case study regions and their watersheds have been summarized in Tables 5.2, 5.3, and 5.4. These summaries of the study area physiography constitute a compromise between the need to condense information and the potential for distortions caused by oversimplifying complex patterns. Nevertheless, this overview of physiographic features of the case study regions underscores the diversity of aquatic resources in the United States.

Northeast

Low ANC surface waters are present throughout the Northeast (Linthurst et al. 1986), but regional case studies are presented only for the Adirondack Mountains, Catskill Mountains, and Maine. Other areas of interest in the Northeast with respect to surface water acidification are the Pocono Mountains of Pennsylvania and central and southern New England, including New Hampshire, Vermont, Massachusetts, Connecticut, and Rhode Island. According to ELS-I results, the three case study regions include more than half of the acidic lakes in the Northeast (Linthurst et al. 1986).

Adirondack Mountains

The Adirondack Mountains constitute a well recognized ecological unit in northern New York, located within the Appalachian Province. The Adirondack Mountains are related topographically and structurally to both the Appalachians and the Canadian Shield (Hunt 1974). Topographically, they are similar to the Appalachians, and the Lower Paleozoic formations circumscribing the region are also similar to the adjacent Valley and Ridge Province in Pennsylvania. However, the Precambrian rocks forming the bulk of the interior of the Adirondack Mountains are clearly a southern extension of the Canadian Shield (Thornbury 1965).

The Adirondack Mountains were glaciated in the Wisconsinan period, during which till of different depth and composition was deposited. Where the till is deep, contact time with the subsurface flow increases and surface water ANC is higher than in watersheds lacking deep till (Chen et al. 1984, Driscoll and Newton 1985, Booty et al. 1988). The climate of the Adirondack Mountains is humid and continental. Annual precipitation is about 125 cm in the southwest portion of the Adirondacks, decreasing to about 95 cm on the northeast side. Atmospheric deposition of H^+ and mineral acid anions is among the highest in the case study regions, particularly on the southwest side, and is exceeded only by deposition in the Catskill Mountains (Table 5.2).

Most of the lakes in the Adirondack Mountains were formed by glacial action, although beaver activity is responsible for creating new ponds and increasing the size of some existing ponds. Beaver dams were present in nearly half of the drainage lakes surveyed by Kretser et al. (1989). Lakes in the Adirondack Mountains typically are larger and deeper than those in other regions of the eastern United States (Overton et al. 1986), although more than 40% of the 1,469 lakes sampled by Kretser et al. (1989) were < 4 ha in area. Most of the lakes in the Adirondack Mountains are drainage lakes with hydraulic residence times < 0.5 yr (Linthurst et al. 1986). The erosion of the structural dome forming the Adirondack Mountains has given rise to a radial drainage pattern and numerous streams.

Maine

Maine is physiographically similar to the Adirondack Mountains and, as a consequence, these two areas are often grouped in common ecological units (Austin 1972, Bailey 1976, Omernik 1987). Together they form the northern portion of the Appalachian Province, and the northwestern portion of Maine is topographically and edaphically similar to the Adirondack Mountains. Both areas were glaciated and the glacial action has resulted in abundant lakes with similar morphometry. Interior Maine and the Adirondack Mountains share similar climate and vegetation types as well.

However, Maine is physiographically more diverse than the Adirondacks. Most geomorphic classifications (e.g., Fenneman 1946, Hunt 1974) recognize a seaboard lowland extending along the coast and an upland inland section. The lakes in the coastal zone of Maine experience a milder climate and receive a substantial contribution of marine aerosols (Mairs 1967, Sullivan et al. 1988). The deposition chemistry shows higher concentrations of SO_4^{2-}, NO_3^-, and H^+ in western Maine compared to sites in the northeastern portion of the state (NADP/NTN 1987, 1988).

The Catskill Mountains

The Catskill Mountains of New York are adjacent to the Poconos of Pennsylvania and border the unglaciated portion of the Appalachian Province. The Catskill Mountains constitute the smallest area represented by a case study in this volume (Chapter 8, Stoddard and Murdoch). Like most of the unglaciated Appalachian Province to the south, the Catskill Mountains have numerous streams and few natural lakes. The climate in the area is also transitional between the cold continental of the Adirondack Mountains and Maine and the warmer, humid climate of the Southeast. Because of proximity to the coast, maritime influence has a moderating effect on the climate, but contribution of marine aerosols to the chemistry of the streams is probably minor. Based on data from the Biscuit Brook precipitation monitoring site, acidic deposition in the Catskill Mountains is greater than that observed in other case study regions, although data from different years and stations would alter these rankings. Much of the Catskill Mountains region is located within a state park, but this area contains approximately 680 km of primary and secondary roads, which are maintained throughout the winter.

Upper Midwest

The Upper Midwest (Chapter 13, Cook and Jager, this volume) is treated as a contiguous group of five subregions. These subregions distinguish continuous, subtle patterns of physiological factors. For this reason, some terrestrially based classifications of the Upper Midwest show the region as largely undifferentiated (Austin 1972, Omernik 1987, Omernik and Gallant 1988). However, there are important variations in geology and climate across the Upper Midwest. For example, the Wisconsinan glacial advances occurred as a series of lobes extending down through the region in a south-southwesterly direction (Thornbury 1965). Differences in lake chemistry have been attributed to differences in properties associated with the glacial lobes in Minnesota (Brugam 1981) and Wisconsin (Simpkins et al. 1978).

Precipitation and evaporation generally vary as a function of longitude, although precipitation increases and evaporation decreases adjacent to Lakes Superior and Michigan. Sulfate and nitrate deposition are greater by nearly a factor of two in Michigan than in Minnesota.

Land use patterns in the region are quite localized. Agricultural influences are greatest in southeastern Minnesota and generally decrease in an easterly and northerly direction. The Boundary Waters of Minnesota and portions of the Upper Peninsula of Michigan are relatively undisturbed by land use activities.

Perhaps the most apparent distinction in the geographic distribution of lake chemistry in the Upper Midwest is the strong contrast between the highly interconnected drainage lakes in the Boundary Waters and the seepage lakes common throughout the remainder of the study area.

Southeast

Only a small portion of the Southeast is represented here in the regional case studies (the mountains of western Virginia and the Southern Blue Ridge) and both these case study regions are highlands of the Appalachian system. Other areas in the Southeast with low ANC surface waters not represented in this book are the Ozarks, the Ouachitas, the Appalachian Plateau, much of the Valley and Ridge Province, the Piedmont, and the Atlantic Coastal Plain. The southeastern case study regions adequately represent the characteristics of the highlands, but because of large differences in geology, topography, soils, and land use, the chemical attributes of streams in the lower elevation areas of the Southeast cannot be inferred on the basis of these case study chapters.

Western Virginia

The western Virginia case study region includes portions of the Valley and Ridge Province and the Blue Ridge Province, which differ from one another with regard to bedrock, soil, and vegetative characteristics (Chapter 10, Cosby et al., this volume). Because the Southeast was not glaciated, the soils are much older than those in the Northeast, and as a consequence, are more highly weathered. The topographic relief is similar to that observed for the low mountains of the Adirondacks and relief is exceeded only in the young mountain systems of the western United States. Because of the lack of glaciation and the presence of a well-developed drainage network in western Virginia, natural lakes are rare.

Precipitation, which occurs primarily as rain, exceeds 140 cm; evapotranspiration is intermediate between that of the Northeast and Florida. Precipitation is highly acidic, and wet deposition of sulfate exceeds 30 kg ha^{-1} yr^{-1} ($\sim$600 eq ha^{-1} yr^{-1}).

Southern Blue Ridge

The Southern Blue Ridge has physical characteristics similar to those of western Virginia, which is to be expected, because the latter also includes the northern portion of the Blue Ridge Province (Chapter 11, Elwood et al., this volume). The Southern Blue Ridge is also dominated by streams, although reservoirs comprise an important surface water resource. The physiography of the Southern Blue Ridge differs from that of the northern section of the Blue Ridge Province in that the former is broader and higher, and displays a prominent frontal scarp on the east side.

Precipitation, evaporation, and temperature for the Southern Blue Ridge are slightly higher than those observed for western Virginia. The lakes in the Southern Blue Ridge sampled during ELS-I show relatively high percentages of urban and agricultural land use (11% and 22%, respectively), which suggests a high potential for anthropogenic contributions from some watersheds. Among the regional case studies, only watersheds in Florida showed a higher percentage of perturbation. Deposition chemistry in the Southern Blue Ridge Province exhibits values similar to values measured in western Virginia.

Florida

Florida is the southernmost extension of the southern coastal plain, and in many respects it is radically different from other case study regions (Chapter 12, Pollman and Canfield, this volume). The calcareous bedrock and highly weathered siliceous sandy soils reflect the marine origins of Florida. The absence of glaciation and minimal recent tectonic activity have resulted in an area with low topographic relief. Lakes in Florida were formed, and continue to be formed, primarily where overlying sands have collapsed into the cavities created by dissolution of the limestone and dolomite bedrock (Fernald and Patton 1984). Lakes are abundant in Florida and most of them are seepage lakes. Although some lakes are technically drainage systems, topographic relief is generally so imperceptible that it is difficult to determine flow paths and areas that contribute to surface flow.

The climate is typical of subtropical conditions in which annual precipitation, temperature, and evaporation are high. Most lakes are within 75 km of the Atlantic Ocean or the Gulf Coast, and the low elevation of the lakes (median = 21 m), combined with minimal topographic relief, results in a strong marine influence on lake chemistry. Precipitation chemistry is mildly acidic in northern Florida, but the high evaporation rate concentrates the deposition to a greater extent than in other regional case study areas.

About 45% of the lakes in Florida sampled in the ELS-I had significant agricultural land use adjacent to the lakes. Most of the agricultural development occurs in the central portion of the state and is a major reason for partitioning the lakes in the peninsula into northcentral and southcentral groups. The third group of low ANC lakes identified in Florida includes those in the panhandle; this group receives considerably more precipitation, has less evaporation, and has far less agricultural influence than lakes in the peninsula.

West

The western Cordilleran Mountains, comprising the Rocky Mountains and the Pacific mountain ranges, are composed largely of intrusive rocks, and to a lesser extent, of volcanic, metamorphic, and sedimentary rocks. Pleistocene glaciation carved abundant glacial cirques in the ridges and formed a smaller number of lakes located on till. Lakes occur from the dense forests of the low elevation montane zone upwards to the barren rocks of the alpine zone. Lakes in the West are typically smaller and deeper than eastern lakes and have watershed area to lake area ratios averaging about twice as large as lakes in the east. Watershed relief is far greater in the West than elsewhere in the United States. This feature, combined with large watershed area to lake area ratios, a high percentage of exposed bedrock, and large snowpacks, often results in hydrologically "flashy" systems.

Nearly all of the areas in the West believed to have low ANC surface waters (as shown in Figure 5.9) are discussed in the three case study chapters on the Rocky Mountains, the Sierra Nevada, and the Cascade Mountains (Chapters 14, 15, and 16, this volume). The areas not addressed are primarily coastal mountain systems from northern California to the Olympic Peninsula. Also excluded are lakes in the Puget Lowlands and coastal dune lakes in Oregon and Washington. Deposition for much of the high-elevation area of the West is either nonacidic or only mildly acidic.

Rocky Mountains

The Rocky Mountains (Chapter 14, Turk and Spahr, this volume) are more of a geographic convenience than a well-defined geomorphic unit. Unlike the Sierra Nevada, which is a discrete geomorphic unit, the Rocky Mountains are a collection of geomorphic units strung together as a discontinuous range. In general, mountain groups within the Rocky Mountains were formed as uplifts of Precambrian rocks, flanked by sedimentary rocks that have been eroded from the Precambrian cores. However, Tertiary volcanic rocks in the central and southern Rocky Mountains and granitic rocks in the northern Rockies are important.

The Rocky Mountain range is typically divided into southern, central, and northern physiographic provinces (Fenneman 1946, Hunt 1974). There are, however, considerable differences among mountain groups within these three divisions that highlight the importance of evaluating the individual uplifts. For example, within the central Rocky Mountains, the Absaroka Mountains (Montana and Wyoming) are composed of Tertiary volcanic rocks, whereas the Uinta Mountains (Utah) are predominantly quartzites and sandstones (Thornbury 1965). The Rocky Mountains receive less precipitation than the Sierra Nevada-Cascade system and a higher proportion as rainfall. The density of lakes in the Rocky Mountains is low, in part because precipitation is relatively low, but the lakes have physical characteristics similar to those of their far western counterparts.

Sierra Nevada

The Sierra Nevada range is a massive fault block, tilted to the west, and contains a large population of extremely dilute lakes (Chapter 15, Melack and Stoddard, this volume). The western portion of the range receives higher precipitation than the east side, which consists of a series of steep escarpments. Most of the lakes occur on the broad western slope of the Sierra Nevada. The almost continuous chain formed by the Sierra Nevada and Cascade ranges provides some justification for evaluating them as a single unit, but the differences in geology, topography, vegetation, and climate warrant separate treatment.

About three-fourths of the Sierra Nevadan rocks are composed of intrusive granites and related rocks; the remainder are largely Paleozoic and Mesozoic meta-sedimentary and meta-igneous rocks. The Sierra Nevadan lakes generally are deeper and are located at much higher elevations than lakes in the

Cascade Mountains. High-elevation watersheds in the Sierra Nevada receive over 90% of their precipitation as snow, a feature that dominates the hydrologic processes of these lakes (Stoddard 1987). A majority (60%) of the land use represented by watersheds from the Western Lake Survey (WLS) in the Sierra Nevada is classified by the U.S. Geological Survey (USGS) as barren/tundra, indicative of an alpine environment.

Cascade Mountains

The southern and middle Cascade Mountains (Chapter 16, Nelson, this volume) are characterized by extrusive volcanic rocks, whereas the northern Cascade Mountains (starting east of Seattle) are formed by granitic material more similar in composition to the Sierra Nevada (Thornbury 1965). Geologically young volcanic peaks protrude up to 2,000 m above the lower elevation mountains. The Cascade Mountains exhibit dramatic variation in precipitation, ranging from 80 cm in the southern Cascade Mountains to about 250 cm in the northern portions. Glaciers are still abundant in the northern Cascade Mountains and continue to shape the range.

High precipitation amounts in the middle and northern Cascade Mountains cause many lakes to have short hydraulic residence times. The typical elevation of lakes in the Cascade Mountains is only half that of lakes in the Sierra Nevada and Rocky Mountains. Most of the Cascade watersheds are situated in montane and subalpine zones where there is greater opportunity for soils to develop.

Conclusions

This book synthesizes our current understanding of low ANC surface waters for major portions of the United States most likely to respond to changes in deposition chemistry. It is one of the first attempts to explicitly organize the information for geographically consistent areas. The key factors we have used to distinguish among regions are geology, soils, climate, hydrology, deposition chemistry, land use, vegetation, and landforms. It is apparent that the physiographic diversity of aquatic resources in the United States is so great that an analysis of the response of these lakes and streams must reflect the fundamentally different properties of the aquatic resources. Classifying these resources is an iterative process, and as we gain a greater understanding of the processes that control surface water chemistry, we anticipate that future attempts to spatially represent areas in the United States susceptible to acidification will better reflect these processes.

Acknowledgments. The authors thank J. M. Omernik for his contributions to the chapter, and acknowledge helpful comments on earlier drafts by D. Charles, R. Cook, W. McFee, S. Norton, D. Schindler, and D. White.

Part II
Case Studies

Northeast Overview

Charles T. Driscoll

The northeastern United States encompasses several subregions with abundant surface water resources characterized by low ionic strength and low ANC (Figure 1). There are 7,096 lakes > 4 ha within lake districts in the Northeast that are potentially sensitive to acidic deposition (Brakke et al. 1988). These areas include the Adirondacks, the Catskills, Maine, central New England, southern New England, and northern Pennsylvania. This book contains chapters on regional analyses of surface waters with respect to acidic deposition for the Adirondacks (Chapter 6, Driscoll et al., this volume) and the Catskills (Chapter 8, Stoddard and Murdoch, this volume) of New York, and Maine (Chapter 7, Kahl et al., this volume). Much of the information in these chapters also applies to those areas not covered.

All three subregions include bedrock and surficial characteristics that result in limited supply of base cations to drainage waters (Chapter 5, Eilers and Selle, this volume). The bedrock geology of the Adirondacks is composed largely of granitic gneisses and metasedimentary rocks. The bedrock material of the Catskills includes sandstone and mudstone. In Maine, there are heterogeneous metasedimentary deposits with granitic intrusions. All three areas were glaciated and are predominantly forested. The soils in the Adirondacks and Maine are largely acidic Spodosols and Histosols, whereas in the Catskills, soils are generally Entisols and Inceptisols.

The climate of the Northeast is generally humid continental, with long cold winters and short cool summers. The coastal areas of Maine, and to a lesser extent the Catskills, experience a milder climate with a marine influence. The Northeast receives relatively high inputs of precipitation (90 to 125 cm yr^{-1}; Figure 2). Much of the precipitation to the region occurs as snow, and snowpack accumulation and melting are an important component of hydrologic dynamics.

Precipitation chemistry varies markedly across the northeastern United States (Figure 3). The highest concentrations and deposition of H^+, SO_4^{2-}, and NO_3^- in the United States occur in the middle Appalachians and extend to the Adirondacks and Catskills. Concentrations of H^+, SO_4^{2-}, and NO_3^- in precipitation decrease across New England to Maine. Seasonal patterns also occur in precipitation chemistry (Figure 4). Sulfate concentrations in precipitation are generally low in the late winter months, increase in the summer, and decline in the fall. This pattern generally coincides with precipitation H^+ concentrations. An exception to this seasonal pattern is a peak in H^+ concentration that occurs during December and January. Concentrations of Ca^{2+} and NH_4^+ are lowest in the winter and peak during the summer months. The low concentrations of these solutes during the winter are partially responsible for the early winter peak in H^+. Nitrate concentrations are generally uniform over the annual cycle.

The variability of geologic, edaphic, and hydrologic characteristics of watersheds in the Adirondacks, Catskills, and Maine results in a range of surface water chemistry and sensitivity to acidic deposition (Figure 1). Generally, the abundance of acidic waters decreases across the Northeast, with the gradient in acidic deposition from the Adirondacks to Maine.

FIGURE 1. Distribution of total lakewater alkalinity (ANC; µeq L^{-1}) in the Northeast. This map represents a more detailed update of an earlier map by Omernik and Powers (1983). The present map is adapted from a national map prepared in 1988 by J.M. Omernik, G.E. Griffith, J.T. Irish, and C.B. Johnson [U.S. EPA Environmental Research Laboratory, Corvallis, Oregon (Chapter 5, Eilers and Selle, this volume)], which used data from the National Surface Water Survey and an additional 36,000 observations.

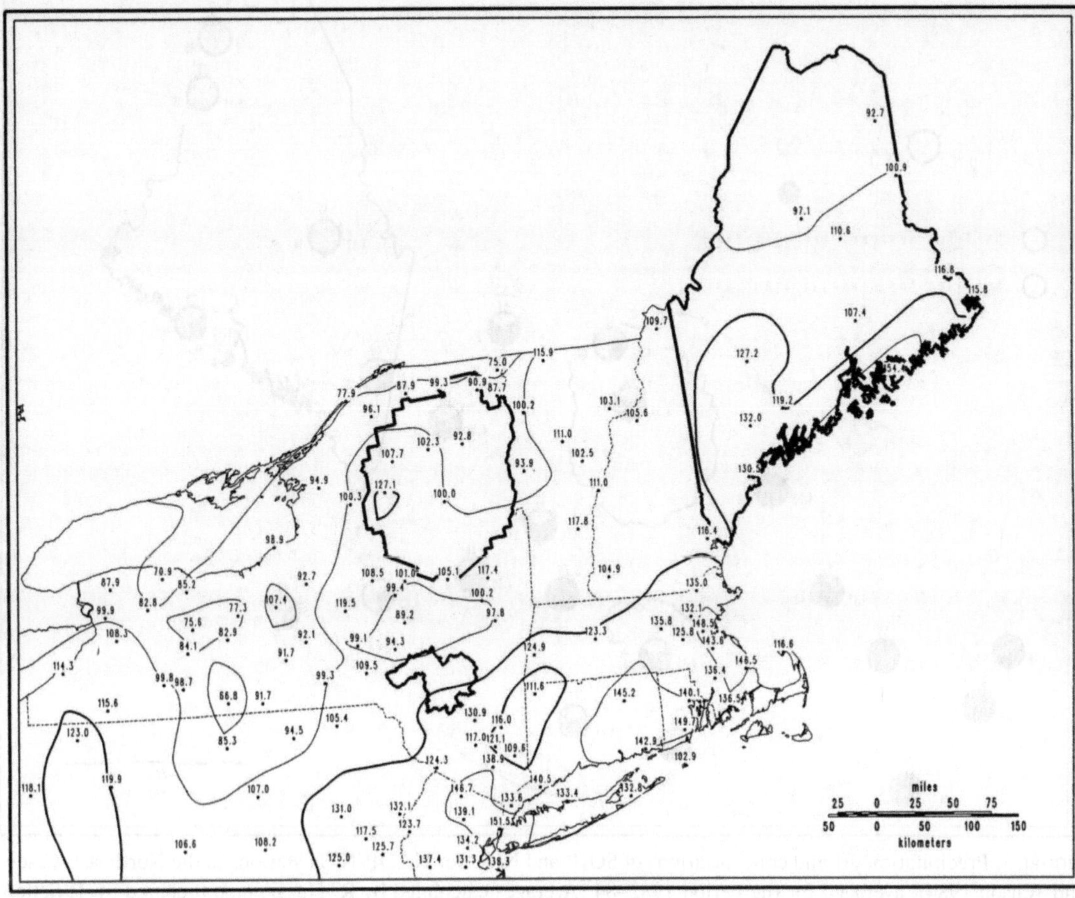

FIGURE 2. Precipitation quantity in the Northeast averaged for the period 1982–84. Contours are in cm yr^{-1} at 20-cm intervals with 60-cm contours in heavier lines; the location of each station in this region that was used in preparing the contour map is indicated, along with its annual precipitation in cm yr^{-1}. Point data calculated by R. Husar; contour map prepared by T. Selle.

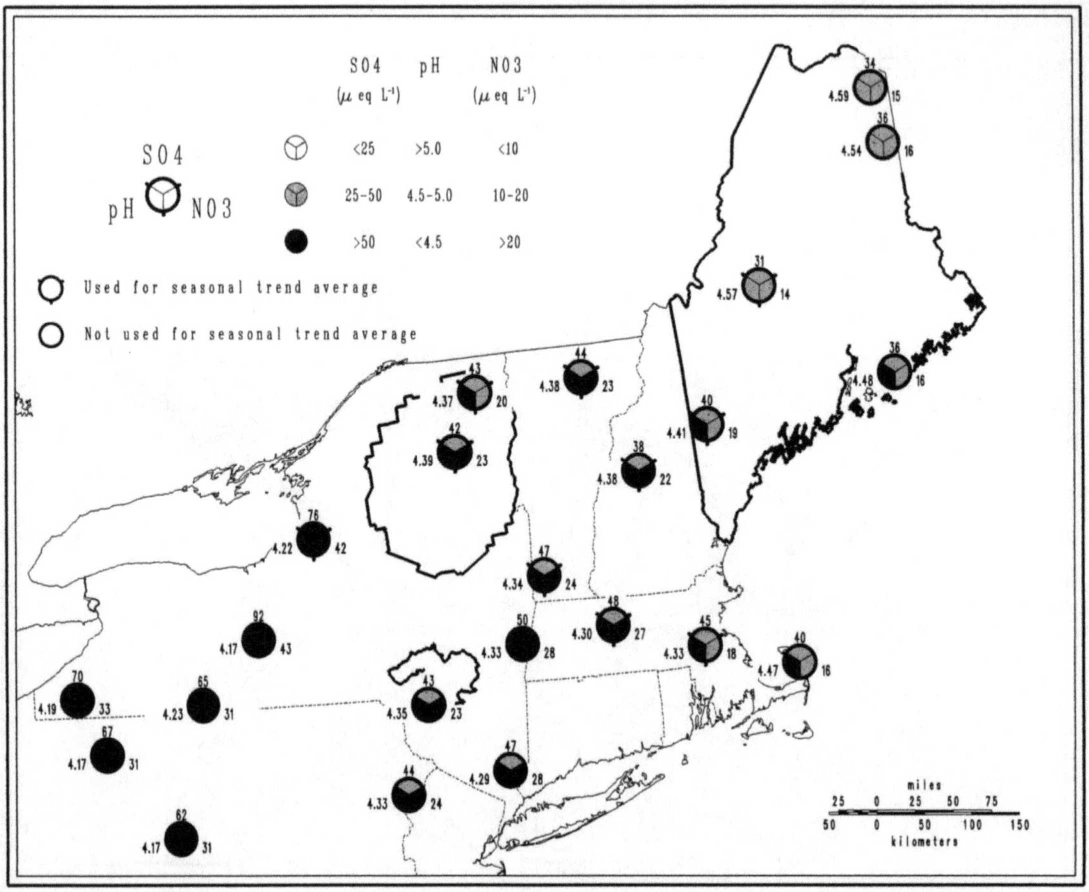

FIGURE 3. Precipitation pH and concentrations of SO_4^{2-} and NO_3^- for NADP/NTN stations in the Northeast (Olsen and Watson 1984), averaged for the period 1982–84. Averages calculated by R. Husar; map prepared by T. Selle.

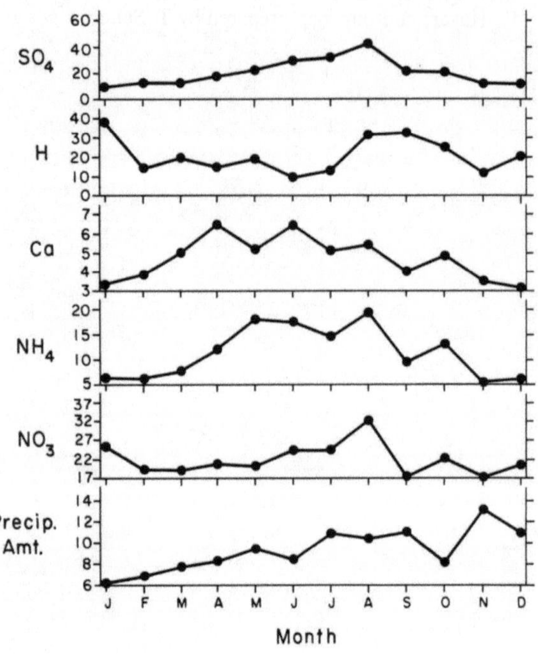

FIGURE 4. Monthly variation in precipitation quantity and chemistry in the Northeast. Chemical values are expressed as $\mu eq \cdot L^{-1}$; precipitation amount is in cm. Data represent 5-year averages for the period 1982–86 for the nine sites indicated in Figure 3. Data are from the NADP (Olsen and Watson 1984) and averages were determined by R. Husar.

6
Adirondack Mountains

Charles T. Driscoll, Robert M. Newton, Chad P. Gubala, Joan P. Baker, and Sigurd W. Christensen

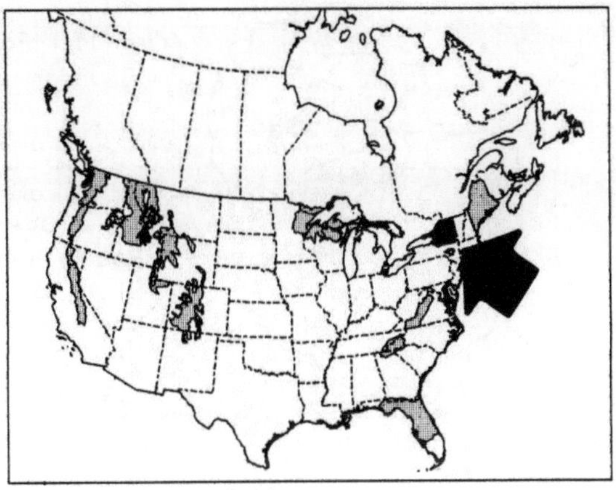

ABSTRACT. The Adirondack Mountain region in northern New York State is a large forested area, underlain by bedrock material primarily composed of granitic gneisses and metasedimentary rocks. The surficial materials covering the region are the result of glacial activity; soils developed from glacial till are typically acidic Spodosols. The Adirondacks receive high deposition of acidic materials (SO_4^{2-} deposition approximately 500 eq ha^{-1} yr^{-1}). The combination of large inputs of strong acids from the atmosphere and geologic/edaphic characteristics that limit base cation supply to drainage waters places the aquatic resources of the Adirondacks at risk from acidic deposition.

Results from the U.S. EPA Eastern Lake Survey (representing a population of 1,290 lakes) suggest that 14% of the lakes > 4 ha in surface area are acidic (acid neutralizing capacity (ANC) ≤ 0 μeq L^{-1}). However, the number and percentage of acidic Adirondack lakes determined from synoptic surveys are influenced by the boundary designating the region as well as by the minimum surface area of the lakes sampled. For example, the Adirondack Lake Survey Corporation (ALSC) survey, which included 1,469 lakes in the Adirondack ecological zone with a surface area > 0.5 ha, determined that 26% of the lakes are acidic.

There is considerable diversity in the acid-base status of Adirondack lakes. Lake-to-lake variation in ANC is largely the result of differences in the supply of base cation concentrations to drainage water, as concentrations of strong acid anions are relatively uniform across the region. Regional patterns in lake chemistry are evident across the Adirondacks. Acidic and low ANC lakes are generally located in the western and southern areas of Adirondack Park. The bedrock geology in this area is principally granitic gneiss and many lakes are situated in high-elevation (> 600 m) basins with shallow glacial till. In the northern and eastern regions of Adirondack Park, lakes occur at lower elevations in watersheds with thick deposits of glacial till or stratified drift. These waters tend to be characterized by higher ANC values. In addition, the bedrock geology of the eastern Adirondacks is calcium-rich anorthosite, which facilitates the supply of base cations to drainage waters. Many lakes along Lake Champlain and outside the Adirondack Park boundary occur at low elevations and are impacted by salt and/or contain carbonate deposits within basins. These waters have high ANC and are generally insensitive to acidic deposition.

Process-level watershed studies indicate that hydrologic flow path is the principal factor regulating the supply of base cations to Adirondack lakes. As a result, a classification system based largely on hydrologic flow paths was developed for Adirondack lakes. Drainage lakes or closed lakes are the predominant hydrologic type (84%) of ponded Adirondack waters. Within the population of drainage lakes, thin till basins (< 5% of the basin area covered by thick till and stratified drift; 28% of Adirondack lakes) have the potential to be chronically acidic under current loadings of acidic deposition.

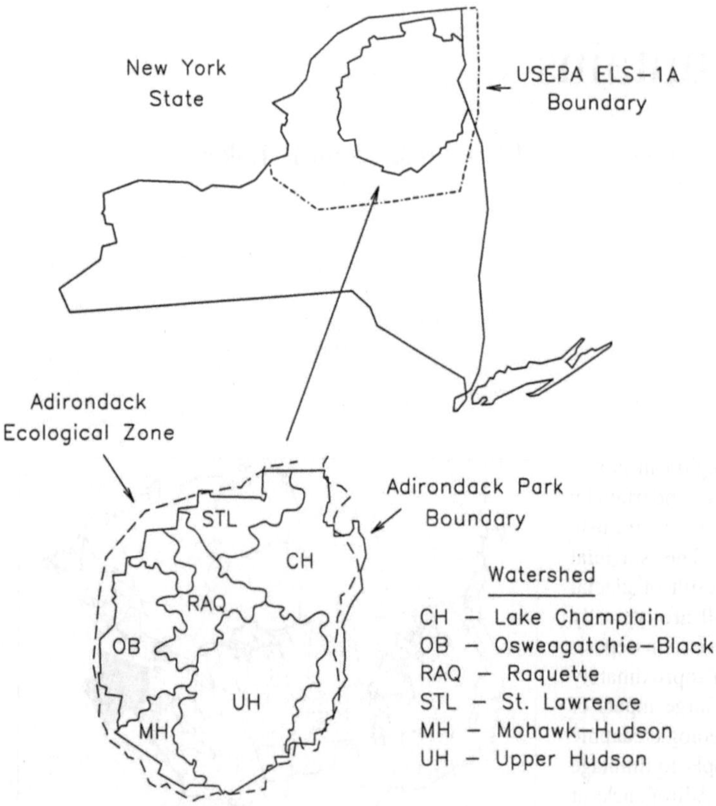

FIGURE 6.1. Location of Adirondack Park, the Adirondack ecological zone, and the boundary of the U.S. EPA Eastern Lake Survey–Phase I (ELS-I) subregion 1A in New York State. The major watersheds within the Adirondack region are indicated.

Intermediate till basins (5% to 25% of the basin area covered by thick till and stratified drift; 18% of Adirondack lakes) generally have a baseflow ANC of < 100 μeq L^{-1}. Surface and outlet waters of these lakes have the potential for substantial episodic acidification (loss of ANC > 50 μeq L^{-1}) during snowmelt. Watersheds with thick till or stratified drift (> 25% of the basin area), or containing deposits of carbonate minerals are characterized by lakes with high ANC and are insensitive to acidic deposition (21% of Adirondack lakes).

Information is available from historical lake surveys, water quality monitoring studies, and sediment records to assess recent changes in the acid-base status of Adirondack waters. Limitations are evident in all individual approaches, which necessitates compositing observations to verify acidification patterns. Most studies indicate that acidification of Adirondack waters has occurred sometime during this century. Although there is considerable uncertainty as to the exact timing of the recent acidification, the limited data suggest that most changes in surface water acidity probably occurred between 1930 and 1960. Sulfate is currently the dominant anion in acidic waters and this SO_4^{2-} can be attributed largely to atmospheric deposition. However, the extent of ANC decreases due to acidic SO_4^{2-} inputs cannot be resolved with existing data. There is limited evidence that base cations and NO_3^- concentrations have also changed in some waters, contributing to trends in ANC. Simulations from the ILWAS acidification model show that there are large lake-to-lake variations in response to decreases in sulfur loading.

Adirondack Lake Survey Corporation observations indicate that 55 species of fish have been collected. There is considerable range in the sensitivity of fish to conditions of acidity and several species of fish are tolerant of acidic conditions. However, the number of fish species caught decreases with decreasing pH in Adirondack lakes. Surveys conducted by the ALSC have shown that no fish were caught in 24% of Adirondack lakes. These waters are generally located in the southwestern Adirondacks. Fishless lakes are significantly correlated with a number of physical and chemical factors, including lake size, lake elevation, pH, and O_2 concentrations. Thus, specific reasons for the absence of fish are difficult to identify, although the absence of fish is correlated with lake pH.

Introduction

The Adirondack region is a large (2,400,000 ha; 6,000,000 acre) predominantly forested area in northern New York State that is endowed with

FIGURE 6.2. Photograph of Clear Pond (43°59′N, 74°50′W) in east-central Adirondack Park.

many lakes and streams (Figures 6.1 and 6.2). Adirondack Park is the largest park in the United States outside Alaska. Geologically, the Adirondack Mountains are among the oldest landforms in eastern North America and are similar in origin to the Laurentian Shield of Canada. The bedrock material is primarily gneisses and metasedimentary rocks. Marble and other calcite bearing bedrock occur in a few scattered locations throughout the region, chiefly to the east. The mountains and uplands are mantled glacial till, thicker in the valleys and becoming progressively thinner upslope. Lower elevation areas may contain glacial meltwater deposits, which are composed primarily of stratified sand and gravel. Soils of the region are generally acidic Spodosols, developed from glacial till.

The Adirondacks receive relatively large inputs of precipitation (annual average approximately 100 cm), accompanied by high stream runoff (annual average approximately 60 cm). The large quantity of precipitation coupled with elevated concentrations of acidic solutes in precipitation results in high deposition of acidic materials to the region (H^+ deposition approximately 500 eq ha^{-1} yr^{-1}; Johannes et al. 1985). The large inputs of strong acids from the atmosphere, coupled with geologic/edaphic characteristics that may limit the neutralization of strong acid inputs, places many of the aquatic resources of the Adirondacks at risk from acidic deposition.

This chapter (1) describes the environmental characteristics of the Adirondack region, (2) summarizes the regional characteristics of Adirondack surface waters sampled as part of the U.S. EPA Eastern Lake Survey (ELS-I), (3) illustrates the major processes regulating the acid-base status of Adirondack surface waters through site-specific investigations, develops a classification system for Adirondack lakes, and determines the distribution of ELS-I lakes in these lake classes, (4) examines evidence for historical acidification trends in the Adirondacks, (5) discusses changes in Adirondack water chemistry that might occur following changes in acidic deposition, and (6) discusses the status of fisheries in the region with respect to the acidity of surface waters.

Regional Characteristics

Climate of the Adirondack Region

There is considerable climatic variability within the Adirondack region, particularly associated

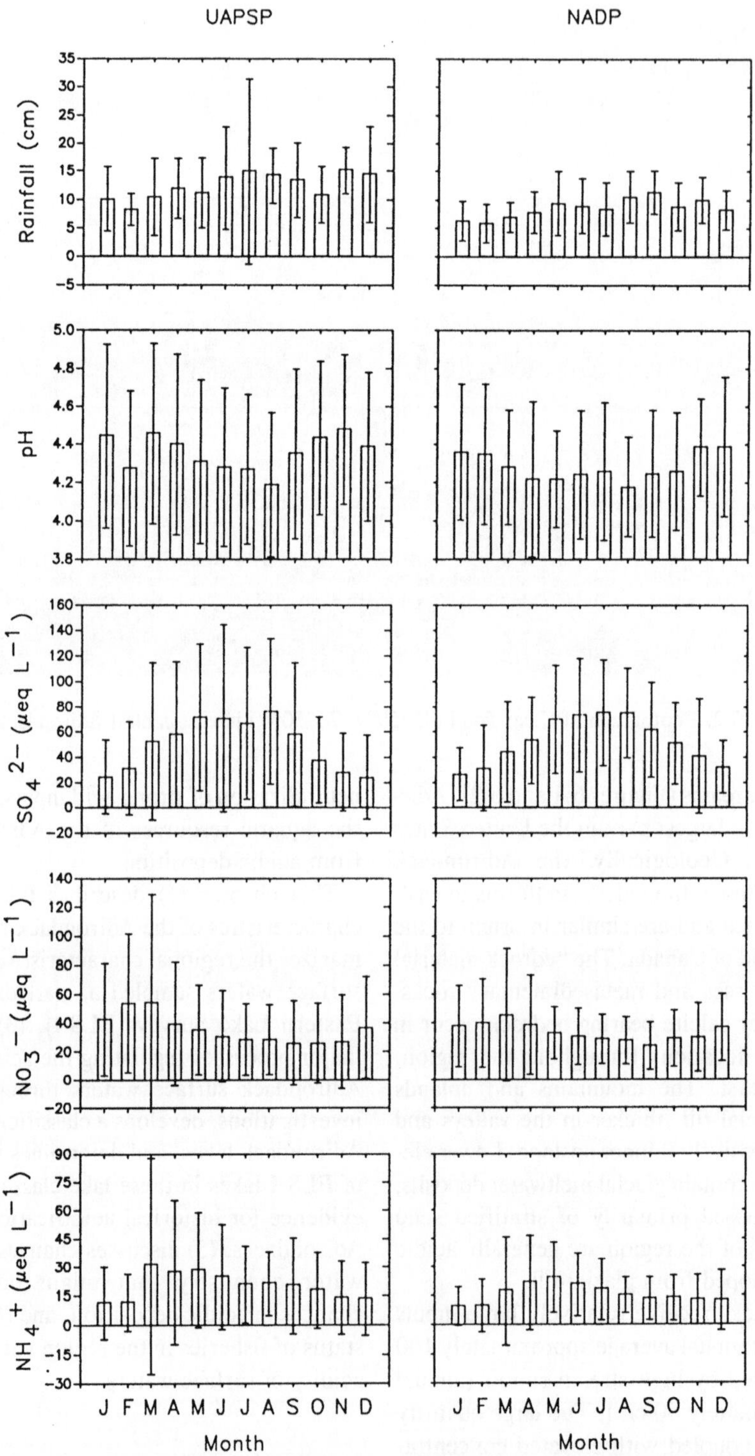

FIGURE 6.3. Monthly trends of precipitation amount and chemical composition for the Utilities Acid Precipitation Sampling Program (UAPSP) site at Big Moose Lake and the National Acid Deposition Program (NADP) site at Huntington Forest in the Adirondacks. Values represent monthly means and standard deviations for the monitoring period (1981–88 for UAPSP and 1978–88 for NADP). The UAPSP monitors precipitation chemistry on an event basis, whereas NADP collections are made at weekly intervals.

TABLE 6.1. Location, elevation, lake type, surficial geology and volume of Adirondack long-term monitoring (ATLM) sites.[a]

Lake	Location	Elevation (m)	Type	Surficial geology	Volume 10^4 m^3
Arbutus Pond (AB)	43°58'N,74°14'W	513	Drainage	Thin/thick till	146
Barnes Lake	43°34'N,75°14'W	397	Seepage	Glacial sand	13
Big Moose Lake (BM)	43°49'N,74°53'W	557	Chain drainage	Thin till	3,600
Black Pond (BL)	44°27'N,74°18'W	494	Drainage	Thick till/stratified drift	180
Bubb Lake	43°46'N,74°15'W	554	Drainage	Thick fill/stratified drift	38
Cascade Lake (CA)	43°48'N,74°48'W	554	Drainage	Thin/thick till	172
Clear Pond (CL)	44°00'N,73°49'W	582	Drainage	Thin/thick till	651
Constable Pond (CO)	43°50'N,74°48'W	584	Chain drainage	Thin till	45
Dart's Lake	43°48'N,74°52'W	536	Chain drainage	Thin till	380
Heart Lake (HE)	44°10'N,73°58'W	662	Drainage	Thin/thick till	54
Lake Rondaxe	43°45'N,74°55'W	525	Chain drainage	Thin/thick till	273
Little Echo Pond	44°18'N,74°22'W	481	Seepage	Spagnum bog	2.8
Moss Lake (MO)	43°47'N,74°51'W	536	Chain drainage	Thin/thick till	260
Otter Lake (OT)	43°11'N,74°30'W	488	Chain drainage	Thin till	22
Squash Pond (SQ)	43°49'N,74°53'W	650	Drainage	Thin till	4.5
West Pond (WE)	43°56'N,74°44'W	579	Drainage	Peat/thin till	56
Windfall Pond	43°48'N,74°50'W	595	Drainage	Thin till	7.8

[a]Eleven ALTM lakes were sampled weekly from January to May in 1986 and 1987 to examine short-term changes in chemistry during snowmelt. The two-letter abbreviations (along with 86 or 87 to indicate sampling year) are used in the text to describe chemical patterns from the snowmelt study. Otter Lake was sampled only during 1986.

with changes in elevation. However, the region as a whole can be characterized as humid continental, with short, cool summers and long, cold winters (Trewartha 1954). The Adirondacks experience considerable variation in weather, with wide ranges in both daily and annual temperature. The 30-year mean air temperature ranges from −10°C in January to 18C in July (U.S. Environmental Data Service 1987).

As prevailing winds travel from west to east across low-lying areas in the midwestern United States and southern Ontario, including the Great Lakes, they accumulate moisture. When moisture-bearing air masses intercept the high-elevation region of the western Adirondacks, a large quantity of precipitation falls (see Figure 2 in Northeast Overview). This orographic effect also results in a rain shadow, so the quantity of precipitation decreases from west to east across the region. Orographic lifting of air masses also explains isolated mountains receiving more than 125 cm of precipitation in the high peaks region of the eastern Adirondacks.

From the Integrated Lake-Watershed Acidification Study (ILWAS), Johannes et al. (1985) studied the precipitation quantity and chemistry of three watersheds in the western Adirondacks over a 4-year period. They found that 95% of the precipitation events ranged from 0.02 to 6.0 cm and that on the average a precipitation event occurred every third or fourth day. Precipitation inputs are distributed evenly over the year (e.g., Figure 6.3). From November until April, snow and mixed snow/rain events dominated the precipitation inputs. The remainder of the year, rainfall was the dominant form of precipitation. During their study, an average of 31% of the total precipitation entering the basins occurred as snow.

The relatively large precipitation input as snow and snowpack accumulation have important hydrologic and water quality considerations. As part of the Adirondack Long-Term Monitoring (ALTM) study of 17 lakes (Table 6.1), snowpack water content and chemistry were monitored weekly at six sites throughout the area during 1986 (Schaefer et al. 1990). Generally, there is an accumulation of snowpack through the winter season (December through April), peaking in mid-to late March (Figure 6.4). Warmer temperatures coupled with rainfall or mixed precipitation events facilitate the loss of snowpack in spring (Rascher et al. 1987). From the ALTM snow survey, it is evident that like precipitation inputs to the region, snowpack accumulation appears to be greatest in the western/southern Adirondacks (e.g., Big Moose, Otter

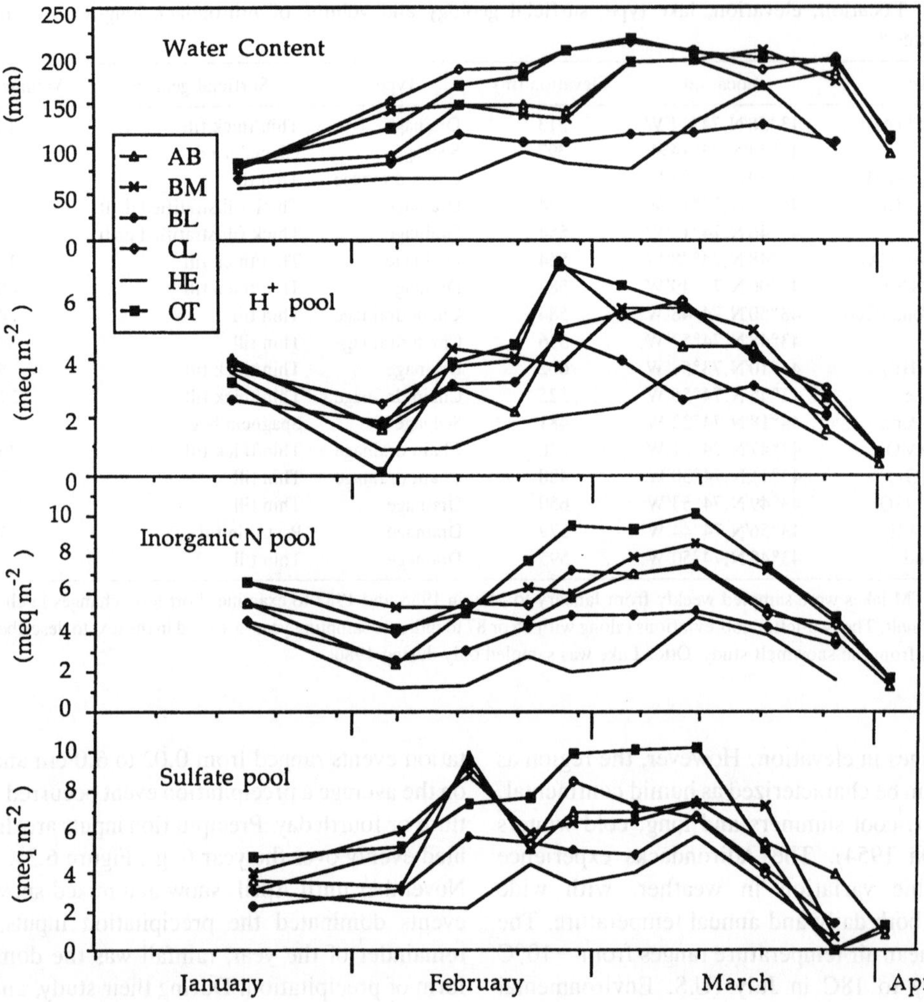

FIGURE 6.4. Snowpack water content and pools of H^+, dissolved inorganic nitrogen ($NH_4^+ + NO_3^-$), and SO_4^{2-} at Adirondack Long-Term Monitoring sites (see Table 6.1) during 1986. (After Schaefer et al. 1990.)

Lake) and diminishes slightly from southwest to northeast (e.g., Black Pond, Heart Lake). Snowpack accumulation and melting, however, are highly variable at different locations, as well as from year to year for any given site. Rainfall events and snowpack loss can occur at any time during the winter season (Johannes et al. 1985; Rascher et al. 1987).

Patterns of Atmospheric Deposition to the Adirondacks

Two precipitation monitoring stations were used in this summary of atmospheric deposition, the Utilities Acid Precipitation Sampling Program (UAPSP) station at Big Moose Lake (43°49′N, 74°53′W) in the westcentral Adirondacks and the National Acid Deposition Program (NADP) site at Huntington Forest (43°58′N, 74°14′W) at Arbutus Pond in the central Adirondacks. Precipitation chemistry for the Adirondack region is acidic, with a mean volume-weighted annual pH value of 4.26 (Figures 6.3 and 6.5). The anion composition at both sites is dominated by SO_4^{2-}. Concentrations of strong acid anions (SO_4^{2-}, NO_3^-, Cl^-) in precipitation are balanced primarily by H^+, and to a lesser extent by NH_4^+ and base cations (SBC: Ca^{2+}, Mg^{2+}, Na^+, K^+).

Seasonal patterns are evident in the chemical characteristics of precipitation for the region

(Figure 6.3), although large year-to-year variations in monthly concentrations occur. Precipitation pH is lowest during the summer months, coinciding with maximum SO_4^{2-} concentrations. Sulfate concentrations have the most distinct seasonal pattern of all major solutes, with low concentrations of about 30 μeq L^{-1} in December and January and peak concentrations of near 80 μeq L^{-1} during the summer months of May through August. Patterns for other solutes are much less distinct. Nitrate and Cl^- appear to have maximum concentrations in the winter and lowest concentrations during the summer months. Ammonium has the highest and most variable concentration in March, and values decline during the rest of the year. There is no apparent seasonal pattern in concentrations of base cations in Adirondack precipitation. As the quantity of precipitation is evenly distributed over the annual cycle, seasonal wet deposition patterns follow trends in solute concentration.

The chemical composition of precipitation is relatively uniform across the Adirondack region (Figure 6.5). Regional patterns in wet deposition are largely due to variations in the amount of precipitation. Isopleths were developed from values of element deposition for NADP stations in the northeastern United States and southern Canada for 1985 (Figure 6.6). Across the Adirondacks, from west to east, SO_4^{2-} wet deposition declines slightly from 625 to 500 eq ha^{-1} yr^{-1}, NO_3^- ranges from about 400 to 350 eq ha^{-1} yr^{-1}, H^+ decreases from about 520 to 470 eq ha^{-1} yr^{-1}, and Ca^{2+} from 25 to 15 eq ha^{-1} yr^{-1}.

Short-term trends in the accumulation and release of solutes from snowpack have important implications for drainage water quality in the Adirondacks (Rascher et al. 1987, Schaefer et al. 1990). Ion pools generally increase in snowpack during the winter season, coinciding with an increase in water content (Figure 6.4). Ion pools often reach maximum values during early to mid-March, and then decline with melting. An important consideration in snowmelt is the preferential elution of solutes from snowpack that occurs during the initial phase of melting (Rascher et al. 1987). This process results in the efflux of high concentrations of elements from snowpack during the early melt period. Ion loss from snowpack generally precedes waters loss by several days (Figure 6.4; Rascher et al. 1987, Schaefer et al. 1990).

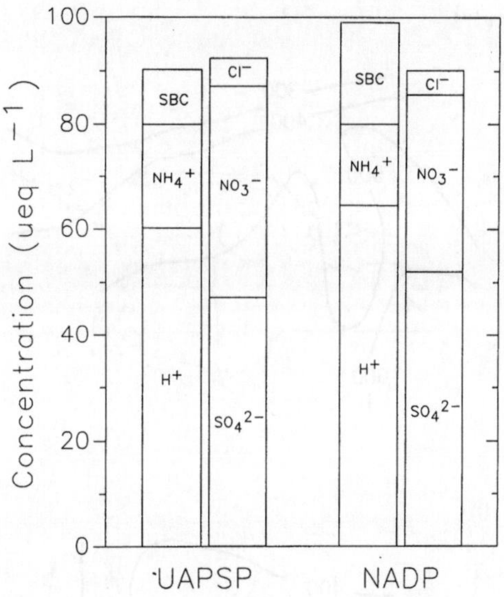

FIGURE 6.5. Ion bar diagrams for mean precipitation chemistry at the Utilities Acid Precipitation Sampling Program (UAPSP) site at Big Moose Lake and the National Acid Deposition Program (NADP) site at Huntington Forest in the Adirondacks.

Nitrate is generally the dominant anion in Adirondack snowpack (Mollitor and Raynal 1982, Rascher et al. 1987, Schaefer et al. 1990), due largely to reduced precipitation concentration of SO_4^{2-} during the winter (Figure 6.3). Hydrogen ion is the dominant cation in snowpack, and NH_4^+ concentrations represent an important fraction of dissolved inorganic nitrogen (NH_4^+ + NO_3^-). As anticipated from trends in snowpack water content, there is a regional gradient in snowpack ion pools. Snow ion pools are generally highest in the western and southern Adirondacks and diminish in a northern and easterly direction (Figure 6.4).

We can have confidence in estimates of wet deposition to the region, but little is known about dry deposition, including atmospheric fluxes of gases and particulate matter. Using a dry bucket approach, Johannes et al. (1985) made annual estimates of dry deposition for Woods Lake in the westcentral Adirondacks for 1978–81. Values ranged from 82 to 118 eq ha^{-1} yr^{-1} for SO_4^{2-}, from 36 to 61 eq ha^{-1} yr^{-1} for NO_3^-, and from 21 to 35 eq ha^{-1} yr^{-1} for NH_4^+. A large discrepancy was evident between their estimates for dry deposition of SO_4^{2-} using dry buckets and the difference between measurements of throughfall and wet deposition

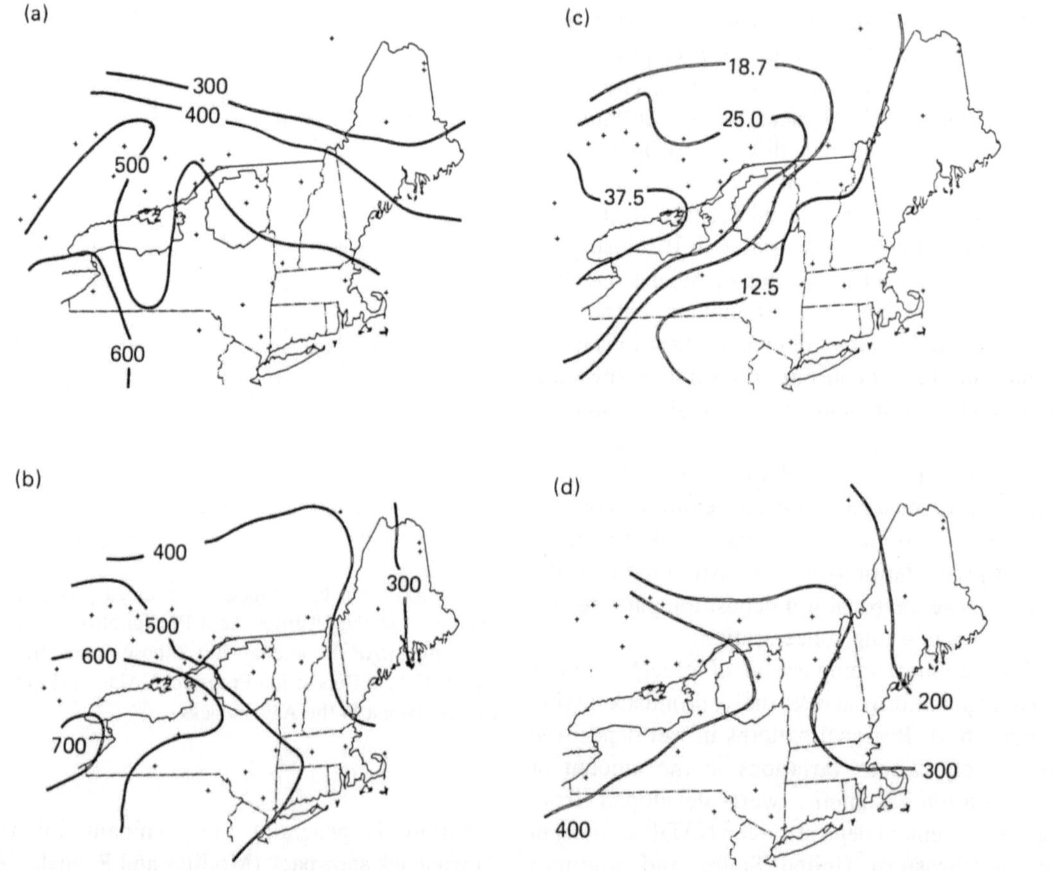

FIGURE 6.6. Isopleths of wet deposition (eq ha^{-1}) of (a) H^+, (b) SO_4^{2-}, (c) NO_3^-, and (d) Ca^{2+} for the northeastern United States and southeastern Canada for 1985. Deposition isopleths were obtained by interpolating between Acidic Deposition System (ADS) stations. The boundary of Adirondack Park is indicated.

(350 eq ha^{-1} yr^{-1}). Johannes et al. (1985) suggested that estimates of dry deposition based on dry bucket measurements probably underestimated true values.

Shepard et al. (1989) conducted a detailed study of dry deposition to the Huntington Forest in the central Adirondacks, which probably represents the best estimate of total atmospheric inputs to the region. In a two-year study, they found that dry inputs of sulfur were 130 eq ha^{-1} yr^{-1}. Dry deposition represented 26% of the total atmospheric deposition of sulfur (490 eq ha^{-1} yr^{-1}) for the study period. Dry inputs were largely attributed to gaseous SO_2 (90 eq ha^{-1} yr^{-1}), with coarse particles contributing 32 eq ha^{-1} yr^{-1}. These values are similar to the estimates of Johannes et al (1985) using the dry bucket approach. Dry deposition of nitrogen for the Huntington Forest was 350 eq ha^{-1} yr^{-1} (230 eq ha^{-1} yr^{-1} for NO_3^- and 18 eq ha^{-1} yr^{-1} for NH_4^+). This input was largely due to deposition of HNO_3 vapor (190 eq ha^{-1} yr^{-1}). Coarse particles contributed 37 eq ha^{-1} yr^{-1} of the NO_3^- loading and 13 eq ha^{-1} yr^{-1} of the NH_4^+ input. Dry deposition was a large fraction (44%) of the total input of nitrogen (570 eq ha^{-1} yr^{-1}) at the site.

Investigators have proposed that measurements of dry/wet deposition ratios for total S (SO_4^{2-} + SO_2, range 0.13 to 1.0; NH_4^+, range 0.11 to 0.39; and NO_3^-, range 0.09 to 1.5) be used to provide regional estimates of dry deposition (Hicks and Matt 1988, Hicks 1989; Appendix B, this volume). Following a critical review of the literature, dry/wet deposition ratios of SO_4^{2-}, NH_4^+, and NO_3^- were selected for the Adirondacks. These values, together with annual wet deposition fluxes, were used to estimate both dry and total atmos-

TABLE 6.2. Estimated annual inputs of SO_4^{2-}, NH_4^+ and NO_3^- (eq ha^{-1} yr^{-1}) to the Adirondacks by wet/dry deposition ratios.

	Wet	Dry	Total	Dry/wet	Reference
SO_4^{2-}	500	250	750	0.5	Appendix B
NH_4^+	150	20	170	0.13	Shepard et al. 1989
NO_3^-	300	290	590	0.97	Hicks 1989
Total *N*			760		

TABLE 6.3. Drainage area, number, and surface area of lakes in major watersheds within the Adirondack ecological zone.[a]

Major watershed	Drainage area (ha)	Number lakes	Lake surface area (ha)
Lake Champlain	598,000	410	24,740
St. Lawrence-Raquette	616,000	643	25,720
Oswegatchie-Black	445,000	844	17,840
Upper Hudson	746,000	635	27,910
Mohawk-Hudson	195,000	264	5,420
Total	2,600,000	2,796	101,630

[a] Based on unpublished data from ALSC.

pheric deposition (Table 6.2). This SO_4^{2-} dry deposition estimate (250 eq ha^{-1} yr^{-1}) is somewhat greater than estimates by Johannes et al. (1985) and Shepard et al. (1989). The dry-deposited nitrogen estimate from wet/dry ratios (Table 6.2; 310 eq ha^{-1} yr^{-1}) is very similar to the 350 eq ha^{-1} yr^{-1} estimated by Shepard et al. (1989). Not considered in these estimates is the potential for nitrogen and sulfur deposition by cloud and fog droplets. Estimates of cloud and fog deposition of nitrogen at individual sites may exceed 2,000 eq ha^{-1} yr^{-1} (Lovett et al. 1982). Within the Adirondacks, cloud and fog deposition is likely to be restricted to montane areas, generally above 1,000 m in elevation (Kinsman and Lovett 1990).

Physiography of the Adirondack Region

The Adirondack region is a unique blend of mountains, uplands, forests-wetlands, and lakes (Figure 6.2). As a result, it is an important recreational area in the northeastern United States. The boundaries of the Adirondack region include the political border of the Park itself and the physiographic boundary of the Adirondack ecological zone (Figure 6.1). The Adirondack ecological zone is defined as the area lying within the 305 m (1,000 ft) elevation contour surrounding the Adirondack upland and mountain area (about 2,600,000 ha; Colquhoun et al. 1984). This includes some minor variations to the 305 m perimeter to accommodate Lake George and the Scanadaga Reservoir. The border of the Adirondack ecological zone generally corresponds to the political boundary of Adirondack Park. Within the Adirondack ecological zone there are 2,796 lakes and ponds > 0.2 ha in surface area, totalling 101,630 ha in surface area (ALSC, unpublished data; Table 6.3).

Elevations in the Adirondack region range from about 30 m near Lake Champlain to 1,630 m (5,344 ft) at Mt. Marcy in the High Peaks area. The area can be subdivided into two physiographic regions, the High Peaks and the highlands. The High Peaks region to the east is characterized by mountains ranging in elevation from 1,200 to 1,600 m above sea level, with up to 1,200 m of relief. Within this area, 46 mountains rise above 1,200 m (4,000 ft). The land surface gradually slopes away from the High Peaks to the north, south, and west, whereas the gradient to the east is much steeper due to faulting in the Lake Champlain area. The highland Adirondacks extend away from the High Peaks region to the north, west, and south. The relief in this area is generally < 600 to 700 m. The entire Adirondack region is criss-crossed by hundreds of deep, linear valleys, some of which are occupied by long, narrow lakes such as Indian Lake and Long Lake. These features are due to glacial erosion along faults (Van Diver 1985).

The large number of lakes in the Adirondacks is a direct result of Pleistocene glaciation, which began approximately 2 million years ago and ended in this region approximately 14,000 to 12,000 years ago. Most of the lakes occur at elevations < 650 m, although a few very small lakes occur at elevations up to 1,200 m. The lakes have been created in a variety of ways. Some lie in deeply eroded bedrock troughs and others are found in bedrock basins dammed by glacial till. Still others occupy depressions entirely within glacial sediments that were created by the melting of buried ice blocks (kettle holes, seepage lakes). The

TABLE 6.4. Major rivers and streams within Adirondack Park.[a]

River	Drainage area (ha)	Length (km)
Upper Hudson River Watershed		
Hudson River	1,198,000	251
Sacandaga River	274,000	129
Schroon River	147,000	85
West Branch, Sacandaga River	59,000	40
East Branch, Sacandaga River	32,000	35
Indian River	52,000	24
Cedar River	42,000	63
Mohawk-Hudson River Watershed		
Mohawk River	897,000	248
West Canada Creek	146,000	111
East Canada Creek	75,000	48
Black Creek	27,000	39
Oswegatchie-Black River Watershed		
Oswegatchie River	415,000	201
East Branch, Oswegatchie River	86,000	82
West Branch, Oswegatchie River	76,000	42
Black River	496,000	203
Moose River	112,000	84
South Branch, Moose River	55,000	56
Beaver River	87,000	87
Woodhull Creek	39,000	26
St. Lawrence River Watershed		
St. Regis River	221,000	135
Grasse River	175,000	175
Indian River	145,000	180
Salmon River	106,000	87
Chateguay River (NY portion only)	42,000	45
Lake Champlain Watershed		
Saranac River	159,000	121
Ausable River	134,000	92
Great Chazy River	78,000	71
Boquet River	72,000	89
Raquette River Watershed		
Raquette River	325,000	245

[a] After Brown 1985.

Adirondacks are subdivided into five large drainage basins (Figure 6.1), each including a large number of lakes (Table 6.3) and important river systems (Table 6.4).

Bedrock Geology of the Adirondack Region

The Adirondacks are part of the Grenville Province of the Canadian Shield, which was the nucleus for the growth of Continental North America. At the center of the Adirondack region is the Anorthosite Massif. This is a large igneous intrusion composed of up to 90% calcium-rich feldspar. The intrusion has undergone extensive metamorphosis since its initial formation.

Surrounding the anorthosite are other igneous and high grade metamorphic rocks. These rocks can be divided into three general types: granite and granite gneiss, quartz syenite, and metasediments (Figure 6.7a). All have undergone extensive metamorphosis. The granite and granite gneiss covers approximately 25% of the region and is characterized by rocks with a granitic composition rich in hornblende (Weiner et al. 1984). In the Adirondacks, the quartz syenite series has been used to define a group of dominantly metamorphic gneisses of mafic syenite composition (with more than 20% to 25% heavy minerals, primarily pyroxenes; Buddington 1968). Metasediment rock covers over 30% of the Adirondack region; composition ranges widely from amphibolites and gneisses to marble. Marble is most abundant in the eastern Adirondacks.

The area of igneous and metamorphic rocks, which geologically defines the Adirondacks, closely corresponds with the boundary of the Adirondack ecological zone. The rocks outside this region are primarily sedimentary rocks (Paleozoic age) and include numerous limestone and shale units.

Surficial Geology of the Adirondack Region

Glaciation of the Adirondack region may have begun as early as 2 million years ago, as the first of a number of Pleistocene glaciers advanced over the region. However, only features of the most recent or Wisconsinan glaciation are preserved. The timing of the retreat of this last glaciation is unclear. Craft (1976) speculated that the continental ice had left the Adirondack region as early as 15,500 years before present (BP), with local alpine glaciers maintained in the High Peaks region until 11,000 BP. The oldest radiocarbon date from the Adirondacks is just over 12,000 BP, suggesting that deglaciation probably occurred between 12,000 and 15,500 BP (Overpeck 1985).

Throughout the Pleistocene, the glaciers covering the Adirondack region eroded the landscape. Sediment deposition from earlier Pleistocene

FIGURE 6.7. Map summarizing the (a) bedrock geology and (b) runoff of the Adirondack region of New York. Runoff values were determined from measured precipitation and an assumed constant rate of evapotranspiration of 38% for the region (Murdoch et al. 1987).

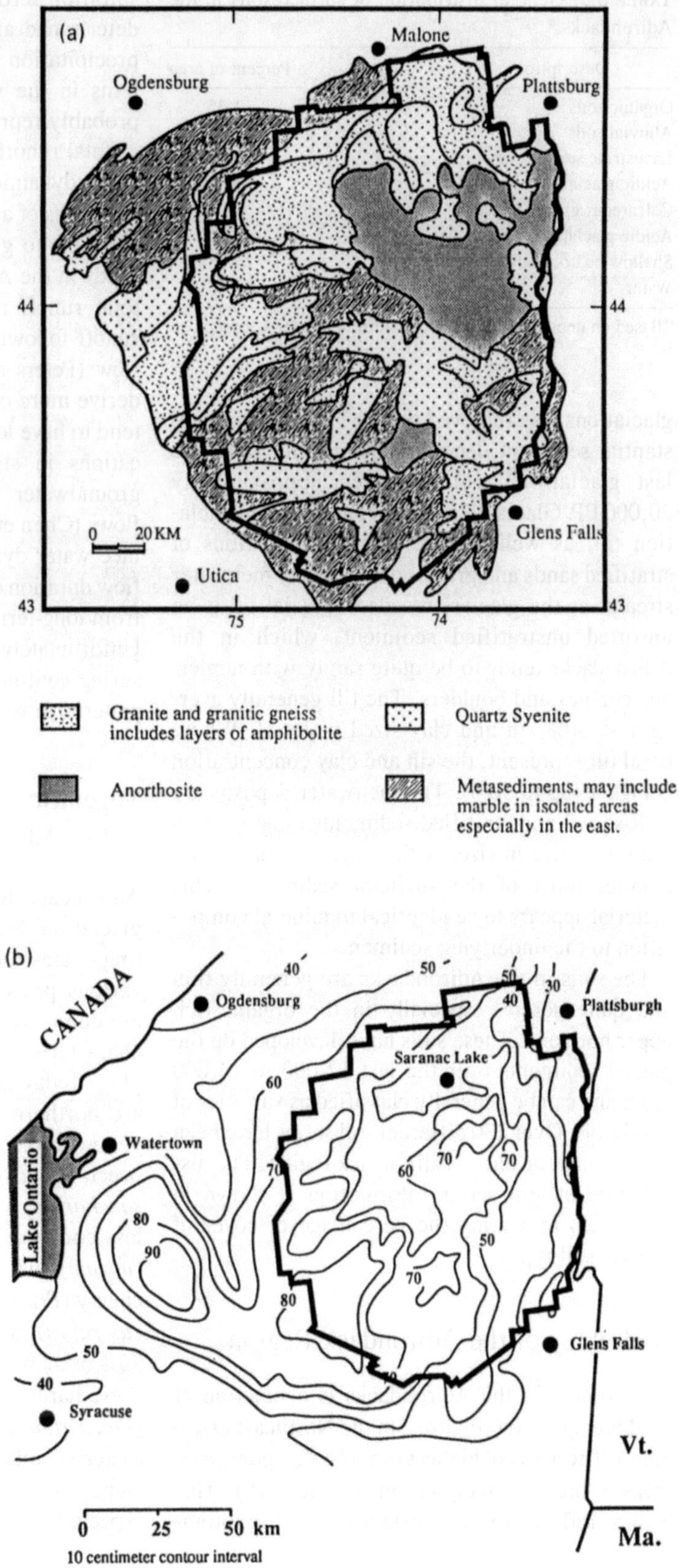

TABLE 6.5. General distribution of surface soils in the Adirondacks.[a]

Description	Percent of area
Organic soils	4.35
Alluvial soils	1.32
Lacustrine soils	1.28
Acidic glacial outwash soils	13.44
Calcareous glacial till soils	1.08
Acidic glacial till soils	48.66
Shallow bedrock soils	24.38
Water	5.47

[a]Based on unpublished data from Adirondack Park Agency.

glaciations was removed during later events. Substantial sediment deposits remain only from the last glaciation, which peaked approximately 20,000 BP. Glacial deposits include basal and ablation till, as well as extensive accumulations of stratified sands and gravels deposited by meltwater streams as the glaciers receded. Glacial till is an unsorted unstratified sediment, which in the Adirondacks tends to be quite sandy with numerous cobbles and boulders. The till generally averages < 30% silt and clay sized material. Where basal till is present, the silt and clay concentration is higher (up to 50%). The meltwater deposits are well-sorted and stratified sediments ranging from sand to gravel in size. A thin layer of aeolian silt mantles most of the surficial sediments. This material appears to be identical in mineral composition to the underlying sediments.

The soils in the Adirondacks are generally thin and quite acidic, especially in the organic-rich upper horizons. These soils have developed on the glacial sediments over the last 12,000 to 14,000 years and can be generally classified as varieties of Spodosols. Over 130 different soil series have been used to map the entire Adirondack region. The distribution of soil series information is shown in Table 6.5, indicating the percentage of cover of general soil types.

Hydrology of the Adirondack Region

Total runoff in the Adirondacks is a function of total precipitation (Figure 2 in the Northeast Overview). The areas of highest runoff correspond with areas of highest precipitation (Figure 6.7b). This pattern reflects a near constant rate of evapotranspiration across the region. Murdoch et al. (1987) determined an evapotranspiration rate of 38% of precipitation inputs for two lake/watershed systems in the western Adirondacks. This value is probably representative for the region as a whole.

Total runoff is not a good measure of the hydrologic dynamics of a watershed. The hydrologic dynamics of a system reflect the proportion of shallow flow to groundwater-derived baseflow. Watersheds in the Adirondacks with the same amount of total runoff may have very different patterns of runoff following hydrologic events or during baseflow (Peters and Driscoll 1987). Watersheds that derive more of their flow through shallow deposits tend to have lower ANC and concentrations of base cations in stream water than those with large groundwater reservoirs and relatively high baseflows (Chen et al. 1984c). The best measure of surface water dynamics is probably obtained through flow duration curves, which can only be constructed from long-term measurements of stream discharge. Unfortunately, there are not enough stations measuring continuous discharge to determine regional patterns in watershed hydrodynamics.

Vegetation and Land Use in the Adirondacks

Adirondack forests have developed since the last glaciation. Many areas have been subjected to cutting, catastrophic blowdowns, severe fires, and various pests. Species distribution and diversity are due to variations in topographic features, soil characteristics, and disturbances, especially logging. Today about 50% of the Adirondack forests are northern hardwoods, composed primarily of yellow birch (*Betula alleghaniensis*), American beech (*Fagus grandifolia*), and sugar maple (*Acer saccharum*; McMartin 1985, Davis 1988). Other, less common hardwoods include red maple (*Acer rubrum*), white ash (*Fraxinus americana*), black cherry (*Prunus serotina*), northern red oak (*Quercus rubra*), and white oak (*Quercus alba*). About 25% of these forests contain various proportions of these hardwoods in combination with red spruce (*Picea rubens*), balsam fir (*Abies balsamea*) and eastern hemlock (*Tsuga canadensis*), depending on slope and soil moisture. On coarse-textured soils, especially following fire or blowdown, stands of eastern white pine (*Pinus strobus*) and other conifers can develop. At higher elevations, above 650 to 800 m, red spruce and balsam fir dominate, with large areas of paper birch (*Betula papyrifera*) where severe fires occurred nearly a century ago. Coniferous stands cover about 10% of the region.

The forest community exhibits a two-tiered age structure (i.e., trees in a stand are typically residuals from logging in the early 1900s or stems that became established after logging). Many of the stands contain 40- to 60-year-old trees. The last major cutting in the late 1800s and early 1900s removed the remaining red spruce and eastern white pine first, then many hardwood species, and lastly eastern hemlock. However, there remain significant tracts of old-growth forests, in which trees are 200 to 400 years old, over 1 m in diameter and 30 or more meters tall.

Wetlands occupy about 5% of all Adirondack vegetative cover types, and are generally found at low elevations and in variable source areas. Nonforested wetlands can be recognized as bogs, fens, and open marshes. Peat mosses (*Sphagnum* spp.), shrubby heath species, and sedges occupy the bog areas, while cattails (*Typha* spp.), various sedges, and grasses dominate the marshes. Forested wetlands account for most of the total wetland area. Wetlands forests are typically characterized by black spruce (*Picea mariana*), balsam fir, tamarack (*Larix larcina*), and northern white cedar (*Thuja occidentalis*), or various hardwood species.

Current land use in Adirondack Park has been largely determined by the classification of state and private lands as designated in the Adirondack Park Master Plan, revised in 1979 and administered by the Adirondack Park Agency (Brown 1985, DiNunzio 1984). About 41% (1,000,000 ha) of the land is State Forest Preserve, with the balance as private holdings (59%; 1,400,000 ha). Most of the state lands have been classified as wilderness or wild forest areas, which are protected from any development. The rest of the state lands are more intensively used for a variety of purposes, from campgrounds and ski areas to administrative facilities. Private land is also regulated by the Adirondack Park Agency. Private lands have been classified according to their ability to withstand use and to preserve open space whenever possible. Most of the private lands have been classified for rural use or resource management. The prime use of resource management lands is for the protection of forest, agricultural, recreation or open space resources. In addition, there are 1,780 km of highway in the park.

Regional Surface Water Chemistry

Previous Surveys of Adirondack Lake Chemistry

A number of synoptic surveys have been conducted to assess the quality of ponded waters in the Adirondack region. A series of biological surveys was initiated by New York State from 1926 through 1939 (New York Department of Conservation 1929–34). In these early surveys, pH was measured using a Hellige color comparitor and ANC was determined by strong acid titration using a methyl orange indicator. Unfortunately, these techniques do not provide accurate analyses of dilute waters (Pfeiffer and Festa 1980, Kramer and Tessier 1982). Moreover, only a fraction of the waters surveyed was actually measured for pH and ANC. Asbury et al. (1989) suggested that historical pH measurements using color comparitors were probably not accurate; however, they found that it may be reasonable to use ANC values after correcting for the bias associated with the methyl orange endpoint. Using an assumed methyl orange endpoint of 4.25 and correcting historical ANC values for an endpoint bias of 54.6 $\mu eq\ L^{-1}$, one can estimate ANC for lakes sampled in the 1929–34 surveys (Figure 6.8). Results of this early study suggest that many Adirondack lakes had low ANC during 1929–34; however, no lakes were acidic (ANC $\leq 0\ \mu eq\ L^{-1}$). Note, however, that lakes surveyed in the early biological surveys were generally larger, located at a lower elevation, and more productive than median characteristics of Adirondack lakes determined from more recent surveys.

The loss of fish populations associated with surface water acidity has been recognized as a problem in the Adirondacks since the 1950s (Schofield 1976a). However, the linkage between acidic deposition and surface water acidification in the region was not established until Schofield (1976a,b) surveyed the water chemistry characteristics of 214 high-elevation lakes (> 610 m) in 1975. He conducted a detailed analysis to elucidate

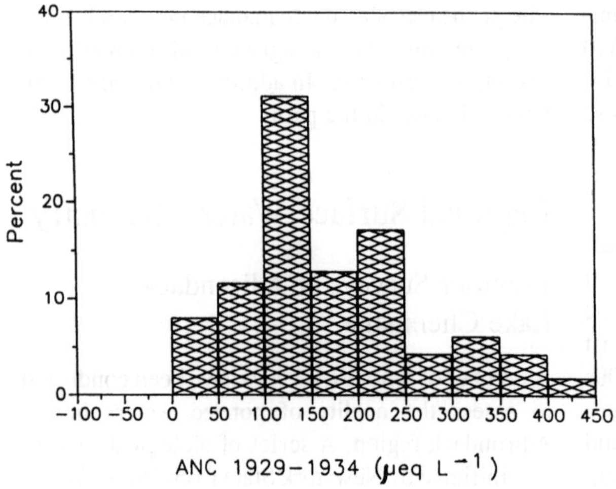

FIGURE 6.8. Percent frequency distribution of ANC measured in 274 Adirondack lakes surveyed between 1929 and 1934. The data were corrected for a 54.6 μeq L^{-1} bias associated with the methyl orange endpoint (Asbury et al. 1989).

factors responsible for regional variations in water chemistry. His conclusions have been verified in recent years by more detailed extensive surveys (e.g., Brakke et al. 1988). Schofield (1976a) reported that 55% of the high-elevation lakes had negative ANC values. The distribution of pH in these waters was skewed, with 52% of the lakes having values < 5.0 and fewer lakes between pH 5.0 and 6.0 (24%) or at circumneutral pH values (24%; > 6.0). Concentrations of SO_4^{2-} were relatively uniform across the region, suggesting a largely atmospheric origin. Variations in ANC were explained largely (77%) by variations in the sum of base cation concentrations (SBC) (with Ca^{2+} alone explaining 74%). It was concluded that lake-to-lake differences in sensitivity to strong acid inputs were due to the heterogeneity in the geological characteristics of the Adirondacks (Schofield 1976a,b).

Schofield (1976a) also showed that ANC is generally lowest in high-elevation lakes in the western Adirondacks, which are typically situated in basins derived from granitic gneiss (Figure 6.7a) and receive large precipitation inputs (Figure 1 in the Northeast Overview). He suggested that these watersheds are characterized by low leaching of base cations in drainage water. Most of the lakes with pH values > 6.0 were situated near Grenville marble outcrops and/or within lower elevation surficial deposits and drift, presumably high in exchangeable Ca^{2+}. As part of this survey, samples were measured for Al, Mn, Fe, Zn, and Cu. Concentrations of Al, Mn, and Zn were found to increase exponentially with decreases in pH.

Wood (1978) surveyed 57 lakes in the Adirondacks in 1975 and found 52% of the lakes with pH < 5.0, 45% with pH values between 5.0 and 6.0 and 2% with pH > 6.0. Unfortunately, ANC was not measured in this survey. Regional patterns reported in this study were similar to the findings of Schofield (1976a,b). One important additional observation was that concentrations of Pb increased with decreasing pH in Adirondack lakes.

The New York State Department of Environmental Conservation (NY DEC) conducted a series of surveys of ponded waters and streams in the 1970s and 1980s to assess the acid-base status of surface waters in the Adirondacks. The results of these studies are summarized in two reports (Pfeiffer and Festa 1980, Colquhoun et al. 1984). From 1975 to 1982, 980 ponds were surveyed: 6% had negative ANC; 18% had ANC between 0 and 40 μeq L^{-1}; 52% had ANC between 40 and 200 μeq L^{-1}; and 25% had ANC above 200 μeq L^{-1} (Colquhoun et al. 1984). The regional patterns of pH and ANC summarized in these studies were similar to those reported by Schofield (1976a,b).

Most recently, the Adirondack Lake Survey Corporation (ALSC) conducted a detailed survey of 1,469 ponded waters in the Adirondack region, generally coinciding with the Adirondack ecological zone (Kretser et al. 1989). They found that 26% of the lakes sampled were acidic. Because an interpretative report of ALSC study is not yet available, a detailed analysis of this comprehensive study is not included in this chapter.

The U.S. EPA Eastern Lake Survey

Description of the U.S. EPA Eastern Lake Survey

In 1984, the U.S. Environmental Protection Agency (U.S. EPA) initiated Phase I of the Eastern Lake Survey (ELS-I), which is part of the National Surface Water Survey (NSWS; Linthurst et al. 1986, Landers et al. 1988; Introduction, Charles, this volume). The objective of the ELS-I was to determine the physical and chemical characteristics of lake populations within regions susceptible to acidic deposition, as well as to estimate the number, percentage, and location of acidic lakes. The target population was generally defined as lakes with a surface area > 4 ha.

As part of the ELS-I, 155 lakes in the "Adirondack region" of New York (subregion 1A) were sampled in order to make population estimates (Brakke et al. 1988). These lakes are thought to represent a target population of 1,290 lakes, which have a total surface area of 118,777 ha. In addition, 49 special interest lakes were sampled in the Adirondacks. The ELS-I Adirondack region (subregion 1A) encompasses a much larger geographic area than that contained within the boundaries of Adirondack Park or the Adirondack ecological zone (Figure 6.1). As a result, many of the lakes sampled in ELS-I subregion 1A would not generally be considered "Adirondack" lakes. Of the target population of lakes sampled in the ELS-I, 1,076 (83%) are located in the Adirondack ecological zone. The balance of the lakes surveyed, representing 214 lakes or 17% of the population, are located outside the Adirondack ecological zone.

Limitations of the Eastern Lake Survey Data for the Adirondacks

Two aspects of the ELS-I design warrant consideration. The first issue is associated with the 4 ha surface area minimum used to select the lakes to be sampled. Baker and Harvey (1984) found that the frequency of acidic lakes in the Adirondacks increased with decreasing surface area. In the ALSC survey, 36% of the drainage lakes and 74% of the seepage lakes have a surface area < 4 ha (Kretser et al. 1989). Johnson et al. (1989) estimated that the Adirondack subregion of ELS-I contains 5,657 lakes > 1 ha and 3,130 (55%) are between 1 and 4 ha. Therefore, small lakes are an important resource in northern New York.

Sullivan et al. (1990) evaluated the influence of the 4 ha surface area cutoff imposed by the design of the ELS-I on estimates of the number of acidic lakes in the Adirondacks. This assessment was accomplished by comparing ELS-I data with data available from the ALSC survey of lakes in the Adirondack ecological zone. The ALSC generally sampled ponded waters > 0.5 ha, but some smaller lakes, identified as having the potential for viable fish populations, were also sampled (n = 151). Although these small lakes (surface area < 0.5 ha) were not sampled randomly and may not be representative of small lakes in the region, Sullivan et al. (1990b) reported that the chemical characteristics of lakes with surface areas < 1 ha were similar to lakes of the size class 1 to 4 ha. Their study indicated that generally there was little difference in the chemical characteristics of ELS-I lakes and ALSC lakes with surface areas > 4 ha. An exception to this pattern was evident for DOC. Adirondack Lake Survey Corporation lakes had higher concentrations of DOC than ELS-I lakes. Distinct differences in chemical characteristics, however, were evident between ALSC lakes < 4 ha and either ALSC or ELS-I lakes with surface areas > 4 ha. Smaller lakes (< 4 ha) tended to have lower pH, ANC, and Ca^{2+} and higher DOC than larger lakes (> 4 ha). The analysis by Sullivan et al. (1990) suggests that increasing the target population to include lakes with surface areas > 1 ha would increase the population estimate of the number of acidic lakes by a factor of eight and double the percentage of acidic lakes.

A second concern about the design of the ELS-I is associated with the geographic definition of the Adirondack region. Although the boundary of Adirondack Park closely coincides with the Adirondack ecological zone (Figure 6.1), the Adirondack subregion (1A) of the ELS-I encompasses a much larger area, including much of northern New York State. Sullivan et al. (1990) suggested that there is little difference in the chemical characteristics of ALSC lakes (within the Adirondack ecological zone) and ELS-I lakes with

TABLE 6.6. Comparison of chemical characteristics of classes of Adirondack ELS-I lakes.[a]

Class	n	pH	ANC (μeq L^{-1})	SO_4^{2-} (μeq L^{-1})	Ca^{2+} (μeq L^{-1})	DOC (μmol L^{-1})	SBC (μeq L^{-1})	H_4SiO_4 (μmol L^{-1})	Al_{ext} (μmol L^{-1})
Total (sample)	155	6.74 (5.94–7.11)	108 (15–230)	120 (107–136)	164 (89–271)	348 (253–452)	277 (162–489)	39 (18–69)	0.28 (0.16–0.81)
Total (population)	1,290	6.74 (5.94–7.11)	108 (15–230)	120 (107–136)	164 (89–271)	348 (253–452)	277 (162–489)	39 (18–69)	0.28 (0.16–0.81)
Adirondack ecological zone	1,076	6.54 (5.69–6.93)	69 (8–168)	117 (105–131)	135 (84–208)	350 (251–452)	226 (144–339)	39 (18–69)	0.37 (0.18–1.98)
Outside Adirondack ecological zone	214	7.45 (7.19–7.74)	483 (343–883)	148 (322–886)	474 (322–886)	276 (63–328)	780 (551–1313)	40 (13–58)	0.26 (0.03–0.28)
Non-salt-impacted	924	6.35 (5.43–6.88)	49 (6–144)	117 (105–131)	118 (76–192)	349 (239–445)	191 (138–306)	37 (1.8–68)	0.41 (0.20–1.32)
Salt impacted	366	7.19 (6.89–7.53)	343 (174–625)	134 (116–157)	322 (221–704)	345 (283–497)	606 (423–1140)	46 (19–69)	0.21 (0.05–0.28)
Acidic[b]	179	4.86 (4.71–5.06)	−15.3 (−24–[−6])	120 (103–130)	51 (45–61)	301 (131–427)	99 (83–113)	19 (11–30)	5.54 (2.59–7.71)

[a]Median values of population estimates are given with 25 and 75% quartiles in parentheses.
[b]ANC ≤ 0 μeq L^{-1}.

surface areas > 4 ha. However, closer examination of the data shows that solute concentrations of ELS-I lakes in the Adirondack ecological zone are distinctly different from those of lakes outside the Adirondack ecological zone (Table 6.6). For example, lakes in the Adirondack ecological zone had a median ANC value of 69 μeq L^{-1}, with 171 acidic (ANC ≤ 0 μeq L^{-1}) lakes (Figure 6.9). Many of the ELS-I lakes located within the Adirondack ecological zone were characterized by low concentrations of SBC and therefore are potentially sensitive to inputs of strong acids. Lakes in the ELS-I subregion 1A, but outside the Adirondack ecological zone, generally had higher pH, ANC, Ca^{2+}, and SBC, and slightly higher SO_4^{2-}, than lakes in the Adirondack ecological zone (Table 6.6). Eight acidic lakes (one lake sampled) were estimated for ELS-I subregion 1A outside the Adirondack ecological zone (0.6% of the total population; Figure 6.9). These differences are consistent with differences in the geological characteristics within and outside the Adirondacks (see sections titled Bedrock Geology of the Adirondack Region and Regional Patterns in Solute Concentrations of the Eastern Lake Survey). The definition of a boundary for the Adirondack subregion of ELS-I that is beyond the border of Adirondack Park or the Adirondack ecological zone influences the interpretation of the physical and chemical characteristics of lakes in the region, as well as population estimates of low pH and acidic lakes in the "Adirondacks" (Table 6.6). For example, if the ELS-I boundary included only the Adirondack ecological zone, the percentage estimate of acidic lakes would increase by 20%.

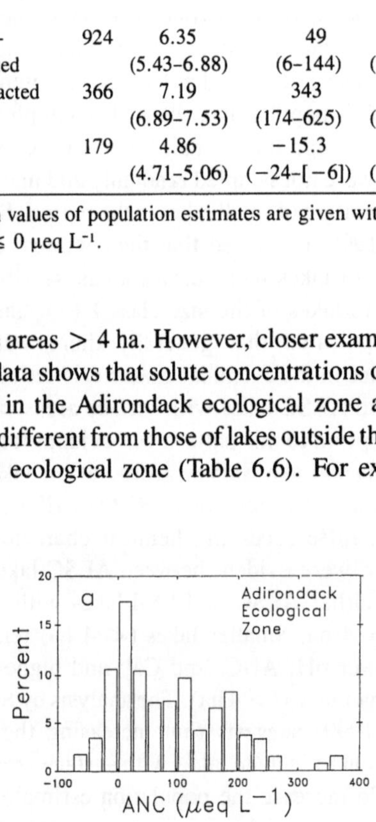

FIGURE 6.9. A comparison of the percent frequency distribution of ANC for ELS-I subregion 1A lakes that are inside and outside the Adirondack ecological zone.

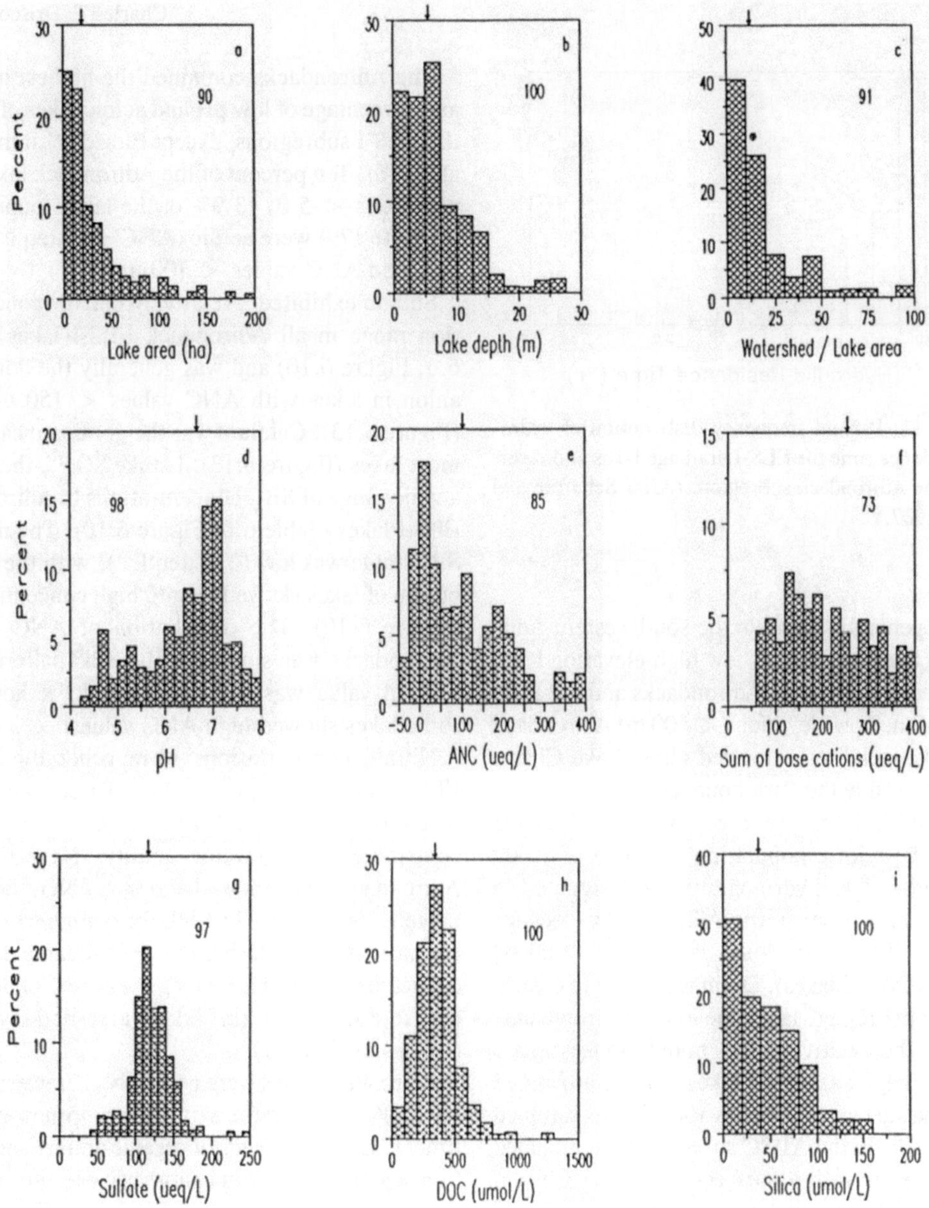

FIGURE 6.10. Frequency histograms of major characteristics of the target population of Adirondack lakes in the EPA ELS-I: (a) lake surface area, (b) sampling depth, (c) watershed area:lake area ratio, (d) air-equilibrated pH, (e) ANC, (f) sum of base cations, (g) sulfate, (h) DOC, and (i) dissolved silica. The arrow at the top of each box indicates the median for the population, and the number in the upper part of each box indicates the percentage of target lakes plotted.

Characteristics of Adirondack Lakes in the Eastern Lake Survey

Lakes sampled as part of the ELS-I in subregion 1A are generally small. The median surface area and depth are 20.9 ha and 5.6 m, respectively (Figure 6.10). The median watershed area is 265 ha, resulting in a median watershed to lake surface area ratio of 12.7. The small lake size coupled with moderate values of watershed/lake surface area causes Adirondack lakes to have generally short hydraulic residence times (median value 0.23 yr). Virtually all lakes in the region have flushing rates of < one year (Figure 6.11). Adirondack lakes of the ELS-I are also located over a wide elevation range, from < 100 to > 800 m. The median elevation of Adirondack lakes is 487 m and 12.5% of the lakes are above 600 m. High elevation (> 600 m) ELS-I Adirondack

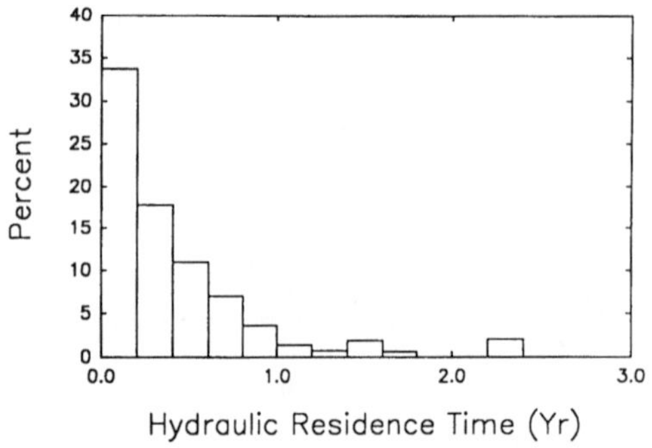

FIGURE 6.11. Percent frequency distribution of hydrologic residence time for ELS-I drainage lakes and reservoirs in the Adirondack subregion. (After Schafran and Driscoll 1987.)

lakes are generally found in the southwestern Adirondacks (Figure 6.12a). A few high-elevation lakes are located in the central Adirondacks and the High Peaks region. Low elevation (< 400 m) Adirondack ELS-I lakes tend to be located along Lake Champlain and outside the Park boundary.

Adirondack lakes are predominantly drainage lakes (78%; with a population estimate of 1,009 lakes). Other lake hydrologic types designated in the ELS-I also occur in the Adirondacks (seepage lakes 8%, 100 lakes; closed lakes 7%, 89 lakes; reservoirs 7%, 92 lakes). Drainage lakes are clearly the major hydrologic lake type in the Adirondacks; however, the relatively low population estimate and percentage of seepage lakes may be attributed to the 4 ha surface area limitation of lakes sampled in the ELS-I. In the ALSC survey, which included ponded waters with a surface area of > 0.5 ha, 14% of the lakes studied were classified as seepage lakes (Kretser et al. 1989). Seepage lakes are a more important resource in the Adirondacks than the ELS-I reflects. In addition, it seems likely that some lakes that were classified as closed lakes in the Adirondack ELS-I were misidentified. For example, of 15 special interest lakes previously studied by Driscoll and Newton (1985), three were classified in the ELS-I as closed lakes, although they are actually drainage lakes. Reservoirs and seepage lakes sampled as part of the Adirondack ELS-I are generally located outside or along the border of the Adirondack Park (Figure 6.12b). In addition, seepage lakes are located in the northern central portion of Adirondack Park.

The Adirondacks contained the highest number and percentage of low pH and acidic lakes of any of the ELS-I subregions, except Florida (Linthurst et al. 1986). Ten percent of the Adirondack lakes had pH values < 5.0; 13.9% of the lakes (population estimate 179) were acidic (ANC ≤ 0 μeq L^{-1}) and 50% had ANC values < 100 μeq L^{-1}.

Sulfate exhibited a relatively narrow concentration range in all Adirondack ELS-I lakes (Table 6.6; Figure 6.10) and was generally the dominant anion in lakes with ANC values < 150 μeq L^{-1} (Figure 6.13). Calcium was the dominant cation in most lakes (Figure 6.13). Unlike SO_4^{2-}, there was a wide range of SBC concentrations in Adirondack ELS-I lakes (Table 6.6; Figure 6.10). The median SBC value was low (277 μeq L^{-1}), with the distribution of lakes skewed toward high concentrations (Figure 6.10). The distribution of ANC in the Adirondacks was similar to the SBC pattern. The median value was low (108 μeq L^{-1}); however, some lakes showed high ANC values.

Nitrate concentrations were generally low in ELS-I samples (Figure 6.14). These values are much lower than SO_4^{2-} and did not significantly contribute to lakewater acidity. Nevertheless, Adirondack lakes showed a range of NO_3^- concentrations. Some lakes had relatively high NO_3^- concentrations, considering that ELS-I samples were collected in the fall following the period of biological retention within lake/watershed systems (Likens et al. 1977).

Adirondack lakes were generally characterized by low DOC concentrations (median concentration 348 μmol C L^{-1}). Estimates of organic anion concentrations by discrepancy in charge balance showed low concentrations relative to SO_4^{2-} (median values 34 μeq L^{-1}; Figure 6.13; Driscoll et al. 1989e). Few waters (0.6% of population) had DOC values >1,000 μmol C L^{-1} or estimated organic anion concentration > 50 μeq L^{-1} (29% of population).

Extractable Al concentrations were higher in the Adirondacks (median concentration 0.28 μmol L^{-1}) than any other subregion of the ELS-I. Concentrations of Al_{ext} were highly variable, with some lakes showing very high concentrations (Table 6.6). In general, elevated concentrations of Al_{ext} (> 2 μmol L^{-1}) were evident only in lakes with pH < 6.0 (Figure 6.15). Like SBC, the median concentration of H_4SiO_4 was low (39 μmol L^{-1}), with a distribution skewed by high values.

FIGURE 6.12. The regional location of (a) elevation classes and (b) hydrological types, of lakes sampled in the ELS-I Adirondack subregion.

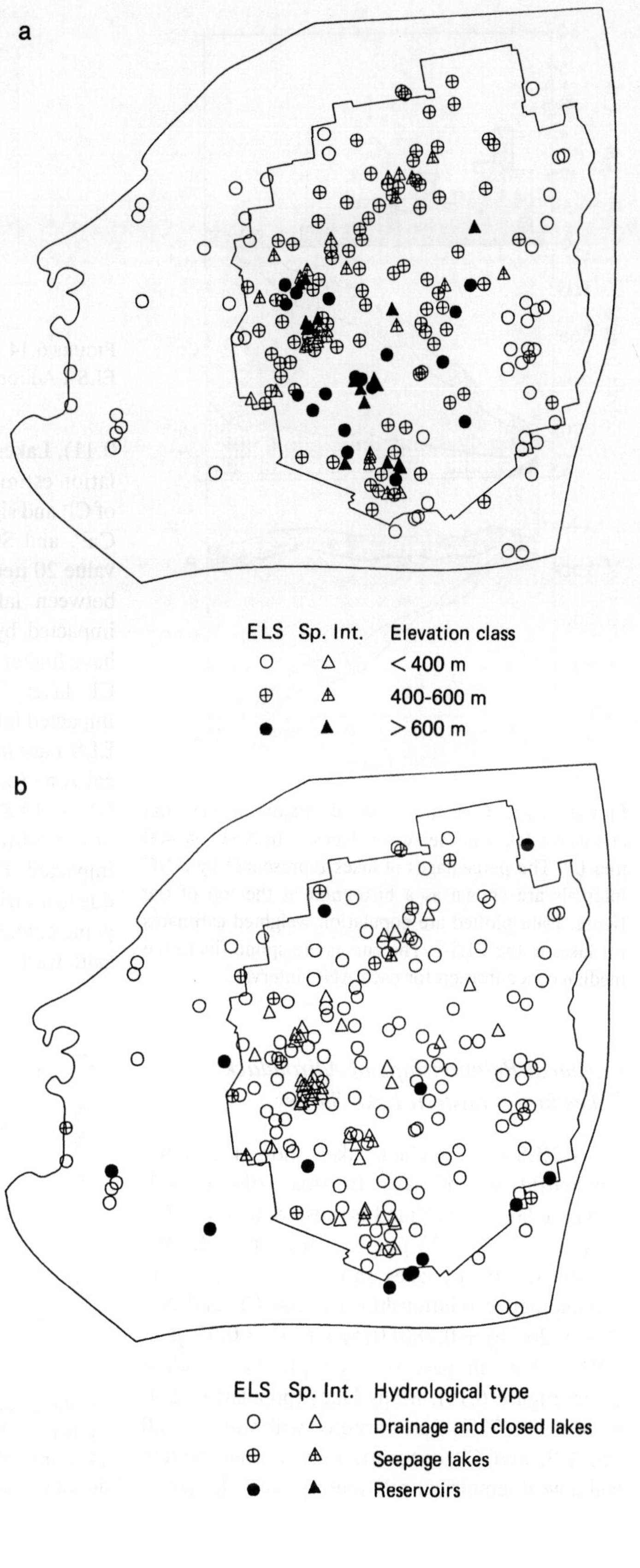

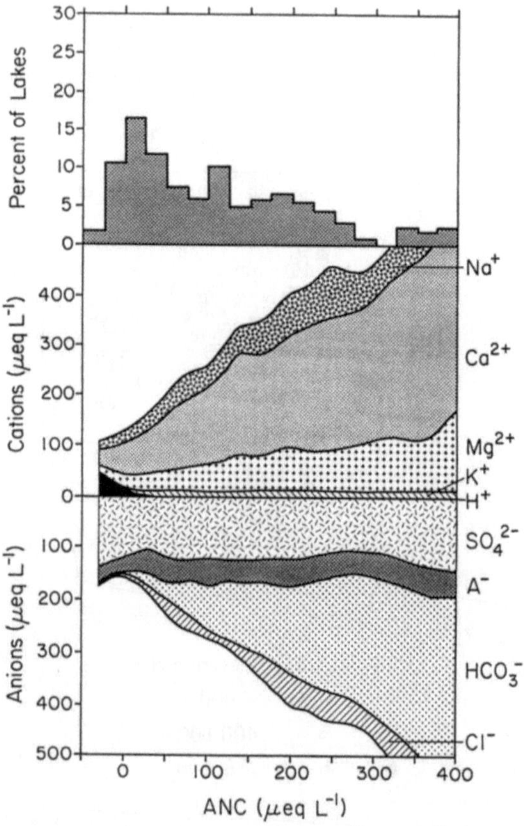

FIGURE 6.13. Concentrations of major cations and anions for lakes in the Adirondacks with ANC $\leq$ 400 μeq L^{-1}. The percentages of lakes represented by ANC intervals are shown as a histogram at the top of the figure. Data plotted are population weighted estimates for lakes in the ELS-I. The curves are spline fits to the median concentration for each ANC interval.

Chemical Relationships of Adirondack Lakes in the Eastern Lake Survey

Many ELS-I Adirondack lakes showed elevated concentrations of salt. In this volume, salt impacted lakes are defined as those waters with Cl^- concentrations > 20 μeq L^{-1} (see Appendix B). Adirondack ELS-I lakes showed a strong, near-stoichiometric relationship between Cl^- and Na^+ ($Na^+ = 26(26) + 0.78(0.015)*Cl^-$; $r^2 = 0.93$; $p <$ 0.01; values in parentheses represent standard error; Figure 6.16). Many lakes (population estimate of 924; 72%) were clustered with values < 20 μeq L^{-1}, and Cl^- concentrations in these waters had a weak empirical relationship with Na^+ (r^2 = 0.11). Lakes with high Cl^- (> 20 μeq L^{-1}) (a population estimate of 366; 18%) showed a wide range of Cl^- and significant linear relationships with Na^+, Ca^{2+}, and SO_4^{2-} (p < 0.01). It appears that the value 20 μeq L^{-1} is appropriate for differentiating between lakes impacted by salt and lakes not impacted by salt. Salt impacted lakes tended to have higher ANC, SO_4^{2-}, SBC, and Ca^{2+} than low Cl^- lakes (Table 6.6). Fifty percent of the salt impacted lakes in the Adirondack subregion of the ELS-I are located outside the Adirondack ecological zone (population estimate 184) (Figure 6.17). Of the ELS-I Adirondack subregion lakes outside of the Adirondack ecological zone, 86% were salt impacted. Enrichment of Cl^- in lake water may be due to a variety of factors. It is likely that anthropogenic contamination largely contributes (e.g., road salt; waste disposal), particularly for those lakes

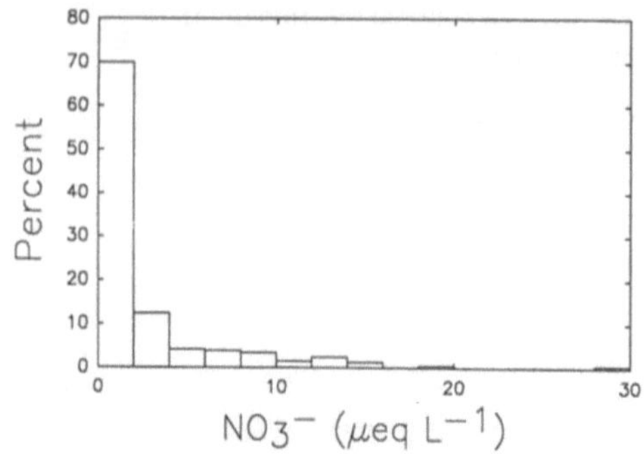

FIGURE 6.14. Percent frequency distribution of NO_3^- for ELS-I Adirondack lakes.

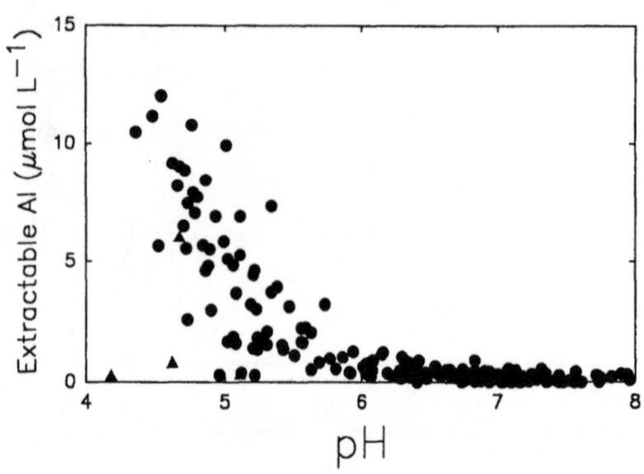

FIGURE 6.15. Concentrations of extractable Al (Al_{ext}) as a function of pH for ELS-I Adirondack lakes. Drainage lakes are indicated by circles, while seepage lakes are shown by triangles.

within the Adirondack ecological zone. Many of the salt impacted Adirondack ELS-I lakes are located near major roads (Figure 6.17). Because salt impacted lakes may be affected by human activity, and thus show the significant empirical relationships between Cl^- and other major solutes (e.g., Ca^{2+}, SBC, SO_4^{2-}), these waters have not been included in the analysis of chemical relationships of ELS-I Adirondack lakes presented in this chapter.

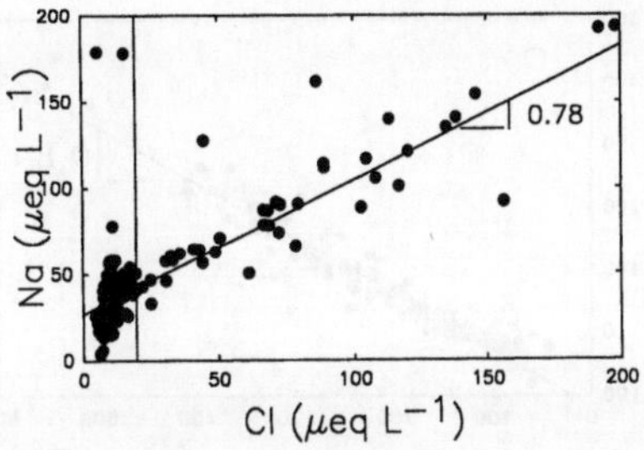

FIGURE 6.16. Concentrations of Na^+ as a function of Cl^- in ELS-I Adirondack lakes. Slope of linear regression analysis is indicated ($r^2 = 0.93$). Concentrations of Cl^- above 20 μeq L^{-1} indicate salt impacted lakes.

Variation in ANC among lakes is due to factors that influence the concentrations of major solutes (Chapters 1 and 2, Munson and Gherini, this volume). In Adirondack lakes, most of the variation in ANC is attributed to variations in the SBC (ANC = −118(42) + 0.89(0.017)*SBC; $r^2 = 0.95$; $p < 0.01$; Figure 6.18). This relationship is due largely to variations in Ca^{2+} (ANC = −94(52) + 1.29(0.031)*Ca^{2+}; $r^2 = 0.92$; $p < 0.01$). No significant relationship ($p > 0.05$) is evident between ANC, and SO_4^{2-} or DOC (Figure 6.18). Estimates of organic anion concentrations show a significant positive relationship with lake ANC (ANC = 30(157) + 1.9(0.25)*A^-; where A^- indicates concentrations of organic anions; $r^2 = 0.27$; $p < 0.01$). Inputs of organic acids do not appear to be a predominant factor contributing to the acidity of ELS-I Adirondack lakes. As the ratio of SO_4^{2-}:SBC increases, ANC decreases (Figure 6.19). Relationships between ANC and the concentrations of major solutes for Adirondack ELS-I data are consistent with the observations of Schofield (1976a,b). Concentrations of SO_4^{2-} were high and relatively uniform in Adirondack lakes

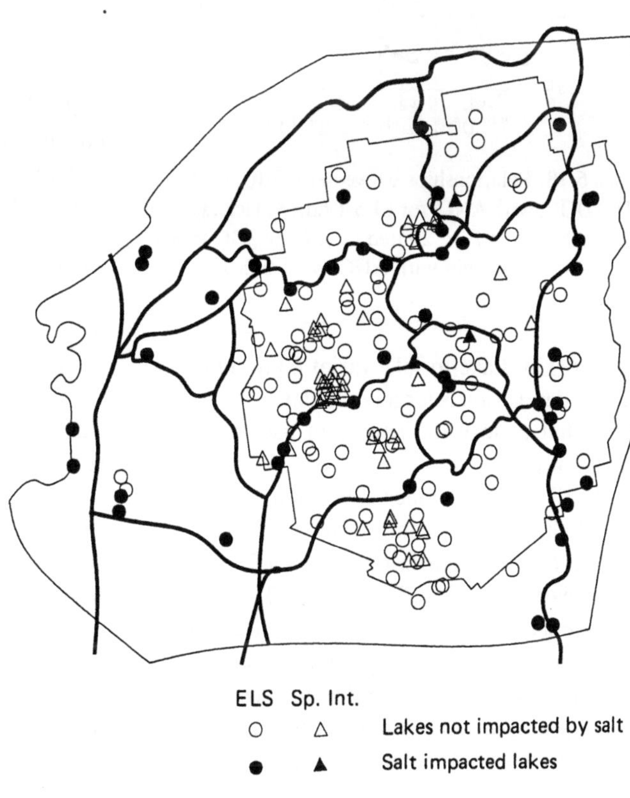

FIGURE 6.17. Location of salt impacted lakes (Cl^- > 20 μeq L^{-1}) in the ELS-I Adirondack subregion. Heavy lines indicate major roads in the region.

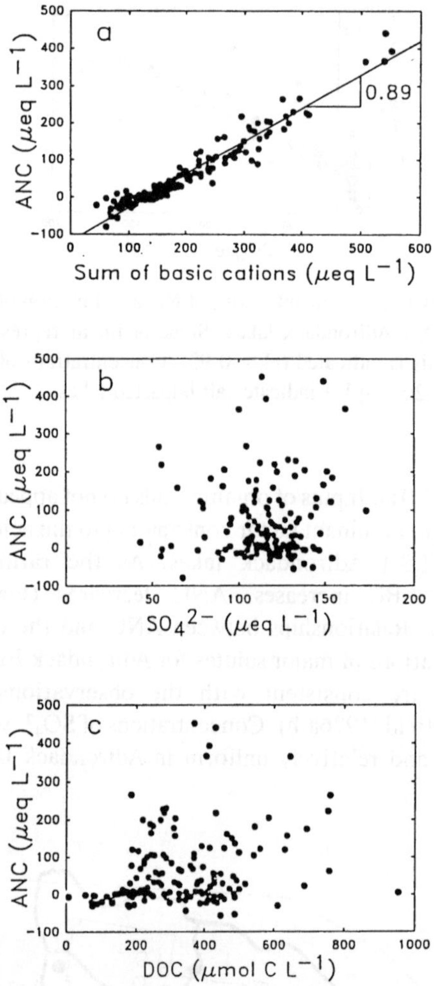

FIGURE 6.18. Relationships between (a) SBC, (b) SO_4^{2-}, and (c) DOC, with ANC, for ELS-I Adirondack lakes not impacted by salt ($Cl^- < 20$ μeq L^{-1}). Only SBC shows a significant relationship with ANC ($p < 0.001$).

(Figure 6.13). Lake-to-lake variation in sensitivity to acidic deposition is largely regulated by geologic factors that control the supply of base cations to drainage water.

In the ELS-I Adirondack subregion, acidic lakes generally had low concentrations of SBC and Ca^{2+} (Table 6.6). Concentrations of SO_4^{2-} in these waters exceeded SBC (Figures 6.13 and 6.19). This characteristic is largely responsible for the low pH and ANC of these systems. Because acidic lakes had low pH, they tended to have elevated concentrations of Al_{ext} (Table 6.6; Figures 6.13 and 6.15). Acidic Adirondack lakes also had low estimated concentrations of organic anions (80% of acidic lakes had concentrations < 10 μeq L^{-1}). In general, organic acids do not supply a large percentage of the anions to acidic Adirondack lakes in ELS-I (Figure 6.13) and are not an important source of acidity to these systems (Figure 6.19).

No significant relationship is evident between H_4SiO_4 and ANC or concentrations of base cations. Dissolved Si is predominantly derived from mineral weathering (Johnson 1984). Because variation in the ANC of Adirondack lakes is largely the result of geologic factors, an empirical relationship between ANC and H_4SiO_4 might be expected. A pattern between H_4SiO_4 and ANC or base cation concentrations may be obscured by a variety of factors, including (1) production of ANC and SBC from weathering reactions that do not release H_4SiO_4, such as dissolution of $CaCO_3$, (2) the importance of watershed processes such as uptake of base cations by vegetation or soil cation exchange, and (3) immobilization of H_4SiO_4 by abiotic or biotic processes, such as secondary mineral formation or diatom assimilation.

Schofield (1976a,b) showed that ANC varied with lake elevation. This trend is also evident for ELS-I Adirondack lakes (Figure 6.20). Adirondack lakes decreased in ANC with increasing elevation (ANC = 479(160) − 0.75(0.11)*Elev.; where elevation (Elev.) is in m; $r^2 = 0.24$; $p < 0.01$; Figure 6.20). This pattern is again due largely to SBC, as concentrations also decreased with increasing lake elevation (SBC = 662(173) − 0.74(0.12)*Elev; where elevation is in m; $r^2 = 0.25$; $p < 0.01$; Figure 6.20). No elevational pattern was evident for SO_4^{2-} or DOC in Adirondack lakes.

Regional Patterns in Solute Concentrations of the Eastern Lake Survey

Regional patterns in the concentration of many solutes are evident for ELS-I Adirondack lakes. Lakes with pH values < 5.0 and ANC values ≤ 0 μeq L^{-1} were generally located in the west and southwest areas of Adirondack Park (Figure 6.21). Many lakes located throughout the Park had ANC values < 100 μeq L^{-1}. As discussed previously, very few acidic (ANC ≤ 0 μeq L^{-1}) or low ANC lakes in the ELS-I Adirondack subregion (1A) were located outside Adirondack Park or the Adirondack ecological zone.

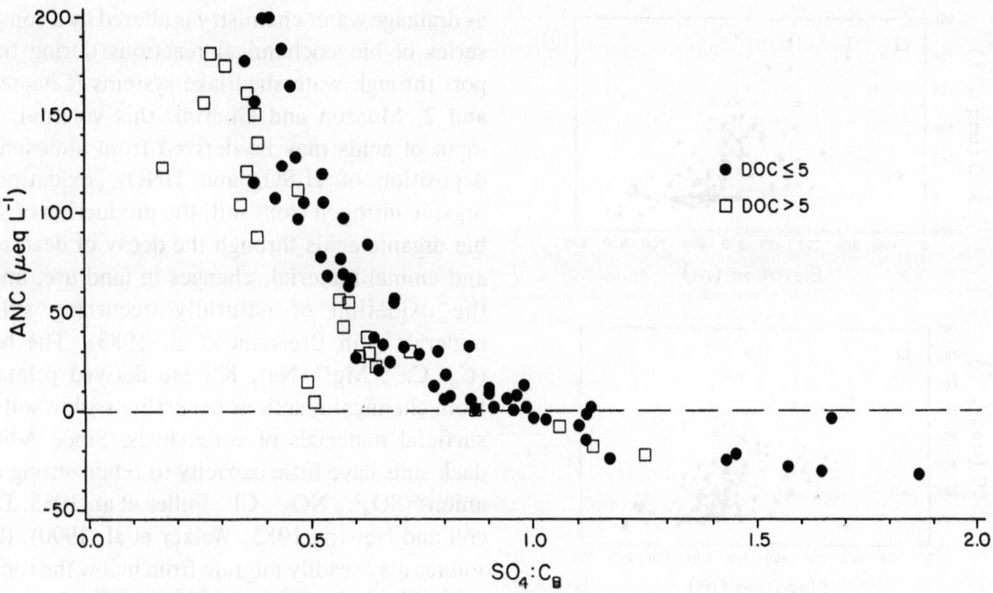

FIGURE 6.19. Variations in ANC with values of SO_4^{2-}:SBC for ELS-I Adirondack lakes. Lakes with DOC concentrations > and < 417 μmol C L^{-1} (5 mg C L^{-1}) are indicated. Values of ANC decrease as SO_4^{2-}:SBC increases in Adirondack lakes. Some loss of ANC may be evident in high DOC lakes, but high DOC lakes tend to have higher ANC values.

Concentrations of SO_4^{2-} were uniform throughout the Adirondacks, particularly within Adirondack Park, as most lakes had concentrations ranging between 100 and 150 μeq L^{-1} (Figure 6.22a). Lakes with concentrations > 150 μeq L^{-1} were generally located near Lake Champlain or outside Adirondack Park. Waters with elevated SO_4^{2-} (> 150 μeq L^{-1}) can probably be explained by internal basin inputs or salt contamination. For example, basins just east of Lake Ontario containing calcite also have deposits of gypsum that may contribute to lake SO_4^{2-}.

Regional variations in Adirondack lake ANC are due to geologic factors that influence base cation supply, rather than to inputs of strong acids (Schofield 1976a,b). Many ELS-I lakes in the western and southern areas of Adirondack Park had low SBC concentrations (Figure 6.22b). This pattern coincides with the location of acidic or low ANC lakes (Figure 6.21) and lakes with low ANC/SBC (Figure 6.23). Lakes within the southwestern section of Adirondack Park are generally found in drainage basins underlain by granitic gneiss (Figure 6.7a). Lakes in the eastern and northern areas of Adirondack Park generally had higher ANC and SBC values. These waters are situated in the bedrock zone of anorthosite that weathers easily and enhances the supply of Ca^{2+} to drainage water (Figure 6.7a). Ponded waters located outside the boundary of Adirondack Park tended to have very high ANC and SBC values (Table 6.6), reflecting the influence of carbonate-bearing bedrock. The bedrock geology in this area includes Ordovician limestones and dolomites. In addition, low ANC lakes are situated in the area of Adirondack Park that experiences the greatest precipitation input (Figure 2 in Northeast Overview). Soils in this area undoubtedly have experienced high rates of leaching, and depletion of exchangeable base cations (Schofield 1976a). Finally, many lakes in the western and southern portions of Adirondack Park are located at high elevations (Figure 6.12a). A distinct trend is evident of decreasing ANC and SBC with increasing elevation in the Adirondacks (Figure 6.20). High-elevation lakes tend to be situated in basins with shallow surficial sediments and have lower rates of base cation supply than lakes in low-elevation basins with thicker surficial deposits.

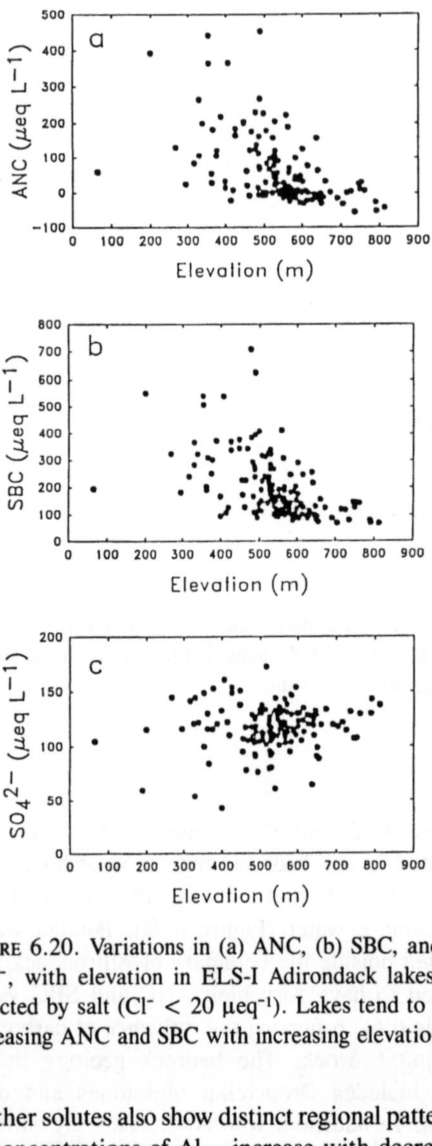

FIGURE 6.20. Variations in (a) ANC, (b) SBC, and (c) SO_4^{2-}, with elevation in ELS-I Adirondack lakes not impacted by salt ($Cl^- < 20$ μeq^{-1}). Lakes tend to have decreasing ANC and SBC with increasing elevation.

Other solutes also show distinct regional patterns. As concentrations of Al_{ext} increase with decreases in pH (Figure 6.15), it is not surprising that lakes with elevated Al_{ext} (> 2 μmol L^{-1}) coincided with low pH lakes in the western and southern areas of Adirondack Park (Figure 6.24a). Lakes showing elevated NO_3^- were also generally found in the southwestern area of Adirondack Park where acidic lakes are relatively abundant (Figure 6.24b).

Processes Influencing Surface Water Chemistry

Introduction

The acid-base status of surface waters is regulated by the input of acids relative to the input of bases, as drainage water chemistry is altered by a complex series of biogeochemical reactions during transport through watershed/lake systems (Chapters 1 and 2, Munson and Gherini, this volume). The input of acids may be derived from atmospheric deposition of H_2SO_4 and HNO_3, oxidation of organic nitrogen from soil, the production of soluble organic acids through the decay of dead plant and animal material, changes in land use, and/or the oxidation of naturally occurring sulfide minerals (van Breemen et al. 1983). The bases (C_B: Ca^{2+}, Mg^{2+}, Na^+, K^+) are derived primarily from chemical reactions occurring within soil and surficial materials of watersheds. Since Adirondack soils have little capacity to retain strong acid anions (SO_4^{2-}, NO_3^-, Cl^-; Fuller et al. 1985, Driscoll and Newton 1985, Walker et al. 1990), these solutes may readily migrate from below the rooting zone, through soil and glacial till, to surface waters. If the quantity of base cations released from the watershed soil/glacial till is not in excess of strong acid inputs, then acidic cations (H^+, Al) will be transported from the soil to the surface water (Driscoll and Newton 1985).

The extent of neutralization of acidic inputs and the ANC of surface waters is determined by the interaction of a complex series of factors, including soil, hydrology, vegetation, geology, climate and atmospheric deposition (Chapters 1 and 2, Munson and Gherini, this volume). The relative contribution of these factors in regulating the acid-base status of surface waters is highly variable, even within very small regions. For example, hydrologic factors may dictate drainage water chemistry in one watershed, while an adjoining watershed may be largely influenced by geologic factors (e.g., the presence of carbonate minerals). Variability in these factors leads to differences in the ability of lake watersheds to neutralize acidic deposition. The major sources of base cations are from cation exchange and mineral weathering reactions occurring in the soils and surficial geologic materials within a watershed (Newton and April 1982, April et al. 1986; Chapter 1, Munson and Gherini, this volume). The rate at which these reactions supply base cations to surface water largely determines its acid-base status.

For most watersheds in the Adirondack region, the relative routing of water through the soils and geologic materials is the major determinant of this base supply rate (Figures 1.6 and 1.7 in Chapter 1,

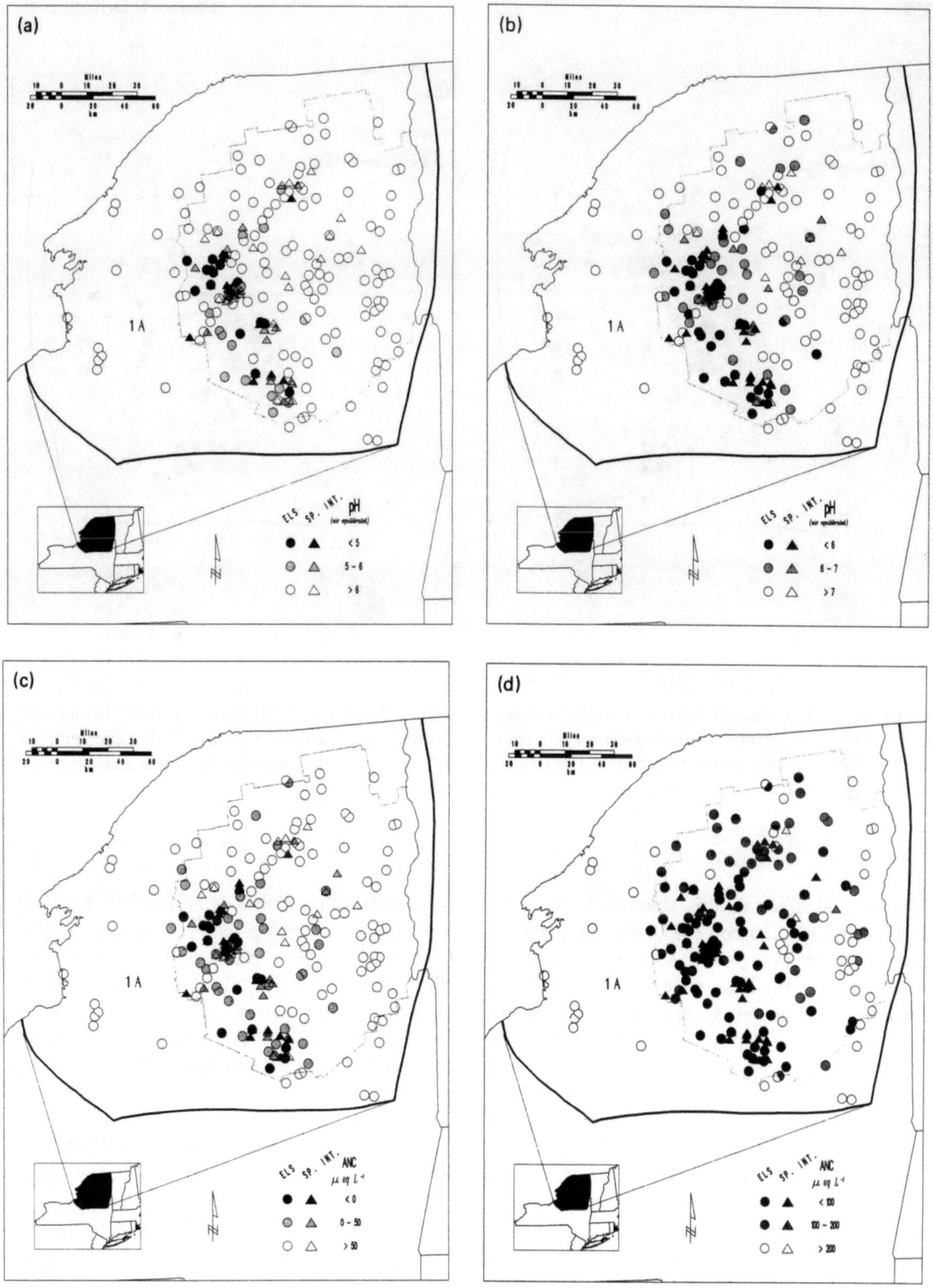

FIGURE 6.21. The regional location of ELS-I and special interest lakes sampled in the Adirondack subregion with (a) pH < 5, 5 to 6, and > 6, (b) pH < 6, 6 to 7, and > 7, (c) ANC < 0, 0 to 50, and > 50 μeq L^{-1}, and (d) ANC < 100, 100 to 200, and > 200 μeq L^{-1}. Lakes in the southwestern portion of Adirondack Park tend to have low pH and ANC values.

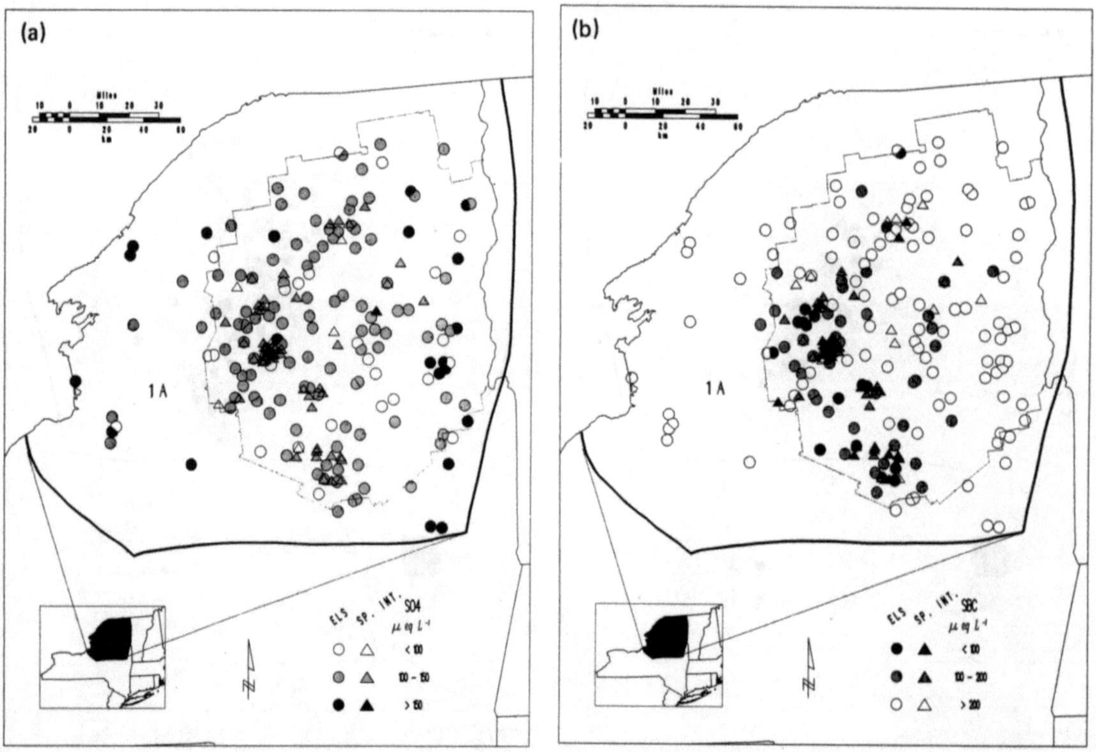

FIGURE 6.22. The regional location of ELS-I and special interest lakes sampled in the Adirondack subregion with (a) SO_4^{2-} < 100, 100 to 150, and > 150 μeq L^{-1}, and (b) SBC < 100, 100 to 200, and > 200 μeq L^{-1}. Lakes with elevated SO_4^{2-} tend to be located along Lake Champlain and outside Adirondack Park. Lakes with low SBC are generally found in the southwestern area of Adirondack Park.

Munson and Gherini, this volume; Chen et al. 1984, Peters and Driscoll 1987). The relative routing of water, or flow path, is a function of both the nature of the surficial material contacting the drainage water, as well as the hydrologic retention time within these deposits. Surficial materials in the Adirondack region may range from highly acidic upper soil horizons to more base rich, relatively unweathered till and stratified drift. Rarely do drainage waters in the Adirondack region contact carbonate minerals in the bedrock. However, when carbonate weathering does occur, the resulting surface water is enriched in base cations (particularly Ca^{2+} and Mg^{2+}) and ANC.

The flow path of water moving through a watershed is a function of a number of lake/watershed characteristics, including thickness of unconsolidated sediments, hydraulic conductivity, and land slope. However, for most Adirondack watersheds, the dominant flow path is determined by the depth of the unconsolidated glacial sediments overlying the bedrock surface (Newton et al. 1987). The depth of surficial deposits is important as it defines the size of the potential groundwater reservoir (Chen et al. 1984c, Peters and Murdoch 1985). Watersheds with thick surficial sediments have a large groundwater storage capacity. During precipitation events, water infiltrates through soil and moves downward to the groundwater table where it is slowly discharged to the surface water. In these basins, deeper flow paths dominate the routing of water and thus, surface waters have higher ANC. In contrast, those watersheds with shallow surficial sediments or bedrock outcrop have only a small groundwater reservoir, which is rapidly filled during the early part of most precipitation and snowmelt events. Consequently, most of the drainage following rainfall or snowmelt is forced to

FIGURE 6.23. The regional location of lakes sampled in the Adirondack subregion, including ELS lakes, with DOC concentrations > and < 417 μmol C L^{-1} (5 mg C L^{-1}) in ANC:SBC classes < .25, .26 to .75, and > .76.

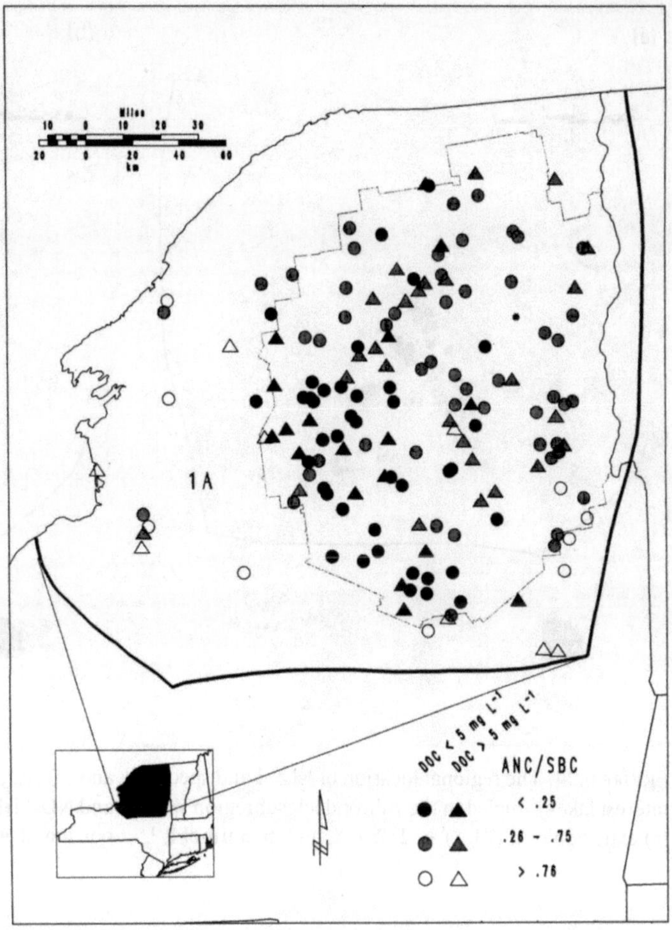

move rapidly as shallow interflow, through the upper acidic soil horizons, or overland flow to surface water, resulting in lower ANC.

In addition to neutralization through the release of base cations from the edaphic environment, ANC production may occur through reduction processes (Kelly et al. 1982) within lake or wetland systems. Many specific reduction processes result in direct production of ANC, but some reactions (e.g., Fe reduction, Mn reduction, NH_4^+ release) are only temporary sources of ANC, as reduced species (e.g., Fe^{2+}, Mn^{2+}, NH_4^+) reoxidize when exposed to O_2 in the surface water column. Sulfate and NO_3^- retention are the only reduction reactions that significantly contribute to permanent production of ANC. Kelly et al. (1987) studied rates of SO_4^{2-} and NO_3^- retention in a series of lakes, including several Adirondack lakes. They found that biological acid neutralization in lakes (AN < μeq L^{-1}) could be predicted from a simple model that requires information only on hydraulic residence time, mean depth, and mass transfer coefficients for SO_4^{2-} and NO_3^-:

$$AN = L[1/(z/t_w) - (1/((z/t_w) + S))] \quad (6\text{-}1)$$

where: AN = ANC production by SO_4^{2-} or NO_3^- retention (μeq m^{-3})
L = areal SO_4^{2-} or NO_3^- loading (μeq m^{-2} yr^{-1})
z = the mean depth of the lake (m)
t_w = the hydraulic residence time (yr)
S = the mass transfer coefficient for SO_4^{2-} (0.4 to 0.54 m yr^{-1}) or NO_3^- (7.4 to 9.2 m yr^{-1})

Since mass transfer coefficients were remarkably constant among study lakes, the extent of ANC

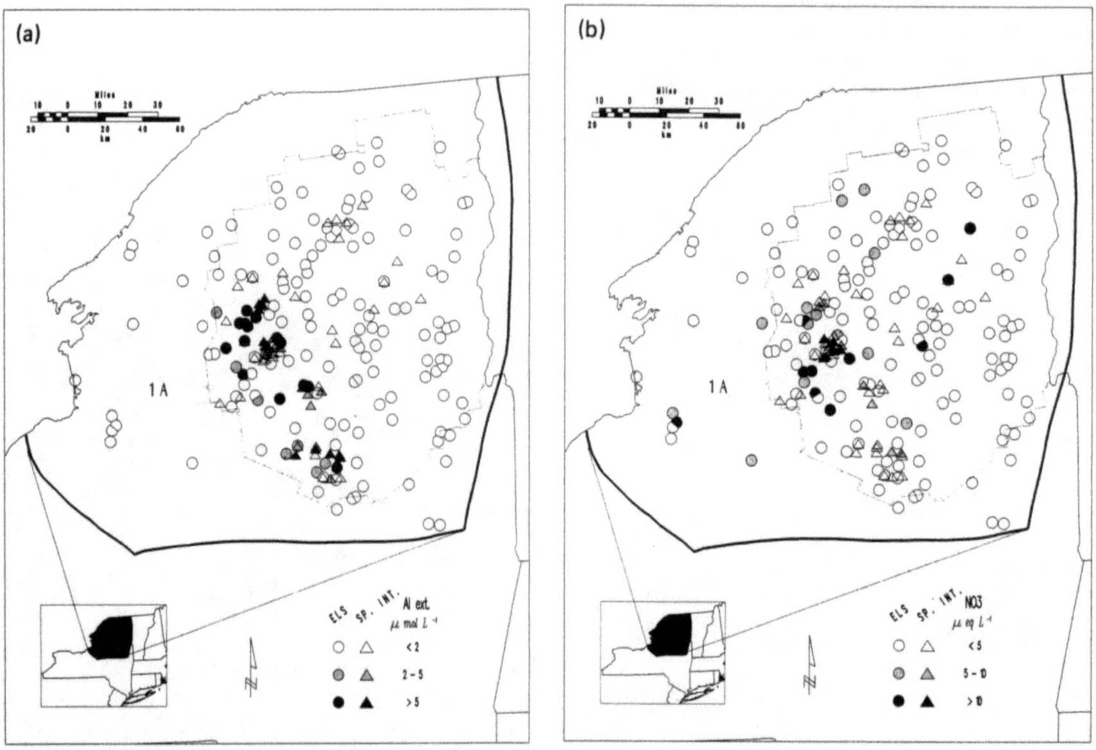

FIGURE 6.24. The regional location of ELS-I and special interest lakes sampled in the Adirondack subregion with (a) extractable Al (Al_{ext}) < 2, 2 to 5, and > 5 μmol L^{-1}, and (b) NO_3^- < 5, 5 to 10, and > 10 μeq L^{-1}. High Al_{ext} and NO_3^- lakes are generally located in the southwestern portion of Adirondack Park.

production by SO_4^{2-} and NO_3^- retention increases with increasing SO_4^{2-} and NO_3^- inputs, and hydraulic residence time. Production of ANC by reduction reactions can also occur within wetlands and saturated soils (Driscoll et al. 1987a).

Lake Classification

Lakes and watersheds can be classified according to lake type and dominant flow path. Drainage and seepage lakes represent the two major hydrologic lake types that occur in the Adirondacks. Drainage lakes are defined as lakes that have one or more distinct surface outlets, whereas seepage lakes are defined as those with no surface outlets (Wetzel 1983). Approximately 86% of the 1,469 Adirondack lakes surveyed by the ALSC are of the drainage type and 14% are seepage lakes (Kretser et al. 1989). Drainage lakes receive water from a variety of flow paths, including surface runoff, tributary streams (surface inflow), direct precipitation, and groundwater inseepage. Water is lost primarily through surface outflow (lake outlet), but evaporation, and in some cases groundwater outseepage, also occurs. The relative importance of these flow paths may differ markedly from lake to lake. Seepage lakes receive water from the same types of flow paths as drainage lakes, but the relative contributions of water from individual flow paths are different for seepage and drainage lakes. For example, many drainage lakes receive much of their water from surface runoff and tributary streams, whereas most seepage lakes have no surface inflow streams and receive only minor amounts of water from surface runoff. Instead, direct precipitation and groundwater inseepage are the most important flow paths for seepage lakes. Since seepage lakes are defined as having no surface outflow, groundwater outseepage and evaporation are the only pathways by which water is lost (Chapter 1, Munson and Gherini, this volume).

Drainage lakes can be subdivided on the basis of the importance of the groundwater inseepage flow path. In watersheds dominated by thick till (> 3 m), groundwater inseepage is important and the lakes generally have neutral pH and high ANC due to the long, deep flow paths of groundwater transported to the lake. In watersheds lacking significant deposits of thick till, groundwater inseepage is small and these lakes have low ANC. The importance of hydrologic flow path in controlling lake chemistry is reduced in watersheds containing carbonate-bearing bedrock. In these systems, all drainage water, even along shallow flow paths, is readily neutralized by dissolution of carbonate minerals (primarily calcite and dolomite).

In some Adirondack watersheds, wetlands are evident adjacent to stream channels (Driscoll et al. 1987a) and along the lake shore (Cronan 1985). The acid-base status of surface waters may be influenced by wetlands through the release of organic acids and/or the reduction of SO_4^{2-} and NO_3^- as drainage water migrates through deposits of peat. Therefore the classes of drainage lakes may each be subdivided into colored or clear waters, depending on whether or not they are significantly influenced by wetlands.

Seepage lakes can also be subdivided according to the dominant input flow path. A groundwater flowthrough seepage lake receives inflow predominantly from the groundwater inseepage flow path (Figure 1.8, Chapter 1, Munson and Gherini, this volume). These lakes, like the comparable drainage lake class, have relatively high ANC, as their water is largely derived from groundwater originating in thick surficial sediments. One might anticipate that groundwater flowthrough seepage lakes would be the dominant type of seepage lake, as most are found in kettle holes surrounded by highly permeable stratified drift. However, many of these kettle hole seepage lakes are actually perched above the regional water table due to the accumulation of organic-rich bottom sediments (Figure 8, Chapter 1, Munson and Gherini, this volume; Driscoll and Newton 1985). Groundwater recharge seepage lakes receive little or no flow from groundwater inseepage. Most of the water entering these lakes comes directly from precipitation, with only minor amounts of surface runoff (Figure 1.8, Chapter 1, Munson and Gherini, this volume). Since little of the incoming water contacts soil or surficial materials, these lakes are almost as acidic as the precipitation inputs (Driscoll and Newton 1985).

Like drainage lakes, both classes of seepage lakes can be influenced by adjacent wetlands. In some basins, seepage lakes are completely isolated from the surrounding sand and gravel aquifers due to the infringement of peat bogs. These lakes are commonly rimmed by a floating mat of sedges and sphagnum, which in time slowly fills the lake with peat. Water input to these lakes primarily occurs through direct precipitation and surface runoff through the uppermost active zone (acrotelm) of the peat. Thus the water tends to be acidic, with elevated concentrations of DOC (Driscoll and Newton 1985).

Case Studies of Adirondack Lakes

Considerable process-level information is available from intensive research on lake-watershed systems in the Adirondacks. This information is summarized in this section as a series of case studies on the different classes of drainage and seepage lake-watershed systems. These studies illustrate the importance of hydrologic flow paths in regulating the acid-base status of surface waters.

Case Studies of Drainage Lakes

Surface Waters Derived from Basins with Thick Deposits of Glacial Till

Panther Lake is a small headwater lake located in the western part of Adirondack Park (43°42'N, 74°56'W) that has physical and chemical characteristics typical of thick till watersheds (Table 6.7; April and Newton 1985). This lake was one of the three ILWAS sites studied intensively from 1977 to 1981 (Goldstein et al. 1985). Panther Lake watershed is underlain by a biotite/hornblende granite-gneiss, which is overlain by thick glacial till with more than 50% of the watershed covered by till > 3 m in thickness. Analysis by seismic refraction indicates a maximum till thickness > 30 m (April and Newton 1985). Numerous soil pits excavated within the watershed during the ILWAS project revealed the presence of an aeolian silt deposit on top of the till. This silt was mineralogically identical to the underlying till. It was presumably deposited during deglaciation approximately

TABLE 6.7. Physical and chemical characteristics of Panther and Woods Lake watersheds.[a]

	Panther Lake	Woods Lake
Physical characteristics		
Basin area (km²)	1.2	2.12
Lake surface elevation (m)	557	607
Relief (m)	170	122
Forest cover (%)	99	96
Mean till depth (m)	24.5	2.3
Lake area (km²)	0.18	0.26
Mean lake depth (m)	3.50	4.0
Maximum lake depth (m)	7	12
Chemical characteristics		
pH	6.2 (4.5 to 7.2)	4.7 (4.4 to 5.9)
ANC	147 (35 to 240)	10 (−60 to 30)
Ca^{2+}	207	73
Mg^{2+}	52	19
Na^+	41	19
K^+	12	6
NH_4^+	2	3
Al	0.5	14
SO_4^{2-}	123	126
No_3^-	23	19
Cl^-	13	9

[a] After Goldstein et al. 1984, Schofield 1984. Average annual concentrations of outlet waters are shown (and range of pH and ANC). All concentrations are expressed in μeq L^{-1}, except Al, which is expressed as μmol L^{-1}, and pH.

14,000 years BP. Soil development since glaciation has resulted in the formation of a Spodosol, consisting of Typic Haplorthods and Fragiorthods (April and Newton 1983).

Atmospheric deposition was monitored in Panther Lake watershed during the period March 1978 through December 1981 (Johannes et al. 1984; summarized in section on Patterns of Atmospheric Deposition to the Adirondacks). A comparison of atmospheric deposition inputs with outputs from the outlet stream (Figure 6.25b) shows a large release of base cations from the watershed and little retention of atmospheric inputs of SO_4^{2-}. There was a large efflux of ANC from the catchment. Atmospheric deposition of dissolved inorganic nitrogen (NO_3^- and NH_4^+), however, was much greater than the efflux from the watershed, due to uptake of nitrogen by vegetation.

The net release of base cations from Panther Lake watershed reflects geochemical reactions occurring along the dominant deep flow path. Hydrologic studies (Murdoch et al. 1987, Newton and April 1982) indicate that most of the water entering the lake initially infiltrates through the soil and then moves through the extensive groundwater system associated with the glacial till before entering the lake. An analysis of base flow recession by Murdoch et al. (1987) indicates a predominance of groundwater input relative to shallow flow, with more than 50% of the observed flow derived from the groundwater system. This large groundwater reservoir attenuates peak stream discharge during precipitation and snowmelt events, since much of the input water infiltrates directly into the groundwater system. Conversely, surficial deposits increase baseflow discharge, as the water in the groundwater reservoir is slowly released to the surface water system during dry periods (Figure 6.26). Approximately 38% of the precipitation input was lost as evapotranspiration over the study period.

The chemical characteristics of water transported through the soil zone were studied during the ILWAS by Cronan (1985). Sulfate was the dominant anion in all soil solutions. Soil solution leachate in the upper soil horizons had higher base cation concentrations and higher concentrations of organic solutes than deeper mineral soil drainage (Table 6.8). This pattern corresponds with higher base saturation and cation exchange capacity (CEC) in the O horizon, than the lower mineral soil (Table 6.9). In general, NO_3^- concentrations in these soil solutions were higher than values in other forested areas of the northeast, where values are near detection limits (Cronan 1984). The NO_3^- concentrations were especially high in winter when vegetative uptake is low. If significant lateral flow occurs through these soil horizons during periods of limited biological activity, the transport of elevated NO_3^- concentrations may contribute to surface water acidity.

Mobilization and transport of aqueous Al was evident through the investigation of soil solution chemistry (Table 6.8). In the O and A horizon leachate, concentrations of Al_{ext} averaged 18–22 μmol L^{-1}, and soil-derived organic acids had a marked effect on Al speciation. In the B horizon solutions, mean Al_{ext} concentrations increased about two times to a range of 37–48 μmol L^{-1} and

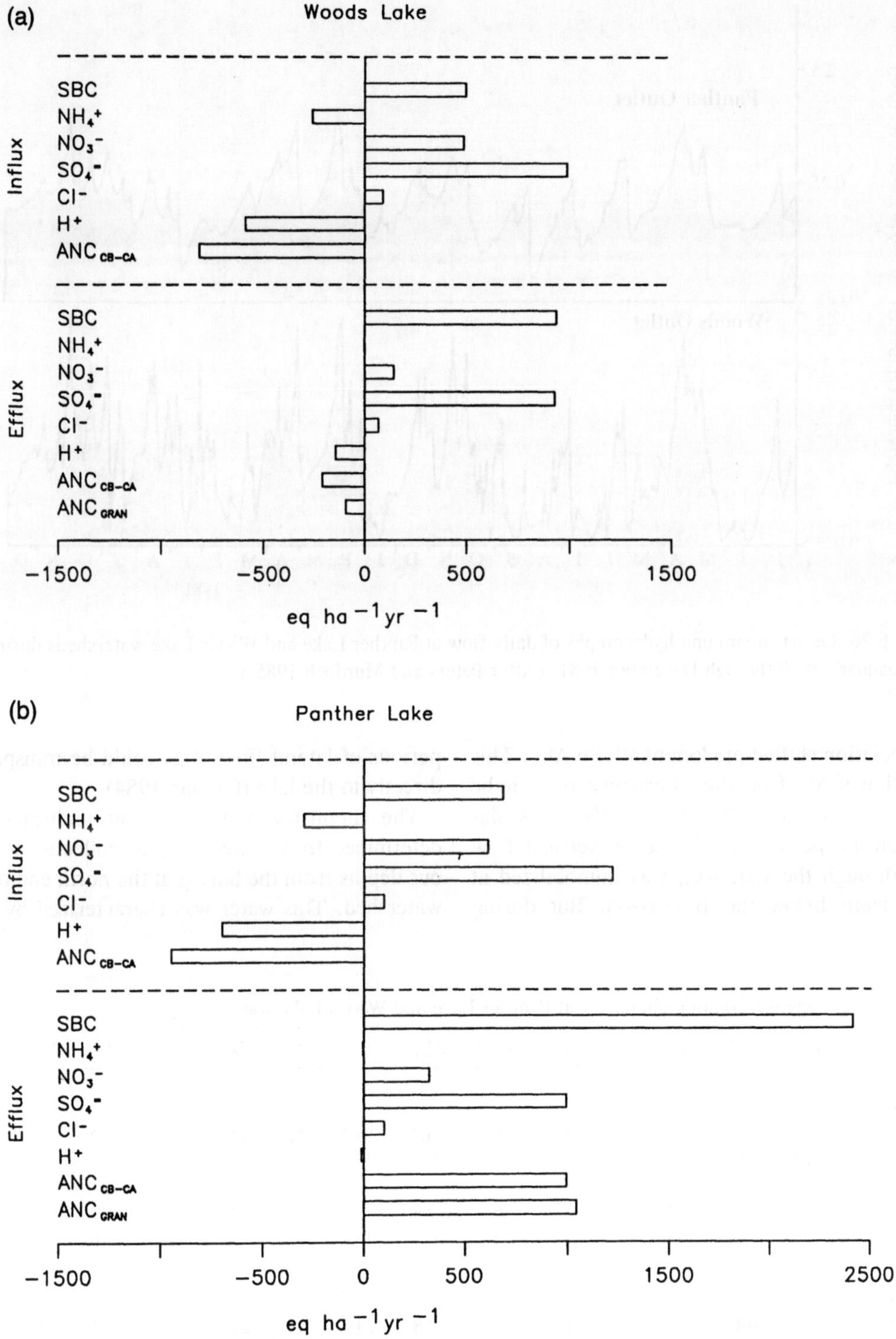

FIGURE 6.25. Summary of atmospheric deposition and biogeochemical processes producing and/or consuming ANC within lake/watershed systems at (a) Woods Lake and (b) Panther Lake. Precipitation solute flux was adjusted for canopy enhanced dry deposition using the ILWAS model. $ANC_{C_B-C_A}$ is the sum of ANC production terms from measured values of major cations and anions.

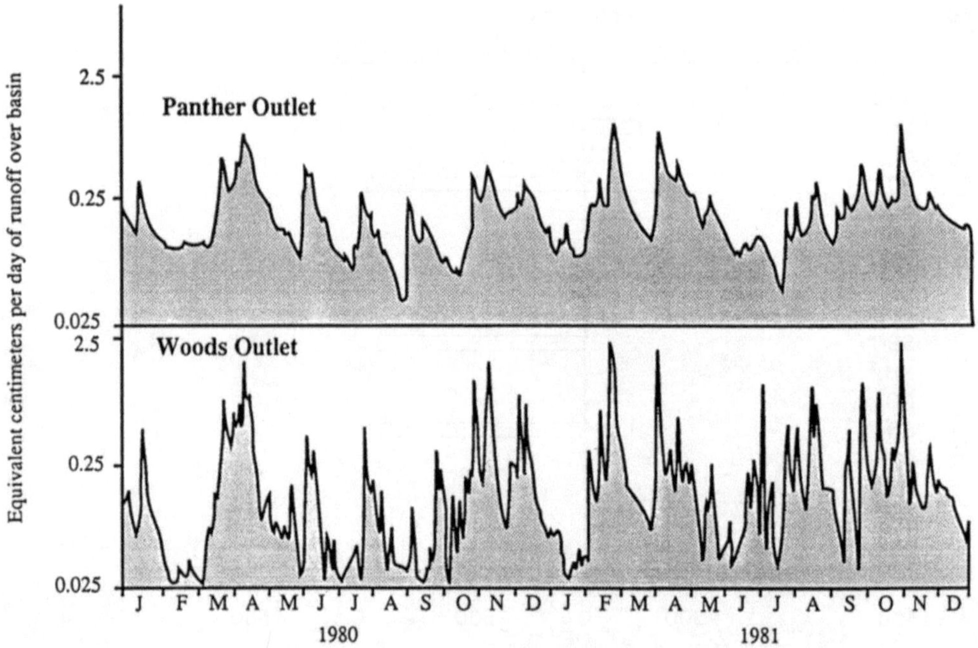

FIGURE 6.26. Outlet stream unit hydrographs of daily flow at Panther Lake and Woods Lake watersheds during the period January 1980 through December 1981. (After Peters and Murdoch 1985.)

the speciation shifted predominantly to Al_{im}. This eluviation of Al_{im} from the B horizon appears to be strongly mediated by H_2SO_4 and HNO_3 dissolution. During periods dominated by vertical flow paths through the soil, Al_{ext} was immobilized at some depth below the B horizon. But during periods of lateral flow, Al_{ext} could be transported directly to the lake (Cronan 1984).

The chemistry of the deep groundwater was determined from water samples collected at various depths from the boring at the north end of the watershed. This water was characterized by high

TABLE 6.8. Average soil solution chemistry at Panther Lake and Woods Lake watersheds.[a]

Site	pH	Ca^{2+}	Mg^{2+}	K^+	Na^+	NH_4^+	Al_{ext}	Al_{om}	Fe	SO_4^{2-}	NO_3^-	Cl^-	DOC	A^-
Panther Lake														
Summer														
20 cm	3.82	139	41	28	23	2	69	58%	12	210	81	32	32	140
50 cm	4.54	98	33	9	27	1	106	13%	–	151	97	20	7	30
Winter														
20 cm	3.90	116	30	36	17	6	64	45%	12	147	153	30	22	75
50 cm	4.55	97	36	9	25	3	129	11%	–	138	163	16	7	5
Woods Lake														
Summer														
20 cm	3.87	108	3i	28	17	2	88	34%	9	210	90	31	27	85
50 cm	4.50	84	21	7	28	2	102	12%	–	191	28	19	7	33
Winter														
20 cm	3.93	103	27	39	15	8	67	29%	9	150	116	26	20	92
50 cm	4.50	83	22	7	27	2	125	11%	–	169	76	17	7	31

[a] After Cronan 1985. All units in μeq L^{-1} except pH and Al_{om}, which is expressed as a percentage of Al_{ext}.
A^- = organic anions = sum of cations minus sum of inorganic anions.

TABLE 6.9. Summary of cation exchange data for ILWAS sites.

Horizon	Exchangeable cations expressed as meq 100 g dry soil^{-1}						
	Ca^{2+}	Mg^{2+}	K^+	Na^+	Acidity[a]	CEC	Percent base saturation
Panther Lake							
O	8.3	2.2	0.83	0.06	16	27	42
E	0.2	0.04	0.04	0.01	3	–	7
Bh	1.0	0.10	0.05	0.01	14	14.75	8
B_{s1}	0.90	0.07	0.07	0.02	16	17.49	6
B_{s2}	0.23	0.04	0.04	0.01	8	7.99	4
B_{s3}	0.17	0.03	0.02	0.01	2	2.50	9
C	0.11	0.02	0.04	0.02	1.0	1.16	16
Woods Lake							
O	8.3	1.7	0.58	0.04	14	25	43
A	0.95	0.19	0.2	0.02	26	27	5
E	0.18	0.04	0.04	0.003	6.2	6	4
B_{s1}	0.29	0.06	0.05	0.009	10	10	4
B_{s2}	0.16	0.03	0.03	0.01	7	7	3
B_{s3}	0.07	0.01	0.01	0.008	4	4	2
C	0.06	0.01	0.004	0.007	2	2	4

[a] Includes exchangeable H^+ and Al.

ANC (1,400 to 2,000 µeq L^{-1}), with Na^+ as the dominant cation (1,000 to 1,600 µeq L^{-1}). The high SO_4^{2-} (290 to 430 µeq L^{-1}) and Cl^- (440 to 980 µeq L^{-1}) concentrations suggest that deep groundwater either has a very long residence time, having been concentrated by evapotranspiration, or has been in contact with sulfur or chloride-bearing minerals. The position of the boring near the hydrologic divide indicates that this area may have a low hydraulic gradient. Twenty till samples collected from the boring were used to determine the cation exchange capacity (1.38 meq 100 g^{-1}) and base saturation (93%). It is evident from the composition of deep groundwater that atmospheric deposition was considerably altered and the acidity was completely neutralized. The mean chemical composition of lake water (Figure 6.25a; Table 6.7) is the result of the mixing of waters entering Panther Lake from the various flow paths. Acidic water enters through direct precipitation and surface runoff, whereas high ANC water enters from the groundwater system. Because the deep groundwater flow path is the dominant source of water, Panther Lake has relatively high ANC.

An analysis of the mineralogy of the soils of Panther Lake basin shows that some of the more weatherable heavy minerals (specific gravity > 2.96), such as hornblende, have been depleted from the upper part of the soil profile (Newton and April 1982). This pattern suggests that primary mineral weathering is important in regulating the base status of these soils.

Panther Lake water has a relatively high concentration of Ca^{2+} compared to Na^+ (Table 6.7). This characteristic may be indicative of the importance of cation exchange reactions. In these reactions, acidity originating from atmospheric deposition or from internal soil/vegetation processes (H^+, Al) is exchanged with Ca^{2+} (or another base cation) on soil exchange sites. Through this process, acidic cations (H^+, Al) are retained in the soils and base cations are released to the surface water. Neutralization by cation exchange reactions can occur only as long as base cations reside on soil exchange sites. With time, pools of base cations are depleted, diminishing pH buffering by cation exchange reactions. The low base saturation of mineral soil horizons (Table 6.9) and acidic soil solution (Table 6.8) in Panther watershed suggests that little neutralization by this process occurs in the upper 1 m of the soil profile. The high base saturation of the till indicates that the deeper materials have a much higher potential for neutralization by cation exchange.

The available data do not allow a quantitative determination of the relative importance of mineral weathering and cation exchange in this watershed. Probably both processes are occurring. However, observations clearly document the importance of

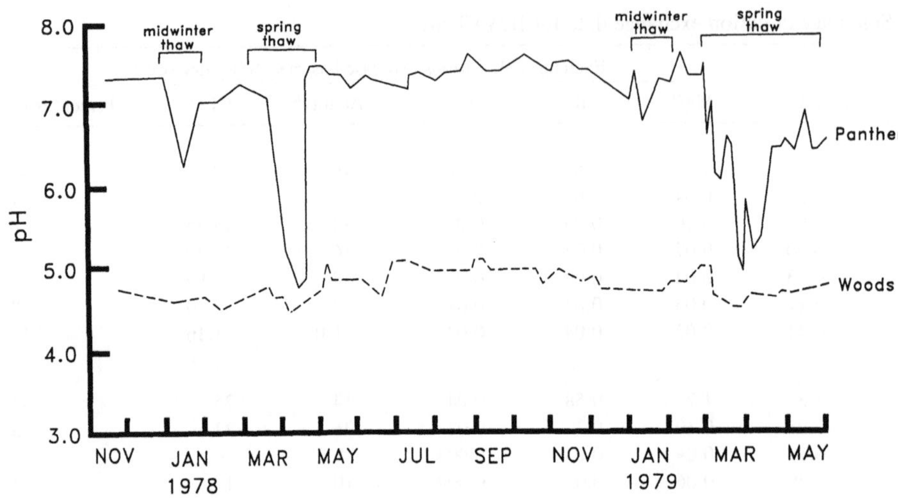

FIGURE 6.27. Temporal trends in air-equilibrated pH of the outlets of Panther Lake and Woods Lake (1978–79). (After Schofield 1984.)

the deep groundwater flow paths in regulating the acid-base chemistry of Panther Lake.

Temporal trends in the chemistry of Panther Lake show short-term acidification, primarily during the spring runoff (Figure 6.27; Schofield 1984). These episodes occur in the upper waters and are characterized by decreased pH and ANC, with corresponding increases in Al_{ext} concentrations. Changes in ANC and pH coincided with decreases in base cation concentrations coupled with increases in NO_3^- (Schofield 1984). These chemical changes reflect lateral flow through the upper soil horizons. This transport occurs during the spring when the groundwater reservoir is filled and additional inputs of water are forced to flow laterally through upper soil horizons. Waters draining these horizons contain low concentrations of base cations and elevated concentrations of SO_4^{2-}, organic acids and Al_{ext}; during spring they have high NO_3^- concentrations.

The hydrology and chemistry of Panther Lake basin are typical of watersheds dominated by thick till. Most water moves along deep flow paths through the till. Even though the till is composed primarily of silicate minerals, the groundwater has sufficient residence time to react with these minerals to neutralize strong acid inputs. This type of watershed is, however, subject to acidification associated with hydrologic events when, for short periods of time, the dominant flow path changes to shallow flow through the upper soil horizons (Figure 6.27). This typically occurs during snowmelt periods when the groundwater capacity is exceeded. During these periods, significant quantities of low pH, Al-rich water enters the lake.

Surface Waters Derived from Basins with Shallow Deposits of Glacial Till

Woods Lake (43°52′N, 71°58′W) was an ILWAS site representing the thin till watershed class. It is similar to Panther Lake watershed in almost every characteristic, including atmospheric deposition, vegetation, soil and soil solution chemistry, and mineralogy (Tables 6.7, 6.8, and 6.9), with the exception of the depth of glacial till. Since the watershed lacks deep surficial deposits for an extensive groundwater reservoir, shallow flow paths are dominant. Because of this characteristic, there is insufficient net internal release of base cations to neutralize the loading of strong acids. Therefore, Woods Lake is acidic (Figure 6.25a; Table 6.7).

Inputs of strong acid anions into Woods Lake watershed were similar to those entering Panther Lake basin (Figure 6.25a). Like Panther Lake basin, atmospheric deposition of SO_4^{2-} was conservatively transported through the watershed. Much of the input of dissolved inorganic nitrogen was retained by the watershed vegetation, but a

FIGURE 6.28. Isopleth of (a) ANC, (b) NO_3^-, and (c) Al_{ext} concentrations in the water column of Woods Lake (1984–85).

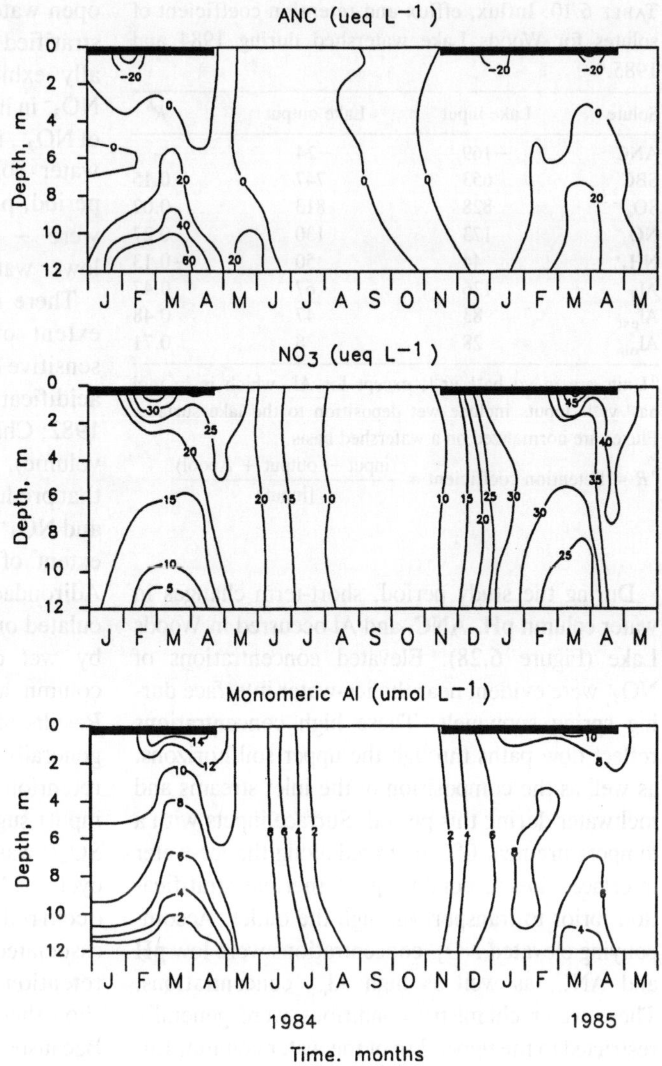

significant fraction was exported as NO_3^-, particularly during spring snowmelt (Schofield et al. 1985). Base cation output from Woods Lake (net release 440 eq ha^{-1} yr^{-1}) watershed was much smaller than at Panther Lake (net release 1,730 eq ha^{-1} yr^{-1}). Clearly, the difference in surface water ANC between these two systems was due to differences in base cation supply rates, rather than the loading of strong acids. The base cation supply rate is, in turn, dependent on the flow path of water through the watershed. Shallow flow paths dominate because the small groundwater reservoir is rapidly filled during precipitation events, forcing most of the water to move laterally as surface runoff and shallow interflow through the upper soil horizons. This shallow interflow water comes in contact with acidic residuum, which has little ability to neutralize acid inputs through the release of base cations.

Within the water column of Woods Lake, spatial and temporal variations in chemistry were evident. The lake experiences ice-cover during the winter and, therefore, exhibited winter temperature stratification (Figure 6.28). Following ice-breakup in the spring, the lake mixed completely and was subsequently weakly stratified during the summer. Woods Lake mixed completely by late fall and ice-cover occurred by late November to early December.

TABLE 6.10. Influx, efflux and retention coefficient of solutes for Woods Lake watershed during 1984 and 1985.[a]

Solute	Lake input	Lake output	R^b
ANC	−169	−24	–
SBC	653	747	−0.15
SO_4^{2-}	828	813	0.02
NO_3^-	173	130	0.27
NH_4^+	44	50	−0.13
Al_T	126	67	0.47
Al_{ext}	83	47	0.48
Al_{om}	28	8	0.71

[a]Units are in eq ha^{-1} yr^{-1}, except for Al, which is in mol ha^{-1} yr^{-1}. Inputs include wet deposition to the lake surface. Fluxes are normalized on a watershed basis.

[b]$R = \text{Retention coefficient} = \dfrac{(\text{input} - \text{output} + \Delta\text{pool})}{[\text{input}]}$

During the study period, short-term changes in water column pH, ANC, and Al occurred in Woods Lake (Figure 6.28). Elevated concentrations of NO_3^- were evident near the ice-water interface during spring snowmelt. These high concentrations reflect flow paths through the upper soil horizons, as well as the composition of the inlet streams and meltwater during this period. Surface inputs with a temperature near 0°C migrated along the ice-water interface, due to conditions of thermal stratification, prior to transport through the outlet. Accompanying elevated NO_3^- concentrations were low pH and ANC, as well as high Al_{ext} concentrations. These water chemistry conditions were generally restricted to the upper 2 m of the water column. Little temporal or spatial variation in water column concentrations of SBC or SO_4^{2-} occurred.

During ice-cover, vertical variations in water chemistry were evident (Figure 6.28). These variations were attributed to inputs of acidic water transported along the ice-water interface, as discussed above, coupled with reduction and mineralization processes that occurred within the sediments. Nitrate depletion was a consequence of denitrification, whereas organic nitrogen mineralization resulted in NH_4^+ release. Both processes served to increase ANC and pH (Chapter 1, Munson and Gherini, this volume) of the lower waters, which in turn coincided with hydrolysis/removal of aqueous Al_{ext}. As a result of these transformations, clinograde profiles in pH, ANC, NO_3^-, NH_4^+, and Al_{ext} were evident during ice-cover. During the open water season, the water column was weakly stratified and solute profiles in Woods Lake generally exhibited an orthograde distribution. Low NO_3^- in inlet waters coupled with in-lake retention of NO_3^-, resulted in low NO_3^- concentrations in the water column during the summer. During this period, pH values were near 5.0 and ANC values were ~ 0. These higher pH conditions caused lower water column Al_{ext} concentrations.

There is considerable interest in assessing the extent of internal ANC production in acid-sensitive lakes, as these processes serve to mitigate acidification by strong acid inputs (Kelly et al. 1982; Chapters 1 and 2, Munson and Gherini, this volume). The most important in-lake processes that produce ANC are biological retention of SO_4^{2-} and NO_3^- (Kelly et al. 1982, 1987). To evaluate the extent of ANC production in an acid-sensitive Adirondack lake, element mass balances were calculated on the inlet streams, the direct lake inputs by wet deposition, the outlet, and the water column at Woods Lake for 1984 (Table 6.10). Results of this analysis indicate that solutes were generally conservative in Woods Lake. Element retention coefficients (element retention/element input) suggest that very little of the mass flux of SO_4^{2-} was retained in Woods Lake over the annual cycle. Moreover, limited production of ANC occurred in the lake, and this was largely associated with the net release of SBC and the retention of NO_3^-. Mass balance calculations also show that Woods Lake was an important sink of Al. Because of chronically acidic conditions, the limited in-lake production of ANC was largely destroyed through the hydrolysis and deposition of Al (Driscoll et al. 1989b). A comparison of terrestrial ANC with in-lake ANC production suggests that in-lake processes are not significant in Woods Lake. Although rates of biological SO_4^{2-} and NO_3^- retention are comparable to values in lakes within regions where these processes are quantitatively important (Kelly et al. 1987), the hydraulic retention time of Woods Lake is short (210 days; Staubitz and Zarriello 1989) and in-lake production of ANC was small relative to watershed production of ANC. Schafran and Driscoll (1987) have conducted a similar analysis for Dart's Lake in the Adirondacks and found comparable results.

The Woods Lake study allows a determination and comparison of the relative importance of ANC generation in terrestrial and aquatic systems. Acid neutralizing capacity generated within the terrestrial system is fundamentally different from that generated within the aquatic system. In the terrestrial system, the dominant processes involve the release of base cations, whereas in the aquatic system, the dominant processes involve uptake of SO_4^{2-} and NO_3^-. The release of base cations can be due either to cation exchange or to mineral weathering reactions. Although the biological uptake of NO_3^- occurs in both the terrestrial and aquatic systems, it is a dominant process in the aquatic environment (Schafran and Driscoll 1987). Acid neutralizing capacity generation in Adirondack drainage lakes is quantitatively unimportant compared to the terrestrial environment.

Adirondack Basins with Carbonate Deposits

Although the Adirondack region is a terrain dominated by igneous and high-grade metamorphic rocks, there are areas with carbonate-bearing bedrock. Carbonate occurs both as deformed strata of relatively massive marble and as disseminated carbonate mineral grains or veins in an otherwise noncarbonate calc-silicate rock. Both types of occurrence can have a profound effect on surface water chemistry if sufficient amounts of water contact these rocks.

The influence of carbonate minerals on surface water quality is illustrated through results from the Regional Integrated Lake-Watershed Acidification Study (RILWAS). One of the goals of RILWAS was to test the hypothesis that hydrologic flow paths regulate surface water quality throughout the Adirondacks. The findings show that for most Adirondack drainage lakes, watershed till depth controls flow path, which in turn regulates surface water chemistry (Driscoll and Newton 1985, Peters and Driscoll 1987). However, there were some exceptions to this pattern. In particular, basins that contain significant quantities of carbonate minerals are characterized by a large supply of base cations to drainage water, regardless of the surficial geology.

Woodruff Pond is located in the central Adirondacks near Newcomb, New York (43°58′N, 74°14′W). The watershed is 8.73 km² in area and is bordered to the north by one of the Adirondack marble belts. Many boulders in the glacial till covering the watershed contain varying amounts of calcite. Most of the watershed is covered by thin till, therefore on the basis of the thickness of surficial deposits, surface waters would be expected to have low ANC. In fact, Woodruff Pond had the highest ANC (480 $\mu eq\ L^{-1}$) observed in any of the RILWAS study lakes (Figure 6.29; Driscoll and Newton 1985).

Basins dominated by thick till are characterized by a large groundwater reservoir and therefore are less responsive to hydrologic events and have a relatively high base flow. Thin till basins are flashier and exhibit lower base flow. The relationship between base flow discharge and SBC is shown in Figure 6.30 for the RILWAS and ILWAS lakes. All the RILWAS lakes except Woodruff Pond fall between the Woods and Panther Lake end-members. Woodruff Pond has a baseflow discharge that is almost as low as Woods Lake, yet it has ANC of 410 $\mu eq\ L^{-1}$ and SBC of 600 $\mu eq\ L^{-1}$. This watershed dramatically shows the influence of carbonate minerals on surface water quality. Although the water moves rapidly along shallow flow paths, there is sufficient time for carbonate minerals to react with H^+ in water. The dissolution rate of calcite is fast and base cations are readily supplied to drainage water, so that flow path is not a determinant in regulating the acid-base status of these waters. Watersheds overlain by thin till may have high ANC surface waters if carbonate minerals are present in the till or are exposed on the bedrock surface.

Carbonate influenced waters can be differentiated from noncarbonate waters by the molar ratio of Ca^{2+} to H_4SiO_4. In the RILWAS project, 22 lake-watershed systems were monitored across the Adirondacks. Of these, only six sites had ANC values > 50 $\mu eq\ L^{-1}$. The Ca^{2+}:H_4SiO_4 values ranged from approximately 0.65 for watersheds underlain by granite gneiss to about 1.15 for watersheds underlain by anorthosite or calc-silicate rocks and to 2.36 for the carbonate influenced Woodruff Pond. Although these ratios may be influenced by the nonconservative behavior of H_4SiO_4 (e.g., secondary mineral formation, diatom activity), it is probably reasonable to suggest that in lakes with molar Ca^{2+}:H_4SiO_4 > 2, ANC values are controlled by carbonate mineral dissolution.

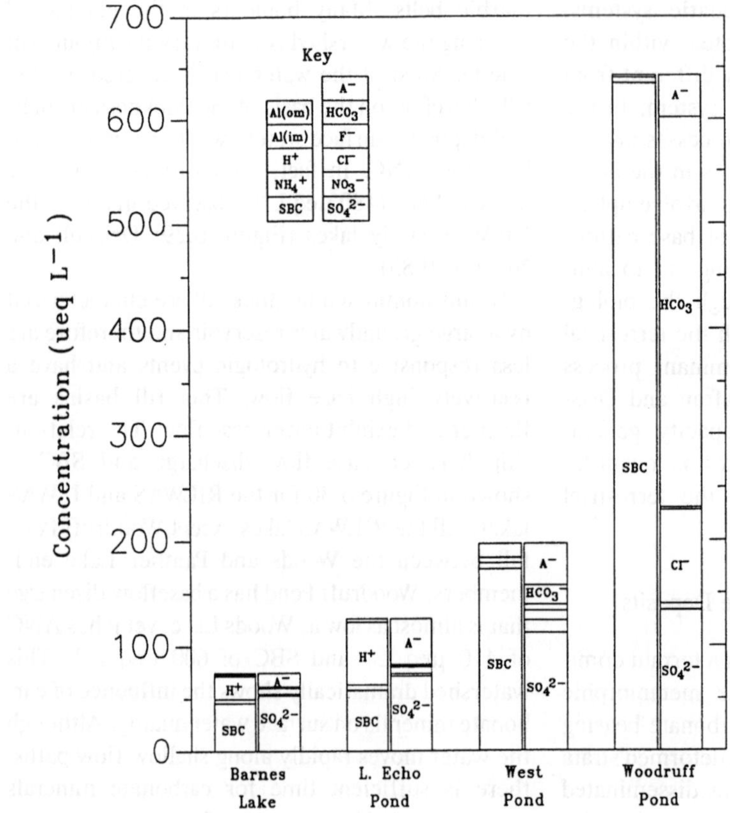

	Barnes Lake	L. Echo Pond	West Pond	Woodruff Pond
pH	4.67	4.23	5.30	6.81
ANC (ueq L^{-1})	−14	−51	28	400
H_4SiO_4 (umol L^{-1})	3.8	4.6	82.3	86.7
DOC (umol C L^{-1})	418	1037	678	714

FIGURE 6.29. Charge distribution in Barnes Lake (a perched seepage lake), Little Echo Pond (a bog seepage lake), West Pond (a wetland dominated drainage lake), and Woodruff Pond (a calcite basin).

Adirondack Basins with Wetlands

Wetlands, primarily swamps, can occur within any lake or stream watershed system and are abundant in the Adirondacks (see section on Vegetation and Land Use in the Adirondacks). Wetlands can greatly influence surface water chemistry as shown by the study of West Pond. West Pond is a small (12 ha) drainage lake located just west of Big Moose Lake in the western Adirondacks (43°49'N, 74°53'W). The lake was investigated during the RILWAS project (Driscoll and Newton 1985, Driscoll et al. 1987b). There is an extensive marsh and bog area immediately surrounding much of the lake. Deposits of peat are over 5 m in depth across much of the wetland.

Water draining the pond had an average of 678 µmol L^{-1} DOC. Moreover, lake water had low ANC (28 µeq L^{-1}), which is expected in a watershed dominated by thin till. However, unlike clearwater lakes, organic acids were an important component of the solution anions composition (Figure 6.29). In West Pond, organic anions (estimated by discrepancy in charge balance) were the second most abundant anion after SO_4^{2-}. In addition, elevated concentrations of organic solutes influence the speciation of Al. Most of the Al appearing in West Pond was associated with organic ligands (Al_{om}) and therefore is probably less toxic to aquatic organisms than inorganic forms (Driscoll et al. 1980).

Extensive Evaluation of Episodic Lake Acidification During Snowmelt

Many synoptic surveys have documented the chronic acidification of Adirondack surface waters (Colquhoun et al. 1984, Driscoll and Newton

1985, Brakke et al. 1988); however, far fewer extensive studies of short-term or episodic acidification have been conducted. Synoptic surveys of Adirondack lake chemistry have typically been conducted during the summer or fall, when ANC is at or near maximum value (Schaefer et. al. 1990). However, rainfall or snowmelt events may result in large hydrologic inputs to surface waters and episodic acidification. Large seasonal changes in acid-base chemistry often occur in Adirondack surface waters, with minimum pH and ANC values, and maximum Al concentrations generally evident during spring snowmelt (Driscoll and Schafran 1984, Driscoll et al. 1987a,b). As a result, there is interest in assessing the extent of episodic acidification and the processes contributing to short-term acidification, and ultimately quantifying a relationship between baseflow ANC measured during lake surveys and minimum ANC occurring during snowmelt.

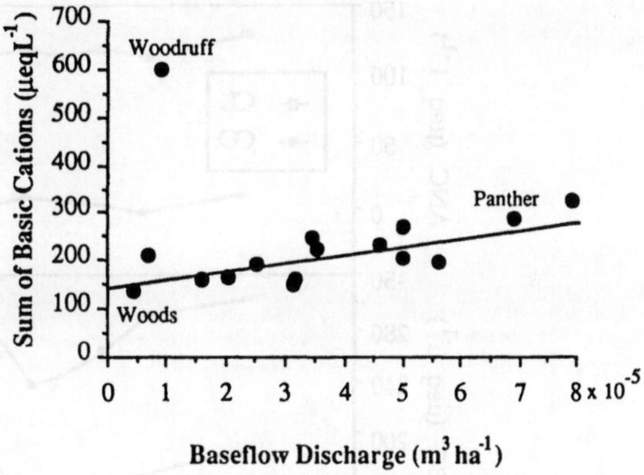

FIGURE 6.30. Average ANC as a function of low-flow discharge for the RILWAS and ILWAS watersheds.

Schaefer et al. (1990) examined changes in the chemistry of outlet water at a subset of the ALTM drainage lakes located throughout the Adirondacks (Table 6.1) during snowmelt in 1986 (11 lakes) and 1987 (10 lakes). They made a quantitative examination of the factors that influence ANC depression during snowmelt for lakes with a range of baseline ANC values. This analysis was accomplished by examining changes in the concentrations of major solutes (e.g., C_B, SO_4^{2-}, NO_3^-) relative to changes in ANC during snowmelt. In this study, baseline ANC was determined by visual inspection of time-series data for the period prior to and following snowmelt ANC depression.

Because of varying water residence times in the soil, groundwater, and lakes, chemical changes associated with snowmelt (and their effects on ANC) are not immediately evident at lake outlets. In contrast to streamwater acidification pulses, which may occur over very short time scales (Sullivan et al. 1986), lakewater ANC depressions tended to persist for 4 to 8 weeks (Figure 6.31). Moreover, peak or minimum solute concentrations are not as extreme in lake water as in stream waters (e.g., Driscoll et al. 1987a). This pattern is undoubtedly due to the attenuation and mixing of acidic water within the lake during snowmelt. During the study years (and indeed most years in the Adirondacks) peak discharge precedes loss of ice cover from the lake surface. As a result, during snowmelt acidic water enters the lake and, due to temperature and ice cover conditions, migrates along the ice-water interface prior to emerging from the outlet. The mixing of melt water with bulk lake water occurs to a depth of about 1 to 2 m (Driscoll et al. 1989b, McAvoy and Driscoll 1989).

To illustrate and compare temporal variations in water chemistry that occur during the spring, time series plots are presented for a high ANC lake, Clear Pond (baseline ANC 100 μeq L^{-1}) and for a low ANC lake, Constable Pond (baseline ANC 10 μeq L^{-1}) (Figure 6.31). Clear Pond lies within a watershed covered by approximately 30% thick till while the Constable Pond watershed has till < 2% thick. Acid neutralizing capacity at both Clear Pond and Constable Pond decreased from baseline values at the onset of snowmelt. The depression in ANC (ΔANC) was much greater in Clear Pond (ΔANC = 70 μeq L^{-1}) than in Constable Pond (ΔANC = 30 μeq L^{-1}), reflecting the general trend that high ANC lakes show a greater depression than low ANC lakes. In Clear Pond, the snowmelt ΔANC closely corresponds with the decrease in base cations (for both study years; ANC = −55 + 0.54SBC; r^2 = 0.72; where ANC and SBC are in μeq L^{-1}). The other major factor that contributes to the short-term acidification of waters is ΔNO_3^-. In Clear Pond, NO_3^- concentrations peaked during snowmelt at 20 μeq L^{-1} from a baseline value of 8 μeq L^{-1}. Therefore, relative to the dilution of base cation concentrations, changes in NO_3^- did not significantly contribute to episodic acidification of Clear Pond (Figure 6.31).

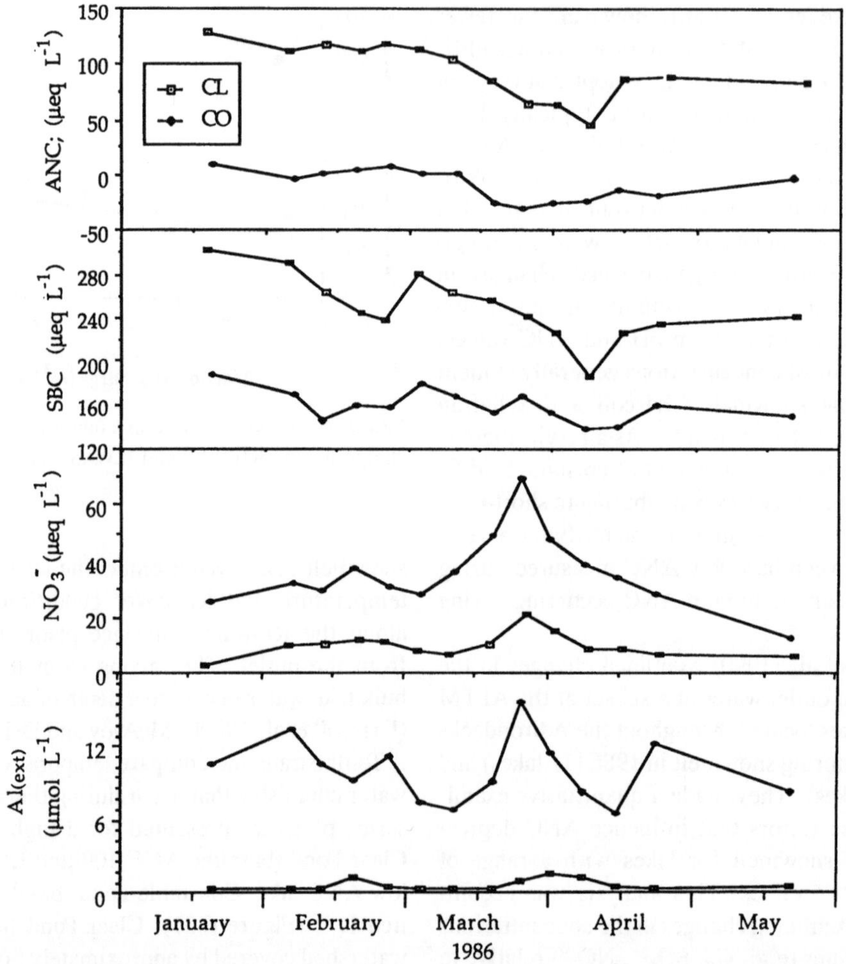

FIGURE 6.31. Weekly concentrations of ANC, SBC, NO_3^-, and Al_{ext} in Clear Pond and Constable Pond from January through May 1986. (After Schaefer et al. 1990.)

In acid-sensitive Constable Pond, the decrease in base cations during snowmelt was much lower (ΔSBC = 20 µeq L^{-1}) than Clear Pond. In Constable Pond, the major acidifying process was an increase in NO_3^- from a baseline value of 30 µeq L^{-1} to a peak of 70 µeq L^{-1}. The increase in NO_3^- concentrations exceeded the depression in ANC during snowmelt. At Constable Pond and other acidic Adirondack waters, the neutralization of acid inputs (base cation dilution and NO_3^- increase) was only partially accomplished by decreases in ANC. The balance was neutralized through the mobilization of Al. In low pH Adirondack waters (pH < 5.5), such as Constable Pond, there is little temporal variation in base cations and SO_4^{2-} concentrations, so a near stoichiometric relationship is often evident between increases in NO_3^- and increases in Al_{ext} (for Constable Pond for both study years; $NO_3^- = 5 + 3.4{*}Al_{ext}$; $r^2 = 0.35$; where NO_3^- is in µeq L^{-1} and Al_{ext} is in µmol L^{-1}) (Driscoll and Schafran 1984, Schofield et al. 1985, Driscoll et al. 1989a).

The magnitude of the springtime ANC depression for all episodic study lakes is not linearly related to either baseline or fall index of ANC; in fact lakes with intermediate baseline ANC exhibited the largest ΔANC (Figure 6.32). In general, the lakes with high baseline ANC have the largest contribution of ΔSBC as a fraction of ΔANC (Figure 6.33).

In Adirondack lakes, changes in SO_4^{2-} concentration did not significantly contribute to snowmelt

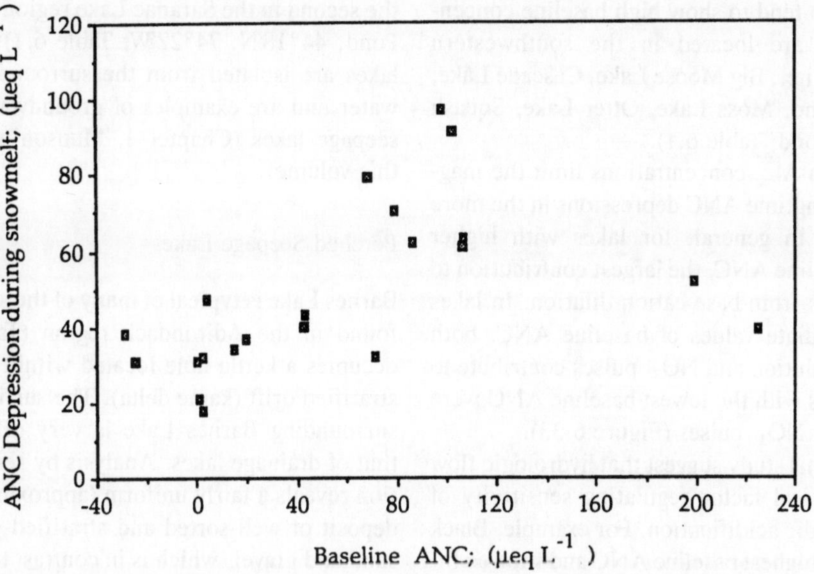

FIGURE 6.32. The maximum depression of springtime ANC values (baseline ANC - minimum ANC) during snowmelt as determined from weekly samples collected from January through May 1986 and 1987. (After Schaefer et al. 1990.)

acidification; as in all lakes, ΔSO_4^{2-} was negative during the ANC depressions (thus ΔSO_4^{2-}:ΔANC was always positive). Unlike SO_4^{2-}, NO_3^- concentrations increased during ANC depression, so the ΔNO_3^-:ΔANC ratios were negative. The proportional contribution of NO_3^- to ANC depression was low in high-baseline ANC lakes, and increased sharply with decreasing baseline ANC (Figure 6.33). Weekly ΔNO_3^- values (e.g., Figure 6.31) indicated that NO_3^- pulses often occurred at the beginning of an ANC depression. In contrast, the latter parts of an ANC depression were generally dominated by dilution of base cations. Lakes that show elevated concentrations of NO_3^- during

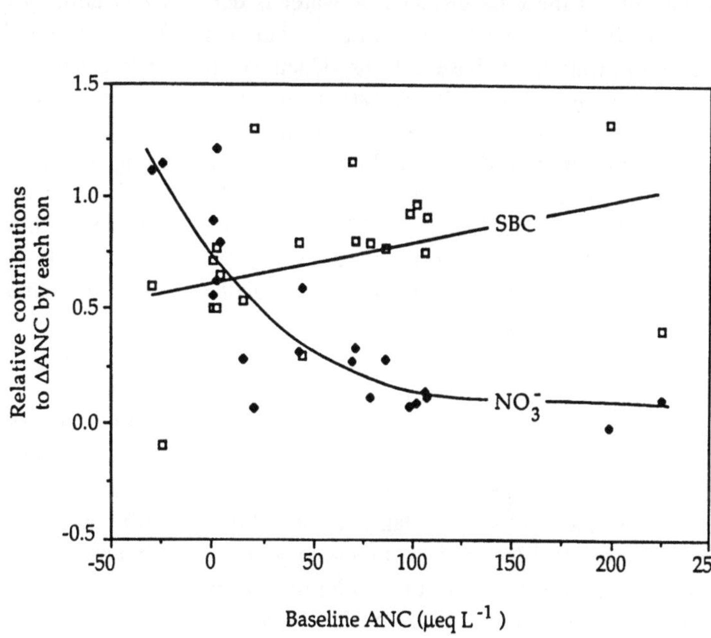

FIGURE 6.33. The relative contributions of SBC and NO_3^- to the springtime ANC depression in Adirondack study lakes. Lakes with lower baseline ANC are more affected by NO_3^- pulses and lakes with higher baseline ANC are more affected by SBC dilution. Solid lines represent a best fit of empirical observations. (After Schaefer et al. 1990.)

snowmelt also tend to show high baseline concentrations and are located in the southwestern Adirondacks (i.e., Big Moose Lake, Cascade Lake, Constable Pond, Moss Lake, Otter Lake, Squash Pond, West Pond; Table 6.1).

Increases in Al_{ext} concentrations limit the magnitude of springtime ANC depressions in the more acidic lakes. In general, for lakes with higher values of baseline ANC, the largest contribution to ΔANC comes from base cation dilution. In lakes with intermediate values of baseline ANC, both base cation dilution and NO_3^- pulses contribute to ΔANC. Lakes with the lowest baseline ANC were dominated by NO_3^- pulses (Figure 6.33).

Results of this study suggest that hydrologic flow path is a critical factor regulating sensitivity of lakes to episodic acidification. For example, Black Pond had the highest baseline ANC and showed little ANC depression during snowmelt (Figure 6.32). The watershed at this site contains extensive stratified drift deposits (Table 6.1). Therefore, surface water in Black Pond is predominantly derived from groundwater. Because of the thick, permeable sand and gravel deposits, snowmelt water moves vertically through the soil to recharge the groundwater reservoir and little lateral flow is generated through the upper soil horizons. In contrast, low ANC sites (Big Moose, Constable Pond, Otter Lake, Squash Pond, West Pond) are generally within watersheds dominated by thin deposits of glacial till. At these basins, surface water is derived largely from shallow soil water, containing low concentrations of base cations (Chen et al. 1984, Peters and Driscoll 1987), rather than from groundwater. Acidic waters show large increases in Al_{ext} concentration during acidification events. Aluminum is an effective pH buffer in acidic waters (Driscoll and Bisogni 1984). Therefore, inputs of acidity at these sites are readily neutralized through the dissolution of soil Al, minimizing the extent of ANC depression.

Case Studies of Seepage Lakes in the Adirondacks

Compared to drainage lakes, there have been far fewer case studies of seepage lakes in the Adirondacks. In the RILWAS project, two seepage lakes were studied, one at the western border of Adirondack Park (Barnes Lake, 43°58'N, 75°14'W) and the second in the Saranac Lake region (Little Echo Pond, 44°18'N, 74°22'W; Table 6.1). Both these lakes are isolated from the surrounding groundwater and are examples of groundwater recharge seepage lakes (Chapter 1, Munson and Gherini, this volume).

Perched Seepage Lakes

Barnes Lake is typical of many of the seepage lakes found in the Adirondack region (Table 6.1). It occupies a kettle hole located within a deposit of stratified drift (kame delta). The surficial geology surrounding Barnes Lake is very different from that of drainage lakes. Analysis by seismic refraction reveals a fairly uniform (approximately 15 m) deposit of well-sorted and stratified glaciofluvial sand and gravel, which is in contrast to the irregular thickness of poorly sorted unstratified till that surrounds most drainage lakes. These glaciofluvial deposits have a much higher hydraulic conductivity than glacial till.

If it is assumed that the thickness of surficial sediments controls flow path and subsequently surface water ANC, as observed for drainage lakes, then Barnes Lake should have a relatively high ANC. However, Driscoll and Newton (1985) found that Barnes Lake was highly acidic (Figure 6.29). A series of wells and borings made around the lake have shown that the lake surface is approximately 10 m above the surrounding groundwater table. This pattern indicates that the lake water is essentially isolated or "perched" above the groundwater. The accumulation of organic-rich bottom sediments together with a mat of sedges and sphagnum along the lake margin have reduced the hydraulic conductivity of the outseepage zone that led to this isolation. This situation is not unique. Recent studies in Wisconsin (Eilers 1984, Lin et al. 1987), a region abundant in seepage lakes, have shown that many of these systems receive most of their inputs from direct atmospheric deposition, indicating that they are also isolated from the surrounding groundwater system.

The chemistry of Barnes Lake reflects these hydrologic flow paths. The water column was acidic with low concentrations of base cations and virtually no H_4SiO_4, but unlike acidic drainage lakes, Al concentrations were low (Figure 6.29). These characteristic are also consistent with the lack of

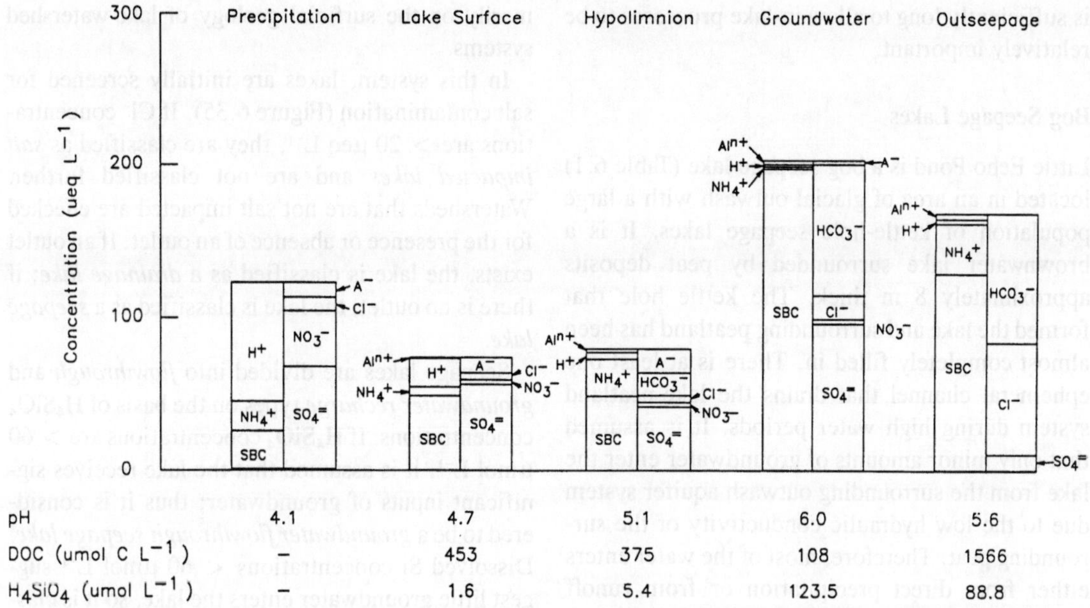

FIGURE 6.34. Ion distribution of waters transported through the Barnes Lake hydrologic system.

significant surface runoff or shallow interflow flow paths, and the dominance of direct precipitation inputs.

Ion charge distribution diagrams of average cation and anion concentrations for precipitation, surface lake samples, hypolimnitic samples, groundwater, and outseepage water summarize drainage patterns for Barnes Lake (Figure 6.34). Precipitation at the site is typical for the Adirondacks, with high concentrations of SO_4^{2-}, NO_3^-, and H^+, reflecting the acidic nature of the input (pH 4.1). Concentrations of base cations in precipitation were low.

The upper waters of Barnes Lake are derived primarily from precipitation, thus they are acidic (pH 4.7). Comparing the composition of Barnes Lake surface waters with precipitation (Figure 6.34), shows that lake water was much more dilute. The upper waters were also significantly higher in base cations, possibly reflecting some minor groundwater inflow from the "fringe zone" surrounding the lake. A major difference between the composition of precipitation and that of lake water is the depletion of inorganic nitrogen. Both NH_4^+ and NO_3^- occurred at lower concentrations in the upper waters than precipitation, undoubtedly due to the consumption of these solutes by bacterial and/or plant activity. In addition, SO_4^{2-} concentrations were somewhat lower in the upper waters suggesting some in-lake retention of SO_4^{2-}.

In the hypolimnion, concentrations of O_2 were depleted in the lower water during periods of stratification. This condition facilitated the retention of SO_4^{2-}, probably from reduction reactions, and the release of NH_4^+, likely from the mineralization of organic nitrogen in sediments. These processes resulted in the hypolimnitic waters having a higher pH and a slightly positive ANC (Figure 6.34).

Outseepage from the bottom of Barnes Lake also suggests reducing conditions. Sulfate concentrations were low and NH_4^+ concentrations were elevated. However, unlike the hypolimnitic waters, concentrations of base cations and H_4SiO_4 were also elevated, although not nearly to the values of the regional groundwater (Figure 6.34). The chemical differences between the outseepage water and the regional groundwater show that the lake is not the major recharge area for the groundwater aquifer.

This case study shows that perched seepage lakes waters are very sensitive to strong acid inputs because they are dominated by the direct precipitation flow path. Watershed flow paths which produce base cations are relatively unimportant in the systems. Unlike drainage lakes, the residence time of water in the perched seepage lake is sufficiently long to allow in-lake processes to be relatively important.

Bog Seepage Lakes

Little Echo Pond is a bog seepage lake (Table 6.1) located in an area of glacial outwash with a large population of kettle-hole seepage lakes. It is a brownwater lake surrounded by peat deposits approximately 8 m thick. The kettle hole that formed the lake and surrounding peatland has been almost completely filled in. There is at least one ephemeral channel that drains the lake-peatland system during high water periods. It is assumed that only minor amounts of groundwater enter the lake from the surrounding outwash aquifer system due to the low hydraulic conductivity of the surrounding peat. Therefore, most of the water enters either from direct precipitation or from runoff through the upper "active" zone of the peat.

The chemistry of Little Echo Pond is similar to Barnes Lake in that both had low concentrations of SBC, Al_{ext}, and H_4SiO_4, and both have very low pH (Driscoll and Newton 1985; Figure 6.29). Although both lakes were acidic, significant differences in their chemical characteristics are evident. The dominant anion in Barnes Lake was SO_4^{2-}; in Little Echo Pond, concentrations of organic anions were comparable to SO_4^{2-} (on an equivalence basis) as demonstrated by the large discrepancy in charge balance. This relatively large concentration of organic anions is consistent with the high concentrations of DOC in Little Echo Pond. Although much of the acidity of the pond can be attributed to atmospheric deposition of H_2SO_4, it appears that the peat deposits and the sphagnum mat surrounding the pond release substantial quantities of organic acids that contribute to the overall acidity.

Criteria for the Classification of Adirondack Lakes

The process-level case studies show that the dominant flow path of water entering a lake is the critical factor in determining the response of Adirondack lake-watershed systems to acidic deposition. Any system developed to classify Adirondack lakes must ultimately be based on factors relating to hydrologic flow paths. The system developed here is based primarily on the surficial geology of lake-watershed systems.

In this system, lakes are initially screened for salt contamination (Figure 6.35). If Cl^- concentrations are $> 20\ \mu eq\ L^{-1}$, they are classified as *salt impacted lakes* and are not classified further. Watersheds that are not salt impacted are checked for the presence or absence of an outlet. If an outlet exists, the lake is classified as a *drainage lake*; if there is no outlet, the lake is classified as a *seepage lake*.

Seepage lakes are divided into *flowthrough* and *groundwater recharge* types on the basis of H_4SiO_4 concentrations. If H_4SiO_4 concentrations are $> 60\ \mu mol\ L^{-1}$, it is assumed that the lake receives significant inputs of groundwater; thus it is considered to be a *groundwater flowthrough seepage lake.* Dissolved Si concentrations $< 60\ \mu mol\ L^{-1}$ suggest little groundwater enters the lake, so it is classified as a *groundwater recharge seepage lake.* These two types of seepage lakes can be subdivided further as *clearwater* ($\leq$ 417 $\mu mol\ C\ L^{-1}$) or *colored* ($>$ 417 $\mu mol\ C\ L^{-1}$), on the basis of DOC concentrations. Dissolved organic carbon is an analog of the amount of organic acidity generated in the lake-watershed system. In addition, in the category of groundwater recharge seepage lakes, clearwater lakes are termed *perched seepage lakes* and colored lakes are termed *bog seepage lakes*.

The process-level studies have shown that flow paths in drainage lakes are primarily controlled by the dominant thickness of glacial till within the watershed. However, a continuum exists between the thick till dominated and thin till dominated endmembers. In the RILWAS project, 23 watersheds of varying size throughout the Adirondacks were studied (Goldstein et al. 1987). The results of this investigation show that in those systems whose water chemistry is controlled by flow path, there is a statistically significant relationship between water chemistry parameters such as ANC, SBC, and Ca^{2+} and the percentage of the watershed covered by thick till and stratified drift (Newton et al. 1987). For the RILWAS watersheds, baseflow Ca^{2+} concentration is strongly correlated with the percentage of area covered by thick till and stratified drift (Figure 6.36a; baseflow $Ca^{2+} = 76 + 3.1*\%$ thick till; $r^2 = 0.87$). Moreover, because of the strong relationship between ANC and Ca^{2+} in

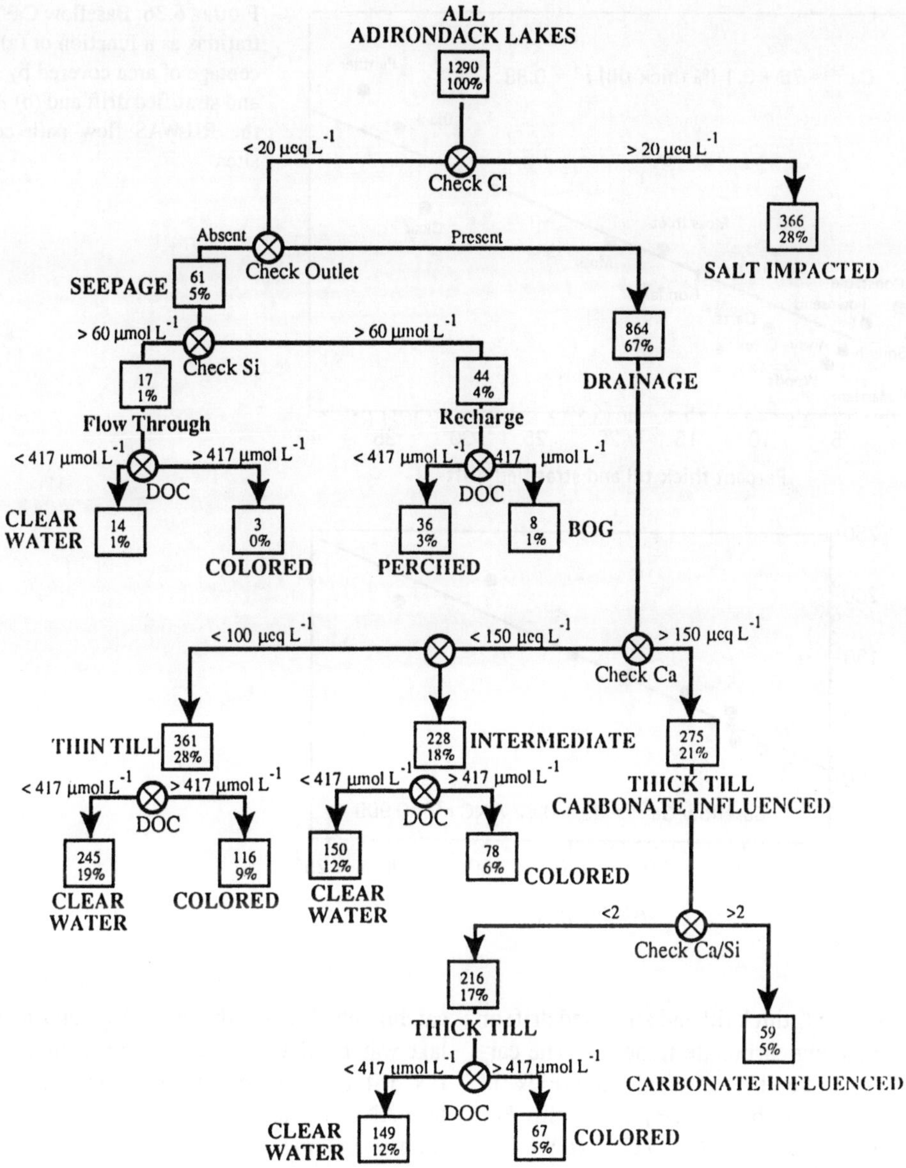

FIGURE 6.35. Flow diagram for a classification system of Adirondack lakes. Number and percentage of ELS-I Adirondack lakes in each class are given. ELS-I lakes are > 4 ha in surface area.

Adirondack lakes, a relationship between baseflow Ca^{2+} and ANC is also expected (for these RILWAS systems, ANC $= -126 + 1.46$ baseflow Ca^{2+}; $r^2 = 0.91$; Figure 6.36b). These relationships were used to develop the criteria for the classification of drainage lakes.

For acidic lakes (ANC ≤ 0 μeq L^{-1}), these empirical relationships suggest baseflow Ca^{2+} concentrations $< \sim 100$ μeq L^{-1}, which corresponds to < 5% of the area covered by thick till and stratified drift. Lakes with an ANC > 100 μeq L^{-1} can be considered relatively insensitive to acidification. They may undergo some spring pH depression (Figure 6.32), but usually ANC values remain positive during hydrologic events. These insensitive lakes have baseflow Ca^{2+} > 150 μeq L^{-1}, and occur in watersheds

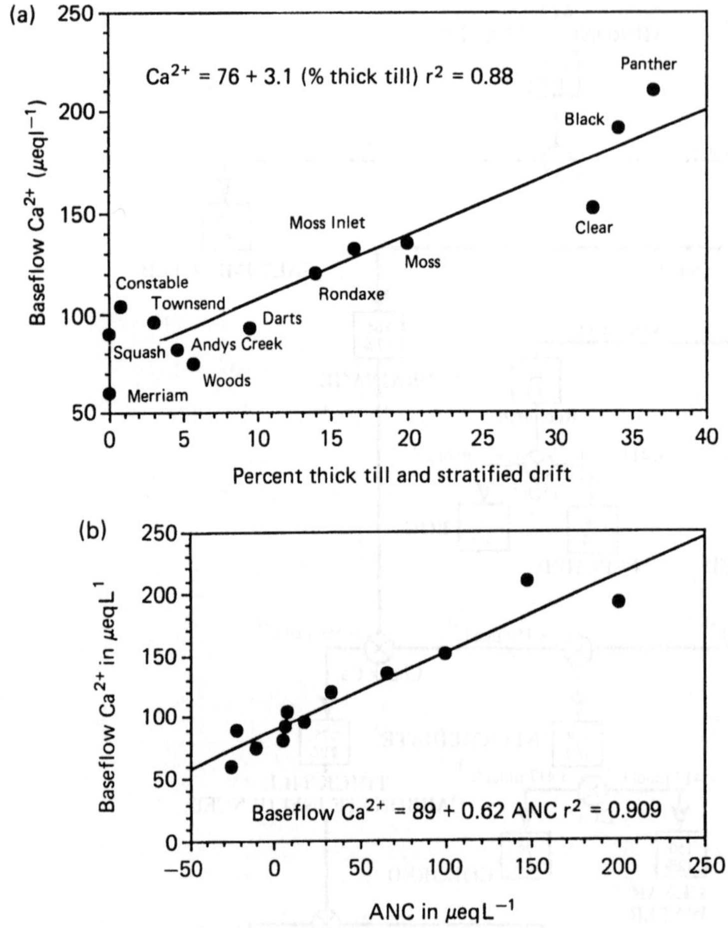

FIGURE 6.36. Baseflow Ca^{2+} concentrations as a function of (a) the percentage of area covered by thick till and stratified drift and (b) ANC, for the RILWAS flow path controlled sites.

containing over 25% thick till and stratified drift or in basins containing carbonate minerals. The carbonate basins can be differentiated from thick till basins by examining the $Ca^{2+}:H_4SiO_4$ ratio. For waters with $Ca^{2+} > 300$ μeq L^{-1}, a $Ca^{2+}:H_4SiO_4 > 2$ suggests that carbonate mineral weathering is the major control of surface water ANC.

Probably the most intriguing Adirondack waters are those with intermediate ANC between 0 and 100 μeq L^{-1}. These lake watersheds are subject to episodic acidification in the spring, causing the ANC of the upper waters to decrease near to, or below zero. These watersheds have thick till covering between 5% and 25% of their watersheds, coinciding with baseflow Ca^{2+} concentrations ranging between 100 and 150 μeq L^{-1}.

Thus three classes of drainage lakes can be defined on the basis of baseflow Ca^{2+}: thin till, intermediate, and thick till/stratified drift or calcite dominated catchments. Most of the thin till systems have acidic lake water (ANC ≤ 0 μeq L^{-1}); the thick till/stratified drift or calcite systems are characterized by high lake ANC and are insensitive to acidification; lake waters draining the intermediate till systems are subject to episodic acidification, which depresses spring ANC values near 0 μeq L^{-1}. Like seepage lakes, all classes of drainage lakes can be subclassified into clearwater and colored lakes on the basis of DOC concentration (≤ or > 417 μmol C L^{-1}, respectively).

Classification of ELS-I Lakes in the Adirondacks

Classification of Drainage Lakes

Drainage lakes are the predominant hydrologic lake type in the Adirondacks (Brakke et al. 1988). Of the 1,009 ELS-I Adirondack (subregion 1A)

TABLE 6.11. Comparison of chemical characteristics of classes of Adirondack ELS-I lakes.[a]

Class	n	pH	ANC (μeq L^{-1})	SO_4^{2-} (μeq L^{-1})	Ca^{2+} (μeq L^{-1})	DOC (μmol L^{-1})	SBC (μeq L^{-1})	H_4SiO_4 (μmol L^{-1})	Al_{ext} (μmol L^{-1})
Total	1,290	6.74 (5.94–7.11)	108 (15–231)	120 (107–136)	164 (89–271)	348 (253–452)	277 (162–489)	39 (18–69)	0.28 (0.16–0.81)
Total drainage	864	6.34 (5.43–6.87)	42 (6–136)	118 (107–132)	119 (77–190)	348 (244–446)	191 (139–312)	38 (19–69)	0.42 (0.21–1.37)
Clearwater drainage	578	6.41 (5.43–6.94)	35 (6–127)	121 (113–134)	110 (78–201)	282 (189–349)	188 (138–314)	31 (18–57)	0.42 (0.18–1.39)
Colored drainage	286	6.26 (5.63–6.74)	55 (14–157)	109 (91–118)	130 (70–177)	486 (446–584)	226 (142–309)	65 (31–84)	0.48 (0.27–1.21)
Thin till	361	5.22 (4.80–5.73)	2 (−15–9)	117 (107–131)	70 (51–86)	347 (176–436)	127 (99–146)	26 (18–45)	2.59 (0.55–6.47)
Intermediate till	228	6.48 (6.16–6.75)	57 (30–83)	116 (105–130)	126 (112–139)	360 (289–445)	207 (189–239)	58 (17.6–79.8)	0.40 (0.19–0.74)
Thick till/stratified drift	199	6.94 (6.78–7.15)	200 (130–315)	121 (111–139)	219 (186–308)	408 280–563)	341 (311–459)	74 (39–89)	0.21 (0.11–0.34)
Calcite	76	7.00 (6.88–7.12)	184 (129–218)	121 (110–145)	221 (203–242)	283 (188–316)	369 (324–396)	20 (15–50)	0.19 (0.12–0.26)
Seepage	61	6.66 (5.11–6.93)	108 (−4–151)	102 (56–110)	115 (38–193)	370 (82–295)	218 (82–295)	5 (1.5–6)	0.27 (0.13–0.35)

[a]Median values of population estimates are given with 25 and 75% quartiles in parentheses. All values do not include salt impacted lakes, except for total.

drainage lakes, 864 are not salt impacted (Figure 6.35; Table 6.11). Drainage waters not impacted by salt can be subdivided into clearwater and colored waters on the basis of DOC concentration. There are 578 clearwater drainage lakes in the ELS-I Adirondack subregion and 286 colored drainage lakes. The chemical characteristics of these waters are generally similar; however, colored drainage lakes have somewhat higher ANC, SBC, Ca^{2+}, and H_4SiO_4 and lower SO_4^{2-} than clearwater drainage lakes (Table 6.11).

Processes regulating the acid-base status of clearwater and colored drainage lakes in the ELS-I can be assessed by determining the atmospheric deposition to individual lakes and, through mass balance calculations, quantifying the biogeochemical processes contributing to the observed lake ANC (Figure 6.37; Chapters 1 and 2, Munson and Gherini, this volume). As discussed in the section entitled Patterns of Atmospheric Deposition to the Adirondacks, ELS-I Adirondack lake-watershed systems receive elevated inputs of atmospheric deposition, resulting in a negative ANC input to watersheds. A number of biogeochemical processes influence drainage water chemistry and lake ANC (Chapters 1 and 2, Munson and Gherini, this volume). In both clearwater and colored lake-watershed systems, the release of base cations from watershed processes is the dominant mechanism for the production of ANC. Lake-watershed systems in the ELS-I Adirondack subregion (1A) also show strong retention of NH_4^+ and NO_3^- inputs from atmospheric deposition. Because atmospheric inputs of NO_3^- exceed NH_4^+, retention of inorganic nitrogen results in some net production of ANC (Figure 6.37).

Clearwater ELS-I Adirondack lake-watershed systems appear to be conservative with respect to atmospheric inputs of SO_4^{2-} (Figure 6.37). Fuller et al. (1985) showed that Adirondack soils contained low concentrations of adsorbed SO_4^{2-} and several researchers have suggested that little adsorption of atmospheric SO_4^{2-} deposition occurred in Adirondack soils (Driscoll and Newton 1985, Walker et al. 1990). Application of the model proposed by Kelly et al. (1987) to ELS-I Adirondack lakes (Equation 1) indicates that in-lake retention of SO_4^{2-} should be small (< 5 μeq L^{-1}), due to the relatively short hydrologic residence times (Figure 6.11). Therefore, lake/watershed processes that act to retain SO_4^{2-} are generally not significant in clearwater systems and lake

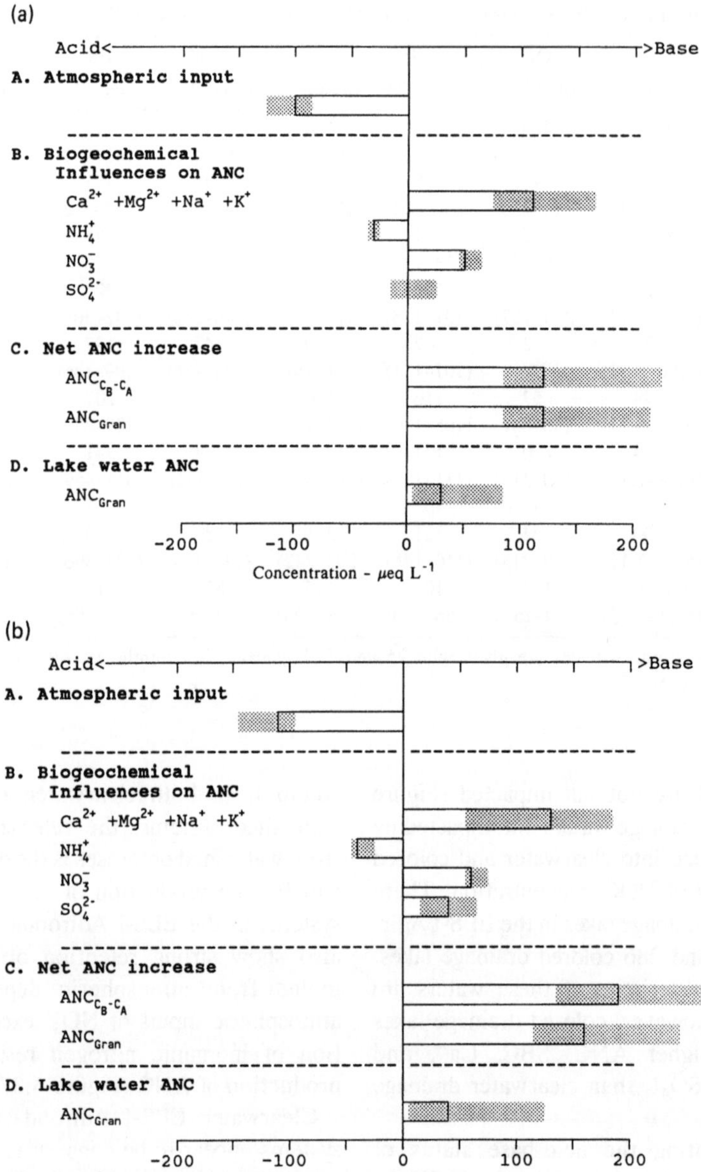

FIGURE 6.37. Modification of atmospheric deposition by watershed/lake biogeochemical processes. A. Precipitation [H^+] adjusted for estimated dry deposition and evapotranspiration (ET). Equivalent to expected lakewater $ANC_{C_B-C_A}$ in the absence of any process other than ET. B. Reactions influencing ANC. Bars to the right indicate ANC production; those to the left show ANC consumption. C. Net ANC production. $ANC_{C_B-C_A}$ is the sum of ANC production terms for measured anions and cations including those shown in B; ANC_{Gran} is determined by difference between input ANC (A.) and lakewater ANC (D.). D. Lakewater ANC represents the net reaction of atmospheric inputs with the lake/watershed system. Shaded bars represent 25% and 75% quartiles. Data are from ELS-I, region 1A. For a further explanation of lake selection, calculation procedures, and figure interpretation, see Appendix A. Figure 6.37a is for clearwater lakes (≤ 417 μmol C L^{-1}). Median DOC = 288.9 μmol C L^{-1} (25% quartile = 189.8 μmol C L^{-1}; 75% quartile = 347.2 μmol C L^{-1}); median A_t = 28.0 μeq L^{-1} (25% quartile = 13.2 μeq L^{-1}; 75% quartile = 46.4 μeq L^{-1}). Figure 6.37b is for colored lakes (> 417 μmol C L^{-1}). Median DOC = 485.4 μmol C L^{-1} (25% quartile = 445.4 μmol C L^{-1}; 75% quartile = 542.0 μmol C L^{-1}); median A_t = 64.7 μeq L^{-1} (25% quartile = 38.3 μeq L^{-1}; 75% quartile = 87.1 μeq L^{-1}).

sensitivity to atmospheric deposition of strong acids is controlled largely by watershed supply of base cations.

Colored Adirondack ELS-I lake-watershed systems (DOC > 417 μmol L^{-1}), like clearwater lake-watershed systems, show production of ANC due largely to the release of base cations (Figure 6.37a). Colored systems also show some net retention of SO_4^{2-}, which probably occurs within wetlands and also serves to increase lake ANC. The greater net release of base cations and retention of SO_4^{2-} in colored lake-watershed systems results in higher production of ANC than clearwater lake-watershed systems (Figure 6.37). However, for colored ELS-I Adirondack waters, the increase in ANC calculated from changes in the concentrations of major solutes is much greater than measured values based on Gran plot analysis. This discrepancy may be attributed to two factors: poor accuracy in Gran titrations of high DOC waters and/or inputs of strong organic acids that depress ANC (Sullivan et al. 1989).

Adirondack drainage lakes sampled in ELS-I can be further subdivided into surficial geology subclasses (based on lake Ca^{2+}). In the target population, there are 361 thin till lakes, representing 42% of the nonsalt impacted drainage lakes (Figure 6.35; Table 6.11). No lakes showing the characteristics of thin till basins (Ca^{2+} < 100 μeq L^{-1}) are salt impacted. As expected, these lakes are characterized by low pH, ANC, SBC, Ca^{2+}, and H_4SiO_4 values and high Al_{ext} concentrations (Table 6.11). Thirty-two percent of the thin till lakes (116 lakes) are colored. These thin till basins are predominantly located in a band along the southwestern portion of Adirondack Park (Figure 6.38a). Of the intermediate till lakes in the ELS-I Adirondack subregion (1A), 228 are not salt impacted. Nineteen salt impacted lakes are classified as intermediate till lakes, which have moderate ANC values (median ANC 57 μeq L^{-1}) and therefore are susceptible to episodic acidification. Thirty-four percent of the intermediate till waters are colored (DOC > 417 μmol C L^{-1}). Intermediate till waters are located generally as a band northeast of the thin till watersheds diagonally across Adirondack Park (Figure 6.38a). Very few intermediate till lakes are located outside Adirondack Park or the Adirondack ecological zone.

Insensitive lakes include those waters with thick glacial till and stratified drift or significant carbonate deposits within the basin. These lakes are scattered throughout the ELS-I Adirondack subregion (Figure 6.38a). In ELS-I Adirondack subregion basins containing thick till and stratified drift, 216 lakes are not impacted by salt. A relatively high percentage (39%) of the lakes in this class are salt impacted. As expected, thick till and stratified drift lakes are characterized by higher ANC, SBC, and Ca^{2+} than thin or intermediate till systems (Table 6.11). Thirty-one percent (67 lakes) of the thick till and stratified drift lakes are colored. Lakes with carbonate deposits within the watershed are also insensitive to strong acid inputs. These waters tend to have very high ANC, SBC, and Ca^{2+} relative to other ELS-I Adirondack lake types (Table 6.11). However, in these basins, H_4SiO_4 concentrations are generally lower than in either intermediate till or thick till and stratified drift systems (Table 6.11). There are 288 carbonate basins in the ELS-I Adirondack subregion (Figure 6.35), generally located outside Adirondack Park, along Lake Champlain and within the marble belt in the central Adirondacks (Figure 6.38b). Most of the calcite lakes are salt impacted (74%); only 59 calcite lakes are not salt impacted. Of the calcite lakes not impacted by salt, 45% are colored.

Classification of Adirondack ELS-I Seepage Lakes

According to the designation by hydrologic lake type, 8% of the lakes in the ELS-I Adirondack subregion are seepage lakes (Brakke et al. 1988). Few differences are obvious in comparisons of the chemical characteristics of seepage lakes with the characteristics of drainage lakes (Table 6.11). The only significant difference is that H_4SiO_4 concentrations were considerably lower in seepage lakes than in drainage lakes. Because of the limited number sampled (population estimate 100 lakes) in the ELS-I Adirondack subregion, it is difficult to make accurate statements concerning the chemical characteristics of seepage lakes. This problem is compounded in attempts to subdivide seepage lakes into lake classes (Figure 6.35). The ALSC survey has shown that seepage lakes are an important component of the aquatic resources in the Adirondacks (14% of the population of lakes > 0.5 ha in the Adirondack ecological zone; Kretser et al. 1989). The low number of seepage lakes in the ELS-I Adirondack sub-

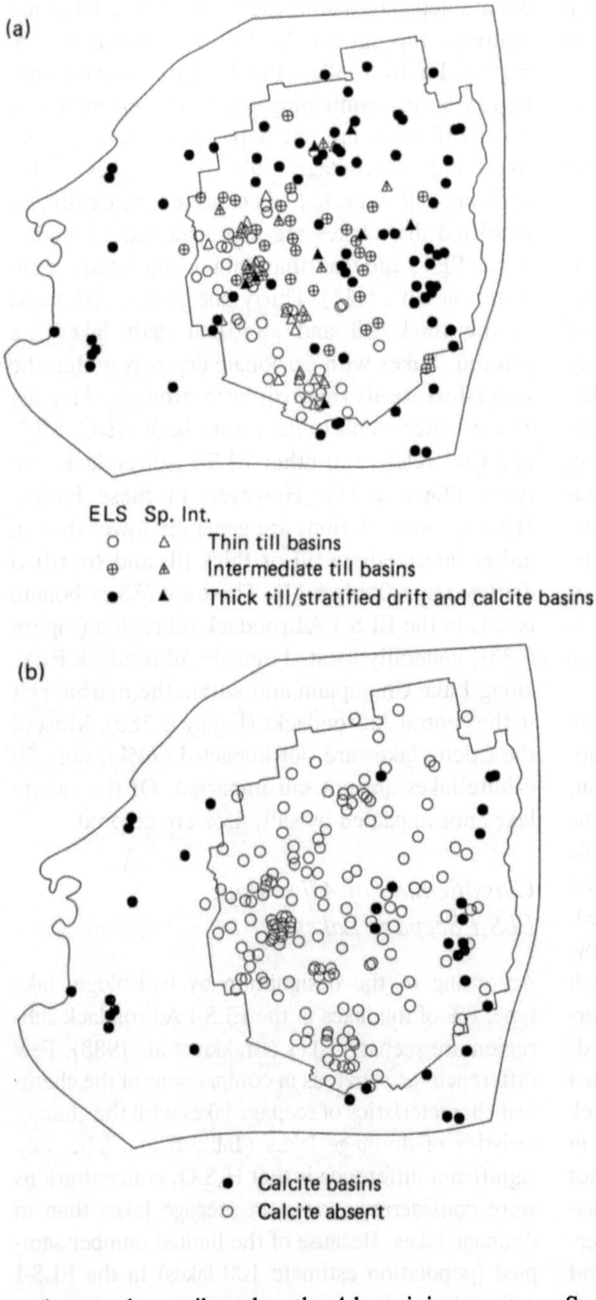

FIGURE 6.38. Location of ELS-I Adirondack drainage lakes by classes of (a) surficial geology and (b) calcite basins. Salt impacted lakes have not been excluded from this figure.

region can be attributed to the 4 ha minimum surface area limitation of lakes sampled.

Evidence of Historical Trends

Introduction

Since studies of acidic deposition were first initiated in the early 1970s, there has been interest in the onset and trends in acidic deposition, and the subsequent acidification of Adirondack waters. Several monitoring studies have collected data to directly assess recent acidification trends. These investigations are relatively short term (< 20 years) and address changes in precipitation, lakewater chemistry, and fisheries.

Given the absence of studies prior to 1970 designed to directly assess acidification, plus the relatively poor analytical methods available for analysis of dilute waters in historical surveys of water quality, several alternative approaches have been used to quantify long-term acidification

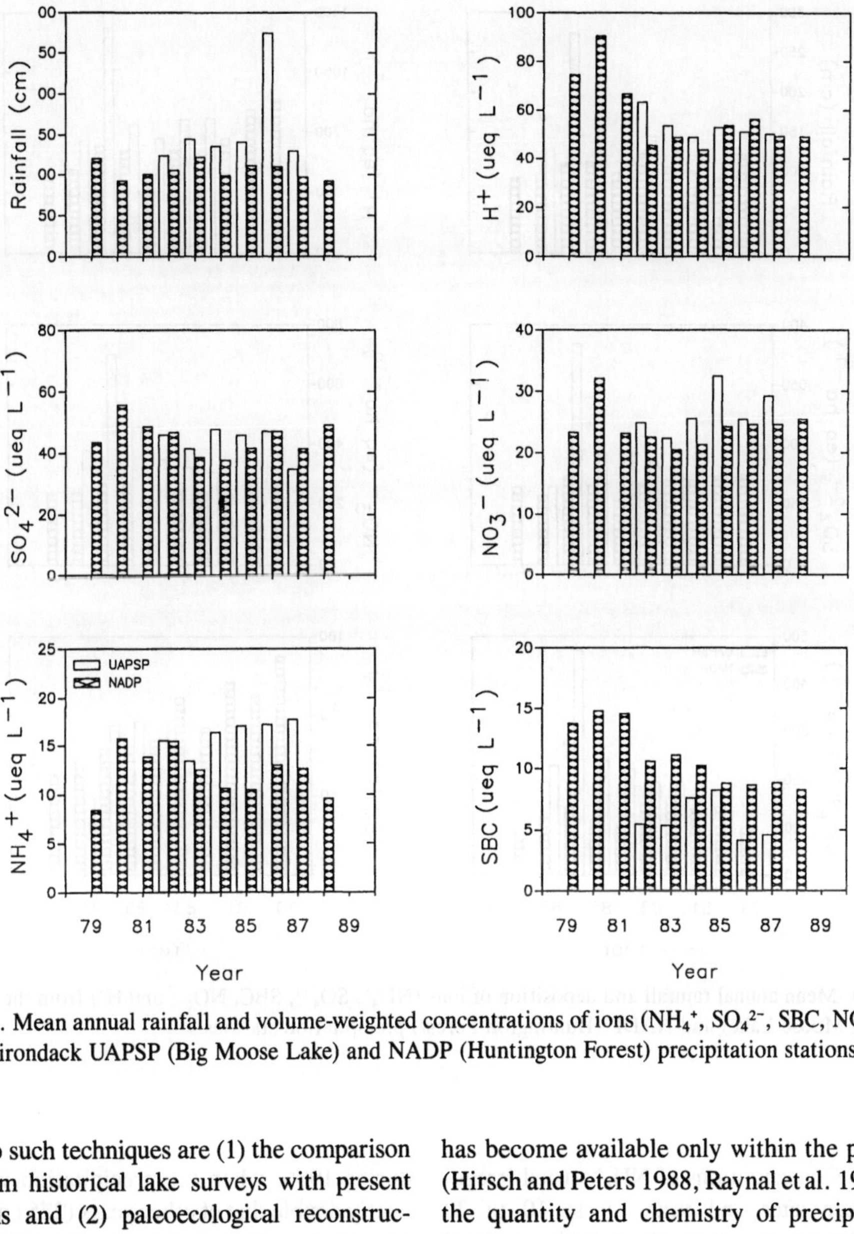

FIGURE 6.39. Mean annual rainfall and volume-weighted concentrations of ions (NH_4^+, SO_4^{2-}, SBC, NO_3^-, and H^+) from the Adirondack UAPSP (Big Moose Lake) and NADP (Huntington Forest) precipitation stations.

trends. Two such techniques are (1) the comparison of data from historical lake surveys with present observations and (2) paleoecological reconstructions of acidification from lake sediment records. Summaries of both short- and long-term studies are presented within this section, giving an overall representation of the temporal trends of acidification for the Adirondack region.

Trends in Precipitation Chemistry in the Adirondack Region

Background

A detailed and uninterrupted 10-year record of precipitation chemistry for the Adirondack region has become available only within the past decade (Hirsch and Peters 1988, Raynal et al. 1987). Since the quantity and chemistry of precipitation are highly variable in the Adirondacks (Figure 6.3), time series trends are not easily resolved with the relatively short records.

UAPSP and NADP Precipitation Stations

The annual volume-weighted concentrations of acidic anions (SO_4^{2-} and NO_3^-) and NH_4^+ from the UAPSP station at Big Moose Lake and the NADP station at Huntington Forest did not exhibit discernible trends over a 10-year period (Figure 6.39). Sulfate and NO_3^- concentrations have remained steady at about 50 and 25 μeq L^{-1}, respectively,

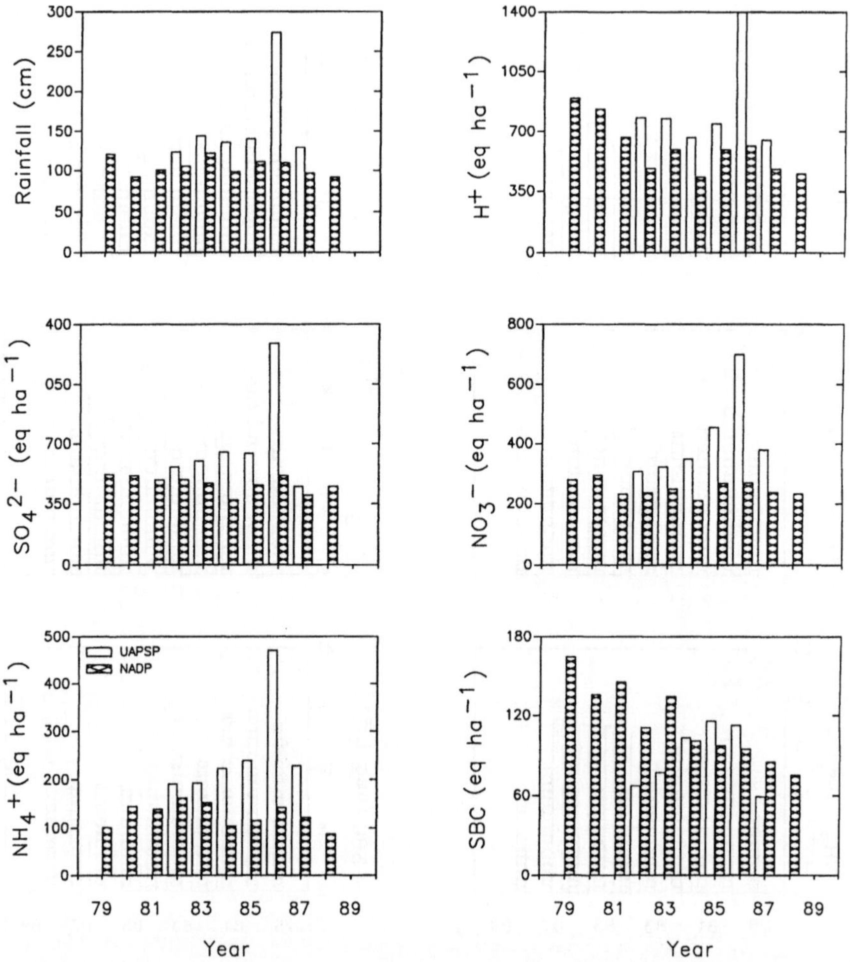

FIGURE 6.40. Mean annual rainfall and deposition of ions (NH_4^+, SO_4^{2-}, SBC, NO_3^-, and H^+) from the Adirondack UAPSP (Big Moose Lake) and NADP (Huntington Forest) precipitation stations.

since 1979. Concentrations of NH_4^+ varied irregularly between sites and years, from 10 to 20 μeq L^{-1}. Concentrations of H^+ at both sites and SBC at the NADP site appear to have declined approximately 25% from 1979 to 1982, remaining relatively constant since 1982. Whether the decline in these two parameters is of statistical significance is currently unresolved (Raynal et al. 1987).

Variations in annual deposition of ions to the UAPSP station at Big Moose Lake and the NADP station at Huntington Forest were controlled primarily by the quantity of precipitation (Figure 6.40). The highest recorded deposition of H^+ (1,400 eq ha^{-1} yr^{-1}) occurred at Big Moose Lake during 1986, when the precipitation volume was nearly double that of other years (275 cm). Deposition of H^+ appeared to be declining at both sites from ~900 eq ha^{-1} yr^{-1} in 1979 to 500 eq ha^{-1} yr^{-1} in 1988, accentuated by the decline in precipitation volume. Base cation trends also declined at the NADP site, from 170 eq ha^{-1} yr^{-1} in 1979 to 80 eq ha^{-1} yr^{-1} in 1988. Base cation deposition trends at the UAPSP station differed from those at the NADP site, ranging from 60 to 120 eq ha^{-1} yr^{-1}. As would be expected from the invariant temporal concentrations of SO_4^{2-} and NO_3^-, deposition rates varied linearly with bulk precipitation rates (Figure 6.40).

Long-Term Synoptic Lake Survey Data

Background

Synoptic surveys of Adirondack lakes have been conducted periodically since the late 1920s and early 1930s (see section entitled Previous Surveys of Adirondack Lake Chemistry). Comparisons between paired historical and contemporary observations of individual lakes provide potentially the best direct evidence for recent (post-industrial revolution) acidification trends in the Adirondacks. Unfortunately, there are two major limitations to this type of comparison. First, the methods for analyzing chemical parameters (e.g., pH and ANC) have improved over the past 60 years, leading to concerns regarding the precision and accuracy of historical data. Second, in most synoptic surveys, only a single surface sample was collected. Hence, within-lake spatial, climatic, and limnological variations are not readily accounted for in comparisons of different surveys. The effects of acidic deposition must then be well pronounced in order to distinguish them from normal ranges of variation.

Comparison of the 1929–34 Biological Surveys with Contemporary Lake Data

During 1929–34, the New York State Conservation Department surveyed a number of lakes within the Adirondacks, reporting methyl orange ANC values. A number of investigators have compared these early water chemistry observations with more recent surveys (Schofield 1976c, Kramer et al. 1986, Asbury et al. 1989). In the most recent analysis, Asbury et al. (1989) determined the bias in methyl orange endpoint relative to modern methods determined by Gran plot analysis (see section on Previous Surveys of Adirondack Lake Chemistry). They matched the ANC values for 274 of the lakes surveyed during 1929–34 with similar analyses from contemporary surveys (1975–85). The authors reported a median loss of 50 μeq L^{-1} ANC from the paired analyses, with 80% of the lakes showing a decline in ANC (Figure 6.41). This is consistent with the findings of Kramer et al. (1986), who reported a median ANC loss of between 0 and 69 μeq L^{-1}, comparing 1929–34 data with the data of Pfeiffer and Festa (1980) and Colquhoun et al. (1984). Higher elevation lakes were most affected. This pattern might be expected, as those waters generally exhibit the lowest ANC (Figure 6.21) and the highest inputs of atmospheric deposition in the Adirondacks.

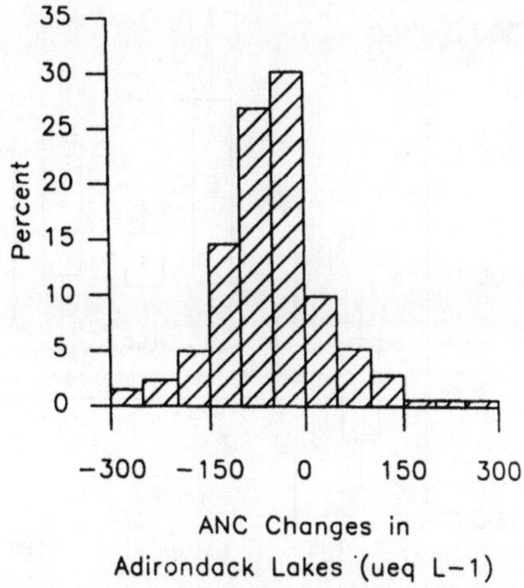

FIGURE 6.41. Frequency distribution of changes in ANC occurring between historical (1929–34) and modern (1975–85) surveys of 269 Adirondack lakes (after Asbury et. al. 1989). Historical ANC values were corrected for 54.6 μeq L^{-1} bias associated with methyl orange endpoint (Asbury et al. 1989).

Comparison of the Schofield (1975) Survey with the ALSC Survey

Another comparison of Adirondack lake synoptic surveys was conducted by Schofield (unpublished data). Lake chemistry data collected from 110 high-elevation lakes not impacted by salt in 1975 (Schofield 1976a; discussed in Previous Surveys of Adirondack Lake Chemistry) were compared to observations from the same lakes from the ALSC survey (1984–87). Although this period (about 10 years) is short, the comparison includes measurements of a large number of chemical parameters made with generally similar analytical techniques.

Generally, no change since 1975 was evident in ANC (mean change ± std. dev. = 4.6 ± 22 μeq L^{-1}) and pH for the study lakes (Figure 6.42). However, SBC and SO_4^{2-} concentrations appeared to have declined through the same period (mean changes −26 ± 34 and −58 ± 72 μeq L^{-1}, respectively). Sulfate was not directly measured during the early survey (Schofield 1976a), but was estimated by SBC − ANC − NO_3^- − Cl^-. Patterns of

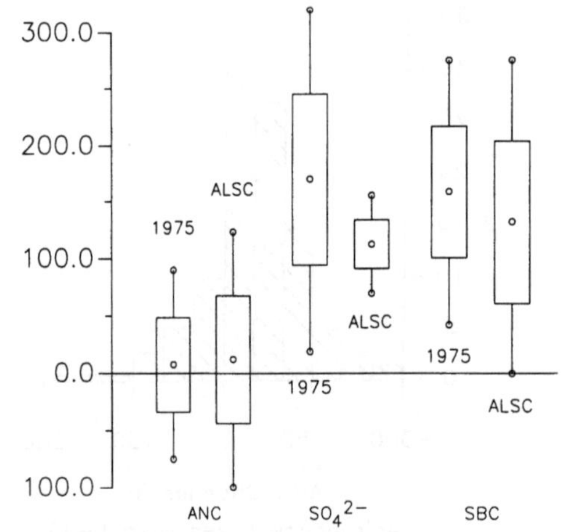

FIGURE 6.42. Changes in ANC, Ca^{2+}, and SO_4^{2-} occurring between the Schofield (1976a) and ALSC (1984–87) surveys (Kretser et al. 1989) for 101 high-elevation (> 610 m) Adirondack lakes (Schofield unpublished data). The line indicates ± 2 std. dev. The box encloses ± 1 std. dev. and the 0 designates the mean value.

declining lake SO_4^{2-} and SBC are qualitatively consistent throughout the Adirondacks (Figure 6.43). Apparent declines in lake SO_4^{2-} parallel decreases in precipitation SO_4^{2-} reported in the northeastern United States in the last 20 to 25 years (Hedin et al. 1987, Driscoll et al. 1989d) and declines in SO_2 emissions (Figure 3.2, Chapter 3, Husar et al., this volume). The decrease in lake SO_4^{2-} appears to have coincided with an equivalent decline in SBC, resulting in little change in lake ANC. Similar findings have been reported for the Hubbard Brook Experimental Forest in New Hampshire (Driscoll et al. 1989d). Most Adirondack lakes also have shown no change or some increase in ANC. Some lakes, largely in the western Adirondacks, have shown recent loss of ANC > 10 μeq L^{-1} (Figure 6.43).

Other Water Quality Trends

Schofield (1972, 1976a) investigated long-term changes in Secchi disk transparency of Adirondack lakes. He noted an increase in transparency coinciding with decreases in pH in 14 lakes since the early 1950s (Schofield 1972) and in two lakes since the late 1930s (Schofield 1976a). In these lakes, the transparency has been most pronounced since the 1950s. Effler et al. (1985) studied the factors responsible for changes in light attenuation in an Adirondack lake. They reported that light was largely attenuated by absorption, due to organic solutes, and noted that short-term increases in transparency were due to loss of DOC from the water column, mediated by coagulation/hydrolysis of Al. One mechanism that may explain the apparent increase in transparency of acidic Adirondack lakes is strong acid leaching of Al from watersheds, followed by hydrolysis/deposition of Al within lakes and associated coagulation of organic solutes and particulates.

Time Series Analysis of Surface Water Data

Long-term chemical records are available for a few sites in the Adirondack region. For example, weekly measurements of pH, ANC, hardness (Ca^{2+} and Mg^{2+}), and Cl^- are available for Hinckley Reservoir on West Canada Creek (1944–75), and monthly observations of pH, ANC, and major solutes are available for the USGS station at the East Branch of the Sacandaga River (1966–75). Unfortunately, these monitoring sites have large drainage areas and solute concentrations are high relative to well-defined headwater systems. As a result, trends for these sites may be difficult to interpret. Regular monitoring of individual Adirondack lakes did not start until the early 1980s with the Adirondack Long-Term Monitoring program (ALTM; Table 6.1). As such, analyses of recent time series of lake data suffer from the same general limitation as precipitation chemistry; the record is not long enough to establish strong trends. However, monthly data from the ALTM program do provide a qualitative indication of changes in Adirondack lake chemistry.

Long-Term Time Series Analysis

Schofield (1976b) summarized the trends for Hinckley Reservoir and the East Branch of the Sacandaga River. Both sites showed a decline in ANC over the period of record. At Hinckley Reservoir, the large decline in ANC (4.6 μeq L^{-1} yr^{-1}) closely coincided with a decline in hardness (5.6 μeq L^{-1} yr^{-1}). In addition to the long-term declines, a cycle of changes in ANC and hardness that may be due to climatic variations was evident. Hard-

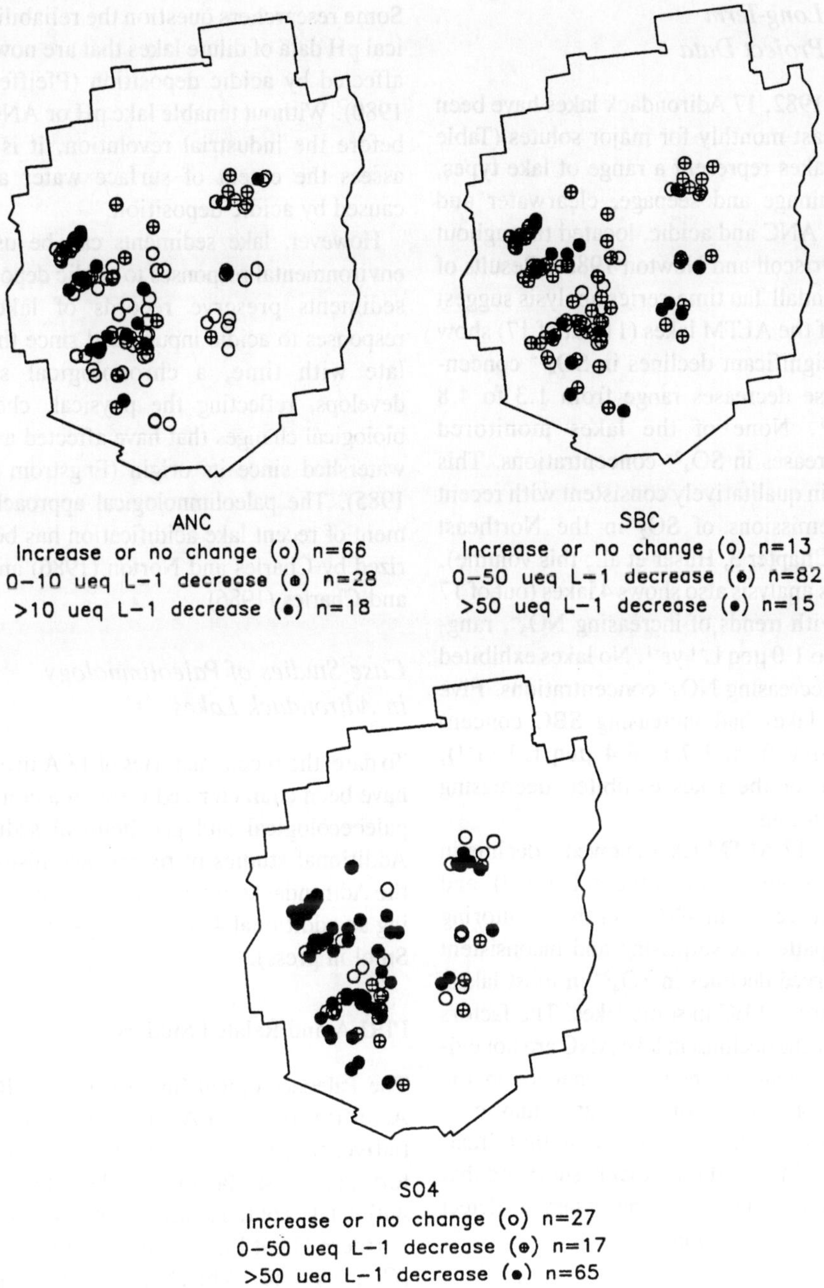

FIGURE 6.43. Geographic distribution of changes in (a) ANC, (b) SBC, and (c) SO_4^{2-} for Adirondack lakes sampled by Schofield (1976a) in 1975 and resampled by the ALSC (1984–87) survey (Kretser et al. 1989).

ness and ANC declined during relatively wet periods in the 1940s and 1970s and increased during extreme drought in the 1960s.

The decline in ANC at the East Branch of the Sacandaga River coincided with an equivalent increase in NO_3^-. Over the record, a decrease in SO_4^{2-} occurred, which was balanced by an equivalent decline in SBC. Schofield (1976b) suggested that increases in NO_3^- were responsible for the decline in ANC at the site.

Adirondack Long-Term Monitoring Project Data

Beginning in 1982, 17 Adirondack lakes have been sampled at least monthly for major solutes (Table 6.1). These lakes represent a range of lake types, including drainage and seepage, clearwater and colored, high ANC and acidic, located throughout the region (Driscoll and Newton 1985). Results of a seasonal Kendall Tau time-series analysis suggest that several of the ALTM lakes (11 out of 17) show statistically significant declines in SO_4^{2-} concentrations. These decreases range from 1.3 to 4.8 $\mu eq\ L^{-1}\ yr^{-1}$. None of the lakes monitored exhibited increases in SO_4^{2-} concentrations. This pattern is again qualitatively consistent with recent declines in emissions of SO_2 in the Northeast (Figure 3.2, Chapter 3, Husar et al., this volume).

Time-series analysis also shows 4 lakes (out of 17 monitored) with trends of increasing NO_3^-, ranging from 0.2 to 1.0 $\mu eq\ L^{-1}\ yr^{-1}$. No lakes exhibited a pattern of decreasing NO_3^- concentrations. Five of the study lakes had increasing SBC concentrations (ranging from 1.7 to 8.4 $\mu eq\ L^{-1}\ yr^{-1}$), whereas none of the lakes exhibited decreasing SBC concentrations.

Seven of the 17 ALTM lakes showed a decline in ANC (ranging from -1.4 to 5.9 $\mu eq\ L^{-1}\ yr^{-1}$), and only one lake increased in ANC over the monitoring period. This pattern is surprising and inconsistent with the observed declines in SO_4^{2-} in most lakes, and the increase in SBC in some lakes. The factors responsible for the declines in lake ANC are not evident from this analysis; however, changes in the concentration of solutes not considered may contribute (e.g., DOC, Al, Cl^-). It is important to realize that the ALTM record is relatively short and that decades of monitoring may be necessary to detect real changes in water chemistry.

Paleoecological Evidence of Recent Acidification of Adirondack Lakes

Background

Combustion of fossil fuels resulting in acidic deposition has occurred since the onset of the industrial revolution (Husar 1986). But an accurate methodology for measuring the pH of low ionic strength lakes has been available only for the last 30 years. Some researchers question the reliability of historical pH data of dilute lakes that are now apparently affected by acidic deposition (Pfeiffer and Festa 1980). Without tenable lake pH or ANC data from before the industrial revolution, it is difficult to assess the extent of surface water acidification caused by acidic deposition.

However, lake sediments can be used to infer environmental responses to acidic deposition. Lake sediments preserve records of lake/watershed responses to acidic inputs, and since they accumulate with time, a chronological stratigraphy develops, reflecting the physical, chemical, and biological changes that have affected a lake and its watershed since its origin (Engstrom and Wright 1985). The paleolimnological approach to assessment of recent lake acidification has been summarized by Charles and Norton (1986) and Battarbee and Charles (1986).

Case Studies of Paleolimnology in Adirondack Lakes

To date, the recent histories of 18 Adirondack lakes have been characterized through a combination of paleoecological and geochemical sediment data. Additional studies of recent lake histories within the Adirondacks are currently in progress, addressing an additional 40 or more systems (Charles and Smol in press).

PIRLA and Related Studies

The Paleoecological Investigation of Recent Lake Acidification (PIRLA) study was a large-scale initiative, designed to determine the acidification history and causal factors of lakewater acidification within lakes of the Adirondacks, New England, the northern Great Lakes states, and northern Florida (Charles and Whitehead 1986). Specifically, PIRLA addressed the impacts of acidic deposition upon 12 primary lakes within the Adirondacks, using 52 additional Adirondack lakes for model calibration.

The PIRLA objectives were met through a combination of paleolimnological techniques, focusing on the analysis of lake sediment cores. Sediment assemblages of diatom and chrysophyte remains were initially characterized within the surficial sediments of the 52 calibration lakes to determine

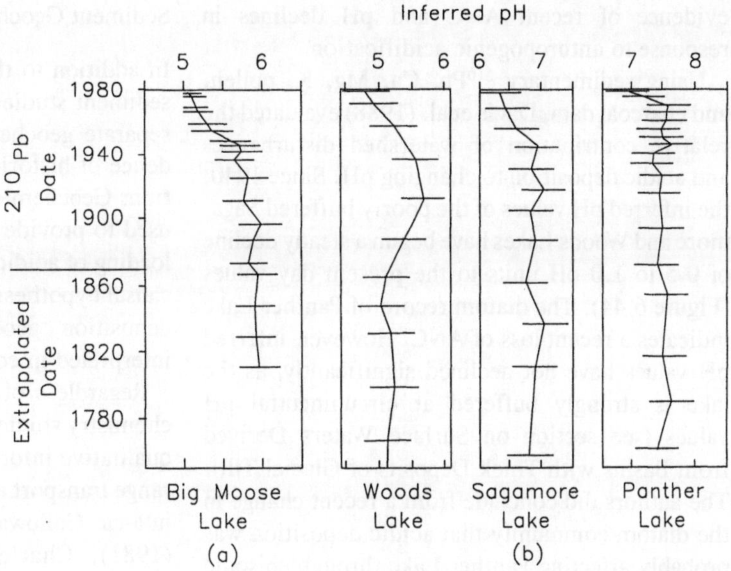

FIGURE 6.44. Diatom-inferred pH trends from (a) Big Moose Lake, and (b) Woods, Sagamore, and Panther Lakes (after Davis et al. 1988), using three different estimation techniques from a single sediment core (after Charles 1984).

their characteristics under current lake conditions. A calibrated model relating sediment assemblages to lakewater chemistry (principally pH) was then developed from the calibration lakes and applied to down-core sediment assemblages. This approach yielded an inference of the historical lakewater pH trends for the 12 primary study lakes in the Adirondacks. Additional analyses, including inorganic elements, sulfur species and isotopes, polycyclic aromatic hydrocarbons, and coal and oil soot were performed upon each core (Chapter 3, Husar et al., this volume). The discrete depth intervals of the lake sediment cores were dated primarily by the analysis of the naturally occurring ^{210}Pb isotope; pollen and charcoal stratigaphies were used to correlate and substantiate the ^{210}Pb dating relationships.

The sediment record of 8 of the 12 PIRLA Adirondack study lakes showed conclusive evidence of a recent (< 100 yr) decline in lakewater pH and ANC values to < 5.5 and 10 μeq L^{-1}, respectively. The PIRLA study concluded that the declines in pH and ANC within the affected lakes were caused primarily by atmospheric deposition of strong acids from the combustion of fossil fuels, though other factors, such as watershed disturbances, may have contributed. Researchers noted correlations between the sediment record of the deposition of combustion by-products (e.g., Pb, Cu, V, Zn, S, PAH, magnetic particles, coal and oil soot) and diatom inferred pH values, which support a linkage between the recent changes in lake chemistry and the increased deposition of atmospheric pollutants.

Big Moose Lake has received considerable attention from PIRLA and other investigators, due to its accessibility, documentation of minimal watershed disturbance, and clarity of the paleoecological and geochemical sediment record (Charles et al. 1987, Charles et al. 1990b). Big Moose Lake has shown signs of recent acidification (since 1950 to present), with an estimated decrease in pH of 0.8 to 1.0 units (Figure 6.44). Diatom reconstructions of lakewater chemistry are consistent with the few colorimetric pH observations made in the 1930s and by current pH values determined potentiometrically (Charles 1984). The full suite of PIRLA analyses indicated that the decline in pH within Big Moose Lake was accompanied by increases in the deposition of combustion by-products (Charles et al. 1987), suggesting that the cause for the recent acidification of Big Moose was atmospheric deposition of strong acids.

Other Diatom Initiatives

Two-hundred-year histories of three Adirondack lakes were carefully described by Davis et al. (1988), using diatom reconstruction techniques. Woods, Sagamore, and Panther Lakes (current pH 4.9, 6.0, and 7.2, respectively), the focus of a number of acidification studies, exhibited evidence of recent ANC and pH declines in response to anthropogenic acidification.

Using sedimentary ^{210}Pb, Ca, Mg, K, pollen, and charcoal data, Davis et al. (1988) evaluated the relative contribution of watershed disturbances and acidic deposition to changing pH. Since 1940, the inferred pH values of the poorly buffered Sagamore and Woods Lakes have begun a steady decline of 0.5 to 1.0 pH units to the present day values (Figure 6.44). The diatom record of Panther Lake indicates a recent loss of ANC. However, inferred pH values have not declined significantly, as the lake is strongly buffered at circumneutral pH values (see section on Surface Waters Derived from Basins with Thick Deposits of Glacial Till). The authors did conclude from a recent change in the diatom community that acidic deposition was probably affecting Panther Lake through episodic pH depressions during the spring snowmelt. The work of Davis et al. (1988) was in agreement with a previous diatom study of Woods, Sagamore, and Panther Lakes associated with the Integrated Lake-Watershed Acidification Study (ILWAS; EPRI 1984). Del Prete and Galloway (1983) were also able to provide conclusive evidence that Woods Lake had recently acidified, but could not substantiate the cause of acidification.

Del Prete and Schofield (1981) and Charles and Norton (1986) used diatom data to infer the historical change in pH for other Adirondack lakes. In Honnedaga Lake (current lakewater pH 4.7 to 4.8), the diatom inferred pH values showed a statistically significant decrease from a background value (pre-1800) of 6.1 to a surface sediment value of 5.2 (Del Prete and Schofield 1981, Charles and Norton 1986). Seventh and Woodhull Lakes (current lakewater pH 6.4 to 7.0 and 5.1 to 5.3, respectively) did not show significant trends in diatom inferred pH through the period of record (~200 years). Charles et al. (1989) used diatom analysis of Deep Lake to disclose a sharp recent (1950–70) decrease in pH from 5.0 to 4.3. Charles et al. (1989) also determined by the same technique that the water of Lake Arnold (current pH 4.5) appeared to have been acidic for some time (pre-1800), but has shown signs of recent increases in the concentration of soluble Al, presumably mobilized from the watershed by inputs of strong acids.

Sediment Geochemical Trends

In addition to the integrated diatom/geochemical sediment studies within the Adirondacks, several separate geochemical studies have provided evidence of historical trends of atmospheric deposition. Geochemical sediment data alone cannot be used to provide an accurate reconstruction of the loading of acidic substances to lakes. To support a causal hypothesis regarding recent trends in acidic deposition, geochemical sediment data must be interpreted in concert with diatom analyses.

Regardless of this shortcoming, sediment geochemistry studies within Adirondack lakes provide qualitative information on the nature of the long-range transport and deposition of atmospheric pollutants. Galloway and Likens (1979), Heit et al. (1981), Charles and Norton (1986), Gubala (1988), and many other researchers have provided evidence for increased deposition of atmospheric pollutants (e.g., trace metals and PAHs) into remote lakes within the Adirondacks. The results derived from these studies routinely conclude that combustion by-products, derived from distant sources, have significantly affected remote lakes within the Adirondacks. However, because of the extreme complexity of mechanisms that govern the ultimate retention of pollutants within lake sediments, a consensus of precise fluxes for most constituents has not yet been achieved.

Fisheries Trends in Adirondack Waters

Several studies of long-terms trends in fish communities in Adirondack lakes have been published (Haines and Baker 1986). These studies have all concluded that fishery declines have occurred in at least some Adirondack lakes and that these fish population losses are generally associated with low present-day pH values and/or pH declines. However, the specific number of fish populations lost, or the number of lakes with fish communities affected by acidification, cannot be determined with confidence due to the lack of historical fisheries data from many Adirondack waters and uncertainties as to specific causes for fishery declines in individual lakes. No analyses of long-term trends are available for fish communities in Adirondack streams.

Schofield (1976a) provided the first evidence for loss of Adirondack fish communities in relation to lake acidity. Forty high-elevation (> 600 m) Adirondack lakes surveyed in the 1930s were resurveyed in 1975. In the 1930s, no fish were caught in only four of these lakes, whereas no fish were caught in 21 lakes in 1975. Thus, 17 of the 40 lakes (42%) had lost their fish communities over the 40-year period. The majority of the lakes in which no fish were caught in 1975 (19 of the 21 lakes) had pH < 5.0. Baker and Harvey (1984) extended this analysis by compiling all available historical fisheries data for Adirondack lakes, collected over the period from the 1920s through 1983. Changes in fish communities over time, and possible explanations for observed trends, were evaluated without references to lake identity or water chemistry data. For those lakes with sufficient survey data, ratings were assigned that defined (1) the strength of the evidence suggesting a fish population decline or loss and (2) the likelihood that any observed decline or loss resulted from factors other than acidification (e.g., shifts in stocking policy, lake reclamation, introductions of competing species, habitat alteration, or potential fish sampling problems). Lakes rated as having lost one or more populations of fish without any obvious alternative explanation (other than acidification) had significantly ($p < - 0.05$) lower present-day pH values than did lakes without fish population declines (Baker and Harvey 1984, Haines and Baker 1986).

Eight of the lakes evaluated for fishery trends by Baker and Harvey (1984) have also been assessed independently using paleocological techniques and sediment diatoms to estimate long-term trends in lake pH. In general, the results from fishery trend analyses and paleoecological analyses are consistent. Those lakes that have lost fish populations without obvious alternative explanations except acidification (5 of the 8 lakes) also have experienced trends in sediment diatoms indicative of lake acidification (Haines and Baker 1986).

Schofield and Driscoll (1987) provided a detailed chronology of changes in fish species distribution for lakes in the North Branch of the Moose River watershed, starting with surveys conducted as early as 1882 and including comprehensive surveys of fish communities in the watershed conducted in 1982–83 as part of the RILWAS project. Dramatic changes in the composition of fish communities inhabiting these lakes first became apparent during the 1940s, following a long period of relative stability from 1882 to 1931. Recruitment failure and declines in lake trout (*Salvalinus namaycush*) condition factors were documented for lake trout populations in Big Moose Lake during the late 1940s and 1950s. Paleoecological analyses of sediment diatoms in Big Moose Lake suggest a rapid decline in lake pH during that time period from pH about 5.8 prior to 1950 to pH < 5.0 in recent years (Charles et al. 1987). The sensitivity of lake trout populations to acidification was confirmed during the experimental acidification of Lake 223 in Ontario; lake trout recruitment failure occurred at pH values 5.0 to 5.6 (Mills et al. 1987). Many minnow and darter species, also expected to be relatively sensitive to acidity (Mills and Schindler 1986), have likewise disappeared from acidic lakes in the North Branch of the Moose River watershed. In contrast, acid-tolerant species such as yellow perch (*Perca flavescens*), central mudminnow (*Umbra limi*), and banded killifish (*Fundulus diaphanus*) are widely distributed in the basin and are typically the dominant fish species in acidic lakes.

Future Considerations

Simulations of Changes in Lake Chemistry Following Decreases in Atmospheric Deposition Using the ILWAS Model

Computer models can be used to simulate the potential response of water chemistry to changes in atmospheric deposition. The ILWAS model, a surface water acidification model, was developed to predict changes in surface water acidity that occur following changes in acidic atmospheric deposition (Goldstein et al. 1984, Gherini et al. 1985). The ILWAS model has been applied to four lake/watershed systems in the Adirondacks (Munson et al. 1987, Davis et al. 1987, Gherini et al. 1989). These include the acidic lake/watershed systems (Woods Lake and Big Moose Lake), as well as higher ANC lakes (Clear Pond and Panther Lake). The model can be used to evaluate the processes responsible for the current lake chemistry. Munson et al. (1987) reported that ILWAS model simulations indicate that cation exchange reactions in soil contribute most of the supply of base cations to both Woods Lake and Panther Lake. The release of base cations by cation exchange occurs to a greater extent in Panther Lake basin because of deeper surficial deposits, which provide a large pool of material rich in base cations and which have a relatively long hydrologic residence time for these reactions to proceed. The importance of cation exchange to the base cation supply rate suggests that future decreases in ANC could occur, even without increases in acidic deposition, if the supply of base cations in a watershed were depleted. The model can be used to test this hypothesis. Simulations, however, suggest that the pool of base cations is large enough to prevent significant changes in Panther Lake ANC under current loading of atmospheric deposition, and any changes in Woods Lake ANC would probably be insignificant because the water is acidic.

An intriguing application of the model involves simulating the effects of changes in acidic deposition to lake/watershed systems. Munson et al. (1987) simulated the response of Woods Lake and Panther Lake basins to a 50% reduction in total sulfur deposition. They found that the largest response occurs at Woods Lake, where pH values increase by as much as a full unit during low flow periods. During other times of the year, the change in lake pH due to decreased sulfur deposition was small (Figure 2.22, Chapter 2, Munson and Gherini, this volume). Overall, pH patterns become more dynamic, as observed for Panther Lake. At Panther Lake, reducing the sulfur load by 50% has very little effect on lake pH (Figure 2.22, Chapter 2, Munson and Gherini, this volume). This response may be attributed, in part, to the higher initial ANC of Panther Lake and the strong pH buffering by the inorganic carbon system (Chapters 1 and 2, Munson and Gherini, this volume). Despite an increase in ANC due to the reduction in sulfur loading, Panther Lake outlet still exhibited episodic acidification in the spring. The episodic acidification is due to the dilution of base cations coupled with an increase in NO_3^- during the spring runoff (see section entitled Extensive Evaluation of Episodic Lake Acidification During Snowmelt).

Unlike the other three lakes, Clear Pond is located in the eastcentral Adirondacks (Table 6.1; Figure 6.2) and receives a lower deposition load than the other watersheds (Gherini et al. 1989). In addition, the mineralogical characteristics of the bedrock and soils are different due to the variations in the underlying bedrock. At Clear Pond, the anorthosite bedrock is characterized by an abundance in calcium-rich feldspars, which weather at a faster rate than the sodium-rich feldspars in the western Adirondacks. The amount of the watershed covered by thick till is similar to that of Panther Lake (33% at Clear Pond, 36% at Panther Lake). Gherini et al. (1989) reported that the ILWAS model predicted that a 50% reduction in total sulfur deposition caused lake outlet SO_4^{2-} to decrease by 30 to 35 μeq L^{-1} over the simulation period. The ANC, however, increased by < 20 μeq L^{-1} because the reduction in sulfur deposition resulted in a decrease in the mobilization of base cations from the terrestrial environment. The change in pH at Clear Pond was < 0.1 unit.

Gherini et al. (1989) also evaluated a scenario of a 100% increase in total sulfur deposition. Under these conditions, the ANC in Clear Pond decreased by about 25 μeq L^{-1} and the maximum change in pH was < 0.5 unit. These simulations suggest that internal buffering processes make this watershed relatively insensitive to changes in atmospheric deposition.

Gherini et al. (1989) simulated the response of Big Moose Lake to changes in atmospheric deposition. Big Moose Lake is very different from the other systems studied. It is not a small headwater lake, but rather a large lake within a 96 km^2 watershed. There are over 20 lakes and ponds and 70 km of stream channel upstream from the lake. It was necessary to consider all the processes occurring in these subsystems in order to make accurate predictions in Big Moose Lake. The ILWAS model was run using a 4-year simulation period. Decreasing the total sulfur loading by 50% caused SO_4^{2-} in Big Moose Lake to decrease from 135 μeq L^{-1} to 80 μeq L^{-1}. The increase in ANC, however, was < 20 μeq L^{-1} and the pH increase was 0.1 unit during the spring melt and 0.5 unit during baseflow periods. Concentrations of Al_{ext} decreased from 50 μeq L^{-1} to 30 μeq L^{-1} during the spring melt. It was evident from the trend in chemical constituents over time that a 4-year simulation was not adequate for the system to achieve steady state. However, from these simulations, it appears that

Big Moose Lake would be more affected by decreases in total sulfur deposition than either Panther Lake or Clear Pond.

Model simulations are consistent with the case studies discussed previously (see Case Studies of Adirondack Lakes) as thin till dominated watersheds are most sensitive to changes in atmospheric deposition, whereas the watersheds dominated by thick till and stratified drift are relatively insensitive. Model results also suggest that changes in total sulfur deposition will increase the minimum ANC associated with episodic acidification, which commonly occurs in the spring in lakes. However, the change in ANC that occurs from baseflow to high flow conditions should remain approximately the same, as episodic acidification is largely due to dilution of SBC and/or increases in NO_3^- (see section entitled Extensive Evaluation of Episodic Lake Acidification during Snowmelt).

Biological Status of the Adirondack Region

Fish Communities in the Adirondacks

The status of fish communities in Adirondack waters with regard to potential effects of acidic deposition has been extensively studied, compared to other regions in the United States. Fish communities in 1,469 Adirondack lakes were surveyed by the ALSC during 1984–87 (Kretser et al. 1989). Based on these and earlier surveys of Adirondack fish communities, regional models of fish population response to lake acidification have been proposed (Reckhow 1987, Reckhow et al. 1987, Baker et al. 1988, Christensen et al. 1988). Intensive site-specific studies of fish population dynamics and effects of lake acidity have also been conducted as components of several major projects, including RILWAS (Schofield and Driscoll 1987), the Aluminum in Streams Study (ALSS; Elwood et al. 1986), and liming experiments in LAMP (Gloss et al. 1989) and the Extensive Liming Study (Schofield et al. 1986). Information on fish in Adirondack streams is more limited. Colquhoun et al. (1981) presented survey data for 42 streams, and Schofield and Driscoll (1987) for 24 streams within the North Branch of the Moose River watershed. Fish bioassays have been conducted in Adirondack lakes and streams by Johnson et al. (1987) and Schofield and Driscoll (1987). Results from each of these studies are reviewed in the following sections.

Regional Surveys

Fifty-three fish species were collected in ALSC surveys of Adirondack lakes (Figure 6.45) (Kretser et al. 1989). The most commonly caught species, in order of frequency of occurrence, were as follows: brown bullhead (*Ictalurus nebulosus*; n = 841 lakes), brook trout (*Salvelinus fontinalis*; n = 579 lakes), white sucker (*Castostomus commersoni*; n = 570 lakes), golden shiner (*Notemigonus crysoleucas*; n = 566 lakes), pumpkinseed sunfish (*Lepomis gibbosus*; n = 502 lakes), creek chub (*Semotilus atromaculatus*; n = 438 lakes), yellow perch (n = 346 lakes), northern redbelly dace (*Phoxinus eos*; n = 218 lakes), and common shiner (*Notropis cornutus*; n = 156 lakes) (Figure 6.45). Brook trout are often the only game species resident in the many small headwater lakes, considered most susceptible to potential effects from acidic deposition (Pfeiffer and Festa 1980). Other important game species, caught in at least 40 of the ALSC lakes, include lake trout, brown trout (*Salmo trutta*), northern pike (*Esox lucius*), largemouth bass (*Micropterus salmoides*), smallmouth bass (*Micropterus dolomieui*) and rock bass (*Ambloplites rupestris*).

The distribution of fish species in ALSC lakes in relation to lake pH is illustrated in Figure 6.45. Four fish species occur in lakes with pH $\leq$ 4.5: central mudminnow, brown bullhead, golden shiner, and yellow perch. Eleven additional species were caught in waters with pH $\leq$ 5.0: pumpkinseed sunfish, brown trout, rock bass, white sucker, brook trout, creek chub, largemouth bass, northern redbelly dace, lake chub (*Couesius plumbeus*), common shiner, and chain pickerel (*Esox niger*). Thus, many of the fish species occurring in Adirondack waters appear quite tolerant of acidic conditions.

No fish were caught in 346 lakes (24% of the lakes surveyed); 61% of the lakes in which no fish were caught occur in the Oswegathchie-Black watershed in the southwestern portion of the Adirondacks (Figures 6.1 and 6.46). Lakes without fish accounted for 7% of the lake area surveyed. Lakes without fish were significantly smaller and shallower and occurred at higher elevations than lakes in which fish were caught. In addition, fishless

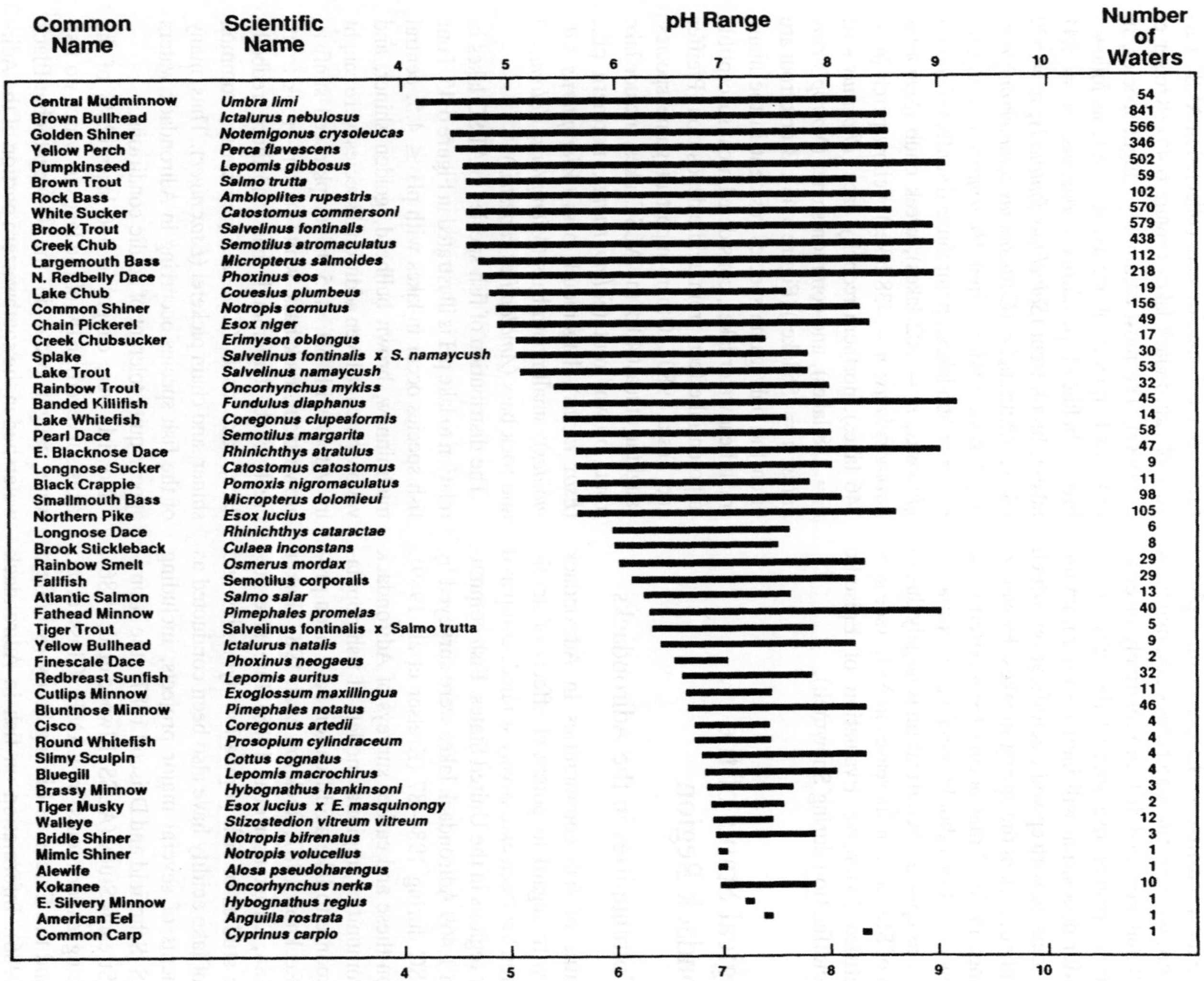

FIGURE 6.45. pH ranges, based on spring and fall chemistry samples, for fish species collected in waters surveyed by the ALSC. (After Kretser et al. 1989.)

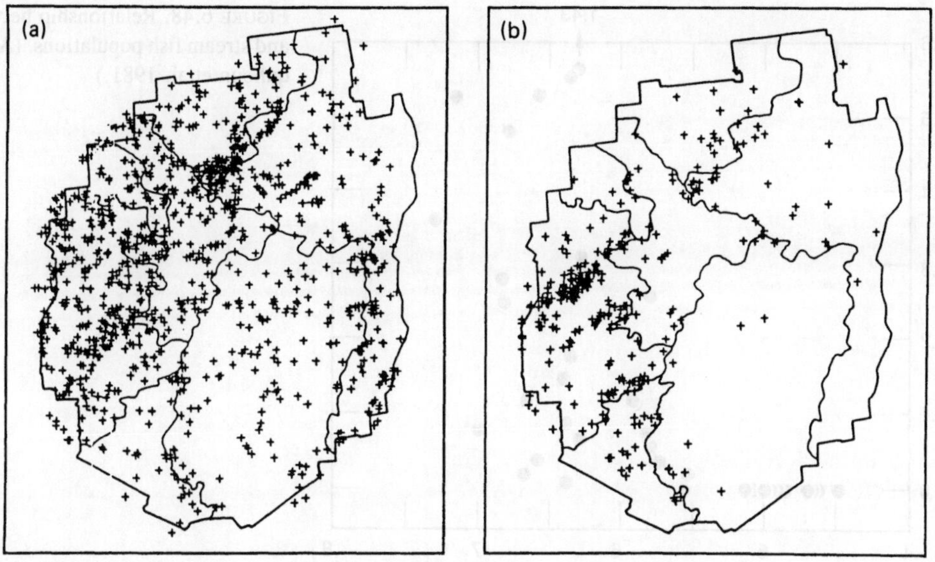

FIGURE 6.46. Geographic distribution of Adirondack waters surveyed by the ALSC in which fish were (a) captured ($n = 1123$) and (b) not captured ($n = 346$). (After Kretser et al. 1989.)

lakes had significantly lower pH, lower ANC, lower Ca^{2+}, higher total Al, and lower concentrations of O_2 than did lakes with fish. Thus, specific causes for the absence of fish are difficult to identify, although as noted in several prior surveys of Adirondack lakes (Schofield 1976d, Colquhoun et al. 1984), the absence of fish was significantly correlated with low pH waters.

The number of fish species caught per lake (species richness) increased with increasing lake pH (Figure 6.47). It is expected that species richness is significantly correlated with other lake characteristics (e.g., lake area and lake type), as observed in surveys for other regions (Rago and Wiener 1986, Haines et al. 1986, Cusimano et al. 1989). Such analyses have not yet been conducted, however, for the ALSC data set.

Colquhoun et al. (1981) and Schofield and Driscoll (1987) observed similar results for Adirondack streams. Numbers of fish caught per unit area were correlated with the minimum pH recorded in spring 1980 in the 42 streams surveyed by Colquhoun et al. (1981) (Figure 6.48). No fish were caught in streams with a minimum pH ≤ 5.2. Possible relationships between fish catch and other stream characteristics were not investigated, however. Schofield and Driscoll (1987) reported a strong correlation ($r = 0.84$) between species richness and baseflow stream pH, but no association ($r = 0.18$; $p > 0.05$) between species richness and stream size, in surveys of streams in the North Branch of the Moose River watershed. Central mudminnow and banded killifish were caught more

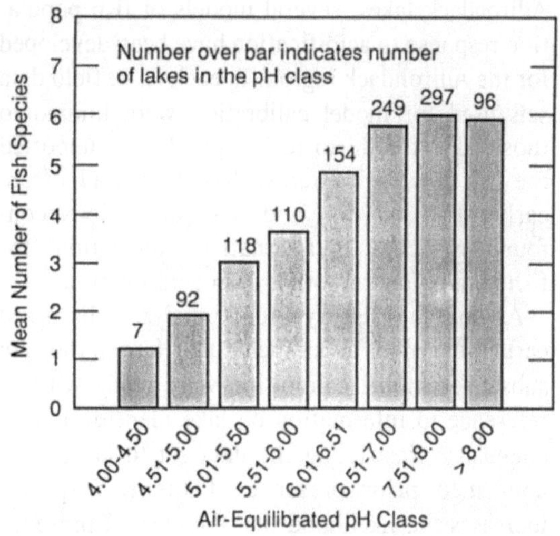

FIGURE 6.47. Distribution of mean number of fish species captured, by air-equilibrated pH class (pH measured at the time of fisheries survey, spring or fall). (After Kretser et al. 1989.)

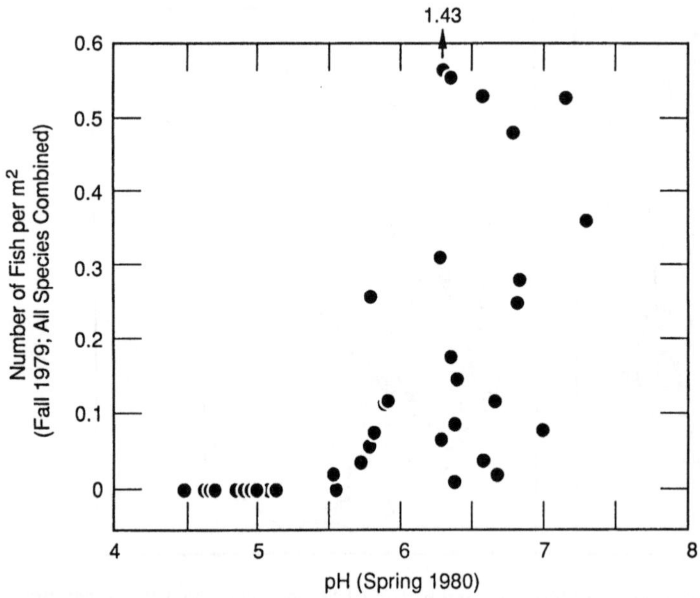

FIGURE 6.48. Relationship between pH and stream fish populations. (After Colquhoun et al. 1981.)

frequently in streams with pH < 6.0 than in streams with pH > 6.0. Minnow species, especially blacknose dace (*Rhinichthys atratulus*), were caught more frequently in streams with pH > 6.0 than in streams with pH < 6.0.

Regional Models of Fish Population Response

Based on the ALSC and other survey data sets for Adirondack lakes, several models of fish population response to acidification have been developed for the Adirondack region. In each case, field data sets used for model calibration were limited to those lakes that historically (pre-1970) supported the fish species of interest, based on results from earlier fish surveys, and without obvious confounding factors or alternative explanations for fish population loss other than acidification.

As for the fishery trend analyses discussed earlier, decisions about including a lake in the data subset for model calibration were made without reference to information on lake identity or lake chemistry. Restricting the data set to lakes with confirmed prior presence of the fish species increases the likelihood that the current presence or absence (defined by catch/no catch) of the species reflects the fish species tolerance to acidity. These models are discussed in greater detail in Chapter 4 (Baker and Christensen, this volume).

Empirical fish response models relate fish species presence/absence to lake chemistry and other characteristics based on logistic regression analysis of field survey data. Reckhow et al. (1987) proposed models for two fish species (brook trout and lake trout) using survey data for Adirondack lakes collected prior to 1984. Baker et al. (1988) have updated these analyses using data collected by the ALSC during the first two years of the ALSC survey (1984 and 1985; $n = 827$ lakes total). Models were developed for brook trout ($n = 193$ lakes with confirmed presence of the species in the past and no obvious confounding factors), white sucker ($n = 170$ lakes), and brown bullhead ($n = 214$ lakes). Lake trout models were developed based on data for Ontario lakes ($n = 192$) and tested with data for Adirondack lakes ($n = 19$) (Baker and Christensen 1990).

A number of lake characteristics were evaluated as potential predictor variables in these empirical fish response models, including pH, ANC, Ca^{2+}, measured total Al, an estimate of Al_{im}, DOC, lake area, depth, and elevation. For brook trout and brown bullhead, based on ALSC fish surveys and summer chemistry data, only lake pH was significant ($p \leq 0.05$) and was selected using both forward and backward stepwise logistic regression analysis:

$$P_{\text{brook trout}} = \frac{1}{1 + \exp[8.62 - 1.76(\text{pH})]}$$

$$P_{\text{brook bullhead}} = \frac{1}{1 + \exp[4.08 - 1.04(\text{pH})]}$$

FIGURE 6.49. Predicted probability of presence of (a) brook trout and (b) brown bullhead, based on a logistic regression model using summer measurements of lake pH in the ALSC survey.

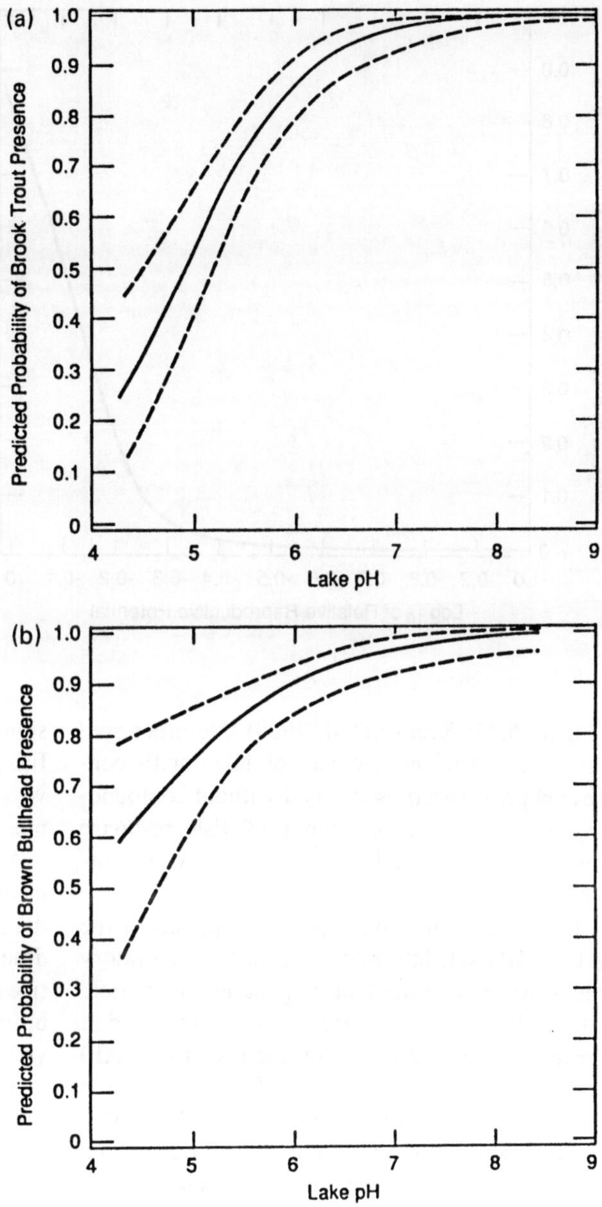

where: P is the predicted probability that the fish species is present in the lake.

The forms of the fitted models for brook trout and brown bullhead are presented in Figure 6.49. As noted above, both of these species are relatively acid tolerant, occurring in lakes with pH as low as 4.49 and 4.64 for brown bullhead and brook trout, respectively (Figure 6.45). This acid tolerance is reflected in the relatively moderate decline in the predicted probability of fish presence at pH values 5.0 to 6.0. Models for white sucker were somewhat more complex, with three variables identified as significant ($p \leq 0.05$) in logistic regression analyses:

$$P_{\text{white sucker}} = \frac{1}{1 + \exp[0.09 - 2.94(\text{pH}) - 3.96(\log \text{DOC}) + 7.99(\log \text{ANC})]}$$

White suckers are also fairly tolerant of low pH, occurring down to pH 4.64 in ALSC surveys

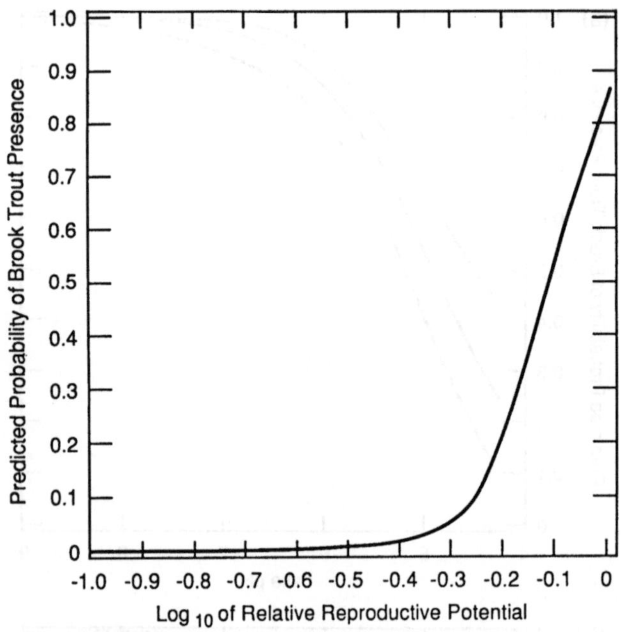

FIGURE 6.50. Predicted probability of brook trout presence as a function of relative recruitment potential from a preliminary calibration of the LAF framework.

(Figure 6.45; Kretser et al. 1989). No other species had sufficient numbers of lakes with confirmed prior fish presence and without confounding factors for development of fish response models based on the first two years of ALSC survey data (Baker et al. 1988).

In a second modeling approach applied to the ALSC data set, information from fish laboratory bioassays was utilized in conjunction with field survey data. A preliminary version of this model, termed the Lake Acidification and Fisheries (LAF) framework, has been developed for brook trout in Adirondack lakes (Christensen et al. 1988, Baker et al. 1988; also see Chapter 4, Baker and Christensen, this volume, for further details on the model structure). Conditional mortality rates as a function of water chemistry are defined for specific fish life stages based on laboratory bioassays, testing a full matrix of combinations of pH, Ca^{2+}, and Al_{im} (Ingersoll 1986, Mount et al. 1989). Predicted effects on individual life stages are integrated within an age-structured, life-cycle model to estimate relative recruitment potential for a population of brook trout exposed to given values of pH, Ca^{2+}, and Al_{im}. The full framework is then "calibrated" for prediction of fish population response in the field (i.e., predicted probability of brook trout presence) using logistic regression analysis and the correlation between observed fish presence/absence (catch/no catch) in field surveys and the estimated brook trout relative recruitment potential for a given fish population (lake).

A preliminary calibration of the LAF framework has been performed using the same subset of ALSC data used for brook trout empirical models described above (n = 193 lakes). Estimates of relative recruitment potential and the predicted probability of fish presence from the LAF framework vary as a function of three parameters (pH, Ca^{2+}, and Al_{im}), making visual presentation of the model results difficult. The association between relative recruitment potential and the predicted probability of brook trout presence is presented in Figure 6.50. These results should be considered preliminary, however, especially due to uncertainties in the estimates of Al_{im} for ALSC lakes (Baker et al. 1988). Only total Al and total dissolved Al were measured in ALSC surveys, whereas the LAF framework requires a value of Al_{im}. In the final application of the LAF framework to brook trout, a number of calibrations will be performed. These will involve consideration of such factors as (1) the appropriate form and means of estimating Al_{im}, (2) the relationship between index water chemistry and the water chemistry to which sensitive life stages are exposed, (3) the expected degree of

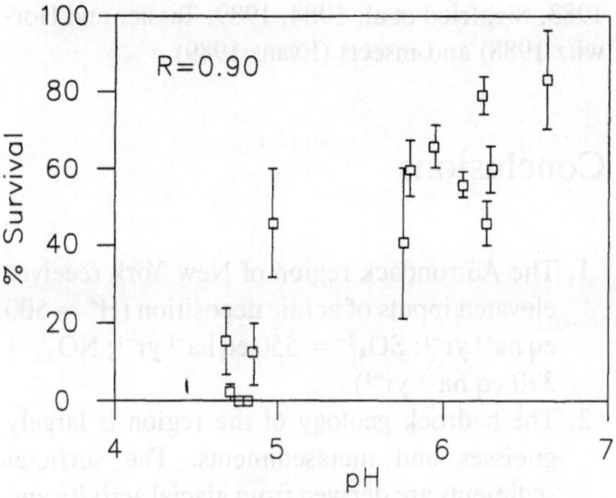

FIGURE 6.51. First year annual survival of 1983 and 1984 year classes of brook trout in lakes included in the Extensive Liming Study, in relation to mean epilimnetic pH in April to May. (After Schofield et al. 1986.)

exposure of different life stages to potentially harmful chemical conditions, and (4) the potential for cumulative effects on growth and fecundity as fish grow older.

Eventually, these and other fish response models are to be applied on a regional scale, using regional chemistry data sets such as the ELS, to estimate numbers of lakes with levels of acidity unsuitable for fish population survival. Further model testing and refinement are needed, however, before such regional assessments can be conducted with confidence.

Intensive Field Studies

Intensive studies of fish communities in selected Adirondack lakes and streams provide insight into the mechanisms of fish population response to acidification. Lake liming experiments, followed by gradual lake reacidification, in particular, afford an opportunity to examine changes in fish population status and condition over time with changes in lake chemistry.

As part of the Extensive Liming Study, 10 small Adirondack lakes were limed in 1983/1984; 3 of these 10 lakes had reacidified by January 1986 (Schofield et al. 1986). Annual brook trout survival rates were significantly correlated with the mean epilimnetic pH during spring turnover (Figure 6.51). Marked population declines followed major

TABLE 6.12. Apparent species tolerance to acidification, based on observed distribution in lakes and streams of the north branch of the Moose River.[a]

Sensitive	Indeterminant	Tolerant
Blacknose dace	Brook trout	Mud minnow
Redbellied dace	White sucker	Killifish
Common shiner	Creek chub	Brown bullhead
Smallmouth bass	Pumpkinseed sunfish	Golden shiner
Rock bass	Finescale dace	Yellow perch

[a] After Schofield and Driscoll, 1987.

hydrologic events that resulted in complete water column reacidification. The toxicity of these episodic incursions of acidic waters was confirmed with in situ bioassays (Schofield et al. 1986). More intensive investigations of two limed lakes as part of the LAMP study indicated, however, that a substantial proportion of this population decline in reacidified waters (25% to 50%) resulted from trout emigration from the lake rather than direct mortality (Gloss et al. 1989).

Decreased fish growth and condition also occurred in some lakes after fish exposure to low over-winter pH values (Schofield et al. 1986). Detailed studies of fish population feeding and growth efficiency in the two LAMP lakes suggest that this decrease results largely from physiological acid stress rather than reductions in food availability (Schofield et al. 1989).

Field bioassays demonstrate the direct toxic effects of surface water chemistry and associated environmental conditions (e.g., temperature) on fish, and these may be used to evaluate relative species sensitivity as well as differences in toxic effects among sites and over time. Based on field surveys, Schofield and Driscoll (1987) proposed a tentative classification of species tolerance to acidification (Table 6.12). "Sensitive" species were collected only at sites with pH > 6.0. "Tolerant" species occurred at sites with pH < 5.0, and "indeterminate" species were caught in both low and high pH waters. This classification was then tested using in situ bioassays, transferring representative species in each category from high to low pH sites within the North Branch of the Moose River watershed. Species in the sensitive (blacknose dace, redbellied dace, and common shiner) and indeterminate (creek chub) groups exhibited high mortality rates when exposed to waters with pH < 5.0. Johnson et al. (1987) also observed high mortality for blacknose dace and creek chub adults and fry tested in situ bioassays in Adirondack waters. Species in the tolerant group (central mudminnow, banded killifish, and yellow perch), on the other hand, experienced relatively low mortality in waters with pH < 5.0. Johnson et al. (1987) conducted toxicity tests at 10 streams and 6 lake sites within the North Branch of the Moose River water-shed. Differences in toxicity among sites and among species (brook trout, lake trout, blacknose dace and creek chub) were consistent with observed patterns of fish distribution in the watershed. Early feeding fry and hatching were identified as the most critical stages for fish survival in acidic waters.

Fish population responses to acidification are complex. The specific mechanisms of effects of fish populations have not yet been defined and are likely to vary among surface waters. Studies in Adirondack waters have contributed the following, however, to our understanding of the biological effects of acidification:

- The importance of Al_{im} as an additional toxic factor in acidic waters (Schofield and Trojnar 1980, Driscoll et al. 1980).
- The role of behavioral avoidance of acidic waters as an important factor influencing the effects of acidification on fish populations (Johnson and Webster 1977, Gloss et al. 1989).
- The importance of relatively small-scale spatial and temporal variations in water chemistry in determining the effects of acidification on fish populations (Schofield et al. 1986, Gloss et al. 1989).
- The effects of acidification on adult fish survival, growth, and production (Schofield et al. 1989).

Other Aquatic Organisms in the Adirondacks

Many studies have been conducted on aquatic organisms other than fish in the Adirondacks (Sutherland 1989). These include investigations on bacteria (Boylen et al. 1983), algae (Conway et al. 1983, Singer et al. 1984, Smol et al. 1984b, Charles 1985, Roberts and Boylen 1988, 1989, Silver 1988, Bukaveckas 1988), macrophytes (Roberts et al. 1985, Bukaveckas 1989), zooplankton (Confer et al. 1983, Siegfried et al. 1984, 1989, Tessier and Horwitz 1988) and insects (Evans 1989).

Conclusions

1. The Adirondack region of New York receives elevated inputs of acidic deposition (H^+ = 500 eq ha^{-1} yr^{-1}; SO_4^{2-} = 550 eq ha^{-1} yr^{-1}; NO_3^- = 370 eq ha^{-1} yr^{-1}).
2. The bedrock geology of the region is largely gneisses and metasediments. The surficial sediments are derived from glacial activity and include glacial till of varying thickness. At lower elevations, glacial outwash is evident, composed of stratified sand and gravel. Soils are generally derived from glacial till and are acidic Spodosols. Rock and soil in the Adirondacks are commonly composed of materials that weather slowly and provide a low supply of base cations.
3. A large number of lakes exists in the Adirondack region. The exact number, as well as the physical and chemical characteristics, is a function of assumptions made on the minimum lake size considered and the boundary of the Adirondack region. The ELS-I estimated 1,290 lakes > 4 ha in surface area for the Adirondack subregion (1A). Subregion 1A extends beyond the boundary identified as the Adirondack ecological zone. There are 1,076 ELS-I lakes in the Adirondack ecological zone. If lakes > 1 ha are considered for the ELS-I subregion 1A, the population estimate increases to 5,657.
4. Adirondack lakes are characterized by small size and short hydrologic residence time. Drainage lakes are the predominant hydrologic lake type (77% of population).
5. There are an estimated 179 acidic lakes (ANC ≤ 0 μeq L^{-1}) in the ELS-I Adirondack subregion (14% of the population). From the Adirondack Lake Survey Corporation survey, which was a more comprehensive study of lakes (> 0.5 ha) in the Adirondack ecological zone, more lakes (26%) were found to be acidic. These waters are acidic due to high concentrations of strong acid anions (largely SO_4^{2-}) relative to concentrations of base cations. Acidic

drainage lakes show elevated concentrations of Al_{ext}.

6. Variations in lake ANC in the Adirondacks is largely attributed to differences in the supply of base cations to drainage waters. Concentrations of SO_4^{2-} are relatively uniform across the region.
7. ELS-I Adirondack subregion waters typically have low concentrations of DOC. Organic acids are not a major source of acidity to low ANC waters.
8. There are distinct geographic patterns in lake chemistry across the Adirondacks. Acidic lakes are generally located at high elevations, in basins with shallow deposits of glacial till in the southwestern area of Adirondack Park. Thin surficial sediments limit the supply of base cations, making waters sensitive to strong acid inputs. In the northeastern portion of the Park, lakes generally are found at lower elevations in watersheds with thick till or stratified drift. Moreover, calcium-rich anorthosite underlies much of the eastern Adirondacks. These characteristics help supply base cations to lake water and increase ANC. Many lakes along Lake Champlain and outside Adirondack Park are impacted by salt and have very high ANC.
9. Episodic acidification associated with hydrologic events is an important process in the Adirondacks. During spring snowmelt, acidification occurs in streams, the upper 2 m of lakes, and lake outlet water. Short-term loss of ANC is due largely to dilution of base cations in high baseflow ANC waters and NO_3^- increases in low baseflow ANC waters. The magnitude of ANC loss during snow melt is most pronounced (60 to 100 μeq L^{-1}) in intermediate ANC waters (80 to 120 μeq L^{-1}).
10. Because of shallow depth and short hydrologic residence time, in-lake processes are generally unimportant in regulating the acid-base status of Adirondack drainage lakes.
11. Most of the SO_4^{2-} in Adirondack lakes can be attributed to atmospheric deposition. Internal sources of SO_4^{2-} are generally insignificant except for salt impacted lakes.
12. It is evident from process-level studies that hydrologic flow paths are the most significant factor regulating the acid-base status of Adirondack waters. A system was developed to classify Adirondack waters on the basis of hydrologic and chemical characteristics. Twenty-eight percent of the ELS-I Adirondack waters were considered salt impacted. The drainage lakes (including drainage, closed, and reservoir hydrologic types) were subdivided on the basis of geologic characteristics. Thin till basins include lakes that could be acidic under current atmospheric deposition (28% of the ELS-I Adirondack lake population). This class generally coincides with the high-elevation lakes in the southwestern portion of Adirondack Park. Intermediate till basins (18% of population) correspond to those waters with baseflow ANC < 100 μeq L^{-1}. These waters are susceptible to episodic acidification (> 50 μeq L^{-1}) during spring snowmelt. Intermediate till basins are generally located as a band from the northwest to the southeast across the center of Adirondack Park. Lakes with basins containing thick glacial till or stratified drift (17% of population) have higher ANC (generally > 100 μeq L^{-1}). These lakes tend to be located in the northeastern area of Adirondack Park, although some thick till basins can be found in the southwestern zone. There are some very high ANC lakes that appear to have deposits of carbonate minerals within their watersheds. These are located in the marble belt in the central Adirondacks, and they also coincide with the salt impacted lakes along Lake Champlain and outside Adirondack Park. Seepage lakes are an important resource in the Adirondacks, representing about 14% of the ponded waters. However, because of their small size they were not adequately characterized in the ELS-I.
13. There is considerable uncertainty in the extent and factors contributing to long-term changes in the acid-base status of Adirondack surface waters. Unfortunately, there are few direct measurements of changes in the ANC or major water chemistry parameters. Observations from historical lakewater quality surveys, time series analysis from monitoring sites, historical fisheries observations, and diatom and geochemical paleoecological data from sediment cores all suggest acidification of some Adirondack waters. The timing of significant

loss of ANC is uncertain, but available data suggests that for some waters acidification occurred between 1930 and 1960. The extent to which this trend is due to increased atmospheric deposition of SO_4^{2-} is not clear, as there is limited evidence that changes in base cations and NO_3^- have occurred and may have contributed to the apparent loss of ANC. Data over the last 20 years suggest a decline in surface water SO_4^{2-}, which is consistent with declines in emissions of SO_2 in the northeastern United States, but large changes in ANC have not occurred.

14. Simulations from the ILWAS model suggest that there is considerable lake-to-lake variation in response to reduction in SO_4^{2-}. Four lakes studied showed a decrease in lake SO_4^{2-} following a 50% reduction in sulfur deposition; however, the changes in pH and ANC were variable, depending on hydrologic and biogeochemical processes occurring within individual basins. Changes in lake pH were most pronounced during low flow summer conditions. Reductions in sulfur loading also increased high flow ANC. However, the change in ANC from baseflow to high flow conditions remained nearly constant as this acidification is largely associated with dilution of SBC and/or increases in NO_3^-.
15. Fifty-three species of fish have been collected in Adirondack lakes from surveys conducted by the ALSC. Brook trout are the only game fish generally caught in small headwater lakes in the region and, therefore, are considered the most susceptible to surface water acidification. Four fish species were found in lakes with pH ≤ 4.5 and 11 additional species were found in waters with pH values ≤ 5.0. Thus, several species of fish occurring in Adirondack waters are tolerant of acidic conditions. However, the number of species caught per lake (species richness) decreases with decreasing pH in Adirondack waters.
16. No fish were caught in 24% of 1,469 Adirondack lakes (1984–86) surveyed by the ALSC. These waters were generally located in the southwestern Adirondacks. Lakes without fish were significantly smaller and shallower and occurred at higher elevations than lakes in which fish were caught. In addition, fishless lakes had significantly lower pH, ANC, and Ca^{2+}, higher Al, and lower O_2 concentrations than lakes with fish. The specific reasons for absence of fish are difficult to identify, although the absence of fish was significantly correlated with low pH waters.

Acknowledgments. We appreciate the help of D. Schaefer, S. Clemence, J. Friedman, H. Booth, D. Long, D. Leopold, J. Gallager, J. Shepard, R. Munson, and S. Gherini in this analysis. Special thanks is extended to R. Munson, K. Roy, W. Kretser, and C. Schofield, who critically reviewed this chapter. We would like to acknowledge precipitation data sets from the Utilities Acid Precipitation Studies Program and the National Acidic Deposition Program (D. Raynal). Support for this analysis was provided in part by the U.S. Environmental Protection Agency (EPA) through the Regional Case Studies Project. S.W. Christensen's contribution was also sponsored by the EPA under Interagency Agreement DW89932112-01 with the U.S. Department of Energy under contract No. DE-AC05-84OR21400 with Martin Marietta Systems, Inc. Finally, we would like to thank the various funding agencies who provided support for the research in Adirondacks discussed in this chapter (Electric Power Research Institute, Empire State Electric Energy Research Corporation, USEPA, New York State Department of Environmental Conservation, U.S. Fish and Wildlife Service, National Science Foundation, and others) and all the researchers who collected data to make this analysis possible. Publication No. 99 of the Upstate Freshwater Institute and No. 3529 of the Environmental Sciences Division, Oak Ridge National Laboratory.

7 Maine

Jeffrey S. Kahl, Stephen A. Norton, Christopher S. Cronan, Ivan J. Fernandez, Linda C. Bacon, and Terry A. Haines

ABSTRACT. The chemical imprint of acidic deposition is evident as increased SO_4^{2-} concentrations in virtually all Maine surface waters, but the effect on acidity status is small. Fewer than 100 nondystrophic lakes are currently estimated to be acidic (ANC ≤ 0), representing < 1.7% of the total number, and < 0.1% of the lake area. Two-thirds of the acidic lakes are smaller than 4 ha, and many are groundwater recharge seepage lakes. Low DOC (< 417 μmol L^{-1}) acidic lakes (about 60% of the total number of acidic lakes) are acidic largely because of atmospheric deposition of acids. High DOC drainage lakes (about 10% of the total) are acidic because of a combination of natural and anthropogenic acids, and high DOC seepage lakes (about 30% of the total) are acidic largely because of organic acidity. High-elevation lake systems are most chemically influenced by acidic deposition. No low DOC lakes are known to have pH < ~5.0.

Base cation release is the most important factor in the generation of ANC. Other factors, such as NO_3^- uptake and SO_4^{2-} retention, also contribute to ANC production. In addition to atmospheric deposition, acidifying processes include the production of organic acids and NH_4^+ uptake. A majority of the region's watersheds are inferred to be presently conservative with respect to SO_4^{2-} on an annual basis. Nitrate loss from watersheds occurs seasonally during snowmelt, or episodically in response to rainfall events.

Historically, nonsignificant decreases in acidity and SO_4^{2-} concentrations in deposition have occurred during the 1980s. Slight, often significant, increases in surface water ANC have occurred during the same period. Few, if any, lakes or streams are expected to become acidic in the next several decades with steady-state atmospheric deposition. A number of lakes and streams spanning the natural pH range are fishless or have depleted populations, but acidic deposition cannot be directly implicated in the status of fisheries.

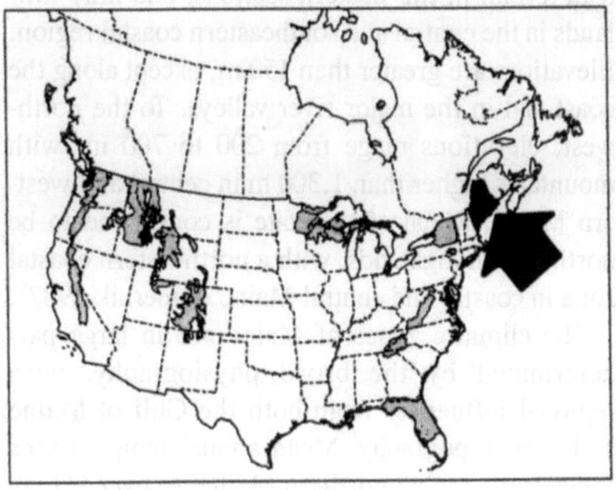

Introduction

Precipitation in the northeastern region of the United States is among the most acidic in the country. Within this region, however, there are depositional gradients of acidic substances that lead to differences in aquatic chemistry among subregions. The Adirondack Mountain region of New York has the greatest percentage of acidic lakes (ANC ≤ 0; 14% of population), whereas Maine has the lowest (1.7%), based on data from the U.S. EPA Eastern Lake Survey–Phase I (ELS-I), conducted in the fall of 1984 (Linthurst et al. 1986a,b, Brakke et al. 1988, Landers et al. 1988a). Maine has a relatively low number of acidic lakes, due in part to its geographic position as the state farthest downwind from emission sources in the industrialized midwest and east coast regions.

This chapter focuses on chemical relationships in the surface waters of Maine. This analysis examines (1) the extent of low ANC surface waters in the state, (2) the relationships between surface water chemistry and acidic deposition, and (3) the major factors governing the effects of acidic deposition on surface waters.

Regional Characteristics

Physiography and Climate

Maine is characterized by hilly to mountainous plateau terrain in the western sections, and hilly lowlands in the central and southeastern coastal region. Elevations are greater than 150 m, except along the coast and in the major river valleys. To the northwest, elevations range from 200 to 700 m, with mountains higher than 1,300 m in central and western Maine. Much of the state is considered to be northeastern highlands, with a northeastern coastal zone in coastal and central Maine (Omernik 1987).

The climatic zones of Maine are in large part determined by the broad physiography, with regional influences from both the Gulf of Maine and local topography. Mean annual temperatures range from 3 °C in northern Maine, to over 5 °C in the southwest (U.S. Department of Commerce 1968). Precipitation distributions are relatively uniform seasonally, but quite variable from year to year. Northern sections receive an average of 90 cm of precipitation, whereas coastal sections average as much as 120 cm. Average total snowfall ranges from about 150 cm in southern and coastal sections, to over 200 cm in inland regions. Winter snow cover in Maine is variable in temporal and spatial extent, especially in southern and coastal regions. The first snow of each year typically falls in November, but at least the southern and coastal areas generally experience multiple significant snowmelts during each winter. A snowpack may persist until early to mid-March in the south, and as late as early May in some years in the northern and upland regions.

Hydrology

Maine has an estimated 5,700 water bodies larger than 0.4 ha (State of Maine Water Quality Assessment 1988), although an unknown number of these are swamps, bogs, and beaver ponds. These lakes cover 4,000 km^2, or about 5% of the state's area. Approximately 3,500 of the lakes are larger than 4 ha. An estimated 91% of the lakes are drainage lakes (Figure 7.1; Linthurst et al. 1986a); approximately 5% are seepage lakes, although many of these are smaller than 4 ha (Kahl 1987). The seepage lakes are typically located in, or confined by, ice-contact stratified sand and gravel. Prior to the late 1960s, timber was transported to lumber and pulp mills via rivers and streams. Consequently, control structures have regulated the water levels in many natural lakes at some time in the past. Some large lakes have been created when rivers were dammed, either for lumbering activity or hydroelectric power. Such lakes are excluded from the discussions presented in this chapter.

The abundance of natural lakes in the state can be attributed to ice activity and periglacial processes during continental glaciation. Most of the lakes lie in bedrock depressions caused or accentuated by glacial activity, or in sand and gravel deposits. A few lakes are apparently younger than the general period of deglaciation, because of delayed hydrologic sealing of incipient lake basins (Caldwell and Davis 1983), or perhaps because of remnant cirque glaciers that delayed lake formation.

Streams are a significant resource in Maine. Approximately 7,300 streams have an estimated length of 52,000 km (State of Maine Water Quality Assessment 1988). The estimated length of low-order streams is nearly 37,000 km. Wetlands are common in areas of low local relief and include swamps, marshes, fens, and bogs. Major wetlands occupy an area of over 1,200 km^2; minor wetlands occupy an estimated three times that area.

Lakes typically freeze over between late October in the northwestern mountains to early January along the southern coast, depending on elevation and latitude. Iceout ranges from early March in the south and along the coast in warm years, to mid-May in the larger, more northerly lakes. Streams normally develop ice somewhat later than lakes, and often become ice-free six to eight weeks earlier than lakes.

Estimates of evapotranspiration (ET) are about 40% (Haines et al. 1990, Knox and Nordenson 1955). Minima in ET occur at high elevations or along often foggy coastal areas. Runoff is typically

FIGURE 7.1. Aerial photo of Mud, Salmon, and Little Long Ponds (left to right), Maine EPA LTM lakes at the Tunk Mountain Watershed, eastern Maine.

between 55% and 65% of measured wet inputs in the region. Runoff minima occur in the late summer, when monthly yields may be as low as 10%. During the spring, monthly yield may exceed 100%, as the decline of the snowpack contributes to discharge.

Bedrock and Surficial Geology

The bedrock of Maine is dominated by igneous and metamorphic rocks (Osberg et al. 1985; Chapter 5, Eilers and Selle, this volume). These rocks are generally resistant to chemical weathering and are in large part responsible for the dilute, oligotrophic nature of many of the surface waters in the state. The northwestern uplands are dominated by chemically and physically resistant noncalcareous lithologies. To the southwest, pelitic and locally calcareous metasedimentary rocks are dominant. These rocks extend through the lowlands of central Maine. The coastal zone consists of complexly deformed rocks of highly variable composition, with large granitic intrusions.

All of Maine was glaciated until 10,000 to 13,000 years ago. During the last glacial advance, compact tills were deposited on flat terrain and on the north or northwest slopes of hills. During glacial wasting, sediments were deposited through a variety of processes. The resultant deposits may be significant aquifers, and may contain or impound lakes and ponds. At modern elevations of < 130 m, glacio-marine silts and clays were deposited locally. These sediments often function as aquitards or aquicludes, and may have important influences on groundwater flow, as well as on groundwater and surface water chemistry.

Soils and Soil Chemistry

Soils in Maine have developed primarily in glacial and marine deposits on bedrock. However, till deposits are commonly similar in composition to the underlying rock. Soils derived from till are typically loamy skeletal, with abundant coarse fragments. Alluvium, lacustrine, and outwash parent materials are also present locally. The dominant

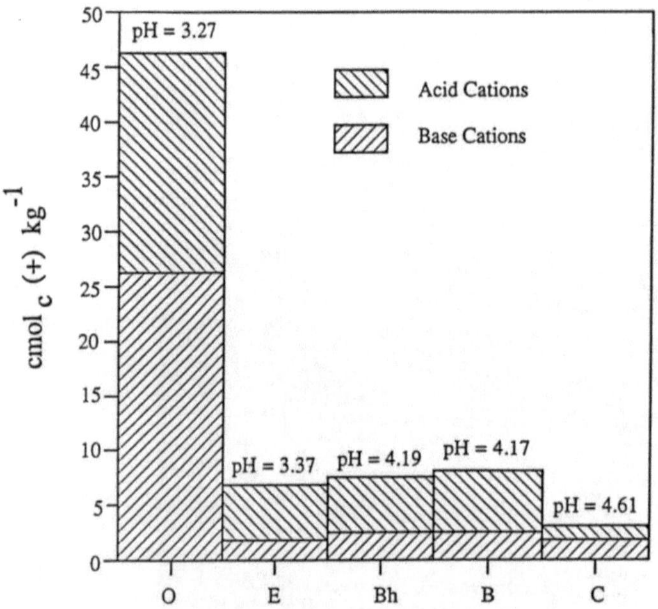

FIGURE 7.2. Representative Maine forest soil effective cation exchange capacity and exchangeable base cations and pH, by horizon. Soils sampled were from a variety of hardwood and softwood sites, at low to mid-elevation in southern and central Maine. Data are from the DDRP pilot survey (Fernandez unpublished).

pedogenic process active in freely draining soils is podzolization. Significant areas of soils with impeded drainage also exist in Maine. Poor drainage conditions in northeastern soils normally result from the presence of basal till, glacio-marine sediments, or shallow bedrock. Weathering generally has developed to depths of 60 to 100 cm where that depth of overburden is present.

Most of the soils are classified as Spodosols, Inceptisols, Entisols, or Histosols, with the first two soil orders accounting for the majority of the land area. These soils principally reflect the development of spodic subsoil horizons through the illuviation of iron, aluminum, and organic materials. Cunningham and Ciolkosz (1984) reported that 59% of Maine soils are in the suborder Orthods, with 24% classified as Aquepts. Further details are provided in Rourke et al. (1978).

The chemical characteristics of soils range considerably. Well-drained soils have low pH, low to moderate cation exchange capacity, low base saturation (Figure 7.2), and low SO_4^{2-} adsorption capacity, typical of soils in glaciated regions (Fernandez and Structemeyer 1985). More than 50% of the surface soils in these regions have < 3 $cmol_c$ kg^{-1} of exchangeable bases (Turner et al. 1986). Although bedrock-based sensitivity maps (Norton et al. 1982, Hendrey et al. 1980) classify substantial portions of Maine as *sensitive*, McFee (1980) classified northern New England soils as *not sensitive* or *slightly sensitive* to long-term atmospheric deposition of acids. With minor exceptions, the soils are believed to allow free leaching of SO_4^{2-}, in conjunction with base cations and Al. Soil Al is normally exported to surface waters only when sub-surface flow is shallow, and soil waters are not neutralized by interactions with aluminosilicate minerals.

Vegetation

Two patterns of natural vegetation predominate in the state. The high-elevation areas ($>$ 600 m) are evergreen boreal forest, dominated by the spruce-fir cover type, similar to the Adirondack province in northern New York. Red spruce (*Picea rubens*), white spruce (*Picea glauca*), black spruce (*Picea mariana*), and balsam fir (*Abies balsamea*) are the dominant species. Low-elevation areas in the northern hardwood zone typically have beech (*Fagus grandifolia*), yellow birch (*Betula alleghen-iensis*), and sugar maple (*Acer saccharum*). These forests typically are a mix of coniferous and deciduous species, with the dominant species being dependent on site characteristics. Poorly drained sites may have black spruce, fir, and northern white cedar (*Thuja occidentalis*), whereas drier sandy sites have white pine (*Pinus strobus*) and oak (*Quercus* sp.) as dominant species. South coastal Maine is broadly defined as the white pine-hemlock (*Tsuga canadensis*) zone, where oak is common.

Land Use

The present distribution and species composition of forests is the result of both natural phenomena and human activity. Insects and disease (e.g., gypsy moth, spruce budworm, chestnut blight, and dutch elm disease) have had major local or regional impact. Fire was the dominant long-term perturbation of forests before the arrival of Europeans. Settlement by Europeans began along the south coastal region during the 1700s. Virtually the entire region has been logged at least once since that time.

Currently, 88% of the state is forested, and approximately 6% of the state is in agriculture (State of Maine Water Quality Assessment 1988). Much of the forestland in central, eastern, and northern Maine is owned by the forest-products industry, and is managed for conifers. Herbicides are commonly used on these lands to control hardwood competition.

Road construction and maintenance has a significant influence on the chemistry of some lakes and streams. State and local governments spread an estimated 55×10^6 kg of NaCl on the state's roads each winter (Maine DEP unpublished). Several lakes were eliminated from analysis in this chapter due to anomalously high Cl^- concentrations.

Atmospheric Deposition

Wet Deposition

Maine has five NADP/NTN precipitation stations. Five additional NADP compatible collectors funded by EPA are located at three sites in eastern and central Maine (Haines et al. 1990, Norton et al. unpublished, Fernandez and Goltz 1988). The spatial distribution of the NADP collectors covers the population distribution of ELS-I lakes. However, the NADP collectors are located at low elevation for each region. Thus, precipitation inputs to higher elevation streams and lakes are not well known. Upland areas are believed to receive significantly elevated inputs of dry and occult deposition of acidic materials (Lovett et al. 1982).

pH and H^+

Unpolluted precipitation in northeastern North America is inferred to have had a pH level of around 5.0 (Charlson and Rodhe 1982, Garland 1981, Newman 1975). This inference is supported by data from remote forested regions of the globe (e.g., Galloway et al. 1984). Southern and coastal areas typically have received precipitation with pH ≤ 4.5 during the 1980s (NADP 1981–88; Figure 7.3). During the same period, precipitation pH in central and northern Maine has been > 4.5. The rain thus has 2 to 5 times the expected acidity. The pH of precipitation tends to be highest during the winter months.

Sulfur and Nitrogen

Sulfate and NO_3^- in precipitation are largely responsible for the pH of precipitation (Figure 7.4). Volume-weighted concentrations of SO_4^{2-} range from about 25 μeq L^{-1} in the north to about 35 μeq L^{-1} in the south. Nitrate concentrations are typically 20 μeq L^{-1} or less. Ammonium concentrations reported by NADP are < 10 μeq L^{-1}. Sea salt corrected sulfur and nitrogen wet deposition range from about 450 and 180 eq ha^{-1} yr^{-1}, respectively, at Acadia National Park, to < 300 and 150 eq ha^{-1} yr^{-1} in northern Maine. Sulfate concentrations are slightly lower during the winter months. Nitrate does not show a seasonal trend, and therefore is relatively more important during the winter.

Base Cations

Concentrations of the four base cations are normally less than a few μeq L^{-1} (Table 7.1), except near the coast, where Mg^{2+} and Na^+ are higher. Inland, the volume-weighted sum of base cations (SBC) is < 10 μeq L^{-1}. A typical value for total base cation wet loading is < 100 eq ha^{-1} yr^{-1}, with markedly higher loading immediately along the coast. Thus cation deposition gradients exist from southeast to northwest, rather than southwest to northeast as for NO_3^-, NH_4^+, and SO_4^{2-}. No seasonal trends are apparent in base cation concentrations.

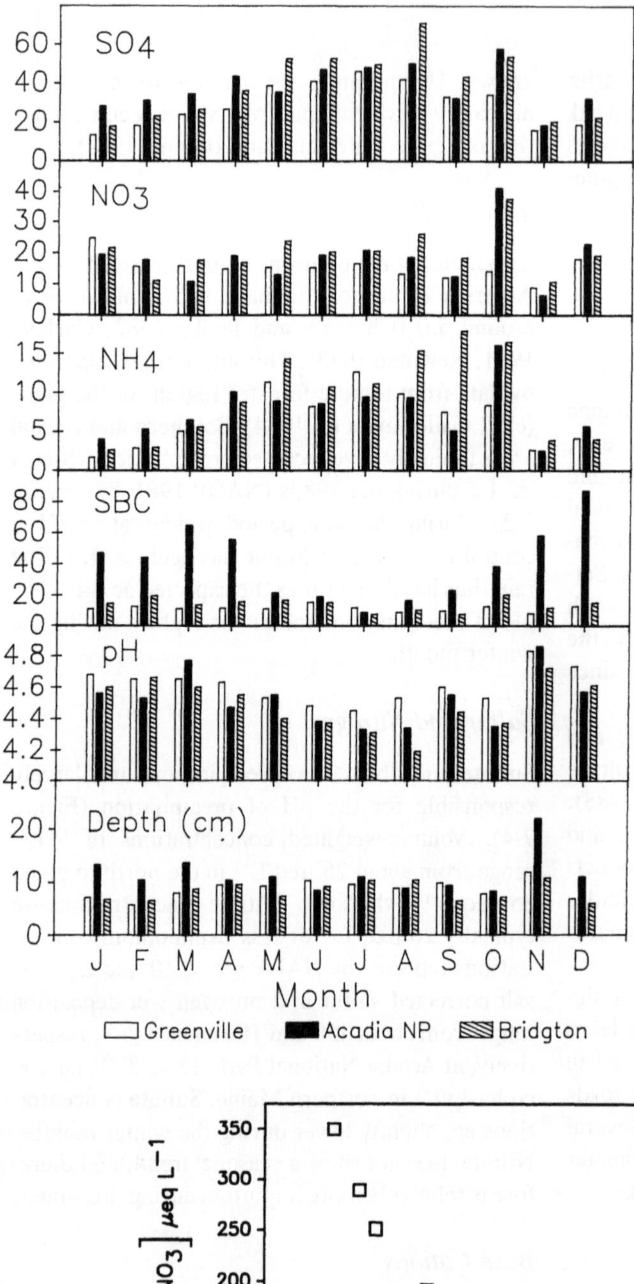

FIGURE 7.3. Histograms of monthly mean volume, SO_4^{2-}, NO_3^-, NH_4^+, SBC, pH, and depth, from NADP stations at Greenville, Acadia National Park, and Bridgton, Maine. Units are μeq L^{-1} unless otherwise indicated. Results calculated from monthly means by NADP, 1980–mid-1989, as available for each station: Greenville, 9 years; Acadia, 7 years; Bridgton, 8 years.

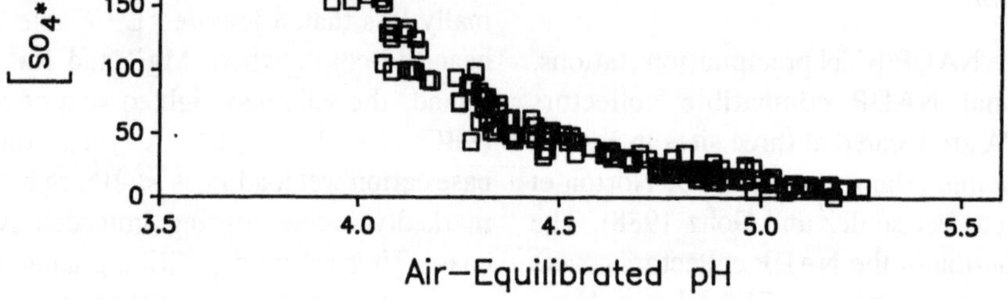

FIGURE 7.4. Precipitation pH versus sum of $SO_4^* + NO_3^-$ during 1985–87 at the Aurora NADP-compatible station (from Haines et al. 1990, and subsequent data). SO_4^* represents the nonmarine SO_4^{2-}, calculated from Cl^- concentrations.

TABLE 7.1. Mean annual wet deposition for NADP and NADP-compatible sites in Maine.[a]

		eq ha^{-1} yr^{-1}								
	pH	H^+	Ca^{2+}	Mg^{2+}	K^+	Na^+	NH_4^+	Cl^-	NO_3^-	SO_4^{2-}
Bridgton	4.46	360	30	25	5	50	72	82	162	370
Acadia	4.49	420	35	115	13	461	89	553	187	458
Aurora	4.51	320	30	33	8	99	105	166	160	289
Greenville	4.55	290	35	25	5	35	78	39	144	301
Caribou	4.56	230	40	16	8	34	61	37	116	260

[a] Data are from 1984 to 1986, the period during which a majority of lake and stream data in this text were collected. All units are converted from units reported in the NADP annual data summaries (NADP, 1984, 1985, 1986). Mean pH is reported as the arithmetic mean of each year of volume weighted data. The Aurora site is not an NADP network station, but is operated by the University of Maine using NADP schedules and protocol. The sites are listed from south to north.

Dry Deposition

Dry deposition inputs are difficult to measure, and little scientific consensus on measurement methods exists. Dry deposition decreases away from its source, due to dispersion and removal (Appendix B, Baker, this volume), and thus is lower in Maine than in areas closer to major industrialization. However, Maine has numerous significant industrial sources of sulfur, resulting in both the highest concentration and highest total emissions of atmospheric sulfur of any New England state (Air Bureau, Maine DEP unpublished).

Dry deposition of sulfur may be comparable to wet inputs in southwestern Maine and at higher elevations, but is inferred to decrease in importance in northern sections (Norton et al. 1988). During the late 1987 growing season, measured (filter pack) dry deposition of SO_2 at the U.S. Forest Service Howland spruce-fir research site in eastcentral Maine was comparable to wet SO_4^{2-} deposition.

Chen et al. (1983a) and Lindberg et al. (1986) have used differences between throughfall and wet chemistry to estimate dry deposition. Several studies have shown enrichment in chemical concentrations in throughfall. At the Howland site, SO_4^{2-} was enriched in spruce-fir throughfall versus wet-only precipitation by a factor of 2 to 3 during 1987 (Fernandez and Goltz 1988). Similar results have been found at the long-term monitoring site at Tunk Mountain (Rustad 1988), and at the Bear Brook Watershed at Lead Mountain (Norton et al. unpublished). These data represent estimates of dry deposition that are higher than those used in this book (Chapter 17, Baker et al., this volume), and by others (e.g., Rochelle and Church 1987).

Regional Surface Water Chemistry

Lake Data Sets

Eastern Lake Survey–Phase I (ELS-I)

The Eastern Lake Survey–Phase I (ELS-I) component of the National Surface Water Survey (NSWS) assessed the chemistry of lakes larger than 4 ha at fall turnover in the eastern United States (Linthurst et al. 1986a, Brakke et al. 1988, Landers et al. 1988a; see Introduction, Charles, this volume). Maine was encompassed by subregions 1C and 1E (Chapter 5, Eilers and Selle, this volume). In Maine, 225 lakes were sampled, representing an inferred population of 1,967 lakes. Data are presented here for fewer than 225 lakes, due to removal of lakes with anomalously high Cl^- and P concentrations.

Data from other studies (Norton et al. 1981, Haines and Akielaszek 1983, Haines and Kahl 1983, Kahl et al. 1985, Longcore and McAuley 1988) indicate that the ELS-I data accurately estimated the ANC distribution and number of acidic lakes in Maine. However, several lake types were under-represented by ELS-I. Therefore, data from special groups, such as high elevation (Kahl and Scott 1988), seepage-input (Kahl 1987), and EPA Long-Term Monitoring (LTM) lakes, complement the ELS-I statistical findings in this chapter. These data were obtained using ELS-I compatible methods for sampling and analysis.

Case Study Projects

High Elevation Lake Monitoring (HELM)

The HELM project sampled all 91 lakes in Maine above 600 m elevation in 1986 and 1987 (Kahl and

Scott 1988, 1990). The 600-m elevation allowed for a reasonably large subpopulation of lakes, and provided for a design parallel to related work on lakes above 600 m in the Adirondack Mountains of New York (e.g., Schofield 1976c). At least one summer and one fall turnover index period sample were taken during 1986 and 1987. The HELM lakes met the following criteria: (1) depth > 1 m, (2) area at least 0.4 ha, and (3) not a beaver impoundment.

Aquifer Lakes Study (ALPS)

The ALPS project sampled lakes that are on, or hydrologically associated with, mapped sand and gravel aquifers. Sampling was conducted in 1986 and 1987, and included at least one fall index sample for each lake. The ALPS population includes 128 of the estimated 150 lakes in Maine that meet the following criteria: (1) location in or on an aquifer or glacial deposit mapped by the U.S. or Maine Geological Surveys, (2) depth > 1 m, and (3) area at least 0.4 ha.

Tunk Mountain Watershed Study (TMWS)

The TMWS is the U.S. EPA Long-Term Monitoring Program in Maine. The site includes five lakes in a watershed of approximately 400 ha. Two lakes are circumneutral, two have pH of ~6.0, and one is acidic. Chemical records exist for a monthly to seasonal sampling schedule beginning in May, 1982.

Eastern Maine Streams Project (EMS)

The EMS was an intensive study of fisheries, stream chemistry, hydrology, and precipitation in six gauged low-order streams in eastern Maine. Hydrology and stream and precipitation chemistry have been summarized in terms of watershed chemical budgets.

Methods

The ELS-I methodology is summarized in Linthurst et al. (1986a) and Landers et al. (1988a), and details are presented in Hillman et al. (1986). In the other Maine projects discussed in this chapter, U.S. EPA NSWS (Hillman et al. 1986) or LTM Project methods were used for all field collection, sample holding, and analytical procedures. The results for the ELS-I were weighted according to EPA population factors to estimate the population results for Maine lakes larger than 4 ha. The sample results for HELM and ALPS data sets (representing essentially entire populations) presented in this chapter are compared to the population estimates of ELS-I.

TABLE 7.2. Average correction for sea salt inputs, by region and project.[a]

Lake population	Ca^{2+}	Mg^{2+}	K^+	Na^+	SO_4^{2-}	Cl^-
ELS-I Maine	1.6	8	0.8	37	5	44
ALPS	3.1	14	1.5	55	9	81
HELM	0.4	2	0.2	10	1	11

[a] Data presented are the means of μeq L^{-1} concentrations subtracted from the analyses of individual lakes.

Comparisons between the 1984 ELS-I Maine data and the 1986–1987 HELM and ALPS populations are complicated by (1) at least a 2-year span between sampling and (2) potential analytical biases between laboratories. However, agreement is good between overlapping lakes in the data sets (Kahl and Scott 1988), even between years that differed markedly in precipitation quantity. Based on these results, comparability is assumed to be good among data sets.

For a majority of the data presentations, Ca^{2+}, Mg^{2+}, K^+, Na^+, Cl^-, and SO_4^{2-} concentrations are mathematically corrected for the presence of sea salts (Chapter 18, Sullivan, this volume). The mean sea salt corrections for lakes in each region or project are summarized in Table 7.2. These data have been screened for road salt contamination using the empirical relationship from the ELS-I for distance from the coast versus Cl^- (Sullivan pers. comm.). All lakes with Cl^- concentrations > 200 μeq L^{-1}, as well as chemical outliers resulting from obvious road salt contamination have been removed from the data sets (Sullivan pers. comm., Linthurst et al. 1986a, Mairs 1967). Comparison of major ion:Cl^- ratios in lake water relative to wet deposition indicates that Cl^- is enriched relative to other chemical constituents. Sea salt corrections of the data based on Cl^- may therefore result in an underestimation of sea salt corrected SO_4^{2-} by as much as 4 or 5 μeq L^{-1}, and by greater amounts for Na^{2+} and Mg^{2+}.

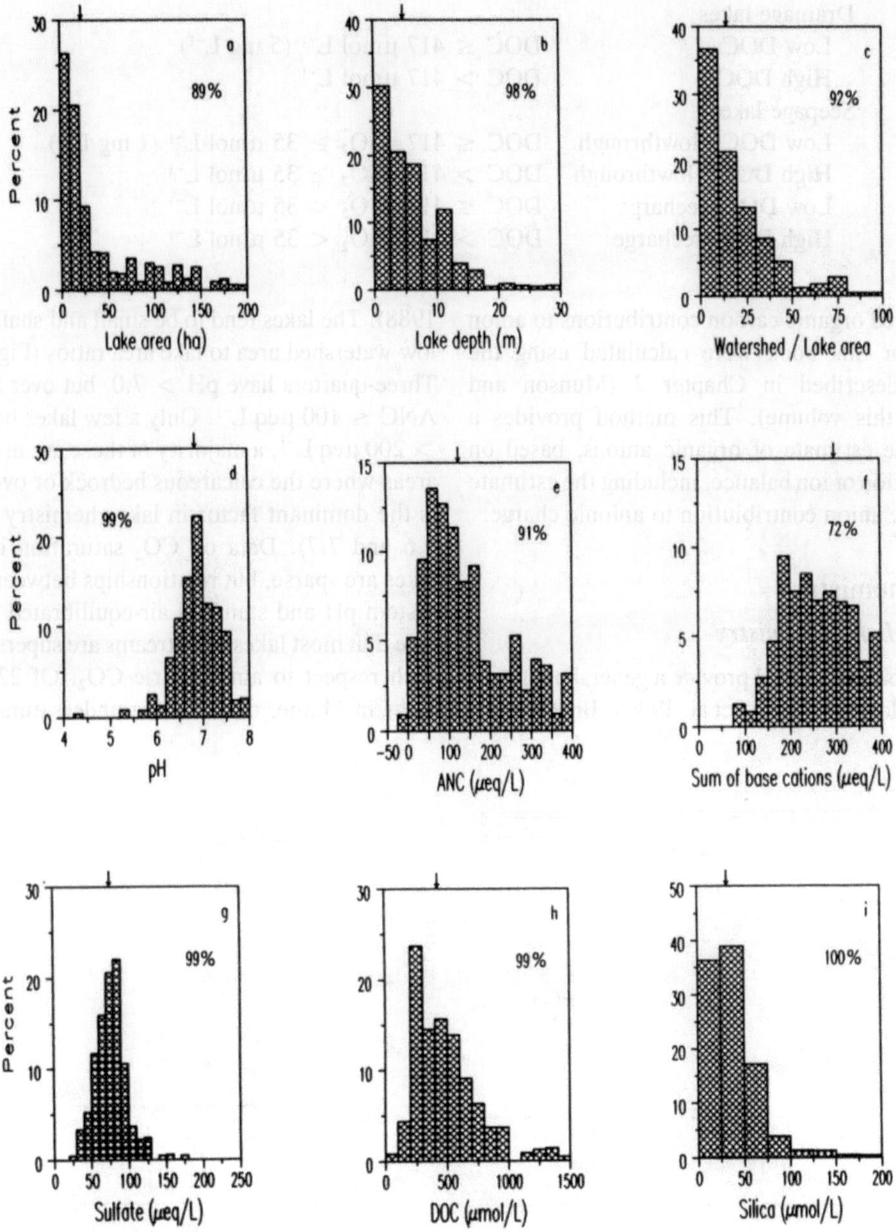

FIGURE 7.5. Frequency histograms of major characteristics of the target population of ELS-I lakes in Maine: (a) lake surface area, (b) lake depth, (c) watershed area:lake area, (d) air-equilibrated pH, (e) ANC, (f) sum of base cations, (g) sulfate, (h) DOC, and (i) silica. The arrow at the top of each box indicates the median for the population, and the number in the upper part of each box indicates the percentage of target lakes plotted.

In several of the following discussions, lake classes are defined as follows (Chapters 1 and 2, Munson and Gherini, this volume):

Drainage lakes	
Low DOC	DOC ≤ 417 μmol L^{-1} (5 mg L^{-1})
High DOC	DOC > 417 μmol L^{-1}
Seepage lakes	
Low DOC, flowthrough	DOC ≤ 417, SiO_2 ≥ 35 μmol L^{-1} (1 mg L^{-1})
High DOC, flowthrough	DOC > 417, SiO_2 ≥ 35 μmol L^{-1}
Low DOC, recharge	DOC ≤ 417, SiO_2 < 35 μmol L^{-1}
High DOC, recharge	DOC > 417, SiO_2 < 35 μmol L^{-1}

Dissolved organic carbon contributions to anion charge for this book were calculated using the method described in Chapter 2 (Munson and Gherini, this volume). This method provides a reasonable estimate of organic anions, based on recalculation of ion balance, including the estimate of organic anion contribiution to anionic charge.

Lake Chemistry

General Lake Chemistry

The results of the ELS-I provide a general picture of lakes in Maine (Linthurst et al. 1986a, Brakke et al. 1988). The lakes tend to be small and shallow, with low watershed area to lake area ratios (Figure 7.5). Three-quarters have pH > 7.0, but over half have ANC ≤ 100 μeq L^{-1}. Only a few lakes have ANC > 200 μeq L^{-1}; a majority of these are in northern areas where the calcareous bedrock or overburden is the dominant factor in lake chemistry (Figures 7.6 and 7.7). Data on CO_2 saturation in Maine lakes are sparse, but relationships between closed-system pH and standard air-equilibrated pH indicate that most lakes and streams are supersaturated with respect to atmospheric CO_2. Of 225 ELS-I lakes in Maine, only 5 were undersaturated with

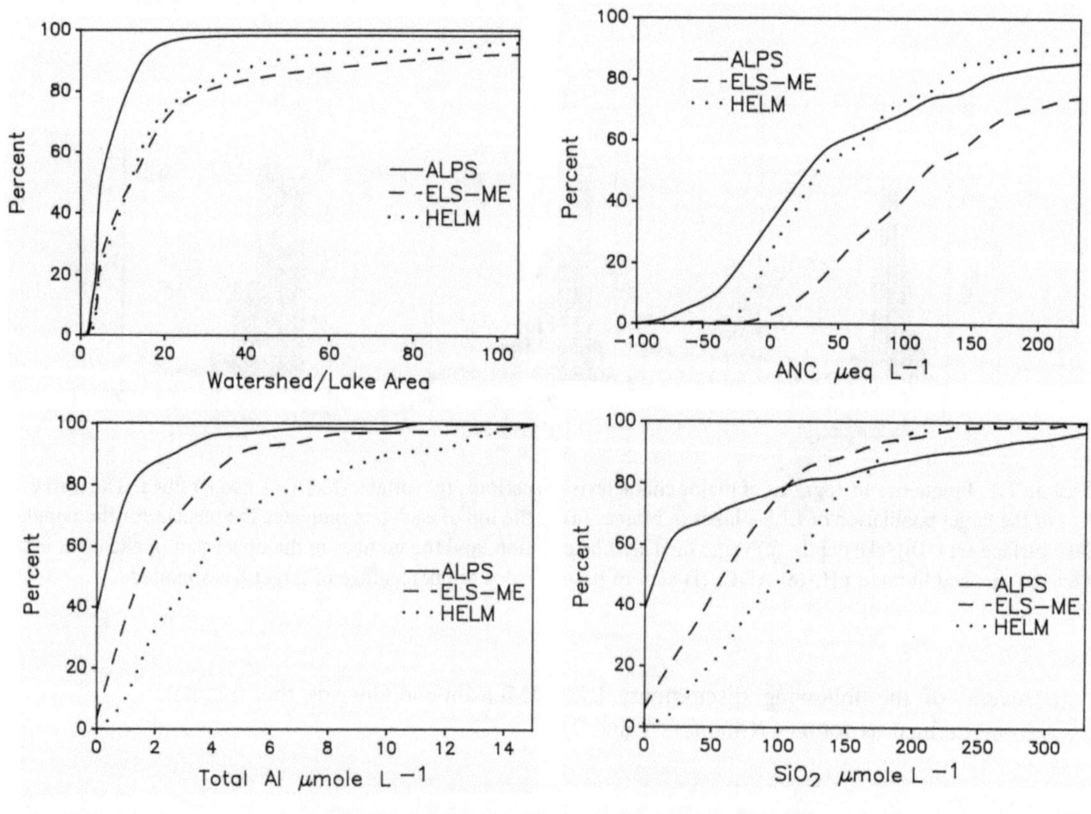

FIGURE 7.6. Cumulative frequency distribution of watershed area:lake area ratio, ANC, total Al, and SiO_2 for ALPS and HELM lakes, with comparison to ELS-ME lakes.

FIGURE 7.7. Location and ANC class of ELS, ALPS, and HELM lakes discussed in this chapter.

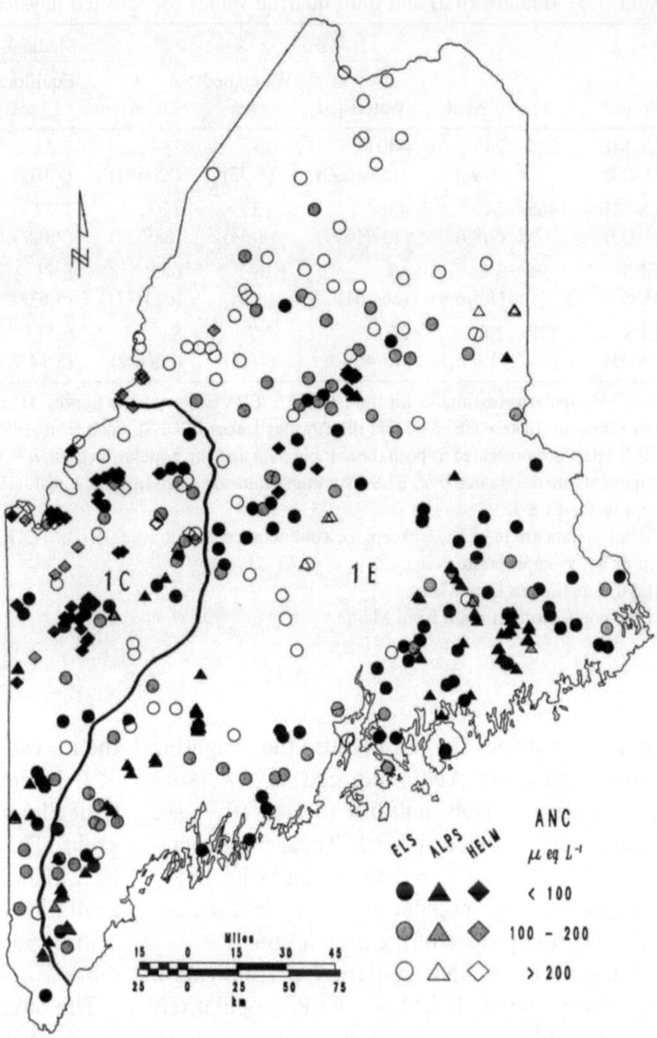

CO_2 at the time of sampling. Generally, Maine lakes with pH < 7.5 are supersaturated.

The sum of sea salt corrected base cations is typically higher than the ANC concentration. Sulfate concentrations average 76 μeq L^{-1}, more than double the 25 to 35 μeq L^{-1} in wet precipitation. Concentration of SO_4^{2-} from wet precipitation and dry deposition by evapotranspiration is responsible for this difference between surface water and precipitation concentrations. Nearly 20% of the lakes have SO_4^{2-} concentrations > 100 μeq L^{-1}, suggesting that either geologic sulfur or dry deposition contribute to lake concentrations of sulfur. An alternative hypothesis is that SO_4^{2-} accumulated in soils during periods of higher SO_4^{2-} deposition from the atmosphere may be leaching to surface waters. Dissolved organic carbon concentrations are low, with few lakes having > 800 μmol C (about 10 mg L^{-1}). However, nearly half the lake population is classified by Linthurst et al. (1986a) as being darkwater (color > 30 PCU). Silica concentrations are also low, with 60% < 75 μmol (2 mg L^{-1}).

Comparative Lake Chemistry

The HELM and ALPS projects (see section on Case Study Projects) were designed to sample the lakes expected to have low ANC. The HELM lakes are small, higher elevation drainage lakes, but they are similar to ELS-I lakes in watershed to lake area ratio (Figure 7.6a). Many of the ALPS lakes are groundwater recharge (perched) seepage lakes, and are thus the most likely to reflect precipitation chemistry.

A summary of selected chemical and physical parameters of ELS-I, HELM, and ALPS lakes is provided in Table 7.3. The ELS-I sample data and

TABLE 7.3. Median, first, and third quartile values for selected physical and chemical parameters.[a]

		Median[b]								
Project	n	Area	Watershed	Watershed area	Elevation	Equilibrated pH	ANC	SO_4^*	Ca*	DOC
ELS-ME[c]	225	24	440	15	164	7.25	119	70	136	430
(Q1/Q3)		(9/98)	(153/1690)	(8/35)	(85/331)	(7.01/7.55)	(69/238)	(57/83)	(92/231)	(269/628)
ELS-ME[d]	1966	24	438	15	170	7.27	138	71	143	433
(Q1/Q3)		(9/99)	(137/1699)	(8/34)	(88/332)	(7.02/7.59)	(71/255)	(58/83)	(92/248)	(269/629)
HELM	91	3.2	58	16	685	6.71	46	83	93	458
(Q1/Q3)		(1.5/6.4)	(26/151)	(8/34)	(624/771)	(5.63/7.16)	(11/125)	(72/94)	(63/128)	(287/558)
ALPS	128	2.8	26	7.7	82	6.51	38	40	50	258
(Q1/Q3)		(1.6/6.1)	(15/44)	(4/13)	(73/102)	(5.14/7.23)	(−6/150)	(28/52)	(20/126)	(170/441)

[a] Data are population estimates for the 1984 U.S. EPA Eastern Lake Survey—Phase I (ELS-I) subregions 1C and 1E in Maine, the High-Elevation Lakes (HELM), and the Aquifer Lakes (ALPS), each from 1986 and 1987.

ELS-I data are presented as both the sample data and the population data; n = population or sample size. The sea salt calculated components are designated by *. ELS-I summary data are from Linthurst et al. (1986) or are calculated from data provided by EPA-Corvallis for ELS-I.

[b] Chemical data are in μeq L^{-1}, except for conductance, which is in μS cm^{-1}, DOC, which is in μmol L^{-1}, and pH; physical data are in m or ha, as appropriate.

[c] ELS-I sample data from Maine.

[d] ELS-I population estimate from Maine.

population data are similar, because the weighting factors for the six ANC map classes (two subregions) are relatively uniform (8.1 to 10.7; see Linthurst et al. 1986a, p. 36). Thus, representations of ELS-I as sample data (of necessity in x-y plots), rather than population data, yield a relatively accurate picture of the population.

Distributions of ANC (Figure 7.6b) indicate a large proportion of HELM and ALPS populations with low ANC, relative to the ELS-I. Thirty percent of the ALPS lakes are acidic (ANC $\leq$ 0), and 21% have air-equilibrated pH < 5.0. Thirteen percent of the HELM lakes are acidic; 9% have pH < 5.0. Only an inferred 1.7% of the ELS-I population was acidic. Including lakes as small as 0.4 ha, 55 lakes in Maine are known to be acidic (37 with pH < 5.0). Sixty percent of these acidic lakes are seepage lakes. Lakes with the lowest pH and ANC in the HELM and ALPS groups appear to occur in clusters (Figures 7.7 and 7.8). These patterns may be related to bedrock geology (Osberg et al. 1985) in the HELM lakes, and to hydrologic control of DOC concentrations in the ALPS lakes.

The HELM lakes have the highest concentrations of total Al, with 50% having Al > 4 μmol L^{-1} (100 μg L^{-1}; Figure 7.6c). The ALPS lakes have the lowest Al concentrations, with 38% having Al < 0.5 μmol L^{-1} (14 μg L^{-1}). The lakes with low Al typically are the most acidic lakes in the ALPS group. This relationship is in contrast to that of the HELM and ELS-I drainage lakes, in which Al concentrations are inversely related to pH. The SiO_2 distributions for the lake groups are similar to those for total Al (Figure 7.6d).

The contributions of major solutes to lake chemistry in the ELS-I population are illustrated in Figure 7.9. The concentration data represent median values for each pH group. In general, the acidic lakes have the lowest solute concentrations. A similar plot for ALPS lakes (not shown) demonstrates that ALPS lakes have markedly lower concentrations of all constituents except Cl^-. Chloride concentrations in the lakes reflect the geographic distribution close to the coast. The HELM lakes have lower concentrations of constituents, except SO_4^{2-}. Base cation concentrations in all groups are similar to the pattern for ANC. Concentrations of base cations tend to be higher in ELS-I lakes than in ALPS or HELM lakes. The sea salt corrected SO_4^{2-} distribution generally reflects variation in ionic strength rather than in pH. In the HELM lakes, however, the relative contribution of SO_4^{2-} to the total anion charge increases at low pH.

FIGURE 7.8. Location and air-equilibrated pH class of ELS, ALPS, and HELM lakes discussed in this chapter.

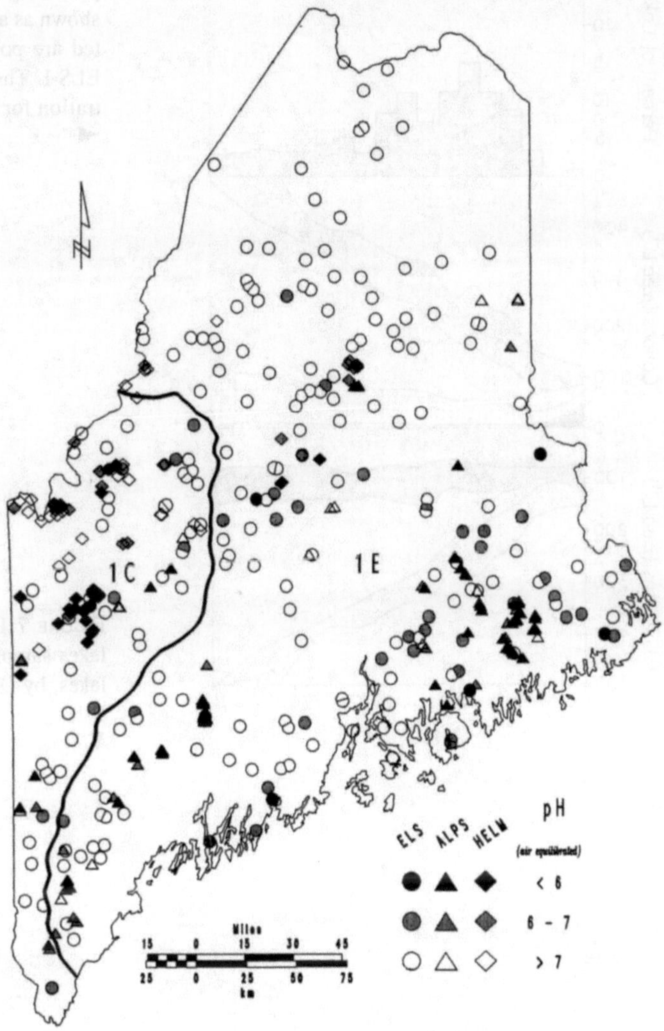

The relationship between ANC and SBC is an indicator of sensitivity to potential acidification from acidic deposition. When expressed as the ratio ANC:SBC, all but one of the ELS-I lakes has a ratio < 0.75 (Figure 7.10a). These ratios are about 1.0 when expressed as ANC:sea salt corrected (Ca^{2+} + Mg^{2+}). There is no apparent geographic pattern to the distribution of the ratio for any lake group within the state (Figure 7.10b). Of the lakes with the lowest ratio, most are low DOC systems, suggesting that acidic deposition is a contributing factor when the ratio is low (Chapter 18, Sullivan, this volume).

Chemistry of Small Streams

A statistical sampling of small streams has not been undertaken in Maine. However, several surveys have been done as part of the site selection process of other studies, and four projects have conducted more extensive chemical characterization and monitoring in low-order streams. Generally, low-order streams are more dilute and more poorly buffered than high-order streams. Sulfate concentrations are highest in upland, first-order streams, and decrease somewhat at lower elevations.

Although chronically acidic perennial streams are unknown in Maine, episodic acidification is common during snowmelt and after heavy rainfall (Figure 7.11). On occasion, pH values < 5.0 have been documented in some systems. ANC, Ca^{2+}, and pH often co-vary during these events, but DOC, SO_4^{2-}, and NO_3^- are less predictable. Rarely can decreases in pH be attributed to absolute increases

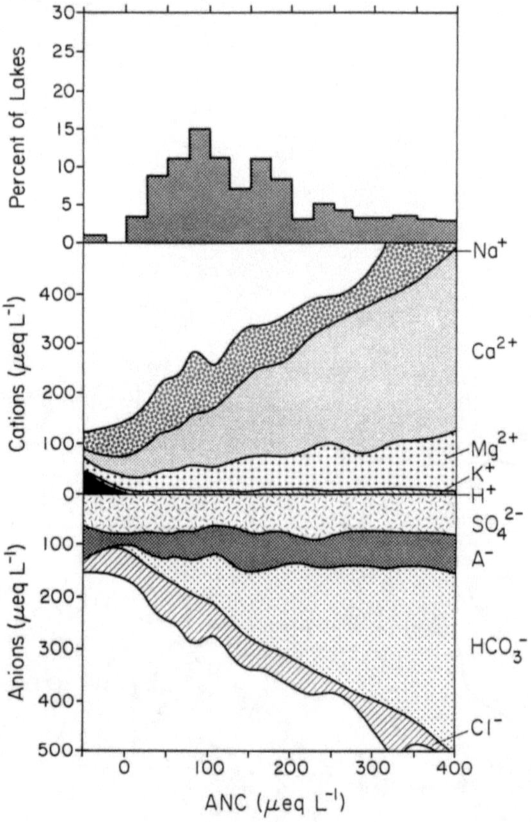

FIGURE 7.9. Concentrations of major cations and anions for ELS-I lakes in Maine with ANC $\leq$ 400 μeq L^{-1}. The percentages of lakes represented by ANC intervals are shown as a histogram at the top of the figure. Data plotted are population-weighted estimates for lakes in the ELS-I. The curves are spline fits to the median concentration for each ANC interval.

◄

FIGURE 7.10. Spatial distribution of ANC:SBC for (a) lakes sampled in the ELS-I, and (b) ALPS and HELM lakes, by DOC class. Low DOC $\leq$ 417 μmol L^{-1} (5 mg L^{-1}).

▼

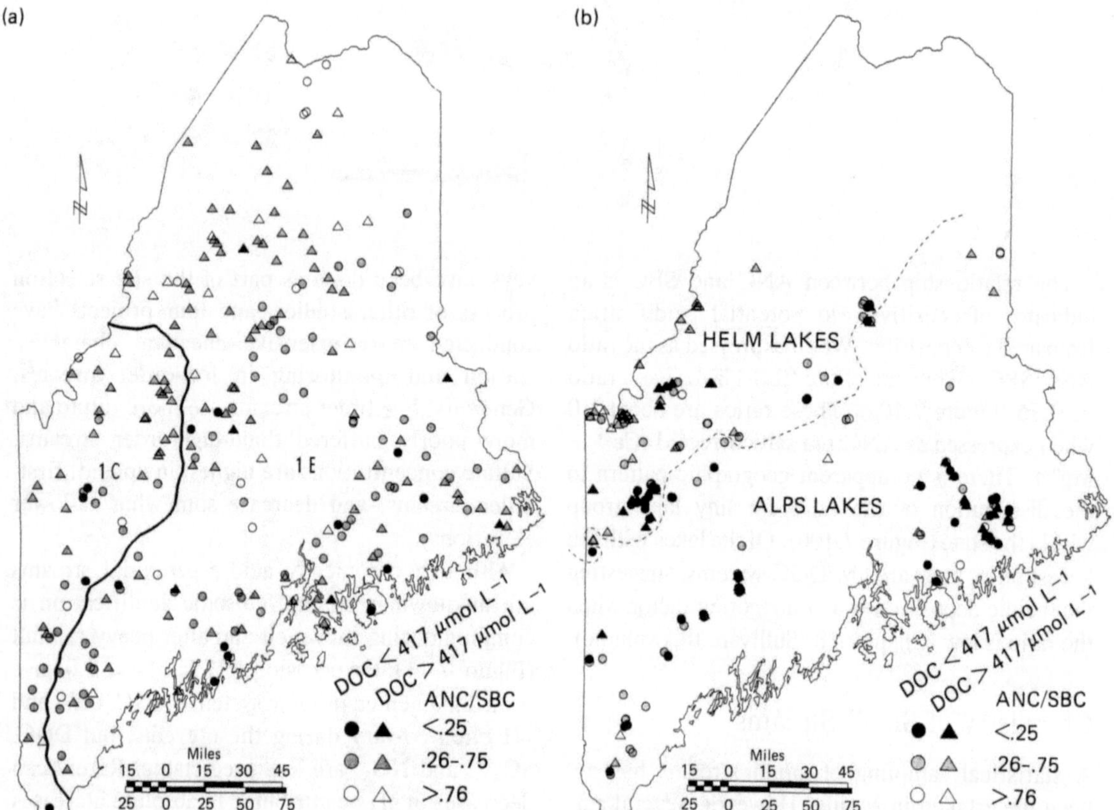

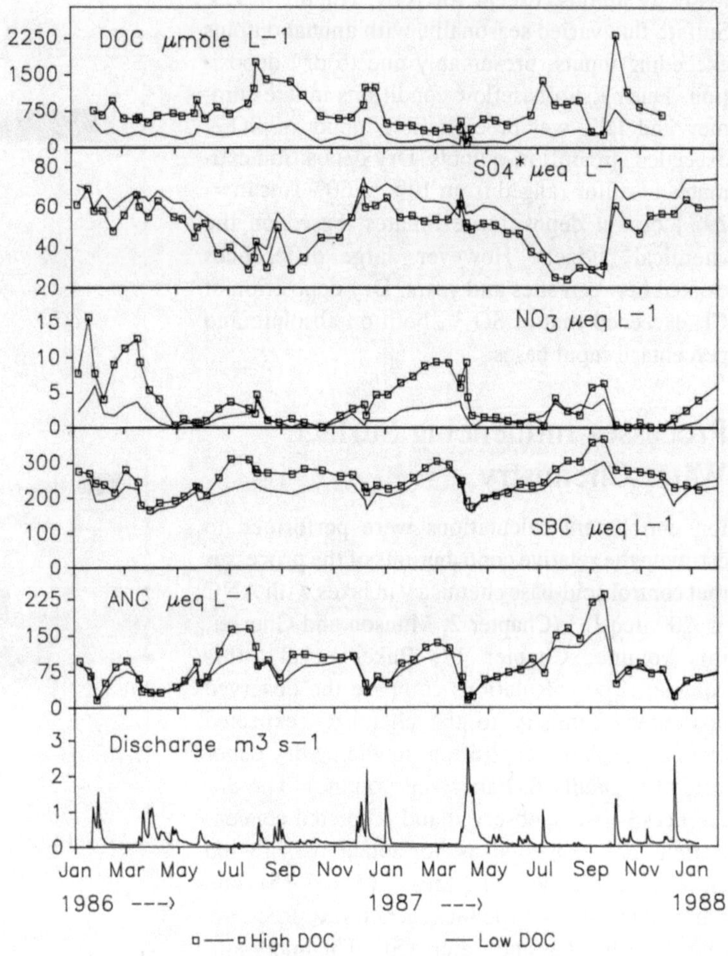

FIGURE 7.11. The relationship between selected chemical parameters and discharge from five low-order streams in eastern Maine (Haines et al. 1990). Data presented as high DOC (> 417 µmol L^{-1}) are the means of three streams in contiguous watersheds; data presented as low DOC (≤ 417 µmol L^{-1}) are the means of two other nearby streams, also in contiguous watersheds. Discharge is from one of the three high DOC streams, presented for reference purposes only.

in SO_4^{2-}. More commonly, episodic loss of ANC is due to base cation dilution coupled with increases in NO_3^- or DOC.

Sulfate exhibits unpredictable response to increasing discharge during episodes (Figure 7.11). Seasonally, SO_4^{2-} concentrations may decline to values less than those in precipitation during low flow in summer and early fall. Typically, total Al increases in response to higher discharge, due to the declining pH. The response is less predictable in streams with higher DOC. Streams with high DOC commonly have the highest total Al. Concentrations of Al are extremely variable among streams. Typical baseflow concentrations are < 2 µmol L^{-1}, whereas peak concentrations at lowest pH have been observed to be as high as 15 µmol L^{-1} (Norton et al. unpublished, Haines et al. 1990, Kahl et al. 1985). Inorganic Al species are usually < about 4 µmol L^{-1}, and quite often < 1 µmol L^{-1}.

Watershed Chemical Budgets

Chemical mass balances were calculated using the midpoint weighting method of Scheider et al. (1978) for five low-order streams studied in eastern Maine during 1985–87 (Haines et al. 1990). Annual water yields averaged between 50% and 60%, and in all months except April, precipitation exceeded discharge. Each watershed exhibited a net export of all cations except H^+ and NH_4^+ due to weathering and ion exchange, or uptake processes (Figure 7.12). Export of SiO_2 and Al were quite consistent among streams. Export of DOC varied markedly among streams. Each watershed had net losses of anions, except for NO_3^- (Figure 7.12). Sulfate flux varied seasonally, with annual outputs exceeding inputs, presumably due to dry deposition. During sub-baseflow conditions in late summer and fall, wet precipitation inputs of SO_4^{2-} exceeded streamflow outputs. Dry deposition estimates of sulfur ranged from 10% to 60% (mean = 29%) of wet deposition estimates, based on the chemical budgets. However, large differences existed between sites and years. Dry deposition of Cl^- exceeded that of SO_4^{2-}, both on absolute and percentage-input bases.

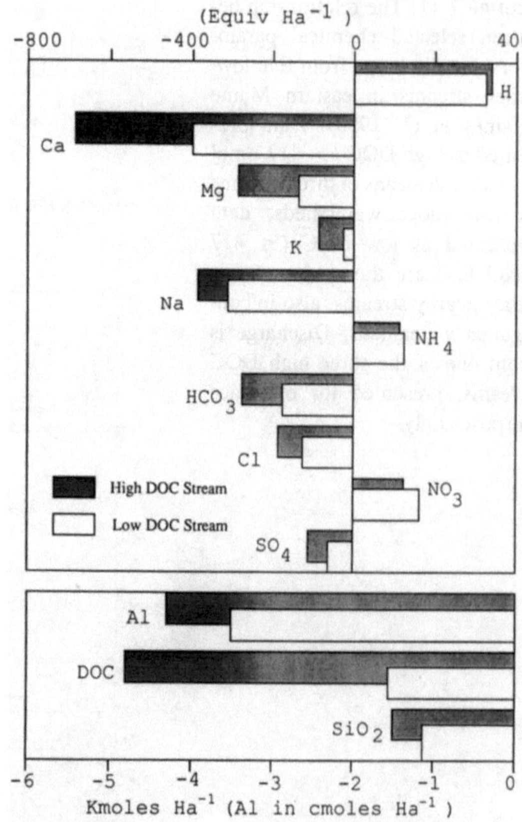

FIGURE 7.12. Mean annual mass balance estimates for major chemical parameters from five watersheds in eastern Maine, 1985–87 (Haines et al. 1990). Horizontal bars to the right indicate net watershed retention (H^+, NH_4^+, NO_3^- only); bars to the left indicate net loss. Data represent the means of three high DOC streams and two low DOC streams.

Processes Influencing Surface Water Chemistry

Ion enrichment calculations were performed to estimate the relative contributions of the processes that control acid-base chemistry in lakes with ANC ≤ 200 μeq L^{-1} (Chapter 2, Munson and Gherini, this volume; Chapter 17, Baker et al., this volume). The calculations compare the observed lakewater chemistry to the chemistry expected from atmospheric deposition, including dry deposition (Appendix B, Baker, this volume). The differences between observed and predicted concentrations are expressed as horizontal bars in ion enrichment diagrams (Figures 7.13 to 7.15), and are used to estimate the biogeochemical processes that occur in lakes and watersheds. The uncertainties in the conclusions are potentially large, due to the uncertainty surrounding dry deposition. However, the expression of these data using the median lake for a given population of lakes minimizes the potential magnitude of error.

Subclasses for hydrologic type and DOC class are shown in Figures 7.13, 7.14, and 7.15, as appropriate. The ELS-I and HELM data are presented for drainage lakes, in two DOC classes. The ALPS lakes are classified as either high or low DOC, and as either groundwater recharge or groundwater flow-through seepage lakes (see section entitled Methods, this chapter, and Chapters 1 and 2, Munson and Gherini, this volume).

Mineral Weathering

In Maine lakes, base cation release is the dominant factor in generating ANC, indicated by horizontal bars to the right in Figures 7.13, 7.14, and 7.15. For the general population of ELS-I lakes, base cation production is responsible for more than 80% of ANC generation (Figure 7.13). Base cation release is slightly lower than 80% in the low DOC lakes, and is higher in the high DOC lakes. Calcium is the dominant cation in each lake class. The data from the high-elevation lakes reflect processes similar to those in the ELS-I lakes, with base cation concentrations being somewhat greater in the high DOC lakes.

Base cation release in the ALPS seepage lakes is also the dominant factor in ANC generation, but with some marked differences between hydrologic types. In the groundwater flow-through lakes, base cations account for as much as 90% of ANC generation. In these lakes, the high DOC systems have

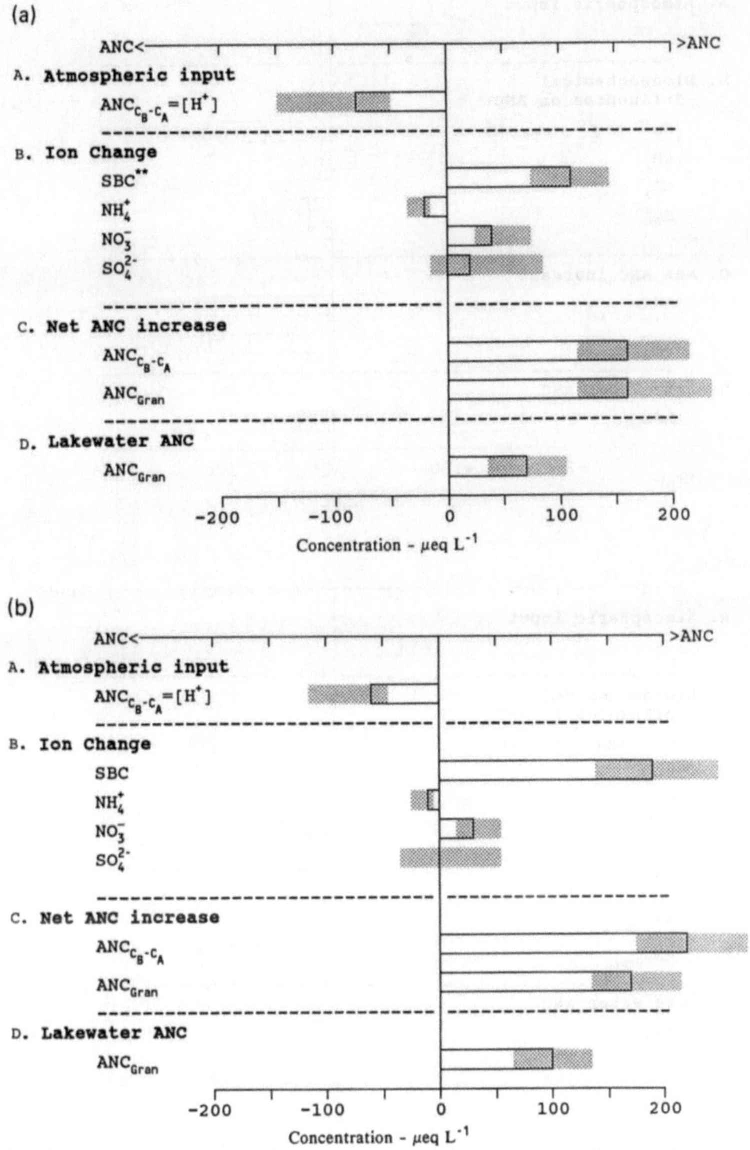

FIGURE 7.13. (a) Modification of atmospheric deposition by watershed and lake biogeochemical processes in the ELS lake population for 62 lakes with low DOC (≤ 417 μmol L^{-1}). A. Precipitation [H^+] adjusted for estimated dry deposition and evapotranspiration (ET); equivalent to expected lakewater $ANC_{C_B-C_A}$ in the absence of any process other than ET. B. Reactions influencing ANC; bars to the right indicate ANC production; those to the left show ANC consumption. C. Net ANC production; $ANC_{C_B-C_A}$ is the sum of ANC production terms for measured anions and cations including those shown in B; ANC_{Gran} is determined by difference between input ANC (A.) and lakewater ANC (D.). D. Lakewater ANC represents the net reaction of atmospheric inputs with the lake/watershed system. Shaded bars represent 25% and 75% quartiles. For a further explanation of calculation procedures and figure interpretation, see Appendix A. Median DOC = 249 μmol L^{-1} (25% quartile = 211 μmol L^{-1}, 75% quartile = 330 μmol L^{-1}); median A_t = 19 μeq L^{-1} (25% quartile = 7 μeq L^{-1}; 75% quartile = 27 μeq L^{-1}). (b) Modification of atmospheric deposition by watershed and lake biogeochemical processes in the ELS lake population for 62 lakes with high DOC (> 417 μmol L^{-1}). For further information, see Figure 7.13a. Median DOC = 596 μmol L^{-1} (25% quartile = 490 μmol L^{-1}, 75% quartile = 872 μmol L^{-1}); median A_t = 68 μeq L^{-1} (25% quartile = 52 μeq L^{-1}; 75% quartile = 94 μeq L^{-1}).

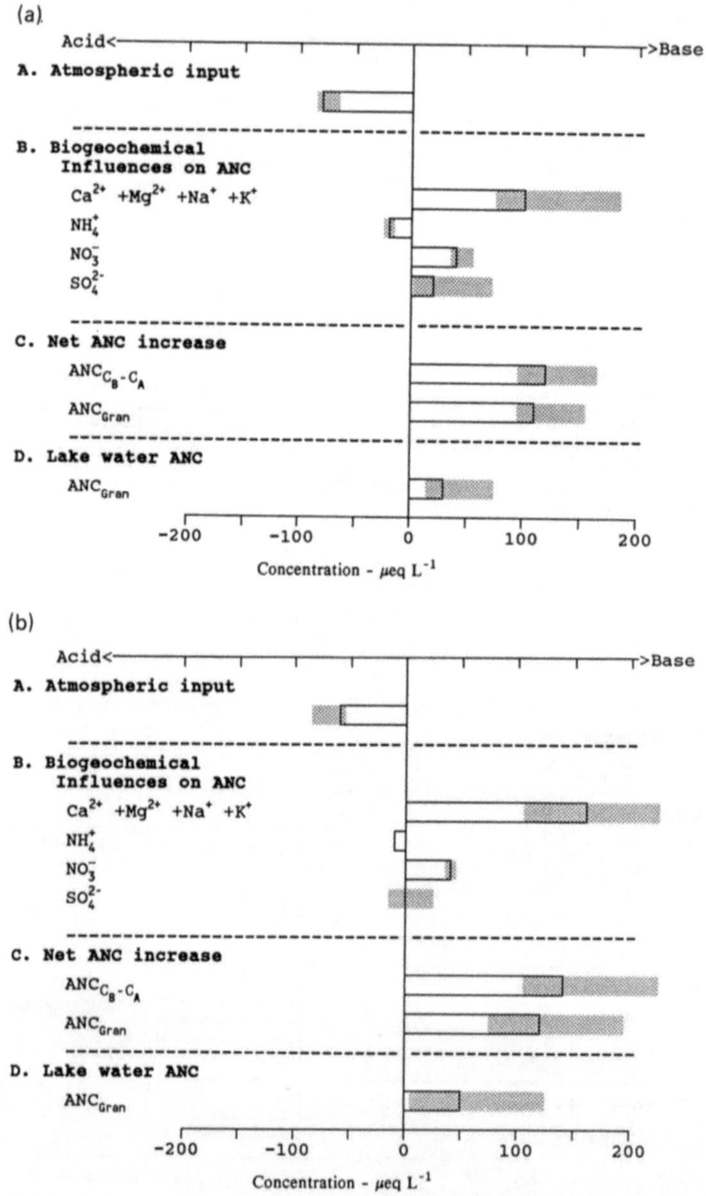

FIGURE 7.14. (a) Modification of atmospheric deposition by watershed and lake biogeochemical processes in the HELM lake population (38 lakes) for lakes with low DOC. A. Precipitation [H^+] adjusted for estimated dry deposition and evaporation (ET); equivalent to expected lakewater $ANC_{C_B\text{-}C_A}$ in the absence of any process other than ET. B. Reactions influencing ANC; bars to the right indicate ANC production; those to the left show ANC consumption. C. Net ANC production; $ANC_{C_B\text{-}C_A}$ is sum of ANC production terms for measured anions and cations including those shown in C; ANC_{Gran} is determined by difference between input ANC (A.) and lakewater ANC (D.). D. Lakewater ANC represents the net reaction of atmospheric inputs with the lake/watershed system. Shaded bars represent 25% and 75% quartiles. For a further explanation of calculation procedures and figure interpretation, see Appendix A. Median DOC = 270 μmol L^{-1} (25% quartile = 171 μmol L^{-1}; 75% quartile = 353 μmol L^{-1}); median A_t = 11 μeq L^{-1} (25% quartile = 2 μeq L^{-1}; 75% quartile = 18 μeq L^{-1}). (b) Modification of atmospheric deposition by watershed and lake biogeochemical processes in the HELM lake population (38 lakes) for lakes with high DOC. For further information, see Figure 7.14a. Median DOC = 510 μmol L^{-1} (25% quartile = 457 μmol L^{-1}; 75% quartile = 643 μmol L^{-1}); median A_t = 31 μeq L^{-1} (25% quartile = 16 μeq L^{-1}; 75% quartile 50 μeq L^{-1}).

slightly less base cation production relative to the low DOC waters. In contrast, base cation production in the groundwater recharge lakes is only 10% to 20% of that in any other lake class. In the colored recharge systems (low ANC, typically acidic lakes) NO_3^- assimilation may be the dominant factor in ANC generation.

Two factors must be considered in these interpretations. First, high DOC drainage lakes necessarily have a significant portion of their water routed through shallow flowpaths, in contact with organic soils. Thus, much of the control on the cation concentration in these lakes results from ion exchange reactions with organic material. In the low DOC lakes, a greater portion of the water is assumed to be derived from deeper flowpaths. Therefore, controls on cation concentrations in these lakes are more likely to be related to primary mineral weathering and exchange reactions on inorganic surfaces. Second, the high DOC seepage lakes are typically situated in or near wetlands. Uptake of nutrient cations, by sphagnum sp. for example, may deplete the base cation concentrations and increase the hydrogen loading to the surface waters. This last factor may confound the interpretation of the ion enrichment diagrams presented here, especially for these latter lake types.

Dissolved Organic Carbon

Seepage lakes have the highest concentrations of organic anions relative to total anion concentration. This type of lake is often transitional into bog lakes. A study of 150 surface waters within ombrotrophic bogs (Davis and Kahl unpublished) found that 91% were acidic (ANC $\leq$ 0) and that 36% had pH $<$ 4.0. Because the pH of wet precipitation is about 4.5, and the mean DOC concentration in these bogs was more than 3,400 μmol L^{-1}, we infer that organic acidity plays an important role in the acid-base status of these waters. High DOC ($>$ 417 μmol L^{-1}) drainage lakes are also influenced by organic acidity. Several lines of evidence support this hypothesis.

If water samples are equilibrated with air of known CO_2 concentration, deviations in pH from calculated pH in the H_2O-CO_2 system can be ascribed largely to the influence of nonvolatile weak acids in solution (Chapter 17, Baker et al., this volume; Chapter 2, Munson and Gherini, this volume). These weak acids are predominantly organic acids in the low Al waters in Maine. The deviations are most evident in the case study lakes (Figure 7.16). Although the pH depression at values $<$ 6.0 is often small (Chapter 17, Baker et al., this volume), the number of μeq L^{-1} represented by the difference at low pH may be considerable. The deviations are highest in high DOC lakes (Figure 7.16). Thus, the influence of dissolved organic matter on pH may be large in low pH, high DOC lakes. The reaction term for organic acidity calculated from the ion enrichment diagram equations is actually larger than the sum of ANC generating terms for the median high DOC recharge lake (Figure 7.15), suggesting that many of these lakes would be acidic even without the additional inputs of acids from the atmosphere.

Calculated organic anions (see Chapter 13, Cook and Jager, this volume, for calculation methods) contribute nearly 60% to the acidity status of the acidic high DOC seepage lakes. This contribution can be compared to 6% in the low DOC ($\leq$ 417 μmol L^{-1}) lakes, and to 19% in the high DOC drainage lakes. Total calculated organic anionic contributions ranged up to 200 μeq L^{-1} in lakes, and were as high as 500 μeq L^{-1} in bogs (Kahl et al. 1989). Thus, high DOC drainage lakes (about 10% of the total number of acidic lakes) are acidic due to a combination of natural and anthropogenic acids, and high DOC seepage lakes (about 30% of the total) are acidic largely due to organic acidity. Contributions from both acidic deposition and organic acidity are necessary but not solely sufficient for the chronically acidic status. No low DOC lakes are known in Maine with pH $<$ 4.9, suggesting that contributions from organic acidity are required to depress pH below about 5.0, under current levels of deposition loading.

Atmospheric Deposition

Anthropogenic Acids

Because acidic precipitation in Maine is dominated by sulfuric acid, correlations should exist between SO_4^{2-} and certain chemical parameters in lake waters. Maine lakes typically have 40 to 120 μeq L^{-1} of SO_4^{2-}, which may represent as much as 20 to 100 μeq L^{-1} more than the amount that existed in the pristine environment several hundred years ago (Brakke et al. 1989, Wright et al. 1980). This excess SO_4^{2-} is evidence of the influence of atmospheric

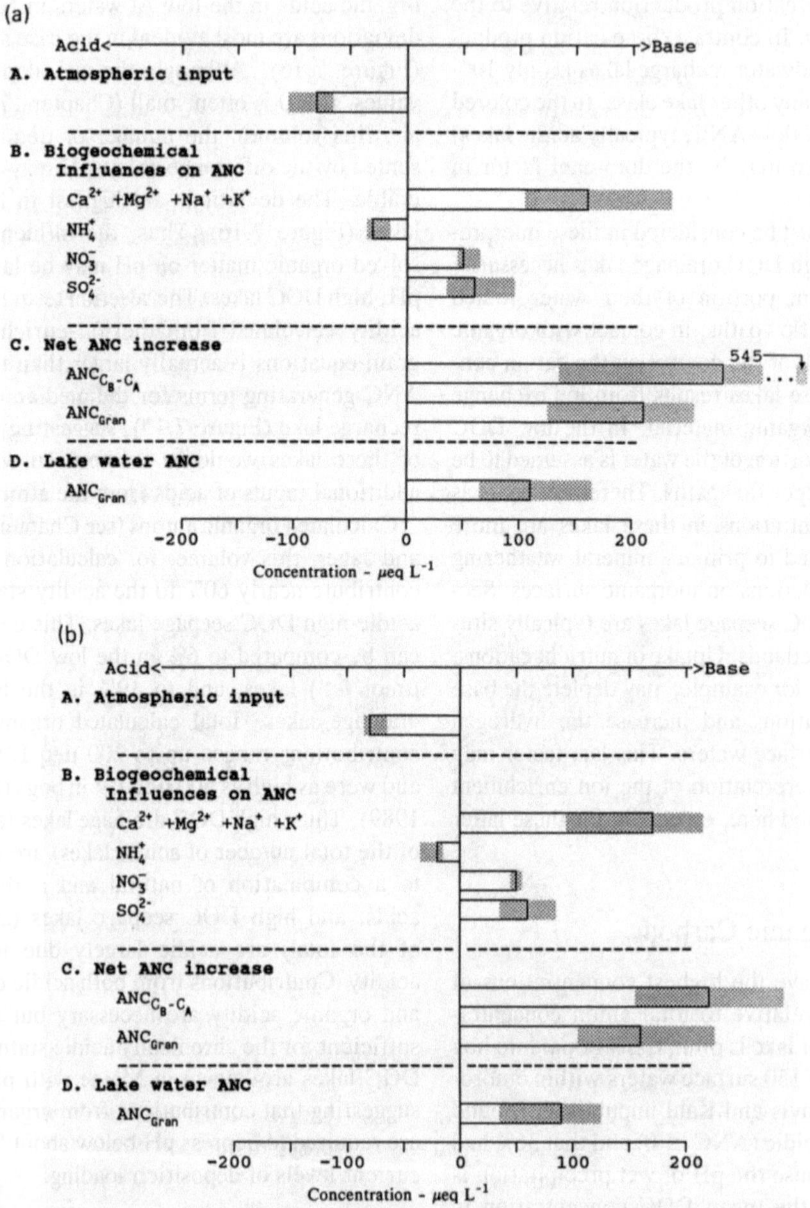

FIGURE 7.15. (a) Modification of atmospheric deposition by watershed and lake biogeochemical processes in the ALPS lake population (17 lakes) for groundwater flow-through lakes with low DOC. A. Precipitation [H^+] adjusted for estimated dry deposition and evapotranspiration (ET); equivalent to expected lakewater $ANC_{C_B\text{-}C_A}$ in the absence of any process other than ET. B. Reactions influencing ANC; bars to the right indicate ANC production; those to the left show ANC consumption. C. Net ANC production; $ANC_{C_B\text{-}C_A}$ is sum of ANC production terms for measured anions and cations including those shown in C; ANC_{Gran} is determined by difference between input ANC (A.) and lakewater ANC (D.). D. Lakewater ANC represents the net reaction of atmospheric inputs with the lake/watershed system. Shaded bars represent 25% and 75% quartiles. For a further explanation of calculation procedures and figure interpretation, see Appendix A. Median DOC = 221 μmol L^{-1} (25% quartile = 75 μmol L^{-1}; 75% quartile = 306 μmol L^{-1}); median A_t = 24 μeq L^{-1} (25% quartile = 8 μeq L^{-1}; 75% quartile = 72 μeq L^{-1}). (b) Modification of atmospheric deposition by watershed and lake biogeochemical processes in the ALPS lake population (12 lakes) for groundwater flowthrough lakes with high DOC. For further information, see Figure 7.15a. Median DOC = 620 μmol L^{-1} (25% quartile = 441 μmol L^{-1}; 75%

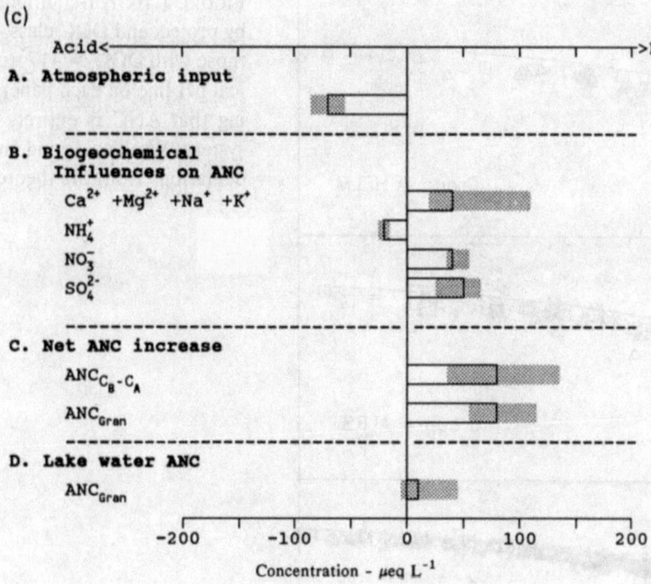

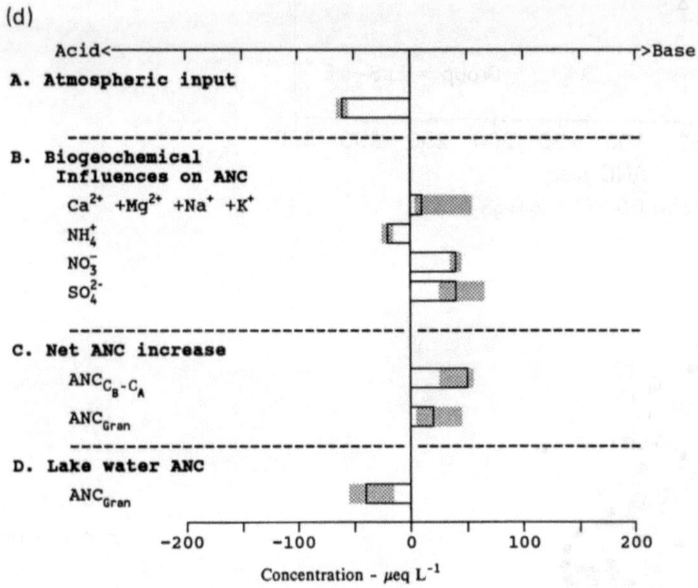

FIGURE 7.15. *(Continued)*

quartile = 1,215 μmol L^{-1}; median A_t 75 μeq L^{-1} (25% quartile = 63 μeq L^{-1}; 75% quartile = 85 μeq L^{-1}). (c) Modification of atmospheric deposition by watershed and lake biogeochemical processes in the ALPS lake population (55 lakes) for groundwater recharge lakes with low DOC. For further information, see Figure 7.15a. Median DOC = 199 μmol L^{-1} (25% quartile = 135 μmol L^{-1}; 75% quartile = 271 μmol L^{-1}); median A_t = 14 μeq L^{-1} (25% quartile = 10 μeq L^{-1}; 75% quartile = 26 μeq L^{-1}). (d) Modification of atmospheric deposition by watershed and lake biogeochemical processes in the ALPS lake population (13 lakes) for groundwater recharge lakes with high DOC. For further information, see Figure 7.15a. Median DOC = 783 μmol L^{-1} (25% quartile = 647 μmol L^{-1}; 75% quartile = 1,267 μmol L^{-1}); median A_t = 40 μeq L^{-1} (25% quartile = 37 μmol L^{-1}; 75% quartile = 50 μeq L^{-1}).

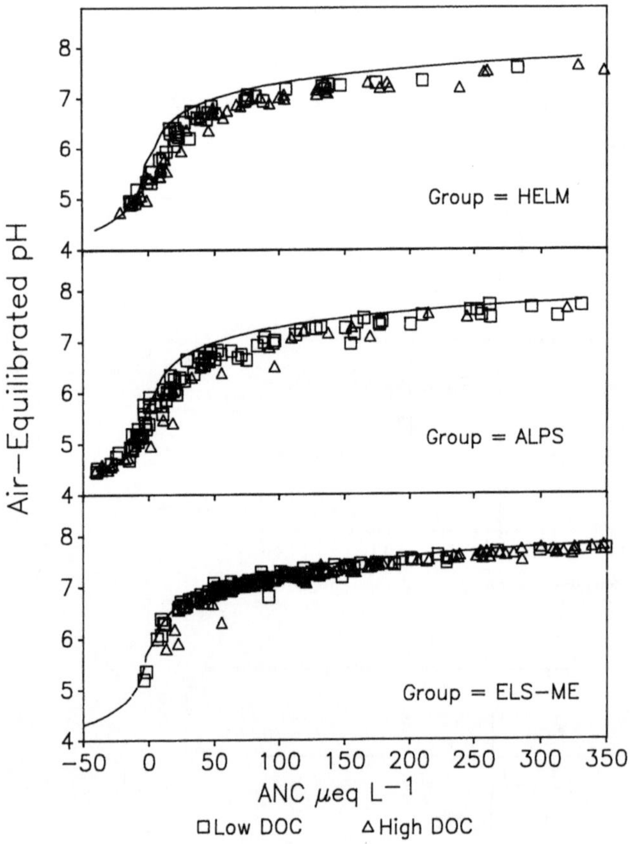

FIGURE 7.16. Air-equilibrated pH versus ANC, by project and DOC class. High DOC lakes are those with DOC > 417 μmol L^{-1}. The theoretical pH line on each panel is calculated assuming that ANC is entirely from the carbonate system. Weak acids and analytical errors cause deviations from the theoretical.

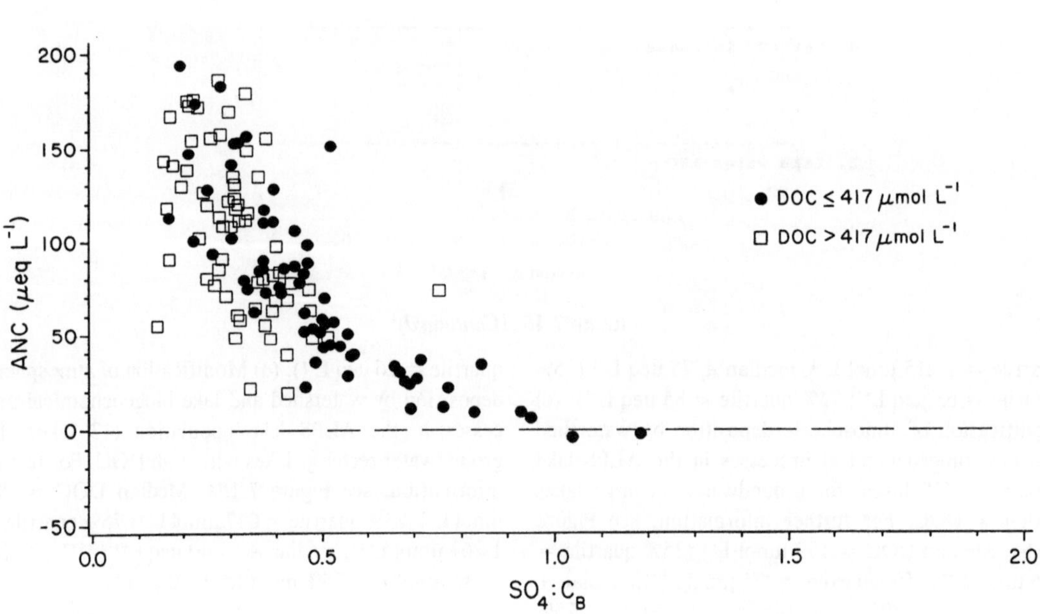

FIGURE 7.17. Relationship between ANC and SO_4^{2-}:C_B for lakes sampled in the ELS-I.

deposition on all Maine lakes, but by itself does not indicate loss of ANC due to atmospheric deposition. A relationship as simple as an inverse correlation between pH and SO_4^{2-} does not occur in population data, because of the diversity of water pathways and the resultant base cation release. However, if Maine lakes once had HCO_3^- waters, then the fraction of the anions now represented by SO_4^{2-} (the sulfate fraction) should indicate an influence of acidic deposition. Twenty-four percent of ELS-I lakes, and 62% and 56% of HELM and ALPS lakes, respectively, have SO_4^{2-} concentrations higher than HCO_3^- concentrations.

Because base cations (C_B) have fewer potential unknown or unmeasured terms, the SO_4^{2-} fraction is expressed as $SO_4^{2-}:C_B$ in this book (Chapter 2, Munson and Gherini, this volume; Chapter 18, Sullivan, this volume). As a group, the HELM lakes have higher SO_4^{2-} fractions than the ELS-I or ALPS lakes. Figure 7.17 presents the data in a standard format that allows for comparison of ELS-I data with data in similar format for other regions discussed in other chapters; Figure 7.18 shows comparisons among Maine chapter data sets. These distributions reflect the relative importance of acidification mechanisms in the three lake groups. The SO_4^{2-} fraction is somewhat lower in ALPS lakes than in either the HELM or ELS-I lakes. For ALPS lakes, the relative contribution of SO_4^{2-} with respect to ANC is more variable than for the other groups (Figure 7.18). Lakes plotting in the lower left of each panel of Figure 7.18 are more chemically influenced by organic acids, whereas lakes in the upper left are interpreted to be more influenced by precipitation chemistry. The low ANC ALPS lakes are more influenced by organic acids than any other group of lakes.

The ion enrichment diagrams (Figures 7.13 to 7.15) can be used to interpret how lake types process nitrogen from atmospheric deposition. For all DOC classes and hydrologic types, NO_3^- concentrations are low or unmeasurable. Thus, the NO_3^- deposited from the atmosphere is assimilated by vegetation, resulting in a gain of ANC that neutralizes a portion of the precipitation acidity. For each lake type, NH_4^+ uptake consumes a small amount of ANC, apparently less than that generated by NO_3^- uptake.

Dry Deposition

There is normally a poor correspondence between wet deposition of SO_4^{2-} and either SO_4^{2-} concentration in lakes or SO_4^{2-} flux from catchments at the local scale. Typically, the flux per unit area of SO_4^{2-} from watersheds exceeds the wet atmospheric flux (NRC 1986). The excess is ascribed to dry deposition of sulfur compounds. Interception of cloud and fog moisture is also a potential source of SO_4^{2-} (Lovett et al. 1982). Sediment sulfur storage, SO_4^{2-} adsorbed by soils, and biological sulfur uptake in the catchment would all decrease lakewater concentrations and lead to an underestimation of dry deposition. Geologic sulfur and biological sulfur release would raise lakewater concentrations, but are considered to be less important than the factors that decrease lakewater SO_4^{2-}.

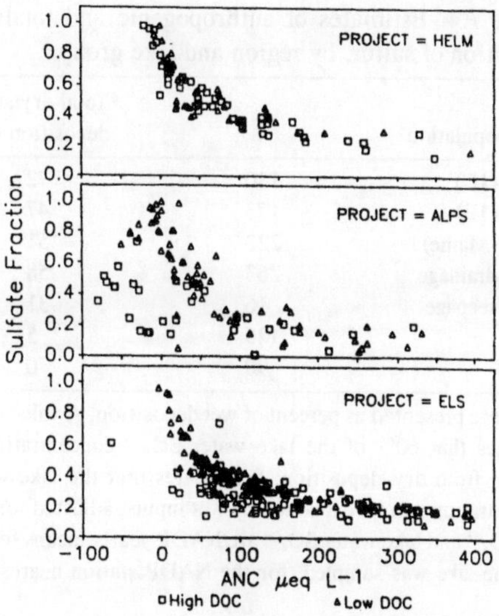

FIGURE 7.18. Relationship of the ratio [nonmarine SO_4^{2-} to sum of nonmarine cations] to ANC by lake population. Lakes plotting toward the lower left of a panel are assumed to be acidic due largely to organic acidity (generally high DOC [> 417 µmol L^{-1}] lakes); lakes plotting toward the upper left of a panel are assumed to be acidic due to SO_4^{2-} acidity from atmospheric deposition (generally low DOC [≤ 417 µmol L^{-1}] lakes).

Estimates of sulfur dry deposition based on lakewater concentrations relative to wet-only deposition indicate that the contribution decreases northeastward (Norton et al. 1988a). These calculations yield values for dry deposition similar to those obtained from watershed budgets (Haines et al. 1990; Table 7.4), and are in general agreement with the data presented in Appendix B (Baker, this volume). The HELM lakes appear to receive dry deposition comparable to that received in ELS regions to the south and west (Table 7.4). In contrast, the ALPS lakes

TABLE 7.4. Estimates of anthropogenic and total dry deposition of sulfur, by region and lake group.[a]

Lake population	n	[b]Total dry sulfur deposition (%)
ELS-I (1C)	130	72
ELS-I (1E)	173	47
ELS-I (Maine)	222	51
ELS-I drainage	263	58
ELS-I seepage	16	31
ALPS	125	5
HELM	90	70

[a] Data are presented as percent of wet deposition. A value of 60 indicates that 60% of the lake water SO_4^{2-} concentration is derived from dry deposition; 0 indicates that the lake water concentration can be explained by wet inputs, adjusted for ET. Wet SO_4^{2-} and lake water SO_4^{2-} are NADP yearly means for the year the lake was sampled, for the NADP station nearest the lake.

[b] $\{(\text{lake water } SO_4^{2-}) - [(\text{wet } SO_4^{2-}) \times 1.6]/(\text{wet } SO_4^{2-})\} \times 100$.

appear to receive an insignificant portion of their sulfur input from dry deposition. The SO_4^{2-} distribution mode in the ALPS seepage lake population is 50 μeq L^{-1}, 25 μeq L^{-1} lower than that of ELS-I lakes in Maine (Figure 7.19). The 50 μeq L^{-1} value could result simply from the mean concentration in precipitation (30 μeq L^{-1}), concentrated by evapotransporation. These data suggest that dry deposition contributions to seepage lakes are small; the lakes have small, low relief watersheds that do not effectively intercept dry deposition. An alternative hypothesis is that SO_4^{2-} reduction is a factor that affects SO_4^{2-} concentrations. This possibility is discussed in the section on sulfur retention.

Marine Aerosols

Maine is unique in the northeastern United States in having many low ANC lakes that receive precipitation laden with a significant marine chemical component. Neutral salts in precipitation may cause episodic decreases in ANC in surface waters, due to an ion exchange salt effect in soils (Kahl et al. 1985, Seip 1980, Rosenqvist 1978a). ANC depressions of up to 30 μeq L^{-1} have been observed in first-order streams at Acadia National Park, related in part to the salt effect (Kahl et al. 1985). There were no increases in either SO_4^{2-} or NO_3^- during the events. The salts are delivered in wet precipitation, or they may accumulate from dry deposition and then be mobilized by rainfall or

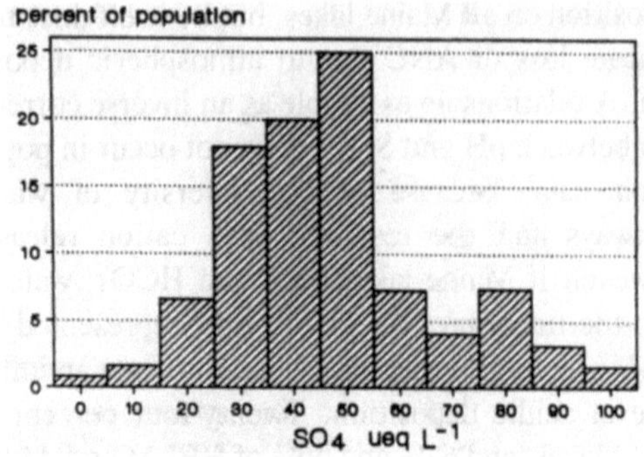

FIGURE 7.19. SO_4^{2-} concentration distribution in the ALPS project seepage lakes.

snowmelt. High concentrations of salt relative to normal values are necessary for the effect to occur. In the soil, H^+ and Ca^{2+} are temporarily displaced by the neutral salts in the percolating soil solution, and Cl^- becomes an important mobile anion. The process requires acidic soils, but it illustrates that acidic precipitation is not necessary for episodic acidification to occur. Chronic acidification by the salt effect alone is not possible, due to the reversible nature of the Na^+ and Mg^{2+} exchange reaction (Seip 1980, Sullivan et al. 1988b).

Sulfur Retention

Watershed sulfur retention may be of two major types, terrestrial or in-lake. Retention by soils is considered of slight importance in the northeast, except on a seasonal basis. Northern Spodosols contain high concentrations of Fe and Al oxides, but most evidence suggests that these forest soils do not presently retain a significant fraction of the atmospheric deposition of SO_4^{2-} (Rochelle and Church 1987, Johnson and Cole 1980). However, SO_4^{2-} adsorption may be important under some special conditions in northeastern watersheds. For example, Nodvin et al. (1988) reported that SO_4^{2-} retention increased significantly in Becket soils at the Hubbard Brook Experimental Forest, following timber harvesting. Cronan (1985) found that soil column leachates from a sandy glacial outwash with Adams series Haplorthod exhibited significant depletion of SO_4^{2-} (up to 95%) compared to precipitation inputs. Chemical budgets from calibrated low-order watersheds illustrate that seasonal sulfur retention is common, probably as a result of retention in upper soil horizons and small wetlands (Haines et al. 1990, Kerekes et al. 1986).

Schindler (1986) established that in-lake generation of alkalinity by sulfur reduction can be a significant fraction of total ANC, if alkalinity generation by the watershed is low. Elevated concentrations of sulfur in recent lake sediments (Mitchell et al. 1985, Nriagu and Coker 1983) are presumably related to acidic deposition. This excess sedimentary sulfur is assumed to have produced ANC.

Using the amount of sulfur storage in sediments of known age, and the maximum net accumulation rate of sulfur, it is possible to establish an upper limit on the alkalinity generation by SO_4^{2-} reduction for specific systems. Evaluation of such data from sediment cores from five low ANC drainage lakes in Maine and Vermont indicates that maximum alkalinity generation from sulfur storage in these lakes is < 10 μeq L^{-1}, and averages < 5 μeq L^{-1} (Norton et al. 1988c). The data are from single hypolimnetic cores, which overestimate sulfur deposition relative to the entire lake. Also, if any portion of the sediment sulfur is geologic, that portion does not contribute to ANC production, but will have been interpreted as such. Sulfur reduction in sediments of drainage lakes appears to contribute only a minor part of the alkalinity generated in northeastern drainage lakes (Norton et al. 1988c, Schafran and Driscoll 1987).

On a unit area basis, alkalinity generation by SO_4^{2-} reduction in lakes may be significant relative to terrestrial alkalinity generation (Schindler 1986, Cook et al. (1987). In-lake alkalinity generation is thus potentially important in Maine seepage lakes, especially hydrologically perched lakes. No data on sulfur reduction in Maine seepage lakes are available, but Baker et al. (1986a) have suggested that sulfur retention in similar systems in Florida and the Upper Midwest may be as much as 40%. In-lake ANC generation as a fraction of total ANC production is substantial in these systems (Baker and Brezonik 1988, Kelly et al. 1987). The ion enrichment diagrams (Figures 7.13 to 7.15) indicate that SO_4^{2-} retention is a net contributor to ANC production in the ALPS seepage lakes, and in the high DOC ELS-I lakes. However, the low relief, tree deficient catchments of the high DOC lakes appear to intercept less dry deposition and aerosols (Norton et al. 1988a) than the general estimate used in this book (Appendix B, Baker et al., this volume). Thus, the contributions to ANC from SO_4^{2-} retention may be overestimated in the seepage lakes by the ion enrichment diagram method, because SO_4^{2-} dry deposition is probably overestimated for seepage lakes.

Evidence of Historical Trends

Historical Water Chemistry

Data on historical water chemistry are sparse in Maine, as they are in all areas of North America. The available data are (1) colorimetric pH measurements from the 1940s (Fuller and Cooper 1946) and (2) often incomplete chemistry on lakes from the late 1970s and early 1980s (Cowing and Scott 1975, 1976, 1977, Davis et al. 1978, Norton et al. 1981, Haines and Akeliazsiek 1983, Kahl et al. 1985). Reliable ANC measurements were not made in Maine until 1982. Early conclusions about rates of pH change in Maine lakes (Davis et al. 1978) appear to be greatly exaggerated, largely due to incomplete data or inappropriate use of colorimetric pH estimates.

During 1985, Haines (unpublished) resurveyed 50 lakes originally studied by Fuller and Cooper (1946). To the extent possible, methods, colorimetric indicators, and sampling dates were duplicated to maximize comparability. Epilimnion pH comparisons indicate no difference between these data collected 40 years apart (Figure 7.20).

Other surveys have provided pH, cation, and some SO_4^{2-} data. Although pH in many lakes appears to have increased by a few tenths to as much as 0.7 units (Kahl unpublished), the apparent trends are probably partly the result of comparing the pH values of field samples supersaturated with CO_2 and measured with slow response gel-filled pH electrodes to standardized air-equilibrated data measured with better equipment in the laboratory after 1984.

At least partial chemical characterization is available for 37 lakes, spanning some portion of the period from 1978 to 1988. Of these lakes, 11 apparently have higher pH than in the past, 8 are essentially the same, and none has declined. ANC increased in 9 lakes, and decreased in 2 lakes during the 6-year period for which there are Gran-plot measurements. Calcium concentrations changed in parallel to ANC changes. Sulfate

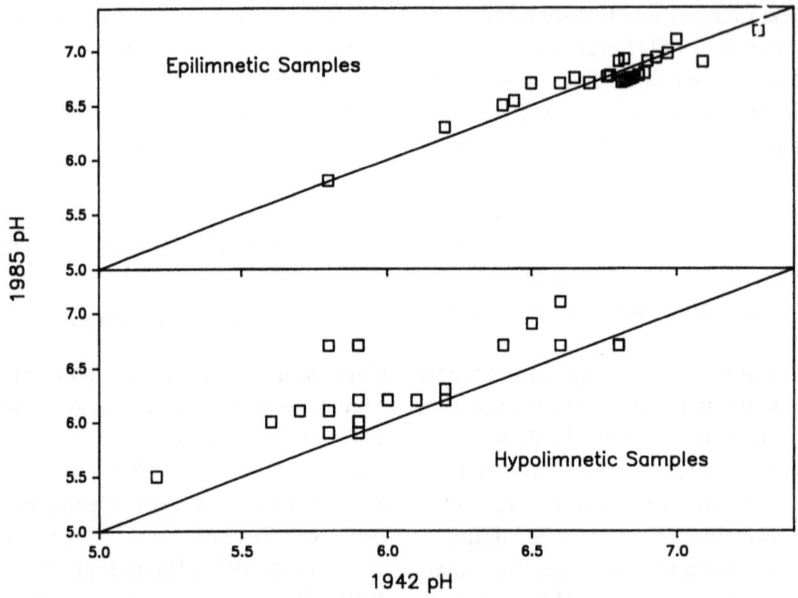

FIGURE 7.20. Haines 1985 resurvey of Fuller and Cooper (1946) lakes. Epilimnion samples were collected in the top meter of the water column. Samples considered hypolimnetic were those taken from depths generally approximating the deep point of the lake. Data were not always available to determine whether the lake was stratified at the time of collection.

increased in five lakes, and decreased in six. The period of record for any lake is at most 10 years, too short to show real trends. However, neither widespread nor rapid acidification in the last decade is indicated by these data.

LTM Lakes at Tunk Mountain

Five ponds located on the Tunk Mountain Granite (Karner 1968) have been studied since 1982, as part of the Tunk Mountain Watershed Study (TMWS) and the EPA Long-Term Monitoring Project (LTM). Selected mean chemistry during fall turnover is listed in Table 7.5. Two ponds are circumneutral (ANC 45 to 60 μeq L^{-1}), two have ANC near zero (0 to 15 μeq L^{-1}), and one is acidic (ANC < −15 μeq L^{-1}). Mud Pond is acidic due to (1) lower base cations, especially Ca^{2+}, and (2) higher SO_4^{2-} concentrations, by a factor of 2, compared to Tilden or Salmon Ponds. The higher SO_4^{2-} is inferred to result from greater interception of dry deposition. The softwood vegetation predominant in the Mud Pond catchment, and the steep southwest watershed aspect, intercept the prevailing winds more than the mixed-forest catchments of the other lakes facing south or east. Sodium, Cl^-,

TABLE 7.5. Mean fall index chemistry for five study lakes, 1985–88, collected during November of each year.

	pH[b]	ANC[b]	Spec. conduct.[b]	DOC[b]	Ca^{2+b}	Total Al[b]	Cl^{-b}	NO_3^{-b}	SO_4^{2-b}
Anderson Pond	6.26	14.5	17.3	150	40	0.7	78	0.0	55
Little Long Pond	6.11	10.8	21.5	137	46	1.3	89	0.0	78
Mud Pond	4.67	−22.8	31.6	362	31	17.5	99	0.0	102
Salmon Pond	6.84	59.4	22.7	290	72	0.8	78	0.0	50
Tilden Pond	6.76	50.3	22.4	237	67	1.0	84	0.0	59

[a] All lakes are part of the Tunk Mountain Watershed Study (TMWS) and the EPA Long-Term Monitoring (LTM) Project. The period for reporting was chosen to correspond with the collection of EPA NSWS methods for air-equilibrated pH for the Maine data.

[b] Data aie in μeq L^{-1} except for pH (pH units), DOC and Al (μmol L^{-1}), and conductivity (μS cm^{-1}).

FIGURE 7.21. ANC and SO_4^{2-} concentrations since 1982 in the Tunk Mountain Watershed lakes (EPA Long-Term Monitoring Project). Data are in μeq L^{-1}.

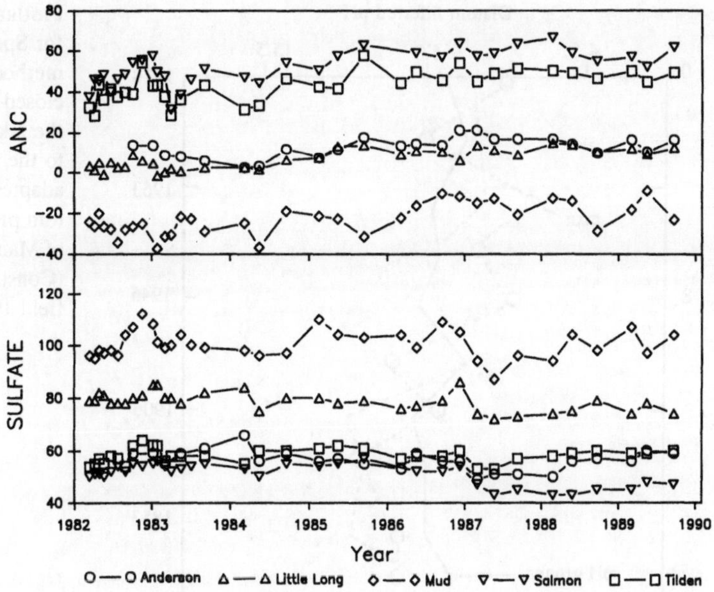

and SO_4^{2-} are highest in Mud Pond compared to the other lakes. These three parameters are also in highest concentration in precipitation and in dry deposition.

Water chemistry data since 1982 are presented in Figure 7.21. ANC values have increased in each of the lakes during the fall index period by as much as 5 to 15 μeq L^{-1} in the six years of study. These increases are significant in four of the five lakes at the $p \leq 0.05$ level (Seasonal Kendall Tau test, Hirsch et al. 1982). Sulfate concentrations have declined slightly, in a pattern parallel to that of precipitation (Chapter 3, Husar et al., this volume; NADP 1982–88), and to the pattern reported by Lins (1986) for northeastern benchmark streams. The decrease in SO_4^{2-} is significant at $p \leq 0.05$ in Little Long Pond. Although eight years is much too short a period to indicate long-term trends, these low ANC lakes clearly are not immediately at risk of further acidification at present levels of deposition.

Paleolimnological Evidence

Diatom pH reconstructions have been performed on ^{210}Pb dated sediment cores from nine lakes in Maine (Davis et al. 1983, Davis et al. 1988). Chrysophyte pH reconstructions have been done on two of these lakes. The conclusions drawn are consistent among lakes, and are consistent with other inferences of acidification. Lakes studied in Maine show little or no acidification during the past 100 years. Speck Pond, the second highest and most southwesterly HELM lake, is typical of low pH lakes (Figure 7.22). The data from diatoms suggest a pH level of about 4.8 in 1978, with a slight decrease in pH from about 5.2 in the last century (R. Davis, cited in Charles and Norton 1986). In 1986–87, Speck Pond had a mean pH level of 5.2 (Kahl and Scott 1990). In general, available paleolimnology data suggest that acidic lakes have always had low pH, but that some have acidified slightly in this century (Charles and Norton 1986). In most instances, pH values recorded in the mid- to late 1980s are the same as the inferred past pH, and are slightly higher than the inferred diatom pH from the early to mid-1980s.

Future Considerations

Existing Low ANC Lakes

Paleolimnological investigations in New England have concluded that many acidic lakes have been low ANC (or acidic) for hundreds of years. Others may have acidified slightly in the past 20 to 50 years. Of the latter, however, all are inferred to have had pH < 6.0 in prehistoric times. Lakes in other pH ranges are either increasing in ANC or show no trend, based on either paleolimnological or existing modern water chemistry data. Therefore, only lakes that currently have pH < 6.0 are considered to be

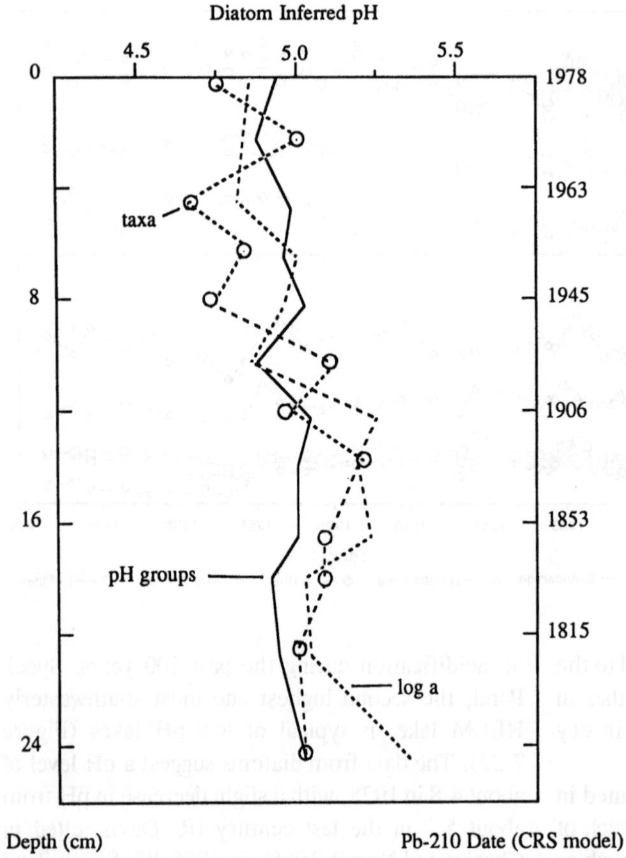

FIGURE 7.22. Downcore diatom inferred pH for Speck Pond, Maine, calculated by several methods (from Battarbee 1984). The 1986–88 closed-system (comparable to in situ) pH of the lake ranged from 5.09 to 5.37, compared to the 1978 inferred pH of 4.5 to 4.9. Figure adapted from Charles and Norton (1986). Diatom pH data are from R.B. Davis, University of Maine; ^{210}Pb dates are from the CRS model (Constant Rate of Supply; Appleby and Oldfield 1978).

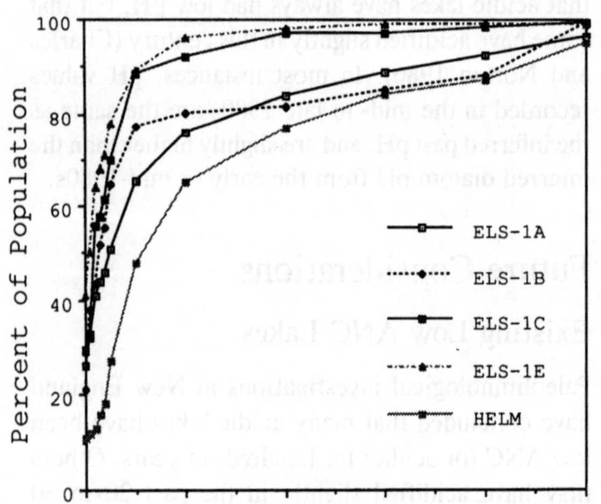

FIGURE 7.23. Cumulative frequency distribution of NO_3^- for the ELS-I populations, and for the HELM project high-elevation lakes. Data are in μeq L^{-1}.

at risk of further acidification at present loadings from the atmosphere. Utilizing the database from which the other data for this chapter are derived, only 41 nondystrophic lakes in Maine are known to have pH between 5.0 and 6.0 and ANC < 20 μeq L^{-1}. Applying the statistical extrapolations from the ELS-I to all Maine lakes, fewer than 100 lakes are thus inferred to be at any risk of further acidification under any reasonable scenario of future emissions. Because existing data for lakes of similar pH suggest recent slight decreases in acidity, rather than increases, Maine surface waters are unlikely to become more acidic in the next few decades at current deposition loadings.

Possible Chemical Changes in Acidic Deposition

Estimates indicate that SO_2 emissions in the northeast have remained constant or declined since

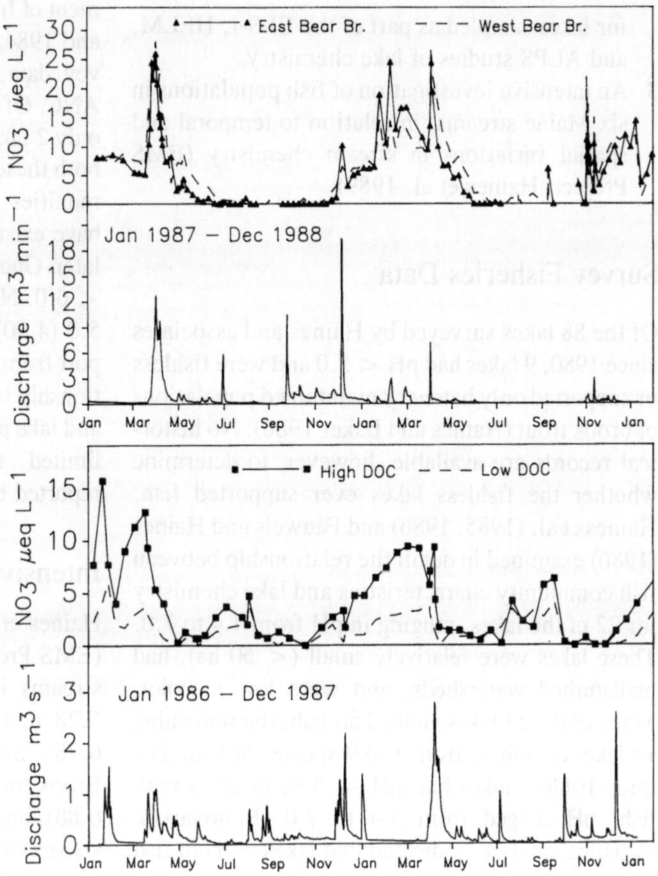

FIGURE 7.24. Nitrate concentrations in two headwater streams, 1987 and 1988 (upper panels), and mean NO_3 concentrations in high DOC and low DOC streams, 1986 and 1987 (lower panels), from Haines et al. (1989). High DOC streams have DOC > 800 µmol L^{-1} (mean = 925); low DOC streams have DOC ≤ 800 µmol L^{-1} (mean = 312).

the 1930s (Chapter 3, Husar et al., this volume). NO_x emissions are inferred to have increased monotonically since the mid-1800s, until leveling off in the mid-1970s. Wet deposition of SO_4^{2-} has declined in northern New England at least since 1970 (Driscoll et al. 1989, NADP 1981–1988), with the decrease being as much as 30% in New England (Chapter 3, Husar et al., this volume). Nitrate in precipitation has shown no recent trend for the region.

These emission and precipitation trends suggest that increases in SO_4^{2-} in either precipitation or surface waters may be less of a concern than increases in NO_3^-. In the Adirondack Mountains, increased NO_3^- in streams has been documented by Driscoll et al. (1989), leading to speculation about forest health, and the potential for surface water acidification due to increases in NO_3^- concentrations and flux. This potential change in NO_3^- cycling is of concern in Maine. Nitrate population data from the high-elevation lakes (Figure 7.23) in comparison to results from the ELS-I and results of recent stream studies (Figure 7.24; Haines et al. 1990, Norton et al. unpublished) show seasonal NO_3^- loss from upland watersheds. No trends can be inferred, due to a lack of historical data.

Biological Status

Fish Communities

Three sources provide data on the present-day status of fish communities in surface waters in Maine:

1. Surveys conducted since 1980 by Haines et al. (1985, 1986), Pauwels and Haines (1986), and Haines and Baker (1986) in 88 lakes selected specifically to evaluate the relationship between fish community status and lake chemistry.
2. Survey data collected by the Maine Department of Inland Fisheries and Wildlife as part of normal fisheries management operations, compiled

for lakes sampled as part of the ELS-I, HELM, and ALPS studies of lake chemistry.

3. An intensive investigation of fish populations in six Maine streams, in relation to temporal and spatial variations in stream chemistry (EMS Project; Haines et al. 1989).

Survey Fisheries Data

Of the 88 lakes surveyed by Haines and associates since 1980, 9 lakes had pH < 5.0 and were fishless or supported only hatchery maintained populations of brook trout (Haines and Baker 1986). No historical records are available, however, to determine whether the fishless lakes ever supported fish. Haines et al. (1985, 1986) and Pauwels and Haines (1986) examined in detail the relationship between fish community characteristics and lake chemistry for 22 of the lakes, ranging in pH from 4.4 to 7.0. These lakes were relatively small (< 50 ha), had undisturbed watersheds, and were low in color. Three of the 22 lakes yielded no fish; the remaining 19 lakes contained from 1 to 9 species of fish. The three fishless lakes had pH < 5.0; in lakes with fish, pH ranged from 5.4 to 7.0. Multivariate statistical analyses indicated that lakes without fish differed from lakes supporting fish primarily with respect to lake chemistry variables related to acidity (i.e., pH, Al, and Ca^{2+}). For the 19 lakes with fish, on the other hand, among-lake variations in species richness and numbers of fish caught were most closely associated with total ionic strength, maximum lake depth, and lake area—factors related to lake productivity and habitat diversity, rather than lake acidity. Several fish species, in particular creek chub (*Semotilus atromaculatus*) and common shiner (*Notropis cornutus*), however, were typically restricted to lakes with pH > 6.0. Other studies have also suggested that minnow species may be especially sensitive to acidic conditions (Mills and Schindler 1986). Thus, the absence of minnow species in lakes with pH < 6.0 may reflect the intolerance of these fish to low pH, although the potential importance of other factors (e.g., biological interactions, lake productivity, and lake habitat characteristics) cannot be dismissed.

Of the lakes surveyed for water chemistry during the ELS-I (subregion 1E), HELM, and ALPS studies, 87 were surveyed for fish by the Maine Department of Inland Fisheries and Wildlife between 1975 and 1984. Most of the lakes with existing fish survey data, however, have relatively high pH and ANC. Of the 62 ELS-I lakes with fish survey data, only 2 lakes had pH < 6.0 (pH 5.73 and pH 5.90); both these lakes presently support diverse fish communities. Twenty-five of the HELM and ALPS lakes have existing fish survey data collected in 1975 or later. One of these 25 lakes had pH < 5.0; 8 had pH < 6.0. No fish were caught in the lake with pH < 5.0 (4.90). The remaining 24 lakes presently support from 1 to 8 fish species, with no apparent relationship between the number of fish species caught and lake pH. Although the data for low pH lakes are limited, these results are consistent with those reported by Haines et al. (1986).

Intensive Stream Fisheries Studies

Haines et al. (1990) conducted an intensive study (EMS Project) of fish populations in six low-order streams in eastern Maine: Halfmile Brook (pH 5.28 to 6.96), Indian Camp Brook (pH 5.28 to 6.78), Spring Brook (pH 5.81 to 6.94), Sinclair Brook (pH 5.07 to 6.88), Rocky Brook (pH 5.23 to 6.88), and Baker Brook (pH 5.33 to 6.98). Of the six streams, Sinclair Brook had the lowest recorded stream pH (5.07), measured in December 1987; pH levels < 5.2 were also recorded in Sinclair Brook in January 1986.

Each of the six study streams supported two or more species of fish, with the lowest number of species occurring in Sinclair Brook (Table 7.6). All streams contained populations of brook trout; three of the six streams also supported Atlantic salmon. In addition, all streams except Sinclair Brook supported one or more species of forage fish; blacknose dace was the most abundant and ubiquitous of the forage species. The absence of forage fish from Sinclair Brook may be the result of unsuitable water chemistry or unsuitable physical stream habitat (e.g., the relatively steep gradient and cold water temperatures in Sinclair Brook), or both.

Among-stream variations in the abundance and production of brook trout and Atlantic salmon were unrelated to stream chemistry. Sinclair Brook, with the lowest measured levels of pH, had the highest annual average values of salmon and trout production. In general, fluctuations in fish abundance over time were also not clearly associated with temporal

TABLE 7.6. Fish species caught in six streams in Eastern Maine.[a]

Fish species	Halfmile Brook	Indian Camp Brook	Spring Brook	Sinclair Brook	Rocky Brook	Baker Brook
Brook trout (*Salvelinus fontinalis*)	x	x	x	x	x	x
Atlantic Salmon (*Salmo salar*)				x	x	x
Brown trout (*Salmo trutta*)			x			
Blacknose dace (*Rhinichthys atratulus*)	x	x	x		x	x
Finescale dace (*Phoxinus neogaeus*)		x	x		x	
Creek chub (*Semotilus atromaculatus*)	x	x	x		x	x
Common shiner (*Notropis cornutus*)		x	x		x	x
Fallfish (*Semotilus corporalis*)		x	x			
Golden shiner (*Notemigonus crysoleucas*)		x				
White sucker (*Catostomus commersoni*)	x	x	x		x	x
Three-spine stickleback (*Gasterosteus aculeatusx*)		x	x			
Ninespine stickleback (*Pungitius pungitius*)		x	x			
Chain pickerel (*Esox niger*)					x	x
Smallmouth bass (*Micropterus dolomieui*)			x			

[a] Data from Haines et al. 1990.

fluctuations in water chemistry, with one possible exception. A marked increase in the mortality of Atlantic salmon presmolts was observed in Sinclair Brook during the period from fall 1985 to spring 1986, perhaps associated with an episodic pH depression to pH < 5.2. The sensitivity of Atlantic salmon to low pH episodes was confirmed during acidification experiments in artificial stream channels. Atlantic salmon fry exposed to pH 5.5 had significantly higher mortality than fish in unaltered stream water. Exposure of Atlantic salmon smolts to pH < 5.5 and exchangeable Al levels > 200 $\mu g\ L^{-1}$ in the artificial stream channels also resulted in some increase in mortality and iono-regulatory stress, reflected by a decline in blood plasma levels of Na^+ and Cl^-, and an increase in hematocrits.

Results from fisheries surveys of Maine surface waters are generally consistent with results and patterns observed in other regions. Brook trout are relatively tolerant of low pH, occurring at pH levels as low as 4.6 to 5.0 in the Adirondack region of New York (Chapter 4, Baker and Christensen, this volume; Chapter 6, Driscoll et al., this volume). Atlantic salmon, on the other hand, may be impacted at somewhat higher pH levels. In Nova Scotia, marked declines in Atlantic salmon catch have been observed for rivers with pH < 5.3 (Watt et al. 1983). Mills and Schindler (1986) suggest that many minnow species are particularly sensitive to acidity, and are often absent at pH levels $<$ 6.0. Thus, it is possible, though not proven, that acidic conditions limit the distribution of some fish species in Maine. The regional extent of such effects cannot be estimated at this time, although the number and area of lakes and streams with low pH levels are small, relative to other regions of the northeastern United States.

Historical Fisheries Data

Haines and Baker (1986) compiled available historical fisheries and chemistry data for 1,459 lakes from the Maine Department of Inland Fisheries and Wildlife. Of these lakes, 312 had been surveyed twice or more and had fisheries survey data suitable for assessment of long-term trends. Twelve of these lakes had lost one or more fish species without obvious cause (i.e., the loss of fish could not be accounted for by reclamation, change in stocking policy, or obvious problems with recent fish sampling procedures). Chemical data were unavailable for 2 of these 12 lakes; 10 lakes had colorimetric pH measurements, although in a number of cases the pH data predated the most recent fish survey. Two lakes had lost lake trout (*Salvelinus namaycush*) populations and were acidic (pH 4.4 to 4.8). The remaining eight lakes had lost brook trout (*Salvelinus fontinalis*), white sucker (*Catostomus commersoni*), or yellow perch (*Perca flavescens*) populations, but had pH > 5.5. Analysis of these data suggests that temporal shifts in fish community composition have occurred in some Maine lakes, although there is no clear association between the decline or loss of fish populations and lake acidity.

Other Aquatic Biota

Few surveys or studies have been conducted on the potential effects of acidification on nonfish biota in Maine surface waters. As part of the HELM project, zooplankton samples were collected from 83 HELM lakes; these data were compared to zooplankton data collected as part of the ELS-II program that sampled a subset of the ELS-I lakes (Tessier and Horwitz 1988a). Although the results of the zooplankton analyses from the HELM lakes generally confirm the data from ELS-II (Tessier and Horwitz 1988b), several important differences were noted. For each data set, the covariance structure indicates that the controlling factors are acidity, dystrophy, and salinity. The body size structure was similar between the two populations, but the response to environmental factors was different. Large zooplankton were not related to any acidity gradients, in contrast to the results of ELS-II. Tessier and Horwitz (1988a) interpret this difference to be related to planktivory. The small, shallow nature of the HELM lakes is inferred to provide less refuge for the larger zooplankton. However, due to decreasing numbers of Daphnia at low pH, the zooplankton as a group showed decreasing abundance in the more acidic lakes. This observation was not made in ELS-II.

Four Maine lakes, two acidic (pH 4.5 to 4.8) and fishless, and two circumneutral (pH 6.3 to 6.8) with fish, were sampled by Hunter et al. (1985, 1986b) to evaluate the potential influence of lake acidity on the growth and behavior of black ducklings (*Anas rubripes*). Ducklings reared on the two acidic lakes grew significantly faster, and spent less time searching for food than did ducklings on the circumneutral lakes. Sweep-net samples collected markedly higher numbers and biomass of invertebrates in the acidic lakes than in the two lakes with higher pH. Fish enclosure experiments in these lakes suggested that the presence of fish decreased invertebrate abundance. Based on this study, Hunter et al. (1985, 1986b) hypothesized that black ducklings and fish compete for food, and that the negative effects of acidification on fish may be beneficial for duckling growth and survival. The overall effects of acidification on black duck populations, or on waterfowl in general, remain uncertain, however. Not all life stages feed primarily on invertebrates or directly compete with fish for food. Some waterfowl species, such as loons, rely on fish as a primary food source.

Conclusions

1. The State of Maine receives precipitation that ranges in mean annual volume-weighted pH from 4.4 to 4.6, south to north.
2. Large portions of the state are underlain by chemically resistant bedrock, and/or covered by thin, acidic soils. Thus, many of the nearly 6,000 lakes and ponds, and over 35,000 km of low-order streams, are poorly buffered and perhaps vulnerable to the effects of acidic deposition.
3. Fewer than an estimated 100 nondystrophic lakes larger than 0.4 ha are presently acidic (ANC $\leq$ 0).
4. Twelve of the 90 lakes above 600 m in elevation were acidic in 1987. Thirty percent of the seepage lakes sampled during 1986–87 were acidic.
5. About 30% of the known acidic lakes are acidic largely due to organic acidity. Some of these lakes would be acidic without the additional acidity from atmospheric deposition. About 60% of the acidic lakes are acidic largely due to acidic deposition.
6. No low DOC ($\leq$ 417 µmol L^{-1}) lakes have pH $<$ 4.9, suggesting that organic acidity is necessary to depress the pH to values lower than about 5.0, at present loading of acids from the atmosphere.
7. Chronically acidic perennial low DOC streams are unknown in Maine. Episodic acidification occurs during snowmelt and after heavy rainfall. Occasional acidification to pH $<$ 5.0 has been documented in some systems, typically in conjunction with increases in DOC or NO_3^-.
8. Several processes exert major control on the chemical status of Maine surface waters. The most important of these are mineral weathering, hydrologic routing, organic acidity, atmospheric deposition, and sulfur retention. Sulfur retention appears to be important only on a localized, often seasonal, scale.
9. Adsorption of sulfur in soils does not appear to be important on a regional scale, although temporary retention during baseflow periods in

summer and fall may be common in many soil types.

10. Available historical data suggest that lake pH may have been relatively constant or may have increased slightly during the past 10 to 40 years. Diatom pH reconstructions from lake sediment cores indicate minor acidification of some lakes during the past century; however, presently acidic lakes have always been low pH or acidic.
11. Existing data suggest that fewer than 100 Maine lakes are potentially at risk of further acidification, at current levels of acidic deposition.
12. Although numerous lakes are fishless, no lakes have been shown to have lost their fish due to acidification. Many of the fishless lakes are small, isolated, and/or at high elevation, with poor breeding habitat.

Acknowledgments. We thank the staff of the Geochemistry Laboratory at the University of Maine, who made much of this work possible. The encouragement, comments, and reviews from Don Charles, Bob Cook, Joe Eilers, John Melack, Matt Scott, and Tim Sullivan both stimulated our thoughts and improved the text. Contributions from Joan Baker (the Biological Status section) and Rudy Husar, and computer assistance from Tony Selle, Jeremy Smith, Bronson Crothers, Jay Clausen, Wayne Persons, and John Scofield are also gratefully acknowledged. The reviews and comments by Dave Brakke, Dave Schindler, Hans Martin Seip, and Bob Stauffer, and the editorial assistance by Susan Christie were invaluable and are much appreciated.

The activities on which this chapter is based were financed in part by the U.S. Environmental Protection Agency (WMP, EMS, TMWS and HELM projects); the State of Maine (HELM project); Department of the Interior, U.S. Geological Survey, through the Maine Land and Water Resources Center (ALPS project); and the Department of the Interior, National Park Service (Acadia National Park project). The contents do not necessarily reflect the views and policies of the Agencies or Institutions, nor does mention of trade names of products constitute their endorsement by any of those organizations.

8
Catskill Mountains

John L. Stoddard and Peter S. Murdoch

Abstract. The Catskill region of New York experiences the highest rates of acidic deposition of any region of the United States with low acid neutralizing capacity (ANC) surface waters. Surface waters are represented almost entirely by streams, whose chemical compositions are strongly affected by discharge. Many streams exhibit ANC $\leq$ 0 μeq L^{-1} during high-flow events; a few are acidic at baseflow. The sum of Ca^{2+} and Mg^{2+} concentrations exceeds ANC in all streams by a consistent 120 to 140 μeq L^{-1}; differences between ANC and [Ca^{2+} + Mg^{2+}] are attributable mainly to high SO_4^{2-} concentrations, which result from acidic deposition. Nitrate concentrations increase during high-flow events. Depressed ANC values during snowmelt and storms are the result of dilution of base cations, relatively consistent SO_4^{2-}, and increased NO_3^- levels.

Site-specific studies indicate that stream SO_4^{2-} is matched very closely by SO_4^{2-} concentrations in precipitation, when adjusted for dry deposition and evapotranspiration. Stream nitrate yield is exceeded by atmospheric deposition on an annual basis. Biological processes may play important roles in controlling the amount of NO_3^- supplied to Catskill streams during high-flow events.

Long-term monitoring data (1922 to the present) indicate that ANC values are increasing, largely as a result of an increasing trend in base cations. Sulfate concentrations are decreasing, but any positive effect this might have on ANC is balanced by a sharply increasing trend in NO_3^-. Limited data indicate that acidic deposition has had no deleterious effects on Catskill stream biota.

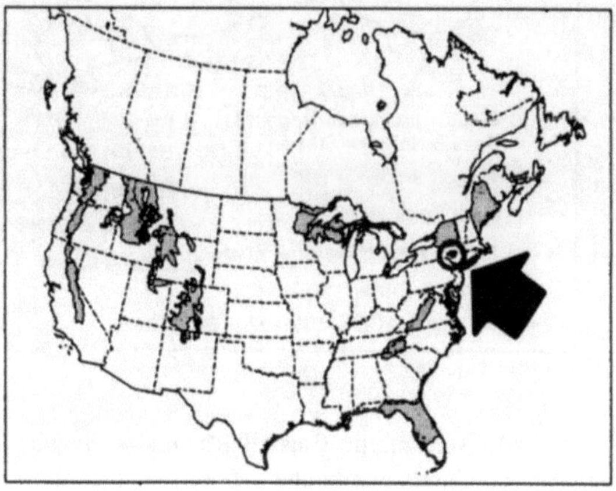

Introduction

The Catskill Mountains of southeastern New York (Chapter 5, Eilers and Selle, this volume) have become a focus of research into the effects of acidic deposition for several reasons. The region experiences higher rates of acidic deposition than any other region in the Northeast (Northeast Overview Section, Driscoll, this volume), and contains several streams with low acid neutralizing capacity (ANC) at baseflow (Murdoch 1988). Streams in the Catskill region supply drinking water for millions of people in the New York City metropolitan area, and Catskill streams have been historically renowned for trout fishing. This chapter summarizes the results of studies conducted by numerous federal, state, and municipal agencies, which indicate that surface waters of the Catskill region have undergone significant chemical changes as a result of acidic deposition.

Surface waters in the Catskill region are represented almost entirely by streams. Only two lakes sampled by the U.S. Environmental Protection Agency (EPA) Eastern Lake Survey (ELS) were

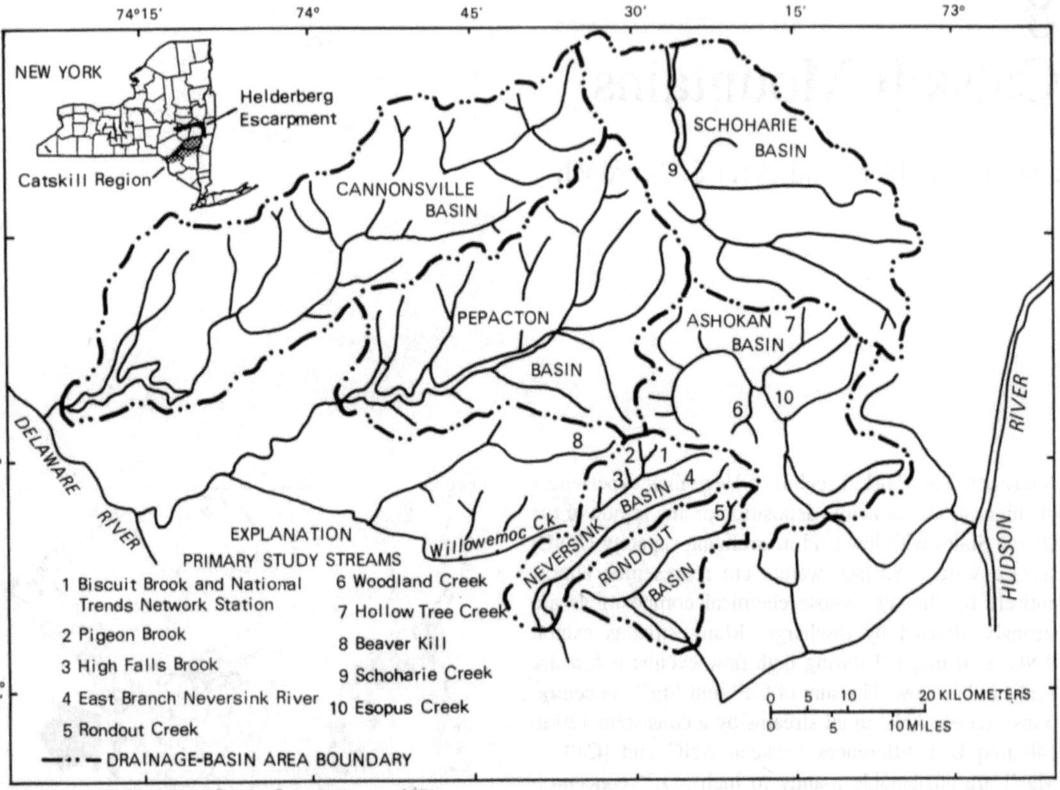

FIGURE 8.1. Map of the Catskill Mountains showing regional boundaries and boundaries of major drainage basins. Also indicated are locations of Long-Term Monitoring (LTM) sites, the Biscuit Brook intensive study and NTN site, and historical data sites on Schoharie and Esopus Creeks.

located within the boundaries of the Catskill region as defined here. As a result, much of the information presented in this chapter is derived from research and monitoring of streams in the region by the U.S. Geological Survey (USGS), the New York City Department of Environmental Protection (NYCDEP), and the EPA.

The USGS has been monitoring stream and precipitation chemistry in the Catskill region since 1983, in cooperation with the NYCDEP during 1985–88 and the EPA during 1983–86. A monitoring station for measuring continuous discharge is being operated at Biscuit Brook in the headwaters of the Neversink River (Figure 8.1). Samples for analyzing stream chemistry are collected at this site weekly, as well as during selected storms and snowmelt. Seven other streams have been sampled and their discharge measured 10 to 12 times per year since 1983 as part of the EPA Long-Term Monitoring (LTM) project. An additional 58 streams (for a total of 66) were sampled seasonally during 1985–87 by USGS and NYCDEP. A National Trends Network (NTN) station adjacent to Biscuit Brook monitors precipitation chemistry and volume weekly. Six Catskill streams were sampled during the EPA's National Stream Survey (NSS) in 1986.

In addition to these recent sampling programs, the NYCDEP has monitored selected chemical variables at approximately 25 stream sites, beginning in 1915 at the earliest site. The USGS has collected water quality data on the Esopus Creek in the Ashokan Basin (Figure 8.1) as part of its Hydrologic Benchmark program since 1967, and also maintains several continuous discharge gauges on large streams throughout the region.

Regional Characteristics

Physiography and Geology

The Catskill Mountains, which form the northeastern end of the Appalachian Plateau (Rich 1934), are the uplifted remnants of a massive Devonian delta that fed into a shallow inland sea to the west. The high ridges are predominantly erosional remnants, and the physiography suggests that the relief is due largely to stream action. The northern, eastern, and western boundaries of the mountain region are defined by the limestone sequences of the Helderberg escarpment, the Hudson River valley, and the Delaware River valley, respectively (Figure 8.1). The southern Catskill boundary is less distinct but can be generally defined by the Delaware River and Rondout Creek valleys.

The Catskill region contains six major watersheds: Esopus Creek (in the Ashokan Basin) and Rondout Creek, which drain east to the Hudson River; Schoharie Creek, which drains north to the Mohawk River; the upper Delaware River and the Willowemoc-Beaverkill systems, which drain west to the main Delaware River; and the Neversink River, which drains south and subsequently to the Delaware River (Figure 8.1). All these streams except the Willowemoc and Beaverkill feed impoundments that provide water for New York City.

Bedrock Composition

Bedrock in the Catskills generally consists of 60% sandstones and interspersed conglomerates and 40% mudstone or siltstone. Quartz makes up approximately 70% of the bedrock detrital material; rock fragments from metamorphic terrane to the east constitute up to 48% of the matrix material (Ethridge 1977). The mineral composition is fairly uniform throughout the region. Muscovite is a common accessory mineral and amphibole is present in small quantities. Calcite and hematite are the primary cement materials, but the amount of interstitial cement is small (Way 1972). Way (1972) reported pyrite and calcite in the shales along the Catskills' southern boundary, and in the area just northeast of the Catskills, but pyrite has not been reported within the Catskill Mountain region.

Surficial Deposits

The surficial deposits of the Catskill Mountains are more varied than the bedrock. Although the last period of continental glaciation (Wisconsinan) ended 14,000 year ago, local alpine-type glaciation continued in many of the headwater valleys. The glacial ice persisted to varying degrees across the area, depending on basin topography, orientation, and the subclimate created by the continental ice margin just to the north. Glacial deposits in headwater valleys are a combination of continental glacier-derived material deposited perpendicular to the direction of regional ice movement and reworked or secondarily scoured alpine glacier-derived material deposited perpendicular to each valley.

Thick drift, deposited and compressed under the glacial ice sheets, and thinner till, deposited as the glaciers melted and retreated, have been reported from repeated glaciations in the main valleys of most Catskill watersheds by Rich (1934). Thin till and exposed bedrock are predominant in the headwater streams, and thick drift is common along valley bottoms. Depth of drift and till varies from a few centimeters along and near ridges, to more than 40 m at the base of the Schoharie basin (Figure 8.1). Holocene alluvial deposits on modern flood plains extend well into many headwater valleys. Glacial striae and stratified deposits indicate that the continental ice sheet flowed downvalley in the Beaverkill and both branches of the Neversink, but upvalley in the Schoharie, Rondout, and Esopus basins. Ice travelled southeastward in the western Catskills, but the valleys containing headwater streams were last occupied by local glaciers moving downslope.

Morainal loops are common in the Catskills (Rich 1934). These consist of unstratified drift deposited at the end of local valley glaciers and are concave upvalley. They have been mapped midway up the Biscuit Brook basin and in the headwaters of the upper East Branch Neversink and Beaverkill basins. Deltaic materials (clays and fine gravels), deposited in ice-impounded lakes by glacial streams, are found in several headwater basins, particularly along the ridge between the Schoharie and Esopus basins (Figure 8.1), where glacial lakes once formed.

Rich (1934) reported that glacial deposits in the lower East Branch Neversink valley (Figure 8.1)

FIGURE 8.2. View of Rondout Creek above Red Brook during snowmelt. Stream channel incised to bedrock, mixed hardwood and hemlock forest, steep slopes, and thin soils are typical of headwater streams in the Catskill Mountains.

are older than Wisconsinan, and dissection of till by streams in this valley is more advanced than would seem possible unless the area had been exposed longer than other valleys to the north. The postglacial flood plains there are wide, and tributary alluvial fans are large. Surficial material in the Catskills consists primarily (> 90%) of local rock and sediment (Parker et al. 1964), but there is a great deal of local variability in the percentage of exotics, particularly carbonate-bearing sediments.

Soils

The soils of the Catskill Mountains are generally categorized as belonging to the Arnot-Oquaga-Lackawanna association; these soils range from excessively well drained to well drained and are on predominantly steep slopes (Tornes 1979). Soils in the basins that have been studied vary locally, but they are predominantly shallow boulder soils on steep slopes conducive to rapid precipitation runoff and are moderately to extremely acidic. The Lackawanna soils in the East Branch Neversink and upper Rondout watersheds contain a fragipan at a depth of 45 to 90 cm. In watersheds containing deltaic deposits, the deposits are generally overlain by Hoosic soils, which are deep and excessively well-drained soils on glacial outwash, and Valois soils, which are deep, well-drained soils developed on till. The soils inherit their character largely from the surficial materials on which they have developed.

Vegetation and Land Use

Vegetation in the Catskill Mountains is primarily hardwood forest that includes mostly American beech (*Fagus granidifolia*), sugar maple (*Acer sacharum*), and yellow birch (*Betula alleghaniensis*). Many watersheds contain first-growth forest at elevations above 730 m (Kudish 1985). Before settlement of the area during 1840–90, the Catskill forests were primarily eastern hemlock (*Tsuga canadensis*). The primary industry was bark peeling of hemlock for tanneries, and hemlock trunks were generally left to rot. Hemlock groves remain throughout the region, primarily along stream banks (Figure 8.2). No major forest fires were reported after the 1840s (Kudish 1985).

The upper watersheds became a state preserve in 1870. Most of the lower watersheds (below 450 m elevation) are privately owned and are in agricultural, recreational, or residential use. Anthropogenic effects are least evident in the Rondout, Beaverkill, and Neversink valleys. The valleys of the east and west branches of the Delaware River and the Schoharie River are primarily agricultural; the valley of the Esopus Creek contains several villages and scattered residential areas.

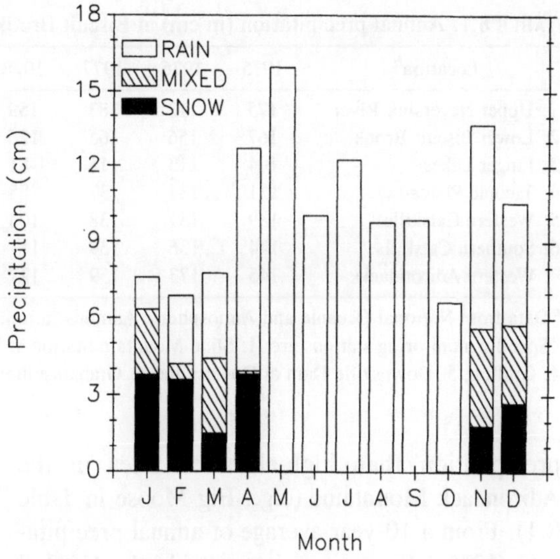

FIGURE 8.3. Mean monthly precipitation at the Biscuit Brook NTN site, 1983–86.

Local Sources of Air Pollution

Sources of air pollution within 50 km are negligible or nonexistent in the Catskill region. Approximately 8.9×10^6 kg of SO_2 and 9.96×10^5 kg of nitrous compounds are emitted annually from sources within a 50 km radius of the NTN station at Biscuit Brook (John Robertson, U.S. Military Academy, 1986, written communication), but most of these sources are to the south and east, and do not contribute significantly to SO_4^{2-} and NO_3^- deposition in the Catskills. The nearest large point source of pollutants is in the area of Kingston, New York, located approximately 40 km due east and downwind of the Catskills. Winds in the Catskills are generally from the west and northwest. The closest potential upwind center for air pollution is Binghamton, New York, about 130 km to the west. Automobile emissions within the region are probably insignificant, but airborne dust from road salt may contribute Na^+ and Cl^- to Catskills watersheds.

Climate

The Catskill Mountains have cold winters and moderately cool to warm summers. Average annual air temperature at Slide Mountain weather station at Winisook Lake is 5°C (National Oceanic and Atmospheric Administration data, 1950–82). Heavy dew is common during the early morning throughout the freeze-free period.

Long-term records of precipitation indicate a very uniform seasonal distribution. Parker et al. (1964) analyzed 30 years (1921–50) of data from Roxbury, New York, in the headwaters of the Pepacton basin (Figure 8.1). Monthly mean precipitation values varied from a minimum of 19 cm in February to a maximum of 25 cm in June. At the Biscuit Brook NTN station, precipitation was greatest during early spring and late fall during 1983–86 (Figure 8.3), but the seasonal distribution of precipitation volume has been variable, at least since the NTN station was established in 1983. Parker et al. (1964) note that yearly minima and maxima are both most likely to occur in late summer and early fall, when thunderstorms and hurricanes can deliver large but unpredictable volumes of precipitation to the region.

Annual volumes of precipitation vary both areally and with elevation. The greatest annual volumes are in the central, high-elevation zones (e.g., Slide Mountain and Frost Valley in Table 8.1). Parker et al. (1964) indicate that maximum precipitation for the region falls in the upper East Branch watershed of the Neversink River (150 to 160 cm yr^{-1} for the period 1921–55), and minimum volumes fall in the northernmost Catskills (90 to 100 cm yr^{-1}). The region receives storms from the south, along the Atlantic coast, and from the west. This characteristic probably is responsible for the greater precipitation volumes at low-elevation Catskill sites (e.g., Downsville and Grahamsville in Table 8.1) than at central New York sites (e.g., Ithaca). High-elevation sites in the central Catskills receive more

TABLE 8.1. Annual precipitation (in cm) at Biscuit Brook and other stations in the state of New York.[a]

Location[b]	1975	1976	1977	1978	1979	1980	1981	1982	1983	1984	Mean
1. Upper Neversink River	175	172	183	159	202	131	168	134	180	149	165
2. Lower Biscuit Brook	167	156	165	126	173	111	–	–	–	135	148
3. Finger Lakes	104	113	118	81	92	80	99	80	90	105	96
4. Taconic Plateau	151	134	137	88	116	98	126	112	135	130	123
5. Western Catskills	139	131	138	104	125	103	97	96	117	131	118
6. Southern Catskills	144	136	139	111	130	88	104	101	–	116	107
7. Western Adirondacks	145	173	159	127	118	120	121	116	125	119	132

[a] Data from National Oceanic and Atmospheric Administration.
[b] Specific monitoring stations are: 1. Slide Mountain Station at Winisook Lake; 2. Frost Valley; 3. Cornell University at Ithaca; 4. Grafton; 5. Downsville Dam at Downsville; 6. Grahamsville; and 7. Big Moose.

precipitation than high-elevation sites in the Adirondack Mountains (e.g., Big Moose in Table 8.1). From a 10-year average of annual precipitation (1975–84), stations just outside the Catskill Mountains to the west and south received 32% and 46% less volume, respectively, than the Slide Mountain station (Figure 8.1; Table 8.1) (Murdoch 1988). Precipitation at the Biscuit Brook NTN station averaged 145.1 cm yr^{-1} for water years (WY) 1984–86 (water year = October 1 through September 30); average annual snowfall at Biscuit Brook was 21.3 cm (snow water equivalence) for the same period (Figure 8.3).

Evapotranspiration (ET) estimates for 1983–86 indicate that about 57 cm of water (40%) leaves the Biscuit Brook watershed annually as ET (ET = total annual precipitation inputs − total annual stream outputs). This value is comparable to estimates for the Catskills and the northeastern United States, based on long-term precipitation and stream discharge data (Knox and Nordenson 1955), and to estimates for the Adirondack Mountains, based on empirical equations (Murdoch et al. 1987).

Chemistry of Atmospheric Deposition

Average concentrations of major constituents in precipitation collected at the NTN station at Biscuit Brook were similar to those observed at other monitoring stations in the Northeast, but precipitation volumes were greater (Table 8.2) (see also Northeast Overview Section). Average volume-weighted precipitation pH at Biscuit Brook was 4.25 during 1983–86 and weekly values ranged from 3.25 to 5.92 (Murdoch 1988).

Data for precipitation at the Biscuit Brook NTN station during 1983–87 may have been affected by the drought that occurred during 1984 and 1985, and these data may be atypical, over the long term, for the Catskill region. Hydrogen ion concentra-

TABLE 8.2. Volume-weighted mean concentrations and pH in weekly precipitation samples from Biscuit Brook and other NADP/NTN stations in northeastern United States, 1984–85.[a]

	Biscuit Brook (New York)		Leading Ridge (Pennsylvania)		Hubbard Brook (New Hampshire)	Huntington (New York)	
	1984	1985	1984	1985	1984	1984	1985
C_B							
$(Ca^{2+}+Mg^{2+}+Na^{+}+K^{+})$	14.3	7.8	15.0	11.5	10.6	10.5	9.6
NH_4^+	8.9	8.9	17.8	13.3	9.4	10.0	10.0
H^+	43.7	50.1	70.7	60.2	38.8	37.5	42.3
SO_4^{2-}	41.2	45.0	69.2	56.4	34.5	36.8	39.7
NO_3^-	21.9	21.9	35.9	28.5	20.6	20.1	23.8
Cl^-	5.4	3.4	5.3	5.1	4.5	3.1	2.8
pH	4.36	4.30	4.15	4.22	4.41	4.43	4.37
Volume	133	122	113	107	125	109	102

[a] Concentrations are in μeq L^{-1}; volumes are in cm. Data are from NADP/NTN Coordinator's Office 1986, 1987.

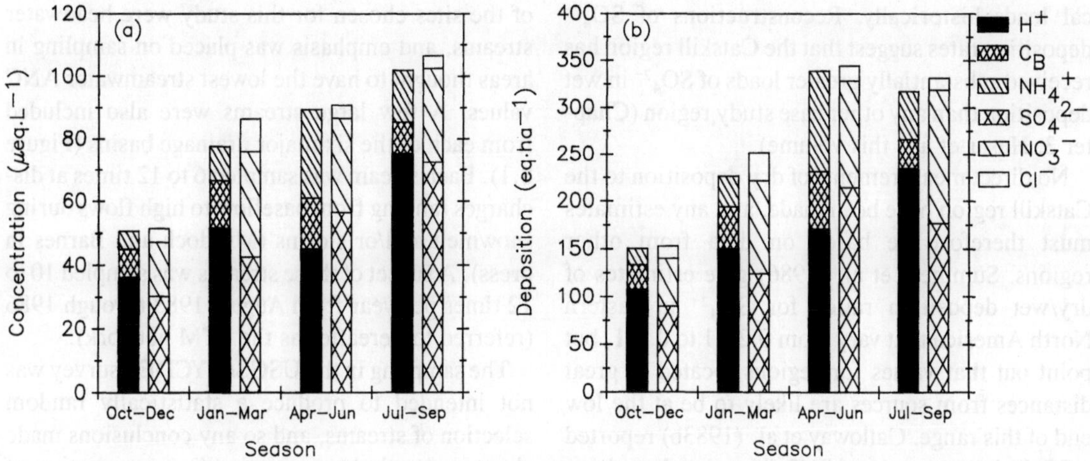

FIGURE 8.4. Mean seasonal concentrations (a), and mean deposition (b), of major ions in precipitation at the Biscuit Brook NTN site, 1983–86.

tions at Biscuit Brook were lower than the 15-year volume-weighted average (1963–77) at Hubbard Brook Experimental Forest (HBEF) in New Hampshire, but higher than those for the dry periods at Hubbard Brook in 1980 and 1981 (Glass and Loucks 1986). Similarly, Cl^- and SO_4^{2-} concentrations at Biscuit Brook during WY 1984–86 were slightly lower than the Hubbard Brook 15-year average, but similar to the 1980–81 concentrations.

The NTN station at Hills Creek in northcentral Pennsylvania, approximately 160 km southwest of the Catskills, recorded slightly higher concentrations of all constituents during WY84 than were recorded at Biscuit Brook (Lynch et al. 1985). The station at Leading Ridge, in southwestern Pennsylvania, received substantially higher SO_4^{2-} concentrations than Biscuit Brook during both WY84 and WY85 (Table 8.2) (Barker and Witt 1990). This pattern of an eastward decrease in SO_4^{2-} concentrations in precipitation is consistent with a hypothesized midwestern source of SO_4^{2-}.

The 1983–86 data on chemical concentrations in precipitation at the Biscuit Brook NTN station show no consistent seasonal trends (Figure 8.4a). Concentrations of H^+ and SO_4^{2-} are highest in summer, whereas NO_3^- concentrations are highest in winter and summer. Wet deposition of acid anions is highest in spring, when volumes are greatest (Figure 8.3); wet deposition of H^+ is highest in summer, when concentrations are highest. Wet deposition of Cl^- is greatest in late winter, which suggests that dust carrying road salt may be entering the watersheds of Catskill headwater streams. Hydrogen accounted for 64% of the cation load at Biscuit Brook during 1983–86, and NH_4^+, the next most abundant cation, accounted for 20%. Of the anion load, SO_4^{2-} accounted for 60% and NO_3^- for 33%.

The chemical loading rates (wet deposition only) observed at Biscuit Brook during 1983–86 are within the range of those observed at Hubbard Brook Experimental Forest in New Hampshire, and at Woods Lake in the Adirondack Mountains. Annual NO_3^- loading during 1963–77 at Hubbard Brook ranged from 108 to 500 eq ha^{-1} (Likens 1985). Nitrate loading at Biscuit Brook was 447 eq ha^{-1} in WY84 and 341 eq ha^{-1} in WY85. The wet deposition of acid anions, NH_4^+, and H^+, was slightly greater at Biscuit Brook than at Hubbard Brook during a period when precipitation volume was similar at both sites (124 cm yr^{-1} from December 1980 through November 1982 at Hubbard Brook and 130 cm yr^{-1} during WY85 at Biscuit Brook), but the base cation concentrations were similar (Glass and Loucks 1986). The ranges in chemical concentrations and deposition rates observed in the central Catskills during the period of record (1983–86) were therefore similar to those at other monitoring stations in the Northeast that have longer periods of record. Long-term annual precipitation averages indicate that the Catskills receive greater precipitation volume, however, and may therefore have received greater annual chemical loads historically. Reconstructions of SO_4^{2-} deposition rates suggest that the Catskill region has received substantially greater loads of SO_4^{2-} in wet deposition than any other case study region (Chapter 3, Husar et al., this volume).

No direct measurements of dry deposition to the Catskill region have been made, and any estimates must therefore be based on data from other regions. Summers et al. (1986) give estimates of dry/wet deposition ratios for SO_4^{2-} in eastern North America that vary from 0.25:1 to 0.8:1, but point out that values for regions located at great distances from sources are likely to be at the low end of this range. Galloway et al. (1983b) reported a SO_4^{2-} dry:wet ratio of 0.42 (i.e., dry deposition is 30% of wet) for three Adirondack watersheds – a ratio that agrees well with an estimated ratio of 0.5 used in ILWAS model simulations (Appendix B). The Catskill region is similar to the Adirondacks region in terms of both elevation and concentrations of acid anions in wet deposition, and should therefore exhibit similar dry/wet deposition ratios. We use a ratio of 0.5 in constructing chemical budgets for Biscuit Brook (see Processes Influencing Surface Water Chemistry), which is sufficient to balance inputs and outputs of SO_4^{2-} and Cl^-.

Inputs of acid anions from clouds and fog represent unknown, but potentially important, inputs to the Catskill region. Weathers et al. (1986) reported on the chemistry of one particularly intense cloud/fog water event at Mohonk Mountain, on the southeastern border of the Catskill Mountains. Sulfate concentrations during this event were 1,100 μeq L^{-1}, and NO_3^- concentrations were 950 μeq L^{-1} (approximately 25 to 40 times the concentrations in wet deposition; Table 8.2). Because these measurements were made on samples from active collectors, it isn't possible to estimate deposition rates for this event. Lovett et al. (1982) suggest that fog may make a significant contribution to rates of acidic deposition in subalpine areas of the eastern United States.

Regional Surface Water Chemistry

Ion Composition of Streams at High and Low Flows

A survey of 66 streams throughout the Catskill Mountains was conducted in a cooperative effort by USGS and NYCDEP during 1985–87 (referred to hereafter as the USGS/NYCDEP survey). Most of the sites chosen for this study were headwater streams, and emphasis was placed on sampling in areas thought to have the lowest streamwater ANC values. A few large streams were also included from each of the six major drainage basins (Figure 8.1). Each stream was sampled 6 to 12 times at discharges ranging from baseflow to high flows during snowmelt and/or storms (Murdoch and Barnes in press). A subset of these streams was sampled 10 to 12 times per year from August 1983 through 1986 (referred to hereafter as the LTM network).

The sampling in the USGS/NYCDEP survey was not intended to produce a statistically random selection of streams, and so any conclusions made about regional characteristics (i.e., proportion of streams with ANC $\leq$ 0) should be considered approximations, at best. Whenever conclusions of this type are made, they are based on a modified dataset. Several sites were dropped from the complete dataset, in order to minimize the bias associated with larger numbers of sampled streams in low ANC basins, leaving an approximately equal number of sites in each of the six basins.

Frequency distributions of key characteristics (watershed area, pH, ANC, sum of base cations (C_B = Ca^{2+} + Mg^{2+} + Na^+ + K^+), SO_4^{2-}, and NO_3^-) in the USGS/NYCDEP survey are shown in Figure 8.5. The plotted chemical data are mean values for each stream, partitioned into high- and low-flow periods to show the effect of stream discharge on the distribution of chemical characteristics in the region. The majority of streams (70%) drain watersheds < 2,000 ha (Figure 8.5a). As mentioned previously, sites were also included at low elevations in each major drainage basin, and some of these larger streams have drainage areas as large as 8,600 ha.

The pH levels showed a strong dependence on stream discharge (Figure 8.5d). At baseflow, 8% of the streams in the modified USGS/NYCDEP dataset had pH values < 5.5. During high flows, this proportion increased to 18%, and the pH distribution became bimodal. The majority of streams exhibited circumneutral pH at high flows, and very few had pH values between 5.0 and 6.0. All the streams sampled had pH values between 4.0 and 8.0 during both baseflow and high-flow periods.

Acid neutralizing capacity is also very dependent on stream discharge. Eight percent of streams had negative ANC at baseflow, and 16% had negative ANC at high flows (Murdoch and Barnes in

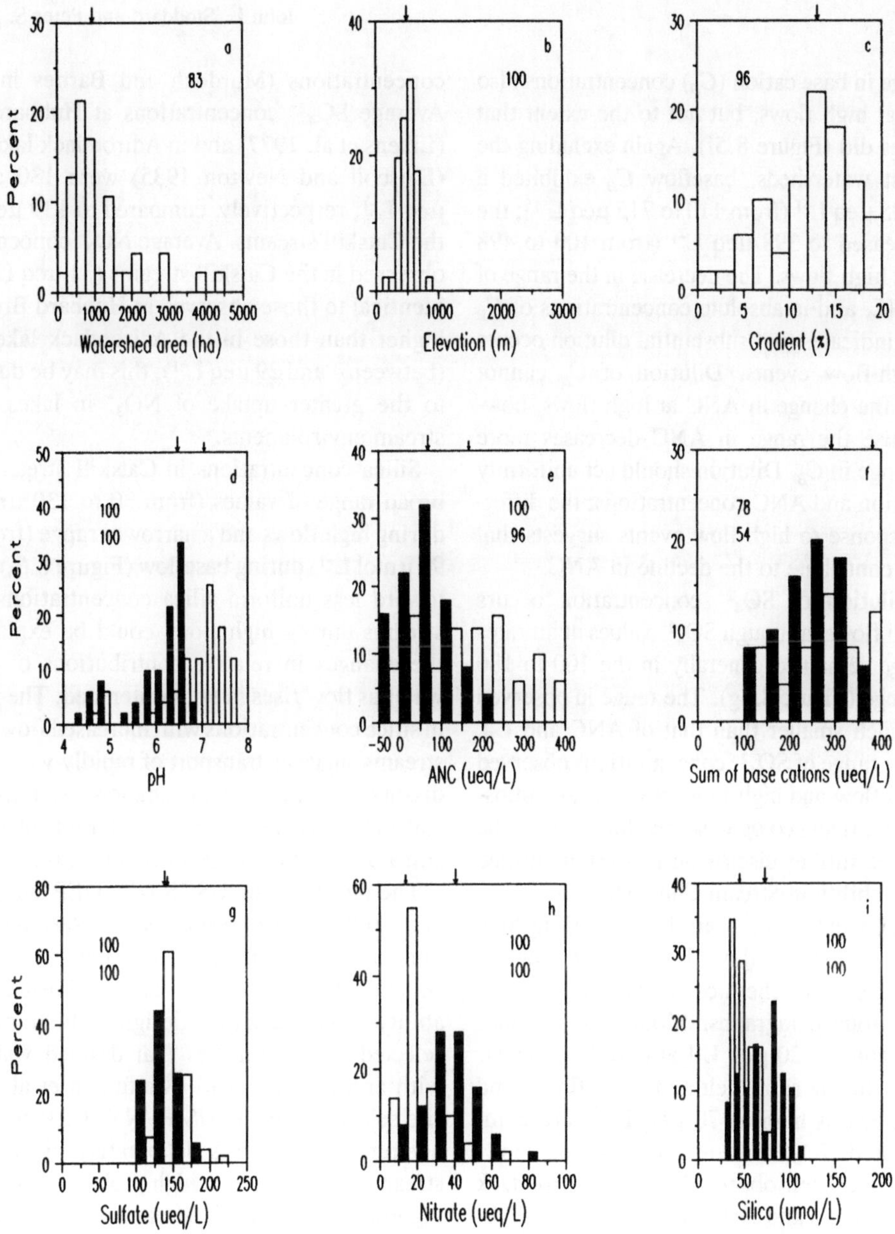

FIGURE 8.5. Histograms of watershed characteristics and streamwater chemistry for the modified USGS/NYCDEP stream survey. Data are from 51 streams, distributed more or less evenly among the 6 major drainage basins. Data for chemical variables are subdivided into results for baseflow (clear bars) and high-flow (solid bars) periods. In order to show both high-flow and baseflow values, bars are only half the width of actual intervals (e.g., 24% of streams had SO_4^{2-} concentrations between 100 and 125 μeq L^{-1} at high flow; at baseflow 8% of streams had SO_4^{2-} between 100 and 125). Values in upper part of box are percentage of stream samples (high flow and baseflow), with values in the graphed range. Arrows indicate median values for each variable; larger arrow is for high-flow values, smaller arrow is for baseflow values.

press; Figure 8.5e). The range of observed ANC values changed at high discharges. At baseflow, the sampled streams (excluding the four largest watersheds with areas > 20,000 ha) exhibited a range in ANC of 408 μeq L^{-1} (from −18 to 390 μeq L^{-1}), with a relatively even distribution. At high flows, the range in ANC decreased to 236 μeq L^{-1} (from −35 to 201 μeq L^{-1}); the distribution shows a very strong mode at a value < 100 μeq L^{-1}.

The range in base cation (C_B) concentrations also decreased at high flows, but not to the extent that ANC values did (Figure 8.5f). Again excluding the four largest watersheds, baseflow C_B exhibited a range of 605 μeq L^{-1} (from 110 to 715 μeq L^{-1}); the range decreased to 398 μeq L^{-1} (from 100 to 498 μeq L^{-1}) at high flows. The decrease in the range of C_B and ANC, and in absolute concentrations of C_B and ANC, indicates that substantial dilution occurs during high-flow events. Dilution of C_B cannot explain all the change in ANC at high flows, however, because the range in ANC decreases more than the range in C_B. Dilution should act uniformly on base cation and ANC concentrations; the difference in response to high-flow events suggests that other ions contribute to the decline in ANC.

Some dilution of SO_4^{2-} concentration occurs during high flows, although SO_4^{2-} values at all rates of discharge remained generally in the 100 to 180 μeq L^{-1} range (Figure 8.5g). The range in observed SO_4^{2-} is much smaller than that of ANC and C_B. The narrow range of SO_4^{2-} concentrations observed at both baseflow and high flows suggests an atmospheric, rather than geologic, source for SO_4^{2-} in the Catskills (see further discussion in section entitled Spatial Variability in Stream Chemistry).

Nitrate concentrations were higher during high flows than during baseflow (Figure 8.5h), unlike SO_4^{2-} and the other chemical variables discussed in the previous paragraphs. Modal NO_3^- values increased from < 20 μeq L^{-1} at baseflow to near 50 μeq L^{-1} during snowmelt and stormflows, and concentrations as high as 70 μeq L^{-1} were common. Elevated NO_3^- concentrations during snowmelt have also been observed in the Adirondack Mountains (Peters and Driscoll 1987) and at Hubbard Brook (Likens et al. 1977), and have been identified as the main cause of episodic ANC declines in chronically acidified Adirondack lakes (Driscoll and Shafran 1984, Driscoll et al. 1987b, Schofield et al. 1985). The source of NO_3^- during stormflows and snowmelt periods is unclear, but some combination of NO_3^- storage in the snowpack (Galloway et al. 1980) and nitrification in the soil (Rascher et al. 1987, Peters and Driscoll 1987) is likely.

Comparisons of the Catskill stream data with data from Hubbard Brook in New Hampshire and from Adirondack lake outlets show the streams of all three regions to have similar SO_4^{2-} and NO_3^- concentrations (Murdoch and Barnes in press). Average SO_4^{2-} concentrations at Hubbard Brook (Likens et al. 1977) and in Adirondack lake outlets (Driscoll and Newton 1985) were 130 and 136 μeq L^{-1}, respectively, compared to 139 μeq L^{-1} in the Catskill streams. Average NO_3^- concentrations observed in the Catskill streams (31 μeq L^{-1}) were identical to those observed at Hubbard Brook, but higher than those in the Adirondack lake outlets (between 4 and 29 μeq L^{-1}); this may be due in part to the greater uptake of NO_3^- in lakes than in stream environments.

Silica concentrations in Catskill streams had a broad range of values (from 30 to 130 μmol L^{-1}) during high flows and a narrower range (from 30 to 90 μmol L^{-1}) during baseflow (Figure 8.5i). A shift toward less uniform silica concentrations among streams during high flows could be explained by the changes in relative contributions of groundwater as flow rises in each watershed. The increase in silica concentrations with increased flow in some streams suggests transport of rapidly weathered or dissolved silica from the shallow groundwater or soil environments, or a contribution of colloidal silica during conditions of rapid runoff.

The results of the USGS/NYCDEP survey compare well with the results of the National Stream Survey (NSS), although the similarity may well be coincidental (Table 8.3). The NSS used a probability-based sampling design, which randomly selected stream reaches that drained watersheds with areas < 155 km² (Kaufmann et al., 1988); site selection for the USGS/NYCDEP survey was not random, and was deliberately biased toward streams in areas with a high probability of exhibiting low ANC. Only six stream reaches in the NSS sample were within the boundaries of the Catskill region, making the calculation of population estimates problematic. Despite these difficulties, NSS population estimates for the Catskills (6% of upstream reach ends with ANC ≤ 0 and 5% with pH ≤ 5 at spring baseflow) and estimates for the modified USGS/NYCDEP survey (8% with ANC ≤ 0 and pH ≤ 5 at baseflow, 16% of ANC values ≤ 0 and 18% of pH values ≤ 5 at high flows) were remarkably close (Kaufmann et al. 1988; Chapter 17, Baker et al., this volume). Sulfate values for the two surveys were also similar (Table 8.3), but SO_4^{2-} was relatively uniform across the Catskill region and site selection is unlikely to have a large

TABLE 8.3. Median values (first and third quartiles in parentheses) of key variables in data sets used in this chapter.[a]

Data set	N	pH units	ANC µeq L⁻¹	SO_4^{2-} µeq L⁻¹	Ca^{2+} µeq L⁻¹	DOC µmol L⁻¹	NO_3^- µeq L⁻¹	$Ca^{2+}+Mg^{2+}$ µeq L⁻¹
USGS/NYCDEP	51	6.60 (6.02–6.74)	119 (28–176)	142 (131–154)	208 (131–267)	–	28 (23–35)	287 (204–366)
NSS (Kaufmann et al. 1988)								
Downstream end								
Sample	6	6.97 (6.81–7.18)	104 (60–204)	149 (134–159)	172 (147–260)	87 (58–113)	16 (15–18)	250 (225–341)
Population	161	7.18 (7.07–7.32)	204 (123–218)	159 (155–166)	259 (183–273)	91 (57–113)	18 (17–18)	341 (263–351)
Upstream end								
Sample	6	6.57 (6.32–6.93)	54 (37–111)	144 (130–158)	164 (100–193)	83 (58–100)	20 (18–23)	234 (163–267)
Population	161	6.59 (6.59–6.93)	58 (58–171)	158 (148–169)	193 (186–267)	95 (49–100)	23 (22–48)	255 (255–341)
LTM	8	5.54 (5.08–5.99)	17 (10–29)	121 (119–126)	116 (87–155)	158 (117–250)	26 (22–31)	158 (130–188)

[a] Values for the USGS/NYCDEP survey are based in survey of 66 streams, later modified to include an even number of sites in each major basin. NSS values are population estimates for upstream and downstream ends of six stream reaches sampled at spring baseflow within the Catskill region boundaries. Long-Term Monitoring (LTM) sites are a low ANC subset of the USGS/NYCDEP survey streams, which were sampled more intensively.

effect on these estimates. Nitrate was substantially lower in the NSS, probably because the samples were intentionally collected at baseflow when NO_3^- concentrations were low (Figure 8.5h); when only baseflow samples were considered, the median NO_3^- concentration for the modified USGS/NYCDEP survey was 14 µeq L⁻¹. The use of spring baseflow samples may be the major weakness of the NSS sampling approach, because it failed to identify the importance of NO_3^- to streamwater chemistry at high flows in the Catskills (Figure 8.5h; see also the section entitled Processes Influencing Surface Water Chemistry).

Spatial Variability in Stream Chemistry

The lowest values for pH and ANC in the Catskills region are found in the two southernmost basins (Figure 8.6a,b), the Neversink and Rondout (Figure 8.1). Acid neutralizing capacity becomes a small fraction of C_B in the low pH, low ANC streams (Figure 8.6c). To investigate these patterns, mean concentrations of major ions were calculated for each of the six major drainage basins of the Catskill region (Murdoch and Barnes in press; Table 8.4). Analysis of these mean concentrations within each of the basins indicates that:

- Concentrations of base cations (Ca^{2+}, Mg^{2+}, Na^+, and K^+) differed markedly among the basins.
- ANC and pH values differed markedly among the basins and generally followed the pattern of Ca^{2+} concentrations.
- Concentrations of SO_4^{2-} and NO_3^- were relatively uniform among the basins.

Streams containing the highest concentrations of all ions except H^+ and aluminum were found in the westernmost basin (Cannonsville). Calcium and Mg^{2+} concentrations there were two to three times higher than in the two southernmost basins (Neversink and Rondout), but SO_4^{2-} concentrations in the Cannonsville basin were only 27% higher than in the Neversink basin and 14% higher than in the Rondout basin. Nitrate, like SO_4^{2-}, was relatively uniform across the region, but the highest concentrations were in the westernmost basins (Murdoch and Barnes in press). Higher Ca^{2+}, Mg^{2+}, SO_4^{2-}, and NO_3^- concentrations in the Cannonsville and Pepacton basins were undoubtedly due, to some degree, to greater agricultural land use in these basins. The higher Na^+ and Cl^- concentrations in the northern and western basins, which are more heavily populated, were at least partly due to greater use of road salt in these than in the southern and eastern basins.

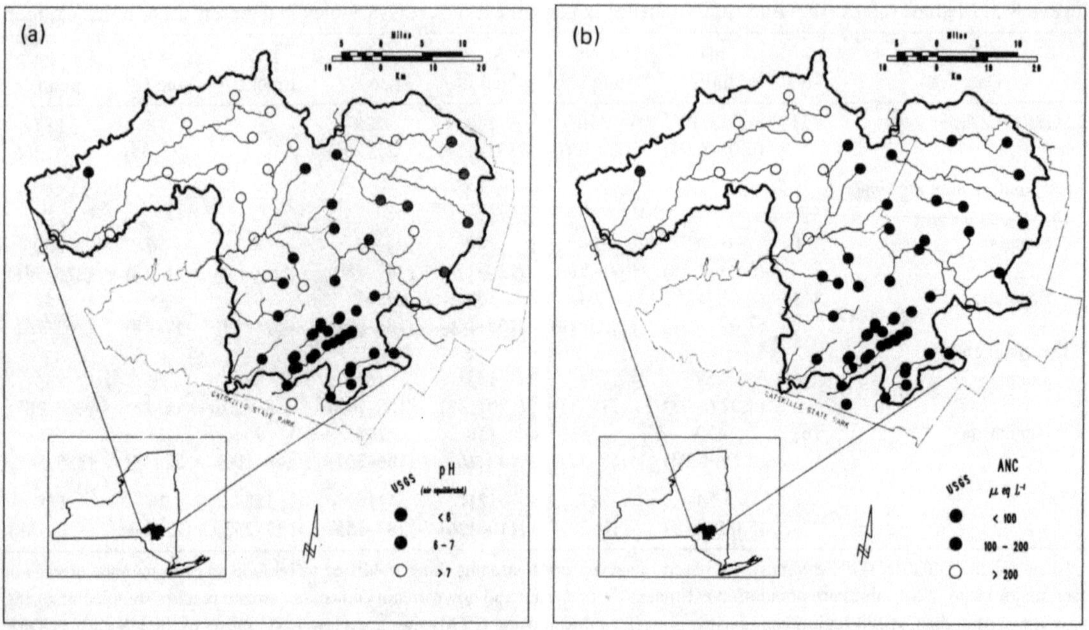

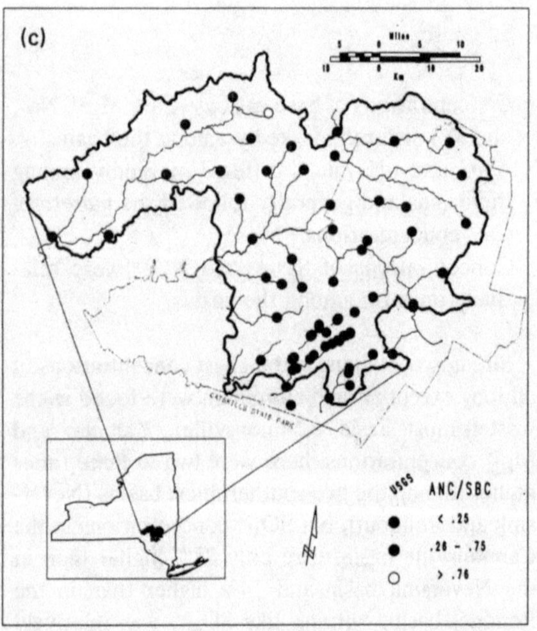

FIGURE 8.6. Mean values (from high-flow and baseflow periods, 1985–87) for (a) pH (calculated from mean hydrogen ion concentration), (b) ANC, and (c) ratio of ANC to base cations (SBC; defined here as the sum of Ca^{2+} + Mg^{2+}) at sites in the USGS/NYCDEP stream survey. Data for all sites are shown; 15 sites in the southern portion of the region were deleted from modified USGS/NYCDEP dataset used in this chapter to characterize regional stream chemistry.

The local differences in ANC among the major drainage basins are clearly related to the base cation concentrations of streams, and not to differences in acid anion concentrations. The source of base cations in each stream, particularly Ca^{2+} and Mg^{2+}, therefore plays an important role in controlling the stream's response to acidic deposition. Three hypothetical causes of local differences in base cation concentration are discussed in the following sections.

Calcium and Magnesium Content of Bedrock

Calcium and Mg^{2+} content of bedrock in the Catskill region may increase westward. The Catskill bedrock materials were deposited in a westward-

TABLE 8.4. Mean concentrations of key chemical variables in six major drainage basins of Catskill region, 1985–87.[a]

Drainage basin	Number of sites	pH	Ca^{2+}	Mg^{2+}	Na^{+}	K^{+}	SO_4^{2-}	NO_3^{-}	Cl^{-}	ANC	Al
Ashokan	10	6.57	224	83	66	8	141	26	58	141	2
Cannonsville	8	6.82	333	131	114	25	161	37	74	275	2
Neversink	23	5.05	114	51	24	8	127	26	24	13	18
Pepacton	9	6.92	268	80	52	11	151	37	47	164	2
Roudout	5	5.74	124	63	41	12	141	26	38	21	10
Schoharie	8	6.54	218	72	74	9	148	24	58	119	3

[a] Basin means are calculated by: (1) calculating mean high-flow and baseflow values for each site, (2) calculating means for each site from high-flow and baseflow means, and (3) calculating means for each basin from overall means at each site. Aluminum (total dissolved) concentrations are in µmol L^{-1}; all others except pH are in µeq L^{-1}.

developing delta, on the edge of a shallow sea that covered much of what is now central New York (Ethridge 1977). The western edge of that delta would therefore have been under marine influences and would have incorporated salts and carbonates. This pattern would be expected to produce greater amounts of carbonate and Na^{+} in the Cannonsville basin, followed by the Pepacton basin. This expectation is borne out by Ca^{2+} and ANC values in these basins (Table 8.4). These marine influences would also be expected to contribute some gypsum to the bedrock and would therefore lead to elevated SO_4^{2-} concentrations (in addition to Ca^{2+}) in the westernmost basins. The highest SO_4^{2-} concentrations in the region are indeed found in the Cannonsville and Pepacton basins (Table 8.4). Differences in ANC among the Neversink, Ashokan, and Schoharie basins are not explained by this hypothesis, however, and data are not available to support any of these suppositions.

Limestone Fragments Deposited by Glaciation

Glaciation may have deposited more limestone fragments in the north Catskills than in the south. Given the glacial deposition patterns previously described, debris from the limestone escarpments north and northeast of the Catskill region could have been deposited by the ice on the leeward side (away from the prevailing direction of ice movement) of any high topographic features the ice encountered (Murdoch and Barnes in press). Deposition would have been greatest at the first ridge encountered and would have decreased over successive ridges. The Neversink and Rondout basins, the farthest from the limestone source, would therefore have received a smaller percentage of the limestone fragments than the northern basins. This is consistent with Ca^{2+} and ANC values in these basins (Table 8.4). Recent studies of till stratigraphy in the western Catskills revealed cobbles of Adirondack origin in the Cannonsville basin but not in the Pepacton basin (D. Ozvath, Lafayette University, written communication). If material from the Adirondack Mountains was transported to the Catskill region, limestone fragments from the closer Helderberg escarpments could also have been transported, and like the Adirondack cobbles, would diminish southward. The control of surface water ANC by small amounts of "exotic" carbonate in till has been documented in the Catoctin Mountains in Maryland (Katz et al. 1985), the Precambrian Shield of Ontario (Hornbrook et al. 1986), and the Cascade Mountains of Washington (Drever and Hurcomb 1986).

Differences in Water Residence Time in Surficial Materials

The thickness, distribution, and permeability of surficial material may also differ among the basins. Because these largely determine the residence time of water in a basin, they may in turn have a significant effect on the base cation concentrations in Catskill streams. The preceding hypothesis suggests that the northernmost basin (Schoharie) should have received the greatest percentage of carbonate; yet its streams have the lowest Ca^{2+} concentrations of the northern basins (Table 8.4). This may be because the Schoharie valley, which is

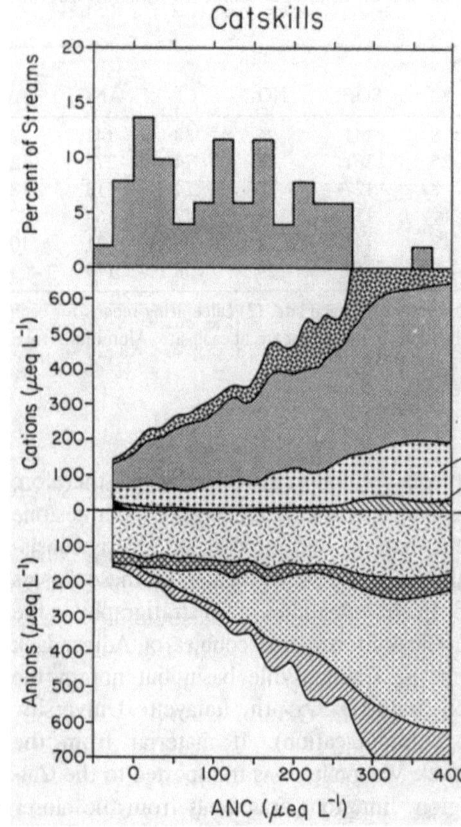

FIGURE 8.7. Spline diagrams of major anion and cation concentrations as a function of ANC for streams in modified USGS/NYCDEP dataset. These plots differ from most others in this book in including NO_3^-, and excluding organic acid anions.

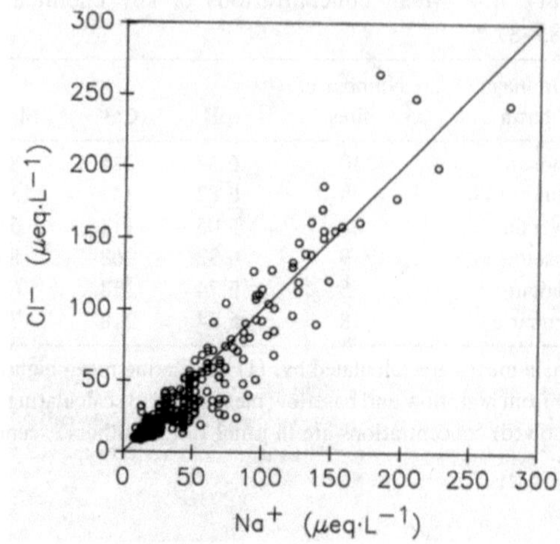

FIGURE 8.8. Relationship between Na^+ and Cl^- in Catskill streams sampled in the USGS/NYCDEP survey. 1:1 line is shown.

oriented toward the north, underwent more glacial scouring and greater deposition of lake clay over the till than did the other northern basins (Murdoch and Barnes in press). Even though the carbonate content of the till may be as high (or higher) in the Schoharie basin as in the southern basins, the carbonate may have been less accessible for weathering during the postglacial period. The thickness and permeability of surficial deposits may therefore differ among the major basins, and the carbonate content of the bedrock and till in the Catskill basins could have less effect on the base cation concentrations of surface water than the thickness, permeability, and transmissivity of the surficial material. Till thickness has been shown to have strongly affected the buffering capacity of surface waters in the Adirondacks (Peters and Murdoch 1985), and could, therefore, be a determinant of ANC in Catskill streams.

Relations Among Chemical Variables

Spline diagrams of major ionic compositions across the range of ANC values observed in the Catskill streams (Figure 8.7) illustrate the uniformity of stream chemistry in the region. Relative cation concentrations are consistent across the entire range of ANC, and the dominance of cations follows the pattern $Ca^{2+} > Mg^{2+} > Na^+ > K^+ > H^+$. Sulfate concentrations are high throughout the observed ANC range, and SO_4^{2-} is the dominant anion at ANC values < 130 μeq L^{-1}. Above 130 μeq L^{-1}, HCO_3^- becomes the dominant anion. Chloride becomes important stoichiometrically above ANC values of 150 μeq L^{-1}, in parallel with increased concentrations of Na^+. In all the streams sampled, Na^+ and Cl^- were found in a nearly 1:1 stoichiometric ratio (Figure 8.8), suggesting a strong influence of road deicing salts on the concentrations of these ions. Because Na^+ in Catskill streamwater results primarily from road salt, and because K^+ is only a minor cation (Figure 8.7), we substitute the sum of Ca^{2+} and Mg^{2+} for C_B in many of the following discussions. However, throughout the chapter we are consistent in using C_B to refer to the sum of all base cations, and we use "$[Ca^{2+} + Mg^{2+}]$" when discussing only Ca^{2+} and Mg^{2+}.

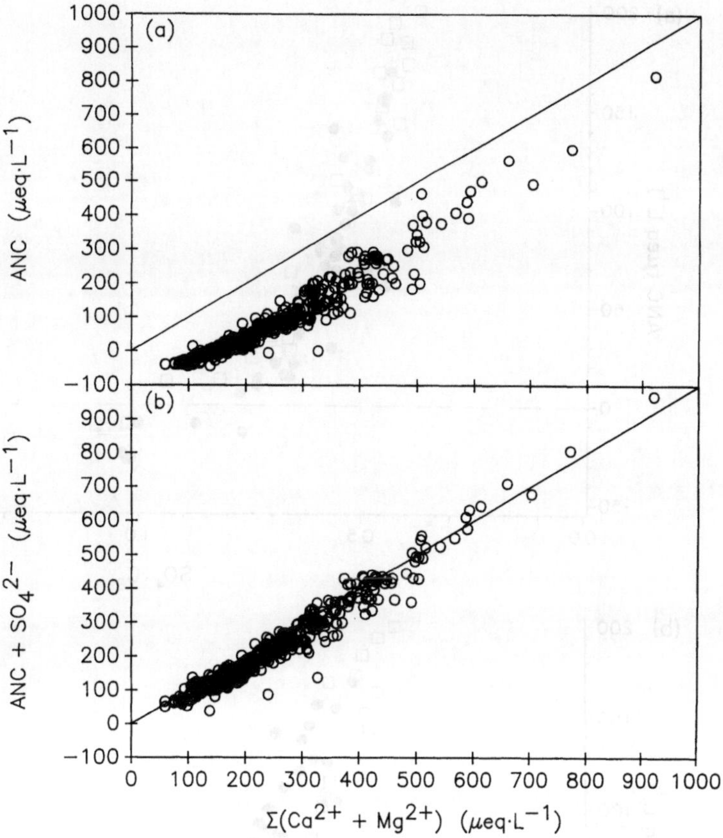

FIGURE 8.9. Plots of (a) ANC versus base cations [Ca^{2+} + Mg^{2+}], and (b) ANC + SO_4^{2-} versus base cations for samples in USGS/NYCDEP stream survey. 1:1 line is plotted to indicate expected relationship.

Calcium and Mg^{2+} are the dominant cations across most of the range of ANC values observed in the Catskills, as would be expected from the weathering of carbonate minerals. Carbonic acid weathering would be expected to produce ANC in concentrations approximately equivalent to the sum of Ca^{2+} and Mg^{2+} concentrations (Chapter 1, Munson and Gherini, this volume). A plot of [Ca^{2+} + Mg^{2+}] against ANC does not indicate a 1:1 relation, however (Murdoch and Barnes in press; Figure 8.9a). Although the scatter clearly indicates a close correlation between ANC and [Ca^{2+} + Mg^{2+}], ANC values fall short of the 1:1 line by a consistent amount across the range of concentrations observed.

When SO_4^{2-} concentration is added to ANC and the relationship with [Ca^{2+} + Mg^{2+}] is re-examined, deviation from the expected relation is decreased dramatically (Figure 8.9b). The difference between ANC and [Ca^{2+} + Mg^{2+}] is closely approximated by SO_4^{2-} concentration, which remains relatively constant at all ionic strengths (Figure 8.7). The difference between observed ANC and expected ANC, based on [Ca^{2+} + Mg^{2+}], is consistent across the Catskill region, as evidenced by the tight cluster in Figure 8.9a,b, and is relatively unaffected by changes in stream discharge (Murdoch and Barnes press). Figure 8.9 includes samples at both baseflow and high flow. This pattern could result if: (1) sulfuric acid deposition decreased ANC by direct titration of HCO_3^- by H^+, (2) sulfuric acid deposition increased base cation concentrations by increasing weathering and/or cation exchange rates (Chapter 1, Munson and Gherini, this volume), (3) some combination of decreased ANC and increased base cation concentrations occurred, or (4) some geologic source of SO_4^{2-} contributed to a natural difference between expected and observed ANC. Of these possibilities, only the last can be ruled out.

Both the spatial (Table 8.4) and seasonal (Figure 8.5g) uniformity in SO_4^{2-} concentrations in the Catskills are inconsistent with a geologic source. Some gypsum may be present in the westernmost basins (see section entitled Spatial Variability in

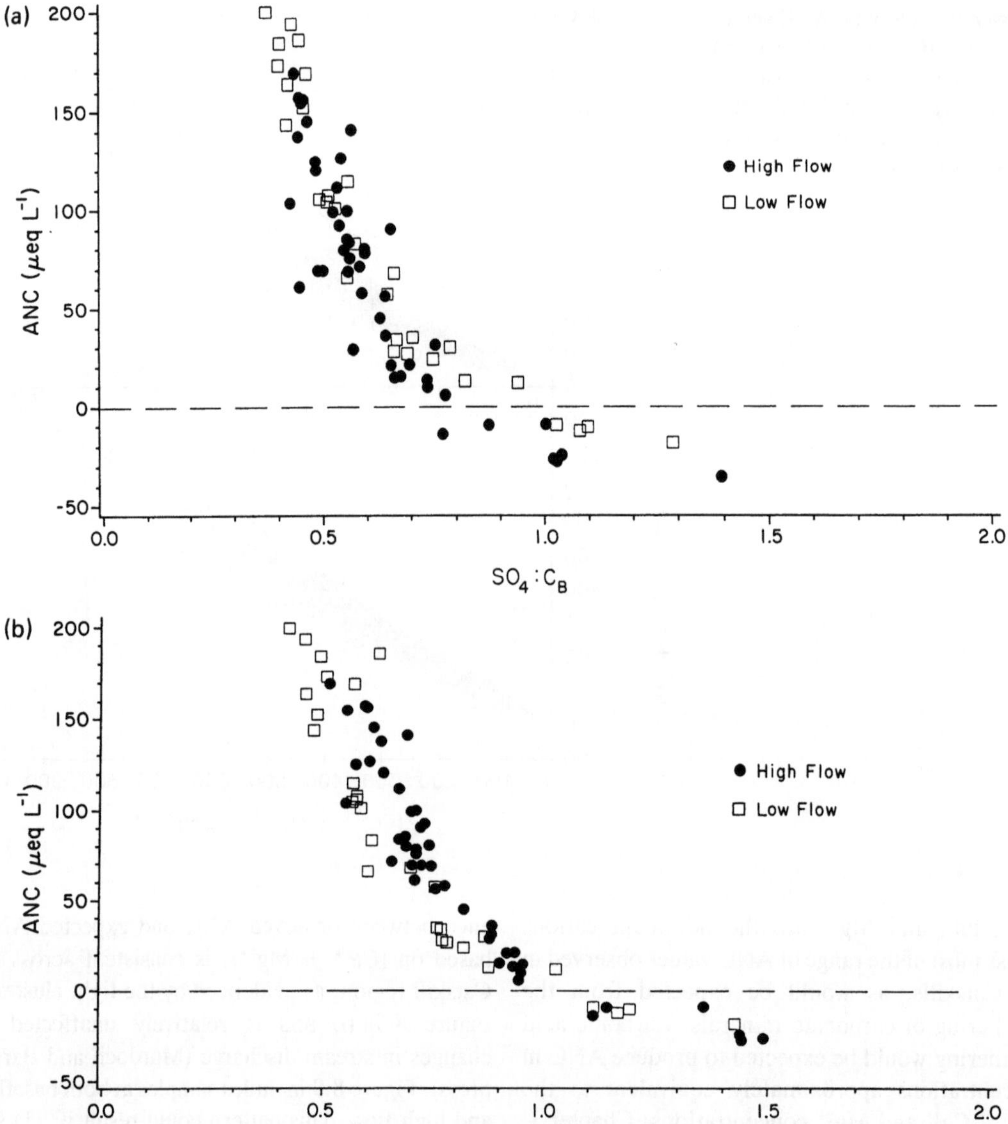

FIGURE 8.10. Plots of (a) ANC versus ion ratio SO_4^{2-}:[Ca^{2+} + Mg^{2+}], and (b) ANC versus ion ratio [SO_4^{2-} + NO_3^-]:[Ca^{2+} + Mg^{2+}] for samples in USGS/NYCDEP stream survey. Nitrate is included in (b) to illustrate its importance as an acid anion in low ANC Catskill streams.

Stream Chemistry), but the difference in SO_4^{2-} concentration between these basins and the remainder of the region is small. If this entire difference were attributed to gypsum dissolution, it would amount to less than 12% of streamwater SO_4^{2-} (i.e., about 15 μeq L^{-1}). Pyrite has been observed in bedrock outcrops to the east and south of the Catskill region (Way 1972), but if it were present within the region boundaries, it would be expected to produce spatially irregular SO_4^{2-} concentrations in streams, reflecting the irregular distribution of pyrite along faults and seams. Perhaps the best evidence against an important geologic source of SO_4^{2-} in the Catskill region comes from input/output budget data from Biscuit Brook (see Comparison of Precipitation and Stream Chemistry), where precipitation inputs of

SO_4^{2-}, adjusted for dry deposition and evapotranspiration, closely match volume-weighted streamwater SO_4^{2-} concentrations (Murdoch 1988).

The preceding discussions imply that a stream's response to acidic deposition depends on rates of base cation release, in addition to the rate of acid anion deposition. Acid neutralizing capacity approaches zero as the ratio of SO_4^{2-} to $[Ca^{2+} + Mg^{2+}]$ approaches 0.8 (Figure 8.10a). Sulfate, although it accounts for a majority of the difference between ANC and $[Ca^{2+} + Mg^{2+}]$, is therefore not the only important acid anion. Nitrate also plays an important role in controlling streamwater ANC. Acid neutralizing capacity approaches zero when the ratio of $NO_3^- + SO_4^{2-}$ to $[Ca^{2+} + Mg^{2+}]$ approaches one (Figure 8.10b). The sum of NO_3^- and SO_4^{2-}, therefore, accounts for nearly 100% of the difference between ANC and $[Ca^{2+} + Mg^{2+}]$ at ANC values above zero.

At ANC values below zero, a deficit in base cation concentrations (relative to anion concentrations) is compensated for by an increase in protolytic cations—H^+ and free aluminum (Driscoll and Newton 1985; Chapter 6, Driscoll et al., this volume). Concentrations of protolytic cations began to increase in several Adirondack lakes when the sum of acid anions ($SO_4^{2-} + NO_3^-$) approached 90% to 100% of the sum of base cations (Driscoll and Newton 1985). A similar pattern is evident in Catskill streams (Figure 8.11), where H^+ and aluminum concentrations increase when the concentration of SO_4^{2-} and NO_3^- exceeds 80% of base cations (defined as $Ca^{2+} + Mg^{2+}$) (Murdoch and Barnes in press). Concentrations of protolytic cations above 60 μeq L^{-1} were found primarily in the Rondout and Neversink basins, where pH and ANC values are lowest (Figure 8.6a,b).

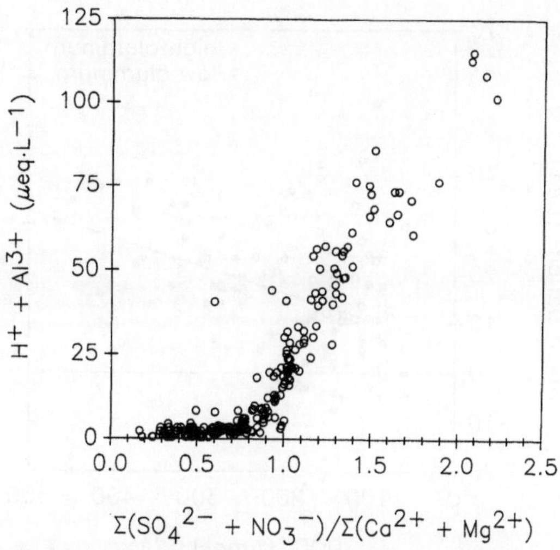

FIGURE 8.11. Plot of $[H^+ + Al^{3+}]$ versus ion ratio $[SO_4^{2-} + NO_3^-]:[Ca^{2+} + Mg^{2+}]$ for USGS/NYCDEP stream survey. Aluminum values are for total dissolved aluminum.

Organic Anions

One of the objectives of this chapter is to assess the relative roles of inorganic acid anions and organic acid anions in controlling Catskill streamwater chemistry. The data necessary to make this assessment are not available for the USGS/NYCDEP dataset, because neither DOC nor aluminum speciation was measured. Data for the LTM sites (a lower ANC subset of the USGS/NYCDEP sample sites; Table 8.1) include DOC measurements, and can be used to make rough estimates of the importance of organic anions.

The total organic acid (A_t) concentration can be estimated by calculating the anion deficit, the difference between the sum of measured cations and the sum of measured anions (LaZerte and Dillon 1984; Chapter 13, Cook and Jager, this volume). The anion deficit theoretically should yield a value equal to A_t (the sum of weak and nonprotonated organic anions), but its calculation has a very large uncertainty due to the propagation of analytical errors associated with the analysis of each major ion. In addition, the inclusion of aluminum in the calculation of anion deficit is problematic when measurements of inorganic monomeric aluminum are unavailable, as is the case for Catskill LTM sites. For the purposes of this chapter, anion deficit is calculated as:

$$\begin{aligned}\text{Deficit} = {} & 2[Ca^{2+}] + 2[Mg^{2+}] + [Na^+] + [K^+] \\ & + 2[Al^{n+}] + 2[Fe^{2+}] + [H^+] \\ & - 2[SO_4^{2-}] - [HCO_3^-] - [NO_3^-] \\ & - [Cl^-] \qquad (8\text{-}1)\end{aligned}$$

where concentrations are in μmol L^{-1}. Aluminum values are for total dissolved aluminum, and a charge of +2 is assumed, consistent with the results of Sullivan et al. (1989). Calculated anion deficits are likely to be overestimates of A_t whenever aluminum concentrations are high (i.e., at

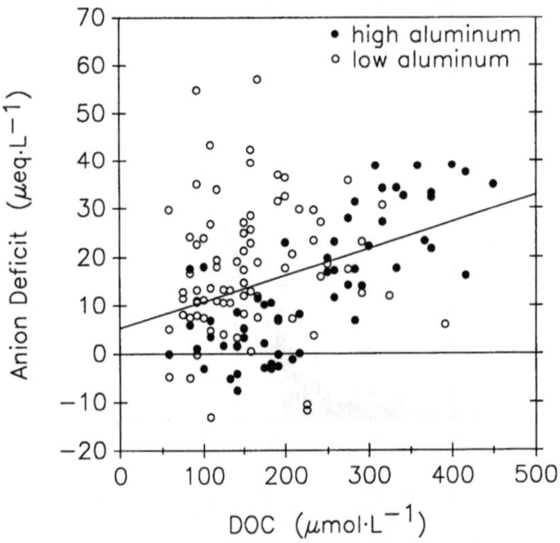

FIGURE 8.12. Plot of anion deficit (sum of measured cations minus sum of measured anions) versus DOC for samples from Catskill Long-Term Monitoring (LTM) sites. Samples with aluminum > 3 μmol L^{-1} are shown as closed circles to illustrate the effect of high aluminum concentrations on anion deficit. Regression line (Anion Deficit = 5.3 + 0.055 × DOC) is based on all samples assuming valence of aluminum (total dissolved) is 2+.

ANC < 0), because total dissolved aluminum values are overestimates of inorganic monomeric aluminum.

If anion deficit is a good measure of A_t, rather than simply a measure of combined analytical error, it should correlate closely with DOC (LaZerte and Dillon 1984, Linthurst et al. 1986). For the Catskill LTM sites the correlation between anion deficit and DOC is very significant, but with high variance ($r^2 = 0.12, p < 0.001$), and suggests that 5.5 μeq of organic acid anions exist per 100 μmol DOC (Figure 8.12). For comparative purposes, Oliver et al. (1983) estimated a charge density of 12.6 μeq per 100 μmol DOC for 20 sites in eastern North America, and Linthurst et al. (1986) estimated 10.1 μeq per 100 μmol DOC for the ELS Catskill/Pocono region. The relationship between anion deficit and DOC in the Catskills, however, is not significant ($r^2 = 0.01, p > 0.3$) if samples with high aluminum concentrations (total dissolved aluminum > 3 μmol L^{-1}) are removed (Figure 8.12). If only high aluminum samples are used, the relationship between anion deficit and DOC is strong ($r^2 = 0.38$) and suggests 9.3 μeq L^{-1} of A_t per 100 μmol L^{-1} of DOC. Interpretation of calculated A_t results are very dependent on aluminum concentrations. Because some organically bound aluminum will necessarily be included in total dissolved aluminum measurements, results for calculated A_t based on these values should be regarded with caution.

When compared with concentrations of inorganic acid anions (SO_4^{2-} and NO_3^-), A_t is relatively unimportant. Median A_t, as calculated above, is 14 μeq L^{-1} (interquartile range is 11 to 19 μeq L^{-1}), or roughly 8% of the total acid anion concentration (Table 8.1). The DOC values increase at high flows, however, and it is possible that organic acid anions could play a larger role during high-flow episodes than is indicated by their central tendency.

Processes Influencing Surface Water Chemistry

Comparison of Precipitation and Stream Chemistry

Differences between average volume-weighted concentrations of major constituents in Biscuit Brook and those in precipitation indicate significant chemical interaction between precipitation and watershed materials (Figure 8.13). The total concentration of dissolved constituents in Biscuit Brook during WY85 and WY86 was 2.5 times that in precipitation, which can be attributed to the concentrating effect of evapotranspiration (52% of precipitation in WY85 and 31% in WY86), and an unknown contribution from dry deposition (estimated as 33% of wet deposition in Figure 8.13). The relative proportions of ions in stream water differ substantially from those in precipitation, however, which indicates that chemical reactions within the watershed also affect stream chemistry. Much of the following discussion is covered in detail by Murdoch (1988).

Calcium in Biscuit Brook replaces hydrogen from precipitation as the dominant cation and accounts for more than 60% of the total cations. The average annual hydrogen concentration in stream water is insignificant except in the most acidic of Catskill streams (1% of total cation concentration at Biscuit Brook). Cations at Biscuit Brook show the same pattern of dominance ($Ca^{2+} > Mg^{2+} > Na^+ > K^+$) as streams in the USGS/NYCDEP survey (Figure

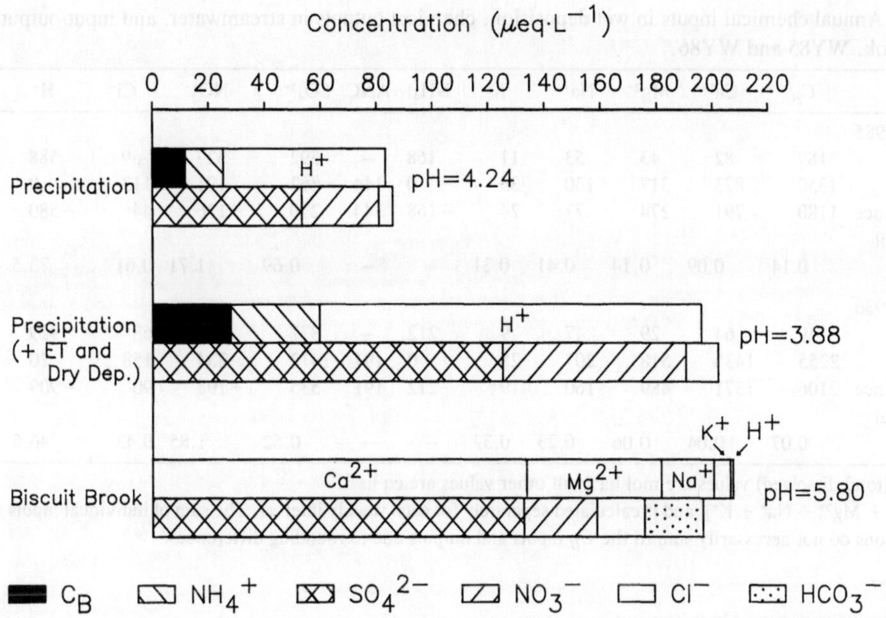

FIGURE 8.13. Ion bar diagrams of precipitation, precipitation modified by dry deposition (assumed to be 33% of wet deposition) and evapotranspiration (ET = 44% of precipitation), and Biscuit Brook stream water for water years 1985 and 1986.

8.7), and their relative concentrations do not change appreciably with increasing discharge. Aluminum and H^+ accounted for as much as 11% and 5%, respectively, of the total cations during peak flows at Biscuit Brook. The general absence of NH_4^+, except at high flows, suggests nearly complete transformation of that ion by biological activity or cation exchange within the watershed.

The dominant anion in both stream water and precipitation is SO_4^{2-}. Sulfate represents 68% of total anion concentrations in Biscuit Brook, and 62% of anionic charge in precipitation (Figure 8.13). Nitrate accounts for approximately one-third of total anion concentrations in precipitation, but only 13% in Biscuit Brook. Chloride makes up a similar proportion of total anions in both stream water and precipitation (8% and 6%, respectively). In Biscuit Brook, HCO_3^- averages approximately 10% of total anion concentrations.

Chemical budgets for Biscuit Brook in WY85 and WY86 suggest a net loss of base cations, SO_4^{2-}, Cl^-, and ANC from the watershed and a net accumulation of NO_3^-, NH_4^+, and H^+ from deposition (Table 8.5; Figure 8.14). Some of the apparent net loss was due to the omission of a dry deposition contribution in the precipitation calculations. As discussed in the section entitled Chemistry of Atmospheric Deposition, estimates of dry deposition for the northeastern United States are quite variable. Both the SO_4^{2-} and Cl^- budgets for Biscuit Brook can be balanced by assuming that dry deposition is 30% to 40% of wet deposition (Figure 8.13). In the case of Cl^-, estimates of dry deposition made from wet deposition chemistry must be made carefully. The close correspondence of Cl^- concentration to Na^+ concentration (Figure 8.8) in the Catskill region suggests strongly that salt (NaCl) is the predominant source of Cl^- in stream water. Appreciable geochemical sources of Cl^- in the Catskills are unlikely, because bedrock is strongly leached along joint fractures and bedding surfaces where most groundwater in bedrock would flow (Murdoch 1988). Biscuit Brook's distance from the ocean makes significant contributions from marine salts unlikely. Road salt is therefore the most likely source of NaCl, and because there are no roads in the Biscuit Brook watershed, redeposition of road salt dust (in both wet and dry deposition) is the only plausible source of salt for the stream. Chloride in dry deposition is therefore

TABLE 8.5. Annual chemical inputs in wet deposition, chemical outputs in streamwater, and input/output ratios for Biscuit Brook, WY85 and WY86.[a]

	C_B[b]	Ca^{2+}	Mg^{2+}	Na^+	K^+	NH_4^+	ANC	SO_4^{2-}	NO_3^-	Cl^-	H^+	Aluminum
Water Year 1985												
Input	189	82	43	53	11	168	–	592	327	69	588	–
Output	1350	873	317	130	35	0	144	862	191	113	9	1.9
Net difference	1180	791	274	77	24	−168	144	270	−136	44	−580	1.9
Input/output ratio	0.14	0.09	0.14	0.41	0.31	–	–	0.69	1.71	0.61	73.5	–
Water Year 1986												
Input	149	64	29	47	9	212	–	877	432	68	929	–
Output	2255	1435	518	207	28	0	191	1412	234	158	20	4.4
Net difference	2106	1371	489	160	19	−212	191	535	−198	90	−909	4.4
Input/output ratio	0.07	0.04	0.06	0.23	0.32	–	–	0.62	1.85	0.43	46.5	–

[a] Aluminum (total dissolved) values are mol ha^{-1}; all other values are eq ha^{-1}.
[b] $C_B = [Ca^{2+} + Mg^{2+} + Na^+ + K^+]$, and is calculated separately for each weekly interval. The sum of individual inputs and outputs of major cations do not necessarily sum to the C_B inputs and outputs due to rounding differences.

very local in origin, and should not be expected to appear as the same proportion of wet deposition that other ions do.

Sulfate yields in Biscuit Brook exceeded SO_4^{2-} deposition in wetfall in both WY85 and WY86 (Table 8.5). Inputs from wetfall accounted for 69% of the stream's SO_4^{2-} yield in WY85 and 62% in WY86. These input-output ratios are similar to those at Hubbard Brook (71%), Panther Lake in the Adirondacks (77%), and Birkenes in Norway (63%), all of which have similar stream SO_4^{2-} concentrations and have had weathering discounted as a significant SO_4^{2-} source (Hemond and Eshleman 1984).

Unlike the Cl^- and SO_4^{2-} budgets, the NO_3^- budget indicates a net retention during both years. The mean annual wetfall input-output ratio for WY85 and WY86 was 1.78, which indicates significant retention and is similar to the ratio calculated for two pristine lake outlets in the Adirondacks during 1978–80 (Galloway et al. 1983b). Also, the input-output ratio for NO_3^- probably represents a minimum estimate because it excludes contributions from dryfall, gaseous nitrogen, or NH_4^+. Most of the NH_4^+ received from atmospheric loading in the Biscuit Brook watershed is retained, as it is in other watersheds studied in the Northeast (e.g., Hemond and Eshleman 1984). If NH_4^+ is included in the total nitrogen input load, the input-output ratio increases to 2.59 for WY85 and 2.75 for WY86. A precise estimate of the nitrogen input-output ratio for the Biscuit Brook watershed is very difficult to make, because dry deposition, nitrification, and denitrification in forest soils, and an unknown export of organic nitrogen, all contribute significant uncertainty to its calculation. Based on budgets of inorganic nitrogen, however, it is clear that substantial nitrogen is retained by the Biscuit Brook watershed (Table 8.5). Both NO_3^- and NH_4^+ are retained, and because their retention has opposite effects on ANC (NH_4^+ retention is an acidifying process, NO_3^- retention produces ANC), the net effect of nitrogen retention on Biscuit Brook ANC is small (+7.6 μeq L^{-1}; Figure 8.14).

A large net loss of base cations occurred in the Biscuit Brook watershed in both WY85 and WY86 (Table 8.5). Weathering and/or cation exchange of C_B in the watershed is the primary mechanism for producing ANC in Biscuit Brook (Figure 8.14). Base cation concentrations and yields are the primary factors that distinguish Biscuit Brook from other Catskill streams, as well as from the Adirondack lake outlets and Hubbard Brook (Murdoch and Barnes in press). Because anion concentrations are similar among all these streams, the differences in their buffering capacities are probably related to the differences in base cation export rates from their watersheds. In the case of Biscuit Brook, the H^+ budget indicates that C_B export is sufficient nearly to balance H^+ inputs (Table 8.5; Figure 8.14).

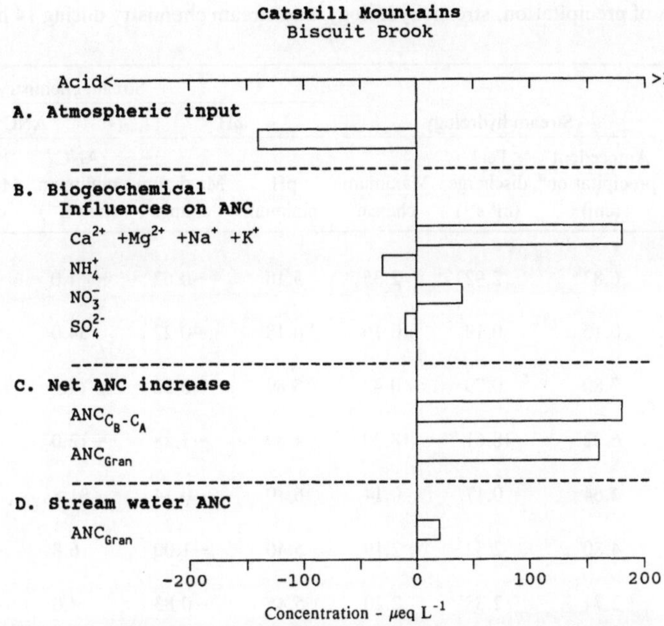

FIGURE 8.14. Modification of atmospheric deposition by watershed and stream biogeochemical processes at Biscuit Brook, WY85 and WY86. A. Precipitation [H^+] adjusted for estimated dry deposition and ET, as in Figure 8.13. Equivalent to expected stream water $ANC_{C_B\text{-}C_A}$ in the absence of any process other than ET. B. Reactions influencing ANC. Bars to the right indicate ANC production; those to the left show ANC consumption. C. Net ANC production. $ANC_{C_B\text{-}C_A}$ is the sum of ANC production terms for measured anions and cations including those shown in B.; ANC_{Gran} is determined by the difference between input ANC (A.) and stream water ANC (D.). D. Stream water ANC represents the net reaction of atmospheric inputs with stream/watershed system. All data are calculated from volume-weighted concentrations in Figure 8.13. For a further explanation of calculation procedures and figure interpretation, see Appendix A. Volume-weighted mean DOC for Biscuit Brook is 303 µmol L^{-1}. Mean A_t = 17 µeq L^{-1}.

Relation Between Stream Ion Concentrations and Discharge

Results of the USGS/NYCDEP survey of Catskill stream chemistry indicated significant changes in concentrations with changes in discharge (Figure 8.5). In order to investigate this pattern further, stream discharge and chemistry were monitored during 14 storms and snowmelt periods between April 1983 and May 1986 at Biscuit Brook (Murdoch 1988). Peak flows ranged from 0.17 to 22.7 m^3 s^{-1} (Table 8.6). Maximum stream acidity occurred at or near peak flows, and peak flow pH ranged from 6.18 to 4.89. The maximum observed change in pH was 1.71 units during Hurricane Gloria on September 27, 1985. Minimum ANC during the 14 high flows ranged from 34 to −16 µeq L^{-1} (net acidity). A maximum observed change in ANC of 60 µeq L^{-1} occurred during Hurricane Gloria.

Stream pH values at Biscuit Brook decrease and approach those of precipitation during very high flow events, but significant buffering occurs in the watershed during most high flows (Table 8.6). Precipitation pH and minimum stream pH were similar only during the large storm of April 3–6, 1984. Minimum stream pH during Hurricane Gloria (September 27–28, 1985) was actually lower than precipitation pH, which indicates a net acidification of the rain water by materials already in the watershed. The minimum stream pH during all other storms exceeded precipitation pH, which indicates a net buffering of rain water acidity.

The relation between stream discharge and precipitation volume during the storms improved with increasing volume but was poorly defined at low volumes because (1) snowmelt contributed to discharge during the spring when precipitation volume was low and discharge was high, and (2)

TABLE 8.6. Characteristics of precipitation, stream hydrology, and stream chemistry during 14 high-flow periods at Biscuit Brook, 1983–86.

	Precipitation		Stream hydrology			Stream chemistry				
						pH		ANC		
Date	Volume (cm)	pH	Antecedent precipitation[a] (cm)	Peak discharge (m^3 s^{-1})	Maximum change[b]	pH minimum	Maximum change[b]	ANC minimum (μeq L^{-1})	Maximum change[b]	Maximum aluminum[c] (μmol L^{-1})
Apr 10–11, 1983	3.04	4.7	6.83	2.92	2.35	5.18	−0.62	1.0	−25.0	8.5
Nov 10–11, 1983	3.96	4.5	0.15	0.19	0.16	6.18	−0.22	34.0	−2.0	1.5
Mar 21, 1984	1.04	4.5	7.80	0.79	0.42	5.89	−0.32	14.0	−4.0	1.9
Apr 3–6, 1984	12.98	4.9	6.02	18.61	18.20	4.89	−1.28	−12.0	−32.0	11.1
Nov 5, 1984	2.08	4.0	4.64	0.17	0.14	6.10	−0.14	31.8	−25.0	1.9
Feb 23–26, 1985	0.46	4.1	4.80	2.32	2.19	5.40	−1.00	6.8	−29.2	5.2
Mar 11–13, 1985	3.83	4.1	2.74	2.38	2.20	5.45	−0.83	4.0	−32.0	5.6
May 17–21, 1985	3.91	4.3	6.85	0.60	0.53	6.02	−0.38	20.0	−22.0	3.7
Jul 31, 1985	4.39	3.9	5.46	1.08	0.79	6.06	−0.84	24.0	−14.0	4.8
Sep 27–28, 1985	12.62	5.3	1.16	22.70	22.60	4.90	−1.71	−10.0	−60.0	11.9
Jan 19–21, 1986	2.94	4.7	0.48	5.04	4.94	5.51	−0.72	8.0	−22.0	5.6
Mar 13–16, 1986	8.99	4.6	0.71	13.90	13.73	4.94	−1.69	−16.0	−38.0	10.0
Mar 18–20, 1986	3.40	4.0	9.70	14.36	13.54	4.92	−0.78	−16.0	−20.0	8.5
May 20–22, 1986	9.01	4.4	2.00	6.95	6.81	5.19	−1.22	−8.0	−36.0	8.5

[a] Precipitation volume during 14 days preceding a storm or snowmelt period.
[b] Difference between initial values and values at peak flow.
[c] Total dissolved aluminum.

antecedent soil-moisture conditions, which affect the amount of precipitation reaching the stream, differed from storm to storm. Adding a surrogate variable for antecedent soil-moisture conditions (precipitation in the 14 days preceding a storm) to the comparison of storm precipitation volume with streamflow volume did not significantly improve the correlation between precipitation and flow (Murdoch 1988). Lynch et al. (1986) have used the hydrologic response of a small stream in central Pennsylvania to precipitation volume as an index of stream sensitivity to acidic deposition. The sensitivity of Biscuit Brook may also be viewed as a function of hydrologic response to precipitation volume.

In general, results of high-flow analyses show dilution of C_B, SO_4^{2-}, and ANC during individual flow peaks, and increased concentrations of H^+ and NO_3^- (Figure 8.15). Concentrations of minor constituents also changed during high flows. Potassium, a minor constituent, showed small increases with increased flow. Hydrogen and aluminum concentrations showed rapid increases with increased flow. Silica decreased with increased flow, which indicates dilution of dissolved weathering products. Values of DOC increased with flow, which suggests flow through the organic soil layer to the stream or bank erosion; seepage from the organic soil layer may provide organic acids to the stream

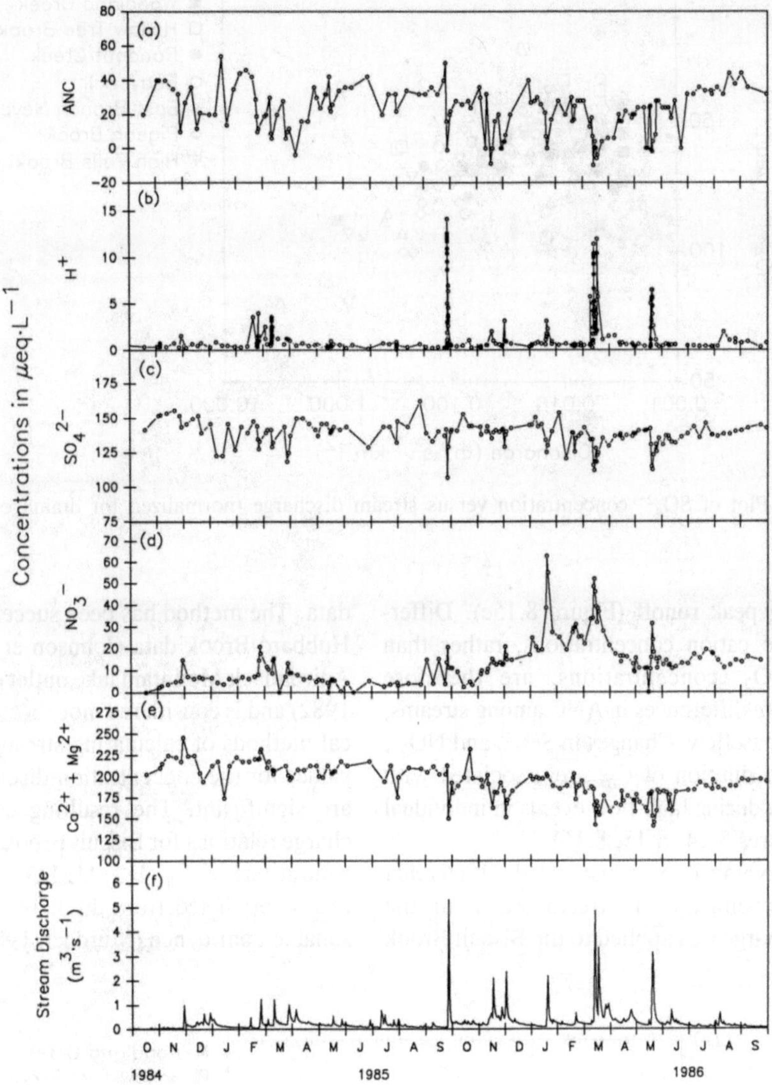

FIGURE 8.15. Temporal patterns in Biscuit Brook stream water chemistry and mean daily stream discharge, WY85 and WY86.

and account for some of the pH decrease observed during storms.

Data from seven LTM streams that together represent a wide range of baseflow ANC values support the concentration-discharge relations observed at Biscuit Brook (Murdoch and Barnes in press). Sulfate concentrations observed in these streams decreased with increased flow and were similar to each other at flows ranging through three orders of magnitude (Figure 8.16). Base cation concentrations, by contrast, differed widely among the streams, and the dilution with increased flow was greatest in the streams with highest concentrations (Figure 8.17). Base cation concentrations in the LTM streams undergo more dilution at high flows than SO_4^{2-} concentrations. This differential dilution causes the $SO_4^{2-}:C_B$ ratio to become higher at high flows, and contributes to low ANC values observed during storm and snowmelt events (Murdoch 1988). Aluminum and H^+ concentrations increased with flow, and NO_3^- increased with flow to values similar to those observed at Biscuit

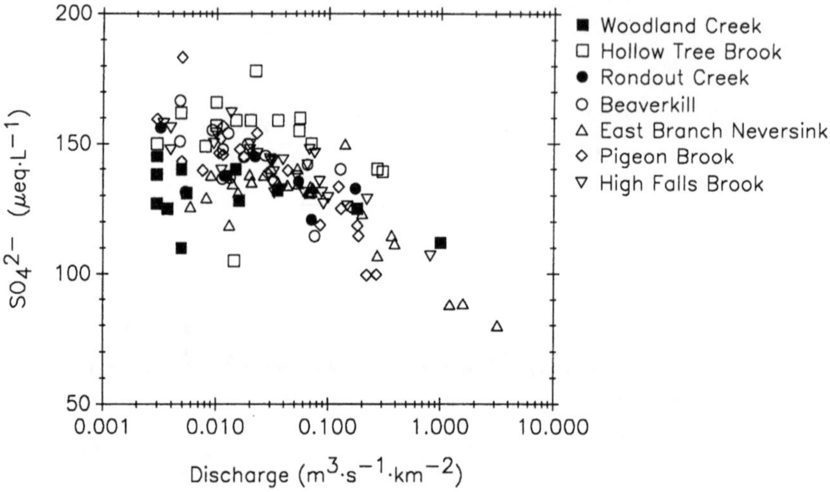

FIGURE 8.16. Plot of SO_4^{2-} concentration versus stream discharge (normalized for drainage area) for Catskill LTM streams.

Brook during peak runoff (Figure 8.15c). Differences in base cation concentrations, rather than SO_4^{2-} or NO_3^- concentrations, are therefore associated with differences in ANC among streams, especially at baseflow. Changes in SO_4^{2-} and NO_3^-, in addition to dilution of C_B, are associated with ANC changes during high-flow events in individual streams (Figures 8.14, 8.15, 8.16).

A model developed by Johnson et al. (1969) that relates the chemistry of stream water to the stream's discharge was applied to the Biscuit Brook data. The method has been successfully applied to Hubbard Brook data (Johnson et al. 1969) and to Adirondack Mountain lake outlet data (Peters et al. 1982) and is considered more accurate than graphical methods of calculating stream yields, if the r^2 values for the concentration-discharge regressions are significant. The resulting concentration-discharge relations for Biscuit Brook suggest that concentrations of C_B, H^+, Al, SO_4^{2-}, Cl^-, and ANC can be predicted from discharge values with reasonable confidence (Murdoch 1988). The r^2 value

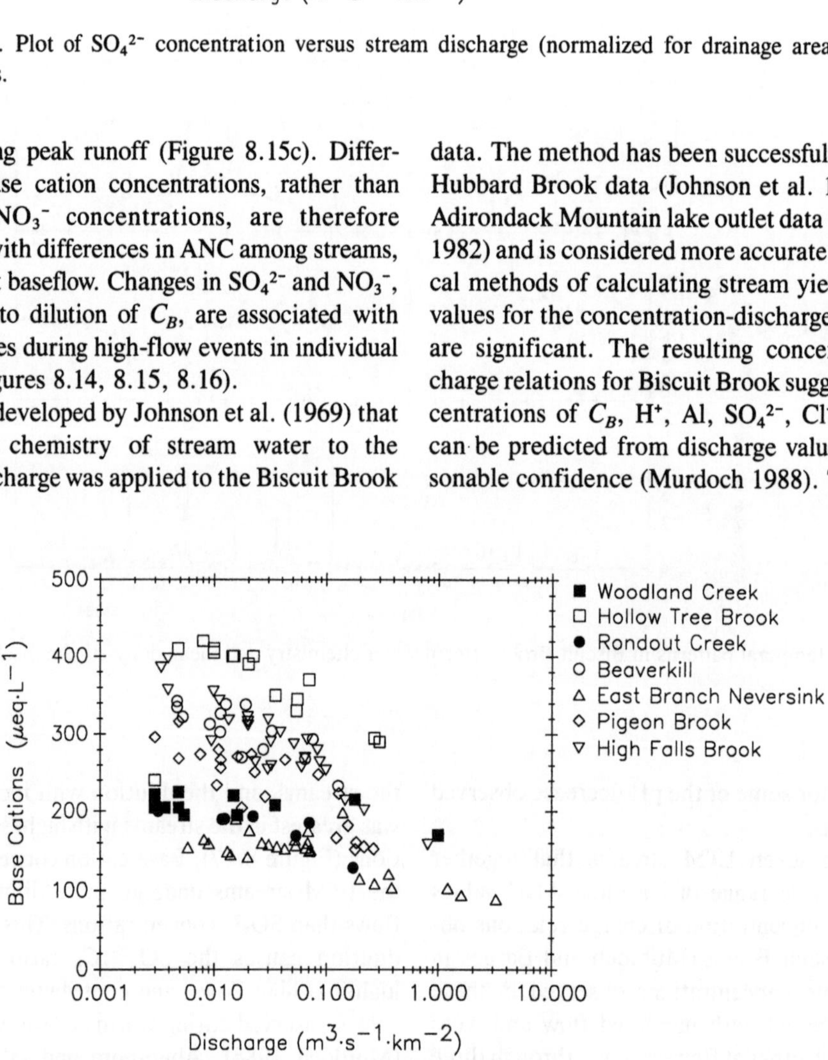

FIGURE 8.17. Plot of C_B (Ca^{2+} + Mg^{2+} + Na^+ + K^+) versus stream discharge (normalized for drainage area) for Catskill LTM streams. Stream symbols are as in Figure 8.16.

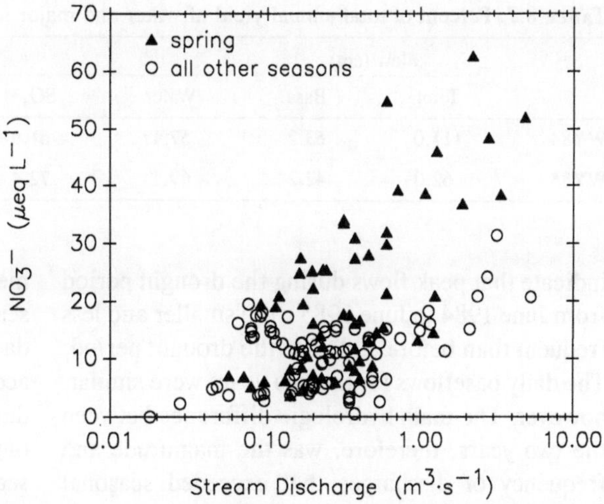

FIGURE 8.18. Plot of NO_3^- concentration versus stream discharge for Biscuit Brook, WY85 and WY86. Spring (March-May) data are plotted separately to illustrate the effect of season on NO_3^- retention by watershed.

for the C_B-discharge relation suggests that 63% of the variability in concentrations can be attributed to changes in discharge. The H^+ and aluminum r^2 values were 0.90 and 0.77, respectively. Sulfate concentration also was strongly correlated with discharge ($r^2 = 0.76$).

Nitrate concentrations were poorly correlated with flow ($r^2 = 0.17$) when data for all seasons were combined (Figure 8.18). The relation was stronger ($r^2 = 0.38$), however, when spring data (March through May) were analyzed separately. Biological uptake of NO_3^- is likely to alter the concentration-discharge relation during the growing season (late spring to early fall), and undoubtedly contributes to the poor relation observed between concentration and discharge on an annual basis. Cessation or slowing of biological activity during the winter months probably causes NO_3^- to be more available to stream runoff during winter and spring, which leads to a steeper increase in NO_3^- concentration at high discharges (Murdoch 1988; Figure 8.18).

Effect of Storm Runoff and Snowmelt on Catskill Stream Chemistry

The data collected at Biscuit Brook during WY84 and WY85 indicate that storm runoff and snowmelt contribute substantially to annual stream acidity in the Catskill Mountains. Plots of daily stream discharge for those years (Figure 8.19)

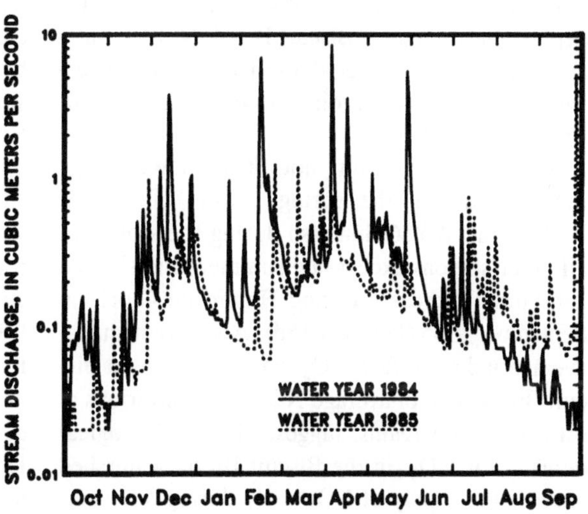

FIGURE 8.19. Hydrographs for Biscuit Brook, WY84 and WY85.

TABLE 8.7. Percent of total annual yield of water and major ions in baseflow at Biscuit Brook, WY84 and WY85.

	Flow (cm)		% in baseflow					
	Total	Base	Water	SO_4^{2-}	NO_3^-	Ca^{2+}	H^+	ANC
WY84	111.0	63.2	57.1	61.9	52.5	63.1	19.0	96.7
WY85	62.0	42.2	69.1	72.4	65.8	72.4	28.5	88.0

indicate that peak flows during the drought period from June 1984 to June 1985 were smaller and less frequent than before and after the drought period. The daily baseflows for the two years were similar, however. The main hydrologic difference between the two years, therefore, was the magnitude and frequency of discharges that exceeded seasonal baseflow values (Murdoch 1988).

The cumulative yields of H^+ and aluminum from Biscuit Brook in WY84 were substantially greater than in WY85 (Murdoch 1988). If chemical yields for each year are divided by graphic flow separation techniques into baseflow and high-flow components, results indicate that the vast majority of the H^+ yield is derived from high flows (Table 8.7). These data suggest that storm runoff and snowmelt provide most of the annual stream acidity and that this acidity develops in the stream during several short episodes over the course of each year.

Another effect of snowmelt and storm runoff is the large increase in stream NO_3^- during the spring (Figure 8.15d). This spring pulse is consistent with results from the Adirondacks and Hubbard Brook, but is contrary to the results at Bickford Reservoir in Massachusetts, where stream NO_3^- concentrations remained low (1 μeq L^{-1}) through the spring (Hemond and Eshleman 1984). The annual wet deposition of NO_3^- at Biscuit Brook (438 eq ha^{-1}) appears slightly greater than at the other watersheds (Bickford Reservoir, 295 eq ha^{-1}; Hubbard Brook, 317 eq ha^{-1}).

Nitrate concentration not only showed large increases during the spring storms of WY84, WY85, and WY86, but also during the storms of all other seasons except late summer, and showed substantial increases during Hurricane Gloria in September 1985 (Figure 8.15d). Analysis of annual average and stormflow NO_3^- concentrations at Biscuit Brook, and of average NO_3^- concentrations at other Catskill streams, suggests that the biological demand for NO_3^- in the Biscuit Brook watershed exceeds the supply of NO_3^- only during late summer. The WY85 net yield of nitrogen showed a seasonal pattern similar to that observed in Adirondack lake outlets (Galloway et al. 1983b), with net accumulation during the summer and fall, net loss during snowmelt periods, and balanced inputs and outputs during winter (Murdoch 1988). This seasonal pattern provides clear evidence of the importance of biological processes in regulating the dynamics of NO_3^- in the watershed.

Evidence of Historical Trends

Evidence of trends in major ion chemistry of Catskill streams comes primarily from historical water quality data gathered by the NYCDEP as part of their routine monitoring network. Because most of New York City's drinking water originates in the Catskill region, laboratories of the NYCDEP have monitored ANC, NO_3^-, water hardness, and other relevant variables since before the construction of the Catskill reservoirs. Data collection in the Schoharie and Ashokan basins began about 1920; data collection in the southern basins began in the 1950s. A second, less extensive, long-term data base has been developed at the USGS hydrologic benchmark station on Esopus Creek in Shandaken, New York (Figure 8.1) where data have been collected since 1964. The following discussion focuses primarily on one site, Schoharie Creek, at Prattsville (Figure 8.1), which has the most complete data record of any of the historical sites. Trends at Schoharie Creek are consistent with trends in about 20 other Catskill streams (Stoddard in review). The USGS has monitored stream discharge at Schoharie Creek since 1907.

One common problem in interpreting historical records of surface water chemistry is that methods of analysis have changed substantially over several decades, and results may be biased as a result of differences between these methods' ability to determine major ion concentrations accurately (Kramer

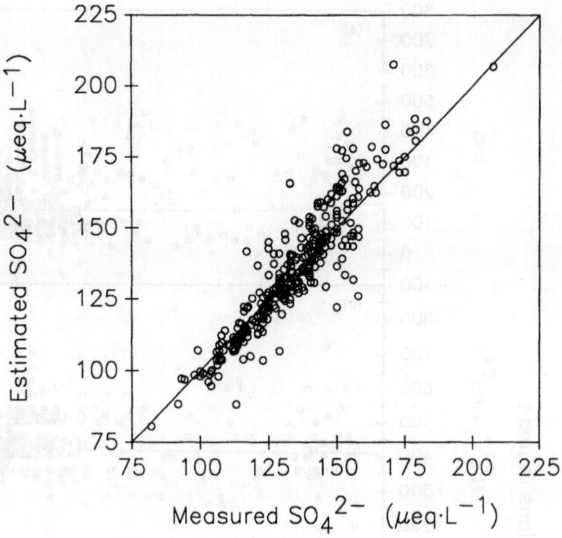

FIGURE 8.20. Plot of estimated SO_4^{2-} vs. measured SO_4^{2-} used as a test of SO_4^{2-} estimation method employed for historical reconstruction of SO_4^{2-} trends. Data for estimated SO_4^{2-} (see text for explanation of estimation method) and measured SO_4^{2-} (by ion chromatography) are for samples in USGS/NYCDEP survey. 1:1 line is included to indicate expected relationship.

and Tessier 1982, Kramer 1985). Acid neutralizing capacity is particularly prone to this problem, and many attempts have been made to determine what effect changes in methods have had on ANC estimates (Kramer and Tessier 1982, Kramer 1985, Pfeiffer and Festa 1980, Kramer et al. 1986). For this chapter, all ANC values were corrected for biases caused by the titration methods used. Correction factors were based on the results of replicated titrations conducted with all of the historical methods, and comparison to Gran titration results. Details on ANC corrections and on outlier tests applied to all chemical data are given in Stoddard (in review). $[Ca^{2+} + Mg^{2+}]$ values are calculated from water hardness measurements; analytical methods for determining hardness were consistent throughout the period of data collection. Nitrate concentrations were determined by two methods; comparisons indicate that accuracy and precision of the two methods were similar.

Sulfate concentrations were not measured directly by NYCDEP laboratories until 1979, and then only infrequently. Sulfate is therefore estimated using the method of Ceraso et al. (1986), modified to utilize the chemical variables available in the NYCDEP dataset. This estimation method is based on the formula for calculated conductivity, which is simplified to remove unmeasured variables (i.e., the major cations), and solved for SO_4^{2-} concentration. For the NYCDEP data, Ca^{2+} and Mg^{2+} concentrations are estimated from water hardness, HCO_3^- is set equal to ANC, and Na^+ is set equal to Cl^-, based on data in Figure 8.8. Potassium, a minor cation in Catskill streams (Figure 8.7) is set equal to 9 $\mu eq\ L^{-1}$, a typical value for the Schoharie basin (Table 8.4). Sulfate is assumed in this method to account for all of the difference between measured and calculated conductivity (i.e., organic acids and minor inorganic ions are assumed to be absent).

The SO_4^{2-} estimation method was tested using data from the USGS/NYCDEP stream survey (Figure 8.20). The results indicate reasonable agreement between estimated and measured values, although estimated values appear to overestimate measured SO_4^{2-} at high values. Estimated SO_4^{2-} concentrations are therefore expected to show greater variability than measured concentrations.

Strong seasonality, which produces the marked scatter among the points in Figure 8.21, complicates the statistical identification of trends in streamwater chemistry. Trends in Schoharie Creek chemical composition were therefore tested using analysis of covariance (ANCOVA) on rank-transformed data (Loftis and Taylor 1989). This method allows the fitting of regression parameters to serially correlated time series data; it has an effect similar to that of deseasonalizing the data through a nonparametric test (e.g., Hirsch et al. 1982, van Belle and Hughes 1984), and has the advantage of detecting nonlinear trends.

Acid neutralizing capacity at Schoharie Creek exhibits a highly significant ($p < 0.0001$) curvi-

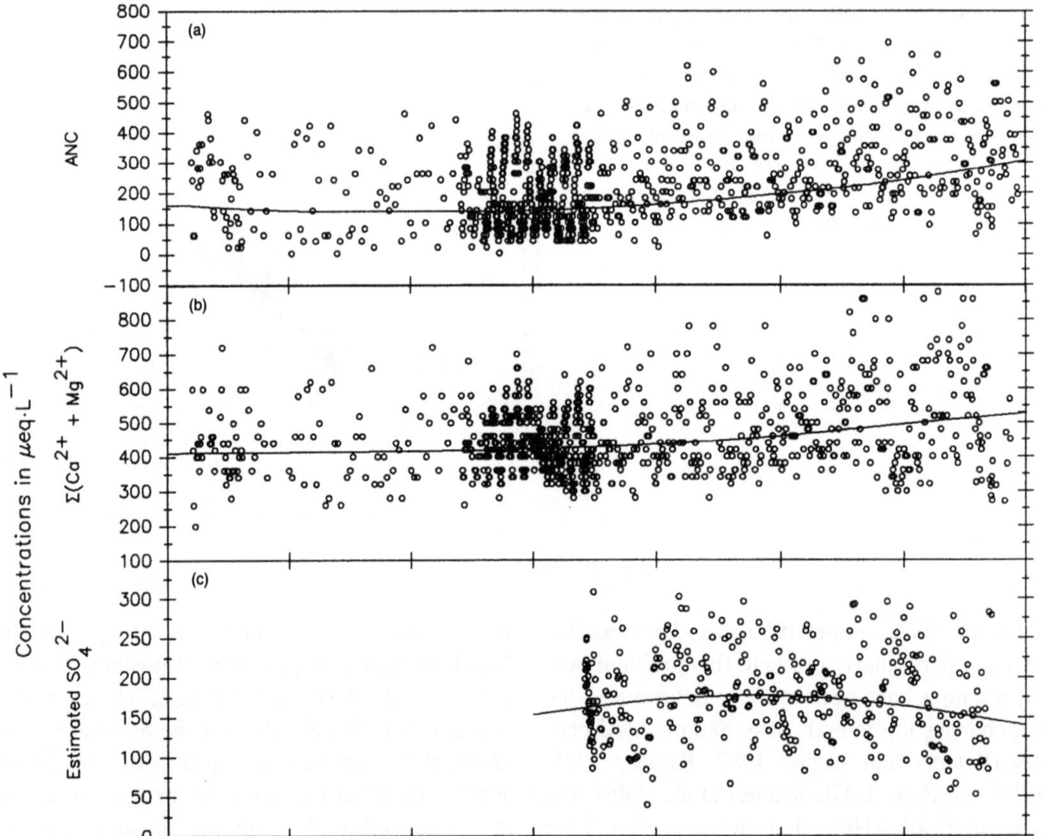

FIGURE 8.21. Temporal trends in streamwater chemistry and stream discharge for Schoharie Creek at Prattsville, 1922–89: (a) ANC values corrected for methods bias; (b) Ca^{2+} + Mg^{2+} concentrations, calculated from water hardness measurements; (c) estimated SO_4^{2-} concentrations (see text for explanation of estimation method); (d) NO_3^- concentrations; and (e) discharge measurements from USGS gauge at site. All concentrations are in μeq L^{-1}. Lines in each box are the results of best-fit regressions, using analysis of covariance (ANCOVA) on rank-transformed data.

linear trend with time (Figure 8.21a); ANC decreased from a value around 160 μeq L^{-1} to a minimum of 140 μeq L^{-1} in the 1940s. After 1950, ANC increased dramatically to a value of approximately 300 μeq L^{-1} in 1989.

The concentration of [Ca^{2+} + Mg^{2+}] at Schoharie Creek also shows a highly significant ($p < 0.01$) curvilinear trend (Figure 8.21b), which increases throughout the period of record. Concentrations have risen from a minimum of 415 μeq L^{-1} to a maximum of almost 500 μeq L^{-1} in 1989, with most of the increase occurring after 1950.

Estimated SO_4^{2-} concentrations (Figure 8.21c) also show a significantly curvilinear trend ($p <$ 0.001) and appear to have reached maximum values during the late 1960s. Estimated SO_4^{2-} decreased from 175 μeq L^{-1} in 1965 to 140 μeq L^{-1} in 1989. A decreasing trend in the SO_4^{2-} concentrations of surface waters since 1970 in the Northeast has been verified by several investigators (Peters et al. 1982, Smith and Alexander 1983) and is consistent with the observed decreases in sulfate deposition recorded in this region since the 1960s (Smith and Alexander 1986, Hedin et al. 1987).

Nitrate concentrations in Schoharie Creek were consistently low (ca. 10 μeq L^{-1}) until approximately 1960 (Figure 8.21d). After 1965, NO_3^- concentrations increased dramatically to levels near 50 μeq L^{-1}. The recent upward trend in NO_3^- concentration in Schoharie Creek is consistent with trends in other surface waters in the Northeast (Smith et al. 1987, Kramer et al. 1986), but not with trends in NO_x emissions. Estimates of NO_x emissions suggest that they have increased monotonically throughout most of the 20th century until about 1970, and have remained more or less constant since then (Husar 1986; Chapter 3, Husar et al., this volume); they certainly do not show any evidence of a precipitous increase beginning around 1965.

Stream discharge has a profound effect on major ion concentrations in Catskill streams (see section entitled Seasonal Variability in Ion Composition of Streams); it is therefore important to determine whether any of the trends shown in Figure 8.21 are attributable to trends in stream discharge. No significant trend ($p > 0.5$) in stream discharge is observable at Schoharie Creek for the period of record (Figure 8.21e).

When combined, the trends illustrated in Figure 8.21 suggest that ANC in Schoharie Creek has increased since 1950, primarily as a result of increasing base cation concentrations. Hypothesized decreases in streamwater SO_4^{2-} (about −35 μeq L^{-1} since 1965), based on estimated concentrations, are nearly offset by increases in NO_3^- concentrations (about +30 μeq L^{-1} since 1965). The mechanisms responsible for the trends in [Ca^{2+} + Mg^{2+}], SO_4^{2-}, and NO_3^- are not obvious and cannot all result from the effects of acidic deposition. Decreasing SO_4^{2-} concentrations are almost certainly the result of decreasing sulfur dioxide emissions in the area upwind of the Catskill region. Increasing [Ca^{2+} + Mg^{2+}] concentrations, on the other hand, cannot be attributed to acidic deposition, because they are accompanied by increases in ANC and they continue during the period when rates of sulfuric acidic deposition are known to be decreasing (i.e., 1970 to the present). The most likely cause of the trend in [Ca^{2+} + Mg^{2+}] is changing land use. The Schoharie basin has undergone substantial development, starting in the period after World War II, and landscape disturbance is likely to have increased base cation concentrations.

Changes in NO_3^- concentrations cannot be readily attributed either to changes in acidic deposition, or to changes in land use. Agricultural activity is minimal in the Schoharie watershed. Other kinds of land use disturbances that would lead to increased streamwater NO_3^- (i.e., inputs from sewage or septic systems) would all produce highest concentrations during baseflow periods. But Schoharie Creek exhibits the same pattern of low NO_3^- in the summer and high NO_3^- in the winter and spring that Biscuit Brook does (Figure 8.15c). One plausible mechanism for increased streamwater NO_3^- in Schoharie Creek is a change in the same kinds of biological processes (e.g., denitrification or nutrient uptake) that are thought to control the seasonal pattern of nitrogen dynamics in the Biscuit Brook watershed (see section entitled Processes Influencing Surface Water Chemistry), but the nature of that change is unknown. Changes in the ability of forests and soils to retain nitrogen may be due to natural causes, such as forest maturation (Melillo et al. 1983), or possibly to the cumulative effects of chronic deposition of sulfuric and nitric acids (Aber et al. 1989). Henriksen et al. (1988) found that NO_3^- concentrations in Norwegian lakes have doubled since 1974, but rates of nitrogen deposition have increased only 10%. Henriksen and Brakke (1988b) suggest that

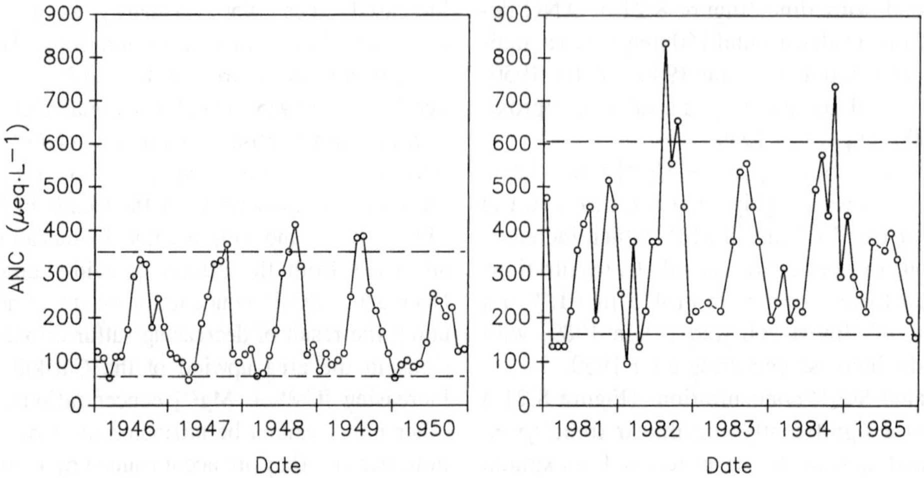

FIGURE 8.22. Short-term temporal patterns in monthly mean ANC during two time periods (1946–50 and 1981–85) in Schoharie Creek at Prattsville. Dashed lines are mean annual maximum (summer) and minimum (spring) values for each 5-year period.

decreases in forest uptake, rather than increases in nitrogen deposition, have caused the increasing trend in NO_3^-. Similar processes may be operating in both the Catskills and in Norway to limit the ability of watersheds to retain nitrogen.

An important pattern in the Schoharie Creek ANC data that warrants further examination is the apparent increase in the annual range of observed values (Figure 8.21a). In the 1940s, yearly ANC minima occurred reliably each spring at around 80 μeq L^{-1} (Figure 8.22), and yearly maxima occurred late each summer at approximately 350 μeq L^{-1}. In the 1980s, the seasonal pattern was still evident, but the minimum and maximum values increased, although to different degrees. Springtime minimum ANC values have increased by approximately 100 μeq L^{-1}, to 180, whereas summer maximum ANC values have increased by more than 250 μeq L^{-1}, to approximately 610 μeq L^{-1}.

The smaller increase in minimum ANC values can be partially explained by the increase in springtime NO_3^- (Figure 8.23). Minimum values for streamwater NO_3^- occur in the late summer, when ANC is maximal, and these minimum NO_3^- concentrations have remained essentially unchanged at approximately 10 μeq L^{-1}. Maximum concentrations of NO_3^- in Schoharie Creek occur in the spring, when stream discharge is high and ANC values are minimal. These maximum NO_3^- concentrations have increased from approximately 30 μeq L^{-1} in the 1940s to near 80 μeq L^{-1} in the 1980s. The increase in springtime NO_3^- is therefore partially responsible for maintaining springtime ANC values at low levels. Nitrate concentrations exert relatively little effect on ANC values in the summer, because NO_3^- levels remain more or less unchanged at this season. Springtime NO_3^- concentrations, however, have increased by as much as 50 μeq L^{-1}, and may control the extent to which springtime ANC values are able to respond to increases in base cations and decreases in SO_4^{2-}.

The only other historical information on Catskill stream chemistry comes from the USGS hydrologic benchmark station on Esopus Creek at Shandaken, where data have been collected since 1964. An increasing trend in ANC (reported increases for the period beginning in 1964 range from +24 to 33 μeq L^{-1}, for a mean change of +1.2 μeq L^{-1} yr^{-1}) and a decrease in SO_4^{2-} concentration (estimated range from −34 to −48 μeq L^{-1} in the same period, or −2.1 μeq L^{-1} yr^{-1}) have been reported for this site (Peters et al. 1982, Smith and Alexander 1983, Kramer et al. 1986). The SO_4^{2-} trend is very similar to the trend in estimated SO_4^{2-} reported for Schoharie Creek (−35 μeq L^{-1} from 1970 to 1989, or −1.9 μeq L^{-1} yr^{-1}), and lends support to the method of SO_4^{2-} calculation used for the NYCDEP data. The ANC trend at Esopus Creek is much smaller than that reported for Schoharie Creek (+140 μeq L^{-1} from 1965 to 1988), but Kramer et al.

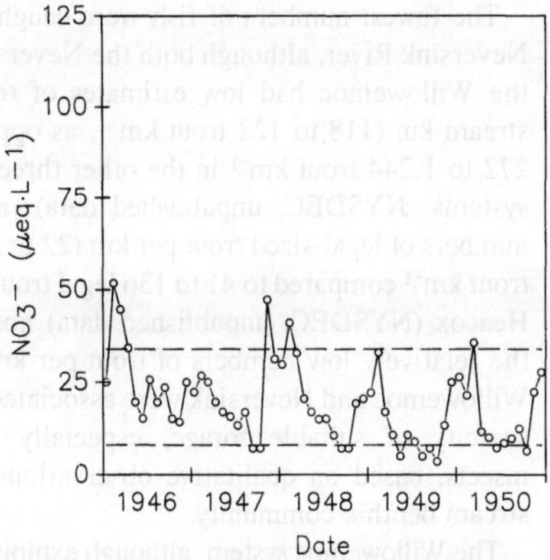

FIGURE 8.23. Short-term temporal patterns in monthly mean NO_3^- during two time periods (1946–50 and 1981–85) in Schoharie Creek at Prattsville. Dashed lines are mean annual maximum (spring) and minimum (summer) concentrations for each 5-year period.

(1986) report no significant trend in Ca^{2+} or Mg^{2+} for the Esopus Creek site, and this is likely to have a strong effect on the ANC trend. In addition, Kramer et al. (1986) have reported a marked increase in NO_3^- concentration at Esopus Creek. In a study of benchmark stations throughout the eastern United States, Smith and Alexander (1986) report that the Esopus Creek at Shandaken site is consistent with sites throughout the Northeast in showing decreasing SO_4^{2-} concentrations at the same time that monitoring stations are reporting decreases in sulfur dioxide emissions densities.

Future Considerations

Analysis of historical streamwater data for the Catskill region suggests that, in the absence of increases in base cation concentrations, ANC values would have remained at their current levels for at least the last 20 years. Rates of SO_4^{2-} deposition are believed to have decreased throughout this time period (Chapter 3, Husar et al., this volume), and conceptual models suggest that this decrease in acid loading should lead to an increase in ANC (Galloway et al. 1983a). In the Catskill region, however, decreases in streamwater SO_4^{2-} concentrations are matched stoichiometrically by increases in streamwater NO_3^-, at least for the time period since 1970.

Predictions of how ANC values will respond to the replacement of SO_4^{2-} by NO_3^- will depend strongly on estimates of the rates of decrease in streamwater SO_4^{2-} and increase in streamwater NO_3^-. Estimation of these rates is complicated by the strong seasonality of streamwater chemistry, and in particular by the association of minimum ANC and maximum NO_3^- during high-flow events in the spring. Seasonal excursions in ANC are much more closely correlated with changes in NO_3^- concentration than with changes in SO_4^{2-} concentration (Figure 8.15) and as a result, an increasing trend in NO_3^- has a greater potential to lower springtime ANC values than a decreasing trend in SO_4^{2-} has to raise them. At our current level of understanding, we cannot make quantitative estimates of future trends in Catskill streamwater ANC. We can say, however, that a continuation of the observed increase in streamwater NO_3^- has the potential to induce a declining trend in springtime ANC values, and to increase the proportion of streams that exhibit negative ANC values during high-flow events.

Biological Status

The Catskill region has long been recognized for its high-quality trout fishing (Kelly 1973, Senectutus 1943). The combination of groundwater inseepage and shading from the forest overstory results in clear, cold waters. Streamwater temperatures rarely exceed 18 to 20°C, even in summer, providing favorable conditions for trout survival and growth.

The following discussions focus on the fish communities in five key headwater streams: the Beaverkill, Willowemoc Creek, Esopus Creek, Rondout Creek, and the Neversink River (Figure 8.1). The Beaverkill, Willowemoc Creek, and Esopus Creek, in particular, are considered among the finest trout streams in the area (Kelly and Gann 1979). Streams in the Neversink and Rondout basins had the lowest levels of ANC, pH, and C_B among these streams (Table 8.4; Figure 8.6a,b).

Brook trout (*Salvelinus fontinalis*) and brown trout (*Salmo trutta*) are the two most important game species in Catskill streams, with brook trout relatively more important in the upstream reaches that may be most sensitive to potential effects from acidic deposition. Brown trout are generally more tolerant of warmer waters and habitat disruption, and thus often dominate in downstream reaches. Brown trout are also more sensitive to acidic conditions; Baker and Christensen (Chapter 4, this volume) report approximate critical pH thresholds of 4.8 to 5.4 and 4.7 to 5.1 for brown trout and brook trout, respectively. Most Catskill headwater streams provide excellent spawning and nursery areas for brown trout and brook trout; the upper reaches of all five stream systems are classified as natural spawning adequate (NSA) by the New York State Department of Environmental Conservation (NYSDEC, unpublished data). In downstream reaches, however, natural trout production is generally supplemented by stocking. Rainbow trout (*Salmo gairdneri*) also occur in some streams, but in general must be maintained by stocking.

Comprehensive surveys of fish communities in the Beaverkill, Willowemoc, Esopus, Rondout, and Neversink systems conducted by Heacox (NYSDEC, unpublished data) in the 1950s provide an index of the relative importance of other fish species in Catskill streams. Minnows, most commonly the blacknose dace (*Rhinichthys atratulus*), and sculpins (*Cottus bairdi* and *C. cognatus*) occur in large numbers; the white sucker (*Catostomus commersoni*) and other nongame species were caught much less frequently and generally only at lower elevations.

The fewest numbers of fish were caught in the Neversink River, although both the Neversink and the Willowemoc had low estimates of trout per stream km (118 to 122 trout km^{-1}, as opposed to 272 to 1,244 trout km^{-1} in the other three stream systems; NYSDEC, unpublished data), and low numbers of legal-sized trout per km (22 to 24 legal trout km^{-1} compared to 41 to 136 legal trout km^{-1}). Heacox (NYSDEC, unpublished data) noted that the relatively low numbers of trout per km in the Willowemoc and Neversink were associated with a scarcity of suitable forage, especially benthic insects, based on qualitative observations of the stream benthic community.

The Willowemoc system, although exhibiting low fish densities, supported a relatively diverse fish fauna. Excluding the Beaverkill, for which individual minnow species were not recorded, the Willowemoc had the highest number of taxa collected: 18, relative to 17 in the Esopus, 13 in the Rondout, and 8 in the Neversink. The relatively low species richness for the Neversink could result from many factors, including lower habitat diversity; numerous investigators, however, have reported lower species richness in waters with lower pH and ANC (e.g., Harvey 1975, Rago and Wiener 1986).

In 1981, Kelly (NYSDEC, unpublished data) duplicated these surveys of fish communities, using similar techniques, at a select group of sites studied in the 1950s by Heacox. For some streams, additional historical data on fish species composition are available for the 1935 and 1936 biological surveys conducted by the NYSDEC (1936, 1937); sampling methods were different for these surveys, however.

Results from the 1950–55 and 1981 surveys are summarized in Table 8.8. No striking pattern of change in the numbers or types of fish caught between the two time periods is evident. Other than brook trout, only sculpins were collected in significant numbers, reflecting the focus in the 1981 surveys on moderate-to-high elevation (335 to 670 m) stream sites that are potentially more sensitive to acidic deposition. In all streams but the East Branch of the Neversink, young-of-the-year trout were collected in 1981, indicating successful natural reproduction and recruitment.

The East Branch of the Neversink had by far the fewest number of fish caught both in 1981 and in

TABLE 8.8. Numbers of fish caught (by species) in 1981 and 1950–55 at selected high-elevation sites in five Catskill stream systems.[a]

Stream	Beaverkill		Willowemoc		Esopus Creek			Rondout Creek		West Branch Neversink		East Branch Neversink			
Elevation (m)	<675	675	590		580			335		395		595		640	
Date	8/9/53	9/81	9/55	9/81	9/50	8/9/51[b]	9/81	8/5	9/81	8/50	9/81	8/51[b]	9/81	8/51[c]	9/81
Temperature (°C)															
Air	26	16	18	14	16	21	19	23	16	17	17	23	17	20	18
Water	18	15	13	13	11	12	12	14	13	12	12	11	11	16	14
Physical characteristics															
Length (m)	90	90	90	107	107	75	112	909	90	90	90	90	150	90	90
Width (m)	7.6	7.0	–	6.1	1.5–4.5	0.6–1.8	2.7	6.1	8.5	6.1	8.2	4.6	4.6	3.6	10.1
Flow (m^3 s^{-1})	–	0.33	–	0.14	0.14	0.85	0.55	–	0.4	–	0.38	–	0.13	–	0.22
Chemical characteristics[d]															
pH	5.9		–		6.5			6.1		5.8		4.8		4.8	
ANC (μeq L^{-1})	31		–		107			50		49		−18		−20	
Fish species															
Brook trout															
Young-of-year	0	8	13	2	32	0	15	15	0	18	15	0	2	0	0
Age > 1 year	14	52	15	22	10	5	22	26	1	42	12	4	6	5	1
Brown trout															
Young-of-year	1	0	3	2	12	0	0	1	1	0	0	0	1	0	0
Age > 1 year	15	0	13	7	15	5	0	1	10	0	3	10	4	0	0
Sculpins	63	28	50	11	0	0	0	80	14	110	28	0	41	0	0
Longnose dace	0	0	0	0	1	0	0	0	0	0	0	0	0	0	0
White sucker	0	0	4	0	0	0	0	0	1	0	0	0	1	0	0

[a] Based on NYSDEC unpublished data.
[b] Stream resurveyed in 1951 to evaluate effects from major floods in November 1950 and spring 1951.
[c] Samples collected following major floods in late 1950/early 1951. No pre-flood data available. Refer to the data for Esopus Creek. Comparison of pre- and post-flood data at a relatively high-elevation site in Esopus Creek suggests potential loss of the trout year-class and adverse effects on sculpins (NYSDEC unpublished data).
[d] Streamwater chemistry data are mean values from USGS/NYCDEP survey, 1985–87.

the 1950s—only one adult brook trout in 1981 and five adult brook trout in 1951. Major floods in late 1950 and early 1951 may be responsible, in part, for the low fish abundance in 1951.

Only limited data are available on nonfish biota in Catskill streams. Qualitative observations of the abundance and types of benthic invertebrates present were recorded as part of both historical and present-day surveys of fish community status. However, these general records (e.g., "aquatic insect life sparse: mayflies and caddis") are not suitable for assessment of potential trends through time. Culp (1980) sampled benthic communities in selected streams in the Schoharie and Ashokan basins during 1975–78. In Esopus Creek, 142 taxa of aquatic macroinvertebrates were collected in total. Aquatic insects, primarily Diptera, Ephemeroptera, and Tricoptera, constituted at least 90%, numerically and gravimetrically, of the total organisms collected. At the uppermost sites in the creek, numerical density ranged between 956 and 6,928 m^{-2} (mean = 1,513 m^{-2}) during 1976–78.

There is no direct evidence from the limited data available that the biological status of Catskill streams has been adversely affected by acidic deposition. Many streams in the Neversink and Rondout basins experience pH < 5.0 on at least an episodic basis during high-flow events (Figure 8.5d); this is within the range of estimated critical pH levels for both brook and brown trout (Chapter 4, Baker and Christensen, this volume), and recruitment failure is likely to occur if early life stages experience these episodic low pH values. The near absence of trout in both the 1951 and 1981 surveys of the East Branch of the Neversink River may indicate the severe effects of poor habitat and flooding, or of acidic conditions that existed at this site during both sampling periods. Data from Schoharie Creek (see Evidence of Historical Trends), a higher ANC stream than the Neversink River, suggest that acidic conditions were as severe, or more severe, in 1951 than they were in 1981. Comparisons of fish data from these time periods are therefore problematic. The lack of difference between fish communities in the 1950s and in 1981 should not be over-interpreted, given the lack of precise information on streamwater chemistry for the earlier sampling period.

Conclusions

1. High rates of acidic deposition in the Catskill region result from concentrations of acid anions and H^+ that are comparable to other regions of the northeastern United States, and volumes that are greater than other regions. Sulfate (605 eq ha^{-1} yr^{-1}), NO_3^- (404 eq ha^{-1} yr^{-1}), and H^+ (663 eq ha^{-1} yr^{-1}) wet deposition rates are higher than in any other Regional Case Studies region.
2. Streamwater chemistry in the Catskill region is strongly affected by stream discharge. A nonrandom sample of 51 streams collected during high-flow and baseflow periods indicates that many streams (16%) are acidic (ANC ≤ 0) during high-flow events. A few (8%) are acidic at baseflow. Base cation concentrations vary with stream discharge in a manner analogous to ANC. Sulfate concentrations are relatively unaffected by discharge, whereas NO_3^- increases from a median of 14 μeq L^{-1} at baseflow to 42 μeq L^{-1} during high-flow events.
3. Spatial variability in streamwater ANC is associated with variability in base cations, rather than variability in acid anion concentrations. Spatial variability in base cation concentrations can be explained by patterns of glacial movement during the last ice age. Till deposition varied with local topography and location, and may have led to spatial variability in the depth, permeability, and carbonate mineral content of surficial materials, which are hypothesized to be controlling factors in streamwater C_B and ANC.
4. ANC values in all Catskill streams at all seasons of the year are lower than expected, based on measured [Ca^{2+} + Mg^{2+}] values and assuming carbonate weathering. The ANC deficit is consistently in the 120 to 140 μeq L^{-1} range, and is closely approximated by streamwater SO_4^{2-} concentrations. The spatial and seasonal uniformity observed in SO_4^{2-} concentrations in the region strongly suggests that atmospheric deposition of SO_4^{2-} is largely responsible for the observed discrepancy between ANC and [Ca^{2+} + Mg^{2+}].
5. Organic acid anions appear to have less potential for affecting ANC than either SO_4^{2-} or

NO_3^- in Catskill streams. The median A_t concentration (12 μeq L^{-1}) represents approximately 8% of total acid anions.

6. Base cation production in stream watersheds is the major process modifying atmospheric deposition and controlling Catskill streamwater ANC. Sulfate is conservative, whereas nitrogen (both NH_4^+ and NO_3^-) is strongly retained.
7. Low or negative ANC values are episodic and associated with high-flow events such as snowmelt or major storms. ANC declines during high-flow events are associated with three processes: (1) dilution of base cations, (2) more or less consistent SO_4^{2-} concentrations (relative to base cations) that increase the SO_4^{2-}:[Ca^{2+} + Mg^{2+}] ratio, and (3) increased NO_3^- concentrations. ANC values drop below zero when the ion ratio (NO_3^- + SO_4^{2-}):[Ca^{2+} + Mg^{2+}] reaches a value of 1.0.
8. Analysis of historical streamwater data for approximately 20 sites in the Catskill region indicates that ANC values in many streams were declining prior to about 1950, while base cation concentrations were rising. After 1950, both ANC and [Ca^{2+} + Mg^{2+}] began to increase and were probably driven by changes in land use. Estimated SO_4^{2-} concentrations are consistent with trends in SO_2 emissions, and have declined approximately 35 μeq L^{-1} since the late 1960s. Nitrate concentrations show a rapidly increasing trend (+30 μeq L^{-1}) since 1965. Because annual peaks in NO_3^- concentration coincide with annual minima in ANC, the increasing trend in NO_3^- has a greater potential to lower springtime ANC than the increases in [Ca^{2+} + Mg^{2+}] and decreases in SO_4^{2-} have to raise springtime ANC.
9. The limited data available for Catskill stream biota show no direct evidence that fish communities have been adversely affected by acidic deposition. This conclusion is based on comparisons of fish survey data collected in the 1950s and in 1981. Historical data on streamwater chemistry, however, suggest that acidic conditions may have been as severe in 1950 as they were in 1981. The lack of difference between 1950s data and 1981 data for fish is therefore not unexpected.

Acknowledgments. The authors would like to acknowledge the considerable assistance of W.H. Kelly (NYSDEC), who provided all of the information on Catskill stream biota for this chapter from unpublished reports. We also thank J.W. Elwood (Oak Ridge National Labaoratory), K.N. Eshleman (University of Virginia), R.B. Cook (Oak Ridge National Laboratory), and D.F. Charles (Indiana University; current address U.S. EPA Environmental Research Laboratory, Corvallis, Oregon) for comments on the manuscript.

9 Southeastern Canada: An Overview of the Effect of Acidic Deposition on Aquatic Resources

Dean S. Jeffries

ABSTRACT. The terrain of a large portion of eastern Canada (with approximately 700,000 lakes) contains lakes that have low acid neutralizing capacity (ANC) and receive > 10 kg wet SO_4^{2-} ha^{-1} yr^{-1} (> 210 eq ha^{-1} yr^{-1}). Subregional variations in lake chemistry are readily explained by the combined influences of terrain geochemistry and atmospheric deposition. In the absence of terrain characteristics that would greatly reduce sensitivity (i.e., $CaCO_3$ in the bedrock or overburden materials, or thick soils) and without elevated deposition (such as occurs in Labrador), most lakes exhibit pH values in the range of 6.0 to 7.0 and ANC < 100 μeq L^{-1}. The presence of easily weathered $CaCO_3$ in the till of northern Ontario yields lakewater pH, ANC, and Ca^{2+} + Mg^{2+} values that are generally higher than those observed in other subregions of eastern Canada. The large emission at Sudbury locally modifies this generalization, however. Sulfate concentrations in lakes are principally controlled by atmospheric deposition. Evaluation of ion ratios provides evidence for significant changes in composition for lakes in southcentral Ontario, southern Quebec, and portions of New Brunswick, Nova Scotia, and Newfoundland. Due to the low base cation content of eastern Canadian waters, the changes are primarily ANC loss in favor of SO_4^{2-}. Paleoecological reconstruction of historical lake pH shows that there are at least some lakes in all subregions receiving elevated deposition that have become more acidic during the past 50–100 years. Episodic acidification often accompanies spring snowmelt. Assessment of the aquatic resource at risk suggests that approximately 12,000 lakes are presently acidic (ANC ≤ 0 μeq L^{-1}) in eastern Canada, and an additional 10,000 to 40,000 will become acidic if present deposition levels persist. Locations presently receiving lower deposition than in the past are exhibiting evidence of improved water quality. Loss of fish populations or reductions in diversity have been documented in Nova Scotia and Ontario. Studies involving experimental lake acidification have defined many of the mechanisms that lead to the biological degradation observed in acidic lakes.

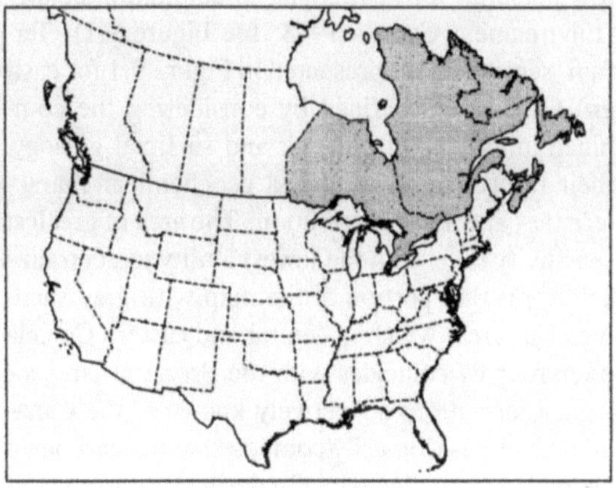

Introduction

Since the mid-1970s, extensive and intensive research has been conducted in Canada to define aquatic effects associated with elevated acidic deposition. This chapter provides an overview of the chemical and biological condition of Canadian aquatic resources as influenced by the long-range transport of air pollutants. It therefore complements the analyses of case study regions in the United States and provides a basis for comparison of effects between the northern and southern portions of North America. The emphasis is on discussion of aquatic chemistry, limited to southeastern

Canada, as explained in the next section, Regional Characteristics. Several other reviews of the effects of acid rain on Canadian waters already exist (Harvey et al. 1981, Canada/United States 1983, RMCC 1986, Cook et al. 1988). The depth of analysis presented in any of these reports is far greater than is possible in this chapter, thus they should be consulted for greater detail.

Regional Characteristics

Canada contains approximately 4 million km² (43% of the land area of the country) classified as having low potential for neutralizing precipitation acidity (Environment Canada 1988; see Figure 9.1). Terrain "sensitivity" as presented in Figure 9.1 for eastern Canada was defined by considering the combined influences of bedrock and surficial geology, their respective physical and geochemical characteristics, and areal distribution. The area of greatest sensitivity (i.e., with the lowest ability to neutralize acidity) is that portion of the map with gray shading. This area, which in central and eastern Canada often roughly coincides with the Precambrian geological formations collectively known as the Canadian Shield, is primarily composed of noncarbonate bedrock overlain by glacially derived, coarsely textured, shallow unconsolidated materials.

Also shown in Figure 9.1 are isopleths of annual wet SO_4^{2-} deposition for 1985 (see also continental deposition maps presented in Chapter 5 (Eilers and Selle, this volume). Although the location of the isopleths varies from year to year, the general pattern of annual deposition has remained constant (i.e., the highest levels (20 to 35 kg SO_4^{2-} ha^{-1} yr^{-1}; 420 to 730 eq ha^{-1} yr^{-1}) occur in southern Ontario, Quebec, and southern New Brunswick, whereas lower values occur to the west, north, and east of these locations. Peaks in wet SO_4^{2-} deposition occurring in western Canada (not shown in Figure 9.1) reflect sources of nonacidifying $CaSO_4$ (Alberta-Saskatchewan) or the combined influence of local sources and high precipitation quantity (southwestern British Columbia). Some large local sources of SO_2 (generally nonferrous smelters) influence water quality in the surrounding region; the most notable of these is the Sudbury nickel mining and smelting complex (Jeffries 1984; see also other papers in Nriagu 1984).

The coincidence of terrain having little ability to neutralize acid inputs and those areas receiving elevated deposition defines the portion of the country of principal concern in Canada, namely, that area east of the Manitoba-Ontario border and roughly south of the 10 kg SO_4^{2-} ha^{-1} yr^{-1} (210 eq ha^{-1} yr^{-1}) isopleth. In order to simplify data analysis and interpretation, the latter isopleth boundary is usually approximated by 52°N latitude (RMCC 1986, Kelso et al. 1986b). Subsequent discussion will focus on this southeastern portion of Canada. Kelso et al. (1986a) have estimated that this area contains more than 700,000 lakes corresponding to approximately 160,000 km² (excluding the Great Lakes). Although the magnitude of error in these estimates is unknown, it is probable that the estimate of the number of lakes is low because the counts were based on interpretation of and extrapolation from existing topographic maps, which generally do not include lakes < 4 ha. The lake area estimates are more reliable. By province in eastern Canada, the percentages of land area ranked in the high sensitivity class (shaded portion of Figure 9.1) are, in order: Quebec, 82%; Newfoundland, 56% (includes Labrador); Nova Scotia, 54%; Prince Edward Island, 46%; Ontario, 34%; New Brunswick, 31%. The remaining western provinces and northern territories range from 6% to 48% (Environment Canada 1988). Comparative information for the United States case study regions is found in Chapter 5 (Eilers and Selle, this volume).

Regional Surface Water Chemistry

Data Base and Subregion Designation

The data base used to describe the water quality of eastern Canadian lakes is an extended version of the one presented by Jeffries et al. (1986). Briefly, it is a compilation of several smaller federal and provincial government data sets and contains information for 7,403 lakes, of which 4,895 are located in Ontario. Lake selection was therefore not based on a single statistical design; for the reasons presented by Jeffries et al. (1986), however, statistical estimates based on the samples reasonably reflect population estimates. Lake samples were collected between 1978 and 1985, generally during the winter, summer, or autumn seasons. Approximately

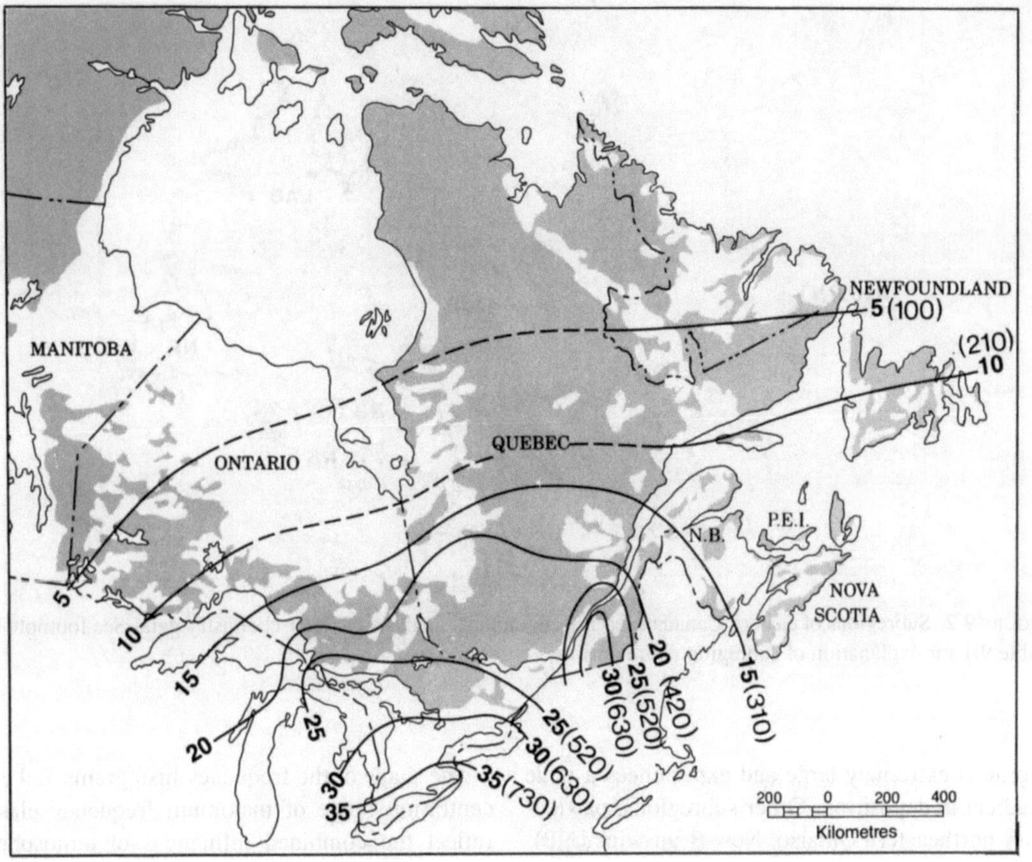

FIGURE 9.1. Sensitivity of Canadian terrain to the atmospheric deposition of acids (after Environment Canada 1988). The gray shaded area is classified as most sensitive and is typified by terrain having noncarbonate bedrock and coarse textured, shallow soils. Isopleths of wet SO_4^{2-} deposition (1985) are also shown in kg ha^{-1} (eq ha^{-1} in brackets).

80% of the samples were "tube" integrated samples, so that errors potentially introduced by seasonal variation in lake composition were minimized. More information on analytical methods, quality assurance, and other details of the sampling may be found in Jeffries et al. (1986) and Jeffries (1986). These same data, plus data from the U.S. EPA Eastern Lake Survey have been used by Schindler (1988b) in his general review of the impact of acid rain on freshwater ecosystems. The data are conservative in nature: first, because they were selected to reflect average lakewater quality, as opposed to, for example, the worst case spring runoff condition and, second, because small lakes that are more likely to be affected are probably underrepresented within the data base.

The data have been separated into eight subregions, as shown in Figure 9.2. Most of the subregions are defined by provincial boundaries, with the exception of the three Ontario subregions. Northwestern Ontario (NW ONT) was arbitrarily separated from northeastern Ontario (NE ONT) at 85° 20′ W longitude, and northeastern Ontario was separated from southcentral Ontario (SC ONT) at 45° 50′ N latitude.

The entire Labrador (LAB) subregion lies north of 52°N and therefore lakes in this area generally receive < 10 kg ha^{-1} yr^{-1} wet SO_4^{2-} (< 210 eq ha^{-1} yr^{-1}); this is also true of many lakes in northwestern Ontario, even though they are commonly south of 52°N (note location of deposition isopleth in Figure 9.1). The Quebec (QUE) sub-

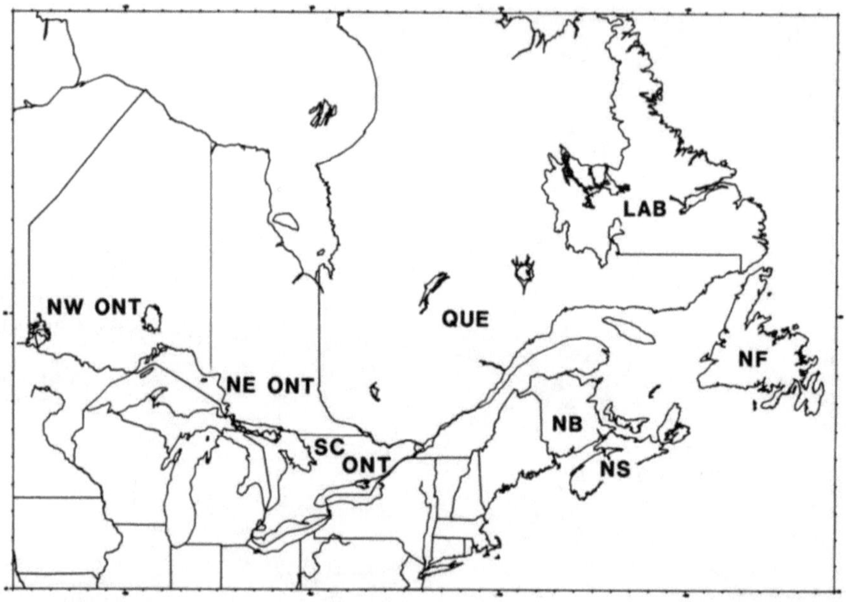

FIGURE 9.2. Subregions of eastern Canada used in geographical analysis of water chemistry data. See footnote b of Table 9.1 for explanation of subregion abbreviations.

region is extremely large and experiences a wide gradient in deposition. Other subregions, particularly northeastern Ontario, New Brunswick (NB), and Nova Scotia (NS) are geologically complex, containing sensitive and insensitive terrain (Figure 9.1); northeastern Ontario also contains the major point source emitter at Sudbury.

Subregional Variability in pH, $Ca^{2+} + Mg^{2+}$, ANC, and SO_4^{2-}

The ions most strongly influenced by acidic deposition are usually H^+ (or pH), base cations, alkalinity or acid neutralizing capacity (ANC), and SO_4^{2-}. For simplicity, the following discussion will be limited to these parameters. Base cations have been approximated by $Ca^{2+} + Mg^{2+}$, since after sea salt correction, Na^+ and K^+ concentrations are comparatively minor. For example, $Na^+ + K^+$ represents only 12% to 18% of the sum of base cations across seven zones in Ontario (Neary and Dillon 1988). Parameter distributions for sampled lakes are illustrated by frequency histograms for the eight subregions of eastern Canada in Figures 9.3 to 9.6, respectively. Corresponding median and quartile values are also presented in Table 9.1.

The shape of the frequency histograms and concentration range of maximum frequency classes reflect the combined influences of atmospheric deposition and terrain geochemistry. The Ontario subregions (bottom three panels of Figures 9.3 to 9.6) are good examples. The glacial tills that cover northern and central Ontario contain a small amount (0 to 10%) of easily weathered $CaCO_3$ derived from a source area in the Hudson Bay lowlands (Shilts 1981). Although the spatial distribution of this material is quite irregular, when present it provides an effective means for neutralizing acidic deposition. Tills closer to the source area (i.e., in northwestern and northeastern Ontario) generally have more $CaCO_3$ than those in south-central Ontario. This regional geochemistry leads to generally higher $Ca^{2+} + Mg^{2+}$ in Ontario lakes (Figure 9.4) compared to the other eastern Canadian subregions; median $Ca^{2+} + Mg^{2+}$ concentrations in Ontario range from 194 to 383 μeq L^{-1} and 73 to 163 μeq L^{-1} for the remaining subregions (Table 9.1). None of the other subregions have a similar, glacially distributed source of $CaCO_3$, although spatially limited occurrences of carbonate bedrock do account for the bimodal nature and/or high concentration tails observed in most of the $Ca^{2+} + Mg^{2+}$ data distributions (Figure 9.4).

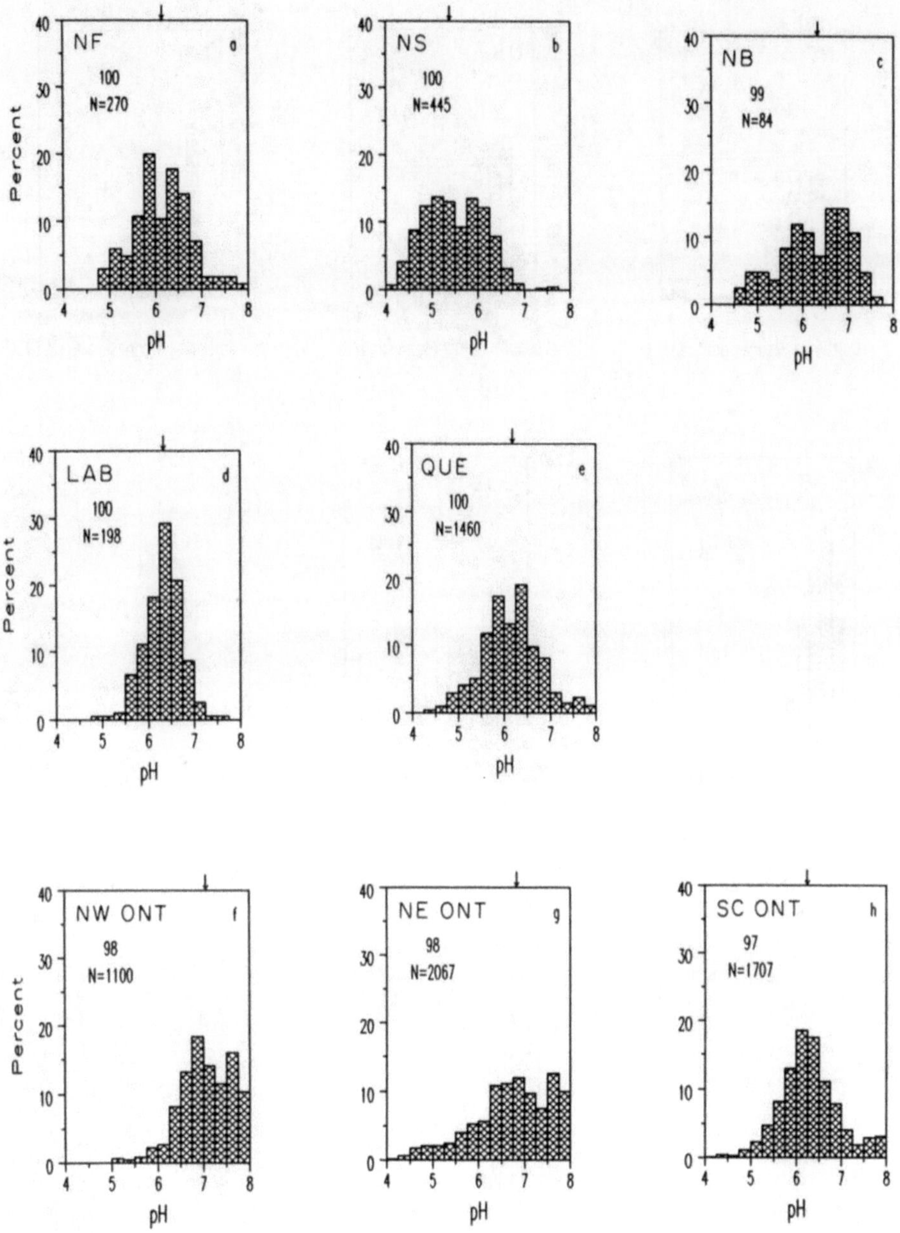

FIGURE 9.3. Frequency histograms for pH for eight subregions of eastern Canada: (a) NF, (b) NS, (c) NB, (d) LAB, (e) QUE, (f) NW ONT, (g) NE ONT, and (h) SC ONT. See Figure 9.2 for subregion location. The arrow at the top of each panel indicates the subregional median value for all data within the concentration range defined by the *x*-axis. *N* indicates the total number of values presented and the number above *N* represents the percentage of lakes in the data set that fall within the range represented by the horizontal axis. Specific quartile values are presented in Table 9.1.

The deposition of marine salts significantly influences the Ca^{2+} + Mg^{2+} concentrations observed in lakes from Newfoundland, Nova Scotia, New Brunswick, and Labrador. Therefore, sea salt corrected quartile values have been included in Table 9.1 for these subregions to demonstrate the shift in data distribution arising from this influence. The magnitude of this effect is primarily related to the lake's distance from the ocean, although local topography may also play a role. As expected, the

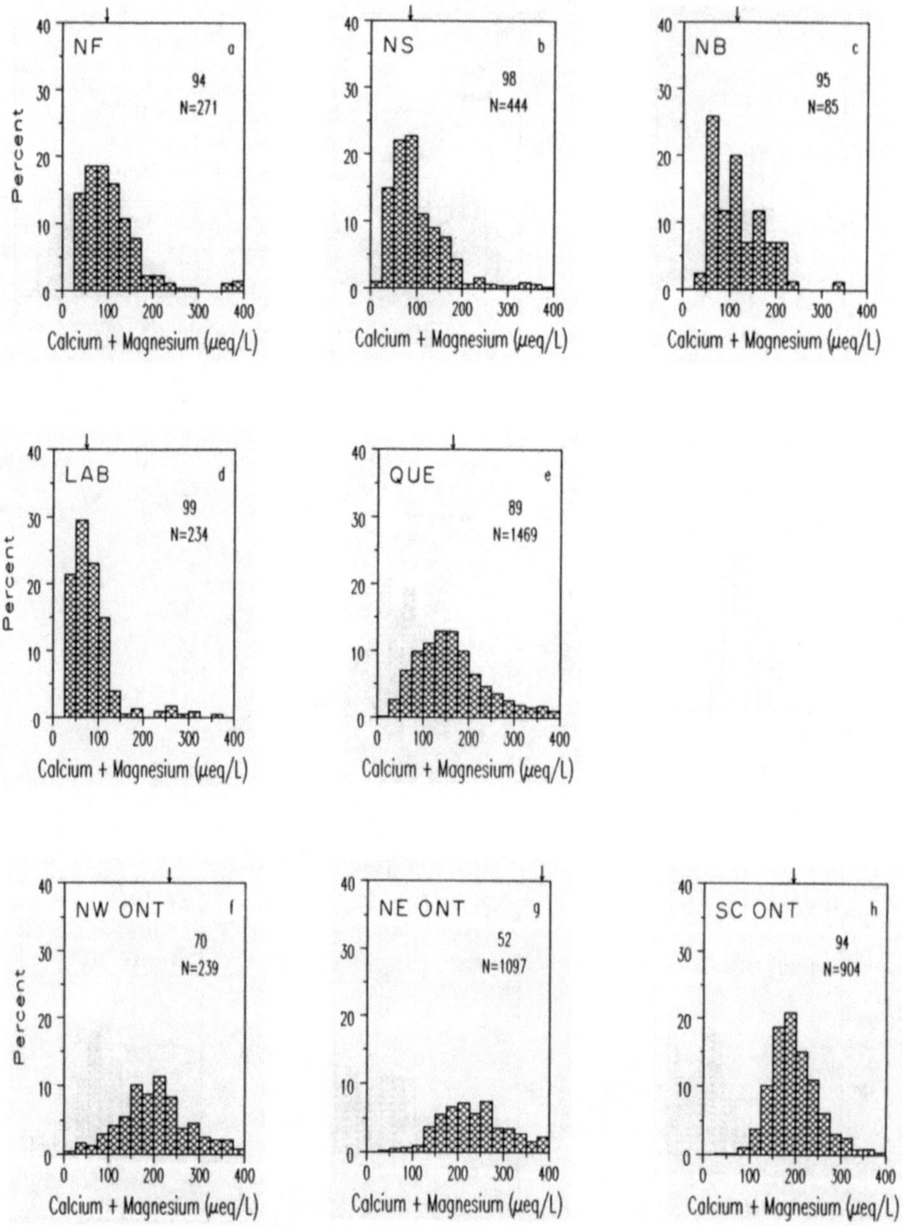

FIGURE 9.4. Frequency histograms for $Ca^{2+} + Mg^{2+}$ (not sea salt corrected) for eight subregions of eastern Canada. Figure preparation and presentation are the same as for Figure 9.3.

chemistry of lakes in Nova Scotia and Newfoundland is most strongly affected by sea salt inputs.

The impact of regional distributions of $CaCO_3$ is also evident in the frequency histograms for pH and ANC (Figures 9.3 and 9.5, respectively). Hence, the Ontario subregions have a large percentage of lakes with high pH (> 6.5), although the presence of a low pH tail in the distributions, particularly for northeastern and southcentral Ontario, reflects the variable distribution of the glacial tills and the impact of acidic deposition originating from both local (i.e., Sudbury in northeastern Ontario) and long range sources. Bimodal distributions in pH are evident for Newfoundland,

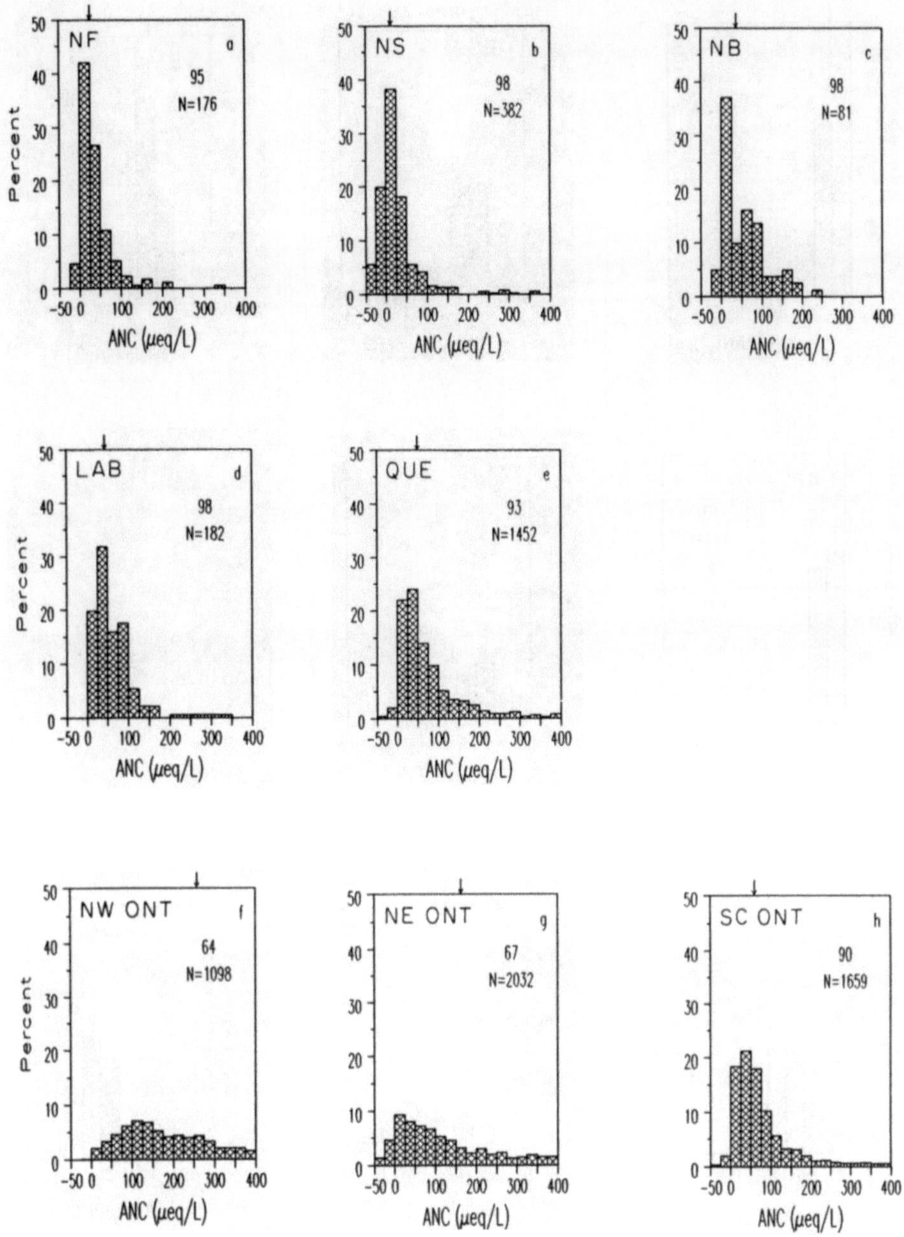

FIGURE 9.5. Frequency histograms for ANC for eight subregions of eastern Canada. Figure preparation and presentation are the same as for Figure 9.3.

Nova Scotia, New Brunswick, and Quebec for the same reasons as the $Ca^{2+} + Mg^{2+}$ distributions discussed in the previous paragraph. Note that a large proportion of lakes in Nova Scotia have pH < 5.5.

A regional distribution map of pH is shown in Figure 9.7. The map was prepared by determining the median lake pH for each cell in a half degree grid overlying the data point coverage. The median pH is represented by a symbol located at the center of the grid cell and appropriately shaded to indicate one of three pH classes (e.g., < 6.0, 6.0 to 7.0, > 7.0). The map clearly shows that median pH > 7.0 (i.e., the open circles) generally follows the occurrence of $CaCO_3$, either in the glacial

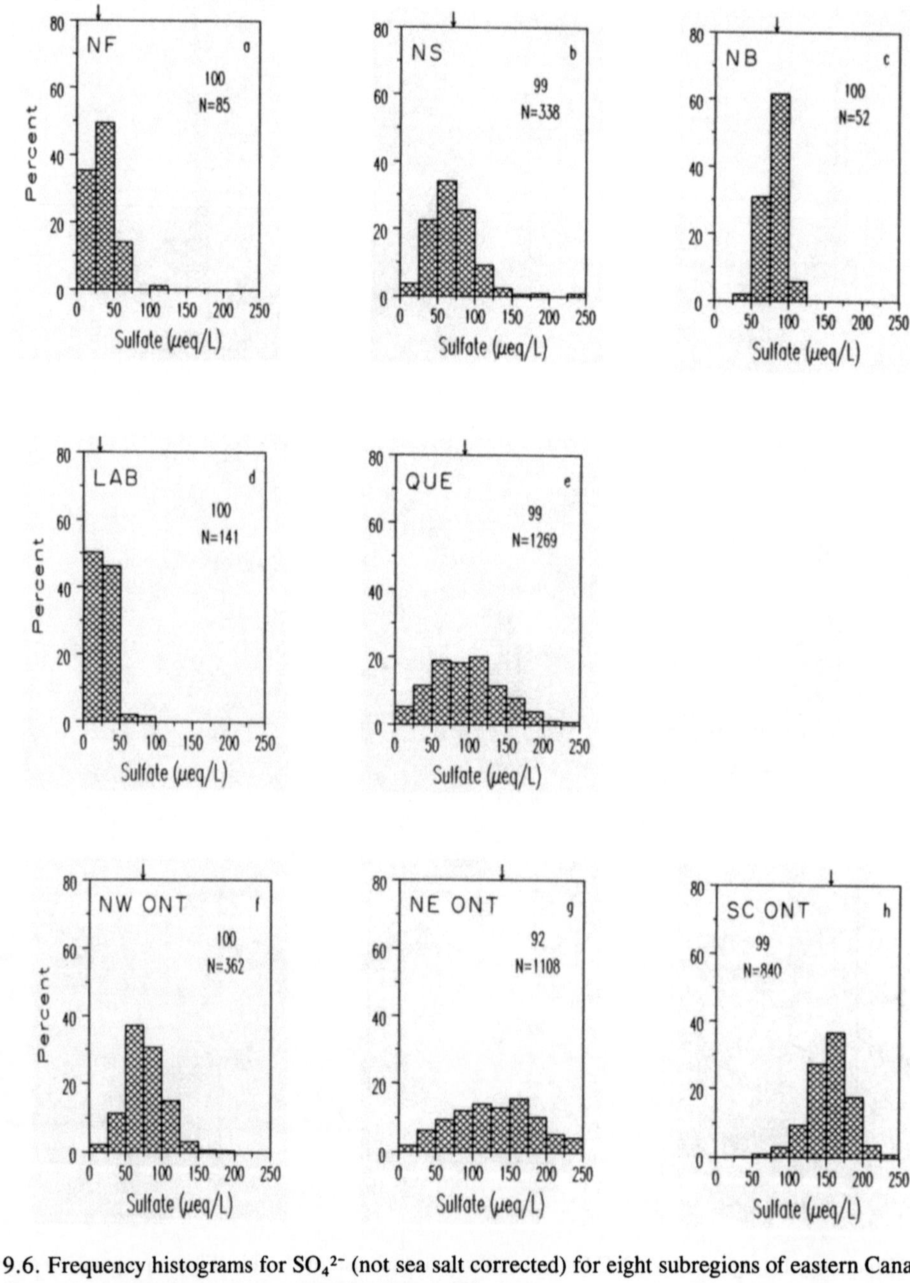

FIGURE 9.6. Frequency histograms for SO_4^{2-} (not sea salt corrected) for eight subregions of eastern Canada. Figure preparation and presentation is the same as for Figure 9.3.

overburden or as bedrock (i.e., northern Ontario, extreme southeastern Ontario, the south shore of the St. Lawrence River, and Anticosti Island). Median pH < 6.0 consistently occurs in Nova Scotia, along the southern shore of Newfoundland, and in portions of southwestern Quebec; the local influence of Sudbury in northeastern Ontario is also visible. The median pH of the majority of cells in Labrador is 6.0 to 7.0, and this is probably the normal range in pH to be expected for Shield lakes that are not influenced by $CaCO_3$ in the tills (yielding higher values) or by acidic deposition (yielding lower values).

The ANC distributions (Figure 9.5) show that all subregions in eastern Canada except Labrador and

TABLE 9.1. Quartile pH and $Ca^{2+} + Mg^{2+}$, ANC, and SO_4^{2-} concentrations ($\mu eq\ L^{-1}$) for the eight subregions of Eastern Canada.[a]

Ion	Percentile	NF[b,c]	NS[b,c]	NB[b,c]	LAB[b,c]	QUE[b]	NW ONT[b]	NE ONT[b]	SC ONT[b]
	25th	5.8	5.0	5.8	6.1	5.8	6.7	6.3	5.9
pH	50th	6.2	5.4	6.3	6.4	6.2	7.1	6.9	6.3
	75th	6.6	6.0	6.9	6.6	6.6	7.5	7.4	6.7
	25th	62	61	70	54	118	173	224	164
	25th[d]	43	31	55	48				
$Ca^{2+} + Mg^{2+}$	50th	98	85	115	73	163	235	383	194
	50th[d]	79	53	90	69				
	75th	141	135	164	100	234	477	1008	236
	75th[d]	118	84	156	100				
	25th	12	−2	6	28	26	129	53	30
ANC	50th	28	14	38	46	52	261	166	61
	75th	50	38	83	82	112	622	615	121
	25th	29	50	65	19	62	62	94	141
	25th[d]	21	36	60	16				
SO_4^{2-}	50th	39	71	83	26	94	75	142	156
	50th[d]	31	50	77	24				
	75th	50	85	94	31	135	94	185	172
	75th[d]	40	65	89	30				

[a] See text for description of data base.
[b] Subregion abbreviations: NF = insular Newfoundland, NS = Nova Scotia, NB = New Brunswick, LAB = Labrador, QUE = Quebec, NW ONT = northwestern Ontario, NE ONT = northeastern Ontario, and SC ONT = southcentral Ontario.
[c] Sea salt corrected data are presented for only NF, NS, NB, and LAB; input of marine salts to the remaining subregions is inconsequential.
[d] Sea salt corrected $Ca^{2+} + Mg^{2+}$ and SO_4^{2-} are given for those subregions experiencing significant deposition of marine salts.

northwestern Ontario contain a significant percentage of acidic lakes (ANC $\leq$ 0), and in particular, > 25% of lakes in Nova Scotia are in this category, reflecting the extreme terrain sensitivity present in certain parts of that province. Figure 9.8 presents the geographical distribution for three classes of ANC (i.e., < 100, 100 to 200, > 200 $\mu eq\ L^{-1}$) in the same manner as for pH as discussed in the previous paragraph. This map demonstrates once again the major influence on lake composition exerted by carbonate bedrock or even a small component of easily weathered $CaCO_3$ in the glacial overburden. Where such geological materials are present (principally subregions in Ontario), median cell ANC is > 200 $\mu eq\ L^{-1}$; where they are not present, the majority of cells have ANC < 100 $\mu eq\ L^{-1}$. The fact that southcentral Ontario has a measureable but relatively smaller amount of $CaCO_3$ in its tills compared to either northwestern or northeastern Ontario and also experiences maximum deposition probably accounts for the extensive occurrence of cell medians < 100 $\mu eq\ L^{-1}$ in this subregion. The widespread occurrence of ANC cell medians < 100 $\mu eq\ L^{-1}$ in subregions outside Ontario, even in Labrador, which receives minimal acidic deposition, highlights the generally high sensitivity that exists throughout this large portion of North America; corresponding subregional median ANC varies from 14 to 52 $\mu eq\ L^{-1}$ (Table 9.1).

The soils of eastern Canada typically do not adsorb SO_4^{2-} (Rochelle et al. 1987); moreover, there is little evidence for widespread geological sources of sulfate to account for the SO_4^{2-} observed in lake waters (Wright 1983, Neary and Dillon 1988). Hence, the SO_4^{2-} frequency histograms (Figure 9.6) best demonstrate the direct influence of acidic deposition on lake chemistry. As can be seen from Figure 9.1, northwestern Ontario, Newfoundland, and particularly Labrador receive the lowest levels of SO_4^{2-} deposition. Consequently, SO_4^{2-} distributions

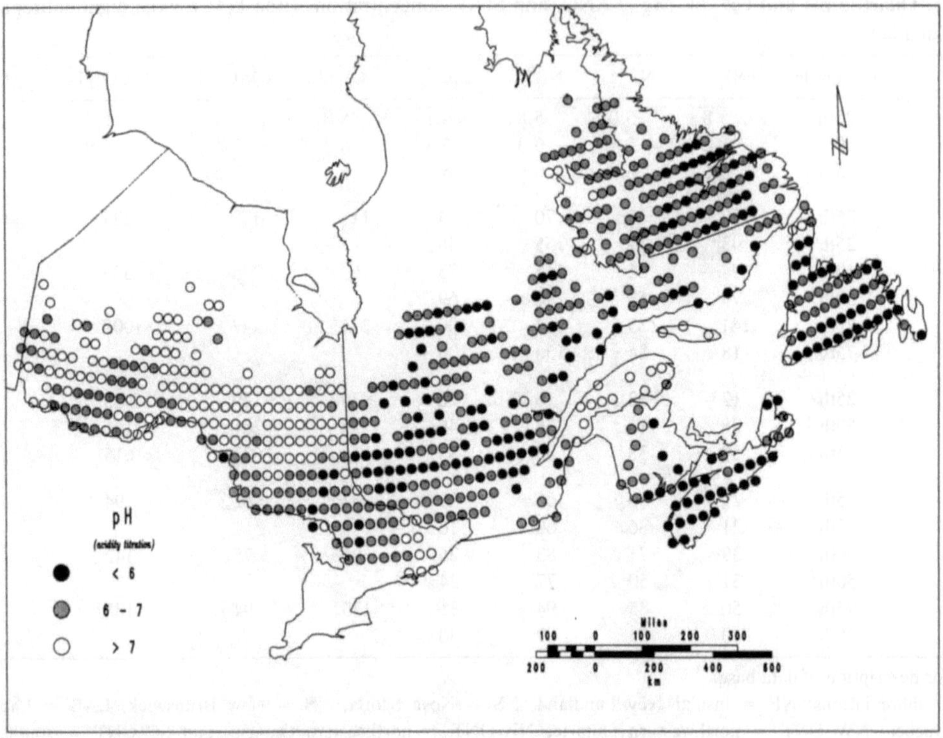

FIGURE 9.7. Geographical distribution of three classes of lakewater pH for 0.5 degree grid cells across eastern Canada. The method of map preparation is explained in the text.

from these regions tend to display a sharp peak at low concentration. In contrast, subregions receiving wider gradients in deposition also display wider distributions (Nova Scotia, Quebec, northeastern Ontario, and southcentral Ontario), although variable SO_4^{2-} reduction in lakes (Schindler 1986a) may also play a partial role. Clearly, the subregion receiving maximum deposition (southcentral Ontario) also exhibits a peak at the upper end of the SO_4^{2-} concentration range, and it is almost a mirror image of the corresponding ANC histogram. Neary and Dillon (1988) have recently demonstrated causal relationships between SO_4^{2-}, pH, and ANC in the Ontario subregions.

Ratio of ANC:Ca^{2+} + Mg^{2+}

Wright (1983) used lake chemistry data from geographically diverse areas of North America and Europe to refine Henriksen's acidification model. As discussed in the last section, lakewater SO_4^{2-} concentrations in eastern Canada generally reflect deposition. In this situation, the acidification model predicts that an increase in lakewater SO_4^{2-} is balanced by a decrease in ANC and/or an increase in base cations, once again approximated by Ca^{2+} + Mg^{2+}. The dynamic interrelationship that exists among these ions means that, in response to elevated SO_4^{2-} deposition, the ratio ANC:Ca^{2+} + Mg^{2+} deviates downwards from an undisturbed background value (see more extended discussion in Chapter 18, Sullivan, this volume). In sensitive terrain receiving minimal deposition (e.g., Labrador), the ANC:Ca^{2+} + Mg^{2+} ratio is typically close to one, implying that the lakewater concentration of SO_4^{2-} and other ions is relatively insignificant. If the terrain is geochemically insensitive due to a ready supply of ANC and Ca^{2+} + Mg^{2+} from carbonate bedrock or the like, the ratio also maintains a value near one, independent of deposition. It is necessary of course to use sea salt corrected Ca^{2+} + Mg^{2+} data so that interpretation of the ratio values is not confused by the influence of neutral marine inputs.

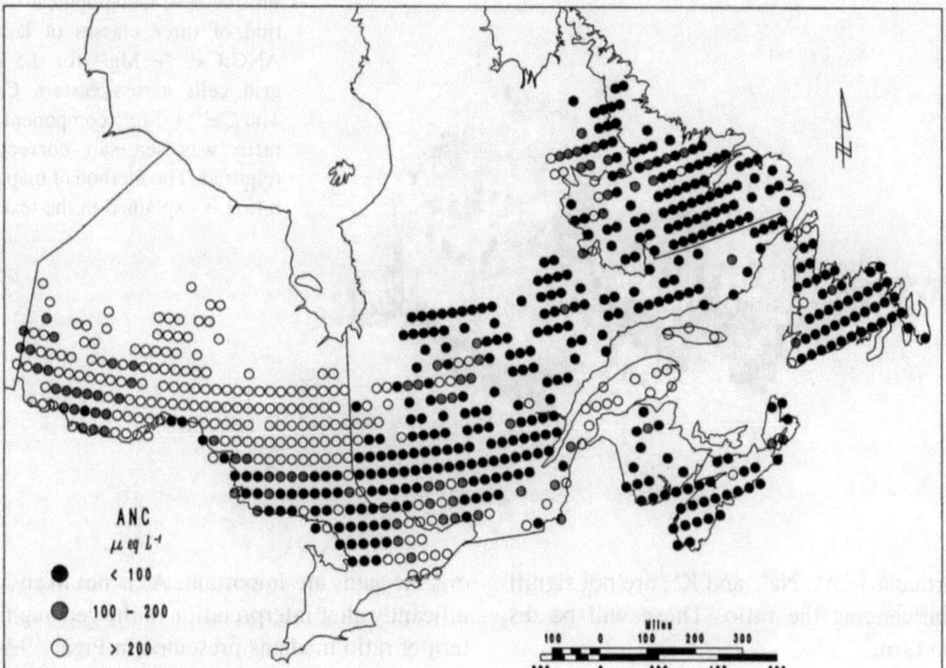

FIGURE 9.8. Geographical distribution of three classes of lakewater ANC for 0.5 degree grid cells across eastern Canada. The method of map preparation is explained in the text.

The spatial variation in ANC:Ca^{2+} + Mg^{2+} was assessed by segregating the data into a 0.5 degree geographic grid in the same manner as explained in the previous section. The median ratio for all lakes observed within each possible grid cell was determined and allocated to three ratio classes (i.e., < 0.4, 0.4 to 0.8, > 0.8). The ratio class is presented by appropriate shading of each cell overlying a Mercator projection of eastern Canada (Figure 9.9). Using this procedure, sufficient data exist to calculate the ratio for 534 cells; the number of lakes included in each cell ranged from 1 to 221 with a median of 3.

Across southeastern Canada, 90% of the median grid ANC:Ca^{2+} + Mg^{2+} fell between 0.13 and 1.02, 70% was < 0.8, and the mean was 0.58, showing that a large proportion of the lakes have experienced changes in composition due to atmospheric SO_4^{2-} deposition. There are, in fact, nine cells in which the median ANC and therefore the median ratio is ≤ 0 (i.e., the median lake for the cell is acidic). These cells most commonly occur in Nova Scotia.

The spatial distribution presented in Figure 9.9 demonstrates that median ANC:Ca^{2+} + Mg^{2+} consistently approaches one in areas receiving low values of SO_4^{2-} deposition. Subregional median ratios for all lakes in northwestern Ontario and Labrador were 0.7 and 0.8 respectively. Furthermore, grid cells located in areas having insensitive terrain, such as the south shore of the St. Lawrence River, also exhibit median ratios near one, despite receiving elevated SO_4^{2-} deposition. The map clearly demonstrates that large areas of southcentral Ontario and southern Quebec, portions of New Brunswick and Newfoundland, and Nova Scotia contain lakes that have experienced significant changes in composition related to SO_4^{2-} deposition (median ratio < 0.4; black grid cells in Figure 9.9).

Interpretation of low ratio values as evidence of acidification depends on the assumption that base cations have not increased in response to SO_4^{2-} deposition, or at least that the absolute magnitude of increase is much smaller than the ANC decrease. It is also necessary to assume that other

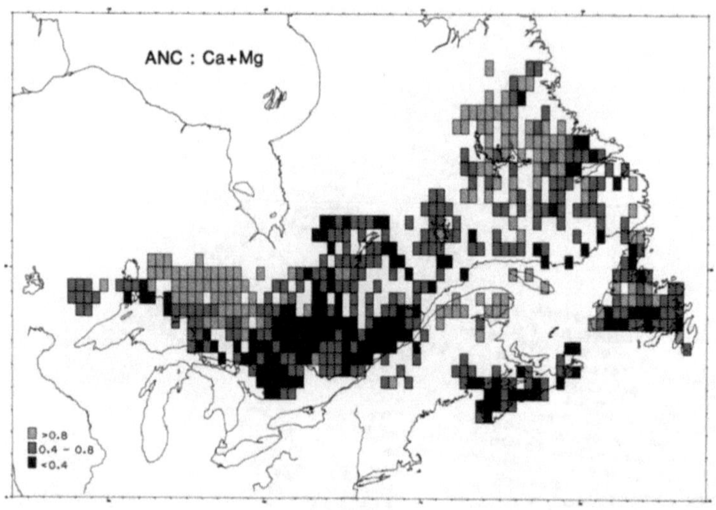

FIGURE 9.9. Geographical distribution of three classes of lakewater ANC:Ca^{2+} + Mg^{2+} for 0.5 degree grid cells across eastern Canada. The Ca^{2+} + Mg^{2+} component of the ratio was sea salt corrected as required. The method of map preparation is explained in the text.

ions, particularly A^-, Na^+, and K^+, are not significantly influencing the ratio. These will be discussed in turn.

Within the acidification model, the magnitude of change in base cations induced by a change in SO_4^{2-} is expressed by the *F*-factor (Wright 1983). As discussed in Chapter 3 (Husar et al. this volume), it is likely that *F* varies as a function of initial base cation concentration. Therefore, terrain with inherently low ion concentrations due to low geochemical weathering rates will have a low value of *F*, and in the absence of other confounding ions (e.g., organic anions), most of the change in ANC:Ca^{2+} + Mg^{2+} can be attributed to loss of ANC rather than increases in cations. As shown in Table 9.1 and in Chapter 17, Table 17.3 (Baker et al., this volume), Ca^{2+} + Mg^{2+} concentrations are particularly low in Labrador, New Brunswick, Nova Scotia, and Newfoundland compared to many other subregions of North America. Wright (1983) examined several Scandanavian and North American data sets and concluded that for sensitive terrain, *F* was generally in the range 0 to 0.4. It is therefore reasonable to expect that waters in southeastern Canada exhibiting very low ratio values, particularly the very dilute waters in the Atlantic provinces, have experienced primarily a loss of ANC rather than an increase in base cations.

Significant concentrations of A^- may influence the ANC:Ca^{2+} + Mg^{2+} ratio also, yielding values < 1 in the absence of SO_4^{2-} deposition. Although there are undoubtedly local situations where organic acids are important, A^- is not likely to significantly alter interpretation of the geographic pattern of ratio medians presented in Figure 9.9. The Nova Scotia subregion contains a large number of grid cells with very low ratio values (subregional median ANC:Ca^{2+} + Mg^{2+} = 0.25) and has an A^- distribution similar to that of Labrador (subregional median = 0.80). Distribution statistics for A^- compiled by Jeffries et al. (1986b) show that there is no relationship between high A^- and low ANC in Canadian lakes, a conclusion subsequently supported by Neary and Dillon (1988) for Ontario lakes. In fact, both studies found that the highest average A^- levels were found in northwestern Ontario, which receives low deposition and generally exhibits ratio values near one. Neary and Dillon (1988) also concluded that only 40% of the spatial gradient in ANC observed for Ontario lakes could be related to the combined influence of A^- and base cations. The remaining 60% was directly linked to replacement of ANC by SO_4^{2-} from atmospheric deposition—a regional finding supportive of earlier site-specific studies that discounted the relative importance of natural acidity as the causal factor explaining observed lake and stream acidification (Lazerte and Dillon 1984). Interpretative errors introduced by failure to consider Na^+ and K^+ and/or natural geological sources of SO_4^{2-} (Kramer and Tessier 1982) are not considered significant for this data analysis for the reasons noted earlier.

Finally, departure of the SO_4^{2-}:Ca^{2+} + Mg^{2+} ratio (sea salt corrected) from a near zero value

(Chapter 3, Husar et al., this volume; Chapter 18, Sullivan, this volume) also suggests that lakes have experienced changes in composition related to SO_4^{2-} deposition. This ratio is less affected by A^- than ANC:Ca^{2+} + Mg^{2+}, and when values exceed one, it is an unequivocal indication of lake acidification by SO_4^{2-} inputs. In Nova Scotia, 35% of the surveyed lakes had SO_4^{2-}:Ca^{2+} + Mg^{2+} > one. Furthermore, similar geographical patterns exhibited by both ratios (Jeffries 1986) lend credence to the acidification interpretation.

Overall, ion ratios can be useful indicators of chemical changes that have occurred in lake waters receiving acidic deposition. The low base cation content of eastern Canadian waters, particularly those in the Atlantic provinces, use of grid cell medians to interpret the regional situation, lack of statistical relationship between ANC and A^-, and a similar geographical pattern for both the ANC:Ca^{2+} + Mg^{2+} and SO_4^{2-}:Ca^{2+} + Mg^{2+} ratios all provide support for interpretation of ratio values in terms of acidification (i.e., replacement of ANC by SO_4^{2-}).

Episodic Acidification

In Canada, pH depressions associated with spring snowmelt constitute the most widely reported occurrences of episodic acidification, although there is no doubt that rain induced episodes also happen. The storage and release of acidic ionic species from the snowpack is a well-documented phenomenon (Jeffries 1990, Jones and Orville-Thomas 1987), and the effects have been reviewed by Marmork et al. (1984). During snowmelt, significant decreases in surface water pH, typically on the order of one unit but sometimes greater, ANC, and base cations have been extensively documented in central Ontario (Jeffries et al. 1979, Jeffries and Semkin 1983, Gunn and Keller 1984, Lazerte and Dillon 1984, Kelso et al. 1986a) and southern Quebec (Jones et al. 1986). Other chemical changes occurring at this time include elevation of Al and sometimes NO_3^-, and decreases in A^- concentrations. These chemical changes are related to simple dilution as well as to alteration of flowpaths resulting in increased transport of acidic chemical species into the surface water environment. It is noteworthy that Lazerte and Dillon (1984) convincingly demonstrated that the springmelt pH depression was more closely related to variations in anthropogenic strong acids than in natural occurring organic acids.

Processes Influencing Surface Water Chemistry

In a region as geographically large as eastern Canada, a broad range of biogeochemical processes operate to control lakewater composition (Chapter 1, Munson and Gherini, this volume), and it is beyond the scope of this overview to address them. Many have been discussed already in the section on regional surface water chemistry, and a comparison of the Canadian and United States case study regions is found in Chapter 17 (Baker et al., this volume). However, four important points summarize the situation.

First, the hydrological characteristics of nearly all eastern Canadian lakes mean that they fall into the drainage lake category as defined in Chapter 2 (Munson and Gherini, this volume). This categorization arises from the predominance of silicate bedrock, thin overburden, and except for portions of northwestern Ontario, annual precipitation near or in excess of 100 cm yr^{-1}. Hence, many processes that dominate in the seepage lakes of Florida and the Upper Midwest find limited expression in Canada.

Second, the glacially derived soils of eastern Canada have a very limited capacity for SO_4^{2-} adsorption. Most investigations of lake ion budgets show that SO_4^{2-} inputs approximately balance with outputs (Jeffries et al. 1988), except in lakes with long residence times, where within-lake SO_4^{2-} reduction can act as a significant source of ANC (Schindler 1986a). Coupled with the lack of evidence for significant geological contributions to SO_4^{2-} in Canadian lakes, this limited adsorption of SO_4^{2-} indicates that SO_4^{2-} levels are principally controlled by atmospheric deposition.

Third, base cation production, primarily Ca^{2+} and Mg^{2+}, either by primary weathering or by cation exchange in the surrounding terrestrial basin, provides most of the ANC supply to Canadian lakes; it is a major component even for the long residence time systems. The variability in mineralogical composition of the glacial overburden is a dominating factor controlling subregional variability in lake chemistry. It is fortuitous that Ontario subregions, which receive some of the highest deposition, also have small quantities of easily weathered $CaCO_3$ in the tills to neutralize the acidity (Shilts 1981). The distribution of till is extremely variable, so that it is commonplace to have lakes with low ANC (< 50 μeq L^{-1}) in proximity to those having higher ANC (> 100 μeq L^{-1}). The rest of eastern Canada does not have the benefit of a glacially dispersed source of $CaCO_3$; thus those subregions generally exhibit a relatively narrower, low concentration ANC distribution.

Fourth, A^- may be an important anion in some waters, particularly those located in the extensive wetland areas of Nova Scotia, Newfoundland, Labrador, and northwestern Ontario; however, there is little evidence that A^- can account for the occurrence of acidic lakes. In Labrador and northwestern Ontario, where many lakes have high A^- but receive only low levels of deposition, there are almost no systems with ANC ≤ 0, and in fact, most lakes have pH > 6.0.

Evidence of Historical Trends

Paleolimnological Evidence

Several investigations have been conducted independently across eastern Canada using diatom, chrysophyte, and metal profiles in lake sediments to infer historical changes in lakewater quality; they have been reviewed previously (Cook et al. 1988, Charles et al. 1990) and only a brief summary is presented here. Many studies have been located near major point sources (e.g., Wawa and Sudbury, Ontario) in order to document localized historical decreases in lake pH and/or increases in heavy metal sedimentation that accompanied operation of the smelters. The paleolimnological records show these effects occurring during the past 50 to 100 years, and recent studies also demonstrate improvements in inferred water quality during the past 10 to 20 years coincident with known reductions in smelter emissions.

Other paleolimnological studies performed at regionally representative sites (i.e., influenced only by long-range transport) suggest a variety of historical responses to acidic deposition. Both statistically significant and insignificant historical decreases in inferred lake pH have been reported; this is reasonable given the variable influence exerted by the surrounding terrain and the numerous biogeochemical processes controlling lake composition as described in the two previous sections and in Chapter 1 (Munson and Gherini, this volume). Nevertheless, in all eastern Canadian subregions receiving acidic deposition, diatom and chrysophyte data suggest that there has been a pH decline during the past 100 to 150 years in at least some lakes.

Recovery of Aquatic Systems

An important test of acidification hypotheses and models is chemical recovery in response to decreased deposition. If SO_4^{2-} deposition decreases, then surface water ANC should increase. Emissions of SO_2 in North America have decreased from peak values in the 1970's (Husar 1986). Furthermore, emission from the large point sources at Sudbury decreased > 50% between 1973 to 1978 and 1979 to 1985. These reductions have permitted observation of several instances of water quality recovery, either increasing pH and ANC or decreasing $Ca^{2+} + Mg^{2+}$.

Thompson (1986) reported decreases in SO_4^{2-} and increases in pH and ANC in Nova Scotia rivers coincident with reductions in SO_4^{2-} deposition. More dramatic evidence has been reported near Sudbury, however, where Lazerte and Dillon (1984), Dillon et al. (1986), Hutchinson and Havas (1986), and Keller and Pitblado (1986) provide a detailed record of water quality improvements (generally increased pH and ANC) in response to deposition reductions. Figure 9.10 taken from Dillon et al. (1986) serves to illustrate this point. Also, paleolimnological (diatom) evidence shows that the downward trend in inferred pH for Sudbury lakes reversed around 1970 (Dixit et al. 1987). Other regions in Ontario not directly influenced by the Sudbury source also demonstrate decreased SO_4^{2-} and increased pH and ANC (e.g., the Algoma region west of Sudbury; Kelso and Jeffries 1988). However, in contrast, Dillon et al. (1987) showed that Plastic Lake in southcentral Ontario responded to decreased deposition by reducing base cation yield from the terrestrial catchment rather than increasing pH and ANC (also observed for certain cases by Kelso and Jeffries 1988).

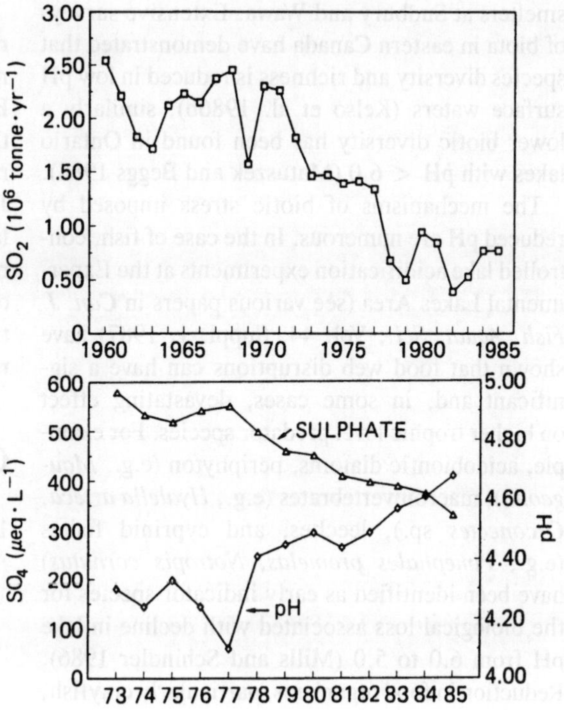

FIGURE 9.10. Sulfur dioxide emissions (10^6 tonne yr^{-1}) in the Sudbury basin from 1960 to 1985, and SO_4^{2-} concentration (µeq L^{-1}) and pH in Clearwater Lake from 1973 to 1985. (After Dillon et al. 1986.)

Future Considerations

Aquatic Resources at Risk

Most of the reviews noted in the introduction to this chapter have estimated the Canadian aquatic resource at risk from atmospheric SO_4^{2-} deposition by using models to predict regional steady-state lake chemistry through extrapolation from the limited survey information. Lake ANC is generally the parameter simulated from which it is possible to calculate lake acidities. The criterion used to define an acidic lake is often slightly different from the one used here; however, predictions reported in the various reviews are very similar. For example, RMCC (1986) used pH < 5.0 as the criterion (also used by Cook et al. 1988) and estimated that approximately 14,000 lakes are presently acidic in eastern Canada, and that if current levels of wet SO_4^{2-} deposition are maintained, 10,000 to 40,000 additional lakes will become acidic as steady-state conditions are achieved. Furthermore, it also notes that the model simulations must be considered as conservative, since they do not account for a possible depletion of ANC within the terrestrial watersheds. While the certainty of these estimates is unknown, the numbers do indicate the large magnitude of the freshwater resource that is at risk in eastern Canada.

Biological Status

There are numerous publications demonstrating the effect of surface water acidification on both fish and nonfish aquatic species in Canada, and it is beyond the scope of this overview to summarize them (RMCC 1986, Cook et al. 1988; see also information in Chapter 4, Baker and Christensen, this volume). The brief comments that follow focus primarily on fish, since they are the most economically important group of biota.

Historical records of fish population densities or harvest are, in most cases, either non-existent or too limited to permit evaluation of acidification impacts. There are two important exceptions, however. Watt et al. (1983) have recorded the decline in angling catch for Atlantic salmon (*Salmo salar*) that was associated with acidification of Nova Scotia rivers. In central Ontario, several authors (e.g., Harvey and Lee 1982, Beggs et al. 1985) have documented fish loss in regions affected by the smelters at Sudbury and Wawa. Extensive surveys of biota in eastern Canada have demonstrated that species diversity and richness is reduced in low pH surface waters (Kelso et al. 1986b); similarly a lower biotic diversity has been found in Ontario lakes with pH < 6.0 (Matuszek and Beggs 1988).

The mechanisms of biotic stress imposed by reduced pH are numerous. In the case of fish, controlled lake acidification experiments at the Experimental Lakes Area (see various papers in *Can. J. Fish. Aquat. Sci.*, Vol. 44, Suppl. 1, 1987) have shown that food web disruptions can have a significant and, in some cases, devastating effect on higher trophic level predator species. For example, acidobiontic diatoms, periphyton (e.g., *Mougeotia*), macroinvertebrates (e.g., *Hyalella azteca*, *Orconectes* sp.), leeches, and cyprinid fishes (e.g., *Pimephales promelas*, *Notropis cornutus*) have been identified as early indicator species for the biological loss associated with decline in lake pH from 6.0 to 5.0 (Mills and Schindler 1986). Reduction in food organisms, particularly crayfish, Mysis relicta, and minnow species, caused gradual emaciation of lake trout (*Salvelinus namaycush*) in experimentally acidified Lake 223; reproductive failure also occurred resulting in eventual catastrophic changes in the fish populations (Mills et al. 1987).

Documentation of biological impacts associated with springmelt chemical episodes (see the section on Episodic Acidification) is extremely limited in Canada, probably due to the great difficulties in making such observations. Harvey and Lee (1982) reported fish kills (primarily *Lepomis gibbosus*) in Plastic Lake in southcentral Ontario during the 1979–81 melt seasons. Gunn and Keller (1984) demonstrated that lake trout sac fry may be affected during spring melt, since spawning beds are characteristically located in shallows within the acidified surface layer present in most lakes during this season. It appears to be generally true that organisms capable of avoiding the adverse chemical conditions can survive the episodic occurrences of low pH and high Al; in contrast, organisms that are trapped in an affected environment or physically incapable of avoiding it may be at far greater risk. For example, Harvey and Whelpdale (1986) have shown that spring runoff at a trout hatchery in southcentral Ontario reduced the pH to 4.0 and caused fish mortality within 28 hours.

There is very meager evidence of biological recovery associated with the cases of chemical recovery noted earlier (see Recovery of Aquatic Ecosystems). Keller and Pitblado (1986) suggest that natural recruitment of some fish species has resumed in some Sudbury area lakes. Kelso and Jeffries (1988) observed the recolonization of a few lakes by white sucker from downstream sources in response to improved pH conditions. The paucity of biological evidence is not surprising, given the time lag expected in biotic response and the recent nature of the chemical observations.

Conclusions

1. Based on consideration of bedrock and surfical geology, Canada contains 4 million km^2 (43% of the country) classified as having low potential to neutralize precipitation acidity. Coincidence of this sensitive terrain with the field of elevated deposition (i.e., > 10 kg wet SO_4^{2-} ha^{-1} yr^{-1}; > 210 eq ha^{-1} yr^{-1}) defines the area of principal concern and is usually approximated by that portion of Canada east of the Ontario-Manitoba border and south of 52°N latitude.
2. A data base containing chemical information for more than 7400 lakes was used to evaluate regional water quality in terms of pH, Ca^{2+} + Mg^{2+}, ANC, and SO_4^{2-} for eight subregions of southeastern Canada. Subregional variation in lake composition is readily explained by a combination of terrain-related and deposition factors. In sensitive areas receiving low levels of deposition (e.g., Labrador), most lakes exhibit pH in the range 6.0 to 7.0 and ANC $\leq$ 100 μeq L^{-1}.
3. The presence of easily weathered carbonate materials (either in the bedrock or in the glacial overburden) in Ontario accounts for the generally higher pH, base cations, and ANC (median ANC = 61 to 261 μeq L^{-1}) observed there compared to the other subregions (median ANC = 14 to 52 μeq L^{-1}).
4. Soil adsorption of SO_4^{2-} is minimal, so that lakewater SO_4^{2-} distributions are principally controlled by atmospheric deposition. Evaluation of the geographic distribution of the ANC:Ca^{2+} + Mg^{2+} ratio showed that southcentral Ontario, southern Quebec, and portions of New Bruns-

wick, Nova Scotia, and Newfoundland all exhibit strong evidence of changes in composition (ratio < 0.4). The low base cation content of these waters suggests that a major component of this change has been loss of ANC (i.e., acidification) in response to SO_4^{2-} deposition.

5. Results of paleolimnological studies conducted to infer historical pH in Canadian lakes reflect the wide variety of conditions that exist. Some lakes demonstrate currently depressed pH compared to the past, other do not. However, in all subregions receiving acidic deposition, there are at least some lakes that have become more acidic during the past 50 to 100 years.
6. Episodic acidification has been observed in many locations and is usually associated with spring snowmelt.
7. Extrapolation of survey information suggests that 14,000 lakes are presently acidic, and models used to assess the aquatic resource at risk in eastern Canada predict that at least an additional 10,000 to 40,000 will become chronically acidic if present deposition levels persist.
8. A few instances of chemical recovery occurring in response to reduced deposition suggest that a regional reduction will yield a general improvement in the condition of the aquatic resource. Documentation of the impact on aquatic biota is necessarily far less complete. Loss of fish populations from some Nova Scotian rivers and either loss or reduction in a diversity of populations in Ontario have been linked to habitat acidification by atmospheric deposition; some losses and reductions are due to local influences, however.
9. Experimental studies have defined many of the mechanisms leading to population loss or diversity reduction. Food chain effects are important.

Southeast Overview

Jerry W. Elwood

The southeastern United States encompasses several subregions containing low ANC surface waters. These areas include portions of the Piedmont in North and South Carolina and Georgia; the coastal plain in Maryland, Virginia, and North and South Carolina; the Valley and Ridge Province in Tennessee, Kentucky, and Virginia; the Ozark/Ouachita Mountains region in Arkansas; and the Blue Ridge Province in Virginia, western North Carolina, eastern Tennessee, western North Carolina, and northern Georgia (Figure 1; Chapter 5, Eilers and Selle, this volume; Kaufmann et al. 1986). The entire region contains a high percentage of dilute, low ANC streams, lakes, and reservoirs.

This book includes chapters on two areas of the inland southeast. Chapter 10 discusses portions of the Blue Ridge Province and the Valley and Ridge Province in Virginia (Cosby et al. this volume). Chapter 11 discusses most of the Southern Blue Ridge Province in western North and South Carolina, eastern Tennessee, southern Virginia, and northern Georgia (Elwood et al. this volume). Both the Southern Blue Ridge and the Valley and Ridge Provinces are unglaciated areas, containing relatively unreactive bedrock and highly weathered residual soils rich in secondary mineral phases. Surface waters consist primarily of small high-gradient streams and a few drainage lakes and reservoirs.

Annual precipitation in the Southeast region ranges from 100 to > 200 cm, depending on elevation (Figure 2). Precipitation maxima occur in the spring and summer and during October and November (Figure 3). Precipitation pH averages < 4.5 and SO_4^{2-} is the dominant anion (Figure 4).

Watershed retention of SO_4^{2-} and NO_3^- accounts for most of the consumption of acidity in acidic deposition in the region; base cation mobilization is the least important source of ANC (Chapter 11, Elwood et al., this volume). Streams, drainage lakes, and reservoirs in the Southern Blue Ridge have similar acid-base chemistry, indicating negligible in-lake production of ANC. Differences in ANC among streams in the Southeast region appear to be related primarily to differences in the capacity of watersheds to mobilize base cations through weathering and cation exchange and to retain SO_4^{2-} by adsorption on soil particles.

Median SO_4^{2-} levels in streams are approximately twice as high in the northern portion of the Blue Ridge Province in Virginia as in the southern portion in North Carolina, Tennessee, and Georgia, despite comparable levels of precipitation in the two regions (Figure 3). Base cation levels are comparable in the two areas, however, resulting in a lower median ANC in streams in the northern portion of the Blue Ridge Province. Sulfate concentrations in streams are higher in the Valley and Ridge Province in Virginia than in the northern portion of the Blue Ridge Province, whereas ANC and base cations are lower. The higher SO_4^{2-} levels in the Valley and Ridge Province may be related to (1) the greater SO_4^{2-} deposition in this area due to its greater proximity to emission sources, (2) greater SO_4^{2-} adsorption capacity by soils in the Blue Ridge Province associated with a greater content of

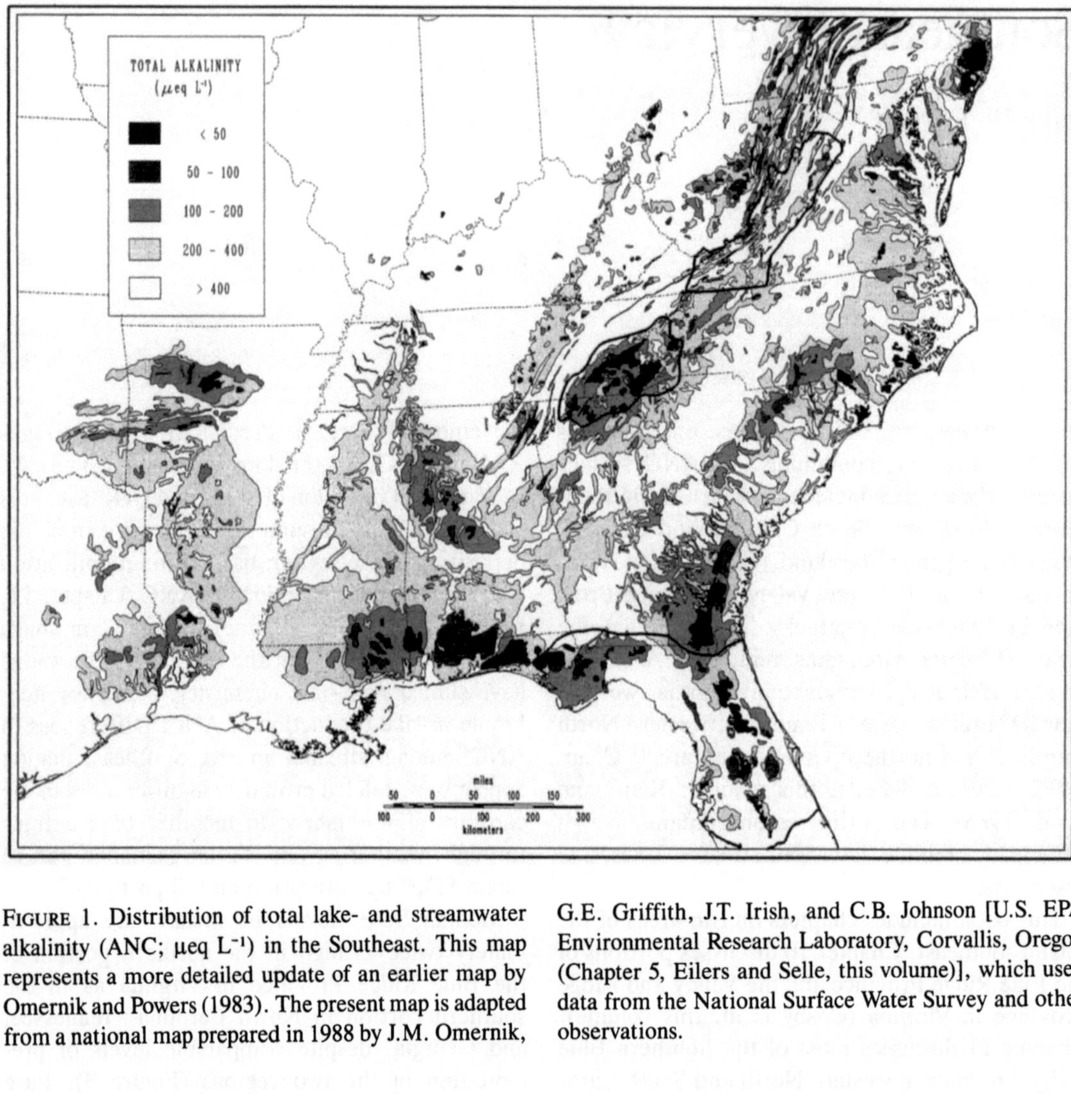

FIGURE 1. Distribution of total lake- and streamwater alkalinity (ANC; µeq L^{-1}) in the Southeast. This map represents a more detailed update of an earlier map by Omernik and Powers (1983). The present map is adapted from a national map prepared in 1988 by J.M. Omernik, G.E. Griffith, J.T. Irish, and C.B. Johnson [U.S. EPA Environmental Research Laboratory, Corvallis, Oregon (Chapter 5, Eilers and Selle, this volume)], which used data from the National Surface Water Survey and other observations.

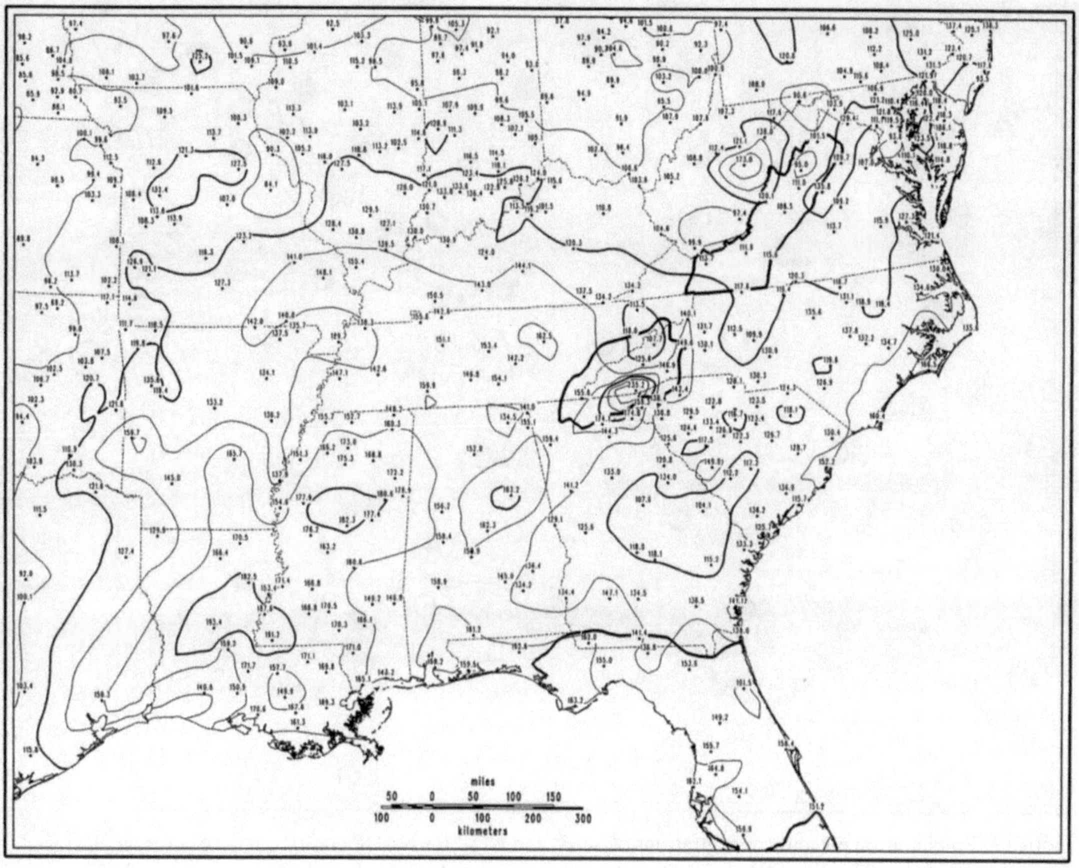

FIGURE 2. Precipitation quantity in the Southeast averaged for the period 1982–84. Contours are in cm yr^{-1} at 20-cm intervals with 60-cm contours in heavier lines; the location of each station in this region that was used in preparing the contour map is indicated, along with its annual precipitation in cm yr^{-1}. Point data calculated by R. Husar; contour map prepared by T. Selle.

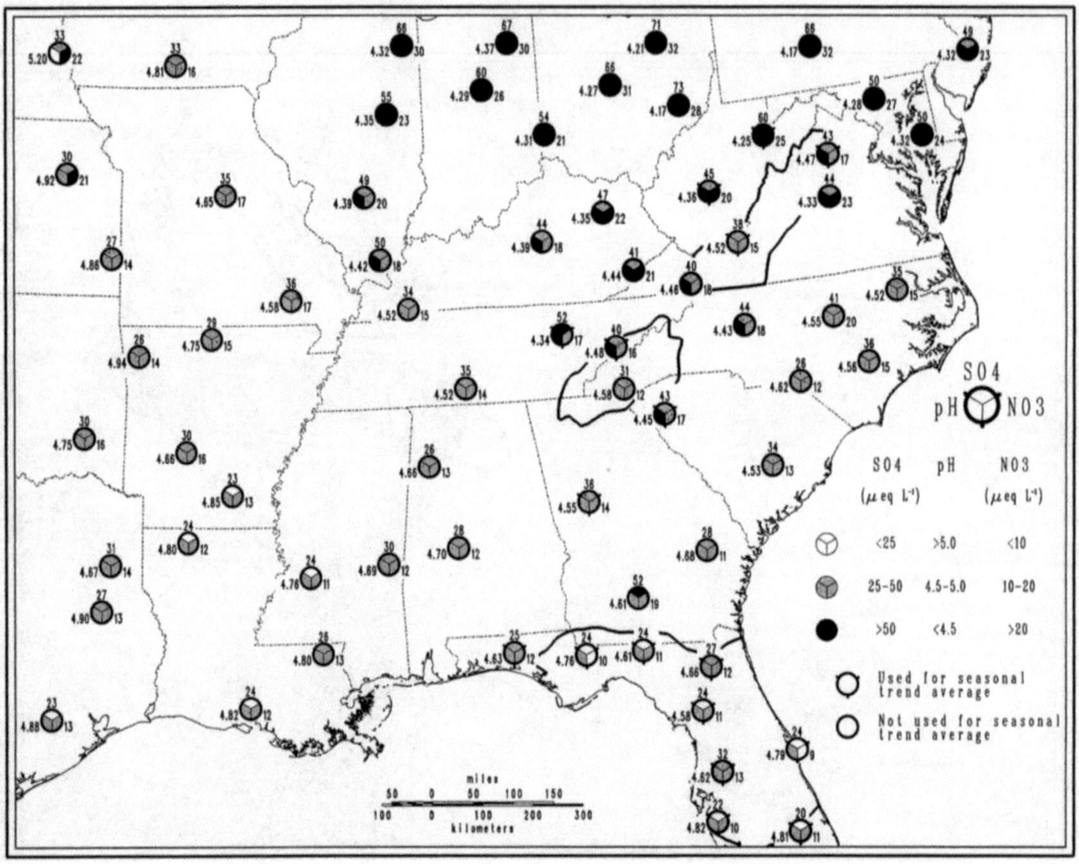

FIGURE 3. Precipitation pH and concentrations of SO_4^{2-} and NO_3^- for NADP stations in the Southeast, averaged for the period 1982–84 (Olsen and Watson 1984). Averages calculated by R. Husar; map prepared by T. Selle.

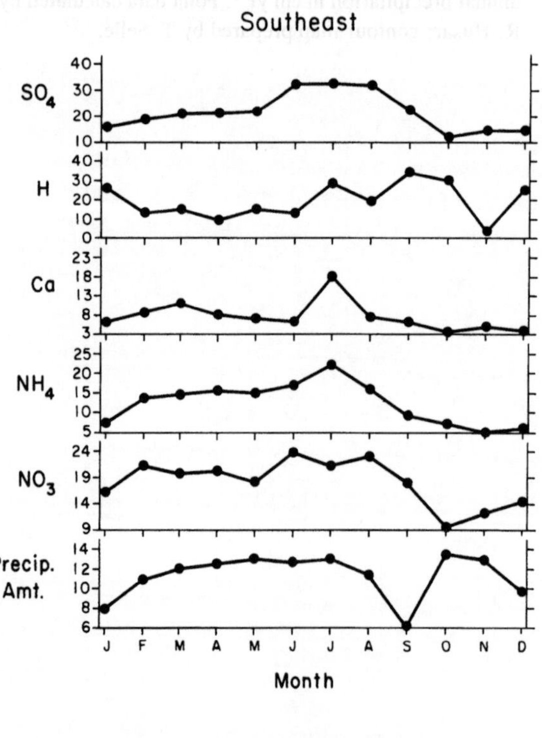

FIGURE 4. Monthly variation in precipitation quantity and chemistry in the Southeast. Chemical values are expressed as μeq L^{-1}; precipitation amount is in cm. Data represent 5-year averages for the period 1982–86 for the sites indicated in Figure 3. Data are from the NADP (Olsen and Watson 1984) and averages were determined by R. Husar.

secondary mineral phases that adsorb anions, or (3) some combination of both.

Ten percent of the streams in the northern portion of the Blue Ridge in Virginia are acidic, compared to < 1% in the southern portion. The differences between the northern and southern portions of the Blue Ridge Province both in ANC and SO_4^{2-} levels in streams and in percentages of acidic streams appear to be a result of lower SO_4^{2-} retention by watersheds in the northern portion of the province.

10
Mountains of Western Virginia

B. Jack Cosby, Patrick F. Ryan, J. Rick Webb, George M. Hornberger, and James N. Galloway

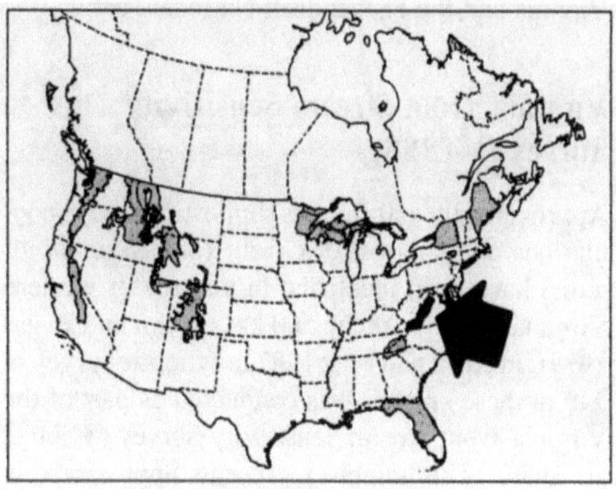

ABSTRACT. Data are presented for surveys of (1) 344 native brook trout streams in the mountains of western Virginia (80% of all such streams in Virginia), (2) 47 streams draining the Shenandoah National Park in west-central Virginia, and (3) 63 headwater streams in the southwestern portion of the Shenandoah National Park. Based on analyses of these survey data and of six years of data from two intensively monitored catchments in the park, we concluded that the mountain streams of Virginia are at considerable risk to damage from acidic deposition. In the largest of the stream surveys, 93% of the streams have ANC concentrations < 200 $\mu eq\ L^{-1}$, 49% of the streams have ANC concentrations < 50 $\mu eq\ L^{-1}$; and 10% of the streams are currently acidic (ANC ≤ 0). Sulfate is the major anion in most streams with low ANC. All catchments surveyed are retaining a significant proportion of atmospherically deposited sulfur; median sulfate retention for all streams surveyed is approximately 60%. Bedrock geology appears to exert the strongest control on the variability of ANC in the streams within the region. Streams in the Valley and Ridge province have lower ANC and higher sulfate concentrations than streams in the Blue Ridge province. The potential for damage to fish populations is large, and at least one instance of decline in fish and invertebrate populations has been reported in the region.

Introduction

Large areas of the southeast, particularly the Appalachian Mountains from Virginia to Georgia, receive acidic deposition at rates equal to or greater than those in the northeast (Chapter 5, Eilers and Selle, this volume). The physical characteristics of catchments in the southeastern United States are different from those in the northeastern United States. Surface waters in the mountains of the southeast are generally limited to low-order or headwater streams; natural lakes are rare. Snow accumulation is hydrologically much less important than it is in more northern climes. The bedrock geology of the southeastern United States is more variable than that of the northeast, including shales and metabasalts as well as granites and quartzites. Soils are typically thin and rocky (usually ultisols or inceptisols). The surficial deposits of the southeast are geologically much older than those of the northeast, since the southeast is below the limit of recent glaciations. Both soils and parent rock in the southeast are highly weathered.

Previous studies (e.g., Galloway and Cowling 1978, Hendrey et al. 1980, Arnold et al. 1985, Kaufmann et al. 1988) have suggested that streams in large areas of the southeast are vulnerable to acidification. The Appalachian Mountains have been identified on national maps as potentially sensitive to acidic deposition (Galloway and Cowling 1978, Omernik and Powers 1983). The Shenandoah National Park in Virginia is located in the Appalachian Mountain chain. Several areas in the park have been designated as wilderness areas and, as such, offer an ideal setting for controlled studies of the effects of acidic deposition on streamwater quality. Other large areas in the Appalachian Mountains of western and southwestern Virginia are included in the National Forest system and offer limited control to facilitate studies of the effects of acidic deposition. This chapter presents a summary and synthesis of several studies of the streams of western Virginia and the Shenandoah National Park.

Virginia Trout Stream Sensitivity Survey (VTSSS)

Approximately 450 streams supporting native populations of eastern brook trout (*Salvelinus fontinalis*) have been identified in an area of western Virginia roughly 100 by 500 km (Mohn and Bugas 1979). In April and May 1987, a synoptic survey of 349 of these streams was conducted as part of the Virginia Trout Stream Sensitivity Survey (Webb et al. 1989). This sample represents approximately 80% of the total number of identified streams with naturally reproducing trout populations in the state. Exclusion of the nonsampled streams was based primarily on watershed disturbance criteria, access difficulty, and watershed size. A small number of streams were excluded due to the presence of carbonate bedrock. Data from 344 of the sampled streams were used in this study.

Shenandoah National Park Synoptic Survey, Phase I (SNP)

The Shenandoah National Park straddles a 100-km segment of the Blue Ridge mountains in northcentral Virginia. Approximately 675 km² are included within the Shenandoah National Park Synoptic Survey, Phase I (SNP-I) boundaries. Fifty-six watersheds representing approximately 70% of the area of the SNP-I were sampled six times each (approximately every other month) from August 1981 through June 1982 (Lynch and Dise 1985). The areas of the sampled watersheds varied from 0.4 to 25 km². Streams were selected for sampling based on (1) presence of one of the major geological formations in the area, (2) accessibility, (3) negligible human impact within the catchment, and (4) fishing or recreational value. The flow-weighted annual average data from 47 of the sampled streams were used in this study.

Shenandoah National Park Synoptic Survey, Phase II (SWSNP)

Since data from the SNP-I indicated that catchments in the southwestern portion of the park had the lowest ANC values, researchers conducted a more detailed survey of the catchments draining that portion of the park (Webb 1988). Ten of the larger catchments of the Phase I survey of the southwestern part of the park were divided into headwater subcatchments with drainage areas from 25 to 180 ha. Sixty-three of these headwater drainages were sampled during the 30-day period from May 10 to June 9, 1985. Data from that survey, the Southwest Shenandoah National Park Synoptic Survey, Phase II (SWSNP), are used in this study.

Shenandoah Watershed Intensively Studied Sites, White Oak Run and Deep Run

Two small watersheds, White Oak Run and Deep Run, in the southwestern portion of Shenandoah National Park, were selected for intensive monitoring. Weekly sampling of precipitation and streamwater chemistry has been carried out at both sites since November 1979 (Ryan et al. 1989, Shaffer and Galloway 1982). Precipitation volume and stream discharge have been continuously monitored throughout the study. Data for the complete calendar years 1980 through 1985 are used in this study.

Regional Characteristics

Climatic, Topographic, and Hydrologic Characteristics

This case study focuses on headwater streams in the mountainous region of western Virginia (Figure 10.1). This area, about 100 by 500 km in size,

FIGURE 10.1. A typical high-gradient upland forested stream in western Virginia. Shown here is an unnamed headwater stream draining forested watershed at an elevation of 280 to 400 m in Frederick County, Virginia, near the town of Mountain Falls.

provides a representative cross-section of the folded Appalachian Mountains of the southeastern United States. In general, the region is made up of parallel mountain ridges ranging from 750 to 1500 m above sea level, with relief ranging up to approximately 750 m above the intervening valley and adjacent Piedmont areas. Forest cover on the mountain ridges is nearly complete and is predominantly mixed deciduous, with limited distributions of transition and boreal forest types on some of the higher crests. Soil cover on the mountain ridges is generally shallow and rocky with occasional talus and bedrock exposure. Climate in the study area is characterized as temperate continental. Air temperature, number of frost-free days, and annual precipitation amounts are fairly uniform across the study region. The mean January temperature is about 2°C and the mean July temperature is about 23°C. Snow falls at higher elevations several times a year, mostly from December to February, but typically melts within two to three days.

Most of the mountain streams in the region are small, usually flowing only 5 to 10 km at high gradient before discharging onto the flatter terrain of adjacent valley or Piedmont areas. The catchment topography associated with these streams is characterized by steep sideslopes with concave upward coves at the head of stream courses. Stream courses typically have alternating rubble and bedrock beds, with narrow alluvial areas and low terraces occurring along higher order reaches. Stream flows vary seasonally with minimum discharge and maximum temperature occurring in late summer and early fall.

The water yields of the catchments draining into these streams are moderate to low. Yields for six small catchments in western Virginia, all at least partially wooded, range from 11% to 43% for the

1958 to 1972 period (data summarized in Haan, 1975). The average yield for White Oak Run, in the Shenandoah National Park, for the years 1980 to 1986 was 40%, ranging from a low of 13% in 1980 to a high of 58% in 1983 (Ryan et al. 1989). The long-term average yield for the James River at Richmond, Virginia (period of record beginning in 1934), is approximately 36% (Prugh et al. 1986; Va. Dept. Conserv. Econ. Develop. 1969). Based on these considerations, the upland streams in the study region can be characterized as having an average water yield of approximately 35%. Precipitation volume for 39 western Virginia climatological stations located at elevations > 300 m averages 104 cm yr^{-1} (NOAA, 1982). Estimated average runoff for catchments in the study area is therefore approximately 36 cm yr^{-1}.

Physiographic, Geologic, and Edaphic Characteristics

Two physiographic (or geomorphic) provinces are recognized in the study region, the Blue Ridge province and the Valley and Ridge province (Fenneman 1928, Dietrich 1970). The Blue Ridge province (Figure 10.2) is situated along the eastern edge of Virginia's mountain region adjacent to the rolling terrain of the Piedmont province. In northern and central Virginia, the Blue Ridge province is represented by a narrow (10 to 25 km wide) range of steep-sided mountains consisting of a single major ridgeline with smaller discontinuous ridges extending to the east and west. In southern Virginia, this province expands to encompass a mountainous plateau area up to 40 km wide. The Valley and Ridge province, consisting of linear mountain ridges and valleys, extends 50 to 80 km west of the Blue Ridge province to the mountainous Allegheny and Cumberland Plateau provinces (Figure 10.2). The mountains of the Valley and Ridge province are separated from the mountains of the Blue Ridge province by a valley area up to 35 km wide (the Great Valley of Virginia). To the west of this broad valley area, the Valley and Ridge province consists of 5–8 major ridges separated by relatively narrow valleys.

The geology of the study region has been described by Butz (1940), Hack (1965), Dietrich (1970), and Gathright (1976). Bedrock types include Precambrian and early Cambrian granites, granulites, and metabasalts along the crest and eastern flank of the Blue Ridge province, with Paleozoic sedimentary rocks along the western flank of the Blue Ridge province and in the Valley and Ridge province (Figure 10.2). Northwestward displacement of these rocks about 250 million years ago produced a pattern of thrust faults and folds reflected in the trend of the modern-day mountain ridges. These ridges, which give rise to most of the streams considered here, are generally underlain by the more weather resistant silicoclastic and crystalline rocks represented in the stratigraphic sequence. The more readily weatherable carbonate and argillaceous rocks generally underlie valleys and topographically low areas within the region.

The distribution of soil types in the study region is closely related to bedrock distribution (see USDA, 1979). The region has not been glaciated, and the typically rocky and patchy mountain soils have formed in residual or colluvial material. In general, ultisols have formed from the crystalline bedrock materials along the crest and eastern flank of the Blue Ridge province. Inceptisols (with poorer base supplies and less clay content than ultisols) have formed from the sedimentary bedrock materials exposed on the western flank of the Blue Ridge province and on the ridges of the Valley and Ridge province to the west. All of these soils are typically shallow (.5 to 2 m to bedrock) and skeletal with a high percentage of rock fragments. The specific depth to bedrock and degree of stoniness appear to be controlled mainly by slope position and orientation of bedrock material.

Vegetation and Land Use Characteristics

The distribution and productivity of forest vegetation in the study region are primarily determined by moisture availability, which in turn is controlled by soil and bedrock conditions as well as topographic features (Hack and Goodlett 1960, Edmunds et al. 1986). The greater moisture holding capacity associated with the ultisols developed on the crystalline rocks of the Blue Ridge is evidenced by the dominance of red oak (*Quercus rubra*), maple (*Acer* spp.), tulip poplar (*Liriodendron tulipifera*), hemlock (*Tsuga canadensis*), rhododendron (*Rhododendron maximum*), and other moisture demanding species. In contrast, the forest associated with the drier inceptisols developed on the sandstone and shale rocks to the west is typically dominated by chestnut oak (*Quercus prinus*), scarlet oak

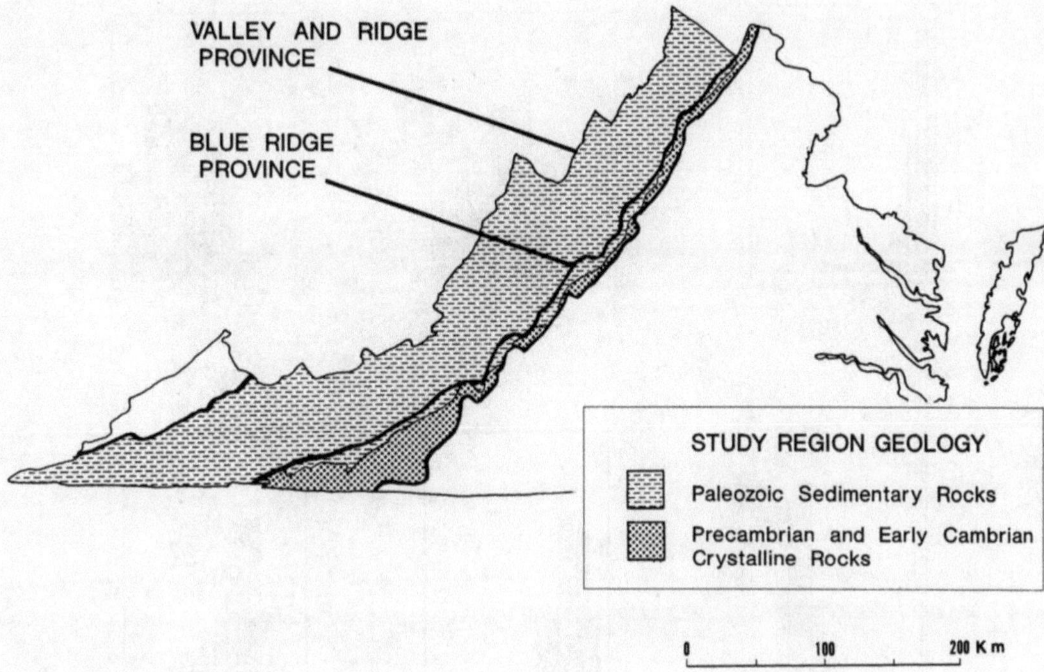

FIGURE 10.2. Physiographic provinces and major geological formations of the western Virginia study region.

(*Quercus coccinea*), black oak (*Quercus velutina*), Virginia pine (*Pinus virginiana*), mountain laurel (*Kalmia latifolia*), azalea (*Rhododendron calendulaceum*), and other less moisture demanding species. Within most mountain watersheds, the distribution and growth of vegetation is further affected by topographic position, as coves and areas adjacent to the stream course typically have relatively greater moisture availability than side slope and ridge crest areas.

Most of the watersheds included in this study have not provided suitable conditions for habitation or agriculture, and watershed disturbance has generally been limited to timber harvest and fire. Indiscriminant and repeated timber harvesting, as well as repeated burning, however, has apparently reduced moisture holding capacity and altered the original distribution of forest species in many watersheds (Moore 1937). Most of this activity ceased 50 to 100 years ago. The majority of the stream/watershed systems included in the study are now protected or managed as national park or national forest land. At present, most of these systems reflect only the effects of natural biogeochemical processes operating in the watersheds, combined with the effects of elevated atmospheric acidic deposition.

Atmospheric Deposition

Precipitation volume is uniformly distributed throughout the year. Prevailing winds are from the west. Sulfate dominates the anionic concentrations in precipitation but NO_3^- concentrations are not insignificant. Neither ammonium ion nor alkaline dust provide significant neutralization of the precipitation. The precipitation of this study region is acidic, consisting primarily of acidic sulfates and nitrates. In this respect, the precipitation chemistry of western Virginia is similar to that of the larger southeastern region of the United States (see Southeast Overview; Chapter 5, Eilers and Selle, this volume).

Wet sulfate deposition in the western portion of Virginia is about 500 eq ha^{-1} yr^{-1} (Buikema et al. 1985, 1986, 1987, 1988, Summers et al. 1986). Regional measurements of sulfur deposition in gaseous and aerosol forms are not available. Although estimates can vary widely, dry deposition is generally estimated to range from 50% to 100% of wet deposition (e.g., Rochelle et al. 1987, Shaffer and Galloway 1982). The sources of the atmospheric sulfur and nitrogen are primarily anthropogenic, from the combustion of fossil fuels

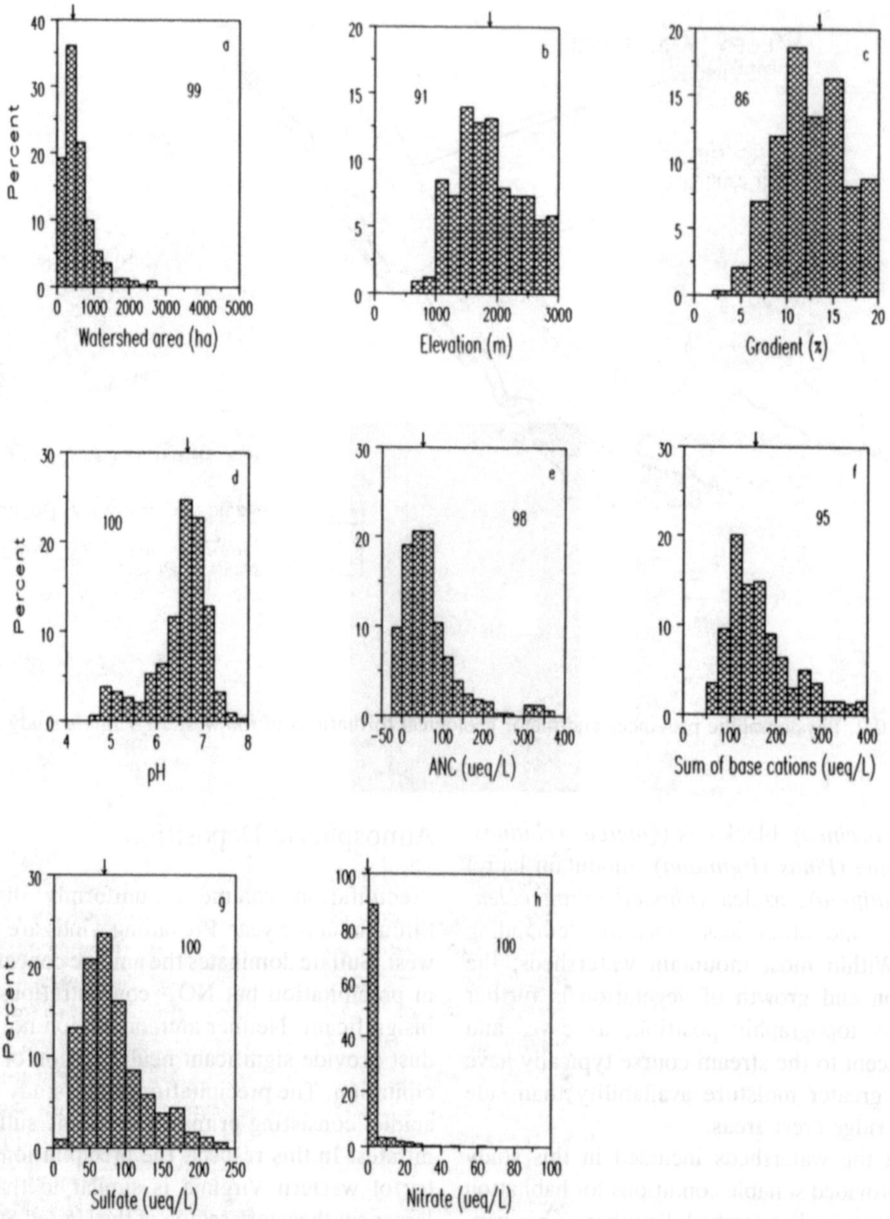

FIGURE 10.3. Frequency histograms of major characteristics of the target population of VTSSS streams: (a) watershed area, (b) elevation, (c) stream gradient, (d) pH, (e) ANC, (f) sum of base cations, (g) sulfate, and (h) Nitrate. The arrow at the top of each box indicates the median for the target population, and the number in the upper right side of each box indicates the percentage of target streams plotted.

and related industrial activities (Galloway and Whelpdale 1980, Galloway et al. 1984). Considering the estimated historical increases in sulfur emissions in the southeastern United States (OTA 1984; Chapter 3, Husar et al. this volume) and sulfur deposition observed in remote, nonindustrial areas of the world (Galloway et al. 1982), current sulfur deposition rates in the study region are estimated to be approximately 10 times the rates that existed before the onset of anthropogenic emissions (i.e., historical deposition equals 10% of current deposition).

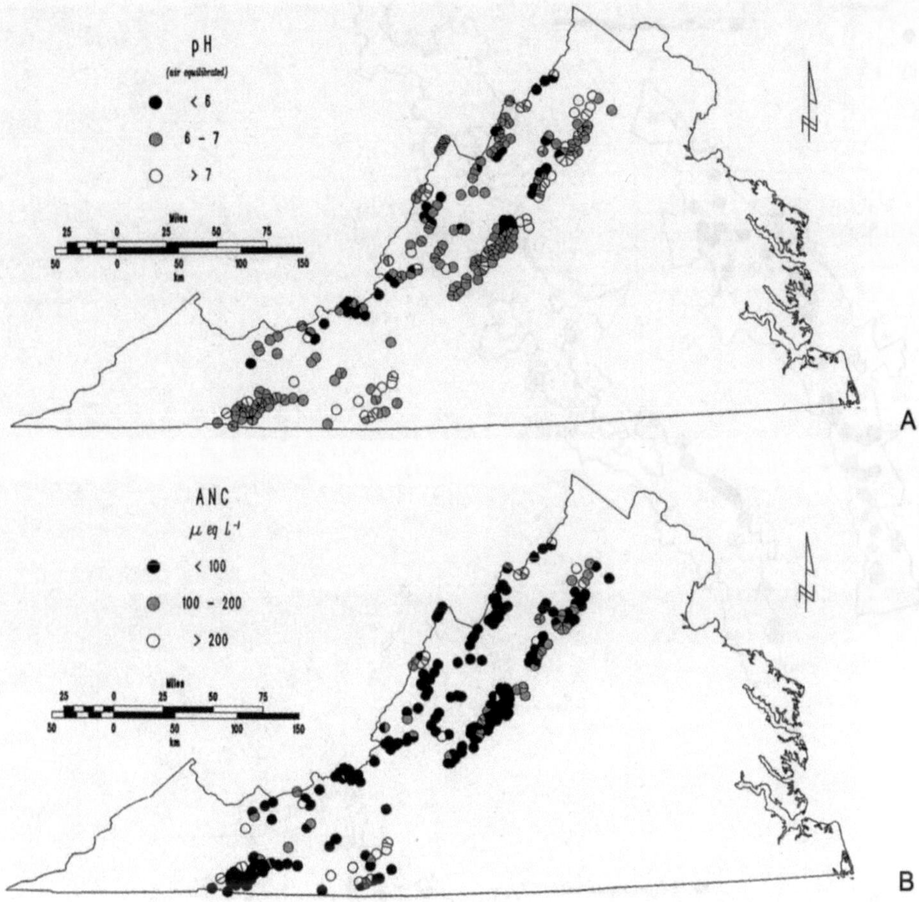

FIGURE 10.4. Geographical distributions of pH, ANC, and ANC:SBC for western Virginia, based on the VTSSS data set (A, B, E) and for Shenandoah National Park based on the SNP-I and SWSNP data sets (C, D, F). (*Continued on following pages.*)

Regional Surface Water Chemistry

Current Status of VTSSS Streams

Concentrations of ANC in the VTSSS streams (Figures 10.3, 10.4, and 10.5; Table 10.1) are low, relative to concentrations commonly cited for the assessment of surface water sensitivity. Only 7% of the VTSSS streams have ANC concentrations > 200 $\mu eq\ L^{-1}$, a value that is perhaps the most commonly cited criterion for identifying sensitive surface waters (Altshuller and Linthurst 1984, Winger et al. 1987, Knapp et al. 1988). The ANC concentrations of 49% of the streams are ≤ 50 $\mu eq\ L^{-1}$, a value that has been cited as a criterion for identifying extremely sensitive surface waters (Gibson et al. 1983, Schindler 1988). A significant proportion (10%) of the sampled streams are currently acidic (ANC ≤ 0). Stream pH values range from 4.7 to 7.0 for those streams with ANC ≤ 100 $\mu eq\ L^{-1}$.

Base cation concentrations of the VTSSS streams are generally low (Figure 10.5; Table 10.1). The ratios of concentrations of pairs of the four base cations are similar to the ratios reported for the major rivers of the the world (Holland 1978). The concentrations in the VTSSS streams, however, are an order of magnitude less than those in the major rivers, emphasizing the low ionic strength of the population of streams under investigation.

Sulfate concentrations (Figure 10.5; Table 10.1) are generally the highest among the measured ions in the VTSSS streams. Nitrate concentrations are

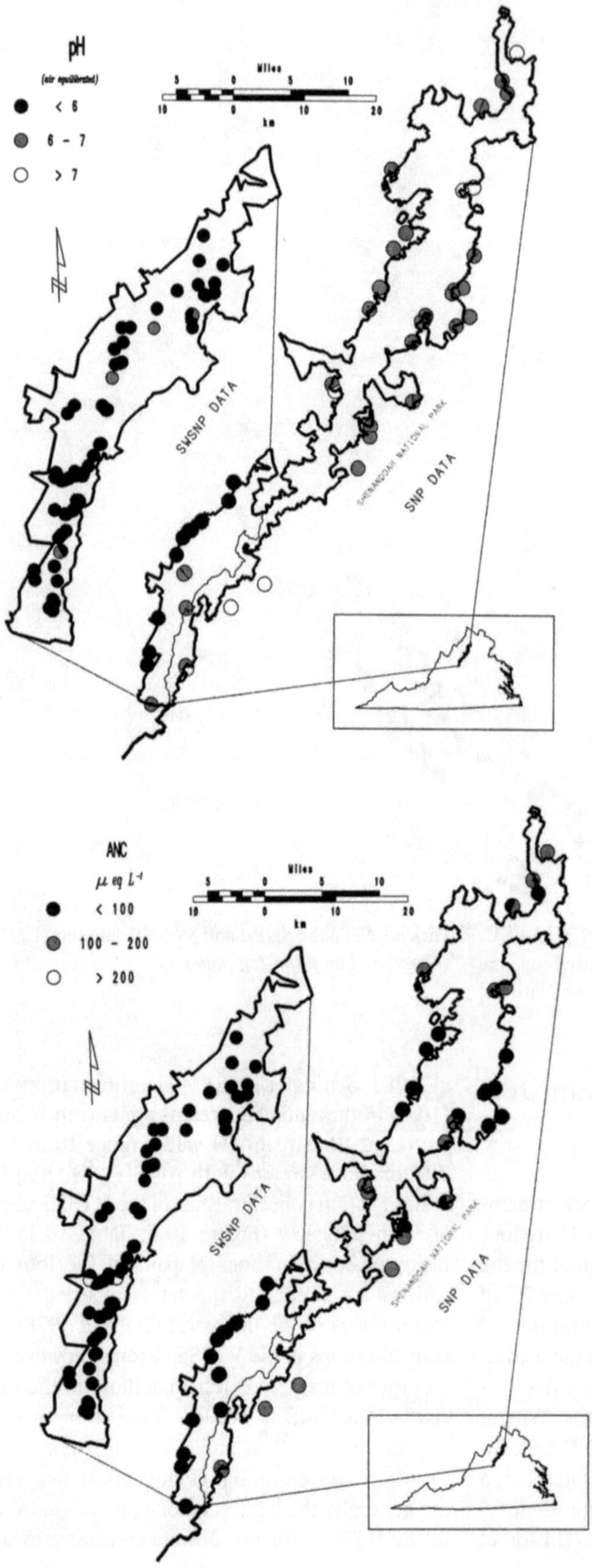

D FIGURE 10.4. (*Continued.*)

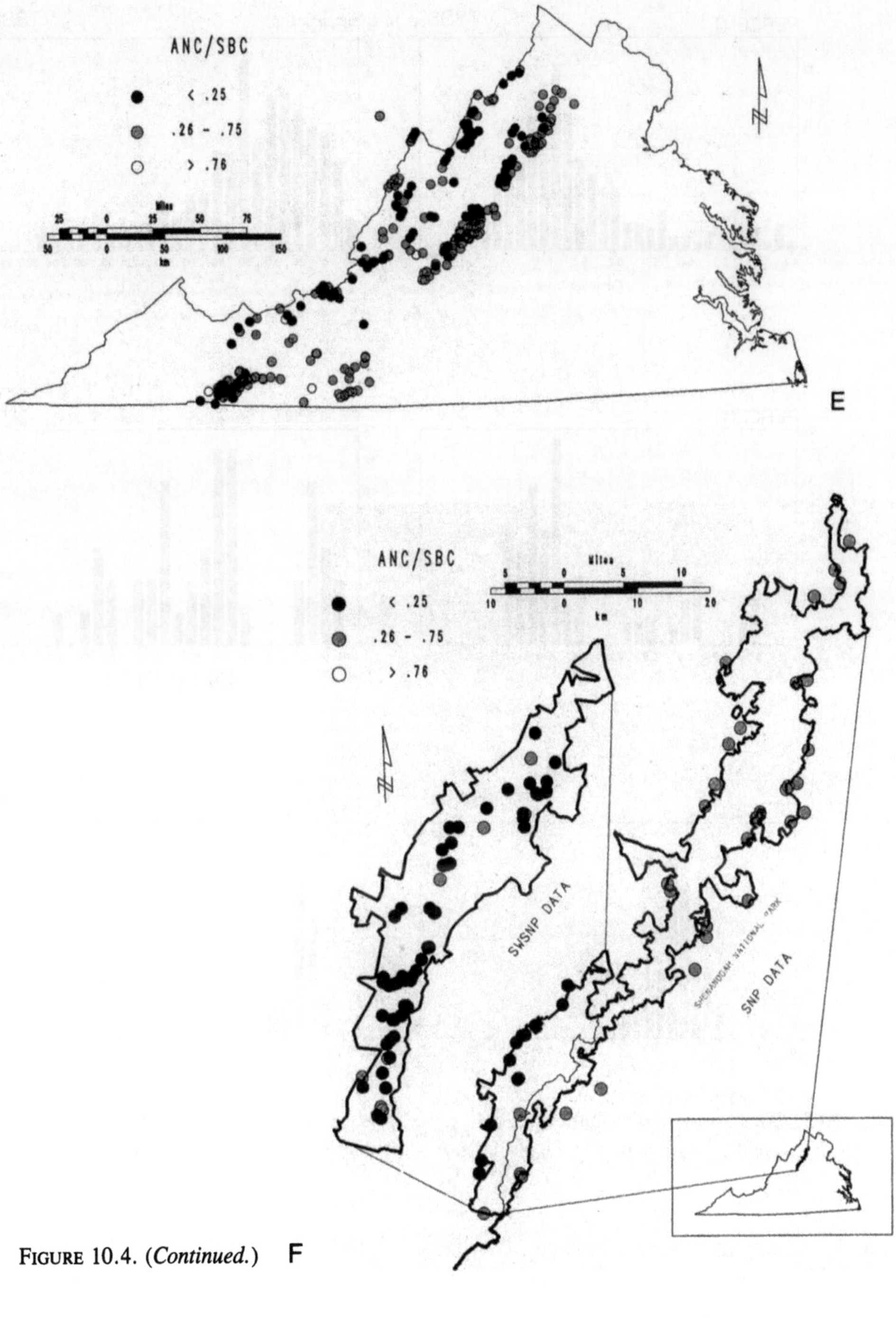

FIGURE 10.4. (*Continued.*)

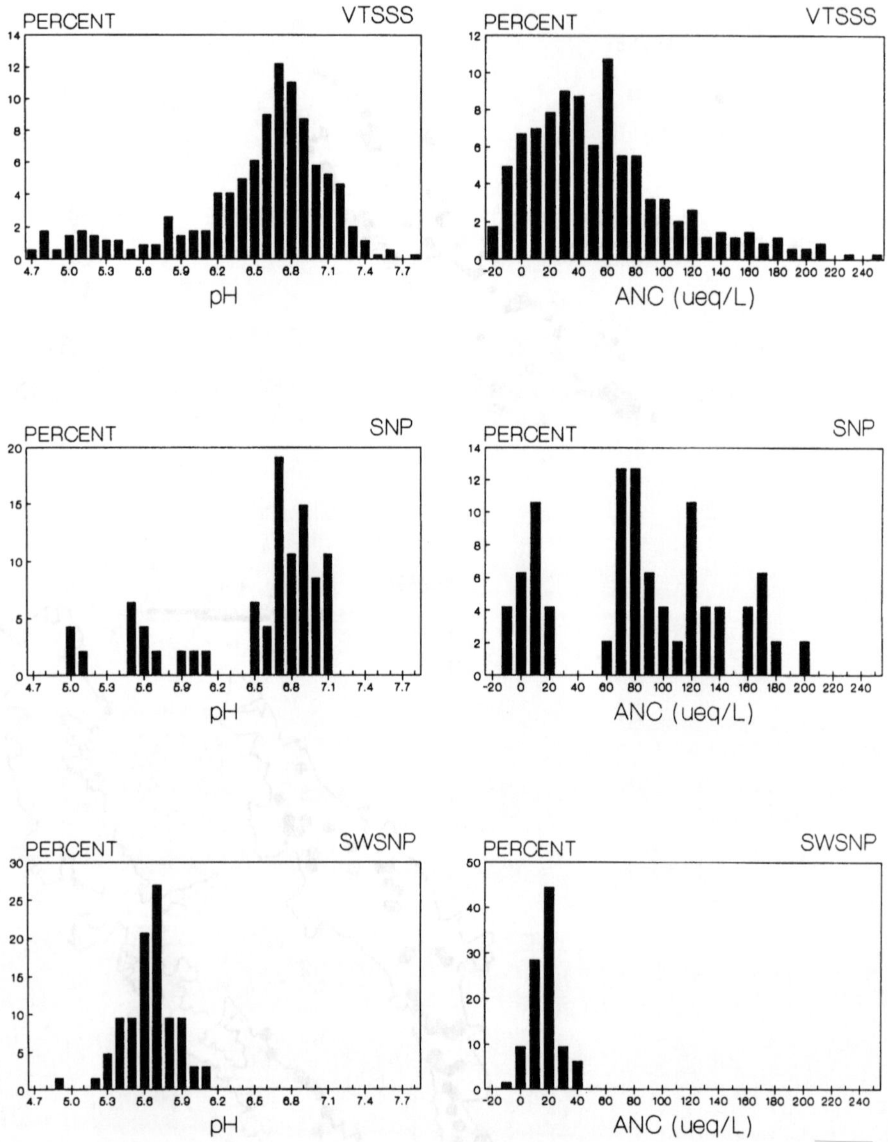

FIGURE 10.5. Comparison of distributions of pH, ANC, base cation concentrations, and SO_4^{2-} concentrations for VTSSS, SNP-I, and SWSNP data sets. Upper tails of some distributions have been truncated.

the lowest among the measured anions. The ratios of the anion concentrations are dissimilar to those reported for major rivers (Holland 1978). Whereas HCO_3^- is the major anion in "average" river water, SO_4^{2-} is a major anion in the sampled native brook trout streams, particularly in those streams with low ANC (Figure 10.6). The VTSSS streams that currently have ANC values ≤ 0 also have $SO_4^{2-}:C_B$ ratios $\sim\geq 1$ (Figure 10.7), which suggests that these streams have lost their ANC as a result of increased mineral acidity derived primarily from acidic SO_4^{2-} deposition from the atmosphere.

Current Status of SNP-I Streams

The ANC concentrations in SNP-I streams are similar to those in the VTSSS (Figures 10.4 and 10.5; Table 10.1). Although the median ANC of streams in the SNP-I (82 μeq L^{-1}) is somewhat higher than that of the VTSSS streams (51

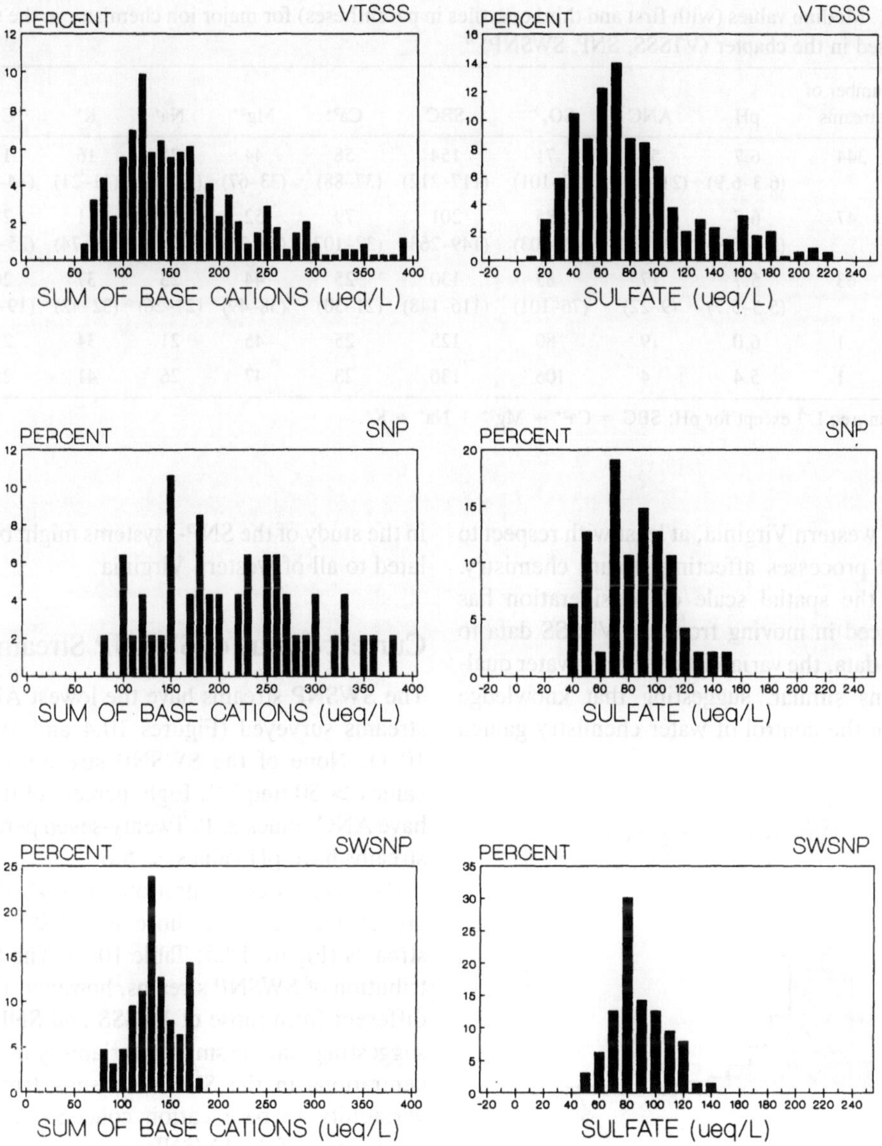

FIGURE 10.5. *(Continued.)*

μeq L^{-1}), all the SNP-I streams have ANC concentrations ≤ 200 μeq L^{-1}. Twenty-five percent of the streams have ANC concentrations ≤ 50 μeq L^{-1}. Six percent of the streams in the SNP-I have ANC ≤ 0. Thirteen percent of the streams have pH values below 5.6.

The distributions of base cations and SO_4^{2-} in the SNP-I data set are very similar to those of the VTSSS data set (Figure 10.5; Table 10.1). The distributions are somewhat "noisier" and the upper tails of the distributions are slightly truncated in the data for the SNP-I, but these differences result mostly from the smaller population of streams in the SNP-I survey; even though the survey is smaller, it still represents all major streams draining the Shenandoah National Park.

The similarity in water quality distributions between the VTSSS streams and the SNP-I streams suggests that the Shenandoah National Park might be considered a microcosm of the mountainous

TABLE 10.1. Median values (with first and third quartiles in parentheses) for major ion chemistry in the survey data sets discussed in the chapter (VTSSS, SNP, SWSNP).[a]

Data set	Number of streams	pH	ANC	SO_4^{2-}	SBC	Ca^{2+}	Mg^{2+}	Na^+	K^+	Cl^-	NO_3^-
VTSS	344	6.7 (6.3–6.9)	51 (21–88)	71 (52–101)	154 (117–212)	58 (37–88)	44 (33–67)	28 (18–48)	16 (11–21)	17 (14–22)	0 (0–0)
SNP	47	6.7 (6.0–6.9)	82 (21–120)	85 (66–103)	201 (149–263)	79 (32–107)	52 (43–76)	60 (28–74)	11 (8–74)	28 (25–32)	7 (3–23)
SWSNP	63	5.7 (5.5–5.7)	17 (9–22)	83 (76–101)	130 (116–148)	25 (21–30)	44 (38–49)	23 (21–26)	37 (32–42)	20 (19–22)	2 (0–5)
WOR	1	6.0	19	80	125	25	45	21	34	21	3
DR	1	5.4	4	106	136	23	47	26	41	25	1

[a] All values in μeq L^{-1} except for pH; SBC = $Ca^{2+} + Mg^{2+} + Na^+ + K^+$.

regions of western Virginia, at least with respect to watershed processes affecting stream chemistry. Although the spatial scale of consideration has been reduced in moving from the VTSSS data to the SNP-I data, the variability of streamwater quality remains similar, suggesting that knowledge concerning the control of water chemistry gained in the study of the SNP-I systems might be extrapolated to all of western Virginia.

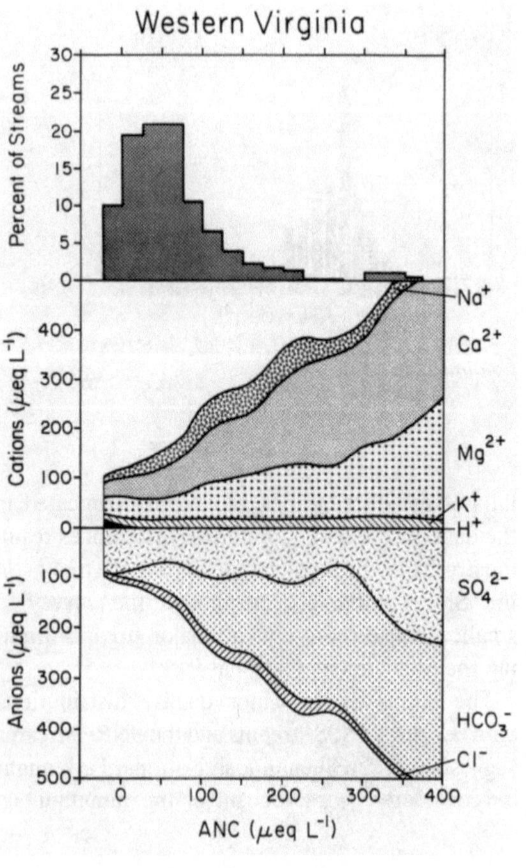

Current Status of SWSNP Streams

The SWSNP streams have the lowest ANC of the streams surveyed (Figures 10.4 and 10.5; Table 10.1). None of the SWSNP streams have ANC values > 50 μeq L^{-1}. Eight percent of the streams have ANC values ≤ 0. Twenty-seven percent of the streams have pH values < 5.6.

Base cation concentrations in SWSNP streams are also lower than those in VTSSS or SNP-I streams (Figure 10.5; Table 10.1). The SO_4^{2-} distribution of SWSNP streams, however, is not very different from those of VTSSS and SNP streams, suggesting that the smaller variability in ANC concentrations in the SWSNP results from smaller variability in base cation concentrations in the catchments of the SWSNP streams. The low ANC concentrations in the SWSNP make these streams ideal for process studies and trend monitoring. If water quality is being affected by acidic deposition, the effects should be most apparent in systems such as these. The long-term monitoring catchments (White Oak Run and Deep Run) are located in this portion of the park.

◄

FIGURE 10.6. Concentrations of major cations and anions for VTSSS streams with ANC ≤ 400 μeq L^{-1}. The percentages of streams represented by ANC intervals are shown as a histogram at the top of the figure. The curves are spline fits to the median concentration for each ANC interval.

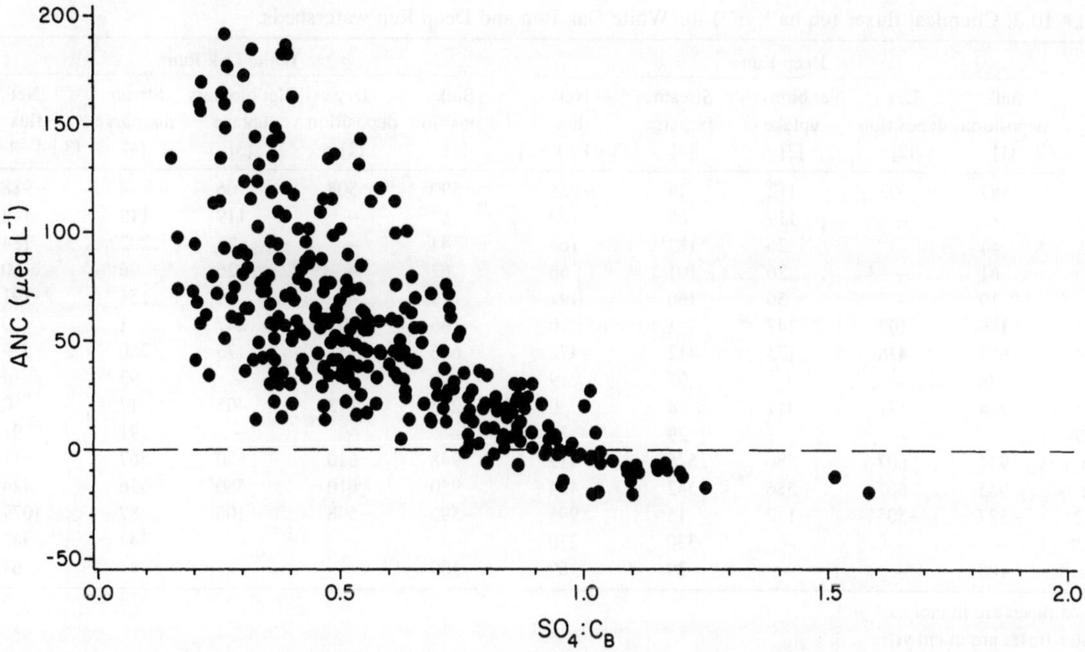

FIGURE 10.7. The relationship between current measured ANC values and the ratio of SO_4^{2-} to total base cation concentrations (both in μeq L^{-1}) for VTSSS streams. All streams have low DOC concentration (≤ 417 μmol L^{-1}).

Current Status of White Oak Run and Deep Run Streams

In White Oak Run and Deep Run streams, the volume-weighted average ANC concentrations from 1980 through 1985 were 19 and 4 μeq L^{-1}, respectively (Table 10.1). These concentrations are consistent with the median ANC concentrations of SWSNP streams (17 μeq L^{-1}) and are typical of the lower quartile ANC concentrations of SNP-I and VTSSS streams (Table 10.1).

The volume-weighted average base cation concentrations of White Oak Run and Deep Run are nearly identical to the median SWSNP base cation concentration (Table 10.1). Although the volume-weighted average SO_4^{2-} concentration of White Oak Run is nearly equal to the median SO_4^{2-} concentrations of all three survey data sets (VTSSS, SNP-I, and SWSNP), that of Deep Run is higher. The input-output budgets of White Oak Run and Deep Run have been monitored since 1979 (Table 10.2). These budgets indicate that H^+, SO_4^{2-}, NO_3^-, and NH_4^+ ions are retained in both watersheds. There are net exports of base cations and SiO_2 from both catchments. White Oak Run exports considerably more HCO_3^- than does Deep Run; Deep Run may be more affected by acidic deposition than White Oak Run.

Comparison of VTSSS to National Stream Survey (NSS) Streams

In this section, data for VTSSS streams are compared to data for a subset of the National Stream Survey (NSS; see Introduction) that encompasses only those reaches on mountain ridges of the Valley and Ridge province and the Blue Ridge Mountains in the states of Virginia, Maryland, and West Virginia (the Mid-Atlantic Ridge subset, NSS-MAR; Chapter 17, Baker et al. this volume). The VTSSS data are also compared to a further subset of NSS reaches found on mountain ridges in the western portion of Virginia (the Virginia Ridge subset, NSS-VR; Chapter 17, Baker et al. this volume). The ridge subsets are used for comparison because VTSSS streams, unlike NSS streams, are primarily headwater and low-order streams; none of the VTSSS streams are located in valleys.

The VTSSS streams showed considerably less regional variation in ANC than did NSS-MAR streams (Figure 10.8). The median ANC of VTSSS streams was less than half of that for either NSS-

TABLE 10.2. Chemical fluxes (eq ha^{-1} yr^{-1}) for White Oak Run and Deep Run watersheds.

	Deep Run					White Oak Run				
	Bulk deposition [1]	Dry deposition [2]	Net biomass uptake [3]	Stream discharge [4]	Net flux [3+4−1−2]	Bulk deposition [1]	Dry deposition [2]	Net biomass uptake [3]	Stream discharge [4]	Net flux [3+4−1−2]
H^+	587	505	112	14	−966	590	508	106	4	−988
Ca^{2+}	87	–	119	88	120	87	–	119	110	142
Mg^{2+}	44	–	26	182	164	44	–	26	202	184
Na^+	61	–	26	101	66	62	–	26	96	60
K^+	19	–	56	160	197	19	–	56	154	191
NH_4^+	146	102	247	1	0	146	102	247	1	0
SO_4^{2-}	623	436	175	412	−472	626	438	175	360	−529
Cl^-	78	–	–	97	19	79	–	–	93	14
NO_3^-	244	171	411	4	0	245	172	405	12	0
HCO_3^-	–	–	–	29	29	–	–	–	91	91
Sum +	944	607	586	546	−419	948	610	580	567	−411
Sum −	945	607	586	542	−424	950	610	580	556	−424
ANC	−587	−505	−112	15	995	−590	−508	−106	87	1079
SiO_2[a]	–	–	–	330	330	–	–	–	342	342
Water[b]	108	–	–	39	−69	109	–	–	45	−64

[a] Silica fluxes are in mol ha^{-1} yr^{-1}.
[b] Water fluxes are in cm yr^{-1}.

MAR or NSS-VR streams. However, the estimated percentage (12%) of acidic streams (ANC $\leq$ 0) in the NSS-VR data set agreed very well with the estimated percentage of VTSSS streams (10% with ANC $\leq$ 0).

The VTSSS streams showed approximately the same median pH, but a narrower distribution of pH, than did NSS-VR streams (Figure 10.8). The percentage of streams with pH < 5.5 is similar in the two data sets, but the NSS-VR population includes more streams with pH > 6.5.

Streams in the VTSSS displayed less regional variation in the sum of base cation concentrations (SBC) than did streams in the NSS-MAR. The median SBC of VTSSS streams was approximately 60% of that estimated for NSS-MAR streams (Figure 10.8). The range of SBC for VTSSS streams showed closer agreement with NSS-VR streams, but the median value of SBC was still considerably lower (VTSSS = 154 μeq L^{-1}; NSS-VR = 255 μeq L^{-1}; Figure 10.8). These contrasts are not surprising in light of the more restricted population of native brook trout streams (mostly headwater reaches) targeted by the VTSSS.

The VTSSS streams also showed considerably less regional variation in SO_4^{2-} concentrations than NSS-MAR streams (Figure 10.8). The median VTSSS SO_4^{2-} concentration was less than half that for NSS-MAR streams. These differences can be expected, given the larger range of atmospheric deposition rates and watershed characteristics represented by NSS-MAR streams. The median VTSSS SO_4^{2-} concentration, however, agreed very well with that of NSS-VR streams (VTSSS = 71 μeq L^{-1}; NSS-VR = 73 μeq L^{-1}; Figure 10.8).

Processes Influencing Surface Water Chemistry

Catchment Input-Output Budgets

The input-output budgets of White Oak Run and Deep Run have been monitored since 1979. An examination of these budgets gives an indication of the magnitude of catchment processes affecting water quality in the streams included in this case study. Results are presented for the six complete years from 1980 through 1985 (Table 10.2).

Bulk deposition can be characterized as an acidic solution of sulfuric and nitric acids containing small amounts of neutral salts (Table 10.2). The bulk deposition measurements include contributions from precipitation and gravitational settling of particles. For sulfur and nitrogen, the deposition of gases and aerosols must also be considered. Based

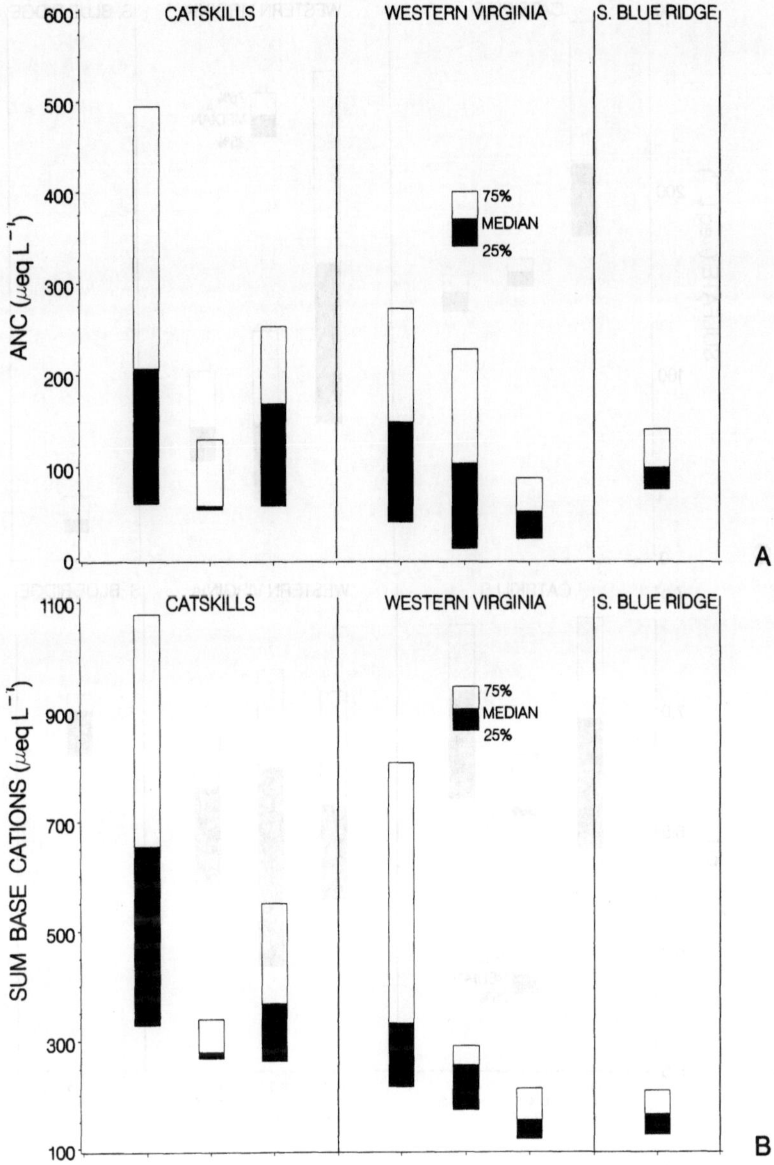

FIGURE 10.8. Comparison of distributions of pH, ANC, base cation concentrations, and SO_4^{2-} concentrations for VTSSS streams and selected subsets of NSS streams. NSS subsets for Western Virginia are described in the text. NSS data for the Catskills and S. Blue Ridge are included for comparison. (*Continued on next page.*)

on measurements of sulfur gases and aerosols and a calculated deposition velocity, dry deposition of sulfur was estimated to be equal to 70% of the bulk deposition of SO_4^{2-} at White Oak Run and Deep Run (Shaffer and Galloway 1982). A similar value is assumed for dry deposition of nitrogen species. For the purpose of estimating the dry deposition contribution to the H^+ budget, it was assumed that dry deposited sulfur is rapidly converted to sulfuric acid. Thus, in order to maintain charge balance, dry deposition of H^+ was set equal to the dry deposition of SO_4^{2-} plus dry deposition of NO_3^- minus dry deposition of NH_4^+. The ANC of bulk and dry deposition was assumed to be equal to the negative of the H^+ deposition (Table 10.2).

A potentially important internal sink for chemical species in these catchments is accumulation in forest biomass. Based on estimates of biomass

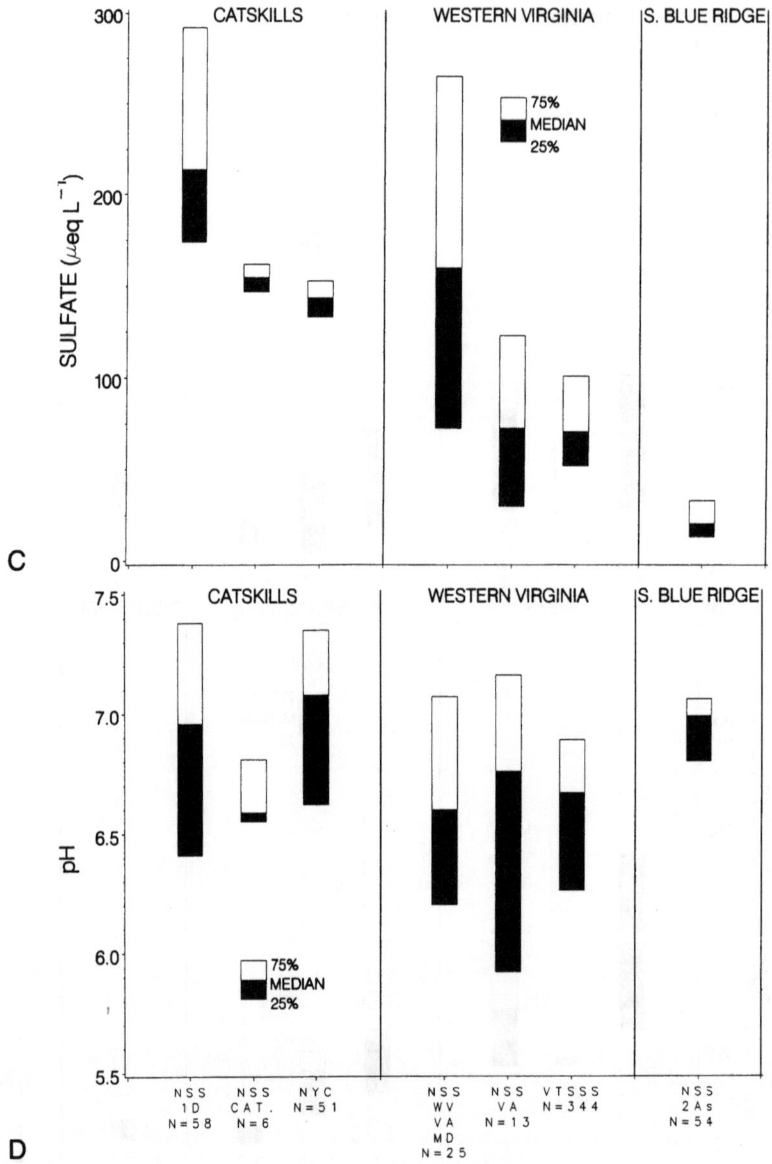

FIGURE 10.8. (*Continued.*)

accumulation for a similar catchment slightly to the north of White Oak Run and Deep Run (Harrison, 1987), the net accumulation of biomass in these catchments is estimated to be approximately 2,000 kg ha^{-1} yr^{-1}. Using this biomass accumulation rate with estimates of the elemental composition of similar forests at the Coweeta Experimental Forest (Weinstein 1982), estimates of net uptake in biomass for the base cations in the catchments were derived (Table 10.2). It was assumed that essentially all retention of nitrogen species in the catchments is in biomass, thus biomass uptake of nitrogen species was set equal to the output flux minus the input flux in Table 10.2). No measure of SO_4^{2-} uptake by plants is available for these sites. Sulfur uptake in biomass at Hubbard Brook is estimated to be approximately 175 eq ha^{-1} yr^{-1} (Likens et al. 1977). This value was assumed for the White Oak Run and Deep Run sites (Table 10.2).

The discharge budgets of the catchments (Table 10.2) show that streams export primarily a mixture of SO_4^{2-} salts and the products of weathering of

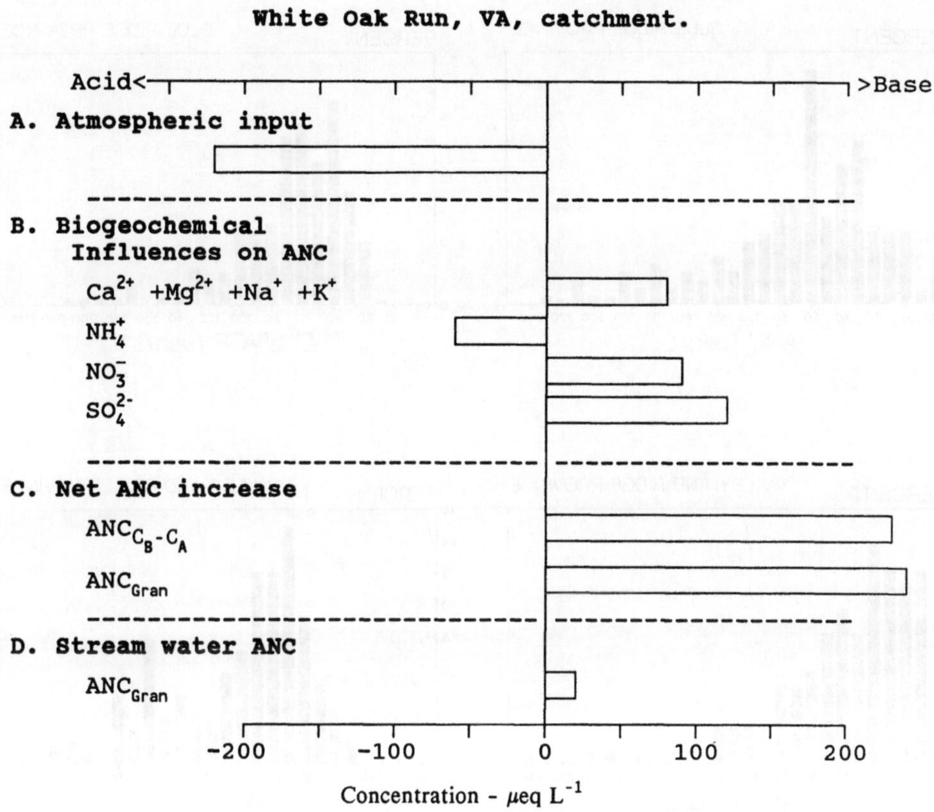

FIGURE 10.9. Modification of atmospheric deposition by watershed processes in the White Oak Run catchment. A. Precipitation [H^+] adjusted for estimated dry deposition and evapotranspiration (ET); equivalent to expected $ANC_{C_B\text{-}C_A}$ in the absence of any process other than ET. B. Reactions influencing ANC; bars to the right indicate ANC production, those to the left show ANC consumption. C. Net ANC production; $ANC_{C_B\text{-}C_A}$ is the sum of ANC production terms for measured anions and cations, including those shown in C; ANC_{Gran} is determined by difference between input ANC (A.) and streamwater ANC (D.). D. Streamwater ANC represents the net reaction of atmospheric inputs with the stream/watershed system. The diagram was constructed using the data in Table 10.2. For a further explanation of calculation procedures and figure interpretation, see Appendix A.

silicate minerals. Bicarbonate fluxes in the streams were calculated as the sum of the ANC and H^+ fluxes (ANC = [HCO_3^-] − [H^+]). The net fluxes for the catchments (Table 10.2) demonstrate that the catchments neutralize essentially all of the incoming atmospheric acidity. The catchments (1) are in approximate steady state with respect to Cl^-, (2) are net sources of base cations, SiO_2, and ANC, and (3) are net sinks for nitrogen species, SO_4^{2-}, and H^+.

The input-output budget for White Oak Run can be used to construct an ion enrichment diagram (Chapter 1, Munson and Gherini, this volume) that summarizes the ANC contributions of the processes operating within the catchment (Figure 10.9). The effects of NO_3^- and NH_4^+ uptake approximately cancel each other, leaving an important observation concerning the processes regulating ANC within these catchments: ANC production from SO_4^{2-} retention is as important as ANC production from those processes generating base cations (i.e., weathering and cation exchange). The total retention of SO_4^{2-} in the catchment is approximately three times greater than can be explained by biological uptake (Table 10.2). The excess retention results from adsorption of SO_4^{2-} by catchment soils (Shaffer and Galloway, 1982). The catchment with higher ANC (White Oak Run) is retaining approximately 50% of deposited SO_4^{2-} by soil adsorption. The catchment with lower ANC (Deep Run) is retaining approximately 45% of deposited SO_4^{2-} by

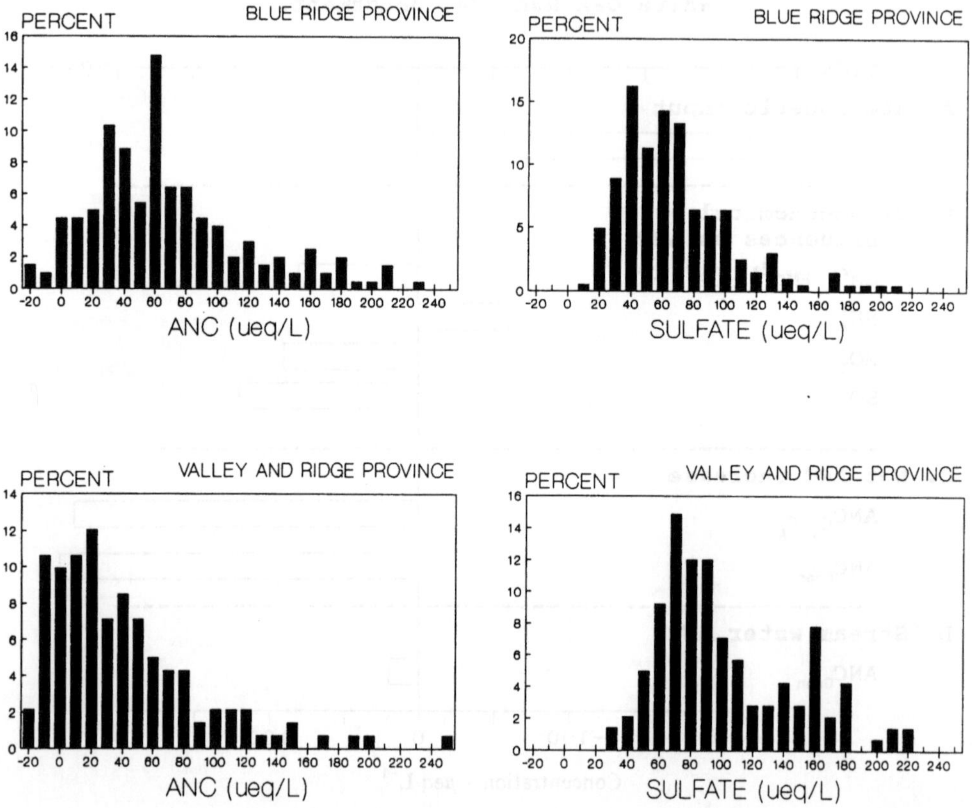

FIGURE 10.10. Distributions of ANC and SO_4^{2-} concentrations for VTSSS streams in the Blue Ridge physiographic province (n = 203) and in the Valley and Ridge physiographic province (n = 141). Upper tails of distributions have been truncated.

soil adsorption. If SO_4^{2-} adsorption were not occurring, the streams would not have positive ANC values (Figure 10.9).

These input-output analyses of two typical catchments in the case study region suggest that an understanding of the distributions of water quality in upland streams of western Virginia might be obtained by examining the survey data with particular attention to those factors controlling base cation availability and SO_4^{2-} adsorption. These factors are most likely to be related to the geologic/edaphic characteristics of the case study region.

ANC and Base Cation Availability

Broad distributions are observed in the streamwater concentrations of ANC and base cations for VTSSS streams (Figure 10.5). There is a strong relationship between ANC and base cation concentrations in these streams (Figure 10.6), suggesting that processes affecting base cation availability exert a strong control on ANC in these streams. The large variations in base cation availability and ANC can be partially understood by examining the differences between two subregions defined by the distinct physiographic provinces in the VTSSS study region (the Blue Ridge and the Valley and Ridge provinces; Figure 10.2).

The ANC values in the Valley and Ridge Province are generally lower than those in the Blue Ridge Province (Figure 10.10). The median ANC concentration for Blue Ridge streams is 60 μeq L^{-1}, compared with 31 μeq L^{-1} for Valley and Ridge streams. This difference is consistent with observations of a strong bedrock-ANC relationship in the northern area of the Blue Ridge Province in Virginia (Hendrey et al. 1980) and with the analysis of Lynch and Dise (1985) for the SNP streams.

In the VTSSS study area, the highest streamwater ANC concentrations are associated with crystalline bedrock types. These bedrock types dominate most of the Blue Ridge Province (Figure 10.2). In contrast, the lowest streamwater ANC concentrations are associated with silico-clastic bedrock types. In the Blue Ridge Province, these rock types are restricted to a band along the western edge, but they occur on most of the ridges in the Valley and Ridge Province (Figure 10.2).

Lynch and Dise (1985) reported for SNP-I streams that streamwater ANC and base cation concentrations are strongly affected by bedrock geology. Streams draining resistant siliceous bedrock have ANC concentrations generally < 20 μeq L^{-1}; streams draining granite and granodiorite generally have ANC concentrations from 20 to 100 μeq L^{-1}; streams draining metamorphosed volcanics generally have ANC concentrations from 100 to 200 μeq L^{-1}. A multiple regression analysis demonstrated that concentrations of ANC and base cations were most strongly related to bedrock type (r^2 values above 0.95 for bedrock type alone; Lynch and Dise 1985).

The SWSNP streams are located on relatively uniform bedrock geology; only two major geological formations are present in the SWSNP catchments. Webb (1988) examined the relationships between ANC and the proportion of the dominant bedrock type in each catchment of the SWSNP streams and found that ANC was correlated with bedrock type ($r^2 = 0.22$). This relationship is weaker than that found by Lynch and Dise (1985) for all SNP-I streams, but it demonstrates that the relationships between ANC and base cation availability and geological factors persist even at small spatial scales.

ANC and Sulfate Retention

Based on wet SO_4^{2-} deposition estimates of 500 eq ha^{-1} yr^{-1} for the case study region, an assumed dry deposition equal to 50% of wet deposition, and an estimated average runoff of 36 cm yr^{-1} for the streams (see section on regional characteristics), the expected average SO_4^{2-} concentration for VTSSS streams is approximately 200 μeq L^{-1}. The median observed SO_4^{2-} concentration for VTSSS streams (71 μeq L^{-1}, Table 10.1) is considerably lower than this value. On average, VTSSS streams appear to be retaining 65% of the total deposited SO_4^{2-}. This retention is consistent with the observed total retention at the intensively monitored sites in this study (White Oak Run and Deep Run) and with values reported for other catchments in the region (Rochelle et al. 1987).

The ANC concentrations in VTSSS streams are relatively independent of SO_4^{2-} concentrations (Figure 10.6), suggesting that the effects of acidic deposition are relatively uniform over the western Virginia mountains. Sulfate concentrations are somewhat higher in the Valley and Ridge Province than in the Blue Ridge Province, although the differences in SO_4^{2-} concentrations between the provinces are not as pronounced as they are for ANC (Figure 10.10). The median SO_4^{2-} concentration for streams in the Blue Ridge Province is 61 μeq L^{-1}, compared with 92 μeq L^{-1} for streams in the Valley and Ridge Province. Geologic/edaphic factors do not appear to exert as strong a control on the variability of SO_4^{2-} concentrations as they do on the variability of ANC and base cation concentrations. Several possible explanations can be advanced, however, for the generally higher Valley and Ridge SO_4^{2-} concentrations.

There may be a difference between the two physiographic provinces in the magnitude of SO_4^{2-} retention. The soils derived from the crystalline rocks of the Blue Ridge Province are predominantly Ultisols; the soils derived from the sedimentary rocks forming the ridges of the Valley and Ridge Province are predominantly Inceptisols. Ultisols have a greater concentration of clays and other secondary minerals associated with SO_4^{2-} adsorption. Greater SO_4^{2-} retention, and thus lower streamwater SO_4^{2-} concentrations, should thus be associated with the Blue Ridge Province relative to the Valley and Ridge Province.

There may be a difference between the two provinces in the magnitude of sulfur deposition. A west-to-east gradient in sulfur deposition might reasonably be expected given both the location of major source areas to the west of the study region and the dominance of airflow from the same direction (Michaels et al. 1988). The Valley and Ridge Province lies to the west of the Blue Ridge and thus closer to deposition sources (Chapter 3, Husar et al. this volume).

There may be a difference between the two provinces in the occurrence of mineral sources of sulfur. Because depositional environments associated with sedimentary rocks are conducive to sulfur mineralization, the effects of sulfur minerals would generally be expected to be greater in the Valley and Ridge Province than in the Blue Ridge Province. Pyritic bedrock has been associated with elevated SO_4^{2-} concentrations in mountain headwater streams to the south of the study area (Winger et al. 1987). To our knowledge, however, similar conditions have not been observed in the case study region.

Lynch and Dise (1985) in their analysis of the SNP-I streams found only a weak relationship between SO_4^{2-} concentration and bedrock type ($r^2 = 0.12$). The best correlation between SO_4^{2-} concentration and any catchment characteristic for SNP-I streams was for geographic location (west slope versus east slope of the Blue Ridge). Webb (1988) examined the relationships between SO_4^{2-} concentrations and the proportion of the regionally dominant bedrock and soil types in each catchment of the SWSNP streams and found that SO_4^{2-} concentration was only weakly correlated with either factor (bedrock type, $r^2 = 0.11$; soil type, $r^2 = 0.22$). Compared to the regional average expected SO_4^{2-} concentration, SNP-I and SWSNP streams are retaining approximately 60% of deposited SO_4^{2-} (Table 10.1).

Evidence of Historical Trends

Levels of atmospheric deposition of acidic compounds are as high in western Virginia as they are in the northeastern United States and southeastern Canada. Yet streams in western Virginia appear to have been protected from increases in acidity because of SO_4^{2-} adsorption by soils. Because the capacity of soils to adsorb SO_4^{2-} is finite, the question of when the streams in this region might show a response to the high levels of atmospheric sulfur deposition is of some importance. Modeling studies (e.g., Cosby et al. 1985, 1986) have suggested that time constants for SO_4^{2-} breakthrough may be on the order of several decades for soils in this region. Given that elevated deposition has been occurring for several decades at White Oak Run and Deep Run, it might be expected that trends in the chemical composition of these streams would be detectable from the monitoring data for these sites.

Ryan et al. (1989) have analyzed the White Oak Run and Deep Run data sets and found increasing concentrations of SO_4^{2-}, H^+, and base cations, and decreasing ANC concentrations in both White Oak Run and Deep Run. Their analyses were based on an examination of both the weekly samples and the volume-weighted annual averages for the period 1980 through 1987 (8 years). Sulfate concentrations were shown to be increasing at a rate of 2 $\mu eq\ L^{-1}\ yr^{-1}$ in both streams. The rates of change of SO_4^{2-} concentration observed at White Oak Run and Deep Run are consistent with values reported for other catchments (Smith and Alexander 1983, Swank and Waide 1988). Ryan et al. (1989) estimated that at current rates of change, the streams would reach steady state with atmospheric deposition in approximately 100 years, at which time the ANC concentrations of the streams could be as low as $-65\ \mu eq\ L^{-1}$.

Future Considerations

Webb et al. (1989) have analyzed the VTSSS data to project streamwater chemistry in the event the streams ultimately reach steady state with respect to atmospheric SO_4^{2-} deposition. Their estimates of potential acidification were made with a simple linear model relating changes in base cation concentrations to changes in concentrations of SO_4^{2-} as the streams approach SO_4^{2-} steady state. Changes in concentrations of base cations were calculated assuming base cation increase factors equal to 0.4 and 0.8 times the expected SO_4^{2-} increase. The median future ANC loss was estimated as 90 and 30 $\mu eq\ L^{-1}$, respectively, for the two assumed factors. The projections indicate that 88% and 32%, respectively, of the streams will have ANC values ≤ 0 when stream SO_4^{2-} concentrations attain steady state. Even with the conservative estimate of ANC loss, the potential for acidification of these stream resources given current SO_4^{2-} deposition is large.

Biological Status

The high-elevation headwater streams in the study region that are most susceptible to acidic deposition are dominated by the eastern brook trout (*Salvelinus fontinalis*), which is the only salmonid native to the region. Despite stocking and establishment of rainbow trout (*Salmo gairdneri*) and brown trout (*Salmo trutta*) in some streams, these species do not appear to have significantly displaced brook trout (Lennon 1961, Sheridan 1971). The uppermost stream reaches may contain only brook trout or no fish at all. Further downstream, mottled sculpins (*Cottus bairdi*) and blacknose dace (*Rhinichthys cataractae*) are also found in association with the trout (Neves and Pardue 1983).

Lennon (1961) listed 21 species of fishes in the SNP-I, of which 11 were considered rare. Owing largely to control of soil erosion and angling restrictions, brook trout populations in the SNP-I increased from scarcity in the 1930s to abundance in the 1950s. Because brook trout populations appear to be unusually dependent upon weather cycles (the short, steep streams are susceptible to drying out during drought and extreme scouring during floods), Lennon (1961) considered SNP-I streams to be marginal for trout. These same characteristics apply to most mountain streams in western Virginia.

A survey of 11 SNP-I streams (Sheridan 1971) revealed as many as 10 nonsalmonid fish species in association with brook trout. Two samples taken above 600 m in elevation collected only brook trout and blacknose dace. Brook trout constituted 100% of the fish biomass at the highest site sampled (850 m) and 17% of the fish biomass at the lowest site (325 m), although there was considerable variation among sites in both the absolute and relative biomass of trout (Sheridan 1971). Smallmouth bass (*Micropterus dolomieui*) were also collected in this survey, but neither rainbow nor brown trout were found to move into SNP-I waters following downstream stocking. Sheridan (1971) attributed much of the variation in brook trout numbers among sampling sites to angler harvest.

Fish species tolerance depends upon a number of environmental factors, but critical pH values range from about 6.0 to 4.7 for a number of fish species commonly present in native brook trout waters (Cooper and Wagner 1973, Haines and Baker 1986, Schofield and Driscoll 1987; Chapter 4, Baker and Christensen, this volume). At present, the median pH for VTSSS streams (6.7) is still well above the range of critical values. However, the observed decreases in ANC and pH reported for the monitored streams in the study region have ecological implications for all streams in the surveys.

Relatively small changes in H^+ can be very important in low ionic strength streams. At presently estimated rates of change (Ryan et al. 1989), an increase in H^+ concentration of 4 μeq L^{-1} would be expected in Deep Run over the next decade. This translates to a change in volume-weighted mean pH from 5.3 to 5.0. Several fish species indigenous to the Shenandoah National Park, such as the common shiner (*Notropus cornutus*) and the fathead minnow (*Pimephales promelas*) lose their ability to reproduce and/or survive in the pH range 5.0 to 6.0 (Chapter 4, Baker and Christensen, this volume). If SO_4^{2-} concentrations continue to increase in the future, the projected decline in pH would greatly enhance the potential for damage to fish populations that are thought to be more acid tolerant, such as small-mouth bass, rock bass (*Ambloplites rupestris*), and white sucker (*Catostomus commersoni*) (Chapter 4, Baker and Christensen, this volume).

There has been at least one report of declining fish populations in the study region. Mohn et al. (1988) surveyed the benthic fauna and fish populations of the St. Mary's River (a stream in the George Washington National Forest near the SNP-I) and compared their results to results of similar surveys conducted in the 1930s (see Mohn et al. 1988 for details of the earlier survey). The comparison revealed declines in most benthic invertebrate genera and acid sensitive fish species, which Mohn et al. (1988) suggested were the results of acidification.

Conclusions

Stream surveys conducted in the Appalachian Mountain region of western Virginia describe a resource at significant risk from the effects of acidic deposition. The potential for deterioration of water quality, and concomitant damage to biological communities, appears to be large. In particular, for the VTSSS survey (the largest sample size covering the greatest geographical area):

1. Relative to commonly applied criteria of stream sensitivity to acidic deposition, 93% of the streams are sensitive (ANC $\leq$ 200 μeq L^{-1}), 49% of the streams are extremely sensitive (ANC $\leq$ 50 μeq L^{-1}), and 10% of the streams are currently acidic (ANC $\leq$ 0).

2. Bedrock geology appears to exert the strongest control on ANC variation of streams within the region.
3. Sulfate is the major anion in the streams; however, all catchments are retaining a significant proportion of atmospherically deposited sulfur (median SO_4^{2-} retention is approximately 60%).
4. The relatively high rate of acidic deposition in the region (coupled with the low buffering capacity of the headwater streams) creates potential problems for the indigenous fish populations in the mountain streams of western Virginia.

Acknowledgments. The research described here was sponsored in part by the National Park Service, the Virginia Department of Game and Inland Fisheries, the National Forest Service and the U.S. Geological Survey. We also acknowledge the generous support and enthusiastic cooperation of the Virginia Council of Trout Unlimited.

11
The Southern Blue Ridge Province

Jerry W. Elwood, Michael J. Sale, Philip R. Kaufmann, and Glenn F. Cada

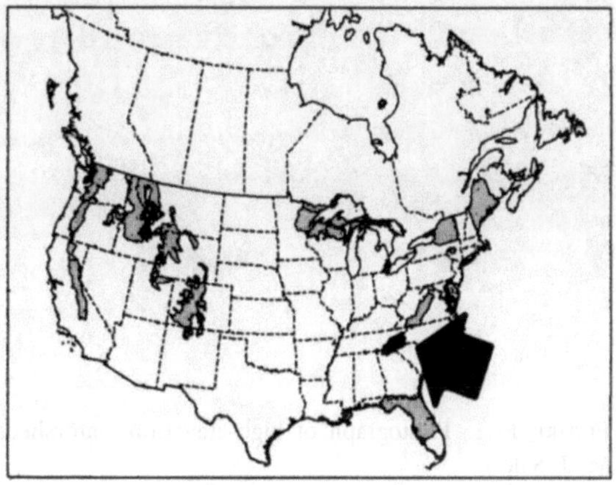

ABSTRACT. The Southern Blue Ridge Province (SBRP) is a steep mountainous region of the southeastern United States characterized by high rainfall, highly weathered base-poor soils, and relatively unreactive bedrock. Streams, drainage lakes, and human-made reservoirs are among the most dilute of any region sampled in the eastern United States (median conductivity < 40 μS cm^{-1} for both streams and lakes). Atmospheric deposition of SO_4^{2-}, NO_3^-, and H^+ in the Southern Blue Ridge Province is among the highest reported in the eastern United States. Although acidic streams are known to occur in the region, no acidic surface waters were found in probability-based regional surveys of 54 low-order streams and 67 lakes and reservoirs.

Watershed retention of SO_4^{2-} plus NO_3^- is the major process generating ANC in drainage waters in the Southern Blue Ridge Province, exceeding base cation mobilization in importance. Among-stream differences in ANC are associated primarily with differences in base cation content. In contrast, SO_4^{2-} and NO_3^- retention is more uniform spatially than base cation production and does not account for much of the regional variation in ANC. Dissolved organic carbon concentrations typically are low in Southern Blue Ridge streams (median DOC = 59 μmol L^{-1}), and do not explain a significant fraction of the ANC variation among streams during spring baseflow.

Sulfate concentrations have increased at an annual rate of approximately 1 μeq L^{-1} over the last 10 years in selected streams draining both high- and low-elevation watersheds in the Southern Blue Ridge Province. In contrast, both ANC and base cation (SBC) concentrations declined in the streams over the same period, indicating that base cation mobilization is not keeping pace with increasing SO_4^{2-} levels in drainage waters in some watersheds in this region. ANC is inversely correlated with SO_4^{2-} only in streams with SBC ≤ 100 μeq L^{-1} (r^2 = 0.27; $p < 0.01$), but is not significantly correlated with SO_4^{2-} in streams with SBC > 100 μeq L^{-1} ($p > 0.20$). Reducing the atmospheric loading of SO_4^{2-} in the Southern Blue Ridge Province thus may have a short-term positive effect on ANC only in those streams draining watersheds with a low capacity for producing base cations, as reflected by their present base cation content (i.e., ≤ 100 μeq L^{-1}). Because of the low capacity of many watersheds in the region to mobilize base cations, the ANC and pH of many surface waters in the Southern Blue Ridge may decline in the future, as watershed retention of acid anions declines.

No adverse biological effects of acidic deposition on streams and lakes have been demonstrated conclusively to date for the Southern Blue Ridge Province, although fish kills have occurred during storms in a trout-holding facility that uses water from a poorly buffered stream experiencing episodic depressions in pH. Biotic and chemical differences between acidic and nonacidic streams in the Southern Blue Ridge Province are similar to those reported for other areas of eastern North America and Europe, indicating that reductions in species richness and diversity of benthic invertebrates,

FIGURE 11.1. Photograph of high-elevation watersheds in the Southern Blue Ridge Province. (Photograph by M. J. Sale.)

declines in decomposition of benthic organic matter, and losses of acid-sensitive fish species such as rainbow trout will occur if streams in the Southern Blue Ridge Province are acidified by atmospheric deposition. The documented decline in the distribution of brook trout into upstream areas of watersheds most likely is due to competition from introduced species (rainbow and brown trout), rather than stream acidification.

Introduction

Rationale

The Southern Blue Ridge Province (SBRP) is a mountainous region (Figure 11.1) in the southeastern United States characterized by high, relatively acidic rainfall, highly weathered base-poor soils, relatively unreactive bedrock, and poorly buffered streams. On the basis of these characteristics, surface waters in this region have been considered potentially vulnerable to acidification from acidic deposition (Hendrey et al. 1980, Olson et al. 1982, Omernik and Powers 1983, Omernik 1985).

Low mobility of acid anions, particularly SO_4^{2-}, in soils in the southeastern United States, however, may be an important factor in reducing the direct effects of acidic deposition on the ANC and pH of surface waters in the region (Cosby et al. 1985a, Reuss and Johnson 1986). It is hypothesized that surface waters in the Southern Blue Ridge, and in other unglaciated regions of the southeast with thick residual soils rich in oxyhydroxides of iron and aluminum, will exhibit less than a direct proportional effect to changes in the deposition of sulfuric acid because of SO_4^{2-} adsorption onto soil particles. As a consequence, much of the SO_4^{2-} is removed from drainage waters, reducing the transport of acidic cations to surface waters (Reuss and Johnson 1985, Cosby et al. 1985).

Evidence from mass balances indicates that watersheds in the Southern Blue Ridge are retaining a significant fraction (70% to 80%) of atmospherically deposited SO_4^{2-} (Rochelle et al. 1987, Rochelle and Church 1987). However, SO_4^{2-} adsorption by soils is a reversible, concentration-dependent process (Johnson and Cole 1980, Reuss and Johnson 1986). Therefore, SO_4^{2-} mobility in watersheds can change with a constant or changing deposition load as the capacity of soils for adsorption of SO_4^{2-} is saturated. Preliminary analysis of the chemistry of selected streams (Burns et al. 1981, Eshleman and Kaufmann 1988) and reservoirs (Talbot and Elzerman 1985) in the Southern Blue Ridge suggests that SO_4^{2-} mobility in watersheds in this region is changing, leading to changes in surface water chemistry.

Nitrate levels in streams, which appear to be mediated primarily by biotic factors in the soil and terrestrial vegetation communities (Vitousek 1977, Martin 1979), may also change as changes occur in the biotic demand for nitrogen by second-growth forests that dominate the region. Water quality surveys in selected areas of the province show significantly higher NO_3^- levels in streams draining watersheds with old-growth forests compared to those with second-growth forests, which are still accumulating biomass (Silsbee and Larson, 1982, 1983, Elwood et al. 1985). As forest succession occurs, NO_3^- levels in streams are expected to increase.

In this chapter, we examine the acid-base chemistry of streams, lakes, and reservoirs in the Southern Blue Ridge Province and the biogeochemical processes that affect their chemistry. The focus is primarily on the spatial and temporal variation in the chemistry among dilute streams and the watershed processes that determine the effects of acidic deposition on stream water. We also include a limited discussion of the acid-base chemistry of natural lakes and human-made reservoirs in this region. Finally, we discuss the biological status of acid-sensitive streams in the Southern Blue Ridge and whether existing data on stream chemistry for this region indicate any short- and/or long-term acidification by acidic deposition.

Description of Region and Subregion Boundaries

The Southern Blue Ridge Province is in the southern Appalachian Mountains in the eastern portion of the Tennessee Valley Drainage (see Figure 5.1, Chapter 5, Eilers and Selle, this volume). It includes the mountainous portions of southern Virginia south of the Roanoke River, western North Carolina, eastern Tennessee, northern Georgia, and a small area in western South Carolina (Figure 11.2).

The Unaka Mountains, Iron Mountains, and Great Smoky Mountains are ranges within the Southern Blue Ridge that straddle the North Carolina-Tennessee border on the western front; the Blue Ridge Mountains define the basin divide on the eastern front. Three major stream valleys, the Hiwassee, Little Tennessee, and French Broad, provide most of the relief in the area, flowing northwest from the Blue Ridge Mountains to the Tennessee Valley. The eastern and southern boundaries of the province are generally defined by the divide separating the Tennessee Valley drainages flowing generally to the west from the Atlantic coast drainages flowing east. Definition of the eastern boundary of the region is somewhat subjective, however, in terms of where the Piedmont ends and the Southern Blue Ridge begins. It can include the higher elevations of the Atlantic Coast drainages in the Blue Ridge Mountains in southern Virginia, North Carolina, South Carolina, and Georgia. The western boundary of the Southern Blue Ridge is determined largely by the limit of overthrust of strongly metamorphosed rocks of the province on the unaltered limestone in the Ridge and Valley Province.

The northern boundary of the Southern Blue Ridge Province is considered to be the Roanoke River in Virginia (Fenneman 1938). The distinction between the northern and southern sections of the province is that the southern section is higher and broader and has greater relief and more transverse divides than the northern section.

Aquatic Resources at Risk

Aquatic resources in the Southern Blue Ridge Province potentially most at risk from acidic deposition consist primarily of high gradient streams and some small upland drainage lakes and reservoirs. The number and length of stream reaches between the upstream and downstream ends in a target population defined for Phase I of the National Stream Survey (NSS-I) have been estimated for a 27,000 km² portion of the Southern Blue Ridge in western North Carolina, eastern Tennessee, northern Georgia, and western South Carolina (Messer et al. 1986, Kaufmann et al.

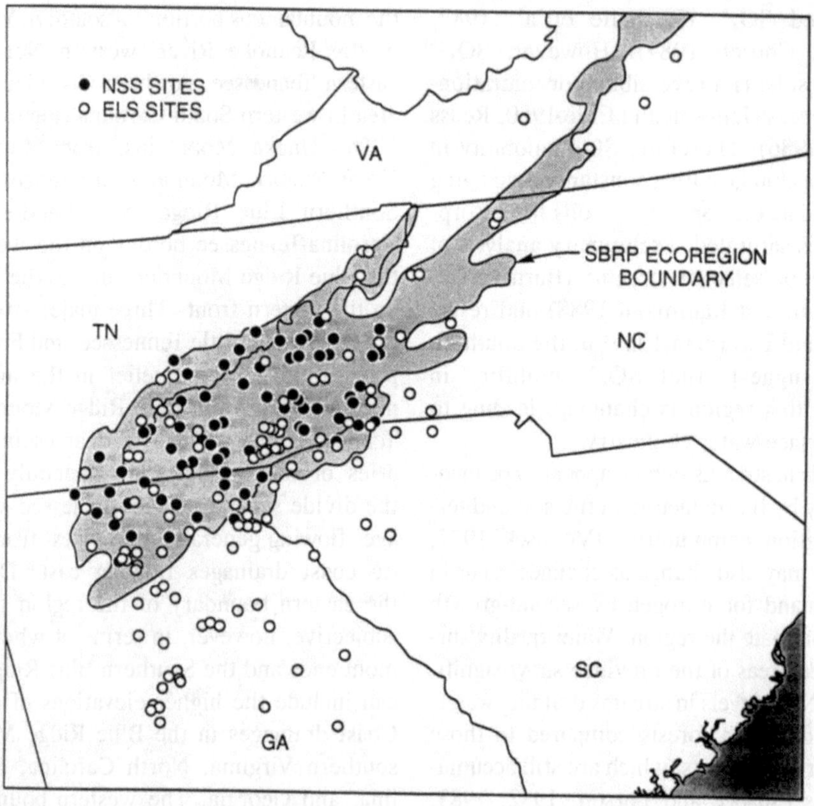

FIGURE 11.2. Map of the ecoregion boundary of the Southern Blue Ridge Province. Also shown are the locations of NSS streams sampled by the EPA during the Pilot Stream Survey, and the locations of ELS lakes and reservoirs sampled in Region 3A of the EPA Eastern Lake Survey. The lakes and reservoirs outside the Southern Blue Ridge ecoregion boundary are in ELS Region 3A.

1988). The NSS target population in this area is estimated to be 2,031 stream reaches, with an interpolated length of 9,036 km (distance between the upstream and downstream ends of reaches; Table 11.1).

These estimates are not of the total population of acid-sensitive streams in the Southern Blue Ridge, because (1) portions of the province were not sampled in the NSS, and (2) the target population does not include low-order reaches that do not occur as blue-line streams on 1:250,000-scale maps. The NSS target population is nonetheless thought to represent a reasonable estimate of the more acid-sensitive streams in the southern portion of the province that are likely to contain fishery resources (e.g., trout populations) of concern.

We are not aware of any estimates of the total stream resource within the Southern Blue Ridge Province. However, the Great Smoky Mountains National Park, which covers an area of 2,300 km² (equal to approximately 8% of the area containing the NSS target population in the Southern Blue Ridge), contains 22 major watersheds with more than 300 separate streams and 1,173 km of stream channel capable of supporting trout and/or smallmouth bass. The Smoky Mountains also have some of the highest quality cold water habitat remaining in the southeastern United States. A hypothetical calculation of the total stream resource in the portion of the Southern Blue Ridge containing the NSS target population, based on the assumption that the number of streams per unit area in the Great Smoky Mountains National Park is typical of the entire Southern Blue Ridge Province, shows that it exceeds 3,500 separate streams and approximately 14,000 km of stream channel. Based on these calculations, the NSS target population of stream reaches in this province equals 58% of the total number of stream reaches

TABLE 11.1. Percentage of the estimated target population of stream reaches in the Southern Blue Ridge Province with the specified index pH and ANC, and the estimated length of stream reaches in the specified reference ranges.[a]

Number of stream reaches[b]

ANC (μeq L⁻¹)									
≤0		>0 to ≤50		>50 to ≤200		>200		Total	
D	U	D	U	D	U	D	U	D	U
<1	<1	5	6	74	78	22	26	2031	2031
pH									
≤5.0		>5.0 to ≤5.5		>5.5 to ≤6.0		>6.0		Total	
D	U	D	U	D	U	D	U	D	U
<1	<1	<1	<1	<1	<1	>99	>99	2031	2031

Interpolated length of stream reaches (km)

ANC (μeq L⁻¹)				
≤0	>0 to ≤50	>50 to ≤200	>200	Total
<1	706 (7.8)	6,378 (70.6)	1,952 (21.6)	9,036
pH				
≤5.0	>5.0 to ≤5.5	>5.5 to ≤6.0	>6.0	Total
<1	<1	<1	9,036 (>99)	9,036

Data from Kaufmann et al. 1988.
[a] Also shown are the number of stream reaches in the target population and the interpolated length (in km) of stream reaches in the target population. Spring was the index sampling period for streams. Values in parentheses next to the interpolated lengths was the percentage of the total interpolated population.
[b] D = downstream end; U = upstream end.

and 65% of the total channel length in the NSS target area of the Southern Blue Ridge Province.

Lakes and reservoirs in a 28,800 km² area of the southern Appalachian Mountains (ELS Region 3A), including the Southern Blue Ridge Province, were also sampled during the Phase I of the Eastern Lake Survey (ELS-I; Linthurst et al. 1986a, Eilers et al. 1988a). Of the 94 lakes and reservoirs sampled in ELS-I Region 3A, 54 were inside the Southern Blue Ridge Province ecoregion boundaries (Figure 11.2). Based on the weighting factor for the 54 lakes and reservoirs sampled, the lake population in that portion of the province sampled in the ESL-I is estimated to be 80 lakes (Table 11.2), consisting of 8 drainage lakes, 3 closed lakes (defined as lakes with surface inlets but not outlets), and 69 reservoirs (Chapter 5, Eilers and Selle, this volume).

Regional Characteristics

Watershed Characteristics

Geology and Physiography

The entire Southern Blue Ridge Province is characterized physiographically as a belt of mountains west of the Piedmont Province that are remnants of a former highland antedating the lower peneplains on either side (Fenneman 1938, Hack 1969). The region consists of uniformly dissected mountains, ranging in elevation from 500 m to over 2,000 m

TABLE 11.2. Percentage of the ELS target population of lakes and reservoirs in the Southern Blue Ridge Province ecoregion with the specified index ANC and pH and the estimated total population of lakes and reservoirs.[a]

ANC (μeq L⁻¹)				Estimated population
≤0	>0 to ≤50	>50 to ≤200	>200	(N)
0	3.9	66.5	29.6	80

	pH			Estimated population
	≤5.0	>5.0 to ≤6.0	>6.0	(N)
Field pH	0.0	1.2	98.8	80
Air-equilibrated pH	0.0	0.0	100.0	80

[a] Data from Linthurst et al. 1986. Fall was the index sampling period for lakes and reservoirs.

above mean sea level (msl). Topography is steep and rugged, with less than 10% of the area having a slope < 10 degrees. The drainage pattern of the area is dendritic and stream valleys are generally V-shaped.

The geology of the area is diverse and complex. The Blue Ridge area in the eastern part of the province is dominated by early Precambrian metamorphic and granitic rocks (granite gneiss, mica gneiss, and mica schist) (Hatcher 1988), and the Great Smoky Mountains in the western part of the province are primarily metamorphosed sedimentary rock, principally sandstone with some limestone, formed during the late Precambrian (King et al. 1968). The sandstones, which cover approximately 60% of the Great Smoky Mountains National Park, are characterized as massive, thick-bedded, and of low porosity, with quartz and potassic feldspars as the dominant minerals.

The geologic formations in the region have undergone complex folding and faulting and are deeply weathered in response to the humid climate. The parent materials weather to form a relatively deep and porous soil mantle, with bare outcrops appearing primarily on the steeper slopes at high elevations. Based on chemical analysis of drainage waters, LeGrande (1958) divided the igneous and metamorphic rocks in the Southern Blue Ridge of North Carolina into two distinct groups. One group (referred to as the granite group), composed of granite, granite gneiss, mica schist, slate, and rhyolite flows and tuffs, yielded soft, slightly acidic water (pH < 7.0), with a low concentration of total dissolved solids (median in springs = 39 mg L^{-1}). The other group (referred to as the diorite group), composed of diorite, gabbro, hornblende gneiss, and andesite flows and tuffs, yielded harder, slightly alkaline water relatively high in total dissolved solids (median in springs = 99 mg L^{-1}). However, only about 10% of that portion of the province located in North Carolina is underlain by the diorite group and very little streamflow originates from the diorite group of rocks (Simmons and Heath 1982). Using a similar approach, Silsbee and Larson (1981, 1982) divided streams in the Great Smoky Mountains National Park, located in the western portion of the Southern Blue Ridge, into different geological groups based on their water quality. Streams draining watersheds in which sandstone was the major rock type had the lowest average conductivity (13.5 $\mu S\ cm^{-1}$) and ANC (52 $\mu eq\ L^{-1}$, respectively), followed in increasing order by watersheds dominated by mixtures of shale, siltstone, sandstone, and quartzite (20.4 $\mu S\ cm^{-1}$; 112 $\mu eq\ L^{-1}$), and shale, siltstone, conglomerate, limestone, and dolomite (25.5 $\mu S\ cm^{-1}$; 134 $\mu eq\ L^{-1}$).

Soils

Residual soils cover almost the entire area of the Southern Blue Ridge Province. Soil depth has not been extensively mapped in the region, but based on road cuts and stream incisions, it appears to be quite variable, depending on slope and elevation. Soil cores in forested watersheds at the Coweeta Hydrologic Laboratory in North Carolina indicate a general decrease in soil depth with elevation (Swift et al. 1988). Consolidated bedrock at Coweeta is typically 5 m to 10 m below the surface, although soil depths on steep slopes and ridges can be < 1 m.

Regional generalizations about soil classification and soil properties in the Southern Blue Ridge cannot be made with certainty because soils have not been extensively mapped or characterized. However, based on limited surveys of a few selected sites, the dominant soil orders on mountain slopes appear to be Inceptisols, Ultisols, and Histosols. Entisols, which are alluvial soils that exhibit little evidence of soil genesis, occur along the floodplain of streams. Inceptisols, which are young soils that have relatively indistinct soil horizonation, appear to be the most common above 1,500 m (Springer 1984). Ultisols appear to be the dominant soil order in the region, however, in terms of their areal extent (Swank and Douglas 1977, Kelly and Mays 1989, Swank and Crossley 1988).

Spodosols or incipient Spodosols also occur occasionally in the Southern Blue Ridge at elevations above 1,800 m under spruce-fir stands (Wolfe 1967, Springer and Elder 1980), and under heath vegetation (e.g., rhododendron) down to elevations as low as 1,000 m (Springer 1984, Kelly and Mays 1989). However, Spodosols do not appear to be common in this region. Histisols also occur in places where organic layers are thick. They generally occur in areas with unstable land forms, such as steep slopes (> 65%) and tree throws, which impede the formation of genetic soil horizons.

The Inceptisols, Ultisols, and Spodosols in the Southern Blue Ridge can be generally characterized as (1) well drained, (2) strongly to extremely acidic (pH 3.6 to 5.5), (3) containing little organic matter (< 5% to 15%), (4) having a low base exchange capacity (< 10%) in the mineral soil horizons, and (5) relatively rich in secondary mineral phases, particularly oxides of iron and aluminum (Johnson et al. 1988, Swank and Douglas 1977, Swank and Crossley 1988, Joslin et al. 1987, Kelly and Mays 1989). Aluminum is likely to be an important, if not the principal, exchangeable cation in most of these soils.

Vegetation, Land Use, and Watershed Disturbance

Most of the region is heavily forested, with a diverse vegetation dominated by a variety of hardwoods: oak (*Quercus* spp.), hickory (*Carya* spp.), maple (*Acer* spp.), birch (*Betula* sp.), beech (*Fagus* sp.), *Rhododendron* spp., and yellow poplar (*Liriodendron tulilpifera*). Pockets of northern hardwoods (e.g., maple, yellow birch, northern red oak) occur at elevations > 700 m and both northern hardwoods and spruce-fir (*Picea* sp., *Abies* sp.) occur at elevations > 1,400 m. Approximately 125 tree species have been identified in the Great Smoky Mountains (Whittaker 1956). Many of the current patterns in vegetation distribution in the Southern Blue Ridge Province are due to logging, farming, and other disturbances, both natural and anthropogenic.

Some of the valley floors and lower slopes, particularly along larger drainages, have been cleared and are under cultivation or are used for pasture. Approximately 80% of the land area in that portion of the province containing the NSS target population of streams is forested, however. The fraction of forested lands in the Southern Blue Ridge with virgin rather than second growth forest is unknown; approximately 20% of the Great Smoky Mountains National Park has little or no record of human disturbance and is considered to be virgin forest (Pyle 1988). Most of the forested land in the Southern Blue Ridge is in federally and state protected and managed lands, consisting of national and state parks and forests.

Climate and Hydrology

The region exhibits considerable climatic variation related primarily to elevation change. The climate can be generally characterized by abundant precipitation and moderate temperatures. Shanks (1954) classified the climate at elevations < 1,500 m in the humid to perhumid mesothermal category; higher elevations (> 1,500 m) were generally classified in the humid microthermal category within Thornthwaite's (1948) modified classification scheme.

Precipitation in the region is of both orographic and cyclonic origin. During summer, thunderstorms caused by the rise of humid air from the valleys advance along the ridges, often discharging large amounts of rain in relatively short periods of time. Weather patterns generally result in precipitation maxima in the late winter–early spring (the primary maximum) and in July–August (the secondary maximum) (Figure 11.3). Fall, particularly in September–October, is typically the driest period in the region. The primary winter maximum in precipitation results from depressions that pick up moisture from the Gulf of Mexico, and then deposit the moisture as they move northeastward. The secondary maximum in midsummer results from showers and thunderstorms generated over Florida interacting with a mid-continent pressure ridge (Dickson 1960).

The frequency of intense rainfall and total precipitation in the Southern Blue Ridge Province increases with increased elevation, with an annual average of 147, 177, 200, 226, and 231 cm at elevations of 445, 762, 1,158, 1,524, and 1,920 m, respectively, in the Great Smoky Mountains National Park (Shanks 1954). There is an additional orographic effect on precipitation, with the eastern portions of the Southern Blue Ridge generally receiving less precipitation at the same elevation than the western portions. Thus, because of the nature of orographic and cyclonic storms, precipitation exhibits high spatial variability in the region, particularly during summer.

The annual mean number of days with snowfall increases from 6.75 days at 445 m elevation to 25.9 days at 1,920 m elevation (Stephens 1969). Snow makes up only 2% to 10% of the annual precipitation at lower elevation (< 700 m) sites (Swift et al. 1988). Average monthly temperatures vary with elevation and range from about 3°C in January to about 22°C in July at Coweeta (Swift et al. 1988) and between −1.8°C and 14°C for February and July, respectively, at 1,524 m elevation in the Great Smoky Mountains National Park (Stephens 1969).

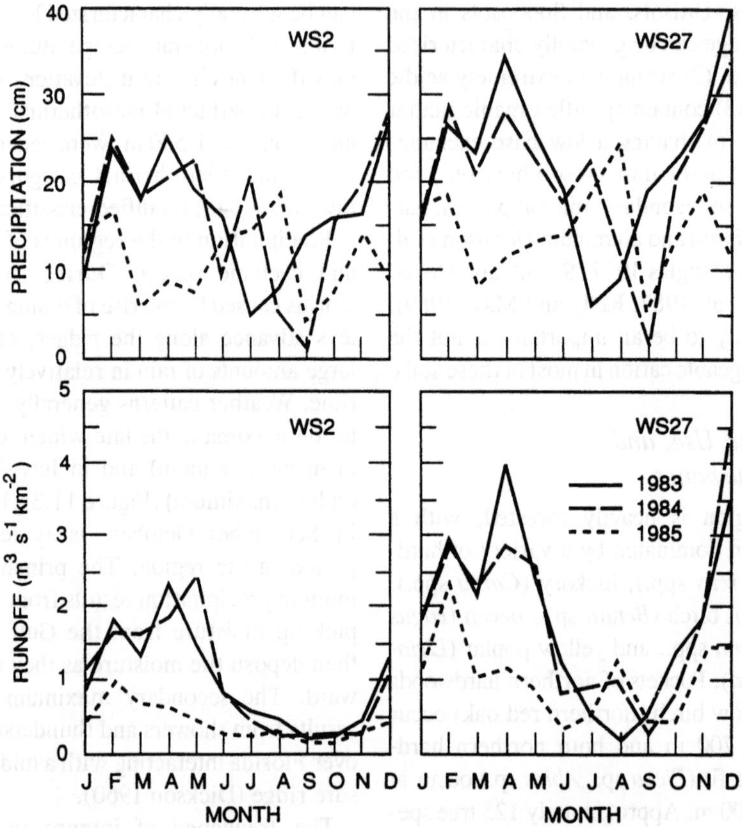

FIGURE 11.3. Monthly precipitation and monthly average streamflow over three years at a low-elevation (WS-2, 709 m at weir) and a high-elevation (WS-27, 1,061 m at weir) watershed at the Coweeta Hydrologic Laboratory, North Carolina. (Data from W. Swank, Coweeta Hydrologic Laboratory.)

Snow cover varies considerably from year to year, particularly at high elevations where snowfall is greater. A snowpack seldom lasts through the winter.

Estimates of evapotranspiration based on the difference between precipitation and runoff in undisturbed forested watersheds at Coweeta range from around 55 cm yr^{-1} for high-elevation catchments (>1,000 m) to 90 cm yr^{-1} for low-elevation catchments (700 m to 800 m) (Swank and Crossley 1988). These water balance estimates of evapotranspiration from control watersheds in the Southern Blue Ridge are in general agreement with those from pan evaporation and empirical and energy balance equations.

The area is drained by clearwater streams with dendritic drainage patterns. Although the range of flow is rather narrow for all months, discharge of most streams is typically highest and most variable during the February–April period when rainfall is greatest, and lowest during the July–November period when rainfall is generally lower (Figure 11.3). Studies at the Coweeta Hydrologic Laboratory in North Carolina indicate that quickflow (or direct runoff) usually is < 10% of the total annual runoff from watersheds (Swank and Douglass 1977). Further, overland flow is rare on undisturbed watersheds in the Southern Blue Ridge, reflecting the high hydraulic conductivity of soils in this region. If overland flow occurs, it is only during intense storms or near streams and along depressions where the water table is at or near the land surface. Lagged streamflow response resulting from snowmelt is a minor factor in the region, because heavy snows and long lasting snowpacks are rare.

On an annual basis, precipitation exceeds evapotranspiration, and most streams flow year round. Studies at Coweeta show that the percentage of precipitation appearing as streamflow increases with elevation because of (1) decreasing soil depth, (2) decreasing evapotranspiration related to lower air temperatures, (3) increasing slope steepness, and (4) greater intensity and amount of precipitation (Swift et al. 1988). The fraction of annual precipitation appearing as streamflow in low-elevation watersheds (< 1,000 m) at Coweeta ranges from 0.48 to 0.58, compared to 0.71 to 0.81 for high-elevation watersheds (> 1,200 m) (Swift et al. 1988). The lower capacity of high-elevation watersheds to retain precipitation also causes a higher percentage of the precipitation to return as quickflow during storms and more variation in streamflow. Such differences may have significant implications on the routing of materials transported with meteoric water through watersheds and thereby on the acid-base chemistry of surface waters, particularly during storms.

Atmospheric Deposition

The longest records of information on atmospheric deposition in the Southern Blue Ridge Province comes from two NADP sites: the Elkmont, Tennessee, site located in the Great Smoky Mountains National Park at 640 m elevation, and the Coweeta site located at the Coweeta Hydrologic Laboratory in southeastern North Carolina at 686 m elevation (NADP 1986, 1987a,b). Deposition data have also been collected at a high-elevation site (1,800 m) located near Clingmans Dome in the Great Smoky Mountains (Lindberg and Johnson 1989, Lindberg et al. 1988), and at two sites located at 646 m and 1,100 m in the Brier Creek Watershed in northern Georgia (Buell and Peters 1988). A Mountain Cloud Chemistry Project (Mohnen 1988) also collects deposition data at two sites in the Southern Blue Ridge, one located on Mt. Mitchell, North Carolina, at an elevation of 1,987 m and the second located on Whitetop Mountain, Virginia, at an elevation of 1,689 m (Mohnen 1988).

Hydrogen Ion, Sulfate, Nitrate, and Chloride

The Southern Blue Ridge Province receives precipitation with estimated annual average pH in the range of 4.5 to 4.3, based on samples collected at five sites. Although the number of sites sampled in the province is too small to determine intraregional patterns in wet deposition, the similarity in volume-weighted average H^+ concentrations in wetfall among these five sites suggests that the amount of precipitation is the primary source of variation in the annual wet deposition of H^+.

Hydrogen and SO_4^{2-} are the two dominant ions in precipitation in the Southern Blue Ridge, constituting more than 60% of the volume-weighted average cationic and anionic composition, respectively. Ammonium and NO_3^- are next most abundant, accounting for about 20% and 27%, respectively, of the composition. Chloride is a minor constituent in rainfall. Precipitation in the province thus can be characterized as a dilute solution of sulfuric and nitric acid.

In contrast to the constant concentration of H^+ among sites in the Southern Blue Ridge, the levels of mineral acid anions in wetfall are more variable. Sulfate and NO_3^- levels in precipitation at Elkmont, for example, are slightly greater than at Coweeta (weighted average of 41 and 31 μeq L^{-1} for SO_4^{2-}, and 16 and 13 μeq L^{-1} for NO_3^-, respectively). The weighted average Cl^- concentration in wetfall is slightly greater at Coweeta (5 μeq L^{-1}) than at Elkmont (3 μeq L^{-1}). This difference in Cl^- is probably due to differences in air trajectories at the two sites. The higher concentrations of SO_4^{2-} and NO_3^- at Elkmont are probably due to its proximity to major point sources of emissions in the Tennessee Valley to the east, whereas the air trajectories at Coweeta are more often from the south.

The ratio of $[SO_4^{2-} + NO_3^-]$ to H^+ in precipitation in the Southern Blue Ridge is approximately 1.5, indicating that some NO_3^- and/or SO_4^{2-} acidity in precipitation has been neutralized by base aerosols or that some of these acid anions were from nonacidic sources. Sulfate contributes over twice as much acidity as NO_3^- to precipitation, based on the ratio of SO_4^{2-} to NO_3^- equivalents in precipitation of approximately 2.5.

The concentrations of acid anions and acidic cations in precipitation in this region vary seasonally, based on data from Elkmont and Coweeta (Figure 11.4). Volume-weighted average SO_4^{2-}, NO_3^-, and H^+ levels in precipitation tended to be slightly higher in summer than in winter at both sites (Figure 11.4), a commonly observed pattern

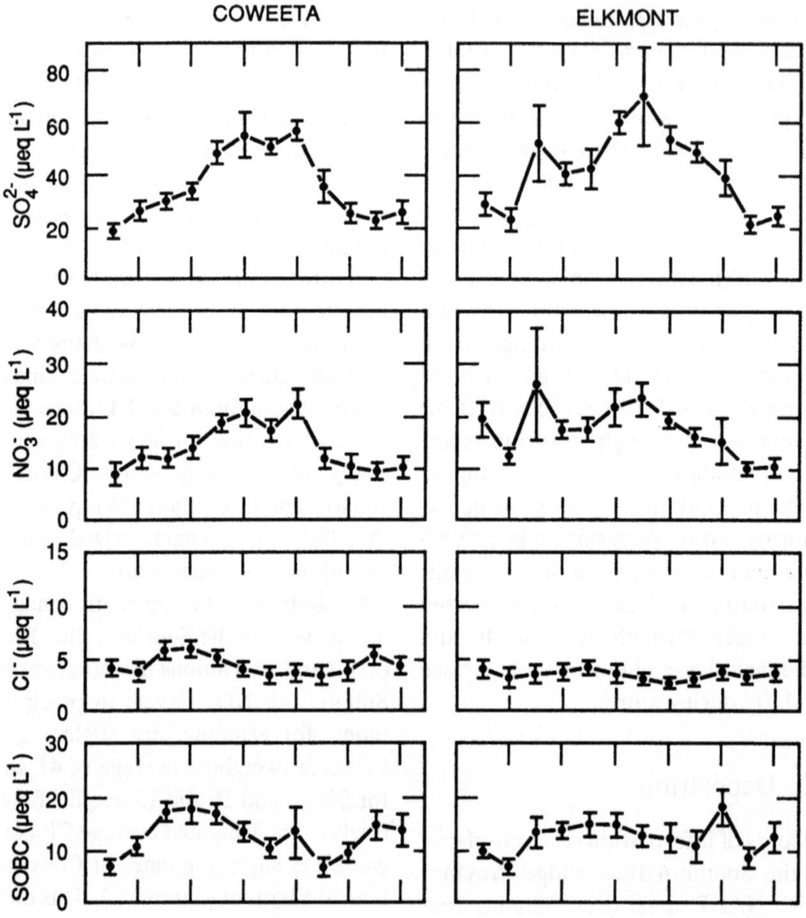

FIGURE 11.4. Seasonal patterns in precipitation chemistry at two sites in the Southern Blue Ridge Province. Values plotted are monthly precipitation-weighted means ± 1 standard error. Means for Coweeta are based on data from 1978–1988; means for Elkmont are based on data from 1980–88. (Data from NADP 1989.)

in the eastern United States (NADP 1986, 1987a,b). Buell and Peters (1988) also report that H^+, SO_4^{2-}, and NO_3^- concentrations in bulk precipitation at the Brier Creek Watrshed in northern Georgia were highest during the spring-summer period. In contrast, the Cl^- concentration in precipitation is relatively constant, exhibiting little if any seasonal pattern (Figure 11.4).

The annual wet deposition rates for H^+ and mineral acid anions at Elkmont and Coweeta appear to be in the range of 40 to 60 meq m^{-2} each for H^+ and SO_4^{2-}, and 16 to 21 meq m^{-2} and 3 to 6 meq m^{-2}, respectively, for NO_3^- and Cl^- (NADP 1986, 1987a,b). The upper value for H^+ deposition in wetfall is close to that for total H^+ deposition (wet + dry) reported by Lindberg and Johnson (1989) for the Clingmans Dome site in the Great Smoky Mountains.

Although there are no long-term records of dry deposition in the Southern Blue Ridge, recent measurements indicate that dry deposition of SO_4^{2-} at higher elevations in this region is comparable to or greater than that of wet deposition (i.e., precipitation). Deposition measurements at the Clingmans Dome site in the Great Smoky Mountains show that dry plus wet deposition of SO_4^{2-} and NO_3^-, respectively, is approximately 2 to 5 times greater than wet only (Lindberg and Johnson 1989). If cloud water is included in the estimates, SO_4^{2-} and NO_3^- deposition, respectively, are 4 to 6 times greater than wet only. The estimates of dry deposition are based on measured air concentrations of SO_2 and NO_x and modeled deposition velocities using meteorological parameters at the site; wet deposition is based on seasonally weighted mean concentrations of SO_4^{2-} and NO_3^- in precipitation and measured water fluxes (Lindberg and Johnson 1989). The wet to dry deposition ratio for Cl^- at the Clingmans Dome site is 1.5, indicating that dry deposition of Cl^- occurs primarily as wetfall. Swank and Waide (1988) report that dryfall accounts for only 10% of the SO_4^{2-} input in bulk precipitation at Coweeta. The lower contribution of dry deposition to SO_4^{2-} inputs at Coweeta, however, is probably not related entirely to intragregional differences in the dry deposition of SO_4^{2-}, but instead is more a function of the method used to collect dry deposition at Coweeta.

Immersion in clouds and fog is an important deposition mechanism in the Southern Blue Ridge, particularly in the high-elevation watersheds where the frequency and duration of cloud immersion are greater. Half the total annual SO_4^{2-} and H^+ deposition, 60% of the Cl^-, and 17% of the NO_3^- deposition to the Clingmans Dome site in the Great Smoky Mountains has been estimated to come from cloudwater interception (Lindberg and Johnson 1989). The annual rates of H^+ and SO_4^{2-} deposition are high at this site, approximately 200 meq m^{-2} y^{-1}; annual deposition rates of these ions at a lower elevation site at Coweeta are approximately half this amount. Somewhat higher cloudwater deposition of SO_4^{2-}, NO_3^-, and H^+ has been reported for other high-elevation sites (Mt. Mitchall, North Carolina, and Whitetop Mt., Virginia) in the Southern Blue Ridge (Mohnen 1988). The higher cloudwater deposition of acidity at these sites is ascribed to their being immersed in clouds and fog for a longer period than Clingmans Dome. The large input of SO_4^{2-}, NO_3^-, and H^+ from cloudwater interception may be somewhat unique to the higher elevation sites in the Southern Blue Ridge, which are in a zone of cloud water and fog immersion for significant periods of time.

Base Cations and Ammonium

The annual deposition rate of base cations and ammonium in the Southern Blue Ridge Province appears to be relatively small. Calcium and Na^+ are the dominant base cations in precipitation in this region (NADP 1986, 1987a,b). The volume-weighted average sum of base cations in wetfall is typically around 15 µeq L^{-1} on an annual basis (NADP 1986, 1987a,b, Buell and Peters 1988), with no consistent seasonal pattern in the concentration (Figure 11.4). The long-term average concentration of base cations in bulk precipitation at Coweeta is approximately 20 µeq L^{-1}; the concentration difference between wetfall and bulk precipitation is due in part to dryfall inputs to the bulk precipitation collected at Coweeta (Swank and Waide 1988).

The low concentrations of base cations in wetfall in the province reflect the lack of significant ground-level sources, such as agriculture and industrial emissions and road dust, which tend to increase the concentrations in precipitation through the resuspension or release of alkaline dust particles. In addition, the humid climate of the region probably plays an important role in regulating base cations in precipitation by reducing the suspension and long-range transport of alkaline dust particles from ground-level sources. As a result, wet deposition would be expected to be a relatively small contributor to the deposition of base cations.

Dry deposition of base cations should also be relatively small in this region. At Coweeta, dryfall contributes about 20% of the Ca^{2+} and K^+ input and 17% and 10%, respectively, of the Na^+ and Mg^{2+} input in bulk precipitation, based on bulk precipitation samples (Swank and Waide 1988). Lindberg and Johnson (1989), however, report that dry deposition of base cations at the Clingmans Dome site in the Smokies was 51 meq m^{-2} y^{-1}, whereas wetfall input was only 14 meq m^{-2} y^{-1}. Total deposition of base cations, which includes cloud and fog water, at this site was 113 meq m^{-2} y^{-1}. Thus, dryfall and fog and cloud water each contribute 40% to 45% of the base cation inputs, and dry deposition contributes the remainder. The high dry deposition of base cations at the Clingmans Dome site, most of which is Ca^{2+}, may be a result of higher wind velocities (3.7 m s^{-1}) than at the other sites and the resulting greater impact on surfaces of aerosols and dust particles containing calcium. Lindberg et al. (1988) reported that the dry deposition rate of Ca^{2+} at the Clingmans Creek site during a six-month period was greater than that at at a low-elevation site located in the Valley and Ridge Province of eastern Tennessee. The lack of data on dry deposition of base cations from other sites in the Southern Blue Ridge obtained using comparable methods, however, precludes drawing any conclusions about spatial variation in dry deposition of base cations in this region.

Of the cations measured in wetfall, NH_4^+ is second only to H^+ in abundance, accounting for about 20% of the cationic equivalents. The volume-weighted average concentration of NH_4^+ at Coweeta, Elkmont, and Brier Creek typically is in the range of 7 to 14 μeq L^{-1}. Ammonium exhibits a seasonal pattern similar to that for SO_4^{2-}, NO_3^-, and H^+, being higher in the spring–summer period and lowest during the winter (Figure 11.4).

Although NH_4^+ is the dominant form of N in the atmosphere, dry deposition of NH_4^+ is relatively unimportant, as it is associated primarily with fine particles with low deposition rates on watershed surfaces (e.g., leaves and stems of canopy vegetation) (Lindberg and Johnson 1989). Dry deposition of NH_4^+ at the Clingmans Dome site in the Smoky Mountains is estimated to be only 21% of the total annual deposition of 51 meq m^{-2} y^{-1} (Lindberg and Johnson 1989). Fog and cloud water account for about 50% of the NH_4^+ input at this site.

Characteristics of Surface Waters in the Southern Blue Ridge Province

Distribution of Data Sets

Data on the acid-base chemistry of surface waters in the Southern Blue Ridge are available from several sources. The U.S. EPA sampled 67 lakes and reservoirs in the Southern Blue Ridge in the fall of 1984 as part of the ELS-I in Region 3A (Linthurst et al. 1986a, Eilers et al. 1988a; Introduction, Charles, this volume) (Figure 11.2). The sample population of 67 lakes and reservoirs represented 76% of the total population estimated to occur in the Southern Blue Ridge Province ecoregion.

The EPA also sampled 61 stream reaches in a 27,000 km^2 area of the Southern Blue Ridge during the spring and summer of 1985 as part of the NSS Pilot Survey (Messer et al. 1986, Kaufmann et al. 1988; Introduction, Charles, this volume) (Figure 11.2). The target population of NSS streams was defined as blue-line streams on 1:250,000-scale USGS topographic maps that (1) occur in areas where low ANC waters would be expected, based on the bedrock geology, (2) drain watersheds of < 155 km^2 (60 mi^2), and (3) do not drain into or out of a lake or reservoir. The basis for using these selection criteria in the NSS was to include streams potentially the most sensitive to acidification from acidic deposition and sufficiently large for fish to inhabit. The target population sampled included both an upstream and a downstream site on each sample reach, with the downstream site determined by the first confluence with another 1:250,000-scale blue-line stream, and the upstream site determined by either a similar upstream confluence or the stream origin, as indicated on the USGS topographic map. Samples for chemical analysis were collected at both the upstream and downstream ends of each reach sampled in the NSS. The probabilistic method of selecting the reaches is described in more detail by Messer et al. (1986).

Region 3A, which was sampled as part of the EPA ELS-I (Linthurst et al. 1986a, Eilers et al. 1988a) included the same area of the Southern Blue Ridge that was sampled in the NSS, as well as a spur of this province extending into southern Virginia. Region 3A also included a portion of the Piedmont in Georgia, South Carolina, North Carolina, and Virgina and a portion of the Valley and Ridge Province in Virginia (Figure 5.1, Chapter 5, Eilers and Selle, this volume). Only one of the four lakes sampled in Virginia actually fell within the Southern Blue Ridge Province boundaries (Figure 11.2). Unless otherwise noted, lakes in ELS Region 3A that fell outside the Southern Blue Ridge ecoregion were excluded from the analysis of lakes in the Southern Blue Ridge Province (Figure 11.2), whereas the characteristics of lakes in the Southern Blue Ridge presented by Eilers et al. (1988a) included data for these lakes.

Both the ELS and the NSS were probability-based samples of streams and lakes drawn from target populations. The resulting statistical samples thus can be used to make quantitative population estimates with known confidence limits for any of the measured parameters of the sampled populations.

Additional data on surface water chemistry in the Southern Blue Ridge Province are available from a study of first- and third-order reaches of 30 streams located in eastern Tennessee, western North Carolina, and northern Georgia (Winger et al. 1987). The study was conducted to determine the chemistry and biology of acid sensitive streams in the Southern Blue Ridge. Streams were sampled during periods when they were considered to be near baseflow. Criteria for selecting streams included elevation of third-order reaches > 600 m, minimal human disturbance in the watershed, and reasonable accessibility. However, the streams were not selected at random from a target population. Hence, the sample data from Winger et al. (1987) should not be used to make quantitative population estimates of the characteristics of first- and third-order stream reaches in the Southern Blue Ridge.

Additional data on the chemistry of selected streams in the Southern Blue Ridge are also available from several studies. These include water quality surveys of selected streams in the Great Smoky Mountains National Park funded by the National Park Service (Silsbee and Larson 1981, 1982, 1983), water quality studies funded by the U.S. Forest Service at the Coweeeta Hydrologic Laboratory located near Franklin, North Carolina (Waide and Swank 1987, Swank and Waide 1988), monitoring and special studies of selected streams by the Tennessee Valley Authority (Meinert and Miller 1981, Jones et al. 1983, Olem 1985, 1986), a study of the Buell Creek Watershed in northern Georgia by the USGS (Buell and Peters 1988), water quality data from USGS benchmark stations in the Southern Blue Ridge (Smith and Alexander 1983, Lins 1986), and a study of the chemistry and biology of streams in the Great Smoky Mountains National Park and the Cherokee National Forest, Tennessee, funded by the Electric Power Research Institute (Elwood et al. 1985).

Data on the chemistry of all major ions are lacking from some of these studies, precluding their use in analyzing the acid-base chemistry of stream water. In addition, some data are from high-order streams and rivers that have been impacted by human disturbances in their watersheds. Thus, they are not regionally representative of low-order, pristine streams in the Southern Blue Ridge. Relevant results from the other studies that contain reasonably complete water chemistry and are thought to be regionally representative of acid sensitive streams in the Southern Blue Ridge are presented or cited where appropriate in the following sections of this chapter.

Current Status of Water Chemistry

Status and Extent

Based on the NSS and ELS, which were conducted during an index period (spring and summer for streams, fall for lakes), no acidic streams or lakes (ANC $\leq$ 0) were identified in the target populations (Messer et al. 1986, Kaufmann et al. 1988, Linthurst et al. 1986a) (Tables 11.1 and 11.2). Because of the uncertainty associated with the population estimates, however, the percentage of NSS stream reaches in the Southern Blue Ridge that were acidic is estimated to be < 1% rather than 0 (Messer et al. 1986), whereas the proportion of lakes estimated to be acidic is 0 (Linthurst et al. 1986a). The results of Winger et al. (1987), in contrast, show that 3% of the first-order streams were acidic over the 1982–1984 period, but none of the third-order reaches had ANC ≤ 0. Because the streams sampled by Winger et al. (1987) were not randomly selected from a target population, the fraction in different ANC and pH classes cannot be extrapolated with confidence to the population of first- and third-order streams in the Southern Blue Ridge as a whole. Nonetheless, their results show that acidic (ANC ≤ 0) streams do exist in the region.

The question arises as to whether the streams sampled by Winger et al. (1987), including those that were acidic, are representative of the Southern Blue Ridge. The two acidic streams drained watersheds containing a pyritic phyllite (Anakeesta formation), which is a potential source of SO_4^{2-} in stream water when the pyrite is exposed to oxidation (Huckabee et al. 1975, Hermann et al. 1976). Sulfate was the dominant anion in these acidic streams, with an average concentration 8 to 9 times higher than that of the streams draining watersheds without a bedrock source of SO_4^{2-} (Winger et al. 1987). Nitrate was also 2 to 3 times higher in the streams affected by the Anakeesta formation. The greater NO_3^- levels in these streams may be due to a co-occurring source of NO_3^- in these watersheds (e.g., organic nitrogen in soil), or to reduced uptake and retention of NO_3^- from atmospheric deposition in these watersheds. Both of the acidic streams (Walker Camp Prong, Beech Flat) drain watersheds that are unlogged at the upper elevations, and previous studies in this region have shown that streams draining watersheds with old-growth (virgin) forests tend to have higher NO_3^- concentrations than those draining watersheds with second-growth forests (Silsbee and Larson 1982, 1983, Elwood et al. 1985). Sources of SO_4^{2-} and NO_3^- in streams in the Southern Blue Ridge are discussed further in the section on watershed processes.

Although no acidic surface waters were sampled in the Southern Blue Ridge during the NSS and ELS, a high proportion of the streams and lakes had low ANC, and thus are potentially sensitive to acidic deposition (Table 11.1). The median ANC values for upstream and downstream reach ends in spring, for example, were 100 and 122 μeq L^{-1}, respectively, whereas the 25th and 75th percentile values for ANC at the upstream reach ends were 75 and 141 μeq L^{-1}, respectively. Hence, on the basis of ANC alone, a large fraction of the NSS stream reaches in the Southern Blue Ridge can be classified as potentially sensitive to acidic inputs.

The fact that no acidic streams were sampled in this province during the NSS-I could be a result of (1) their being outside the boundaries and selection criteria used to define the NSS target population in the Southern Blue Ridge, (2) a low probability of encountering them due to a low abundance, or (3) some combination of (1) and (2). The sampling frame for the NSS-I included only blue-line stream reaches on 250,000-scale USGS topographic maps. Stream reaches with perennial flow are known to extend upstream of some blue-line reaches in the Southern Blue Ridge. Although these upstream reaches may not be an important habitat for fish because of their shallow depth and small size, they may have a lower ANC than the blue-line reaches, based on the finding that the ANC of stream water, including that in the Southern Blue Ridge, typically declines from downstream to upstream (Table 11.3; Johnson 1979, Messer et al. 1986, Elwood et al. 1985, Kaufmann et al. 1988). Thus, it is likely that stream reaches with the lowest ANC are located primarily upstream of the blue-line reaches. As such, they are outside the NSS target population of interest.

The ANC distribution of ELS lakes in this province is similar to that for streams, with none of the lakes sampled in the region found to be acidic (ANC ≤ 0) (Table 11.2). About 3.9% of the ELS lakes in this region, however, are estimated to have ANC ≤ 50 μeq L^{-1} (Table 11.2).

Frequency histograms of selected characteristics of Southern Blue Ridge stream and lake populations sampled during the NSS and the ELS illustrate the distribution of the various physiographic and water quality factors in the sampled populations (Figures 11.5 and 11.6). Most of the NSS watersheds were small and low order (first to fourth), with a median area of < 25 km^2 (Figure 11.5a). When classified according to their Strahler order, the stream reaches in the NSS target population for this region range from first to fourth order on 250,000-scale maps, with most of the upstream ends being first order. However, maps of this scale appear to provide an underestimate of the true Strahler order of blue-line streams in the Southern Blue Ridge by at least one order (e.g., streams appearing as first order on 250,000-scale maps are

TABLE 11.3. Sample and population statistics for major ion chemistry of streams in the Southern Blue Ridge Province.[a]

	Node	$N(n)$	pH_{ae}[b]	Field pH	ANC	SO_4^{2-}	Ca^{2+}	DOC	NO_3^-	Cl^-	Na^+	SBC	Specific conductance
NSS Sample	U	(54)	7.32 (7.08–7.60)	6.96 (6.78–7.16)	100 (60–169)	21.9 (14.9–33.5)	46.0 (34.0–100.2)	47.4 (39.6–58.2)	8.0 (2.0–18.5)	19.7 (15.4–26.8)	59.7 (49.6–70.2)	149 (121–257)	15.5 (12.9–23.2)
NSS Sample	L	(54)	7.28 (7.14–7.54)	7.05 (6.91–7.23)	117 (77–215)	28.1 (21.3–47.1)	67.5 (42.3–131.7)	60.2 (46.6–77.9)	8.6 (3.7–16.8)	23.7 (18.9–31.2)	59.2 (44.8–81.3)	185 (138–305)	19.9 (15.1–31.6)
NSS Population	U	2031	7.28 (7.10–7.62)	6.98 (6.80–7.06)	100 (75–141)	20.7 (13.1–33.1)	44.7 (37.0–81.5)	48.0 (42.6–58.2)	6.9 (1.7–15.8)	20.7 (15.8–23.7)	60.5 (51.2–67.5)	164 (125–208)	15.9 (12.8–19.5)
NSS Population	L	2031	7.28 (7.15–7.46)	7.03 (6.87–7.15)	122 (89–178)	22.9 (19.3–40.8)	64.7 (44.0–105.7)	59.6 (49.2–83.8)	7.6 (3.4–16.4)	23.4 (19.1–31.1)	59.1 (47.4–81.3)	188 (137–245)	20.6 (14.7–26.0)
Winger et al. Sample	U	(30)	NR[c]	6.62 (6.46–6.82)	61.3 (48.4–87.9)	22.8	NR[c]	NR[c]	5.5	20.1	NR[c]	178.8	NR[c]
Winger et al. Sample	L	(30)	NR[c]	6.69 (6.59–6.82)	74.6 (46.0–80.3)	22.6	NR[c]	NR[c]	6.2	18.7	NR[c]	198.0	NR[c]

[a] Data are based on two data sets discussed in this chapter. Values presented are medians, with first and third quartiles in parentheses. Node refers to the upstream (U) and downstream or lower (L) samling location. All values are in μeq L^{-1}, except for DOC (μmol L^{-1}), specific conductance (μS cm^{-1}), and pH.
[b] pH_{ae} = air-equilibrated pH.
[c] NR = not reported.

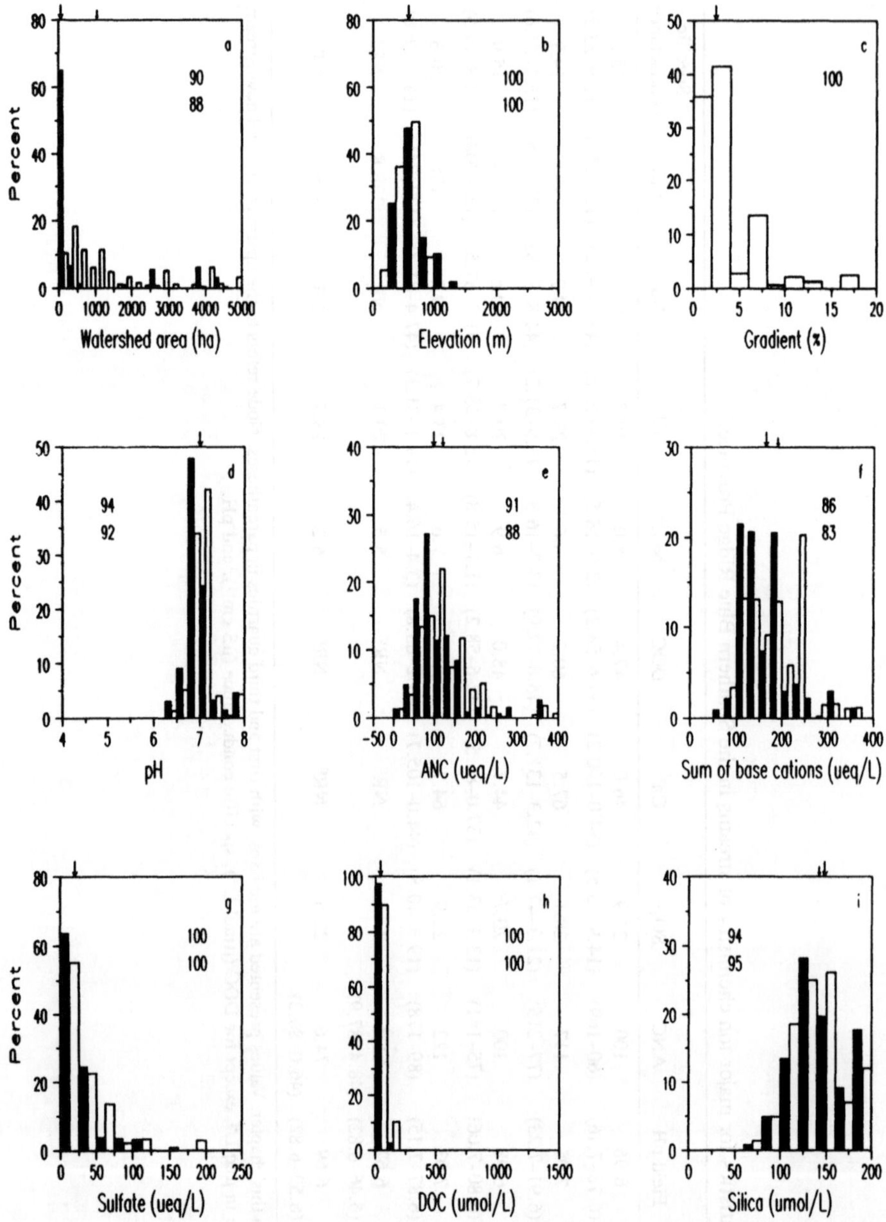

FIGURE 11.5. Frequency histograms of major characteristics of the target population of NSS streams in the Southern Blue Ridge Province: (a) watershed area, (b) site elevation, (c) stream gradient between the upstream and downstream reaches, (d) pH, (e) ANC, (f) sum of base cations, (g) SO_4^{2-}, (h) DOC, and (i) silica. The solid bars are the upstream ends and the open bars are the downstream ends. The long and short arrows indicate the median values for the upstream and downstream ends of reaches, respectively. The numbers in the histograms are the percentages of the NSS target population included in the range plotted (i.e., upper number is for the upstream reach end).

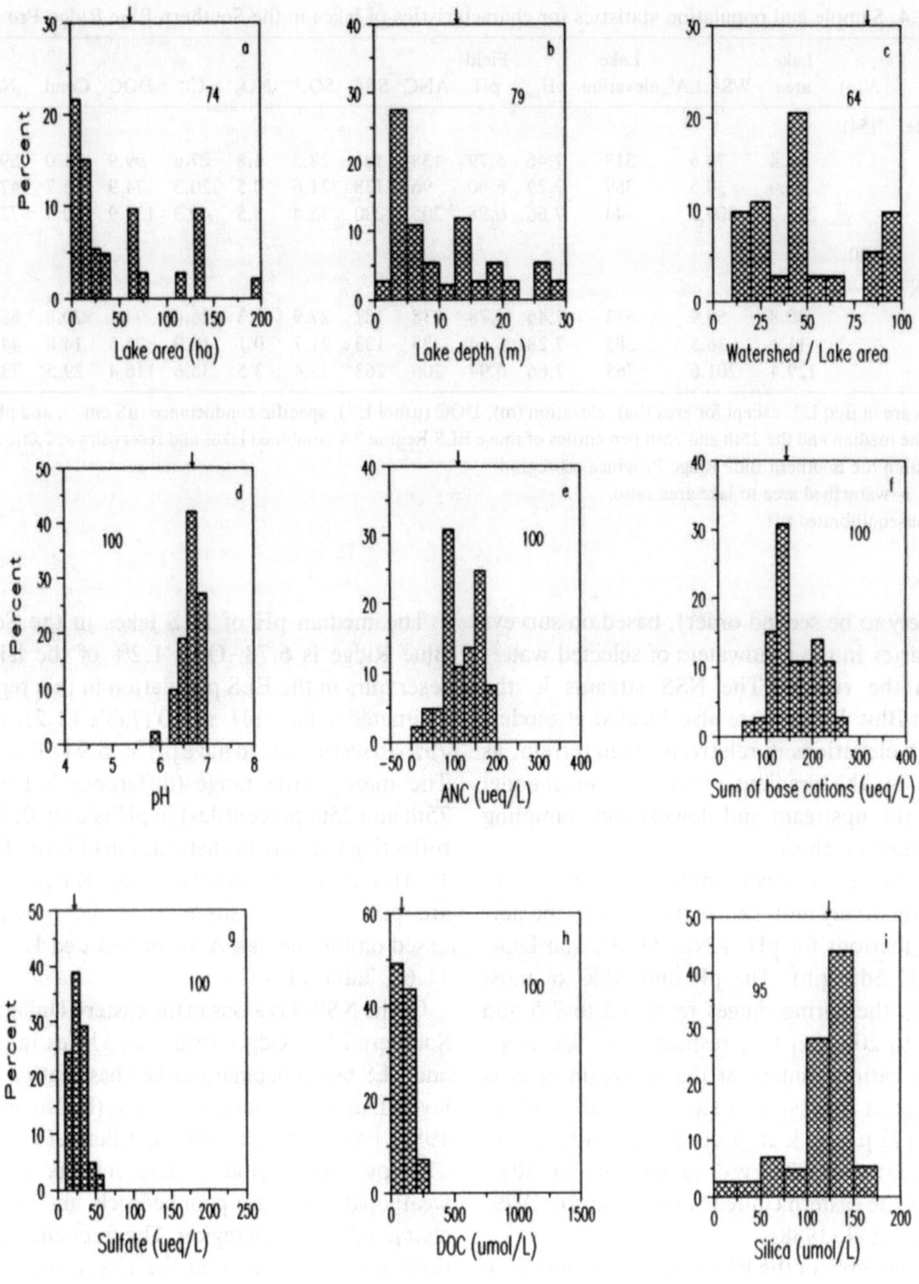

FIGURE 11.6. Frequency histograms of major characteristics of the target population of ELS Region 3A lakes and reservoirs in the Southern Blue Ridge Province (see Figure 11.2): (a) lake surface area, (b) sampling depth, (c) watershed area:lake area, (d) pH, (e) ANC, (f) sum of base cations, (g) sulfate, (h) DOC, and (i) silica. The arrow at the top of each box indicates the median value; the number in each histogram is the percentage of target lakes plotted.

TABLE 11.4. Sample and population statistics for characteristics of lakes in the Southern Blue Ridge Province.[a]

	$N(n)$	Lake area	WSA:LA[b]	Lake elevation	pH_{ae}[c]	Field pH	ANC	SBC	SO_4^{2-}	NO_3^-	Cl^-	DOC	Cond.	Na^+	Ca^{2+}
ELS sample	(54)														
Median		30.4	74.6	518	7.46	6.79	138	191	28.3	1.8	27.6	99.9	21.0	59.2	68.1
25th		11.0	30.5	389	7.29	6.60	96	138	21.6	0.5	20.3	74.9	15.7	47.4	47.9
75th		231.0	209.5	844	7.66	6.98	205	280	35.4	5.5	40.3	124.9	30.5	72.6	99.3
ELS population	80														
Median		30.8	56.9	513	7.46	6.78	138	187	27.9	1.3	26.4	94.1	20.8	58.3	67.6
25th		11.1	36.5	383	7.28	6.64	88	133	21.7	0.1	19.0	76.5	14.8	44.8	46.6
75th		129.4	201.6	765	7.66	6.94	206	268	35.4	3.5	35.6	116.4	29.5	73.3	92.6

[a] All values are in μeq L^{-1}, except for area (ha), elevation (m), DOC (μmol L^{-1}), specific conductance (μS cm^{-1}), and pH. Values given are the median and the 25th and 75th percentiles of those ELS Region 3A combined lakes and reservoirs <2,000 ha in size that are within the Southern Blue Ridge Province ecoregion.
[b] WSA:LA = watershed area to lake area ratio.
[c] pH_{ae} = air-equilibrated pH.

more likely to be second order), based on surveys of tributaries in the headwaters of selected watersheds in the region. The NSS streams in the Southern Blue Ridge were also located at moderately high elevations in relatively steep terrain, as reflected by the gradient of the stream channel between the upstream and downstream sampling sites (Figure 11.5b,c).

The chemistry of streams in the NSS target population is relatively uniform, as reflected by the narrow distributions for pH, ANC, SO_4^{2-}, and DOC (Figure 11.5d,e,g,h). The pH and ANC of most streams in the spring ranges from 6.5 to 7.5 and from 50 to 200 μeq L^{-1}, respectively. The ANC and base cation content at the upstream sites is lower than at downstream sites (median ANC of 100 and 122 μeq L^{-1}, respectively), a pattern consistent with that observed in streams in other regions of the eastern United States (Johnson 1979, Kaufmann et al. 1988).

More than 90% of the ELS lakes in this province are human-made reservoirs that are small (median area = 27.1 ha^2) and relatively deep (median depth = 11.0 m) (Figure 11.6a,b,c). Drainage lakes and reservoirs in this region have large watershed areas (median = 966 ha^2 and 1,474 ha^2, respectively), with the lakes themselves making up < 2% of the area. A few of the larger reservoirs with surface areas > 150 ha^2 were sampled; these account for a small percentage of the total ELS lake population in the region.

The median pH of ELS lakes in the Southern Blue Ridge is 6.78. Only 1.2% of the lakes and reservoirs in the ELS population in this region are estimated to have pH ≤ 6.0 (Table 11.2), whereas 75% are estimated to have pH < 6.94 (Table 11.4). The interquartile range (difference between the 75th and 25th percentiles) in pH is only 0.24 units, reflecting the narrow distribution of lake pH (Table 11.4). Lakes in the Southern Blue Ridge, however, are potentially sensitive to acidic precipitation based on the median ANC of 138 μeq L^{-1} (Figure 11.6e, Table 11.4).

Of the NSWS regions in the eastern United States, Southern Blue Ridge streams and lakes in the NSS and ELS target populations had base cation content levels that were among the lowest (Kaufmann, et al. 1988, Linthurst et al. 1986a, Eilers et al. 1988a). The low base cation content reflects the highly weathered soils, base-poor bedrock, and high runoff characteristic of the region. The frequency distributions for base cations in the target population of streams and lakes is similar to but somewhat broader than that for ANC (Figures 11.5f and 11.6f). Most streams and lakes in the Southern Blue Ridge have a base cation content < 200 μeq L^{-1} (medians at the upstream and downstream ends = 164 and 188 μeq L^{-1}, respectively) (Tables 11.3 and 11.4).

Sulfate levels in ELS lakes in this province are low (Table 11.4, Figure 11.6g), probably due primarily to the adsorption of SO_4^{2-} by soils. The median (27.9 μeq L^{-1}) and 75th percentile (35.4

μeq L^{-1}) SO_4^{2-} values in Southern Blue Ridge lakes were the lowest of any among the lake regions sampled in the eastern United States (Linthurst et al. 1986a). Sulfate levels in approximately 4% of the lakes, however, exceeded 115 μeq L^{-1}, suggesting a watershed source of SO_4^{2-} in these catchments.

Sulfate and DOC concentrations are also low in the NSS target population of streams, with medians of 21 μeq L^{-1} and 48 μmol L^{-1}, respectively, at upstream reach ends in the spring (Table 11.3, Figures 11.5g,h). Stream SO_4^{2-} levels are similar, but DOC levels are lower than in lakes and reservoirs, which have median SO_4^{2-} and DOC concentrations of 28 μeq L^{-1} and 94.1 μmol L^{-1}, respectively (Tables 11.3 and 11.4). The higher DOC levels in lakes and reservoirs compared to streams may result from algal growth and decomposition or from additional contact with soil and vegetation encountered by lower elevation reservoir runoff. However, the DOC levels in these lakes and reservoirs are relatively low, and these waters can generally be classified as moderately clearwater lakes, based on the median color of 25 platinum cobalt units (PCU) and a median Secchi disc transparency of 2.4 m.

Silica concentrations in the Southern Blue Ridge target population of streams and lakes are relatively high (Figures 11.5i and 11.6i), especially in reference to the major ions in stream and lake water. Silica levels in lakes (median Si = 268 μmol L^{-1}) rank higher than those for any of the other subregions sampled (Linthurst et al. 1986a). The high Si levels in this region reflect the weathering of silacious, base-poor saprolites and bedrock (e.g., sandstones, quartzite) that dominate the geology.

The relative importance of the ions that determine the ANC of water in streams can be evaluated using spline diagrams (Chapter 2, Munson and Gherini, this volume). Figure 11.7, for example, illustrates how ion dominance and ionic strength change with decreasing ANC in the NSS target population of streams in this province. The ionic composition is relatively stable, with the ionic strength and the concentration of most ions decreasing proportionally with decreasing ANC. The decline in ionic strength is associated primarily with decreases in Ca^{2+} and HCO_3^-. In high ANC streams, Ca^{2+} is the dominant cation, whereas in low ANC streams, which dominate the NSS target population, Na^+ is the dominant cation (Figure 11.7).

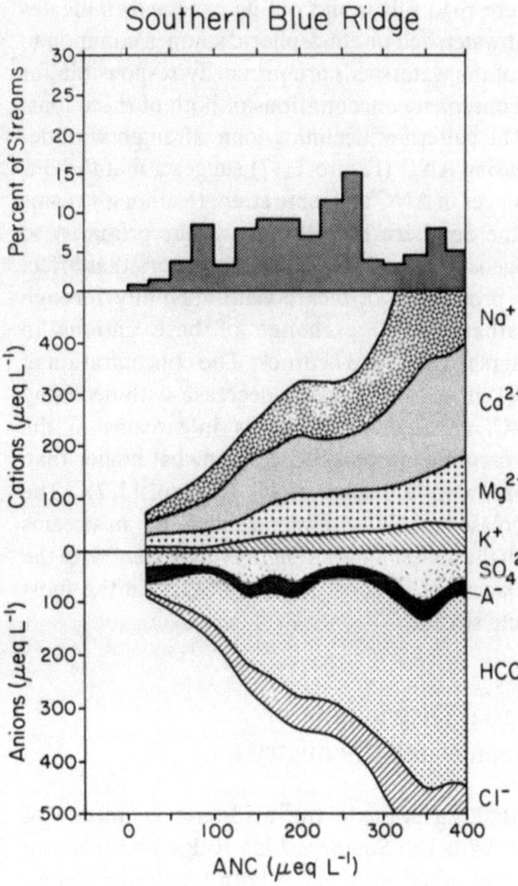

FIGURE 11.7. Concentrations of major cations and anions in NSS streams in the Southern Blue Ridge Province with ANC ≤ 400 μeq L^{-1}. The percentages of streams represented by ANC intervals are shown as a histogram at the top of the Figure. Data plotted are population-weighted estimates for streams in the NSS. The curves are spline fits to the median concentration for each ANC interval. Contributions of cations such as H^+, Al^{n+}, Fe^{n+}, Mn^{n+}, and NH_4^+, and anions such as NO_3^- and Cl^-, which are important in some streams, are exluded here to increase the clarity of presentation.

The high levels of Na^+ could be due to weathering of albite, a sodic plagioclase feldspar, which is present in the parent geologic material in the region (Velbel 1985a,b). Marine aerosols carried aloft and intercepted by these high-elevation watersheds could also be a source of some of the Na^+ and Cl^- in stream water. Although road salt is used in the Southern Blue Ridge, the relatively stable proportion of Na^+ and Cl^- in streams, including those where road salt would not be expected, indicates that watershed or atmospheric sources common to all of the watersheds are primarily responsible for the observed concentrations of both of these ions.

The pattern of declining ionic strength with decreasing ANC (Figure 11.7) suggests that the differences in ANC and ionic strength among streams in the Southern Blue Ridge are due primarily to biogeochemical and hydrologic factors that affect the production of bicarbonate alkalinity through weathering and exchange of base cations in watershed soils and bedrock. The concentration of SO_4^{2-} in streams tends to decrease with declining ANC, although SO_4^{2-} levels in streams in the lowest ANC groupings are somewhat higher than those with a higher ANC (Figure 11.7). The decrease in ANC with increasing SO_4^{2-} in streams with the lowest ionic strength is consistent with the replacement of some ANC by SO_4^{2-} in the most dilute streams.

Spatial Distribution of Streamwater Chemistry

Spatial variation in the acid-base chemistry of streams in the Southern Blue Ridge Province can be considered in terms of both longitudinal (i.e., upstream-downstream) changes within the same watershed and differences among streams in different watersheds. Longitudinal differences in the acid-base chemistry of streams in the NSS target population in this province are clearly evident. Upstream reaches have lower pH and ANC than downstream reaches, whereas strong mineral acid anions in the upstream reaches are either approximately the same as (SO_4^{2-}) or lower (NO_3^-) than those in the downstream reaches (Table 11.3, Figure 11.5). This indicates that the lower ANC and pH in the upstream reaches are generally due to a lower base cation content rather than to a greater concentration of mineral acids.

Similar longitudinal (i.e., upstream-downstream) patterns in the acid-base chemistry of streams have been reported for the Southern Blue Ridge (Elwood et al. 1985, Winger et al. 1987) and for other areas of the eastern United States (Vitousek 1977, Johnson 1979, Kaufmann et al. 1988). Although Johnson (1979) presented no data on mineral acid anions, he suggested that the longitudinal pattern in acidic cations and base cations in Falls Brook, New Hampshire, was a result of the neutralization of acidic deposition containing strong mineral acids by weathering reactions, with the concurrent gain of strong bases and loss of acidic cations from upstream to downstream. Johnson implied that this pattern was due, at least in part, to the characteristics of watershed hydrology, wherein acidic inputs falling on upstream areas of a watershed have less opportunity to be neutralized by weathering and cation exchange reactions compared to those falling on lower portions of the watershed. The extent to which these upstream-downstream differences in base cation levels result from elevational differences in hydrology (e.g., greater precipitation and less evapotranspiration at higher elevations) versus differences in soils and geology (e.g., differences in base saturation of soil and weatherability of bedrock geology) is not known. Similar results for the Southern Blue Ridge, however, suggest that these longitudinal patterns in stream chemistry are common and probably represent a general phenomenon that is characteristic of montane watersheds.

The regional pattern of water chemistry of NSS streams in the Southern Blue Ridge is illustrated with ANC and pH data for the downstream reach ends (Figures 11.8 and 11.9). The spatial pattern in ANC and pH appears to be related to a combination of land use, geology, and elevational differences among watersheds. The approximately contiguous geographic distribution of the major ANC classes (Figure 11.8) suggests that the acid-base chemistry of streams within a grouping is a result of factors common to the watersheds within that area. For example, the NSS stream reaches with the highest ANC (> 200 μeq L^{-1}) and pH are located along the northern and western edges of the province, where limestone from the adjacent Valley and Ridge Province frequently is mixed with the felsic saprolites of the Southern Blue Ridge. The high ANC of streams in this area thus is probably due to the presence of limestone in these watersheds. A second high ANC group is located in the French Broad and Hiawassee River valleys where agriculture and other watershed disturbances are more prevalent. The high ANC of streams in this grouping thus may be related to liming or to other anthropogenic inputs to, or disturbances of, the watersheds within the grouping.

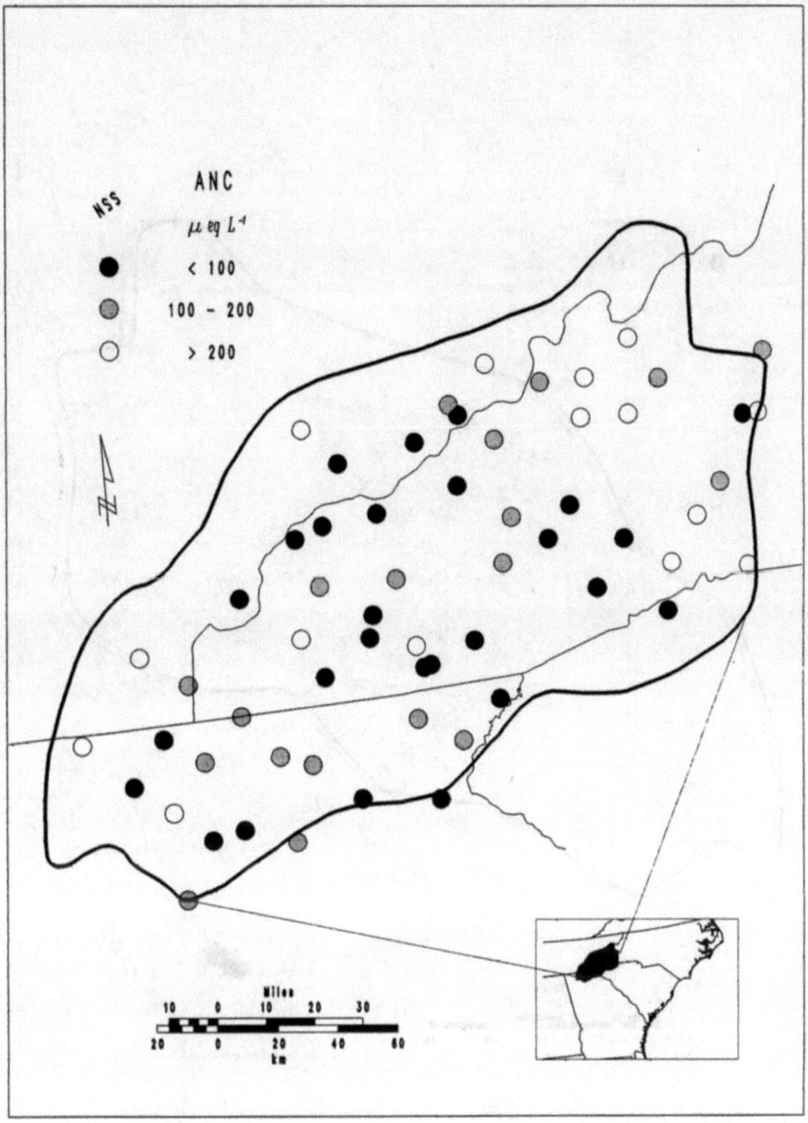

FIGURE 11.8. Geographic distribution of NSS streams in different ANC classes in the Southern Blue Ridge Province in the spring.

Stream reaches in the NSS target population with the lowest ANC ($\leq$ 100 μeq L^{-1}) are located in the highlands in the northcentral part of the Southern Blue Ridge, whereas those in the next highest ANC class (100 to 200 μeq L^{-1}) are located at somewhat lower elevations (Figure 11.8). Log ANC exhibits a significant inverse correlation with stream elevation ($p < 0.001$) in the Southern Blue Ridge, accounting for approximately 27% of the variation in ANC among stream reaches in the NSS target population. The effect of elevation on ANC is probably related to hydrologic and soil factors that regulate the production of alkalinity through their control of weathering and exchange reactions in watershed soils and bedrock. This hypothesis is supported by the fact that the base cation concentration of streams in the NSS target population in this province is also inversely correlated with elevation ($r = -0.43$; $p < 0.001$). The influence of watershed processes on the acid-

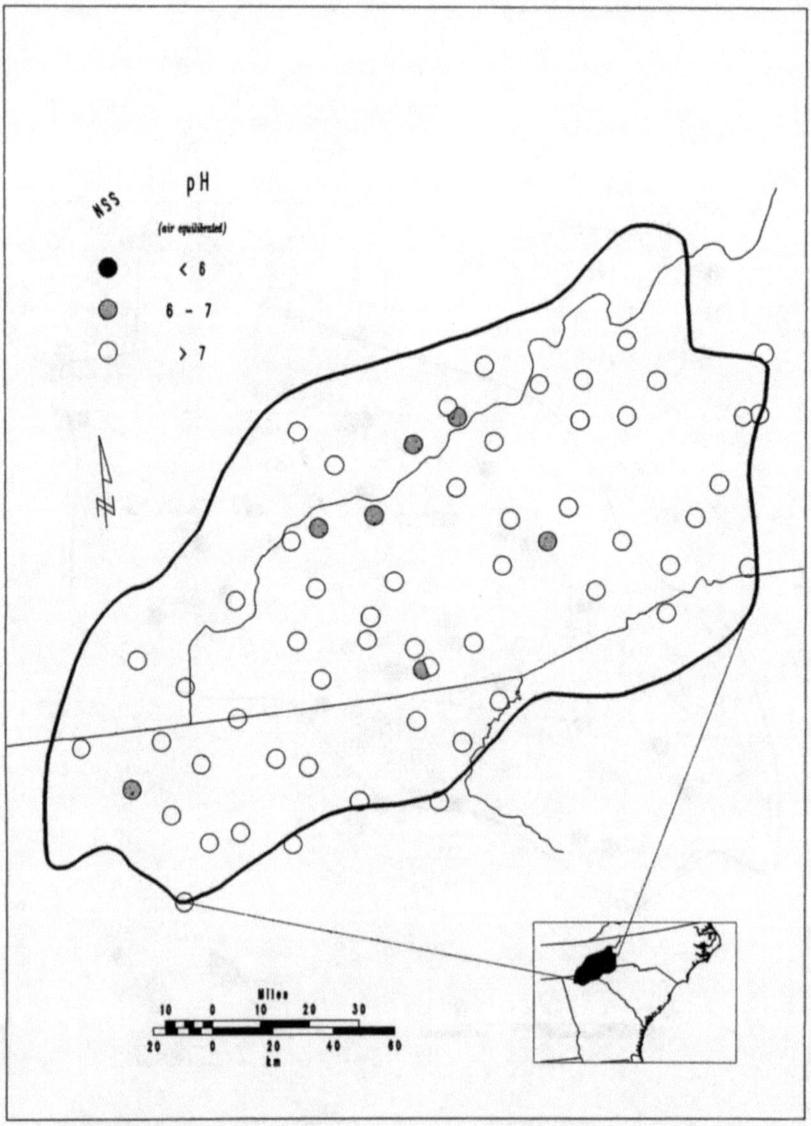

FIGURE 11.9. Geographic distribution of NSS streams in different pH classes in the Southern Blue Ridge Province in the spring.

base chemistry of surface waters in this region is examined in more detail in the section on watershed processes.

Association Among Chemical Variables

Effects of Weak Acids on pH

As discussed by Munson and Gherini (Chapter 1, this volume), the pH of surface waters is controlled primarily by a combination of carbonate species, weakly ionizable organic acids, and carbonic acid (pCO_2). The effect of a weak volatile acid (carbonic acid) on stream pH in the Southern Blue Ridge was determined by comparing the pH of samples isolated from the atmosphere, in order to prevent loss of CO_2 (so-called closed-system pH), with the pH of samples equilibrated with 300 ppm of CO_2 (i.e., $pCO_2 = 10^{-3.5}$ atm.), which approximates equilibrium with atmospheric CO_2. The

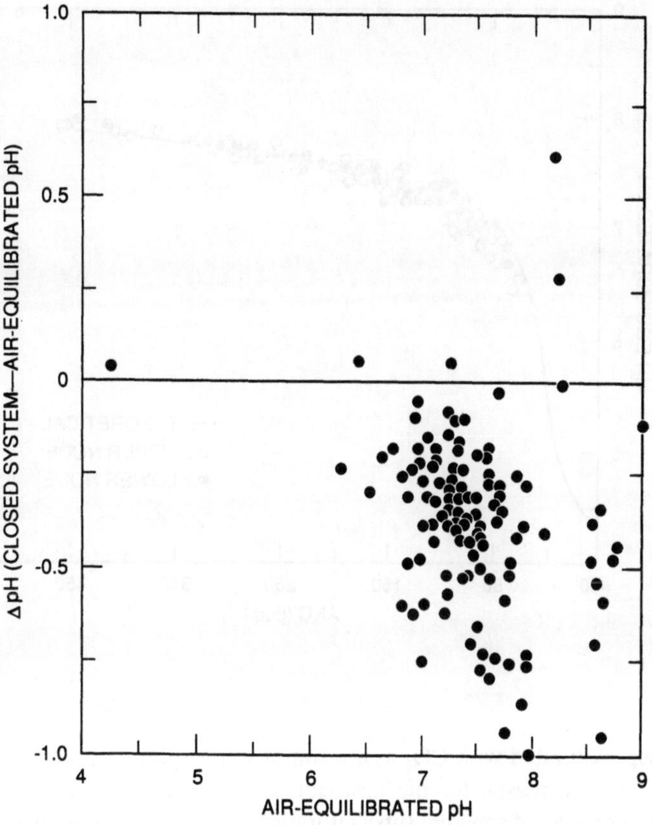

FIGURE 11.10. ΔpH plotted as a function of the air-equilibrated pH of streamwater samples from upstream and downstream ends of NSS stream reaches in the Southern Blue Ridge Province. Negative values indicate that the sample was supersaturated with respect to the pCO_2 of the atmosphere, whereas positive values indicate that the sample was undersaturated.

ΔpH, defined as the difference between the closed-system pH and the air-equilibrated pH, of nearly all samples was negative (Figure 11.10), indicating that the pH of stream water from the NSS reaches in the Southern Blue Ridge increased when equilibrated with 300 ppm of CO_2. A similar increase in pH was also noted for lakewater samples (Table 11.4), which indicates that the samples were either supersaturated with CO_2 or that they contained some other volatile weak acid that decreased in concentration when the samples were air equilibrated with CO_2.

To assess whether the change in pH was due to a systematic error in the pH measurement, the theoretical pH of NSS stream samples in equilibrium with atmospheric CO_2, as calculated from the dissolved inorganic carbon (DIC) data, was compared to the air-equilibrated pH (Figure 11.11). These data show that all the samples plot on or very near the expected pH curve, demonstrating that the air-equilibrated pH is consistent with the theoretically expected value. These results indicate that there is very little if any noncarbonate ANC affecting the acid-base chemistry of these streams and that the increase in pH when samples are air equilibrated is due to a volatile weak acid, which is most likely carbonic acid.

The effect of volatile weak acids on stream pH amounts to a depression of 0.75 units in approximately 20% of the upstream and downstream ends of stream reaches in the Southern Blue Ridge (Figure 11.10). The ΔpH is somewhat greater at the upstream sites than at the downstream sites, indicating that the upstream sites contain more of the volatile weak acid, presumably because of greater supersaturation with CO_2 than at the downstream sites. The hypothesis that carbonic acid is the volatile weak acid causing the change in pH in these streams is partially confirmed by the fact that the median DIC concentrations in stream water declined by approximately 34 μmol L^{-1} when the samples were air equilibrated with CO_2. The cause of the CO_2 supersaturation is assumed to be primarily soil water and groundwater entering the stream

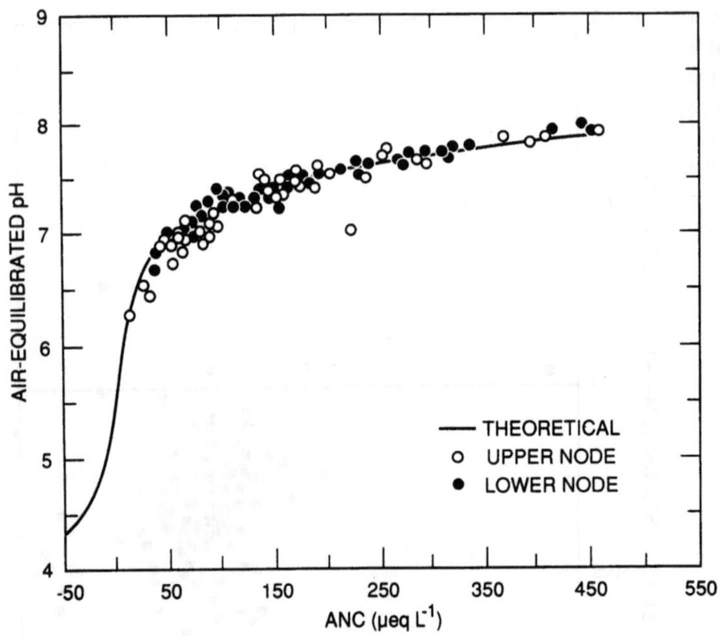

FIGURE 11.11. Theoretical versus observed pH of stream water equilibrated with 300 ppm CO_2 as a function of the ANC at the upstream and downstream ends of NSS stream reaches in the Southern Blue Ridge Province.

supersaturated with CO_2 as a result of soil respiration. The reason for the upstream sites having a higher level of apparent supersaturation is unknown but may be due to greater groundwater inputs in the upper reaches and less time for the excess CO_2 in stream water to equilibrate with the atmosphere.

The fact that the streams were still supersaturated with CO_2 during the day when the samples were collected indicates that CO_2 inputs in groundwater and soil water exceed losses from plant uptake and outgassing throughout the 24-hour period in these high gradient, turbulent streams. The pCO_2 of most streams in this province thus probably remains above air saturation throughout the day and night, resulting in a continuous depression of pH. Although excess CO_2 would be expected to degas rapidly from the water once groundwater reached the open channel of a turbulent stream, depression of stream pH due to excess CO_2 appears to be a common phenomenon (Kaufmann et al. 1988). Neal (1988) reported that the pCO_2 of water samples, based on calculated values from pH and alkalinity data from two streams in Wales, commonly exceeded the atmospheric level by a factor of 2 to 3, and some samples were more than 10 times the atmospheric level. Together these results indicate that carbonic acid is an important source of acidity even in high-gradient turbulent streams, such as those in the Southern Blue Ridge, and that in measuring pH, caution must be used to minimize or correct for CO_2 loss from samples.

Organic Anions

Weak Organic Anions (A_w^-)

As discussed by Munson and Gherini (Chapter 1, this volume), weak organic anions react with hydrogen ions during a Gran ANC titration. Although the definition is operational, it provides information on the concentration of the weak organic anions that may buffer H^+ for surface waters with pH values between 4.5 and 7.0. In Southern Blue Ridge streams, the measured air-equilibrated pH agrees well with that expected for a pure carbonate system (Figure 11.11), based on the measured DIC in stream water. This indicates that at least during the spring baseflow, there is very little if any noncarbonate ANC in streams in this province to buffer H^+ over a pH range of 4.5 to 7.0. Hence, weak organic anions do not appear to be important as a source of ANC in streams in this region, a result that is not surprising given the low concentrations of DOC (medians = 55 and 64 μmol L^{-1} during spring at the upstream and downstream sites, respectively).

Nonprotonated Organic Anions (An^-)

Nonprotonated organic anions do not react with H^+ during a Gran ANC titration and hence do not appreciably neutralize H^+ or exert control on pH in waters with pH between 4.5 and 7.0. The concentration of An^- in the NSS streams in the Southern Blue Ridge was estimated from the difference in the charge balance:

$$An^- = [cations] - [anions]$$

where:

$[cations] = Ca^{2+} + Mg^{2+} + Na^+ + K^+ + NH_4^+ + H^+$

$[anions] = SO_4^{2-} + NO_3^- + Cl^- + [ANC+H^+] + F^-$

ANC encompasses both the carbonate and noncarbonate protolytes, including the weak organic anions. Aluminum and manganese were not included in the charge balance because of the low concentrations of these metals in the circumneutral streams in this region (median extractable aluminum (Al_{ext}) and dissolved Mn of approximately 4 and 6 μg L^{-1}, respectively, at both upstream and downstream sites). The calculated anion deficit was regressed against the DOC concentration at upstream and downstream sites combined. Although the regression was significant ($p < 0.01$), it was relatively weak, with DOC accounting for only 12% of the variation in the anion deficit among NSS stream reaches in the region. Based on this, we conclude that most of the anion deficit among streams in the Southern Blue Ridge target population is due to analytical error rather than to unprotonated organic anions, and that organic acids are a minor source of acidity in streams in this region during baseflow.

Geochemical Relationships Between ANC, Base Cations, and Strong Acid Anions

To assess the relative importance of strong mineral acids, organic anions, and strong bases as sources of variation in ANC among streams in the Southern Blue Ridge, standardized multiple regression analysis was conducted using data for (1) all streams except those with Cl^- concentration ≥ 40 μeq L^{-1} and (2) streams with Cl^- < 40 μeq L^{-1} and ANC ≤ 200 μeq L^{-1}. Streams with Cl^- ≥ 40 μeq L^{-1} were excluded from the analysis to eliminate the potentially confounding effect of anthropogenic and watershed sources of Cl^- in some watersheds. A Cl^- concentration of 40 μeq L^{-1} is the maximum that could reasonably be expected in stream water from the evapoconcentration of Cl^- in wetfall in watersheds in this province.

Comparing the absolute value of the standardized regression coefficients provides a means of ranking the relative importance of each independent variable in explaining the among-stream variation in ANC within the region. Thus, a greater absolute value of the standardized regression coefficient for SO_4^{2-} compared to that for NO_3^-, for example, indicates that SO_4^{2-} is relatively more important than NO_3^- in explaining the among-stream variation in ANC. The results of these standardized regression analyses describe only the statistical associations among the chemical variables; they do not prove causality.

The five independent variables used in the standardized regression model were the sum of base cations (SBC = $Ca^{2+} + Mg^{2+} + K^+ + Na^+$), SO_4^{2-}, NO_3^-, Cl^-, and DOC. DOC was used as a surrogate for organic acids to avoid using estimates of organic acidity derived from the charge imbalances or from the difference between the calculated and measured alkalinity. Such calculated values of organic acidity may not be completely independent of the other independent variables used in the regression model because the estimated organic acidity is derived using the other independent variables in the model.

The standardized coefficients show that SBC ranks as the most important factor in explaining the variation in ANC among streams in the Southern Blue Ridge, regardless of whether all streams are considered or just those with ANC ≤ 200 μeq L^{-1} (Figure 11.12a,b). The five independent variables together account for ≥ 98% of the among-stream variation at both the upstream and downstream reach ends, confirming the charge balance.

When the other independent variables are ranked in order of importance, SO_4^{2-} is second to base cations when all streams are considered (Figure 11.12a). Sulfate, however, is substantially less important than base cations in explaining the variation in ANC among streams in this region, and NO_3^- and Cl^- are less important than SO_4^{2-}. The standardized regression coefficient for NO_3^- is only marginally significant ($p = 0.10$) at the upstream

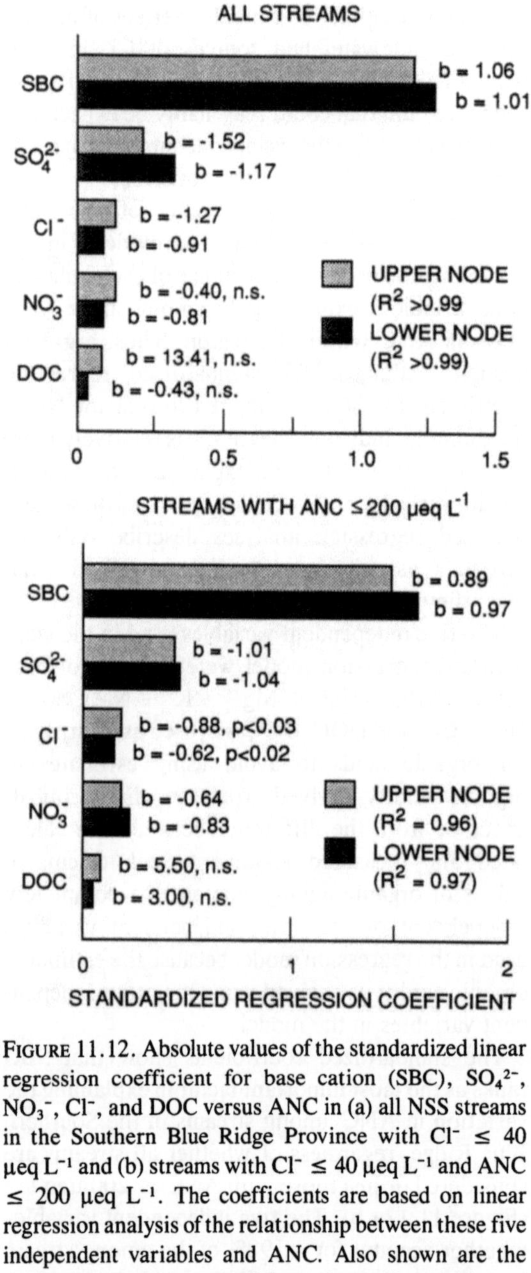

FIGURE 11.12. Absolute values of the standardized linear regression coefficient for base cation (SBC), SO_4^{2-}, NO_3^-, Cl^-, and DOC versus ANC in (a) all NSS streams in the Southern Blue Ridge Province with $Cl^- \leq 40$ µeq L^{-1} and (b) streams with $Cl^- \leq 40$ µeq L^{-1} and ANC ≤ 200 µeq L^{-1}. The coefficients are based on linear regression analysis of the relationship between these five independent variables and ANC. Also shown are the estimated regression coefficients for each variable and the coefficient of determination (R^2) for the full model (all variables included). Unless otherwise noted, the standardized regression coefficients are highly significant ($p < 0.001$).

sites, and for DOC it is not statistically significant at either the upstream or the downstream sites ($p > 0.23$) (Figure 11.12a). This supports earlier findings that organic acids do not contribute significantly to the acid-base chemistry of streams in the NSS target population in the Southern Blue Ridge during baseflow periods.

For streams with ANC ≤ 200 µeq L^{-1} and Cl^- < 40 µeq L^{-1}, SO_4^{2-} and NO_3^- are highly significant ($p < 0.001$) sources of among-stream variation in ANC, whereas Cl^- is only marginally significant ($p < 0.05$) (Figure 11.12b). All these acid anions, however, rank well below base cations in importance. Again, DOC is nonsignificant at both the upstream and down-stream sites, indicating that organic acids are not a source of variation in ANC among streams in this group during spring baseflow. These results thus indicate that differences in ANC among poorly buffered streams in the Southern Blue Ridge are due primarily to variation in the base cation levels. The fact that SO_4^{2-} and NO_3^- account for much less of the variation in ANC than base cations among Southern Blue Ridge streams is evidence that watershed retention of acid anions in this region is more uniform spatially than base cation production.

Evidence of Acidification in the Southern Blue Ridge Province

In areas where SO_4^{2-} was the major acid anion in precipitation and there was little sulfur supplied by weathering, Henriksen (1979, 1980) and others have shown that the difference (in equivalents) between the divalent base cations and alkalinity was directly proportional to the nonmarine SO_4^{2-} concentration. Differences between the nonmarine base cation equivalents and the ANC or alkalinity were assumed to provide a quantitative measure of acidification. This approach for estimating acidification, however, is based on the assumption that the discrepancy between ANC and base cation equivalents is caused entirely by the input of strong acids that titrate some of the ANC and that the concentration of base cations has remained constant over time (i.e., no increase in the export of base cations) (Chapter 3, Husar et al., this volume).

The relationship between nonmarine base cations (C_B) and ANC was examined to determine if the base cation content alone could account for the differences in ANC among streams. Figure 11.13a shows that the ANC values are displaced from those expected if an equivalent of alkalinity is produced for every equivalent of base cations released in the watershed (i.e., SBC exceeds measured ANC in almost all Southern Blue Ridge streams). Although the difference is relatively small for most streams, the fact that most of the streams plot below the one-to-one line indicates a relatively consistent, region-wide difference, resulting from a significant contribution of strong acid anions to the charge balance. As a result, there is an excess of base cations relative to that expected from carbonic acid weathering alone. This excess may result from (1) mineral acid weathering and exchange, (2) a decline in ANC due to inputs of strong acids, (3) weathering of neutral salts of SO_4^{2-} (e.g., gypsum), or (4) some combination of these.

A plot of [ANC plus nonmarine SO_4^{2-}] versus SBC in stream water shows that most of the streams plot on or near the one-to-one line (Figure 11.13b). Stepwise regression analysis of the difference between the observed ANC and SBC versus the individual acid anions shows that the nonmarine SO_4^{2-} alone can explain 94% ($p < 0.01$) and 53% ($p < 0.01$) of the difference at the downstream and upstream sites, respectively.

Because watersheds differ in their capacity for neutralizing acidity, as a result of differences in their capacity for mobilizing base cations and retaining acid anions, the concentration of strong mineral acid anions such as SO_4^{2-} is expected to correlate better with the loss of ANC than with ANC itself (Sullivan et al. 1988a). As a consequence, data on historical changes in ANC are necessary to unequivocally prove that acidification has occurred. However, when the capacity of watersheds for mobilizing base cations, as reflected by their current SBC, is taken into account, correlations of mineral acid anions and ANC may exist for streams that drain catchments with a limited capacity for neutralizing acidity by releasing base cations.

To examine whether base cation content provides an indication of the susceptibility of streams to acidification, streams were stratified according to their [Ca^{2+} + Mg^{2+}] concentration. Acid neutralizing capacity was then regressed against SO_4^{2-} concentration for each class (Figure 11.14). In the lowest base cation class ($\leq$ 100 μeq L^{-1}), there was a significant inverse relationship between ANC and SO_4^{2-} concentration ($p < 0.001$), with SO_4^{2-} explaining 25% of the variation in ANC among streams (Figure 11.14). In contrast, the regression coefficients of ANC versus SO_4^{2-} for the higher base cation classes (100 to 200, 200 to 300, and 300 to 400 μeq L^{-1}) were not significantly different from zero, indicating that there was no significant relationship between ANC and SO_4^{2-} in any of these groups of streams with a higher base cation content.

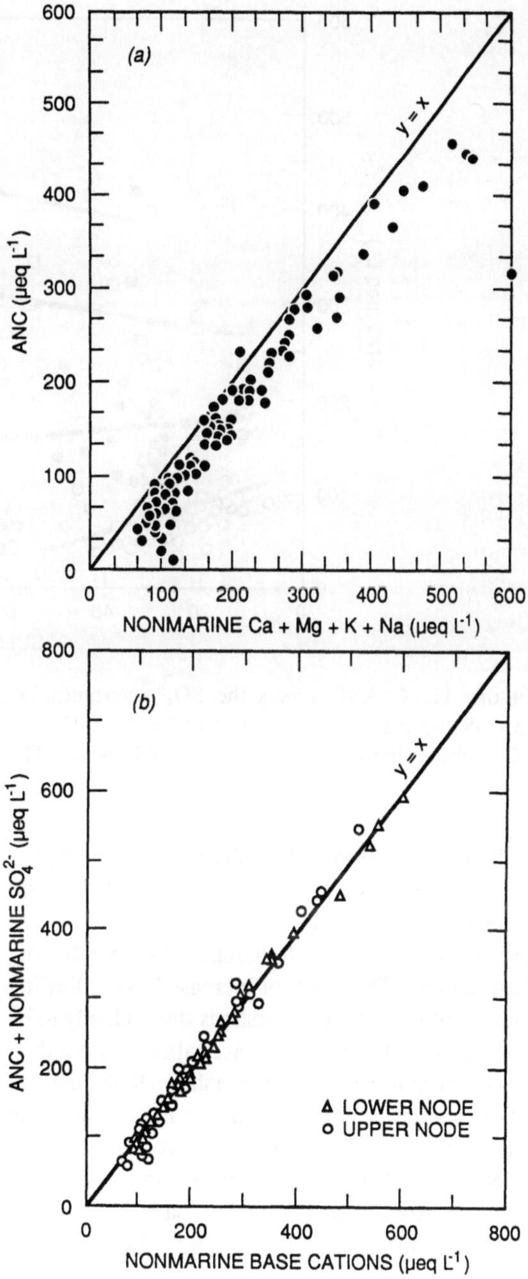

FIGURE 11.13. (a) ANC versus the sum of nonmarine base cations, and (b) ANC + nonmarine SO_4^{2-} versus the sum of nonmarine base cation concentration in NSS streams in the Southern Blue Ridge Province.

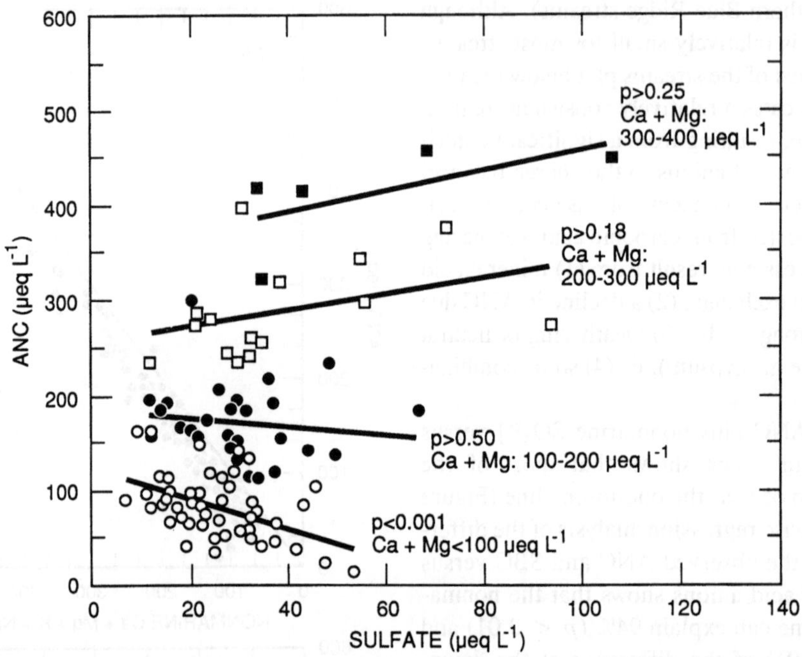

FIGURE 11.14. ANC versus the SO_4^{2-} concentration in NSS streams in different base cation (i.e., Ca^{2+} + Mg^{2+}) classes in the Southern Blue Ridge Province. The regression coefficient is significantly different from zero ($p < 0.05$) only in the lowest base cation class (≤ 100 µeq L^{-1}).

This pattern suggests that streams with the higher base cation content (i.e., > 100 µeq L^{-1}) drain watersheds containing soils and bedrock with a greater capacity for neutralizing acidity by releasing base cations. The effect of increased SO_4^{2-} loading and mobility in these watersheds thus is likely to be an increase in the export of base cations, rather than acidic cations from the watershed. Reducing the SO_4^{2-} loading to these streams might reduce the export of base cations but would probably have no effect on the ANC of stream water. In contrast, the ANC of streams in the lowest base cation class (≤ 100 µeq L^{-1}) might increase if the SO_4^{2-} loading were lower.

Additional evidence of the effects of SO_4^{2-} on streams in the Southern Blue Ridge was presented by Eshleman and Kaufmann (1988). They used a chemical model similar to Henriksen's (1979), which assumes that stream waters in the Southern Blue Ridge have been titrated by an amount of sulfuric acid equivalent to the present SO_4^{2-} concentration in stream water, and that historical pH and ANC can be calculated from current chemical data. They further assumed that all of the increased SO_4^{2-} was compensated for by a decrease in the ANC (i.e., $F = 0$), rather than an increase in the production of base cations from cation exchange or accelerated mineral weathering. Based on these worst case assumptions, the median historical decline in ANC and pH of streams in the Southern Blue Ridge attributable to atmospheric deposition of sulfuric acid was estimated to be 23 µeq L^{-1} and 0.09 units, respectively. The relatively small predicted changes are consistent with the finding of very few acidic streams in this region and the hypothesized delayed response of these watersheds to SO_4^{2-} inputs (Galloway et al. 1983). Nonetheless, there is an inverse correlation between the ratio of SO_4^{2-} to base cations and ANC (Figure 11.15a; Eshleman and Kaufmann 1988), indicating that the importance of SO_4^{2-} to the charge balance increases with declining ANC. This increase in the SO_4^{2-}:C_B ratio with decreasing ANC is due to lower concentrations of base cations for relatively constant concentrations of SO_4^{2-} in low ANC waters. A similar pattern exists for lakes and reservoirs in the region (Figure 11.15b). Although this pattern alone does not prove that there has been a loss of ANC in streams in the Southern Blue Ridge due to SO_4^{2-} inputs, it is con-

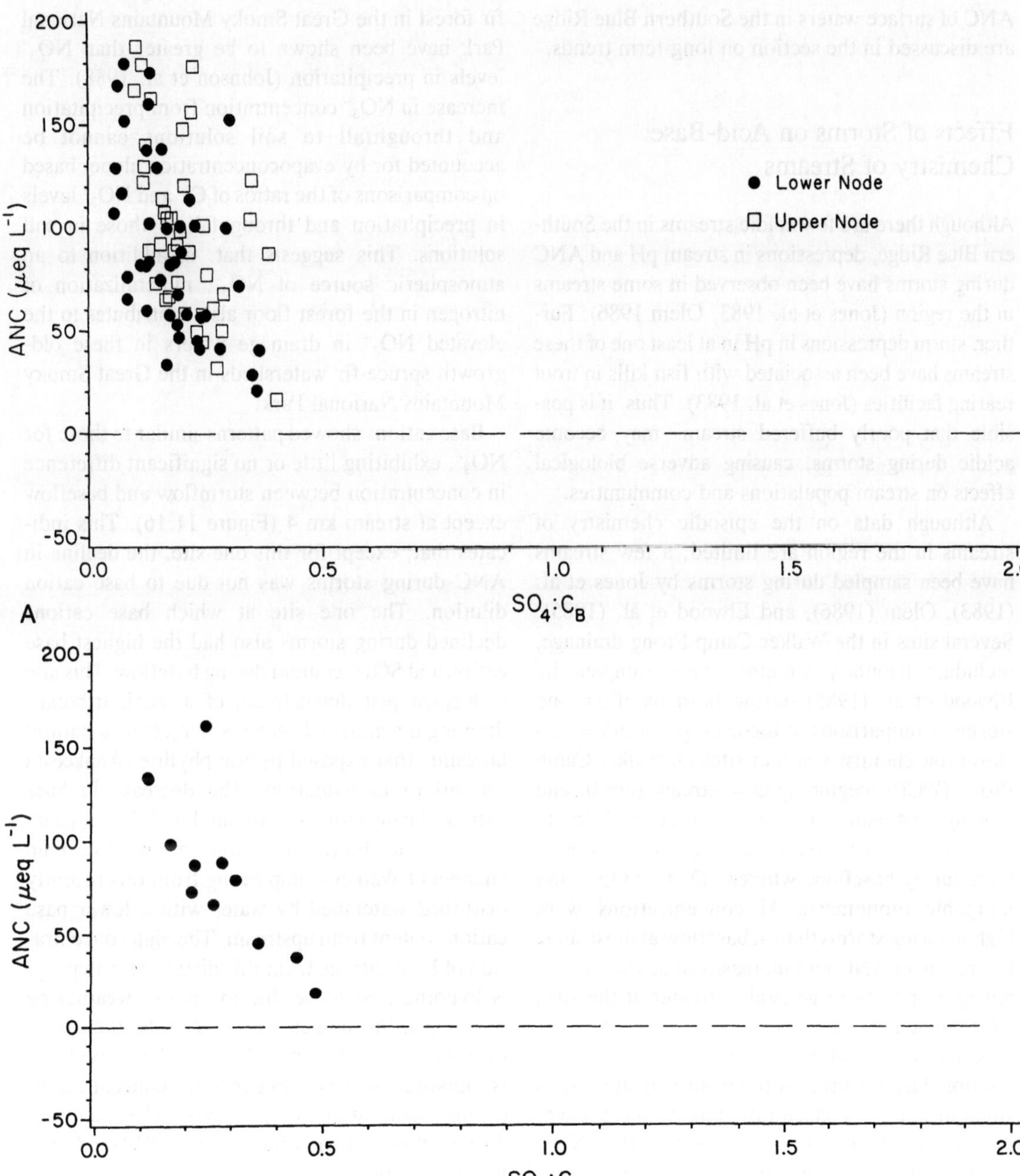

FIGURE 11.15. ANC versus the ratio of SO_4^{2-} to the sum of base cations in (a) low DOC (≤ 417 μmol L^{-1}) NSS streams in the Southern Blue Ridge Province and (b) low DOC (≤ 417 μmol L^{-1}) ELS lakes in the Southern Blue Ridge Province with Cl^- ≤ 40 μeq L^{-1}.

sistent with the results of the sulfuric acid titration model, suggesting that low ANC streams in this region have experienced some acidification.

Thus, surface waters in the Southern Blue Ridge, particularly those with a low base cation content, may be exhibiting a small response to SO_4^{2-} deposition on the watersheds. Such a response would be expected from a reduction in the SO_4^{2-} adsorption capacity of the soils as they approach a new steady state. Evidence of long-term changes in the pH and

ANC of surface waters in the Southern Blue Ridge are discussed in the section on long-term trends.

Effects of Storms on Acid-Base Chemistry of Streams

Although there are few acidic streams in the Southern Blue Ridge, depressions in stream pH and ANC during storms have been observed in some streams in the region (Jones et al. 1983, Olem 1986). Further, storm depressions in pH in at least one of these streams have been associated with fish kills in trout rearing facilities (Jones et al. 1983). Thus, it is possible that poorly buffered streams may become acidic during storms, causing adverse biological effects on stream populations and communities.

Although data on the episodic chemistry of streams in the region are limited, a few streams have been sampled during storms by Jones et al. (1983), Olem (1986), and Elwood et al. (1985). Several sites in the Walker Camp Prong drainage, including tributary streams, were sampled by Elwood et al. (1985) during both baseflow and storms. Comparisons of the average baseflow and stormflow chemistry at four sites on Walker Camp Prong (WCP), beginning downstream (km 0) and moving upstream, are shown in Figure 11.16. In general, pH and ANC were lower during storms than during baseflow, whereas DOC, SO_4^{2-}, and inorganic monomeric Al concentrations were higher during storms than at baseflow at most sites. Decreases in ANC and increases in acidic cations during storms were generally greater at the sites with the highest ANC.

Nitrate exhibited either no change or a slight dilution during storms at the upstream sites and a slight increase in concentration at the most downstream site (Figure 11.16). The fact that NO_3^- levels in stream water during these storms were greater than NO_3^- levels in precipitation in the region (Figure 11.4) implies that the watershed has a large capacity for supplying NO_3^- to drainage waters and maintaining a relatively constant NO_3^- concentration in stream water. Vegetation in the upper portions of this watershed is old-growth forest, in which nitrogen uptake would be expected to be low (Vitousek 1977). In addition, NO_3^- levels in soil solutions from the A and B horizons from a high-elevation watershed with old-growth spruce-fir forest in the Great Smoky Mountains National Park have been shown to be greater than NO_3^- levels in precipitation (Johnson et al. 1988). The increase in NO_3^- concentration from precipitation and throughfall to soil solutions cannot be accounted for by evapoconcentration alone, based on comparisons of the ratios of Cl^- and NO_3^- levels in precipitation and throughfall to those in soil solutions. This suggests that, in addition to an atmospheric source of NO_3^-, mineralization of nitrogen in the forest floor also contributes to the elevated NO_3^- in drainage waters in these old-growth spruce-fir watersheds in the Great Smoky Mountains National Park.

Base cations showed patterns similar to those for NO_3^-, exhibiting little or no significant difference in concentration between stormflow and baseflow except at stream km 4 (Figure 11.16). This indicates that, except for this one site, the decline in ANC during storms was not due to base cation dilution. The one site at which base cations declined during storms also had the highest base cation and SO_4^{2-} content during baseflow. This site is located just downstream of a small tributary draining a watershed recently subject to a natural landslide that exposed pyritic phyllite (Anakeesta formation) to oxidation. The decrease in base cations during storms at stream km 4 thus appears to be due to dilution of water entering the main channel of Walker Camp Prong from this recently disturbed watershed by water with a lower base cation content from upstream. The high concentration of base cations from this disturbed watershed is hypothesized to be due to intense weathering resulting from the generation of H^+ by the oxidation of the recently exposed pyrite. This hypothesis is supported by the decline in SO_4^{2-} concentrations during storms at this site, whereas SO_4^{2-} increases during storms at the other sites on Walker Camp Prong (Figure 11.16).

The higher levels of SO_4^{2-} in Walker Camp Prong during storms indicates that at least a fraction of the stream acidification during storms is due to increased SO_4^{2-} loading to the stream. Increases in SO_4^{2-} during storms have been reported for other streams in the southeastern United States and are attributed to the lateral flow of water through upper soil horizons containing a higher proportion of water-soluble SO_4^{2-} and having a lower SO_4^{2-} adsorption capacity (Johnson and

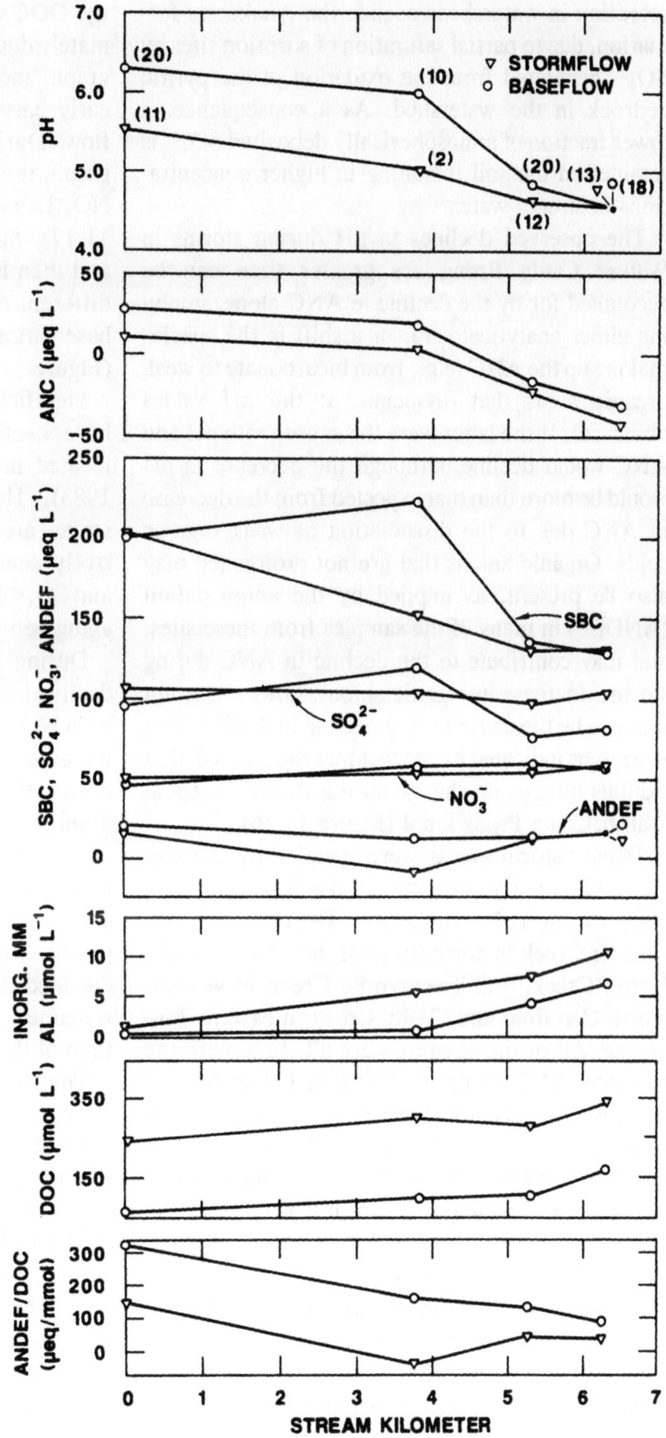

FIGURE 11.16. Stormflow versus baseflow chemistry of stream water at several sites in Walker Camp Prong located in the Great Smoky Mountains National Park. Stream kilometers are plotted from downstream (stream km = 0) to the most upstream site (stream km = 6.3).

Henderson 1979, Joslin et al. 1987, Swank and Waide, 1988). The higher levels of SO_4^{2-} in Walker Camp Prong during baseflow compared to those in other streams in the Southern Blue Ridge are most likely to be caused by the bedrock source of SO_4^{2-} in this watershed. The retention of atmospherically deposited SO_4^{2-} by soils in the upland portions of this watershed may also be lower than retention in watersheds outside the Anakeesta formation, due to partial saturation of sorption sites by SO_4^{2-} generated from the oxidation of the pyritic bedrock in the watershed. As a consequence, a lower fraction of atmospherically deposited SO_4^{2-} is retained in the soil, resulting in higher concentrations in stream water.

The observed declines in pH during storms in Walker Camp Prong are greater than can be accounted for by the decline in ANC alone, implying either analytical error or a shift in the species making up the ANC (e.g., from bicarbonate to weak organic acids that dissociate at the pH values observed). If the latter were the cause, both pH and ANC would decline, although the decrease in pH would be more than that expected from the decrease in ANC due to the dissociation of weak organic acids. Organic anions that are not protonated may also be present, as implied by the anion deficit (ANDEF) in many of the samples from these sites, and may contribute to the decline in ANC during storms. Increase in organic anions, however, is not likely to be the cause for the decline in ANC during storms, as indicated by the fact that the anion deficit exhibits little or no change during storms except at Walker Camp Prong km 4 (Figure 11.16).

Twenty storm events were sampled by Olem et al. (1986) in five streams in the Southern Blue Ridge during a one-year period. The five sites were Chester Creek in northern Georgia; Moses Creek, Jarrett Creek, and Twentymile Creek in western North Carolina; and Cosby Creek in eastern Tennessee. All of the streams were alkaline, with the baseflow ANC being < 100 μeq L^{-1} in four of the five streams; the lowest was 28 μeq L^{-1} in Chester Creek, Georgia, and the highest was 134 μeq L^{-1} in Moses Creek, North Carolina. When sampled during a major storm in February, Cosby Creek had an average baseflow ANC of approximately 36 μeq L^{-1}. Samples were not collected immediately before the increase in stage height, but ANC and pH remained relatively unchanged over the rising limb of the storm hydrograph, when stage height increased by approximately 1 m (Figure 11.17). During this same period, combined SO_4^{2-} and NO_3^- increased by 57 μeq L^{-1}, whereas base cations increased by 43 μeq L^{-1}. The lack of a decrease in ANC apparently was due to an increase in weak (partially dissociated) organic acids that contribute to the noncarbonate alkalinity. DOC concentrations in stream water approximately doubled during the rising limb of the hydrograph, increasing from 100 μmol L^{-1} during the early part of the storm to 175 μmol L^{-1} at peak flow. During the descending limb of the hydrograph, the concentrations of acid anions [SO_4^{2-} + NO_3^-], base cations, and DOC declined (Figure 11.17). Surprisingly, ANC and pH also declined and then increased slightly, apparently due to the different rates of change in the concentrations of base cations, acid anions, and weak organic acids (Figure 11.17).

Significant increases in H^+ and Al during storms have been reported for Raven Fork, a stream located in western North Carolina (Jones et al. 1983). The baseflow pH and alkalinity of this stream are 5.7 and 20 μeq L^{-1}, indicating it is relatively sensitive to acidic inputs. Baseflow NO_3^- and SO_4^{2-} levels are approximately the same, averaging between 22 and 33 μeq L^{-1}.

During a major storm in Ravens Fork, pH declined to 4.3 (Figure 11.18). Concentrations of both NO_3^- and SO_4^{2-} increased to levels exceeding the base cation equivalents. Nitrate remained high during the descending limb of the storm hydrograph, whereas SO_4^{2-} declined. Base cations increased during the rising limb of the storm hydrograph, then declined following the peak discharge, a pattern similar to that observed for Cosby Creek (Figure 11.18). The pH of the stream declined during the early part of the storm and remained below 5.0 throughout the descending limb of the hydrograph (Figure 11.18).

The mineral acid anion equivalents (SO_4^{2-} + NO_3^-) exceeded those for base cations throughout most of the storm, suggesting that mineral acidity was the source of the low pH during much of the storm. At certain times during the descending limb of the hydrograph, however, the pH was < 5.0 when the ratio of base cations to acid anions was > 1. This suggests that other sources of acidity (e.g., carbonic acid, Al, organic acids) also contributed to the low pH during the descending limb of the hydrograph. Unfortunately, the potential contribution of weak acids cannot be assessed, since no data on ANC, inorganic Al, and DOC are available for this storm. Increases in total organic carbon during storms have been reported for other storms in this watershed (Jones et al. 1983). Total Al levels also increased during the storm, reaching

FIGURE 11.17. Changes in stage height and stream chemistry during a major storm in Cosby Creek located in the Great Smoky Mountains National Park. (Data from Olem 1986.)

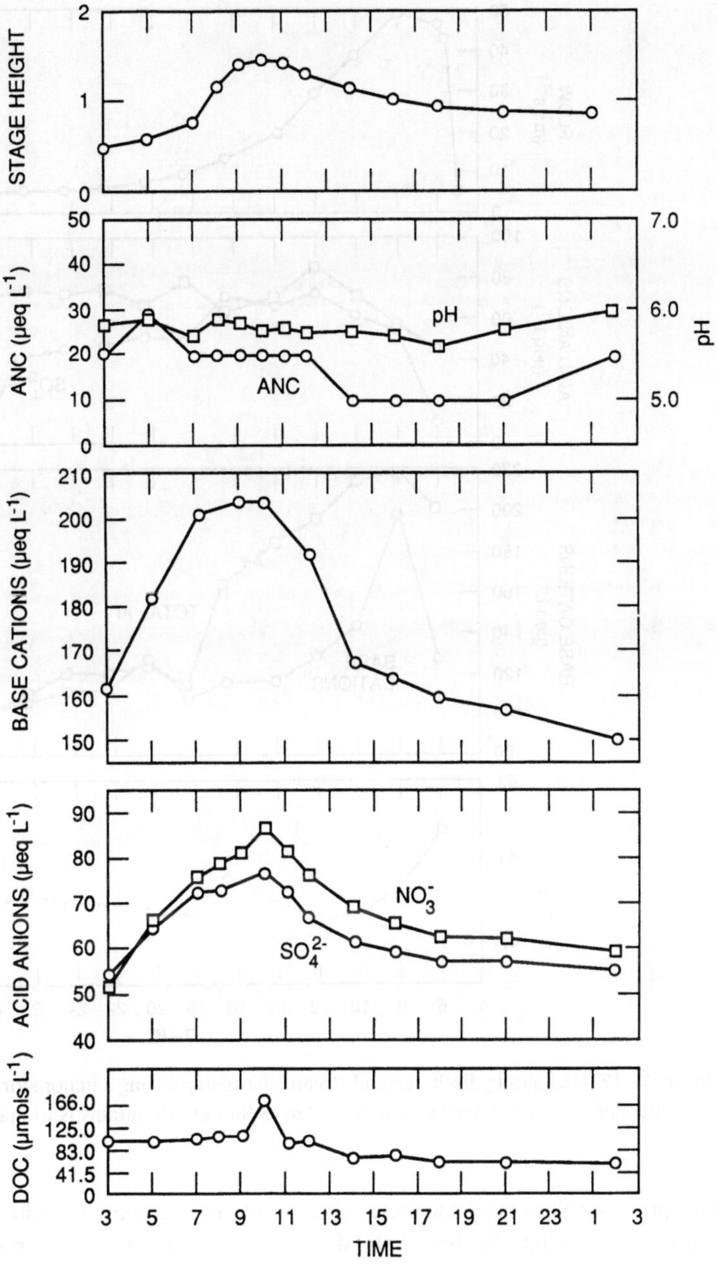

a maximum at peak flow and then declining (Figure 11.18). The fraction of the total Al that was soluble however is unknown.

Vegetation type on watersheds and changes in the flow routing of water through soils during storms may have an important effect on the chemical response of Southern Blue Ridge streams to increased discharge. Joslin et al. (1987) reported that concentrations of SO_4^{2-}, Al, H^+, and undissociated organic acids in soil solutions collected from the O horizon of soils in the Raven Fork watershed were greater than in throughfall samples, indicating that the O horizon was a source of some of the acidity in soil water. Concentrations of these ions in soil solutions were also significantly greater at sites dominated by spruce (*Picea rubens*) and birch (*Betula* sp.) than at other sites dominated by hardwoods in the same watershed, indicating

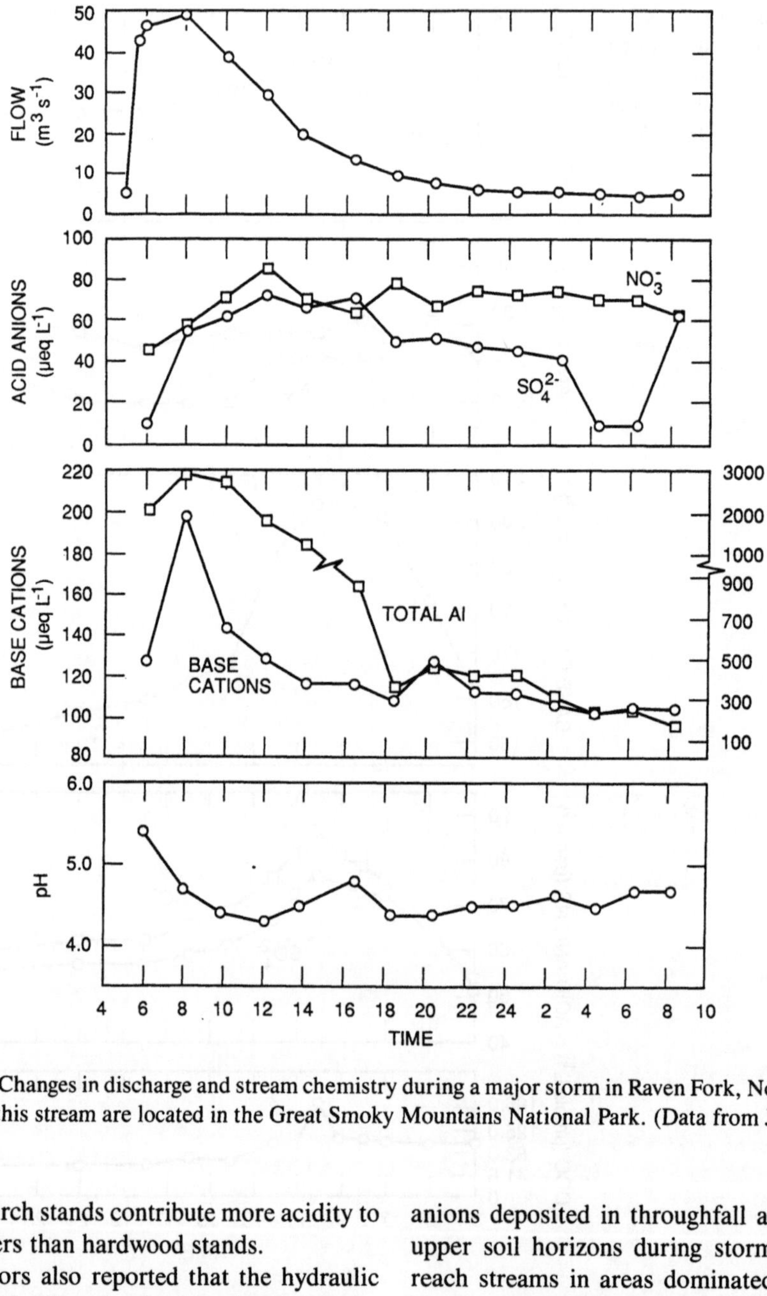

FIGURE 11.18. Changes in discharge and stream chemistry during a major storm in Raven Fork, North Carolina. The headwaters of this stream are located in the Great Smoky Mountains National Park. (Data from Jones et al. 1983.)

that spruce-birch stands contribute more acidity to drainage waters than hardwood stands.

These authors also reported that the hydraulic conductivity of soils under hardwood stands in the Raven Fork watershed was significantly greater than that of soils under spruce-birch stands. They demonstrated that SO_4^{2-} and NO_3^- retention by the mineral soil at spruce-birch sites was not as effective in removing these anions from drainages waters as at the hardwood sites because of the more rapid flow of water through macropores in and below the mineral soil. Joslin et al. (1987) concluded that, as a consequence, more of the acid anions deposited in throughfall and mobilized in upper soil horizons during storms was likely to reach streams in areas dominated by spruce and birch compared to that in areas dominated by hardwoods. The pH depressions observed in Raven Fork during storms thus may be related to rapid subsurface flow and interflow of water emanating from the acid spruce-birch sites that dominate the higher (> 1,700 m) elevations in this watershed.

In summary, storm sampling of streams in the Southern Blue Ridge indicates that depressions in ANC and pH do occur, with the largest changes occurring in streams draining watersheds dominated

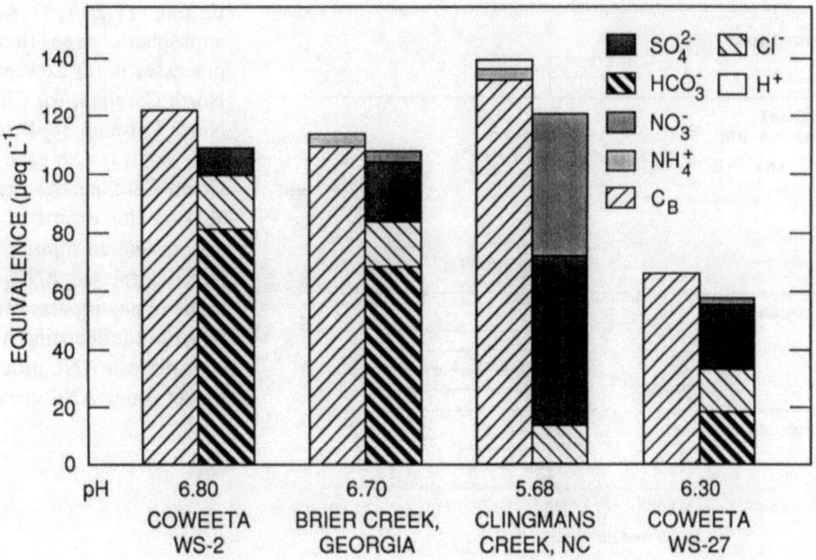

FIGURE 11.19. Ion composition of four streams in the Southern Blue Ridge Province used in the ion enrichment analysis to estimate reaction terms for processes that generate ANC (see Figure 11.20).

by spruce-fir vegetation. The ANC depressions are associated mainly with increases in mineral acid anions, primarily SO_4^{2-}, with little or no change in base cations. Increases in dissociated (unprotononated) organic anions may also contribute to the depressions in pH and ANC during storms.

Processes Influencing Surface Water Chemistry

Ion Enrichment Analysis

The relative importance of watershed processes in buffering acidic inputs from atmospheric deposition and thereby affecting the acid-base chemistry of surface waters can be assessed using ion enrichment analysis (Appendix A, Baker, this volume; Chapter 2, Munson and Gherini, this volume). This analysis, which is based on the charge-balance definition of alkalinity, illustrates the relative importance of different watershed processes that increase and decrease alkalinity in drainage waters. Thus, on a net basis, any process that increases the concentration of base cations without increasing the concentration of acid anions will increase alkalinity; conversely, any process that increases acid anions without increasing base cations will decrease alkalinity. Calculation of the contribution of various processes and the basis for their inter-pretation are discussed by Baker (Appendix A, this volume).

Ion enrichments were calculated for four stream sites in the Southern Blue Ridge using the measured or calculated ion concentrations in precipitation and the average annual concentration of ions in stream water. Because deposition was measured differently at the different sites, the concentration of ions in precipitation was based on either the total annual ion deposition (wet plus dry) rate divided by the total annual precipitation or the average annual concentration of ions in bulk precipitation.

Only four streams were selected for this analysis because data were available for stream chemistry and deposition chemistry from the same watershed or an adjacent watershed. Also, the four sites have contrasting H^+ deposition rates and ANC and thus provide a range of conditions for assessing the relative importance of different watershed processes in buffering acidic inputs in this region. The four sites are Clingmans Creek, a high-elevation stream (1,567 m) near Clingmans Dome in the Great Smoky Mountains National Park; Brier Creek, a lower elevation (646 to 1,267 m) site in northern Georgia; Coweeta Watershed 2 (WS-2), an even lower elevation site (709 to 1,004 m); and Coweeta Watershed 27 (WS-27), a high-elevation site (1,061 to 1,455 m). The last two sites are located at the

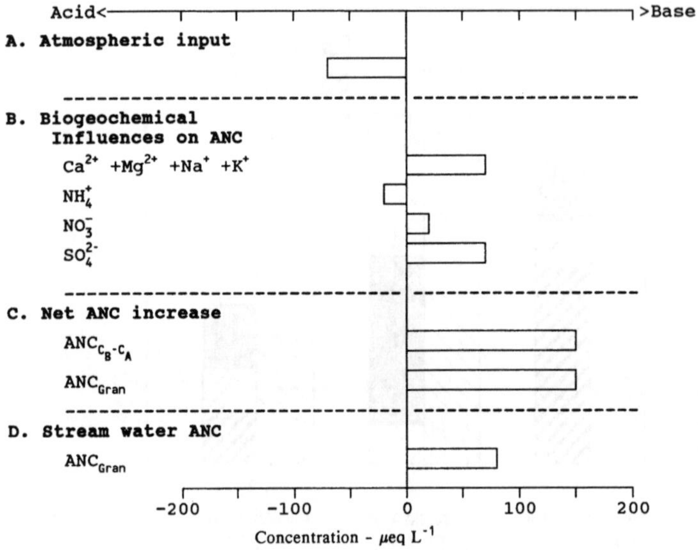

A

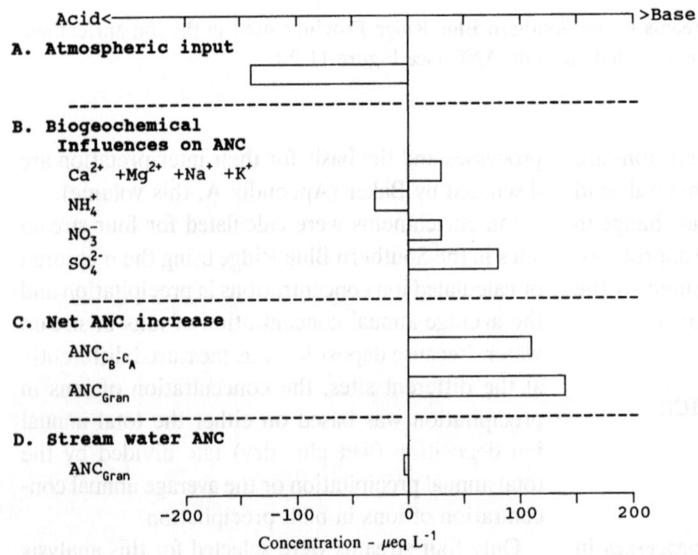

B

FIGURE 11.20A,B. Modification of atmospheric deposition by watershed processes in (a) Coweeta Watershed 2, North Carolina, (b) Clingmans Creek, North Carolina, (c) Brier Creek, Georgia, and (d) Coweeta Watershed 27, North Carolina. A. Precipitation [H^+] adjusted for estimated dry deposition and evapotranspiration (ET); equivalent to expected $ANC_{C_B-C_A}$ in the absence of any process other than ET. B. Reactions influencing ANC; bars to the right indicate ANC production, those to the left show ANC consumption.

Coweeta Hydrologic Laboratory in southwestern North Carolina.

Clingmans Creek is a slightly acidic stream (average ANC −4 μeq L^{-1}) with a high deposition rate of H^+ (> 200 meq m^{-2} y^{-1}, Lindberg and Johnson 1988). In contrast, Brier Creek and Coweeta WS-2 and WS-27 have positive ANC (68, 81, and 19 μeq L^{-1}, respectively). Coweeta WS-2 and WS-27 are both alkaline streams (ANC 81.5 and 22.7 μeq L^{-1}, respectively) (Swank and Waide 1988), and have the lowest annual deposition rate of H^+ in bulk precipitation of these three sites. The H^+ deposition in bulk precipitation at Brier Creek is comparable to the total H^+ deposition at Clingmans Creek.

The average ion composition for the four streams is shown in Figure 11.19. Three of the four streams are somewhat similar in terms of their average ionic strength, with WS-27 the lowest (approximately half that of the other three sites) and Clingmans Creek the highest. The difference in ionic strength between Clingmans Creek and WS-27 is surprising given that both are high-elevation sites and probably receive more precipitation (> 200 cm y^{-1}) than the other two. Swank and Waide (1988) report that streams draining high-elevation watersheds at Coweeta typically have a lower ionic strength than those draining low-elevation catchments, despite the higher

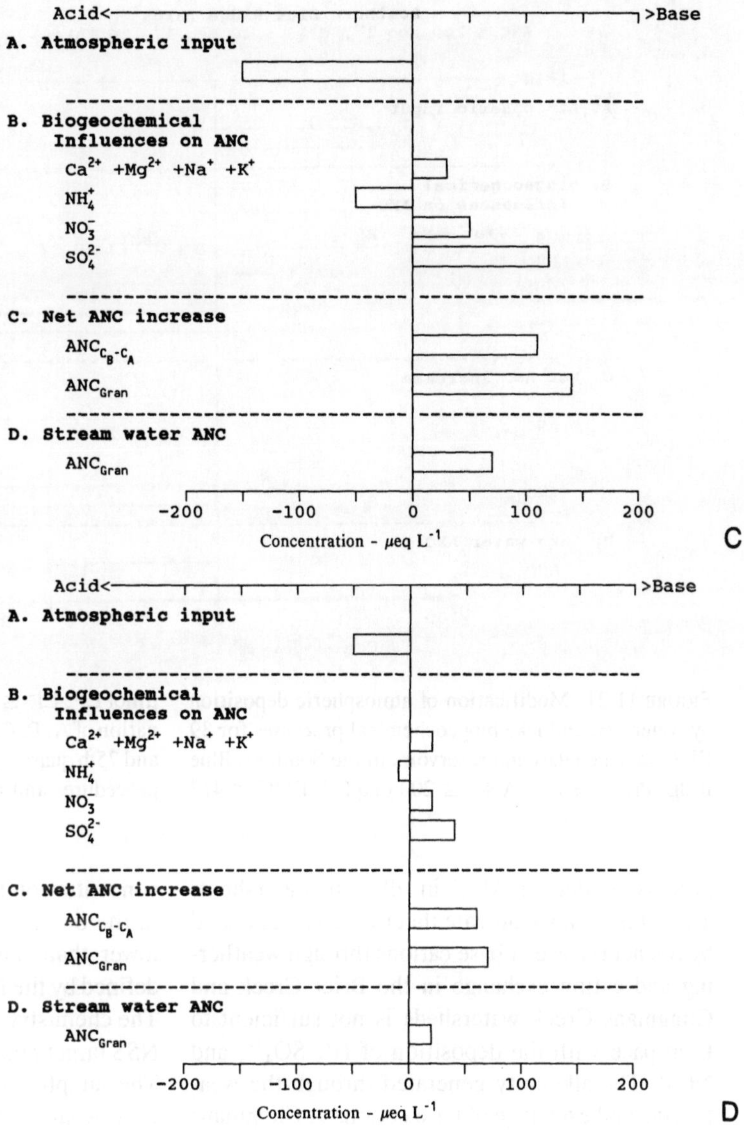

FIGURE 11.20C,D. Net ANC production. $ANC_{C_B\text{-}C_A}$ is sum of ANC production terms for measured anions and cations, including those shown in C; ANC_{Gran} is determined by difference between input ANC (A.) and streamwater ANC (D.). D. Streamwater ANC represents the net reaction of atmospheric inputs with the stream/ watershed system. Data for WS-2 and WS-27 from Swank and Waide, 1988, and Waide and Swank, 1987; data for Brier Creek from Buell and Peters, 1988; data for Clingmans Creek from Elwood et al. 1989 and Johnson et al. 1988. Deposition data for WS-2, WS-27, and Brier Creek based on bulk deposition only; data for Clingmans Creek based on wetfall and dryfall chemistry. DOC concentrations for WS-2, WS-27, and Brier Creek are unknown but are probably < 150 μmol L^{-1}; average DOC concentration in Clingmans Creek is 51 μmol L^{-1}. For a further explanation of calculation procedures and figure interpretation, see Appendix A.

SO_4^{2-} levels in the high-elevation streams. The difference in ionic strength between Clingmans Creek and WS-27 is due primarily to the higher concentrations of SO_4^{2-}, NO_3^-, and H^+ in Clingmans Creek.

The reason for the anion deficits of the four streams (Figure 11.19) is unknown but it is most likely a statistical artifact associated with calculating the average concentration of ions in samples collected at different times, or it may be an analytical error. Dissociation of organic acids is an unlikely cause of the discrepancies because of the low concentrations of DOC (< 100 μmol L^{-1}) in the three streams.

The calculated reactions of the various processes affecting ANC at the four sites are somewhat similar in terms of their sign and ranking in relative importance (Figure 11.20). Base cations, SO_4^{2-}, and NO_3^- exhibit positive terms, whereas that for NH_4^+ is negative at all the sites. Thus, these watersheds exhibit a net release of base cations and a net retention of SO_4^{2-}, NO_3^-, and NH_4^+.

The reaction term for base cations is low at all sites, and less than that for acid anion retention. Further, the reaction term for base cations is lower than that for H^+ inputs at Brier Creek, WS-27, and Clingmans Creek (Figure 11.20b,c,d). Thus, acid anion (SO_4^{2-} and NO_3^-) retention is the dominant

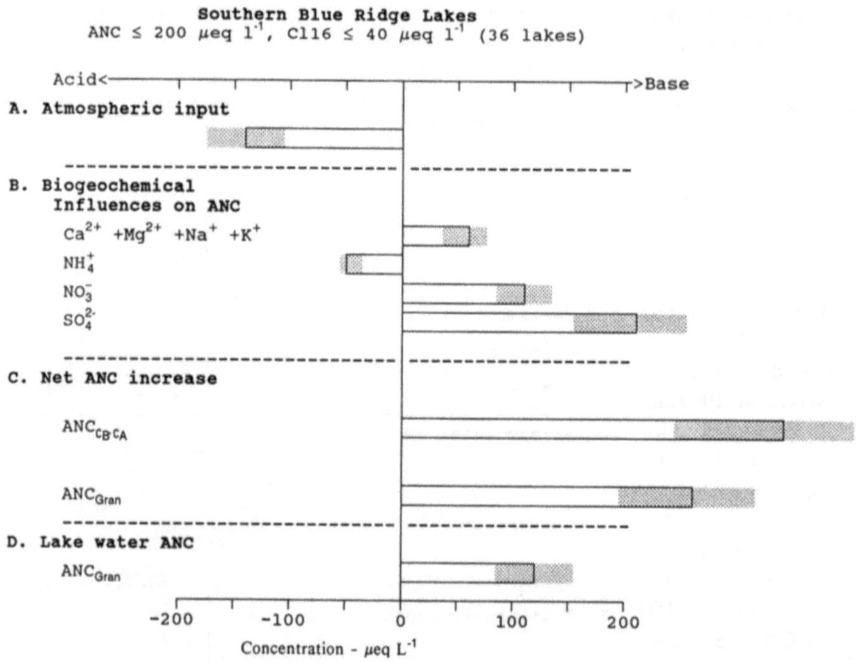

FIGURE 11.21. Modification of atmospheric deposition by watershed and lake biogeochemical processes for 39 ELS drainage lakes and reservoirs in the Southern Blue Ridge Province with ANC ≤ 200 μeq L^{-1}, DOC ≤ 417 μmol L^{-1}, Cl^- ≤ 40 μeq L^{-1} (see Figure 11.20 for explanation of A, B, C, and D). Shaded bar represents the 25th and 75th quartiles. For further explanation of calculation procedures and figure interpretation, see Appendix A.

process producing ANC in all four watersheds. These results also indicate that the ANC generated by the net release of base cations through weathering and cation exchange in the Brier Creek and Clingmans Creek watersheds is not sufficient to keep pace with the deposition of H^+, SO_4^{2-}, and NO_3^-. The alkalinity generated through the weathering and exchange of base cations at Clingmans Creek, which has the highest base cation content in stream water of the four sites, is balanced by the release of acidic cations resulting from the net retention of NH_4^+ in the watershed.

At Brier Creek, WS-2, and WS-27, acid anion retention and base cation production combined exceed H^+ inputs and NH_4^+ uptake, resulting in a positive ANC at these three sites. At Clingmans Creek, however, the input of H^+ combined with the net uptake of NH_4^+ exceeds base cation production and acid anion retention. Thus, even though there is a net retention of both acid anions and a net release of base cations in this watershed, Clingmans Creek is slightly acidic.

The higher concentrations of SO_4^{2-} and NO_3^- in Clingmans Creek approach the upper end of the concentration range for these ions in NSS streams in the Southern Blue Ridge, whereas the ANC is lower than that measured in any of the streams defined by the NSS target population in this region. The chemistry of this stream thus is atypical of the NSS target population in the Southern Blue Ridge. The sampling location on Clingmans Creek, however, would not have been included in the NSS target population because it is a first-order reach that does not appear as a blue-line stream on the 250,000-scale map.

The watershed from which Clingmans Creek drains may also be somewhat atypical of the NSS target population in the Southern Blue Ridge. Vegetation on the watershed is dominated by an old-growth spruce-fir forest that is degrading, whereas the other three streams drain watersheds with second-growth aggrading forests. Although the proximate cause of these elevated NO_3^- levels in streams draining unlogged watersheds is unknown, reduced uptake by the old degrading forest of nitrogen deposited on the watershed and/or increased mineralization of organic N in soil are potential sources (Vitousek and Reiners 1975). In addition, average

SO_4^{2-} levels in soil solutions from the A and B horizons in this watershed exceed 250 μeq L^{-1} (Johnson et al. 1988), indicating that soil SO_4^{2-} adsorption is lower at this site.

Ion enrichment analysis was also conducted for lakes in the Southern Blue Ridge using data from the ELS and interpolated deposition data from the NADP. Population-weighted estimates of ANC production for Southern Blue Ridge watersheds (Figure 11.21) show that clear drainage lakes and reservoirs combined have DOC concentrations ≤ 420 μmol L^{-1}, ANC ≤ 200 μeq L^{-1}, and Cl^- ≤ 40 μeq L^{-1}. In the absence of watershed processes other than evapotranspiration, median inputs of acidity to lakes with ANC < 200 μeq L^{-1} and Cl^- < 40 μq L^{-1} would be 136 μeq L^{-1} (Figure 11.21). Median production of ANC by watershed processes was 264 μeq L^{-1}, however, resulting in a median ANC in lake water of 128 μeq L^{-1}. Most of the ANC production was due to anion retention (97%), whereas base cation production accounted for only 16.9% (Table 11.5). Ammonium retention consumed an amount of ANC approximately equal to the base cation production. The relative importance of the different processes in these watersheds thus is consistent with that for the four watersheds containing streams.

TABLE 11.5. Median percent contribution of the various processes to the Gran ANC of drainage lakes and reservoirs in the Southern Blue Ridge Province ecoregion.[a]

Process	Percent contribution to ANC
Base cation mobilization	16.8 (8.2–25.5)
Sulfate retention	63.2 (56.4–69.1)
Nitrate retention	33.8 (31.7–36.0)
Ammonium retention	−14.1 (−15.0– −13.1)

[a] Results are based on ion enrichment analysis of 36 drainage lakes and reservoirs in the Southern Blue Ridge Reservoir ecoregion with ANC ≤200 μeq L^{-1} and Cl^- ≤40 μeq L^{-1} that were sampled in the ELS in the fall. Values in parentheses are the interquartile ranges.

Weathering of parent material in the watersheds appears to be the primary source of base cations in stream water in this region, based on the geochemistry of stream water, parent material, and soils in the region. For example, data on K^+ and SiO_2 for most streams in the NSS target population have been shown to fit a mineral stability curve for weathering of potassium feldspars (Messer et al. 1986), lending support to the hypothesis that the stream concentrations of these elements, and perhaps other elements, such as Na^+, associated with feldspars, are controlled primarily by the weathering of feldspars such as albite.

The relatively high concentrations of SiO_2 in stream water, and the fact that Na^+ is one of the dominant base cations, also suggests that weathering of parent material is a more important source of base cations to drainage waters in the Southern Blue Ridge than cation exchange. This is because the base saturation of soils in this province is low (typically < 10% in mineral soil horizons) (Johnson et al. 1988), and Na^+ typically is lower than any of the other exchangeable cations in highly weathered soils in the Southern Blue Ridge (Johnson et al. 1988, Swank and Douglas 1977). In addition, Na^+ is generally the least replaceable of the base cations on soil exchange sites, making it unlikely that much of the Na^+ in stream water is coming from soil exchange (Wiklander 1964). Thus, available evidence of soil chemistry, bedrock geology, and stream chemistry suggest that weathering of parent material is more important than cation exchange as a source of base cations to streams in the Southern Blue Ridge.

Velbel (1985a,b) also hypothesized that the geochemistry of streams in the Southern Blue Ridge is controlled primarily by weathering, with the concentration of solutes in streams controlled kinetically by (1) the residence time of drainage water in contact with subsoils and bedrock and (2) the abundance of readily weatherable mineral phases. The high rainfall and warm temperatures in the region provide an ideal environment for intense weathering as evidenced by the well-weathered soil mantle in the region. The fact that SiO_2 levels are relatively high in surface waters in ths region, whereas base cations are low, indicates that the parent material being weathered in many watersheds is relatively low in bases.

The question arises as to whether in-lake processes such as SO_4^{2-} and NO_3^- reduction and denitrification contribute substantially to the ANC of lakes and reservoirs in this region. Based on the large ratio of mean lake depth to hydrologic residence time of water in drainage lakes and reservoirs in the Southern Blue Ridge (median = 48; Linthurst et al. 1986a), and the inverse relationship between SO_4^{2-} retention in lakes and this ratio (Baker et al. 1986a), within-lake processes are unlikely to be a major source of ANC in most drainage lakes and reservoirs in this region. Shaffer and Church (1989) concluded that in-lake alkalinity generation in the Southern Blue Ridge is a minor contributor to the net basin alkalinity production, contributing < 3% of total basin alkalinity production in typical watersheds in this region. Their result is further supported by the fact that the median SO_4^{2-} level in Southern Blue Ridge drainage lakes and reservoirs in the fall (27.9 μeq L^{-1}; Table 11.4) is approximately the same as that at the downstream sites in the spring (22.9 μeq L^{-1}, Table 11.3; Messer et al. 1986).

Because SO_4^{2-} retention in drainage lakes and reservoirs should reduce the concentration in lake water relative to that in inflowing streams, the similarity in median SO_4^{2-} concentration among Southern Blue Ridge lakes and reservoirs and streams indicates that in-lake alkalinity generation due to SO_4^{2-} reduction cannot be a signficant source of ANC in this region.

Evidence of Historical Trends

Although results of surface water surveys in the Southern Blue Ridge show that there are few acidic streams and no acidic lakes or reservoirs in this region, the region contains a large number of relatively low ANC streams and lakes. The question arises as to whether there has been a long-term decline in the ANC and pH of these systems due to acidic deposition. This is particularly important given that (1) acid anion retention is a major source of ANC in watersheds in the Southern Blue Ridge, typically exceeding base cation production, (2) watershed retention of SO_4^{2-} and NO_3^- in this province is variable, (3) SO_4^{2-} adsorption is a reversible, concentration-dependent process, (4) sulfur and nitrogen oxide emissions are increasing in the region (Chapter 3, Husar et al., this volume), and (5) an increasing trend in stream SO_4^{2-} levels in the southeastern United States over a 13-year period (1969–81) has been documented (Lins 1986).

Burns et al. (1981) reported that the alkalinity of streams in the Blue Ridge Mountains of North Carolina was significantly lower in 1979 than in the early 1960s, suggesting a long-term decline in the ANC of streams in this area of the province. In addition, the average ratio of bicarbonate alkalinity to the sum of base cations in 1979 was approximately 0.6, a value significantly less than that expected for streams draining watersheds in which carbonic acid weathering is the dominant process controlling the release of base cations to drainage waters. Differences in analytical methodologies used to measure ANC, however, place the results of Burns et al. in doubt (Church 1983).

Smith and Alexander (1983) reported statistically significant increases in SO_4^{2-} emissions during a 10- to 15-year period over a broad area of portions of the eastern United States, including the southeast. Sulfate concentrations in streams at USGS hydrologic benchmark stations in this same area showed statistically significant increases, whereas alkalinity decreased (Smith and Alexander 1983, Lins 1986). The only site located in the Southern Blue Ridge, however, showed no significant temporal trends in alkalinity, SO_4^{2-}, or the ratio of alkalinity to major cations. The site, Cataloochee Creek, located near Cataloochee, North Carolina, had a mean alkalinity of 110 μeq L^{-1}. However, because this stream drains a relatively large (127 km^2) watershed, it is less likely that acidic deposition would cause a temporal decline in alkalinity and increase in SO_4^{2-}. Large watersheds offer a greater opportunity for buffering of acidic inputs, as a result of the longer residence time of drainage waters in contact with soil and bedrock.

At the Coweeta Hydrologic Laboratory in North Carolina, significant increases over a 10-year period in the volume-weighted concentration of SO_4^{2-} have been documented in streams draining smaller undisturbed (control) watersheds (Figure 11.22; Swank and Waide 1988, Waide and Swank 1987). Some of the streams drain low-elevation watersheds and others drain medium-or high-elevation catchments. Waide and Swank (1987) report that the intercept of regressions of stream SO_4^{2-} concentration and time is greater for the high-elevation streams, reflecting the lower retention of SO_4^{2-} in high-elevation catchments. However, the regression coefficient (slope) does not appear to differ between the low- and high-elevation streams, averaging about +0.7 μeq L^{-1}, which indicates that high- and low-elevation watersheds in this area of the Southern Blue Ridge behave similarly in terms of the release of SO_4^{2-}.

Swank and Waide (1988) also report statistically significant decreases in alkalinity and base cations and significant increases in H^+ and anion deficit

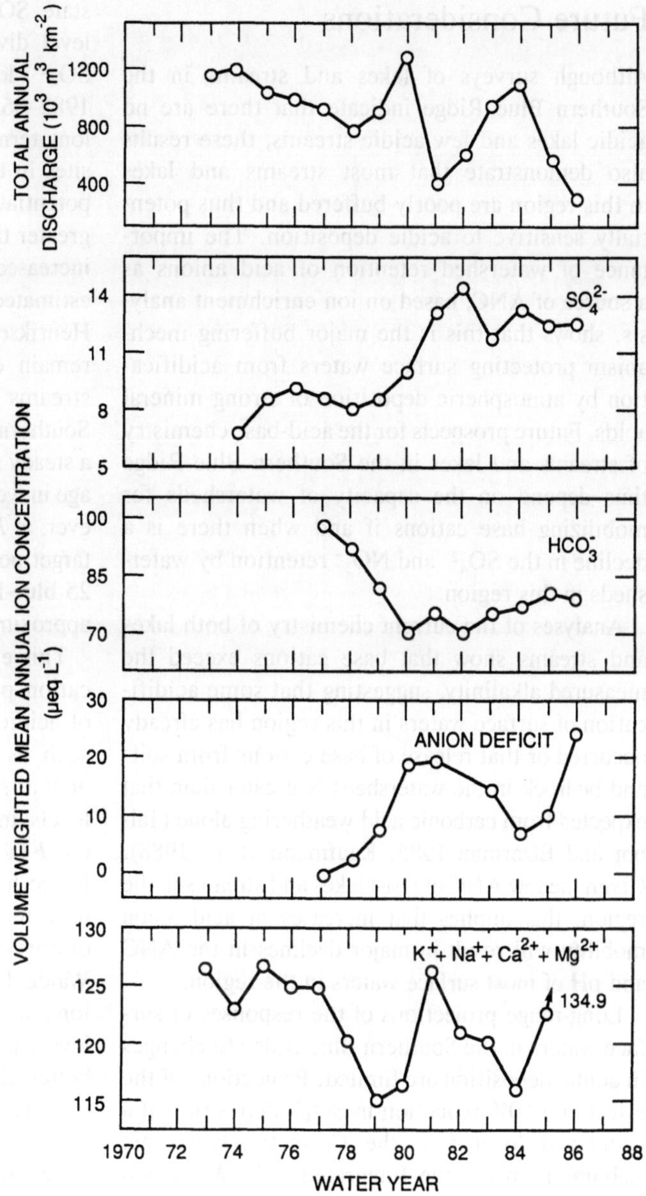

FIGURE 11.22. Temporal trends in discharge and the concentration of selected solutes in the stream draining Watershed 2 at the Coweeta Hydrologic Laboratory, North Carolina. (Data from Waide and Swank 1987.)

over the same period (Figure 11.22). Concentrations of NO_3^- and H^+ in precipitation over this period showed no significant temporal trends, whereas SO_4^{2-} exhibited a weak positive correlation with time; base cations in bulk precipitation exhibited a slight decline over time. Net SO_4^{2-} budgets show that all these control or reference watersheds retained a large fraction of atmospherically deposited SO_4^{2-}, with the net retention lower on high-elevation than low-elevation watersheds. These results are significant because they show that declines in both ANC and base cations in stream waters at Coweeta occurred over a period of increasing SO_4^{2-} levels in stream waters. This suggests that the capacity for releasing base cations when SO_4^{2-} retention declines and acid anion mobility increases is limited in some watersheds in the Southern Blue Ridge.

Future Considerations

Although surveys of lakes and streams in the Southern Blue Ridge indicate that there are no acidic lakes and few acidic streams, these results also demonstrate that most streams and lakes in this region are poorly buffered and thus potentially sensitive to acidic deposition. The importance of watershed retention of acid anions as a source of ANC, based on ion enrichment analysis, shows that this is the major buffering mechanism protecting surface waters from acidification by atmospheric deposition of strong mineral acids. Future prospects for the acid-base chemistry of streams and lakes in the Southern Blue Ridge thus depend on the capacity of watersheds for mobilizing base cations if and when there is a decline in the SO_4^{2-} and NO_3^- retention by watersheds in this region.

Analyses of the current chemistry of both lakes and streams show that base cations exceed the measured alkalinity, suggesting that some acidification of surface waters in this region has already occurred or that release of base cations from soils and bedrock in the watersheds is greater than that expected from carbonic acid weathering alone (Talbot and Elzerman 1985, Kaufmann et al. 1988). Given the low ANC of most lakes and streams in the region, this implies that increases in acid anion mobility will result in major declines in the ANC and pH of most surface waters in the region.

Long-range projections of the responses of surface waters in the Southern Blue Ridge to changes in acidic deposition are limited. Projections of the effect of a 50% reduction in SO_4^{2-} deposition at a watershed located at the Coweeta Hydrologic Laboratory using the Integrated Lake-Watershed Acidification Model (ILWAS) (Munson et al. 1987; Chapter 2, Munson and Gherini, this volume) suggest that there would be no effect on stream pH. This lack of an effect was attributed to the fact that SO_4^{2-} adsorption is the dominant source of ANC in the watershed and SO_4^{2-} adsorption is modeled as a reversible equilibrium process. Thus, when SO_4^{2-} deposition declines, it desorbs from the soil, resulting in no net change in ANC and pH.

For streams in the NSS target population in the Southern Blue Ridge, Church et al. (in press) computed an index of Potential Future Acidification (PFA), defined as the difference between the steady state SO_4^{2-} concentration and the current SO_4^{2-} level divided by the current ANC. Steady state SO_4^{2-} levels in streams were computed using 1982–86 wet deposition chemistry and average long-term normal precipitation for 1951–80 for sites in the Southern Blue Ridge. The greater the potential increase in SO_4^{2-} relative to ANC, the greater the potential for future acidification due to increased SO_4^{2-} flux. The change in ANC was then estimated by the PFA, using different values for Henriksen's (1984) *F*-factor, which was assumed to remain constant. For $F = 0$, over 40% of the streams in the NSS target population in the Southern Blue Ridge would be acidic (ANC $\leq$ 0) at a steady state SO_4^{2-} concentration equal to the average in wet deposition for the 1982–86 period. However, at $F = 0.75$, only 2.6% of the streams in the target population would be acidic. This represents 25 blue-line streams on a 1:250,000-scale map, or approximately 100 km of stream channel.

This exercise points out the importance of base cation production in assessing the future effects of acidic deposition on surface waters in regions such as the Southern Blue Ridge that are not now at steady state with respect to current SO_4^{2-} levels in wet deposition. Although estimates of the *F*-factor for watersheds in the Southern Blue Ridge have not been determined, results of long-term monitoring at the Coweeta Hydrologic Laboratory suggest that it may be < 0 (Swank and Waide 1988), based on the finding that base cations in streams are decreasing, whereas SO_4^{2-} levels are increasing. There is clearly a need to better define the change in base cation levels occurring with changing acid anion mobility in drainage waters in representative Southern Blue Ridge watersheds in order to narrow the range of uncertainty in the projected effects of acidic deposition on surface waters in this region.

Biological Status

Fish Communities of Acid Sensitive Streams

The low-order, high-elevation streams in the Southern Blue Ridge that are expected to be most sensitive to acidic deposition are generally dominated by trout species, especially the eastern brook trout or brook char (*Salvelinus fontinalis*). The brook trout is the only salmonid native to the Southern Blue Ridge, but because of stocking, this species has been replaced over much of its original range by rainbow trout (*Onchorhynchus mykiss*) and brown trout (*Salmo trutta*).

Kelly et al. (1980) summarized changes that have occurred in trout distributions in the Great Smoky Mountains National Park, based on surveys conducted periodically since the 1930s. Brook trout have always been limited to higher elevation streams in the Park; in the early 1900s, they were generally found above 600 m msl, and sections of streams between the lower limits of brook trout and the appearance of smallmouth bass below 490 m msl contained no gamefishes. Since that time, the brook trout range has steadily decreased. Surveys indicate that the average minimum elevation of brook trout distribution was over 900 m by the 1930s, and over 1,000 m by the 1970s. Brook trout once occupied an estimated 680 km of the 1,173 km of fishable trout streams in the Great Smoky Mountains National Park, but now are relegated to 197 km of mostly small headwater streams (Moore et al. 1983). The ranges of rainbow trout, and to a lesser extent brown trout, have increased correspondingly, and recent management efforts have been aimed at controlling or eradicating rainbow trout from selected streams formerly occupied by allopatric brook trout (Moore et al. 1983, 1986).

The decline of brook trout in the Great Smoky Mountains National Park, which is also representative of changes in the distribution of this species throughout the Southern Blue Ridge, has been attributed to a number of causes (Larson and Moore, 1985). Brook trout losses prior to the establishment of the Park are thought to have resulted from logging and overfishing. Because neither logging nor harvest of brook trout is now allowed in the Park, continuing declines have been most commonly attributed to competition from rainbow trout. During the period following creation of the Park in the mid-1930s, rainbow trout began to move upstream from where they were stocked and invaded unstocked brook trout waters. Kelly et al. (1980) noted that where the two species occur together, significant differences existed in the fecundity, age at initial reproduction of females, number of spawnings, and longevity, all of which favor the rainbow trout. Larson and Moore (1985), on the other hand, could find no differences between brook and rainbow trout in longevity and age at first spawning, but noted size and sex ratio advantages for rainbow trout. The only other fish whose range overlaps with brook trout in the Great Smoky Mountains National Park is the blacknose dace; although the two species have similar food habits, the dace is not found in enough brook trout streams to be regarded as a significant competitor (Kelly et al. 1980).

Because brook trout are more tolerant of acidic conditions than many other fish species, and are still found in many high-elevation, low pH streams in the Southern Blue Ridge, Kelly et al. (1980) believed that low pH was not a major factor in the decline of brook trout in the Great Smoky Mountains National Park. They speculated, however, that low pH may limit the spawning and upstream movement of rainbow trout into brook trout streams. This is supported by surveys of Walker Camp Prong, a high-elevation, acidified stream in the Park (Elwood et al. 1985). A lower section of Walker Camp Prong that had a mean pH of 6.18 contained both rainbow and brook trout, whereas sections up to 325 m higher in elevation had lower mean pH (5.74 to 5.90) and only brook trout. That Southern Blue Ridge streams can become too acidic to support even brook trout is evidenced by the absence of fish in nearby stream reaches with mean pH < 5.0 (Elwood et al. 1985). The exposure and subsequent oxidation of pyritic bedrock in the Great Smoky Mountains National Park have been shown to depress stream pH below levels suitable for survival of brook trout (Huckabee et al. 1975) and native shovel-nosed salamanders (Huckabee et al. 1975, Mathews and Morgan 1982). The naturally occurring acidic conditions brought about by weathering of pyritic rocks may cause fishkills directly or increase the susceptibility of fish populations to atmospheric acidic deposition in some watersheds.

The highest elevation streams in the Southern Blue Ridge may contain only brook trout, or sympatric populations of brook and rainbow trout. Nonsalmonid fishes whose ranges overlap with the brook trout in first-order (headwater) streams include the blacknose dace (TVA 1971, Kelly et al. 1980, Whitworth and Strange 1983) and mottled sculpin (Fowler 1985, Fowler et al. 1987). Larger brook trout streams in the Southern Blue Ridge still support relatively simple fish communities, but may also include other minnows such as the creek chub and longnose dace (Fowler et al. 1987). At lower elevations, Southern Blue Ridge streams are warmer and less rapid, and have a lower gradient. The lower elevation streams may contain a mixed assemblage of salmonids, minnows, suckers, sculpins, and darters (Lennon 1962, TVA 1971, Mathews 1978, Loar 1985).

Fowler et al. (1987) examined the relationships among fish communities and the pH and alkalinity of 12 stream systems in the Southern Blue Ridge. All 12 streams had low alkalinity and would be highly sensitive to acidic deposition, but they had circumneutral pH (measured pH ranged from 6.25 to 7.00). No consistent, statistically significant relationships were found among fish assemblage characteristics (occurrence, density, biomass) and either pH or alkalinity, which the authors have interpreted as indicating that acidification had not significantly impacted fish communities in streams of the Southern Blue Ridge. The only documented cases of fish kills that have been attributed to acidic deposition in the Southern Blue Ridge occurred in fish-rearing facilities that utilized stream water from Raven Fork, North Carolina. Several fishkills of brown trout and rainbow trout in holding tanks supplied with water from Raven Fork have occurred during storms (Jones et al. 1983). Stream acidification was implicated as the cause of the fishkills based on (1) observed decreases in pH and increases in Al to levels known to be acutely toxic to rainbow trout and (2) a lack of trout mortality in the rearing facilities after the incoming stream water was limed (Jones et al. 1983). Mortality of native trout in Raven Fork during storms has not been documented, suggesting that the trout in the holding tanks were more sensitive than native trout to changes in pH because of the stress of being confined in holding tanks. Alternatively, mortality of native trout during storms may not occur because of avoidance behavior to acidic conditions or because the fish have become acclimated to acidic conditions and hence are less sensitive to pH changes during these storm events. It is also possible that the apparent lack of fish mortality in Raven Fork during storms was due to problems in observing dead fish and stressed and dying fish during periods of high flows in turbid, turbulent waters.

Benthic Invertebrate Communities of Acid Sensitive Streams

Benthic invertebrate communities of streams in the Southern Blue Ridge are characterized by relatively high diversity (Lasier 1986, Silsbee and Larson 1983, Rosemond et al. submitted) and low standing crops (TVA 1971, Loar 1985, Rosemond et al. submitted). Because of the complex and interacting influences of habitat and physicochemical variables on the distribution of aquatic invertebrates, studies have often shown considerable variation in benthic community parameters among nearby sites. For example, Stoneburner (1977) observed distinct differences in the aquatic macroinvertebrate fauna along a 370-m altitudinal gradient in first-order streams draining a single ridge in the Great Smoky Mountains. With increasing altitude, there were significant decreases in water temperature and pH, decreases in the frequency of Ephemeroptera, and increases in taxonomic richness, number of dipteran taxa, and abundance of Plecoptera. Sheldon (1985) also found elevation to be an important variable affecting the distribution of Plecoptera at 56 sites in the Great Smoky Mountains.

Lasier (1986) related information about benthic macroinvertebrate communities to alkalinity and pH in 30 streams representing all the major drainages in the Southern Blue Ridge. With the exception of two streams whose pH and ANC were greatly reduced by oxidation of pyritic bedrock of the Anakeesta formation, community characteristics (taxonomic richness, total density and biomass, diversity, evenness, and the relative abundance of functional feeding groups) were not reliable in determining acid sensitivity. However, the abundances of five species commonly found throughout the Southern Blue Ridge were significantly correlated with pH and alkalinity and were considered to be reliable indicators of sensitivity to acidification. Lasier (1986) noted that certain common, but acid intolerant taxa were already absent from some low pH stream reaches and might be eliminated from other streams if anthropogenic acidification occurs.

The macroinvertebrates of Walker Camp Prong were studied by Rosemond et al. (submitted). The species richness, density, and diversity of the benthic invertebrates were significantly lower at baseflow pH values as low as 4.5, compared to circumneutral sites on the same stream. However, functional feeding group composition was not strongly affected by pH. Two of the genera that were regarded as indicative of sensitivity to acidification by Lasier (1986), that is, the mayflies *Epeorus* and *Stenonema*, were rare or absent from the most acidic Walker Camp Prong sites (Rosemond et al. submitted). In addition, both Lasier (1986) and Rosemond et al. (submitted) found the stoneflies *Leuctra* and *Isoperla* to be tolerant of acidic conditions; these findings are consistent with studies of other low pH systems (Chapter 4, Baker and Christensen, this volume).

Conclusions

1. Streams and lakes (90% of which are estimated to be reservoirs) in the Southern Blue Ridge Province are relatively dilute systems that are potentially sensitive to acidic deposition. Based on a random sampling of 54 low-order streams in the spring and 67 lakes and reservoirs in the fall, no acidic streams, lakes, or reservoirs were found in this region.
2. Sulfate and NO_3^- are the dominant anions in the few acidic streams known to occur in the Southern Blue Ridge. These streams drain unlogged old-growth watersheds with a bedrock source of SO_4^{2-} (pyritic phyllite) that becomes exposed to oxidation through natural landslides and road building.
3. Carbonic acid has a major effect on the pH of most NSS streams in the Southern Blue Ridge due to the fact that they are supersaturated with CO_2, which causes a depression in pH of more than half a unit in many streams.
4. In terms of explaining the variation in ANC among streams in the Southern Blue Ridge, base cation content is most important, followed in decreasing importance by SO_4^{2-}, NO_3^-, and Cl^- concentration. Sulfate is substantially less important than base cation content as a source of variation.
5. Dissolved organic carbon concentrations are very low (median < 60 µmol L^{-1}) in Southern Blue Ridge streams during spring baseflow and do not explain a significant fraction of the variation in ANC among streams. This indicates that organic acids are not important contributors to either weak or strong acidity of streams in this region during baseflow.
6. Acid anion (SO_4^{2-} + NO_3^-) retention is more important in generating alkalinity than base cation release in watersheds in the Southern Blue Ridge, accounting for approximately 97% of the Gran ANC in drainage lakes in this region. Increases in acid anion mobility in soils in the Southern Blue Ridge without concomitant increases in base cation release from weathering and cation exchanges thus will cause substantial reductions in the ANC and pH of streams and lakes in the region.
7. The present chemistry of streams and lakes suggests that mineral acidity, primarily SO_4^{2-}, is having an effect on their acid-base chemistry. The ANC of most lakes and streams in the region is less than that predicted from the carbonic acid weathering of base cations. In addition, the ANC of streams with a base cation content of < 100 µeq L^{-1} is inversely related to SO_4^{2-} concentration, suggesting that reducing the SO_4^{2-} loading to these systems will result in an increase in ANC.
8. Although data on stream chemistry during storms for the Southern Blue Ridge are limited, results suggest that SO_4^{2-}, NO_3^-, and base cations tend to increase during the rising limb of hydrographs of major storms and then decline, with base cations declining at a faster rate than any of the acid anions, resulting in a depression in ANC and pH during the falling limb of the hydrograph. Increases in DOC during storms may help buffer streams from major pH depressions by providing some noncarbonate ANC.
9. Selected streams draining undisturbed watersheds exhibit increasing trends in SO_4^{2-} concentration and decreasing trends in alkalinity and base cation content over the last decade or more, suggesting that the capacity of watersheds in the Southern Blue Ridge for retaining SO_4^{2-} and mobilizing base cations is declining.
10. Documented declines in the distribution of brook trout populations in Southern Blue Ridge streams are probably not a result of

acidification, because brook trout populations are now found primarily at the higher elevations where acidic conditions are more likely to exist. The upstream distribution of rainbow trout in drainages with acidic headwaters, however, appears to be limited by water quality.

11. Studies of benthic invertebrate communities in acidic (ANC $\leq$ 0) and circumneutral streams in the Southern Blue Ridge indicate that both direct effects of pH and Al and indirect effects of food availability account for differences in macroinvertebrate community composition at the different sites. Species richness and diversity of benthic macroinvertebrates are positively related to pH in streams in this province. The composition of functional feeding groups shifts from dominance by shredders at low pH sites to fewer shredders, fewer collectors and gatherers, and increased abundance of grazers and predators at sites higher in pH. Species richness of mayflies, stoneflies, and caddisflies are positively related to pH and inversely related to inorganic monomeric Al concentration.

Acknowledgments. The authors acknowledge Drs. Wayne Swank and Jack Waide for providing the precipitation, flow, and chemistry data from sites at the Coweeta Hydrologic Laboratory, North Carolina; Dr. Harvey Olem, TVA, for providing data on Ravens Fork and Cosby Creek; and Dr. Jake Peters, USGS, for providing data on Brier Creek, Georgia. We also wish to acknowledge Dr. Steve Lindberg, ORNL, for providing deposition data on the Clingmans Creek site. Drs. Joan Baker, Steve Nodvin, Dave Schindler, John Stoddard, Bob Cook, Don Charles, and Robbins Church offered constructive comments and suggestions for improvement on an earlier version of this manuscript. Jeremy Smith provided the ion enrichment analysis of lakes in the Southern Blue Ridge Province and Tony Selle furnished the frequency histograms of lake and stream characteristics.

This research was sponsored in part by the U.S. Environmental Protection Agency under Interagency Agreement DW89932112-01-7 with the U.S. Department of Energy and by the Electric Power Research Institute under contract No. RP-2336-1 with Martin Marietta Energy Systems, Inc. The Oak Ridge National Laboratory (ORNL) is operated by Martin Marietta Energy Systems, Inc., under contract DE-AC05-84OR21400 with the U.S. Department of Energy. Environmental Sciences Division, ORNL Publication No. 3402.

Florida Overview

Curtis D. Pollman

Florida is relatively flat and featureless, compared to the landforms of other major regions containing lakes in the United States. Nonetheless, subtle differences in topography help to define the occurrence of major lake types and their geochemistry. Most acidic or low ANC lakes in Florida are seepage lakes that lie in ridge and highland regions where a veneer of deep acid sands isolates the lake basin from the underlying limestone bedrock (cf. Figure 12.1, Chapter 12, Pollman and Canfield, this volume, and Figure 1, Southeast Overview, this volume). These Pleistocene sands are highly weathered, and groundwater seepage to these lakes provides little in the way of base cations and ANC. Acidic lakes in Florida are divided almost equally into two categories: (1) colored lakes, in which acidity is related to high concentrations of organic acids and (2) clearwater lakes, in which acidity is due to strong acids, principally H_2SO_4.

Florida is a region in transition, like much of the Southeast. Since 1940, Florida's population has increased approximately sixfold, from nearly 2 million to about 12 million in 1988. The effects of Florida's burgeoning population on its aquatic resources are pervasive: of the approximately 6,500 lakes > 4 ha, about two-thirds were excluded from consideration in the EPA's Eastern Lake Survey (ELS) because of obvious disturbances in their watersheds (e.g., intense agricultural, urban, or industrial land use, point source discharges). In addition, of the target population of 2,049 lakes that were included in the ELS, about 57% show geochemical signs of disturbance related primarily to citrus agriculture. Nonetheless, relatively pristine regions persist, and among these regions lies the largest number of acidic lakes in the United States (Linthurst et al., 1986).

Florida can be divided into three distinct subregions based on lakewater chemistry: the Panhandle, the Northern Peninsula, and the Central Peninsula (Figure 12.1, Chapter 12, Pollman and Canfield, this volume). These subregions also correspond to distinct climatic regimes. Precipitation is highest in the Panhandle and lowest in the Central Peninsula (Figure 2, Southeast Overview, this volume). Conversely, lake evaporation is highest in the Central Peninsula and lowest in the Panhandle. Precipitation chemistry also varies somewhat spatially (Figure 3, Southeast Overview, this volume); however, the overriding determinant of differences in major ion chemistry of Florida's seepage lakes across the subregions appears to be a combination of climatic and localized hydrologic factors (Chapter 12, Pollman and Canfield, this volume). Florida has essentially two seasons, and precipitation chemistry varies consistently between the two. Summertime concentrations (April 1–September 30) invariably are higher for all major constituents than wintertime concentrations (Figure 1; see also Chapter 12, Pollman and Canfield, this volume).

The sheer number of acidic lakes in Florida (approximately 460 lakes with ANC $\leq$ 0) give this region special significance. Moreover, whereas both anthropogenic emissions of sulfur and acidic deposition in the Northeast and Upper Midwest appear

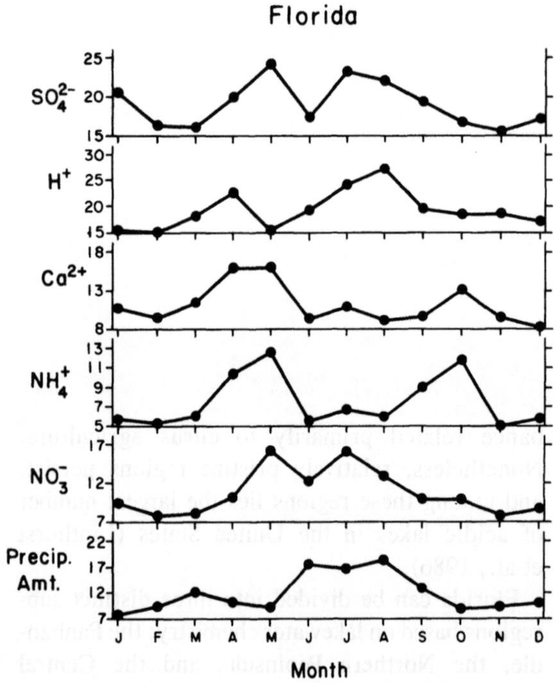

FIGURE 1. Monthly variation in precipitation quantity and chemistry in Florida. Data represent 5-year averages for the period 1982–86 for the sites indicated in Figure 3 in the Southeast Overview (this volume). Data are from the NADP (Olsen and Watson 1984) and averages were determined by R. Husar.

to be decreasing (Chapter 3, Husar et al., this volume), emissions of sulfur in Florida are likely to increase over the next two decades as power generation keeps pace with the increased demands imposed by Florida's expanding population. The effects of increased acidic deposition in Florida are uncertain.

12
Florida

Curtis D. Pollman and Daniel E. Canfield, Jr.

ABSTRACT. Florida lakes are extremely diverse. Although many are influenced by agriculture and other aspects of cultural land use, Florida also contains more acidic (ANC $\leq$ 0) lakes than any other region in the United States. Approximately 460 lakes are acidic (22% of the Eastern Lake Survey – Phase I population) despite receiving only moderate rates of wet deposition of H^+ and SO_4^{2-}. About half the acidic lakes appear to be naturally so because of organic acids. The remaining clearwater systems derive their acidity from mineral acids. As in the Upper Midwest and Maine, the occurrence of acidic lakes is strongly related to hydrologic type; over 80% of the acidic lakes in Florida are seepage lakes.

Unlike other case study regions, neutralization of acidic deposition in dilute seepage lakes is due largely to anion retention (SO_4^{2-} and NO_3^-). Base cation supply accounts for only 12% of ANC generation. Low base cation supply reflects the depleted base status of the Plio-Pleistocene sands mantling the karst terrane characteristic of Florida seepage lake districts. Within a particular subregion, differences in ANC among undisturbed, softwater seepage lakes reflect hydrologic factors, and the proximity of clay and carbonate deposits to the lake bed.

Major ion chemistry varies longitudinally across the state, primarily because of differences in relative amounts of precipitation and evaporation. Regionally, precipitation:evaporation ratios appear to exert competitive effects on lakewater ANC through (1) direct evapoconcentration of atmospheric deposition, and (2) controls on the relative importance of groundwater inseepage on hydrologic and mineral budgets of seepage lakes. The most sensitive lakes are located in the Panhandle, where precipitation: evaporation ratios are high and inseepage fluxes of ANC are believed to be small.

Diatom reconstructions of historical pH in the Trail Ridge lake district indicate that some lakes have become more acidic since the 1950s. Chemical evidence indicative of recent acidification is more ambiguous. Increased acidic deposition and changes in groundwater hydrology

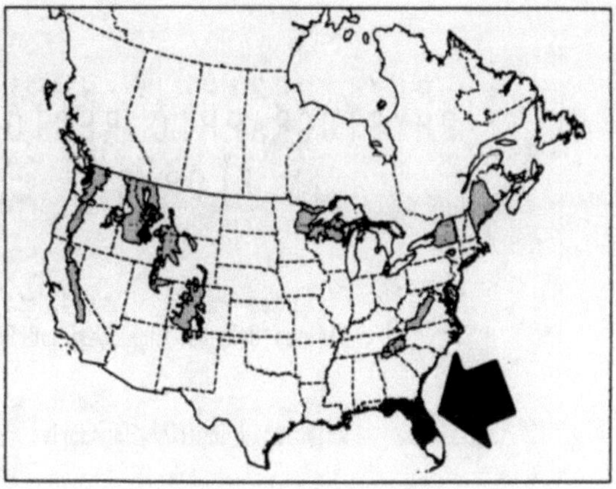

(with associated changes in base cation supply) have been hypothesized to account for the paleolimnological record. Regional increases in SO_2 emissions and progressive declines in the artesian aquifer potentiometric surface near the Trail Ridge have been documented since 1940; model hindcasting suggests that both phenomena may account for observed changes in pH. Depending on the mechanism, different responses in other major ion components are predicted and insufficient data are available to distinguish between the two hypotheses.

Introduction

Florida has the distinction of containing the largest number (453) and frequency (22%) of acidic (ANC $\leq$ 0) lakes of any major region in the United States, despite receiving only moderate amounts of acidic deposition compared to heavily impacted areas of the Northeast (ESE 1987, Stensland et al. 1986). A number of hypotheses, both anthropogenic and natural, may be invoked to explain the acidic status of Florida: (1) acidic deposition, (2) modifications

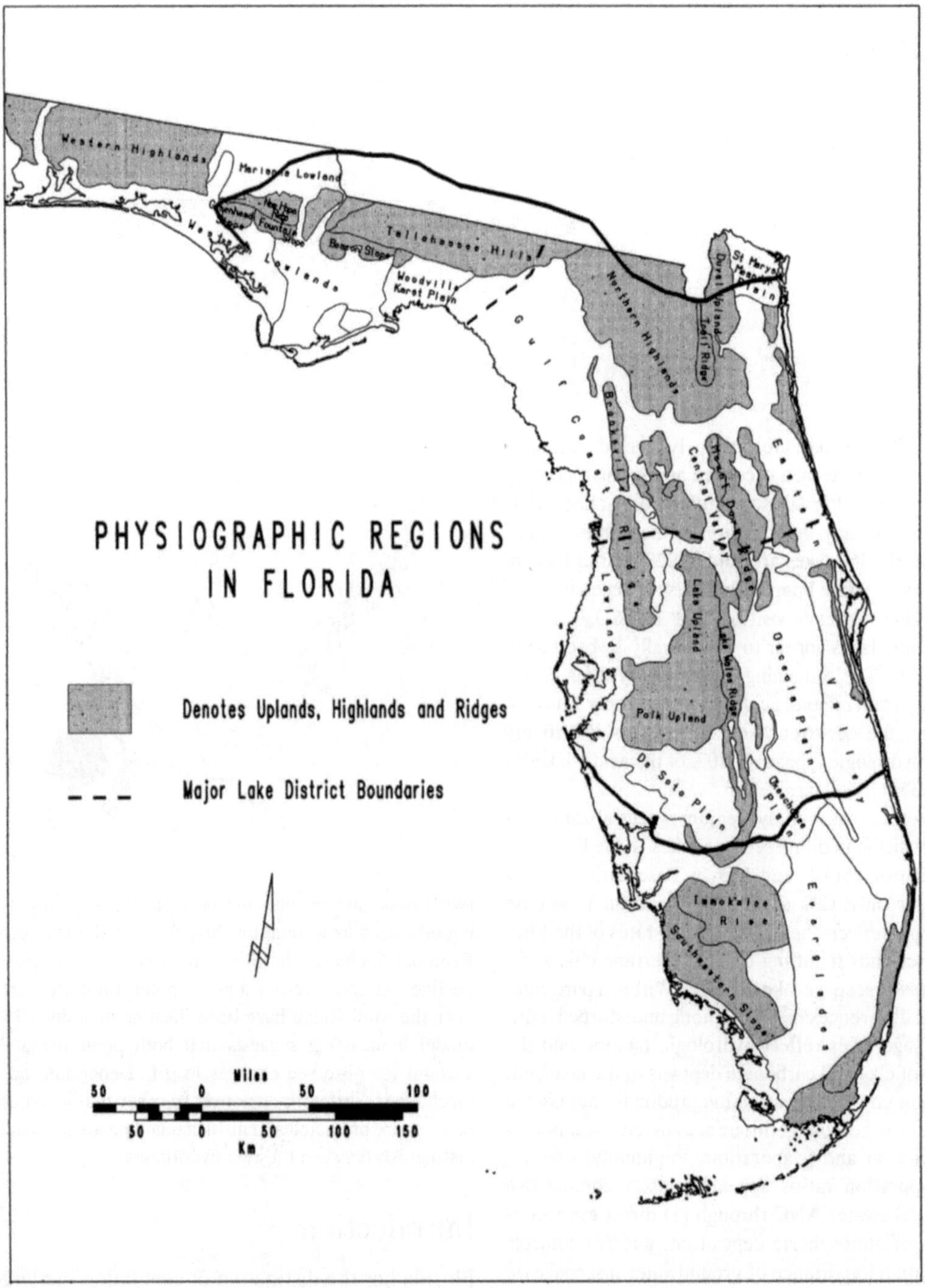

FIGURE 12.1. General physiography and location of major lake districts within the Florida case study region, denoted by the heavy solid line. Dashed lines delineate the three lake regions: the Florida Panhandle, the Florida Northern Peninsula, and the Florida Central Peninsula. (After Puri and Vernon 1964.)

in lake hydrology because of increasing demands for water supply, (3) naturally occurring organic acids, and (4) acidification due to deposition of sea salt and release of H^+ via cation exchange. Nearly 84% of acidic lakes in Florida are seepage lakes that receive nearly all their hydraulic income from direct rainfall; nearly half of these acidic seepage lakes have SO_4^{2-} concentrations that exceed [Ca^{2+} +

Mg^{2+}], suggesting that SO_4^{2-} inputs rather than organic acids are responsible for the acidic status of clearwater lakes in Florida (Sullivan et al. 1989, Eilers et al. 1988a). In the Southeast, strong increases in SO_2 emissions observed since 1960 have persisted into this decade (Husar 1986; Chapter 3, Husar et al., this volume). Estimates of future emissions in Florida project increases of perhaps 30% by the year 2000 (FDER 1984). Paleolimnological reconstructions of lakewater pH for several lakes in the Florida Northern Peninsula indicate that acidity in two lakes recently has increased (post-1950), and may be related to acidic deposition (Sweets et al. 1987). Modeling suggests that increases in SO_4^{2-} deposition of this magnitude will result in further decreases in lakewater ANC (Pollman and Dickinson in preparation).

The Regional Case Study (RCS) boundaries for Florida encompass all the major softwater lake districts in Florida (Figure 12.1). The western boundary marks the edge of the New Hope Ridge and Greenhead Slope of the Western Highlands, which contains the most dilute lakes in Florida. The southern boundary is marked by the southern limit of the Lake Wales lake district. Other major lake districts include the Trail Ridge and the lakes within the Ocala National Forest, both located in the Northern Peninsula. Within these bounds, soils are highly weathered and acidic, with limited ability to supply base cations. Chemical characteristics of Florida lakes are nonetheless extremely diverse (Canfield 1981, Eilers et al. 1988a, Stauffer in review), and reflect a mixture of hydrologic and geochemical factors, as well as landuse patterns. Seepage lakes are clearly the most important class of acidic lakes in Florida, and they are the focal point of discussion.

Regional Characteristics

Watershed Characteristics

Bedrock and Surficial Geology

The geology of Florida is primarily marine in origin. Bedrock geology is dominated by the Floridan aquifer, an extensive lithologic series of limestone and dolomite of Eocene to Miocene age that contains minor amounts of clay, sand, and marl (Heath and Conover 1981). The Floridan underlies virtually all of Florida, although its proximity to the land surface varies. It is near the surface or exposed along parts of the Gulf Coast and along the northern Florida-Alabama border, but it can be more than 100 m below the surface in the mid-Peninsula (Conover et al. 1984).

In many places, surficial deposits of sands from the Pliocene to principally Pleistocene ages are isolated hydraulically from the Floridan aquifer by the Miocene age Hawthorn Formation. The Hawthorn Formation is extremely variable in composition, consisting of varying amounts of clay and clayey sands, quartz sand, carbonate and dolomite deposits, and phosphatic granules (Scott and McGill 1981, Scott 1983). Kaolinite is the predominant clay in the more weathered or highly leached sections, including the major lake districts in the Panhandle and the Trail Ridge; otherwise, palygorskite or montmorillonite predominates. Within the Trail Ridge and other northern Florida locations, fossiliferous limestone is present just above the Hawthorn (Pirkle et al. 1977).

Topographic features in Florida are largely the result of wave and stream activity on the land surface over the past 10 to 15 million years. A number of marine terraces resulting from eustatic changes in sea level are present in Florida, resulting in progressively higher and older terrace slopes moving inland, both in the Panhandle and towards the central axis of peninsular Florida (Vernon 1942, Healy 1975b, Fernald and Patton 1984). Surface features are dominated by karst terrane (Figure 12.2). Major districts of softwater lakes in northwest Florida are associated with New Hope Ridge-Greenhead Slope and Woodville Karst Plain, which range in elevation from 15 m to about 80 m above sea level. Major softwater lake districts in peninsular Florida include the Trail Ridge, Mount Dora Ridge and Sumter Uplands, and the Lake Wales Ridge, all of which are part of the central Florida ridge extending along the spine of peninsular Florida (Figure 12.1). Paleo dunes are characteristic features of the Central Ridge, which, because of the xeric or excessively drained nature of the sand hills, is the principal recharge area of the Floridan aquifer (Brooks 1982).

Origin of Lakes

Florida contains an estimated 6,530 lakes $\geq$ 10 ha in size (Shafer et al. 1986), of which nearly all are natural in origin. Traditionally, the majority of

FIGURE 12.2. Oblique aerial photo of two doline ELS-I lakes lying in karst terrane: Crystal Lake (foreground) and Compass Lake (background), in the Florida Panhandle on Greenhead Slope.

these lakes have been viewed as solution lakes formed by dissolution of underlying limestone by percolating surface or groundwater followed by collapse of the overlying deposits into the cavity (Vernon 1942, Musgrove et al. 1965, Schmidt and Clark 1980, Palmer 1984). Indeed, Hutchinson (1957) cites Florida as containing perhaps the most important group of solution lakes in the Western Hemisphere. However, Arrington and Lindquist (1987) argue that there is no evidence supporting sinkhole formation by collapse of large caverns within the limestone—at least within the Florida Northern Peninsula. Unlike other regions where karst is as highly developed, the surface karst features of the Florida Northern Peninsula are formed entirely within the clastic deposits overlying the limestone, as noted by Arrington and Lindquist. The limestone bedrock is not exposed, and only some of the deepest sinks extend into the Hawthorn Formation. According to Arrington and Lindquist (1987), piping of mantle sediments into an extended network of solution channels within the limestone ultimately leads to collapse and sinkhole formation at the land surface.

Two types of lakes related to solution processes occur in Florida karst: steep-walled, conical basins with nearly circular surfaces, and polje basins. The origin of the second form is controversial. Palmer (1984) suggests that poljes form in valleys when drainage into a sinkhole within the valley causes erosion, eventually plugging the sinkhole and giving rise to broad, irregularly shaped lakes that may be interconnected with other basins. White (1970) suggests that poljes were formed during periods when the potentiometric surface of the Floridan was above the basin elevation (see also Hendry and Sproul 1966). Lateral movement of groundwater charged with CO_2 through insoluble surficial clastic deposits dissolved the uppermost limestone, giving rise to broad shallow lake basins. Sinkholes characteristically drain these lakes, indicating that the systems are in the latter stage of lake development.

Examples of the steep-walled basin type include Sheelar Lake in the Trail Ridge (conductivity 19 $\mu S\ cm^{-1}$, pH 5.14; Hendry 1983) and Open Lake in the Panhandle (conductivity 14.9 $\mu S\ cm^{-1}$, pH 4.96; Kanciruk et al. 1986). Perhaps the most striking example of a Florida polje is Lake Jackson near Tallahassee. Similar to other large lakes in the Tallahassee Hills lake district, Lake Jackson periodically disappears, most recently in fall 1982. The lake is quite shallow and the existence of the lake follows long-term trends in precipitation coupled with periodic opening of sinkholes within the lake basin (Hendry and Sproul 1966, Jensen 1987).

Most solution lakes apparently formed when sea levels declined, resulting in corresponding declines in local groundwater tables and formation of funnel-shaped dolines, which in turn filled as the sea level rose in post-glacial times (Vernon 1942, Hutchinson 1957). The preponderance of solution lakes in Florida occurs along the Central Ridge and in the central Panhandle (Holmes and Washington counties), where the limestone is covered by 10 to 60 m of cohesive, relatively impermeable clayey sediments (Sinclair and Stewart 1985).

Soils

Four major orders of soil predominate in Florida: Entisols, Histosols, Spodosols, and Ultisols (Figure 12.3). Softwater lakes in Florida are clustered in portions of the Panhandle (Lakeland-Troup-Alpin association) and along the central axis of the peninsula (Candler-Apopka-Astatula association), where thick deposits of deeply weathered entisols with high rates of infiltration dominate (SCS hydrologic group A; Caldwell and Johnson 1982). Principally quartz sands, the soils are quite acidic ($pH \leq 5.0$), with low cation exchange capacity (< 5 meq $100\ g^{-1}$). Base saturation levels generally are $< 10\%$ and sometimes $< 2\%$.

Typic quartzipsamment and paleudult soils dominate in highland and ridge area watersheds. Pollman and Hendry (1983) have calculated average exchangeable base quantities of only 54 keq ha^{-1} and 52 keq ha^{-1} in the surficial 25 cm of paleudult and quartzipsamment soils respectively; values < 10 keq ha^{-1} have been calculated for paleudult and quartzipsamment soils in the Apalachicola National Forest in the Panhandle (Pollman unpublished data). Using cation exchange capacity (CEC) as the criterion, McFee (1980) classified the ridge and highland soils as sensitive to acidic deposition (i.e., the total equivalents of acid deposited by 100 cm rainfall at pH 3.7 for 25 years exceeds the CEC in the surficial 25 cm soil).

Vegetation

Major natural plant communities within the Florida RCS region differentiate and develop in response to prevailing surface hydrologic and associated fire regimes (Clewell 1981, Monk 1960, 1968, Laessle 1942, 1958, Duever 1981, Veno 1976). Dominant plant communities include hammock, sandhill, flatwoods, baygall, seepage bog, swamp, and marsh communities. Most acidic seepage lakes in Florida are found in sandhills, which normally burn every few years. As fires sweep through the xeric, rolling sandhill terrain, invading turkey oaks (*Quercus laevis*), shrubs, and other hardwoods are pruned back, leaving an open canopy of longleaf pines (*Pinus palustris*) and an understory of wiregrass (*Aristida stricta*) and wildflowers.

The other major community type characteristic of softwater lake districts is scrub. Similar to sandhills, scrub communities are xeric, but have adapted to much more severe fires occurring at irregular 30- to 120-year intervals (Myers 1985, Abrahamson 1984, Laessle 1984, Duever 1981). Sand pine (*Pinus clausa*), sand live oak (*Quercus geminata*), scrub oak (*Q. inopina*), myrtle oak (*Q. myrtifolia*), rosemary (*Ceratiola ericoides*), fetterbush (*Lyonia ferruginea*), and other leathery leaved shrubs characterize this habitat.

Other major plant community types are found in wetter regimes where vertical infiltration of precipitation is not as excessive or the water table is higher. At lower elevations where the water table is close to the land surface, flatwoods are the dominant community. Flatwoods are fire-maintained pinelands characterized by a sparse canopy of longleaf pines or slash pines (*Pinus elliottii*), a shrub layer of saw palmetto (*Serenoa repens*), and a groundcover of wiregrass, other grasses, and forbs. On saturated organic soils, baygall plant communities consisting of dense forests of sweetbay (*Magnolia virginiana*), loblolly bay (*Gordonia lasianthus*), swampbay (*Persea palustris*), and dahoon holly (*Ilex cassine*) develop. Seepage bogs with shrubby or herbaceous vegetation occur along slopes where sweeping fires prevent forest development.

Swamps typically dominated by cypress (*Taxodium distichum* and/or *T. ascendens*), blackgum (*Nyssa biflora*), and red maple (*Acer rubrum*) develop where periodic inundation rather than infiltration dominates surface hydrology. Fires still play a major role in wetland plant community structure, and where fires occur too often for trees to become established, marshes are the usual wetland vegetation. On sites that burn every one to two years, marshes consist of herbaceous emergents such as

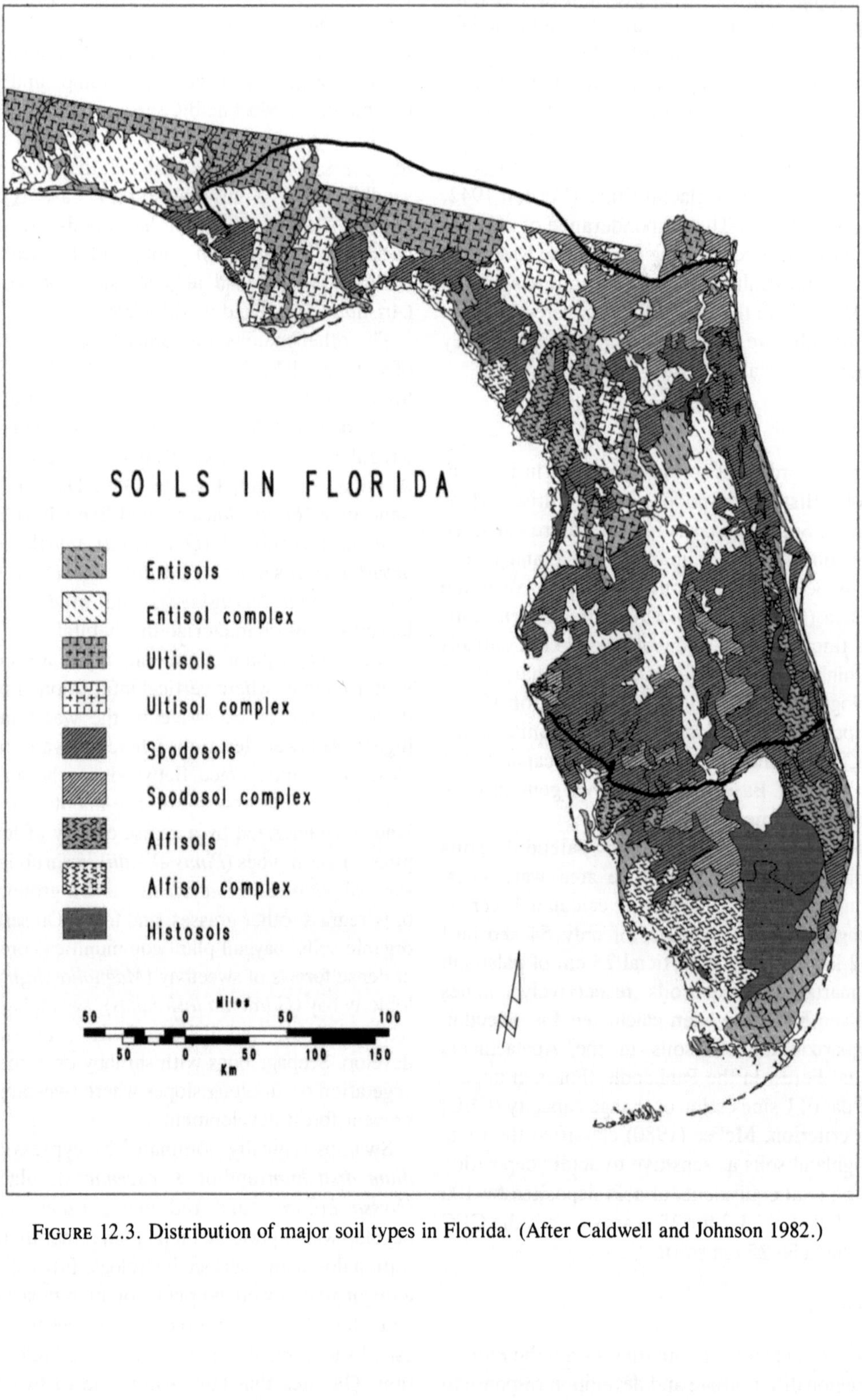

FIGURE 12.3. Distribution of major soil types in Florida. (After Caldwell and Johnson 1982.)

maidencane (*Panicum hemitomon*). Where fires are less frequent, shrubs such as coastal plain willow (*Salix caroliniana*) and buttonbush (*Cephalanthus occidentalis*) dominate.

The sandhill and scrub communities that dominated the original vegetation of Florida softwater lake districts have undergone striking changes over the past century (Means and Grow 1985, Noss 1988, Myers and White 1987, Means and White 1987, Means 1987, Tebo 1985, Peroni and Abrahamson 1989). Most of the longleaf pine was cut in the early 1900s, starting with logging of the Panhandle in the late 1800s and extending to the Lake Wales Ridge in the Florida Central Peninsula by the 1920s and 1930s. Logging and fire suppression have fostered the development of scruffy stands of oaks in former longleaf forests with little remaining of the original light-dependent groundcover. Over half of this xeric landscape has been converted to agricultural and residential uses. Below the 29° parallel, citrus groves have replaced most of peninsular Florida's scrub/sandhill systems, starting in the late 1800s through the present, although freezes in the last decade have led to complete losses of most groves north of Orlando. Because wiregrass groundcover of the sandhills does not recolonize after disturbance and fires are much less extensive than in the original landscape, such abandoned agricultural lands are developing into oak forests rather than reverting to grassy pinelands.

Climate/Hydrology

Precipitation and Evaporation

Florida's climate is subtropical, with relatively high temperatures and rates of rainfall. Intraregional climatic differences within Florida are greatest in winter, due to the alternating influence of warm maritime and cold continental air masses. Invading cold air masses typically originate from the northwest, and usually weaken or stagnate before reaching southern Florida. Low pressure areas associated with fronts over the Gulf of Mexico contribute additional rainfall during late winter. Rainfall during the winter months decreases monotonically across the state from the northwest (ca. 12.5 cm month^{-1}) to the southcentral part of the state (ca. 6 cm month^{-1}) (Heath and Conover 1981, Jordan 1984).

Summer climatic differences by comparison are much smaller. Summer storms are largely small-scale, convective disturbances, and the distribution of rainfall is relatively uniform along the Central Florida Ridge and the Panhandle where most softwater lakes are located. Rainfall is highest throughout the region in the summer, although the ratios of summer to winter rainfall vary from about 1.4 (northwest) to 4 (south).

The aggregated result of frontal and convective storms over the course of a year is decreasing precipitation from the western Panhandle to the Central Peninsula. Average annual rainfall ranges from about 152 cm (60 inches) in northwest Florida to 131 cm (52 inches) in the Central Peninsula (NOAA 1986; Figure 12.4). In contrast, annual lake evaporation and evapotranspiration rates are highest in south Florida and lowest in the Panhandle where temperatures, particularly in the winter, are comparatively cooler. Superimposed on Figure 12.4 is the difference between rainfall and potential evaporation, the latter defined as the amount of evaporation that would occur annually if the region were continuously wet. So defined, potential evaporation always exceeds actual evaporation, and the difference between precipitation and potential evaporation (surplus precipitation) defines the lower limit of regional outflow rates (runoff and recharge).

Seepage Lake Hydrology

The existence of gradients in precipitation and evaporation across Florida has implications for similar gradients in seepage lake hydrology. Over time scales of moderate length (> 6 to 12 years), lakes typically maintain stage or volume about some mean value. Oscillations about this mean value in general are tied directly to precipitation patterns. In order to conserve volume over the time scales cited above, gains and losses from inseepage and outseepage must balance precipitation inputs and evaporative losses:

$$\Delta S = P * A + Q_i - E * A - Q_o = 0 \quad (12\text{-}1)$$

where: ΔS = change in storage or volume (m^3) of the lake
P = precipitation (m)
E = evaporation (m)
A = lake area (m^2)
Q_i = inseepage (m^3)
Q_o = outseepage (m^3)

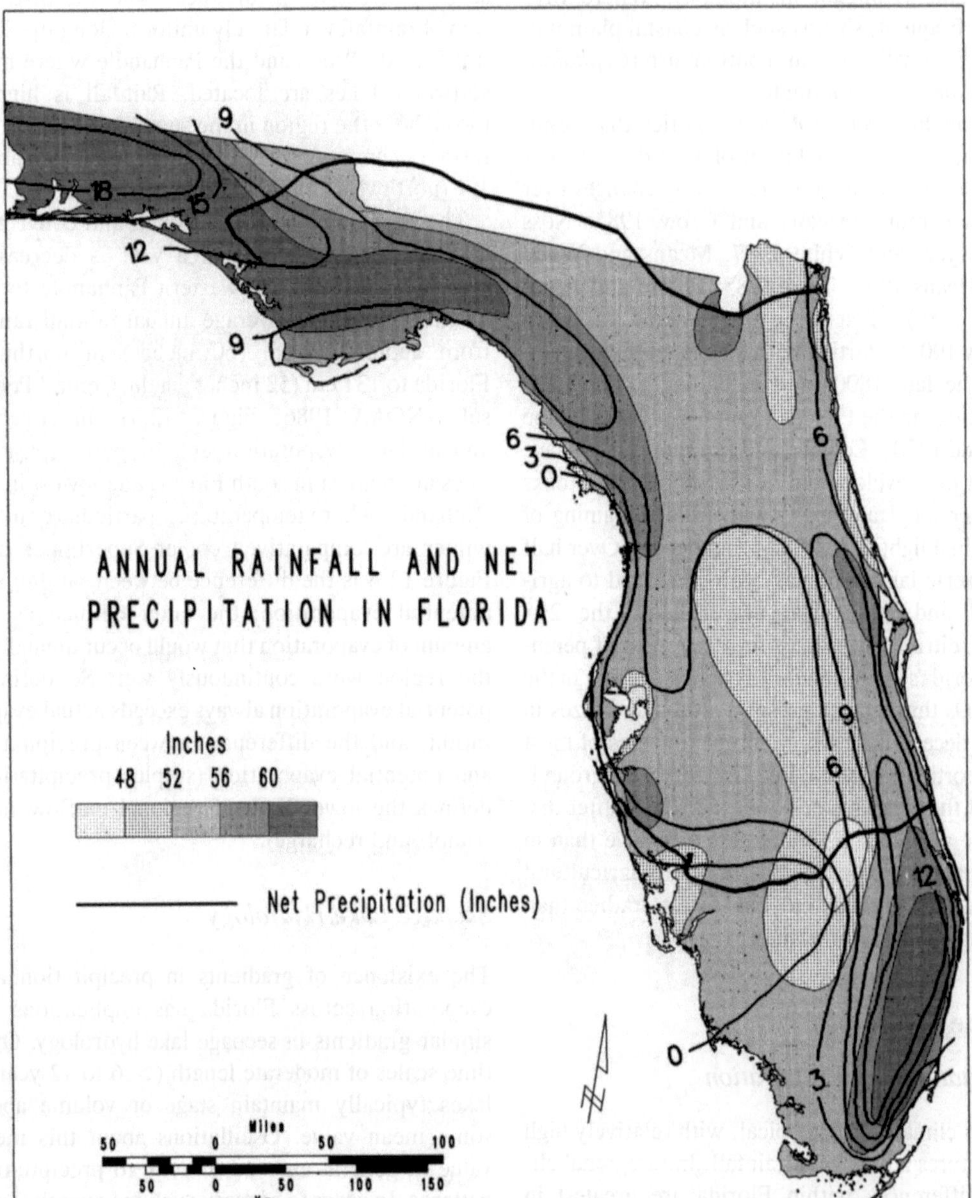

FIGURE 12.4. Annual rainfall (cm) in Florida. Superimposed on the figure are isolines showing the difference between rainfall and potential evaporation. Potential evaporation is defined as evaporation that would occur if the region were continuously wet and thus defines the maximum evaporation that could occur. (After Jordan 1984, and Heath and Conover 1981.)

TABLE 12.1. Hydrological conditions in study area and estimated water budgets for seepage lakes with an assumed mean depth of 3 m.[a]

Location	Precipitation[b] (cm)	Pan evaporation[b] (cm)	Groundwater[c] Inflow (cm)	Groundwater[c] Outflow (cm)	Percent of input Precipitation	Percent of input Ground water	τ_w[d] (yr)
Milton Experimental Station	164	151	1	60	99.0	1.0	5.0
Gainesville	136	166	16	35	89.3	10.7	8.6
Lisbon	119	149	10	23	92.2	7.8	13.0
Archbold Biological Station	128	175	26	25	83.3	16.7	12.0

[a]From Baker et al. 1988a.
[b]Precipitation and pan evaporation are 20-year means, except at the Archbold Biological Station, which includes precipitation data from the discontinued Lake Placid station and evaporation data from the Lake Alfred station (100 km north).
[c]Groundwater flow was estimated for seepage lakes from mean monthly $P - E$, where lake $E = 0.70 \times$ pan evaporation. The analysis assumes that if $P - E > 0$, then flow is out; conversely, if $P - E < 0$, then flow is in. Annual inflows and outflows are the sums of monthly values.
[d]Conservative solute residence time = groundwater outflow/mean depth. Mean depth was assumed to be 3 m.

Assuming that lake volumes approximate steady state over a 20-year period, Baker et al. (1988a) applied the continuity equation (Equation 12-1) to examine subregional patterns in Florida seepage lake hydrology. Monthly net groundwater flow was estimated from 20-year means of monthly precipitation and pan evaporation data collected at NOAA climatological stations throughout Florida. Pan evaporation data were corrected to lake evaporation using a pan coefficient of 0.70 (Linsley et al. 1975) and net inseepage and outseepage were computed by summing monthly precipitation and lake evaporation values. From this approach, Baker et al. calculate that seepage lakes in Florida receive 83% to 99% of their water from direct precipitation (Table 12.1). Inferred relative inputs of seepage are lowest in the Panhandle because of the general surplus of net precipitation; τ_w* values correspondingly are lower in the Panhandle because of greater rates of outflow. By holding lake volume constant, this analysis probably overestimates the amount of seepage (in and out) that occurs; τ_w values presented in Table 12.1 are probably higher. This analysis is regional and is of course not strictly valid for individual systems; nevertheless, limited hydrologic data for Lakes McCloud (Trail Ridge, τ_w = 9.6 years; inseepage equivalent to 10% total hydraulic inputs; Baker 1984) and Lucerne (Central Peninsula; T.M. Lee, pers. comm.) suggest that the estimates of Baker et al. are reasonable.

*In this chapter, we use the outflow based residence time (volume/outflow or τ_w; Chapter 17, Baker et al., this volume). This parameter explictly defines the residence time of conservative solutes and accordingly defines more clearly the response time of solutes to perturbations in mass loading, as opposed to the true hydraulic residence time.

Interactions with the Deep Aquifer

Because the Floridan aquifer is largely limestone and dolomite, the nature and direction of flow between surface waters and the Floridan can be very important in defining the geochemistry and ANC regimes of Florida lakes. Artesian flow from the confined aquifer is restricted to areas where the potentiometric surface of the aquifer exceeds the elevation of the lake or land surface. Conversely, recharge to the Floridan occurs only in areas where the elevation of the shallow water table is above the Floridan potentiometric surface. Figure 12.5 shows generalized areas of recharge. Artesian flow can occur in about 45% of Florida, principally in coastal areas and south Florida south of Lake Okeechobee (Conover et al. 1984, Healy 1975). High rates of recharge (25 to 50 cm yr^{-1}) occur in the very porous, well-drained sand ridges of central and westcentral Florida, as well as in the Trail Ridge lake district where internal drainage through sink holes connects the land surface to the Floridan aquifer. High recharge areas also include the major seepage lake districts of the Panhandle (cf. Stewart 1980). Areas of very low recharge (< 5 cm yr^{-1}) occur where the

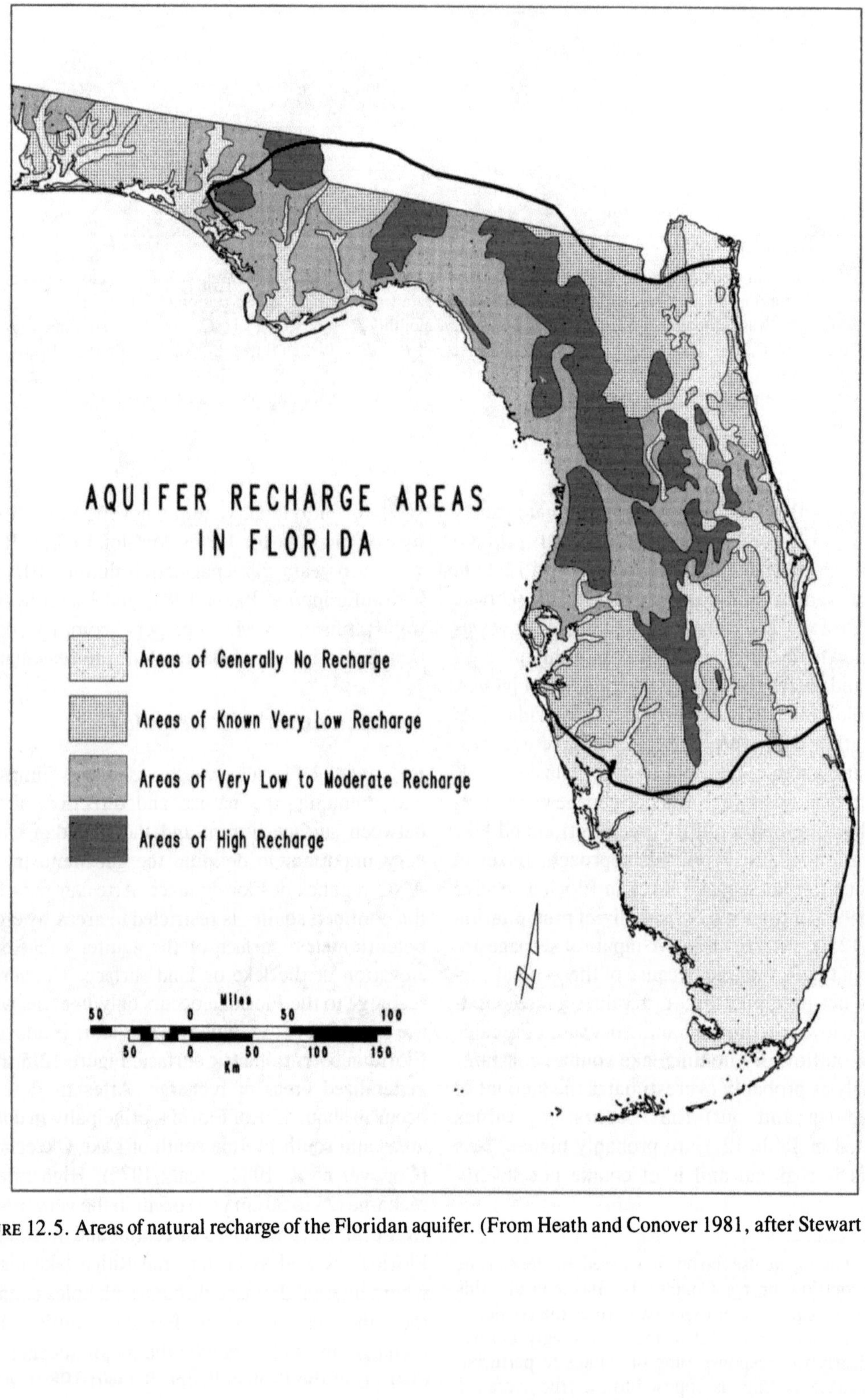

FIGURE 12.5. Areas of natural recharge of the Floridan aquifer. (From Heath and Conover 1981, after Stewart 1980.)

Floridan is overlain by relatively impermeable confining beds 8 m or more thick. Areas of very low to moderate recharge (5 to 25 cm yr^{-1}) represented in Figure 12.5 are defined by areas where the confining beds are less than 8 m thick or breached. This category also includes areas where confining beds are absent but where the water table and the potentiometric surface both lie close to the land surface, resulting in little recharge.

The principal source of potable water in Florida is the Floridan aquifer (Heath and Conover 1981). Between 1960 and 1980, total consumptive use of freshwater nearly doubled from 14.2×10^6 to 27.7×10^6 m^3 day^{-1}, paralleling similar increases in population (Leach 1982). Heavy consumptive use of the Floridan has resulted in modifications of natural recharge patterns; as a result, recharge of the Floridan and other confined aquifers now occurs in regions that formerly were discharge areas (Conover et al. 1984).

Atmospheric Deposition

The most comprehensive precipitation monitoring network in Florida is the Florida Acid Deposition Study (FADS) network established in June 1981 by the Florida Electric Power Coordinating Group (FCG), a consortium of all major utilities in Florida (ESE 1983, 1984, 1985, 1987, FCG 1986). Originally established as a 14-station network, the network currently maintains 7 wet/dry collectors throughout the state, including 6 collectors in the RCS study area (Figure 12.6). Depending on the site, collectors are sampled either daily or weekly. By comparison, National Atmospheric Deposition Program (NADP) network coverage is less dense (2 stations within the Florida RCS region) and sample intensity and the scope is not nearly as well developed as that of the FADS network. In September 1982, the FADS network began sampling ambient air concentrations of SO_4^{2-}, SO_2, HNO_3, and NO_2 every three days. Ambient air measurements now extend to NO_3^-, NH_4^+, Na^+, and NH_3. In addition, FADS dry buckets were sampled monthly and analyzed for major ions between September 15, 1982, and September 14, 1983.

Precipitation Chemistry

Volume-weighted mean concentrations of H^+, nonmarine SO_4^{2-}, and NO_3^- in Florida wet deposition between October 1981 and September 1986 are presented in Figure 12.6. In general, H^+ concentrations decrease nearly 40% from north to south, ranging from 15.2 μeq L^{-1} in Highlands County to 28.1 μeq L^{-1} in the Northern Peninsula (Gainesville). H^+ concentrations in the Panhandle averaged 20.8 to 23.4 μeq L^{-1}. Average annual H^+ wet deposition rates over the same period show the same fundamental gradient, with rates ranging from 184 eq ha^{-1} yr^{-1} in the Central Peninsula to 331 eq ha^{-1} yr^{-1} in the western Panhandle. Deposition was somewhat higher in the Northern Peninsula (373 eq ha^{-1} yr^{-1}), apparently reflecting localized sources superimposed upon the general north-south gradient (ESE 1987).

Nonmarine SO_4^{2-} concentrations (using Na^+ to correct for marine contributions) show a similar regional gradient as H^+, although the effect of localized sources on SO_4^{2-} is more pronounced near Tampa (site 8). Average annual deposition of nonmarine SO_4^{2-} ranged from 188 to 369 eq ha^{-1} yr^{-1}, with similar rates of deposition (ca. 300 eq ha^{-1} yr^{-1}) observed in the Panhandle and the Northern Peninsula. Multiple linear regression shows that variations in excess SO_4^{2-} concentrations can account for 58% to 77% of the variability of H^+ in wet precipitation within the Florida RCS region (ESE 1987). Secondary determinants include nonmarine Cl^- and Ca^{2+} as well as NH_4^+; NO_3^- typically ranks third or fourth in importance as a predictor of H^+.

Nitrate concentrations average about 47% of the concentration of SO_4^{2-} in wet deposition. Nitrate concentrations were relatively uniform, ranging from 9.5 to 13.3 μeq L^{-1}; deposition ranged from 120 to 161 eq ha^{-1} yr^{-1}. Ammonium concentrations vary considerably among sites and show little evidence of a statewide trend, apparently because of localized factors (ESE 1987). The lowest average annual concentration of NH_4^+ was observed at coastal site 4 (4.2 μeq L^{-1}); maximal concentrations were observed in central Florida (site 8; 9.7 μeq L^{-1}).

Major constituent concentrations in Florida precipitation vary in definable patterns. Analysis of variance (ANOVA) of differences in volume-weighted mean concentrations of H^+, excess SO_4^{2-}, NO_3^-, NH_4^+ excess Ca^{2+}, and excess Cl^- between October 1981 and September 1986 across FADS sites are presented in Table 12.2 (ESE 1987). Depending on the constituent, three to four

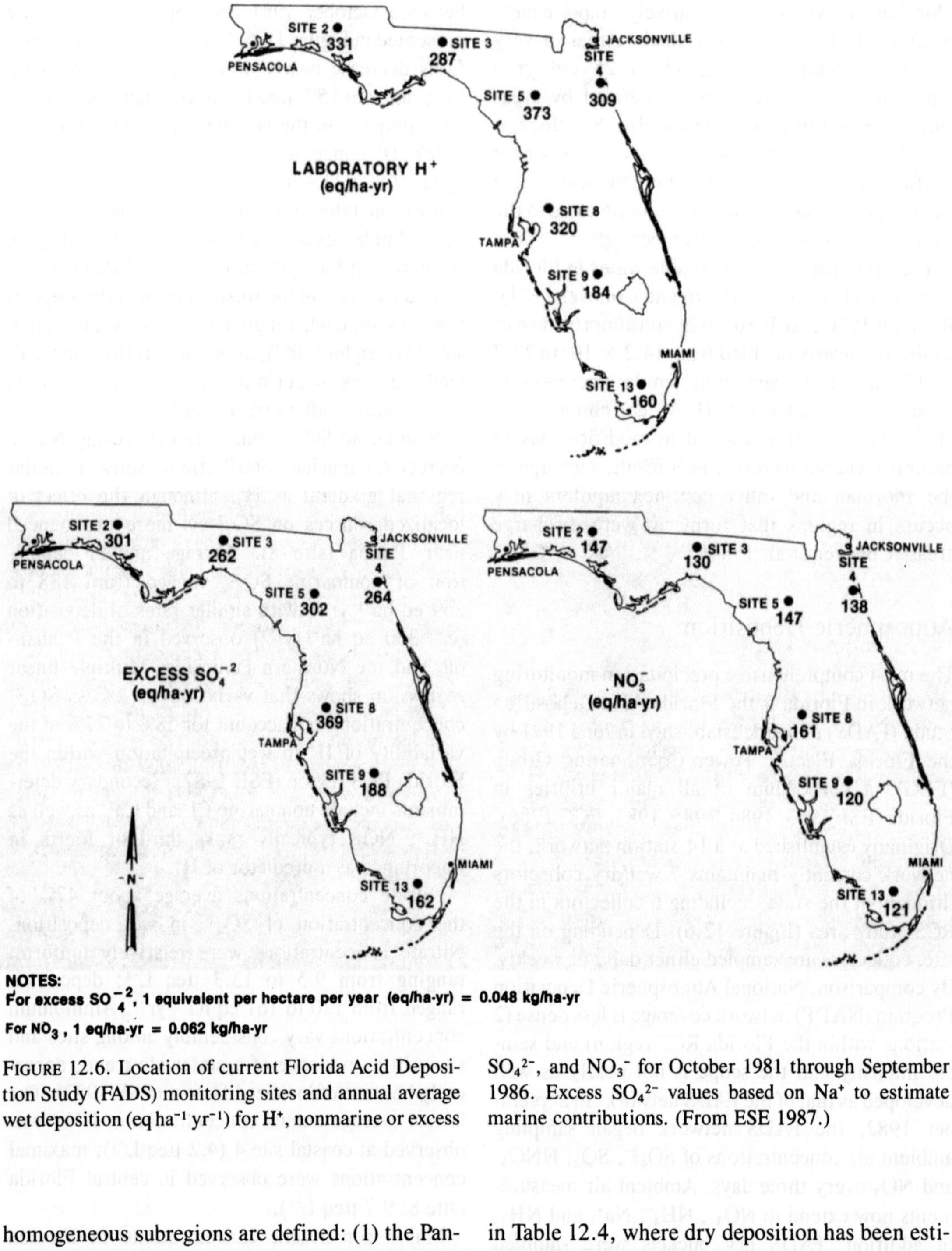

FIGURE 12.6. Location of current Florida Acid Deposition Study (FADS) monitoring sites and annual average wet deposition (eq ha^{-1} yr^{-1}) for H^+, nonmarine or excess SO_4^{2-}, and NO_3^- for October 1981 through September 1986. Excess SO_4^{2-} values based on Na^+ to estimate marine contributions. (From ESE 1987.)

homogeneous subregions are defined: (1) the Panhandle near Tallahassee, (2) the western Panhandle and the Northern Peninsula, (3) central Florida near Tampa, and (4) the Central Peninsula.

Table 12.3 summarizes average wet deposition rates by FADS site for calendar years 1982 through 1986. Total (wet plus dry) deposition is presented in Table 12.4, where dry deposition has been estimated from dry bucket data and ambient air measurements (cf. Baker et al. 1988a). Deposition fluxes of gases and SO_4^{2-} aerosols were calculated using ambient air concentrations measured as part of the FADS program between September 1982 through 1986 and applying deposition velocities

TABLE 12.2. ANOVA results for volume-weighted mean concentrations[a] during the FADS 5-year study period.

	Analyte[b]											
	Lab H^+		NH_4^+		xs-Ca^{2+}		xs-SO_4^{2-}		NO_3^-		xs-Cl^-	
Site	VWM	Rank	VWM	Rank	VWM	Rank	VWM	Rank	VWM	Rank	VWM	Rank
2	23.	A/B	6.3	B	4.1	C	21.3	B	10.4	A/B	1.1	B
3	20.8	B	5.6	B	4.4	B	19.2	B	9.5	B	1.3	B
4	22.9	B	4.2	B	4.5	B	19.8	B	10.4	A/B	2.9	A/B
5	28.1	A	6.1	B	5.0	B	22.9	B	11.1	A/B	1.1	B
8	26.2	A	9.7	A	7.3	A	30.3	A	13.3	A	1.6	A/B
9	15.2	C	5.6	B	6.0	A	15.9	C	10.1	B	1.2	B

[a] From ESE 1987. See Figure 12.6 for locations of sites. All concentrations are $\mu eq\ L^{-1}$.
[b] Ranks refer to significantly different concentrations at the 95% confidence level. Concentrations are ranked in descending order from A to D. A/B indicates volume-weighted mean (VWM) concentrations that fall between two ranks.
xs = Excess or nonmarine concentrations using Na^+ as a marker.

derived from the literature to give fluxes (Appendix B). Disregarding marine influences at site 4, wet deposition of Ca^{2+} increases from north to south, reflecting subregional trends in surficial geology. Calcareous outcrops are more prevalent in south Florida (Fernald and Patton 1984) and windblown $CaCO_3$ probably contributes to the observed gradient in Ca^{2+} deposition, resulting in some neutralization of precipitation acidity.

Seasonal Variability

Precipitation chemistry in Florida shows moderate to strong seasonal variability for all major constituents (ESE 1987). Summertime (April 1 through September 30) concentrations invariably are higher for all ions throughout the Florida RCS region; these seasonal differences range from 50% to 220% higher for H^+ concentrations and 25% to 75% higher for nonmarine SO_4^{2-} concentrations in the summer. Brezonik et al. (1980) conjectured that these trends may reflect either seasonally higher rates of power generation during the summer (see section on Emissions Inventory and Source Attribution) or differences in scavenging efficiencies by summer convective storms compared to winter frontal events. Multiple linear regression analysis conducted across seasons also shows that excess SO_4^{2-} is a comparatively more important predictor of H^+ in the summer (ESE 1987).

Acid precursor (SO_2 and NO_2) and oxidation product (particulate SO_4^{2-} and HNO_3) concentrations also show distinct seasonal fluctuations, although there are marked differences between the two types of ambient air species. Typically higher in winter, SO_2 and NO_2 exhibit seasonal differences averaging about 25% and 65%, respectively (ESE 1987). Conversely, particulate SO_4^{2-} and HNO_3 are typically higher in the summer, with seasonal differences averaging about 13% and 22%, respectively.

TABLE 12.3. Average wet deposition rates by FADS site, calendar years 1982–86.[a]

	$eq\ ha^{-}\ yr^{-1}$									
Site	H^+	Na^+	K^+	NH_4^+	Ca^{2+}	Mg^{2+}	Cl^-	NO_3^-	SO_4^{2-}	Cat/Anion
2	332.0	165.4	11.6	91.0	65.0	56.0	201.6	148.0	322.6	1.073
3	287.8	173.0	11.8	77.0	66.4	56.4	218.2	131.2	283.4	1.063
4	309.8	519.0	16.6	57.0	80.4	135.2	626.0	138.4	327.2	1.024
5	375.8	143.0	10.6	81.8	72.4	45.8	180.2	148.8	322.0	1.120
8	320.0	159.0	11.4	120.2	95.2	53.2	200.8	162.6	371.6	1.033
9	186.2	161.2	8.4	68.2	77.8	56.6	198.6	121.2	210.0	1.054

[a] From ESE 1987. See Figure 12.6 for locations of sites.

TABLE 12.4. Total (wet plus dry) deposition in eq ha^{-1} yr^{-1}.[a]

Site	H^+	Na^+	K^+	NH_4^+	Ca^{2+}	Mg^{2+}	Cl^-	NO_3^-	SO_4^{2-}	Cat/Anion
2	564.5	213.7	26.0	124.0	111.2	78.2	257.8	261.7	549.4	1.045
3	347.9	221.1	41.3	104.5	221.0	92.5	274.2	211.6	487.2	1.057
4	513.1	716.5	32.2	76.2	158.4	183.9	856.0	247.4	528.1	1.030
5	582.4	218.5	21.9	91.8	186.5	75.9	268.1	275.7	540.1	1.086
8	474.0	267.8	46.3	168.3	314.0	95.1	327.5	301.3	654.0	1.064
9	303.8	243.7	20.3	98.2	168.7	91.3	294.7	205.6	386.6	1.044

[a] Values derived from wet deposition presented in Table 12.3 and estimated dry deposition from ambient air concentrations and wet bucket measurements. See Figure 12.6 for locations of sites. See text and Baker (Appendix B, this volume), for details.

Emissions Inventory and Source Attribution

An emissions inventory of anthropogenic SO_2 and NO_x for Florida compiled by FCG (1986) for the period September 1982 through September 1983 is presented in Table 12.5. Statewide SO_2 and NO_x emissions totaled 850,000 and 735,000 tonnes yr^{-1} respectively. Emissions of SO_2 are dominated by point source contributions, with utility sources accounting for 68% of the total. Emissions of NO_x are more evenly distributed among point (52%) and area (48%) sources. Source-receptor relationships evaluated with the Eastern North America Model of Air Pollution (ENAMAP) suggest that instate sources contributed approximately 66% of the total deposition of sulfur (S) and approximately 45% of wet deposition of SO_4^{2-} in Florida in 1983 (FCG 1986).

TABLE 12.5. Summary of estimated anthropogenic SO_2 and NO_x emissions for Florida, September 15, 1982, through September 15, 1983.[a]

	SO_2		NO_x	
Source category	Emissions	Percent total	Emissions	Percent total
Point sources				
Utility	580,256	68.3	232,257	31.6
Nonutility	194,440	22.9	152,675	20.8
Total for point sources	774,696	91.2	384,931	52.4
Area sources				
Fuel consumption	35,864	4.2	22,982	3.1
Solid waste	236	0.03	1,468	0.2
Transportation	38,649	4.5	314,966	42.9
Miscellaneous	278	0.04	10,479	1.4
Total for area sources	75,027	8.8	349,895	47.6
Total for point and area sources	849,723	100.0	734,826	100.0

[a] All emissions in tonnes yr^{-1}. After FCG 1986.

Monthly inventories for utility emissions of SO_2 and NO_x compiled by FCG (1986) for the same period show that maximum utility emission rates for both contaminants occur in July; minimum emissions occur in November for SO_2 and February for NO_x. SO_2 emissions for area and nonutility point sources are relatively constant during the year. NO_x emissions from area and nonutility point sources are distinctly more seasonal, with lowest emissions in the spring and fall (FCG 1986).

Estimated and projected SO_2 emissions for Florida between 1976 and 2000 have been developed by the Florida Department of Environmental Regulation (FDER) in conjunction with the Florida Public Service Commission (PSC) and the Florida Electric Power Coordinating Group (FCG) (Figure 12.7; FDER 1984). Data for 1976 through 1981 are based on reported fuel consumption. Estimates for 1982 through 2000 were based on anticipated population growth, changes in electrical consumption, changes in fuel mix, industrial development, changes in transportation, and application of existing state and federal air pollution regulations with no further controls. In the absence of additional regulatory controls, SO_2 emissions from Florida are predicted to increase over current levels by nearly 30% by the year 2000 (FDER 1984).

Regional Surface Water Chemistry

Overview of Softwater Lake Chemistry Data Sets for Florida

The U.S. EPA Eastern Lake Survey–Phase I (ELS-I) conducted in fall 1984 is the only chemical survey

FIGURE 12.7. Estimated and predicted SO_2 emissions (tonnes) for Florida: 1976–2000. (Data from FDER 1984.)

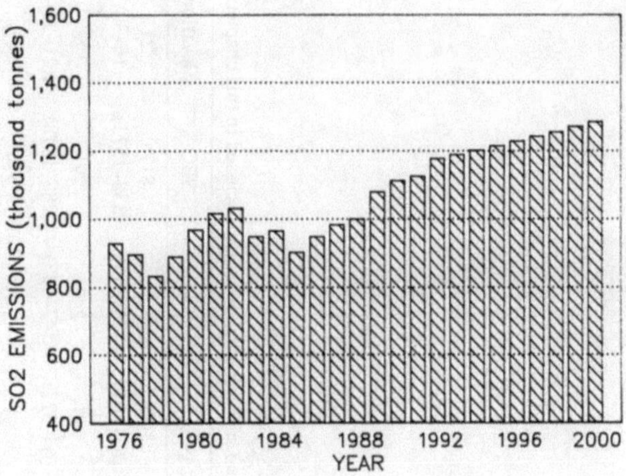

of Florida lakes that allows for unbiased estimation of regional characteristics. This inferential ability to estimate population characteristics owes to a stratified random sampling scheme incorporated as a fundamental feature of the ELS-I (Linthurst et al. 1986; Introduction, Charles, and Chapter 1, Munson and Gherini, this volume). A number of surveys had been conducted in Florida previous to the ELS-I, notably studies by Shannon and Brezonik (1972), Shannon (1970), Brezonik et al. (1983), Hendry and Brezonik (1984), and Canfield (1981, 1983). These surveys, however, were biased toward easily accessible lakes where water chemistry was influenced to varying degrees by cultural eutrophication (Pollman 1986).

The ELS-I used a fall index to assess chemical status—an approach that was useful for interregional comparisons but provided no information on temporal variations (see Introduction, this volume, for discussions of the ELS-I sample design and the fall index approach). For this chapter, we have used selected data sets to supplement the ELS-I data base for Florida, principally to examine historical changes in water chemistry, as well to provide more information on dilute lakes. The following discussion briefly describes the supplemental data sets used for Florida.

Between 1958 and 1960, Clark et al. (1962) sampled a number of lakes in the Florida Northern Peninsula, including six softwater lakes in the Trail Ridge lake district. Sampling was conducted at approximately quarterly intervals, with the number of sampling events ranging from four to six. The softwater lakes sampled by Clark et al. were resurveyed in 1968 and 1969 by Shannon and Brezonik (1972) as part of a 55 lake survey in north and central Florida examining factors controlling eutrophication. Shannon and Brezonik sampled 12 softwater lakes quarterly in the Trail Ridge. Ten years later (1978, 1979), Brezonik and co-workers (Brezonik et al. 1983, Hendry and Brezonik 1984, Hendry 1983) returned to the Trail Ridge as part of a study on the effects of acidic deposition. Thirteen softwater lakes were sampled along or near the Trail Ridge, including nine of Shannon and Brezonik's survey lakes; an additional seven lakes were located in the Highlands Ridge in Florida.

Garren et al. (1989) conducted a quarterly survey of 60 lakes in the Ocala National Forest in the Florida Northern Peninsula between November 1982 and February 1984. Twenty-seven of these lakes were resampled in 1985 and 1986 as part of the Paleolimnological Investigation of Recent Lake Acidification (PIRLA) project funded by the Electric Power Research Institute (EPRI) and the National Science Foundation (NSF) (Sweets et al. 1990). Thirty-three lakes were included in PIRLA, with four lakes in the Trail Ridge and one lake in the Panhandle.

Table 12.6 summarizes the median, upper, and lower quartile distributions of selected major constituents for the ELS-I and the supplemental data sets. As in other chapters, ELS-I distributions are presented both as weighted population estimates, using weighting factors derived from the statistical design of the sampling conducted for the region, and as unweighted sample distributions with no inference toward population characteristics. All the supplemental sets were combined to give a larger data base from which to develop quartile

TABLE 12.6. Median values[a] (with first and third quartiles in parentheses) for major ion chemistry and elevation in the ELS-I and supplemental data sets discussed in this chapter.

Data set	*n*	pH	ANC	SO_4^{2-}	DOC	NO_3^-	Cl^-	Ca^{2+}	Mg^{2+}	Na^+	SBC	Elevation
ELS-I population estimate	150	6.95 (6.01–7.65)	86 (5–591)	92 (32–244)	730 (420–1010)	1.0 (0.2–4.8)	224 (141–274)	246 (89–661)	162 (64–401)	202 (126–311)	817 (338–1573)	21 (16–30)
ELS-I sample lakes	150	6.71 (5.30–7.39)	65 (−5–382)	153 (59–475)	500 (250–990)	1.3 (0.2–6.8)	281 (160–485)	229 (57–486)	169 (64–522)	243 (147–333)	791 (288–1573)	28 (21–32)
Supplemental[b]	68	4.88 (4.59–5.37)	30 (12–46)	100 (62–131)	ND	<1.0 (<1.0)	221 (185–248)	45 (34–89)	63 (51–82)	155 (137–181)	279 (241–368)	ND

[a] All values in μeq L^{-1} except for pH, DOC (μmol L^{-1}), and elevation (m). SBC = $Ca^{2+} + Na^+ + K^+$. ND = no data.

[b] Supplemental data include Clark et al. (1964), Shannon and Brezonik (1972), Brezonik et al. (1983), Garren et al. (1989), and Sweets and Kahl (unpublished data). Supplemental data sets have been combined, giving a total of 68 lakes not included in the ELS-I.

estimates. Comparison between the ELS-I and supplemental data sets shows that the supplemental lakes are more dilute and acidic, with tighter distributions of ANC and base cations. Differences between the two major data sets relate to the directed selection of lakes for the supplemental data sets. The PIRLA survey and the surveys conducted by Brezonik et al. (1983) and Garren et al. (1989) sampled strictly softwater lakes with low ANC, and only softwater lakes from the surveys conducted by Clark et al. (1962) and Shannon and Brezonik (1972) were included in our analysis.

Regional Overview

Within its RCS boundaries, Florida has a total of 2,049 lakes satisfying the ELS-I definition of target lakes. Frequency distribution histograms derived from the ELS-I results for Florida are shown in Figure 12.8 for major lake physical (lake surface area, watershed area:lake area, and sample depth) and chemical (pH, ANC, SBC, SO_4^{2-}, DOC, and Si) characteristics.

The preponderance of lakes in Florida are rather small (median area 17.3 ha; Figure 12.8a) and quite shallow (median ELS-I sample depth 2.7 m; Figure 12.8b). Most Florida lakes are seepage lakes (66%) in which the relationship between topographically determined watershed areas and actual contributing watershed areas is uncertain; as a result, the watershed area:lake area ratio has no quantitative hydrologic significance in Florida. As might be expected of seepage lakes in karst terrane, nearly 68% of the ELS-I target population had watershed area:lake area ratios between 1 and 10 (Figure 12.8c).

Distribution of pH in Florida lakes is bimodal, with most lakes clustered about pH 7.0 (67.8% between pH 6.0 and 8.0) and a smaller acidic subset (21.9%) with pH levels < 5.5 (Figure 12.8d). These results are nearly identical with earlier survey results of Canfield et al. (1983) (see also Pollman and Hendry 1983), who found 16 lakes from a survey population of 165 lakes with pH < 5.0 (compared to an ELS-I population estimate of 12.2%). The small number of lakes in the transitional pH range 5.5 to 6.0 reflects the coincident minimum in buffering intensity characteristic of natural waters in equilibrium with atmospheric p_{CO_2} but with no other source of bicarbonate (Stumm and Morgan 1981, Henriksen 1980). Separation of Florida lakes into two distinct subpopulations related to ANC also is evident (Figure 12.8e; Eilers et al. 1988a). Lakes with ANC ≤ 50 μeq L^{-1} represent 33.8% of the population, compared with 54.0 percent with ANC > 200 μeq L^{-1}.

Base cation concentrations are high in Florida relative to low ANC lakes found in other RCS regions. Figure 12.9 is a spline diagram detailing ionic composition as a function of ANC. At ANC ≤ 0, the dominance of marine Na^+ is clear and generally exceeds the sum of other base cations. Because of marine Na^+, most Florida lakes with ANC ≤ 50 μeq L^{-1} have base cation (C_B) concentrations > 400 μeq L^{-1}. In other RCS regions where marine inputs are low, lakes with ANC ≤ 50 μeq L^{-1} typically have C_B < 200 μeq L^{-1} (cf. Eilers et al. 1988b; Brakke et al. 1988; Landers et al. 1987). Only in southern New England and, to a lesser degree, Maine, is the influence of marine inputs on total ionic strength as prevalent as it is in Florida.

Overall, the dominant cations are Na^+, Mg^{2+}, and Ca^{2+} (Figure 12.9). Potassium is a relatively minor base cation constituent, but it has been shown to be a tracer of inputs associated with citrus agriculture (Stauffer in review). Most of the lakes in the Florida ELS-I target population show signs of cultural activity within their watersheds. In fact, comparison of the ELS target population of 2,049 lakes with a state-wide total of 6,529 lakes > 4 ha (Shafer et al. 1986) hints at the widespread problem of cultural eutrophication in Florida. Using aerial photographs to identify nearshore agricultural and residential activity, Baker et al. (1988b) estimated that 29%, or 601, of the lakes in the original ELS-I population are undisturbed. Stauffer (in review) cites K^+ concentrations exceeding 15 μeq L^{-1} as a signature for significant cultural inputs and in particular, citrus growing. Using 15 μeq L^{-1} K^+ alone as a marker, we estimate an undisturbed population of 886 lakes (43%). At ANC ≤ 200 μeq L^{-1}, Mg^{2+} concentrations typically exceed Ca^{2+} and there is some evidence of Ca^{2+} depletion in acidic Florida lakes, apparently due to algal uptake and subsequent deposition and burial in the sediments (Baker et al. 1988b; cf. Baker et al. 1988c). The ANC increases are related in part to weathering of limestone and dolomite; for disturbed lakes, increased base cation supply related

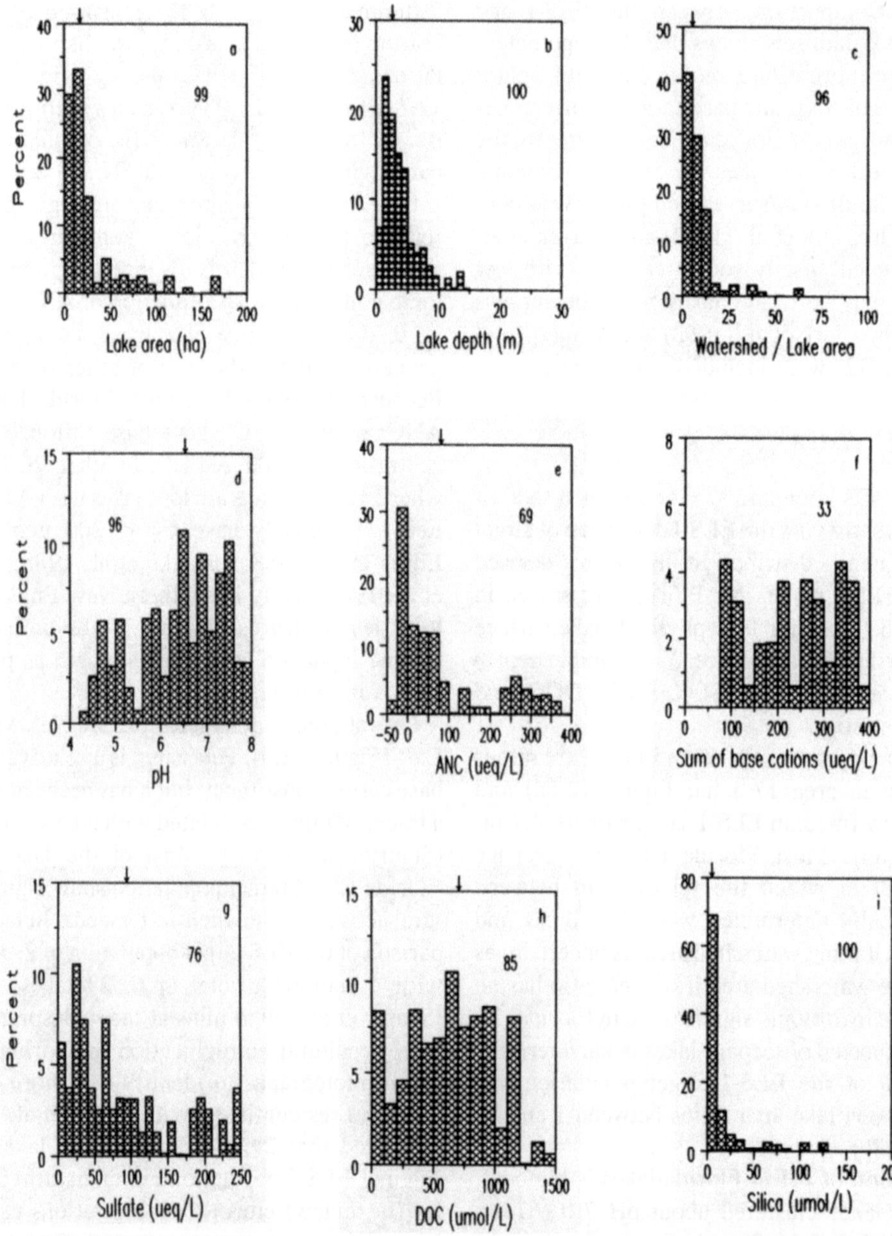

FIGURE 12.8. Frequency histograms of major characteristics of the ELS-I target population of lakes in Florida: (a) lake surface area, (b) sampling depth, (c) watershed area:lake area, (d) air-equilibrated pH, (e) ANC, (f) sum of base cations, (g) sulfate, (h) DOC, and (i) silica. The arrow at the top of each box indicates the median for the target population, and the number in the upper right side of each box indicates the percentage of target lakes plotted.

to soil fertilization is also quite important (Stauffer in review).

Sulfate concentrations are relatively high in Florida, with a median concentration of 94 μeq L^{-1}. The distribution is heterogeneous with some clustering of lakes below 70 μeq L^{-1} superimposed upon a continuum of occurrence frequencies of 60 to 80 lakes (2.9% to 3.9%) between 0 and 160 μeq L^{-1} (10 μeq L^{-1} frequency intervals; Figure 12.8f). Nearly 24% of the Florida population has SO_4^{2-} concentrations > 250 μeq L^{-1}.

Dissolved organic carbon is a relatively important constituent for the overall population of Florida lakes. The median concentration is 720

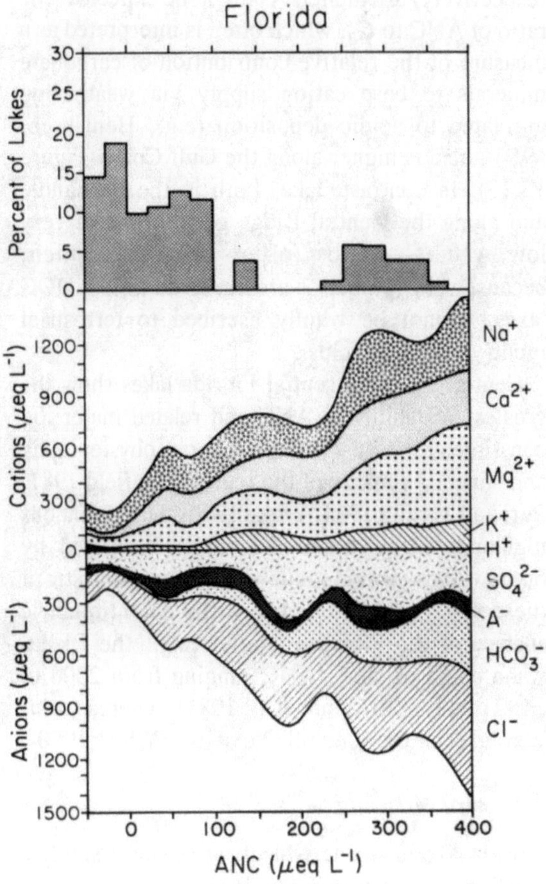

FIGURE 12.9. Concentrations of major cations and anions for lakes in Florida with ANC ≤ 400 μeq L⁻¹. The percentages of lakes represented by ANC intervals are shown as a histogram at the top of the figure. Data plotted are population-weighted estimates for lakes in the ELS-I. The curves are spline fits to the median concentration for each ANC interval.

TABLE 12.7. Subregional median chemical characteristics of Florida ELS-I lakes.

Parameter*	Panhandle (N = 260)	North Peninsula (N = 845)	South Peninsula (N = 945)
pH	4.87	6.44	6.79
Conductivity	17.6	59.2	119.9
ANC	−23.7	79.4	247.8
Ca^{2+}	18.4	177	282
Mg^{2+}	28.4	100	216
Na^{+}	55.4	157	278
K^{+}	3.2	9.8	84.0
SO_4^{2-}	31.9	83.8	199.0
Cl^{-}	64.2	166.6	374.5
DOC	327	577	797
SiO_2	1.2	5.1	5.9
Organic acids[b]	29	67.9	93.9
Total Al	1.4	2.1	1.5

[a] All concentrations in μeq L⁻¹ except Al and SiO_2 (μmol L⁻¹) and conductivity (μS cm⁻¹).
[b] Calculated by charge deficit.

μmol L⁻¹, and, based on calculated charge deficit, DOC contributes 98 μeq L⁻¹ to the overall ion balance (Eilers et al. 1988a). Of the Florida lakes, 85.7% have moderate amounts of DOC, defined as concentrations > 250 μmol L⁻¹ (cf. Kramer and Davies 1988). Despite the prevalence of DOC in the overall population, DOC is generally less important than SO_4^{2-} as an acid anion in very low ANC and acidic lakes (Figure 12.9).

Silica concentrations are interesting in Florida because concentrations are extremely low; 76% of the population has SiO_2 concentrations below 16.6 μmol L⁻¹, with a median concentration of only 5.0 μmol L⁻¹ (Figure 12.8i). These concentrations are lower than any other RCS region or subregion, including northcentral Wisconsin (median concentration 10 μmol L⁻¹; Eilers et al. 1988b), which of all the ELS-I study regions has surficial geological characteristics most similar to Florida. Soils in Florida are highly weathered, acidic quartz sands. Low SiO_2 concentrations directly reflect the absence of primary weatherable minerals (cf. Johnson 1984; Stumm and Furrer 1987), thus limiting the supply of SiO_2 via inseepage. Yearlong diatom uptake and losses to bottom sediments likely contribute to SiO_2 depletion as well.

Spatial Patterns of Distribution

The Florida region can be separated into three subregions on the basis of water chemistry: the Panhandle, the Florida Northern Peninsula, and the Florida Central Peninsula (Figure 12.3). Table 12.7 summarizes median chemical characteristics of selected parameters. Hendry and Brezonik (1984) related differences in lakewater chemistry, including ANC, between seepage lakes in the Northern Peninsula (Trail Ridge) and the Central Peninsula of Florida (Highlands Ridge) primarily to gradients in atmospheric deposition. We propose here that differences in major ionic constituents for undisturbed lakes relate principally to differences in evaporative concentration of atmospheric deposition. That is, hydrologic differences across subregions are the predominant determinants of major ion chemistry even though acidity in clearwater seepage lakes ultimately appears to be derived from atmospheric deposition (Baker et al. 1988b). In addition, subregional differences in land use practices further contribute to gradients in major ion concentrations across the subregions (Stauffer in review). These concepts are developed more fully in the section entitled Groundwater Hydrology/Weathering.

Striking differences in major ion chemistry occur across Florida with the most dilute lakes occurring in the Panhandle. Median pH, ANC, and base cations all increase from the Panhandle to Florida Central Peninsula (Table 12.7). Base cation increases are related to higher rates of evapoconcentration and increased agricultural inputs, particularly below 29° latitude. Atmospheric loadings of nonmarine Ca^{2+} also are higher in the Central Peninsula, apparently due to increased deposition of weathered calcareous material from the southern Peninsula where calcareous soils and limestone outcrops are more common.

Acidic lakes are clustered along the New Hope Ridge-Greenhead Slope and Woodville Karst regions of the Panhandle and the Central Ridge of the peninsula (Figures 12.10, 12.11). Seventy-five percent of the Panhandle population of lakes are acidic, declining to 26% in the Northern Peninsula and only 3.7% in the Central Peninsula. Most of the lakes in the Panhandle are clearwater (DOC < 420 μmol L^{-1}) seepage lakes (Figure 12.12), where SO_4^{2-} is the dominant anion (excluding Cl^-). The Trail Ridge also contains a number of clearwater seepage lakes. The DOC concentrations increase toward the south and, within the Lake Wales Ridge, commercially mineable deposits of peat occur (Fernald 1981).

To the west of the lakes of the Central Ridge lies a series of lakes along the Gulf Coast. These lakes are almost exclusively drainage and closed basin lakes (Figure 12.12), with ANC typically > 200 μeq L^{-1} (Figure 12.10). Unlike the xeric sandhills of the Panhandle and Central Ridge, surficial drainage is poorer along along the Gulf Coast because the Floridan aquifer (and the Hawthorn Formation when present) lies close to the ground surface and in some areas, discharges (Healy 1975a). Accordingly, both DOC and Ca^{2+} (median concentrations 825 μmol L^{-1} and 463 μeq L^{-1}, respectively) are higher. As might be expected, the ratio of ANC to C_B, which often is interpreted as a measure of the relative contribution of carbonate minerals to base cation supply via weathering unrelated to acidic deposition (e.g., Henriksen, 1980), also is higher along the Gulf Coast (Figure 12.13). In contrast, lakes both in the Panhandle and along the Central Ridge generally have very low ANC:C_B ratios (often < 0.25), which, because very low ratios are observed in low DOC lakes, cannot be wholly ascribed to terrestrial inputs of organic acids.

Central and southcentral Florida lakes show the greatest variability in ANC and related major ion constituents because of the diverse physiography and surficial geology of the region (Canfield 1981, Puri and Vernon 1964, White 1970). Discontinuous highlands form subparallel ridges separated by broad valleys. Broad shallow lakes characteristic of these valleys tend to lie below the potentiometric surface of the Floridan aquifer (e.g., the Okalawaha chain of lakes, ANC ranging from 2000 to 2460 μeq L^{-1}; Brezonik et al. 1981), whereas solution lakes in the sandhills lie above (White 1970).

Temporal Variability

Florida lakes characteristically are shallow and few stably stratify (Shannon and Brezonik 1972, Beaver et al. 1981, Garren et al. 1989, Brenner et al. 1990). Lakes that have sufficient depth to stratify may be classified as warm monomictic systems, mixing only in winter. The effects of seasonal dynamics on major ion chemistry in softwater lakes are not well-studied in Florida. The most intensive study published to date is an ion budget conducted by Baker (1984) of McCloud Lake in the Trail Ridge of the Florida Northern Peninsula. Baker conducted monthly sampling for all major ions for 22 months. Monthly variations in pH, SO_4^{2-}, Na^+, and Cl^- occurred and have been related to evapoconcentration rather than seasonal differences in loadings or biological activity (Pollman 1986). Lakewater pH varied by 0.42 units over the study (standard error 0.090); Cl^- and SO_4^{2-} ranged from 137 to 205 μeq L^{-1} and 111 to 231 μeq L^{-1}, respectively.

Associations Among Variables

As discussed previously, marine inputs of Na^+ and Cl^- dominate the ionic balance of softwater Florida

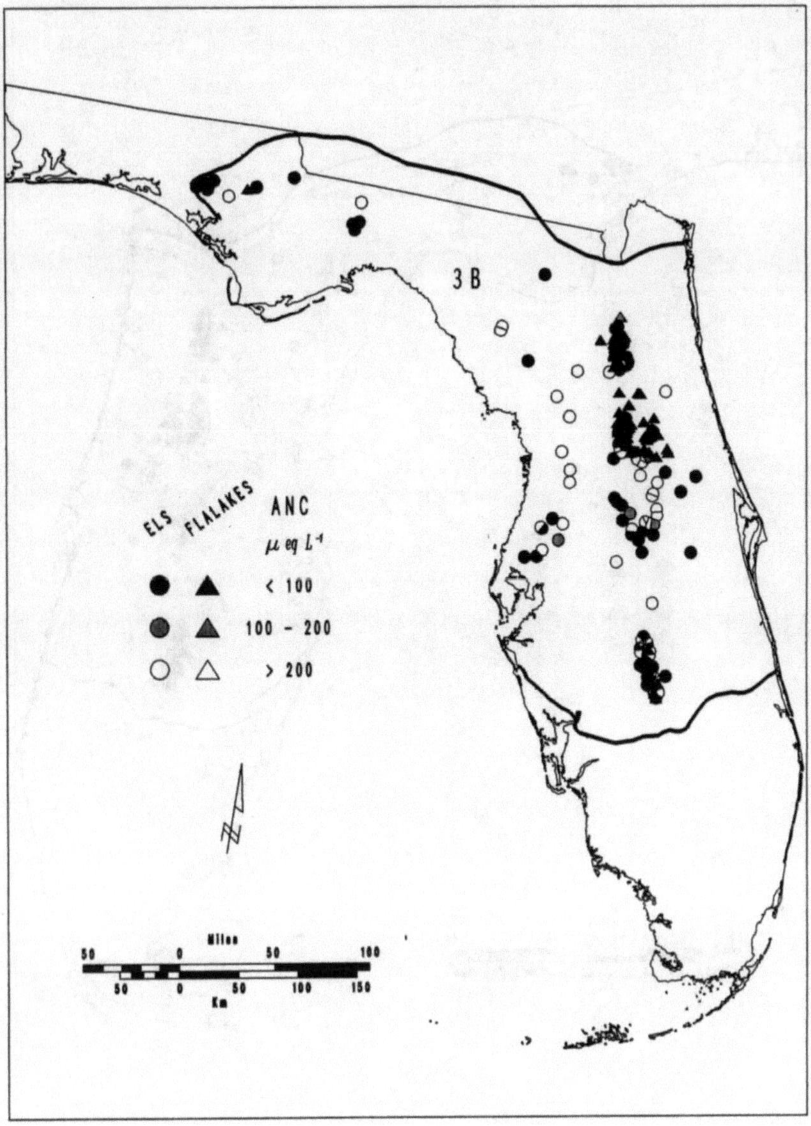

FIGURE 12.10. Areal distribution of ANC by concentration category. Two data sets are represented in the figure: ELS-I lakes (ELS) and lakes from other surveys not included in the ELS (FLALAKES).

lakes (Figure 12.9). In the following discussion of interparameter relationships, SO_4^{2-} and base cation concentrations have been corrected for sea salt influences using Cl^- as the marker for marine associated inputs. The focus of this discussion is on lakes with ANC $\leq$ 200 μeq L^{-1}.

In waters in which base cation supply is derived solely from weathering of carbonate materials or primary minerals such as K-feldspars, and the acid driving weathering reactions is H_2CO_3, ANC should be equivalent to SBC (e.g., Henriksen 1980). Figure 12.14a shows ANC in relation to C_B (SBC) for Florida ELS-I lakes for both low and high DOC lakes. For ELS-I lakes in Florida, the correlation between ANC and C_B is poor, with considerable scatter, and all ANC values fall well below the 1:1 line (slope = 0.098; r^2 = .25). This apparent charge imbalance or ANC deficit must be satisfied by other anions, notably SO_4^{2-} and DOC. A number of factors may account for the ANC

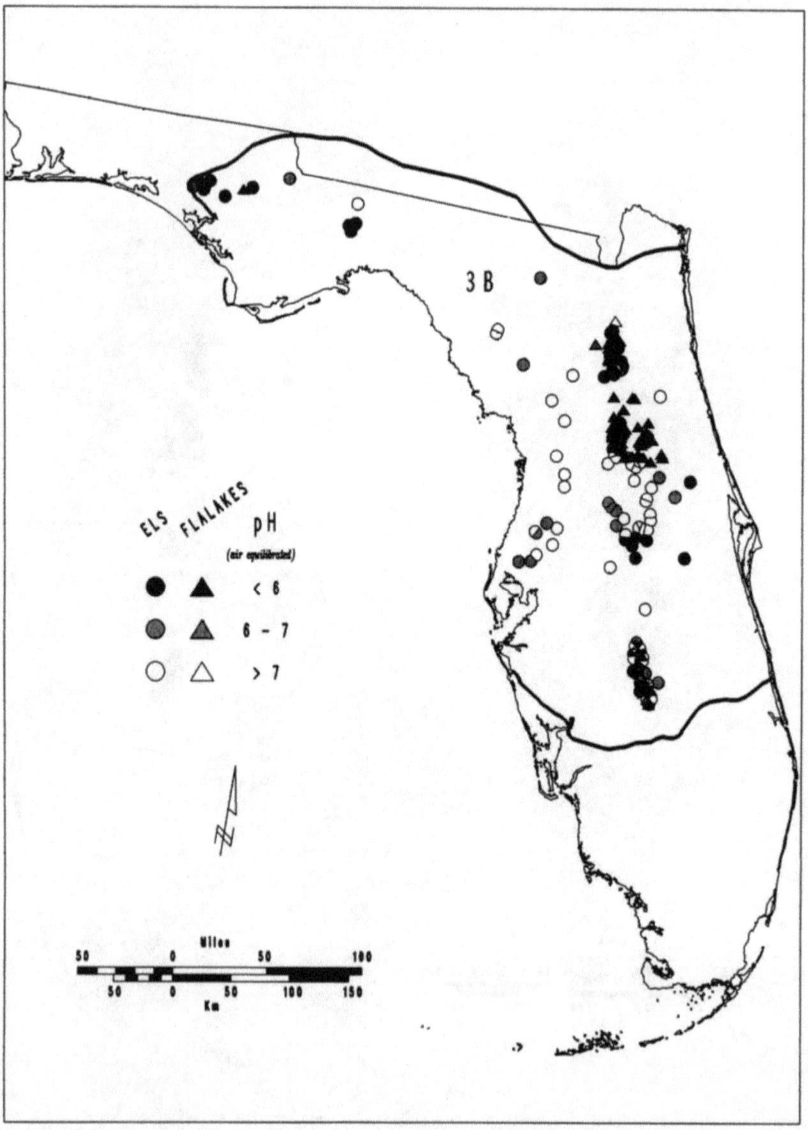

FIGURE 12.11. Areal distribution of pH by category. Two data sets are represented in the figure: ELS-I lakes (ELS) and lakes from other surveys not included in the ELS-I (FLALAKES).

deficit, including (1) base cation supply dominated by weathering promoted by strong organic (A_s) or mineral (H_2SO_4) acids, (2) terrestrial sources of SO_4^{2-}, and (3) lake acidification by atmospheric deposition of SO_4^{2-} (Chapter 13, Cook and Jager, this volume).

The effect of nonmarine SO_4^{2-} on resolving apparent charge imbalances is shown in Figure 12.14b. Inclusion of nonmarine SO_4^{2-} brings the relationship for all lakes close to the expected relationship (slope = 0.97; r^2 = .91), but the intercept is −38 μeq L^{-1}. Evaluating the relationship for low and high DOC lakes separately shows the effect of DOC on the ANC deficit in colored lakes. For clearwater lakes only, the intercept is closer to zero (−4.1 μeq L^{-1}) and the fit to the relationship is better (r^2 = 0.98); conversely, the intercept for high DOC lakes is −54.3 μeq L^{-1} and the fit

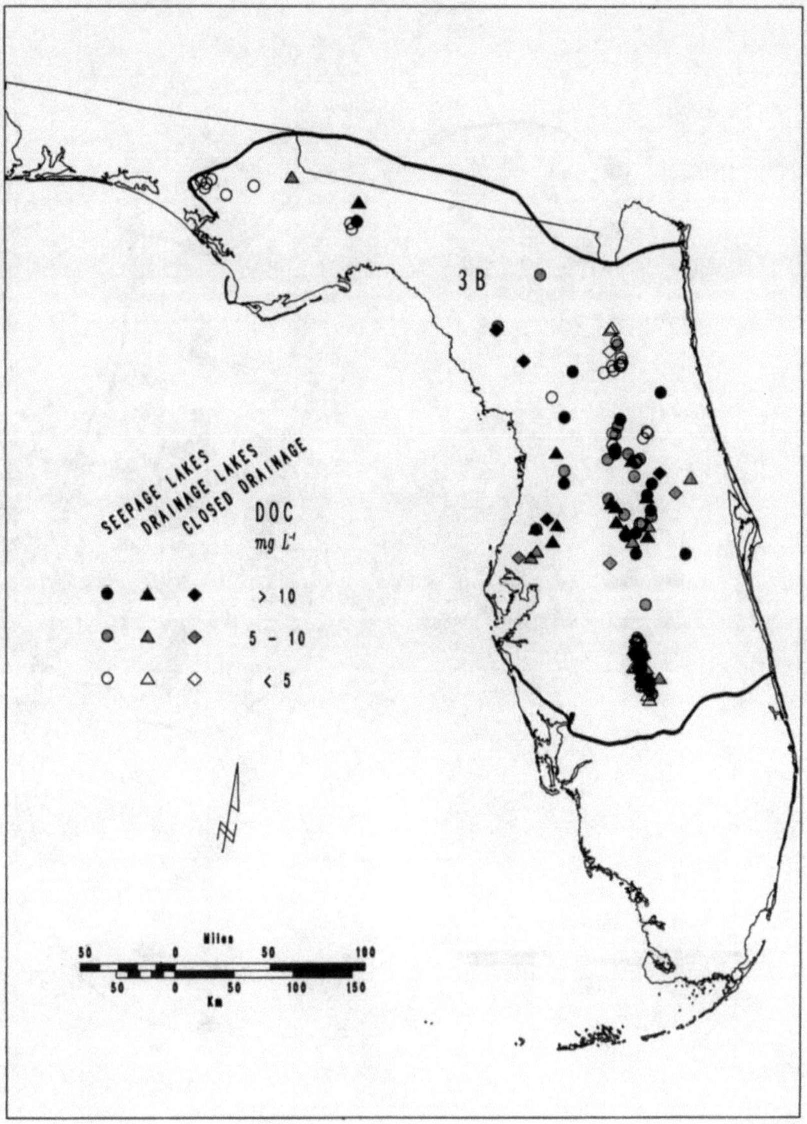

FIGURE 12.12. Areal distribution of DOC by concentration category and lake hydrologic type. Seepage lakes have no defined inlets or outlets, closed basin lakes have inlets but no outlets, and drainage lakes have both inlets and outlets (Chapter 1, Munson and Gherini, this volume).

(r^2 = .87) is poorer, as might be expected because of the broader range of concentrations of DOC included among the high DOC lakes. For high DOC lakes, the magnitude of the organic acid effect on the ANC deficit is particularly important at C_B concentrations between 200 and 400 μeq L^{-1}.

Sensitivity of surface waters to acidification depends in part on the ability of the watershed to release base cations relative to the influx of acid anions (Driscoll and Newton 1985). This is illustrated by plotting the ratio of SO_4^{2-}:C_B as a function of ANC (Figure 12.15). For clearwater systems receiving mineral acids, ratios < 1 imply the presence of ANC; at ratios > 1, SO_4^{2-} exceeds base cations and ANC should be negative. In Florida, most acidic clearwater lakes indeed have

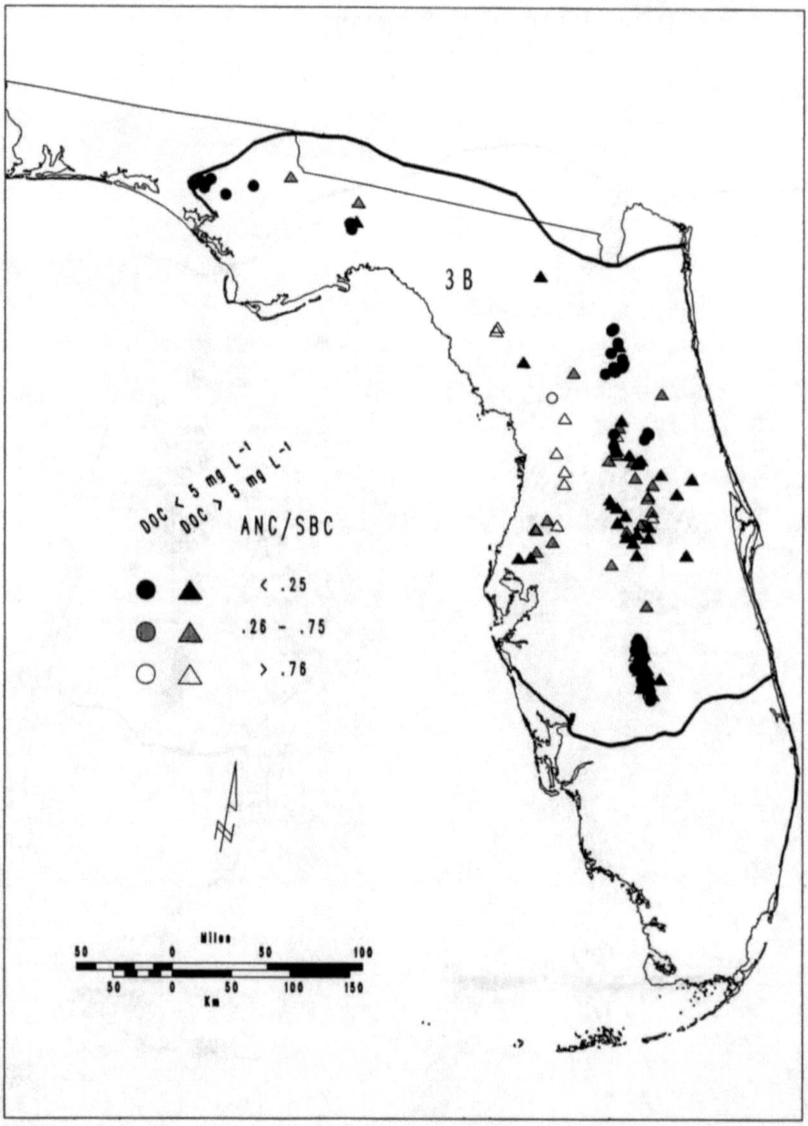

FIGURE 12.13. Areal distribution of the ratio of ANC to SBC by DOC category for low (≤ 5 mg L^{-1}; ≤ 417 µmol L^{-1}) and high (> 5 mg L^{-1}; > 417 µmol L^{-1}) DOC lakes. Only ELS lakes are shown in the plot.

$SO_4^{2-}:C_B > 1$, whereas $SO_4^{2-}:C_B$ in colored lakes is largely < 1, even in acidic lakes. The low ratios for some acidic, colored lakes arise because organic acids contribute to base cation supply through weathering and cation exchange, phenomena related to surficial geologic features restricting groundwater infiltration and resulting in high DOC systems. Unlike most other regions, Florida also has a number of lakes with moderate ANC (> 100 µeq L^{-1}) with $SO_4^{2-}:C_B > 0.25$ (cf. Chapter 13, Cook and Jager, this volume). This divergence with other regions owes to the high ionic strength of Florida lakes. More explicitly, when C_B and SO_4^{2-} are large, relatively small differences between the two parameters (i.e., $SO_4^{2-}:C_B$ approaching 1 but < 1) still translate to moderate amounts of ANC.

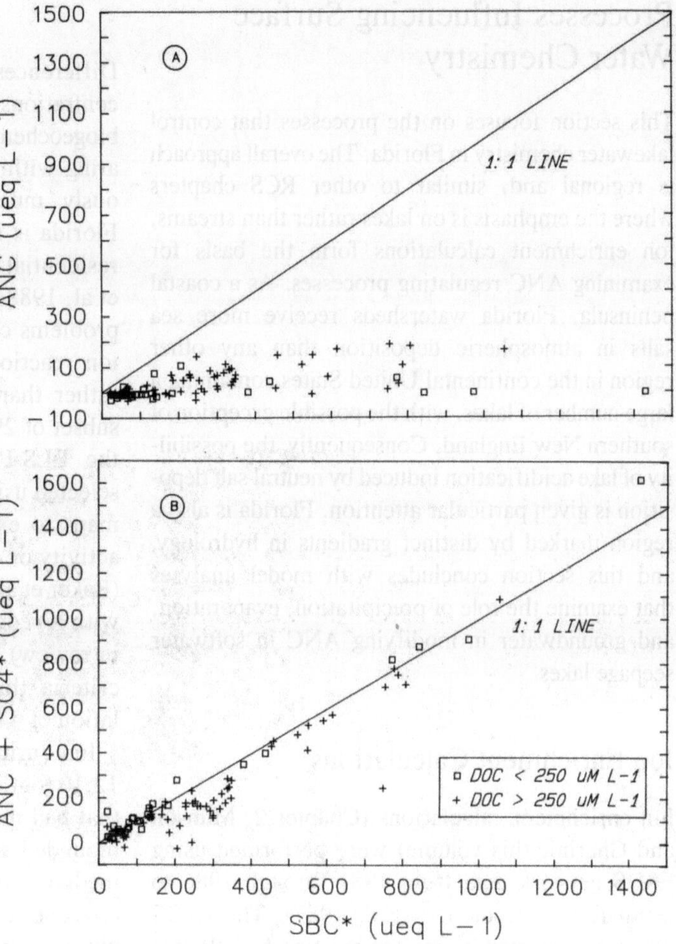

FIGURE 12.14. Ion equivalency relationships for high (> 250 µmol L⁻¹) and low DOC (< 250 µmol L⁻¹) lakes: (a) relationship between ANC and sum of base cations corrected for sea salt contributions (SBC*), and (b) relationship between ANC + SO_4^{2-} and SBC*. Line represents 1:1 equivalency between SBC* and dependent variable.

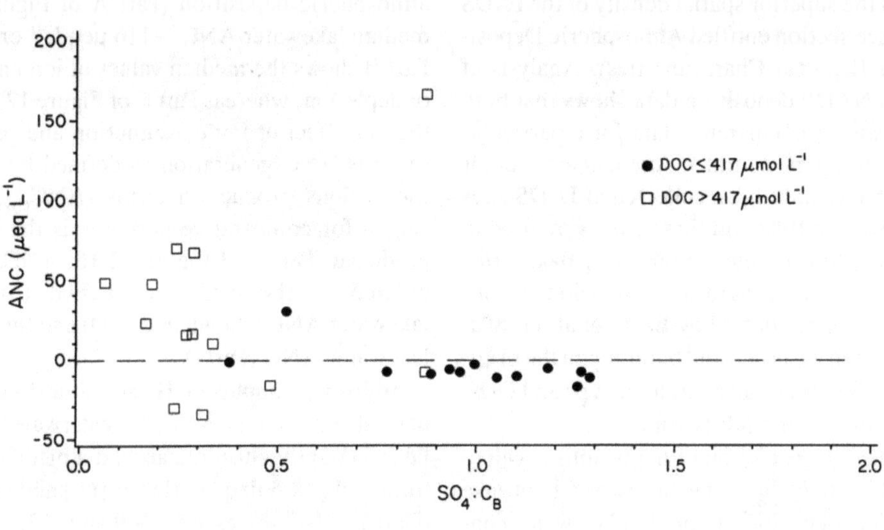

FIGURE 12.15. Relationship between ANC and nonmarine SO_4^{2-}:sum of base cations (SO_4:C_B) for high and low DOC lakes sampled during the ELS-I in Florida. DOC categorization as in Figure 12.13.

Processes Influencing Surface Water Chemistry

This section focuses on the processes that control lakewater chemistry in Florida. The overall approach is regional and, similar to other RCS chapters where the emphasis is on lakes rather than streams, ion enrichment calculations form the basis for examining ANC regulating processes. As a coastal peninsula, Florida watersheds receive more sea salts in atmospheric deposition than any other region in the continental United States containing a large number of lakes, with the possible exception of southern New England. Consequently, the possibility of lake acidification induced by neutral salt deposition is given particular attention. Florida is also a region marked by distinct gradients in hydrology, and this section concludes with model analyses that examine the role of precipitation, evaporation, and groundwater in modifying ANC in softwater seepage lakes.

Ion Enrichment Calculations

Ion enrichment calculations (Chapter 2, Munson and Gherini, this volume) were performed using FADS network data from 1981 through 1986 to estimate wet deposition (ESE 1987). The FADS data were selected in favor of existing NADP data because of the superior spatial density of the FADS network (see section entitled Atmospheric Deposition under Regional Characteristics). Analysis of FADS and NADP deposition data shows that both networks collect comparable data for a particular site (R. Husar, pers. comm.). Ambient air data for gaseous species have been collected at FADS sites since September 1982 and these data were used to infer dry deposition rates (Appendix, Baker, this volume). Our analysis here is an extension of similar calculations performed by Baker et al. (1988a) on Florida seepage lakes, and builds upon the original analysis by using lake hydrologic type and DOC content as discriminating factors.

Enrichment or depletion of major ions is calculated in this study by using ratios of Cl^- in lake water to total deposition to predict lakewater concentrations for a particular constituent, i, as a function of atmsopheric loadings (Appendix A, Baker, this volume):

$$[i]_{exp} = [i]_d \, [Cl^-]_m \, / [Cl^-]_d \qquad (12\text{-}2)$$

Differences between predicted and measured concentrations are a measure of the magnitude of biogeochemical and anthropogenic processes operating within the watershed. As mentioned previously, much of the ELS-I target population for Florida is influenced directly by agricultural and residential activities within the watersheds (Baker et al. 1988a, Stauffer in review). In order to avoid problems caused by trying to distinguish between ion reactions related to anthropogenic activities rather than natural biogeochemical processes, a subset of 29 undeveloped lakes was selected from the ELS-I data base for Florida. Lakes were selected using aerial photographs and topographic maps to exclude lakes with obvious agricultural activity or other nearshore land use modifications (Baker et al. 1988a), followed by eliminating lakes with K^+ concentrations > 15 μeq L^{-1} (cf. Stauffer in review). Based on ELS-I population weighting criteria, this undisturbed subset represents a population of 444 lakes.

Ion enrichment results are presented in Figure 12.16 for the only hydrologic/DOC class in Florida that had more than 10 lakes present—clearwater, mounded recharge lakes. The figure depicts the median concentration of expected lakewater H^+ concentrations if evapoconcentration was the only physical or biogeochemical process operating on atmospheric deposition (Part A of Figure 12.16; median lakewater ANC −116 μeq L^{-1} or pH 3.9). Part B shows the median values of ion enrichment or depletion, whereas Part C of Figure 12.16 shows the net effect of ion consumption and production towards ANC generation as defined by summing the various production terms ($ANC_{C_B\text{-}C_A}$). Also shown for comparative purposes is the net ANC produced (Part C of Figure 12.16; ANC_{Gran}), calculated as the difference between measured lakewater ANC (shown in Part D) and the expected lakewater ANC (Part A).

Atmosperic inputs of H^+ are almost completely neutralized. Actual ANC for clearwater recharge lakes is tightly clustered around a median concentration of −8.5 μeq L^{-1} (lower [Q_1] and upper [Q_3] quartile ANC values are −9.9 and −5.6 μeq L^{-1}). Anion reactions are largely responsible for neutralizing acid inputs, and unlike seepage lakes of similar hydrologic/DOC type in the Upper Midwest

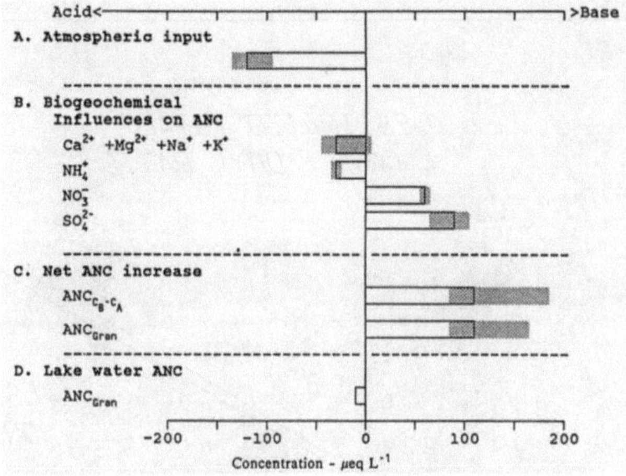

FIGURE 12.16. Modification of atmospheric deposition by watershed and lake biogeochemical processes in low DOC Florida lakes (≤ 417 μmol L^{-1}; ≤ 5 mg L^{-1}). A. Precipitation [H^+] adjusted for estimated dry deposition and evapotranspiration (ET); equivalent to expected lakewater $ANC_{C_B-C_A}$ in the absence of any process other than ET. B. Reactions influencing ANC; bars to the right indicate ANC production, those to the left show ANC consumption. C. Net ANC production; $ANC_{C_B-C_A}$ is sum of ANC production terms for measured anions and cations including those shown in C; ANC_{Gran} is determined by difference between input ANC (A.) and lakewater ANC (D.). D. Lakewater ANC represents the net reaction of atmospheric inputs with the lake/watershed system. Shaded bars represent 25% and 75% quartiles. For a further explanation of calculation procedures and figure interpretation, see Appendix A. Median DOC = 275.6 μmol L^{-1} (25% quartile = 112.4 μmol L^{-1}; 75% quartile = 327.2 μmol L^{-1}; median A_t = 20.0 μeq L^{-1} (25% quartile = 19.5 μeq L^{-1}; 75% quartile = 24.0 μeq L^{-1}).

(Chapter 13, Cook and Jager, this volume; Chapter 17, Baker et al., this volume), there is little or no indication of base cation enrichment. The Ca^{2+} losses are striking, presumably through seston deposition (Baker et al. 1988c) or incorporation into macrophytes (Baker et al. 1988b), with a median loss of 31 μeq L^{-1} (Q_1 = 4.5 and Q_3 = 43.4 μeq L^{-1}). Baker et al. (1989) estimated annual lake-wide sedimentation rates (gross) for Ca^{2+} of 14.1 meq m^{-2} yr^{-1} in Little Rock Lake in northern Wisconsin, which are comparable to estimated total atmospheric deposition rates in Florida (11 to 31 meq m^{-2} yr^{-1}; Table 12.4). K^+ also shows modest losses, apparently via biogenic pathways (Hutchinson 1957, Ruttner 1963), with median losses of 5.5 μeq L^{-1}.

Skartveit (1980) demonstrated that lakes in coastal Norway can undergo short-term acidification due to retention of marine sodium within the watershed (as a result of ion exchange of H^+), followed by transport of H^+ with mobile Cl^- to the lake basin. Because Florida lakes have long τ_w, short-term pulses in acidity related to neutral salt deposition are unlikely. Long-term acidification in clear-water recharge lakes related to neutral salt deposition is discounted by ion enrichment calculations as well. Previous researchers working with the ELS-I data set also have discounted the neutral salt hypothesis for Florida, based on the observation that lakewater Na^+ concentrations increase with distance from the coast (Sullivan et al. 1988). Sullivan et al. (1988) suggested that this relationship reflects more substantial contributions from the watershed for inland Florida lakes, particularly from groundwater sources.

Figure 12.17 shows Na^+ plotted versus Cl^- for all 29 undisturbed lakes. Most of the lakes closely match concentrations expected from either total or wet deposition only, and the Sullivan et al. hypothesis holds only if weathering releases Na^+ and Cl^- in stoichiometric ratios consistent with estimated atmospheric deposition ratios (0.807 for wet only precipitation, and 0.829 for estimated total deposition). Primary minerals are almost completely absent from surficial Florida soils, and there is no evidence of chloride bearing minerals such as halite (NaCl). Thus, higher Na^+ and Cl^-

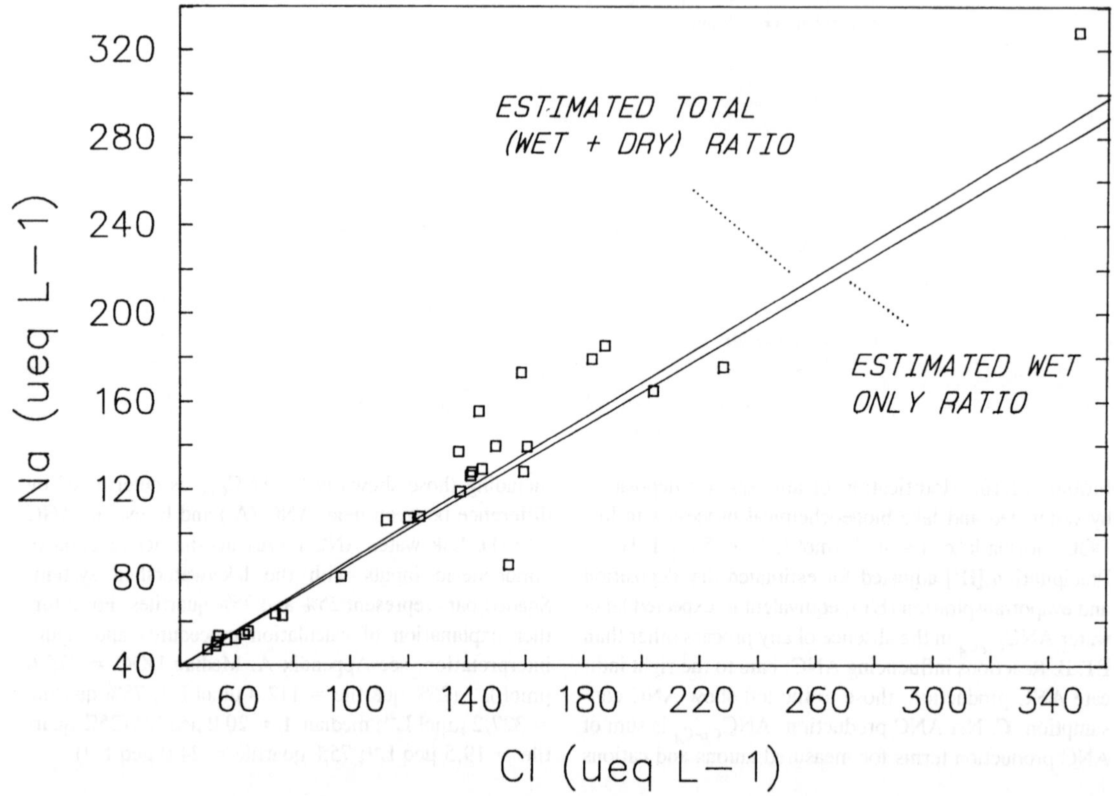

FIGURE 12.17. Relationship between Na^+ and Cl^- concentrations in undisturbed Florida lakes. Upper line represents predicted Na^+ concentrations at estimated Na^+:Cl^- ratios in total (wet + dry) deposition (0.829). Lower line is same as upper line but represents the measured wet only deposition Na^+:Cl^- ratio (0.807).

concentrations in inland Florida lakes cannot be ascribed to weathering inputs. Rather, the relationship between Na^+ and distance from the coast simply reflects fundamental hydrologic differences between inland and coastal lakes (Figure 12.12). Drainage and closed basin systems represent 59% of the hydrologic types for coastal lakes, whereas seepage lakes are the predominant lake type for inland lakes in the Florida Northern Peninsula and the Panhandle (69% and 83%, respectively). As a general rule, drainage and closed basin lakes receive higher rates of hydraulic loading than seepage lakes. Accordingly, drainage and closed basin lakes expectedly have shorter hydraulic retention times. Lower Na^+ and Cl^- concentrations in drainage and closed basin lakes compared to seepage lakes would result.*

Despite relatively high acidities in Florida clearwater recharge lakes, aluminum mobilization is low, with 75% of the lakes releasing less than 1.0 µeq L^{-1} extractable aluminum into the water column (0.3 µmoles L^{-1}). Aluminum mobilization is lower compared to drainage lakes of similar acidity (e.g., Driscoll and Newton 1985), apparently because (1) aluminum mobilized in surficial Florida soils is rapidly removed from solution by precipitation at still shallow depths (75 cm) within the soil profile (Graetz et al. 1985), and (2) the bot-

*Other marine-associated cations, particularly Mg^{2+} and, to a lesser degree, Ca^{2+}, may be more important than Na^+ toward neutral salt induced acidification because clays tend to selectively adsorb divalent cations, displacing monovalent cations such as H^+. As mentioned previously, our ion enrichment calculations show that Ca^{2+} is depleted, although Mg^{2+} is not. Based on measurements of inseepage chemistry of McCloud lake, depletion of Ca^{2+} appears to occur within the lake and not in the watershed soils (cf. Baker 1984, Baker et al. 1988a).

FIGURE 12.18. SO_4^{2-} enrichment factors (dimensionless) plotted against pH in undisturbed Florida lakes. SO_4^{2-} enrichment factor is the ratio of observed to predicted lake water SO_4^{2-} concentrations, using estimated total SO_4^{2-} and Cl^- atmospheric loadings to calculate predicted SO_4^{2-} from Cl^- deposition:lake water ratios.

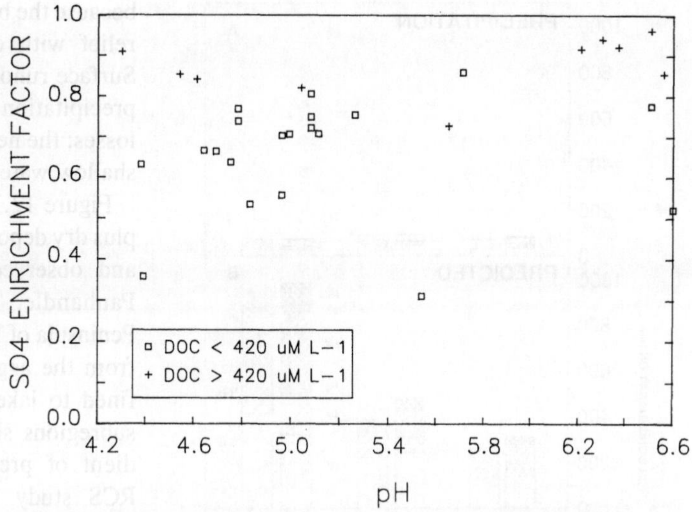

tom sediments of acidic lakes are buffered at higher pH levels than the overlying lake water because of biogeochemical processes occurring within the sediments (Baker 1984b). A number of studies have shown that base cation supply increases from sediments with increasing H^+ (Chapter 13, Cook and Jager, this volume); however, Perry et al. (1986) demonstrated that for 13 Florida softwater lake sediments, mineral dissolution and cation exchange of base cations for added H^+ occurred only when SO_4^{2-} reduction was prevented by poisoning the sediments with formalin.

Sulfate consumption is likely the principal reaction generating ANC in clearwater recharge lakes. Median production was 91 μeq L^{-1} and Baker et al. (1988b) have calculated that an average of 69% in input SO_4^{2-} is retained in undisturbed Florida softwater lakes. Laboratory sediment-water microcosm studies show losses of added SO_4^{2-} ranging from 32% to 94% in Florida lakes (Perry et al. 1986). Sulfate retention and the concomitant production of ANC can occur through dissimilatory SO_4^{2-} reduction (Cook et al. 1986, Rudd et al. 1986a, 1986b, Baker et al. 1989) and specific anion adsorption (Rajan 1978, Johnson and Cole 1980). For Florida lakes, sediment adsorption of SO_4^{2-} is negligible and SO_4^{2-} reduction appears to be most important (Perry et al. 1986). Measured laboratory rates for Florida lakes range from 96 to 303 meq m^{-2} yr^{-1} (Perry et al. 1986) compared to rates of nearly 60 meq m^{-2} yr^{-1} for Lac du Chevreuil in southwestern Quebec based on ^{210}Pb-ated sulfur profiles (Carignan and Tessier 1988). Higher rates of reduction in Florida may reflect higher temperatures driving microbial activity. Sulfate reduction appears to be limited by rates of diffusion across the sediment-water interface, leading to first-order kinetics (Baker et al. 1986a, Cook and Schindler 1983). Baker et al. (1988c) also indicate that algal uptake and subsequent losses to sediments is an important sink for SO_4^{2-} in low SO_4^{2-} seepage lakes, although the relative role of this process in Florida lakes presently is unknown. The degree to which SO_4^{2-} is retained is variable in response to τ_w (Baker et al. 1986a), and appears to be independent of pH (Figure 12.18).

Nitrate assimilation is also quite important in neutralizing acid inputs in Florida lakes. Nitrate assimilation in Florida lakes is virtually complete (ca. 99%; Baker et al. 1988b), and the median contribution of 84 μeq L^{-1} to ANC production is second only to SO_4^{2-} reduction. Algal uptake and denitrification both may contribute to NO consumption. Baker (1984) measured 90% loss rates of NO_3^- in eluate collected from laboratory columns of sediment from McCloud Lake. Algal and terrestrial assimilation is also efficient at removing NH_4^+, generating 56 μeq L^{-1} acidity. Consistent with the definition of clearwater recharge lakes, terrestrial production of both weak and strong organic acids is minor compared to acid inputs from precipitation and acidity generated by base cation and NH_4^+ losses.

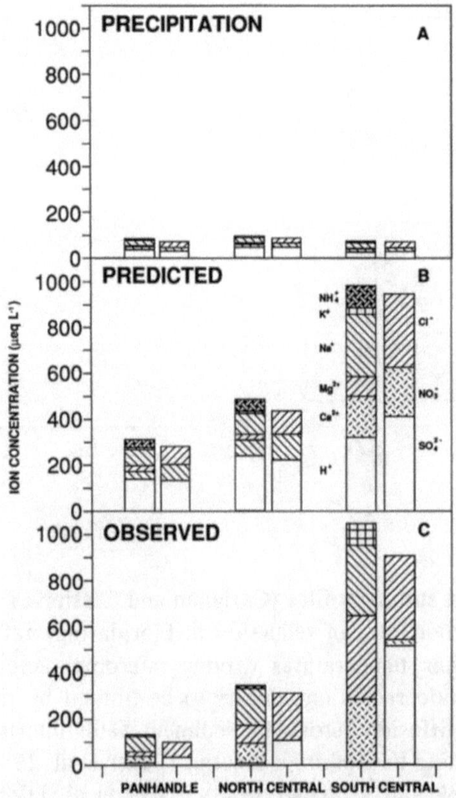

FIGURE 12.19. Comparison of major ion concentrations in (a) wet plus dry precipitation, with (b) predicted and (c) observed lakewater concentrations. Predicted concentrations derived from "Teflon-lake" model described in text.

Precipitation/Evaporation Effects

The influence of precipitation/evaporation gradients in Florida (see section on climate/hydrology) on overall ionic composition can be examined independently of the use of Cl^- to account for evapoconcentration as water moves through the watershed. This alternative approach is based on calculating concentration factors based on the ratio of total precipitation to net precipitation, where net precipitation is calculated as the difference between precipitation and pan evaporation (corrected to represent lake evaporation). For this analysis, runoff is neglected. Instead, lakes are considered to be inert, unreactive basins essentially isolated from their watersheds. For many seepage lakes in Florida, this "Teflon-lake" model appears to be a reasonable first approximation because the basins of many of these lakes lie in low relief with expectedly low hydraulic gradients. Surface runoff also does not occur. Finally, direct precipitation inputs typically exceed evaporative losses; the net flow of water is from the lake to the shallow water table rather than the reverse.

Figure 12.19 shows volume-weighted mean wet plus dry deposition, predicted or model lake water, and observed lakewater concentrations for the Panhandle, Northern Peninsula, and Central Peninsula of Florida. Observed concentrations are from the aggregated lake data base and are confined to lakes with ANC $\leq$ 400 μeq L^{-1}. The subregions shown in Figure 12.19 span the gradient of precipitation/evaporation ratios in the RCS study area. Predicted concentrations are based on 20-year average precipitation and pan evaporation, applying a pan coefficient of 0.70 to estimate lake evaporation (Linsley et al. 1975).

The model correctly predicts the overall trend of increasing ionic content for lakes located in the southern part of the region. For the Panhandle and the Central Peninsula, the model predictions are reasonable for Na^+ and Cl^- (within 8% and 13% for the Panhandle and 13% and 18% for the Florida Central Peninsula), given the sensitivity of the model to estimates for precipitation, evaporation, and the pan coefficient. The model underpredicts evapoconcentration in the Florida Northern Peninsula by 46% (Na^+) to 48% (Cl^-). Estimated concentration factors range from 3.8 in the Panhandle to 14.1 in the Central Peninsula. Based solely on evapoconcentration and precipitation chemistry, the lower limit for predicted lakewater pH in the Panhandle is 3.8; for the Northern Peninsula and Central Peninsula, lower limits of 3.6 and 3.5, respectively, are obtained for predicted pH. If the model estimate for lakewater pH in the Florida Northern Peninsula is modified to account for underprediction of evapoconcentration, the predicted pH is 3.2.

Comparison of measured versus predicted pH shows that between 93% and 99% or more of the H^+ inputs are neutralized. The extent of neutralization is greatest in the Florida Central Peninsula (pH 5.6). Two mechanisms may account for this pattern. First, residence times generally increase from north to south within the case study region, and in-lake processes and SO_4^{2-} reduction should be more complete in slower flushing systems

(Baker et al. 1986b, Kelly et al. 1987). Second, inseepage appears to constitute a greater fraction of the hydrologic budget for seepage lakes as the P/E ratio decreases (Baker et al. 1988a). Although soils in the major highland lake districts in Florida are quite acidic, the limited data available for McCloud Lake in the Florida Northern Peninsula indicates that shallow groundwater and inseepage are net sources of ANC (Baker 1984, Graetz et al. 1985). McCloud Lake watershed soils are typical of most highland and ridge soils: a typic quartzipsamment soil, pH 4.8, CEC 3.44 meq 100 gm^{-1} and 3.73% base saturation (Volk and Byers 1983). Using McCloud Lake as a qualitative model, inseepage ANC fluxes should be higher in the Central Peninsula compared to the Panhandle and the Northern Peninsula.

The regional-wide effect of evapoconcentration on the ANC regime of Florida lakes is thus perceived as twofold and antagonistic. Concentrating an acidic solution in an unreactive system clearly makes the solution more acidic and reduces the concentration of ANC. However, the magnifying effect of high concentration factors on lakewater acidity may be offset by accompanying changes in lake hydrology and residence time. As precipitation:evaporation ratios decline across the region, inseepage assumes greater importance in the hydrologic budget. The contribution of inseepage to the ANC budget is greater than its contribution to the hydrologic budget. Moreover, as mentioned above, low precipitation/evaporation values imply long τ_w and greater ability for a lake to internally regulate ANC via SO_4^{2-} reduction. For lakes in the Central Peninsula of Florida, residence time effects and higher rates of inseepage appear to outweigh the concentrative effect of very low precipitation:evaporation.

Groundwater Hydrology/Weathering

Ion enrichment calculations presented in the previous section for clearwater, groundwater recharge lakes show that production of base cations via weathering and ion exchange are minor processes neutralizing atmospheric inputs of acidity. Baker et al. (1988a) have shown that base cation inputs are more substantive in seepage lakes with ANC > 30 $\mu eq\ L^{-1}$ and, because in-lake processes such as SO_4^{2-} reduction and NO_3^- assimilation so closely balance atmospheric acidity, small increases in base cation supply related to hydrologic effects or surficial geologic differences among lakes can dictate whether a lake is acidic or has modest levels of ANC. This section examines the role of the watershed on base cation supply.

Within Florida, lakes may be lumped into two broad hydrologic types: lakes that recharge the Floridan aquifer and lakes to which the Floridan aquifer discharges (i.e., artesian areas). Low ANC lakes generally are not found in discharge areas because hydraulic contact with the calcareous Floridan aquifer provides a source of very hard water. Notable examples of major lakes in discharge areas include Lake Okeechobee (ANC = 1,300 to 3,310 $\mu eq\ L^{-1}$; Joyner 1972) south of the Lake Wales Ridge and Lake Apopka (ANC = 2,380 $\mu eq\ L^{-1}$), which lies in the Central Valley below the Mount Dora Ridge in central Florida. Gourd Neck Spring supplies an average of about 2,400 $\mu eq\ L^{-1}$ ANC to Lake Apopka (Pollman et al. 1980 Roseneau et al. 1977).

For lakes within recharge areas of Florida, the interaction of laterally moving shallow groundwater with carbonates and clays relates in part to the proximity of the Hawthorn Formation to the ground surface (Baker et al. 1988a). If the Hawthorn lies particularly close to the surface, vertical infiltration and drainage is impeded and runoff and seepage inputs assume greater hydrologic importance. Base cation and DOC concentrations are higher as well.

Geochemical Factors Affecting Shallow Groundwater Chemistry

Easily weathered primary minerals are virtually nonexistent in the deep acid sands of the Florida sandhills. Kaolinite and gibbsite are typically formed under conditions of excessive leaching (Bolt and Bruggenwert 1976), and quartz, kaolinite, and to a lesser degree, gibbsite are the principal clay-sized particles found in the sandhills of the Panhandle and the Florida Northern Peninsula (Carlisle et al. 1978). Neutralization of acidic deposition within the soil profile thus proceeds via ion exchange, anion retention, or dissolution of kaolinite or gibbsite.

Few empirical data are available on the geochemistry of shallow groundwater in Florida. The

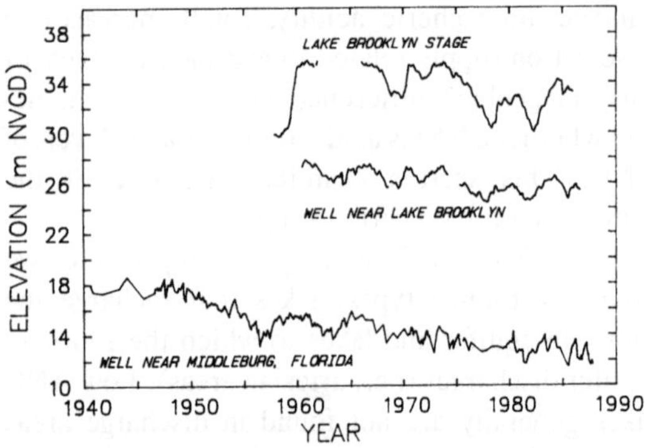

FIGURE 12.20. Long-term declines in the potentiometric surface of the Florida aquifer. Lower curve: well near Middleburg, Clay County, Florida, 1940–87. Middle curve: well near Lake Brooklyn in the Trail Ridge, Clay County, Florida, 1960–87. Upper curve: stage of Lake Brooklyn, 1957–87.

only field research within Florida to evaluate the effects of acidic deposition on soil and groundwater geochemistry was a series of three studies conducted within the McCloud Lake watershed (Volk and Byers 1983, Byers 1985, Graetz et al. 1985). In these experiments, a series of transects were irrigated with acidified lake water (pH 3.0 to ambient, 4.6) to simulate the effects of acidic deposition on soil acidification and nutrient flux. The following sequence shows the relative importance of major ion mobilization (cations) and retention (anions) to neutralization of added acidity at current levels of deposition (Graetz et al. 1985):

$$Ca^{2+} >> Na^+ > K^+, Al^{3+} > Mg^{2+}, SO_4^{2-}$$

This sequence is for 15-cm depth within the soil profile. Sulfate retention was minor compared to inputs (5% of inputs). The SO_4^{2-} adsorption isotherm studies using the same typic quartzipsamment soil show that SO_4^{2-} adsorption is minor in the surficial 25 cm (Volk and Pollman 1985), although adsorption may still be significant deeper within the profile (D. Johnson, pers. comm.). The pH 3.6 treatment gave the following sequence of contributions to H^+ neutralization:

$$Ca^{2+} >> Na^+ > SO_4^{2-} > K^+, Mg^{2+}, Al^{3+}$$

Mobilization of Al^{3+} and SO_4^{2-} retention did not dominate H^+ neutralization until the irrigation pH was reduced from 3.6 to 3.0:

$$Al^{3+} > SO_4^{2-} > Ca^{2+} > NO_3 > Na^+ > Mg^{2+}$$

The McCloud Lake studies indicate that Al^{3+} mobilized in surficial soils by high levels of acid loading is removed to some extent from solution via precipitation and ion exchange as water moves through the soil profile. Between 15 and 75 cm depth, 50% of the Al^{3+} mobilized at 15 cm was removed from solution at 75 cm. SO_4^{2-} retention at pH 3.6 and 3.0 averaged 36% and 60%, respectively. Soil solution pH at 75 cm depth at the conclusion of the study ranged from 6.4 to 6.6 for the plots receiving pH 3.6 and pH 4.6 treatments; soil solution pH for the pH 3.0 plots averaged 5.8 (Graetz et al. 1985).

Measurements of water collected in situ with seepage meters for McCloud Lake (Baker 1984) suggest that ion exchange is the principal mechanism contributing to terrestrial or paralimnetic neutralization of acid inputs. Using Baker's data, calculated concentrations of ANC ($C_B - C_A$) in inseepage average 116 μeq L^{-1} with base cation enrichment (relative to expected concentrations based on Cl^-) contributing 118 μeq L^{-1}. Inseepage SO_4^{2-} and NO_3^- are depleted by 45 (55%) and 38 μeq L^{-1} (89%); NH_4^+ uptake generates 10 μeq L^{-1} acidity. Sulfate adsorption in a variety of Florida softwater lake sediments, including McCloud, appears unlikely, based on laboratory studies (Baker 1984, Perry et al. 1986), and SO_4^{2-} reduction may account for SO_4^{2-} losses in inseepage.

Recent changes in groundwater head may have a significant effect on seepage lake ANC. Most seepage lakes in Florida are located in higher elevation areas that historically have recharged the Floridan aquifer. Reducing the potentiometric surface due to consumptive use sharpens the hydraulic gradient between the water table and the Floridan aquifer. Relatively small declines in the potentiometric surface may translate to relatively large increases in the vertical gradient. For example, the average head difference between lakes in the Trail Ridge and the Floridan aquifer is about 3 to 6 m (Clark et al. 1964, Yobbi and Chappell 1979); a sustained decline of 1 m is equivalent to a 17% to 33% increase in the vertical gradient and, according to Darcy's Law, is equivalent to a corresponding increase in the vertical flow of shallow groundwater recharging the Floridan. Reduction in lateral flow or inseepage to lakes may result, affecting the supply of base cations and ANC. For Lake Five-O in the Florida Panhandle, the head difference between the lake and the Floridan aquifer is less than 2 m (W. Aucott and J. Oliveros, pers. comm.).

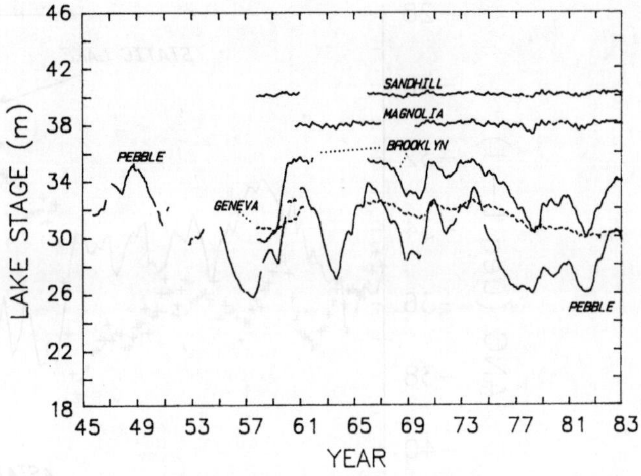

FIGURE 12.21. Long-term stage relationships for lakes in the Trail Ridge, Florida. Lakes Sandhill and Magnolia are classified as drainage type lakes, but typically behave as seepage lakes, except under during periods of high rainfall.

Elevational data for a well located 25 to 30 km from the Trail Ridge indicates that the potentiometric surface of the Floridan aquifer has declined about 6 m since 1940 (USGS unpublished data; Figure 12.20). The potentiometric surface of the Floridan aquifer near Lake Brooklyn has declined by about 2 m since 1960 (Figure 12.20). Hypothesized effects of aquifer drawdown on historical changes in seepage lakewater chemistry are modeled in the section entitled Model Hindcasting Analysis of Historical Trends.

Model Evaluation of Hydrologic Effects on Lakewater Chemistry

Seepage lakes by definition have very limited interaction between the basin proper and the surrounding watershed soils. In-lake biogeochemical processes tend to dominate ANC regulation in these systems; for example, Baker et al. (1986b) estimate that inseepage accounts for only 33% to 37% of ANC generation in McCloud Lake. The In-Lake Alkalinity Generation (IAG) model developed by Baker et al. (1985c, 1986a) focuses explicitly on internal processes to predict lake acidification. Originally developed to analyze acidification of Florida seepage lakes, the IAG model has since been calibrated and verified using ion budget data for 14 seepage and drainage lakes, including 3 seepage lakes in Florida (Baker and Brezonik 1988). The IAG model is used here to illustrate the effects of hydrology and changing atmospheric inputs on ANC and other major ion constituents for a hypothetical clearwater seepage lake. The model presently is based on the continuity equation and uses the principle of electroneutrality to simulate ANC as a function of base cation, SO_4^{2-}, NO_3^-, and NH_4^+ inputs and reactions through a series of coupled first-order differential equations (Pollman and Sweets in review).

The simulated lake is based on the chemistry of Lake Barco (ANC = −35 μeq L^{-1}, SO_4^{2-} = 131 μeq L^{-1}, Cl^- = 125 μeq L^{-1}, C_B = 222 μeq L^{-1}), located in the Trail Ridge. Simulations are begun by assuming that the lake is at steady state with current levels of wet plus deposition (see atmospheric deposition section). Model coefficients were initially calibrated by assuming that inseepage from the shallow water table averaged 11.5% of direct precipitation and that inseepage had the following chemical characteristics: ANC = 115 μeq L^{-1}; SO_4^{2-} = 38 μeq L^{-1}; SBC = 195 μeq L^{-1} (cf. Baker 1984, for McCloud Lake). Average depth is 4.3 m and τ_w is 15.5 years.

Lake stage regulation may be viewed to varying degrees as a homeostatic process, where stage at any point in time reflects a balance between antecedent rainfall and inflow and outflow rates dictated by both the stage of the lake and conditions in the water table. The extreme of self-regulation is a lake in which lake stage remains constant by compensating instantaneously for excess precipitation or evaporation with corresponding increases in outseepage or inseepage, respectively. Examples of lakes approaching this hydrologic type include Sandhill Lake in the Trail Ridge, which undergoes stage fluctuations of < 0.3 m (Figure 12.21). Alternatively, seepage lake stage varies significantly with long-

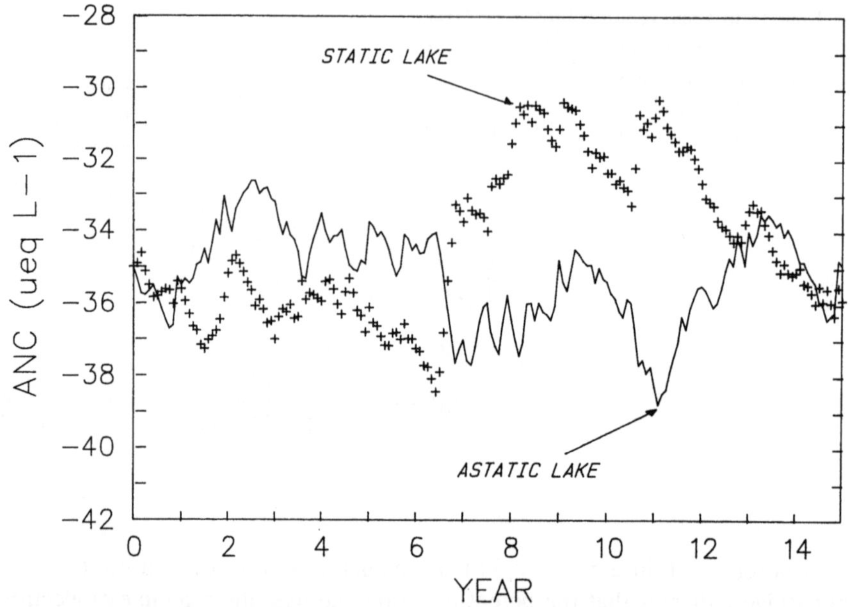

FIGURE 12.22. IAG model time series predictions for ANC using steady state model coefficients and atmospheric inputs. For static lake, stage was held constant by adjusting seepage flows according to antecedent rainfall (see text). For astatic lake, stage was allowed to vary with antecedent rainfall.

term patterns of precipitation and evaporation (e.g., Pebble, Brooklyn, and Geneva lakes; Figure 12.21; Deevey 1989). Pebble Lake is an extreme example of an astatic lake. For the Trail Ridge, which has more stage data than any other seepage lake district in Florida, this latter type of lake appears to include most seepage lakes.

For the purpose of modeling, we have included both hydrologic types. For the astatic lake, inseepage (Q_i) and outseepage (Q_o) were assumed to vary as first order functions of stage:*

$$Q_i = (z_{max} - z_t) * \alpha$$

$$Q_o = (z_t - z_0) * \beta$$

where: z_t = lake stage at any time, t
z_{max} = maximum lake stage
z_0 = elevation of the potentiometric surface
α,β = calibration constants

A 15-year period of monthly precipitation and evaporation data was compiled for Gainesville from October 1970 through September 1985 and used as input to examine the effects of drought and wet periods on ANC.

Simulated ANC in response to variable precipitation inputs are presented for the static and astatic model lakes in Figure 12.22. Although precipitation was allowed to vary, atmospheric deposition of H^+ and other major ions was held constant throughout the simulation. As might be expected, both lake types give similar ranges of response for ANC and other major constituents, owing to the same long-term hydrologic and chemical inputs. Simulated ANC concentrations oscillated about -35 µeq L^{-1}, ranging as high as -32 µeq L^{-1} and as low as -39 µeq L^{-1}. Concentrations of other major consituents were more variable: SO_4^{2-} ranged from 121 to 148 µeq L^{-1}; C_B ranged from 204 to 250 µeq L^{-1}; and Cl^- ranged from 115 to 141

*The intent of this simulation is to examine how seepage lakes might respond chemically during extreme periods of rainfall and drought in Florida. Intuitively we suspect (but do not know for certain) that outseepage rates fall with lake stage. Similarly, we also suspect that inseepage reaches a minimum at maximum lake stage. Certainly this is true for *net* inseepage but may be incorrect for *gross* inseepage since head levels in porous media (because porosity is less than unity) will increase more rapidly than lake stage during a rain event. The hydrologic algorithm is simply a vehicle to account for this perceived overall behavior.

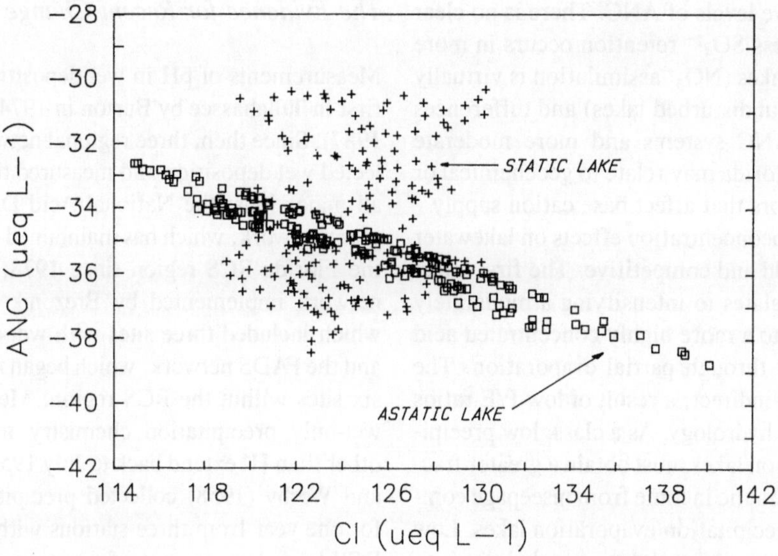

FIGURE 12.23. Predicted ANC as a function of Cl^- for the astatic and static stage lakes. Cl^- is used as a tracer of evaporative concentration.

μeq L^{-1}. That ANC does not show the same relative variability in concentration as other major ions is related to the homeostatic nature of in-lake ANC generation; as evapoconcentration drives constituent concentrations higher, net rates of internal production of ANC increase as well. NO_3^- and NH_4^+ respond very rapidly to changing net precipitation and vary by less than $\pm$ 1 μeq L^{-1}.

Our model results are consistent with the hypothesis that evapoconcentration exerts competitive or antagonistic effects on ANC in Florida seepage lakes. Figure 12.23 shows the relationship between ANC and changing Cl^- concentrations, with Cl^- used as a surrogate for evapoconcentration. The ANC for the astatic lake is clearly related to Cl^-; as drought progresses, evapoconcentration drives ANC down at rates faster than the increased fluxes of inseepage can compensate for it. This response is consistent with changes in H^+ observed in McCloud Lake, where H^+ is inversely related to lake volume (Pollman 1986). For the static lake, inseepage rates increase instantaneously as net negative P/E exerts pressure to reduce lake stage. During drought periods, this enhanced flux of seepage can negate the acidifying effects of evapoconcentration. As a result, no correlation is seen between ANC and Cl^- for the model static lake (Figure 12.23).

Insofar as long-term, steady state concentrations are concerned, the critical response parameters are τ_w and the SO_4^{2-} loss term (cf. Imboden and Lerman 1978, Sonzogni et al. 1976). In addition, it is clear that inseepage associated fluxes may play a pivotal role in the ANC regime of seepage lakes. Baker and Brezonik (1988) have shown that the IAG model is insensitive to changes in k_{NO3} and k_{NH4}; this directly reflects the rapid rates at which NO_3^- and NH_4^+ are consumed in dilute, nutrient-poor seepage lakes. Reasonable deviations for either coefficient still give rapid loss rates. Long residence times yield higher retention of SO_4^- and greater ANC generation production as well as a slower rate of response to perturbations.

Summary of Processes

Regulation of ANC in acidic softwater lakes in Florida is due largely to anion retention (SO_4^{2-} and NO_3^-) processes. Both processes appear to dominate in dilute, clearwater recharge lakes and Baker et al. (1988a) have shown that net ANC generation nearly balances atmospheric inputs of H^+. Base cation production is comparatively small but, because of the close balance between ANC generated by anion retention with acid inputs, small changes in base cation supply may be sufficent to yield net positive levels of ANC. There is no clear evidence that less SO_4^{2-} retention occurs in more acidic Florida lakes (NO_3^- assimilation is virtually complete in all undisturbed lakes) and differences between low ANC systems and more moderate ANC lakes in Florida may relate to geochemical or hydrologic factors that affect base cation supply.

Finally, evapoconcentration effects on lakewater ANC are twofold and competitive. The first effect is direct, and relates to intensifying a moderately acidic solution to a more highly concentrated acid solution simply through partial evaporation. The second effect is indirect, a result of low P/E ratios on seepage lake hydrology. As a class, low precipitation/evaporation lakes must obtain a greater fraction of their hydraulic income from inseepage compared to high precipitation/evaporation lakes. Low precipitation/evaporation lakes are believed to receive greater inputs of base cations accordingly, and this flux of ANC helps to offset the acidifying effect of evapoconcentration. Also, for lakes that maintain relatively constant stage, compensatory increases of inseepage during periods of drought may increase base cation supply and ANC, thereby counteracting evaporative concentration. Astatic lakes do not appear to receive the degree of inseepage that static lakes do during drought and ANC regulation cannot keep up with evaporative acidification.

Evidence of Long-Term Trends

Atmospheric Deposition Trends

Trends in atmospheric deposition can be evaluated in a number of ways. For periods when chemical measurements actually exist, deposition differences in time can be quantified directly. Otherwise, inferential assessments may be made based on reconstructions of historical emission inventories (Chapter 3, Husar et al., this volume). Changes in the accumulation rate in lake sediments of sufur and other tracers of anthropogenic emissions to the atmosphere (e.g., V, Zn, and Cu, among other elements) also provide an indirect record of changes in deposition of acidity to lakes. In this section, direct and inferential evidence for changes in SO_4^{2-} deposition are examined. The evidence for Florida is by no means consistent.

The Evidence for Recent Change

Measurements of pH in wet deposition were made first in Tallahassee by Burton in 1974 (cited in ESE 1987). Since then, three regional networks have collected wet deposition and measured the samples for all major ions: the National Acid Deposition Program network, which has maintained two stations in the Florida RCS region since 1978; the statewide network implemented by Brezonik et al. (1983), which included three sites with wet-dry collectors, and the FADS network, which began in 1981 and has six sites within the RCS region. Measurements of wet-only precipitation chemistry for major ions other than H^+ extend back to July 1955, when Junge and Werby (1958) collected precipitation samples for one year from three stations within the Florida RCS boundary as part of a nationwide precipitation chemistry network. Samples were collected and analyzed on a monthly basis. Between 1960 and 1966, the National Center for Atmospheric Research (NCAR) collected wet-only precipitation on an event basis from a site in Tampa as part of a 33 station nationwide network.

Comparing Junge and Werby's data for non-marine SO_4^{2-} with bulk collector chemistry collected from five sites in 1978–79, Brezonik et al. (1980) cited average increases of 60% and 350% for volume-weighted mean concentrations of SO_4^{2-} and NO_3^-, respectively. Precipitation H^+ (inferred from ionic balance) increased from < 2.5 μeq L^{-1} in 1955–56 to 17 to 24 μeq L^{-1} in the Panhandle and Northern Peninsula in 1978–79 (Brezonik et al. 1980). The comparison across the two studies is confounded in part because Brezonik et al. collected bulk deposition and Junge and Werby wet-only deposition. When correction factors empirically derived by Brezonik et al. to account for differences in collector type are applied to their data, the estimated increases in SO_4^{2-} and NO_3^- deposition in the Panhandle and Northern Peninsula average 38% and 320% (volume-weighted mean differences of 7 and 9 μeq L^{-1}, respectively).

Husar (Chapter 3 this volume) has estimated historical wet deposition of SO_4^{2-} to Florida using reconstructed emissions inventories. His calculations indicate that wet deposition of SO_4^{2-} has increased most rapidly since 1950. Between 1955–56 and 1978–79, estimated SO_4^{2-} deposition increased from 124 to 226 eq ha^{-1} yr^{-1} (Figure 3.3, Chapter 3, Husar et al., this volume); current depo-

FIGURE 12.24. Sulfur profiles in three northcentral Florida lakes (Barco, Fore, and Mary) and one Panhandle lake (Mirrow). From Mitchell et al. (1988).

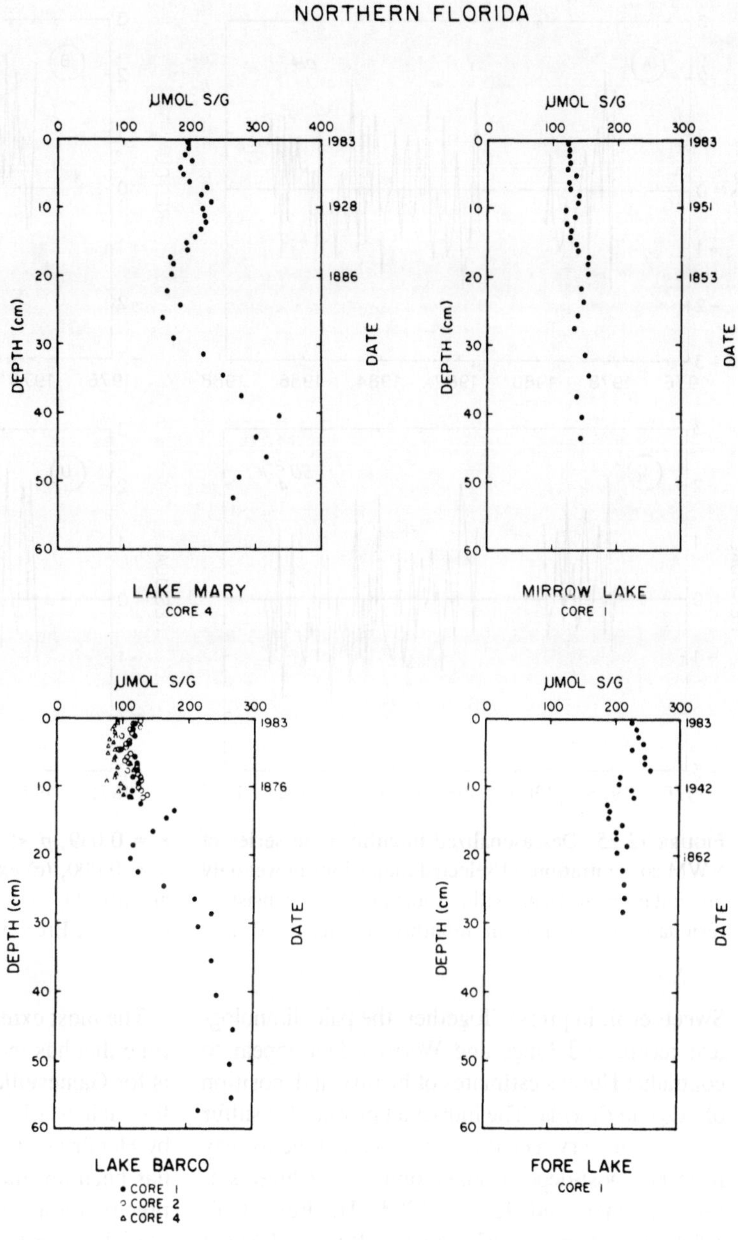

sition is approximately 350% higher than preindustrial (ca. 1860) levels.

The Evidence Against Recent Change

Retrospective comparison of Junge and Werby's data with more recent rainfall chemistry does not clearly support the notion that SO_4^{2-} deposition has increased significantly since the 1950s. Estimated nonmarine SO_4^{2-} deposition near Tallahassee* between July 1955 and June 1956 was virtually identical to wet deposition measured at nearby Monticello between 1982 and 1986 (ESE 1987): 250 to 260 eq ha^{-1} yr^{-1} in 1955–56 compared to 262 eq ha^{-1} yr^{-1} in 1982–86.

The paleolimnological record also suggests that SO_4^{2-} increases since 1950, if anything, have been quite modest (Figure 12.24; Mitchell et al. 1988,

*Calculations based on Junge and Werby's data and rainfall volume measured during the same period at Monticello (NOAA 1956 and 1957).

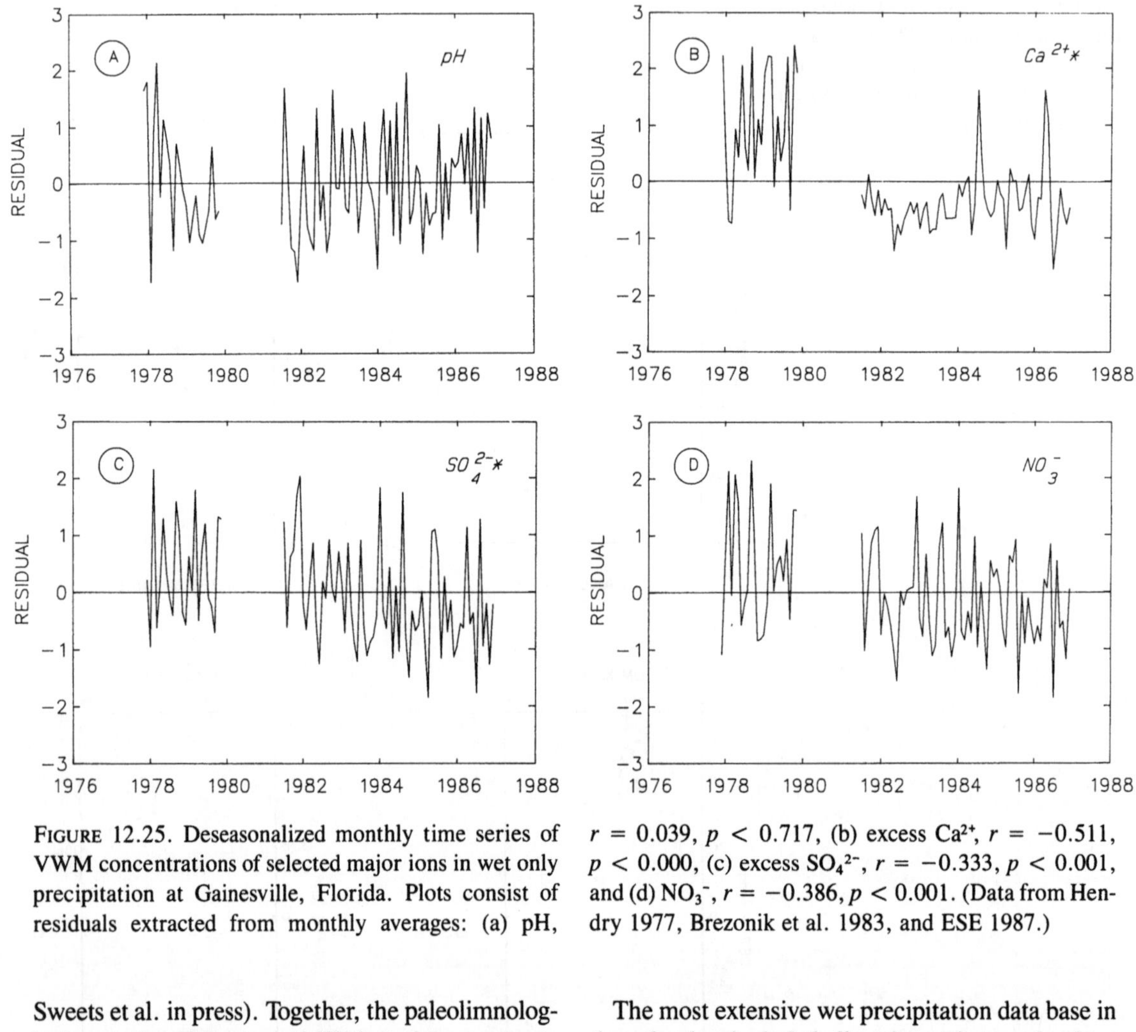

FIGURE 12.25. Deseasonalized monthly time series of VWM concentrations of selected major ions in wet only precipitation at Gainesville, Florida. Plots consist of residuals extracted from monthly averages: (a) pH, $r = 0.039$, $p < 0.717$, (b) excess Ca^{2+}, $r = -0.511$, $p < 0.000$, (c) excess SO_4^{2-}, $r = -0.333$, $p < 0.001$, and (d) NO_3^-, $r = -0.386$, $p < 0.001$. (Data from Hendry 1977, Brezonik et al. 1983, and ESE 1987.)

Sweets et al. in press). Together, the paleolimnological record and Junge and Werby's data appear to contradict Husar's estimates of historical deposition of SO_4^{2-} in Florida. The indistinct profiles for sulfur accretion in very recent Florida lake sediments may reflect (1) post-depositional mobility of reduced sulfur (Carignan and Tessier 1985, Holdren et al. 1984) or (2) degassing of reduced sulfur (as H_2S) not sequestered by iron in the sediments and subsequent loss to the atmosphere because of the low solubility and slow oxidation kinetics of H_2S in the water column (Morel 1983). Reconciliation of the Junge and Werby data with the emissions record is problematic. The quality of precipitation data collected from the 1950s and 1960s has been debated (e.g., Hansen and Hidy 1982) and there is concern that evaporation and possibly dust leakage rendered the early measurements more like bulk rather than wet deposition (Stensland et al. 1986).

The most extensive wet precipitation data base in time that has included all major cations and anions is for Gainesville. Wet precipitation chemistry data for Gainesville includes measurements conducted by Hendry (1977) in 1976 to 1977, and Gainesville has been an integral site of all the major wet-only precipitation regional networks operating in Florida. Temporal trends for pH, nonmarine SO_4^{2-}, NO_3^-, and nonmarine Ca^{2+} are presented in Figure 12.25 for the period 1977 through 1986. Seasonality from each parameter has been removed by analyzing the residuals from overall monthly averages and testing for autocorrelation. Results show that pH has remained essentially constant, although both nonmarine SO_4^{2-} and NO_3^- have declined significantly ($p < 0.001$). The trend for SO_4^{2-} is corroborated by emission inventories for the southeast, which show a similar decline from 1977 through 1984 (Figure 3.2, Chapter 3, Husar et al., this

TABLE 12.8. Summary of lakewater chemistry surveys and analytical techniques conducted in the Trail Ridge Lake district.[a]

Study	Time frame	Analytical methods: pH	Analytical methods: SO_4^{2-}	Analytical methods: ANC[b]	ANC method[c] correction (μeq L^{-1}): Low	ANC method[c] correction (μeq L^{-1}): High
Clark et al. (1964)	1958–60	Electrometric	Turbidimetric	4.5	−28	−31
Shannon (1970)	1968–69	Electrometric	Turbidimetric	4.5–5.1	+4	−31
Hendry and Brezonik (1984)	1968–69	Electrometric	Methylthymol blue	BCG-MR[d] (4.9–5.0)	0	−12
Baker (1984)	1980–81	Electrometric	Methylthymol blue	NA[e]		
ELS-I	1984	Electrometric	Ion chromatography	Gran		
PIRLA	1985	Electrometric	Ion chromatography	Gran		

[a] ANC corrections based on C_T and titration method endpoint after Kramer (1986). See text.
[b] Titration pH endpoint.
[c] Range in ANC correction based on stated range of titration endpoints and estimated range of carbonate C_T (see text).
[d] Bromcresol green-methyl red mixed indicator.
[e] NA = not analyzed.

volume). Emissions of NO_x in the southeast essentially peaked in 1977 and remained constant through 1981 (Figure 3.4, Chapter 3, Husar et al., this volume). That rainfall pH levels did not decrease along with SO_4^{2-} and NO_3^- may be related to reduced concentrations of base inputs. Between 1977 and 1986, Ca^{2+} decreased significantly; similar decreasing trends for base cations and chloride have been observed at Hubbard Brook (Stensland et al. 1986).

Historical Water Chemistry Comparisons

Historical measurements of major ion chemistry in softwater Florida lakes are limited to the Trail Ridge Lake District. Several lakes in the Trail Ridge have been surveyed periodically by a number of investigators since the late 1950s, including studies by Clark et al. (1962) from 1957 to 1960, Shannon (1970) from 1968 to 1969, Brezonik et al. (1983) from 1978 to 1979, the ELS in 1984, and the PIRLA study in the Florida Northern Peninsula (Sweets et al. 1990) in 1985. In addition, Baker (1984) collected detailed chemistry from McCloud Lake, also in the Trail Ridge, as part of an ion budget study of that softwater system. Table 12.8 summarizes the various studies and the analytical techniques used in each for pH, ANC, and SO_4^{2-}.

Two published analyses of historical trends in the chemistry of the Trail Ridge lakes predate this chapter. Crisman et al. (1980) compared pH, ANC, and SO_4^{2-} concentrations from the 1960s to 1978–79 levels and reported an average decline in pH of 0.5 units. Increased concentrations of SO_4^{2-} as well as the complete titration of low levels of ANC originally present were also noted.

Hendry and Brezonik (1984) examined historical trends in acidification of the Trail Ridge using data collected by Clark et al. (1964) and Shannon (1970), plus their own data. Hendry and Brezonik apparently used the same data set as Crisman et al., and Hendry and Brezonik's analysis also appears more rigorous. Alkalinity measured by Clark et al. (1964) was corrected to account for diffences in analytical technique by subtracting 20 μeq L^{-1} from measured values; this correction gives a titration endpoint of about pH 4.9 (neglecting CT effects). The Shannon data were not corrected. There is no indication that Crisman et al. made similar attempts to correct historical values of ANC as part of their analysis. Of the 10 lakes included by Hendry and Brezonik in their analysis, ANC declined or remained unchanged in 8 lakes; 2 lakes, Geneva and Kingsley, showed large increases in ANC, apparently because of extensive home development (including fertilizer and septic tank use) along the shorelines of both lakes. For lakes that had data extending back to 1957, nonmarine SO_4^{2-} concentrations showed increases ranging from essentially undetectable (16 μeq L^{-1}) to 73 μeq L^{-1}.

Although Clark et al. sampled approximately every six months, Hendry and Brezonik had access only to the average results and were unable to exam-

TABLE 12.9. Summary of historical changes in pH, ANC, nonmarine SO_4^{2-}, nonmarine Ca^{2+}, and Cl^- in Trail Ridge Lakes.

Lake	Time frame	Slope[a]					ANC[b]	
		H^+	SO_4^{2-}	Ca^{2+}	Conductivity	Cl^-	Low	High
Brooklyn	1957–84	+0.29	–	–	–	−1.75	−1.53	−0.91
Geneva	1957–79	−0.19	3.18	3.36	0.51	–	−0.80	–
Kingsley	1957–85	−0.02	1.75	5.32	0.40	−2.15	–	–
Magnolia	1957–84	+0.22	–	–	–	–	−1.58	−1.42
McCloud	1968–85	+1.25	5.30	–	0.57	–	+0.80	−1.42
Sandhill	1957–84	–	1.39	–	–	–	−1.41	−1.10
Santa Rosa	1969–84	+0.28	–	–	–	–	–	−1.64

[a] Slope μeq L^{-1} yr^{-1}, except for conductivity (μS cm^{-1} yr^{-1}). Slopes significant at $p < 0.05$.
[b] ANC corrected for differences in analytical methodology between data sets (see Table 12.8 and text). Low and high ANC slope estimates based on the range of corrected ANC values.

ine the range in variability of pH, ANC, or SO_4^{2-}, nor did Hendry and Brezonik include any of the 1957–60 data in any statistical analysis of trends. Extreme variability in lake stage and hydrology during this period also confounds the analysis—a point that Hendry and Brezonik took care to include. As Clark et al. began collecting their data in 1957, north Florida was in a record drought, and stage elevations for a number of seepage lakes were at or near record low levels (Figure 12.21). The effects of long-term drought on lakewater ANC are believed to be antagonistic (see section on groundwater hydrology/weathering) and are explored from a conceptual perspective in the section entitled Model Hindcasting Analysis of Historical Trends.

We have revisited the analysis of long-term acidification of the Trail Ridge lakes, using a more extensive data set compared to Hendry and Brezonik's original analysis. Our data set includes all the raw data collected by Clark et al., as well as ELS-I and PIRLA analyses as available for some of the lakes. Both the ELS-I and PIRLA surveys determined ANC by Gran titration, and earlier surveys were corrected to the appropriate titration endpoint using Kramer's (1986) method. Because C_T values were not available for all the lakes, a range of correction values was used based on the range of possible C_T values and the reported range of titration endpoints for each of the early surveys (Table 12.8). Reported ELS-I data for air-equilibrated DIC concentrations for Trail Ridge lakes ranged from 17 to 21 μmol L^{-1} and Hendry and Brezonik report inorganic carbon concentrations as high as 223 μmol L^{-1} in Kingsley Lake. Since Kingsley Lake contains the greatest amount of ANC of all the Trail Ridge lakes in the data set, it defines the upper end of reasonable C_T concentrations. A C_T range of 17 to 223 μmol L^{-1} consequently was used to correct all early survey ANC measurements and, depending on the method endpoint, corresponding ANC corrections ranged from −31 to +4 μeq L^{-1} (Table 12.8).

Table 12.9 summarizes slope estimates of annual changes in H^+, nonmarine SO_4^{2-}, ANC, nonmarine Ca^{2+}, conductivity, and Cl^-. Of the seven lakes analyzed, four showed significant increases of H^+ with time. The other three lakes either showed significant declines in H^+ (Geneva and Kingsley) or no trend (Sandhill). Nonmarine Ca^{2+} also increased in Lakes Geneva and Kingsley, but not in the other lakes. Trends of increasing nonmarine SO_4^{2-} were observed in four of the lakes analyzed, but only one of these lakes (McCloud) also showed increasing H^+. The differences in SO_4^{2-} concentrations approach analytical precision limits of the colorometric and turbidimetric procedures used historically (ca. 10 to 20 μeq L^{-1}; cf. APHA 1976 and U.S. EPA 1983) and although consistent, there is sufficient analytic variability such that these trends are not statistically significant for Lakes Brooklyn, Magnolia, and Santa Rosa.

The most extensive data base available to infer historical changes in lakewater chemistry of any lake in the Trail Ridge is McCloud Lake, an undeveloped seepage lake. In addition to the surveys conducted by Shannon (1970) and Hendry and Brezonik (1984), Baker (1984b) collected major ion chemistry monthly between October 1980 to August 1982. Data from 1985 are also available as part of the PIRLA water chemistry

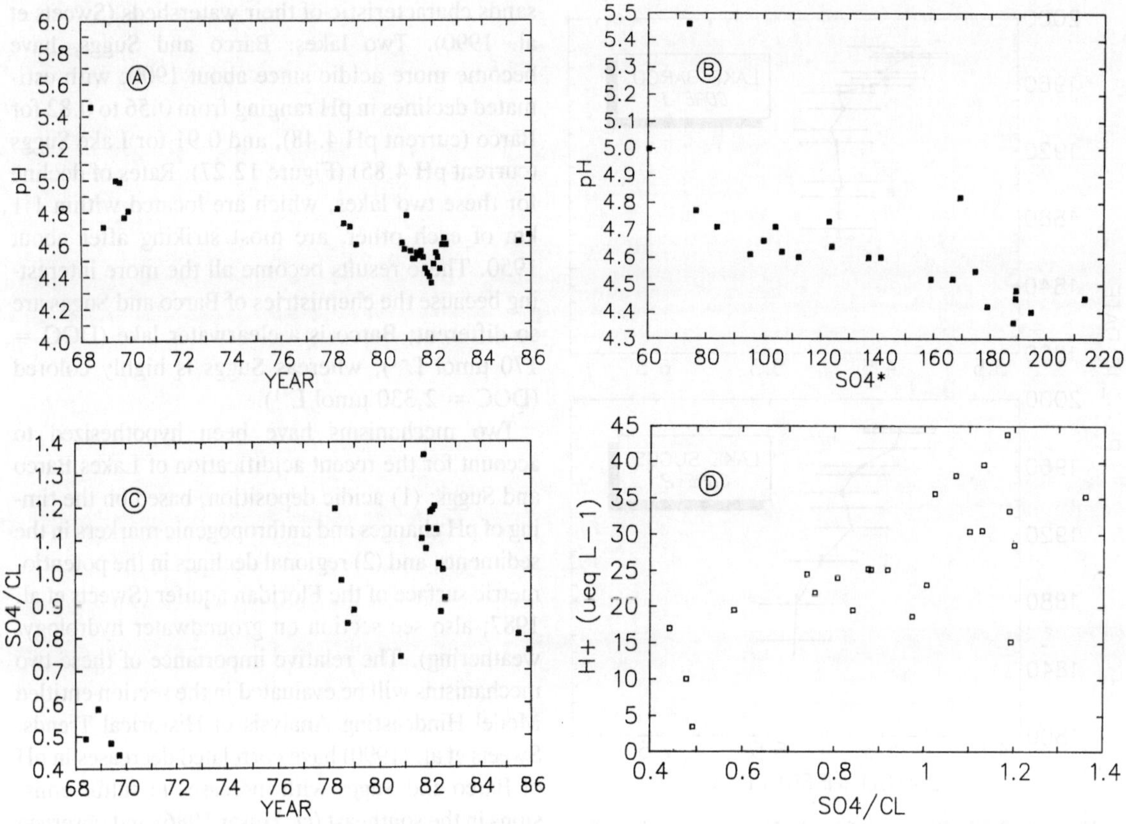

FIGURE 12.26. Long-term trends and interparameter relationships in McCloud Lake, Florida: (a) pH as a function of time, (b) pH as a function of nonmarine SO_4^{2-}, (c) SO_4^{2-}/Cl^- as a function of time, and (d) H^+ as a function of SO_4^{2-}/Cl^-.

sampling program. Regression analysis indicates that H^+ has increased about 1.25 μeq L^{-1} yr^{-1} since 1968, essentially doubling acidity (Table 12.9; Figure 12.26a). The resultant change in pH is about 0.3 units, from 4.90 in 1968–69 to 4.64 in 1985. Whether this trend is real, much less related to acidic deposition, is not clear. The trend in Figure 12.26a is driven by Shannon's data collected in 1968–69; later surveys give relatively consistent values (with some short-term variability; see section on ion enrichment calculations) for pH after 1978. Shannon used soft glass dissolved oxygen bottles to collect his samples for pH; the bottles themselves may be a source of 20 to 100 μeq L^{-1} of ANC, depending upon the condition of the glass surface and sample storage temperature and holding times (Kramer and Tessier 1982). There is evidence that holding times were minimized by Shannon because of his stated concern to prevent CO_2 transfer with the atmosphere, which in turn would reduce the magnitude of ANC contributed by the glass bottle surfaces.

Comparison of nonmarine SO_4^{2-} concentrations and pH shows clearly that pH in McCloud Lake is inversely related to nonmarine SO_4^{2-} (Figure 12.26b). This alone does not confirm increasing acidic deposition as a causal mechanism driving changes in pH in the lake. Drought and evapoconcentration can produce the same effect. For example, increasing the concentration of a dilute solution with initial concentrations of pH 4.8, Cl^- 145 μeq L^{-1}, and nonmarine SO_4^{2-} 127 μeq L^{-1} by 25% via evapoconcentration results in a more acidic solution of pH 4.7, Cl^- 181 μeq L^{-1}, and nonmarine SO_4^{2-} 159 μeq L^{-1}. Nonmarine SO_4^{2-} and H^+ have both increased because of evapoconcentration, but the ratio of H^+ to nonmarine SO_4^{2-} has remained unchanged. Figure 12.26d shows

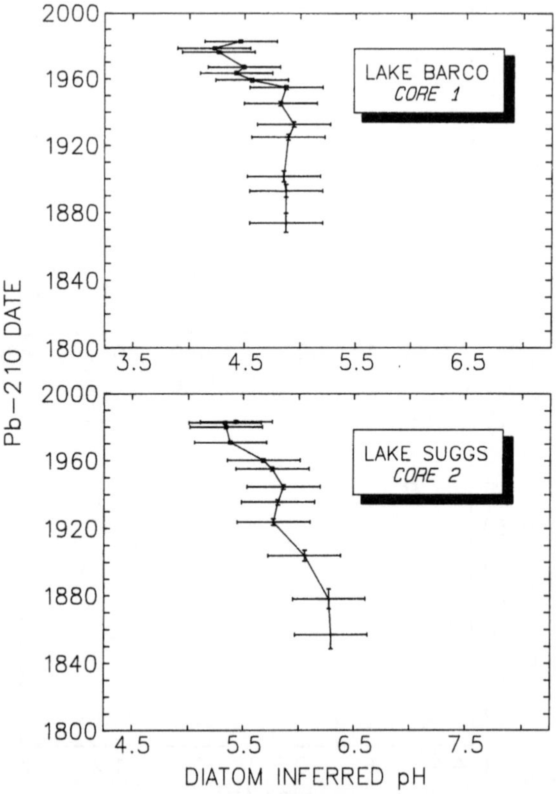

FIGURE 12.27. Diatom inferred pH reconstructions for (a) Lake Barco and (b) Lake Suggs in the Trail Ridge, Florida. (From Sweets et al. 1990.)

H^+ as a function of the ratio of nonmarine SO_4^{2-} to Cl^-, filtering out evapoconcentrative effects on pH. Lakewater acidity is clearly correlated with SO_4^{2-}/Cl^-. Moreover, since 1968–69, SO_4^{2-}/Cl^- ratios in McCloud Lake have increased from about 0.5 to 0.8, suggesting that increased deposition of nonmarine SO_4^{2-} has promoted higher lakewater concentrations of nonmarine SO_4^{2-} and H^+ (Figure 12.26c and d).

Paleolimnological Evidence of Historical Trends

Paleolimnological analysis of changes in diatom assemblages in the sedimentary record of six lakes in north Florida has been used to infer recent historical change in pH and ANC as part of the PIRLA study (Sweets et al. 1990). Four of the six lakes had low pH values predating 1850, at least in part reflecting the low base saturation of the deep acid sands characteristic of their watersheds (Sweets et al. 1990). Two lakes, Barco and Suggs, have become more acidic since about 1900, with estimated declines in pH ranging from 0.56 to 0.82 for Barco (current pH 4.48), and 0.91 for Lake Suggs (current pH 4.85) (Figure 12.27). Rates of decline for these two lakes, which are located within 1.1 km of each other, are most striking after about 1950. These results become all the more interesting because the chemistries of Barco and Suggs are so different; Barco is a clearwater lake (DOC = 170 μmol L^{-1}), whereas Suggs is highly colored (DOC = 2,330 μmol L^{-1}).

Two mechanisms have been hypothesized to account for the recent acidification of Lakes Barco and Suggs: (1) acidic deposition, based on the timing of pH changes and anthropogenic markers in the sediments, and (2) regional declines in the potentiometric surface of the Floridan aquifer (Sweets et al. 1987; also see section on groundwater hydrology/weathering). The relative importance of these two mechanisms will be evaluated in the section entitled Model Hindcasting Analysis of Historical Trends. Sweets et al. (1990) have correlated decreases in pH of Barco and Suggs with increases in sulfur emissions in the southeast (cf. Husar 1986) and observed similar increases in Pb and polynuclear aromatic hydrocarbons (PAH) within the core profiles. In addition, sedimentary sulfur appears enriched by 10% to 138% in post- 1900 sediments when compared to pre-1900 levels (Mitchell et al. 1988).

Model Hindcasting Analysis of Historical Trends

We use the IAG model here to analyze paleolimnological results for Lake Barco in order to evaluate possible mechanisms contributing to its inferred, recent acidification. The model was calibrated to existing major ion chemistry in the lake, assuming steady state conditions (Table 12.10). Once calibrated, the model was used to hindcast steady state historical major ion chemistry prior to cultural impacts.

Although no hydrologic or chemical inseepage data are currently available for Lake Barco, Baker (1984) measured inseepage chemistry and calculated seepage flow rates (in and out) for McCloud Lake. The two lakes lie within 2.2 km of each other in the Trail Ridge and the geologic and hydrologic

TABLE 12.10. Chemical composition (μeq L^{-1}) of Lakes Barco and McCloud in 1985.[a]

	McCloud (n = 2)[b]	Barco (n = 3)[b]
H^+	23	32
Ca^{2+}	39	39
Mg^{2+}	59	56
K^+	5	5
Na^+	144	136
SO_4^{2-}	119	132
Cl^-	152	125
Cations	270	267
Anions	271	257

[a] Data from Sweets et al. (in press).
[b] n = number of observations.

TABLE 12.11. IAG model hindcasting results for historical water chemistry in Lake Barco.[a]

Constituent[b]	Scenario I C_o	Scenario II C_o	Present C_t
pH	4.78	4.72	4.45
ANC	−15.9	−17.9	−35.0
SO_4^{2-}	122.3	100.3	131.7
NO_3^-	4.6	3.5	**
NH_4^+	4.4	4.4	**
C_B	218.6	206.5	234.8
Cl^-	112.0	125.0	125.0

[a] Results are based on two scenarios: Scenario I holds deposition constant at current levels but assumes that seepage fluxes have declined by 25% in response to sustained declines in the potentiometric surface of the Floridan aquifer; Scenario II assumes that H^+ deposition has increased 60% since the 1950s.
[b] Concentrations in μeq L^{-1}. C_O is initial or historical concentration (i.e., concentration observed at time zero of simulation). C_t is concentration of constituent at end of simulation. ** = not detected.

settings for both lakes are almost identical. The chemistry of both lakes also is quite similar (Table 12.10) and model calibration was performed using McCloud seepage fluxes.

Two effects of two model scenarios were evaluated: (1) an assumed 25% decrease in inseepage flow rates because of changes in the downward hydraulic gradient, and (2) an increase in H^+ deposition of 60% since the 1950s. The first scenario is based on long-term declines in the potentiometric surface of the Floridan aquifer and its potential effect on groundwater hydrology (see section on groundwater hydrology/weathering). Within the past 20 years, the potentiometric surface of the Floridan has declined about 2 m and there is evidence suggesting that this decline may be part of a longer decline related to heavy withdrawal from the aquifer extending back to 1940. By increasing the hydraulic gradient between the water table and the Floridan aquifer, rates of downward leakage are believed to be enhanced, and lake inseepage rates may be reduced.

The second scenario is based on analysis of historical precipitation chemistry for Florida (Brezonik et al. 1980). For simulation purposes, it was assumed that two-thirds of the increase in H^+ deposition is attributable to increased SO_4^{2-} deposition; the remaining increase in H^+ deposition was attributed to NO_3^-. This scenario is consistent with current SO_4^{2-}/NO_3^- ratios in Florida precipitation. However, since emissions of SO_2 have increased by 70% to 80% in the southeast since 1950 (Figure 3.2, Chapter 3, Husar et al., this volume), our analysis probably underestimates the effect of recent increases in SO_4^{2-} deposition on Lake Barco.

Model hindcasting estimates of historical chemical conditions in Lake Barco as a function of each scenario are presented in Table 12.11. Both scenarios predict essentially the same historical pH and are consistent with pre-acidification levels inferred by diatom analysis. However, subtle differences in other major ion components are predicted by the model. The increased atmospheric deposition scenario predicts that Cl^- concentrations have remained relatively constant, whereas increases of 30% for SO_4^{2-} and 14% for base cations are predicted. Effects on SO_4^{2-} and base cations are less pronounced for the reduced inseepage scenario, but Cl^- is predicted to have increased by about 12% and the effect of increased SO_4^{2-} and Cl^- concentrations on ANC is essentially equivalent.

No historical water chemistry are available for Lake Barco to discriminate between the two hypotheses. Insofar as the effects of groundwater flow path alteration are concerned, there is no indication of increasing Cl^- concentrations for other lakes in the Trail Ridge to support the hypothesis (Table 12.11). Data for Sandhill and Magnolia lakes show that Cl^- levels have remained essentially constant since the mid 1950s, whereas SO_4^{2-} concentrations have increased around 10 to 20 μeq L^{-1}. McCloud Lake SO_4^{2-} concentrations have increased by nearly 70 μeq L^{-1} since 1968–69, whereas average concentrations of Cl^- for 1968–69 and 1981–82 are within 3 μeq L^{-1} (Baker 1984). These results, in conjunction

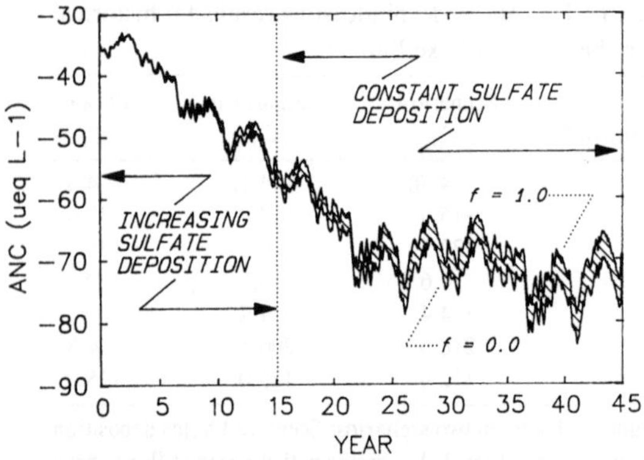

FIGURE 12.28. IAG model time series predictions for ANC for hypothetical seepage lake, assuming a 30% increase in SO_4^{2-} deposition. SO_4^{2-} deposition is stepped in equal increments to yield a total increase of 30% 15 years after t_0; years 15 through 45 maintain SO_4^{2-} deposition at 30% above initial level. Upper curve represents predicted lake response if watershed *F*-factor is 1; lower curve represents predicted response if *F*-factor is 0.

with the hindcasting model analysis, suggest that atmospheric deposition is contributing to progressive acidification of the Trail Ridge lakes. However, further chemistry data on inseepage chemistry in the region must also be obtained before aquifer effects on long-term changes in ANC and other major ions can be verified or discounted. Also, it must be emphasized that these results are simply model simulations and are not conclusive at this point. Differences in lake water SO_4^{2-} concentrations predicted by the model lie within the limits of laboratory analytical precision.

Future Considerations

The astatic type of seepage lake used in the section on Summary of Processes to examine hydrologic effects on short-term changes in ANC is employed here to conduct a hypothetical analysis of the effects of increased SO_4^{2-} deposition on lakewater ANC. FDER (1984) predicts that SO_2 emissions will increase by approximately 30% by 2000 as the population of Florida continues to grow rapidly. The effects of a similar increase in total SO_4^{2-} deposition (30%, 182 eq ha^{-1} yr^{-1}) was modeled with an equivalent increase in H^+ deposition. The increase was stepped incrementally during the simulation such that, after 15 years, SO_4^{-2} and H^+ deposition had increased by 182 eq ha^{-1} yr^{-1}. The simulation was then continued for another 30 years at the new deposition levels. Effects on groundwater ANC of increased SO_4^{2-} and H^+ deposition on ANC were simulated by allowing the *F*-factor describing the efficiency of weathering and cation exchange to respond to increases in SO_4^{2-} concentrations (Chapter 18, Sullivan, this volume) to range from 0 to 1.

Simulated results for the astatic lake are presented in Figure 12.28. Note that by definition, both model lake types, static and astatic, give the same steady state response. Depending on whether base cation supply is enhanced, steady state ANC is predicted to decline to between -70 ($F = 1$) to -75 μeq L^{-1} ($F = 0$), whereas SO_4^{2-} concentrations are predicted to increase by 30% from 132 to 170 μeq L^{-1}. The simulation shows that lakewater ANC in seepage lakes responds more rapidly than predicted by τ_w alone. The ANC response time is limited by SO_4^{2-} uptake kinetics, which is much slower than corresponding NO_3^- and NH_4^+ rates (Baker and Brezonik 1988). In comparison with the τ_w of the lake (15.5 years), the SO_4^{2-} residence time is nearly 50% shorter (8.5 years) and ANC concentrations approach steady state within 25 years of when atmospheric loadings become constant (year 40 of the simulation). Little difference is seen between predicted ANC for the two different *F*-factor scenarios.

This simulation illustrates an important aspect of lake response time to perturbations in atmospheric deposition: the timeframe of response by the lake is governed not strictly by τ_w but by turnover times of the most important ions contributing to ANC production. For lakes in which SO_4^{2-} adsorption is the principal acid neutralizing mechanism, response time is controlled largely by the time lag for SO_4^{2-} breakthrough through the soil profile to occur. For groundwater recharge seepage lakes, in-lake processes dominate, and lakewater ANC response approaches the SO_4^{2-} residence time, viz. $V/(kA + Q_o)$, where V is the lake volume (m^3), A is the lake area (m^2), and k is the SO_4^{2-} loss constant (m yr^{-1}).

Biological Status

Fish

Historical data on the fisheries of aquatic systems located in the Florida RCS region are sparse (e.g., Meehean 1941, Dickinson 1948). The available data, however, suggest that Florida fishes may have evolved to tolerate extremely low pH (ESE 1984). Species distributional studies have collected 66 species of fish from waters having a pH of 5.0 or less and 37 species from waters having pH values as low as 4.0 (ESE 1984). Although fish abundance is lower in acidic compared to nonacidic lakes (Keller 1984), Meehean (1941) reported fish standing crops ranging from 25 to 118 kg ha^{-1} in five Ocala National Forest lakes with pH values between 5.4 and 5.6. Meehean (1941), however, attributed these differences to the ecological maturity of the lakes, with the younger, least productive lakes supporting the lowest fish standing crops.

In the early 1980s, two fisheries surveys were initiated to determine the current status of the fisheries in Florida lakes potentially sensitive to acidic deposition (Keller 1984, Canfield et al. 1985). Using data from the Florida Game and Fresh Water Fish Commission and other sources, Keller (1984) correlated fish species numbers with lake area and pH. Canfield et al. (1985) sampled 11 lakes in the Northern Peninsula and found that plasma osmotic concentrations, growth, and coefficients of condition of largemouth bass were significantly lower in acidic clearwater lakes than in acidic colored and circumneutral lakes. These studies, however, cautioned that reductions in fish growth, condition, and plasma osmotic concentrations, as well as reduced species richness in the surveyed acidic lakes, could be related to either acidic conditions or other environmental factors such as lake trophic status (Keller 1984, Canfield et al. 1985).

Lamia (1987), therefore, initiated in 1984 an intensive 2-year study of the limnology and fishery of Cue Lake. Cue Lake was chosen for study because (1) the fisheries survey by Canfield et al. (1985) did not collect adult or young-of-the-year bluegill (*Lepomis macrochirus*), and (2) largemouth bass (*Micropterus salmoides*) collected from the lake had low plasma electrolyte and osmotic concentrations, reduced condition factors (KTL), and deformed otoliths. Cue Lake also had the lowest pH (average and minimum pH 4.1 and 3.7, respectively) and highest Al concentration (5.93 μmol L^{-1}) of any lake sampled. Lamia (1987), however, collected 12 fish species, including bluegill, from Cue Lake with the most commonly collected species being the lake chubsucker (*Erimyzon sucetta*), yellow bullhead (*Ictalurus natalis*), warmouth (*Lepomis gulosus*), and largemouth bass (*Micropterus salmoides*). Cue Lake's total fish biomass, as estimated by mark-recapture, was 14 kg ha^{-1} in 1984 and 11 kg ha^{-1} in 1985. The largemouth bass population (fish > 150 mm total length [TL]) was 6 fish ha^{-1} (2 kg ha^{-1}), but the harvestable (> 250 mm TL) large-mouth bass population was only 2 fish ha^{-1} (1 kg ha^{-1}). Population estimates for lake chubsuckers, yellow bullheads, and warmouth were 42 fish ha^{-1} (6 kg ha^{-1}), 10 fish ha^{-1} (4 kg ha^{-1}), and 5 fish ha^{-1} (0.2 kg ha^{-1}), respectively. Reproduction by these four species was observed, however, and fry and young-of-year were collected, indicating that reproduction and recruitment were occurring during the study period. Lamia (1987), therefore, concluded that the fishery of Cue Lake was not yet sufficiently stressed by its low pH to cause reproductive and recruitment failure. She further concluded that the fishery was probably influenced primarily by factors related to lake trophic status. Cue Lake was highly oligotrophic and had Secchi disc measurements to the lake bottom (9 m) throughout her study. Total phosphorus and total nitrogen concentrations averaged 4.9 μg L^{-1} and 110 μg L^{-1}, respectively, and chlorophyll a concentrations averaged 2.1 μg L^{-1}.

In 1987, Canfield (unpublished data) began intensive fisheries studies for 14 lakes located in the Florida RCS region, including 5 lakes originally sampled as part of the ELS-I. Preliminary results indicate that the relationship between pH and the number of fish species present in the acidic lakes is weak (r = .46, $p < 0.10$; Table 12.12). The lowest numbers of fish species, however, have been collected in the most unproductive lakes. Total fish standing crops have ranged from 8 to 95 kg ha^{-1} (Table 12.12) and show no relationship with pH (r = .02), although there is a strong relationship with lake trophic status (r = .82, $p <$ 0.05; Tables 12.12 and 12.13).

TABLE 12.12. Average water chemistry and fisheries values for some acidic lakes in Florida.[a]

Lake TL	pH	Aluminum µg L⁻¹	Total *P* µg L⁻¹	Total *N* µg L⁻¹	Chlorophyll *a* µg L⁻¹	No. fish species	Fish standing crops (kg ha⁻¹) Total	Fish standing crops (kg ha⁻¹) Largemouth bass sportfish	>250 mm TL	Largemouth bass (fish ha⁻¹) 150–249 mm TL	Largemouth bass (fish ha⁻¹) >250 mm
Barco	4.5	90	3	110	0.8	5	8	5	0	24	4
Lawbreaker	4.4	270	5	130	0.7	4	22	1	0	0	0
Cue	4.6	110	5	110	1.9	5	22	14	6	4	2
Turkey Pen	4.7	70	2	160	0.7	6	–	–	–	6	6
Keys	5.4	20	1	250	0.9	8	27	14	1	–	–
McCloud	4.6	60	4	180	1.7	5	–	–	–	35	30
Moore	5.6	40	2	390	1.5	15	45	19	2	5	31
Lofton Pond	4.8	30	3	450	1.0	11	56	37	21	6	14
Deep	4.6	60	3	120	0.9	14	59	20	14	11	14
Tomahawk	4.9	50	8	220	1.7	12	59	34	11	209	48
Gobbler	4.0	100	7	680	4.8	8	–	–	–	0	0
Suggs	5.0	70	57	820	5.4	24	17	12	0	1	2
Brock	4.9	80	51	–	4.6	13	–	–	–	6	20
Crooked	4.6	80	11	320	10	10	95	45	12	33	51

[a] From Canfield (unpublished data). Bass estimates obtained from mark-recapture studies.

Sport fish populations in the acidic lakes are currently estimated at 5 to 45 kg ha⁻¹ (Table 12.12). Comparing these estimates with similar estimates from less acidic lakes (Table 12.13) indicates that sport fish biomass is strongly influenced by lake trophic status. Reproduction and recruitment of sport fish has been documented in nearly all the acidic lakes (Table 12.12), but two of the study lakes, Lawbreaker and Gobbler, currently support no largemouth bass. Lakes Lawbreaker and Gobbler are located next to each other in the Ocala National Forest. Lake Lawbreaker is a clearwater lake, with an average total aluminum concentration of 10 µmol L⁻¹ (Table 12.12). Lake Gobbler is highly colored, but has an average aluminum concentration of only 3.7 µmol L⁻¹ (Table 12.12). The reason for the lack of largemouth bass in these lakes, however, is currently unknown, but may also be related to lake trophic status or competition between different fish species.

Phytoplankton/Macrophytes

Phytoplankton

Phytoplankton diversity and abundance are generally related to pH, although the relationship may reflect covariation with nutrients rather than direct effects of pH (cf. Canfield et al. 1983, Pollman et al. 1985). In a survey of 20 softwater lakes in the

TABLE 12.13. Average water chemistry and fisheries values for some Florida lakes having pH values >6.0.[a]

Lake	pH	Aluminum mg L⁻¹	Total *P* mg L⁻¹	Total *N* mg L⁻¹	Chlorophyll *a* mg L⁻¹	No. fish Species	Fish standing crops (kg ha⁻¹) Total	Fish standing crops (kg ha⁻¹) Sportfish
Carr	6.3	15	22	820	14	15	120	36
Baldwin	8.0	19	22	630	16	19	206	57
Susannah	7.7	22	27	780	17	22	339	78
Wauberg	7.6	21	160	1500	110	21	600	130
Hollingsworth	8.9	15	110	2400	120	15	600	140
Bivens Arm	9.6	11	400	3300	250	11	500	230

[a] From Canfield (unpublished data). Fish standing crops estimated by rotenone sampling.

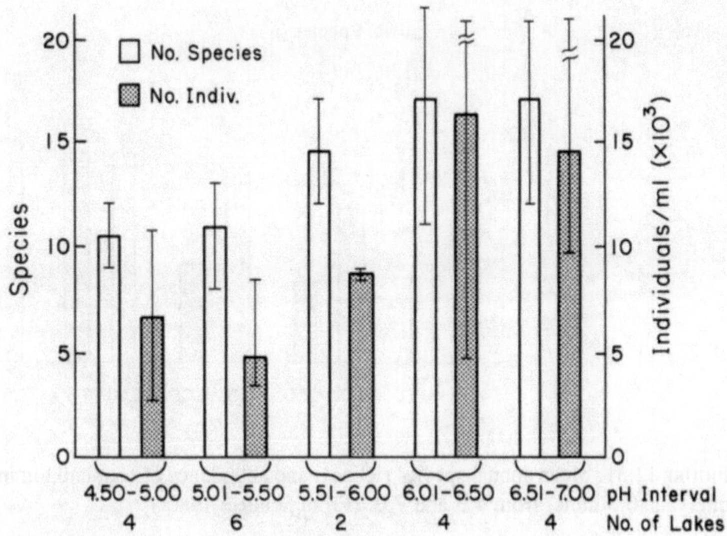

FIGURE 12.29. Mean annual species richness and abundance of phytoplankton in 20 softwater lakes grouped across pH intervals (0.5 units) from 4.5 and 7.0. (From Schulze 1980.)

Trail and Highlands ridges, Schulze (1980, Brezonik et al. 1984) correlated phytoplankton species numbers with pH (Figure 12.29; $p < 0.01$). Scatter in the relationship at pH > 6.0 was ascribed to differences in total phosphorus among the nonacidic lakes (Brezonik et al. 1984). Algal species composition also changed along the gradient in pH defined by Schulze's survey lakes (Figure 12.30). Bluegreen algae, which were dominant above pH 6.0, were replaced by green algae in lakes with pH < 6.0. With the exception of an increase in pyrrophytes at pH < 5.0, subdominant algal taxa showed little relationship with pH.

The significance of pH dependent correlations with algal biomass and diversity are unclear. Multivariate analysis conducted by Canfield et al. (1983) of pH, chlorophyll *a*, and total phosphorus concentrations in 165 Florida lakes indicates that phosphorus concentrations are more strongly correlated with algal standing crop than is pH.

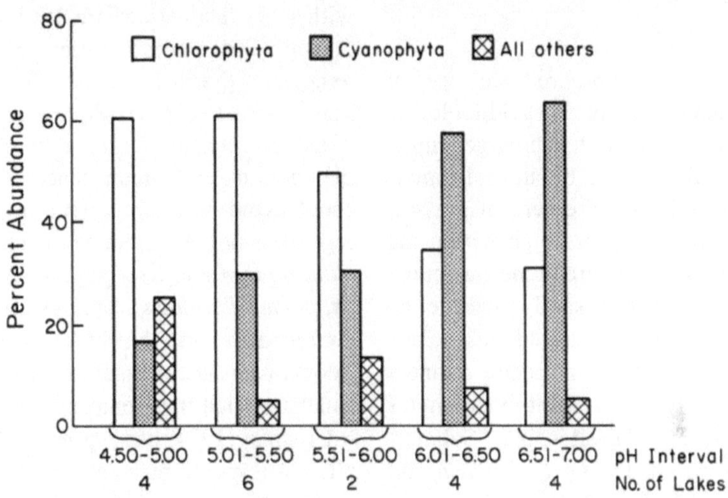

FIGURE 12.30. Relative contribution of major taxonomic groups to total abundance of phytoplankton in 20 softwater Florida lakes. (From Schulze 1980.)

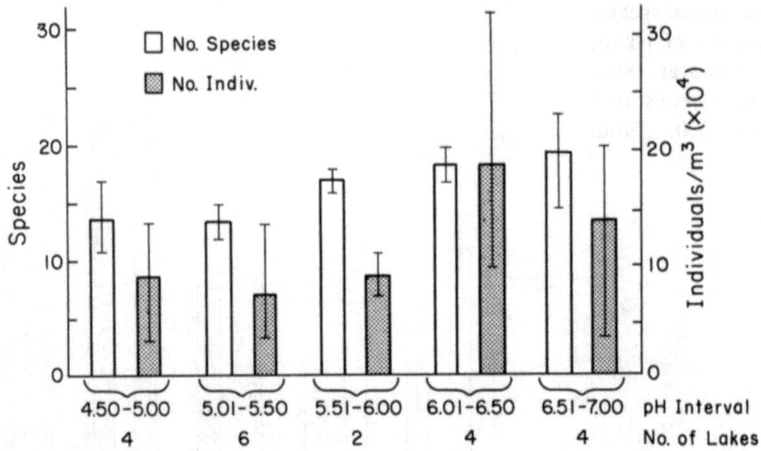

FIGURE 12.31. Mean annual species richness and abundance of zooplankton in 20 softwater lakes grouped across pH intervals (0.5 units) from 4.5 and 7.0. (From Schulze 1980.)

Pollman et al. (1985) analyzed the watershed characteristics of Schulze's (1980) survey lakes and found that chlorophyll *a* was correlated at least as strongly with watershed factors such as land use as it was with pH. In situ mesocosm (limnobag) studies in McCloud Lake also suggest that nutrient availability is a stronger determinant of algal abundance than is pH. Ogburn (1984) acidified a series of bag enclosures in McCloud Lake and detected no response in chlorophyll *a* for pH levels down to 3.7, whereas nutrient additions to the enclosures promoted increases in chlorophyll *a*.

Macrophytes

Crisman et al. (1986) have observed two types of macrophyte community structure in acidic Florida lakes (pH 4.0 to 5.5). In the first, submergent macrophytes are totally absent, and the littoral zone is confined to a narrow fringe of emergent grasses. Epiphytic algal biomass appears high within the grass zone, and benthic algae are numerous in the granular interstices of the sandy sediments, although algal mats do not develop in these lakes. Other lakes show no reduction in species richness or extent of coverage by submergent macrophytes associated with lower lakewater pH. Submergent vegetation is dominated by *Utricularia* and *Websteria*. Epiphytic algae are lush and Crisman et al. (1986) note the occurrence of dense clouds of epiphytic algae during summer.

Zooplankton

The survey results of Schulze (1980, Brezonik et al. 1984) show that zooplankton species richness is lower in acidic lakes (Figure 12.31). Numbers of zooplankton species reported by Schulze (mean of 14 species for lakes with pH < 5.5; range 11 to 17) were much higher than reported for temperate lakes of comparable pH (Lievestad 1976, Raddum et al. 1980, Schindler et al. 1985). Brezonik et al. (1984) attribute the relative richness of Florida zooplankton to lower concentrations of aluminum and toxic trace elements in clearwater Florida lakes compared with acidic lakes elsewhere. Brezonik et al. also cite climatic differences between the subtropical and temperate thermal regimes (i.e., less profound seasonal fluctuations in Florida), and the possibility of natural selection for a regime predisposed to low pH under natural circumstances. Several species of zooplankton were important in all survey lakes regardless of pH, including the cladocerans *Eubosmina tubicen* and *Daphnia ambigua*, the copepods *Diaptomus floridanus*, *Cyclops varicans*, and *Mesocyclops edax*, and the rotifer *Keratella cochlearis*. Zooplankton assemblages in Florida lakes tend to be smaller bodied than temperate lake assemblages.

Crisman (1984) has suggested that the most significant response by zooplankton to reduced pH in Florida lakes is given by ciliated protozoans. In a study of 21 lakes ranging in pH from 4.7 to 7.0, Beaver and Crisman (1981) observed declines in ciliated protozoan abundance and biomass with declining pH. These results were initially attributed to inferred reductions of bacterial and algal populations corresponding to pH, although trophic state is now believed to be the controlling factor (Beaver and Crisman 1982). Reductions in pH also were accompanied by a compositional shift from small-bodied to large-bodied assemblages.

In situ enclosure studies in McCloud Lake indicate that major changes in zooplankton community structure can occur with incipient acidification. Comparison of acidified (pH 3.6) and control (pH 4.6) littoral zone enclosures with an additional enclosure receiving base additions (> pH 5.1) showed decline and replacement of copepods with rotifers in the acidified enclosure (Crisman et al. 1982). Total zooplankton abundance in the acidified enclosure was greatly reduced. Subsequent acidification of pelagic enclosures confirmed the almost complete elimination of copepods at pH 3.7 (Ogburn 1984).

Summary

Available data indicate that there currently has been no widespread biological damage in the Florida RCS region due to acidic precipitation. There definitely have been no widespread fisheries losses in Florida similar to those reported for the northeastern United States, despite the extremely low pH values that occur in many Florida lakes. Florida's acidic lakes, however, support fewer plant and animal species and are generally less productive than its nonacidic lakes, but it is difficult to separate the direct effects of pH from covarying factors related to lake nutrient status. Lake trophic status, however, currently seems to be the dominant factor influencing the biotic communities in Florida's acidic lakes.

Conclusions

1. More acidic lakes are found in Florida than any other region in the United States—even though atmospheric deposition of H^+ and SO_4^{2-} is relatively moderate. The vast majority of acidic lakes in Florida are seepage lakes and, although the acidity in half of these lakes may be ascribed to organic acids, the remainder are clearwater systems where strong acid inputs associated with SO_4^{2-} appear to be the major source of acidity.
2. Control of ANC in undisturbed, clearwater seepage lakes appears to be dominated by in-lake biogeochemical processes. Anion reduction processes are the major sources of ANC production; in comparison to the amount of ANC generated by SO_4^{2-} and NO_3^- consumption, base cation supply is quite small. Nevertheless, because ANC generated by anion reduction very nearly balances acid inputs, small amounts of base cation supply may be sufficent to maintain positive levels of ANC in some lakes.
3. Because (1) the extent of SO_4^{2-} consumption appears to be independent of pH, and (2) NO_3^- assimilation is virtually complete in all seepage lakes, differences between low and more moderate ANC lakes in Florida appear to be related to geochemical or hydrologic factors that affect base cation supply. These factors include the relative balance between precipitation and evaporation, which in turn not only controls evapoconcentration effects on solute concentrations but influences the role of inseepage on the overall hydrologic budget as well. Localized variations in the proximity of the Hawthorn Formation to the lake bed also appear to influence ANC through base cation and possibly organic acid supply.
4. Analysis of long-term chemistry in the Trail Ridge Lake district suggests recent acidification of some seepage lakes. Depending on the correction factor used for historical ANC measurements, four to five of the seven lakes analyzed showed significant declines in ANC; however, corresponding increases in nonmarine SO_4^{2-} only accompanied two of the lakes. The strongest direct chemical evidence of recent lake acidification is for McCloud Lake, where increasing H^+ has been correlated with increasing SO_4^{2-}/Cl^- ratios, suggesting that increased deposition of nonmarine SO_4^{2-} has promoted higher lakewater concentrations of nonmarine SO_4^{2-} and H^+.
5. Paleolimnological reconstructions of diatom-inferred pH for Lakes Barco and Suggs also indicate progressive acidification since about 1950, although the mechanisms are not clearly understood. Model hindcasting suggests that

either of two hypotheses may account for recent trends in ANC: increased atmospheric deposition of SO_4^{2-} and H^+, or inferred changes in groundwater hydrology related to consumptive use of the underlying confined aquifer. Analysis of model hindcasts coupled with observed trends in SO_4^{2-} and Cl^- for other Trail Ridge lakes suggests that atmospheric deposition is contributing to progressive acidification of the Trail Ridge lakes. However, further data on inseepage chemistry in the region must be obtained before aquifer effects on long-term changes in ANC and other major ions can be verified or discounted.

6. Available data indicate that currently there has been no widespread biological damage in the Florida RCS region due to acidic precipitation. In contrast to the Adirondacks, no fisheries losses have been verified in Florida, despite the extremely low pH values (< 5.0) observed in many Florida lakes. Florida's acidic lakes, however, support fewer plant and animal species and are generally less productive than its nonacidic lakes, but it is difficult to separate the direct effects of pH from covarying factors related to lake nutrient status. Nutrient status currently appears to be the dominant factor influencing the biotic communities in Florida's acidic lakes. The reasons for the success of fisheries in acidic Florida lakes are not known, and the apparent differences between the fisheries responses in northern regions and Florida suggest the need for additional research.

Acknowledgments. Many of the ideas presented in this chapter evolved from extended conversations with Joe Eilers and Larry Baker on the nature of Florida seepage lake chemistry. Linda Duever wrote much of the material on vegetation and P. Roger Sweets provided unpublished diatom reconstructions. Tony Selle and Jeremy Smith provided numerous iterations of graphic and computer analyses. Bob Cook and Don Charles reviewed various drafts and provided much insight and critical input. Finally the authors wish to acknowledge painstaking reviews by Mark Brenner and the late Edward S. Deevey, whose comments provoked careful reconsideration of our views of Florida limnology. This work was funded in part by a grant from the Florida Department of Environmental Regulation.

Upper Midwest Overview

Robert B. Cook

Northern Minnesota, Wisconsin, and Michigan—the Upper Midwest—contain about 10,000 lakes with a broad range of chemical, physical, and hydrologic characteristics (Chapter 5, Eilers and Selle, this volume). As a consequence of the geology and hydrology of the region, many lakes in the

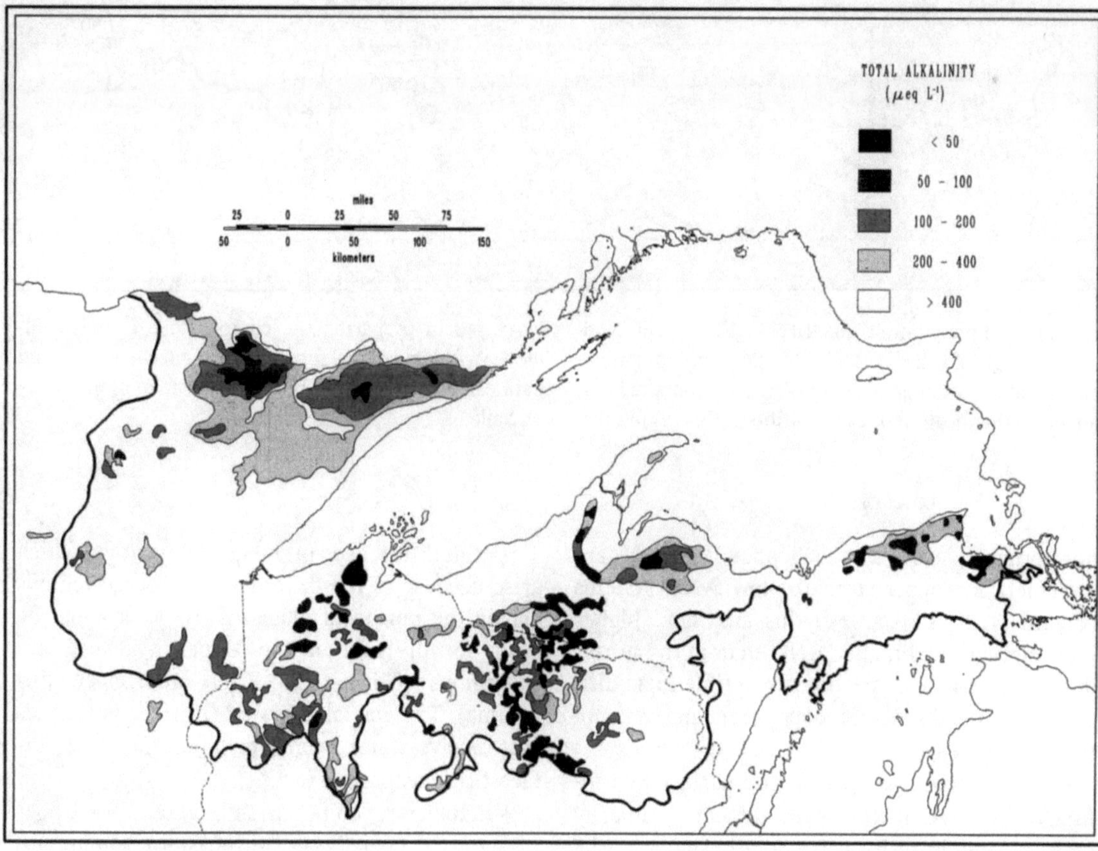

FIGURE 1. Distribution of total lakewater alkalinity (ANC; μeq L^{-1}) in the Upper Midwest. This map represents a more detailed update of an earlier map by Omernik and Powers (1983). The present map is adapted from a national map prepared in 1988 by J.M. Omernik, G.E. Griffith, J.T. Irish, and C.B. Johnson [U.S. EPA Environmental Research Laboratory, Corvallis, Oregon (Chapter 5, Eilers and Selle, this volume)], which used data from the National Surface Water Survey and other observations in the Upper Midwest.

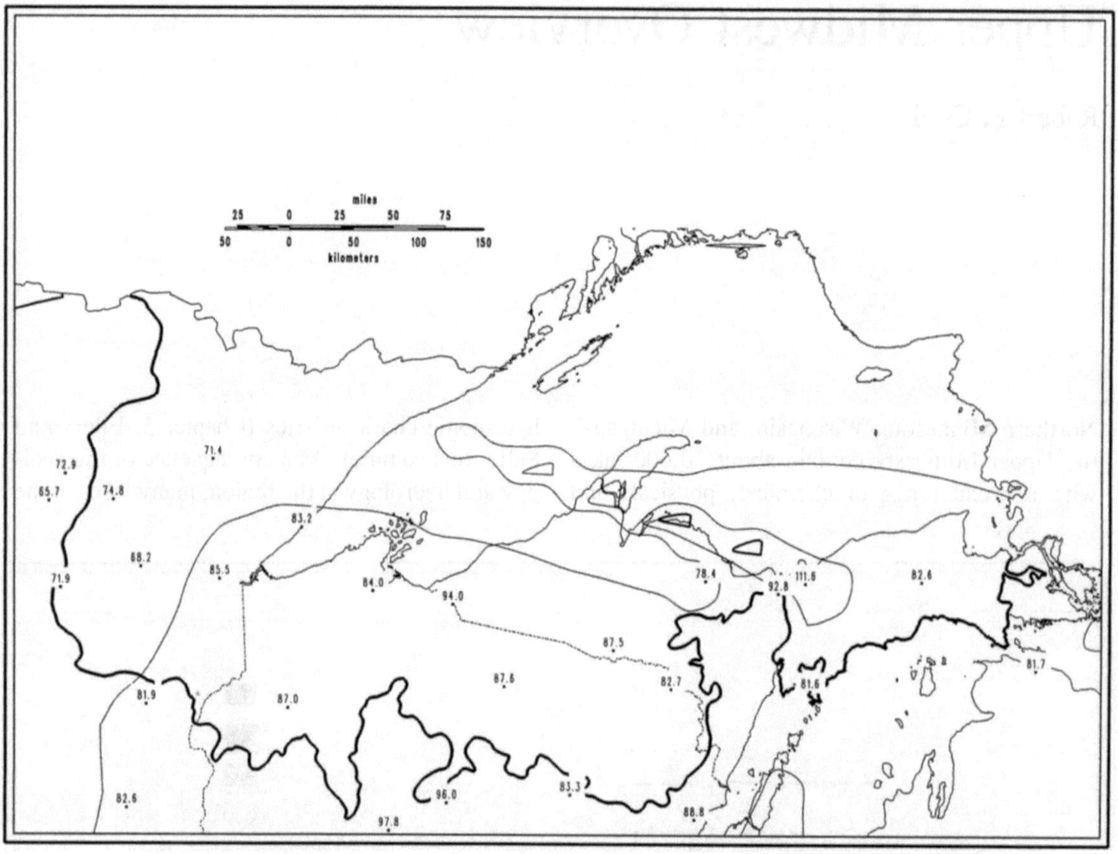

FIGURE 2. Precipitation quantity in the Upper Midwest averaged for the period 1982–84. Contours are in cm yr^{-1} at 20-cm intervals with 60-cm contours in heavier lines; the location of each station in this region that was used in preparing the contour map is indicated, along with its annual precipitation in cm yr^{-1}. Point data calculated by R. Husar; contour map prepared by T. Selle.

northernmost part of the Upper Midwest are extremely low ionic strength, low ANC systems (Figure 1). Lakes south of this area have higher ionic strength and higher ANC than in the north; these southern lakes are not susceptible to acidification caused by acidic deposition and are not treated in this section.

Where bedrock is close to the surface (northeastern Minnesota), it is Precambrian, primarily igneous, and is resistant to weathering and the production of acid neutralizing capacity. Lakes in this part of the Upper Midwest are typically interconnected drainage lakes. Thick glacial deposits in the remainder of the region are derived from this Precambrian bedrock and provide few solutes to lake water.

Spatial differences in the glacial geology, which arise because of the differences in source regions and in the period of glacial activity, provide the basis for division of the Upper Midwest into subregions (Chapter 13, Cook and Jager, this volume). The glacial deposits are dotted with lakes that receive water from precipitation and subsurface flow (seepage lakes).

Wetlands are common throughout the Upper Midwest, especially in northeastern Minnesota, and may be partially responsible for the high lakewater concentrations of dissolved organic carbon (DOC).

Precipitation quantity in the region exhibits little spatial pattern (Figure 2); evapotranspiration is greater in the western portion of the region, with

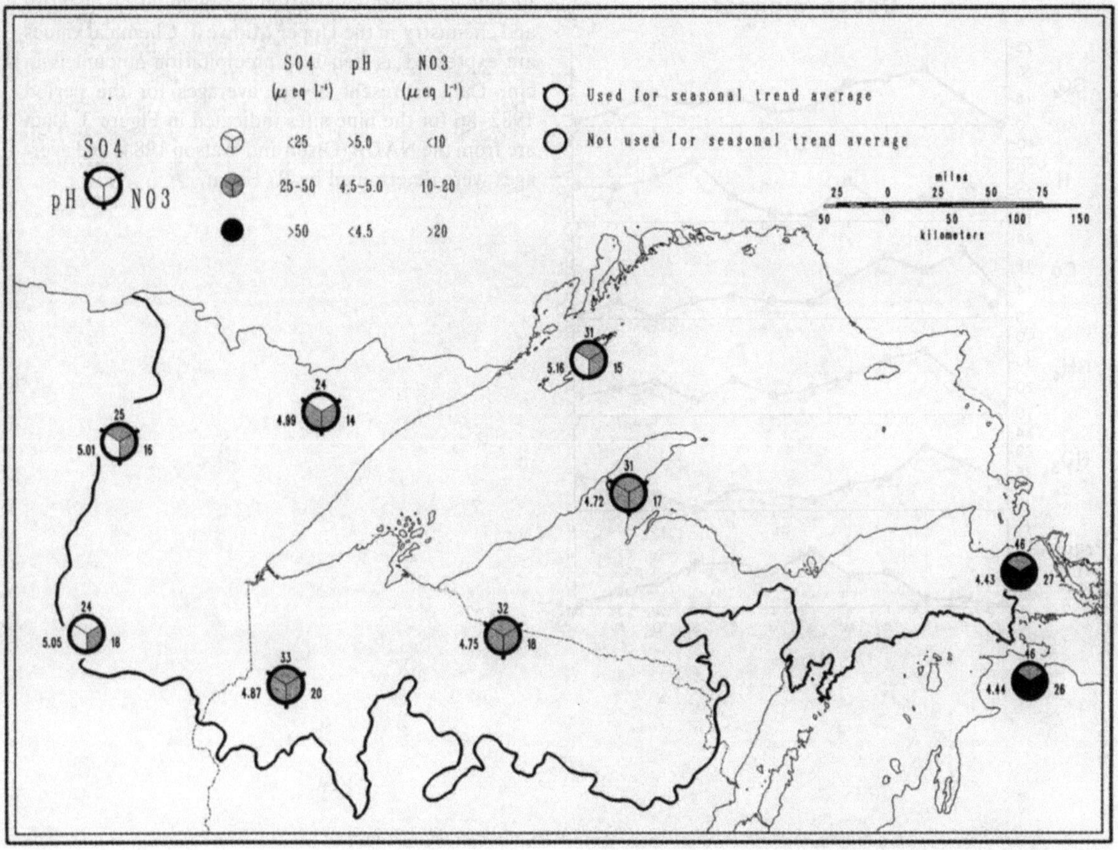

FIGURE 3. Precipitation pH and concentrations of SO_4^{2-} and NO_3^- for NADP stations in the Upper Midwest, averaged for the period 1982–84 (Olsen and Watson 1984). Averages calculated by R. Husar; map prepared by T. Selle.

the result that runoff declines in an east-to-west direction. The late spring and summer months receive the most precipitation, averaging about 1 cm each month, whereas the winter months typically receive about 0.5 cm each. Precipitation quality displays a marked zonal gradient, with lower pH and higher SO_4^{2-} and NO_3^- in the eastern part of the Upper Midwest (Figure 3). For most individual solutes, the greatest precipitation concentrations occur during the spring period, with lowest concentrations in the fall and winter (Figure 4).

This combination of edaphic and atmospheric deposition characteristics leads to the large number of low ANC lakes in the Upper Midwest (Figure 1). Lakes with ANC < 50 μeq L^{-1} are found in northeastern Minnesota and northern Wisconsin, and in the northern portion of upper Michigan. Spatial differences in lake water chemistry in the Upper Midwest are related to spatial differences in hydrologic lake type and atmospheric deposition, as discussed in detail in Cook and Jager (Chapter 13, this volume).

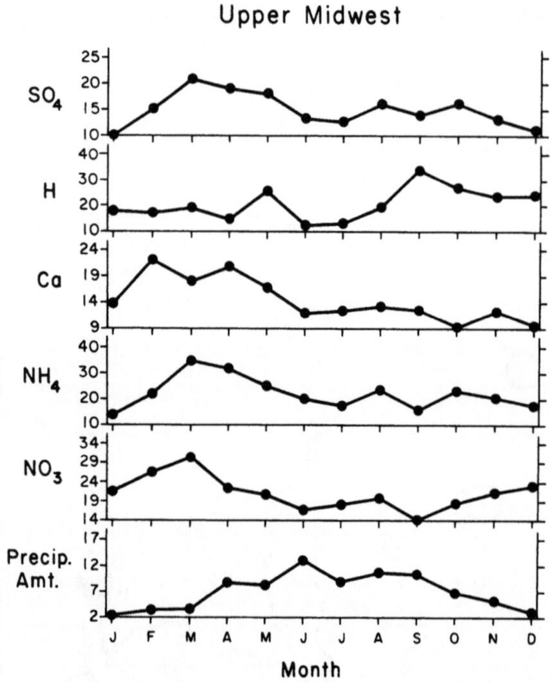

FIGURE 4. Monthly variation in precipitation quantity and chemistry in the Upper Midwest. Chemical values are expressed as $\mu eq\ L^{-1}$; precipitation amount is in cm. Data represent 5-year averages for the period 1982–86 for the nine sites indicated in Figure 3. Data are from the NADP (Olsen and Watson 1984) and averages were determined by R. Husar.

13
Upper Midwest

Robert B. Cook and Henriette I. Jager

ABSTRACT. The glaciated terrain in the Upper Midwest contains the largest density and diversity of lakes in North America outside Florida and southeastern Canada. The region receives a gradient in atmospheric deposition, with higher deposition of SO_4^{2-}, NO_3^-, and H^+ in upper Michigan. However, the rates of deposition are less than 50% of those observed in eastern North America.

The effects of acidic deposition on softwater lakes in northern Minnesota, Wisconsin, and Michigan were assessed using data collected in large-scale synoptic surveys of lakes along with data collected during intensive studies of individual lakes. Of the estimated 7,557 lakes in the Upper Midwest with ANC ≤ 400 μeq L^{-1}, 2,511 have ANC ≤ 200 μeq L^{-1}, and 128 have ANC ≤ 0 μeq L^{-1}. Of those lakes with ANC ≤ 0, 80% are seepage lakes.

Factors affecting lakewater chemistry in the Upper Midwest include atmospheric deposition, hydrologic lake type, watershed production of cations and dissolved organic matter, and watershed and in-lake retention of nitrate, ammonium, and sulfate. Key among these are base cation production and hydrologic lake type. For example, lakes with ANC ≤ 200 μeq L^{-1} exhibit a west to east decline in ANC and pH that is the direct result of the west-to-east decline in base cation concentration and the west-to-east increase in proportion of seepage lakes.

Because the rates of atmospheric deposition are lower in the Upper Midwest relative to other regions in eastern North America, the acidification of surface waters in the Upper Midwest is not as extensive as in other regions of eastern North America. Based on historical water chemistry and diatom-inferred pH, the acidification has amounted to ANC decreases of 10 μeq L^{-1} for the most sensitive lakes.

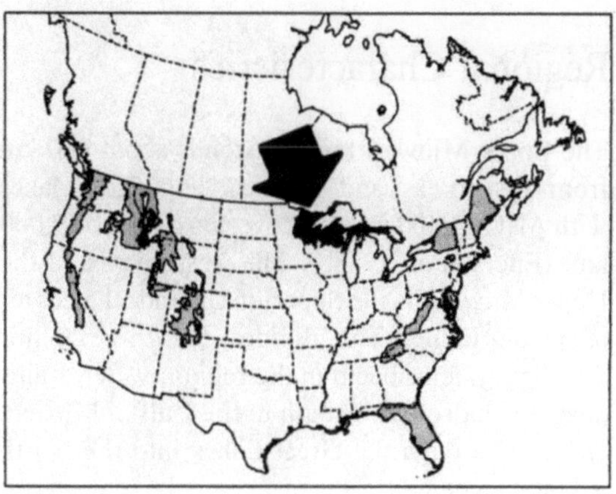

Introduction

The glaciated terrain in the Upper Midwest—northern Minnesota, Wisconsin, and Michigan—contains the largest density and diversity of lakes in North America outside Florida and southeastern Canada (Wright 1972). As a consequence of the resistance of the bedrock and glacial till in the region to weathering, many of the lakes in the Upper Midwest are softwater, have low acid neutralizing capacity (ANC), and are vulnerable to the effects of acidic deposition (Eilers et al. 1983, 1988b, Gorham et al. 1983, Rapp et al. 1985, 1987). The region receives atmospheric H^+ deposition that is 50% or less of rates observed in eastern North America (Glass and Loucks 1986; Chapter 3, Husar et al., this volume; Chapter 5, Eilers et al., this volume). However, paleolimnological studies suggest that atmospheric deposition has caused acidification of some lakes in the Upper Midwest (Kingston et al. in press). It is important to understand the biogeochemical controls of acid-base chemistry for these lakes, because future increases in acidic deposition may cause acidification of some of the low ANC lakes. In fact, modeling results suggest that future increases in acidic deposition will cause acidification of the most sensitive lakes in the Upper Midwest (Schnoor et al. 1986).

This chapter focuses on softwater lakes. Information on the effects of acidic deposition on streams in the Upper Midwest is more limited and has been reviewed by Wiener and Eilers (1987).

Regional Characteristics

The Upper Midwest region extends about 900 km from west to east and includes about 7,500 lakes with ANC ≤ 400 μeq L^{-1} covering a total of 5,000 km^2 (Eilers et al. 1988b). The western part of the Upper Midwest is the Superior Upland, the center of an anticline. The divides for three major drainage systems begin in the region: water drains north to Hudson Bay, south to the Gulf of Mexico, and east through the Great Lakes into the North Atlantic Ocean.

For the analysis presented in this chapter, the Upper Midwest has been divided into five subregions (A, B, C, D, E, F; Figure 13.1), based on physiographic differences within the Upper Midwest (Eilers and Bernert in review). This regional stratification is different from that used in previous analyses of lakes in the Upper Midwest. In the new subregions, lakewater chemistry and atmospheric deposition are more homogeneous than in the subregions used in previous studies (Eilers et al. 1988b).

Physical Characteristics of Watersheds

Surficial Geology

The physiography of the Upper Midwest reflects both the bedrock topography and the late Pleistocene glaciation (Figure 5.1, Chapter 5, Eilers and Selle, this volume). In northern Minnesota and in most of northern Wisconsin, the bedrock is the southern extension of the Precambrian Shield. In northern Minnesota, the main bedrock types are gabbro and granite, with some small areas of economically important deposits of iron and sulfide minerals in the mafic bedrock (Sims and Morey 1972). In most of northern Wisconsin, the bedrock is mafic, intermediate, and felsic igneous rocks (Mudrey et al. 1982). The contact between the Precambrian shield and Paleozoic sedimentary bedrock extends northward from the area near Green Bay through northwestern Wisconsin and into upper Michigan. Upper Michigan is underlain by Cambrian sandstone in the north and by Silurian and Ordovician carbonate rocks in the south (Rapp et al. 1985).

The entire Upper Midwest is mantled by Wisconsin age glacial features: lake-dotted moraines, outwash plains, till, and drumlin fields (Wright 1972). Glacial lobes entering northern Minnesota from the northeast brought noncalcareous sandy till, and lobes entering from the northwest brought calcareous clayey till. Near the Canadian border in northeastern Minnesota, the late Wisconsin age glaciation scoured the area leaving thin (≤ 5 m) till. Lake basins in this area were excavated in weak zones (faults, dikes) of the underlying bedrock, which causes the lakes to be elongated in an east-west direction (Wright 1972).

South of the Minnesota-Canada border region and into northern Wisconsin the glacial drift is thicker, commonly ranging from 30 to 190 m in thickness, and lake basins are kettles or in the pitted outwash plains, ground moraines, end moraines, and drumlins. The morphometry of lakes in till is highly variable, ranging from perfectly round to irregular shapes with complicated bathymetry (Figure 13.2; Hutchinson 1957). The central and eastern part of the Upper Midwest was transected by at least six distinct glacial lobes in the late Pleistocene (Fenneman 1938, Simkins 1978, Gorham et al. 1984). The general direction of glacial movement was from north-northeast to south-southwest. Differences in the till deposited by these lobes provide the basis for dividing the Upper Midwest into subregions (Figure 13.1; Eilers and Bernert in review). The Green Bay lobe was the only one that contained appreciable amounts of carbonate material, derived from bedrock in the southern half of upper Michigan (Simkins 1978).

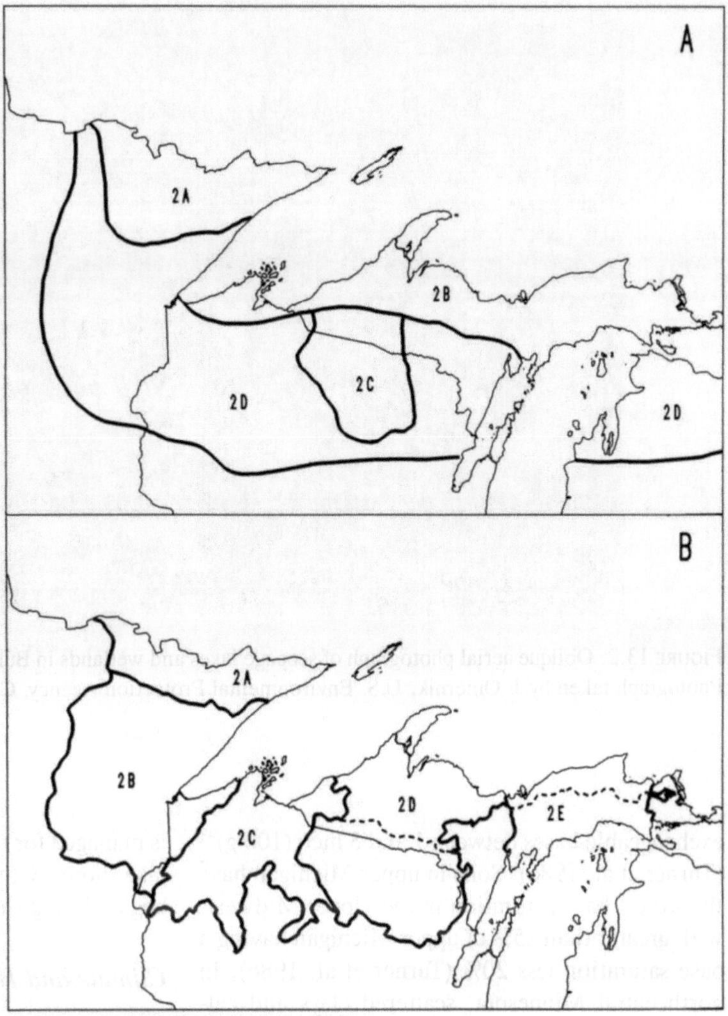

FIGURE 13.1. Subregions in the Upper Midwest. (a) Original Eastern Lake Survey subregions (Eilers et al. 1988b). (b) Subregions based on physiography (Eilers and Bernert in review) that will be used in the remainder of this chapter. In the new division of the Upper Midwest, lakes in the lower peninsula of Michigan are not incorporated.

In upper Michigan, deposits of glacial lacustrine sediments occur near the Lake Superior shoreline and quartz-rich glacial drift up to 35 m thick occurs in the central portion of the peninsula (Rapp et al. 1987). The till in the southern part of upper Michigan is calcareous and thin ($\leq$ 5 m). Outcrops of bedrock are more frequent in the western portion of upper Michigan and are commonplace near the Keweenaw Peninsula and in the southern upper Michigan along the Lake Michigan and Lake Huron shoreline.

Bog and muskeg deposits are scattered throughout eastern upper Michigan, northern Wisconsin, and, particularly, northcentral Minnesota (Hunt 1986).

Soils

The soils in the Upper Midwest, which derive primarily from the sandy glacial drift, are poorly developed and thin. The predominant suborders are orthods and boralfs in upland areas, and saprists and aqualfs in low lying areas (U.S. Department of Agriculture 1981). Within the entire Upper Midwest, between 50% and 70% of the forested area has soils with cation exchange capacity < 10 meq $(100\ g)^{-1}$ and base saturation is 20% to 60% for over 50% of forested soils (Turner et al. 1986). Greater than 50% of the area in upper Michigan has exchangeable bases between 1 and 3 meq $(100\ g)^{-1}$, and 30% to 50% of the remainder of the Upper Midwest has

FIGURE 13.2. Oblique aerial photograph of seepage lakes and wetlands in Burnett County, northwestern Wisconsin. Photograph taken by J. Omernik, U.S. Environmental Protection Agency, Corvallis, Oregon.

exchangeable bases between 1 and 3 meq $(100\ g)^{-1}$ (Turner et al. 1986). Soils in upper Michigan have the lowest base saturation in the Upper Midwest, with greater than 25% of upper Michigan having a base saturation less 20% (Turner et al. 1986). In northcentral Minnesota, scattered clays and calcareous till from glacial Lake Agassiz may cause locally high base saturation content and alkaline drainage water (Wright 1972).

Vegetation and Land Use

Much of the Upper Midwest contains second-growth mixed deciduous-conifer forest. The area was extensively logged of white pine (*Pinus strobus*) and to a lesser extent red pine (*P. resinosa*), hemlock (*Tsuga canadensis*), and hardwoods in the late 1800s and early 1900s. In addition, fires occurred throughout the region during and following logging until the implementation of fire control programs in the 1940s. There are many national forests and parks in the region, for which human influence during the last several decades is minimal. Much of the privately owned forest land is managed for timber production. Development of lake shores with seasonal and year-long residences began during the past 50 years.

Climate and Hydrology

The area has a continental climate, with average annual temperatures of 2°C to 8°C (U.S. Department of Agriculture 1981). Annual precipitation averages from 57 cm in northcentral Minnesota to 75 cm in eastern upper Michigan (U.S. Department of Agriculture 1981), with 20% to 40% of the annual precipitation falling as snow. In recent years (1970–82), the coefficient of variation for the amount of precipitation was 20% for a station in northwestern Ontario (Linsey et al. 1987). The distribution of precipitation through three water years at Vandercook Lake, northcentral Wisconsin, is presented in Figure 13.3 (Wentz and Rose 1989). The summer months have the largest precipitation amounts and the winter months the smallest amount.

Evapotranspiration, estimated from regional precipitation and regional runoff (Geraghty et al.

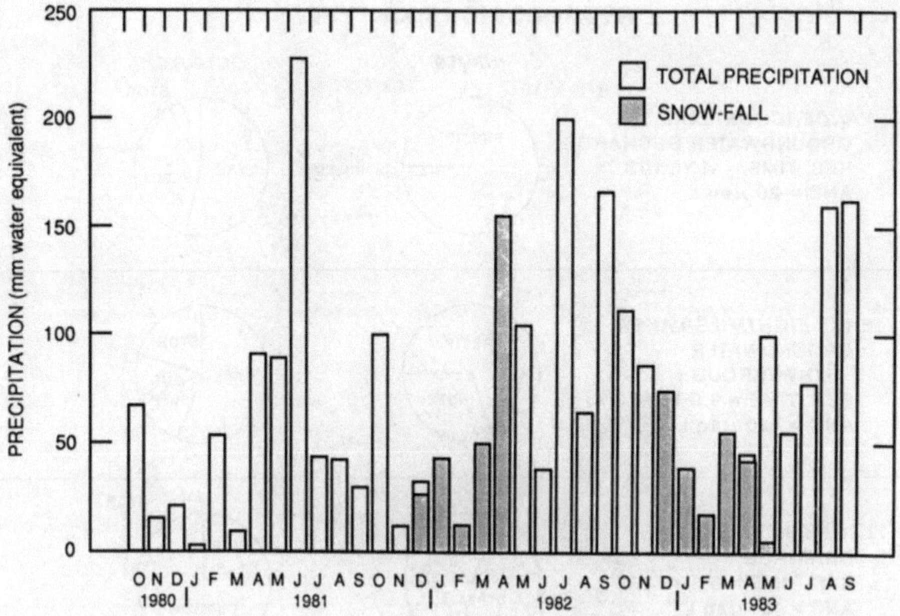

FIGURE 13.3. Total precipitation at Vandercook Lake during water years 1981–83 (redrawn from Wentz and Rose 1989, with additional data from K. Webster, Wisconsin Department of Natural Resources, Fitchburg, Wisconsin, pers. comm.). Values presented in water year 1981 were estimated based on regression from nearby stations; water equivalents of snow are not indicated for this water year. Reproduced with permission.

1973), averages between 31 cm and 49 cm, with higher evapotranspiration in the west. The estimate of actual evaporation from open-water surfaces (estimated from Class A pans) is about 60 cm each year in the region (Geraghty et al. 1973). As a result of the gradients in precipitation and evaporation, annual runoff exhibits a spatial gradient from 46 cm yr^{-1} in eastern upper Michigan to 20 cm yr^{-1} in northcentral Minnesota (Gebert et al. 1987). However, northern Wisconsin and adjacent portions of upper Michigan have little spatial variation and an estimated runoff between 30 cm yr^{-1} and 38 cm $year^{-1}$ (Gebert et al. 1987). In areas of thin till, runoff usually is maximum during the snowmelt period and is low during the remainder of the year (Schindler et al. 1976, Siegel 1981). In areas where the till is thick, snowmelt typically enters the soil to become part of interflow and the local and regional groundwater, rather than direct runoff (Kratz et al. 1987, Wentz and Rose 1989).

Lakes in the Upper Midwest are ice-covered from approximately late November to late April. As a consequence of warm air temperatures when ice-off occurs, thermal stratification is established quickly and spring overturn is sometimes incomplete. Fall overturn typically results in complete mixing of the water column (Schindler 1971, Cook and Schindler 1983).

Hydrologic and Physical Characteristics of Lakes

Surficial deposits govern the physical characteristics and the water balance for lakes in the Upper Midwest. Where Pleistocene deposits are thin, along the Minnesota-Canada border, southern part of upper Michigan, lakes are commonly drainage lakes, in which nonevaporative loss of water occurs by flow from an outlet (Wetzel 1975; Chapter 1, Munson and Gherini, this volume). Lakes in outwash plains or in kettles are commonly seepage lakes that receive water primarily from precipitation falling directly on the lake surface and lose water through evaporation from the lake surface and through groundwater recharge from the littoral zone (Wetzel 1975, Winter 1977, Born et al. 1979, Anderson and Bowser 1986; Chapter 1, Munson and Gherini, this volume). The characteristic

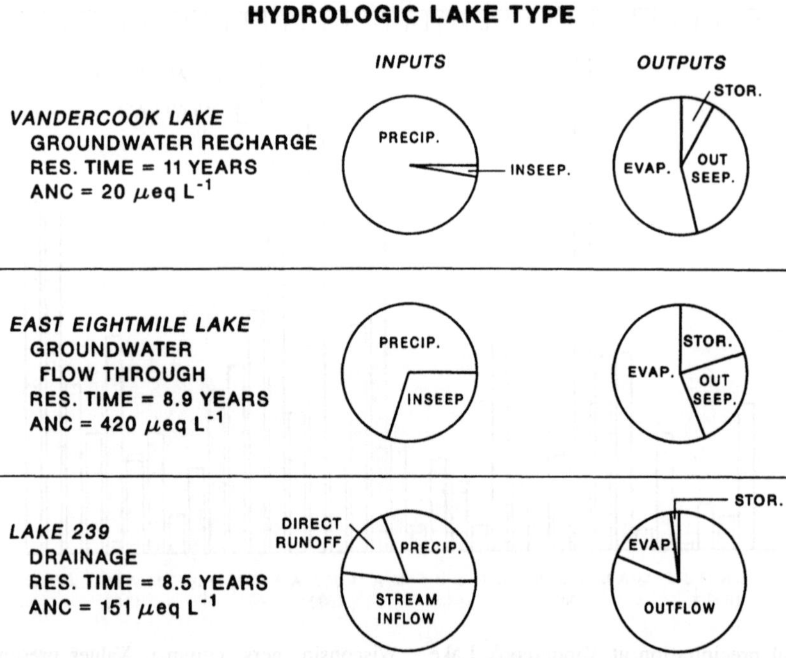

FIGURE 13.4. Relationship between hydrologic lake type, hydrologic budgets, and ANC for Vandercook Lake, northcentral Wisconsin (Lin and Schnoor 1986), East Eightmile Lake, northwestern Wisconsin (Wentz et al. 1987) and Lake 239, northwestern Ontario (Schindler et al. 1976, Schindler 1986). Change in storage is represented as an output so that the sum of all outputs equals the total inputs for the budget years.

morphometry of seepage lakes and low relief of the Upper Midwest is shown in Figure 13.2. In this chapter, seepage lakes are further classified into groundwater recharge lakes (i.e., outseepage from the lake replenishes the groundwater) with lakewater silica < 1 mg L^{-1}, and groundwater flow-through lakes with lakewater silica ≥ 1 mg L^{-1} (Chapter 1, Munson and Gherini, this volume).

To illustrate the differences in water balances for drainage and seepage lakes, annual water inputs, outputs, and change in storage are presented in Figure 13.4 for Vandercook Lake, a groundwater recharge lake in northcentral Wisconsin (Lin and Schnoor 1986), East Eightmile Lake, a groundwater flow-through lake in northwestern Wisconsin (Wentz et al. 1987), and Lake 239, a drainage lake in northwestern Ontario about 300 km north of the Minnesota border (Schindler et al. 1976). Seepage inflow in Vandercook Lake accounts for approximately 3% of the water input, with precipitation directly on the lake surface accounting for the balance. This value for groundwater input is based on both hydrologic and chemical data, and is considerably smaller than the estimate of 16.3% made by Wentz and Rose (1989) based solely on hydrologic information. Vandercook Lake can have fluctuations in lake volume, corresponding to changes in lake stage of 0.25 m, resulting primarily from storm inputs of precipitation and evaporation during dry periods (Schnoor et al. 1986). Evaporation (54%) and groundwater recharge (38%) are the primary means of water loss for Vandercook (Lin and Schnoor 1986, Wentz and Rose 1989). Hydrologic budgets for two groundwater recharge lakes in the Upper Midwest (Crum Lake, Minnesota, and Crystal Lake, Wisconsin) are presented in Chapter 2 (Munson and Gherini, this volume). East Eightmile Lake has a larger inseepage (30%) than does Vandercook, resulting in a shorter residence time (8.9 vs. 11 years) and lakewater chemistry more influenced by groundwater chemistry than that of Vandercook. Lake 239 receives 70% of its water from stream inflow and direct runoff and loses 80% of its water via stream outflow. Lakewater chemistry, particularly ANC, for drainage lakes and seepage lakes is greatly

TABLE 13.1. Estimated percentage of lakes by hydrologic type in the Upper Midwest.

Subregion	Subpopulation[a]	n^b	N^b	Lake type (%) Drainage	Seepage	Closed	Reservoir
NE Minnesota	All	136	1302	80	10	8	1
(2A)	ANC ≤ 200	81	591	73	11	13	3
WC Minnesota	All	41	1810	53	40	1	5
(2B)	ANC ≤ 200	14	310	8	93	0	0
NW Wisconsin	All	117	1520	32	53	15	0
(2C)	ANC ≤ 200	73	570	18	78	4	0
NC Wisconsin &	All	218	2388	39	55	3	3
NW Michigan	ANC ≤ 200	116	848	26	70	4	0
(2D)							
NE Michigan	All	62	539	54	38	7	0
(2E)	ANC ≤ 200	33	192	17	71	12	0
Upper Midwest	All	573	7557	49	42	6	2
	ANC ≤ 200	317	2511	32	61	6	1

[a] All is the subpopulation of lakes in the Upper Midwest with surface area less than 2,000 ha. ANC ≤ 200 is the subpopulation of lakes with ANC ≤ 200 μeq L^{-1}; Cl^- ≤ 20 μeq L^{-1}; total P ≤ 30 μg L^{-1}; and surface area ≤ 2000 ha.
[b] n is the number of sampled lakes in the subpopulation and N is the estimated total number of lakes in the subpopulation.

influenced by the sources of water (Anderson and Bowser 1986; Chapter 2, Munson and Gherini, this volume; also see section entitled Processes Influencing Surface Water Chemistry).

The hydrologic lake type for all lakes with a surface area between 4 and 2,000 ha in the Upper Midwest is evenly divided between seepage and drainage lakes, although the distribution of hydrologic type varies among subregions. From Eastern Lake Survey–Phase I (ELS-I) data, 49% of the lakes in the Upper Midwest were estimated to be drainage lakes, and 42% of the lakes were estimated to be seepage lakes (Table 13.1). In northeastern Minnesota (2A) where the glacial deposits are thin (see section entitled Hydrologic and Physical Characteristics of Lakes), 80% of the lakes were classified as drainage lakes. In the pitted outwash plain of northwestern and northcentral Wisconsin (2C and 2D), 53% and 55%, respectively, of the lakes were estimated to be seepage lakes (Table 13.1).

In the Upper Midwest, seepage lakes tend to have lower ANC than drainage lakes. Here, 80% of the 128 lakes with ANC ≤ 0 μeq L^{-1} are seepage lakes. Of the subset of lakes in the Upper Midwest with ANC ≤ 200 μeq L^{-1} that are not influenced by human activity (Appendix A, Baker, this volume), as determined by Cl^- (≤ 10 μeq L^{-1}) and total (dissolved and particulate) phosphorus concentrations (≤ 30 μg L^{-1}), 61% are seepage lakes and 32% are drainage lakes. Because of the restrictions on ANC and Cl^-, these seepage lakes are predominantly groundwater recharge lakes. In subregions 2B through 2E, 70% to 93% of the lakes with ANC ≤ 200 μeq L^{-1} are seepage lakes, whereas in northeastern Minnesota (2A), 73% of the lakes with ANC ≤ 200 μeq L^{-1} are drainage lakes. The spatial distribution of drainage and seepage lakes with ANC ≤ 200 μeq L^{-1} in the Upper Midwest is shown in Figure 13.5.

Lakes in northeastern Minnesota with ANC ≤ 200 μeq L^{-1} are larger and somewhat shallower and have larger watersheds than lakes in the other subregions in the Upper Midwest (Table 13.2). The watershed areas of lakes in northcentral Wisconsin (2C) and upper Michigan (2E), as determined by watershed topography, are smaller than those in northeastern Minnesota. However, watershed area determined from topographic divides may have little correspondence with the terrestrial contributing area of a seepage lake, which receives no water from the surface and small amounts of water from the regional and local groundwater.

Atmospheric Deposition

Wet Deposition of pH, Sulfate, and Nitrate

The region receives a spatial gradient of acidic and ionic wet atmospheric deposition (Glass and

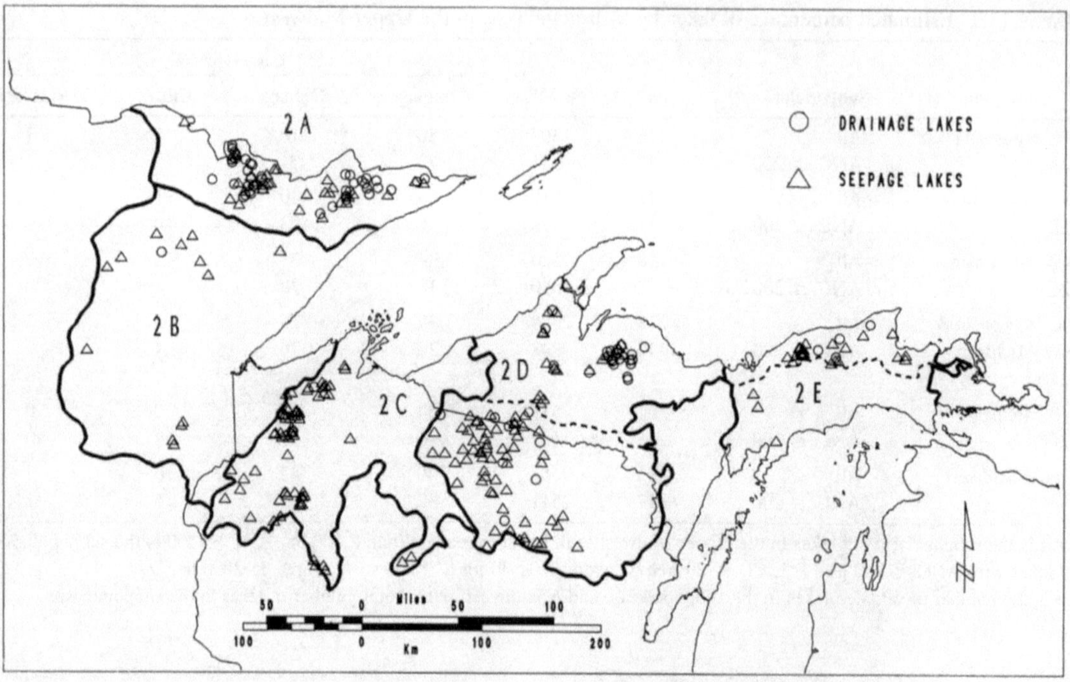

FIGURE 13.5. Location of drainage and seepage lakes sampled in the Upper Midwest as part of the Eastern Lake Survey. The plotted lakes have ANC $\leq$ 200 µeq L^{-1}, surface area $\leq$ 2,000 ha, Cl^- $\leq$ 20 µeq L^{-1}, and total phosphorus $\leq$ 30 µg L^{-1}.

Loucks 1986; Upper Midwest Overview). Precipitation has an annual average pH level of 4.4 in upper Michigan, 4.8 in northern Wisconsin, and 5.0 in northern Minnesota (Table 13.3). Sulfate in precipitation exhibits a similar spatial gradient, from 46 µeq L^{-1} in upper Michigan to 24 µeq L^{-1} in Minnesota (Table 13.3; Munger and Eisenreich 1983, Glass and Loucks 1986). The concentration and deposition gradients of SO_4^{2-} arise in part directly from anthropogenic emissions from eastern North America (Galloway et al. 1984) and also indirectly from re-emission of sulfur from wetlands, some fraction of which is of anthropogenic origin (Nriagu et al. 1987); soil dust in the western portions of the Upper Midwest may also contribute SO_4^{2-} to these wet deposition values (Munger 1982, Munger and Eisenreich 1983, Verry and Harris 1988). The poor correlation observed between H^+ and SO_4^{2-} in the Upper Midwest for all wet precipitation data has been attributed to soil sources of neutral SO_4^{2-} salts (Sievering 1987).

Back-trajectory analysis for particulate and gaseous sulfur measured in precipitation at the Experimental Lakes Area, northwestern Ontario, suggests that the source region for these forms of sulfur was the lower Great Lakes region (Barrie et al. 1980). In northern Wisconsin during 1982–84, an estimated 60% to 75% of the SO_4^{2-} in precipitation in 30 storms originated outside the state (Wisconsin Department of Natural Resources unpublished data).

NO_3^- in precipitation exhibits a slight decreasing gradient from east to west from 26 µeq L^{-1} to 14 µeq L^{-1} (Table 13.3). The equivalence ratio of SO_4^{2-}:NO_3^- in wet deposition indicates that SO_4^{2-} provides about twice as many equivalents as NO_3^- on an annual basis (Table 13.3).

Wet Deposition of Base Cations and Ammonium

The concentrations of ammonium and base cations in precipitation are relatively uniform across the study region (Glass and Loucks 1986). Prairie dust appears to be a major contributor of base cations in precipitation throughout the Upper Midwest, causing high concentrations of base cations in wet

TABLE 13.2. Median physical characteristics of lakes in the Upper Midwest.

Subregion	Subpopulation[a]	Lake area (ha)	Watershed area (ha)	Watershed/ lake ratio	Site depth[b] (m)	Residence time[c] (years)
NE Minnesota	All	19	222	8.9	5.2	3.7
(2A)	ANC ≤ 200	15	169	8.4	4.8	0.8
WC Minnesota	All	24	376	12.0	6.6	–
(2B)	ANC ≤ 200	10	73	7.8	2.8	–
NW Wisconsin	All	12	105	6.5	5.6	–
(2C)	ANC ≤ 200	8	48	4.6	6.8	–
NE Wisconsin &	All	14	162	7.1	5.4	–
NW Michigan (2D)	ANC ≤ 200	9	65	6.3	5.9	0.8
NE Michigan (2E)	All	10	92	8.4	1.9	–
	ANC ≤ 200	6	45	4.5	3.5	–
Upper Midwest	All	16	162	8.5	5.6	0.6
	ANC ≤ 200	10	71	6.8	5.5	6.2

[a] The lake populations are either the entire population of lakes in the region or the subpopulation of lakes that have ANC ≤ 200 μeq L^{-1}; Cl$^-$ ≤ 20 μeq L^{-1}; surface area < 2,000 ha; and total P ≤ 30 μg L^{-1}.
[b] Site depth is the depth at the location of sample collection and is an estimate of the maximum depth of the lake.
[c] Residence time is determined from estimates of regional runoff, watershed area, and lake volume. This residence time is calculated with respect to inflows to the lake, and thus should not be compared to outflow-based residence times.

deposition (Munger 1982, Thornton and Eisenreich 1982, Linsey et al. 1987). The seasonal pattern of base cation concentrations in precipitation follows the pattern of agricultural activity, with maximum concentrations occurring in the spring and autumn periods of tilling and minimum concentrations occurring when the ground is frozen or snow covered (Verry 1983, Gorham et al. 1984, Linsey et al. 1987). During times when concentrations of soil-derived base cations are high (e.g., during periods of soil tilling), H^+ is not significantly correlated with SO_4^{2-} and NO_3^- (Verry and Harris 1988). During these periods, SO_4^{2-} is strongly related to base cation concentrations (Sievering 1987). In contrast, when soil-derived base cation concentrations are low, H^+ varies with the sum of NO_3^- and SO_4^{2-} (Verry and Harris 1988). Although the proportions of base cations, NH_4^+, SO_4^{2-}, and NO_3^- control the value of pH in precipitation (Thornton and Eisenreich 1982, Verry 1983), in general, annual concentrations of base cations and NH_4^+ in wet deposition have little influence on the regional precipitation pH gradient, because of the uniformity in these cations in the Upper Midwest.

Dry Deposition

Estimates of the dry deposition of major solutes are few and the uncertainties associated with these estimates are considerable. The estimates given here should be viewed with an appreciation of how difficult dry deposition is to measure on a regional basis and also with the understanding that the

TABLE 13.3. Volume-weighted annual average precipitation chemistry for the study area for the period 1982–84.[a]

Study area	Annual precipitation	H^+	NH_4^+	Ca^{2+}	Mg^{2+}	Na^+	K^+	SO_4^{2-}	Cl^-	NO_3^-
Fernberg, Minnesota	83	10	14	12	4	3	1	24	2	14
Spooner, Wisconsin	91	14	23	14	4	4	1	32	4	20
Trout Lake, Wisconsin	92	14	23	14	4	2	1	32	2	18
Douglas Lake, Michigan	94	36	18	10	4	2	1	46	3	26

[a] NADP data with volume-weighted annual means computed by R.W. Husar. Concentrations given in μeq L^{-1}. Precipitation is in cm.

values given are highly uncertain. Furthermore, estimates of dry deposition of gases or of particles to water surfaces, which are important for the chemistry of seepage lakes, are even more poorly known than estimates of dry deposition to terrestrial systems (L.A. Baker pers. comm).

Measurements of dry deposition in lower Michigan, near Pellston, during a 5-month period in late winter (1982) showed that dry deposition of H^+, SO_4^{2-}, and NO_3^- was less than 14% of the wet deposition (Cadle et al. 1984a,b). Dry deposition of base cations and ammonium ranged from 17% to 35% of the wet deposition. These measurements indicate that during the winter months, dry deposition was not a major input, which is in agreement with the findings of Munger (1982) and Thornton and Eisenreich (1982).

Barrie and Sirios (1986) determined, on the basis of daily air and precipitation measurements, that dry deposition of sulfur was 26% of total deposition in northwestern Ontario and 16% of total deposition in northern Ontario (Algoma) near eastern upper Michigan. Dry deposition of nitrogen species was between 22% and 44% of total deposition in northwestern Ontario and 28% to 36% near Algoma; the range is caused by uncertainty in the contribution of NO_2 to dry deposition. Baker (Appendix B, this volume) has estimated average dry deposition in the Upper Midwest from NADP dry bucket data and air concentrations to be 27% of total sulfur deposition and 23% of total NO_3^- deposition. Baker (Appendix B, this volume) estimated the dry deposition of other major ions using only NADP dry bucket data. Values Baker obtained were: dry Ca^{2+} deposition was 35% of total Ca^{2+} deposition, dry Mg^{2+} deposition was 40% of total Mg^{2+} deposition, dry K^+ deposition was 60% of total K^+ deposition, dry Na^+ deposition was 20% of total Na^+ deposition, dry Cl^- deposition was 16% of total Cl^- deposition, and dry NH_4^+ was 16% of total NH_4^+ deposition.

Characteristics of Lakewater Chemistry in the Upper Midwest

This section presents regional lakewater chemistry data sets, describes the acid-base chemistry of the lakes, and examines the relationships among water quality parameters. The biogeochemical processes responsible for lakewater chemistry in the Upper Midwest are discussed in the section entitled Processes Influencing Surface Water Chemistry.

Distribution of Data Sets in the Region

The areas of interest in Minnesota, Wisconsin, and Michigan are defined as those regions expected to contain the most lakes with ANC $\leq$ 400 μeq L^{-1} (Omernik and Griffith 1986); the boundaries of the Upper Midwest region in the ELS-I are based on this criterion (Figure 13.1). Other high quality lakewater chemistry data sets for lakes located within the ELS-I boundaries were also obtained to provide a more thorough spatial coverage for the Upper Midwest; these other data bases are from earlier work by the U.S. Environmental Protection Agency (EPA) and from the U.S. Forest Service (USFS).

The ELS-I was designed using a statistical approach to estimate characteristics and total numbers of acidic and low ANC lakes (Eilers et al. 1988b; Introduction, Charles, this volume). Lakes were selected using three stratification factors: region, subregion, and alkalinity map class. A total of 625 lakes (573 lakes in the probability sampling and 52 special interest lakes) in the Upper Midwest were sampled in October and November 1984 and analyzed for major solutes and physical attributes (Tables 13.1, 13.2, and 13.4).

In 1984, the USFS sampled 73 low ANC lakes – from 9 to 13 lakes in each of the 7 national forests in northern Minnesota, northern Wisconsin, and northern and central Michigan (Nichols and McRoberts 1986). Lakes in these national forests were not sampled in the ELS-I or the earlier EPA surveys. To obtain better spatial coverage, USFS data are included in some of the analyses that follow. Characteristics for the lakes sampled by the USFS are presented in Table 13.4.

During late 1970s and early 1980s, the EPA commissioned a broad-scale regional survey of low ANC lakes in the Upper Midwest. This survey was not statistically designed to estimate the total population of lakes in the Upper Midwest, yet the survey provides additional lakewater chemistry data characterizing the lakes in the region. The data set includes a total of 523 lakes – 214 in northeastern Minnesota (Rapp et al. 1985), 270 in north-

TABLE 13.4. Median population characteristics of lakes in the Upper Midwest sampled in recent surveys.[a]

	n	N	pH	ANC (μeq L^{-1})	SO_4^{2-} (μeq L^{-1})	Ca^{2+} (μeq L^{-1})	DOC (μmol L^{-1})
ELS-I sampled lakes	625	–	6.68 (6.4–7.15)	108 (36–344)	60 (45–78)	102 (62–234)	533 (349–816)
ELS-I population estimate, all lakes	573	7,557	7.02 (6.58–7.52)	295 (90–928)	53 (36–74)	214 (87–606)	632 (416–966)
ELS-I population estimate, lakes with ANC ≤ 200	317	2,511	6.36 (6.00–6.74)	69 (28–108)	60 (45–75)	74 (49–102)	558 (383–833)
USFS	73	–	7.28 (6.46–7.71)	234 (42–636)	78 (59–100)	195 (71–451)	NA NA
USEPA	595	–	7.01 (6.45–7.44)	187 (65–444)	73 (61–90)	168 (87–357)	487 (306–726)

[a] Values in parentheses are the 25th and 75th percentiles. n = number of lakes sampled. N = population estimate.

central Wisconsin (Eilers et al. 1983), and 39 in upper Michigan (Rapp et al. 1987). Chemical characteristics of the lakes in this data base are presented in Table 13.4. Only 35 of the lakes sampled in this early EPA study were also sampled in the ELS-I probability sampling and 38 of the lakes sampled by the EPA were sampled as special interest lakes during ELS-I. Comparisons of (1) the 73 lakes sampled in common between ELS-I and the earlier EPA survey and (2) the distribution of water quality parameters (Table 13.4) revealed that both data sets yield essentially the same information concerning water quality. Because of the similarity between the ELS-I and the earlier EPA data sets, as well as the large number of samples contained in both, data analysis in this chapter is primarily restricted to ELS-I and USFS data. Furthermore, the ELS-I sampling was population based, in contrast to the earlier EPA survey, which was nonrandom.

Regional Maps of Acid-Base Chemistry

The pH and ANC of lakes in the ELS-I and USFS data sets are mapped in Figures 13.6 and 13.7. These maps show the high density of lakes in the Upper Midwest and the large number of low ANC lakes, particularly in subregions 2A and 2D. Furthermore, a range of lakewater ANC and pH are found in each of the five subregions. Examination of lakes with ANC ≤ 200 μeq L^{-1} (Figure 13.8) illustrates the east-west gradient in lakewater ANC. A few lakes in subregions 2A, 2B, and 2C have ANC values between 0 and 50 μeq L^{-1}; none with ANC < 0 μeq L^{-1} was sampled. In subregions 2D and 2E, lakes with ANC < 0 μeq L^{-1} were present and lakes with ANC between 0 and 50 μeq L^{-1} were more abundant (Figure 13.8).

Physical and Chemical Characteristics of Lakes

The ELS-I data provide a means of estimating characteristics of the entire target population of lakes in the Upper Midwest. Lakes in the Upper Midwest have an estimated median surface area of 16 ha, a median sampling depth of 5.6 m, and a median watershed area:lake area ratio of 8.5 (Figure 13.9, Table 13.2). For the entire population of lakes in the Upper Midwest, median pH is 7.02 and median ANC is 295 μeq L^{-1} (Figure 13.9, Table 13.4). Concentrations of SO_4^{2-} are relatively uniform across the Upper Midwest, with a median value of 53 μeq L^{-1}. The sum of base cations (SBC) has a skewed distribution with only an estimated 47% of the lakes having values < 400 μeq L^{-1}. DOC concentrations in the Upper Midwest have a median value of 632 μmol L^{-1} (7.6 mg L^{-1}). Silica concentration exhibits a wide range, yet most of the lakes have concentrations < 50 μmol L^{-1} (3 mg L^{-1}).

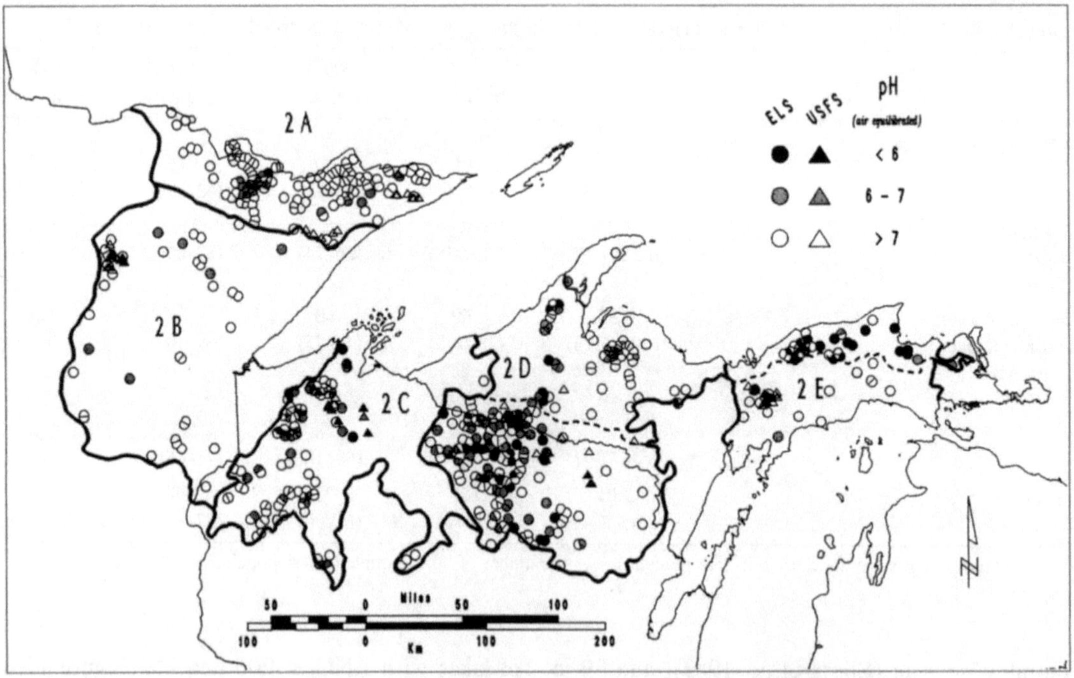

FIGURE 13.6. Location of lakes sampled in Eastern Lake Survey and in the U.S. Forest Service survey, coded by pH measurements. No subsetting was performed on the survey data presented in this plot.

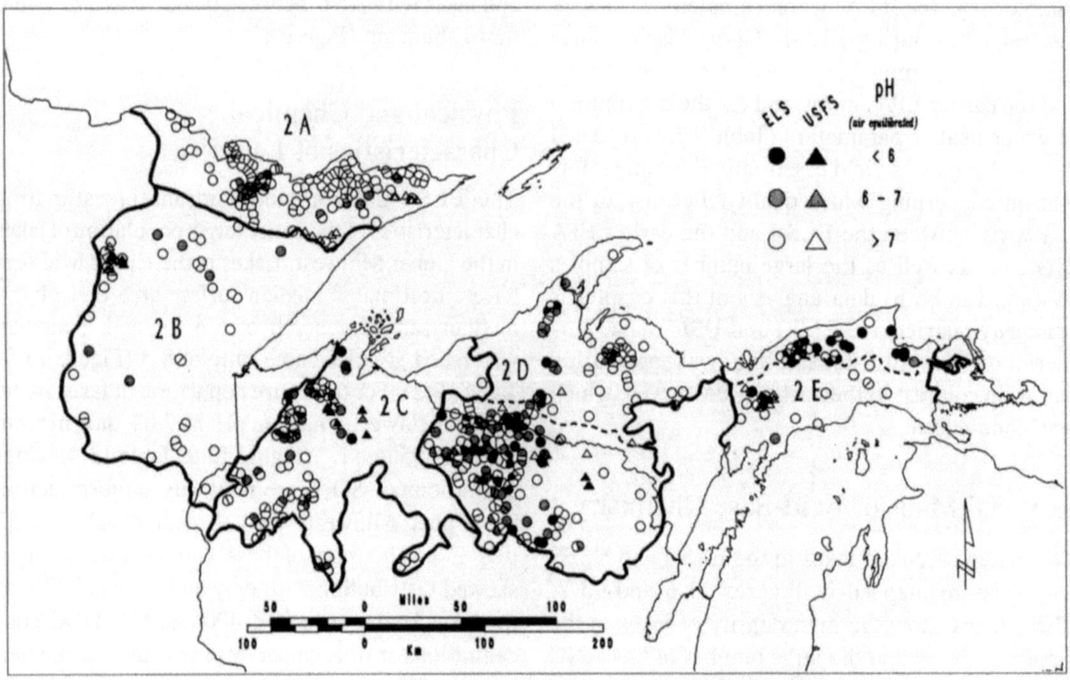

FIGURE 13.7. Location of lakes sampled in Eastern Lake Survey and in the U.S. Forest Service survey, coded by ANC measurements. No subsetting was performed on the survey data presented in this plot.

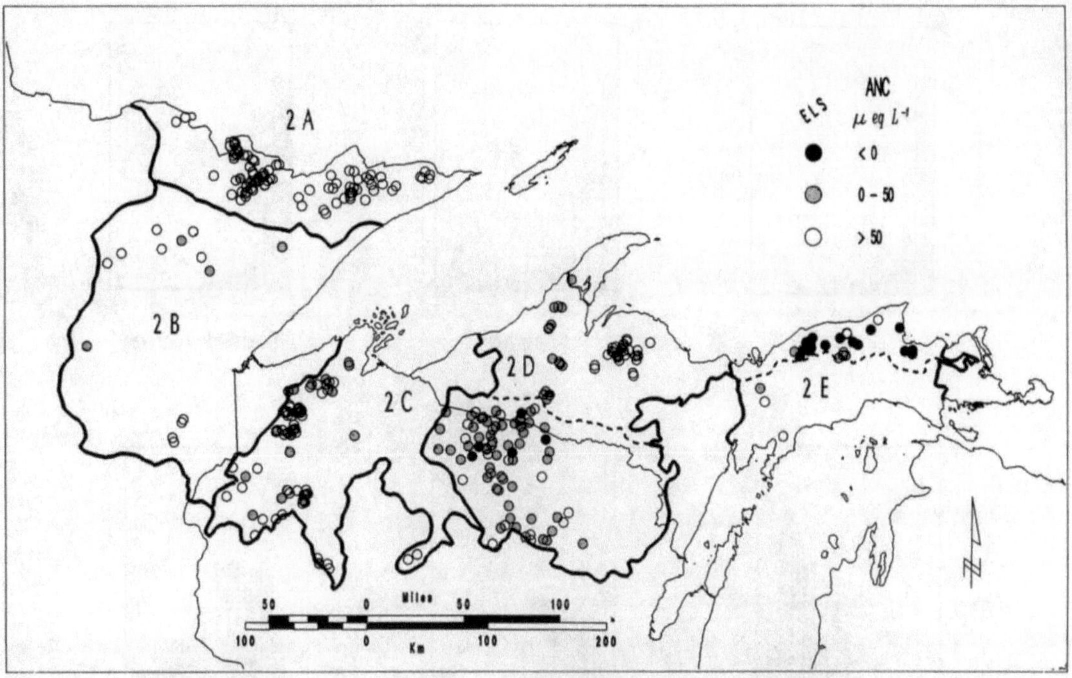

FIGURE 13.8. Map with subregion boundaries in the Upper Midwest, showing location of lakes in the Eastern Lake Survey with ANC $\leq$ 200 μeq L^{-1}, surface area $\leq$ 2,000 ha, Cl^- $\leq$ 20 μeq L^{-1}, and total phosphorus $\leq$ 30 μg L^{-1}.

The lowest ANC lakes in the Upper Midwest also have the lowest concentrations of solutes, whereas higher solute concentrations occur at the higher ANC values (Figure 13.10). This pattern is due in part to lower concentrations of base cations and HCO_3^- at more acidic ANC values. These trends suggest that the watershed-lake systems of the more acidic lakes (ANC < 50 μeq L^{-1}) have a lower capacity to produce base cations from weathering and cation exchange reactions than do those for lakes having ANC > 200 μeq L^{-1}.

The dominant cations are Ca^{2+}, Mg^{2+}, and Na^+; the concentration of K^+ is relatively low and K^+ is a minor contributor to the total cations (Figure 13.10). The primary anions are SO_4^{2-}, HCO_3^-, and organic anions, which were estimated from the model of Oliver et al. (1983). HCO_3^- and organic anions decrease in proportion with decreasing ANC, leaving SO_4^{2-} as the dominant anion in lakes with ANC $\leq$ 0 μeq L^{-1}. Cl^- is generally present in low concentrations, with the higher concentrations probably caused by road de-icing salts.

In the remainder of this chapter, only the subset of ELS-I lakes with low ANC are examined. This subset includes lakes that have ANC $\leq$ 200 μeq L^{-1} and a surface area $\leq$ 2,000 ha. Lakes with ANC $\geq$ 200 μeq L^{-1} have sufficient acid neutralizing capacity to buffer pH against changes in atmospheric deposition. Five lakes with surface area $\geq$ 2,000 ha are not included in these population estimates because of the disproportionately large influence of these lakes on population estimates for lake surface area. To remove lakes that may have Cl^- concentrations influenced by road de-icing salts, lakes with Cl^- > 20 μeq L^{-1} are excluded from this analysis. This Cl^- concentration was determined from estimates of the Cl^- expected from atmospheric deposition (3 to 4 μeq L^{-1}; Table 13.3) and an evapotranspiration factor characteristic for the Upper Midwest (Appendix A, Baker, this volume). Approximately 90% of the entire population of lakes in the Upper Midwest have Cl^- values < 20 μeq L^{-1}. Also, to remove lakes with human disturbance, only lakes that have a total (dissolved and particulate) phosphorus $\leq$ 30 μg L^{-1} are considered (Appendix A, Baker, this volume). This total phosphorus criterion removed about 10% of the entire population of lakes in the Upper

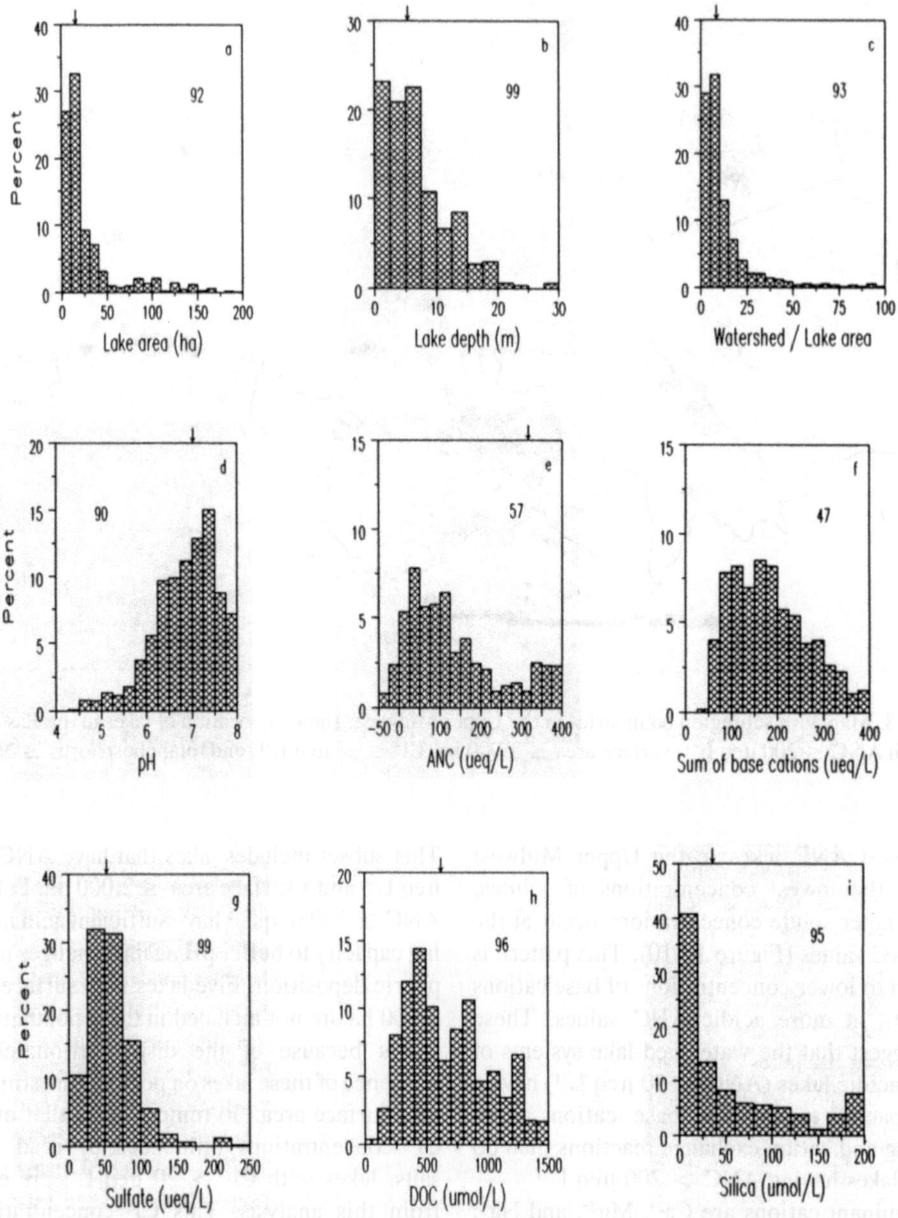

FIGURE 13.9. Frequency histograms of major characteristics of the target population of Upper Midwest lakes in the EPA Eastern Lake Survey: (a) lake surface area, (b) sampling depth, (c) watershed area:lake area, (d) air-equilibrated pH, (e) ANC, (f) sum of base cations, (g) sulfate, (h) DOC, and (i) silica. The arrow at the top of each box indicates the median target population, and the number in the upper right corner of each box indicates the percentage of target lakes plotted.

Midwest. The subset of lakes with ANC ≤ 200 μeq L^{-1}, no human disturbances, based on Cl^- and total phosphorus, and surface area ≤ 2,000 ha are referred to in the discussion that follows as the low ANC subset.

Spatial Distribution of Water Chemistry

The median pH, ANC, base cations (SBC), and DOC for the low ANC subset of lakes decreases from west to east across Region 2 (Figure 13.11).

The decrease in base cations is due in part to a shift from predominantly drainage lakes in northeastern Minnesota to predominantly seepage lakes in upper Michigan (Table 13.1, Figure 13.5). Median SO_4^{2-} concentrations increase slightly from central Minnesota to upper Michigan. However, the range within each subregion is quite large, reflecting variation in biogeochemical cycling of SO_4^{2-}, and there is considerable overlap of the ranges. Greater lakewater SO_4^{2-} concentrations in the eastern part of Region 2 have been attributed to the greater rates of atmospheric deposition of SO_4^{2-} there (Rapp et al. 1987, Eilers et al. 1988). Bedrock containing minor to trace amounts of sulfur may contribute to the high SO_4^{2-} concentrations in northeastern Minnesota (subregion 2A) (Rapp et al. 1985).

The DOC decline from west to east in the region (Figure 13.11) is probably caused by the west-to-east decline in peatlands and marshes and the west-to-east decline in the abundance of drainage lakes. Drainage lakes are expected to have greater inputs of water flowing through organic soil horizons into the lake without losing DOC than seepage lakes. Baker et al. (Chapter 17, this volume) showed that higher DOC concentrations in Minnesota drainage lakes are associated with larger ratios of terrestrial watershed area:lake area.

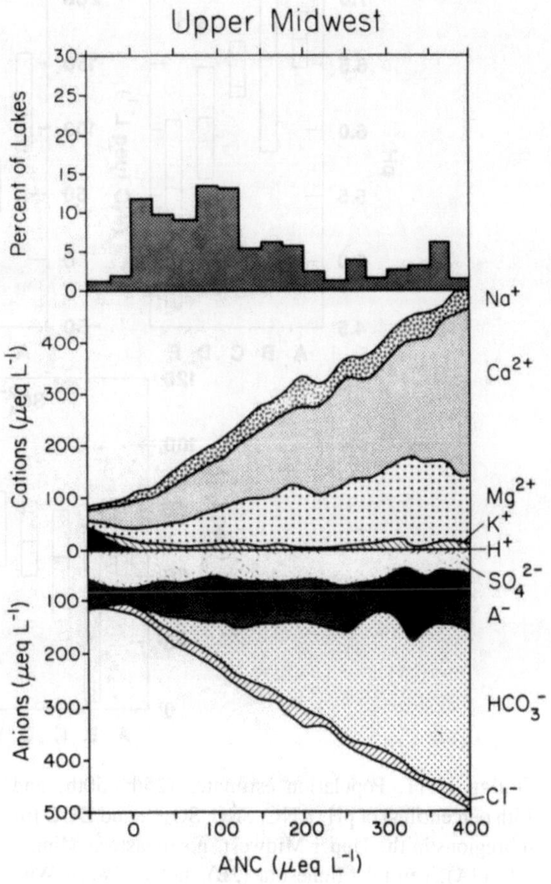

FIGURE 13.10. Concentrations of major cations and anions for lakes in the Upper Midwest with ANC ≤ 400 μeq L^{-1}. The percentages of lakes represented by ANC intervals are shown as a histogram across the top of the panel. Data plotted are population weighted estimates for lakes in the ELS-I. The curves are spline fits to the median concentrations for each ANC interval.

Spatial Relationships Between Lakewater Sulfate and Bedrock Type in Northeastern Minnesota

The spatial pattern of lakewater SO_4^{2-} in northeastern Minnesota may be influenced by the presence of bedrock that contains sulfur (Rapp et al. 1985, Nichols and McRoberts 1986). To investigate whether a relationship exists between bedrock sulfur and lakewater SO_4^{2-}, lakes in the ELS-I, the earlier EPA survey, and the USFS survey were assigned to one of three bedrock types: (1) granitic/felsic, (2) mineralized volcanic and sedimentary, or (3) mafic/volcanic–based on bedrock geology maps and the coordinates of the lakes. Granitic/felsic bedrock (type 1) contains low concentrations of sulfur and low abundances of sulfur-bearing minerals (Green 1982) and is located in the western part of subregion 2A. Mineralized volcanic and sedimentary bedrock (type 2) is associated with the iron formations and is located in a thin band running northeast to southwest in the center of Region 2A. This bedrock type contains high concentrations of sulfur, has high abundances of sulfur-bearing minerals (Green 1982), and is expected to be associated with high lakewater SO_4^{2-} concentrations. Mafic/volcanic bedrock (type 3) may contain sulfur and is located in the eastern portion of northeastern Minnesota (Green 1982).

The median and 25th and 75th percentiles for lakewater SO_4^{2-} concentrations in the sampled lakes were determined for lakes in each bedrock type and for subregion 2A (Table 13.5). Although the lakes underlain by granitic and felsic rocks exhibited median SO_4^{2-} concentrations that were 20 μeq L^{-1} lower than those of the other two areas, the range in SO_4^{2-} concentrations in the three areas

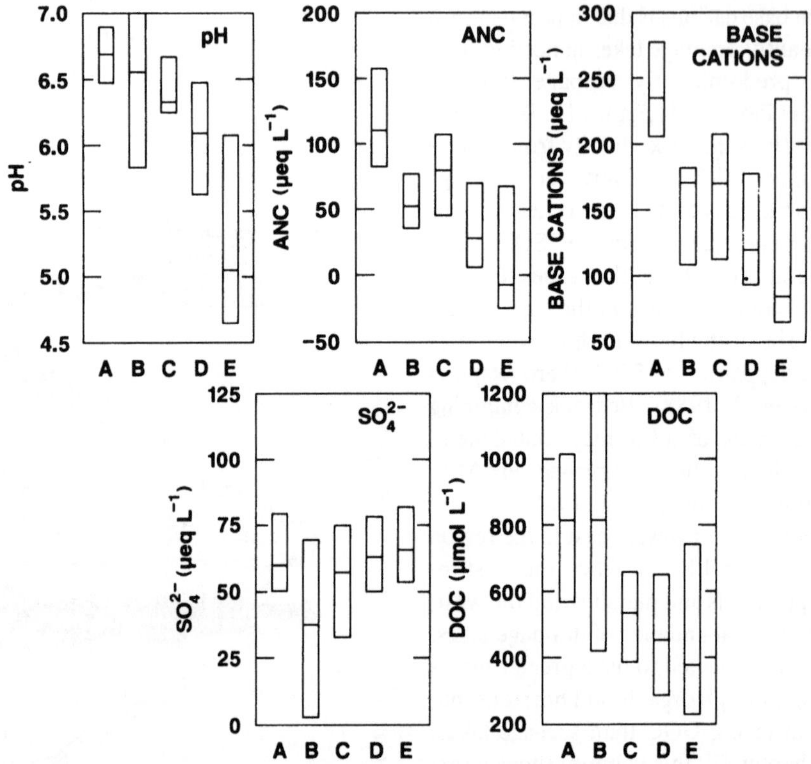

FIGURE 13.11. Population estimates (25th, 50th, and 75th percentiles) of pH, ANC, SBC, SO_4^{2-}, and DOC for subregions in the Upper Midwest: northeastern Minnesota (2A); central Minnesota (2B); northwestern Wisconsin (2C); northeastern Wisconsin (2D); and upper Michigan (2E). Data shown are for the subpopulation of all lakes with ANC ≤ 200 μeq L^{-1}, Cl^- ≤ 20 μeq L^{-1}, surface area < 2,000 ha, and total phosphorus ≤ 30 μg L^{-1}.

overlapped. This overlap suggests that bedrock SO_4^{2-} supply is not the most important factor controlling lakewater SO_4^{2-}. Furthermore, this overlap of lakewater SO_4^{2-} concentrations suggests that the supply of geologic SO_4^{2-} derived from the mineralized and mafic/volcanic zones does contribute to the lakewater SO_4^{2-} concentration, along with atmospheric deposition and biogeochemical cycling in the lakes.

TABLE 13.5. Quartiles of SO_4^{2-} concentrations in lakes sampled in northeastern Minnesota in the Eastern Lake Survey, the earlier EPA Survey, and the U.S. Forest Service Survey.

		SO_4^{2-} (μeq L^{-1})		
Bedrock geology	*n*	25th	50th	75th
Granite/felsic	146	54	64	75
Mineralized volcanic and sedimentary	87	65	83	96
Mafic/volcanic	165	65	83	100
Region 2a	398	60	73	91

Even though the spatial distribution of lakewater SO_4^{2-} in subregion 2A is influenced by bedrock sulfur, the interregional differences in lakewater SO_4^{2-} in the Upper Midwest are not explained solely by bedrock sulfur sources (Figure 13.11). The median SO_4^{2-} of the 58 ELS-I lakes in the low ANC subset in subregion 2A underlain by granite is 60 μeq L^{-1} and the interquartile range is 50 to 69 μeq L^{-1}. This SO_4^{2-} concentration distribution does not differ significantly from that for all ELS-I lakes in the low ANC subset in subregion 2A (Figure 13.11).

The slight differences in lakewater SO_4^{2-} concentration between subregion 2A and the other Upper Midwest subregions (2B, 2C, 2D, 2E; Figure 13.11) are likely to be related to a combination of interregional differences in atmospheric deposition, in the extent of biogeochemical processes retaining SO_4^{2-}, and in bedrock sulfur sources. These factors are discussed in the section entitled Processes Influencing Lakewater Chemistry.

Spatial Relationships Between Water Chemistry and Wet Atmospheric Deposition

The relationship between wet atmospheric deposition and lakewater chemistry for the low ANC subset of ELS-I lakes in the Upper Midwest was examined using interpolated estimates of wet SO_4^{2-} deposition (Chapter 3, Husar et al., this volume). Lakes were assigned to six classes of wet SO_4^{2-} deposition (< 300, 300 to 325, 325 to 350, 350 to 375, 375 to 410, and > 410 eq ha^{-1} yr^{-1}). This classification scheme resulted in roughly a zonal distribution of lakes related to the east-to-west gradient in wet SO_4^{2-} deposition. Population characteristics for each deposition class reveal that ANC and pH decline with increasing wet SO_4^{2-} deposition (Figure 13.12). Lakewater SO_4^{2-} remains relatively constant across the twofold range of SO_4^{2-} deposition, implying that zonal differences in SO_4^{2-} retention associated with biogeochemical reactions are responsible for this uniform pattern of lakewater SO_4^{2-}. The ANC trend results from west-to-east decreases in base cation concentrations and the relatively constant lakewater SO_4^{2-}. The base cation trend reflects the spatial trend in hydrologic lake type, with seepage lakes that have lower base cation concentrations predominantly occurring in the eastern portion of the region (Table 13.1, Figures 13.5 and 13.11).

Eilers et al. (1988b) examined the relationship between atmospheric deposition and the ratio ANC:(Ca^{2+} + Mg^{2+}) for lakes with conductivity < 30 μS cm^{-1}. The observed spatial decrease in this ratio with increasing H^+ deposition was found to be consistent with the spatial trend in acidic deposition. An alternative explanation for this decreasing ratio is that ANC and (Ca^{2+} + Mg^{2+}) are decreasing because of a shift in hydrologic lake type. Seepage lakes are more prevalent in the eastern part of the Upper Midwest and seepage lakes have lower (Ca^{2+} + Mg^{2+}) than drainage lakes. In addition, ANC, which may be approximated by Ca^{2+} + Mg^{2+} − SO_4^{2-}, is affected by the west-to-east decline in (Ca^{2+} + Mg^{2+}), because SO_4^{2-} is uniform in this region.

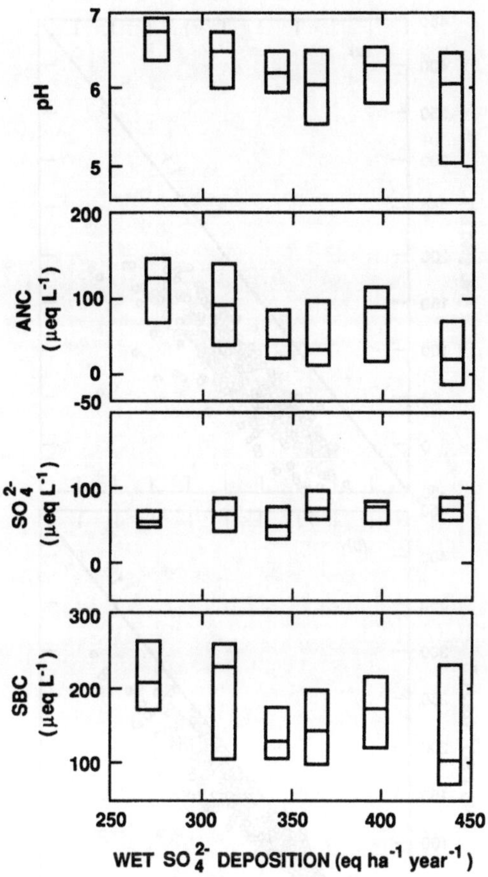

FIGURE 13.12. Relationship between wet deposition of SO_4^{2-} and lakewater chemistry for ELS-I lakes in the Upper Midwest with ANC ≤ 200 μeq L^{-1}, Cl^- ≤ 20 μeq L^{-1}, surface area < 2,000 ha, and total phosphorus ≤ 30 μg L^{-1}. Bars represent the 25th, 50th, and 75th percentiles.

Associations Among Variables

The variable controlling the pH of softwater lakes is ANC, which in turn is a function of major ions (Chapter 1, Munson and Gherini, this volume):

$$ANC = SBC + NH_4^+ + 2Al - SO_4^{2-} - Cl^- - NO_3^- - F^- \quad (13\text{-}1)$$

where each of the terms is the total analytical concentration. For most waters in the Upper Midwest, NH_4^+, Al, Cl^-, NO_3^-, and F^- do not contribute significantly to the charge balance (Figure 13.10) and the following definition can be used:

$$ANC = SBC - SO_4^{2-} \quad (13\text{-}2)$$

This simplified definition of ANC will be used to examine the ions responsible for the acid-base chemistry of lakes.

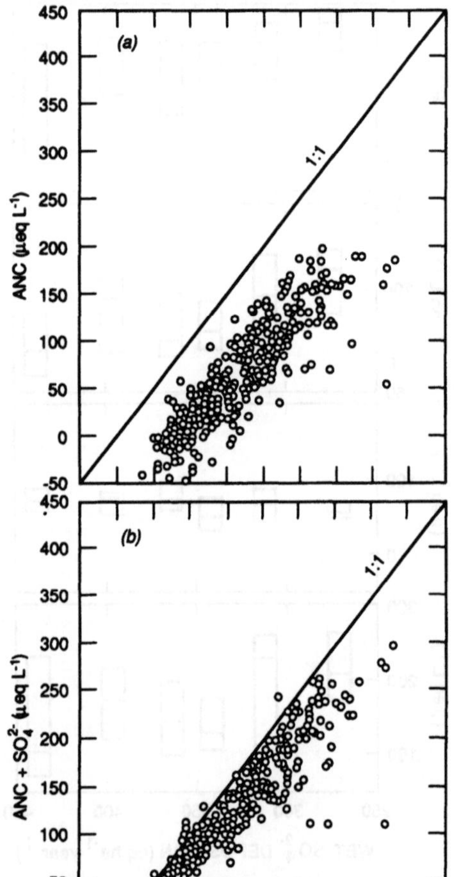

FIGURE 13.13. Relationship between major ions in ELS-I lakes in the Upper Midwest with ANC ≤ 200 μeq L^{-1}, Cl^- ≤ 20 μeq L^{-1}, surface area < 2,000 ha, and total phosphorus ≤ 30 μg L^{-1}: (a) relationship between sum of base cations (SBC = Ca^{2+} + Mg^{2+} + Na^+ + K^+) and ANC; (b) relationship between SBC and the sum of ANC and SO_4^{2-}. In both panels, the 1:1 line is shown for comparison.

In areas where ANC is the primary anion in surface waters and the concentrations of other ions (SO_4^{2-}, Cl^-, NO_3^-, F^-, NH_4^+, and Al), are low, a plot of ANC versus SBC follows the SBC = ANC line, as in the western United States, for example (Chapter 15, Melack and Stoddard, this volume; Chapter 16, Nelson, this volume; Chapter 14, Turk and Spahr, this volume). Base cation production via weathering and cation exchange in these western regions is driven by carbonic acid, with HCO_3^- and base cations as the major products. In the Upper Midwest, ELS-I data fall below the ANC = SBC line (Figure 13.13a), indicating that other anions contribute to the charge balance. The spatial distribution of these two variables, as the ANC:SBC ratio, is shown in Figure 13.14. This plot illustrates that other anions besides weak acids (HCO_3^- and organic acids) contribute to the anionic charge throughout the Upper Midwest.

The contribution of SO_4^{2-} to the anionic charge may be shown with a plot of ANC plus SO_4^{2-} versus SBC (Figure 13.13b), which yields a better fit to the 1:1 line than ANC vs. SBC. The observation that SO_4^{2-} contributes to the anionic charge suggests that either: (1) sulfuric acid weathering reactions contribute to base cation production, with the source of sulfuric acid being either atmospheric deposition or reduced sulfur in bedrock, (2) there are sources of neutral salts of SO_4^{2-} (e.g., $CaSO_4$) in the region, (3) acidification has occurred, with ANC lost and SO_4^{2-} replacing the lost HCO_3^-, or (4) some combination of these three. Sources of sulfur occur in the mafic bedrock in northeastern Minnesota and may contribute to high levels of sulfur there (Rapp et al. 1985; see section entitled Spatial Distribution of Water Chemistry). However, for the other subregions of the Upper Midwest, the amount of sulfur in surficial deposits, particularly in glacial till (see section entitled Regional Characteristics), does not contribute to the SO_4^{2-} in surface waters (Rapp et al. 1987, Eilers et al. 1988). Atmospheric deposition in the region is low relative to areas in eastern North America (Table 13.3; Chapter 3, Husar et al., this volume; Chapter 5, Eilers and Selle, this volume) and paleolimnological results and comparison of historical and recent lakewater chemistry data suggest that acidification attributable to atmospheric deposition of the magnitude 50 to 100 μeq L^{-1}—the deviation from the ANC = SBC line (Figure 13.13a)—has not occurred in the Upper Midwest (see section entitled Evidence for Long-term Trends). Thus, the concentrations of SO_4^{2-} in the Upper Midwest may not have been negligible in the past and may have contributed to the charge balance, causing historical lakewater chemistry in the Upper Midwest to fall below the ANC = SBC line. Therefore, in the Upper Midwest, deviation

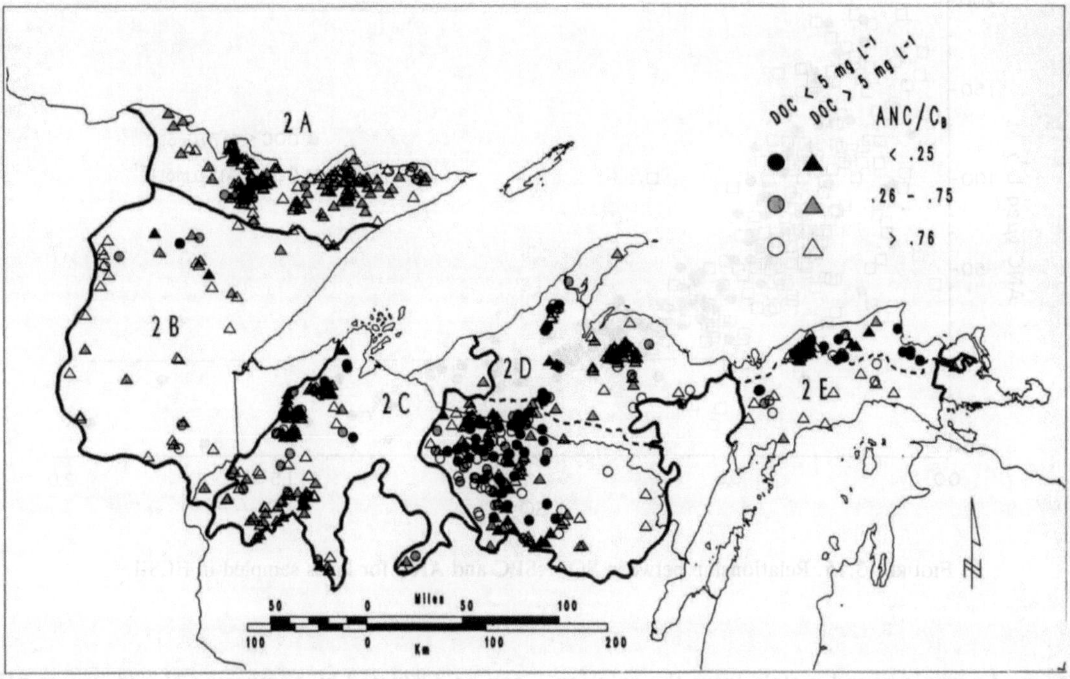

FIGURE 13.14. Map showing ANC:SBC (SBC = Ca^{2+} + Mg^{2+} + Na^{+} + K^{+}) for lakes with DOC $\leq$ 5 mg L^{-1} ($\leq$ 417 μmol L^{-1}) sampled in ELS-I.

from the line does not necessarily indicate acidification has taken place.

The ratio SO_4^{2-}:SBC, which provides an estimate of the relative contribution of base cations and SO_4^{2-} in surface waters (Chapter 3, Husar et al., this volume), increases in the Upper Midwest as ANC decreases (Figure 13.15). This plot suggests that at high ANC, SO_4^{2-} is not the primary anion, but at low ANC the importance of SO_4^{2-} in the charge balance increases. The observed increase in this ratio at low ANC is due to lower concentrations of base cations for relatively constant concentrations of SO_4^{2-} in low ANC waters (Figures 13.11 and 13.13); these lakes appear in the extreme lower left part of the ANC versus SBC plot (Figure 13.13a). Low ANC lakes with SO_4^{2-} as the dominant anion are likely to have experienced some acidification; for these lakes, low rates of base cation production in the watershed-lake system are insufficient to neutralize the acidity of atmospheric deposition.

The role of organic anions in surface waters can be estimated from total organic ligand (A_t) present in surface waters (LaZerte and Dillon 1984; Chapter 1, Munson and Gherini, this volume). A_t is based on the discrepancy in charge balance.

$$A_t = \text{SBC} + NH_4^{+} + H^{+} + Al^{3+} + Fe^{2+} + Mn^{2+} - HCO_3^{-} - SO_4^{2-} - Cl^{-} - F^{-} \qquad (13\text{-}3)$$

where all terms are the total analytical concentration. This method of estimating organic anions includes both free organic anions and organic ligands bound to Al, Fe, Mn, and base cations. A_t represents the organic acids that were produced in the watershed and lake and contributed H^{+} to surface waters (LaZerte and Dillon 1984).

Organic ligands can exist as either metal-bound organic anions or free anions. The complexed anions are predominantly associated with iron (LaZerte and Dillon 1984) and aluminum (Driscoll et al. 1984), and to a lesser extent with manganese and with alkali and alkaline earth cations (Perdue 1985). During an ANC titration, metal-bound organic anions may take up H^{+} and contribute to measured ANC. The definition of A_t (Equation 13-3) differs from the estimate of organic anions made by Oliver et al. (1983) because Oliver et al.

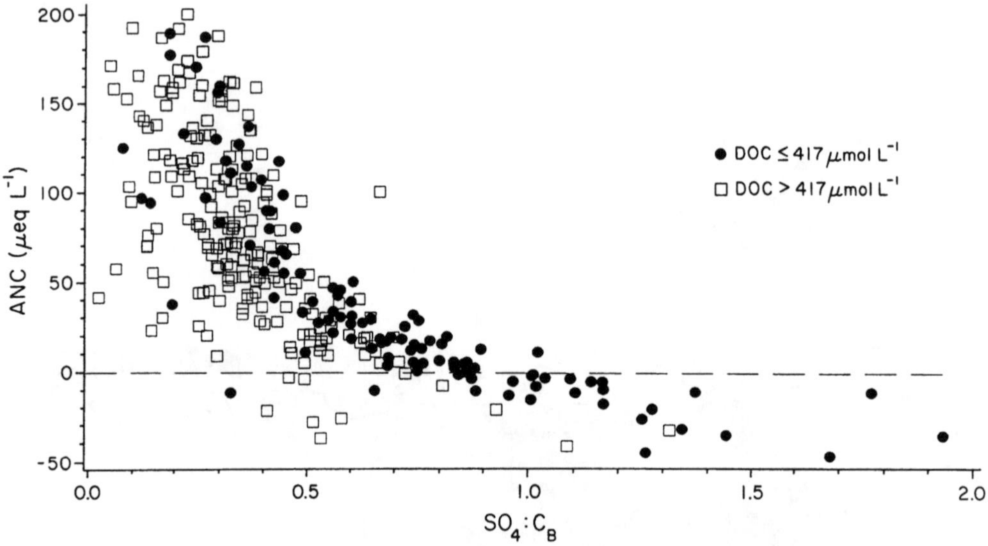

FIGURE 13.15. Relationship between SO_4^{2-}:SBC and ANC for lakes sampled in ELS-I.

removed organic-bound metals from the solution prior to their titrametric analysis.

The ratio A_t:SBC in the Upper Midwest is low for all values of ANC, except for a few sites in northeastern Minnesota, northcentral Wisconsin, and upper Michigan (Figure 13.16). For the majority of lakes in the Upper Midwest, organic anions play a minor role ($\leq$ 40%) in the charge balance (Figure 13.9) and in the acid-base status across the ANC range considered (−50 to 200 μeq L^{-1}). For lakes in northeastern Minnesota, organic anions contribute a large part (> 40%) of the total anionic charge, and organic ligands contribute to the acid-base chemistry of the lake water. However, lakes in northeastern Minnesota have ANC > 50 μeq L^{-1}. Thus, most lakes in the Upper Midwest with ANC $\leq$ 50 μeq L^{-1} do not have low ANC values as a result of organic anions, but rather as a result of SO_4^{2-}.

Relationships Among pH (In situ and Air Equilibrated) and pCO_2

In surface waters, pH is primarily controlled by a combination of carbonate species, weakly ionizable organic acids, and the partial pressure of carbon dioxide (pCO_2) (Chapter 1, Munson and Gherini, this volume). The effects of pCO_2 on pH can be evaluated by comparing closed-system pH to air-equilibrated pH. An increase in pH upon aeration indicates that the lakewater was supersaturated with respect to atmospheric pCO_2 and a decrease in pH indicates undersaturation.

The ELS-I lakes with pH > 5.5 experienced as much as a 1 unit increase in pH upon air equilibration, indicating supersaturation with respect to atmospheric pCO_2. A pH change of 1 unit corresponds to a CO_2 supersaturation of 10 times the partial pressure in equilibrium with atmospheric CO_2 (Stumm and Morgan 1980). Thus, carbonic acid has a large influence on pH for these lakes (Figure 13.17). ELS-I lakes with pH < 5.5 exhibited a smaller difference between air-equilibrated pH and closed-system pH; this is expected because the effect of supersaturation or undersaturation of CO_2 is diminished at pH < 5.0, where carbonic acid does not efficiently dissociate (Stumm and Morgan 1981).

The degree of supersaturation is a function of gas exchange, photosynthesis and respiration, groundwater and stream water input, and pH (Chapters 1 and 2, Munson and Gherini, this volume). Supersaturated conditions indicate that the input of CO_2 into the water column by respiration and water inflow is greater than the output of CO_2 via water outflow, gas exchange, or uptake by photosynthesis. Supersaturated conditions are expected at fall

FIGURE 13.16. Relationship between A_t/SBC and ANC (A_t = total organic ligand; SBC = Ca^{2+} + Mg^{2+} + Na^+ + K^+) for lakes sampled in ELS-I.

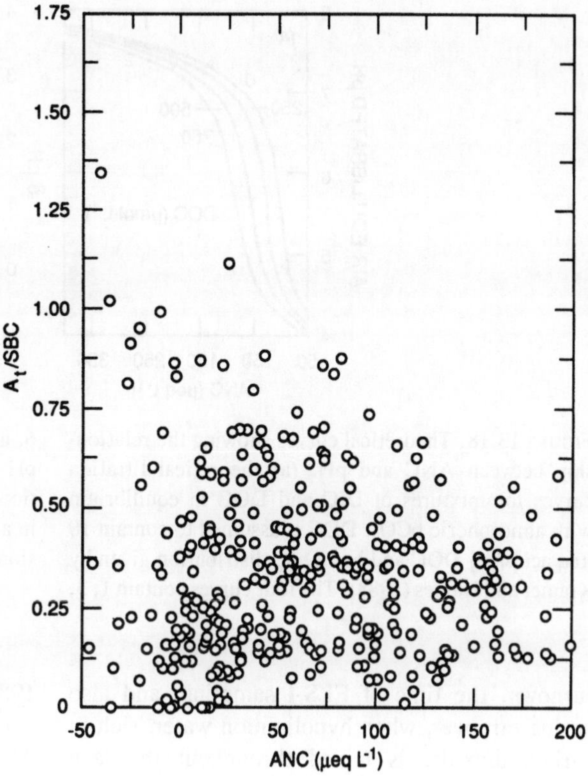

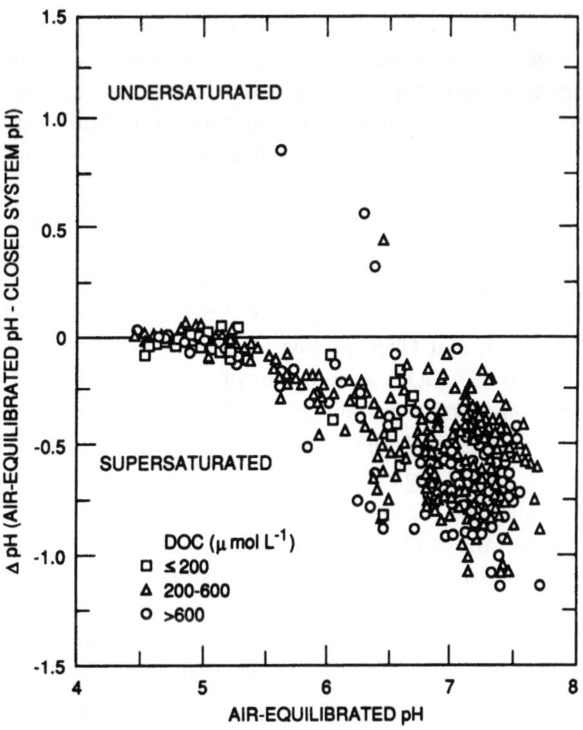

FIGURE 13.17. Change in pH upon equilibration with 300 μatm CO_2 in ELS-I lakes in the Upper Midwest with ANC ≤ 200 μeq L^{-1}, Cl^- ≤ 20 μeq L^{-1}, surface area < 2,000 ha, and total phosphorus ≤ 30 μg L^{-1}. The change in pH is between closed-system pH and air-equilibrated pH. Positive changes indicate undersaturation, negative values indicate supersaturation, and values of 0 indicate equilibrium with respect to 300 μatm CO_2.

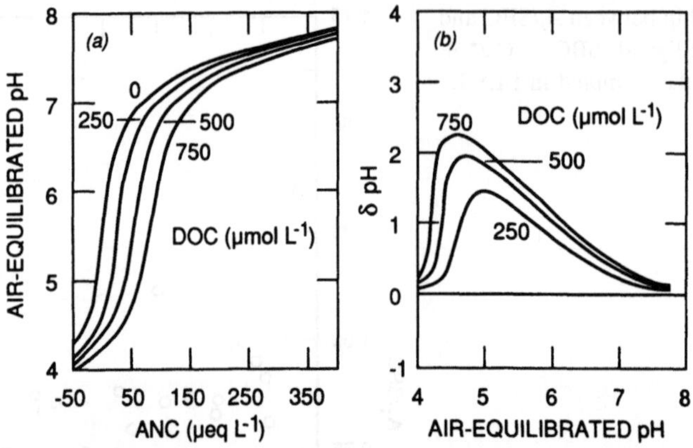

FIGURE 13.18. Theoretical curves showing the relationship between ANC and pH: (a) theoretical titration curves for mixtures of DIC and DOC in equilibrium with atmospheric pCO_2. DOC is assumed to contain 10 μeq acids/mg DOC and has a pK_a distribution given by Kramer and Davies (1988). The four curves contain 1, 3, 6, and 9 mg L^{-1} DOC. (b) difference in air-equilibrated pH between the solution containing DOC and the solution containing no DOC (Figure 13.18a). The difference in air-equilibrated pH (δpH) is a measure of the depression in pH due to weak organic acids.

turnover, the time of ELS-I sampling, and also spring turnover, when hypolimnion water, rich in carbon dioxide, is mixed throughout the lake (Schindler et al. 1980, Herczeg et al. 1985, Kratz et al. 1987, see section entitled Episodic Changes in Lakewater Chemistry). In softwater lakes where patterns in pCO_2 have been measured or estimated for the ice-free season, saturation (R.H. Hesslein, Freshwater Institute, Winnipeg, pers. comm.), supersaturation (up to threefold) (Herczeg et al. 1985, Cook et al. in press), and undersaturation (Schindler et al. 1972) have been reported.

Organic Acids: Effects on pH

Organic acids can be an important source of natural acidity in softwater lake systems (Chapters 1 and 2, Munson and Gherini, this volume). The concentrations of DOC, an indicator of organic acids, vary considerably in the Upper Midwest (Figure 13.11), because of the variable occurrence of DOC source areas–organic-rich surficial deposits–in the region and because of the spatial distribution of drainage and seepage lakes (see section entitled Spatial Distribution of Water Chemistry).

Naturally occurring organic acids are characterized as having a continuum of dissociation constants (pK_a), many of which are less than that of carbonic acid (Kramer and Davies 1988, Perdue 1985). Organic anions may exert pH buffering control in the pH range 4.5 to 7.0 (i.e., contribute to ANC). However, the buffer intensity of organic anions is a function of the concentration of organic anions (DOC) and the distribution of pK_a of the organic acids present in the lake water. As a consequence, even though two samples have the same DOC, the acid-base chemistry (ANC, titration curve, and buffer intensity) of the samples can be quite different. Because of these characteristics of the organic acid functional groups, the presence of organic acids in surface waters causes the air-equilibrated pH (pH-ae) to be less than that expected if the only pH buffering species present were dissolved inorganic carbon (DIC) (Chapter 1, Munson and Gherini, this volume).

To illustrate the role of DOC in controlling pH-ae, the theoretical relationship between pH-ae and ANC for carbonate species alone and for several DOC concentrations is shown in Figure 13.18a. The DOC used in this relationship contained 0.120 μeq organic acids (μmol $DOC)^{-1}$ (Oliver et al. 1983, Cook et al. 1987). The pK_a distribution for the organic acids employed was determined for a subset of high DOC lakes in the ELS-I by Kramer and Davies (1988). The difference between theoretical pH-ae and measured pH-ae, defined as δpH, is plotted in Figure 13.18b. The maximum value of δpH caused by DOC occurs in the pH-ae

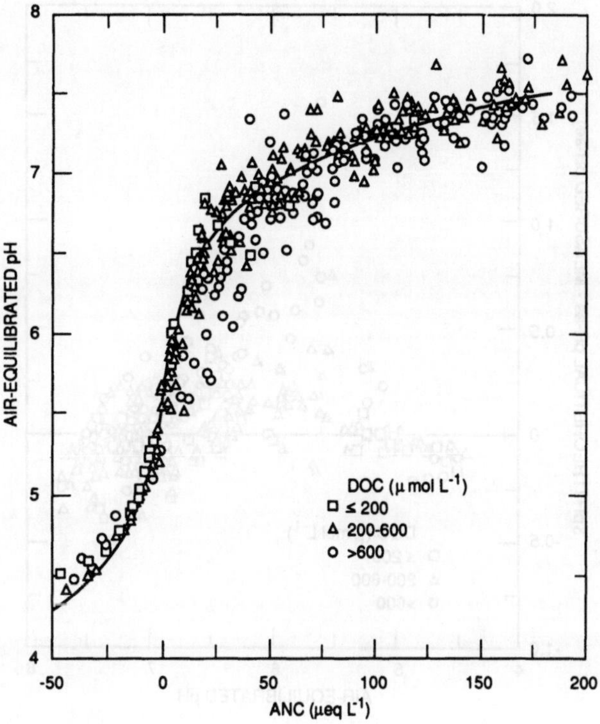

FIGURE 13.19. Relationship between air-equilibrated pH and ANC for ELS-I lakes with ANC $\leq 200\ \mu eq\ L^{-1}$ in Upper Midwest. Solid line indicates theoretical relationship for $pCO_2 = 340$ µatm assuming all ANC is due to carbonate species. Analytical uncertainties and weak acids other than carbonate species cause the lake values to be different from the theoretical relationship.

4.0 to 7.0 range and increases with increasing DOC concentration to 2 pH units for a DOC of 750 $\mu mol\ L^{-1}$ (Figure 13.18b). For DOC concentrations $\geq 750\ \mu mol\ L^{-1}$, greater values of δpH may occur at lower pH-ae values.

A plot of pH-ae versus ANC in the Upper Midwest reveals that few lakes fall on the theoretical curve for carbonate species in equilibrium with atmospheric pCO_2 (Figure 13.19). The difference between theoretical pH-ae and measured pH-ae (δpH) is greatest between pH-ae 5.0 and 6.5 (Figure 13.20). In general, higher DOC concentrations are associated with higher values of δpH and lower DOC values are associated with lower values of δpH. Even though DOC can contribute to pH-ae for pH-ae < 5.0 (Figure 13.18a), observed δpH in this portion of the curve is small, suggesting that for most acidic lakes in the Upper Midwest, organic acids do not contribute significantly to the acid-base status (δpH ≤ 0.2).

The δpH may also be influenced by the presence of other weak acids (e.g., aluminum). However, in the Upper Midwest, aluminum concentrations are relatively low (typically $< 50\ \mu g\ L^{-1}$; Eilers et al. 1988b) and, thus, Al does not contribute to δpH. The large scatter in δpH for each DOC class between pH-ae 5.0 and 6.0 may occur in part because δpH is most sensitive to analytical uncertainties in ANC and pH in this pH range, which is in the steepest portion of the titration curve (Figure 13.18a).

Seasonal Variability in Lakewater Chemistry

The ELS-I provides lakewater chemical characteristics for the fall index period, a time when the water column should be well mixed (Landers et al. 1988). Seasonal data from intensively studied sites can be used to gauge how well the fall index conditions represent the lakewater concentrations and also to determine if the acid-base chemistry fluctuates markedly from the fall index conditions. Additional work is underway to determine the relationship between the annual average water quality and that of the fall index sample (Eastern Lake Survey–Phase II).

A number of processes can cause seasonal variation in the acid-base chemistry of lakes, as illustrated by data from Vandercook Lake (Figure

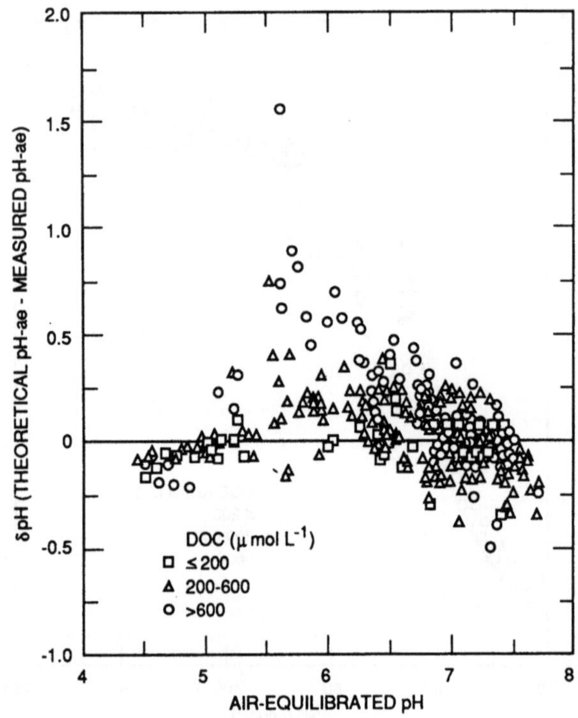

FIGURE 13.20. Effect of noncarbonate weak acids on pH as measured by deviation (δpH) as a function of air-equilibrated pH for lakewater chemistry plotted in Figure 13.19.

13.21; K. Webster, Wisconsin Department of Natural Resources, unpublished data). Highest pH values were observed during the ice-free season, not as a result of high ANC, but probably as a consequence of photosynthetic uptake of CO_2 (see section entitled Relationships Among pH and pCO_2). The lowest pH values occurred under the ice in April 1981, 1982, and 1983 and appeared to be related to changes in pCO_2, rather than to ANC, which was relatively high in April 1982 and low in April 1983. This pH depression during late winter and early spring was also observed at six lakes in northcentral Wisconsin (see section entitled Episodic Changes in Lakewater Chemistry; Kratz et al. 1987). ANC in the surface water of Vandercook increased in one of the two winter periods sampled, perhaps in part by in-lake SO_4^{2-} retention (Figure 13.21; Garrison et al. 1987). After ice-out in 1982 and 1983, the pH rose and ANC, Ca^{2+}, and SO_4^{2-} decreased by dilution with water from snow- and ice-melt. The temporal lakewater chemistry data from Vandercook Lake shows that acid-base chemistry fluctuates during the year and that at certain times of the year (e.g., late winter-early spring), ANC and pH may be dramatically different from the fall index sample.

Episodic Changes in Lakewater Chemistry

Episodic pH and ANC excursions due to snowmelt are not as severe in the Upper Midwest as in the northeastern United States, in part because the snowpack accumulations are smaller in the Upper Midwest than in the northeast and atmospheric deposition is less (see section entitled Regional Characteristics; Chapter 5, Eilers et al., this volume; Chapter 6, Driscoll et al., this volume). For seepage lakes, most of the meltwater flows into the thick till, which is frozen only to shallow depths as a result of insulation by the snowpack or not frozen at all (Kratz et al. 1987). Nevertheless, episodic pH depressions do occur in the Upper Midwest.

During spring snowmelt, Siegel (1981) and Schnoor et al. (1984) observed pH depressions from 6.6 to 5.5 in Filson Creek, northeastern Minnesota, and attributed them primarily to dilution of

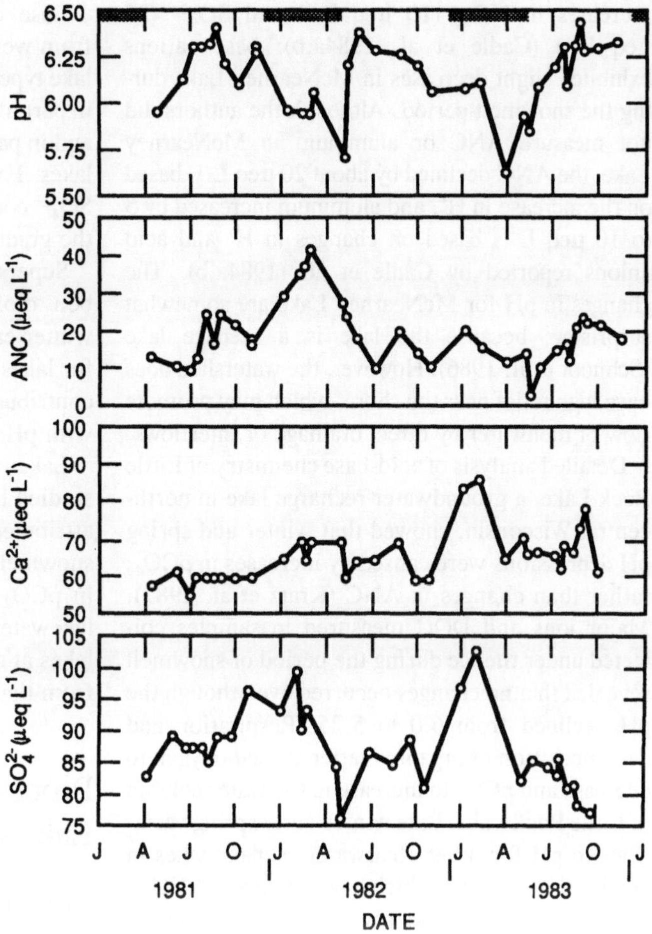

FIGURE 13.21. Temporal trend in surface lakewater chemistry for Vandercook Lake: (a) pH, (b) ANC, (c) Ca^{2+}, (d) SO_4^{2-}. Bars indicate ice cover. Data from K. Webster, Wisconsin Department of Natural Resources, unpublished.

base cations and, secondarily, to increases in SO_4^{2-} and NO_3^-. Schnoor et al. (1984) measured episodic ANC declines of 400 μeq L^{-1} (from 458 to 67 μeq L^{-1}), of which approximately 25 μeq L^{-1} was caused by increases in SO_4^{2-}, 50 μeq L^{-1} by increases in NO_3^-, and 300 μeq L^{-1} by base cation dilution. Although DOC was not measured in either study, the stream water is colored due to humic substances. Anion deficits were noted for Filson Creek, although no information on temporal variation was given (Siegel 1981, Schnoor et al. 1984). Decreases in DOC caused by snowmelt may also cause changes in the relative importance of organic acids in the acid-base balance of Filson Creek (Schnoor et al. 1984).

ANC and pH depressions during snowmelt were observed in two lakes in northeastern Minnesota located near the Filson Creek site (Wonson-Liukkönen 1987). For both lakes, the snowmelt changes in pH were caused by base cation dilution, with small or no increase in SO_4^{2-}.

Detailed studies of snowmelt in northern Michigan revealed that acidity in the first 50% of the melt was largely associated with SO_4^{2-} and that NO_3^- concentrations were smaller than expected from the snowpack concentrations, probably due to interaction of NO_3^- with the forest floor (Cadle et al. 1984a,b, Cadle et al. 1987). Changes in the pH of streams draining the study areas were attributed to base cation dilution, with changes in SO_4^{2-} accounting for no more than 10% of the total change in ANC, as estimated from data in Cadle et al. (1984a,b). The pH excursion during the snowmelt period in one of the study lakes (McNearney Lake) from 4.4 to 4.2, which corresponds to an increase in H^+ from 40 to 60 μeq L^{-1}, was attributed to increases in NO_3^- (10 μeq L^{-1}) and SO_4^{2-} (25 μeq L^{-1}) (Cadle et al. 1984a,b); base cations exhibited slight decreases in McNearney Lake during the snowmelt period. Although the authors did not measure ANC or aluminum in McNearney Lake, the ANC declined by about 20 μeq L^{-1}, based on the increase in H^+, and aluminum increased by 5 to 10 μeq L^{-1}, based on changes in H^+ and acid anions reported by Cadle et al. (1984a,b). The changes in pH for McNearney Lake are somewhat surprising, because the lake is a seepage lake (Schnoor et al. 1986). However, the watershed does have high relief near the shore, which may promote flow of meltwater by direct drainage or interflow.

Detailed analysis of acid-base chemistry of Little Rock Lake, a groundwater recharge lake in north-central Wisconsin, showed that winter and spring pH depressions were caused by increases in pCO_2, rather than changes in ANC (Kratz et al. 1987). Major ions and DOC measured in samples collected under the ice during the period of snowmelt revealed that no changes occurred even though the pH declined from 6.0 to 5.25. Respiration and decomposition of organic matter caused oxygen to decrease and pCO_2 to increase in the water column of Little Rock Lake. Late winter-early spring variations in pH for other clearwater seepage lakes in northern Wisconsin, which have a range in ANC of 10 to 800 μeq L^{-1}, were also attributed to changes in pCO_2 (Kratz et al. 1987). Similar pH depressions, probably caused by increases in pCO_2, were observed during late winter-early spring in Vandercook Lake as well (Figure 13.21).

Summary of the Characteristics of Lakes

Out of an estimated 7,557 lakes within the ELS-I target population in the Upper Midwest, 2,511 (33.2%) have ANC $\leq$ 200 μeq L^{-1}, and 128 (1.7%) have ANC $\leq$ 0 μeq L^{-1}. The lakes with ANC $\leq$ 0 μeq L^{-1} are located primarily in northcentral Wisconsin and upper Michigan (subregions 2D and 2E) and are predominantly (80%) groundwater recharge lakes. The major ions in the ANC $\leq$ 200 μeq L^{-1} subset of lakes are base cations, SO_4^{2-}, and ANC (HCO_3^-, and organic anions). In acidic waters, base cations, SO_4^{2-}, and H^+ are the primary ions. This observation suggests that in acidic waters, SO_4^{2-} and not organic anions are related to the low ANC.

Base cations concentrations decline markedly from west to east, due to differences in hydrologic lake type. DOC also declines from west to east, due in part to the decrease in the abundance of wetlands and in part to a shift from drainage lakes to seepage lakes. For the low ANC subset of lakes, lakewater SO_4^{2-} concentrations are relatively uniform, despite the gradient in atmospheric deposition of SO_4^{2-}.

Supersaturation with respect to atmospheric carbon dioxide during the fall turnover and late winter-early spring periods may cause reduced pH for lakes with pH $>$ 5.0. Weak organic acids also contribute to reduced pH, but primarily for lakes with pH $>$ 5.0.

Lakewater chemistry for the few intensively studied lakes exhibits annual variation that can be attributed to dilution by snowmelt, inputs of snowmelt-associated NO_3^- and SO_4^{2-}, and changes in pCO_2. Because of the seasonal fluctuation in lakewater chemistry, the acid-base chemistry of lakes at certain times of the year may be different from that measured for the fall index sample.

Processes Influencing Lakewater Chemistry

Lakewater chemistry is the product of atmospheric deposition, hydrology, and biogeochemical reactions that occur in the terrestrial watershed and in the lake. Because there are only a few process-level studies examining biogeochemical reactions in the Upper Midwest, the relative importance of these biogeochemical reactions on a regional scale will be evaluated in this section by using ion enrichment calculations. In addition, the few site-specific studies on biogeochemical processes will be examined.

Processes as Estimated by Ion Enrichment Calculations

Ion enrichment calculations for major lakewater solutes in the Upper Midwest were performed to estimate the watershed and lake biogeochemical processes that control the acid-base chemistry of lakes. The calculations were made for the low ANC subset of lakes ($\leq$ 200 μeq L^{-1}), which are relatively undisturbed, based on total phosphorus concentrations ($< 30\ \mu g\ L^{-1}$). Ion enrichment values are calculated from the difference between the observed lakewater chemistry and that predicted from atmospheric deposition and an evapoconcentration factor (Cook et al. 1987, Lee and Schnoor 1988; Chapter 2, Munson and Gherini, this volume). These differences provide an estimate of the net biogeochemical reactions that occurred in the watershed-lake system. The net reactions are expressed in terms of their contribution to ANC generation (Chapter 2, Munson and Gherini, this volume; Appendix A, Baker, this volume).

The evapoconcentration factor is the total hydrologic input to the lake divided by the outflow from the lake. In the Upper Midwest, mapped runoff and precipitation data (see section entitled Regional Characteristics) yield a range in evapoconcentration factors of 1.5 to 3.5, with higher values in Minnesota and lower values in upper Michigan. These runoff-based values of the evapoconcentration factors assume that all of the precipitation incident on the watershed flows into the lake via direct runoff or interflow (i.e., no loss of precipitation to the local or regional groundwater table) and that groundwater inputs and outputs are negligible. This is a valid assumption for drainage lakes, but is not an appropriate means of calculating the evapoconcentration factor for seepage lakes.

For individual lakes in the Upper Midwest, evapoconcentration factors were calculated from the ratio of Cl^- in lake water to Cl^- in atmospheric deposition. The assumption made in using this value of evapoconcentration is that Cl^- is supplied only by atmospheric deposition and that it is geochemically conservative. To remove lakes that have other significant sources of Cl^- (e.g., road deicing salts, geologic deposits) from the data set, we excluded those lakes having $Cl^- > 20\ \mu eq\ L^{-1}$ (see section entitled Regional Characteristics); about 10% of the lakes sampled in the Upper Midwest in the ELS-I had Cl^- concentrations $> 20\ \mu eq\ L^{-1}$. Median evapoconcentration factors determined from Cl^- range from 1.5 to 1.7 for the class of lakes with $Cl^- \leq 20\ \mu eq\ L^{-1}$ (Table 13.6), in general agreement with the evapoconcentration factors determined from mapped runoff.

Wet atmospheric deposition to each lake was interpolated by Husar (Appendix C, this volume). Dry deposition, as estimated by Baker (Appendix B, this volume; see section entitled Regional Characteristics), was added to the wet deposition values, to obtain an effective concentration for wet plus dry deposition.

For lakes with ANC $\leq 200\ \mu eq\ L^{-1}$ in the entire Upper Midwest, median inputs of acidity amounted to $50\ \mu eq\ L^{-1}$, median production of ANC was about $125\ \mu eq\ L^{-1}$, and median lakewater ANC was $75\ \mu eq\ L^{-1}$ (Figure 13.22). The production of ANC was derived from primarily base cation production (75%), with smaller contributions from NO_3^- uptake and SO_4^{2-} retention; uptake of NH_4^+ consumed an amount of ANC roughly equal to that produced by NO_3^- uptake (Figure 13.22). Median organic ligand production was $60\ \mu eq\ L^{-1}$ and median DOC production was $550\ \mu mol\ L^{-1}$.

To examine the spatial variation in processes affecting lakewater chemistry, ion enrichment calculations were also made for combinations of hydrologic lake type (drainage and closed lakes; groundwater recharge lakes; and flow-through seepage lakes) for each of two DOC groups: $\leq 400\ \mu mol\ L^{-1}$ ($5\ mg\ L^{-1}$) and $> 400\ \mu mol\ L^{-1}$ (Table 13.6). The two seepage lake types were distinguished based on a silica concentration of $17\ \mu mol\ L^{-1}$ ($1\ mg\ L^{-1}$).

The high DOC ($> 400\ \mu mol\ L^{-1}$) drainage lake class in northeastern Minnesota (2A), upper Michigan (2D), and the entire Upper Midwest have similar median lakewater ANC concentrations (Table 13.6). For these lake classes, ANC is primarily produced by base cation production. Median base cation production in $\mu eq\ L^{-1}$ in these two high DOC drainage lake classes is greater than that for all lakes in the Upper Midwest with ANC $\leq 200\ \mu eq\ L^{-1}$ (Figure 13.22). The median ANC produced from SO_4^{2-} reactions is not different from $0\ \mu eq\ L^{-1}$ for the high DOC drainage lakes in these two subregions. This low rate of ANC production from SO_4^{2-} retention is probably related to the short lakewater residence times for low ANC, high DOC drainage lakes (Table 13.2; Chapter 17, Baker et al., this volume). The low DOC drainage lakes have a lower proportion of ANC produced by base cation reactions and a higher proportion of ANC contributed by SO_4^{2-} reactions than the high DOC drainage lakes (Table 13.6). As expected,

TABLE 13.6. Ion enrichment data for lakes in the Upper Midwest.[a]

Region	Hydrologic lake type	DOC	n	ET	R_{SBC}	$ANC_{SBC\%}$	$R_{SO_4^{2-}}$	$R_{SO_4^{2-}\%}$	$ANC_{SO_4^{2-}\%}$	A_t	ANCGEN Gran	ANCGEN C_B-C_A
2	Drainage	≤ 5	23	1.92	89	73	−22	−21	12	24	182	164
				(1.07, 2.86)	(89, 108)	(40, 97)	(−83, 4)	(−52, 5)	(−3, 37)	(17, 37)	(94, 220)	(104, 178)
2	Drainage	> 5	107	1.68	184	109	−3	−4	2.5	95	170	189
				(1.27, 2.14)	(129, 214)	(95, 130)	(−29, 9)	(−33, 13)	(−7, 14)	(68, 136)	(129, 200)	(143, 227)
2A	Drainage	> 5	67	1.68	188	113	−1	−1	1	104	166	179
				(1.32, 2.14)	(142, 214)	(99, 133)	(−23, 13)	(−26, 22)	(−8, 13)	(73, 138)	(121, 200)	(147, 224)
2D	Drainage	> 5	24	1.53	180	116	−3	−4	2	88	182	187
				(1.26, 1.80)	(122, 234)	(94, 128)	(−23, 8)	(−35, 16)	(−6.7, 18)	(68, 140)	(130, 197)	(143, 239)
2	GWR	≤ 5	86	1.56	47	66	−16	−22	17	20	76	62
				(1.12, 2.55)	(33, 68)	(35, 98)	(−58, 9)	(−50, 14)	(−14, 50)	(12, 25)	(52, 133)	(47, 132)
2	GWR	> 5	73	1.67	88	80	−31	−55	38	66	108	115
				(0.94, 2.42)	(49, 101)	(40, 107)	(−80, −18)	(−78, −24)	(21, 49)	(49, 85)	(58, 160)	(101, 171)
2C	GWR	≤ 5	18	1.70	68	75	−20	−25	15	26	90	83
				(1.4, 2.0)	(59, 110)	(62, 80)	(−27, −7)	(−29, −11)	(12, 26)	(26, 30)	(76, 159)	(70, 148)
2D	GWR	≤ 5	46	1.33	40	72	−10	−19	11	17	68	53
				(1.11, 1.73)	(28, 62)	(40, 107)	(−32, 14)	(−37, 23)	(−22, 35)	(8, 21)	(41, 96)	(37, 73)
2E	GWR	≤ 5	18	1.08	35	86	−7	−12	5	25	55	51
				(0.86, 1.58)	(26, 63)	(52, 93)	(−12, 9)	(−13, 18)	(−28, 13)	(17, 39)	(32, 128)	(46, 61)
2	GWFT	> 5	25	1.53	131	96	−19	−24	22	85	105	140
				(1.22, 1.95)	(66, 152)	(80, 136)	(−38, 0)	(−47, 0)	(0, 46)	(64, 130)	(63, 172)	(85, 230)

[a] ET represents evapotranspiration as estimated by Cl^- (lake)/Cl^- (deposition). R is the reaction, in μeq L^{-1} or percentage, for SBC ($Ca^{2+} + Mg^{2+} + Na^+ + K^+$) or SO_4^{2-}. The contribution of SBC or SO_4^{2-} reactions to the total generation of ANC (ANCGEN [C_B-C_A]) is indicated by ANC_{SBC} or $ANC_{SO_4^{2-}}$. Values presented are medians estimated for the target population; the 25th and 75th percentiles are enclosed in parentheses. GWR is groundwater recharge; GWFT is groundwater flow-through.

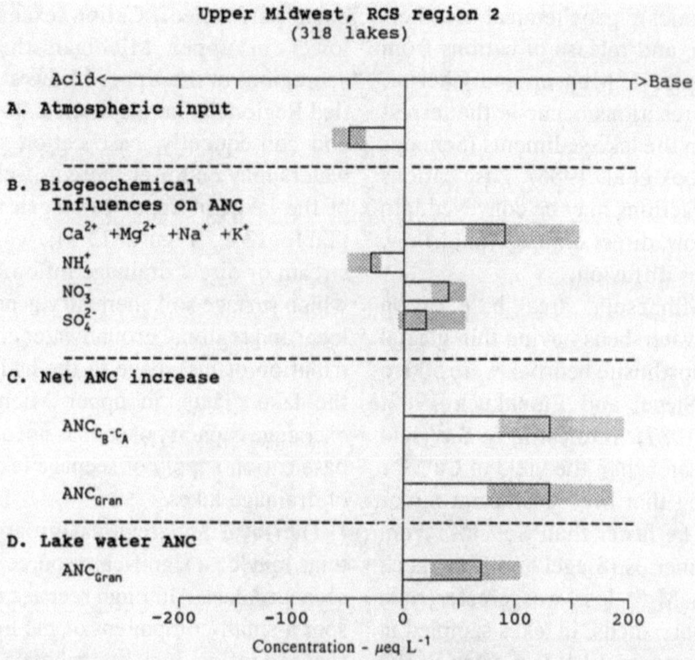

FIGURE 13.22. Modification of atmospheric deposition by watershed and lake biogeochemical processes. A. Precipitation [H^+] adjusted for estimated dry deposition and evapotranspiration (ET); equivalent to expected lakewater $ANC_{C_B\text{-}C_A}$ in the absence of any process other than ET. B. Reactions influencing ANC; bars to the right indicate ANC production, those to the left show ANC consumption. C. Net ANC production; $ANC_{C_B\text{-}C_A}$ is the sum of ANC production terms for measured anions and cations, including those shown in C; ANC_{Gran} is determined by the difference between input ANC (A.) and lakewater ANC (D.). D. Lakewater ANC represents the net reaction of atmospheric inputs within the lake/watershed system. Shaded bars represent 25% and 75% quartiles. For a further explanation of calculation procedures, see Appendix A. Median DOC = 549.5 µmol L^{-1} (25% quartile = 383.0 µmol L^{-1}; 75% quartile = 832.6 µmol L^{-1}); median A_t = 59.5 µeq L^{-1} (25% quartile = 24.5 µeq L^{-1}; 75% quartile = 91.6 µeq L^{-1}).

organic acids, as estimated by A_t, have a higher concentration in the high DOC drainage lakes than in the low DOC drainage lakes.

Low DOC groundwater recharge lakes in subregions 2C, 2D, and 2E have the lowest lakewater ANC of any of the hydrologic lake types, thus this is the lake type most vulnerable to the chemical effects of acidic deposition (Table 13.6). Base cation production is relatively low for groundwater recharge lakes (about 68 to 35 µeq L^{-1} from west-to-east), yet it still produces 72% to 86% of the total ANC production. This spatial gradient in base cation production causes a spatial gradient in median ANC for groundwater recharge lakes (Table 13.6, Figure 13.11). Sulfate reactions contribute a median ANC production of 15 µeq L^{-1}, or 5% to 15% of the total ANC production in the low DOC groundwater recharge lakes.

Groundwater flowthrough lakes in the high DOC class have a median ANC production similar to that for the groundwater recharge lakes, but lower than that for the drainage lake classes. ANC in groundwater flow-through lakes is produced primarily by base cation reactions, with SO_4^{2-} reactions contributing 22% of the total ANC production. The greatest contribution of SO_4^{2-} reactions to ANC production occurs in the high DOC groundwater recharge lakes (Table 13.6).

In the next section, results from intensively studied sites are used to evaluate the ion enrichment analysis.

Base Cations

Base cations can be supplied to surface water by atmospheric deposition (see section entitled

Regional Characteristics), geochemical weathering, cation exchange, and release of cations from organic matter (Chapter 1, Munson and Gherini, this volume). These reactions occur in the terrestrial watershed and in the lake sediments (Schnoor and Stumm 1984, Cook et al. 1986). Base cations produced by these reactions may be conveyed into the lake by stream flow, direct drainage, interflow, seepage, or molecular diffusion.

In northeastern Minnesota, areal base cation yields from several watersheds having thin glacial till, wetlands, and anorthosite bedrock were 500 to 600 eq ha^{-1} yr^{-1} (Siegel and Pfannkuch 1984, Wonson-Liukkonen 1987). Interestingly, the yield of Mg^{2+} was 50% greater than the yield of Ca^{2+} for these sites, indicating that Mg^{2+} loss from minor mafic minerals may be faster than Ca^{2+} loss from major plagioclase minerals (Siegel and Pfannkuch 1984). Even though Mg^{2+} loss was greater than Ca^{2+} loss for these watersheds, in lakes sampled in northeastern Minnesota in ELS-I, Ca^{2+} was the predominant cation and was always present in greater concentrations than Mg^{2+} (Figure 13.10). The yield of base cations determined by Siegel and Pfankuch (1984) and Wonson-Liukkonen (1987) is considerably larger than atmospheric deposition of hydrogen ion and ammonium to northeastern Minnesota (200 to 275 eq ha^{-1} yr^{-1}; Table 13.3), suggesting that base cation production caused by reactions of natural acids (i.e., carbonic acid and organic acid) in the watershed accounts for at least about 300 to 400 eq ha^{-1} yr^{-1} (about 60%) of the net yield of base cations.

Bockheim et al. (1987) measured base cation yields of 500 to 730 eq ha^{-1} yr^{-1} from forest and meadow plots in northwestern Wisconsin, based on increases in base cations as atmospheric deposition percolated to the base of the rooting zone (60 cm). The soil was sandy, well-drained Haplorthod derived from glacial outwash. Although some buffering occurred as atmospheric deposition interacted with the forest canopy of these plots, most of the hydrogen ion neutralization and base cation production occurred in the soil, primarily by cation exchange. The presence of weatherable minerals (feldspars, mica, olivine, hornblende, and augite) within the rooting zone suggests that, in addition to cation exchange, weathering is also responsible for some of the base cation production.

Detailed studies of base cation supply from soils and surficial deposits in upper Michigan have not been performed. Cation exchange capacity is lower in upper Michigan than in the other subregions of the Upper Midwest (see section entitled Regional Characteristics; Turner et al. 1986), and consequently, base cation supply to surface waters may be lower there as well. However, 80% of the lakes in upper Michigan are seepage lakes (Table 13.1, Figure 13.5), which do not have stream or direct drainage inflows of water and for which surface soil chemistry is not as important as local and regional groundwater chemistry and contribution of inseepage to the hydrologic budget of the lake. Thus, in upper Michigan, the cation exchange capacity of soils is not as important to the base cation supply of seepage lakes, as it is to that of drainage lakes.

The local and regional groundwater flow systems may be a significant source of base cations to seepage lakes. Although seepage inflow may represent a small component of the hydrologic budget, seepage inflow may contribute a large component of base cation input, because groundwater is typically enriched in base cations due to longer term weathering reactions (Hurley et al. 1985, Stauffer 1985, Lin et al. 1987, Garrison et al. 1987). For example, the groundwater inflow to Vandercook Lake was 3% of the total water input (Figure 13.4), yet it supplied 42% of the total base cation input and 24% of the ANC input to the lake (Lin and Schnoor 1986, Lin et al. 1987). Base cation input to East Eightmile Lake from seepage inflow was 91% of the total base cation input and resulted from a seepage input of about 30% of the hydrologic input (Figure 13.4; Wentz et al. 1987). The difference in ANC concentrations between Vandercook Lake (20 μeq L^{-1}) and East Eight Mile Lake (420 μeq L^{-1}) arises because of differences in the supply of base cations from seepage inflow.

Reduction or elimination of groundwater flow would have a dramatic effect on the base cation economy of seepage lakes. Reductions of groundwater flow may result from a decrease in the hydrologic head, due to consumptive use of groundwater or due to climate variations (e.g., Chapter 12, Pollman and Canfield, this volume). The new steady state achieved in response to changes in base cation supply will occur after 2 to 3 lakewater residence times, which amounts to 10 to 20 years for the groundwater recharge lakes in the Upper Midwest. Garrison et al. (1987) modeled Vandercook Lake with the ILWAS model (Gherini et al. 1985) and found that with no groundwater input of base cations and associated ANC, the lakewater pH would drop from 6.1 to 5.4 in 20 years. Lin et al. (1987) showed that increasing groundwater inseepage from 2% to 5% of the total hydrologic flow to Vandercook causes the base cation input from inseepage to increase from 42% to 62% of the total base cation input. These results demonstrate the importance of knowing the groundwater inputs accurately to evaluate the role of inseepage on the acid-base chemistry of lake water.

The acidity of atmospheric deposition would probably not influence the inseepage supply of base cations to seepage lakes, because groundwater has relatively high concentrations of base cations and ANC caused by base cation production in the aquifer (Anderson and Bowser 1986, Lin et al. 1987, Wentz et al. 1987). This is in contrast to drainage lakes in the northeastern United States, for which terrestrial base cation production can be relatively low and atmospheric deposition acting over long time periods may modify this capacity of the watershed to supply base cations.

Base cations are also supplied to lakes from sediments (Lerman and Brunskill 1971, Carignan 1985, Cook et al. 1986), through cation exchange and weathering reactions. The flux of base cations into the water column determined for a number of low ANC lakes from input-output budgets and pore water gradients is 10 to 50 meq m^{-2} $year^{-1}$ (Cook et al. 1987, Lin et al. 1987, Wentz et al. 1987, Brezonik et al. 1987). The production of base cations by the sediments is likely to increase with increasing concentration of hydrogen ion, as has been shown in experimental acidification of lakes (Cook et al. 1986), in microcosms (Schiff and Anderson 1987), and in intact sediment cores (Baker et al. 1985a,b). In an experimentally acidified lake, Cook et al. (1986) found that the increase in base cation concentration was 40% of the increase in SO_4^{2-} concentration. However, the capacity of the sediments to produce base cations is limited and may be decreased with time by elevated hydrogen ion concentrations, so that the capacity of the lake sediments to release base cations is diminished. For example, Baker et al. (1985a) measured a base cation exchange capacity (to pH 5.0) for the upper 1 m of sediments for three northern Wisconsin seepage lakes of 1 eq m^{-3}. This exchange capacity would be depleted in about a decade of atmospheric deposition in the region (40 meq m^{-2} yr^{-1}; Table 13.3) in the absence of other neutralization processes.

The role of sediment-water exchange in influencing water column chemistry is limited to slowly flushing lakes and to lakes with shallow mean depths (Kelly et al. 1987, Baker and Brezonik 1988). In the Upper Midwest, drainage lakes have a median residence time of 1 year and a mean depth of about 2 to 3 m (corresponding to a maximum depth of about 5 m, Table 13.2). A base cation flux from the sediments of 25 meq m^{-2} yr^{-1} would result in a base cation production of 8 to 12 μeq L^{-1} (flux × res. time/mean depth; Kelly et al. 1987). Faster flushing drainage lakes would have a correspondingly lower contribution to the water column from the production of base cations by sediments. In comparison to drainage lakes, seepage lakes have longer water residence times (5 to 10 years; Schnoor et al. 1986, Wentz et al. 1987) and mean depths of 2 to 5 m (corresponding to sampling depths of 3 to 7 m, Table 13.2). For seepage lakes, a base cation flux of 25 meq m^{-2} $year^{-1}$ would generate about 50 μeq L^{-1} of base cations to the water column. This sediment-derived concentration of base cations is about half of the base cation production estimated from ion enrichment calculations for groundwater recharge lakes (see section entitled Processes as Estimated by Ion Enrichment).

Two studies have concluded that base cations are consumed within the lake (Wentz et al. 1987, Baker et al. 1988), with attendant consumption of ANC. Wentz et al. (1987) calculated an in-lake retention of base cations in East Eightmile Lake of 6% of the total inflow of base cations; this retention is within the uncertainties of hydrologic and geochemical budgets. Preliminary results, based largely on sediment trap collections, suggest that base cations are being retained by Little Rock Lake (Baker et al. 1988), which has a base cation concentration of 85 μeq L^{-1}. However, the complete ion input-output budgets for this groundwater recharge lake have not been determined and the relative roles of seepage inflow, in-lake production (Cook et al. 1987, Brezonik et al. 1987), and in-lake consumption have not been assessed.

Sulfate

The median production of ANC from SO_4^{2-} reactions as estimated from ion enrichment calculations is approximately 10 to 50 μeq L^{-1} and is greater for low DOC groundwater recharge lakes than for high DOC drainage lakes (see section entitled Processes as Estimated by Ion Enrichment). For all the lakes with ANC $\leq$ 200 μeq L^{-1}, SO_4^{2-} reactions produce a small but significant component of lakewater ANC. In the absence of sulfate reactions, lakewater ANC would be lower and the lakes would have a lower pH.

Siegel and Pfannkuch (1984) measured input-output budgets for SO_4^{2-} on a small catchment in northeastern Minnesota that contained appreciable wetlands and determined that the watershed was retaining 185 eq ha^{-1} yr^{-1}; similar values for this watershed were obtained by Schnoor et al. (1984). Retention was believed to be the result of a combination of uptake in the wetlands and by soils. The budgets were based on bulk precipitation, which may underestimate the contribution of gases to dry deposition, thereby causing this areal retention to be a lower limit. A small portion of the bedrock in the study watershed contains iron, nickel, and copper sulfide minerals and supplies SO_4^{2-} to some of the groundwater samples in a limited portion of the watershed. Because of the discontinuous extent of the thick till in the watershed, the authors state that the SO_4^{2-}-rich groundwater did not contribute to SO_4^{2-} in surface water. Supply of SO_4^{2-} from the bedrock, particularly from the gabbro in the northeastern part of Minnesota and from contact zones, has been cited as a cause of higher SO_4^{2-} concentrations in this area (Rapp et al. 1985, Nichols and McRoberts 1986, cf. Figure 13.11). However, stable sulfur and oxygen isotope measurements made on a number of lakes in the region suggests that the influence of bedrock sulfur on lakewater SO_4^{2-} is highly localized and limited in regional extent (Malander 1983).

Input-output mass balances of SO_4^{2-} for forested watersheds in the Upper Midwest and adjacent parts of Canada suggest that SO_4^{2-} is retained by SO_4^{2-} reduction in wetlands (Schindler et al. 1976, 1986) and by plant uptake (Pastor and Bockheim 1984), but not by SO_4^{2-} sorption onto mineral soils (Rochelle et al. 1987). Bockheim et al. (1988) observed that SO_4^{2-} was leached from soil plots forested by pine in northwestern Wisconsin, although deciduous and meadow soil plots exhibited retention of between 7% and 35% of the input SO_4^{2-}; the authors did not ascribe this uptake to any specific process.

Sulfate may be retained in the lake by dissimilatory SO_4^{2-} reduction followed by incorporation into the sediments in an organic or inorganic form (Nriagu and Soon 1985, Cook et al. 1986, Rudd et al. 1986a,b, Brezonik et al. 1987). Whole-lake input-output budgets for SO_4^{2-} for four seepage lakes in northern Wisconsin indicate that 40 to 80% of the input SO_4^{2-} was retained by in-lake processes (Lin and Schnoor 1986, Greb et al. 1987). Rates of SO_4^{2-} retention determined by pore water gradients, limnocorrals, and whole-lake budgets are 20 to 50 meq m^{-2} yr^{-1} for low ANC lakes (Cook et al. 1987). The retention of SO_4^{2-} in lakes by dissimilatory SO_4^{2-} reduction is a function of mean depth and water residence times, with slowly flushing, shallow lakes likely to retain a larger percentage of input SO_4^{2-} than fast flushing lakes (Kelly et al. 1987, Baker and Brezonik 1988). For drainage lakes in the Upper Midwest (res. time = 1 year, mean depth = 2.5 m, Table 13.2), SO_4^{2-} retention of 25 meq m^{-2} yr^{-1} corresponds to a loss of SO_4^{2-} of 10 μeq L^{-1}, which is in close agreement with the ion enrichment SO_4^{2-} reaction term (Table 13.6). For seepage lakes in the Upper Midwest (res. time = 7.5 years, mean depth = 3 m), this areal SO_4^{2-} retention rate corresponds to a loss in SO_4^{2-} from the water column of 50 μeq L^{-1}. This loss in SO_4^{2-} concentration for seepage lakes is within the 25th and 75th percentiles calculated by ion enrichment (Table 13.6).

Changes in SO_4^{2-} concentrations are likely to cause changes in the rates of SO_4^{2-} retention, because of the first-order dependence of this process on SO_4^{2-} concentrations (Cook and Schindler 1983, Kelly and Rudd 1984, Cook et al. 1986, Lovely and Klug 1986). However, the capacity of lakes, particularly seepage lakes, to retain SO_4^{2-} may be limited due to the low rate of supply of allochthonous iron and of metabolizable organic carbon (Rudd et al. 1986a,b, Brezonik et al. 1987, Cook in press). For example, Mitchell et al. (1988) observed low SO_4^{2-} retention (12 meq m^{-2} yr^{-1}) in McNearney Lake, a low pH, high SO_4^{2-} lake in upper Michigan, for which the accumulation rate of iron and organic carbon is also quite low (Cook et al. 1990).

Sulfate may also be retained during photosynthesis by assimilatory SO_4^{2-} reduction. This process should be a relatively minor retention process, given the large supply of sulfur in precipitation relative to the small sulfur requirements of terrestrial (Likens et al. 1977) and aquatic (Wetzel 1975) plants. However, for a lake in the Adirondacks, David and Mitchell (1985) found that deposition of sulfur associated with seston, as determined by sediment trap collections, was a more important sink for SO_4^{2-} than was bacterial sulfate reduction.

Nitrate and Ammonium

The concentrations of NO_3^- and NH_4^+ are low and do not contribute significantly to the ionic charge of lake water in the Upper Midwest (Figure 13.10). Because the concentrations of these ions are significant in atmospheric deposition (Table 13.3) and low in the lakes, biogeochemical processes (e.g., uptake during plant growth, denitrification) are efficiently removing these nutrients. NO_3^- and NH_4^+ are efficiently retained in nitrogen-limited terrestrial ecosystems (Bormann et al. 1977). In aquatic ecosystems studied in the Upper Midwest, NO_3^- supplied in excess of that required for growth is effectively taken up by denitrifying communities and NH_4^+ supplied in excess of growth requirements is nitrified (Schindler et al. 1985, Kelly et al. 1987).

Summary

The acid-base status of lakes in the Upper Midwest is controlled largely by base cation production; other ions do not contribute significant amounts of ANC to the majority of lakes in the Upper Midwest. Low ANC lakes, particularly mounded seepage lakes, have a lower base cation production and other ions contribute significantly to the acid-base status. Of particular importance are the reactions of NO_3^-, NH_4^+, and SO_4^{2-}, in addition to base cation production, in controlling ANC in low ANC lakes.

Evidence of Long-Term Trends

Changes in Anthropogenic Emissions and Atmospheric Deposition

Because the majority of sulfur and nitrogen in atmospheric deposition is derived from anthropogenic emissions, estimates of the magnitude and the timing of changes in the amount of anthropogenic emissions can provide some perspective on changes in the sulfur and nitrogen content of atmospheric deposition. Husar (1986) and Husar et al. (Chapter 3, this volume) estimated the anthropogenic emissions of nitrogen and sulfur derived from coal and oil combustion and copper and zinc smelting for Minnesota, Wisconsin, and upper Michigan. The sulfur emissions in this region, currently < 1 Tg S yr^{-1}, have been relatively constant or perhaps increasing slightly over the past 100 years. Nitrogen oxide emissions have been moderately increasing in the Upper Midwest states during the past 100 years, and presently are about 1 Tg N yr^{-1}. Sulfur and nitrogen emissions are currently low relative to emissions in other sectors of eastern North America (about 12% to 25% of sulfur emissions and 5% to 25% of the nitrogen oxide emissions in the other sectors) (Husar 1986; Chapter 3, Husar et al., this volume).

Husar et al. (Chapter 3, this volume) have also estimated historical atmospheric SO_4^{2-} deposition in the Upper Midwest based on estimates of historical emission rates and atmospheric transport. For the Upper Midwest, atmospheric deposition increased from the 1900s to 1920s values up to fourfold estimated background rates. Since that time, estimated historical deposition declined somewhat to rates about twice the estimated background SO_4^{2-} deposition (Chapter 3, Husar et al., this volume).

Even though emissions are lower in the Upper Midwest and the wind patterns are generally from west to east, the region does receive deposition of sulfur and nitrogen that originated in the industrialized sectors of eastern North America, as determined by calculated back trajectories (Barrie et al. 1980, Wisconsin Department of Natural Resources, unpublished data). Furthermore, industrial activity to the south of the Upper Midwest may have contributed to higher atmospheric deposition in the summer, when wind patterns are from south to north. Sources such as paper mills and smelters operating in the region, although not contributing significantly to emissions to the region, may have contributed substantially to local deposition. The Upper Midwest receives relatively small amounts of anthropogenic material via the atmosphere compared to other areas in eastern North America, based upon the low accumulation rates of atmospherically transported anthropogenic material in lake sediments, for example, PAHs (Furlong et al. 1987), V, and Pb (Cook et al. 1990, Kingston et al. in press).

Historical Water Chemistry

Comparison of the acid-base chemistry of surface waters using historical and current data provides an additional means of evaluating changes in the acidity of surface waters. In many regions, historical studies were poorly documented or used unreliable analytical methods, and are therefore not suitable for historical comparisons. However, in their historical studies in northcentral Wisconsin, Birge and Juday (Juday et al. 1938) used methods that are directly comparable to present day methods or that can be adjusted so that they are comparable (Eilers et al. 1989, cited in Kramer et al. 1986).

During the period 1925 to 1941, Birge, Juday, and coworkers collected and analyzed samples from about 600 lakes in northern Wisconsin (Juday et al. 1938). Eilers et al. (1989, cited in Kramer et al. 1986) selected a consistent set of data from 145 lakes sampled by Birge and Juday and rigorously corrected pH and ANC values so that historical data could be compared to the present chemistry from those lakes determined by Eilers et al. (1983). Eilers et al. (1989) concluded that, on average, conductivity and calcium, sulfate, alkalinity, and pH increased over the past 50 years. The median ANC increase was estimated to be 38 $\mu eq\ L^{-1}$ and the median pH increase was estimated to be 0.5 pH units; SO_4^{2-} exhibited a mean increase of 10 $\mu eq\ L^{-1}$. The lakes with the largest increases in pH, ANC, Ca^{2+}, and conductivity had increases in watershed development. Two undeveloped lakes experienced significant ANC increases, apparently associated with major fires in their respective watersheds during the decade preceding recent surveys. When only the 19 lakes whose watersheds have not been developed are considered, 7 lakes experienced a decline in pH. However, because of the modest levels of atmospheric deposition in northcentral Wisconsin, the uncertainty in historical methods, and the complexity of interacting natural and anthropogenic factors that affect lakewater chemistry, Eilers et al. (1989) could not attribute the changes in lakewater chemistry to acidic deposition alone.

Paleolimnological Evidence for pH Changes and for Changes in Atmospheric Deposition

Lacustrine sediments can provide information useful for assessing modification of the watershed-lake system caused by natural changes and anthropogenic activity, including changing atmospheric deposition. Key among the paleolimnological evidence are diatom species abundance, trace metal concentrations, and the concentrations of human-produced organic compounds (e.g., PAH).

Past and present diatom-inferred pH values from three recent studies in the Upper Midwest are collected in Table 13.7. The lakes selected for this analysis have a range of lakewater pH, ANC, and DOC and were selected because they have had minimal watershed disturbance (Kingston et al. in press). Of the 15 lakes for which pH has been inferred, 5 lakes showed a decline of 0.2 to 0.4 pH units during the past 50 to 100 years: Andrus, Brown, Denton, Hillis, and McGrath lakes (Table 13.7). The pH changes cannot be attributed to changes in atmospheric deposition alone, and probably resulted from some combination of land-use changes and changes in atmospheric deposition. Diatom-inferred pH for Andrus Lake was highly variable, with a minimum value in the 1920s, a maximum in the 1940s, and a decline since the 1940s of 0.2 pH units (Kingston et al. in press). One of the 15 lakes (Otto Mielke Lake, Table 13.7) showed an increase of 0.2 pH units, whereas the 9 remaining lakes did not show any recent pH changes. McNearney Lake, Michigan, is a naturally acidic lake that is buffered by aluminum and H^+ against large pH changes (Cook et al. 1990). Organic acids do not appear to have been a factor in the past acidity because of the seepage hydrology and because the diatom taxa do not indicate significant amounts of DOC and organic acids in the past.

The trace elements Pb, Zn, Cu, and V are introduced into lake sediments by atmospheric deposition and inputs from the watershed. Through use of trace element to titanium ratios in lake sediment cores, the terrestrially derived trace metal burden can be factored out of the inventory of trace elements, thereby allowing determination of the component derived from atmospheric deposition (Norton et al. in press, Kingston et al. in press). These results indicate that the atmospheric deposition of Pb, Zn, Cu, and V increased in concentration and accumulation rate around the 1900s and that the magnitude of the increase was smaller than that observed in the industrialized sectors of eastern North America (Charles and Norton 1986). For some lakes, particularly McNearney and Brown, the accumulation rate of Zn is smaller than expected

TABLE 13.7. Inferred trends in the acid-base chemistry of lakes in the Upper Midwest obtained by paleolimnological reconstruction.

State	Lake	Current pH	Inferred-pH trend and time frame	Reference
Michigan	Andrus	5.6	From about 1940 to the present, pH declined from 6.2 to 6.0. Diatom-inferred pH record quite variable.	Kingston et al. in press
	McNearney	4.4	Over the past 2,000 years, pH remained constant at 4.9. Poor agreement between diatom-inferred current pH and current pH is due to uncertainties in regression.[a]	Cook et al. 1990; Kingston et al. in press
Wisconsin	Brown	5.7	From 1860 to 1900, pH was 6.2. After 1900, pH declined to 5.9.	Kingston et al. in press
	Camp 12	5.4	From about 1850 to present, pH was 5.7	Kingston et al. in press
	Denton	5.2	Between 1850 and the present, pH fluctuated by 0.6 pH units from 5.8 to 5.2. pH highs of 5.8 in the 1950s may be related to watershed logging. pH declined from 5.8 to 5.4 from 1950s to 1984.	Kingston et al. in press
	Otto Mielke	5.3	From 1850 to present, pH increased from 5.5 to 5.7.	Kingston et al. in press
Wisconsin	Bastile	5.0	Over the period 1880–1985 pH has remained constant at 5.6. Poor agreement between diatom-inferred current pH and current pH is due to uncertainties in regression.[b]	Kingston, unpublished data
	Hillis	5.6	Over the period 1880–1985, pH has declined from 5.8 to 5.5	Kingston, unpublished data
Wisconsin	McGrath	5.2	Over the period 1880–1985, pH declined from 5.6 to 5.4	Kingston, unpublished data
	Morgan	4.6	Over the period 1880–1985, pH remained constant at 5.4. Poor agreement between diatom-inferred current pH and current pH is due to uncertainties in regression.[b]	Kingston, unpublished data
	Sugar Camp	5.4	Over the period 1880–1985, pH remained constant at 5.7.	Kingston et al. in press
Minnesota	August	6.8	Over the past 5,000 years, pH remained constant at pH 6.9	Wonson-Liukkonen 1987
	Dunnigan	6.7	During the period 1850–1984, pH remained constant at 6.5.	Kingston et al. in press
	Nels	6.4	During the period 1840 to 1984, pH remained constant at 6.6	Kingston et al. in press
	Hustler	6.8	During the period 1850 to 1984, pH remained constant at 6.8.	Kingston et al. in press

[a] The calibration data set on which pH inference equations are based includes only one lake with pH < 5.1 (McNearny). Therefore, pH inferences for lakes with pH < 5.1 are not well constrained and have a large uncertainty.

[b] The calibration data set on which pH inference equations are based includes only three lakes with pH < 5.1. Therefore, pH inferences for lakes with pH < 5.1 are not well constrained and have a large uncertainty.

based on the other trace elements, perhaps due to a decrease in the rate of removal from the low pH water column or to uncertainties in the estimates of Zn contributed from the watershed (Norton et al. in press, Kingston et al. in press).

PAH accumulation rates were determined in several Upper Midwest lakes by Hites and coworkers. The accumulation rates at the sediment surface are higher than those from horizons corresponding to the 1800s in these cores and are similar in magnitude to those observed at the sediment surface in other remote areas and also northern Florida (Gschwend and Hites 1981, Furlong et al. 1987, McVeety and Hites 1988). However, the integrated flux is an order of magnitude less than those observed for sediments in the Adirondack Mountains and other lakes receiving high rates of atmospheric deposition (Hites 1981, Furlong et al. 1987), indicating that the levels of atmospheric deposition of anthropogenic elements in the Upper Midwest are smaller than those in other regions of North America.

Dynamic Models – Hindcasts

Hindcasts of dynamic watershed models provide an indication of how surface water chemistry may have varied in the past. Schnoor et al. (1986) have made hindcasts of lakewater chemistry for three Upper Midwest seepage lakes using time-variable and steady-state versions of the Trickle Down model. The three lakes – Clara (Wisconsin), Vandercook (Wisconsin), and McNearney (Michigan) – are similar in surface area, mean depth, and water residence times, but McNearney, which is in eastern upper Michigan, receives 30% higher levels of acidic deposition. Weathering rates for the Trickle Down model were determined from the present day concentration of base cations along with hydrologic information determined from field studies and from Cl^- concentrations (Schnoor et al. 1986). Due to relatively long water residence times, the three lakes may not be at steady state with respect to the current level of deposition.

The time-variable model was used in conjunction with estimates of past levels of atmospheric deposition to simulate the time-trend of acid-base chemistry; the authors acknowledge that, because of the uncertainty in the past changes in atmospheric deposition and uncertainties in the response of weathering to changes in atmospheric deposition, the model results should be viewed as indicators of the types of changes the lake may have experienced rather than a firm reconstruction of past acid-base chemistry. Schnoor et al. (1986) used a fivefold increase in the acidity of deposition for Clara and Vandercook and a sevenfold increase for McNearney over the past 80 years. These levels are high relative to changes in emissions of sulfur and nitrogen in the region as estimated by Husar (1986) and Husar et al. (Chapter 3, this volume).

The acidification projected to have occurred in Clara and Vandercook lakes by Schnoor et al. (1986) is about 10 to 15 $\mu eq\ L^{-1}$ of ANC, and in McNearney is 60 $\mu eq\ L^{-1}$ of ANC. This acidification of Clara and Vandercook is similar to the variability in present day field data ($\pm$ 10 $\mu eq\ L^{-1}$), which suggests that, according to the Trickle Down model, the lakes have not acidified to a measurable extent. The model-projected acidification of McNearney is inconsistent with the paleolimnological analysis for this lake, which suggested that no change in pH occurred (see section on Historical Water Chemistry). Causes of disagreement between the diatom-inferred pH reconstructions and the model results may be due to some combination of the estimated time trend and magnitude of changes in atmospheric deposition and the response of weathering mechanisms to changes in atmospheric deposition. Also the levels of aluminum in McNearney provided some buffering against pH changes, so that increases in atmospheric deposition (i.e., SO_4^{2-}) may have been partially offset by increases in aluminum, with no change in pH. The Trickle Down model did not treat aluminum, which may also explain the apparent discrepancy between the model and paleolimnological results.

These model runs provide a perspective on the changes in acid-base chemistry that may have occurred in the Upper Midwest. Although low ANC seepage lakes in Wisconsin appear to have limited resistance to changes in atmospheric deposition due to low base cation production, the possible losses of ANC have been small and within the seasonal to annual variability in lakewater ANC.

Summary of Trend Data

The emissions of nitrogen and sulfur in the Upper Midwest have not increased greatly over the past 100 years and are considerably lower than emissions estimated for other regions in eastern North America. Changes in estimated historical deposition are also lower in the Upper Midwest, compared to those for other regions (Chapter 3, Husar et al., this volume). Paleolimnological analysis and comparison of recent and historical lakewater chemistry data indicate a decrease of 0.2 to 0.3 pH units for some lakes in the Upper Midwest. Dynamic watershed modeling suggests that historical changes in deposition caused relatively small changes in ANC for two of the three lakes modeled. Model hindcasts for the third lake simulated a large ANC decline, in part because the model does not consider buffering by aluminum reactions. The changes indicated by these lines of evidence are less than those observed in the Adirondacks (Chapter 6, Driscoll et al., this volume) and were probably caused by a number of factors, including changes in atmospheric deposition and land-use. The changes cannot be attributed to changes in atmospheric deposition alone.

Future Considerations

Dynamic watershed models have been developed to project future changes in acid-base status of lakes that may occur in response to changes in environmental parameters, such as atmospheric deposition. The models that have been applied to the Upper Midwest are ILWAS and Trickle Down; these models are described in Munson and Gherini (Chapter 2, this volume), Gherini et al. (1985), and Schnoor et al. (1986). The models incorporate a number of processes that affect acid-base chemistry, with formulation that allows the rates of processes to change in response to changes in environmental parameters. For example, the models have base cation weathering rates that are functions of the H^+ concentration. The models are calibrated using field measurements made over a relatively short period. Intrinsic limitations of models of this type are (1) the short time period of calibration and (2) the lack of validation that the model formulation and parameterization are adequate to predict with certainty the future changes in acid-base chemistry of surface waters. Although the models do include many of the processes affecting lakewater chemistry, they are necessary simplifications of these processes and are based on many assumptions concerning spatial, temporal, and process aggregation. For example, Trickle Down does not incorporate the behavior of aluminum and DOC, and ILWAS does not treat the production and consumption of DOC as a function of H^+ (Chapter 2, Munson and Gherini, this volume).

This section discusses published model projections of changes in the acid-base chemistry of lakes in response to changes in atmospheric deposition. Projections are based on several alternative atmospheric deposition scenarios. The projections are meant to give an indication of possible responses to future changes in atmospheric deposition and should be viewed as experimental simulations made using models that have not been adequately tested under the range of possible future conditions.

The ILWAS Model

The ILWAS model has been used to project the changes in acid-base chemistry for five seepage lakes in northcentral Wisconsin and one drainage lake in northeastern Minnesota (Chen et al. 1988, Garrison et al. 1987, Greb et al. 1987). The results of model forecasts after a 20-year period, which is long enough for a new steady state to be established, are presented in Table 13.8. Even though the forcing function of the models is an increasing atmospheric deposition of SO_4^{2-}, the present loading and the projected future loading are still smaller than SO_4^{2-} loading today in the northeastern United States (Chapter 5, Eilers et al., this volume; Table 13.3).

A 50% increase in atmospheric deposition of strong acids would result in a drop of 0.75 pH units for Crystal Lake in northcentral Wisconsin and a decline of 18 $\mu eq\ L^{-1}$ ANC (Table 13.8). The simulated response to a 25% reduction in atmospheric deposition was projected to be an increase of 0.1 to 0.2 pH units and ANC increases of 10 to 13 $\mu eq\ L^{-1}$ for Crystal (Table 13.8). This simulation shows the sensitivity of this seepage lake to changes in atmospheric deposition.

A twofold increase in atmospheric deposition to Meander Lake, northeastern Minnesota, would cause the pH to decrease by 0.5 pH units and ANC to decrease by 32 $\mu eq\ L^{-1}$. A 50% decrease in atmospheric deposition for Meander Lake caused projected ANC to increase by 16 $\mu eq\ L^{-1}$ and projected pH to increase by 0.1 pH units (Table 13.9).

Changes in pH and ANC for increased rates of atmospheric deposition were smaller for Meander than those for Crystal, primarily because the rates

TABLE 13.8. Summary of results of RILWAS model scenarios at steady state.[a]

Scenario	Parameter[b]	Crystal Lake Present	Crystal Lake Predicted	Vandercook Lake Present	Vandercook Lake Predicted
25% reduction of atmospheric sulfur loading	pH	6.1	6.25	6.1	6.3
	ANC	18	28	28	41
	SO_4^{2-}	75	65	75	62
50% reduction of atmospheric sulfur loading	pH	6.1	6.4	–	–
	ANC	18	37	–	–
	SO_4^{2-}	75	56	–	–
50% increase of atmospheric sulfur loading	pH	6.1	5.35	–	–
	ANC	18	0	–	–
	SO_4^{2-}	75	94	–	–
Elimination of sulfate reduction	pH	6.1	5.5	6.1	4.7
	ANC	18	6	28	−18
	SO_4^{2-}	75	91	75	121
Elimination of groundwater input	pH	6.1	5.4	6.1	5.4
	ANC	18	0	28	0
	SO_4^{2-}	75	75	–	–
	Ca^{2+}	55	45	–	–
50% groundwater increase	pH	6.1	6.3	–	–
	ANC	18	25	–	–
	SO_4^{2-}	75	75	–	–
	Ca^{2+}	55	60	–	–

[a] Greb et al. (1987). Reprinted with permission.
[b] ANC, SO_4^{2-}, and Ca^{2+} expressed as μeq L^{-1}.

of atmospheric deposition both today and in the two future scenarios are different. Also, the capacity of the soil and bedrock to supply base cations to Meander Lake by geochemical weathering and by cation exchange is quite high. Meander Lake also has much higher DOC concentrations than Crystal Lake, which contributes both weak and strong organic acids, thereby contributing to the lake-water ANC (Sullivan et al. 1989).

TABLE 13.9. Summary of RLWAS model scenarios and results at steady state for Meander Lake, Minnesota.[a]

	Parameter	Meander Lake Present	Meander Lake Predicted
Deposition increased by 50%	pH	6.1	5.9
	ANC	70	52
	SO_4^{2-}	79	[97][b]
Deposition increased by 100%	pH	6.1	5.6
	ANC	70	37
	SO_4^{2-}	79	[111][b]
Deposition decreased by 50%	pH	6.1	6.2
	ANC	70	86
	SO_4^{2-}	79	[63][b]

[a] Data from Chen et al. 1988.
[b] Predicted SO_4^{2-} is estimated from the change in ANC, assuming that changes in ANC are determined solely from changes in SO_4^{2-}.

The Trickle Down Model

The Trickle Down model predicts that Vandercook Lake is not at steady state with the current level of deposition and that after about 20 years the ANC will decline from current levels of 15 μeq L^{-1} to 5 μeq L^{-1} (from pH 6.1 to 5.8) (Schnoor et al. 1986). A 100% increase in the atmospheric deposition of SO_4^{2-} simulated an ANC loss of 49 μeq L^{-1} to an ANC of −34 μeq L^{-1} (pH 4.5). Differences between ILWAS and Trickle Down projections for Vandercook Lake are due to differences in model algorithms, to differences in today's hydrologic budget, and differences in the changes in atmospheric deposition. The ILWAS model used an inseepage that was 16% of the total hydrologic input, and Trickle Down used a seepage input of 3% of the total; these differences were the result of different groundwater wells used to determine the water chemistry of inseepage to Vandercook Lake. Trickle Down used a zero-order rate expression for ANC production by SO_4^{2-} retention in the sediments. Research on SO_4^{2-} reduction rates has shown that the process is first order with respect to SO_4^{2-} concentrations (Cook and Schindler 1983, Lovley and Klug 1986). For future elevated rates of atmospheric deposition, the ANC production rate

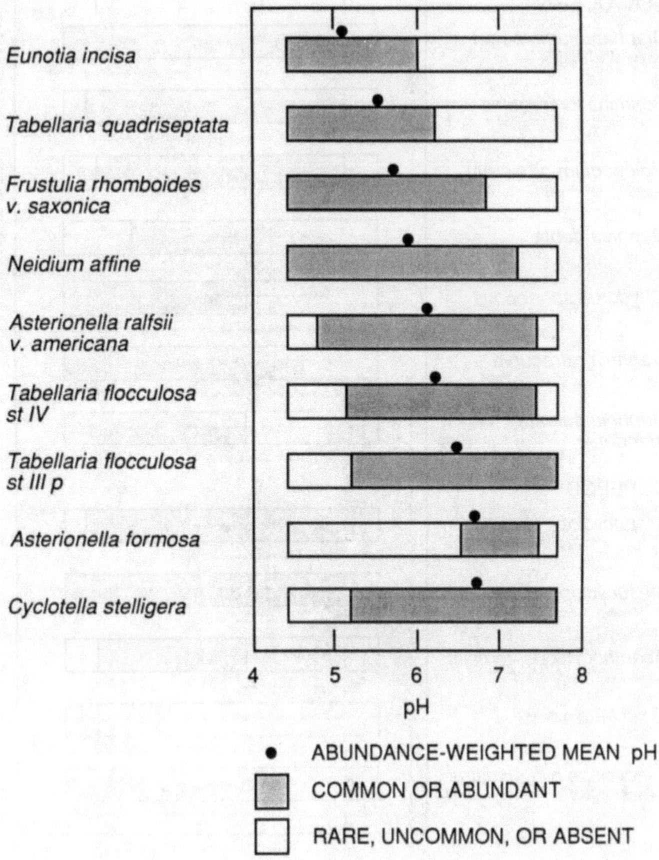

FIGURE 13.23. pH distribution of selected diatom taxa from 36 lakes in the Upper Midwest. Drawn from data in Kingston et al. (in press). "Common or abundant" corresponds to an abundance of > 10% of the maximum for a species in this data set. "Rare, uncommon, or absent" corresponds to an abundance of < 10% of the maximum for a particular species. Abundance weighted mean pH is an estimate of the species optimum along the pH gradient.

in the lake is likely to increase, thereby mitigating some of the increase in acidity. ILWAS and an enhanced version of Trickle Down incorporate this first-order dependence on SO_4^{2-} concentrations (Chapter 2, Munson and Gherini, this volume).

Biological Status

Plankton

Studies of the regional association between non-fish biota and lake acidity have not been published for the Upper Midwest region, although several data sets are currently being processed and analyzed. As part of a project to use diatom assemblages in sediments to reconstruct the pH history of lakes in the Upper Midwest, the distribution, abundance, and association of over 400 diatom taxa were investigated in surface sediment samples from 36 lakes in the Upper Midwest (Kingston et al. in press). The distribution of most fossil diatom taxa were closely related to lakewater pH (Figure 13.23), indicating that many diatom taxa have specific pH preferences and tolerances within the region. Assemblage diversity and richness were low in the more alkaline, plankton dominated lakes and in two acidic lakes dominated by Melosira species (Kingston et al. in press). Shallow acidic lakes generally contained higher abundances of benthic species than deep acidic lakes (Kingston et al. in press). Statistical analysis suggests that, although pH and pH correlated parameters explain most of the variance in diatom assemblage composition, other factors (e.g., DOC, nutrient content, lake morphometry, human disturbance, fish population manipulations) may also influence diatom assemblage composition (Kingston et al. in press).

The Wisconsin Department of Natural Resources (WDNR) has compiled data on zooplankton samples for 120 lakes in northern Wisconsin. Preliminary results from analysis of this data set suggest that many of the zooplankton species that commonly occur in lakes in the area are relatively acid

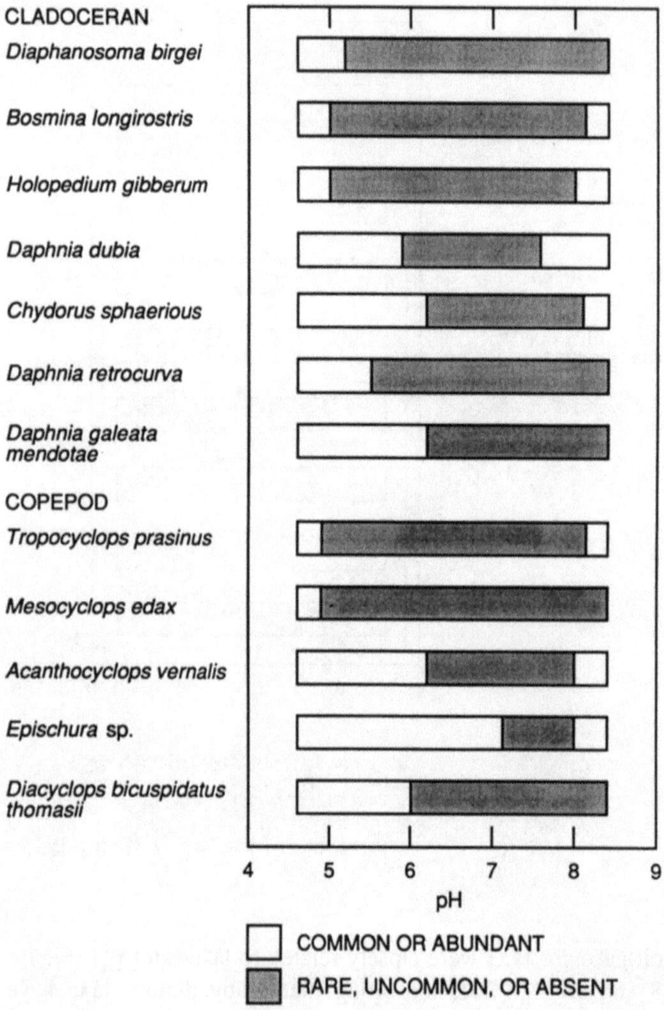

FIGURE 13.24. pH distribution of selected zooplankton taxa from 120 lakes in northcentral Wisconsin. "Common or abundant" represents greater than 51 individuals. "Uncommon, rare, or absent" represents 50 individuals or fewer. Absolute abundances were not determined. Data from K. Webster, Wisconsin Department of Natural Resources, unpublished.

tolerant (K. Webster, WDNR, pers. comm.; Figure 13.24). Only 1 of the 120 lakes examined, Morgan Lake with pH 4.6, had a zooplankton community considered typical for an acidic lake (i.e., dominated by a single calanoid copepod species, *Diaptomus (Leptodiaptomus) minutus*; (Hodgson et al. 1987). Species distributions and species richness decrease at pH levels below about 5.2 to 5.5, although several species occur at pH levels as low as 5.0 (K. Webster, WDNR, pers. comm.; Figure 13.24).

Fish Communities

Fish communities in nonwinterkill, low ANC lakes of northeastern Minnesota, northern Wisconsin, and upper Michigan typically are dominated by yellow perch (*Perca flavescens*), largemouth bass (*Micropterus salmoides*), sunfishes [chiefly bluegill (*Lepomis macrochirus*) and pumpkinseed (*L. gibbosus*)], bullheads (*Ictalurus* spp.), and white sucker (*Catostomus commersoni*) (Payer 1985, Rahel and Magnuson 1983, Wiener 1983, Cusimano et al. 1988). Other sport fishes common in low ANC lakes in the region include northern pike (*Esox lucius*), walleye (*Stizostedion vitreum vitreum*), smallmouth bass (*M. dolomieui*), black crappie (*Pomoxis nigromaculatus*), and, to a lesser degree, salmonids such as brook trout (*Salvelinus fontinalis*) and lake trout (*S. namaycush*).

Several investigators have observed a spatial association between fish community composition and lake acidity in the Upper Midwest. Low pH lakes (pH < 6.0) in both northcentral Wisconsin (Wiener et al. 1984, Rago and Wiener 1986) and

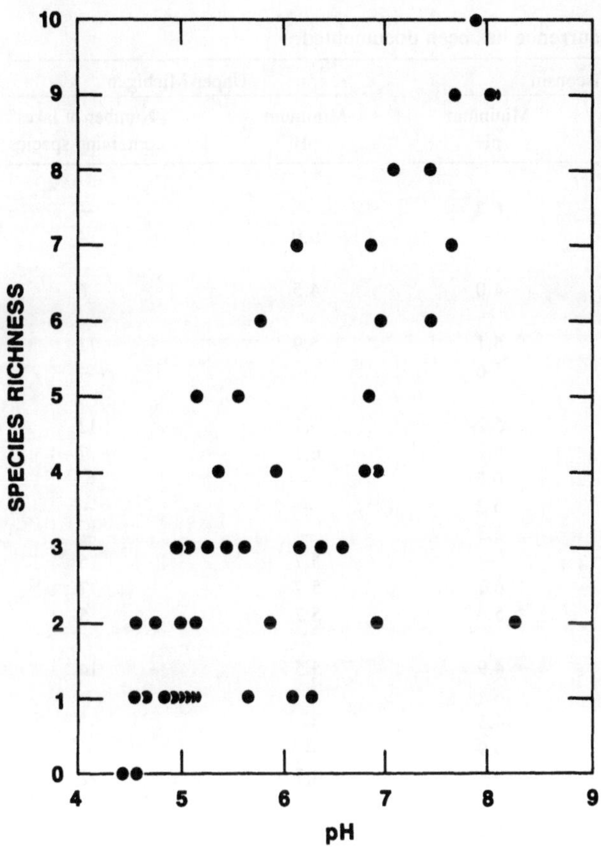

FIGURE 13.25. Species richness (number of fish species caught) as a function of lake pH for the 49 lakes surveyed in Upper Michigan as part of Phase II of the Eastern Lake Survey. (From Cusimano et al. 1990.)

upper Michigan (Cusimano et al. 1990; Figure 13.25) tend to have significantly fewer fish species than do higher pH lakes of comparable size and depth. Wiener and Eilers (1987) compiled information on fish species distribution relative to lake pH for 150 Wisconsin lakes. Twenty-nine fish species occurred in at least 10 lakes; minimum pH values for species occurrence are tabulated in Table 13.10. Ten species–four cyprinids (*Cyprinidae*), three percids (*Percidae*), and the cisco (*Corgonus artedii*), trout-perch (*Percopsis omiscomaycus*), and burbot (*Lota lota*)–had a pH minimum ≥ 6.0. Five species had pH minima in the range of pH 5.5 to 5.9, six had minima in the range 5.0 to 5.4, and eight were found at pH < 5.0. The most acid-tolerant fishes were the central mudminnow (*Umbra limi*) and yellow perch. Results for 49 lakes in upper Michigan were generally quite similar (Cusimano et al. 1990; Table 13.10).

Caution must be applied, however, in assuming that the observed associations between fish community characteristics and lake chemistry imply a causal relation to acidification. The Upper Midwest is used extensively for recreation and many of the fisheries in the region's lakes have been manipulated through fish poisoning and stocking.

Rahel (1986) noted that several other factors are also highly correlated with lake pH and may be responsible, in part or entirely, for patterns of fish species distribution in the region. Most of the clearwater acidic lakes in the Upper Midwest are seepage systems that have sandy substrates in littoral areas (Eilers et al. 1988). Smallmouth bass and walleye require wave-washed, hard-bottomed substrate as spawning sites; such areas are often lacking in acidic seepage systems. Northern pike spawn in vegetated littoral zones, a habitat also uncommon in small, low-nutrient waters. The absence of certain acid-sensitive cyprinids and darters (*Etheostoma*) from small oligotrophic acidic lakes reflects an intolerance of low pH. For more tolerant species, biotic interactions, for example, predation and competition from centrachids (*Centrarchidae*) and yellow perch, may assume greater importance (Rahel 1986, Tonn and

TABLE 13.10. Minimum pH at which fish species occurrence has been documented.[a]

Family and species	Northern Wisconsin: Number of lakes[b] containing species	Northern Wisconsin: Minimum pH	Upper Michigan: Minimum pH	Upper Michigan: Number of lakes[b] containing species
Salmonidae				
Cisco	17	6.2	–	–
Brook trout	–	–	5.0	4
Umbridae				
Central mudminnow	52	4.0	4.5	6
Esocidae				
Northern pike	71	5.1	5.9	11
Muskellunge	44	5.6	–	–
Cyprinidae				
Golden shiner	69	5.2	5.1	12
Common shiner	32	6.2	6.1	7
Blacknose shiner	17	6.5	–	–
Mimic shiner	22	6.2	–	–
Northern redbelly dace	13	5.3	–	–
Finescale dace	–	–	5.7	6
Bluntnose minnow	50	6.2	5.7	7
Creek chub	14	5.6	5.7	5
Catostomidae				
White sucker	93	4.9	5.5	14
Ictaluridae				
Black bullhead	65	4.5	–	–
Yellow bullhead	43	4.9	–	–
Brown bullhead	–	–	4.7	12
Percopsidae				
Trout-perch	10	6.2	–	–
Gadidae				
Burbot	25	6.0	–	–
Gasterosteidae				
Brook stickleback	11	5.4	(4.6)	(3)
Centrarchidae				
Rock bass	63	5.2	7.0	4
Pumpkinseed	88	4.9	4.9	15
Bluegill	95	4.5	4.5	16
Smallmouth bass	49	5.2	7.0	5
Largemouth bass	92	4.6	5.0	10
Black crappie	66	5.6	–	–
Percidae				
Iowa darter	25	6.2	5.7	7
Johnny darter	40	6.2	–	–
Yellow perch	126	4.4	4.5	31
Walleye	60	5.5	–	–
Logperch	15	6.3	–	–
Cottidae				
Mottled sculpin	30	5.6	–	–

[a] Data for northern Wisconsin lakes ($n = 150$) were compiled by Wiener and Eilers (1987) from Rahel and Magnuson (1983) and Wiener (1983). Data for lakes in upper Michigan ($n = 49$) from Cusimano et al. (1990). Scientific names corresponding to the common names given in this table may be found in Robins et al. 1980.

[b] Only fish species caught in at least 10 lakes in northern Wisconsin or 4 lakes in upper Michigan are included in the table.

Magnuson 1982). The specific environmental factors that limit fish species distribution are difficult to define based on survey data alone. However, acid sensitivity for many cyprinids has been recorded in the ample toxicological data from laboratory and field studies (Chapter 4, Baker and Christensen, this volume). The survey data are thus consistent with the acid sensitivity for certain of these taxa.

Surveys of fish communities in upper Michigan (Cusimano et al. 1990) were conducted as part of Phase II of the ELS (ELS-II). Lakes were selected based on a variable probability sampling design from waters surveyed during ELS-I (Linthurst et al. 1986, Hillman et al. in preparation); thus, the survey results can be used to establish regional estimates of the current status of fish communities in the upper Michigan subregion (subregions 2C, 2D, and 2E in Figure 13.1b). ELS-I indicated that an estimated 11.3% of the lakes in this subregion had ANC ≤ 0 μeq L^{-1} in fall 1984; 9.4% had pH < 5.0; 17.7% had pH < 6.0. Despite the relatively large numbers of acidic and low pH lakes in upper Michigan, most lakes in the area (over 99% of the target population of lakes for ELS-II, by number and by lake area) supported at least one fish species. Sport fish, defined as yellow perch, walleye, largemouth and smallmouth bass, brook trout, northern pike, and lake trout, occurred in 84% of the lakes, or 96% by lake area. The most common fish in upper Michigan was yellow perch, an acid-tolerant species that occurred in an estimated 70% of the lakes, or 89% by lake area. Only six fishes – all considered tolerant of low pH – were caught in lakes with pH < 5.0: yellow perch, central mudminnow, brook stickleback (*Culaea inconstans*), brown bullhead (*Ictalurus nebulosus*), bluegill, and pumpkinseed (Table 13.10) (Wiener and Eilers 1987; Chapter 4, Baker and Christensen, this volume). Similar regional probability samples of fish communities are not available for other areas in the Upper Midwest.

Historical Fisheries

Wiener and Eilers (1987) examined the available data on temporal fishery trends for low ANC lakes in the Upper Midwest. Changes in fish community status have occurred in recent years in several low ANC lakes in northern Wisconsin (Table 13.11), consistent with the hypothesis of lake acidification, but natural causes of fishery declines could not be conclusively ruled out. For example, in Morgan Lake, the density and diversity of the lake's fish fauna have apparently declined substantially in the past 15 to 20 years (Hodgson et al. 1987). The low pH (4.6) and elevated metal concentrations in Morgan Lake suggest toxic conditions for fish. Wiener and Eilers (1987) concluded that the available data were not sufficient to determine the degree to which changes in lakewater chemistry (pH and Al) may have affected fishery resources in the area.

Biological Communities and Processes

The possible effects of acidification on biological communities and processes in lakes in the Upper Midwest are being evaluated as part of the experimental acidification of Little Rock Lake, a clearwater seepage lake in northcentral Wisconsin (Brezonik et al. 1986). Frost et al. (1988) summarized results from the first 3 years of the experimental acidification, first to pH 5.6 (2 years) and then to pH 5.1 (1 year). The major responses observed (relative to the reference basin at pH 6.1) were:

- An increased prevalence of benthic algae (principally *Mougeotia*)
- Increased lake transparency
- No overall change in epilimnetic chlorophyll-*a* or ^{14}C measures of primary productivity
- A shift in the relative abundances of major algal groups during the second year of acidification
- Shifts in zooplankton species composition
- No change in abundance or species composition of benthic invertebrate communities
- Reduced emergence of one of four chironomid subfamilies in the lake
- Reduced rates of leaf decomposition for some species of leaves
- No decrease in the reproductive activity of yellow perch, largemouth bass, black crappie, and rock bass

These preliminary results from the Little Rock Lake experiment are consistent with the types and magnitude of biological effects from acidification observed in other regions and for other clearwater

TABLE 13.11. Partial listing of clearwater lakes in Northern Wisconsin showing chemical conditions and biological symptoms consistent with recent acidification.[a]

Lake	County	Depth (m)	Maximum area (ha)	pH	ANC (μeq L⁻¹)	Ca^{2+} (μeq L⁻¹)	Total Al (mg L⁻¹)	Dissolved organic carbon (μmol L⁻¹)	SO_4^{2-} (μeq L⁻¹)	Biological reports
Morgan	Florence	3.7	17.2	4.60[b,c]	−27	63	–	–[d]	–	Nearly fishless; apparent decline in density and diversity of fish fauna; no recruitment of fishes evident.[e]
				4.59[f]	−17	56	172	–	150	
Silverbass	Oneida	3.4	12.0	4.93[b,c]	−6	55	85	33	130	Reported loss of largemouth bass.
Sand	Oneida	5.9	15.0	5.01[b]	−9	87	70	75	168	Recent loss of largemouth bass; yellow perch still present.
				5.17[g]	9	81	35	92	148	Recent extensive growth of filamentous gree algae.[h]
McGrath	Oneida	7.4	21.4	5.22[b,c]	−5	49	41	233	100	Recent stocking of smallmouth bass failed.
				5.31[f]	2	56	39	–	106	Recent extensive growths of filamentous gree algae.[h]
				5.45[g]	1	47	29	275	84	
Sugar Camp	Oneida	11.7	220.6	5.40[b]	0	77	48	75	135	White suckers had poor condition factors; other fish species including smallmouth bass, reported in good condition.[h]
				5.50[g]	4	71	37	92	118	

[a] Wiener and Eilers (1987). Reprinted from the *Journal of Lake Management* with permission of the North American Lake Management Society.
[b] Chemical data in row, unless noted otherwise, are unpublished data from J.M. Eilers and G.E. Glass (U.S. E.P.Ag, Duluth, MN).
[c] Possible historical decline in pH.
[d] Mean apparent color is 2.5 PCU; Secchi disk transparency exceeded maximum depth (Schmidt, 1985).
[e] Data from J.R. Hodgson (St. Norbert College, De Pere, WI, pers. commun.).
[f] Chemical data in row are annual mean values for samples collected at 0.5-m depth (Schmidt, 1985).
[g] Chemical data in row are from the Eastern Lake Survey (Kanciruk et al. 1986).
[h] K.E. Webster and M.D. Johnson (Wisconsin Department of Natural Resources, Rhinelander, WI, pers. commun.).

lakes with low ionic strength (Chapter 4, Baker and Christensen, this volume).

Conclusions

1. The majority of the Upper Midwest is underlain by surficial deposits that provide few solutes to lakes, resulting in low ionic strength and low ANC systems. The region receives a gradient in atmospheric deposition, but SO_4^{2-} and NO_3^- deposition rates in the Upper Midwest are less than 50% of those observed in the eastern United States.
2. Hydrologic lake type provides some of the chemical diversity exhibited by lakes in the Upper Midwest, with seepage lakes accounting for 61% to 78% of the lakes with ANC $\leq$ 200 $\mu eq\ L^{-1}$ in northern Wisconsin and upper Michigan. In contrast, seepage lakes are only 11% of the lakes with ANC $\leq$ 200 $\mu eq\ L^{-1}$ in northeastern Minnesota. Within the entire Upper Midwest region, an estimated 80% of the 128 lakes with ANC $\leq$ 0 $\mu eq\ L^{-1}$ are seepage lakes.
3. In the Upper Midwest, the lowest ANC systems also have the lowest concentration of solutes, suggesting that solute production for the low ANC lakes is lower than for the high ANC lakes. The primary cations are Ca^{2+}, Mg^{2+}, and Na^+, HCO_3^-, SO_4^{2-}, and organic anions are the predominant anions at ANC > 50 $\mu eq\ L^{-1}$, whereas SO_4^{2-} is the dominant anion in the lakes with ANC $\leq$ 0 $\mu eq\ L^{-1}$.
4. The key process regulating the acid-base chemistry of lakes with ANC > 50 $\mu eq\ L^{-1}$ in the Upper Midwest is base cation production, from weathering and base cation exchange. Watershed-lake systems, particularly seepage lakes, that have low rates of base cation production typically have low ANC. The spatial gradient in ANC for lakes with ANC $\leq$ 200 $\mu eq\ L^{-1}$ within the Upper Midwest results from the gradient in base cation production. Groundwater recharge seepage lakes in upper Michigan have the lowest watershed-lake production of base cations of any lake type within the Upper Midwest.
5. Although wet SO_4^{2-} deposition increases from west-to-east by a factor of 1.6, lakewater SO_4^{2-} concentrations in lakes with ANC $\leq$ 200 $\mu eq\ L^{-1}$ do not exhibit a corresponding spatial gradient. This observation signifies that SO_4^{2-} concentrations in low and negative ANC lakes (ANC $\leq$ 50 $\mu eq\ L^{-1}$) are controlled by a combination of atmospheric deposition, biogeochemical reactions within the lakes, and, in northeastern Minnesota, by bedrock sources. Lakes in northern Minnesota underlain by bedrock that contains appreciable sulfur have median SO_4^{2-} concentrations 20 $\mu eq\ L^{-1}$ higher than lakes in that subregion underlain by bedrock that does not contain sulfur. However, the variability of lakewater SO_4^{2-} is such that the concentrations are not significantly different from those in other subregions of the Upper Midwest.
6. Lakewater DOC concentrations exhibit a west-to-east gradient as a result of a shift from predominantly drainage lakes in Minnesota to seepage lakes in the eastern part of the region. Also, wetlands are more prevalent in the western part of the Upper Midwest. DOC and organic acids are not responsible for causing the pattern of low and negative ANC lakes.
7. Analysis of historical evidence suggests that emissions and atmospheric deposition have not increased greatly over the past 100 years, in contrast to trends estimated for the eastern United States. Comparisons of recent and historical lakewater chemistry data, and also analysis of paleolimnological data suggest that some lakes have experienced a pH loss of 0.2 to 0.3 units, which corresponds to an ANC loss of 10 $\mu eq\ L^{-1}$. However, atmospheric deposition cannot be singled out as the cause of these pH losses.
8. Forecasts from the Trickle Down dynamic watershed model predict that with no change in atmospheric deposition, about 10 $\mu eq\ L^{-1}$ of ANC will be lost from some of the lakes, because the lakes are currently not at steady state with today's deposition. The ILWAS and Trickle Down models both predict that future increases in atmospheric deposition are likely to cause further declines in ANC. However, the magnitude of the acidification would be smaller for every percentage increase in deposition than similar percentage changes in the northeastern United States, because the current levels of deposition are low and percentage increases would also be low.

9. The distribution of phytoplankton, zooplankton, and fishes is closely related to lakewater pH, indicating that many species have specific pH preferences and tolerances within the Upper Midwest. In shallow acidic lakes, benthic diatom species prevail over planktonic species. In northern Wisconsin, many zooplankters are acid tolerant, yet zooplankton species distribution and species richness decline for the lower pH lakes. Low pH lakes (pH < 6.0) in both northcentral Wisconsin and upper Michigan tend to have significantly fewer fish species than do high pH lakes of comparable size and depth. Changes in fish community status have occurred in recent years in several low ANC lakes in northern Wisconsin, consistent with the hypothesis of lake acidification, but natural causes of fishery declines could not be conclusively ruled out.

Acknowledgments. Joan Baker, Kathy Webster, and Jim Wiener contributed information on aquatic biota for the section on Biological Communities in Low ANC Lakes in the Upper Midwest. The assistance of the following people is acknowledged: Don Charles and Joe Eilers for offering support and encouragement; Dale Nichols for supplying data tapes; Jim Wiener, Steve Gherini, Ron Munson, Kathy Webster, and Paul Garrison for providing copies of unpublished manuscripts; John Kingston for access to unpublished data and reports on paleolimnology in the Upper Midwest; John Beauchamp for statistical advice; and Jeremy Smith for assistance with ion enrichment calculations. D.F. Charles, J.M. Eilers, J.W. Elwood, P. Garrison, R.S. Turner, D.W. Schindler, J. Stoddard, K. Webster, and J. Wiener critically reviewed this manuscript and provided many helpful comments. Martha Kirby assisted with manuscript preparation. This work was supported by the U.S. Environmental Protection Agency (EPA) under Interagency Agreement DW89932112-01-7 with the U.S. Department of Energy. This chapter does not necessarily reflect the opinions of the EPA, and no official endorsement should be inferred. Oak Ridge National Laboratory (ORNL) is managed by Martin Marietta Energy Systems, Inc., under contract DE-AC05-84OR21400 with the U.S. Department of Energy. Publication No. 3390, Environmental Sciences Division, ORNL.

West Overview

John M. Melack

The lakes of the western United States (Figure 1) are mostly of glacial origin and contain the most dilute waters in the country. The lowest median ANC values occur in the Pacific mountain ranges,

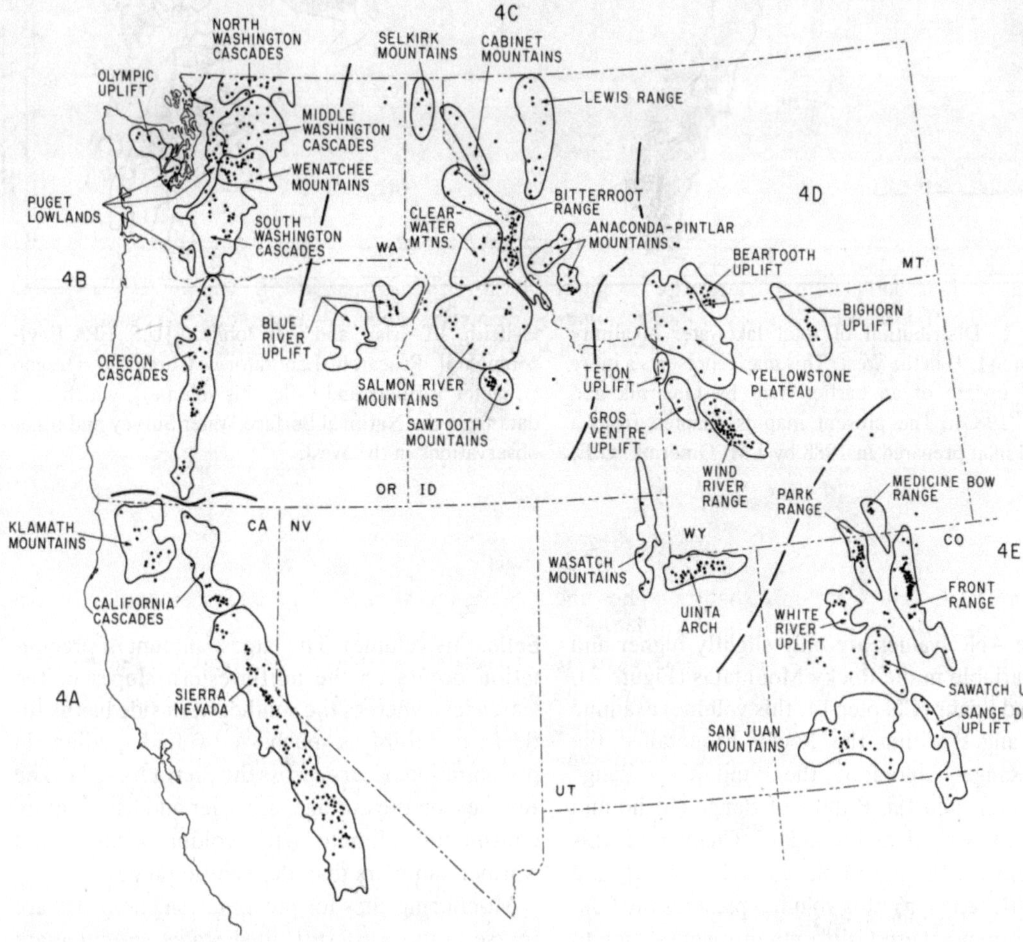

FIGURE 1. Locations and names of geomorphic units that represent physiographic areas of common geological origin and that generally coincide with major mountain ranges. Dashed lines delineate WLS subregions 4A to 4E. (From Landers et al. 1987.)

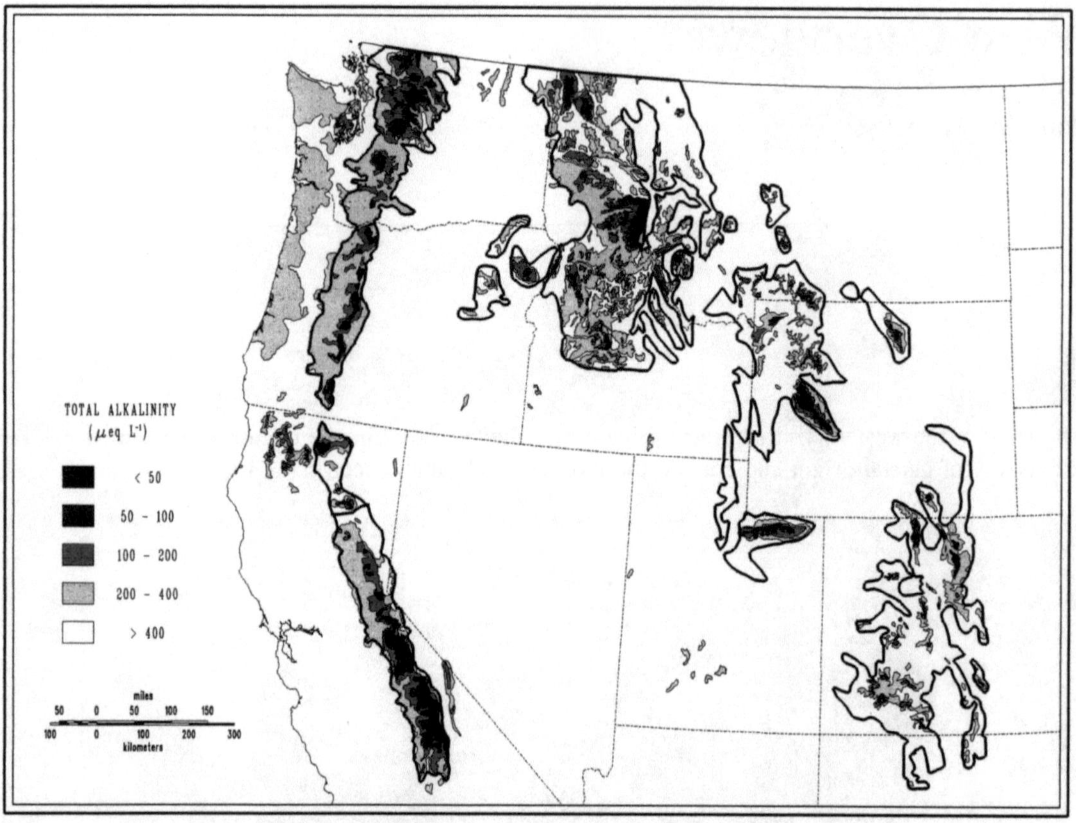

FIGURE 2. Distribution of total lakewater alkalinity (ANC; $\mu eq\ L^{-1}$) in the West. This map represents a more detailed update of an earlier map by Omernik and Powers (1983). The present map is adapted from a national map prepared in 1988 by J.M. Omernik, G.E. Griffith, J.T. Irish, and C.B. Johnson [U.S. EPA Environmental Research Laboratory, Corvallis, Oregon (Chapter 5, Eilers and Selle, this volume)], which used data from the National Surface Water Survey and other observations in the West.

and the ANC values are only slightly higher and more variable in the Rocky Mountains (Figure 2). Turk and Spahr (Chapter 14, this volume) examine three ranges within the Rocky Mountains: the Front Range (Colorado), the Wind River Range (Wyoming), and the Bitterroot Range (Idaho and Montana). Melack and Stoddard (Chapter 15, this volume) treat the Sierra Nevada (California), and Nelson (Chapter 16, this volume) presents the Cascades from northern California through Oregon to northern Washington as six subunits (Figure 1).

Large variations in precipitation quantity occur within and among subregions across the western United States (Figure 3; Chapter 5, Eilers and Selle, this volume). The largest amount of precipitation occurs on the northwestern slopes of the Cascades, whereas the southeastern side has as little as one-third as much. A twofold gradient in precipitation occurs across the Sierra Nevada. The Rockies, on the average, are drier and have a more continental climate with colder winters and warmer summers than the Pacific ranges.

Monitoring sites for precipitation chemistry are sparse in the western United States and none are located in the high mountains (Figure 4). Differences in precipitation chemistry are relatively small among the three mountain regions addressed in this book. Values of pH are higher and concen-

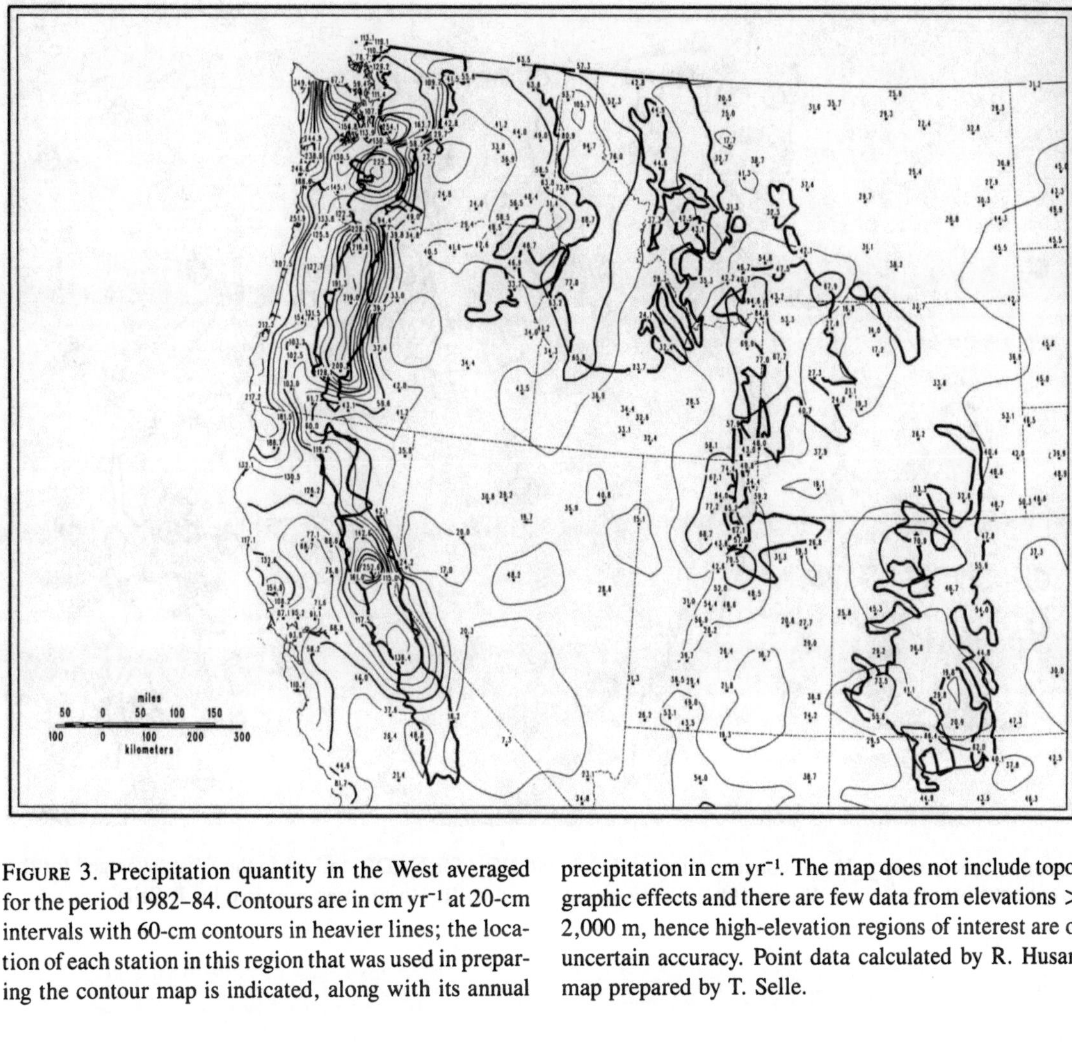

FIGURE 3. Precipitation quantity in the West averaged for the period 1982–84. Contours are in cm yr^{-1} at 20-cm intervals with 60-cm contours in heavier lines; the location of each station in this region that was used in preparing the contour map is indicated, along with its annual precipitation in cm yr^{-1}. The map does not include topographic effects and there are few data from elevations > 2,000 m, hence high-elevation regions of interest are of uncertain accuracy. Point data calculated by R. Husar; map prepared by T. Selle.

trations of SO_4^{2-} and NO_3^- are much lower than in the eastern United States.

Throughout the western mountains, snow accumulates water and solutes during the winter, and these are released during a brief period of snowmelt. This annual episodic event flushes most of the lakes with very dilute water that may contain elevated concentrations of SO_4^{2-} and NO_3^-. Hence, an episodic depression of ANC attributable to dilution and perhaps titration is an important general characteristic of the western montane and alpine lakes.

Weathering reactions within the watersheds of high-elevation western lakes are the principal source of ANC. However, the stoichiometry of the reactions implicates rocks that may be a negligible portion of the watersheds in all regions, but especially in the Rocky Mountains. The large watershed area to lake volume ratio, the steep relief, and the deep snowpacks reduce the role of within-lake processes as ANC sources, at least during snowmelt.

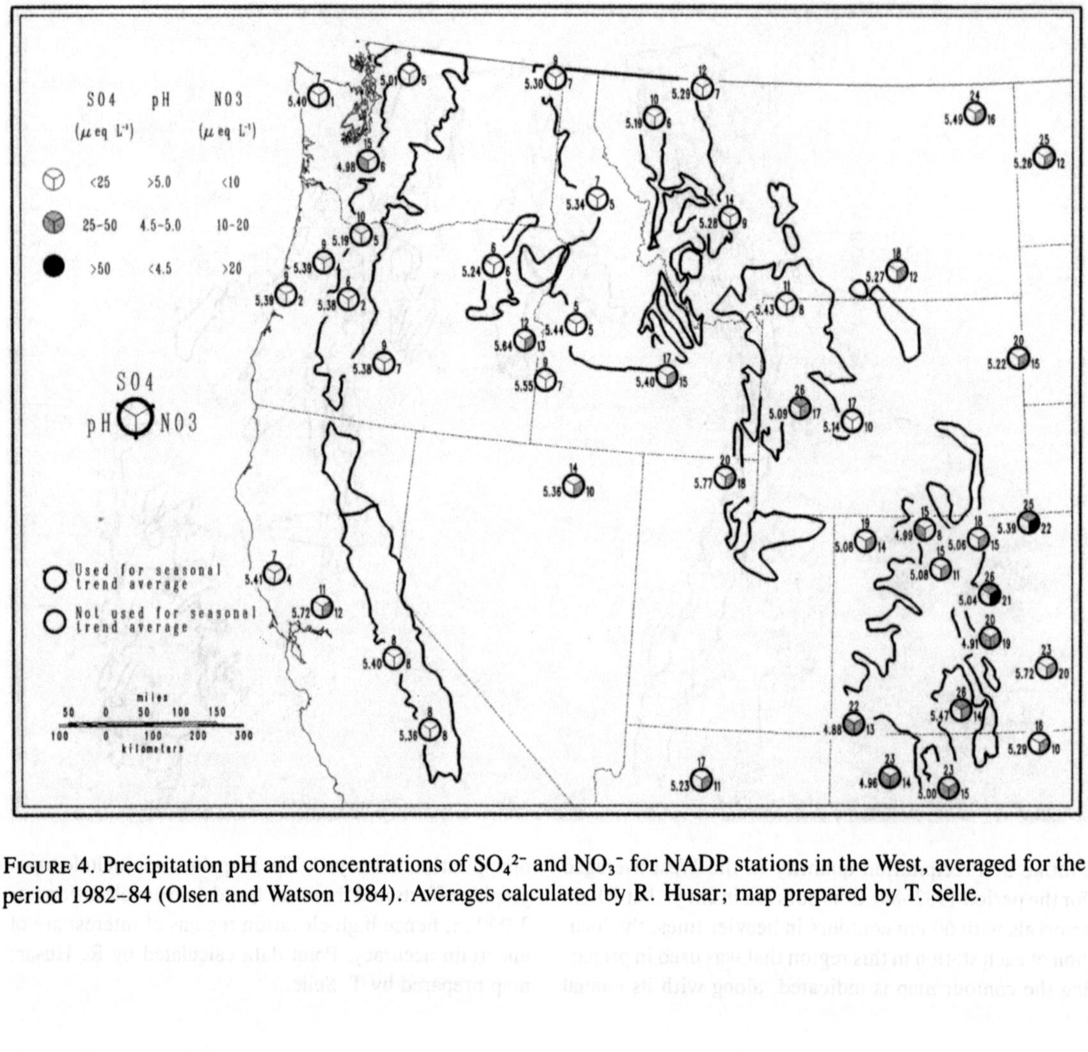

FIGURE 4. Precipitation pH and concentrations of SO_4^{2-} and NO_3^- for NADP stations in the West, averaged for the period 1982–84 (Olsen and Watson 1984). Averages calculated by R. Husar; map prepared by T. Selle.

14
Rocky Mountains

John T. Turk and Norman E. Spahr

ABSTRACT. Most Rocky Mountain lakes sampled by the Western Lake Survey are not acidic, although they are very sensitive to acidification. Because of a lack of historic data, indirect methods must be used to estimate past acidification. Lakes having ANC of ≤ 200 μeq L^{-1} are estimated to have lost no more than 5 μeq L^{-1} in the Bitterroot Range, 10 μeq L^{-1} in the Colorado Front Range, and 12 μeq L^{-1} in the Wind River Range. Present (1989) data are inadequate for determining whether episodic acidification occurs.

Concentrations of SO_4^{2-} and Cl^- in dilute Rocky Mountain lakes are controlled primarily by the concentration of these ions in wetfall. Evapotranspiration, dryfall, anion adsorption, SO_4^{2-} reduction, and mineral weathering rarely are half as important as wetfall in controlling the lake concentration of these ions.

Lake concentrations of major cations, silica, and ANC are not consistent with a simple stoichiometric weathering of the granitic minerals common to the lake watersheds. In particular, lake concentrations of sodium and silica are smaller than would be expected from such stoichiometric weathering. Preferential dissolution of common granitic minerals, atmospheric sources of readily weathered minerals, or trace amounts of readily weathered minerals within the watersheds may be the source of most of the major cations within the lakes.

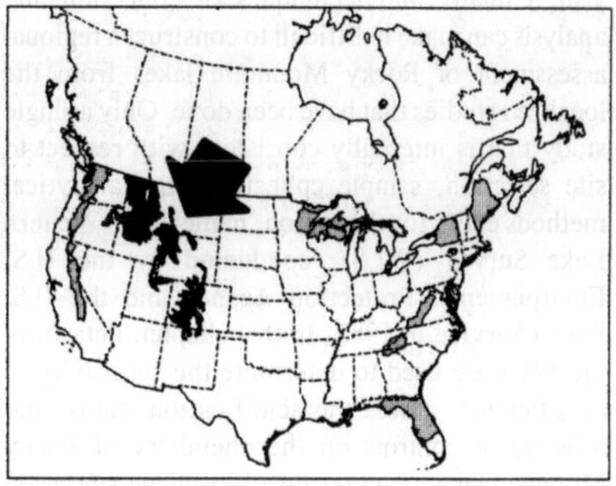

Introduction

The Rocky Mountain region contains many thousands of lakes that are important habitat for wildlife, an integral part of many mountain watersheds, and an essential part of the recreational experience sought by those who tour the backcountry. Many of these lakes are in wilderness areas, where direct anthropogenic effects on the watersheds are minimal. Thus, the chemistry of these wilderness lakes is controlled primarily by natural watershed and in-lake processes that have been altered in many lakes in other parts of the nation. Such pristine lakes and watersheds provide an opportunity to study natural systems. The need for special protection of these scarce wilderness systems was incorporated into the Wilderness Act and the Clean Air Act, which require an assessment of the effects of new sources of atmospheric emissions on wilderness areas. In spite of the remote setting and legal protection of most Rocky Mountain lakes, some research has indicated possible acidification (e.g., Lewis 1982).

Assessment of the effects of emissions on wilderness lakes requires knowledge of the processes that control lake chemistry and biology. However, study of these controlling processes is made difficult by physical and legal restrictions on access. Many wilderness lakes are accessible only during two or three months of the year; in addition, wilderness areas generally may not be disturbed or instrumented, as is normally necessary for detailed study of processes. Thus, much of our knowledge of wilderness lakes must be inferred from simple data on lake characteristics, rather than from complex studies in which each component and process is individually quantified.

Problems can develop in comparing data from different studies, even with the most basic of measurements of lake chemistry. Random and systematic variations caused by differences in season, year, climate, and techniques of collection and analysis can make it difficult to construct a regional assessment of Rocky Mountain lakes from the localized studies that have been done. Only a single study that is internally consistent with respect to site selection, sample collection, and analytical methods exists for the region, namely, the Western Lake Survey (WLS), conducted by the U.S. Environmental Protection Agency and the U.S. Forest Service in 1985. In this chapter, data from the WLS are used to determine the sensitivity to acidification, assess the acidification status, and infer some controls on the chemistry of Rocky Mountain lakes. The results of localized studies in the region have been summarized previously (Turk 1983, Turk and Spahr 1989). Because the Rocky Mountain region is so large and its watershed characteristics are so diverse, it may not always be appropriate to consider the lakes as part of a single group. By comparing subsets from different areas within the Rocky Mountain region, it is possible to test for patterns of similarity or difference within the region. Such tests can aid in determining the controls on lake chemistry.

Regional Characteristics

The Rocky Mountains are a discontinuous series of mountain ranges extending from New Mexico into Canada. The bedrock geology varies greatly. Although the dominant bedrock of most of the Rocky Mountains tends to be granite or gneiss, many areas are composed of quartzite, basalt, volcanic breccia, monzonite, or other igneous and metamorphic bedrock. Thus, bedrock geology is best described for local areas rather for than the Rocky Mountains as a whole. In this chapter, background characteristics are provided in the section entitled Processes Affecting Surface Water Chemistry.

Topographic relief is great, with many of the mountain ranges forming the Continental Divide. Many individual peaks are greater than 4,000 m in elevation and rise above lowlands of less than 2,000 m. Because of the great elevational gradient, several vegetative zones are common. Dense montane forests grade into transitional subalpine forests and, commonly, into alpine zones. Vegetative characteristics are discussed in more detail in the sections on the individual areas.

The Rocky Mountains have a severe continental climate. Snow can occur during any month, and is common from September through May. Snow tends to accumulate continuously from early fall, until a single, major snowmelt occurs from about May through July. This snowmelt supplies most of the annual runoff of most lakes and streams in the region. The volume of water from snowmelt commonly is sufficient to completely replace the water stored in lakes several times during the snowmelt period. Thus, lake chemistry often is greatly influenced by the chemistry of the preceding year's snowpack.

Figure 14.1 is a photograph of King Lake, Colorado, sampled during the Western Lake Survey. Its watershed characteristics are common to many of the Rocky Mountain lakes that are most sensitive to acidification. Its chemical characteristics are typical of sensitive lakes in the Colorado Front Range.

The large geographic extent of the Rocky Mountains results in greatly varying amounts of precipitation. Differences in source areas of moisture, orography, latitude, extent of persistent subtropical moisture flow "monsoonal patterns," and extent of late summer cold fronts can cause differences in timing and amounts of precipitation. Assessment of precipitation is complicated by the fact that stations that measure precipitation amount are uncommon at the elevations of lakes discussed in this chapter. Those stations existing at such elevations generally have missing records, which makes it difficult to compute temporal variation in precipitation amount. Figure 14.2 shows the monthly variation in precipitation amount at Fraser, Colorado, and at

FIGURE 14.1. Aerial oblique photo of King Lake, Colorado, sampled during the Western Lake Survey.

Estes Park, Colorado (Hansen et al. 1978). These two stations are located within 50 km of each other on opposite sides of the Colorado Front Range. Figure 14.2a indicates somewhat uniform distribution of precipitation throughout the year at the Fraser station, with slightly increased amounts in January through May. In comparison, extreme variation occurs at the nearby Estes Park station (Figure 14.2b). Many precipitation stations in the Colorado Front Range receive five times the average monthly precipitation in late spring or early summer that they receive in winter. Thus, the amount and seasonal distribution of precipitation in the Rocky Mountains seems to be site specific. Any generalized statement about distribution of precipitation amount is likely to be misleading, even for small areas within the Rocky Mountains. To generalize for the entire Rocky Mountain region would risk misleading the reader. Similarly, to generalize temporal patterns of deposition of solutes in precipitation would be misleading.

Regional Surface Water Chemistry

Recent data collected during the WLS (Eilers et al. 1987, Landers et al. 1987) provide an internally consistent regional assessment of the acidification status of lakes. The data were collected using a stratified random sampling design that allows estimation of characteristics of all lakes with surface areas > ~1 ha. Weighted population estimates are used for the areas discussed in this chapter.

Geographic distributions of pH, ANC, and ANC/SBC for Rocky Mountain lakes (Figures 14.3, 14.4, and 14.5) indicate that a wide range of values for each constituent occurs throughout the Rocky Mountains. Areas that consistently contain smaller values of these constituents and thus are likely to be sensitive to acidification are the Colorado Front Range, the Wind River Range of Wyoming, and the Bitterroot Range of Idaho and Montana.

The physical characteristics estimated for Rocky Mountain lakes are shown in Figures 14.6(a-c).

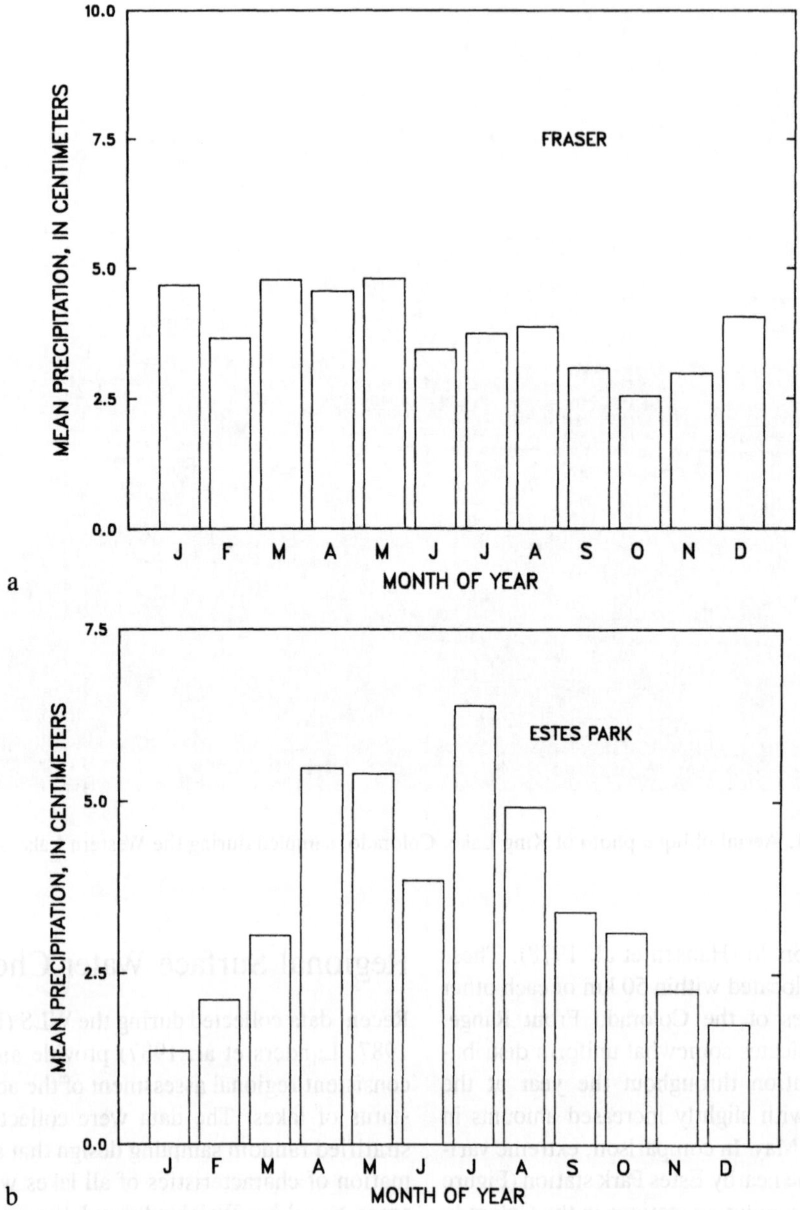

FIGURE 14.2. Average monthly precipitation amount at (a) Fraser, Colorado, and (b) Estes Park, Colorado.

These lakes are small, with surface areas commonly $< \sim 10$ ha. Although maximum depth varies greatly, it most commonly is $< \sim 10$ m. Many lakes have maximum depths of only a few meters; thus, these lakes may completely freeze during winter and be unsuitable habitat for fish. The ratio of watershed area to lake surface area commonly is ≥ 15; thus, even with a snowpack water content of 1 m, there is sufficient snowmelt to equal lake volume for most lakes.

The distributions of pH and ANC estimated for Rocky Mountain lakes are shown in Figures 14.6(d–e). Of the 424 lakes sampled in this region, only Fern Lake, Wyoming, had pH < 5.6 (pH $=$ 4.7). The SO_4^{2-} concentration of Fern Lake was 818 μeq L^{-1}, whereas the concentration in wetfall

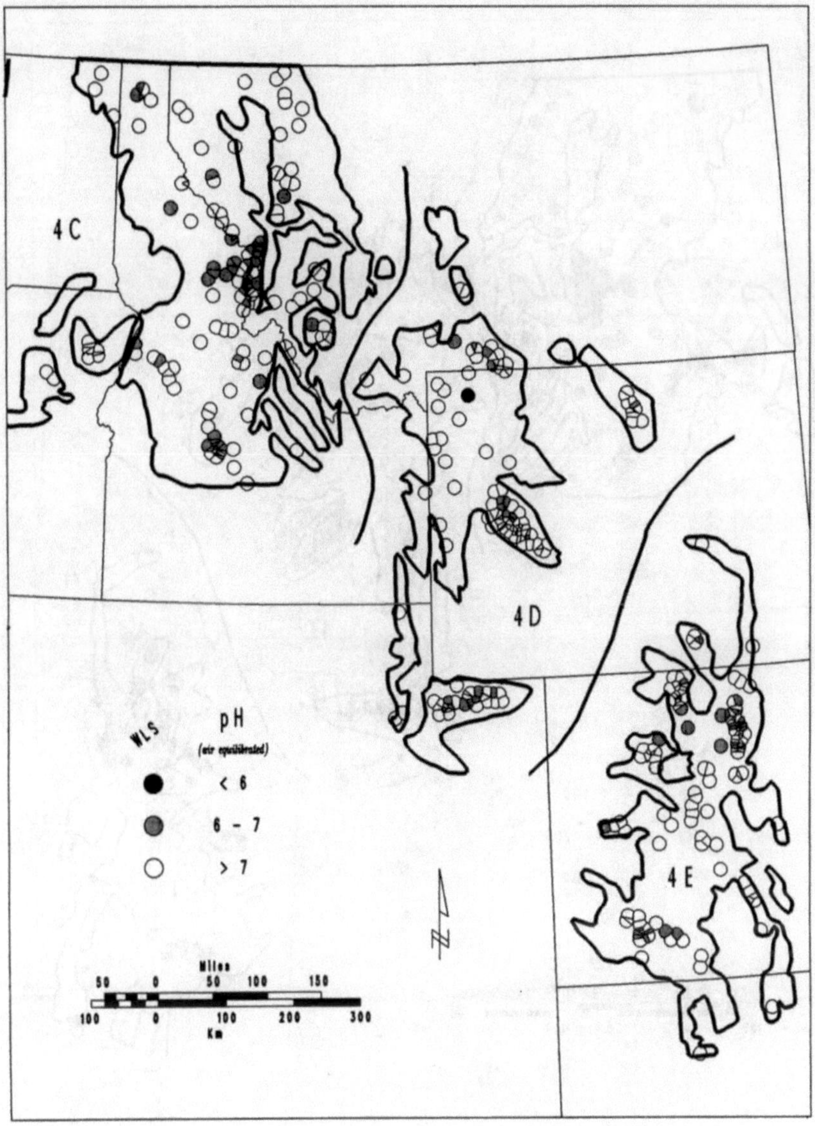

FIGURE 14.3. Distribution of lakes in three pH categories.

is about 2% of this value, indicating that watershed sources of acidity, rather than atmospheric deposition, caused the low pH of 4.7. Most of the lakes sampled have pH $\geq$ 7.0, and thus are not acidic. However, most of the lakes sampled have ANC < 200 $\mu eq\ L^{-1}$ and are sensitive to acidification (Hendrey et al. 1980, Turk and Adams 1983). Historical data that would allow a direct determination of whether ANC has decreased in these sensitive lakes are not available. Thus, indirect methods are necessary to estimate past acidification.

The SBC estimated for Rocky Mountain lakes (Figure 14.6f) has a distribution similar to that of ANC (Figure 14.6e), although SBC is slightly greater than ANC. Anions other than ANC are thus significant, although not dominant, in these lakes. Sulfate concentrations estimated for Rocky Mountain lakes (Figure 14.6g) are most commonly 25 $\mu eq\ L^{-1}$ or less.

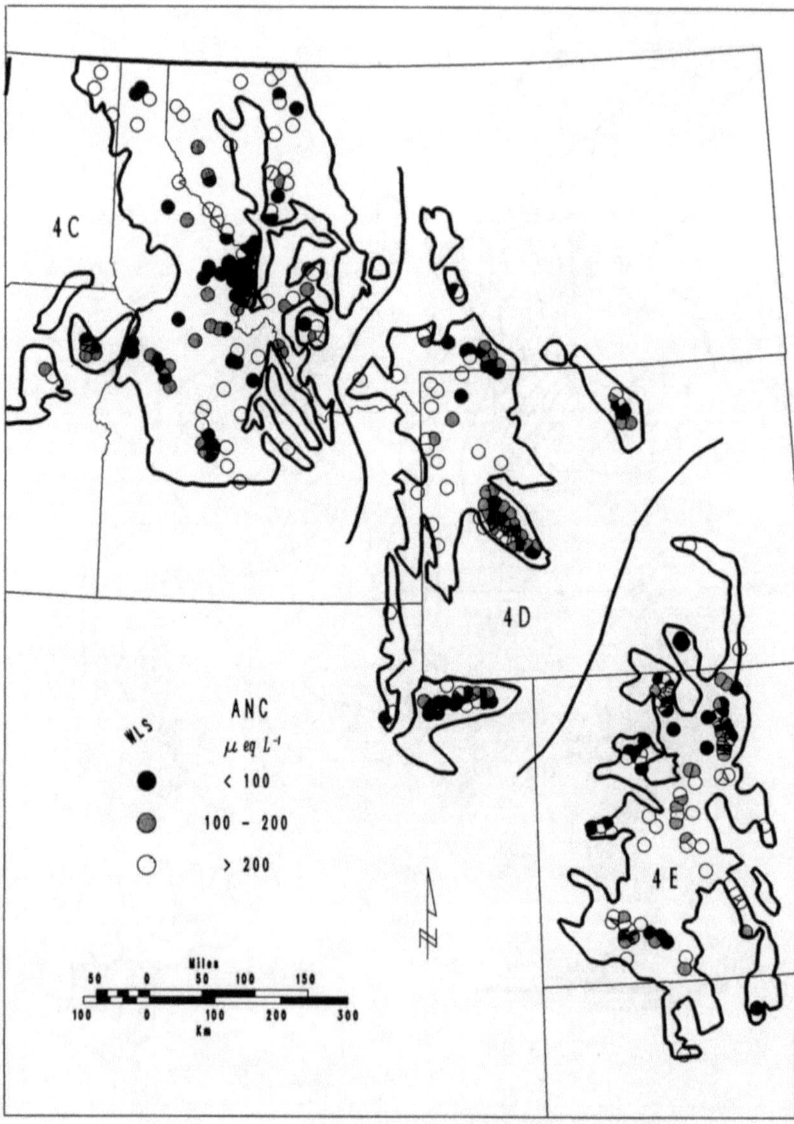

FIGURE 14.4. Distribution of lakes in three ANC categories.

Dissolved organic carbon concentrations estimated for Rocky Mountain lakes (Figure 14.6h) are usually $< \sim 50\ \mu mol\ L^{-1}$. Thus, DOC commonly is of minor importance in the charge balance of Rocky Mountain lakes, assuming a carboxyl content similar to that determined by Oliver et al. (1983).

Silica concentrations estimated for Rocky Mountain lakes (Figure 14.6i) most commonly are $< \sim 50\ \mu mol\ L^{-1}$. Thus, SiO_2 concentrations are small compared to other major weathering products, such as SBC and ANC. Silica concentrations may be small because of removal by processes such as uptake by diatoms.

Variation in ANC has little relation to variation in $SO_4{:}C_B$, as indicated by the near vertical distribution of most of the data points in Figure 14.7. Thus, acidification by SO_4^{2-} and associated H^+ is minimal in most of these lakes.

Among cations, the most variation with respect to changes in ANC (Figure 14.8) occurs in Ca^{2+}

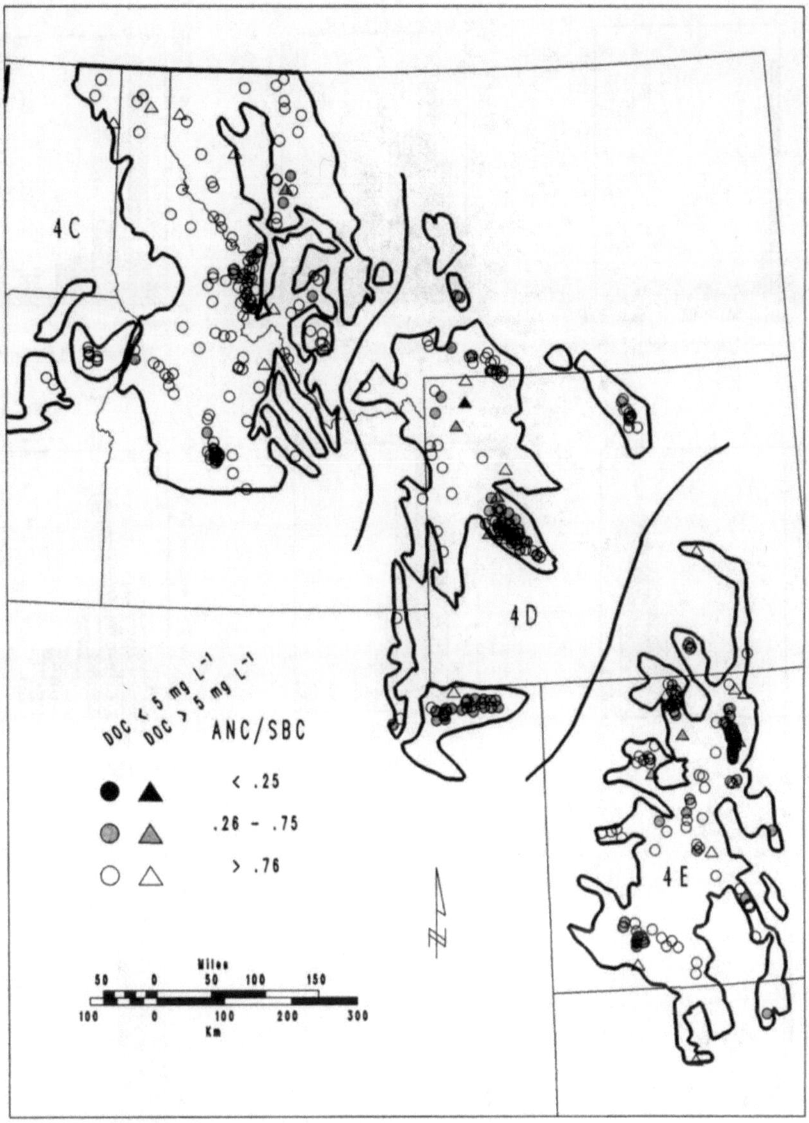

FIGURE 14.5. Distribution of lakes in ANC/SBC classes. Low DOC lakes have DOC values ≤ 417 μmol L^{-1} (5 mg L^{-1}).

and Mg^{2+}, which increase consistently with increase in ANC. Because anions other than ANC are comparatively invariant, their concentrations, compared to ANC, increase as ANC decreases.

Although no data set other than the WLS covers the entire Rocky Mountains, several studies have sampled lakes within the area. Table 14.1 lists data from the WLS along with a 1987 sampling of lakes in the Colorado Front Range done cooperatively by the U.S. Geological Survey, the U.S. Environmental Protection Agency, and the Colorado Department of Health. Whereas the WLS emphasized a stratified random sampling of lakes, the 1987 study sampled Front Range lakes, hypothesized to have acidified, for which historical data were available (Lewis 1982). The 1987 study used sampling and analysis protocols consistent with those of the WLS. Median concentrations for most constituents in

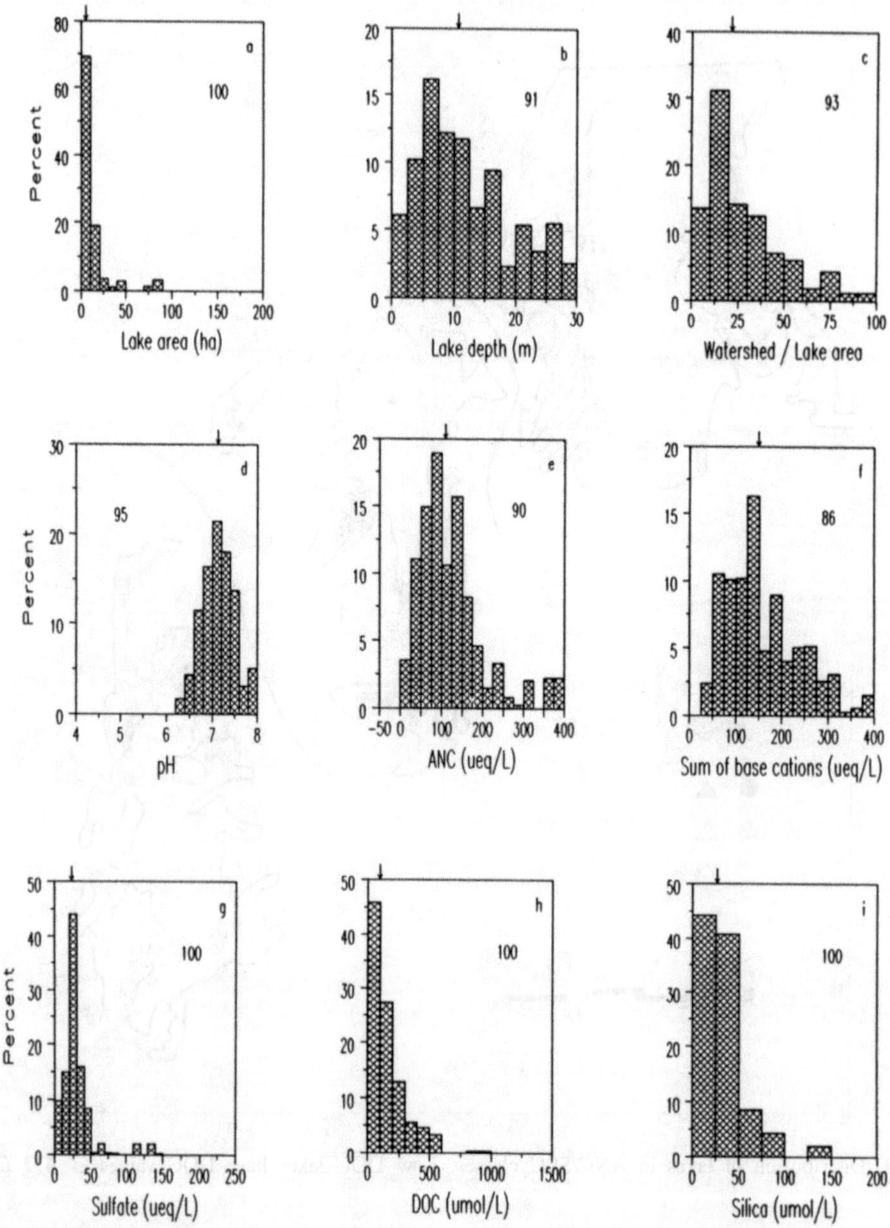

FIGURE 14.6. Frequency histograms of major characteristics of the target population of lakes in the WLS: (a) lake surface area, (b) sampling depth, (c) watershed area:lake area, (d) air-equilibrated pH, (e) ANC, (f) sum of base cations, (g) sulfate, (h) DOC, and (i) silica for Rocky Mountain region. The arrow at the top of each box indicates the median for the target population and the number in the upper right side of each box indicates the percentage of target lakes plotted.

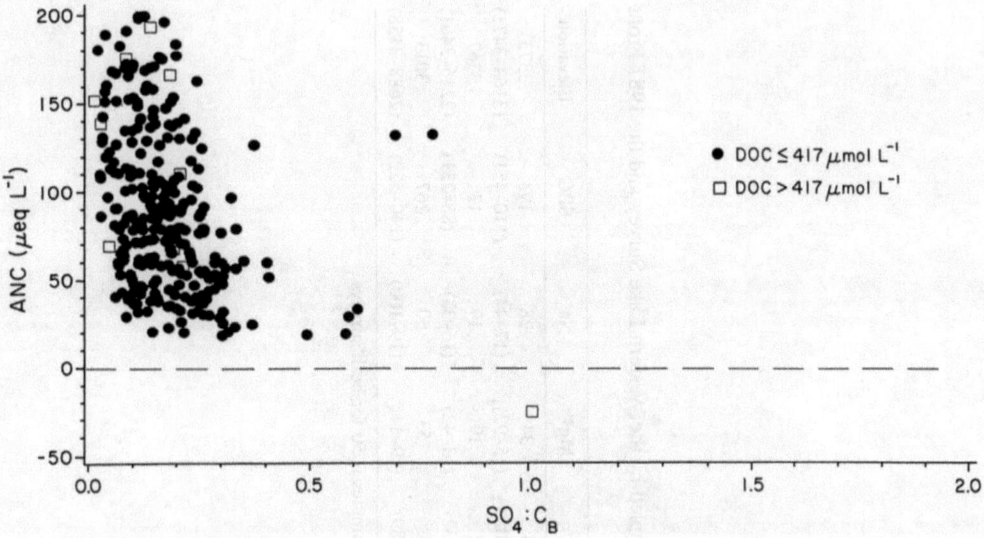

FIGURE 14.7. Relationship between ANC and SO_4^{2-}:C_B for lakes sampled in the WLS.

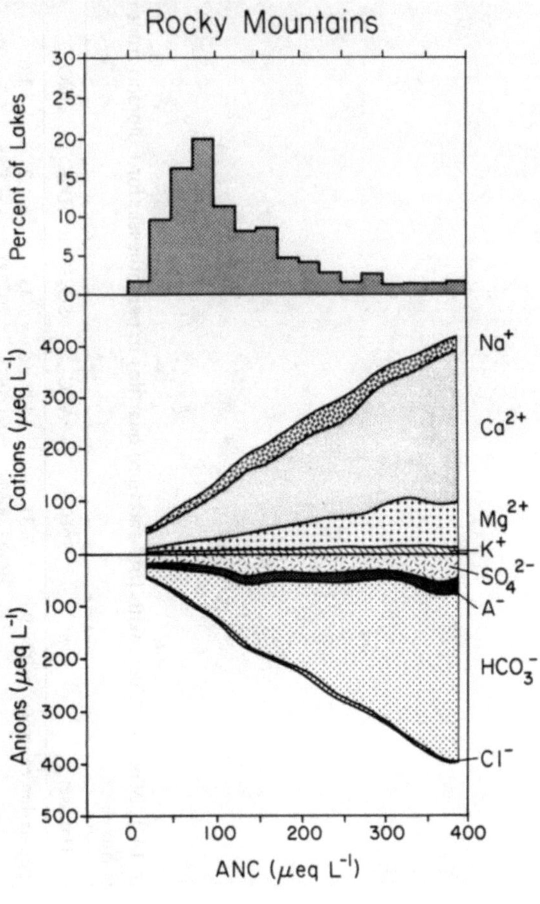

FIGURE 14.8. Concentrations of major cations and anions for lakes in the Rocky Mountain region with ANC ≤ 400 μeq L^{-1}. The percentages of lakes represented by ANC intervals are shown as a histogram at the top of the figure. Data plotted are population-weighted estimates for lakes in the WLS. The curves are spline fits to the median concentration for each ANC interval.

TABLE 14.1. Median values, with first and third quartiles in parentheses, for Colorado front range lakes sampled by the Western Lake Survey and the 1987 Front Range Survey.[a]

Data set	*n*	pH	ANC	SO_4^{2-}	DOC	NO_3^-	Cl^-	Ca^{2+}	Mg^{2+}	Na^+	SBC	Elevation
WLS population estimate	228.9	7.31	127	33	1.2	2.6	3	104	34	28	193	3373
		(6.97–7.67)	(60–363)	(20–38)	(.8–1.8)	(.5–6.3)	(2–4)	(64–205)	(15–92)	(17–44)	(110–451)	(3163–3471)
WLS sample lakes	50	7.10	80	24	1.0	3.2	2	82	19	19	131	3367
		(6.90–7.35)	(44–158)	(19–36)	(.7–1.5)	(.8–7.7)	(2–4)	(53–132)	(13–42)	(16–33)	(88–231)	(3236–3461)
1987 survey	50	7.51	160	46	–	0.2	6	157	53	50	267	3013
		(7.18–8.11)	(88–365)	(24–73)		(.2–.9)	(5–20)	(80–289)	(19–115)	(19–100)	(140–522)	(2688–3158)

[a] All concentrations are in µeq L^{-1}, except for pH in standard units, DOC in µmol L^{-1}, and elevation in m. Number of samples is 50 for both surveys.

samples from the 1987 survey were about twice those of the WLS. The 1987 survey included many lakes at lower elevations than those sampled by the WLS. The lower elevation lakes consistently were more concentrated than lakes at higher elevations. Thus, much of the difference in chemistry between the two surveys is attributable to differences in sampling design.

Acidification of Rocky Mountain Lakes

Acidification of a lake can be either chronic or episodic. The WLS data indicate that in the Rocky Mountain region only one of the sampled lakes was acidic during September–October 1985. Thus, lakes of the region are not chronically acidic. However, they do receive much of their water during the brief period of snowmelt, and intense thunderstorms in steeply sloping watersheds may produce pulses of runoff that have not had an opportunity to react with soil or rock and become neutralized. If such snowmelt and runoff sufficiently dilute or physically displace lake water that is more alkaline with undiluted snowmelt or runoff, an episode of decreased ANC and pH occurs. If the snowmelt and runoff are sufficiently acidic, further episodic decreases in alkalinity will occur, since this acidity neutralizes the diluted ANC of the lake water.

There are no region-wide data that address the question of episodic acidification. Thus, it is not possible to determine whether such acidification is widespread or localized. Problems of access make it difficult to collect samples during snowmelt. In the few lakes that are being studied during snowmelt, dilution or total physical displacement of lake water with snowmelt often occurs. However, because of the pH of wetfall (Figure 14.9), such snowmelt and runoff are unlikely to be acidic enough to neutralize much ANC and still produce large decreases in pH. Episodic decreases in ANC and pH have been measured in a pond in western Colorado (Blanchard et al. 1988). Some of this decrease was attributed to dilution of the surface 1 m of the pond with snowmelt, and some decrease was attributed to neutralization reactions in that layer. Data do not exist to determine whether snowmelt in the region is now more acidic than in previous periods of lesser emissions. Thus, decreases in ANC and pH that do occur may be natural phenomena or natural phenomena accentuated by an unknown amount by anthropogenic emissions.

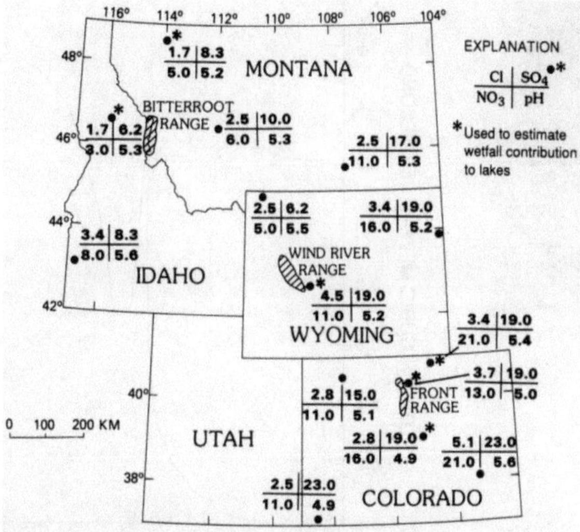

FIGURE 14.9. Annual precipitation-weighted mean Cl^-, SO_4^{2-}, and NO_3^- concentration (μeq L^{-1}) and pH for 1985.

Very little published information is available that describes the response of Rocky Mountain lakes to seasonal changes in hydrology. However, data are available for the Loch Vale watershed in Rocky Mountain National Park (Baron and Bricker 1987). Figure 14.10 shows the variation of SO_4^{2-} with season and flow during a hypothetical year, based on data from 1982–85. Data presented (Baron and Bricker 1987) indicate that major weathering products, such as Ca^{2+} and ANC, as well as atmospherically derived components, such as Cl^-, vary similarly to the illustrated SO_4^{2-} concentration. Baron and Bricker (1987) attribute increased ion concentrations during winter to lack of flushing, which presumably allows groundwater that is more concentrated to affect lake chemistry, and to the exclusion of ions from the developing ice layer. Initial snowmelt causes a minor increase in SO_4^{2-} concentration as SO_4^{2-} is preferentially released in the early phase of snowmelt. Subsequent snowmelt results in dilution of all ions.

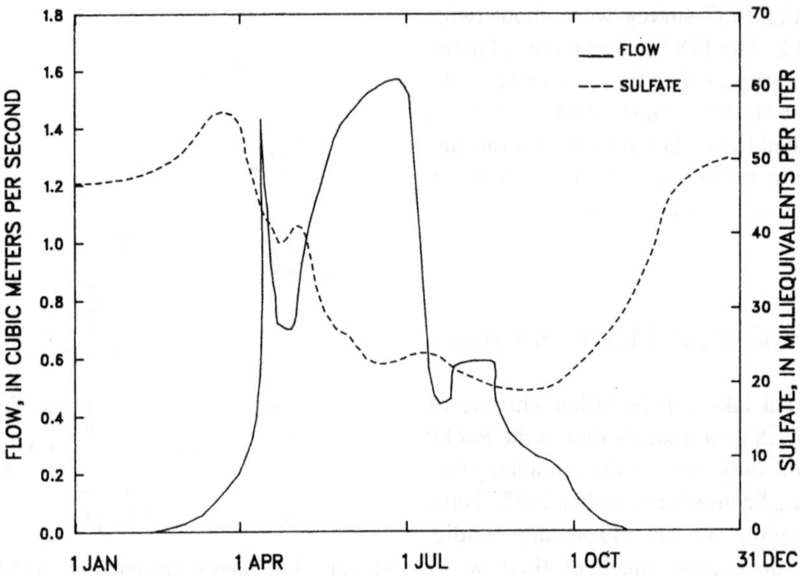

FIGURE 14.10. Change in SO_4^{2-} concentration and discharge in Loch Vale watershed, Colorado.

Processes Affecting Surface Water Chemistry

Western Lake Survey data indicate that many lakes in the region are sensitive to acidification but that they are not presently (1985) acidic. In the absence of historical data, it is difficult to determine whether some acidification has occurred. It is possible to estimate how much acidification could have occurred if we know the amount of acids being deposited by wetfall and dryfall, the extent to which these deposited acids are concentrated or added to by evapotranspiration or watershed sources of acidity, and the extent to which these acids are removed by reduction and sorption within the watershed or lake. Unfortunately, direct measurement of these sources, sinks, and processes is impossible in most wilderness watersheds. Thus, it is necessary to estimate the magnitude of these controls on lake chemistry by indirect methods. Indirect methods are necessary not only because of the difficulty of direct measurement of controls on lake chemistry, but also because of the difficulty of extrapolating direct measurements made in an intensively studied watershed to a region as large as the Rocky Mountains. For example, determining the importance of dryfall contributions of SO_4^{2-} requires measuring the site-specific characteristics affecting dryfall of the intensively studied watershed. Then the distribution of all these characteristics, as well as the distribution of SO_4^{2-} in the air masses of the region, must be determined. In an intensively studied catchment, such an extrapolation is difficult; in a region as large as the Rocky Mountains, it is impractical.

Indirect methods have been used to estimate acidification of lakes in the Rocky Mountains (Gibson et al. 1983, Turk and Campbell 1987). These indirect methods use the concentration of SO_4^{2-} in lakes to estimate how much acidity could have been deposited along with the SO_4^{2-}. However, to provide a realistic estimate of acidification, such methods must be able to determine how much SO_4^{2-} in a lake results from the various possible sources and processes, some of which contribute acidity and some of which do not. Also, acidity may be deposited along with NO_3^-, although the biogeochemistry of nitrogen often is more complex than that of sulfur, and thus more difficult to quantify either directly or indirectly.

Effects of Processes on a Single Lake

The potential contribution of wetfall to chemistry at Mirror Lake, Colorado, is illustrated in Figure

FIGURE 14.11. Contribution by wetfall to the composition of Mirror Lake, Colorado.

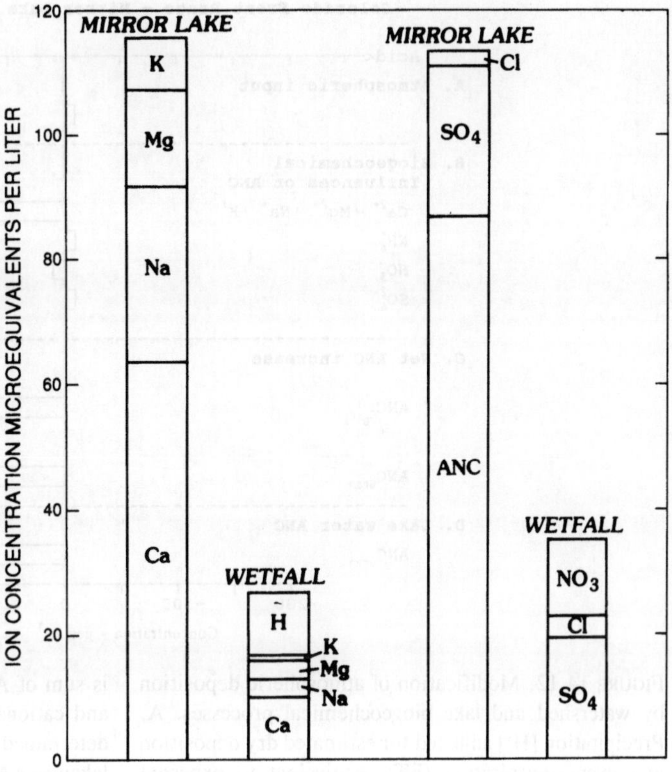

14.11. The wetfall concentrations are volume-weighted mean concentrations, in μeq L^{-1}, for 1985 from the Beaver Meadows site in Rocky Mountain National Park. Mirror Lake is near the Beaver Meadows site and its major ion chemistry is representative of other Colorado Front Range lakes having ANC < 200 μeq L^{-1}. As with other lakes in this area, lake Cl^- closely approximates wetfall Cl^-; thus, evapotranspiration probably can be ignored. Lake SO_4^{2-} is about 25% greater than wetfall SO_4^{2-}; thus, little contribution from dryfall, weathering, or evapotranspiration can have occurred. The major difference in anion composition is the large increase in lake ANC and decrease in lake NO_3^- compared to wetfall. Because net SO_4^{2-} reduction or retention does not occur, these processes cannot have added to lake ANC, and NO_3^- reduction or retention is insufficient to account for the large increase in lake ANC.

Lake cation concentrations also are much greater than wetfall cation concentrations. Further, the increase in lake cation concentrations is sufficiently large to account for the increase in lake ANC compared to wetfall. Neutralization of wetfall H^+ is insufficient to account for all the lake cation increase; thus, weathering by other sources of acid, such as soil carbon dioxide, probably causes most of the increase in cations and ANC.

The net contribution of several sources and sinks for ANC is illustrated in Figure 14.12, using Mirror Lake as an example. In Mirror Lake, only a minor amount of ANC generation or consumption is likely to be caused by processes involving NH_4^+, NO_3^-, SO_4^{2-}, and H^+; most net increase in ANC can be accounted for only by increased lake cation concentrations. Thus, weathering, rather than in lake processes or watershed transformations of ions derived from wetfall, is likely to be responsible for ANC generation.

Effects of Processes on Regional Distribution of Chloride and Sulfate – Development of a Conceptual Model

The lake SO_4^{2-} concentration integrates the effects of all sources, sinks, and processes within the watershed and lake that control the biogeochemistry of sulfur. If there is a large degree of variation

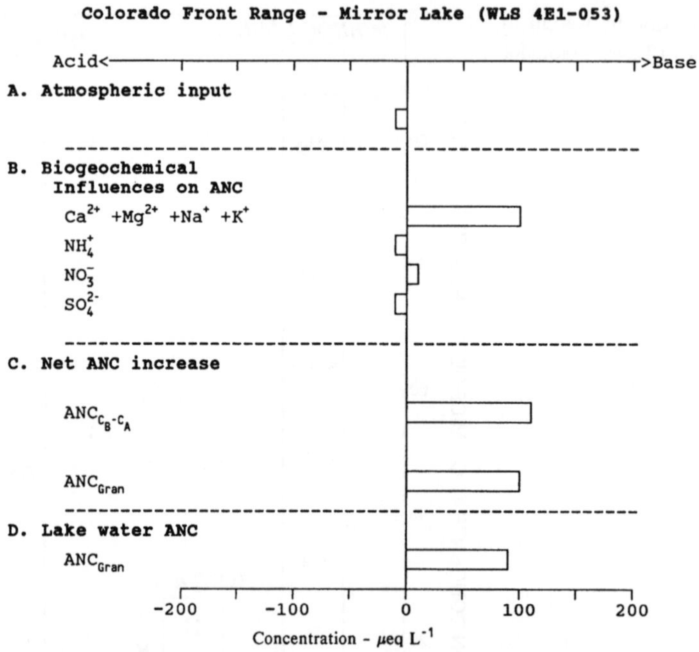

FIGURE 14.12. Modification of atmospheric deposition by watershed and lake biogeochemical processes. A. Precipitation $[H^+]$ adjusted for estimated dry deposition and evapotranspiration (ET); equivalent to expected lakewater $ANC_{C_B\text{-}C_A}$ in the absence of any process other than ET. B. Reactions affecting ANC; bars to the right indicate ANC production; those to the left show ANC consumption. C. Net ANC production; $ANC_{C_B\text{-}C_A}$ is sum of ANC production terms for measured anions and cations, including those shown in C; ANC_{Gran} is determined by difference between input ANC (A.) and lakewater ANC (D.). D. Lakewater ANC represents the net reaction of atmospheric inputs with the lake/watershed system. For a further explanation of calculation procedures and figure interpretation, see Appendix A. DOC = 74.1 μmol L^{-1}.

in the importance of these controls among the lakes of a region, the lake SO_4^{2-} concentration will also have a large degree of variation. However, if only a few of these controls are important, especially if only one control commonly is dominant, the lake SO_4^{2-} concentration will tend to be near some regional average. This can be illustrated with a simple conceptual model of the effects of common sulfur biogeochemical controls on the frequency distribution of lake SO_4^{2-} concentration. Figure 14.13 shows a hypothetical frequency distribution of lake SO_4^{2-} concentration that might occur in a region with many controls on sulfur biogeochemistry. Each labeled bar represents a subset of lakes for which some individual control on sulfur biogeochemistry is important relative to the other controls. In Figure 14.13, the lakes can be divided into two modes. The smallest concentration mode includes lakes in which the wetfall source of SO_4^{2-} is the most common dominant control on lake SO_4^{2-} concentration. Lakes in which all other controls are unimportant will be in the concentration range represented by the bar labeled *wetfall*, which will be the concentration of SO_4^{2-} in wetfall. However, in some lakes, the SO_4^{2-} contributed by wetfall is enhanced by the concentrating effect of evapotranspiration or the additional source of SO_4^{2-} from dryfall. Such lakes will have lake SO_4^{2-} concentrations in the concentration range represented by the bars labeled *evapotranspiration* and *dryfall*. Depending on the relative proportion of lakes in each concentration range, the shape of the distribution may change markedly from the hypothetical distribution shown in Figure 14.13, which assumes that wetfall is the only important control in most lakes. Similar effects, from processes that

FIGURE 14.13. Hypothetical frequency distribution of lake-sulfate concentration as affected by several processes.

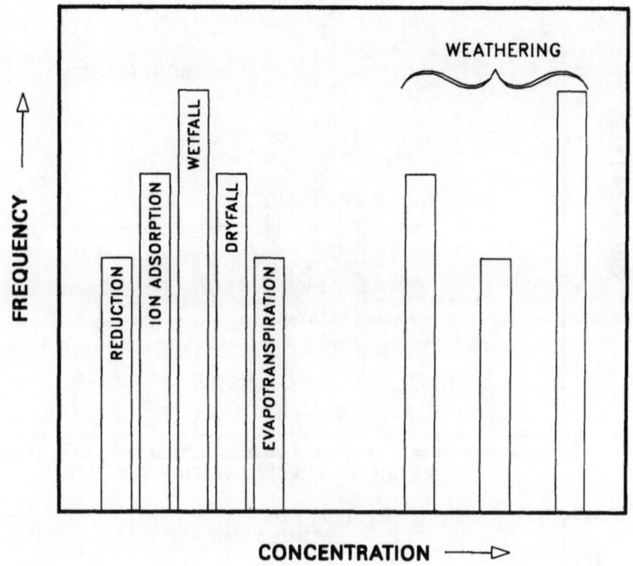

remove SO_4^{2-} from wetfall contributions, are illustrated by the bars labeled *adsorption* and *reduction*. Finally, a second, broader and greater concentration mode is labeled *weathering*. This concentration range represents lakes for which watershed sources of sulfur are assumed to be readily available to weathering, which greatly exceeds the incremental effects of the previously discussed controls. This assumption is likely to be true in the case of watersheds with readily weatherable deposits of an element, such as SO_4^{2-} contributed by weathering of ore deposits or sulfur-rich sedimentary rock. This assumption is less likely to be true if sulfur is disseminated widely in trace amounts of slowly weathering minerals.

It is possible to prepare a frequency distribution of lake SO_4^{2-} concentration for a region of any size; however, as the size of the region increases, the likelihood lessens that controls will remain constant. Thus, in a region as large as the Rocky Mountains, it can be useful to consider subsets representing areas of consistent wetfall composition, geology, or other factors that may vary greatly from area to area in the region. In this chapter, emphasis is on the Colorado Front Range, the Wind River Range of Wyoming, and the Bitterroot Range of Montana and Idaho. These areas provide good geographic coverage of the entire Rocky Mountain region and help to reduce variance in some factors. For example, the concentration of Cl^- and SO_4^{2-} in wetfall varies among these areas (Figure 14.9).

The Colorado Front Range

The Colorado Front Range is west of and near the urbanized corridor between Ft. Collins and Colorado Springs. Many studies have investigated the effects of anthropogenic emissions from the urban area on the chemistry of precipitation and surface water and these have been summarized previously (Turk 1983, Turk and Spahr 1989). The bedrock of the Colorado Front Range is primarily Precambrian granite, biotitic gneiss, schist, and migmatite. Quaternary glacial drift and Precambrian felsic and hornblendic gneiss are present in substantial, but lesser amounts (Tweto 1979). Soils and vegetation were classified for eight representative watersheds (J. Gibson, National Atmospheric Deposition Program, Ft. Collins, CO, pers. comm.). Dominant soils were Cryoboralfs, Cryumbrepts, Cryochrepts, Cryaquepts, and Cryorthents. Montane forests of ponderosa pine, lodgepole pine, and aspen grade into subalpine forests of Englemann spruce and subalpine fir. Willow is a common shrub of the alpine zone.

The WLS sampled 50 lakes in the Colorado Front Range during September–October 1985. Data from 41 lakes with ANC $\leq$ 200 μeq L^{-1} were

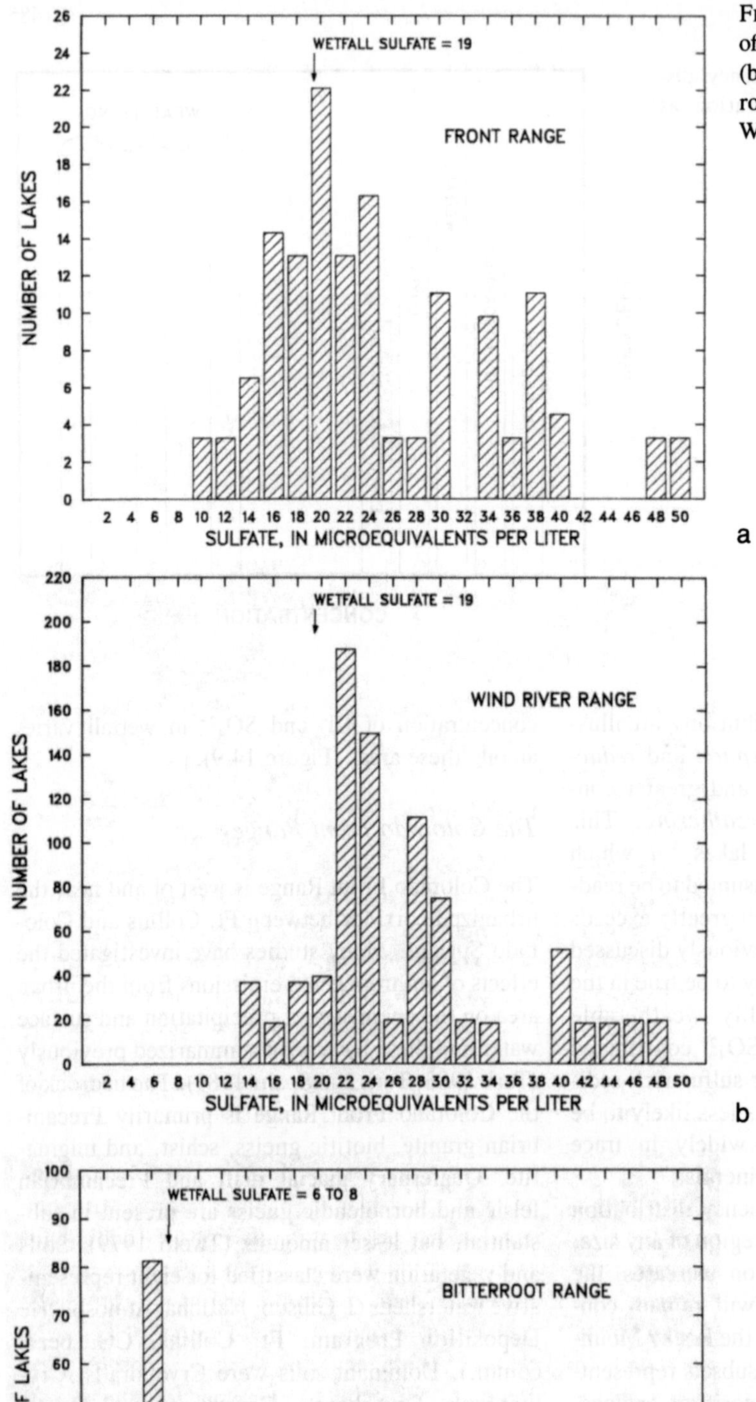

FIGURE 14.14. Frequency distribution of SO_4^{2-} for (a) Colorado Front Range, (b) Wind River Range, and (c) Bitterroot Range lakes. (Data are from the WLS.)

used to compute the weighted population estimates considered here. The frequency distribution of SO_4^{2-} in these lakes is presented in Figure 14.14. The SO_4^{2-} concentration of most of these lakes is about 20 μeq L^{-1}, compared to a wetfall concentration in 1985 of about 19 μeq L^{-1} (Figure 14.9). This similarity in SO_4^{2-} concentrations for lakes and wetfall indicates either that no processes affecting the wetfall contribution of SO_4^{2-} are significant in these watersheds or that a fortuitous balance exists between processes adding and removing SO_4^{2-}.

It is possible to quantify the process of evapotranspiration by the use of Cl^- concentrations in lakes (Figure 14.15) and wetfall (Figure 14.9), assuming that Cl^- behaves as a conservative ion in the watershed and lake. The lake Cl^- concentration most commonly is about 2 μeq L^{-1}; the wetfall Cl^- concentration for this area was about 3 μeq L^{-1} in 1985. The smaller concentration of Cl^- in some lakes compared to wetfall probably results from slightly decreased concentrations in wetfall at the elevation of the lakes compared to wetfall at the measurement stations. Although most lakes have Cl^- concentrations equal to or less than wetfall concentrations, only a few have Cl^- concentrations greater than those in in wetfall. Thus, evapotranspiration can be ignored for most of these lakes.

Direct measurement of lake evaporation at Ned Wilson Lake, located about 150 km west of the Front Range and at about the same elevation as the lakes discussed here, indicates that < 25 cm of water, or about 8% of the lake volume, evaporates from the lake surface during the summer open-water period (Spahr and Turk 1985). An adjustment factor for the effect of evapotranspiration is calculated as the ratio of lake Cl^- concentration to wetfall Cl^- concentration for lakes with Cl^- concentration greater than wetfall Cl^- concentration; the factor is 1 if lake Cl^- is less than or equal to wetfall Cl^- concentration. If lake SO_4^{2-} concentration is multiplied by this factor, the distribution obtained (Figure 14.16) indicates that wetfall SO_4^{2-} concentration is the dominant control on lake SO_4^{2-} concentration.

The frequency of occurrence of lakes with lake SO_4^{2-} concentration exceeding wetfall SO_4^{2-} concentration by more than about 25% becomes very small (Figure 14.16). Thus, processes that add SO_4^{2-}, such as mineral weathering and dryfall, rarely contribute more than 25% of the SO_4^{2-} provided by wetfall in the sampled lakes. Some lake SO_4^{2-} concentrations are less than the wetfall SO_4^{2-} concentration. These lakes probably have significant SO_4^{2-} removal processes, such as SO_4^{2-} reduction in the lake or watershed; however, such lakes are less common than lakes controlled only by wetfall SO_4^{2-} concentration. This conclusion is especially significant because of its implications for the importance of dryfall contributions of SO_4^{2-}. Dryfall is difficult to measure directly, even in the most ideal conditions. However, to predict the response of these watersheds to changes in atmospheric deposition of acidity, it is necessary to know how important dryfall is, compared to wetfall, which is measured more readily. The results discussed here indicate that dryfall is a minor source of SO_4^{2-} compared to wetfall. If true, this considerably simplifies the prediction of watershed response to changes in acid deposition, since it implies that only wetfall must be considered.

The adjusted SO_4^{2-} concentrations (Figure 14.16) indicate that most lakes having ANC ≤ 200 μeq L^{-1} sampled by the WLS along the Colorado Front Range have a very simple biogeochemistry of sulfur. In these lakes, SO_4^{2-} is contributed almost entirely by wetfall and moves as a conservative constituent through the watershed, as does Cl^-. This conservative or steady-state behavior has been observed for lakes on granitic terrain in Norway (Wright and Henriksen 1978). If all this SO_4^{2-} occurred in the form of anthropogenic sulfuric acid, then lakes in this area may have undergone ANC reduction of as much as 20 μeq L^{-1} as a result of acidification by sulfuric acid, assuming H^+ is not retained or neutralized within the watershed.

In addition to sulfuric acid, anthropogenic acidification of lakes along the Colorado Front Range may result from nitric acid (Lewis and Grant 1980). Unfortunately, NO_3^- is less likely than SO_4^{2-} to behave conservatively in watersheds. Thus, frequency distributions of lake NO_3^- concentrations are unlikely to indicate NO_3^- contributions from wetfall. Also, biological uptake within the watershed may neutralize much of the nitric acid from wetfall (Hemond and Eshleman 1984). As a worst case, assume that all the NO_3^- in wetfall is in the form of nitric acid, and that any biological uptake within the watershed does not neutralize this acidity. Also assume that any nitric acid neutralization results in an equivalent loss of ANC. Thus, the maximum loss of ANC attributable to

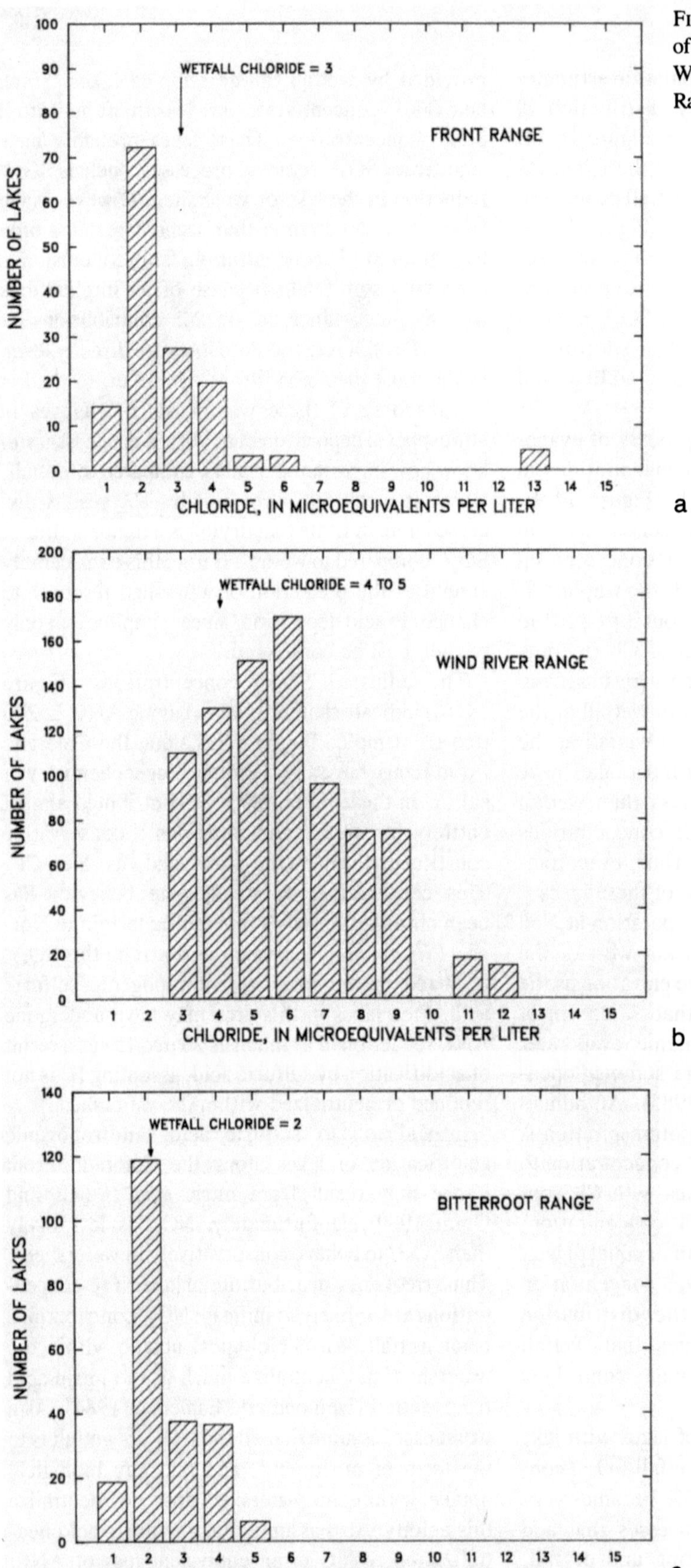

FIGURE 14.15. Frequency distribution of Cl^- for (a) Colorado Front Range, (b) Wind River Range, and (c) Bitterroot Range lakes. (Data are from the WLS.)

FIGURE 14.16. Frequency distribution of SO_4^{2-} adjusted for evapotranspiration, for (a) Colorado Front Range, (b) Wind River Range, and (c) Bitterroot Range lakes. (Data are from the WLS.)

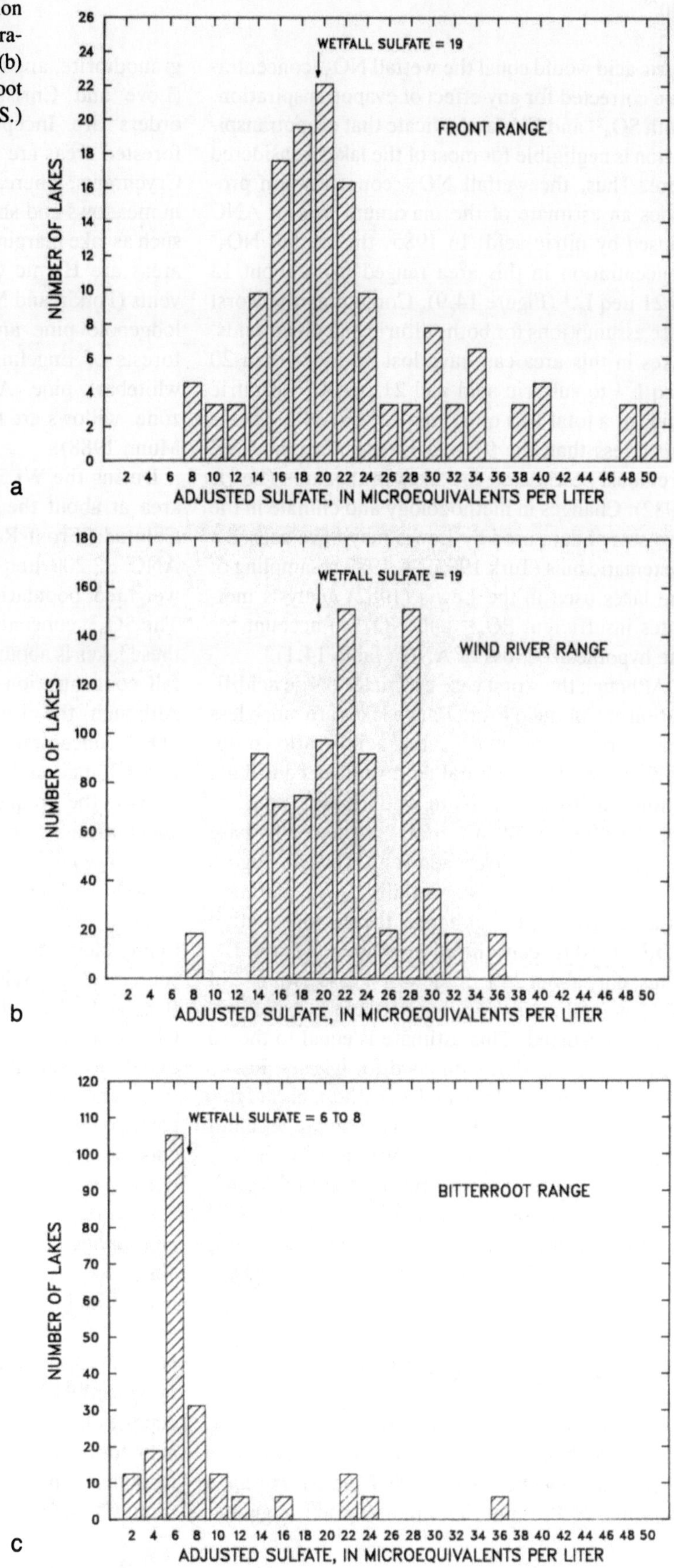

nitric acid would equal the wetfall NO_3^- concentration corrected for any effect of evapotranspiration. Both SO_4^{2-} and Cl^- data indicate that evapotranspiration is negligible for most of the lakes considered here. Thus, the wetfall NO_3^- concentration provides an estimate of the maximum loss of ANC caused by nitric acid. In 1985, the wetfall NO_3^- concentration in this area ranged from about 13 to 21 μeq L^{-1} (Figure 14.9). Combining the worst case assumptions for both sulfuric and nitric acids, lakes in this area can have lost no more than 20 μeq L^{-1} to sulfuric acid and 21 μeq L^{-1} to nitric acid, or a total loss of 41 μeq L^{-1} of ANC. This is much less than the 97 μeq L^{-1} decrease in ANC previously estimated for lakes in this area (Lewis 1982). Changes in methodology and climate in the two sets of data used by Lewis may have caused a systematic bias (Turk 1983). A 1987 resampling of the lakes used in the Lewis (1982) analysis indicates insufficient SO_4^{2-} and NO_3^- to account for the hypothesized loss of ANC (Table 14.1).

Although the worst case estimates of the acidification of Colorado Front Range lakes are much less than a previous estimate, actual acidification probably is even less. Natural acid sources contribute sulfuric and nitric acids to wetfall; also, some of the SO_4^{2-} and NO_3^- in wetfall is balanced by base cations instead of hydrogen ions. Data in Figure 14.9 indicate an average wetfall H^+ concentration of about 10 μeq L^{-1}, whereas the average wetfall $SO_4^{2-} + NO_3^-$ concentration is about 41 μeq L^{-1}. Thus, only about 24% of the worst case estimate of ANC neutralization, or about 10 μeq L^{-1}, is likely to have occurred. This estimate is equal to the 10 μeq L^{-1} loss of ANC estimated for lakes in Rocky Mountain National Park in the northern part of this area (Gibson et al. 1983). This estimate assumes that all of the acidity now in wetfall is anthropogenic. Estimates of the pH range of natural wetfall (Charlson and Rodhe 1982, Galloway et al. 1982) overlap the present pH range for this area. Thus, actual ANC neutralization in these lakes may have been much less than 10 μeq L^{-1}.

The Wind River Range of Wyoming

The Wind River Range of Wyoming is located downwind from the urbanized area of Salt Lake City, Utah. The bedrock of the Wind River Range is primarily Precambrian granite, granitic gneiss, granodiorite, and metasediments or metavolcanics (Love and Christiansen 1985). Dominant soil orders are Inceptisols and Entisols. Soils in forested areas are Lithic Cryochrepts and Dystric Crychrepts, whereas Typic Cryumbrepts are found in meadows and shrubland. Soils in riparian areas such as lake margins, stream channels, and seepage areas are Humic Cryaquepts and Typic Cryofluvents (Potkin and Munn 1988). Montane forests of lodgepole pine and aspen grade into subalpine forests of Engelmann spruce, subalpine fir, and whitebark pine. At timberline and in the alpine zone, willows are the dominant shrub (Potkin and Munn 1988).

During the WLS, 53 lakes were sampled in this area at about the same time as sampling in the Colorado Front Range. Data from 44 lakes with ANC ≤ 200 μeq L^{-1} were used to compute the weighted population estimates considered here. The SO_4^{2-} concentration (Figure 14.14) in most of these lakes is about 22 μeq L^{-1}, compared to a wetfall concentration in 1985 of about 19 μeq L^{-1}. Although the frequency distributions for lake SO_4^{2-} concentration are similar for the Colorado Front Range and the Wind River Range (Figure 14.14), the frequency distributions for lake Cl^- concentrations are very different.

Average lake Cl^- concentrations are greater in the Wind River Range than in the Colorado Front Range (Figure 14.15) and concentrations also are less tightly clustered in the Wind River Range. Upwind sources of Cl^- salts such as Great Salt Lake, salt flats, and salty soils may explain the increased lake Cl^- concentrations in the absence of increased lake SO_4^{2-} concentrations. The large size of these resuspended particles of Cl^- salts and dust, compared to the much smaller SO_4^{2-} particles formed by condensation, could cause a more variable deposition rate than might be expected for long-range sources of Cl^- affecting the Colorado Front Range. Decoupling of deposition of Cl^- and SO_4^{2-} has been observed, in which large seaspray particles contribute Cl^- and smaller particles contribute anthropogenic SO_4^{2-} (Wright and Henriksen 1978). Use of lake Cl^- concentrations to calculate evapotranspiration is problematic for the Wind River Range. If the deposition rate of Cl^- is quite variable, then the assumption that wetfall Cl^- concentrations from the National Atmospheric Deposition Network sites (Figure 14.9) represent wetfall Cl^- concentrations at the lake watersheds is false. The nonnormal distribution of the lake SO_4^{2-} concentrations, adjusted for evapotranspiration (Figure 14.16), indicates that Cl^- deposition is much more variable than SO_4^{2-} deposition in the Wind River Range. Thus, the use of Cl^- to estimate evapotranspiration probably is not valid in this area.

Although it would be preferable to have an independent measure of evapotranspiration, apparent variability in Cl^- deposition makes this impossible for the Wind River Range. However, evapotranspiration in the Wind River Range is probably not greater than in the Colorado Front Range. The Wind River Range is farther north and the sampled lakes are at about the same elevation as those sampled in the Colorado Front Range. Also, the similarity in wetfall SO_4^{2-} concentration and lake SO_4^{2-} concentration indicates that evapotranspiration is unlikely to be large in most of these lakes. In the few lakes that have lake SO_4^{2-} concentrations greater than wetfall SO_4^{2-} concentration (Figures 14.9 and 14.14), evapotranspiration, dryfall, and weathering may increase lake SO_4^{2-} concentration by up to 100% relative to wetfall. For simplification, this effect also can be assumed to apply to NO_3^- and any other wetfall source of acidity. Thus, maxiumum decreases in ANC would fall between the value of wetfall H^+ concentration and a value 100% greater than that concentration, or about 6 to 12 $\mu eq\ L^{-1}$, if all present wetfall acidity were anthropogenic.

The Bitterroot Range of Montana and Idaho

The Bitterroot Range is located in an area remote from major urban and obvious natural sources of SO_4^{2-} and Cl^-. The bedrock of the Bitterroot range is primarily Lower Cretaceous granite, granitic gneiss, and gneissic or quartz monzonite associated with the Idaho Batholith. Sedimentary rock impregnated with granitic material is also present (Ross et al. 1955). Dominant soils in the lower elevation montane zone are Ustochepts, Xerolls, and Boralfs, whereas in the subalpine and alpine, Cryochrepts are most common. Montane forests of ponderosa pine, lodgepole pine, and Douglas fir grade into subalpine forests of alpine fir and whitebark pine (N. M. Davis, U.S. Forest Service, written comm. 1988).

During the WLS, 36 lakes were sampled in this area. Data from 33 lakes with ANC $\leq$ 200 $\mu eq\ L^{-1}$ were used to compute the weighted population estimates considered here. The lake SO_4^{2-} concentration (Figure 14.14) is most commonly about 6 $\mu eq\ L^{-1}$, compared to a wetfall SO_4^{2-} concentration of about 6 to 8 $\mu eq\ L^{-1}$ in 1985 (Figure 14.9). Thus, both wetfall SO_4^{2-} concentration and lake SO_4^{2-} concentration are much less in the Bitterroot Range than in the Wind River Range and the Colorado Front Range. The lake Cl^- concentration in this area is very similar to that in the Colorado Front Range (Figure 14.15). The most common lake Cl^- concentration of 2 $\mu eq\ L^{-1}$ equals the wetfall Cl^- concentration in 1985. Thus, evapotranspiration can be ignored for most of these lakes. If lake SO_4^{2-} concentration is adjusted for the effects of evapotranspiration (Figure 14.16), few lakes may have sources of SO_4^{2-}, such as dryfall or weathering, that would increase lake SO_4^{2-} concentrations to more than the 6 to 8 $\mu eq\ L^{-1}$ from wetfall. Thus, maximum decreases in ANC would be about equal to the wetfall H^+ concentration of about 5 $\mu eq\ L^{-1}$, if all present acidity is anthropogenic.

Summary

The wetfall Cl^- and wetfall SO_4^{2-} concentrations vary by about 300% among the lakes in the mountain ranges discussed in this chapter. Lake Cl^- and lake SO_4^{2-} concentrations vary in a monotonic pattern with wetfall concentrations. This close agreement between wetfall and lake concentrations throughout such a great range of concentration is unlikely unless wetfall is the major control on lake SO_4^{2-} and Cl^-. Direct measurement of most processes that may affect Cl^- and SO_4^{2-} is generally lacking in Rocky Mountain watersheds and lakes. Direct measurement of evaporation confirms that this process is likely to have only very minor effects on lake SO_4^{2-} and lake Cl^- concentrations.

Major Weathering Products of Rocky Mountain Lakes

Preceding sections of this chapter indicate that lakes in the Rocky Mountain region are sensitive to acidification and not chronically acidic, and that the maximum amount of chronic acidification that could have occurred is 5 to 12 μeq L^{-1}. The degree of protection of these lakes from future acidification is a function of the mineral weathering and other processes that produce the major cations and ANC. Lake ANC is often used as a measure of sensitivity to acidification (Hendrey et al. 1980, Turk and Adams 1983). Frequency distributions for lake ANC, such as the one shown in Figure 14.6e, can povide a simple first approximation of the number or fraction of lakes at risk from acidification. Unfortunately, such estimates do not provide much knowledge of the processes that produce ANC and base cations. The response of these processes to changes in the chemistry of atmospheric deposition will determine whether a lake acidifies. These processes can be primarily terrestrial, such as mineral weathering, or primarily, in-lake, such as SO_4^{2-} reduction in lake sediments.

The net production of major cations, ANC, and SiO_2 is primarily a result of mineral weathering. For example, constituents such as K^+ and SiO_2 may be significantly involved in biological processes; however, such biological processes are not net sources of these constituents. Thus, knowledge of the mineral weathering process is vital to predicting future acidification. In wilderness areas, it is impractical to directly measure characteristics that control weathering, for example, mineralogy, surface area of each mineral per unit volume of soil and bedrock, and rate of flow of water through soil and bedrock. Thus, indirect techniques are necessary to understand weathering processes in these wilderness areas.

One approach to determining the controlling mineral weathering processes is to examine the lake concentration data for patterns in the relative concentrations of each constituent as a function of some measure of degree of mineral weathering. Any product of mineral weathering can be used as the independent variable in this approach, as can a measure of solution concentration such as ionic strength or specific conductance. In this chapter, Ca^{2+} and ANC are used as the independent variables for measuring degree of weathering because

- Calcium and ANC, as evidenced by their large concentrations in lakes, are major products of mineral weathering for the igneous, metamorphic, and sedimentary rocks that are common throughout the region.
- In the concentration range of the lakes with ANC $\leq$ 200 μeq L^{-1}, Ca^{2+} and ANC are unlikely to be affected by solubility limits.
- Analytical precision is better for Ca^{2+} and ANC than for many other constituents in the concentration ranges observed for these samples.
- Calcium and ANC are not as likely as many other constituents to be significantly affected by biological processes, even though some biological effects are likely.

The approach used here is (1) to determine whether a net weathering reaction seems plausible, based on lake chemistry, and (2) to ascertain the stoichiometry of any apparent net weathering reaction. To determine the plausibility and stoichiometry of a net weathering reaction, each weathering product is regressed against Ca^{2+} or ANC. Correlation between Ca^{2+} or ANC and other weathering products is assumed to indicate that a common weathering reaction or a series of weathering reactions is the principal source. The slope of the regression is assumed to be an indication of stoichiometry; however, the net reaction inferred from the regression slope may be composed of several independent reactions. Thus, it is possible for the data to be consistent with a net reaction stoichiometry for a reaction that does not actually occur. Reactions consistent with the data may be verified by examining changes in mineralogy between fresh and weathered rock or soil; however, it is often possible only to eliminate hypothesized reactions involving minerals that do not occur in the area. Lack of correlation between Ca^{2+} or ANC and other weathering products is assumed to indicate (1) prevalence of differing weathering reactions among lake watersheds, (2) processes that preferentially affect the removal of the weathering product relative to Ca^{2+}, or (3) sources other than weathering of the weathering product.

Because the wetfall concentration of these weathering products typically is only about 10% of the mean lake concentration, and because evapotranspiration has been shown to be unimportant in these lakes, actual lake concentrations have been used without subtracting the wetfall contribution

TABLE 14.2. Regression characteristics, by area, of weathering products on calcium (ANC).[a]

Area	Product	n	R^2	$P > F$[b]	Slope[c]
Front Range	ANC	41	0.73(−)	0.0001(−)	1.14(−)
	Mg	41	0.48(0.80)	0.0001(0.0001)	0.30(0.29)
	Na	41	0.08(0.27)	0.0743(0.0005)	0.13(0.18)
	K	41	0.05(0.10)	0.1472(0.0376)	0.02(0.02)
	SiO_2	41	0.04(0.19)	0.2339(0.0044)	0.01(0.01)
Wind River Range	ANC	44	0.85(−)	0.0001(−)	1.13(−)
	Mg	44	0.47(0.69)	0.0001(0.0001)	0.29(0.29)
	Na	44	0.38(0.66)	0.0001(0.0001)	0.22(0.24)
	K	44	0.34(0.32)	0.0001(0.0001)	0.05(0.04)
	SiO_2	44	0.09(0.24)	0.0438(0.0007)	0.01(0.01)
Bitterroot Range	ANC	33	0.92(−)	0.0001(−)	1.07(−)
	Mg	33	0.65(0.69)	0.0001(0.0001)	0.20(0.19)
	Na	33	0.04(0.09)	0.2666(0.0774)	0.05(0.07)
	K	33	0.10(0.06)	0.0616(0.1727)	0.03(0.02)
	SiO_2	33	<0.01(<0.01)	0.7500(0.6714)	<0.01(<0.01)

[a] Data are from WLS.
[b] Probability of a greater F statistic.
[c] Slope values are in μeq/μeq, except for SiO_2, which is in mg/μeq.

to lake concentration. Because of the small concentration of H^+ in wetfall (< 10 μeq L^{-1}) and lack of evapotranspiration, net release of cations by ion exchange within the soil is an unlikely source of cations compared to mineral weathering.

A linear regression model of ANC as a function of Ca^{2+} (Table 14.2) accounts for 73% of the variance in ANC for Colorado Front Range lakes, 85% for Wind River Range lakes, and 92% for Bitterroot Range lakes. The slope of the regressions indicates 1.07 to 1.14 μeq of ANC per μeq of Ca^{2+}. The high correlation and a slope near unity indicate that the mineral source of both ANC and Ca^{2+} is the same and that the mineralogy actually weathering is rich in Ca^{2+}, or that Ca^{2+} preferentially weathers compared to other major cations.

A linear regression model of Mg^{2+} as a function of Ca^{2+} (Table 14.2) accounts for 47% to 65% of the variance in Mg^{2+}. The slope of the regressions is 0.2 to 0.3 μeq of Mg^{2+} per μeq of Ca^{2+}. Thus, if a single mineral source supplies both Ca^{2+} and Mg^{2+}, it involves a fairly constant proportion of Mg^{2+} to Ca^{2+}; if separate reactions are involved for Ca^{2+} and Mg^{2+}, they proceed at a constant rate relative to each other. The decrease in variance of Mg^{2+}, compared to variance in ANC, explained by Ca^{2+}, indicates that some source or sink exists for Mg^{2+} in addition to those for Ca^{2+}. Acid neutralizing capacity accounts for more of the variance in Mg^{2+} (69% to 80%) than does Ca^{2+}. Thus, a weathering reaction that produces ANC but is different from the reaction producing Ca^{2+} is likely to supply some of the Mg^{2+}.

A linear regression model of Na^+ as a function of Ca^{2+} (Table 14.2) accounts for only 38% of the variance in Na^+ for the Wind River Range and no more than 8% for the other areas. The variable and small correlation of Na^+ with Ca^{2+} suggests that there are probably different sources for each constituent. Thus, plagioclase weathering, a likely source of Na^+, may not be the source of most of the Ca^{2+}. A linear regression model of Na^+ as a function of ANC accounts for 66% of the variance in Na^+ for the Wind River Range but only 27% for the Colorado Front Range and 9% for the Bitterroot Range. Thus, except for lakes in the Wind River Range, Na^+ is supplied by weathering or other sources at a variable rate compared to Ca^{2+} and ANC. Sodium may also be contributed by wetfall. Because wetfall seems to be the major source of Cl^-, Na^+ was regressed against Cl^- to determine whether wetfall is a likely source of Na^+. A linear regression model of Na^+ as a function of Cl^- accounts for 38% of the variance in Na^+ for the

Colorado Front Range and for about 20% in the other areas. Thus, variations in the atmospheric deposition of Cl^-, or minor evapotranspiration that occurs in some lake watersheds, accounts for more variance in Na^+ among the three areas than does weathering of a common mineral source for Ca^{2+} and Na^+. Within the Wind River Range, however, mineral weathering of a common source may account for more variance in Na^+ than does wetfall and evapotranspiration.

A linear regression model of K^+ as a function of Ca^{2+} (Table 14.2) accounts for 34% of the variance in K^+ for lakes in the Wind River Range, but no more than 10% for the other areas. Potassium is likely to come from a different mineral source than Ca^{2+}; it may come from atmospheric deposition, or it may be very much affected by processes such as biologic uptake and release. A linear regression model of K^+ as a function of Cl^- accounts for 10% of the variance in K^+ for lakes in the Wind River Range and 15% for the Bitterroot Range. Thus, in these two areas, variations in the atmospheric deposition of Cl^-, or minor evapotranspiration that occurs in some lake watersheds, accounts for about as much variance in K^+ as does mineral weathering. A linear regression of K^+ as a function of ANC accounts for no more variance in K^+ than does the regression against Ca^{2+}. Thus, K^+ is supplied at a variable rate compared to ANC or is selectively affected by processes such as biological uptake.

A linear regression model of SiO_2 as a function of Ca^{2+} (Table 14.2) accounts for only 9% of the variance in SiO_2 for lakes in the Wind River Range, and no more than 4% for the other areas. Regression against ANC accounts for only slightly more of the variance in SiO_2 than does Ca^{2+}. The mean concentration of SiO_2 is about 2 mg L^{-1} in each area. The small degree of correlation with Ca^{2+} and ANC, and the fairly constant concentration among areas, suggests that SiO_2 may be limited by some process such as (1) the solubility of some secondary mineral, (2) biologic uptake, for example, diatoms, or (3) adsorption.

Linear regression (Table 14.2) indicates that ANC and Mg^{2+} are consistently related to Ca^{2+} and, presumably, to primary mineral weathering reactions in all three areas. Regressions of Na^+, K^+, and SiO_2 are significantly ($p < 0.05$) related to Ca^{2+} only in the Wind River Range and to ANC only in the Wind River Range and Colorado Front Range. Potassium often is affected by vegetative uptake and release; perhaps the alpine conditions common to the Wind River Range minimize the biological complications that result in poor correlation between K^+ and Ca^{2+} or ANC in the other areas.

Thus, regional consistency among the three areas is restricted to the interdependence of Ca^{2+}, Mg^{2+}, and ANC. Such a relationship would be expected from the dissolution of a mixture of calcite and dolomite rather than the granitic mineralogy of the Rocky Mountain region. This relationship is similar to that reported for dilute water in the Cascade Mountains (Drever and Hurcomb 1986). There, calcite occurring in trace amounts in veins, on joint surfaces, and as a subglacial surficial deposit accounts for most of the mineral weathering in a watershed underlain by igneous and high-grade metamorphic rocks. Whether calcite and dolomite actually exist even in trace amounts in the Rocky Mountain watersheds reported here is unknown. Trace amounts of calcite have been identified in granite and gneiss of the Loch Vale watershed in the Colorado Front Range (Mast 1989). In addition to weathering of carbonate minerals, weathering of silicates rich in Ca^{2+} and Mg^{2+}, followed by precipitation or biologic uptake of SiO_2 could explain the relationships among the major ions. Augite and plagioclase were proposed as the major mineralogic sources of weathering products at a site in southwestern Colorado (Claassen et al. 1983), even though augite comprised < 3% of the mineral content.

Aeolian deposition of readily weatherable minerals or salts rich in Ca^{2+} and Mg^{2+} also could account for observed lake chemistry. For example, during infrequent periods of drought and strong winds, soil might be transported from low-elevation, sedimentary areas in the Great Basin to high-elevation watersheds. Such aeolian material should contain readily weatherable calcite, dolomite, and gypsum. Some of the deposited minerals would be incorporated into the soil and be available for future weathering. Further, the cations released upon dissolution of these minerals would increase the ratio $[Ca^{2+} + Mg^{2+}]:Na^+$ on exchange sites within the soil. Thus, slow release of cations from cation exchange in the soil could maintain a large ratio of

$[Ca^{2+} + Mg^{2+}]:Na^+$ long after the episodic source had disappeared. Aeolian transport of materials from lowland sedimentary terrane to high-elevation granitic terrane has been documented in the Tesuque watersheds of New Mexico (Gosz et al. 1983) and in the Colorado Front Range (Litaor 1987).

Preferential weathering of major minerals within the watersheds also could account for observed lake chemistry. For example, studies in watersheds of the Idaho Batholith indicate preferential weathering of anorthite-rich feldspar within zoned plagioclase feldspars (Clayton 1988). There, plagioclase is about An_{19} in composition; however, stream chemistry is consistent with weathering of An_{45} plagioclase. Balance of calculated and measured dissolved SiO_2 fluxes indicates that cations are being released by silicate mineral weathering. Examination of thin sections indicates preferential weathering of anorthite within the zoned plagioclase. Rapid physical removal of weathered material and exposure of fresh mineral surfaces maintain a large ratio of $Ca^{2+}:Na^+$ in stream chemistry.

The relationships among major ions in dilute Rocky Mountain lakes is not consistent with a simple, stoichiometric weathering of the major minerals within the watersheds. The mineral weathering reactions controlling ANC, Ca^{2+}, and Mg^{2+} are uncertain; thus, it is difficult to predict how these reactions will respond to any future change in the chemistry of atmospheric deposition. Determining the mineral weathering reactions and making predictions regarding their future rates will require intensive research in regionally representative watersheds.

Evidence of Historical Trends

Precipitation chemistry of the Rocky Mountain region is documented at a few sites, but only for the period since about 1979. Similarly, quality assured data about the chemistry of lakes only recently have become available on a regional basis. Few lakes or streams have been sampled sufficiently to enable determination of the presence or absence of trends related to acidification.

Precipitation Chemistry Data

The first documentation of trends in the chemistry of precipitation for the region is for the Como Creek watershed, 6 km east of the Continental Divide in Boulder County, Colorado (Lewis and Grant 1980). During 150 weeks beginning in 1975, analysis of weekly bulk deposition samples indicated a linear decrease in pH of about 0.8 unit. The magnitude of the increase in H^+ loading was balanced by an equivalent increase in the NO_3^- loading. Sulfate concentration did not have a statistically significant trend with time; SO_4^{2-} was only slightly larger in loading rate than NO_3^-: $SO_4^{2-} = 196$ eq ha^{-1} yr^{-1}; $NO_3^- = 129$ eq ha^{-1} yr^{-1} (Grant and Lewis 1982).

The Como Creek precipitation data are considered to be affected by the densely urbanized Denver Front Range metropolitan area. From January to July, when easterly winds are the most prevalent, a hundredfold increase in concentrations of pollutants in air masses from the metropolitan area is the main factor that affects average pollutant concentration during collection periods of about a week, such as used in the work of Lewis and Grant (1980) on Como Creek (Kelly and Stedman 1980). The large effect of the metropolitan area on precipitation chemistry along the Front Range was documented further by Huebert et al. (1983), who concluded that more than 60% of the total concentration of NO_x and HNO_3 was attributable to the nearby metropolitan area.

Bulk deposition samples were collected from a statewide network of 42 sites in Colorado between May 1982 and May 1983 (Lewis et al. 1984). The calculated deposition rate of oxides of nitrogen and sulfur was about equal to emissions from power plants in and near Colorado. A negative correlation of pH, as a function of elevation, was attributed to the preferential removal of alkaline crustal materials relative to acidic particles from air masses by orographic effects. Five stations had an average pH < 5.0; the minimum average pH was 4.7. Measurements of wetfall at a site in westcentral Colorado indicate an average pH of 4.9 for 1981–84 (Harte et al., 1985).

Atmospheric deposition of SO_4^{2-} in wetfall has been correlated with SO_2 emissions from nonferrous smelters. The annual, volume-weighted concentration of SO_4^{2-} at eight wetfall sites in Arizona, Colorado, Idaho, and Montana correlated with annual SO_2 emissions from smelters in Arizona, New Mexico, Nevada, and Utah for 1980–83

(Oppenheimer et al. 1985). This correlation approach was modified to examine the correlation between monthly volume-weighted concentrations of SO_4^{2-} at five wetfall sites in Colorado and monthly SO_2 emissions from smelters during 1980–84 (Epstein and Oppenheimer 1987). Little correlation occurred between monthly smelter emissions of SO_2 and monthly volume-weighted concentrations of SO_4^{2-}. Modification of the data set to eliminate months with incomplete sample collection and to deseasonalize the data resulted in an explanation of 41% of the variance in wetfall SO_4^{2-} concentration caused by the smelter emissions ($p < 0.001$).

Wetfall chemistry data are available for the Rocky Mountain region from the National Atmospheric Deposition Program (Figure 14.9). These data provide an internally consistent regional assessment of the acidity status of wetfall. Although most of the stations are located at low elevations, a few stations are located at elevations typical of many sensitive aquatic systems. The sparsity of high-elevation sites makes regional contour maps ambiguous, and the short period of record makes determination of trends difficult.

Surface Water Data

At the Como Creek site discussed by Lewis and Grant (1980), the increase in H^+ loading to the watershed corresponds to an approximately equivalent decrease in stream transport of HCO_3^-. The stream loading of SO_4^{2-}, NO_3^-, NH_4^+, and dissolved organic matter also indicated significant increases. The decrease in base cation loading in the stream accounted for only about one-third of the measured decrease in ANC, indicating that much of the decrease in ANC was attributable to neutralization by H^+ (Lewis and Grant 1979).

In a 1979 reconnaissance of Colorado lakes initally sampled during 1938–42 and 1949–52, a 17% decrease in ANC and a 0.2-unit decrease in pH occurred. Both decreases were attributed to acidification by precipitation (Lewis 1982). Differences in methods and a lack of data about major ion chemistry have prevented determination of whether this apparent acidification is plausible.

The ANC of six ponds in westcentral Colorado, initially measured in 1972, was remeasured in 1981–83; a trend toward decreased ANC in the more recent data was evident. Seasonal variance of ANC in these systems and differences in methods hamper determination of whether a significant trend exists (Schneider 1984, Harte et al. 1985).

A reconnaissance of lakes in Rocky Mountain National Park, Colorado, and Yellowstone National Park, Wyoming, indicated no evidence of chronic acidification in either area (Gibson et al. 1983). Maximum possible acidification, estimated from the concentrations of NO_3^- and SO_4^{2-}, was < 10 $\mu eq\ L^{-1}$ in seven of eight watersheds studied. One watershed was calculated to have lost a maximum of 18 $\mu eq\ L^{-1}$.

A 1983 reconnaissance of 92 lakes in the Wind River Range of Wyoming determined that although the area contained lakes with low ANC, the anion chemistry was dominated by HCO_3^-. Four lakes were monitored routinely during 1983; they showed no evidence of acidification (Stuart 1984).

Monitoring of 13 lakes in Glacier National Park, Montana, began in 1984 (Ellis et al. 1986). During 1984–85, the minimum ANC was 80 $\mu eq\ L^{-1}$ and the minimum pH was 6.5. In the lakes with the lowest ANC, SO_4^{2-} concentrations were only about 10 $\mu eq\ L^{-1}$, and NO_3^- concentrations were only a few $\mu eq\ L^{-1}$. Thus, little ANC could have been lost by neutralization of acidity, which possibly was associated with SO_4^{2-} and NO_3^-.

A 1983 reconnaissance of 70 lakes in the Mt. Zirkel Wilderness Area of Colorado indicated that, although the lakes had little ANC (about one-half the lakes had ANC < 100 $\mu eq\ L^{-1}$), there was no evidence of acidification. Maximum possible acidification, estimated from concentrations of NO_3^-, SO_4^{2-}, Cl^-, and H^+ in wetfall and concentrations of SO_4^{2-} and Cl^- in lakes, was no more than 9 $\mu eq\ L^{-1}$ (Turk and Campbell 1987).

A statewide reconnaissance of 175 Colorado lakes in 1984 was used to develop a classification system for all Colorado lakes (Chappell et al. 1985, Chappell et al. 1986). The minimum ANC for samples collected during the spring was 22 $\mu eq\ L^{-1}$; whereas that for samples collected during the fall was 130 $\mu eq\ L^{-1}$. Minimum pH during the spring was 6.4; minimum pH during the fall was 6.9. Factor analysis explained the differences in lakewater chemistry with factors interpreted to be related to: (1) bedrock geology, (2) vegetation, (3) hydrologic setting, and (4) human use. No factors associated with atmospheric deposition were determined.

The presence of long-term trends in acidification was investigated by diatom and metal stratigraphy for four lakes in Rocky Mountain National Park, Colorado (Baron et al. 1986). The stratigraphic data indicated no historical effect on pH attributable to atmospheric deposition; however, there was evidence of enhanced lead deposition between 1855 and 1985.

Stream chemistry data are available for several Rocky Mountain region headwater streams in the National Hydrologic Benchmark Network (Smith and Alexander 1983). Of the nine streams in the region, only four in Colorado and northern New Mexico have ANC low enough to be likely to respond to acidification. Data from three of the four streams indicate an increasing trend in SO_4^{2-}; data from all four indicate a decreasing trend in ANC between the mid-1960s and 1981.

Future Considerations

Potential for Future Acidification

Most Rocky Mountain lakes are not acidic, but would be very sensitive to acidification if atmospheric deposition of acids should increase. The region is rich in natural resources such as oil, gas, oil shale, coal, and nonferrous metals. Development of these resources, as well as population growth, may result in increased atmospheric emissions of acids. Many of these resources are located nearby and upwind of sensitive Rocky Mountain lakes. Thus, increased atmospheric deposition of acids to lakes and their watersheds may occur.

Presently (1990), increases in atmospheric emissions in much of the Rocky Mountain region are regulated by the Clean Air Act and the Wilderness Act. Those administering these acts have commonly used models of atmospheric transport and deposition, in conjunction with simple lake titration models, to estimate the effects of proposed new sources. Depending on the model predictions, applications for new emissions sources are disapproved, approved, or approved with specifications of levels of emissions control and possible mitigation procedures. Making such decisions effectively requires knowledge of processes that control lake chemistry. Such knowledge is generally lacking for Rocky Mountain lakes. Thus, although legal mechanisms exist to protect these lakes, the scientific understanding necessary to effectively utilize these legal mechanisms is incomplete. This lack of knowlege may hamper the application of regulatory processes, either by failing to protect wilderness lakes or by overly restricting emissions because worst-case estimates of effects of emission are used in the absence of more definitive knowledge.

In addition to regulation of new emissions sources, monitoring of lake and watershed characteristics is likely to be important in determining the potential for future acidification. Careful selection of monitoring systems could allow detection of changes in chemistry or other characteristics before damage can occur. Presently (1990), no coordinated monitoring of lakes and watersheds on a regional scale exists. Also, commitments to long-term monitoring at existing sites are rare. Thus, coordinated design, funding, and operation of a regional monitoring network are needed to ensure that future acidification can be reversed before damage occurs.

Selection of Regionally Representative Watersheds

If we are to more fully understand watershed processes that control the present chemistry of Rocky Mountain lakes, and the probable response of these processes and lakes to future changes in atmospheric deposition, regionally representative research and monitoring watersheds must be selected. Because of the large range of geographic, geologic, and hydrologic conditions throughout the region, only a small number of lakes and watersheds can be effectively studied; therefore, special care needs to be used in selecting these few systems.

The discussion presented in this chapter uses regional patterns in lake chemistry to infer likely mechanisms controlling lake chemistry. This approach is very useful in generating hypotheses about which sources, sinks, and processes are important to lake chemistry and which are not. This approach, by starting at the regional scale, allows determination of regionally common lake chemistry and, perhaps, controlling mechanisms. However, verification of the hypothesized controlling mechanisms must be done at a smaller scale in systems representative of the regional population of lakes. Considering the number of lakes in the Rocky Mountain region, and the number of important watershed characteristics that may affect watershed response to acidity, very few systems are monitored routinely or are intensively studied to determine processes that control lake chemistry. Effective testing of hypothesized controls requires intensively studied watersheds representing the range of geology, soil, vegetation, climate, atmospheric deposition, and hydrologic characteristics common to the region.

In selecting a regionally representative watershed, it is possible to select for appropriate geology, watershed area, elevation, vegetation, and other characteristics and still obtain a nonrepresentative watershed. Minor amounts of readily weathering minerals are not included on most geologic maps and are easily missed in field verifications. Hydrology, which controls contact and reaction between water and minerals, often is complex and cannot be known without direct measurement of flow paths. One approach to minimizing these sources of variance from regionally normal characteristics is to use lake chemistry as an indicator of controlling processes. Illustrations in this chapter presenting the frequency distribution of lake SO_4^{2-} and Cl^- concentrations demonstrate that some concentration ranges are common, whereas other concentration ranges seldom occur. Further, the various concentration ranges within these frequency distributions indicate which controlling processes may be important. A regionally representative watershed thus should have lake SO_4^{2-} and Cl^- concentrations within a readily defined range. If it is necessary to study a process that produces lake concentrations outside the regional normal range, such as SO_4^{2-} reduction, lakes within that part of the frequency distribution should be considered as candidate sites.

Biological Status

Due to natural barriers to fish migration, most high-elevation lakes throughout the Rocky Mountain region were originally fishless. By the late 1800s, trappers, miners, and settlers, and in later years, sportsmen, began transplanting fish into these waters. For example, lakes in the Bridger Wilderness of the Wind River Mountains were mostly devoid of fish before 1910; trout populations existed only in some lower elevation lakes and streams adjacent to the present wilderness areas (Hudelson et al. 1980). The widespread development of hatcheries in the late 1800s and early 1900s accelerated introductions of trout into barren lakes.

The earliest transplants used indigenous species (Baxter and Simon 1970, Brown 1971), especially cutthroat trout (*Oncorhynchus clarki*). But, they also included grayling (*Thymallus arcticus*) in the upper Missouri River system of northern Wyoming and Montana and rainbow trout (*Salmo gairdneri*) and bull trout (*Salvelinus confluentus*) in the west slope drainages of the northern Rockies. Later, along with the above species, other trout species were introduced, including brown trout (*Salmo trutta*), golden trout (*Oncorhynchus aquabonita*), brook trout (*Salvelinus fontinalis*), lake trout (*Salvelinus namaycush*), and Sunapee trout (*Salvelinus alpinus aureolis*). Some nongame species, including cyprinids and catostomids, have also been introduced into some high-elevation lakes, mostly by fishermen. All 10 of these trout species now occur in at least some low alkalinity lakes in the Rocky Mountains. However, the species of principal concern in relation to the effects of acidification are cutthroat, rainbow, golden, and brook trout, as these are now the species most prevalent in high-elevation lakes.

Natural reproduction by trout in many high-elevation Rocky Mountain lakes is limited by lack of spawning habitat, especially spawning streams, since many of the lakes are headwater lakes (Hudelson et al. 1980). This is especially critical for obligate stream spawners, such as rainbow, cutthroat, and golden trout. Brook and lake trout often find suitable spawning habitat within high-elevation lakes, and are not dependent on streams for spawning. In Colorado, relatively few high-elevation lakes have self-sustaining trout populations, except for a few brook trout, lake trout, and cutthroat trout lakes. Thus, most high-elevation lakes in Colorado depend on regular stocking to maintain their fisheries (Nelson 1987).

The first stockings of high-elevation lakes attempted only to establish new fisheries. But current management focuses on providing diverse, quality fisheries. In the Bridger Wilderness, for example, stocking is primarily limited to maintaining existing populations and to introducing new species; golden and cutthroat trout dominate present stocking efforts (Hudelson et al. 1980). Throughout the Rockies, rainbow, cutthroat, and golden trout are the predominant species now stocked in high-elevation lakes. Angler usage in these lakes ranges from only a few to over a thousand fishing days per year, depending largely on ease of access.

Limited data exist for assessing the sensitivity to acidification of the native Western fish species inhabiting high-elevation Rocky Mountain lakes. As discussed in Chapter 4, survival of fish in acidifying surface waters depends primarily on levels of pH, inorganic monomeric Al, and Ca^{2+}. Briefly, high acid and inorganic monomeric Al concentrations and low Ca^{2+} concentrations all can cause stress and reduce survival. At concentrations typically found in acidifying surface waters, Al can generally produce the greatest reduction in survival. But this relationship can vary with life stage (Mount and Marcus 1988). Therefore, evaluating potentially toxic effects in acidifying surface waters requires that both the water quality conditions and the life stages present be carefully defined.

Figure 14.17 shows survival during laboratory toxicity tests using two life stages of brook trout and rainbow trout exposed to 5.2 pH, 50 μg L^{-1} total dissolved Al, and 1 mg L^{-1} of Ca^{2+}, conditions similar to those that could occur during potential acidification of sensitive lakes in the Rocky Mountains. These plots indicate that rainbow trout are potentially more sensitive to acidification than brook trout. Other recent toxicity data indicate that cutthroat trout may be more sensitive than rainbow trout (Farag 1988, Woodward et al. 1989). Although not yet tested, golden trout, another *Oncorhynchus* species related to rainbow trout, may also be more sensitive to acidification than brook trout.

Thus, western trout species appear to be more sensitive to acidic water than brook trout, the trout species of greatest concern in the East. Since, as noted earlier in this chapter, many high-elevation

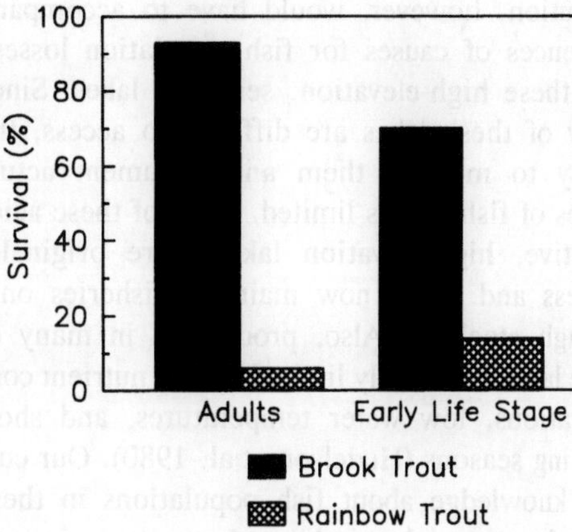

FIGURE 14.17. Survival by brook trout and rainbow trout exposed to 5.1 pH, 50 μg L^{-1} total dissolved Al and 1 mg L^{-1} Ca^{2+}. Exposures for adults extended for approximately 5 to 6 months, while early life stage exposures extended for approximately 3 months following fertilization (Unpublished data from the Lake Acidification and Fisheries project, University of Wyoming, funded by Electric Power Research Institute, used by permission.)

Rocky Mountain lakes have very high sensitivities to potential acidification, the relatively high sensitivities of western trout species imply that a large proportion of this fishery resource is potentially at high risk. Since several of these native Rocky Mountain fish species and subspecies are considered to be threatened or endangered by both federal and state authorities, concern for their protection is increased (Behnke and Zarn 1976, Johnson 1987). It is important to emphasize here, however, that presently there is no known evidence of impact from acidification on the fishery resources of high-elevation Rocky Mountain lakes.

Where impacts to fisheries by acid depositions have been reported, the first observed impacts have accompanied snowmelt and heavy rainfall events, when runoff waters contain acutely toxic acid and inorganic monomeric Al concentrations (e.g., Leivestad and Muniz 1976, Schofield 1980). Many high-elevation streams and lakes in the Rocky Mountains have high flushing rates. Therefore, in the event of acidification of lakes and streams in the Rocky Mountains, fish populations would be most likely at highest risk during such episodic runoff of acutely toxic waters. Also, the first impacts to fisheries would accompany such runoff events.

Caution, however, would have to accompany inferences of causes for fish population losses rom these high-elevation, sensitive lakes. Since many of these lakes are difficult to access, the ability to monitor them and document actual causes of fish loss is limited. Most of these acid-sensitive, high-elevation lakes were originally fishless and many now maintain fisheries only through stocking. Also, production in many of these lakes is severely limited by low nutrient concentrations, low water temperatures, and short growing seasons (Hudelson et al. 1980). Our current knowledge about fish populations in these high-elevation lakes is limited mostly to incomplete stocking records, angling surveys, and occasional gill net samplings. Until this knowledge is substantially improved, great care must be exercised in interpreting causes of fish population losses or changes in Rocky Mountain lakes sensitive to potential acidification.

Conclusions

1. Most of the Rocky Mountain lakes sampled during the Western Lake Survey have pH $\geq$ 7.0, but ANC $\leq$ 200 μeq L^{-1}. Thus, these lakes are not acidic, but are very sensitive to acidification. Lack of historic data prevents a direct determination of whether ANC has decreased in these sensitive lakes. Thus, indirect methods are necessary to estimate past acidification. Present data are inadequate for determining whether episodic acidification occurs; however, these data are adequate for determining whether the lakes have acidified during September–October.
2. Based on lake concentrations of SO_4^{2-} and Cl^-, and on wetfall concentrations of SO_4^{2-}, NO_3^-, and H^+, lakes with ANC $\leq$ 200 μeq L^{-1} are estimated to have lost no more than 5 μeq L^{-1} in the Bitterroot Range, 10 μeq L^{-1} in the Colorado Front Range, and 12 μeq L^{-1} in the Wind River Range.
3. Indirect methods used to estimate acidification also suggest which processes are likely to be most important in controlling concentrations of SO_4^{2-} and Cl^- in these dilute lakes. The wetfall concentration of SO_4^{2-} and Cl^- is the most important control in the Colorado Front Range, the Wind River Range, and the Bitterroot Range.
4. Evapotranspiration, dryfall, anion adsorption, SO_4^{2-} reduction, and mineral weathering rarely are half as important as wetfall in controlling the concentration of SO_4^{2-} and Cl^-. In most instances, these processes are insignificant compared to wetfall for the most dilute lakes.
5. The watershed mineralogy of the lakes discussed in this chapter is mostly granitic in composition; however, the concentrations of major cations and SiO_2 in the lakes are not consistent with a simple stoichiometric weathering of dominant granitic minerals. Stoichiometry of lake chemistry is consistent with the mineral weathering reactions being dominated by dissolution of calcite and dolomite. In particular, lake concentrations of Na^+ and SiO_2, compared to Ca^{2+}, indicate that stoichiometric weathering of the common silicate minerals cannot account for lake chemistry.
6. Many possible explanations exist for the apparent inconsistency between bedrock and lake chemistry. For example, SiO_2 produced in weathering of silicate minerals may be removed by precipitation of silica-rich clays or other minerals, by uptake by diatoms, or by adsorption. Atmospheric sources of readily weathered minerals may be the source of much of the lake Ca^{2+}, Mg^{2+}, and ANC, as may trace amounts of readily weathered minerals, such as calcite, within the watersheds. Also, preferential chemical weathering of silicates, in conjunction with rapid physical weathering of the regolith, may account for the unexpectedly great ratios of $[Ca^{2+} + Mg^{2+}]:Na^+$. Unfortunately, data are insufficient for determining which explanation is most important.
7. This chapter presents hypotheses of processes that control the chemistry of dilute lakes in the Rocky Mountain region. These hypotheses are developed using regional data on lake chemistry and a conceptual model of the effects of various processes on lake chemistry. However, these hypotheses need to be verified by intensive work on regionally representative watersheds. Representativeness can be achieved by selection of watershed characteristics common

to the region. If particular processes need to be studied, lakes in which that process is important to the chemistry can be selected based on frequency distributions of lake chemistry, as interpreted by the conceptual model presented here.

Acknowledgments. The section on biological status was written by Dr. Michael D. Marcus of Western Aquatics, Inc., in Laramie, Wyoming. Data from the Lake Acidification and Fisheries Project were used with permission from the Electric Power Research Institute.

15
Sierra Nevada, California

John M. Melack and John L. Stoddard

ABSTRACT. The lakes and streams of the Sierra Nevada are vulnerable to acidic deposition because of the predominance of granitic rocks and thin acidic soils in their catchments, and the large quantity of precipitation in the region. Most of the precipitation to the Sierra Nevada falls as very dilute snow. When this is combined with the more acidic spring, summer, and autumn rain or wet snow, annual volume-weighted mean pH values of precipitation are between 5.2 and 5.5. Annual deposition (meq m^{-2}) ranges from 2 to 14 for H^+, 2 to 12 for NO_3^- and 1.5 to 13 for SO_4^{2-}.

Characteristics of Sierra Nevada lakes obtained from the EPA's Western Lake Survey can be summarized as follows: 70% were < 10 ha in area and < 10 m deep; about 75% had pH values between 6.5 and 7.5 and only one had a pH value < 6.0; 65% had ANC values < 100 μeq L^{-1} and 75% had SO_4^{2-} values < 10 μeq L^{-1}. Calcium is the dominant cation in all the lakes. Weathering of plagioclase feldspars in intrusive igneous rocks produces the observed strong correlation of Ca^{2+} and Na^+ with ANC for most lakes. An R-mode factor analysis confirmed this and identified associations of Ca^{2+} and SO_4^{2-} or Ca^{2+} and Mg^{2+} in moderate-to-high ANC waters. Samples from a nonrandom subset of 15 lakes indicated that pH values are lower by 0.2 to 1.0 units in the spring or early summer, compared to autumn values.

Processes influencing lacustrine chemistry are examined with data from intensively studied Emerald Lake and Gem Lake and from five other lakes sampled year round. Rock weathering in the catchment is the major source of ANC; sediments within the lakes contribute slightly to ANC accumulation during stratified periods under ice and in mid-summer. Brief episodes of near zero ANC can occur during exceptionally heavy rains and during snowmelt. The main influence of the large volume of water released during snowmelt is to flush the lakes with very dilute water. A slight acidifying effect is also associated with snowmelt and rains, but the effect is episodic, not chronic.

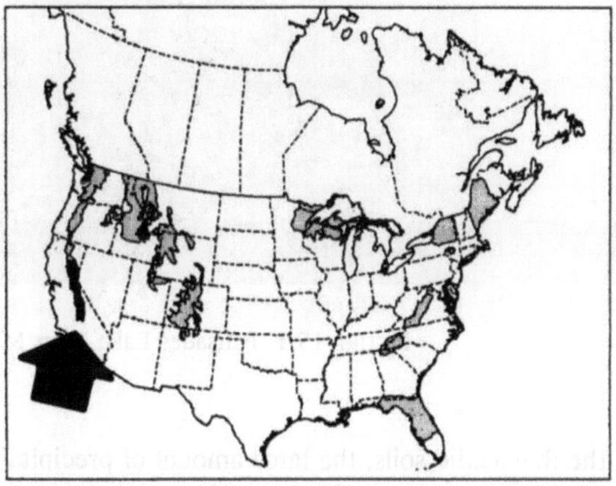

Zooplankton and benthic macroinvertebrates known to be sensitive to acidic inputs are common and abundant in Sierra Nevada waters, hence it is unlikely that acidic deposition has affected the biota. Furthermore, paleolimnological data derived from diatom microfossils indicate no evidence for regional acidification during the last century. Lakes of the Sierra Nevada are not now acidified but are extremely vulnerable to increases in acidity of atmospheric precipitation.

Introduction

As evidence for acidic deposition in California and its potential consequences has increased (California Air Resources Board 1988), the chemical and biological conditions in the lakes and streams of the Sierra Nevada have become special concerns (Figure 15.1). The waters of the Sierra Nevada are vulnerable to acidic deposition because of the predominance of intrusive igneous rocks that weather slowly,

FIGURE 15.1. Palisades Lake, near Mather Pass, Kings Canyon National Park.

the thin acidic soils, the large amount of precipitation, and the low ANC of the lakes (Melack et al. 1985; Landers et al. 1987). Runoff from these mountains supplies most of the fresh water used by the cities of San Francisco and Los Angeles and the rich agricultural lands in the Central Valley. Recreational activities in Yosemite, Kings Canyon, and Sequoia National Parks, and in National Forest lands, much of which are designated wilderness areas, are focused on the lakes and streams.

The longest continuous mountain range in the conterminous United States, the Sierra Nevada extends about 700 km with a width of 100 to 130 km. The northern end (ca. 40°15'N) of the range is located near Lake Almanor, just south of the volcanic rocks associated with Mt. Lassen. Tehachapi Pass (ca. 35°N) marks the southern extent. To the west lies the Central Valley and to the east the Great Basin. The tilted fault block forming the Sierra Nevada rises from the west as a band of foothills 50 to 80 km wide and reaches to the High Sierra, the high elevation region above 2,700 m and 16 to 32 km wide (Whitney 1979). Spectacular river gorges course to the west and include the 900-m glacier carved walls of Yosemite Valley and 2,450-m deep Kings Canyon. Peaks along the Sierra crest exceed 4,250 m in the Palisades and Mt. Whitney areas; north of Lake Tahoe, few peaks reach 2,400 m and at the southern end, the highest ridges are near 2,150 m. The steep, eastern escarpment of the Sierra is especially impressive above Owens Valley where the crest rises 2,700 to 3,300 m above the valley floor.

Eleven major rivers flow west into the Central Valley from the Sierra Nevada. The eastern side is drained by several rivers and many streams, all of which terminate in the Great Basin, except those diverted to the City of Los Angeles. About 2,200 lakes > 1 ha and < 2,000 ha are located in the Sierra Nevada (Landers et al. 1987).

Regional Characteristics

Watershed Characteristics

Geology

The Sierra Nevada is the product of depositional, erosional, volcanic, and tectonic processes of the past 500 million years (Matthes 1930, Blackwelder 1931, Bateman 1961, Hill 1975, Norris and Webb 1976). The range is made up largely of igneous and metamorphic rocks of diverse composition and age. Shallow sea floor sediments and submarine extrusive igneous rocks deposited in the Paleozoic and Mesozoic Eras were metamorphosed and uplifted. Beginning in the early Triassic period, the uplifted rocks were intruded by molten igneous rock, and by the late Cretaceous period, about 70 million years ago, the granitic rocks that form the Sierran batholith were emplaced. The period between about 80 and 40 million years ago was one of erosion, during which much of the ancient metamorphosed rock was removed.

Extensive eruptions of basic volcanic material buried much of the northern Sierra during the mid- to late Tertiary period. Discontinuous but regionally important basaltic volcanism has continued through the Pleistocene to the present throughout the Sierra. Increased tectonic activity rejuvenated the Sierran rise and contributed to accelerated downcutting of Sierran drainages during the last 10 million years.

The glacial epoch of the Pleistocene began about 3 million years ago. It was during this period that much of the magnificent sculpting of the Sierran rock faces occurred. The glacial ice retreated between 10,000 and 25,000 years ago, leaving behind depressions present today as debris filled meadows and alpine lakes (Figure 15.1).

Soils

High-elevation Sierra Nevada soils are characterized by a frigid or cryic soil temperature regime, that is, the mean annual soil temperature ranges from 0°C to 8°C (Huntington and Akeson 1987, Reilly and Zasoski 1989). In the frigid regime, the difference between mean summer and mean winter soil temperature is more than 5°C. The lower elevational boundary of the frigid regime moves downslope with increasing latitude. Accordingly, frigid soils occur at approximately 2,300 m in the southern Sierra, and at approximately 1,800 m in the northern Sierra.

In the Sierra Nevada, soil development is greater at lower elevations. Montmorillonitic and kaolinitic clays are present in limited amounts and are more abundant in the frigid than cryic soil temperature regime. Soil organic matter tends to accumulate at the higher elevations where cool temperatures reduce microbial decomposition rates. Vegetation enhances the rate at which soil parent material weathers by the action of organic acids. Coniferous litter, particularly red fir litter, is low in base content, and soils that develop under conifers have low pH and very low ionic content in soil solutions.

Soils that develop from the granitic rocks of the Sierra are higher in sand and lower in clay than are soils that develop from the volcanic deposits or metamorphic rocks. The coarse textured soils developed from granitic parent materials tend to be permeable and acidic (pH 5.0 to 6.5), and to have low base saturation (5% to 15%) and cation exchange capacity (average about 10 meq per 100 g) and limited moisture retention capacity (Huntington and Akeson 1987, Lund et al. 1987, Reilly and Zasoski 1989).

Soils that develop from basic volcanic and metamorphic rocks are typically finer textured and contain larger amounts of iron-and magnesium-rich minerals and calcic plagioclase feldspars than granitic soils. Because of their finer texture, these soils have higher moisture retention capacity and cation exchange capacity than granitic soils. Base saturation is higher because of the greater abundance of bases in the parent material. High amounts of oxidized iron impart a striking red color to many of these soils.

Vegetation

Above 3,200 m, a variety of perennial plants occurs in alpine fell fields, but the most notable characteristic of this region is the scarcity of vegetation. Alpine meadow and meadow-like vegetation is often found near lakes above 3,200 m. Whitebark pines are scattered between 3,000 and 3,350 m. Open, subalpine forest occurs between 2,800 and 3,350 m, and upper montane/red fir forest grows between 1,800 and 2,800 m. More detailed descriptions of the vegetation found in Sierra Nevada watersheds are provided in Munz and Keck (1959), Bennett (1965), Klikoff (1965), Chabot and Billings (1972), Majors and Taylor (1977), Burke (1977), Rundel et al. (1977), and Melack et al. (1983).

Climate

The Sierra Nevada climate is characterized by warm, dry summers and cool, wet winters; the climate varies strongly with altitude and topography (Miller 1955). The paucity of meteorological stations at the higher elevations limits the detail and accuracy of descriptions of Sierra Nevada climatology. A compilation of available data for the southern Sierra is available in Stephenson (1988), the source for the following summary. The lapse rate is about 5°C km^{-1}, and the tree line, which corresponds approximately to the elevation with a mean temperature of 10°C during the warmest months, occurs at about 3,200 to 3,400 m. Mean annual precipitation on the western side of the crest increases from less than 400 mm in the foothills (below 250 m) to about 1,400 mm at mid-elevations (2,000 m) and then remains about the same at higher altitudes. The eastern side of the crest receives half to two-thirds as much precipitation as the western side. The proportion of precipitation that falls as snow is negligible below 500 m, rises to about half at 2,100 m, and continues to increase to between 80% and 100% above 3,000 m. Almost all the precipitation falls from October through May; convective summer rains are more common at higher than lower elevations. Runoff from mid- to high-elevations comes predominately during snowmelt from April through July.

California receives prevailing westerly winds, with the North Pacific high-pressure cell having the dominant influence on the low-level winds, especially during the summer. As the Pacific high weakens and moves south in the winter, its influence is lessened. Differential heating of the land surface as compared to the adjacent Pacific Ocean enhances the westerly flow during the warm portion of the year. On a regional or local scale, winds are channeled by valleys with a tendency for upvalley flow during warm months and downvalley flow during cool months. Upslope winds from the Central Valley into the Sierra Nevada are most intense during hot, summer afternoons. Conversely, downvalley flow is strongest at night in winter. The air flows also have a diurnal pattern (Ewell et al. 1989) with daytime, upslope and nighttime, downslope flows. The climatology of surface winds in California is summarized and diagrammed in Hayes et al. (1984).

Land Use

Most human activities in the High Sierra are now recreational, with about 1.5 million ha lying within national parks and wilderness areas (Whitney 1979). Mining operations are limited today and hydroelectric reservoirs are found largely at lower elevations. Visitation frequency varies considerably among high-elevation Sierra Nevada lakes, but very few, if any, are likely to have had their nutrient or major ion chemistry influenced by human activity within their basins. Further information about tourist use and possible impacts on Sierra Nevada lakes is provided in Baas et al. (1976), Ghirelli et al. (1977), Silverman and Erman (1979), Melack et al. (1983), Parsons (1983), and Parsons and Stohlgren (1987).

Atmospheric Deposition

Regional patterns of atmospheric deposition in the Sierra Nevada are not well understood because of the inadequate number of monitoring sites and their location at moderate elevations, the strong influence of topographic relief on the amount of precipitation and the possible importance of dry deposition during the summer. At high elevations (> 2,500 m), snowfall accounts for almost all the input of water (Sickman and Melack 1989, Dozier et al. 1989), and the snowpack accumulates through the winter without much melt until March or April. Hence, regional surveys of snow depth and density in late winter permit reliable estimation of the contribution of snow to the annual deposition of water (e.g., California Cooperative Snow Survey 1986).

Very few sites in the Sierra Nevada are monitored for chemical composition of atmospheric deposition (Table 15.1; Figure 15.2; Western Region Overview). The National Atmospheric Deposition Program (NADP) currently operates two sites, and the California Acid Deposition Monitoring Program (CADMP) now has six year-round sites, two of which are colocated with NADP collectors. All these sites are at low-to-moderate elevations on the periphery of the High Sierra. Data from high elevations are available in the southwestern Sierra and eastern-central Sierra.

Snow cores collected at 25-km intervals in a transect along the Sierra Nevada crest in March 1983 provide another valuable description of Sierra Nevada snowpack chemistry, albeit from a single sampling period (Laird et al. 1986). Synoptic surveys of Sierra Nevada snow date to the late 1950s (Feth et al. 1964a), and samples were collected intermittently during the 1970s (Brown and Skau 1975, McColl 1980, Leonard et al. 1981, Powers

TABLE 15.1. Annual wet atmospheric deposition and volume-weighted mean concentrations in Sierra Nevada, California.

Location (Altitude, m)	Dates	Precipitation[a] cm	H^+ μeq L^{-1}	H^+ meq m^{-2}	NO_3^- μeq L^{-1}	NO_3^- meq m^{-2}	SO_4^{2-} μeq L^{-1}	SO_4^{2-} meq m^{-2}	Reference or source
Giant Forest, Sequoia N.P. (1,856)	1981	64	4.79 (5.32)	3.1	8.9	5.7	12.1	7.7	NADP[b]
	1982	203	5.0 (5.3)	10.2	3.5	7.1	6.0	12.2	
	1983	148	3.63 (5.44)	5.4	6.1	9.0	7.1	10.5	
	1984	69	3.63 (5.44)	2.5	7.1	4.9	8.7	6.0	
	1985	72	4.48 (5.35)	3.2	7.7	5.5	6.9	5.0	
Hodgson Meadows, Yosemite N.P. (1,408)	1982	212	4.17 (5.38)	8.8	3.2	6.8	4.2	8.9	NADP
	1983	232	3.39 (5.47)	7.9	5.0	11.6	5.6	13.0	
	1984	82	3.39 (5.47)	2.8	7.4	6.1	8.9	7.3	
	1985	91	4.57 (5.34)	4.2	8.4	7.6	6.2	5.6	
Quincy (1,000)	July 85–June 87	91[d]	5.6 (5.25)	5.1	5.1	4.6	4.1	3.7	CADMP[c]
Soda Springs, Norden (2,200)	July 85–June 87	112[d]	4.9 (5.31)	5.5	4.4	5.0	3.4	3.8	CADMP
S. Lake Tahoe (1,880)	July 84–June 87	40[d]	5.7 (5.25)	2.3	6.8	2.7	5.5	2.2	CADMP
Hodgson Meadows, Yosemite N.P. (1,408)	July 85–June 87	77[d]	4.9 (5.31)	3.8	8.9	6.8	6.2	4.8	CADMP
Giant Forest Sequoia N.P. (1,856)	July 85–June 87	85[d]	5.9 (5.23)	5.1	9.3	8.0	5.9	5.0	CADMP
Lake Isabella (1,200)	July 85–June 87	24[d]	5.1 (5.29)	3.1	8.9	2.2	6.6	1.6	CADMP
Mammoth Mtn. (3,000)	June 81–April 82	150	3.2 (5.5)	4.8	2.8	4.2	–	–	Melack et al. (1982)
Gem Lake (3,340)	Dec 83–Oct 84	174	6.6 (5.18)	11.6	5.8	10.2	5.7	10.0	Stoddard (1986, 1987a)
Emerald Lake (2,800)	Oct 85–Sept 86	276	4.86 (5.31)	13.4	3.2	8.9	2.3	6.5	Dozier et al. (1987), CADMP
	Oct 84–Sept 87	158[e]	5.3 (5.28)	8.4	5.0	7.9	4.3	6.7	Dozier et al. (1989)

[a] Precipitation is the depth of water deposited per year from the atmosphere.
[b] National Atmospheric Deposition Program.
[c] California Acid Deposition Monitoring Program.
[d] Mean annual precipitation for 2 or 3 years.
[e] Mean for water years 1985 through 1987.

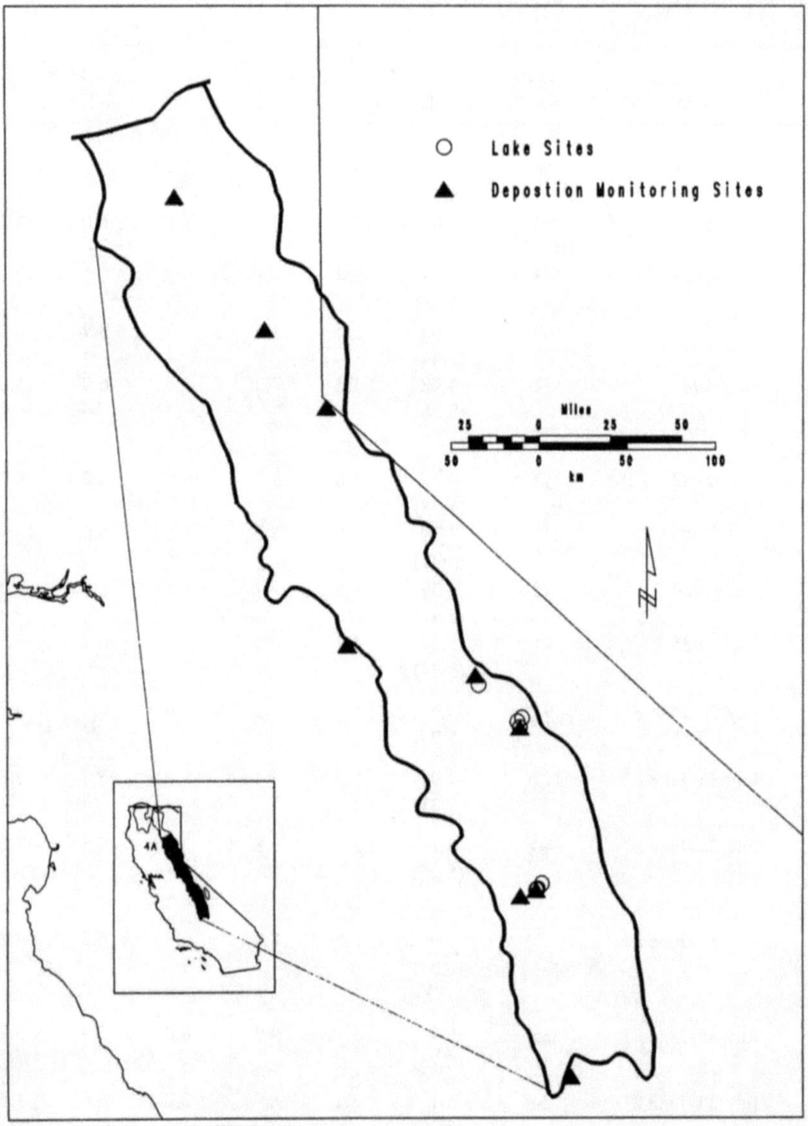

FIGURE 15.2. Locations of deposition monitoring stations and intensively studied lakes; see Tables 15.1 and 15.3.

and Rambo 1981). However, sampling and analytical differences do not permit use of these data to establish temporal trends.

A compilation of volume-weighted mean concentrations and annual wet deposition for several Sierra Nevada sites with data in the 1980s (Table 15.1) indicates that, despite the low solute concentrations, the large volume of precipitation results in considerable chemical loading. Annual deposition (meq m^{-2}) ranges from about 2 to 14 for hydrogen, 2 to 12 for nitrate and 1.5 to 13 for sulfate. These values are low in comparison to most regions of the United States (Chapter 3, Husar et al., this volume).

Although a small proportion of the precipitation falls as rain or nonaccumulating snow, this deposition makes a disproportionate contribution to the total chemical inputs because of the much higher concentrations of solutes (Melack et al. 1982, Sickman and Melack 1989, Dozier et al. 1989). Furthermore, dry deposition increases total chemical loading during summer and early autumn

(Cahill et al. 1986, Bytnerowicz and Olszyk 1988, Ewell et al. 1989). During periods with snow cover, dry deposition is sampled with the snowpack. Cloudwater can also contribute acidity, and the limited data available for Giant Forest, Sequoia National Park, are variable; pH values range from 4.4 to 5.5 (Hoffman et al. 1989).

Regional Surface Water Chemistry

Distribution and Description of Data Sets

The only available statistical sampling of lakes in the Sierra Nevada is the Western Lake Survey (WLS) conducted by the U.S. Environmental Protection Agency (EPA) in 1985 (Landers et al. 1987; Eilers et al. 1987), and this data set is used throughout this chapter to describe regional characteristics of lakes. Four additional data sets, which were not collected with probability based surveys, exist for the Sierra Nevada and are useful for determining temporal variation among Sierra Nevada lakes, as well as providing a more detailed description of the subset of lakes they were designed to sample.

Melack et al. (1985) sampled 73 lakes along the Sierra crest in 1981 and 1982. Data from nine additional lakes sampled in 1983 (Stoddard unpublished data), using field and analytical methods identical to those of Melack et al. (1985), are included here. This data set includes multiple sampling (up to four dates) of several lakes, and the sampling was designed to include lakes whose watersheds cover the range of bedrock types found in the Sierra Nevada.

Stoddard and Holmes sampled 29 lakes in 1985 to relate pH to diatom distribution in Sierra Nevada lakes (Holmes 1986). Most of these lakes were the same as those sampled by Melack et al. (1985); this subset of lakes was selected to cover a range of pH.

Melack and Setaro (1986) sampled 17 Sierra Nevada lakes in 1985 and 1986. Most lakes were sampled twice during the ice-free seasons of 1985 and 1986, and during ice cover in 1986. Many lakes were sampled at more than one station and at several depths. Some lakes were the same as those sampled by Melack et al. (1985). McCleneghan et al. (1985, 1987) sampled 34 lakes in the Sierra Nevada during spring and autumn of 1985 and 1986 as part of a statewide survey of lakes vulnerable to acidic deposition. Six of these lakes coincide with those sampled by Melack and Setaro (1986). Several other collections are available, but lack complete major ion chemistry (Tonnessen 1983, Bradford et al. 1968, Bradford et al. 1981). These are not discussed here.

The range of characteristics exhibited by the WLS and nonprobability based data sets is shown in Table 15.2. We include values for both the sample of WLS lakes and the target population of WLS lakes, in order to illustrate the importance of considering the entire population in describing the results of the WLS. Median values for many variables (especially pH, ANC, Mg^{2+}, and Na^{+}) are higher, and median elevation is lower, in the WLS sample than in the three nonprobability based data sets that are biased toward high-elevation lakes (Melack et al. 1985, Holmes 1986, Melack and Setaro 1986).

Values for the WLS target population are well within the range of values for these data sets. A major factor in the difference between WLS sample and population characteristics is illustrated in Figure 15.3, where histograms of elevation are shown for the lakes in all of the data sets. The WLS sampled an even distribution of lakes between 1,800 m and 3,800 m, but the target population has a distinct cluster of lakes at high elevations (3,000 m to 3,600 m). This cluster of high-elevation lakes was sampled preferentially in the three nonprobability based data sets listed in the previous paragraph. The McCleneghan et al. (1985, 1987) data set resembles the WLS sample distribution more closely than the WLS target population or the other nonprobability based data sets.

Description of Surface Water Chemistry

Histograms of major characteristics (lake surface area, sample depth, lake area:watershed area ratio, pH, ANC, sum of basic cations (SBC), and SO_4^{2-}, DOC, and SiO_2 concentrations) for the WLS target population are illustrated in Figure 15.4. Maps of the distribution of ANC, pH, and ratio of ANC to SBC are also provided (Figures 15.5, 15.6, and 15.7).

The vast majority of lakes in the Sierra Nevada are small (Figure 15.4a) and shallow (Figure

TABLE 15.2. Median values, with first and third quartiles in parentheses, for elevation and major ion chemistry in five data sets discussed in this chapter.[a]

Data set	*n*	pH	ANC	SO_4^{2-}	DOC	NO_3^-	Cl^-	Ca^{2+}	Mg^{2+}	Na^+	SBC	Elevation
Western Lake Survey	2119	6.93	60	7	67	0.4	2	43	6	19	76	3008
(population)		(6.65–7.30)	(37–130)	(4–7)	(42–158)	(0.1–1.7)	(1–3)	(29–104)	(3–9)	(11–21)	(51–120)	(2150–3140)
WLS (sample)	114	6.95	71	6	106	0.3	3	45	10	23	86	2626
		(6.71–7.29)	(45–137)	(4–11)	(58–183)	(0.1–1.2)	(2–6)	(29–91)	(4–20)	(14–40)	(58–164)	(2131–3159)
Melack et al.	82	6.95	48	15	—	0.4	6	37	5	19	73	3255
(1985)		(6.51–7.26)	(25–98)	(9–31)		(0.1–1.5)	(4–10)	(24–83)	(3–10)	(11–31)	(44–137)	(2999–3424)
Holmes (1986)	29	6.93	67	12	—	0.7	2	58	9	12	88	3194
		(6.44–7.42)	(27–258)	(5–83)	—	(0–3.3)	(2–4)	(42–345)	(4–18)	(9–29)	(60–413)	(2999–3365)
Melack and Setaro	17	6.64	57	10	—	2.6	3	46	6	17	74	3170
(1986)		(6.50–6.72)	(53–101)	(8–14)		(1.3–3.9)	(2–4)	(40–92)	(4–9)	(10–19)	(64–113)	(2899–3365)
McCleneghan et al.	34	6.56	66	22	—	0.4	3	47	11	12	88	2664
(1985, 1987)		(6.19–6.73)	(30–97)	(16–39)		(0.1–1.2)	(2–4)	(22–79)	(6–19)	(18–26)	(52–121)	(2195–3146)

[a] DOC is in $\mu mol\ L^{-1}$; elevations are in meters; all other values, except pH, are in $\mu eq\ L^{-1}$. SBC is sum of base cations. WLS pH values are closed-system pH; all others are partially air equilibrated.

15.4b), and drain watersheds substantially larger than the lakes themselves (Figure 15.4c). Approximately 70% of the lakes have surface areas less than 10 ha, and are less than 10 m in depth. They typically represent 5% to 15% of their watersheds. Several larger, lower elevation lakes were included in the WLS sampling; these range up to 500 ha in surface area and 90 m in depth, and include up to 40% of the watershed area.

Approximately 75% of the sampled lakes had pH values between 6.5 and 7.5 (Figure 15.4d). Only one sampled lake had closed-system pH < 6.0 (its air-equilibrated pH was 6.2), and fewer than 5% had pH above 8.0.

Most (65%) Sierra Nevada lakes have ANC values $\leq$ 100 μeq L^{-1} (Figure 15.4e), and more than 80% are considered sensitive to acidic deposition (ANC $\leq$ 200 μeq L^{-1}). A small number of high ANC lakes also were sampled; these have ANC values as high as 1,300 μeq L^{-1}.

The distribution of the sum of basic cations in Sierra Nevada lakes is similar to that for ANC; a vast majority (about 80%) have concentrations $\leq$ 200 μeq L^{-1} (Figure 15.4f). As with ANC, a few lakes have high SBC concentrations, ranging up to 1,400 μeq L^{-1}.

Sulfate concentrations are very low in the Sierra (Figure 15.4g); 75% of the sampled lakes have concentrations $\leq$ 7 μeq L^{-1}. Given the lack of low pH lakes in this region, the existence of a few high SO_4^{2-} lakes suggests catchment sources.

Dissolved organic carbon (DOC) concentrations are low in Sierra Nevada lakes; 75% have concentrations $\leq$ 160 μmol L^{-1} (Figure 15.4h), and only 10% have values $>$ 300 μmol L^{-1}. All the lakes sampled were identified as clearwater lakes (Landers et al. 1987), with a median color of 0.0 platinum cobalt units (PCU).

Silica concentrations in Sierra Nevada lakes are moderately high relative to the major ion concentrations (Figure 15.4i). Most lakes have SiO_2 concentrations from 50 to 150 μmol L^{-1}.

Other ions of interest that are not illustrated in Figure 15.4 are consistent with the pattern of low ionic strength waters in the Sierra Nevada. Extractable Al concentrations are consistently $\leq$ 2 μmol L^{-1}, the median NO_3^- concentration is 0.4 μeq L^{-1}, and Cl^- concentrations are typically $\leq$ 5 μeq L^{-1} (Landers et al. 1987).

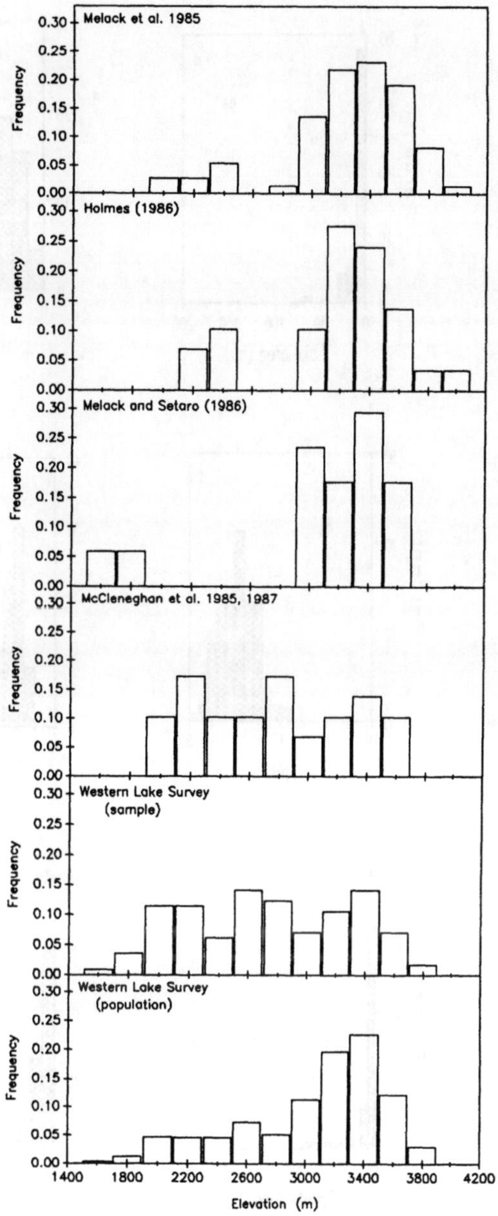

FIGURE 15.3. Frequency distribution of elevations of lakes sampled in five surveys.

There is little evidence that organic acid anions play a significant role in Sierra Nevada lakes. Total organic acid (A_t) concentrations can be estimated from anion deficit (Oliver et al. 1983, LaZerte and Dillon 1984; see also Chapter 13, Cook and Jager, this volume), which should in turn be closely

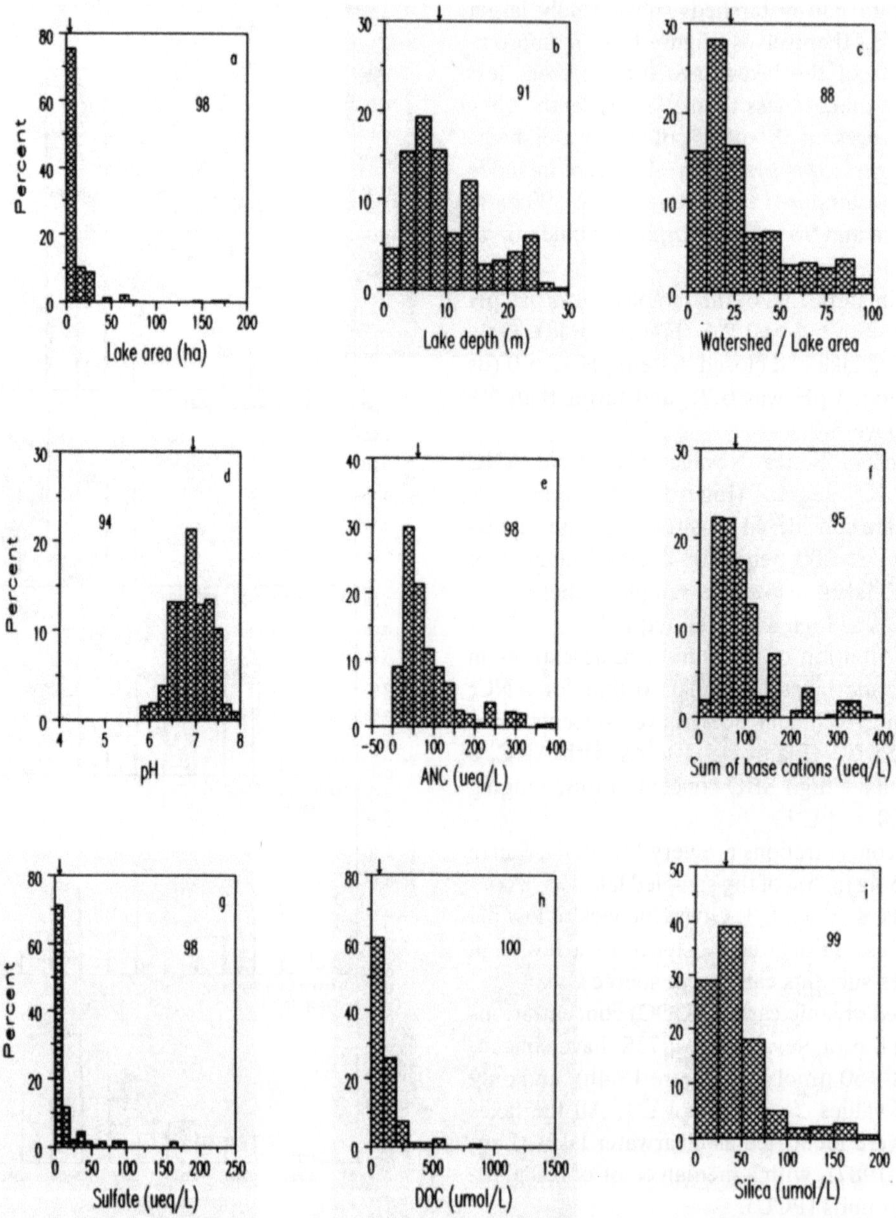

FIGURE 15.4. Frequency histograms of major characteristics of the target population of lakes in the EPA Western Lake Survey: (a) lake surface area, (b) sampling depth, (c) watershed area:lake area, (d) air-equilibrated pH, (e) ANC, (f) sum of base cations ([Ca^{2+} + Mg^{2+} + Na^{+} + K^{+}]), (g) sulfate, (h) DOC, and (i) silica. The arrow at the top of each box indicates the median of the target population, and the number in the upper right side of each box indicates the percentage of target lakes plotted.

correlated with DOC concentrations (Figure 15.8). Many of the problems with this method of estimating A_t, especially problems dealing with the influence of Al and iron on anion deficit, are minimal in the Sierra Nevada because metal concentrations are universally low (Landers et al. 1987). Measurement uncertainties are large, however, both for DOC concentrations, where values are often near detection limits, and for anion deficit, where analytical errors associated with the

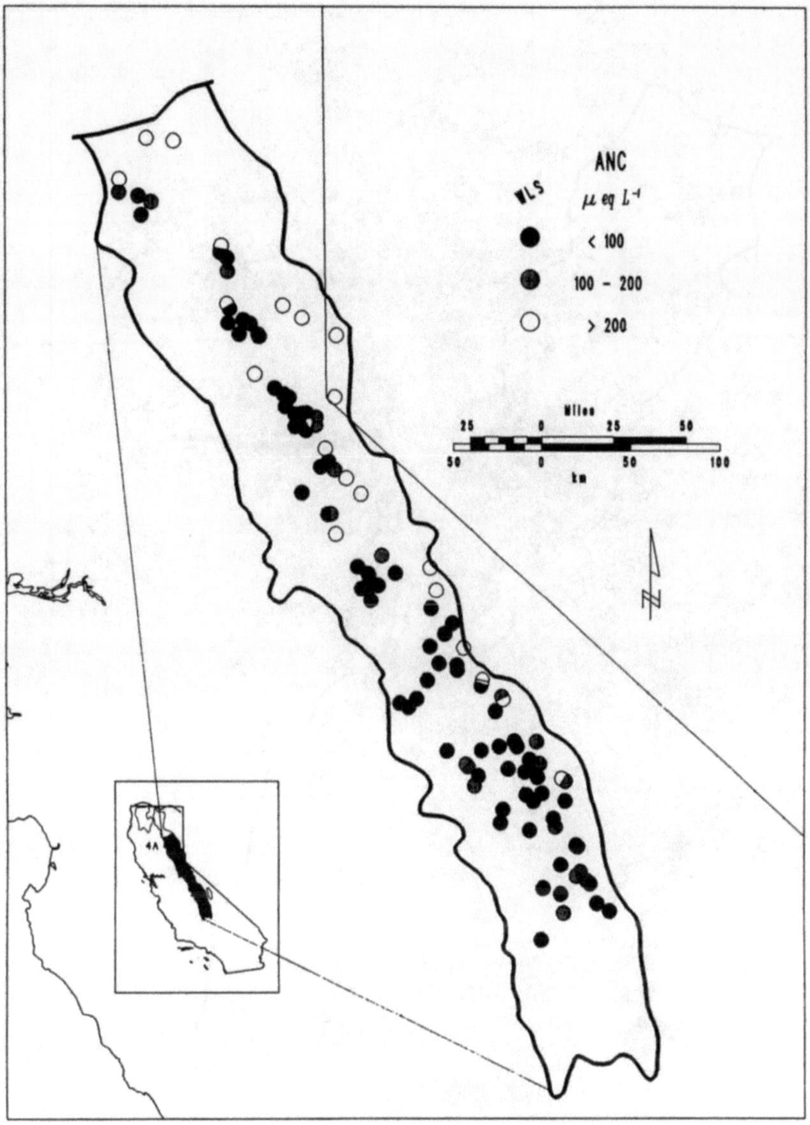

FIGURE 15.5. Distribution of lakes in three ANC categories.

measurement of small concentrations of major ions are propagated in the calculation of ion sums. The result of these uncertainties is a weak ($r^2 = 0.19$) but significant ($p < 0.001$) linear relationship between anion deficit and DOC for Sierra Nevada lakes (Figure 15.8), which suggests that lakes with median DOC (Table 15.2) contain < 5 μeq L^{-1} of organic acid anions. Nelson (Chapter 16, this volume) reports similar results for the Cascades.

Overall, the characteristics shown in Figure 15.4 indicate a region with a large number of small, very dilute lakes. Landers et al. (1987) identify them as the most dilute group of lakes sampled to date in the United States. They are comparable to lakes of very low ionic strength in Norway and Sweden (Henriksen and Brakke 1988a). ANC values are always above zero during autumn, and their distribution closely mirrors that of basic cation concentrations. Concentrations of strong acidic anions are usually very low (NO_3^- values characteristically < 1 μeq L^{-1} and $SO_4^{2-} < 10$ μeq L^{-1}), and the few cases of high concentrations are not associated with acidic lakes.

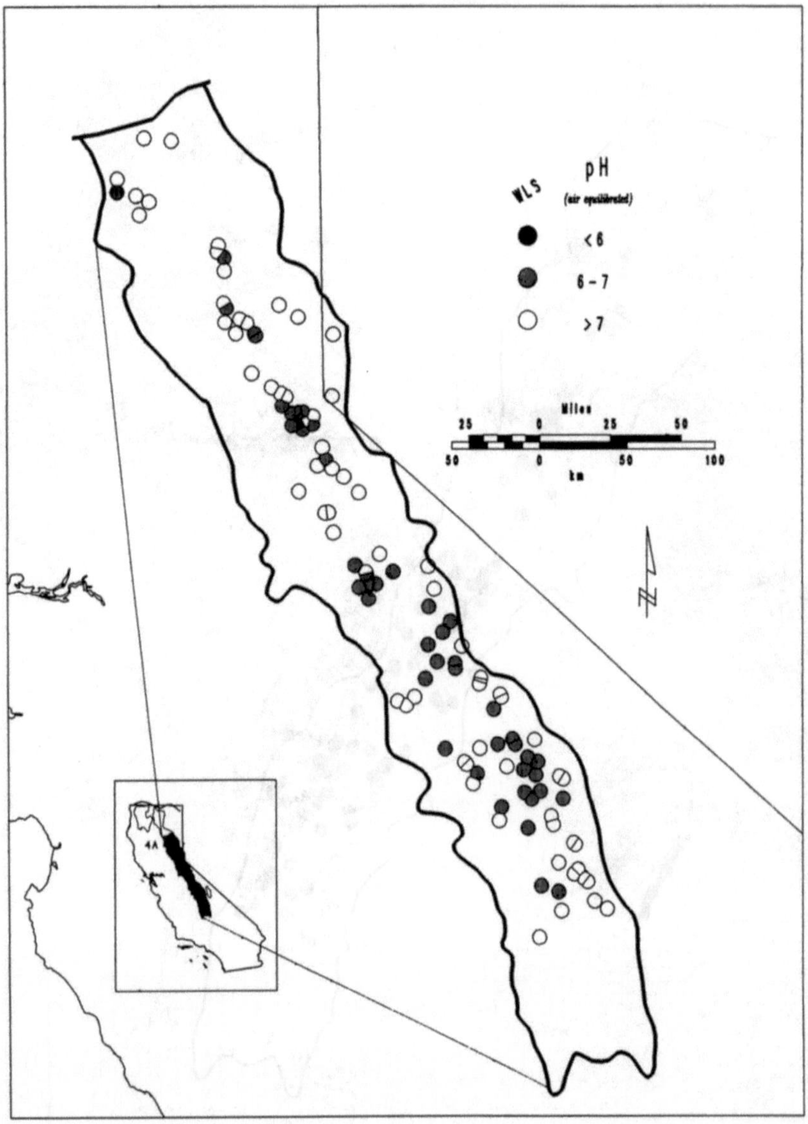

FIGURE 15.6. Distribution of lakes in three pH categories. pH value is air equilibrated and differs from other WLS pH values (closed-system pH) in text.

A spline diagram of major ion composition across the range of ANC observed in Sierra Nevada lakes (Figure 15.9) illustrates how ion dominance changes with increasing ionic strength. At all ANC values, Ca^{2+} is the dominant cation. At low ANC values, Na^+ is second in dominance, but Mg^{2+} concentration is as great or greater than Na^+ at higher ANC. Potassium is consistently low in concentration, as is H^+. Bicarbonate exceeds the concentration of all anions at all levels, with SO_4^{2-} second in dominance. Large increases in SO_4^{2-} concentration at intermediate ANC coincide with peaks in Ca^{2+} concentration, and are disjunct with the ANC ranges where Mg^{2+} becomes the second most dominant cation. The most dilute Sierra Nevada lakes, therefore, are $Ca^{2+} - Na^+ - HCO_3^-$ waters, whereas those of higher ionic strengths are either $Ca^{2+} - HCO_3^- - SO_4^{2-}$ waters or $Ca^{2+} - Mg^{2+} - HCO_3^-$ waters. These patterns are the result of differences in the bedrock weathering regimes that

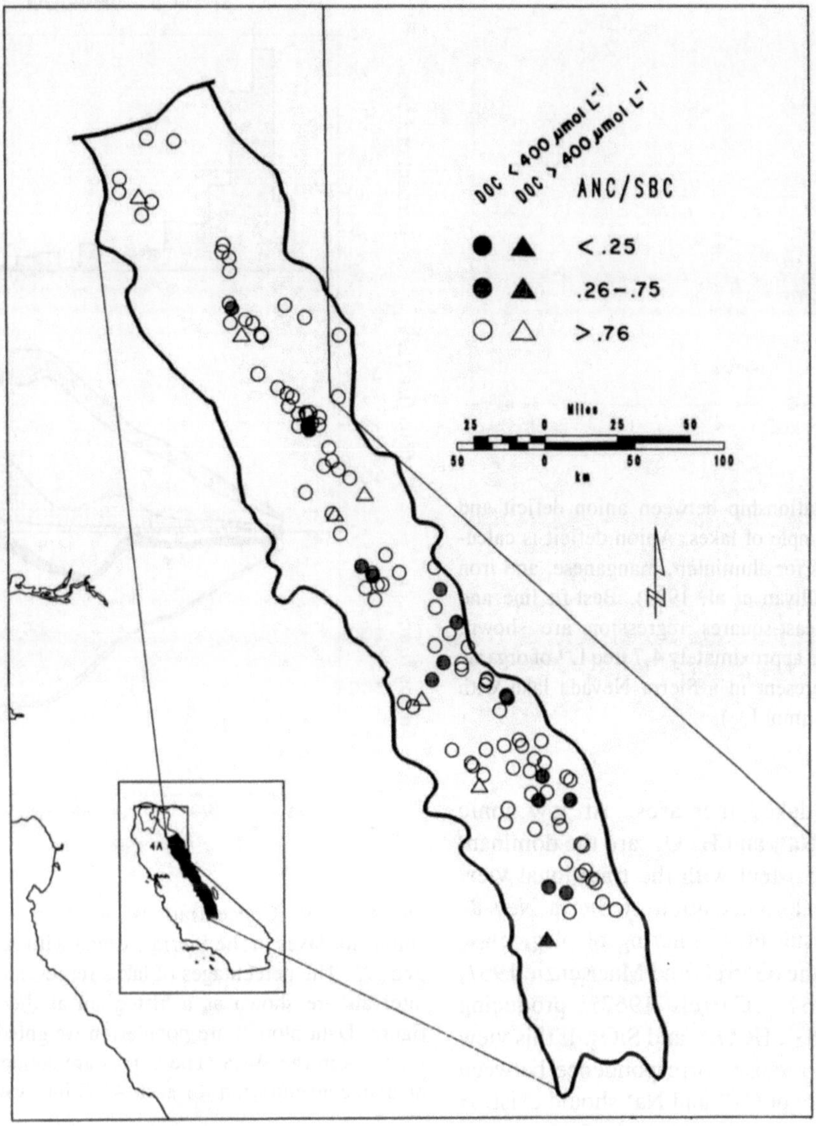

FIGURE 15.7. Distribution of lakes in ANC/SBC classes.

produce the water in each ANC class (see section on associations among variables).

Temporal variability on seasonal or interannual scales can be evaluated for few Sierra Nevada lakes. Melack and Setaro (1986) compared mid-summer to early autumn pH and ANC in 33 lakes over a 2- to 5-year period and reported standard errors of 0.01 to 0.62 for pH and 0 to 68 μeq L^{-1} for ANC. No interannual temporal trend was apparent. In contrast, when pH in autumn was compared with pH in spring or early summer for 15 lakes, Melack and Setaro (1986) found a reduction of 0.2 to 1.0 pH units in all the lakes. Further discussion of seasonal variations is presented in the context of processes influencing solute chemistry (see section on processes influencing surface water chemistry).

Associations Among Variables

The spline diagram (Figure 15.9) discussed in the previous section shows a progression in the dominance of major anions and cations as the ANC of

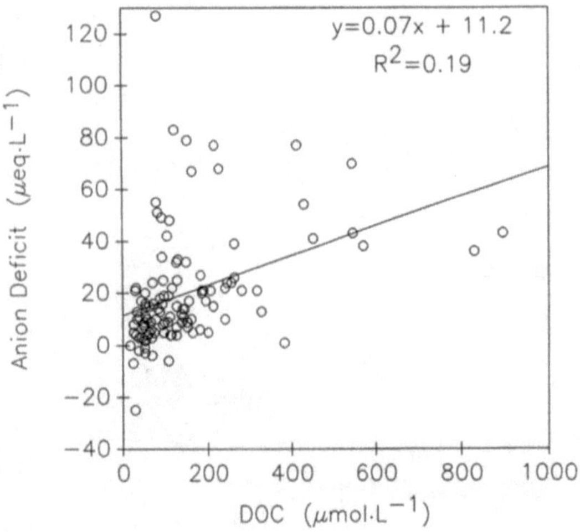

FIGURE 15.8. Relationship between anion deficit and DOC for WLS sample of lakes. Anion deficit is calculated with charges for aluminum, manganese, and iron all set to +2 (Sullivan et al. 1989). Best-fit line and parameters for least-squares regression are shown; regression suggests approximately 4.7 μeq L^{-1} of organic acid anions are present in a Sierra Nevada lake with median DOC (67 μmol L^{-1}).

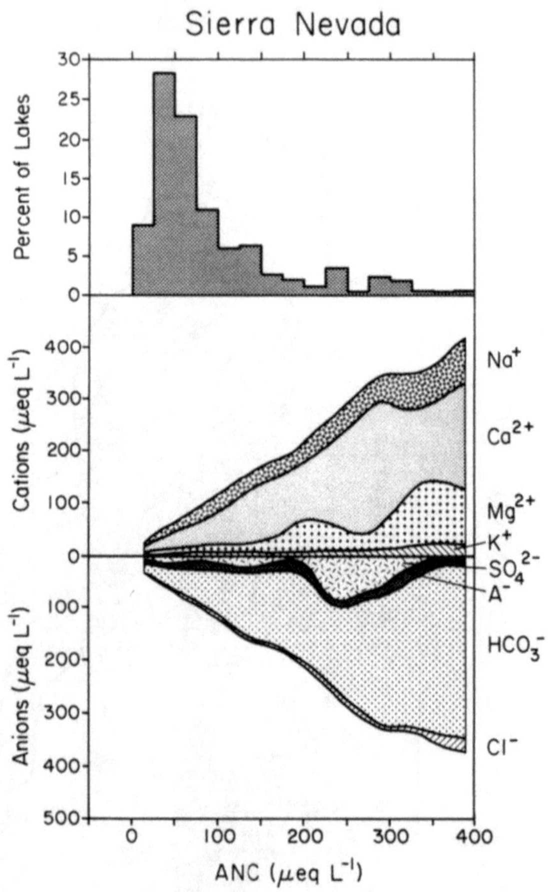

FIGURE 15.9. Concentrations of major cations and anions for lakes in the Sierra Nevada with ANC ≤ 400 μeq L^{-1}. The percentages of lakes represented by ANC intervals are shown as a histogram at the top of the figure. Data plotted are population weighted estimates for lakes in the WLS. The curves are spline fits to the median concentration for each ANC interval.

Sierra Nevada lakes increases. At low ionic strengths, Ca^{2+}, Na^{+}, and HCO_3^- are the dominant ions. This is consistent with the traditional view that the chemical composition of Sierra Nevada waters is the result of weathering of plagioclase feldspars in granite (Garrels and MacKenzie 1967, Feth et al. 1964b, Garrels 1967), producing primarily Ca^{2+}, Na^{+}, HCO_3^-, and SiO_2. If this view is correct, then a close correspondence between ANC and the sum of Ca^{2+} and Na^{+} should exist. A plot of this relationship (Figure 15.10) shows reasonably good agreement at low ionic strengths, but considerable scatter, especially at ANC values > 150 μeq L^{-1}. Estimates of analytical error (among-batch precision estimates from WLS audit samples; Landers et al. 1987) can be used to determine which points in Figure 15.10 are within acceptable limits (their scatter can be attributed to error) and those where the presumed 1:1 relationship between ANC and sum of Ca^{2+} and Na^{+} cannot be confirmed. This error analysis suggests that many lakes do not exhibit chemistries that are consistent with the plagioclase weathering model. Turk and Spahr (Chapter 14, this volume) present evidence for the Rocky Mountains that major cations, SiO_2, and ANC are not consistent with simple stoichiometric weathering of the granitic minerals common to the watersheds.

In order to test whether geochemical processes other than plagioclase weathering are responsible for producing the chemistry of some Sierra Nevada lakes, the major ion data from the WLS were subjected to an *R*-mode factor analysis. Factor analysis is often used to simplify complex relationships among chemical variables; the goal of factor analysis is to reduce a large number of interrelated and/or uninterpretable variables to a small number of factors that explain (statistically) a large proportion of the variance in the data set (Reeder et al. 1972, Drever 1982). The relationships between single variables and the factors are expressed in terms of loadings, which vary from −1 to +1, and which are interpretable in terms of known geochemical processes if the factor analysis is successful.

Results of the factor analysis for the WLS data set are shown in Figure 15.11. Three factors explain 99% of the variance in major ion composition of the sampled lakes. The first factor has very high loadings for ANC and for all of the major cations. Silica has a somewhat lower, but still very significant loading for factor #1. This factor, by itself, explains 90% of the variance in composition, and is related to weathering of minerals in granitic bedrocks (plagioclase and K-feldspars, hornblende and biotite). Near zero loadings for SO_4^{2-} and extractable Al further indicate that ANC is not closely related to concentrations of strong acid anions and that ANC therefore remains high enough to limit dissolved Al concentrations. Factor #2 has high loadings only for SO_4^{2-} and Ca^{2+}. This factor is associated with the peaks of SO_4^{2-} and Ca^{2+} observable at moderate ANC values in the spline diagram (Figure 15.9), and is interpreted as an indicator of small amounts of gypsum ($CaSO_4$ - H_2O) or pyrite (FeS_2, which oxidizes to form sulfuric acid and attacks associated calcite, liberating Ca^{2+}) in the drainage basins of some lakes. Factor #2 is responsible for the points in Figure 15.10 that fall well below the expected 1:1 line, due to the presence of substantial Ca^{2+}, which is not associated with ANC. The third factor has high loadings for Ca^{2+}, Mg^{2+}, and ANC, and is associated with areas in the spline diagram (Figure 15.9) where Mg^{2+} becomes the cation second in rank of dominance. It is interpreted as evidence of the presence of small amounts of magnesium-bearing minerals (e.g., dolomite, pyroxenes, or amphiboles) in the

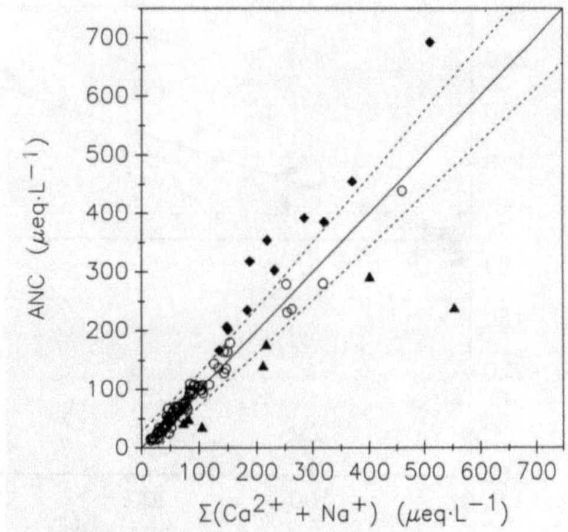

FIGURE 15.10. Relation between ANC and [Ca^{2+} + Na^+] for WLS sample of lakes. 1:1 line is shown to indicate expected relationship if plagioclase weathering is responsible for ANC production. Dashed lines are bounds of variability attributable to analytical error. 95% confidence interval = 2 × (Variance [Ca^{2+}] + Variance [Na^+] + Variance [ANC])$^{1/2}$, where variances are based on range of variability observed in WLS audit samples (Landers et al. 1987). Lakes falling significantly below these bounds are plotted as filled triangles. Lakes falling significantly above these bounds are plotted as filled diamonds. Four lakes with ANC values > 750 μeq L^{-1} are not shown; all fall significantly above the allowable error bounds.

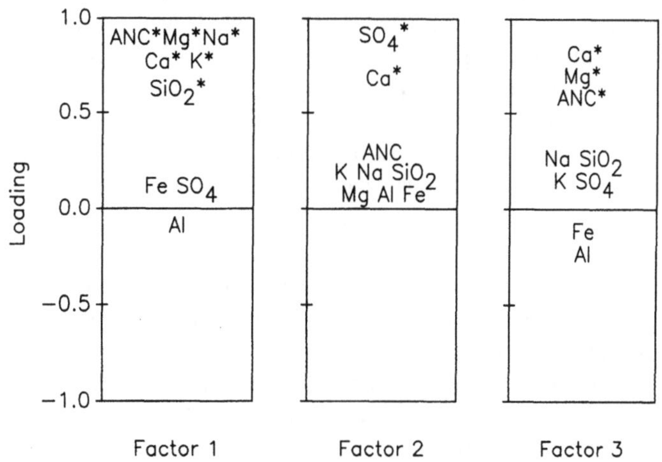

FIGURE 15.11. Results of factor analysis for WLS sample of lakes. Variables with significant loadings for each factor are labeled with asterisks.

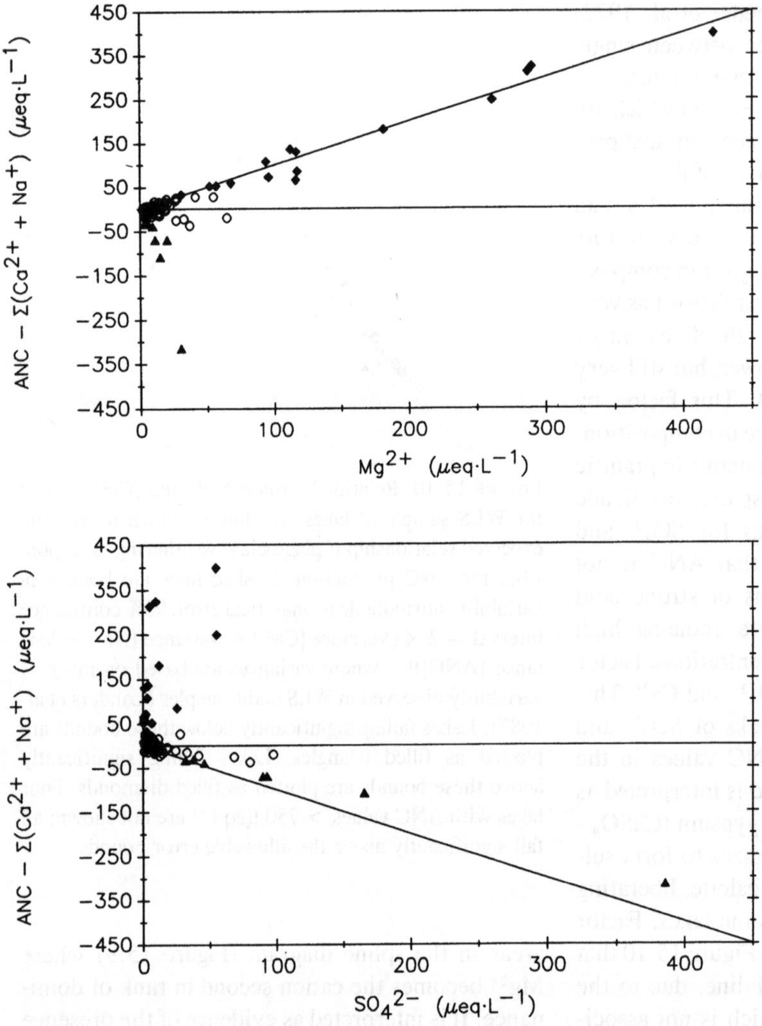

FIGURE 15.12. Deviation from plagioclase-weathering model [deviation is defined as: ANC − $\Sigma(Ca^{2+} + Na^{+})$; see Figure 15.10] plotted as a function of (a) Mg^{2+} and (b) SO_4^{2-}. 1:1 lines are shown to illustrate expected relationship for high Mg^{2+} and high SO_4^{2-} lakes. Symbols are consistent with those in Figure 15.10.

watersheds of moderate to high ANC lakes. Factor #3 produces the points in Figure 15.10 that lie well above the expected 1:1 line.

The importance of SO_4^{2-} and Mg^{2+} in producing surface waters that do not fit a plagioclase weathering model can be further tested by plotting the deviation of individual lakes from the expected 1:1 line in Figure 15.10, against either SO_4^{2-} or Mg^{2+} (Figure 15.12). In each case, the *y*-axis represents the amount of ANC unaccounted for by plagioclase weathering. Excess ANC is explained by high Mg^{2+} concentrations (diamonds in Figure 15.12); the deviations of points lying above the 1:1 line in Figure 15.10 are correlated with Mg^{2+}. Deficit ANC is explained by high SO_4^{2-} concentrations (triangles in Figure 15.12); ANC values that fall much lower than expected do so because a large proportion of the Ca^{2+} present did not produce ANC when weathered.

If the above interpretations of the factor analysis results are correct, then the same factors should have high loadings for the types of geological formations that produce the described patterns in major ion composition – granodiorite or diorite for factor #1, gypsum or pyrite for factor #2, pyroxenes or dolomite for factor #3. Unfortunately, the number of lakes for which there is detailed geologic information (approximately half of the lakes in the WLS data set) is too small to produce valid factor analysis results; a nonstatistical examination of geological patterns is still possible, however. All the lakes in the high Mg^{2+} class (diamonds in

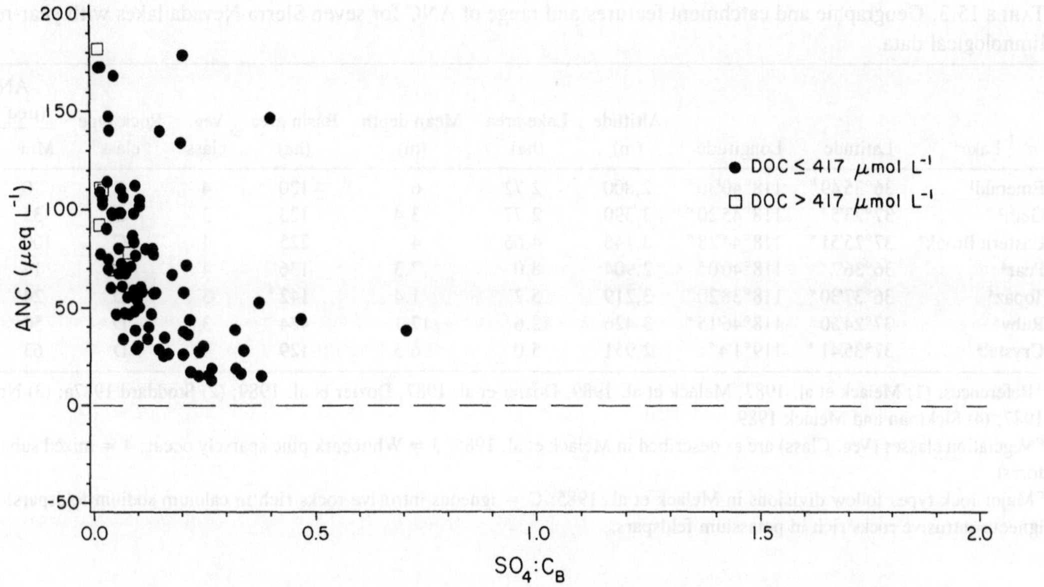

FIGURE 15.13. Relationship between ANC and SO_4^{2-}:C_B for lakes sampled in the WLS.

Figures 15.10 and 15.12), for which there is detailed geological information available, are located in basins with substantial metavolcanic bedrock, especially andesite and basalt. These rocks contain significant amounts of hornblende and pyroxenes (and therefore Mg^{2+}) and weather more easily than granites (Schnoor and Stumm 1986). They can therefore contribute substantially to the chemistry of surface waters, even when they are found in minor amounts in watersheds. Some of the lakes in the high SO_4^{2-} class (triangles in Figures 15.10 and 15.12) are associated with metasedimentary bedrocks, especially marbles, in the Convict Lake area of the eastern Sierra Nevada; gypsum is a likely source for both SO_4^{2-} and Ca^{2+} in these lakes. The remainder of the high SO_4^{2-} lakes, however, are located in areas of largely unbroken granodiorite and quartz monzonite. Small inclusions of pyrite (and presumably calcite) seem likely to be present, especially along faults, within the bedrock in these basins. Although small in mass percent, these inclusions can have very significant effects on surface water chemistry. It is important to note that high SO_4^{2-} concentrations turn up in low ANC lakes as well as high ANC lakes, and that their presence is not predictable from surface geology.

For many regions experiencing high inputs of acidic deposition, useful information can be gained by plotting the ratio SO_4^{2-}:C_B against ANC. Sierra Nevada WLS data are plotted in this form in Figure 15.13 for comparative purposes; lakes with unusual patterns of major ion composition, such as those discussed in the previous paragraphs, confuse the interpretation of this plot. Because strong acid anion concentrations are generally very low in the Sierra Nevada, high values of SO_4^{2-}:C_B arise only when nonatmospheric sources of SO_4^{2-} are present. The five lakes with highest SO_4^{2-}:C_B ratios in Figure 15.13 (values > 0.35) correspond with the high SO_4^{2-} lakes shown in Figures 15.10 and 15.12 (two high SO_4^{2-} lakes have ANC > 200 μeq L^{-1} and do not appear in Figure 15.12).

Processes Influencing Surface Water Chemistry

Description of Study Sites

Seven high-altitude Sierra Nevada lakes and watersheds have received sufficient study to allow examination of seasonal differences in processes influencing their aquatic chemistry (Figure 15.2). Emerald Lake is the most thoroughly studied; Eastern Brook Lake and Gem Lake, in addition to four others, have sufficient data to consider some

TABLE 15.3. Geographic and catchment features and range of ANC for seven Sierra Nevada lakes with year-round limnological data.

Lake[a]	Latitude	Longitude	Altitude (m)	Lake area (ha)	Mean depth (m)	Basin area (ha)	Veg. class[b]	Rock type class[c]	ANC ($\mu eq\ L^{-1}$) Min	Max
Emerald[1]	36°35′49″	118°40′30″	2,800	2.72	6	120	4	C	0	63
Gem[2]	37°23′5″	118°45′20″	3,390	2.77	3.4	123	3	C	35	210
Eastern Brook[3]	37°25′51″	118°44′28″	3,146	4.66	4	225	4	C	100	380
Pear[4]	36°36′2″	118°40′0″	2,904	8.0	7.3	136	4	D	14	628
Topaz[4]	36°37′30″	118°38′20″	3,219	5.2	1.4	142	3	D	24	85
Ruby[4]	37°24′50″	118°46′15″	3,426	12.6	17.1	424	3	D	50	80
Crystal[4]	37°35′41″	119°1′4″	2,951	5.0	6.5	129	4	D	63	107

[a]References: (1) Melack et al. 1987, Melack et al. 1989, Dozier et al. 1987, Dozier et al. 1989; (2) Stoddard 1987a; (3) Nodvin 1987; (4) Sickman and Melack 1989.
[b]Vegetation classes (Veg. Class) are as described in Melack et al. 1985: 3 = Whitebark pine sparsely occur; 4 = mixed subalpine forest.
[c]Major rock types follow divisions in Melack et al. 1985: C = igneous intrusive rocks rich in calcium sodium feldspars; D = igneous intrusive rocks rich in potassium feldspars.

processes (Table 15.3). Before discussing specific processes, a brief description of Emerald Lake and Gem Lake is provided to help characterize the conditions in high-altitude Sierra Nevada catchments. This material complements the general remarks in the section on Watershed Characteristics.

Emerald Lake

The Emerald Lake watershed in Sequoia National Park is located in the upper Marble Fork of the Kaweah River drainage, approximately 8 km from and 800 m above the nearest trailhead. The basin is a small high-altitude cirque, with streams that drain into Emerald Lake. The highest point in the watershed is Alta Peak (3,416 m). The watershed faces north and is steep, with half the area containing slopes > 30° (Dozier et al. 1987).

The bedrock of the Emerald Lake basin consists mainly of granite with some mafic inclusions, aplite dikes, and pegmatite veins (Sisson and Moore 1984, Clow 1987). Granodiorite occurs in the western part of the basin and surrounds and underlies the lake. The basin contains numerous joints where the rock has fractured (Moore and Wahrhaftig 1984). Conversion of plagioclase to kaolinite is the dominant weathering reaction at Emerald Lake (Clow 1987). The primary clay minerals found in the basin are illite and several types of vermiculite (Weintraub 1986).

More than one-third of the basin is exposed bedrock (Huntington and Akeson 1986, Lund et al. 1987), and the remainder has a thin mantle of talus, colluvium, or poorly developed soils covering about 20% of the basin and classified as Entisols or Inceptisols. Vegetation is sparse (Rundel et al. 1988), with scattered coniferous trees, but shrubs and herbaceous perennials are abundant where soils occur. Mean daily air temperatures are usually between 6°C and 13°C in summer and between −4°C and +4°C in winter (Dozier et al. 1987).

Precipitation is highly seasonal and variable. About half the annual total falls in the winter months and very little in the summer. Snowfall accounts for most of the input of water (Dozier et al. 1987). Almost 3 m of precipitation were recorded at Emerald Lake in water year 1986, but less than 90 cm fell in water year 1987. The snowpack accumulates from November through March and melts from April through the summer. Sublimation from the snow is the principal loss of water to the atmosphere, which totals about 20% of the annual precipitation. Groundwater storage and release is minor. Most of the precipitation runs off via the outflow during snowmelt (Dozier et al. 1989).

Most of the data for Emerald Lake used here for illustration of processes influencing water chemistry are for water year 1986. During this period, snow accounted for 98% of the total precipitation, and the Kaweah basin received 1.8 times its 50-year mean snowfall (California Cooperative Snow Survey 1986). Heavy snowfall and avalanches accumulated on the frozen surface of the lake, creating a white-ice layer of 6-m thickness;

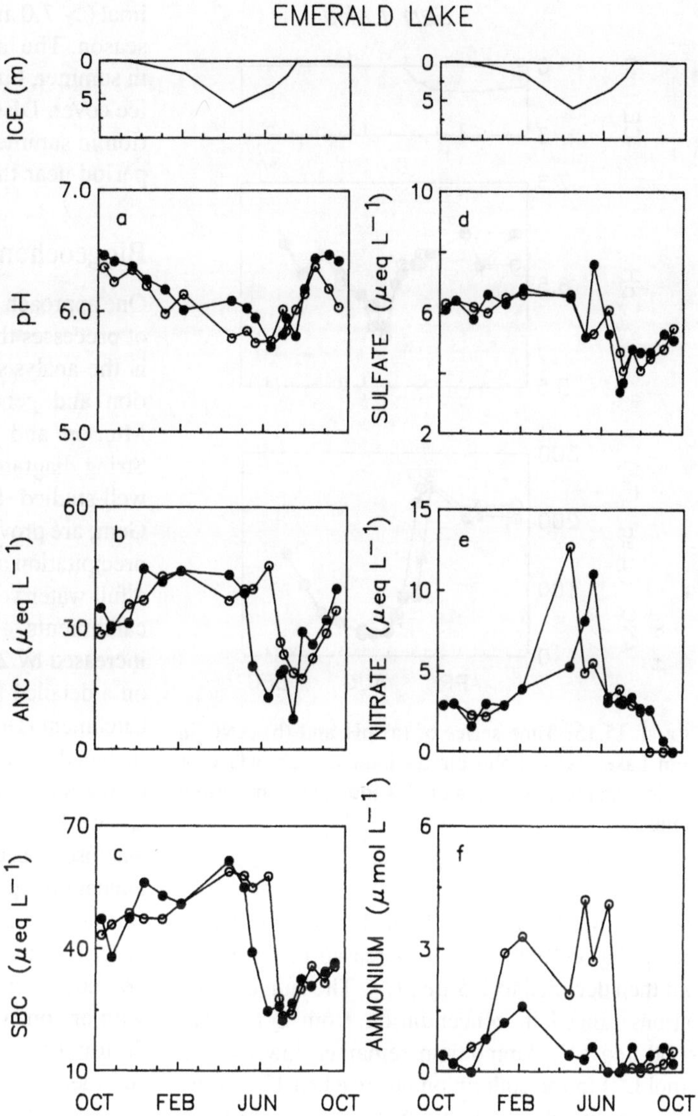

FIGURE 15.14. Time series of water chemistry at mid-lake station in Emerald Lake, October 1985 through September 1986: (a) pH, (b) ANC, (c) sum of base cations, (d) sulfate, (e) nitrate, and (f) ammonium. Solid circles indicate subsurface or sub-ice samples, and open circles indicate near bottom samples.

lake water was displaced, and the volume of water below the ice decreased to 25% of the ice-free amount. The winter of 1986 was abnormally severe, with unusually thick ice and snow accumulation on the lake (Melack et al. 1989).

Emerald Lake was dimictic; it was thermally stratified under ice, mixed from top to bottom at iceout, restratified in midsummer, and mixed again in autumn (Melack et al. 1987). Under ice, dissolved oxygen was depleted near the sediments. During holomixis in spring and autumn, dissolved oxygen was near saturation throughout the water column. In midsummer, near-anoxia developed in the hypolimnion.

Times-series plots of important chemical species at subsurface and 9.5-m depths (Figure 15.14) illustrate seasonal differences related to stratification and snowmelt (Melack et al. 1987). In subsurface waters, pH ranged from 5.6 to 6.5, with low values associated with snowmelt; near bottom pH followed a similar pattern. Acid neutralizing capacity reached a minimum of about 10 μeq L^{-1} during snowmelt in 1986; a maximum of 45 μeq L^{-1} occurred under ice. In general, ANC was higher in the deeper water than in the near-surface water when the lake was thermally stratified. Nitrate concentrations ranged from 1 to 14 μeq L^{-1}; the highest values occurred during snowmelt in 1986.

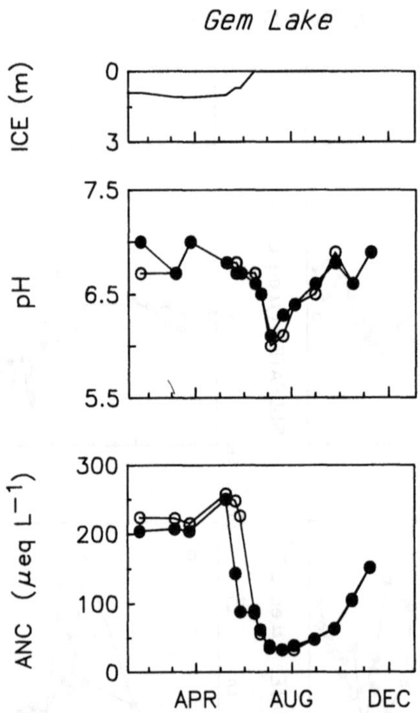

FIGURE 15.15. Time series of (a) pH and (b) ANC in Gem Lake, 1984. Solid circles indicate subsurface or sub-ice samples, and open circles indicate near bottom samples.

Sulfate ranged from 5 to 7 μeq L^{-1}, except during the spring of 1986 when it increased to 9 μeq L^{-1} and then declined to 1.5 μeq L^{-1}. The sum of basic cations ranged in concentration from 8 μeq L^{-1} to 60 μeq L^{-1}. Ammonium remained low (< 1.5 μmol L^{-1}) in the epilimnion but reached 12.5 μmol L^{-1} in the hypolimnium during periods of stratification. Data for major solutes from 1983 to 1988 indicate similar seasonal patterns (Melack et al. 1989).

Gem Lake

Gem lake is a small, shallow, high-elevation lake on the eastern slope of the Sierra Nevada. It lies at timberline in a basin composed almost entirely of granodiorite talus (Stoddard 1987a). Time-series plots for pH and ANC are shown in Figure 15.15. The timing of decreases in pH and ANC varied from year to year, but minimum values were evident during snowmelt (pH 6.0, ANC between 30 and 40 μeq L^{-1}). Both pH and ANC increased gradually during the ice-free season and were maximal (> 7.0 and > 200 μeq L^{-1}) in the ice-covered season. The lake was stratified weakly or not at all in summer, but showed strong stratification during ice cover. Dissolved oxygen remained near saturation in summer; near-anoxia developed for a brief period near the bottom in late winter.

Biogeochemical and Hydrologic Factors

One approach to elucidating the relative importance of processes that influence surface water chemistry is the analysis of string diagrams. Their construction and general interpretation are provided in Munson and Gherini (Chapter 2, this volume). String diagrams and related ion bar charts for two well-studied Sierra Nevada lakes, Emerald and Gem, are provided in Figures 15.16 and 15.17. The precipitation amount and chemistry presented span a full water year and were obtained for the specific catchments. Precipitation concentrations were increased by 20% to account for evaporation based on a detailed hydrological study of Emerald Lake's catchment (Dozier et al. 1989). Dry deposition was included with the snow chemistry but not calculated or measured during periods with nonaccumulating snow. Mean lake chemistry for mid- to late summer was used because this represents the result of the annual precipitation near the end of the water year. At this time in Emerald Lake, total Al was about 1 μmol L^{-1} (inorganic monomeric Al concentrations are not available), fluoride was below detection with an ion specific electrode and DOC was about 80 μmol L^{-1}. Aluminum, fluoride, and DOC measurements are not available for Gem Lake.

It is clear that the dominant mechanisms of ANC generation are associated with release of basic cations. Weathering reactions, with some contribution from cation exchange, are the presumed mechanisms. A minor source of ANC is associated with NO_3^- consumption, which may be attributed to biological assimilation or denitrification. Slight losses of NH_4^+ and gains of SO_4^{2-} represent minor ANC sinks.

Biogeochemical studies of the soils in the Emerald Lake basin corroborated some inferences made from the string diagrams (Lund and Brown 1989). Under laboratory conditions, plagioclase feldspar, the mineral most important in the weathering of the granite or granodiorite bedrock, generated ANC at a rate of 10 meq m^{-2} d^{-1} per top 10 cm of

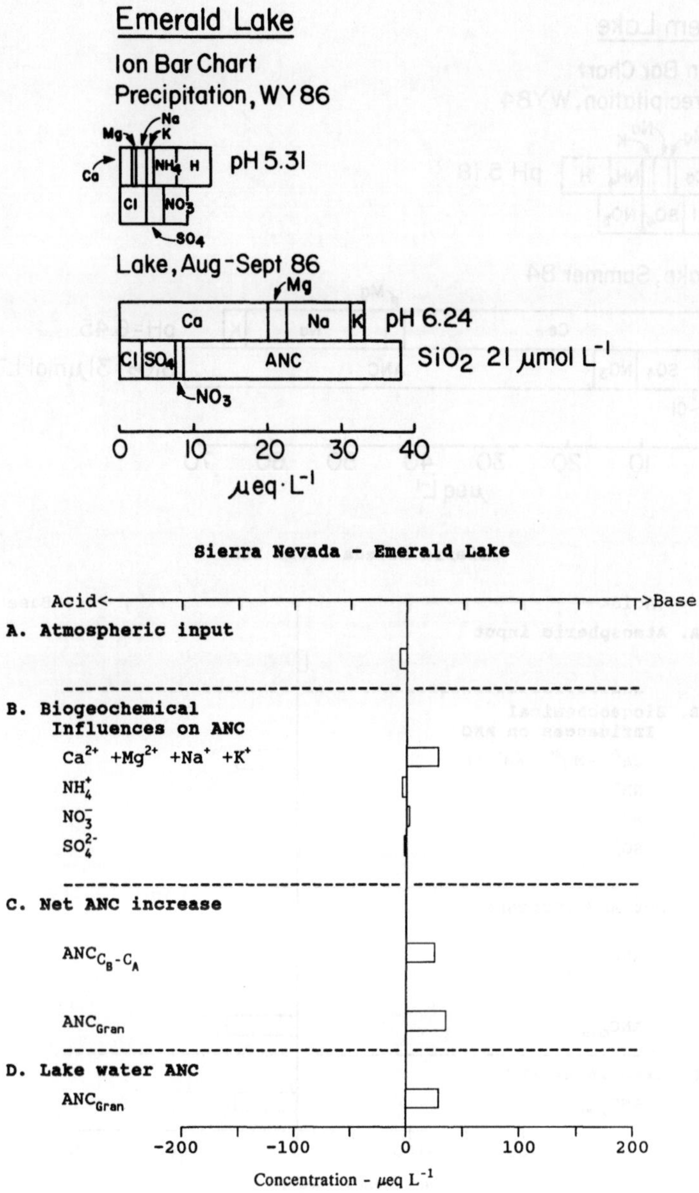

FIGURE 15.16. Ion bar chart for precipitation and lake (*top*), and string diagram of watershed processes (*bottom*), for Emerald Lake, WY86. String diagram illustrates the modification of atmospheric deposition by watershed and lake biogeochemical processes. A. Precipitation $[H^+]$ adjusted for evapotranspiration (ET); equivalent to expected lakewater $ANC_{C_B\text{-}C_A}$ in the absence of any process other than ET. B. Reactions influencing ANC; bars to the right indicate ANC production; those to the left show ANC consumption. C. Net ANC production; $ANC_{C_B\text{-}C_A}$ is sum of ANC production terms for measured anions and cations, including those shown in (B); ANC_{Gran} is determined by difference between input ANC (A.) and lakewater ANC (D.). D. Lakewater ANC represents the net reaction of atmospheric inputs with the lake/watershed system. For a further explanation of calculation procedures, see Appendix A.

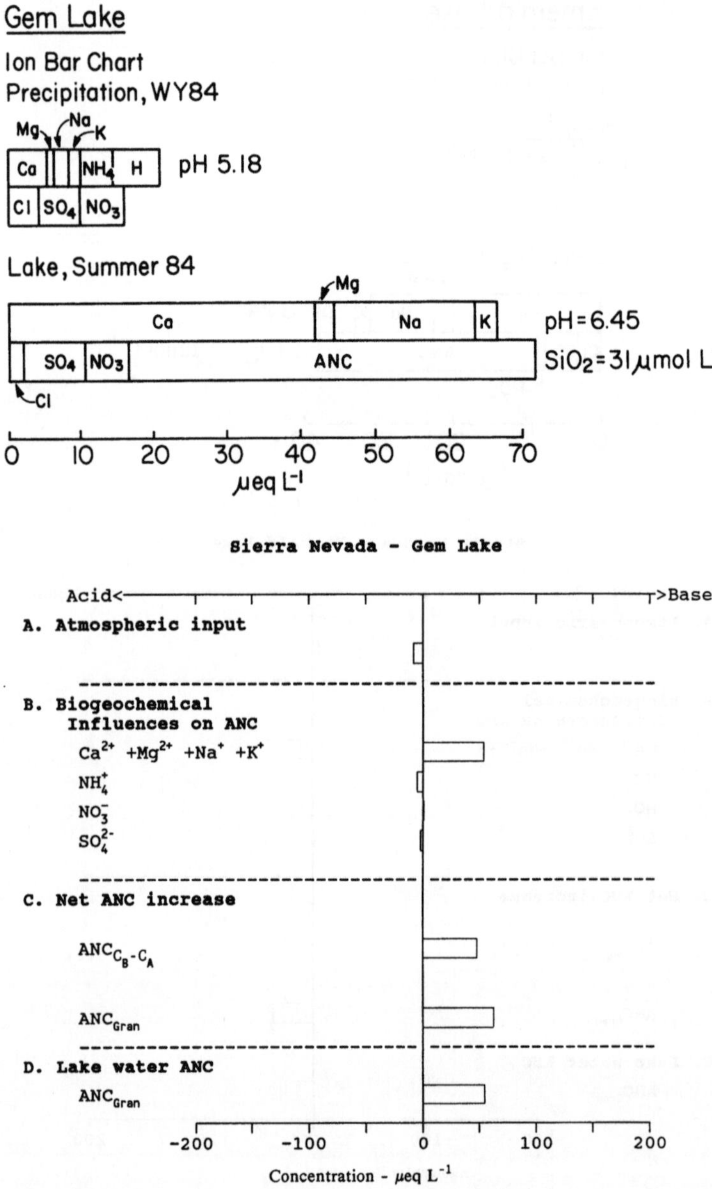

FIGURE 15.17. Ion bar chart for precipitation and lake (*top*), and string diagram of watershed processes (*bottom*), for Gem Lake, WY 84. String diagram illustrates the modification of atmospheric deposition by watershed and lake biogeochemical processes. A. Precipitation [H^+] adjusted for evapotranspiration (ET); equivalent to expected lakewater $ANC_{C_B\text{-}C_A}$ in the absence of any process other than ET. B. Reactions influencing ANC; bars to the right indicate ANC production; those to the left show ANC consumption. C. Net ANC production; $ANC_{C_B\text{-}C_A}$ is sum of ANC production terms for measured anions and cations, including those shown in (B.); ANC_{Gran} is determined by difference between input ANC (A.) and lakewater ANC (D.). D. Lakewater ANC represents the net reaction of atmospheric inputs with the lake/watershed system. For a further explanation of calculation procedures, see Appendix A.

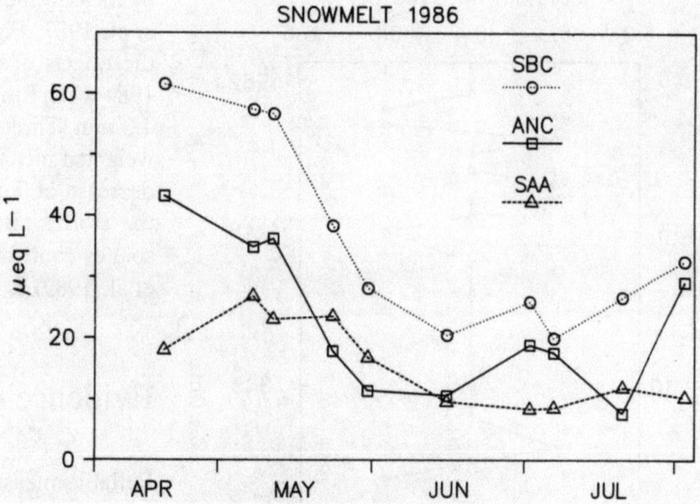

FIGURE 15.18. Time series of subsurface water chemistry in Emerald Lake during snowmelt 1986.

soil; this rate is expected to be at least 100 times less under field conditions. Another acid neutralizing weathering reaction identified in the laboratory was the release of Al^{3+} from gibbsite, aluminum hydroxide trapped in vermiculite or associated with organic matter. Furthermore, exchangeable basic cations, mainly Ca^{2+}, in the soils represented a large pool in comparison to the current atmospheric loading of H^+. Direct measurements of denitrification were variable; rates were zero when soils were dry or frozen and were on the order of 5 meq NO_3^- lost m^{-2} d^{-1} at other times.

Further insight into controls of Sierra Nevada lake chemistry is provided by examination of seasonal changes in amounts and vertical distributions of ANC. Both Emerald and Gem lakes have increased ANC under ice. A similar pattern occurred in Crystal, Ruby, Pear, Topaz, and Eastern Brook lakes (Figure 15.2; Table 15.3) and in most lakes sampled (during ice cover) by Melack and Setaro (1986). Hypolimnetic ANC increases also occurred in these lakes during periods of mid-summer stratification, which were associated with depletion of dissolved oxygen. In contrast, ANC declined in Sierra Nevada lakes during snowmelt. Hence, two issues are raised: what is the source of elevated ANC and what causes its subsequent decline?

The conditions in Gem Lake and its watershed that are responsible for its elevated ANC under ice and low ANC during snowmelt are discussed in detail by Stoddard (1987a). Dilution rather than acidification accounted for the snowmelt related decrease in ANC, and ANC increases in autumn and winter were attributed almost entirely to weathering reactions in the watershed. Cryoconcentration made a trival contribution to ANC increases under ice (Stoddard 1988). Subsequent analysis of winter profiles from Gem Lake (Stoddard 1988) indicated that $<$ 1% of the annual ANC flux was attributable to in-lake production; of this 1%, half was due to Ca^{2+} exchange in the sediments, and the remainder to sulfate reduction and denitrification, which were limited by the low SO_4^{2-} and NO_3^- concentrations and fast flushing times.

Snowmelt at Emerald Lake was sampled on 10 dates in 1986 (Figure 15.18; Melack et al. 1987). During the initial stages (April and early May), the acidic anions (SO_4^{2-}, Cl^-, and NO_3^-) increased, whereas basic cations and ANC declined in the subsurface water collected at from one to four stations. Minima in all constituents occurred on 9 June 1986. These data indicate that a combination of dilution and strong acid titration contributed to the ANC decline. Further evidence for these factors is provided by examination of the main inflowing streams. During the initial phase of snowmelt in 1986, ANC (calculated as the difference between basic cations and strong acid anions) decreased 4 µeq L^{-1} because of dilution and 10 µeq L^{-1} because of strong acid anions (Williams and Melack unpublished).

The accumulation of ANC near the bottom of Emerald Lake during stratified periods indicates a probable role of the sediments as a source. Further evidence for within-lake ANC generation in Sierra

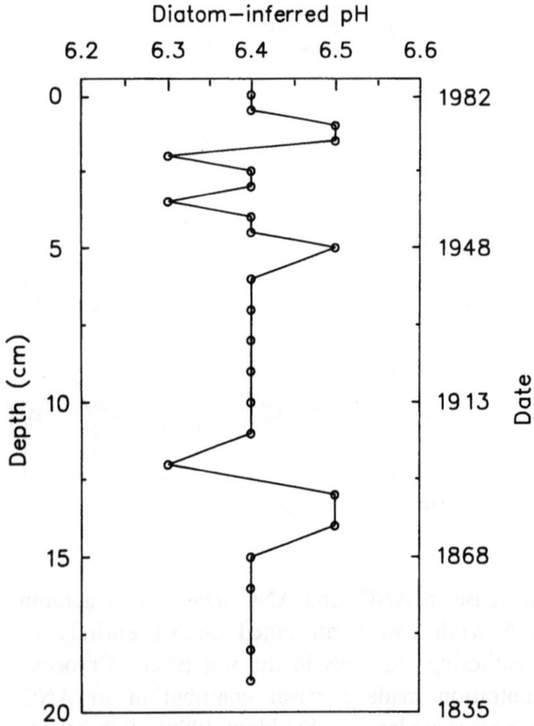

FIGURE 15.19. Diatom inferred pH, Emerald Lake, 1835–present.

Nevada lakes is provided by the direct measurements of ANC flux (Melack et al. 1987) and by the chemical gradients in the interstitial water measured in Mosquito, Eastern Brook, and Emerald lakes (Amundson et al. 1988). These data also indicate a minor role for SO_4^{2-} or NO_3^- reduction and major roles for mineral weathering, NH_4^+ production, and perhaps cation exchange as ANC sources. Although much less important on an annual basis than the catchments, because of the high watershed/lake volume ratio typical of western montane lakes, the sediments can be a minor source, at least in Emerald Lake, during stratified periods with low flow, such as mid-winter and late summer. The greatly reduced lake volume during mid-winter in years with thick ice cover can enhance the role of the sediments.

Although no chronically acidic lakes occur in the Sierra Nevada, episodic reductions in ANC can occur during snowmelt and during unusually intense summer rains, which are quite acidic. An ionic pulse caused by the preferential elution of solutes during the initial phases of snowmelt was indicated by measurements at Emerald Lake in 1986 (Dozier et al. 1987; Figure 15.18). An extreme example of the effects of summer rains occurred in mid-July 1984 when Emerald Lake experienced 5.6 cm of pH 4.3 rain. Three days after these storms, the volume-weighted mean ANC for the lake was 2 µeq L^{-1}, a decrease of 7 µeq L^{-1} from the ANC level before the storms. Both dilution and titration by acidic solutes contributed to the decline in ANC (Melack et al. 1989).

Evidence of Long-Term Trends

Reliable measurements of ANC in waters of the Sierra Nevada date only to 1979, hence it is not possible to recognize directly if changes in ANC have occurred during the last few decades. Although pH records for two reservoirs in the western foothills of the Sierra Nevada (Pardee and Hetch Hetchy) indicate pH declines of about 0.4 and 0.2 over the period 1954–79 (McColl 1981a), interpretation of these data is equivocal on methodological grounds and extrapolation to the high-elevation lakes is of uncertain validity.

Paleolimnological evidence for past pH and ANC values in lakes can be derived from microfossils such as diatom frustules preserved in lake sediments (Whitehead et al. 1986, Charles 1985, Davis and Anderson 1985). Holmes et al. (1989) have applied this technique to sediment cores taken from Emerald Lake. From 1825 to the present, the period spanned by these cores, neither pH (Figure 15.19) nor ANC had a consistent, time related trend; only minor fluctuations are evident. Diatom assemblages in the surficial sediments of Emerald Lake accurately confirm the current lake pH, corroborating the effectiveness of the predictive equations developed to predict pH in Sierra Nevada lakes. Thus, this analysis does not provide evidence of significant or permanent changes in pH or ANC in Emerald Lake during recent increases in agricultural and industrial activities in the western United States. Analyses from three additional high-elevation Sierra Nevada lakes corroborate the lack of recent acidification; one lake showed no trends, one lake showed very slight long-term pH increase, and one decreased very slightly in pH (Whiting et al. 1989).

Future Considerations

Predicting the impact, or potential impact, of acidic atmospheric deposition on a lake requires measurements of the areal loading to the drainage basin, and subsequently to the lake, as modified by buffering within the basin and the lake. In the Sierra Nevada, most precipitation accumulates as snow and then more or less abruptly enters the lake during snowmelt. Laboratory and field examinations of snowmelt chemistry indicate the solutes are concentrated in the first stages of runoff (Johannessen and Henriksen 1978, Colbeck 1981).

As a hypothetical exercise, if a range of snowpack depths (0.8 to 1.6 m of water) with different pH values (5.1 to 5.4) is applied to a range of basins with a ratio of basin surface area to lake volume from 3 to 200, an estimate of the impact of spring snowmelt on the pH of the lakes can be made (Melack et al. 1982a). The results of such calculations for Sierra Nevada lakes indicate that most of the lakes will attain the pH of runoff water. The model assumes that (1) 80% of the solutes enter the runoff in the first 50% of the snowmelt, (2) snow ANC is 2 $\mu eq\ L^{-1}$, (3) no buffering of the runoff occurs before it reaches the lake, (4) no thermal stratification occurs in the lakes, and (5) there is no contribution from the sediments to the ANC of the lake during the brief period of snowmelt. If a greater proportion of the solutes leave the snow earlier in the melt or if the lake is stratified and thus the runoff is added to a smaller volume of lake water, the effective acidic loading would be greater. These two possibilities are likely. The major uncertainty in this model is the extent of buffering and contributions of additional ANC from weathering in the basin and from the lake sediments.

In the alpine basins of the Sierra Nevada, thick snowpacks with ice layers interspersed tend to route some portion of the runoff through the snow itself. The underlying ground is largely bare granite with little soil. These factors reduce the extent of the buffering in the basin. However, experimental additions of $^{45}Ca^{2+}$ to Norwegian snowpacks overlying an area with thin soils indicated considerable interaction between melt water and soil (Seip 1980). Moreover, the runoff during snowmelt in the Emerald Lake basin contains solutes derived from soils (Williams and Melack unpublished). The lake sediments can also contribute to the ANC of the overlying water (Tonnessen and Harte 1982, Melack et al. 1987, Amundson et al. 1988). However, during the short period when there are large inflows of snowmelt runoff into a lake that may be stratified, the influence of the sediments is likely to be very small.

Recent efforts to model responses of Sierra Nevada waters to increases in acidic atmospheric deposition provide further indication of the susceptibility of these waters to acidification. Schnoor and Nikolaidis (1989) have formulated a model of episodic snowmelt and summer runoff from rain based on mass balance for ANC; the model assumes no acid neutralizing reactions in the terrestrial watershed. A Monte-Carlo simulation technique was used for each of 168 lakes in the Sierra Nevada to provide an estimate of the uncertainty of the predictions. The model predicts that the percentage of lakes with ANC $< 40\ \mu eq\ L^{-1}$ will increase from 29% before snowmelt to 79% $\pm$ 9% after 20 days of early spring snowmelt with present acidic loading. Doubling the acidic loading has only a slight effect, because current loading is so low.

Hooper et al. (1989) developed a model of solute concentrations during snowmelt in the Emerald Lake basin; the model includes a pseudo-kinetic representation of primary mineral weathering as the only ANC generating mechanism. If deposition of NO_3^- and SO_4^{2-} doubled, the model predicts a decline in minimum ANC of 2 to 5 $\mu eq\ L^{-1}$. Depending on the depth of water in the snowpack and the proportion of solutes released during the first phase of snowmelt, current atmospheric deposition of strong acids is predicted to require increases of two to eight times to acidify Emerald Lake during snowmelt.

Nishida and Schnoor (1989) have applied a steady state, charge balance model of ANC to 198 Sierra Nevada lakes under different atmospheric loading scenarios. Their model used a Henrickson *F*-factor, or cation replenishment rate, of 0.4, which means that 40% of an increase in acid would be neutralized and result in release of an equivalent amount of basic cations. A doubling of annual loading that combined equivalent increases in sulfuric acid and ammonium nitrate was predicted to increase by 9% the number of lakes with ANC $\leq 40\ \mu eq\ L^{-1}$, with 2.5% of these lakes having an ANC ≤ 0. The results were sensitive to the ratio of $NH_4^+:NO_3^-$ in deposition, because NH_4^+ uptake is acidifying.

Biological Status

The distribution of fish species in the Sierra Nevada is only partially documented. Fish stocking records are irregular, and those that are available do not usually match the species present in high-elevation lakes many years later. A few lake surveys, conducted by the California Department of Fish and Game, and Sequoia and Kings Canyon National Parks (Zardus et al. 1977), provide more reliable data. High-elevation lakes either lack fish or contain golden trout (*Salmo aguabonita*), rainbow trout (*Oncorhynkus mykiss*), or eastern brook trout (*Salvelinus fontinalis*).

Survival of fish in response to acidification is largely determined by pH, dissolved monomeric Al, and Ca^{2+}. Physiological and ecological responses to acidification of species in the genera *Salmo* and *Salvelinus* are well studied because of the decline of these fish in acidified waters of Europe and eastern North America (Chapter 4, Baker and Christensen, this volume). In the Sierra Nevada, many lakes are naturally fishless. In those with fish, food is often very limited, and eggs and young face difficult conditions such as frozen stream bottoms and severe floods. Although pH can briefly decline to values near 5.5, where acidic stress may be expected, the generally unfavorable environment in high-elevation Sierra Nevada lakes is currently the major limitation, rather than acidification (Cooper et al. 1988a).

A thorough analysis of zooplankton assemblages and distributional patterns in the Sierra Nevada is provided by Stoddard (1987b). As in other mountainous regions, fishless lakes in the Sierra are dominated by large, highly pigmented zooplankton (Anderson 1971, 1974, Patalas 1964, Sprules 1972, Williams 1976). In the Sierra Nevada, these communities are always made up of *Daphnia middendorffiana* and *Diaptomus shosone* or *Diaptomus eiseni* (but never both), and occasionally the anostracan *Branchinecta dissimilis*.

A typical community in a lake with fish in the Sierra Nevada consists of *Daphnia rosea* (the most common species found), and *Diaptomus signicauda* and *Chydorus sphaericus* (the next two most common species). As species richness increases (e.g., at lower elevations or in more productive lakes), species such as *Bosmina longirostris*, *Holopedium gibberum*, *Ceriodaphnia lacustris*, or *Alona affinis* become common.

Zooplankton species richness in the Sierra Nevada correlates strongly with elevation, but shows no relationship to pH. Despite a range in H^+ concentration of nearly three orders of magnitude (pH 5.7 to 9.4), neither species richness nor the distribution of any individual species showed a relationship with pH that was significantly nonrandom (Stoddard 1987b).

It is very difficult to take a census of macroinvertebrates in the zoobenthos of streams and lakes because of their large spatial variability. Only Emerald Lake and several neighboring lakes and streams in the Kaweah drainage have received sufficient sampling for the abundance of these animals in high-elevation Sierra Nevada waters to be characterized. The qualitative results from Emerald Lake are in agreement with surveys of other regions in the Sierra Nevada (e.g., Taylor and Erman 1980); thus results from Emerald Lake are summarized here based on Melack et al. (1987) and Melack et al. (1989).

The dominant macroinvertebrates in inlet and outlet streams were chironomid larvae. Simuliid larvae (*Simulium* sp., *Prosimulium* sp.), *Baetis* mayfly nymphs, and trichopteran larvae *Ecclisomyia* and *Rhyacophila* were commonly collected in both inlet and outlet streams. In addition, larvae of ephemerellid and leptophlebid mayflies (*Drunella spinifera*, *Caudatella hystrix*, *Serratella* sp., *Paraleptophlebia* sp.), nemourid and perlodid stonefly nymphs (*Malenka californica*, *Zapada* spp., *Isoperla quinquepunctata*), tipulid larvae (*Dicranota* sp., *Hexatoma* sp.), water mites (Acaridae), oligochaetes, and sphaeriid clams (*Pisidium* sp.) were often collected in the outflow stream. Low densities of a sponge (*Spongilla lacustris*) were found on or under hard substrata in the lake. The dominant benthic invertebrates in the lake included chironomids, water mites, sphaeriid clams (*Pisidium* sp.), and chydorid cladocerans.

The inlet and outlet streams and lake contain taxa known to be sensitive to acidic inputs. The mayflies (*Baetis*, *Serratella*, *Drunella*) and sphaeriid clams, in particular, are often rare or absent in acidic waters (Hall and Ide 1987, Okland 1980, Bell 1971, Sutcliffe and Carrick 1973, Friberg et al. 1980, Singer 1982). Because these taxa are common in the Emerald Lake system, it is unlikely that acidic deposition has had a long-term effect on the biota.

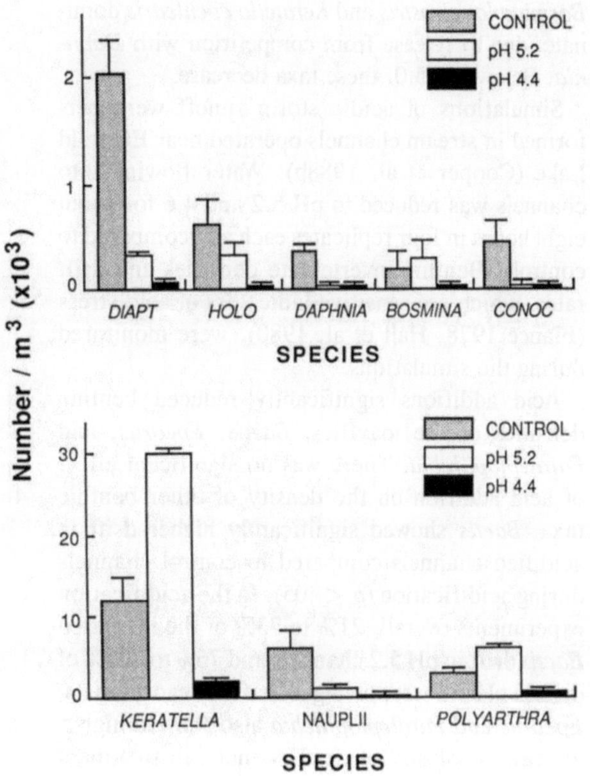

FIGURE 15.20. Zooplankton responses to experimental acidification of enclosures in Emerald Lake. Mean density ± one standard error indicated. Abbreviations indicate species as follows: Diapt–*Diaptomus signicauda*, Holo–*Holopedium gibberum*, Daphnia–*Daphnia rosea*, Bosmina–*Bosmina longirostris*, Conoc–*Conochilus unicornis*, Keratella–*Keratella conchlearis*, nauplii–*Diaptomus nauplii*, Polyarthra–*Polyarthra vulgaris*.

To evaluate the sensitivity of Sierra Nevada plankton and stream invertebrates to increased acidity, Melack et al. (1987), Melack et al. (1989), Cooper et al. (1988b) and Hopkins et al. (1989) conducted experimental manipulations in large tubes in Emerald Lake and in channels along a neighboring stream. Experimental stream channels and large enclosures in the lake constitute good compromises between the naturalness of whole system manipulations and the control and replication of laboratory experiments; they approximate natural conditions while providing the replication needed for statistical analyses. Artificial stream channels have been successfully used in past studies to examine the effects of increased acidity on stream systems (Zischke et al. 1983, Burton et al. 1985, Servos and Mackie 1986). Large bags in lakes have been used to examine the effects of acidic inputs on plankton (Marmorek 1983, 1984, Yan and Stokes 1978) and periphyton (Müller 1980).

Phytoplankton responses during the course of one to four weeks of treatments with pH values maintained at about 5.2 and 4.4 were minor. No significant differences in chlorophyll-*a* concentrations or total cell numbers occurred. Slight shifts in the relative abundance of dominant morphotypes and in total number of morphotypes were observed.

Zooplankton responses to two acid levels in Emerald Lake enclosures are illustrated in Figure 15.20 (Melack et al. 1987). *Conochilus*, *Daphnia*, and *Diaptomus* were very sensitive to acidic inputs; *Holopedium* and *Bosmina* were significantly reduced only at low pH values (pH < 5), and the rotifers *Keratella* and *Polyarthra* increased relative to controls in bags reduced to intermediate pH values (pH 5.2), then declined at low pH values (pH < 4.8).

Overall, the experiments conducted by Melack et al. (1987) and Melack et al. (1989) indicate that the effects of acidification depend on the pH involved, the species assemblage present, and the biological interactions among zooplankton species. Few changes occur as pH declines from 6.3 to 5.6; however, at pH levels between 5.6 and 5.5, a number of zooplankton taxa, including the competitive dominant (*Daphnia rosea*), abruptly decline, owing to direct toxic effects. As pH declines from 5.5 to 5.1,

Bosmina longirostris and *Keratella cochlearis* dominate, due to release from competition with *Daphnia*. Below pH 5.0, these taxa decrease.

Simulations of acidic storm runoff were performed in stream channels operated near Emerald Lake (Cooper et al. 1988b). Water flowing into channels was reduced to pH 5.2 and 4.6 for about eight hours in four replicates each and compared to controls. Benthic invertebrate densities and drift rates, which are sensitive indicators of acid stress (Fiance 1978, Hall et al. 1980), were monitored during the simulations.

Acid additions significantly reduced benthic densities of the mayflies, *Baetis*, *Epeorus*, and *Paraleptophlebia*. There was no significant effect of acid addition on the density of other benthic taxa. *Baetis* showed significantly higher drift in acidified channels compared to control channels during acidification ($p < .05$). In the acidification experiments overall, 21% to 73% of the increased *Baetis* drift in pH 5.2 channels and 76% to 100% of drift in pH 4.6 were owing to drift of dead animals. *Epeorus* and *Paraleptophlebia* also showed higher drift rates, mostly of dead animals, in treatment channels during acidification, but both taxa were less sensitive to acid inputs than *Baetis*.

Conclusions

1. Surface waters of the Sierra Nevada currently show no signs of chronic acidification, but are very dilute and highly susceptible to the impacts of acidic deposition. Episodic snowmelt events of lowered pH and ANC in intensively studied lakes are associated with dilution by large volumes of dilute runoff and slightly elevated levels of strong acid anions relative to basic cation concentrations.
2. Weathering of granitic minerals is the major source of ANC to Sierra Nevada lakes, although in situ processes can make minor contributions when lakes are stratified. Weathering of non-granitic bedrocks, low in mass percent in all watersheds, can have significant effects on surface water chemistry and lead to concentrations of SO_4^{2-} and Mg^{2+} much higher than would be expected from granite weathering.
3. Concentrations of strong acid anions in precipitation are low compared to acid sensitive areas of the eastern United States, but high amounts of precipitation lead to considerable rates of acidic deposition. These rates are low to moderate by standards of the eastern United States, but should be considered significant when the dilute nature of Sierra Nevada waters is taken into account.
4. Species of fish, zooplankton, and stream invertebrates that are considered sensitive to acidification are common throughout the Sierra. It is unlikely that their current distribution has been influenced by acidic deposition.

Acknowledgments. We acknowledge T. Reilly for help with the description of soils. S. Cooper and K. Kratz made available information on biological responses to acidification, and N. Stephenson offered chapters of his Ph.D. thesis that aided our description of regional climate. D. Parsons and A. Esperanza provided the photograph for Figure 15.1. We thank J. Sickman for assistance with graphics. Many people contributed to the California Air Resources Board's Integrated Watershed Study at Emerald Lake, and special thanks are owed J. Sickman, Michael Williams, R. Kattelmann, Mark Williams, C. Soiseth, S. Hamilton, S. Sippel, D. Lucero-Sickman, and F. Setaro for technical assistance. We thank D. Charles, D.W. Schindler, H.M. Seip, S.D. Cooper, C.T. Driscoll, K.A. Tonnessen, R.B. Cook, J.M. Eilers, S.K. Hamilton, S. Nodvin, and an anonymous reviewer for useful comments on the manuscript.

16
Cascade Mountains

Peter O. Nelson

ABSTRACT. Lakes in the Cascade Mountains of the Pacific Northwest are among the most dilute in the world and second only to Sierra Nevada lakes in the United States. High precipitation quantities, unreactive bedrock minerals, and generally thin soil development contribute to the dilute chemical composition of Cascade lakes. The principal data set used as the basis for evaluating the status of the lakes with regard to acidic deposition was the Western Lake Survey (WLS), a statistically designed survey of low ANC lakes conducted by the U.S. EPA as part of the National Acid Precipitation Assessment Program (NAPAP). Other data sets contributed either to the baseline chemical data set or to the evaluation of processes affecting lakewater composition.

Based on the WLS data set, Cascade lakes have the following overall population characteristics. Median values of key chemical parameters are: pH, 7.2; ANC, 102 μeq L^{-1}; SO_4^{2-}, 10.0 μeq L^{-1}; sum of base cations, 127 μeq L^{-1}; Ca^{2+}, 70.5 μeq L^{-1}, and DOC, 101 μmol L^{-1}. Eighty-two percent of Cascade lakes have ANC ≤ 200 μeq L^{-1}, 94% have SO_4^{2-} ≤ 50 μeq L^{-1}, and 78% have pH < 7.5.

Six geomorphic subregions of the Cascade Mountains along a 1,000-km north-south transect were considered. Based on population medians, California Cascade lakes had the most dilute ionic composition and the lowest ANC, and northern Washington Cascade lakes the highest values of the two parameters, with a generally increasing gradient of most chemical parameters proceeding from south to north. Sulfate was very low in Oregon Cascade lakes and highest in the middle Washington Cascades, which receive deposition from anthropogenic sources in the Puget Sound urban complex.

Seasonal trends in Cascade lake chemical composition are dominated by the relatively rapid dilution effect of snowmelt in the late spring to early summer period, followed by a gradual increase in concentrations throughout the remainder of the year. Variations in composition range from 10% to 20% of the mean for lakes with long hydrologic residence times to 100% to 300% of the mean for lakes with short hydrologic residence times. Total dissolved Al, although very low, can double from July to October, with the major increase attributable to the organically complexed fraction. Minor pulses of acid anions may occur during the snowmelt period, but changes in ANC due to acidification have not been clearly demonstrated.

The ANC and base cation composition of Cascade lakes is acquired predominantly from weathering reactions of basic minerals in lake watersheds. Precipitation accounts for 15% to 20% of lake ionic composition. Watershed and in-lake biochemical processes for important redox components (SO_4^{2-}, NO_3^-, NH_4^+, Fe^{3+}) do not appear to be generally important for Cascade lakes. Sulfate is geochemically derived in the watersheds of some Washington Cascade lakes but not at detectable levels in Oregon and California Cascades lakes.

Although the aquatic biota of Cascade lakes have generally not been definitively characterized, species diversities of phytoplankton and zooplankton are high, whereas abundances of organisms are low, indicative of the gener-

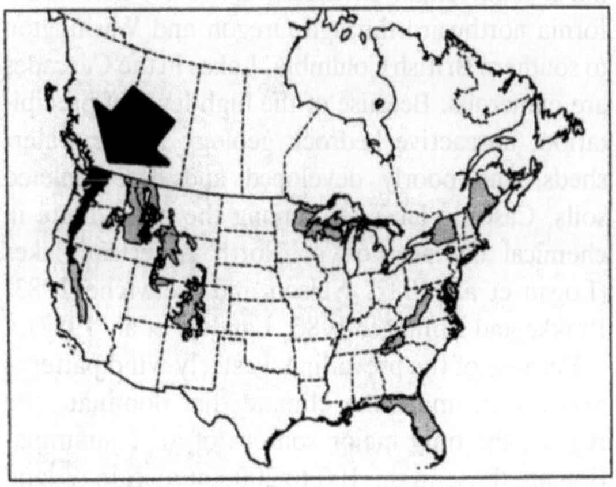

ally unproductive, low nutrient waters. No discernible impacts from acidic deposition are expected under present conditions in Cascade lakes.

In general, Cascade Mountain lakes are highly susceptible to acidification, but there is no indication of acidification at present, based on current atmospheric deposition loading levels, lake chemistry, and assessments using empirical chemical models.

Introduction

The Cascade Mountains are the major mountain range of the Pacific Northwest, forming a 1,100-km, north-south chain extending from northern California northward through Oregon and Washington to southern British Columbia. Lakes in the Cascades are numerous. Because of the high level of precipitation, unreactive bedrock geology in the watersheds, and poorly developed and base-depleted soils, Cascade lakes are among the most dilute in chemical composition of North American lakes (Logan et al. 1982, Nelson and Delwiche 1983, Brakke and Loranger 1986, Landers et al. 1987).

Because of the prevailing westerly wind patterns and Pacific maritime climate that dominate the region, the only major sources of air contamination are those in the 100 to 200 km margin of land lying between the Cascade Range and the Pacific Ocean. Anthropogenic sources of air pollution, particularly sulfur and nitrogen oxides associated with acidic deposition, are relatively minor compared to those in sensitive regions of the eastern and midwestern United States.

Cascade mountain lakes are perhaps unique in the continental United States in that they are among the most dilute, and therefore most sensitive to acidic deposition, yet receive only minor amounts. In this sense, they may serve as a benchmark for comparison to other areas that are sensitive but receive much greater loadings of acidic deposition.

Regional Characteristics

Watershed Characteristics

Physiography

The Cascade Mountains consist of uplifted lava plateaus surmounted by more recent volcanic peaks that form a north-south oriented backbone to the range. Figure 16.1 shows an oblique aerial photograph of the Cascades region. The Cascade range crest lies approximately 200 to 250 km inland from the Pacific coast, and the range varies in width from about 50 km in southern Oregon to 130 km in northern Washington. The range connects with the Sierra Nevada near Mount Lassen in California and with the Coast Mountains of British Columbia.

Cascade Range physiography is dominated by the high volcanic peaks that generally lie on the eastern half of the north-south range. The high elevations are characterized by a rugged alpine zone with craggy peaks and permanent glaciers. The Columbia River forms a deep gorge that breaches the Cascade range at the Oregon–Washington border. The Klamath River, near the Oregon–California border, forms the only other breach in the 1,100-km Cascade chain of mountains.

The Cascade range separates the mild, maritime climate on the west side, dominated by the Pacific Ocean, from the colder, arid, high-elevation Columbia-Snake River Plateau in eastern Washington and Oregon, and from the Great Basin in California. Western slopes of the Cascades receive high precipitation amounts and are heavily vegetated, whereas the eastern flank transcends to arid and sparsely vegetated high desert regions.

Geomorphic Subregions

The Cascade Mountain Range can be divided into subregions based on geology, physiography, and climate. Hunt (1974) divides the Cascade Range into three sections. The Southern Cascades extend from the lava-covered structural sag in the Sierra-Cascade uplift north to the Klamath River. The Middle Cascades consist of an uplift of Middle Tertiary lavas and extend from the Klamath River north past Mount Rainier to the Wenatchee Mountains of Washington. The Northern Cascades, which extend from Snoqualmie Pass in the Wenatchee Mountains north to Canada, are similar structurally to the Sierra Nevada in that they are dominated by granitic batholiths and older metamorphosed rocks (see Chapter 15, Melack and Stoddard, this volume).

For this study, in order to emphasize the changing character of Cascade lakes along the north-south transect, six geomorphic subregions were delineated in accordance with those of Landers et

FIGURE 16.1. Oblique aerial photograph of Cascades region (Big Lake, Oregon).

al. (1987) (see map of subregions in Western Region Overview, this volume):

- North Washington Cascades, including lakes from the Canadian border south to Rainy Pass
- Middle Washington Cascades, including lakes between Rainy Pass and Stevens Pass
- Wenatchee Mountains, including lakes between Stevens Pass and Snoqualmie Pass
- South Washington Cascades, including lakes between Snoqualmie Pass and the Columbia River (Washington-Oregon border)
- Oregon Cascades, including all Cascade lakes in Oregon
- California Cascades, including lakes from the Klamath River (approximate California-Oregon border) south to just south of Mount Lassen near Lake Alamor (northern extent of Sierra Nevada)

Geology

The principal geological features of the Cascade Mountain range are relatively recent, developing during the Cenozoic era. Older igneous rocks from the Paleozoic era and metamorphic rocks from the Mesozoic era are found only in the North Cascades of Washington. The relatively young major Cascade volcanic peaks developed during the Pleistocene and are classic examples of stratovolcanoes, consisting of alternating layers of lava (typically andesite) and pyroclastic material. Glaciers that formed on the volcanic cones during the recent Wisconsinan glaciation have extensively eroded their flanks, producing the deeply gouged valleys of the present mountains.

McKee (1972) defines the North Cascades on a geological basis as that region of the Cascades bounded on the north by the Frazier River in British Columbia and on the south by Snoqualmie Pass east of Seattle. This corresponds to the northern Cascades section of Hunt (1974). The North Cascades thus encompass three geomorphic subregions delineated in the previous section, North Washington Cascades, Middle Washington Cascades, and Wenatchee Mountains. The geology of the North

Cascades is more complex than that of other Cascades subregions; it is a heterogeneous mixture of older metamorphic rocks intruded with batholiths of granite, granodiorite, and quartz diorite (Welch et al. 1987). Schists, gneisses, and granites are the dominant bedrock minerals (McKee 1972). Individual watersheds in the North Cascades may also contain significant percentages of sedimentary rocks and glacial till formed from the extensive glacial erosion of volcanic and metamorphic rocks (Brakke and Loranger 1986, Duncan and Lillquist 1986). Trace amounts of exposed reactive minerals such as calcite may dominate weathering reactions in some watersheds (Drever and Hurcomb 1986).

The Middle Cascades section (Hunt 1974) encompasses the South Washington Cascades and Oregon Cascades geomorphic subregions (delineated in Geomorphic Subregions). This section is geologically distinct from and more uniform than the North Cascades, lacking major exposures of intrusive igneous and metamorphic rocks. Bedrock geology is dominated by extrusive igneous rocks, primarily andesite and basalt. Vents of dacite and rhyodacite are also common. Intrusions of granodiorite, quartz monzonite, diorite, and quartz diorite are exposed in areas of the Quaternary volcanic cones. Batholiths of granite and granodiorite are infrequent. The Middle Cascades have a lower average elevation and lesser amounts of exposed rock than the North Cascades.

The Southern Cascades section of the Cascade Range is identified by the lava-covered structural sag in the Sierra-Cascade uplift (Hunt 1974). This section corresponds to the California Cascades geomorphic subregion. Bedrock geology is similar to that of the Oregon Cascades, being dominated by extrusive igneous rock types. The Western Cascades of Oregon extend southward to about Mount Shasta in California, south of which older sedimentary rocks are found (MacDonald 1966).

Soils

Soils in the Cascade Mountains are given the general classification of Udic soils of high rainfall mountains (Omernik and Gallant 1986). According to the soil classification system of the U.S. Department of Agriculture, soils in the Cascade Range are predominantly Inceptisols and Ultisols, which are found primarily in alpine and subalpine areas (see soils map, Chapter 5, Eilers and Selle, this volume). These soils are naturally acidic and are the only two soil orders in the western United States that are considered susceptible to acidic deposition (Hidy et al. 1986).

Soil development in the Cascade Mountain Range has resulted from two major processes, weathering and glaciation. As such, soils generally reflect the mineral composition of the parent materials, including breccias and tuffs of basic igneous rock types, mainly andesite, basalt, and dacite. Residual deposits formed from weathering are most prominent on the western wet slopes of the mountains and at lower elevations. Deep, sandy loams are found in valley bottoms; these are older soils dating from the Tertiary period and are generally classed with the Red and Yellow Podzols. At higher elevations and more directly relevant to the Cascade lakes of this study, soils formed by weathering date from the Quaternary period and are Gray-Brown Podzols. These soils are often medium to very shallow in depth with medium to coarse grain sizes and good permeability. Glacially derived soils were formed during the Wisconsinan and Holocene ages and are relatively recent. These soils are generally found at valley bottoms or sideslopes and are often deep. Glacial soils consist of glacial drift and outwash that has been modified by colluvial action. On the eastern slopes of the Cascades, soils are more glacially influenced. Variable soils exist at lower elevations, graduating to moderately coarse soils in the forest zone, and fine soils at higher elevations (Duncan and Lillquist 1986).

Vegetation

Most of the Cascade Mountain Range is densely forested below about 1,500 m elevation. Above this elevation, subalpine and alpine vegetation becomes dominant. Marked changes in vegetation accompany the pronounced climatic change (see Climate) that occurs eastward across the range. Timberline occurs at 1,400 to 1,900 m in the North Cascades, rising to 2,100 to 2,500 m in the southern Cascades (Arno and Hammerly 1984). Only ice-covered areas and the craggiest peaks of the Cascades are completely barren (Rasmussen and Tangborn 1976).

The dominant forest species on the western slopes of the Cascades in Oregon and Washington is the Douglas fir (*Pseudotsuga menziesii*), with lesser numbers of western hemlock (*Tsuga heterophylla*) and western red cedar (*Thuja plicata*). Forest understories are also dense in the Western Cascades, grading toward sparse in the high Cascades and east of the Cascade crest. The California Cascades lie in the rain shadow of the Klamath Mountains to the west. Due to the lower precipitation and warmer climate, open forests of sugar pine (Pinus lambertiana) dominate western slopes in this subregion.

Eastern slopes of the Cascades lie in the rain shadow of the high Cascades. East slope forests consist of open stands of ponderosa pine (*Pinus ponderosa*) at high elevations, where precipitation amounts are greater, and lodgepole pine (*Pinus contorta*) at lower elevations and in areas where precipitation amounts are down to 35 cm per year. Where annual precipitation averages < 30 cm, sagebrush grass or western juniper open woodlands dominate.

At high elevations in the Cascades, above 1,200 to 1,500 m in the north and above 2,000 to 2,300 m in the south, a transition occurs to subalpine and alpine vegetation. Important forest species are Pacific silver fir (*Abies amabilis*), subalpine fir (*Abies lasiocarpa*), mountain hemlock (*Tsuga mertensiana*), and Alaska cedar (*Chamaecyparis nootkatensis*) on the wetter western slopes, and whitebark pine (*Pinus albicaulis*) and Englemann spruce (*Picea engelmannii*) on the drier eastern slopes. Subalpine fir dwindles in importance in the southern Oregon Cascades and is nearly absent in the California Cascades. In southern Oregon, the Pacific silver fir is replaced by the Shasta red fir (*Abies magnifica shastensis*), which is prominent in the California Cascades. Open areas above timberline are occupied by alpine meadows.

Land Use

In the Cascade Mountain Range, the major land cover is evergreen forest land and the major use of this land in unrestricted areas is silviculture (Franklin and Dyrness 1973, Omernik and Gallant 1986). Approximately 101 of the 146-lake sample population of the WLS lakes included in this study were located in wilderness areas or national parks (Landers et al. 1987). An estimated 70% of the total lake population lies in watersheds physically undisturbed by current human activities other than recreation. Of the remaining 30% of the estimated lake population, most is likely to be in evergreen forest land managed for silvicultural activities. Infrequently, alpine meadows provide range for livestock grazing, and mining for metals may occur in some areas.

Climate

The climate of the Cascade Mountain Range is influenced by variations in latitude, altitude, and proximity to the Pacific Ocean. Prevailing winds are westerly, and weather patterns are mostly developed from frontal systems evolving out of the Pacific "low" regime (Hidy and Young 1986). West of the range, the climate has a strong maritime influence, with mild, wet winters and dry, moderate summers. East of the range, in a prominent rain shadow, the climate is that of the arid inland plateau, much dryer with hot summers and cold winters. The Cascade Range itself is an orographically influenced transition zone between these climate types.

Mean annual temperatures for the Cascade Mountains range from 4 to 10°C (Hunt 1974). Gradients in temperatures and precipitation are great. Winter temperature lows may reach −18°C (0°F) and −35°C (−30°F) west and east of the Cascade crest, respectively (Arno and Hammerly 1984).

Precipitation in the Cascade Mountain Range exhibits marked variation seasonally and spatially. Average annual precipitation amounts range between 250 cm yr^{-1} to about 120 cm yr^{-1} from north to south along the Cascade crest (Gerahty et al. 1973, Hunt 1974; see also precipitation map in Western Region Overview, this volume). About 80% of annual precipitation in Cascade lake watersheds (above 600 m) falls during the winter months (October to April), mostly as snow, whereas 20% falls during the summer months (May to September), mostly as rain (Rasmussen and Tangborn 1976). Thunderstorms are rather common in the summer, occurring an average of 5 to 10 days per year (Hunt 1974). Dethier (1979) has summarized annual precipitation data along a transect from Everett, Washington, near Puget Sound (sea level, 100 cm yr^{-1} precipitation) to Copper Lake (1,850 m mean watershed elevation, 470 cm yr^{-1} precipitation) in the Middle Washington Cascades, showing a nearly linear dependence on elevation with the gradient reversing on the eastern slopes to about 50 cm yr^{-1} annual precipitation. The coefficient of variation in annual precipitation amount is high (between 20% and 25%) in the Pacific Northwest (Hidy and Young 1986) and may be greater in the Cascade Mountain Range.

Hydrology

Stream density is typically 0.9 to 1.2 km of perennial streams per km^2 in the Cascades, determined from 7.5- and 15-minute topographic maps. Watersheds of perennial streams may be as small as 3 to 5 km^2 and commonly exceed 1,300 km^2 (Omernik and Gallant 1986). Runoff hydrographs for stream and river systems draining the Cascades are strongly influenced by a precipitation pattern in which most of the precipitation falls as snow during the winter months (see previous section, Climate). Runoff is relatively low and constant during the winter months when precipitation is stored as snowpack. A steep rise in the hydrograph from snowmelt begins in late April to May with peak runoff occurring in June. The hydrograph drops rapidly in July through August as the snow is depleted and precipitation amounts are very low. Minimum runoff is reached in September–October, until winter rains or snows begin. Representative mean annual precipitation and runoff hydrographs for west and east sides of the Cascade crest are shown in Figure 16.2 (Rasmussen and Tangborn 1976).

Cascade lakes from the WLS were classified by hydrologic type as seepage (no inlet or outlet), drainage (outlet), closed (inlet but no outlet), or reservoir (human-made impoundment). Reservoirs were not considered in the Cascade lakes analyzed in this chapter. Of the 146 sample lakes, 104 were classified as drainage, 41 as seepage, and 1 as closed. These determinations were based on USGS topographic maps (Eilers et al. 1987a). Inlet and outlet streams in the Cascades are often ephemeral, occurring primarily as snowpack meltwater; they may occur as subsurface or fracture flow in the volcanic terrain. Western Lake Survey hydrologic classifications based on mapped information may not represent substantial differences in Cascade lakes hydrologic regimes (Landers et al. 1987).

Few data appear to be available to directly estimate evapotranspiration (ET) for Cascade lakes and watersheds. The Water Atlas of Gerahty et al. (1973) permits only general estimation due to insufficient scale in the Cascade Mountains. For the North Washington Cascades, potential evapotranspiration is less than about 30%, whereas for the California Cascades, it may be greater than 60% of annual precipitation. Evapotranspiration varies between these extremes along the Cascade north-south transect (see also map of pan evaporation data, Chapter 5, Eilers and Selle, this volume). From the data of Rasmussen and Tangborn (1976), Dethier (1979) states that evapotranspiration is < 30 cm for watersheds in the Middle Washington Cascades, which equals about 6% to 15% of mean annual precipitation. Evaporation measured from a land pan at Odell Lake (elevation = 1,459 m) in the central Oregon Cascades was estimated at 61 cm, or about 39% of annual precipitation (Phillips 1968). At Crater Lake (1,882 m) in the southern Oregon Cascades, Phillips (1968) estimates the annual evaporation at 58 cm, or 34% of annual precipitation, although recent hydrologic calculations by Redmond (1989) indicate that evapotranspiration may be as much as 60% of annual precipitation, because the lake remains unfrozen in most winters.

Atmospheric Deposition

A major source of precipitation chemistry data in the United States is the National Atmospheric Deposition Program/National Trends Network (NADP/NTN) monitoring sites program originally established in 1978. At present, there are eight sites in California, seven in Oregon, and five in Washington (NADP 1987a,b). No NADP/NTN monitoring sites are located in the watersheds of the Cascade lake population considered in this study. Compared to the Cascade study lakes, all the NADP/NTN monitoring sites located in the vicinity of the Cascade Mountain Range are situated at lower elevations and most are located nearer to the Pacific coast. Precipitation chemical parameters of the NADP/NTN sites are variably influenced by sea salts. Thus there are no long-term NADP/NTN precipitation chemistry data directly applicable to the study of individual lake chemistry in the Cascades.

Independent research studies have reported precipitation, deposition, and snow core chemistry data for selected Cascade lakes and watersheds. Table 16.1 presents a summary of these studies, indicating the Cascades subregion, period of study,

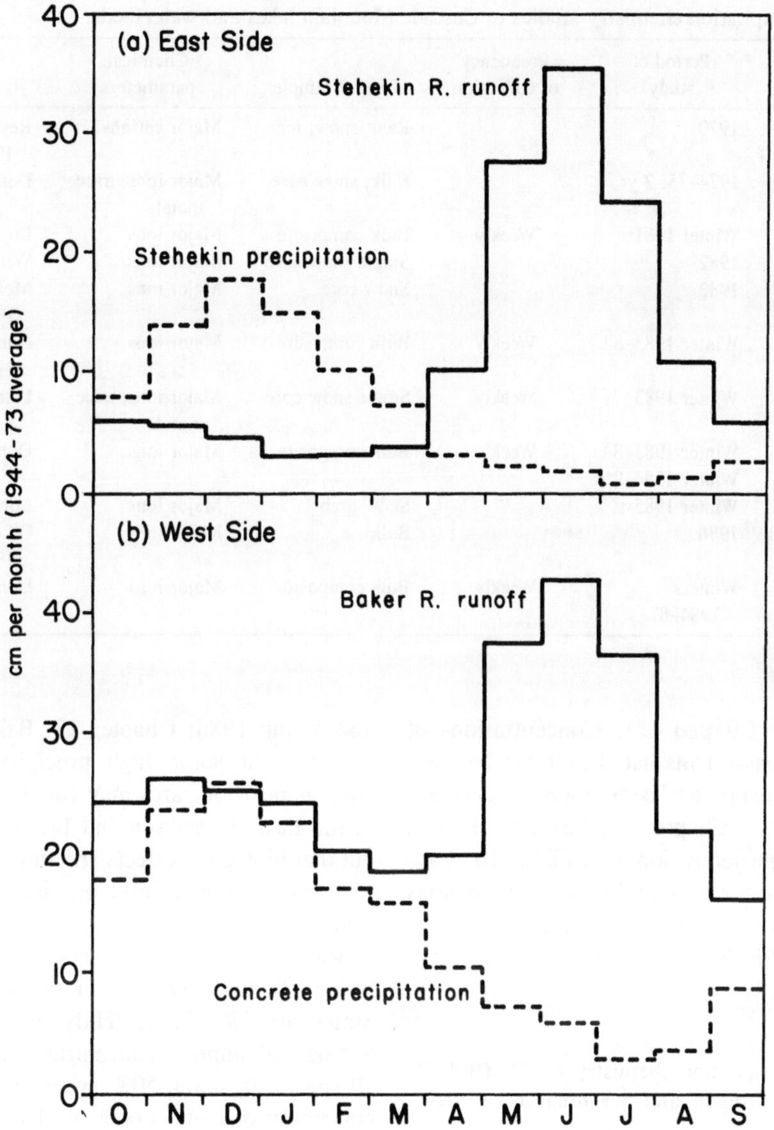

FIGURE 16.2. Representative mean annual precipitation and runoff distributions for (a) east and (b) west sides of the Cascade crest. (From Rasmussen and Tangborn 1976.)

frequency of sample collection, type of sample, and chemical parameters determined.

Hidy and Young (1986) have prepared a comprehensive analysis of precipitation chemistry in the western United States based on NADP/NTN network sites for the years 1981–84. Although not directly applicable to Cascade lakes chemistry, the study results do elucidate general characteristics of precipitation chemistry in western Oregon and Washington. Table 16.2 was prepared from the results of Hidy and Young (1986) for NADP/NTN sites in western Oregon and Washington. It shows ranges of annual volume-weighted mean concentrations for pH, SO_4^{2-}, NO_3^-, NH_4^+, and Ca^{2+}, and annual deposition rates of H^+, SO_4^{2-}, NO_3^-, and NH_4^+ during the four years 1981–84.

Eilers and Selle (Chapter 5, this volume), in comparison, list volume-weighted means from selected NADP stations for the Cascades in the year 1987 of: pH = 5.31, $[SO_4^{2-}]$ = 5.0 μeq L^{-1},

TABLE 16.1. Precipitation chemistry studies of Cascade Mountain lakes and watersheds.

Cascades subregion	Period of study	Frequency of collection	Type of sample	Chemical parameters	Reference
North Washington	1970		Rain; snow; ice	Major cations	Reynolds and Johnson 1972
Mid. Washington	1974–75: 2 yr		Bulk; snow core	Major ions, trace metals	Dethier 1979
Wenatchee	Winter 1981	Weekly	Bulk composite	Major ions	Logan et al. 1982
Wenatchee	1982		Single snow core	Major ions	Welch et al. 1984
British Columbia	1982		Snow core	Major ions	McBean and Nikleva 1986
Wenatchee (2 sites)	Winter 1983–84	Weekly	Bulk composite	Major ions	Duncan and Ausserer 1984
All	Winter 1983	Weekly	Single snow core	Major ions, trace metals	Laird et al. 1986
Wenatchee (2 sites)	Winter 1983–84 Winter 1984–85	Weekly	Bulk composite	Major ions	Duncan et al. 1986
North Washington	Winter 1985		Snow grab	Major ions	Loranger 1986
Wenatchee/Mid. Washington	1986		Bulk	Major ions	O'Loughlin et al. 1986
Wenatchee (2 sites)	Winters 1984–87	Weekly	Bulk composite	Major ions	Duncan et al. 1988

and $[NO_3^-]$ = 1.9 μeq L^{-1}. Concentrations of precipitation constituents listed in Table 16.2 are probably at or near natural background levels (Galloway et al. 1982; Chapter 17, Baker et al., this volume). Average deposition rates (Table 16.2) are lower than those for comparable sensitive regions in the eastern United States, ranging up to about 25% of those in the East for most parameters (Hidy and Young 1986; Chapter 17, Baker et al., this volume). At some high precipitation sampling sites, deposition rates may equal or exceed those in the East. Henriksen and Brakke (1988a) point out that biological effects are more closely related to concentrations in precipitation than to deposition rates.

TABLE 16.2. Precipitation chemistry for NADP/NTN sites[a], Western Oregon and Washington, 1981–84 annual means.[b]

	Number of sites[c]	Range of concentration (mean) (μeq L^{-1}, except pH)	Range of wet deposition rate (kg ha^{-1} y^{-1})
pH	8	4.95–5.50	0.012–0.201[d]
SO_4^{2-}	6	4.2–12.5	2.4–13.8
NO_3^-	6	1.6–6.3	1.9–6.4
NH_4^+	6	0.6–6.1	0.2–4.1
Ca^+	6	2.0–6.0	

[a] Site locations: Oregon (Alsea, H.J. Andrews, Lost Creek, Pendleton, Schmidt Farm, Bull Run); Washington (Olympic National Park, Marblemount).
[b] Data from Hidy and Young (1986).
[c] Sites meeting primary or secondary data quality criteria. Number may vary from year to year. Number reported is maximum.
[d] Expressed as H^+.

Seasonal variations in concentrations of precipitation chemical species in the western United States are significant (Hidy and Young 1986). Spring and summer concentrations are higher for all species by about 50%, whereas fall and winter concentrations are lower by 10% to 20% than annual volume-weighted means. Seasonal variations in areal deposition rates are not significant for most species except those of H^+ and NO_3^-, which are about 20% greater in the summer, and NH_4^+, which is about 20% less in the fall.

Few direct measurements have been reported for dry deposition in the western United States. Estimates of rural background rates of dry deposition based on ambient concentrations of chemical species and characteristic deposition velocities for SO_4^{2-} and NO_3^- are 7 and 5 kg ha^{-1} yr^{-1}, respectively (Hidy and Young 1986). These are comparable to the wet deposition rates reported in Table 16.2, but may be high estimates for the Cascades, where wet deposition is expected to predominate

TABLE 16.3. Median values[a] (with first and third quartiles in parentheses) for major ion chemistry in primary Cascade lakes data sets – summary.

Data set	Number of lakes	pH	ANC	SO_4^{2-}	Ca^{2+}	DOC	C_B
Cascade Lakes – WLS							
Raw data	144	7.17	97.2	9.2	68.4	104	125.2
		(6.85-7.47)	(47.3-176.0)	(2.1-22.9)	(30.7-121.9)	(48-169)	(66.1-199.9)
Population estimate	1450	7.21	102.2	10.0	70.5	101	126.9
		(6.91-7.48)	(52.3-184.5)	(2.3-24.6)	(30.1-126.3)	(48-168)	(70.9-206.5)
Oregon Cascades – Nelson and Delwiche (1983)							
Raw data	59	7.14	113.5	4.8	49.2	–	141.6
		(6.37-7.46)	(16.9-208.8)	(2.4-7.1)	(12.2-86.6)		(52.7-237.6)
North Washington Cascades – Brakke (1984)							
Raw data	33	6.50	91.5	14.6	96.1		132.7
		(6.40-6.67)	(73.9-115.3)	(12.5-31.7)	(59.5-118.1)	–	(89.2-157.9)

[a] All units μeq L^{-1} except DOC and (μmol L^{-1}) and pH.

(Dethier 1979). Recent mass deposition rate data for Crater Lake in the southern Oregon Cascades, using a bulk collector, indicates that dry deposition contributes 5% to 10% of major cations and 1% to 5% of acid anions relative to wet deposition (Reilly 1989). Another deposition mechanism that is difficult to quantify is termed occult deposition, or deposition associated with clouds, fog, and rime (frozen cloudwater). Estimates of occult deposition suggest that it may be comparable in magnitude to wet deposition in the Cascades (Hidy and Young 1986).

Regional Surface Water Chemistry of the Cascades

Distribution of Data Sets

Three principal data sets were used to analyze the status of Cascade Mountain lakes with respect to acidic deposition: (1) the Western Lake Survey (WLS) (Landers et al. 1987, Eilers et al. 1987a), (2) Nelson and Delwiche (1983), and (3) Brakke (1984). Principal data sets (1) included a complete determination of major cations and anions such that a cation/anion balance could be computed and (2) encompassed a representative number of lakes for the Cascade Mountain region or a subregion. Only the EPA's WLS covers the entire Cascade lakes study area. Table 16.3 summarizes five major chemical parameters for these data sets. In a stratified data distribution for three key parameters (ANC, pH, and SO_4^{2-}), numbers of lakes sampled and estimated lake population are quite evenly distributed over ranges of parameter values appropriate for Cascade lakes (Table 16.4).

Trophic Status and Physical Limnology

The chemical and physical limnology of Cascade lakes indicates ultra-oligotrophic conditions (Wetzel 1975). Low nutrient concentrations are reported in the WLS (Landers et al. 1987) and numerous other studies (EPA National Eutrophication Survey 1978, Wismar et al. 1982, Bortleson et al. 1974). The median total phosphorus concentration for the WLS Cascade lakes population is 2.8 μg L^{-1}, with subregional medians ranging from 1.9 μg L^{-1} in the North Washington Cascades to 4.7 μg L^{-1} in the California Cascades. Estimated NO_3^--N and NH_3-N population medians for the WLS Cascade lakes are also very low, with a concentration of 1.5 μg L^{-1} (0.11 μeq L^{-1}) for NO_3^--N and below detection (10 μg L^{-1} or 0.71 μeq L^{-1}) for NH_3-N.

Estimated median secchi disk transparency, turbidity, and color for the WLS Cascade lakes population are 5.84 m (Q1 = 3.50 m, Q3 = 10.00 m), 0.40 NTU (Q1 = 0.20 NTU, Q3 = 0.40 NTU), and 5 APHA color units (Q1 = 3 color units, Q3 = 9 color units), respectively. These values are indicative of very clear lakes with low organic content.

Temperature and dissolved oxygen profiles for selected Cascade lakes are in most cases typical of

TABLE 16.4. Lake distribution over selected parameter data ranges for WLS Cascade lakes and subregions.

Parameter	Data set[a]	Number of lakes[b]	Parameter range				
ANC (μeq L^{-1})			<0	0–50	50–200	200–400	>400
	A	*n*	0	37	81	14	10
		N	0	351	812	151	106
	B	*n*	0	7	7	1	1
		N	0	41	47	8	5
	C	*n*	0	11	26	4	0
		N	0	113	275	43	0
	D	*n*	0	7	5	4	4
		N	0	86	53	45	45
	E	*n*	0	6	22	1	2
		N	0	60	226	10	22
	F	*n*	0	4	16	1	0
		N	0	42	166	11	0
	G	*n*	0	2	4	3	3
		N	0	20	45	34	33
	H	*n*	0	14	20	14	3
	I	*n*	0	3	20	4	6
AEpH[c]			<5.5	5.5–6.5	6.5–7.5	7.5–8.5	>8.5
	A	*n*	0	11	101	29	0
		N	0	109	1002	308	0
	B	*n*	0	1	13	2	0
		N	0	5	82	14	0
	C	*n*	0	4	32	5	0
		N	0	40	338	53	0
	D	*n*	0	3	9	8	0
		N	0	32	96	90	0
	E	*n*	0	1	23	7	0
		N	0	10	234	73	0
	F	*n*	0	1	19	1	0
		N	0	11	197	11	0
	G	*n*	0	1	5	6	0
		N	0	10	55	68	0
	H	*n*	0	16	29	14	0
	I	*n*	1	8	20	4	0
SO_4^{2-}, μeq L^{-1}			0–2	2–5	5–10	10–50	>50
	A	*n*	33	23	18	59	8
		N	320	213	183	614	89
	B	*n*	7	7	1	1	0
		N	47	44	5	5	0
	C	*n*	24	11	3	2	1
		N	251	116	32	21	11
	D	*n*	0	1	1	15	3
		N	0	11	11	162	34
	E	*n*	1	3	8	19	0
		N	12	30	82	194	0
	F	*n*	1	0	5	13	2
		N	11	0	53	134	21
	G	*n*	0	1	0	9	2
		N	0	12	0	97	24
	H	*n*	7	23	26	3	0
	I	*n*	0	0	2	22	9

[a] Data sets: (A) WLS, Cascade lakes combined. (B) WLS, California Cascades. (C) WLS, Oregon Cascades. (D) WLS, Southern Washington Cascades. (E) WLS, Wenatchee Cascades. (F) WLS, Mid-Washington Cascades. (G) WLS, Northern Washington Cascades. (H) Nelson et al. (1983), Oregon Cascades. (I) Brakke (1984), Northern Washington Cascades.

[b] Number of lakes: *n* = lakes sampled; *N* = estimated lake population.

[c] Air-equilibrated pH.

oligotrophic, dimictic lakes. Temperature and dissolved oxygen profiles for selected Cascade lakes have been determined for the following subregions by various investigators:

- Oregon Cascades, Nelson and Delwiche (1983) and Johnson et al. (1985)
- South Washington Cascades (Mount Rainier National Park), Turney et al. (1986) and Nelson and Baumgartner (1988)
- Wenatchee and Middle Washington Cascades, Bortelson et al. (1974), Bortelson et al. (1976), and McConnel et al. (1976)

For deep (> 10 m) Cascade Lakes, iceout typically occurs from May to early July, depending on elevation, with stratification commencing almost immediately thereafter. Stratified conditions remain until mid-September to early October. Epilimnia of Cascade lakes generally remain well oxygenated and near saturation. Hypolimnetic oxygen depletion is often relatively small (< 25% dissolved oxygen depletion) during stratified conditions, although deep lakes with high DOC concentrations may experience greater dissolved oxygen depletion and concomitant CO_2 enrichment, as for example, Golden Lake, Mt. Rainier National Park (Turney et al. 1986, Nelson and Baumgartner 1988).

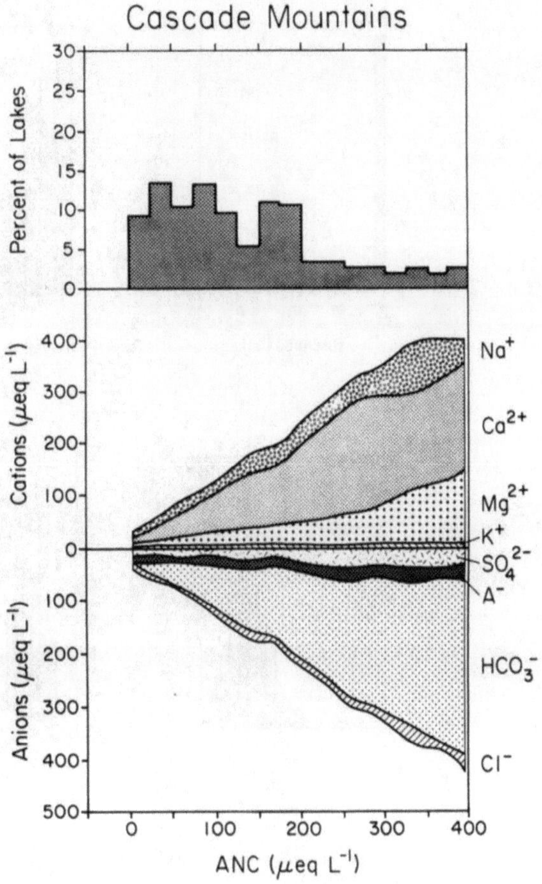

FIGURE 16.3. Concentrations of major cations and anions versus ANC for Cascade lakes with ANC ≤ 400 μeq L^{-1}. The percentages of lakes represented by ANC intervals are shown as a histogram at the top of the figure. Data plotted are population-weighted estimates for lakes in the WLS. The curves are spline fits to the median concentration for each ANC interval.

Water Chemistry Status

The spline diagram in Figure 16.3 shows the distribution of major (base) cations and major anions as a function of ANC (lakes with ANC ≤ 400 μeq L^{-1}) for WLS Cascade lake sample data. For the base cations (Ca^{2+}, Mg^{2+}, Na^+, and K^+), relative proportions remain approximately constant over the ANC distribution range, except for Na^+ at low ANC. Inputs of Na^+ from sea salt occur for all ANC classes, but are only seen out of proportion for low ANC lakes. The overall constancy of the lake cation distribution reflects the uniform volcanic geologic origin of the bedrock and soils in the Cascade Mountain Range. However, considerable subregional heterogeneity may exist in base cation distributions, as a result of localized differences in geology (see sections on Geology and Distribution of Base Cations).

For anions, there is evidence of a linear correspondence between the bicarbonate ion and ANC (Figure 16.3). Sulfate increases somewhat with lake ANC, but Cl^- and A^- (organic anions) remain relatively constant in concentration and therefore represent decreasing fractions of the anions with increasing lake ANC.

Figures 16.4a through 16.4i show the estimated population frequency distributions of selected physical and chemical parameters for the WLS Cascade lake population. The variations of eight lake water chemistry parameters (ANC, C_B [sum of Ca^{2+}, Mg^{2+}, Na^+, and K^+], pH, SO_4^{2-}, conductivity, Cl^-, DOC, and SiO_2) with three lake physical parameters (watershed area, watershed area to lake area (WSA:LA) ratio, and lake elevation) for WLS lakes are summarized in Table 16.5. Most Cascade lakes are small with surface areas < 5 ha (Figure 16.4a).

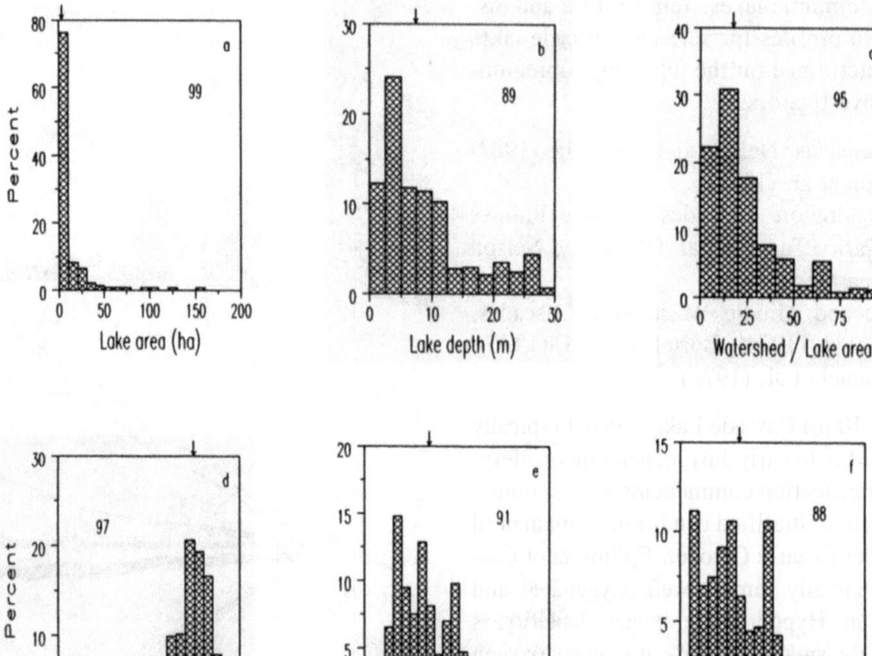

FIGURE 16.4. Frequency histograms of major characteristics of lakes in the entire WLS Cascade lake population: (a) lake surface area, (b) sampling depth, (c) watershed area:lake area, (d) air-equilibrated pH, (e) ANC, (f) sum of base cations, (g) sulfate, (h) DOC, and (i) silica. The arrow at the top of each box indicates the median target population, and the number in the upper right side of each box indicates the percentage of target lakes plotted.

The watershed area to lake area ratios (Figure 16.4c) are generally < 20. Most sampling site lake depths are < 10 m (Figure 16.4b).

Closed-system pH values (Figure 16.4d; Table 16.3) are approximately normally distributed around pH = 7.0, with observed minimum and maximum values of 5.8 and 9.3, respectively. Air-equilibrated pH values (Table 16.3) are about 0.2 pH units higher than those for closed-system pH, but with a narrower range (minimum and maximum

TABLE 16.5. WLS Cascade lake population data distribution variation with selected watershed parameters.

Parameter	Data distribution				
Watershed area (ha)	0–50	50–100	100–200	200–500	>500
n, number of lakes	41	33	29	21	17
N, estimated population	420	324	283	214	178
Population medians					
ANC, μeq L^{-1}	86.9	119.0	89.8	66.0	191.0
C_B, μeq L^{-1}	104.0	153.8	104.7	90.7	204.1
pH	7.1	7.2	7.0	7.0	7.5
SO_4^{2-}, μeq L^{-1}	9.1	7.9	16.3	10.7	9.1
Cond., μS cm^{-1}	10.9	16.0	11.3	9.5	21.8
Cl^-, μeq L^{-1}	7.4	8.7	5.6	5.8	6.6
DOC, μmol L^{-1}	98.8	119.1	89.0	96.1	117.7
SiO_2, μmol L^{-1}	35.0	51.7	46.7	56.7	68.3
Watershed area:lake area ratio	0–10	10–20	20–50	>50	
n, number of lakes	33	45	41	22	
N, estimated population	335	445	424	215	
Population medians					
ANC, μeq L^{-1}	94.0	116.2	102.5	105.1	
C_B, μeq L^{-1}	110.8	122.4	153.1	147.9	
pH	7.2	7.1	7.2	7.3	
SO_4^{2-}, μeq L^{-1}	8.6	14.5	7.4	8.0	
Cond., μS cm^{-1}	11.1	13.0	15.6	16.2	
Cl^-, μeq L^{-1}	9.3	6.7	6.4	5.5	
DOC, μmol L^{-1}	103.5	90.0	96.9	154.6	
SiO_2, μmol L^{-1}	35.0	35.0	58.3	60.0	
Elevation (m)	600–1,200	1,200–1,600	1,600–2,000	>2,000	
n, number of lakes	22	61	42	16	
N, estimated population	239	630	425	126	
Population medians					
ANC, μeq L^{-1}	99.5	126.6	86.0	60.1	
C_B, μeq L^{-1}	127.4	152.3	94.4	64.6	
pH	7.18	7.22	7.15	7.00	
SO_4^{2-}, μeq L^{-1}	14.1	10.9	4.8	3.3	
Cond., μS cm^{-1}	13.7	15.6	9.5	7.0	
Cl^-, μeq L^{-1}	15.2	9.1	4.4	2.8	
DOC, μmol L^{-1}	120.3	110.7	87.5	65.0	
SiO_2, μmol L^{-1}	60.0	48.3	48.3	23.3	

values of 5.8 and 8.5, respectively). Differences in closed-system and air-equilibrated pH values result from the highly variable dissolved CO_2 content of lake waters and indicate that in situ (closed-system pH) samples are generally equilibrated with CO_2(g) concentrations higher than the 300 ppm CO_2 standard for air-equilibrated values.

Approximately 78% of Cascade lakes have air-equilibrated pH values < 7.5. Subregional pH differences are notable (Table 16.4), with greater proportions of lower pH lakes in the southern Cascades (California and Oregon subregions) and an increasing pH trend in subregions to the north along the Cascade transect (Figure 16.5). Very few lakes had pH values < 6.0. Lake pH values were somewhat higher in lakes with watershed areas > 500 ha, varied only slightly with WSA:LA ratio, and were somewhat lower in lakes with elevations > 2,000 m.

Approximately 82% of Cascade lakes are estimated to have ANC values ≤ 200 μeq L^{-1} and 93% ≤ 400 μeq L^{-1} (Figure 16.4e; Table 16.3). In

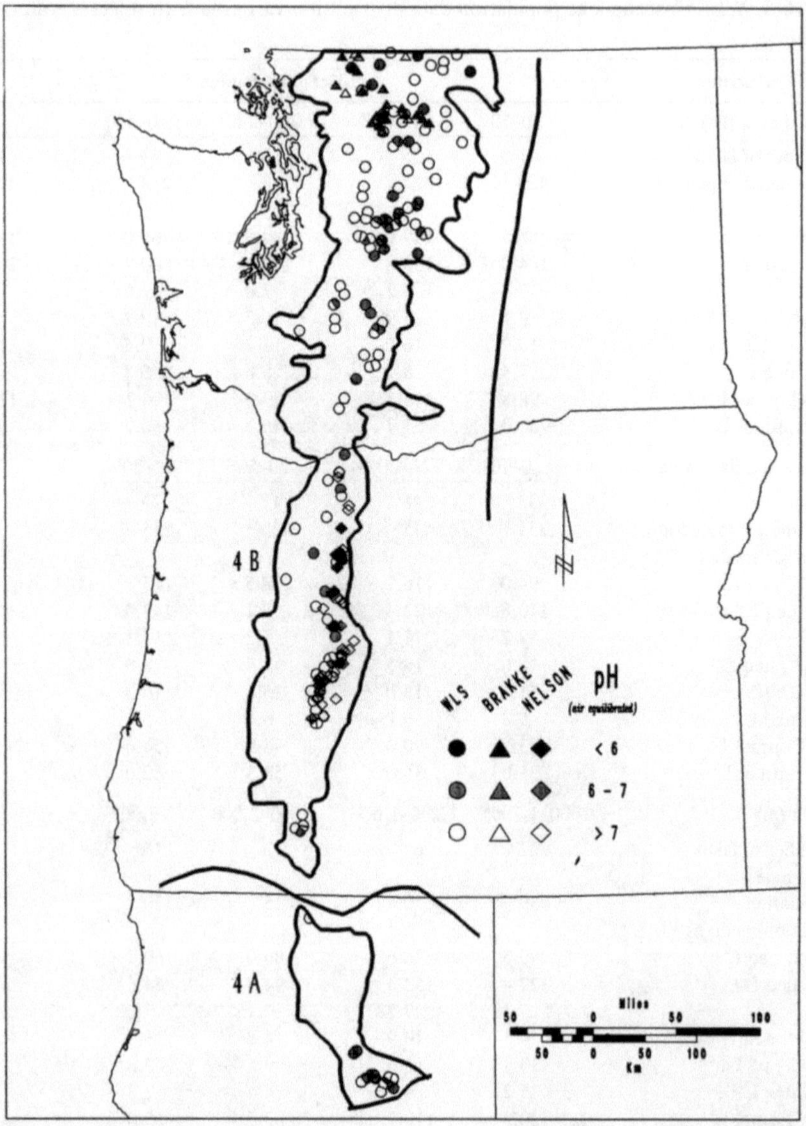

FIGURE 16.5. Distributions of Cascade lakes for three data sets in three pH categories.

the United States, only the Sierra Nevada region has a greater percentage of low ANC lakes (Landers et al. 1987; Chapter 15, Melack and Stoddard, this volume). None of the lakes sampled in the WLS had negative ANC values (minimum ANC = 3.9 μeq L^{-1}).

Lake ANC distributions varied by subregion (Table 16.4; Figure 16.6), with greater proportions of lower ANC lakes in the southern Cascades (California and Oregon subregions) and greater proportions of higher ANC lakes in the northern Cascades (especially North Washington Cascades subregion). Lake ANC values were higher for watershed areas > 500 ha, somewhat lower for WSA:LA ratios < 10, and lower for lake elevations > 2,000 m (Table 16.5).

The frequency distribution for sum of base cations (Figure 16.4f) is very similar to that of ANC, but with base cation concentrations typically about 20 μeq L^{-1} greater (Table 16.3). No Cascade lakes sampled had ANC:C_B ratios < 0.25 (Figure 16.7). Virtually no Oregon or California Cascade lakes

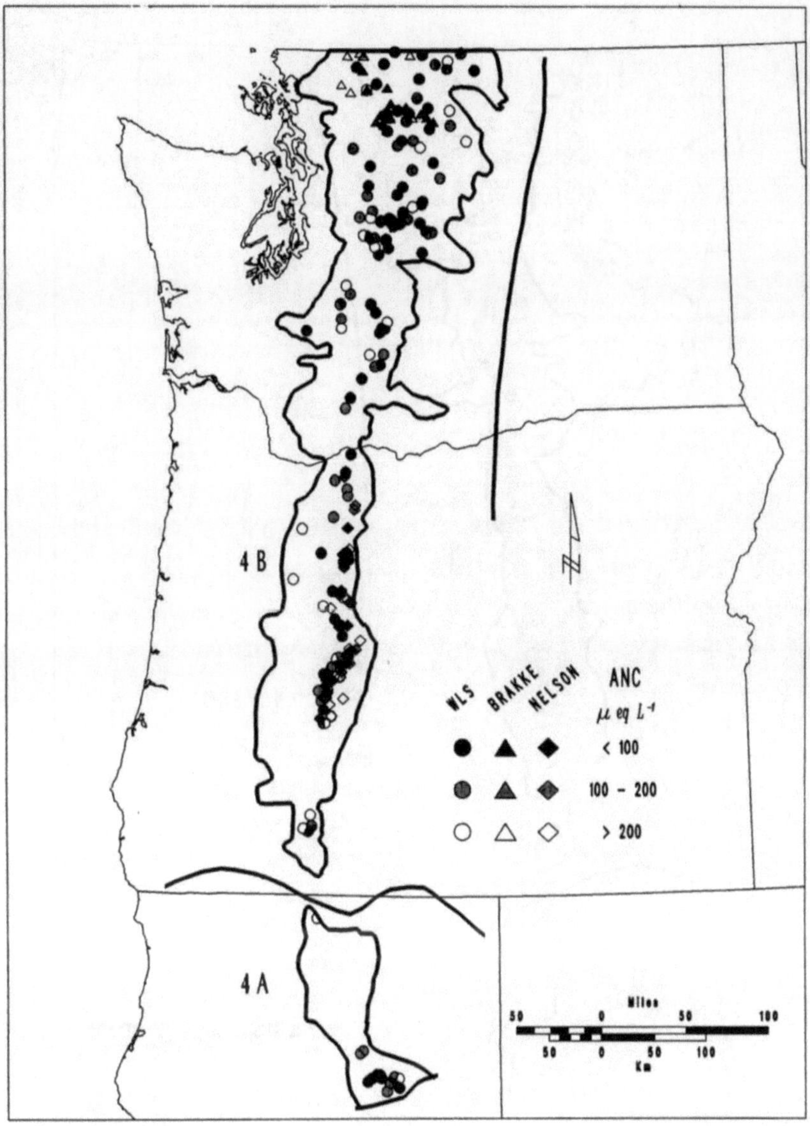

FIGURE 16.6. Distributions of Cascade lakes for three data sets in three ANC categories.

had ANC:C_B ratios < 0.75, but about 25% of Washington Cascade lakes did, which is attributable to their greater anion fraction of SO_4^{2-} (Figure 16.8; Table 16.4). Median base cation sums ranged from a low of 76 µeq L^{-1} in the California Cascades to 250 µeq L^{-1} in the North Washington Cascades. Lake C_B values were higher for watershed areas > 500 ha, somewhat lower for WSA:LA ratios < 10, and lower for high-elevation lakes (1,600 to 2,000 m and > 2,000 m) (Table 16.5).

Calcium is the dominant cation for the Cascade Region, accounting for an estimated median of 58.6% of the base cation equivalent fraction for the WLS lake population. Median fractions for other base cations are Na^+, 20.8%, Mg^{2+}, 17.2%, and K^+, 3.5%. When corrected for sea salts, the base cation sum decreased by about 13% (M = 109.8 µeq L^{-1}), and the relative fractions of Ca^{2+} and K^+ increased, whereas those of Na^+ and Mg^{2+} decreased, as follows: Ca^{2+}, 64.2%, Na^+, 15.4%, Mg^{2+}, 16.7%, and K^+, 3.7%. Subregional variations in base cation

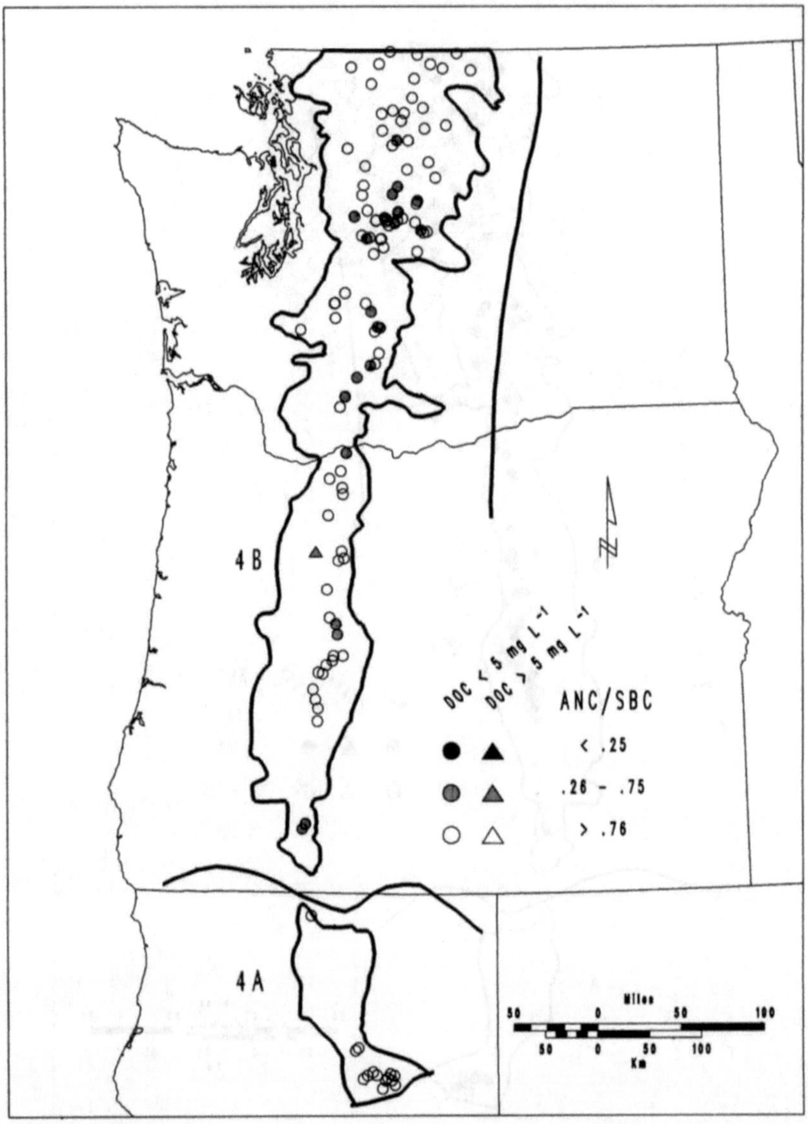

FIGURE 16.7. Distributions of WLS Cascade lakes in three categories of ANC:SBC. Low DOC lakes have DOC values $\leq$ 417 μmol L^{-1} (5 mg L^{-1}).

equivalent fractions are substantial. Calcium was the dominant cation in all subregions, except California, where Mg^{2+} was the dominant cation, with a base cation fraction of 41.5%, compared to 31.4% for Ca^{2+}. Sodium dominated over Mg^{2+} as the second highest base cation in the Oregon and South Washington Cascades, but was less than or equal to Mg^{2+} when corrected for sea salts. Calcium accounted for 80.8% of the base cations in the North Washington Cascades. It is apparent that despite the common volcanic origin of the Cascades' prominent geological formations, considerable heterogeneity in bedrock geology at the watershed level exists along the transect from Northern California to the Canadian border.

Sulfate concentrations are generally very low in Cascade lakes (Table 16.3), and the median SO_4^{2-} fraction comprises about 9% of the anion sum. About 50% of Cascade lakes have SO_4^{2-} concentrations < 10 μeq L^{-1} and 94% have SO_4^{2-} < 50

FIGURE 16.8. Variation of lake SO_4^{2-} concentrations in Cascade subregions.

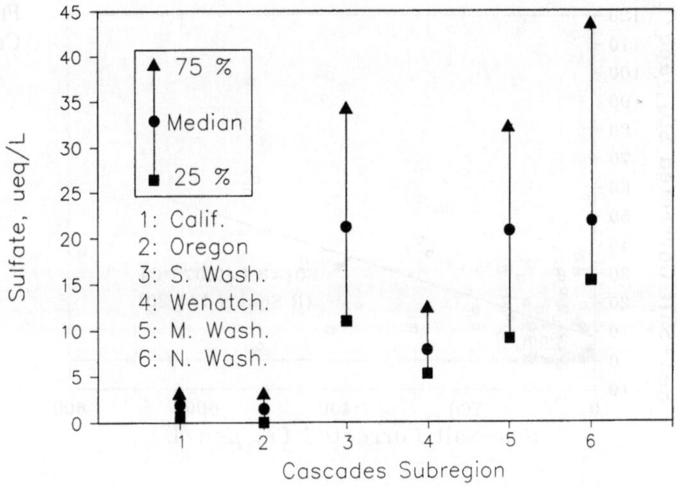

μeq L[-1] (Figure 16.4g). The distribution of lake SO_4^{2-} concentrations among subregions is markedly different (Figure 16.8; Table 16.4). Southern Cascade medians are extremely low (California Cascades, M = 1.9 μeq L^{-1}; Oregon Cascades, M = 1.5 μeq L^{-1}), and Washington Cascade subregional medians range from 8.0 μeq L^{-1} (Wenatchee) to 21.6 μeq L^{-1} (North Washington). Lake SO_4^{2-} concentrations appear to be higher for watershed areas between 100 and 200 ha and for WSA:LA ratios between 10 and 20, and to decrease with increasing lake elevation (Table 16.5).

Sulfate attributable to sea salts equals about 0.9 μeq L^{-1}, or about 9% to 10% of lake SO_4^{2-} concentrations, with substantial subregional variation. Lakes with higher SO_4^{2-} (sea salt corrected) concentrations tend to be those with higher pH values and higher Ca^{2+} concentrations. The weak (r^2 = 0.20) but significant ($p < 0.001$) correlation of sea salt corrected lake SO_4^{2-} with Ca^{2+} (Figure 16.9) suggests that watershed sources of SO_4^{2-} from mineral weathering are important for some Cascade lakes.

Nitrate concentrations in the WLS Cascade lake population are typically very low (M = 0.11 μeq L^{-1}; Q1 = 0.05 μeq L^{-1}; Q3 = 0.36 μeq L^{-1}), and thus NO_3^- constitutes a negligible fraction of the equivalent anion sum (about 0.1%). The highest subregional median concentration was 0.23 μeq L^{-1} in the Wenatchee Cascades, showing the apparent influence of the urbanized Puget Sound area upwind.

The estimated median Cl^- concentration for the WLS lake population is 7.2 μeq L^{-1} (Q1 = 3.8 μeq L^{-1}; Q3 = 13.0 μeq L^{-1}). Subregional population estimated median Cl^- concentrations were highest in the South Washington Cascades (M = 14.5 μeq L^{-1}) and decreased both northward and southward. The median Cl^- fraction was 8% of the anion sum, ranging from 13% for the Oregon Cascades to 1% for North Washington Cascade subregions. Sea salt effects are thus important for much of the Cascade Mountain Range. Chloride showed no significant correlation with sum of base cations (r^2 = 0.01; $p = 0.17$), indicating no general mineralogical source. There was a tendency for lake Cl^- concentrations to be higher at lower lake elevations (Table 16.5). This corresponds to the significant negative correlation of precipitation Cl^- with elevation (see in section entitled Processes Influencing Surface Water Chemistry). For Pacific Northwest lakes (subregion 4B) of the WLS, lake Cl^- concentrations decrease with distance from the Pacific coast up to about 50 km inland, beyond which they remain relatively constant up to 100 km inland (Landers et al. 1987). The diphasic behavior of Cl^- in precipitation, first decreasing with distance from the coast and then increasing with elevation in the Cascade Mountains farther inland, may be explained by an initial rainout of lower altitude, larger diameter aerosols, followed by the scavenging of higher altitude, smaller diameter aerosols by orographic effects.

Dissolved organic carbon (DOC) concentrations are low in Cascade lakes (Figure 16.4h; Table

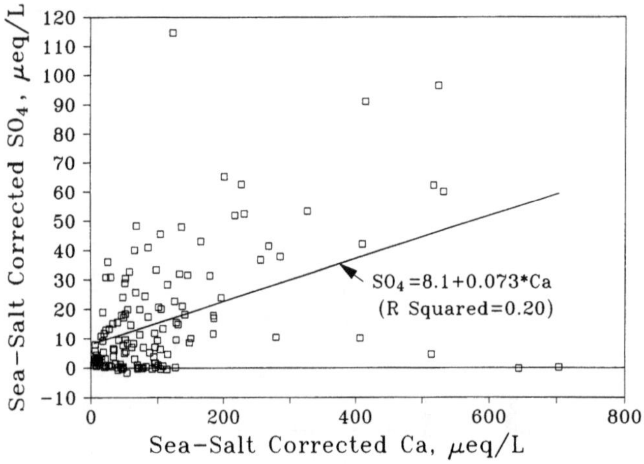

FIGURE 16.9. Relationship between SO_4^* and Ca* for WLS Cascade lakes.

16.3). Estimated lake population median concentrations decrease in Cascade subregions from south to north, with highest values in the California Cascades (M = 170 µmol L^{-1}) and lowest values in the Middle Washington Cascades (M = 30.2 µmol L^{-1}) and North Washington Cascades (M = 38.3 µmol L^{-1}). Lake DOC concentrations do not vary greatly with watershed area, are higher for WSA:LA ratios > 50, and decrease with increasing lake elevation (Table 16.5). Overall, 75% of the estimated Cascade lakes population had DOC < 2 mg L^{-1} compared to 83% for the Sierra Nevada, the lowest DOC in WLS subregions.

There is a weak but significant ($p < 0.001$) correlation of anion deficit with DOC. The median anion deficit is 18.0 µeq L^{-1}, yielding a relationship of 0.18 µeq anion deficit/µmol organic carbon (15.0 µeq/mg C). This value is within the range of total organic acid contents of DOC summarized by Cook et al. (1987) for surface waters, groundwaters, and soils (0.08 to 0.21 µeq/µmol C). The anion deficit from the inorganic ion balance is attributable to both ionized organic acids (see section entitled Dissolved Organic Acid Anions for further discussion) and cumulative analytical errors (Landers et al. 1987).

Low dissolved SiO_2 concentrations (M = 49.0 µmol L^{-1}; Q1 = 27.2 µmol L^{-1}; Q3 = 77.2 µmol L^{-1}) are consistent with the low ionic solute concentrations of Cascade lakes (Figure 16.4i). Weak though significant ($p < 0.001$) positive correlations exist between dissolved SiO_2 and pH, ANC, and C_B. This is explained by the weathering of plagioclase feldspar minerals that account primarily for lake solute composition (Reynolds and Johnson 1972, Nelson 1985, Drever and Hurcomb 1986; see section entitled Processes Influencing Surface Water Chemistry). Lake SiO_2 concentrations generally increase with increasing watershed area, and with increasing WSA:LA ratio, and decrease with increasing lake elevation (Table 16.5).

Dissolved Al concentrations are generally very low for Cascade lakes. For the WLS lakes data, the estimated population median total Al concentration was 1.05 µmol L^{-1} (Q1 = 0.70 µmol L^{-1}; Q3 = 1.67 µmol L^{-1}) and the median extractable Al was 0.23 µmol L^{-1} (Q1 = 0.11 µmol L^{-1}; Q3 = 0.35 µmol L^{-1}). Subregional variations were not great, ranging from population medians of 0.67 µmol L^{-1} for the North Washington Cascades to 1.48 µmol L^{-1} for the California Cascades. There were no significant correlations of total Al with pH or with DOC. All lakes with total Al > 2.0 µmol L^{-1} had ANC values ≤ 200 µeq L^{-1}. The correlation between total Al and ANC for Cascade lakes was not significant.

Seasonal Variations in Lake Chemical Composition

Seasonal variations in the chemical composition of Cascade lakes have generally been confined to a relatively short ice-free period of about four months from July to October. Limited data also exist for the other seasons, and one North Cascade lake was sampled over an annual cycle (Loranger

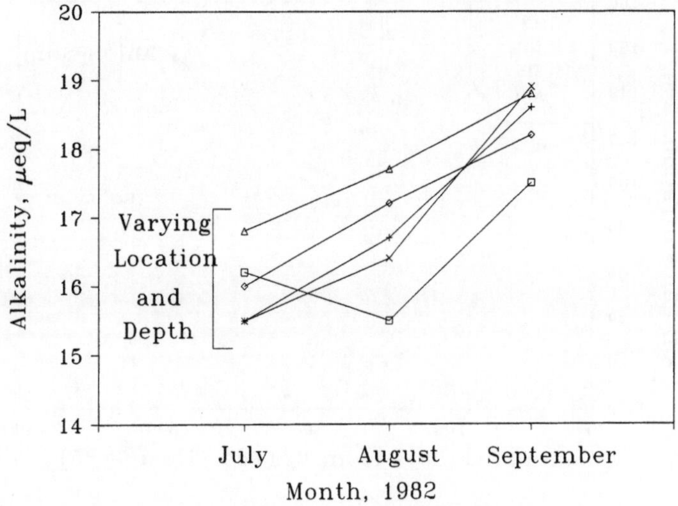

FIGURE 16.10. Open water seasonal alkalinity variation in Waldo Lake, Oregon.

1986). The WLS (Landers et al. 1987, Eilers et al. 1987a) does not include seasonal data.

Nelson and Delwiche (1983) sampled eight Oregon Cascade lakes three times each at approximately monthly intervals from July through October, 1982. The July sampling period occurred near the end of snowmelt in the watersheds of these lakes. Three low ANC lakes with small surface areas, shallow depths, and estimated short hydrologic residence times exhibited temporal fluctuations in ANC up to 60% of the mean. For the other lakes, temporal ANC fluctuations were < 20% of the mean. Open-water seasonal variations in alkalinity at different sampling locations and depths in Waldo Lake, which is deep (max. depth 128 m) and has a long hydrologic residence time (32 yr; Johnson et al. 1985), are representative of lakes with larger surface area and greater depth (Figure 16.10). In general, alkalinity was lowest in the first sampling period in July, which was attributed to dilution from snowmelt. Alkalinity increased in the second sampling period in August and in the third sampling period in September and October as well, for most sampling locations. In July, surface samples were generally higher in alkalinity than depth samples, but for August and September–October samples, the opposite was generally observed.

Welch et al. (1986) observed the seasonal variation in alkalinity and acid anions for seven lakes in the Alpine Lakes Wilderness of the Wenatchee Cascades subregion. Alkalinity increased over the July through September sampling period for three of four lakes with short hydrologic residence times (fast-flush), but remained relatively constant for three lakes with long hydrologic residence times (slow-flush). The alkalinity change in lakes with short residence times was attributed to initial dilution by snowmelt water followed by slow replacement over the summer season with higher alkalinity groundwater. The variability in alkalinity ranged from < 10% to > 300% increase over the first (July) sample. Sulfate remained relatively constant and NO_3^- decreased during the sampling period.

Loranger et al. (1986) sampled five lakes in the North Washington Cascades over the open water period (August–October) of 1984. Initially, surface water alkalinity was depressed and acid anion concentrations were increased from snowmelt. As summer and fall progressed, alkalinity increased and acid anions decreased. Surface sample alkalinity in shallow lakes experienced the greatest variations, ranging up to about 75% increase from the initial (early summer) value after snowmelt.

Loranger (1986) followed the annual cycle of major ions in one North Cascades lake with a low hydrologic residence time, from summer 1984 to summer 1985 (Figures 16.11 and 16.12). Total anion (Figure 16.11) and total cation (Figure 16.12) sums varied by nearly 50% of the mean during the annual cycle, but anion and cation fractions remained approximately constant, showing that dilution and concentration were causing the seasonal pattern. Ion concentrations were most dilute in the early summer (July) after snowmelt,

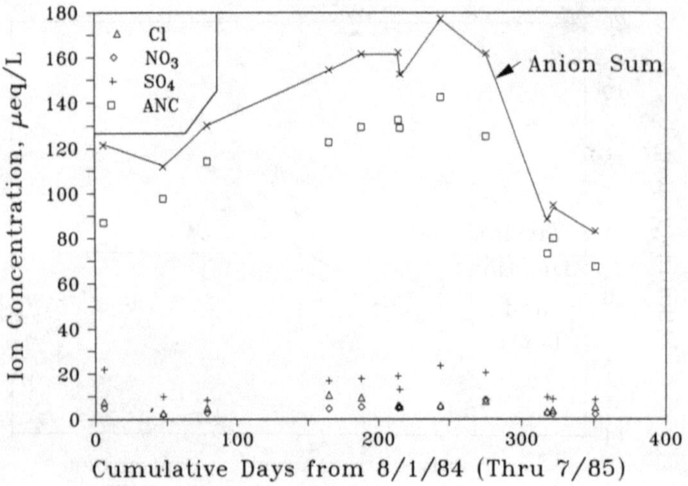

FIGURE 16.11. Annual cycle of major anions, Bagley Lake, Washington.

and then increased for about 9 to 10 months throughout the fall, winter, and early spring, reaching a maximum in April. For two to three months during the spring and early summer snowmelt period, ion concentrations decreased due to snowmelt dilution.

Nelson and Baumgartner (1988) reported variations of major ions and Al species for selected lakes in Mt. Rainier National Park of the South Washington Cascades during the period from July to September 1985. Alkalinity increased during the sampling period for most lakes, but for some lakes sampled after fall turnover, alkalinity decreased slightly between August and late September. Ratios of acid anions to alkalinity did not change appreciably. Average total Al concentration, although low, increased by more than 100% during the sampling period (Figure 16.13). Exchangeable (inorganic) Al species decreased from about 85% to about 65% of total Al during this period, whereas organically complexed Al species increased proportionately. Organically complexed Al was positively correlated ($r^2 = 0.80$) with DOC, and multiple regression analysis indicated that more than 65% of the variation in total Al was accounted for by DOC and pH. Dissolved Al was also predicted to be in equilibrium with an aluminum trihydroxide phase similar to natural gibbsite, as Driscoll et al. (1984) also found for lakes receiving substantially greater acidic deposition.

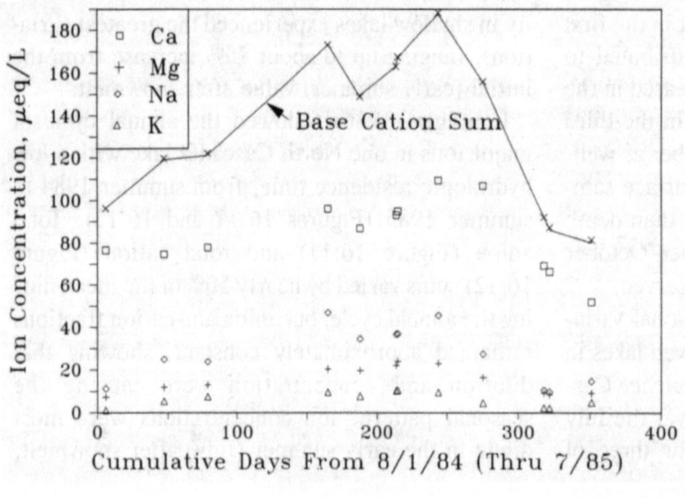

FIGURE 16.12. Annual cycle of base cations, Bagley Lake, Washington.

FIGURE 16.13. Temporal variation of lake dissolved Al species. Mean values for 16 lakes in Mount Rainier National Park.

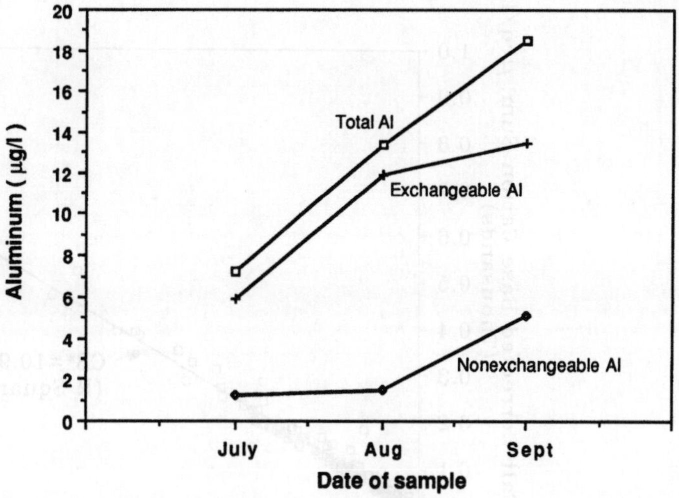

Associations Among Variables

Analysis of the relationships among ions provides insight into processes controlling lake chemical composition and facilitates comparisons among data sets and regions. Table 16.6 presents linear regressions of sea salt corrected base cations (Ca*, Mg*, Na*, K*) with ANC for WLS Cascade lakes. Calcium and Mg^{2+} are the dominant cations but Na^+ is also important. Only K^+ is weakly correlated with ANC, which may be due in part to its low concentration in most lakes (median = 4.1 µeq L^{-1}) and higher relative analytical error (Landers et al. 1987). All base cation correlations with ANC were highly significant ($p < 0.001$).

The relationship between sum of base cations (sea salt corrected) and ANC (Figure 16.14) is very near equivalency for WLS Cascade lakes. The high correlation of sum of base cations with ANC ($r^2 = 0.99$), the slope near unity (1.07), and the significance level ($p < 0.001$) are strong evidence that the same geochemical processes control both ANC and the base cation content of Cascade lake waters. The correlation between calculated ANC (difference between sum of base cations and sum of acid anions, excluding organic anions) and measured ANC (Gran titration) is also highly significant ($r^2 = 1.00$; $p < 0.001$), with a slope near unity (1.04; Figure 16.15). It can be inferred from these two strong relationships with ANC that the influence on ANC of nonprotonated (strong) organic acids and of watershed (weathering) sources of SO_4^{2-} are of only minor importance for Cascade lakes in aggregate. In low ANC lakes, the influences of organic acids and weathering sources of SO_4^{2-} are proportionally greater (see section entitled Water Chemistry Status).

The relationship between the ratio $SO_4^{2-}:C_B$ and ANC is used to estimate the relative contribution of SO_4^{2-} to the charge balance for changing lake ANC (Figure 16.16). For Cascade lakes, higher values of this ratio are found only at low ANC values. It cannot be inferred that lakes with high ANC:C_B ratios and low ANC have experienced acidification without corroborating evidence, since SO_4^{2-} correlates positively with both C_B and ANC (see Empirical Models). Clearly for some low ANC Cascade lakes, SO_4^{2-} is the primary anion.

TABLE 16.6. Correlations of base cations (sea salt corrected) with ANC.

Base cation[a] (C_B)	Intercept (a)	Slope (b)	Correlation coefficient r^2
Ca*	27.6	0.54	0.76
Mg*	−22.3	0.37	0.78
Na*	1.6	0.15	0.68
K*	4.4	0.009	0.11
C_B	11.8‘	1.07	0.88

[a] Equation: $C_B = a + b * \text{ANC}$.

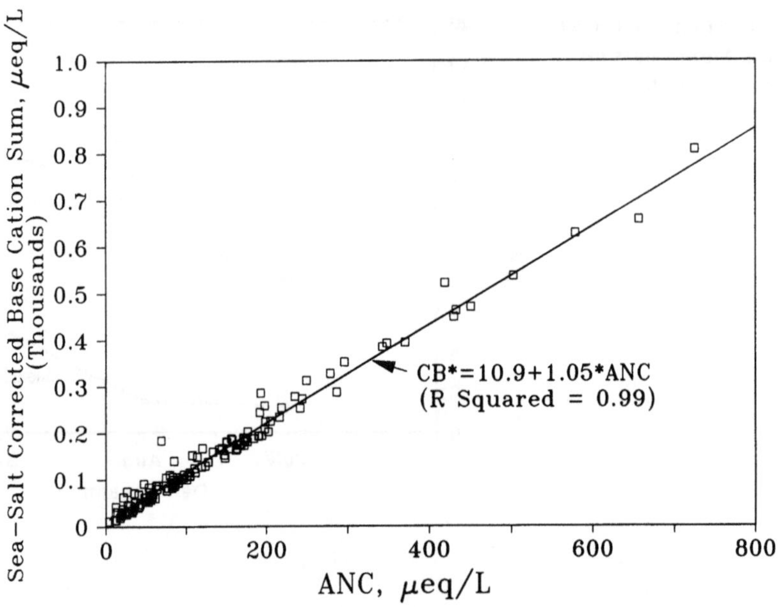

FIGURE 16.14. Relationship between C_B* and ANC for WLS Cascade lakes.

Processes Influencing Surface Water Chemistry

Precipitation and Deposition

Chemical components in precipitation and dry deposition, corrected for evapotranspiration, form the baseline composition of lakes from which watershed and in-lake processes affecting lake chemical composition can be assessed. This baseline lake composition formed from precipitation alone has been termed a model lake by Cook et al. (1987).

The amount of evapotranspiration in a lake watershed can be estimated by assuming that the Cl^- is conserved in watersheds (Cook et al. 1987), although this assumption is subject to several limitations (Lee and Schnoor 1988). The ratio of the Cl^- concentration in lake water to that in precipitation represents the increase in concentration that is due to evapotranspiration. The Cl^- ratio is thus a concentration factor that can be applied to all

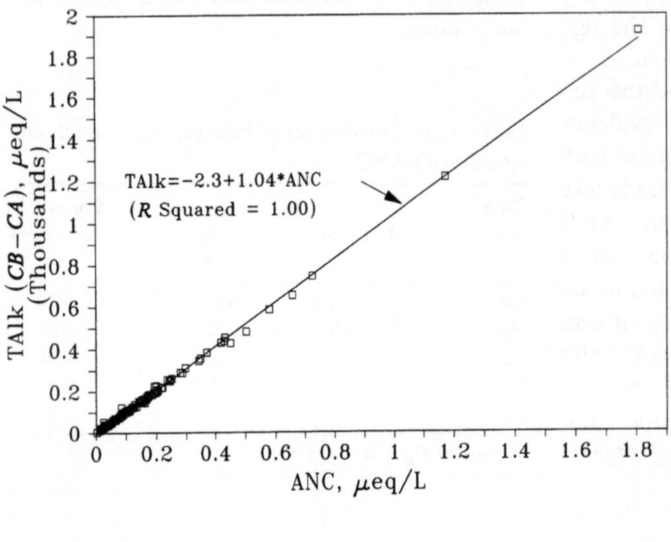

FIGURE 16.15. Relationship between calculated total alkalinity ($C_B - C_A$) and ANC_{Gran} for WLS Cascade lakes. All lakes have DOC ≤ 417 μmol L^{-1} (5 mg L^{-1}).

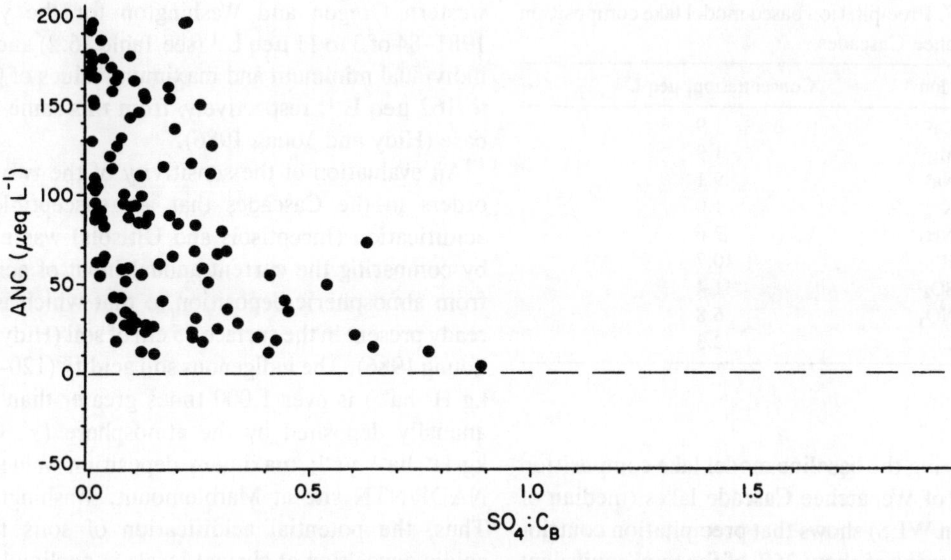

FIGURE 16.16. Relationship between ANC and SO_4^*:C_B for Cascade lakes sampled in the WLS.

solutes in precipitation to calculate model lake composition. It is preferable, however, to use volume-weighted mean annual precipitation data to perform this calculation (Chapter 17, Baker et al., this volume).

The precipitation data of Logan et al. (1982), Duncan et al. (1986), and Duncan and McGinty (1988) for the Central Washington Cascades (Wenatchee Cascades subregion) appear to be the most complete record of precipitation chemistry for any Cascade subregion. The data are volume-weighted and represent precipitation mainly as snow during the late fall to early spring period of about six months. Synoptic lake chemistry data to enable direct calculation of Cl^- concentration factors were not obtained. A linear regression fit calculated for the Cl^- data of Logan et al. (1982) and Duncan et al. (1986) along a west-east transect across the Cascade crest shows that Cl^- concentrations in precipitation depend on elevation, according to the relationship: Cl^- (μeq L^{-1}) = 17.1 − 0.0073 * elev (m) (r^2 = 0.78). A similar dependence of precipitation Cl^- on elevation is obtained from the data of O'Loughlin et al. (1986) along a west-east transect directly east of Seattle, yielding the relationship: Cl^- (μeq L^{-1}) = 16.8 − 0.0080 * elev (m) (r^2 = 0.74). Chloride concentrations in Cascade lakes also depend on elevation (see section entitled Water Chemistry Status).

The regression relationships for Cl^- dependence on elevation were used to estimate precipitation Cl^- concentrations for Wenatchee Cascade lakes based on each lake's elevation, from which Cl^- concentration factors were determined, as described in the previous paragraph. Median values of 1.2 to 1.4 were determined for the concentration factor and a value of 1.3 was selected. Synoptic snow core and lake Cl^- data of Loranger (1986) were used to determine concentration factors of 1.1 to 1.4 for two North Washington Cascades lakes, although considerable uncertainty existed in the data. Dethier (1979) assumed a concentration factor of 1.15 for stream chemistry runoff in the North Washington Cascades. Overall, evapotranspiration factors for central and northern Washington Cascade lakes are low, indicating relatively low evapoconcentration compared to higher values in southern Cascade lakes (see sections on Climate and Hydrologyy) and other case study regions (Chapter 17, Baker et al., this volume).

The Cl^- based concentration factors presented in this section are consistent with evapotranspiration data summarized in the sections on Climate and Hydrology. Using the mean concentration factor of 1.3 for the Wenatchee Cascades, a model (precipitation based) lake composition was calculated from the precipitation data of Duncan and McGinty (1988) (Table 16.7). The cation:anion ratio calculated for the model lake is 0.89.

TABLE 16.7. Precipitation based model lake composition for Wenatchee Cascades.

Ion	Concentration, μeq L^{-1}
Ca^{2+}	1.9
Mg^{2+}	1.9
Na^{+}	9.1
K^{+}	1.0
NH_4^{+}	2.6
H^{+}	10.7
SO_4^{2-}	11.4
NO_3^{-}	6.8
Cl^{-}	12.3

Comparing the baseline model lake composition with that of Wenatchee Cascade lakes (median of 31 lakes in WLS) shows that precipitation contributes an average of about 25% of the total equivalent ionic chemical composition. The remainder is produced by watershed and in-lake processes, as shown in the following discussion. Precipitation is the major source of acid anions, although mineral weathering may also be an important source of SO_4^{2-} (see Water Chemistry Status). NADP/NTN precipitation data (Table 16.2) provides a range of mean annual SO_4^{2-} concentrations from about 4 to 12.5 μeq L^{-1}. Hidy and Young (1986) estimate the background SO_4^{2-} at 4–8 μeq L^{-1} in precipitation in the western United States, thus implying that anthropogenic SO_4^{2-} is in the range of 4 to 8 μeq L^{-1}. In the Cascades, a generally increasing gradient in lake SO_4^{2-} has been shown from south to north (see Water Chemistry Status). The same gradient may hold for precipitation SO_4^{2-}, both background and anthropogenic, as inferred by comparison of subregional lake SO_4^{2-} distributions (Table 16.4) to the estimated precipitation background concentration. Based on the median sea salt contribution to lake SO_4^{2-} of 0.9 μeq L^{-1} (see Water Chemistry Status) and assuming an evapotranspiration factor of 1.3 and no watershed uptake of SO_4^{2-} (see section entitled Acid Neutralizing Capacity), the median sea salt contribution to precipitation SO_4^{2-} is about 0.7 μeq L^{-1}.

The H^{+} contribution from precipitation (calculated from pH), based on the mainly wintertime data of Duncan and McGinty (1988), is variable and falls in the approximate range of 5 to 20 μeq L^{-1} for the Cascades. This contrasts with the range of means from NADP/NTN stations in western Oregon and Washington for the years 1981–84 of 3 to 11 μeq L^{-1} (see Table 16.2) and the individual minimum and maximum values of 0.03 to 162 μeq L^{-1}, respectively, from this same data base (Hidy and Young 1986).

An evaluation of the sensitivity of the two soil orders in the Cascades that are susceptible to acidification (Inceptisols and Ultisols) was made by comparing the current annual input of acidity from atmospheric deposition to that which is already present in the surface 15 cm of soil (Hidy and Young 1986). The indigenous soil acidity (120–370 kg H^{+} ha^{-1}) is over 1,000 times greater than that annually deposited by the atmosphere (< 0.20 kg H^{+} ha^{-1} yr^{-1}; maximum deposition value for NADP/NTN site at Marblemount, Washington). Thus, the potential acidification of soils from acidic deposition at current levels is negligible.

Acid Neutralizing Capacity (ANC)

The ANC of natural lakes represents the net modification of meteoric and telluric water by watershed and in-lake processes and by evapotranspiration. Acid neutralizing capacity is acquired primarily through the weathering of basic minerals or from cation exchange reactions in watershed soils and exposed rock surfaces. It is also influenced by watershed and in-lake changes of biogeochemically important redox components (SO_4^{2-}, NO_3^{-}, NH_4^{+}, DOC, Fe^{3+}, Fe^{2+}). Redox processes that increase ANC include in-lake reduction reactions and biological assimilation of oxidized chemical forms (e.g., SO_4^{2-}, NO_3^{-}), whereas those that decrease ANC include in-lake oxidation reactions and biological assimilation of reduced chemical forms (e.g., NH_4^{+}, Fe^{2+}). ANC is further decreased by organic acids leached from soils. Fluxes of ANC components into and from the sediments affect the lakewater chemistry of some lakes; the magnitude of the effect depends on their hydrologic residence times (Cook et al. 1987, Schafran and Driscoll 1987). A more detailed discussion of these processes is presented by Munson and Gherini (Chapter 1, this volume).

The relative importance of the above processes in determining lake ANC can be assessed by examining the change of lake cation and anion concentrations relative to those in the precipitation-derived, or model, lake (see section entitled Enrichment

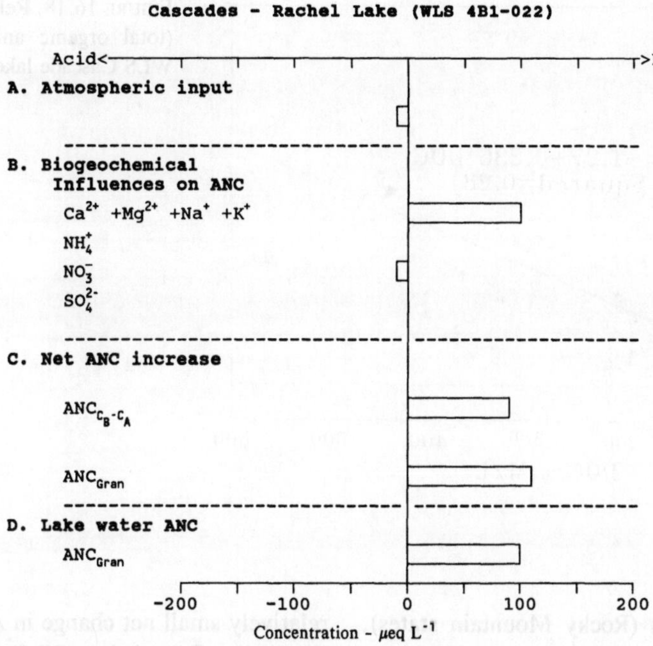

FIGURE 16.17. Modification of atmospheric deposition by watershed and lake biogeochemical processes. A. Precipitation [H^+] adjusted for estimated dry deposition and evapotranspiration (ET); equivalent to expected lakewater $ANC_{C_B-C_A}$ in the absence of any process other than ET. B. Reactions influencing ANC; bars to the right indicate ANC production, those to the left show ANC consumption. C. Net ANC production; $ANC_{C_B-C_A}$ is the sum of ANC production terms for measured anions and cations including those shown in B; ANC_{Gran} is determined by difference between input ANC (A.) and lakewater ANC (D.). D. Lakewater ANC represents the net reaction of atmospheric inputs with the lake/watershed system. For a further explanation of calculation procedures and figure interpretation, see Appendix A.

Factor Analysis in Chapter 2, Munson and Gherini, this volume). The model lake incorporates concentrated acidic precipitation and typically has negative ANC due to the presence of anthropogenically derived strong acid anions (SO_4^{2-} and NO_3^-). The negative ANC is estimated by the hydrogen ion concentration as determined by pH. Weathering and ion exchange processes increase base cation concentrations and lake ANC. Ammonium ion is removed by nitrification or assimilation processes that decrease lake ANC. Conversely, NO_3^- is removed by denitrification or assimilation processes that increase lake ANC. Sulfate reduction to sulfide (with subsequent precipitation of metal sulfides and removal from solution) decreases lake SO_4^{2-} concentration and increases lake ANC, whereas weathering reactions that increase lake SO_4^{2-} concentration decrease (sulfide-bearing minerals) or have no net effect (SO_4^{2-}-bearing minerals, such as gypsum) on lake ANC.

Processes influencing ANC are illustrated in an ion enrichment diagram (Figure 16.17) for Rachel Lake, which is approximately median in ANC for Wenatchee Cascade lakes and typical of central Washington Cascade lakes. Effects of organic acids are not included in the diagram but are calculated separately and discussed in the section on Dissolved Organic Acid Anions.

Precipitation contains about 11 μeq L^{-1} of H^+ ions. This is about 60% of the acid anion sum (corrected for neutral sea salts), and is also approximately equal to base cations (sea salt corrected and excluding NH_4^+). Thus, about 60% of SO_4^{2-} and NO_3^- in precipitation occurs as strong acid equivalents and 40% as neutral salts. Neutral salts can be formed either from strong acids that are subsequently neutralized by base minerals in terrestrial dusts, or directly from dissolution of minerals in terrestrial dusts such as gypsum ($CaSO_4$). Creager (1985) has shown that for precipitation in the

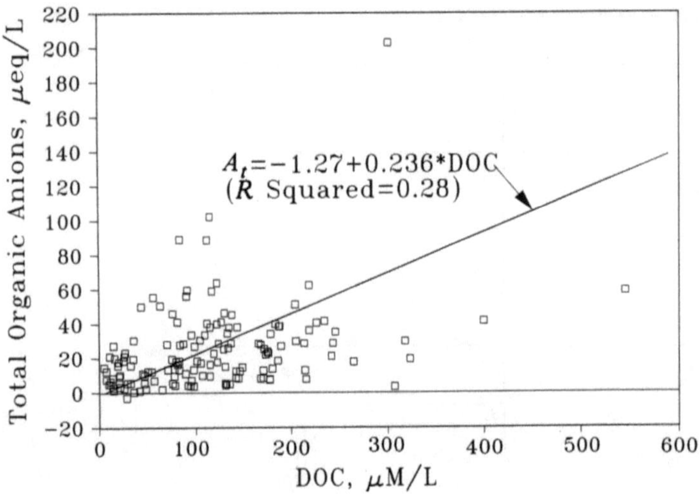

FIGURE 16.18. Relationship between A_t (total organic anions) and DOC for WLS Cascade lakes.

western United States (Rocky Mountain states), SO_4^{2-} is significantly correlated with Ca^{2+} and not with H^+, indicating the importance of terrestrial dusts as a source of SO_4^{2-}.

It is clear that processes associated with changes in base cations are the principal source of ANC for Cascade lakes. For Rachel Lake (Figure 16.17), the increase in sum of base cations over that in precipitation is about 98 μeq L^{-1}, which is about 94% of the lake's ANC production. The sum of base cations, and all individual base cations except K^+, were previously shown to be highly correlated with ANC. Soils in the Cascades are generally low in pH and depleted in base cations, with resultant low cation exchange capacity (Hidy and Young 1986). It is unlikely that cation exchange is a major contributor to the base cation increase. Weathering of basic minerals in soils and exposed rock surfaces is thus the likely principal source of base cations and ANC to Cascade lakes.

Biological assimilation, and in-lake oxidation and reduction processes, do not appear to be generally important for Cascade lakes, largely because of the low concentrations of nitrogen species and SO_4^{2-} in precipitation. The model lake SO_4^{2-} concentration (11.4 μeq L^{-1}) is about 10% of the ANC for Rachel Lake, which places an upper bound on ANC that could be generated from SO_4^{2-} reduction. Furthermore, the SO_4^{2-} concentration for Rachel Lake is 12.9 μeq L^{-1}, showing that SO_4^{2-} is approximately conserved in this watershed with relatively small net change in ANC. The range of Wenatchee Cascade lake SO_4^{2-} concentrations (1.3 to 43.7 μeq L^{-1}) is much greater than the amount that could be attributed to variation in precipitation SO_4^{2-}. It can thus be surmised that in some lakes, SO_4^{2-} reduction is occurring, whereas in the watersheds of others, SO_4^{2-} is being supplied by weathering of pyritic and sulfidic minerals, as discussed earlier.

Nitrate and NH_4^+ concentrations in the model (precipitation-derived) lake for the Wenatchee Cascades (6.8 and 2.6 μeq L^{-1}, respectively), although low, are considerably greater than Rachel Lake concentrations of these ions (0.2 and 0.5 μeq L^{-1}, respectively). Inorganic nitrogen in precipitation is thus efficiently retained or assimilated in the watershed and lake. Removal of NO_3^- and NH_4^+ is most likely from assimilation by watershed and lake biota but could also involve denitrification or nitrification processes. The alkalinity and acidity produced by uptake processes for NO_3^- and NH_4^+ approximately cancel each other, and considering their small magnitude relative to lake ANC, the overall influence on lake acid-base chemistry is very minor (also see Figure 16.3).

Organic anion contributions to lakewater acid-base chemistry can be estimated by measured inorganic cation and anion concentrations and measured ANC (see section on empirical models). For Rachel Lake, the calculated total organic anion concentration of 4.4 μeq L^{-1} comprises weak

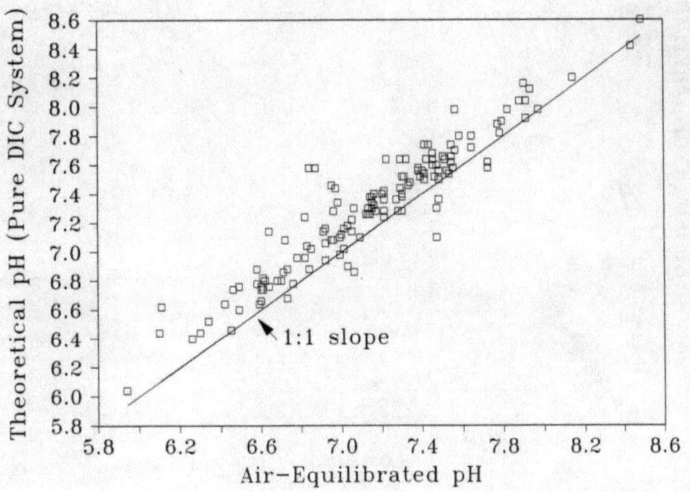

FIGURE 16.19. Relationship between calculated pure DIC air-equilibrated pH (AEpH) and measured air-equilibrated pH for WLS Cascade lakes.

organic acid anions only. Thus, about 5% of measured ANC is noncarbonate alkalinity from weak organic acid anions and 95% is carbonate alkalinity from carbonate species. Metal hydroxide complexes of Al and Fe have negligible influence on lake ANC values in Cascade lakes.

Dissolved Organic Acid Anions

The difference between the sums of measured inorganic cations and anions, or anion deficit, represents the fraction of the charge balance attributable to dissolved (total) organic anions. Organic anions may derive from weak organic acids that are largely titrated during the ANC determination and from stronger organic acids that are not protonated during ANC titrations. For Cascade lakes, the calculated population median total organic anion concentration is 18.4 $\mu eq\ L^{-1}$. Using the methods of Cook et al. (1987), it is apparent that weak organic acids compose most of the charge, or contribute most of the H^+ ions, attributable to dissolved organic compounds. Total organic anion concentration is weakly though significantly ($r^2 = 0.27$, $p < 0.001$) correlated with lakewater DOC (Figure 16.18). The weakness of this correlation may be due to the heterogeneous nature of natural organic compounds with varying ratios of net charge, or acidity, to carbon (Cook et al. 1987).

The major component of DOC in surface waters is fulvic acids (humic materials), for which the dominant functional protolyte influencing pH at pH levels < 7.0 is the carboxyl group (Oliver et al. 1983). Values for pK_a range between 3.5 and 5.5 for carboxyl groups, whereas pK_{a1} for the carbonate system is about 6.4 (Stumm and Morgan 1981, Kramer and Davies 1988). The net effect of weak organic acids in lake water buffered by the carbonate system is thus to change the form of ANC from carbonate to noncarbonate with resultant lowering of pH from that controlled by a pure dissolved inorganic carbonate (DIC) system. Figure 16.19 shows the relationship between calculated pure DIC air-equilibrated pH (AEpH) and measured air-equilibrated pH for WLS Cascade lakes. Most AEpH values are higher than corresponding measured values, indicating that weak acids in addition to the carbonate system are influencing pH. The weak acids are likely to be organic.

Noncarbonate ANC (ANC due to weak acids other than pure DIC) is calculated as the difference between Gran (measured) ANC and pure DIC ANC. Noncarbonate ANC is significantly correlated with ANC ($p < 0.001$) with a slope of 0.14 μeq noncarbonate ANC/μeq ANC, indicating that noncarbonate ANC comprises about 14% of ANC (Figure 16.20). Noncarbonate ANC corresponds closely to anion deficit, corroborating that both are primarily attributable to weak organic acids.

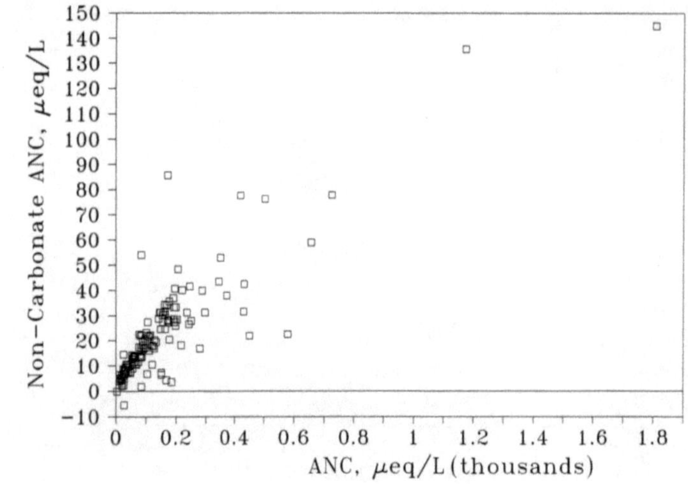

FIGURE 16.20. Relationship between noncarbonate ANC and ANC for WLS Cascade lakes.

Evidence for Long-Term Trends

Atmospheric Deposition

Hidy and Young (1986) have summarized historical data (1950–80) and estimated future trends (1980–2000) in anthropogenic emissions of sulfur dioxide (SO_2) and nitrogen oxides (NO_x) in the western United States. Sulfur dioxide emissions increased from about 3.3 million tons yr^{-1} in the 1950s to a peak of 4.8 million tons yr^{-1} in 1970 and then decreased to about 3.0 million tons yr^{-1} in 1980. The latter decrease was attributed to a reduction in emissions from nonferrous smelters, which accounted for about 70% of SO_2 emissions in the 1970's. Nitrogen oxide emissions have risen at an increasing rate from 1950 to 1980, from 1.4 million tons yr^{-1} in 1950 to 3.6 million tons yr^{-1} in 1980. Transportation and stationary combustion sources accounted for about equal proportions of the emissions and increase in emissions.

Sulfur dioxide emissions are projected to decrease during the 1980's due to further reductions in smelter emissions and to reach a level of 2.2 million tons yr^{-1} by 1990. Increasing emissions from stationary combustion sources will counteract the continuing decline in smelter emissions after 1990, resulting in a projected SO_2 emission rate of 2.6 million tons yr^{-1} by the year 2000. Nitrogen oxide emissions are projected to remain constant at about 3.5 million tons yr^{-1} during the 1980's due to emission controls. Gradual increases will occur after 1990, with the projected NO_x emissions reaching 4.6 million tons yr^{-1} by the year 2000. Overall, the 50-year trend from 1950 to 2000 shows a marked change, with nitrogen oxide emissions predominating over sulfur oxide emissions.

It is uncertain whether changes in atmospheric deposition comparable to those projected for emissions of sulfur and nitrogen oxides will occur. Source-receptor relationships have not been clearly demonstrated for sulfur and nitrogen oxides in sensitive areas of the western United States (Hidy and Young 1986). Some investigators have suggested that linearity exists between smelter SO_2 emissions and SO_4^{2-} deposition, although results are conflicting (National Research Council 1983). Vong et al. (1988) demonstrated a significant decrease in mean H^+ and excess SO_4^{2-} concentrations for precipitation in the region immediately downwind of a large copper smelter in the Puget Sound area after its closure. However, changes in winter precipitation chemistry in the Cascades farther downwind of this same source were not detectable (Duncan et al. 1986, Duncan and McGinty 1988). Sulfur to nitrogen ratios in atmospheric deposition are lower in the western United States than in the east and should continue to decrease (Hidy and Young 1986).

Historical Water Chemistry Comparisons

Major surveys of chemical data for Cascade lakes were first conducted in the early 1970s. Prior to this, only scattered data appear to exist for selected study lakes. No Cascade lakes appear to have been sampled over an extended period of time (25 to 50 years or more) for water chemistry parameters to enable comparison of historical trends. In addition, difficulties arise in comparing historical data with current data because of differences in the chemical methods used for determining pH, alkalinity, and other key parameters (Kramer and Tessier 1982).

The U.S. Geological Survey, in conjunction with the Washington Department of Ecology, began surveying the chemical composition of Washington lakes in 1971 (Bortleson et al. 1974, Bortleson et al. 1976, McConnell et al. 1976). Unfortunately, that survey did not include any of the WLS subregion 4B Cascade lakes sampled for this study, although it did include three other WLS subregion 4B lakes excluded from this study. Determination of changes in water chemistry from these relatively recent data is thus precluded.

Duncan and Lillquist (1986) reported six years of chemical data for one Wenatchee Mountain (Alpine Lakes Wilderness) lake. No trends in chemical composition were evident. Nelson and Delwiche (1983) compared ANC data determined by Gran's method (Stumm and Morgan 1981) with ANC data of Shulters (1976) determined by fixed endpoint at pH = 4.5 for 16 Oregon Cascade lakes common to both studies. The ANC data were found to be highly correlated ($r^2 = 0.98$) and in agreement with the theoretical expected difference, confirming that Gran and fixed endpoint methods can be intercompared for historical data (see also Kramer and Tessier 1982). The comparison thus indicated that no detectable change in ANC had occurred over the intervening seven years.

Welch et al. (1987) have concluded that no discernible changes in alkalinity have occurred in selected Wenatchee Cascade lakes as a result of decreases in SO_2 emissions after the closure of a major copper smelter downwind. Decreases in SO_2 emissions from Mount St. Helens volcano nearby are similar in magnitude and complicate clear interpretation of trends.

Overall, historical chemical data for Cascade lakes are limited to the recent past and do not reveal any consistent trends in lake chemical composition.

Empirical Models

Empirical equations used to estimate quantitative changes in lake solute chemistry from anthropogenic acidification are based on the charge balance equation (Chapter 3, Husar et al., this volume; Chapter 18, Sullivan, this volume). Qualitative empirical equations based on ion ratios are used to infer the direction of changes that may have occurred (Chapter 1, Munson and Gherini, this volume).

Henriksen (1979) proposed that the change in ANC due to acidification be estimated as the difference between Ca^{2+}, or the sum of Ca^{2+} and Mg^{2+} (sea salt corrected), and ANC, a method that has inherent limitations and that has been reformulated to include other important factors that determine the extent of ANC change (Chapter 18, Sullivan, this volume). Several previous investigators have applied Henriksen's empirical approach to Cascade lakes chemical data and have found that it predicts negligible acidification to have occurred (Welch and Chamberlain 1981, Logan et al. 1982, Nelson and Delwiche 1983, Turney et al. 1986). A greater fraction of base cations is attributable to Na^+ and K^+ in many Cascade lakes than in other sensitive regions studied by Henriksen (Nelson and Delwiche 1983) due to the greater importance of Na-feldspar and K-feldspar weathering (Nelson 1985). The same is true of Sierra Nevada lakes (see Chapter 15, Melack and Stoddard, this volume). Total base cation models provide a more suitable model for testing acidification since they have greater theoretical justification from the charge balance (Kramer and Tessier 1982). The sum of base cations is more highly correlated with ANC than is Ca^{2+} (Table 16.6). Application of the total base cation model to Cascade lakes similarly predicts negligible acidification to have occurred (Nelson and Delwiche 1983).

Wright (1983) developed an electroneutrality-based acidification equation that incorporates certain empirical simplifications to relate SO_4^{2-} from acid deposition (net SO_4^{2-}) to reduction in alkalinity. Duncan and Ausserer (1984) used this index to predict that acidification of up to 40 $\mu eq\ L^{-1}$ has occurred for some Wenatchee Cascade lakes, but further stated that this result may be attributable to oxidative weathering of minerals containing SO_4^{2-} in the watershed. Nelson and Baumgartner (1988) predicted negligible acidification of selected Mt. Rainier lakes (South Washington Cascades) using Wright's equation.

The estimated historical change in ANC that has occurred due to anthropogenic acidification can be calculated as the algebraic sum of the increases in base cations plus the increase in inorganic Al minus the increase in SO_4^{2-}, using sea salt corrected data (see Chapter 3, Husar et al., this volume). A worst case analysis for decrease in ANC assumes that base cations and Al have not changed, equivalent to assuming that no neutralization reactions have occurred in watershed or lake processes. The change in SO_4^{2-} can be calculated if the background SO_4^{2-} is known or estimated. For Cascade lakes, both background SO_4^{2-} and anthropogenic SO_4^{2-} are estimated to range from 4 to 8 μeq L^{-1} (see section entitled Precipitation and Deposition). Thus, the estimated historical maximum change in ANC is calculated to be 4 to 8 μeq L^{-1} for the population of Cascade lakes.

The dependency of lake ANC on the ratio of $SO_4^{2-}:C_B^*$ (Figure 16.16) can be used as an indicator of lake acidification (see Chapter 1, Munson and Gherini, this volume). Cascade lakes with ANC values > 100 μeq L^{-1} have $SO_4^{2-}:C_B^*$ ratios ranging from about 0.0 to 0.3, whereas those with ANC values ≤ 100 μeq L^{-1} have ratios ranging from 0.0 to about 0.75. The higher $SO_4^{2-}:C_B$ ratio for some low ANC lakes has several possible causes, including acidification of lake water by acid SO_4^{2-} without corresponding release of base cations, the presence of natural organic acids that tend to lower Gran ANC relative to base cation concentrations, and watershed sources of SO_4^{2-} (Chapter 3, Husar et al, this volume). For Cascade lakes, SO_4^{2-} is generated in some watersheds (1) by weathering reactions of sulfidic and pyritic minerals that decrease ANC (from H^+ ions produced) with concomitant production of SO_4^{2-}, and (2) by gypsum. Evidence for these weathering sources of SO_4^{2-} in Middle Washington and North Washington Cascade subregions has been cited (see section entitled Water Chemistry Status) and has been documented by other investigators (Reynolds and Johnson 1972, Duncan and Ausserer 1984, Loranger 1986, Drever and Hurcomb 1986).

Figure 16.14 shows the relationship between the sum of base cations (sea salt corrected) and ANC for WLS Cascade lakes. Burns et al. (1981) have contended that in lakes unaffected by acidic deposition, the slope of the ANC versus C_B relationship (linear plot) exceeds 0.75 (slope of C_B versus ANC < 1.33). All WLS Cascade lakes satisfied this criterion for being unaffected by acidic deposition, providing further evidence that Cascade lakes, as a group, have not experienced chronic acidification.

Future Considerations

Atmospheric Deposition

Sulfur dioxide emissions in the western United States are projected to follow a decreasing pattern during the 1980s and reach a level of 2.2 million tons yr^{-1} by 1990, followed by slow increases in emissions, resulting in a projected SO_2 emission rate of 2.6 million tons yr^{-1} by the year 2000 (see section entitled Evidence for Long-Term Trends). Nitrogen oxide emissions are projected to remain constant at about 3.5 million tons yr^{-1} during the 1980s followed by gradual increases after 1990, with the projected NO_x emissions reaching 4.6 million tons yr^{-1} by the year 2000. The 50-year trend from 1950 to 2000 shows a change in emissions patterns, with nitrogen oxide emissions predominating over sulfur oxide emissions. Although it is not clear whether linearity exists between emissions of sulfur and nitrogen oxides and their atmospheric deposition, it is apparent from emissions trends that deposition of nitrogen oxides will continue to increase in importance in the next 20 years, whereas deposition of sulfur oxides will not change greatly.

Lake Chemistry

For many regions of the United States sensitive to acidic deposition, dynamic watershed models have been developed to predict future changes in lake chemistry in response to changes in lake chemical inputs, particularly input from acidic deposition (see Chapter 2, Munson and Gherini, this volume). The models have been used to predict changes for increases as well as recovery times for decreases in deposition loadings. No such modeling effort has been applied to Cascade Lakes, so hypothetical future changes must be predicted from changes that have been observed in acidified regions.

The main question to be addressed is whether Cascade Lakes, which have been shown in aggregate to be largely unimpacted from the present low level of acidic atmospheric deposition in the Pacific Northwest, are at risk for possible future changes in depositional loadings. Sulfur oxide acidic loadings are predicted to remain about constant and those of nitrogen oxides may increase by up to 40% over the next 20 or more years (see section on atmospheric deposition). At present, loadings of these acidic precipitation constituents range up to about 25% of those for impacted regions of the eastern United States, whereas corresponding concentrations (near background) range from 5% to 15% of those for impacted regions of the east (see Atmospheric Deposition in section entitled Regional Characteristics). Thus, for the immediate future (20 years), there appears to be no imminent risk to Cascade lakes as a whole from acidic deposition. However, approximately 11% of the Cascade lake population has ANC values $\leq$ 25 µeq L^{-1}. These lakes are generally at high elevations with small watersheds, sparse vegetation, and poorly developed soils. Some of these lakes are also directly downwind of major metropolitan areas, such as those in the Wenatchee Cascades east of Puget Sound. This subpopulation of Cascade lakes may be at considerably greater risk than the general population, particularly during episodic inputs of acidic constituents during snowmelt events.

Lack of long-term atmospheric deposition and lake chemistry data has severely limited assessments of past conditions and future trends in Cascade lake chemistry. The data bases for deposition chemistry and lake chemistry are presently inadequate for accurately estimating the magnitudes of either past changes that may have occurred or future changes that may occur. Because of their uniqueness as low ANC lakes receiving close to background concentrations of acidic deposition, much of the Cascade lakes region may provide valuable information as a benchmark against which other regions impacted by acidic deposition can be compared. A long-term atmospheric deposition and lake chemistry monitoring network should be established in the Cascade Mountains to enable such future assessments.

Biological Status

Phytoplankton and Zooplankton Status

No comprehensive references appear to exist that characterize the flora and fauna of Cascade lakes. Definitive studies on invertebrates in Cascade lakes have not been made (Anderson 1989). The biological status of a few Cascade lakes has been studied intensively, including Crater Lake, Oregon (Hasler 1938, Larson 1972, Larson 1988), three lakes in Mt. Rainier National Park (Larson 1973, Hall 1973, Funk et al. 1985), and Findley Lake, Washington (Hendrey and Welch 1974, Pederson et al. 1976).

Generally, conditions of low temperature, low nutrient concentrations, extended winter ice-snow cover up to nine months, and, in shallow lakes, intense solar radiation during open-water periods inhibitory to photosynthesis, combine to adversely affect algal growth. Three studies of lakes in Mt. Rainier National Park found very low productivities and standing crops typical of ultra-oligotrophic conditions (Larson 1973, Hall 1973, Funk et al. 1985). Phytoplankton are characterized by low population density and relatively high species diversity. Hall (1973) reported net plankton density for subalpine Shadow Lake to be typically < 1,000 cells L^{-1}, although 55 algal species were identified. Planktonic species were dominated by Cyanophyceae with Bacillariophyceae (diatoms) becoming important in late August to September. Phytoplankton biomass peaked in late July in Mowich Lake (Larson 1973).

The higher solute concentrations in Crater Lake are atypical of Cascade Lakes, owing to probable inputs from subsurface springs (Phillips and Van Denburgh 1968). Despite Crater Lake's extreme clarity, its productivity is higher than that of most ultra-oligotrophic Cascade lakes (Larson 1972), and 142 phytoplankton taxa were found (McIntire and Debacon 1988).

The zooplankton communities of the Cascade lakes are severely limited in development by the low primary productivities and by abiotic factors that also inhibit phytoplankton growth. At Mt. Rainier National Park, in deep, ultra-oligotrophic Mowich Lake, the zooplankton community consisted mainly of five species of rotifers, although eight were identified along with one crustacean (Larson 1973); in intermediate depth Shadow Lake, five species of rotifers, four crustaceans, and one diptera larva constituted the zooplankton (Hall 1973). In shallow Reflection Lake, eight species of rotifers, five species of cladocerans, and two species of eucopepoda were reported (Funk et al. 1985). In Crater Lake, two species of cladocerans and eleven species of rotifers were collected in studies from 1985 to 1987 (Karnaugh 1988).

Zooplankton are the major primary consumers in many naturally fishless Cascade lakes, most of which are too low in primary productivity to sustain fish populations (see Fisheries Sensitivity to Acidification). Zooplankton numbers are greatest in the fall season due to the short open water seasonal development of primary producers (Larson 1973).

Effects of acidification on phytoplankton and zooplankton communities in sensitive, poorly buffered Cascade lakes should be similar to those summarized by Baker and Christensen (Chapter 4, this volume) for similar lakes in artificial acidification studies and in documented studies of acidified regions of the world.

Fisheries Sensitivity to Acidification

Most high-elevation lakes of the northwestern United States, including those of the Cascade Mountains, were naturally fishless until stocking efforts were initiated in the mid- to late 1800s, although isolated lakes had indigenous populations of rainbow trout (*Salmo gairdneri*), cutthroat trout (*Salmo clarki*), and kokanee salmon (*Oncorhynchus nerka*). Subsequently, stocking spread these species through the high-elevation lakes of the region and introduced several non-native species, including eastern brook trout (*Salvelinus fontinalis*), golden trout (*Salmo aguabonita*), brown trout (*Salmo trutta*), Atlantic salmon (*Salmo salar*), and grayling (*Thymallus articus*).

Of these species, brook trout are most successful at establishing self-sustaining populations; they can overpopulate some lakes and streams, leading to stunting in some populations. Rainbow trout, cutthroat trout, kokanee, and grayling also maintain naturally reproducing populations in some of the high-elevation lakes of the Northwest. Relatively few of these lakes maintain naturally reproducing fish populations (Larmie 1955, Johnston 1977, Johnson et al. 1985). Approximately 75% of the high-elevation lakes of Oregon and Washington are stocked to maintain fisheries (Jim Griggs, Oregon Department of Fish and Wildlife, pers. comm.; Johnston 1977.)

Angling pressure in these lakes can be relatively high; an estimated 800,000 fishing trips were made into the alpine lakes of Washington during 1986 (Paul Mongillo, Washington Department of Game, pers. comm.). To maintain the populations necessary to satisfy anglers, the high-elevation lakes are generally stocked every second year in Oregon and on 3- to 10-year cycles in Washington. To permit recovery of invertebrate populations, many of these lakes in Washington are intentionally left fishless for one or more years before restocking. Currently, brook and rainbow trout are stocked primarily in the high-elevation lakes of Oregon, and rainbow and cutthroat trout are introduced in Washington. Limited stocking of other trout and salmon species continues in both states. It should be noted, however, that National Park policy excludes all fish stocking in lakes and streams of the national parks, except for North Cascades National Park, Washington, where limited stocking still occurs. This policy is based on a goal of allowing many aquatic ecosystems in the parks to return to their naturally fishless conditions.

No existing studies directly address the sensitivity to acidification of many of the fish species inhabiting the high-elevation lakes of the Northwest. But, as discussed in Chapter 14 on the Rocky Mountains (Turk and Spahr, this volume), some laboratory toxicology data are available that can be extrapolated to estimate the relative sensitivity of these fish species. These data indicate that rainbow and cutthroat trout appear to be more sensitive to elevated acidity and dissolved monomeric Al concentrations than are brook trout. Chapter 4 (Baker and Christensen, this volume) contains a discussion of how acidity, dissolved Al, and Ca^{2+} interact to affect the survival of fish in acidifying lakes and streams. Based on this relationship, populations of rainbow trout, cutthroat trout, and perhaps other Salmo species are potentially at greater risk from acidification, if lake acidities increase, than are populations of brook trout.

Earlier studies suggest that in regions experiencing acidification, initial impacts to fisheries may result from episodic runoffs of snowmelt or rainstorm waters containing acutely toxic concentrations of acidity or dissolved Al (Leivestad and Muniz 1976, Schofield 1980). These events can result in the mortality of adults, or more frequently, of earlier life stages, which can become apparent in the population as missing year classes. But, as this section emphasizes, natural conditions in these lakes often severely limit fish populations, with many lakes maintaining populations only through regular stocking. And, because topography and climate generally restrict access to these lakes, monitoring of the lakes and documentation of actual causes of fish mortality are severely limited. Therefore, sound cause-effect relationships should be established before causes of fish mortality in high-elevation lakes are attributed to extrinsic causes.

Conclusions

1. The Cascade Mountain Range forms a transect 1,100 km long through the Pacific Northwest from northern California to southern British Columbia. Along this transect, varying biogeochemical and atmospheric influences cause substantial subregional differences in lake chemical composition.
2. Cascade Mountain lakes as a group were confirmed to be extremely dilute in chemical composition and thus sensitive to possible inputs of acid deposition.
3. No lakes in the Cascade lake population were found to be acidic (ANC $\leq$ 0 μeq L^{-1}) or to experience chronic acidification (depletion of ANC) based on empirical chemical assessment methods. The pH values (air equilibrated) were generally > 7.0 and the lowest value determined was 5.94.
4. Seasonal changes in chemical composition, determined from comparing early summer post-snowmelt samples to late summer samples, vary from less than 10% for lakes with high hydrologic residence times to more than 300% for lakes with low hydrologic residence times. Evidence for episodic acidification during the snowmelt period indicates that slight changes in base chemistry may occur in the Middle Washington Cascades.
5. Winter precipitation is slightly acidic but concentrations of acid anions are very low in comparison to sensitive areas of the eastern United States. Anthropogenic influences are noticeable, especially east of the Puget Sound metropolitan area.
6. The principal process controlling the ANC and base cation composition of Cascade lakes is mineral weathering. Watershed and in-lake biogeochemical processes are generally unimportant. Organic acid anions are minor components of lake acid-base chemistry.
7. The aquatic biota of Cascade lakes have generally not been definitively characterized or studied for impacts from acidification, but no discernable impacts from acidic deposition are expected under present conditions.

Acknowledgments. The section entitled Fisheries Sensitivity to Acidification was prepared by Michael D. Marcus, Western Aquatics, Inc., Environmental Consultants, Laramie, Wyoming. I thank Gary L. Larson, Cooperative Park Studies Unit, College of Forestry, Oregon State University, and Joseph M. Eilers, who provided copies of unpublished manuscripts and general technical assistance. I have learned much from former graduate students who were involved in earlier lake studies, including Gregory K. Delwiche, Robert P. Baumgartner, and Joseph F. Reilly.

Part III
Synthesis and Integration

17
Interregional Comparisons of Surface Water Chemistry and Biogeochemical Processes

Lawrence A. Baker, Joseph M. Eilers, Robert B. Cook, Philip R. Kaufmann, and Alan T. Herlihy

ABSTRACT. The case study regional populations are a heterogeneous assemblage of surface waters that include streams, rapidly flushed drainage lakes, and seepage lakes located across a range of landscape and climatic conditions. Surface water SO_4^{2-} concentrations vary by an order of magnitude among the study regions, paralleling variations in wet SO_4^{2-} deposition. Surface waters with mineral acidity ($SO_4^{2-}:C_B$ ratios >1.0) are found only in moderate and high deposition areas: the Adirondacks, Florida, the eastern Upper Midwest, the western Virginia mountains, and the Catskills in the United States, and southcentral and northeastern Ontario, Quebec, New Brunswick, and Nova Scotia in Canada. Lakes dominated by organic acidity ($A^-:C_B$ ratios > 1.0) are found in the Upper Midwest, Florida, and the Canadian maritime provinces. Lakes in the latter group are probably naturally acidic.

The heterogeneity of surface water chemistry is reflected in the diversity of biogeochemical processes that regulate ANC. Overall, base cation production is the dominant source of ANC throughout the case study regions; an exception is base cation retention in some Florida lakes, a process that consumes ANC. Among seepage lakes, C_B concentrations are linearly related to groundwater inputs. Sulfate adsorption retains the majority of SO_4^{2-} deposited in the Southern Blue Ridge Province and the western Virginia mountains and is the dominant source of ANC for surface waters in the Southern Blue Ridge. Sulfate reduction is an important source of ANC in seepage lakes in the Upper Midwest and Florida. Elsewhere, SO_4^{2-} inputs appear to closely match outputs on a watershed scale. Geologic sources of SO_4^{2-} appear to be relatively unimportant compared with atmospheric inputs. Both NO_3^- and NH_4^+ are efficiently retained by most case study watersheds. In the east, where NO_3^- inputs exceed NH_4^+ inputs, efficient retention of both results in net ANC production.

Organic acids play an important role in the acid-base status of many lakes in the Upper Midwest, Florida, and Maine. Overall, virtually all case study lakes included in the ion enrichment analysis produce ANC. Production of ANC appears to be highest in drainage lakes, intermediate in groundwater flowthrough lakes, and lowest in groundwater recharge lakes.

Introduction

There is considerable heterogeneity in the physical and chemical nature of aquatic resources among the case study regions, as Chapters 6 to 16 demonstrate. For example, Sierra Nevada lakes have inherently low ANC, even though they receive little atmospheric input of anthropogenically derived acids. Streams in the Southern Blue Ridge Province also have low ANC, but very few are acidic (ANC ≤ 0), even though precipitation [H^+] is nearly 10 times higher in this region than in the Sierra Nevada. By contrast, precipitation pH in the Adirondack Mountains and Florida case study regions is acidic and many lakes in both regions are acidic. However, acidic lakes in the Adirondack Mountains typically are rapidly flushed drainage lakes, whereas most acidic lakes in Florida are seepage lakes with long water residence times. The processes that regulate ANC in these systems are very different and we would expect system responses to changes in atmospheric loadings to be different.

In this chapter, we synthesize our knowledge of low ANC systems throughout the case study regions and address the general question: "What are the characteristics of low ANC and acidic surface waters in North America?" More specifically:

- What are the hydrologic characteristics and landscape features associated with low ANC surface waters?
- What are their chemical characteristics?
- What biogeochemical processes regulate ANC and what does this imply about future acidification or recovery of these systems?

In the first part of this chapter we compare the diverse landscape, climatic, and hydrologic conditions that characterize low ANC systems among the case study regions and briefly summarize variations in deposition chemistry among these regions. We then examine interregional variations in surface water acid-base chemistry and the nature of the biogeochemical processes that regulate alkalinity in lakes and streams in the case study regions. We synthesize these findings in a region-by-region synopsis of surface water acid-base chemistry and the processes that influence acidification. Chapter 18 addresses the subject of how the chemistry of these surface waters has changed in response to acidic deposition and how they might respond to further changes in deposition chemistry.

Many analyses throughout this book and in this chapter, particularly ion enrichment analysis, are restricted to low ANC systems (ANC $\leq$ 200 μeq L^{-1}) because low ANC lakes and streams are more likely to exhibit chemical response to changes in acid loadings and are therefore of greater interest than lakes with ANC > 200 μeq L^{-1}. The low ANC systems used for these comparisons include lakes with ANC $\leq$ 200 μeq L^{-1} that have been screened to remove the apparent influences of road salt or watershed development (see Appendix A). For many analyses, lakes were subdivided into three hydrologic classes (drainage, groundwater flow-through, and groundwater recharge) and two DOC classes (DOC $\leq$ 5 mg L^{-1} and DOC > 5 mg L^{-1}; 417 μmol). This classification scheme makes it possible to compare geochemically similar systems among regions.

Regional Landscape, Climatic, and Hydrologic Characteristics

Landscape Characteristics

Among the physiographic provinces, there are two distinct glaciated regions, the western alpine and the continental. Each of these has two major subgroups: the Cascade/Sierra Nevada range and the Rocky Mountain range in the West, and the Upper Midwest and Northeast in the East. Other case study regions are in the unglaciated Southeast, represented by the western Virginia mountains and the Southern Blue Ridge Province. Lakes in the Florida karst district are the final case study group.

The regional case study areas have been classified on the basis of bedrock geology, surficial geology, soils, topography, vegetation, and land use (Chapter 5; Table 17.1). These factors, combined with the climatic characteristics, determine the hydrologic properties of lakes and streams discussed later in this chapter.

Bedrock Geology

Bedrock geology has long been considered a key terrigenous factor used to define regions containing low ANC surface waters. Sensitive lakes and streams have been associated with bedrock highly resistant to weathering (Norton et al. 1982, Bricker and Rice 1989). This is largely the case for many lakes in the West and in Norway, which typically have little overburden or dense vegetative cover to modify the weathering products of granitic bedrock. In the Cascade Mountains, where the predominant bedrock types are moderately weatherable andesite, basalt, and mixed metamorphics, base cation concentrations in lakes are among the lowest anywhere. In fact, Lake Notasha in the Cascade Mountains has been identified as the most dilute lake in the world (Eilers et al. 1990). However, overlying till and soils can substantially modify regional sensitivity or even reverse the sensitivity status inferred from bedrock geology alone. For example, many lakes in northeastern Wisconsin have high ANC, even though the underlying Precambrian shield is highly resistant to weathering, because they are located in thick till containing highly weatherable calcite (Simpkins et al. 1978). In contrast, the bedrock in Florida and in parts of northern Michigan is limestone and dolomite, yet many lakes have low ANC as a consequence of being situated on highly weathered quartzite sand (Chapter 12, Pollman and Canfield, this volume; Chapter 13, Cook and Jager, this volume). We conclude that bedrock geology, although an important determinant of ANC, is a poor predictor of low ANC surface

TABLE 17.1. Summary of regional landscape and climatic characteristics.

Case study area	Topography (watershed relief)[a]	Geology (glaciated = g)	Area of region (km² × 10³)	Density[b] of lakes (no./100 km²)	Precipitation[c] in cm/yr (% rain/snow)	Major soil types (orders)	Vegetation/ land use	Climate	Dominant low ANC resources
Adirondack Mountains	Low mountains (2.4)	Granite (g)	51	3	101–123 (50)	Spodosols; inceptisols	Mixed deciduous	Warm wet summer Cold wet winter	Drainage lakes[d]
Maine	Low mountains; plains (2.4)	Metasedimentary; granite intrusive (g)	84	4	92–109 (50)	Spodosols; inceptisols	Mixed deciduous	Warm wet summer Cold wet winter	Drainage and seepage lakes, streams
Catskill Mountains	Low mountains (NA)	Sandstone; mudstone (g)	3	1	107 (50)	Inceptisols	Mixed deciduous (originally hemlock)	Warm wet summer Cool wet winter	Streams
Upper Midwest Boundary waters	Irregular plains (1.3–1.6)	Gabbro, granite (g)	13	12	61–81 (30)	Boralfs and orthods in uplands; qualfs in lowland (spodosols)	Mixed coniferous/ deciduous (originally dominated by white pine)	Warm wet summer Cold dry winter	Seepage lakes
NE Minnesota		Mafic; intermediate and felsic igneous (g)	38	7					
NW Wisconsin			22	10					
NE Wisconsin			36	8					
UP Michigan			13	6					
Sierra Nevada	High mountains (4.4)	Granite (g)	45	3	93–153 (90)	Frigid, cryic ultisols	Alpine/subalpine	Cool arid summer Cold wet winter	Drainage lakes[d]
Cascade Mountains	High mountains (4.4)	Andesite, basalt mixed metamorphic (g)	81	1	80–250 (80)	Inceptisols; ultisols	Alpine/subalpine	Cool arid summer Cold wet winter	Drainage lakes[d]
Rocky Mountains	High mountains (4.4)	Mixed metamorphic; granite (g)	353	<1	60–120 (50)	Frigid, cryic alfisols	Alpine/subalpine	Cool summer Cold wet winter	Drainage lakes[d]
Florida	Coastal plains (~1)	Limestone bedrock; ancient marine surficial sands; karst	102	8	131–152 (0)	Entisols	Deciduous; conifer plantations; citrus groves	Hot wet summer Warm dry winter	Seepage lakes
Western Virginia	Low mountains (NA)	Granite; metabasalts; sedimentary	33	<1	101–142 (10)	Ultisols; inceptisols; entisols	Mixed deciduous	Warm wet summer Cool wet winter	Streams
S. Blue Ridge	Low mountains (NA)	Granite; sandstone, limestone	29	<1	142–162 (10)	Inceptisols; ultisols; spodosols	Mixed deciduous	Warm wet summer Cool wet winter	Streams; reservoirs

[a] Watershed relief expressed on an ordinal scale measured from the center of the lake where 1 = plain (<100′), 2 = rolling hills (100–500′), 3 = low hills (500–1,000′), 4 = high hills/low mountains (1,000–2,000′), 5 = mountains (>2,000′).

[b] Data are derived from the EPA's National Lake Survey and are computed using the weighting factors. The statistics for the western regions (Sierra Nevada, Cascades, Rockies) are for lakes with surface area >4 ha.

[c] Total precipitation represents long-term (30 yr) annual average. The estimated relative proportions of rain and snow are shown in parentheses.

[d] Streams have not been studied extensively in this region, but they are expected to have chemical characteristics similar to lakes.

waters in North America when other factors are not considered.

Surficial Geology

On a local scale, glacial history and soil types are also important factors in determining the location of low ANC surface waters, but again, on a regional scale, their association with low ANC systems is highly variable. Low ANC lakes in North America are present where glacial retreats have left deep overburdens of material derived from bedrock resistant to weathering (e.g., northern Wisconsin) and where the glacial action has left a high proportion of exposed, resistant bedrock (e.g., Sierra Nevada). In some areas such as the Adirondack Mountains, the depth of till appears to exert strong controls on lake chemistry (Chen et al. 1984a, Driscoll and Newton 1985, Booty et al. 1988; Chapter 6, Driscoll et al., this volume). In the Upper Midwest, the composition of the till appears to be far more important than depth of till in influencing lake chemistry (Simpkins et al. 1978). However, glaciation is not a necessary condition for low ANC surface waters, as many low ANC lakes and streams are found in the Southeast Highlands and Florida. The Florida case study region has the highest proportion and number of acidic lakes, yet the lakes were commonly formed from karst activity, which is often associated with extremely high ANC waters. Most of the soils in the regional case study areas have low cation exchange capacity and low base saturation but vary greatly in texture, organic content, horizon development, and other characteristics. Whereas the influence of till on lake chemistry is highly variable on a regional scale, low ANC lakes and streams are present only where soils have low cation exchange capacity and low base saturation (e.g., McFee 1980).

Topography

Low ANC systems are typically found at high elevations in the West (Chapter 14, Turk and Spahr, Chapter 15, Melack and Stoddard, Chapter 16, Nelson, all this volume), the Southeast Highlands (Chapter 10, Cosby et al., Chapter 11, Elwood et al., this volume), Maine (Kahl et al., this volume), and the Adirondack Mountains (Driscoll et al., this volume), but elevation alone may not be a good predictor of ANC within a region. Lake ANC was inversely related to elevation in Colorado (Turk and Adams 1983), but elsewhere in the West elevation was a poor predictor of lake ANC (Eilers et al. 1987). Lake elevation is a reasonably good predictor of ANC in the Adirondacks (Hunsaker et al. 1986). In Florida and the Upper Midwest, elevation is a poor predictor of lake chemistry. In these regions, concentrations of ANC in seepage lakes are strongly influenced by position within a local flow system, such that ANC is higher in lakes that intersect the groundwater flow path than in lakes located above the general groundwater system (Kenoyer 1986, Wentz et al. 1987). This was also observed for drainage lakes in Minnesota (Rapp et al. 1985) and southern Ontario (Booty and Kramer 1984). Thus, the degree to which lake chemistry reflects topography or elevation appears to be determined largely by how strongly the measures of topography reflect lithological and hydrologic characteristics.

Vegetation and Land Use

Vegetation can alter the acid-base status of lakes and streams through enhancement of dry deposition, ion uptake and exchange, and organic acid production during decay. The type, age, and density of vegetation are important determinants of its geochemical influence. For example, coniferous forests can enhance acidification through more efficient retention of dry deposition (Johannes and Altwicker 1983). Furthermore, the age of the forest has a strong influence on the rate of cation and nutrient uptake (Likens et al. 1977). Removal of natural vegetation by storms, logging, fire, disease, agriculture, or urbanization are also important in modifying mineral uptake or release from the watershed (Likens et al. 1977, Bayley and Schindler 1987, Wright 1976, Krug and Frink 1983). The case study regions display great variability in vegetation types, ranging from the coniferous forest of the West to the xeric vegetation of some parts of Florida. However, associations among vegetation types and low ANC surface waters in the United States are not evident or are difficult to separate from other factors. For example, most lakes in the Sierra Nevada exhibit a pronounced difference in chemistry between the montane and the alpine/sub-alpine groups, but differences in elevation, soils, and precipitation volume also contribute to the differences in lake chemistry.

Cultural influences can greatly alter the chemical and biological conditions of surface waters, as seen in the widespread association between urbanization and eutrophication (Hasler 1975). Many land use changes are perceived to cause increased mobilization of base cations from the watershed through accelerated erosion, exposure of weatherable subsoil, and modifications of hydrologic flowpaths. In northern Wisconsin, cultural development was associated with increase in ANC over a 50-year period (Eilers et al. 1989). Agricultural land use is positively correlated with ANC (Chapter 5, Eilers and Selle, this volume), an association that may reflect the higher fertility of agricultural soils rather than an impact of changing land use on water chemistry. However, some land use changes may favor acidification. The export of base cations may decrease, for example, when groundwater levels are lowered (Chapter 12, Pollman and Canfield, this volume) or when dirt roads are paved (Gatz et al. 1986). The export of mineral acids may increase as a result of acid mine drainage (Herlihy et al. in press). And neutralization of acidic deposition may decrease when the loss of wetland habitat causes a decrease in watershed SO_4^{2-} reduction. Change in forest cover, because of logging or other reasons, has received much attention as a mechanism for causing changes in the acid-base status of surface waters, primarily because so much of eastern North America has experienced forest removal at one time or another. The regrowth of forests following logging has been invoked as an alternative explanation for recent acidification (Krug and Frink 1983), and recent acidification in the Adirondacks has been attributed to a massive forest blowdown. These explanations have been disputed (Havas et al. 1984; Charles et al. 1987; Norton et al. 1989; Johnson et al. 1987), and it is unlikely that changes in forest cover alone have caused the widespread occurrence of acidic lakes in the Adirondacks or elsewhere. Changes in forest cover undoubtedly alter the acid-base chemistry of surface waters to some extent, but at present we are not able to assess the magnitude of that effect.

Climatic Characteristics

The climatic zones for the regional case studies correspond closely with differences in physiography. Three major climatic zones are represented: the cool continental climate of the Northeast and the Upper Midwest, the humid subtropical climate of the Southeast, and the alpine climate of the mountainous West. Total precipitation is generally greatest in the Southeast and in the high elevations of the West, but the seasonal distribution is juxtaposed in these two regions (Table 17.1). Precipitation is highly variable in areas with substantial topographic relief, and precipitation in the West is greatly dependent on elevation. Only in the West does the majority of precipitation occur as snow during the winter. Evaporation is typically less variable than precipitation, spatially and temporally. Only in Florida does evaporation approach 80 cm yr^{-1}, and for most of the case study regions it ranges from 40 to 60 cm^{-1}. Most lakes in Canada, and in the Northeast, Upper Midwest, and West, freeze in the winter, whereas those in the Southeast and Florida do not freeze. One consequence of these climatic differences among regions is that interregional comparisons of loadings or concentrations of acidic constituents in atmospheric deposition can be misleading. In Florida, where evaporation and deposition rates are high, the potential for acidification of aquatic resources may be better represented by total loading than by threshold precipitation pH values developed for temperate lakes. In the northern Cascade Mountains, in contrast, there is little evapoconcentration and the potential for acidification of lakes may be better reflected by H^+ or SO_4^{2-} concentrations in precipitation rather than deposition loadings.

Hydrologic Characteristics

The susceptibility of lakes and streams to acidification is largely influenced by the hydrologic characteristics of their watersheds. The amount, timing, and form of precipitation play a major role in determining susceptibility to acidification, as do evapotranspiration, flowpath length through the soil, groundwater inputs, lake flushing rates, and mixing characteristics (Eilers et al. 1983, Chen et al. 1984a, Marmorek et al. 1986, Baker et al. 1988b; Chapter 1, Munson and Gherini, this volume).

Hydrologic characteristics of low ANC surface waters range from flashy first-order streams to seepage lakes with residence times > 10 years (Figure 17.1). In the western Virginia mountains and Southern Blue Ridge Province, precipitation is relatively well distributed throughout the year. This is illustrated by the hydrograph for White Oak Run,

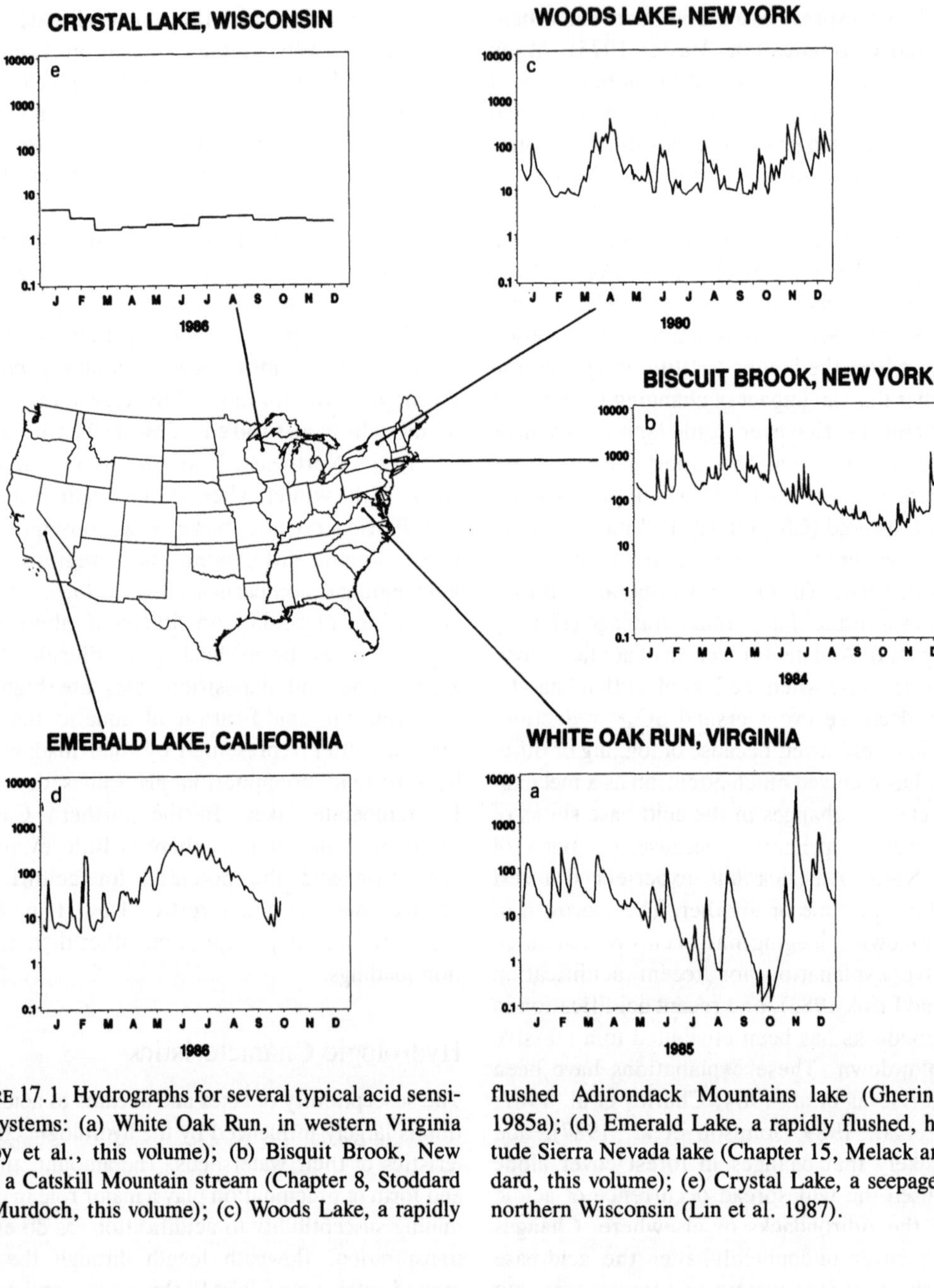

FIGURE 17.1. Hydrographs for several typical acid sensitive systems: (a) White Oak Run, in western Virginia (Cosby et al., this volume); (b) Bisquit Brook, New York, a Catskill Mountain stream (Chapter 8, Stoddard and Murdoch, this volume); (c) Woods Lake, a rapidly flushed Adirondack Mountains lake (Gherini et al. 1985a); (d) Emerald Lake, a rapidly flushed, high altitude Sierra Nevada lake (Chapter 15, Melack and Stoddard, this volume); (e) Crystal Lake, a seepage lake in northern Wisconsin (Lin et al. 1987).

Virginia, in which peaks represent storm events and very low flows result from high evaporation rates during summer (Figure 17.1a). By comparison, hydrographs for two northeastern systems, Bisquit Brook (Figure 17.1b) and Woods Lake (Figure 17.1c), show broad peaks during spring snowmelt with additional peaks caused by storms during the spring and summer.

The influence of snowmelt is even more pronounced in high-elevation western lakes, illustrated by Emerald Lake in Figure 17.1d, where about 90% of the precipitation occurs as snowfall. Snowmelt causes sustained high flow from March through July and is followed by low flows for the rest of the year, with few summer storm events.

The other end of the hydrologic spectrum is represented by Crystal Lake, a Wisconsin seepage lake, illustrated in Figure 17.1e. The hydrology of seepage lakes is characterized by the following:

- Direct precipitation to the lake surface is the dominant source of water.
- Hydraulic residence times are longer than for most drainage lakes.*
- Evaporation from the lake surface is a major loss term from the lake hydrologic budget.

In this book, we have segregated seepage lakes into two groups: (1) groundwater recharge lakes, which receive very little groundwater input, and for this reason have low base cation concentrations, and (2) groundwater flow-through lakes, which receive a greater fraction of their water input from groundwater and generally have high base cation concentrations (Born et al. 1979, Eilers et al. 1983, Baker et al. 1988a; Chapter 12, Pollman and Canfield, this volume; Chapter 13, Cook and Jager, this volume). The distinction between the two groups is arbitrary, in large part because hydrologic budgets have been developed for only a small number of seepage systems. However, from a standpoint of alkalinity contributions from lakes and catchments, the available evidence suggests that seepage lakes receiving < 10% of their total water input might be considered groundwater recharge lakes in which in-lake processes are the dominant source of ANC. Seepage lakes receiving larger fractions of their water input would be considered groundwater flowthrough systems in which terrestrial ANC inputs are the dominant source of ANC (Baker and Brezonik 1988). Operationally, we have classified seepage lakes into *groundwater recharge* or *groundwater flowthrough* lakes on the basis of their SiO_2 or K^+ concentrations, with seepage lakes having $SiO_2 \leq 1$ mg L^{-1} (or $K^+ \leq 15$ μeq L^{-1} in the case of Florida) assigned to the groundwater recharge category. All nonseepage lakes were classified as *drainage* lakes (Appendix A, Baker, this volume).

Interregional Differences in Hydrology

Sixty-five percent of the NSWS lakes in the case study regions were drainage lakes and 35% were seepage lakes. This apportionment varied considerably with region. Drainage lakes made up over 90% of the NSWS lakes in the West, the Adirondack Mountains, Maine, and the Southern Blue Ridge Province. By contrast, seepage lakes constituted 71% of the the NSWS lakes in Florida and 53% of the NSWS lakes in the Upper Midwest. In Maine and the Adirondack Mountains, seepage lakes were relatively more numerous among the lakes smaller than 4 ha that were not included in the ELS (Chapter 7, Kahl et al., this volume; Sullivan et al. 1990).

There has been no formal hydrologic categorization of Canadian lakes, but the clear impression of Canadian limnologists is that the vast majority of Canadian lakes are drainage systems (D. Jeffries, Canadian Center for Inland Waters, pers. comm.). For some Canadian drainage lakes, inflow may be intermittent, but even these lakes nearly always experience inflow during the snowmelt period. Outflow-based residence times of several years are apparently common for Canadian drainage lakes (Kelly et al. 1987).

In two RCS regions, the western Virginia mountains and Southern Blue Ridge Province, streams are the dominant aquatic resource and there are few natural lakes. Nearly all NSWS lakes in the Southern Blue Ridge Province are human-made reservoirs (Chapter 11, Elwood et al., this volume).

Drainage Lakes

Interregional differences in the hydrology of drainage lakes are related to differences in catchment characteristics, precipitation, and evapotranspiration. Median *RT* values are < 0.5 yr for low ANC drainage lakes in Maine, the Adirondack Mountains, and the West, and somewhat longer

*We note that residence time (*RT*) has been defined in the ELS and WLS data base as lake volume (m^3)/inflow (m^3 yr^{-1}). We would like to point out the difference between this definition and the alternative definition based on lake *outflow* that is commonly used in mass balance models (e.g., Vollenweider 1975). To distinguish between the two definitions, we have used the symbol t_w to identify outflow-based residence times, where t_w = lake volume/outflow. For rapidly flushed lakes, inflow ~ outflow and $RT \sim t_w$. The difference between the two values becomes greater for less rapidly flushed systems, particularly seepage lakes. For these lakes, evaporation losses are significant, hence outflow < inflow and $t_w > RT$. For example, *RT* for Crystal Lake, Wisconsin, is 10 yr, whereas t_w is 30 yr (data from Armstrong et al. 1987).

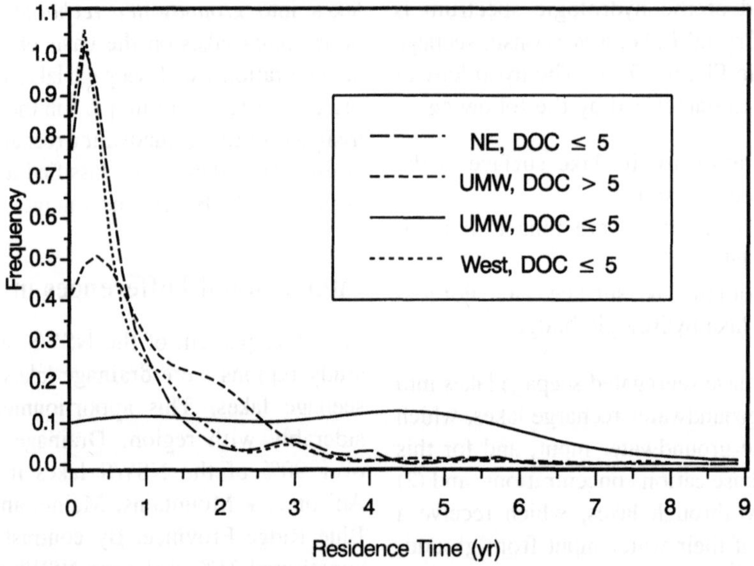

FIGURE 17.2. Density function of residence time (*RT*) values for several groups of low ANC (≤ 200 μeq L^{-1}) drainage lakes throughout the case study regions; low DOC (< 5 mg L^{-1}; 417 μmol) Northeast lakes, low DOC and high DOC Upper Midwest lakes, and low DOC Western lakes. Data are from the National Lake Survey, screened to remove lakes with watershed disturbances. *RT* = lake volume ÷ annual inflow.

= 0.8 yr) for low ANC drainage lakes in the Upper Midwest (Figure 17.2; Table 17.2). The most rapidly flushed lakes are found in the West and the Southern Blue Ridge Province, where median *RT* values are 0.3 yr and 0.1 yr, respectively. Low *RT* values for the western lakes can be related to small lake sizes, high terrestrial watershed area to lake area (WSA:LA) ratios, high precipitation (at least for the Cascade Mountains and Sierra Nevada), and impervious watersheds. Low *RT* values for the Southern Blue Ridge lakes occur because most of these lakes are reservoirs with very large terrestrial catchments and high WSA:LA ratios (Table 17.2).

There are no obvious differences in lake size or depth (as reflected in sampling site depth) or WSA:LA ratios that can account for the relatively long *RT* values in Upper Midwest drainage lakes compared to *RT* values for Maine and the Adirondack Mountains. However, precipitation is lower in the Upper Midwest (50 cm yr^{-1} < P < 80 cm yr^{-1}), compared with the Northeast (100 cm yr^{-1} < P < 130 cm yr^{-1}) and evaporation rates are similar. As a result, runoff rates are lower in the Upper Midwest (20 to 50 cm yr^{-1}, versus 40 to 100 cm yr^{-1} for the Northeast), resulting in higher *RT* values for lakes of similar size in the Upper Midwest.

In the three eastern RCS regions with significant numbers of low DOC and high DOC drainage lakes (Maine, Adirondack Mountains, and Upper Midwest), median *RT* values for the high DOC lakes are shorter than those for low DOC lakes (Table 17.2). The discrepancy is most pronounced in the Upper Midwest, where the median residence time of low DOC systems is five times longer than that of high DOC systems (3.1 yr versus 0.6 yr, respectively). To a large extent, lower *RT* values for high DOC lakes reflect their larger WSA:LA ratios (Brakke et al. 1987). For example, the median WSA:LA ratio for high DOC drainage lakes in the Upper Midwest was twice that of the low DOC drainage lakes (M = 10:1 and 5:1, respectively). All other factors being equal, the higher WSA:LA ratios for the high DOC systems would result in more total runoff and lower *RT* values. High DOC systems in the Adirondack Mountains and Maine also have higher WSA:LA ratios than do low DOC lakes, but the discrepancy is smaller than in the Upper Midwest (Table 17.2).

Seepage Lakes

Water budgets have been compiled for a number of seepage lakes in Florida and the Upper Midwest,

TABLE 17.2. Hydrologic characteristics of drainage lakes in the low ANC data set.[a]

	RT			Watershed area (ha)			Lake area (ha)			Watershed area:lake area		
	Q_1	M	Q_3	Q_1	M	Q_3	Q_1	M	Q_3	Q_1	M	Q_3
A. All DOC levels	0.1	0.4	1.1	117	256	728	10	18	49	7	11	26
Maine	0.1	0.2	0.7	158	404	1,476	10	25	101	8	13	28
Adirondacks	0.1	0.2	0.6	145	249	702	9	20	40	7	13	29
Upper Midwest	0.3	0.8	2.0	73	148	342	10	14	27	5	8	7
West	0.1	0.3	0.8	58	145	371	2	5	11	14	22	46
B. DOC ≤5 mg L⁻¹	0.2	0.6	1.6	101	249	803	10	21	75	5	8	19
Maine	0.2	0.6	1.2	122	347	1,147	10	33	129	6	8	18
Adirondacks	0.2	0.3	0.6	135	238	671	10	21	41	7	11	24
Upper Midwest	0.5	3.1	4.7	31	67	132	8	10	15	3	5	7
Southern Blue Ridge Province	0.1	0.1	0.3	1,116	5,765	20,518	22	130	212	26	44	109
West	0.1	0.3	0.8	60	152	380	2	4	9	14	22	46
C. DOC >5 mg L⁻¹	0.1	0.2	0.8	124	262	728	9	15	40	8	15	33
Maine	0.1	0.1	0.3	174	438	2,580	10	18	92	12	22	35
Adirondacks	0.1	0.2	0.3	145	280	702	7	15	30	9	17	36
Upper Midwest	0.2	0.6	1.6	98	202	596	10	17	29	6	10	23

[a] Data from National Surface Water Survey data base. *RT* = inflow-based water residence time (yr).; Q_1 = first quartile; M = median; Q_3 = third quartile.

but regional estimates of median hydrologic characteristics have not been made. Published studies show that seepage lakes typically have outflow-based residence times of 9 to 113 years (Baker et al. 1985c, Baker et al. 1986b, Lin et al. 1987, Wentz et al. 1987, Armstrong et al. 1987). Compared to seepage lakes in the Upper Midwest, seepage lakes in Florida receive more precipitation but also have higher evaporation rates. Because of this, and because of sea salt influences, the Florida lakes are more concentrated with respect to conservative constituents (see the discussion of evapoconcentration), but there is no clear regional difference in *RT* or t_w between the two regions.

Streams

In the three RCS regions where streams are the dominant surface water resource, watersheds are fairly steep with thin soils, and most headwater streams have steep gradients. Water yields appear to be substantially higher in the Southern Blue Ridge Province, where yields range from 50% to 80%, depending upon elevation, than in the western Virginia mountains or the Catskill Mountains, where yields are 35% to 40% (Chapter 8, Stoddard and Murdoch, this volume; Chapter 10, Cosby et al., this volume; Chapter 11, Elwood et al., this volume). The timing of high flows also varies. In the Southern Blue Ridge Province, peak monthly discharges occur in February to April, during the period of highest rainfall, and are at a minimum during July to November (Chapter 11, Elwood et al., this volume). Although precipitation in the western Virginia mountains is relatively uniform throughout the year, peak monthly discharges occur in late winter, reflecting decreased evapotranspiration (Webb et al. 1989). Seasonal patterns are variable in the Catskills, with peak flows occurring in early spring or late fall. Among the three stream regions, only in the Catskills does snowmelt contribute to increased spring flow, and even in this region runoff from snowmelt accounts for only 14% of annual precipitation.

Atmospheric Deposition

Atmospheric deposition throughout the Regional Case Studies areas varies from pristine to highly polluted (Figure 17.3). Precipitation in the Sierra Nevada Mountains (Emerald Lake) and the Cascade Mountains (H.J. Andrews National Forest) has pH ≥ 5.3, $[SO_4^{2-}]^* \leq 5$ µeq L^{-1}, and $[NO_3^-] \leq 5$ µeq L^{-1} (volume-weighted wet-only means) and is comparable in composition with precipitation at remote sites (Galloway et al. 1984). Wet SO_4^{2-} loadings at the Sierra Nevada and Cascade sites (65 and 138 eq ha^{-1} yr^{-1}, respectively) are higher than one might expect on the basis of the very low $[SO_4^{2-}]^*$ at these sites, because precipitation volumes are very high (> 250 cm yr^{-1}) compared to the other regions.

Precipitation quality at most Rocky Mountain sites is also typical of remote areas (see Chapter 14, Turk and Spahr, this volume), but the Rocky Mountain Beaver Meadow NADP site (Figure 17.3) has relatively high $[SO_4^{2-}]^*$ (19 µeq L^{-1}) compared with worldwide background concentrations. This may reflect a higher natural background SO_4^{2-} level resulting from Great Basin dust inputs (Young et al. 1988) or it may be a result of anthropogenic activities.

Upper Midwest sites, illustrated by Trout Lake, Wisconsin, and the Experimental Lakes Area (ELA) of Canada (Figure 17.3), show evidence of anthropogenic influence on precipitation chemistry, with pH ≤ 5.0 and $[SO_4^{2-}]^* \sim 30$ µeq L^{-1}. There is a distinct east-west gradient in SO_4^{2-} and wet H^+ deposition across the Upper Midwest, with wet SO_4^{2-} deposition ranging from > 400 eq ha^{-1} yr^{-1} in the eastern Upper Penninsula of Michigan to 200 eq ha^{-1} yr^{-1} in northeastern Minnesota.

Throughout much of the eastern United States and southeastern Canada, wet precipitation is acidic and has pH < 4.7, with average minimum values as low as 4.2 to 4.3 (e.g., 4.26 at Dorset, Ontario). Wet H^+ loadings exceed 400 eq ha^{-1} yr^{-1} in several RCS regions, including the Adirondack Mountains, southern Ontario, the western Virginia mountains, and the Southern Blue Ridge Province. Sulfate is the dominant anion in precipitation throughout the eastern United States and southeastern Canada. Excess (sea salt corrected) wet SO_4^{2-} loadings exceed 300 eq ha^{-1} yr^{-1} at all eastern United States and southeastern Canada stations in Figure 17.3 except Kejimkujik, Nova Scotia; they exceed 400 eq ha^{-1} yr^{-1} at the Shenandoah National Forest in Virginia, at Coweeta, South Carolina, at Huntington, New York, and at Dorset, Ontario. Wet NO_3^- loadings at representative sites throughout this area range from 100 to 300 eq ha^{-1} yr^{-1}, with a maximum of 320 eq ha^{-1} yr^{-1} at Dorset.

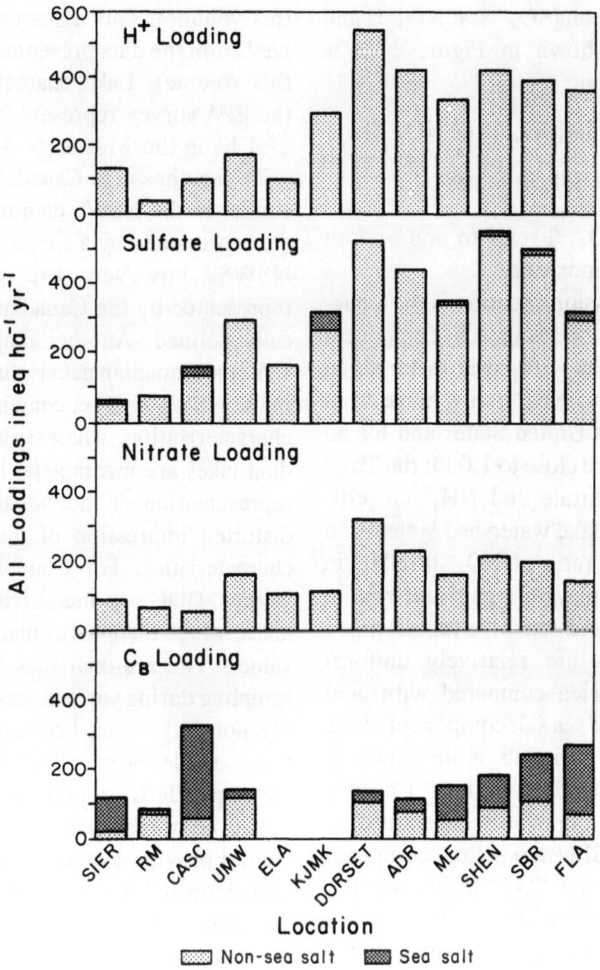

FIGURE 17.3. Wet deposition loadings throughout the Regional Case Study regions. Wet deposition of (a) H^+, (b) sulfate, (c) nitrate, and (d) C_B at representative sites throughout the Regional Case Studies area. Data sources: Sierra Nevada (SIER) = J. Melack, U. of Calif. Santa Barbara, pers. comm.; Experimental Lakes Area (ELA) and Kejimkujik (KJMK) = R.J. Vett, Environment Canada, pers. comm.); Dorset (DORSET) = Ontario Ministry of Environment Atmospheric Deposition in Ontario Study (APIOS 1982–84); Gainesville, Florida (FLA) = ES&E (1985) and C.D. Pollman, KBN Engineering and Applied Sciences, Gainesville, Florida, pers. comm. All others are NADP stations: Maine (ME) = Bridgeton, Maine; Adirondack Mountains (ADR) = Huntington, New York; Shenandoahs (SHEN) = Shenandoah Big Meadows, Virginia; Upper Midwest (UMW) = Trout Lake, Wisconsin; Cascade Mountains (CASC) = A.J. Andrews, Oregon; Rocky Mountains (RM) = Rocky Mt. National Park, Beaver Meadows, Colorado. For the ELA site, data for Cl^-, K^+, and SO_4^{2-} were considered unreliable (R.J. Vett, Environment Canada, pers. comm.) and C_B loadings are not shown. The non-sea salt concentration of sulfate and C_B was calculated by subtracting the sea salt contribution from the total concentration by using Cl^- as a sea salt tracer. For the ELA site, we assumed that the relationship between excess and sea salt sulfate was the same as that for Trout Lake, Wisconsin.

Throughout the eastern United States and southeastern Canada, excess SO_4^{2-}:NO_3^- ratios in wet deposition are close to 2:1 (range = 1.7 to 2.6); this ratio drops slightly at the two Upper Midwest sites (1.8 at Trout Lake and 1.6 at ELA). There is no clear relationship between excess wet SO_4^{2-} and NO_3^- loadings at the remote western sites (Young et al. 1988).

The relationship between $[SO_4^{2-*} + NO_3^{-*}]$ and $[H^+]$ for the stations shown in Figure 17.3 is described by the equation:

$$[H^+] = -0.91 + 0.59\,[SO_4^{2-*} + NO_3^{-*}]$$

$$(r^2 = 0.80;\ SE_Y = 6.9;\ n = 10)$$

where $[SO_4^{2-*}]$ and $[NO_3^{-*}]$ refer to non-sea salt concentrations in wet deposition.

The slope of the regression equation is < 1, suggesting that a portion of the strong acid component is neutralized by reactions with basic materials in the atmosphere. Nitrate to NH_4^+ ratios range from 1.2 to 1.9 in the eastern United States and for all three western sites, but are close to 1.0 for the Trout Lake and ELA sites. Nitrate and NH_4^+ are efficiently retained in most lake/watershed systems on an annual basis and the ratio of NO_3^- to NH_4^+ in atmospheric deposition plays a significant role in alkalinity regulation in acid-sensitive ecosystems.

Base cation loadings are relatively uniform throughout North America compared with acid anion or H^+ loadings. The sea salt component of the sum of base cation (SBC) inputs is important in near-coastal areas (Florida, western Cascade Mountains, and Maine), but even so, there is only a threefold range in wet SBC deposition, compared with a tenfold range in wet SO_4^{2-} loadings and a fifteenfold range in wet H^+ loading.

In summary, precipitation at most sites in the western RCS regions shows little or no evidence of anthropogenic influence. The extent of acidic precipitation increases significantly from west to east across the Upper Midwest. Precipitation throughout the eastern United States and southeastern Canada RCS regions is clearly acidic as the result of strong acids (H_2SO_4 and HNO_3) derived from human acitivities (Chapter 3, Husar et al., this volume).

Current Status of Surface Water Chemistry

Lakes

Nature and Limitation of Data

Our assessment of the current chemical status of lakes in the United States is based on the EPA's National Lake Survey (NLS) (Linthurst et al. 1986a, Landers et al. 1987; Introduction, Charles, this volume). Our assessment for Canada is derived from the data presented in Chapter 9 (Jeffries, this volume). Lake characterizations based upon the EPA survey represent fall conditions for lakes > 1 ha in the west or > 4 ha in the east.

The southeastern Canada lake chemistry data are similar to the NSWS data in that most of the lakes are represented by a single observation. Unlike the NSWS, however, the population of lakes represented by the Canadian survey is not statistically defined. Another major difference between U.S. and Canadian data is that each lake observation in the NSWS data set contains a complete chemical characterization, whereas the data for many Canadian lakes are incomplete. It is likely that unequal representation of individual variables results in a distorted impression of the true lake population characteristics. For example, in some Canadian studies, DOC was measured only in highly colored lakes, which obviously creates bias in median DOC values. The seasonal bias that might occur from sampling during various seasons and years is probably not very large, because a large proportion of these samples were collected using a composite tube technique (Jeffries, pers. comm.).

The combined data sets for the regional case studies provide a strong basis for assessing the current status of lake chemistry. The summary of lake chemistry for 6 regions from the United States is based on observations on more than 1,100 lakes from the ELS; the 8 regional data sets from southeastern Canada comprise observations on more than 7,400 lakes. Nevertheless, for the reasons discussed in the previous paragraph, we caution against direct comparisons between results from the U.S. and Canadian lake surveys.

Interregional Comparisons

Important aspects of the acid-base chemistry of regional case study lakes can be evaluated on the basis of several key variables: pH, ANC, DOC, SO_4^{2-}, and $[Ca^{2+} + Mg^{2+}]$ (Table 17.3). Nitrate and NH_4^+, although important in the acid-base status of lakes in other areas such as Norway (Henriksen et al. 1988) and The Netherlands (Schuurkes 1986), are generally found at concentrations near 1 $\mu eq\ L^{-1}$ for North American lakes sampled in the fall (Linthurst et al. 1986, Landers et al. 1987). We have assumed that Cl^- is conservative and not

TABLE 17.3. Quartile (25th, 75th percentiles) values of pH, ANC, DOC, SO_4^{2-}, and $[Ca^{2+} + Mg^{2+}]$ for lakes in case study regions of the United States and Southeastern Canada.[a]

Case study region	Sample size	pH	ANC (μeq L^{-1})	DOC (μM)	SO_4^{2-} (μeq L^{-1})	$Ca^{2+} + Mg^{2+}$ (μeq L^{-1})
United States[b]						
Adirondack Mountains	104	5.4–6.9	5–136	244–445	107–132	103–264
Maine	185	6.6–7.2	69–252	272–653	62–86*	132–327*
Upper Midwest	426	6.3–7.2	55–404	400–908	43–75	117–482
Southern Blue Ridge[c]	54	6.6–6.9	89–220	83–117	22–36	80–158
Florida[d]	31	4.8–6.4	−10–49	328–1130	28–94*	65–233*
	(138)	(5.8–7.2)	(5–585)	(417–1076)	(23–206)*	(116–919)*
West	393	6.8–7.3	48–160	51–159	5–27	42–156
Southeastern Canada[e]						
Northwestern Ontario	69–1100	6.7–7.5	129–622	477–925	62–94	173–477
Northeastern Ontario	903–2068	6.3–7.4	53–615	267–625	94–185	224–1008
Southcentral Ontario	839–1707	5.9–6.7	30–121	267–450	141–172	164–236
Quebec	1071–1460	5.8–6.6	26–112	333–667	62–135	118–234
Labrador	180–302	6.1–6.6	28–82	292–542	19–30*	54–100*
New Brunswick	52–84	5.8–6.9	6–83	275–533	65–94*	70–164*
Nova Scotia	275–445	5.0–6.0	−2–38	317–631	50–85*	61–135*
Newfoundland	176–270	5.8–6.6	12–50	49–792	29–50*	62–141*

[a] Values adjusted for sea salt contributions are indicated with an asterisk.

[b] Lakes with significant land use disturbances have been excluded. Values represent weighted population estimates for lakes with surface area >4 and ≤2,000 ha. Data from National Surface Water Survey.

[c] As defined in the NSWS (subregion 3A).

[d] Values for lakes unscreened for land use disturbance are shown in parentheses.

[e] Not all samples from Canadian lakes have complete chemical characterization, therefore the number of lakes represented by the interquartile range varies depending upon parameter. The range of lakes represented is shown in column 2. See Jeffries et al., this volume, for discussion of Canadian data bases.

associated with chronic acidification of lakes in relatively undisturbed lakes in North America (Sullivan et al. 1988b, Baker et al. 1988b).

Unlike southern Norway, where about 70% of the sampled lakes are acidic (Henriksen et al. 1988), relatively small percentages of lakes in the case study regions have pH ≤ 5.0 or ANC values ≤ 0 (Table 17.4). Florida has the highest percentage of acidic lakes (46% of lakes in the case study region), but this reflects the exclusion of lakes with substantial watershed disturbances. Including these lakes lowers the estimated percentage of acidic lakes in Florida to 22% (Linthurst et al. 1986).

The inclusion of small lakes (< 4 ha) in the assessment of lake chemistry substantially increases the number of acidic lakes in the Adirondack Mountains (Kretser et al. 1989) and Maine (Chapter 7, Kahl et al., this volume), but does not alter the assessment for lakes in the western United States, where the lake chemistry is independent of lake area (Landers et al. 1987). The acid-base status of lakes in the northeastern and western United States and southeastern Canada is expected to be very sensitive to temporal variation associated with snowmelt and storm events (see section on episodic acidification).

The enumeration of acidic waters is highly sensitive to the criteria used to define "acidic." For example, an ANC value of ≤ 0 has commonly been used in the United States to describe acidic waters. However, from a biological perspective, aquatic biota may exhibit detrimental effects as pH decreases to values below 6.0 (Schindler 1988b). The percentage of lakes considered acidic based on ANC ≤ 0 and pH ≤ 6.0 for many areas differs by a factor of two to four. However, the percentages of acidic lakes defined on the basis of $SO_4^*:C_B^* > 1$ are in closer agreement with estimates based on ANC ≤ 0 (Table 17.4). One major exception occurs with lakes in Florida, where the estimated percentage of acidic lakes defined on the basis of $SO_4^*:C_B^* > 1$ is far lower than that based on either pH or ANC criteria. This occurs because organic acids contribute to the acidity of many Florida lakes.

TABLE 17.4. Percentage of lakes in the case study regions above or below specified reference values.[a]

	pH		ANC (μeq L^{-1})		DOC (μm)	SO_4:C_B	SO_4:[Ca+Mg]	A^-:C_B	A^-:[Ca+Mg]
	≤5.0	≤6.0	<0	≤200	(>417)	(>1)	(>1)	(>1)	(>1)
United States[b]									
Adirondack Mountains	14.4	35.8	18.4	85.9	33.2	16.6	32.7	0	1.1
Maine	0.5	5.1	1.5	69.4	57.1	2.6*	4.1*	1.5*	2.6*
Upper Midwest	2.9	14.5	5.4	57.8	73.9	1.7	4.7	2.8	6.5
Southern Blue Ridge	0	1.2	0	70.4	1.7	0	0	0	0
Florida[c]	34.8	67.7	46.0	87.1	62.7	19.7*	19.7*	15.1*	17.3*
	(11.7)	(31.8)	(21.9)	(53.9)	(69.3)	(9.6)*	(16.7)*	(6.8)*	(10.0)*
West	0	1.2	0	83.0	4.5	0	0	0	1.9
Southeastern Canada[d]									
Northwestern Ontario	0	4.0	0	39.7	81.2	[e]	0	[e]	0
Northeastern Ontario	4.5	18.1	6.9	53.2	48.0	[e]	6.9	[e]	0
Southcentral Ontario	1.7	29.8	2.8	83.9	31.9	[e]	10.1	[e]	0.1
Quebec	4.1	42.0	3.1	87.0	64.2	[e]	7.7	[e]	4.4
Labrador	0.4	16.6	0	95.1	52.5	[e]	0.6*	[e]	22.7*
New Brunswick	7.1	35.3	11.1	96.3	47.4	[e]	32.7*	[e]	1.4*
Nova Scotia	25.8	75.1	27.5	97.1	54.6	[e]	17.0*	[e]	17.9*
Newfoundland	3.0	44.3	6.8	94.3	52.6	[e]	7.4*	[e]	17.4*

[a] Variables with an asterisk have been seasalt corrected.
[b] Lakes with significant land use disturbances have been excluded. Values represent weighted population estimates for lakes with surface area >4 ha and ≤2,000 ha. Data from National Surface Water Survey data base.
[c] Percentages for the population of lakes not screened for land use disturbances are shown in parentheses.
[d] Not all Canadian lakes have complete chemical characterization; therefore, percentages of lakes represent values based on unequal numbers of lakes for each variable. Caution should be used in interpreting the significance of the percentage of Canadian lakes above or below reference values because it is uncertain how the sample represents the total population of Canadian lakes. See Jeffries et al., this volume, for description of Canadian data bases.
[e] Relatively few Canadian lakes have complete analyses on all base cations; therefore, percentages are computed only for lakes with [Ca+Mg].

Figure 17.4a depicts the location and hydrologic characterization of low ANC lakes for regional case studies in the United States. Most of the low ANC lakes in the United States are drainage lakes in the West, the Northeast, and the Upper Midwest. Most of the high DOC lakes are in the eastern United States, particularly in Minnesota, Maine, and Florida, where wetlands are common.

The distribution of *acidic* lakes (ANC ≤ 0) in the United States is radically different from the distribution of low ANC lakes (Figure 17.4). Within the screened RCS data set of lakes > 4 ha, an estimated 247 acidic lakes are in Florida, exceeding the number of acidic lakes in the Adirondack Mountains and Maine. The West, which has the largest number of low ANC lakes in the United States, is virtually devoid of acidic lakes. Of the acidic lakes > 4 ha in the case study regions, 63% are characterized as groundwater recharge lakes; only 32% are drainage lakes. The majority (72%) of the acidic lakes are low DOC systems and a high proportion (68%) of the acidic, low DOC lakes are groundwater recharge systems.

Multivariate View of Acid-Base Chemistry

Using [Ca^{2+} + Mg^{2+}] to represent lake sensitivity, SO_4^{2-} as the mineral acid component, and DOC as the organic acid surrogate, we can view the distribution of selected groups of low ANC lakes in three dimensions (Figure 17.5). The Adirondack Mountains have the highest [SO_4^{2-}] and are intermediate with respect to [Ca^{2+} + Mg^{2+}] and DOC concentrations (Figure 17.5a). Sierra Nevada lakes have the lowest concentrations of all three constituents. Characteristics of the Cascade Mountain lakes (not shown) are intermediate between those of Sierra Nevada lakes and Rocky Mountain lakes. Sulfate-retaining soils

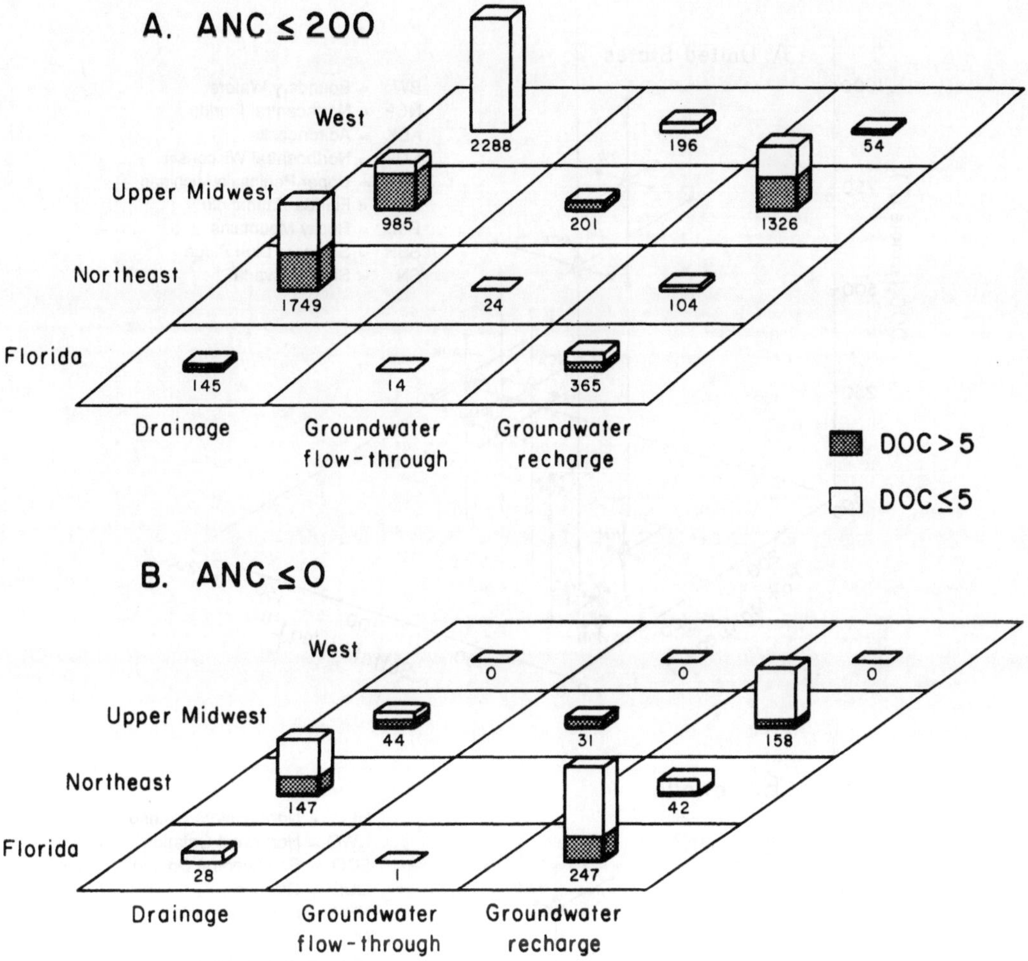

FIGURE 17.4. Frequency of low DOC (≤ 5 mg L^{-1}; 417 μmol) and high DOC lakes for case study regions. The West includes the Cascade Mountains, Sierra Nevada, and Rocky Mountains; the Upper Midwest includes subregions 2a–2e; the Northeast represents the Adirondack Mountains and Maine only; and Florida includes subregions 3a and 3b. The frequencies are based on the estimated number of lakes > 4 ha and ≤ 2,000 ha from EPA's National Lake Survey, screened to exclude lakes perceived to have substantial anthropogenic land use disturbance; (a) is limited to lakes with ANC ≤ 200 μeq L^{-1}; (b) is restricted to lakes with ANC ≤ 0. Hydrologic and DOC classifications are described in Appendix A.

in the Southern Blue Ridge Province maintain chemical characteristics in the streams and reservoirs similar to lakes in the West, despite a fivefold difference in SO_4^{2-} deposition between the regions. Weathering of Na^+ from watersheds in the West and Southeast contributes substantially to total cation production, and if included in the cation axis, would place these lake groups near those in the Upper Midwest and Northeast. The Florida case study region can be divided into three distinct groups. Lakes in the Florida panhandle have low DOC, are extremely dilute, and are notable in their homogeneity, whereas lakes in the northcentral peninsula have higher DOC and SBC and are more variable in composition. Lakes in the southcentral group (not shown) are extremely heterogeneous (Chapter 12, Pollman and Canfield, this volume). Lakes in the Boundary Waters area of northern Minnesota have the highest median concentrations of DOC and $[Ca^{2+} + Mg^{2+}]$, and low $[SO_4^{2-}]$ (Figure 17.5a). There is a downward gradient in $[Ca^{2+} + Mg^{2+}]$ and DOC moving eastward across the Upper Midwest,

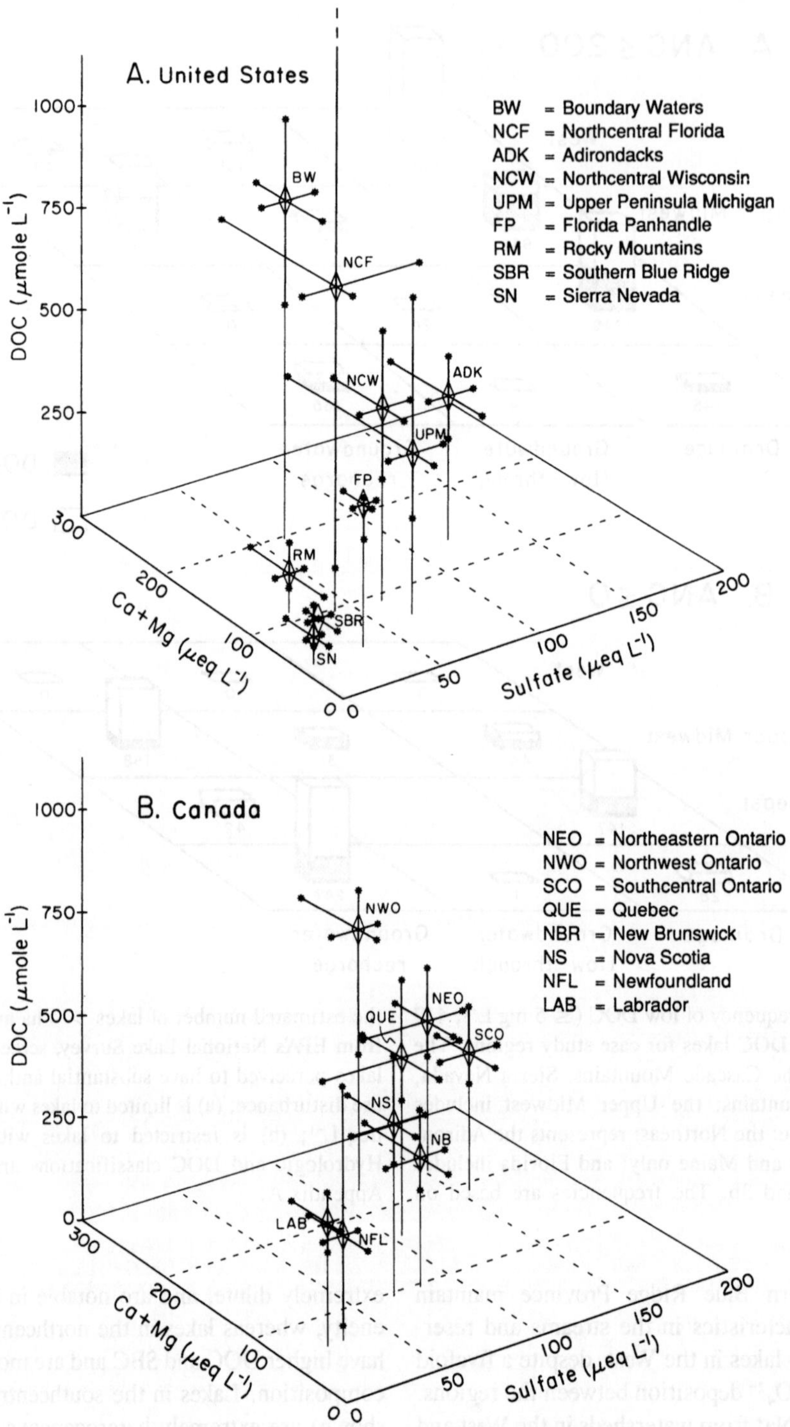

FIGURE 17.5. Representation of selected case study lake populations in three-dimensions represented by DOC, Ca^{2+} + Mg^{2+}, and SO_4^{2-}. The midpoint of each lake group is the estimated population median; the quartiles are represented by the terminal points of the intersecting lines. (A) Low ANC ($\leq$ 200 μeq L^{-1}) lakes in the United States, screened to remove anthropogenic disturbances. (Data from National Lake Survey.) (B) Low ANC lakes in Canada. (Data from Jeffries et al., Chapter 9, this volume.)

reflecting an increasing predominance of seepage lakes. Although there is a twofold increase in wet SO_4^{2-} deposition across the Upper Midwest, median lakewater [SO_4^{2-}] does not follow a clear parallel trend.

A plot of these three variables for low ANC Canadian lakes (Figure 17.5b) depicts a pattern quite different from that observed for lakes in the United States. Lakes in northwestern Ontario have high concentrations of DOC, divalent base cations, and SO_4^{2-}, a combination observed in the United States only in Florida's southcentral ridge. Because northwestern Ontario includes lakes from a wide range of deposition zones, it is difficult to evaluate the primary source of SO_4^{2-} to this group of lakes. However, the sample size of DOC for northwestern Ontario lakes is small ($n = 69$) and it is uncertain how representative these few samples are for the area. Lakes in northeastern Ontario and southcentral Ontario have lower concentrations of base cations and DOC and higher SO_4^{2-} than northwestern Ontario. Canada's four maritime provinces are distinguished primarily on the basis of SO_4^{2-}. Concentrations are typically threefold greater in Nova Scotia and New Brunswick than in Labrador and Newfoundland. Lakes in Quebec have concentrations of divalent base cations and SO_4^{2-} intermediate between those in the maritime provinces and Ontario, although the interquartile range for SO_4^{2-} in Quebec is notably greatest among all provinces. This may reflect a high degree of scatter associated with a small sample size for Quebec. The moderate to high concentrations of DOC in the southeastern Canadian lakes are comparable to those in the Northeast and Upper Midwest, but the validity of this comparison is limited by the questionable representativeness of the Canadian data.

TABLE 17.5. Aluminum concentrations and percentages of lakes and streams with high Al in the case study regions.

Regional case study area	n^a	Al_{MIBK}[b] (µg L^{-1}) Q_1 - Q_3	Al_{MIBK} > 100 µg L^{-1} (%)
Lakes			
Adirondack Mountains	104	5 - 34	18.1
Maine	185	4 - 16	0.5
Upper Midwest	426	2 - 17	1.0
Southern Blue Ridge	54	2 - 3	0
Florida	31	0 - 22	4.5
	(138)	(0 - 18)	(2.6)
West	393	1 - 6	0
Streams			
Catskill Mountains			
NSS subregion 1D	58	2 - 20	5.1
NSS Catskill sites	6	–	1.5
NYC/USGS study[c]	51	15 - 30	10.2
Western Virginia[d]			
NSS WV/MD/VA ridges	25	2 - 13	5.6
NSS VA ridges	13	3 - 11	12.2
NSS Southern Blue Ridge	54	2 - 4	0

[a] Sample size for same lakes presented in Tables 17.3 and 17.4.
[b] Methyl-isobutyl-ketone (MIBK) extractable Al (Hillman et al. 1986).
[c] Total dissolved aluminum.
[d] Aluminum was not measured in the VTSSS.

Aluminum

Aluminum plays a minor role in the acid-base chemistry of most case study lakes, but Al has a major effect on the acid-base chemistry and the aquatic biota in acidic lakes (Driscoll et al. 1980). Aluminum measurements in Canadian lakes have generally been restricted to localized research efforts and this summary of Al is limited to lakes in the United States. Aluminum exists in a variety of forms (Chapter 2, Munson and Gherini, this volume), but only inorganic monomeric Al (Al_i) is believed to be associated with fish toxicity. Methyl-isobutyl ketone aluminum (Al_{MIBK}) measured in the National Lake Survey generally overestimates Al_i because it includes organically bound Al (Al_o). In highly acidic, low DOC waters, $Al_{MIBK} \sim Al_i$, but the deviation between the two parameters increases with increasing pH and DOC. High concentrations of Al_{MIBK} occur in acidic Adirondack lakes, presumably because of high rates of mobilization from the watersheds to these short residence time lakes (Table 17.5). Aluminum concentrations in drainage lakes in the Northeast probably increase substantially during the spring as a consequence of increased mobilization from upper soil horizons, enhanced by elevated NO_3^-, and increased hydrologic throughput (Driscoll and Newton 1985). Aluminum concentrations in most acidic seepage lakes are low because these lakes receive most of their water input from direct precipitation. Since there is limited contact between acidic deposition and the soil, there is little opportunity for Al solubilization. In other groups of RCS lakes and streams in the United States, there are few acidic systems, and consequently few systems with high Al concentrations (Table 17.5).

Streams

Nature and Limitation of Data

The primary data base for the analysis of lakes in the RCS project was the National Lake Survey (NLS), which included the ELS and WLS. Thus, a statistically based sampling approach and uniform methods were used to analyze a defined target population of lakes in all regions. In contrast, most of the information about streams in the mountains of western Virginia and the Catskill Mountains presented in this book (Chapter 10, Cosby et al., this volume; Chapter 8, Stoddard and Murdoch, this volume) was based on smaller regional surveys. These surveys were not based upon a probability sampling design, nor were sites selected in the same way as in the National Stream Survey (NSS). The NSS was not completed until the RCS project was well underway; thus, the NSS was the primary data base for only one case studies region, the Southern Blue Ridge Province (Chapter 11, Elwood et al., this volume).

The primary data base used by Stoddard and Murdoch (Chapter 8, this volume) in the analysis of Catskill streams was a study by the City of New York and the U.S. Geological Survey. The 51 streams in this survey were chosen because their watersheds provide drinking water for the City of New York; site selection was deliberately biased toward low ANC systems (Chapter 8, Stoddard and Murdoch, this volume). The NSS included 58 sampled reaches, representing a population of 3,244 $\pm$ 347 reaches, in NSS subregion 1D, which also included the Pocono Mountains and the glaciated plateau region of northeast Pennsylvania, in addition to the Catskills. For comparison to the NYC/USGS survey, we also present data for the Catskills area of NSS subregion 1D, but we emphasize that this data base comprises only 6 sampled reaches, representing a population of 161 $\pm$ 86 reaches located in the Catskill Mountains.

Chapter 10 (Cosby et al., this volume), on the western Virginia mountains, is based largely on the Virginia Trout Stream Sensitivity Study (VTSSS) and two smaller surveys in the Shenandoah National Park. Again, these surveys are not directly comparable with the NSS. The VTSSS included 344 stream sites, believed to include about 78% of Virginia streams with naturally reproducing brook trout populations, whereas the Shenandoah Park surveys were limited to the confines of the Shenandoah National Park. The VTSSS is compared with (1) ridge sites in the West Virginia/Maryland/Virginia portion of the NSS (25 sampled sites representing a population of 6,330 $\pm$ 1,182 stream reaches), and (2) ridge sites in Virginia only (13 sampled reaches representing a population of 3,231 $\pm$ 884 reaches).

Interregional Comparisons

Although the designs among the stream surveys differ and none have the same design as the NSS, the composite picture (Figure 17.6[a–d]) shows clear interregional variations in stream chemistry. Streams in all three regions are essentially dilute solutions of Ca^{2+} and Mg^{2+}, with SO_4^{2-} as the dominant strong mineral acid anion (Chapter 8, Stoddard and Murdoch, this volume; Chapter 10, Cosby et al., this volume; Chapter 11, Elwood et al., this volume; Kaufmann et al. 1988). In all three stream regions, baseflow DOC concentrations are generally < 2 mg L^{-1} and virtually all streams have < 4 mg L^{-1} (330 μmol) DOC. Thus, unlike the situation for many case study lakes, organic acids are a relatively unimportant component of the acid-base chemistry in the large majority of streams in the three stream case study regions. We note, however, that organic anions play an important role in the acid-base status in many NSS streams that were not included in the Regional Case Studies, notably those in the Mid-Atlantic Coastal Plain and Florida (Kaufmann et al. 1988).

Base cation concentrations are generally highest in the Catskills region, intermediate in the western Virginia mountains, and lowest in the Southern Blue Ridge Province (Figure 17.6a). Southern Blue Ridge streams also have much less regional variation in base cation concentrations than do the other two NSS subregions (Figure 17-6a).

The north–south trend in stream SO_4^{2-} concentrations is even more dramatic (Figure 17.6b). Median SO_4^{2-} concentrations are well over 100

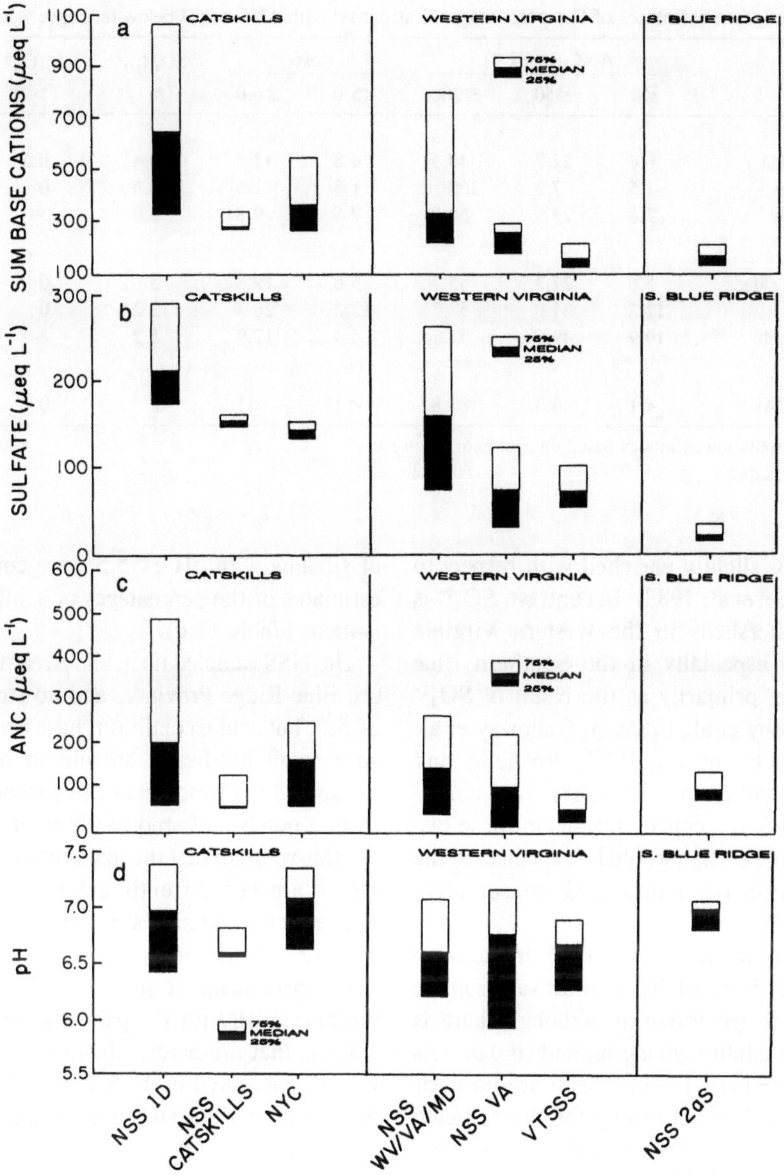

FIGURE 17.6. Comparison of regional chemical status and RCS sample characteristics during spring baseflow in streams of the Catskill Mountains, western Virginia, and the Southern Blue Ridge Province: (a) C_B, (b) SO_4^{2-}, (c) Gran ANC, and (d) pH (closed headspace). Sample codes are as follows: NSS-1D = upstream ends of reaches in NSS subregion 1D (n = 58); NSS Catskills = NSS upstream ends of reaches in the Catskill Mountains (n = 6); NYC = New York City/USGS study (n = 51); NSS WV/VA/MD = NSS sites in the ridges of West Virginia, Virginia, and Maryland (n = 25); NSS-VA = NSS ridge sites in Virginia (n = 13); VTSSS = Virginia Trout Stream Sensitivity Survey (n = 344); NSS 2As = upstream ends of reaches in NSS subregion 2As (n = 54).

μeq L⁻¹ in the Catskill region and in all of NSS subregion 1D. Sulfate concentrations are lower in the the western Virginia mountains and in the ridge sites of West Virginia, Maryland, and Virginia, and lower yet, typically < 30 μeq L⁻¹, in the Southern Blue Ridge streams. Streams in the Poconos/Catskills region (NSS subregion 1D) are apparently near steady state (e.g., watershed inputs = water-

TABLE 17.6. Population estimates of the percentage of streams[a] with ANC or pH below reference values.

	ANC (μeq L^{-1})			pH		SO_4^*:C_B^*	A^-:C_B^*	DOC
	≤0	≤50	≤200	≤5.0	≤6.0	(>1)	(>1)	(>416 μM)
Catskills								
NSS Subregion 1D	6.4	22.7	48.3	4.8	12.8	4.4	0.5	6.1
NSS Catskill sites	1.5	7.3	100	1.5	1.5	1.5	0	0
NYC/USGS study	7.8	23.5	60.8	7.8	9.8	2.0	—[b]	—[b]
Western Virginia								
NSS WV/V/MD ridges	5.6	27.3	55.7	5.6	19.8	5.6	0	0
NSS VA ridges	12.2	31.0	57.7	12.2	26.3	12.2	0	0
RCS VTSSS sample	10.9	49.9	92.7	4.4	17.9	1.2	—[b]	—[b]
Southern Blue Ridge								
NSS Subregion 2As	<1	6.3	83.8	<1	<1	0	0	0

[a] NSS values are population estimates based on upstream reach ends.
[b] DOC was not measured.

shed outputs) or slightly enriched with respect to SO_4^{2-} (Kaufmann et al. 1988). In contrast, SO_4^{2-} is retained in watersheds in the western Virginia mountains and especially in the Southern Blue Ridge Province, primarily as the result of SO_4^{2-} adsorption (Cosby et al. 1985a,b, Galloway et al. 1983a,b, Rochelle et al. 1987, Rochelle and Church 1987, Kaufmann et al. 1988). As a consequence, stream SO_4^{2-} concentrations in these two regions are far lower than would be expected on the basis of atmospheric inputs and conservative behavior in watersheds.

Concentrations of ANC generally decrease from highest values in the Catskills to lowest values in the Southern Blue Ridge Province, although there is considerable variability among individual data sets within regions (Figure 17.6c). There are no clear differences in median pH among the three stream regions (Figure 17.6d). In the Catskills case study region, Stoddard and Murdoch (Chapter 8, this volume) report that 8% of the streams were acidic during baseflow, compared with a population estimate of 6.4% for stream reaches in NSS subregion 1D (Table 17.6). In the Catskills area of the NSS, an estimated 1.5% of the streams were acidic, but this estimate is probably not reliable since it is based on only six sampled sites. In the western Virginia mountains, estimates of the percentage of acidic systems were consistent between the VTSSS study (10.7%) and the NSS Virginia ridge sites (12.2%); these estimates are higher than those for the entire NSS West Virginia/Maryland/Virginia subset (5.6%). For both of these regions, the percentages of streams with pH ≤ 5.5 were comparable with estimates of the percentages of acidic (ANC ≤ 0) systems (Table 17.6).

The NSS sampled no acidic streams in the Southern Blue Ridge Province, and no streams with pH ≤ 5.5, but acidic conditions have been reported for very small headwater streams at high elevations (Chapter 11, Elwood et al., this volume). It is clear from a number of studies (Chapter 11, Elwood et al., this volume) that the main reason streams in this region are not currently acidic is that watershed soils effectively retain atmospheric inputs of SO_4^{2-} by SO_4^{2-} adsorption.

The percentage of streams with Al_{MIBK} concentrations > 100 μg L^{-1} parallels the percentage of streams that are acidic (Table 17.5). Although Al was not measured in the VTSSS, about 12% of the NSS Virginia ridge sites had Al_{MIBK} concentrations > 100 μg L^{-1}. This percentage is somewhat higher that that estimated for ridge sites in the West Virginia/Maryland/Virginia area combined (Table 17.5). The NYC/USGS study indicates that total dissolved Al was > 100 μg L^{-1} in 10.2% of the Catskill streams; < 2% of the Catskills NSS streams have Al_{MIBK} > 100 μg L^{-1}. This discrepancy parallels the discrepancy in estimates of acidic systems between the two studies discussed earlier. Aluminum concentrations in the Southern Blue Ridge Province are universally < 100 μg L^{-1}, reflecting the absence of acidic systems in the sampled NSS population. Because most of the case study region streams have low DOC, Al_i generally constitutes the dominant fraction of total monomeric Al.

Acidic streams in the Catskill and western Virginia mountain case study regions are almost certainly acidic as a result of acidic deposition rather than other causes. This assertion is based upon three lines of evidence:

1. Sulfate is the dominant anion in acidic streams in these regions.
2. Ratios of $SO_4^{2-}:C_B$ exceed 1.0 in most acidic streams in these regions, strongly supporting the contention that acidification has occurred because H_2SO_4 inputs have exceeded base cation production (Table 17.6).
3. Input-output budget calculations show that atmospheric deposition is the dominant source of SO_4^{2-} (i.e., oxidation of reduced sulfur minerals is not an important source of SO_4^{2-} for streams in these regions). In fact, there is strong retention of SO_4^{2-} in watersheds in the western Virginia mountains (Chapter 10, Cosby et al., this volume) and in the Southern Blue Ridge Province (Chapter 11, Elwood et al., this volume).

In making these conclusions, we note that regional surveys avoided sampling streams believed to be impacted by acid mine drainage and that the NSS population estimates are based on a sample that has been screened to remove acid mine drainage streams (Kaufmann et al. 1988, Herlihy et al. in press).

Relationship of Surface Water Chemistry to Acidic Deposition

If all lakes and streams had similar chemistry prior to the onset of acidic deposition, the relationship between acidic deposition and surface water chemistry would be clear: SO_4^{2-} and C_B would increase and ANC would decrease, along a deposition gradient. In reality, surface water chemistry is highly variable, even within a small geographic area, as discussed in Chapters 6 through 16. Despite this within-region heterogeneity, spatial patterns of surface water chemistry across the continent show a clear relationship to acidic deposition.

First, the relationship between atmospheric wet SO_4^{2-} loadings and surface water SO_4^{2-} concentrations is more or less linear across the United States case study regions. Surface water SO_4^{2-} concentrations are very low, typically < 25 μeq L^{-1} throughout the western case study regions (Figure 17.7), corresponding to low atmospheric deposition rates (< 10 kg ha^{-1} yr^{-1}; 20 eq ha^{-1} yr^{-1}). Surface water SO_4^{2-} concentrations increase across the Upper Midwest and are typically highest in the mid-Appalachian area, approaching 200 μeq L^{-1} in the Poconos/Catskills region. The only departure from this general trend is in the Southeast, where SO_4^{2-} adsorption results in much lower concentrations than one would expect on the basis of atmospheric inputs and steady-state conditions. A similar relationship exists among the Canadian case study regions, with lowest SO_4^{2-} concentrations occurring in the low deposition regions (northwest Ontario, Labrador, Newfoundland, and New Brunswick) and much higher concentrations occurring in southcentral Ontario, north-eastern Ontario, and Quebec.

The relationship between the occurrence of acidic lakes and wet SO_4^{2-} loadings is certainly not linear (Figure 17.7), because many factors in addition to acidic deposition control ANC, including hydrologic flow path, soil characteristics, lake hydraulic residence times, and bedrock geology (Chapter 2, Munson and Gherini, this volume). A relatively small percentage of lakes and streams in high deposition areas are acidic. The fact remains, however, that acidic surface waters in the case study regions are found almost exclusively in areas receiving wet SO_4^{2-} loadings > 20 eq ha^{-1} yr^{-1}, with the exception of a number of acidic, high DOC lakes in the Canadian maritime provinces (Table 17.4). In low deposition areas within the United States, including the western case study regions and Minnesota, there are virtually no acidic lakes, nor were any sampled in northwestern Ontario. There are very few acidic lakes or streams in the Southern Blue Ridge Province (Chapter 11, Elwood et al., this volume), even though wet SO_4^{2-} deposition is very high, reflecting the importance of SO_4^{2-} adsorption, which confers an additional degree of buffering to surface waters beyond that provided by base cation production.

The relationship between acidic deposition and surface water acidification is made clearer by examining ratios of acid anions (SO_4^{2-} and A^-) to base cations (C_B) in acidic lakes along a SO_4^{2-} deposition gradient (Figure 17.7), for U.S. case study regions only. As discussed in Chapter 2 (Munson and Gherini, this volume), a $SO_4^{2-}:C_B$

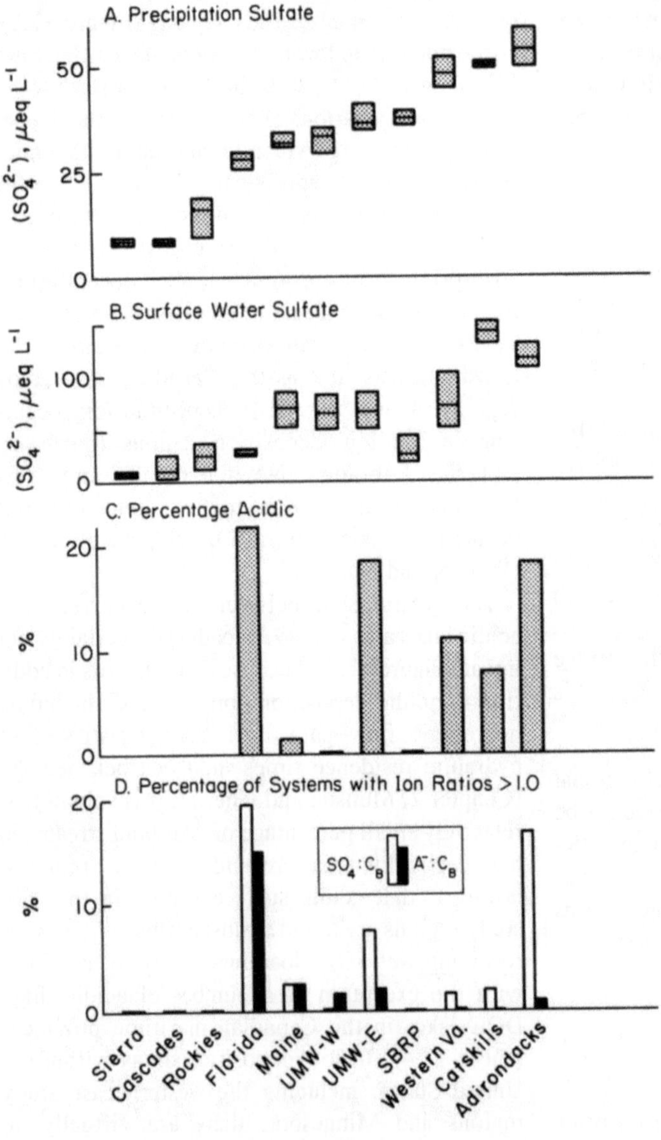

FIGURE 17.7. Estimated median and quartile values for: (a) precipitation [SO_4^{2-}*], (b) surface water [SO_4^{2-}*], (c) percent acidic systems, and (d) percentages of acidic systems with ion ratios > 1.0, in case study regions plotted in order of increasing estimated [SO_4^{2-}] in wet deposition. Lake data are derived from EPA's National Lake Survey, screened to remove sites with watershed disturbances. Stream data represented are as follows: Southern Blue Ridge Province (SBRP), National Stream Survey; Catskill mountains, NYC/USGS study (Stoddard and Murdoch, Chapter 8, this volume); western Virginia mountains, Virginia Trout Stream Sensitivity Study (Cosby et al., Chapter 10, this volume).

ratio > 1.0 is fairly clear evidence of anthropogenic acidification, assuming that watershed SO_4^{2-} production from sulfide mineral weathering is not important. Similarly, $A^-:C_B$ ratios > 1.0 suggest that organic acids alone are sufficient to cause acidic conditions. In the U.S. case study regions, lakes and streams with $SO_4^{2-}:C_B$ ratios > 1.0 are found primarily in the Adirondacks, Florida, the eastern Upper Midwest, the western Virginia mountains, and the Catskills, but are absent in the western Upper Midwest and the West. In Canada, $SO_4^{2-}:(Ca^{2+} + Mg^{2+})$ ratios > 1.0 are found primarily in southcentral and northeastern Ontario, Quebec, New Brunswick, and Nova Scotia. Thus, ratios of $SO_4^{2-}:C_B$ > 1.0 are found only in areas with elevated SO_4^{2-} deposition (Figure 17.7). In the United States, ratios of $A^-:C_B$ > 1.0 are found only in the Upper Midwest and Florida, and ratios of $A^-:(Ca^{2+} + Mg^{2+})$ > 1.0 are fairly common in the Canadian maritime provinces (Table 17.4). Flat topography and extensive wetland area is a ubiquitous feature of regions where $A^-:C_B$ ratios

> 1 occur; these features appear to be important in defining the distribution of naturally acidic, organic surface waters.

Episodic Acidification

Most regional assessments of lake and stream acid-base status have focused on chronic acidification, as indicated by summer or fall index samples for lakes (Linthurst et al. 1986a, Jeffries et al. 1986a, Cook et al. 1988), or on spring baseflow sampling for streams (Kaufmann et al. 1988). These assessments underestimate the extent of acidification with respect to damage to aquatic organisms because they do not consider the pH and ANC depressions and increases in dissolved Al that are associated with episodic events (snowmelt or storms).

Episodic Processes

Episodic excursions from baseline pH and ANC occur during storm events, but the greatest potential for episodic pH depressions is associated with snowmelt events. Several mechanisms contribute to episodic events (Marmorek et al. 1986).

Dilution of Baseflow

During snowmelt or major storm events, there is often less contact between runoff and soils than at other times of the year, because the soils are either saturated or frozen. As a result, chemical modification of runoff by weathering and cation exchange reactions is reduced and stream concentrations of base cations are lower than during baseflow periods (Galloway et al. 1987, Driscoll et al. 1987a,b).

Concentration of Solutes During Snowmelt

Much of the solute loading that accumulates during snow cover is released during the early melt period, which concentrates acidic materials (Semkin and Jeffries 1986, Cadle et al. 1987) and results in an acid pulse during early snowmelt.

Watershed Biogeochemical Processes

In watersheds that contain bogs or other reducing environments, oxidation of reduced sulfur can result in releases of sulfuric acid during storms following dry periods (Kerekes et al. 1982, Bayley et al. 1986). Pulses of organic acids during storm events may contribute to episodic acidification of streams in the Northeast (Hemond and Eshleman 1984; Chapter 7, Kahl et al., this volume). Retention of NO_3^- from biological uptake is generally thought to ameliorate episodic pH and ANC depressions associated with nitric acid inputs (Cadle et al. 1987), but mineralization of organic nitrogen followed by nitrification may contribute to acidification of some Adirondack Mountains lakes (Rascher et al. 1987, Galloway et al. 1987).

Degassing During Ice-out and Vernal Mixing

In addition to the foregoing mechanisms that cause changes in ANC, wintertime respiration during ice cover can also result in carbonation of lake water, which decreases pH but has no effect on ANC. In several northern Wisconsin lakes, this mechanism has been shown to cause gradual decreases in pH during the winter, with late-winter (before ice-out) pH values as much as 1 pH unit or more lower than summertime averages. These lakes degas rapidly following ice melt, with a concomitant increase in pH to summertime values (but no change in ANC) over a period of a few days (Kratz et al. 1987).

Extent and Magnitude of Episodic Acidification

There have been no synoptic studies of episodic acidification in any of the case study regions except the Catskill Mountains (Chapter 8, Stoddard and Murdoch, this volume), and the extent and severity of episodic events in other areas are uncertain. Nevertheless, episodic events have been studied in one or more systems in nearly every case study region and these studies have been augmented by a model-based regional assessment (Eshleman 1988).

Episodic pH and ANC depressions associated with snowmelt have been reported for a number of short residence time lakes in the Adirondack Mountains and are associated with pulses of HNO_3^- and dilution of base cations (Galloway et al. 1980, 1987, Driscoll and Schafran 1984, Driscoll et al. 1987a,b). The importance of NO_3^- in episodic acidification has been ascribed to the relatively high ratio of NO_3^- to SO_4^{2-} in snowpack (Johannes et al. 1980) and nitrification in the forest floor (Rascher et al. 1987). Panther Lake probably represents an extreme case of episodic lake acidification: pH declined from 6.6 to 4.8 and ANC decreased from 162 to −18 μeq L^{-1} during the 1978 snowmelt (Galloway et al. 1980). Eshleman (1988) developed a regression relationship between fall and spring ANC for lakes in the Adirondack Mountains. He concluded that median spring ANC values were 65 μeq L^{-1} lower than median fall ANC values and that 35% of the lakes in the Adirondacks would have ANC ≤ 0 during spring snowmelt, compared with 11% for the fall index sample.

A survey of 51 streams showed that episodic acidification is fairly extensive throughout the Catskills (Chapter 8, Stoddard and Murdoch, this volume). At base flow, only 8% of the streams had mean pH ≤ 5.5 and 8% had mean ANC ≤ 0. At high flow, 18% had pH ≤ 5.5 and 16% had ANC ≤ 0. These results are similar to model predictions reported by Eshleman (1988) indicating that 23% of the upstream ends and 4% of the downstream ends of NSS stream reaches in the larger Catskills/Poconos region (NSS 1D) would have ANC ≤ 0 during episodic events. Episodic acidification of streams in the Catskill Mountains is associated with both snowmelt and storm events, and results from dilution of base cations and increased NO_3^-, even though there is some dilution of SO_4^{2-}. Because base cations are diluted to a greater extent than SO_4^{2-}, the $SO_4^{2-}:C_B$ ratio increases during episodes (Chapter 8, Stoddard and Murdoch, this volume).

Episodic acidification is well documented in central Ontario and southern Quebec (Chapter 9, Jeffries, this volume). In the Muskoka-Haliburton area, spring pH depressions are commonly in the range of 1 pH unit (Jeffries et al. 1979). In this region, SO_4^{2-} is the dominant anion during snowmelt.

There is relatively little information on episodic pH and ANC depression in the Upper Midwest. Available evidence suggests that ANC depressions occur in many streams, but severe pH depressions are not common. The apparent lack of severe snowmelt episodes in this region has been attributed to smaller snowpack (relative to the Northeast), lower atmospheric deposition of acid anions, and the predominance of seepage lakes that receive little water input from their watersheds (Chapter 13, Cook and Jager, this volume). In several streams that have been studied in detail, ANC depressions appear to be more closely related to dilution of base cations and to a lesser extent, increases in organic acids, than to increases in mineral acid anions (Siegel 1981, Schnoor et al. 1984, Cadle et al. 1987, Eilers and Bernert 1990; Chapter 13, Cook and Jager, this volume).

Episodic events in the western Virginia mountains and Southern Blue Ridge Province are associated with storm events rather than snowmelt. Model predictions show that 9% of the upstream ends of reaches in the Valley and Ridge Province and 3% of the upstream ends in the Southern Blue Ridge Province should experience episodic acidification severe enough to cause ANC ≤ 0 (Eshelman 1988, Kaufmann unpublished). Cases of substantial pH or ANC depressions have been documented for individual streams in these regions, for example, Raven Fork, North Carolina (Chapter 11, Elwood et al., this volume), and Deep Run, Virginia (Chapter 10, Cosby et al., this volume).

Large declines in ANC during snowmelt have been observed in lakes throughout the western case study regions, primarily as a result of dilution (Baron and Bricker 1987, Stoddard 1987; Chapter 15, Melack and Stoddard, this volume), but acid pulsing may contribute to episodic ANC depressions in the North Cascade Range (Welch et al. 1986, Loranger et al. 1986). Blanchard et al. (1988) report a case of episodic acidification in the southern Rocky Mountains in which the surface (1 m) ANC of a small pond (Pond L12) dropped from > 30 μeq L^{-1} (summertime peak) to ≤ 0 μeq L^{-1}. They attributed a portion of this ANC depression to snowmelt containing acid anions from atmospheric deposition. To our knowledge, this is the only published case of episodic acidification to ANC ≤ 0 μeq L^{-1} in the West. Storm events may also cause significant ANC depressions, at least in the Sierra Nevada (Chapter 15, Melack and Stoddard, this volume).

Processes that Regulate ANC in Lakes and Streams

Approaches to Evaluating the Importance of Biogeochemical Processes

This book presents a variety of approaches to evaluating processes that regulate alkalinity in lakes and streams throughout the case study regions. For individual systems such as whole lakes and watersheds, ion budgets have been widely used to quantify ANC production and determine which ions are involved in ANC regulation. Watershed ion budgets that have been compiled for the case study regions are summarized in Table 17.7; ion budgets for lakes have been compiled by Brezonik et al. (1987) and Schindler (1986a). Modeling studies have provided useful insights into processes that regulate alkalinity, particularly for intensively studied lakes and watersheds in the Adirondack Mountains and the Upper Midwest (Chapter 1, Munson and Gherini, this volume; Schnoor et al. 1986, Gherini et al. 1985b). Finally, detailed biogeochemical studies (e.g., adsorption isotherms, lysimeter studies, isotope tracer studies) provide the basis for inferences about the specific processes that cause changes in the ionic composition of water passing through ecosystems.

The objective of ion enrichment analysis, as used here, is to calculate the net reaction of each ion in a watershed and then to assess net ANC production. First, we calculate the chemical composition of lake water that would occur in the absence of any biogeochemical reactions, that is, the composition that would occur if lake water were merely precipitation concentrated by evaporation, as in a "Teflon" lake. The "expected" concentration of ion i (i_{exp}) is calculated as:

$$[i]_{exp} = [i]_d*[Cl^-]_L/[Cl^-]_D \qquad (17.1)$$

where: $[Cl]_L$ = lakewater Cl^- concentration
$[i]_D$ = volume-weighted atmospheric deposition of ion i (wet + dry deposition divided by annual precipitation volume)

Enrichment or depletion reactions (R_i) are then calculated by comparing $[i]_{exp}$ with measured concentrations of $[i]_m$:

$$R_i\ (\mu eq/L) = ([i]_m - [i]_{exp}) \qquad (17\text{-}2)$$

$$R_i\ (\%) = \frac{([i]_m - [i]_{exp}) \times 100}{[i]_{exp}} \qquad (17\text{-}3)$$

Net ANC production is calculated in two ways, first, by the sum of ionic reactions ($ANCGEN_{C_B\text{-}C_A}$ where:

$$ANCGEN_{C_B\text{-}C_A} = R_{Ca} + R_{Mg} + R_K + R_{Na} + R_{NH4} + R_{Al} + R_{Mn} - R_{SO4} - R_{NO3} - R_F \qquad (17\text{-}4)$$

and second, on the basis of the Gran ANC reaction ($ANCGEN_{Gran}$), with the assumption that atmospheric inputs are entirely strong acids. Details of this analysis, including justification of assumptions, aluminum speciation, hydrologic classification, and land use screening, appear in Appendix A, and dry deposition estimates are developed in Appendix B. Table 17.7 summarizes this analyis.

Most watershed and lake ion budgets support our assumption that Cl^- is conservative in dilute systems and therefore a suitable ion for calculating evapoconcentration in ion enrichment analysis (Table 17.8). In several watersheds in the Southeast and in the Adirondack Mountains that clearly show watershed Cl^- production (Brier Creek, Buell and Peters 1988; Panther and Sagamore Lakes, Galloway et al. 1983b), the Cl^- source was attributed to Cl^- bearing minerals. In several other cases where watershed ion budgets indicate Cl^- production, there may have been watershed disturbances (e.g., road salt) or problems with Cl^- measurements (Jeffries et al. 1984). To a large extent, watersheds with major Cl^- sources would have been excluded from ion enrichment analysis by the Cl^- screening step (Appendix A).

Evapoconcentration

Calculated Cl^- ratios (lakewater Cl^- ÷ [volume-weighted wet + dry Cl^- deposition]) used in ion enrichment analysis are in general agreement with evapoconcentration estimates derived from regional hydrologic characteristics. We examined these relationships closely for a group of 104 lakes in the Adirondack Mountains of New York. Evapoconcentration estimates from Cl^- ratios were compared directly with evapoconcentration estimates derived by four other methods (Table 17.9). The mean Cl^- ratio (1.50) was somewhat lower than the mean evapoconcentration value computed from regional runoff estimates, but it was in very close agreement with the mean value (1.44) computed from an elevation-adjusted runoff model. Elsewhere there was general agreement between regional precipitation/runoff estimates and median Cl^- ratios

TABLE 17.7. Ion budgets for watersheds in regions susceptible to acidic deposition within the case study regions.

Lake	Number per year	Deposition[a]	Ca^{+} net flux	Mg^{2+} net flux	Na^{+} net flux	K^{+} net flux	C_B net flux	NH_4^+ Net flux	NH_4^+ R (%)	SO_4^{2-} Net flux	SO_4^{2-} R (%)	NO_3^- Net flux	NO_3^- R (%)	Cl^- Net flux	Cl^- R (%)	ANC production Measured[b]	ANC production C_B-C_A
Northeast																	
Darts[d]	1	W	564	164	156	70	954	−146	−83	356	69	−105	−35	54	159	621 G	503
Panther[e]	2	B; W+D					2078	−278		0		−176		36		1944 (G)	1940
Woods[e]	2	B; W+D					450	−238		15		−226		0		833 (G)	423
Sagamore[e]	2	B; W+D					1433	−257		252		−180		48		888 G	1056
Hubbard Brook[f]	11	B	575	209	246	25	1055	−142	−88	321	40	−42	−13	−57	−29	991 (?)	694
Bickford Res.[g]	2	B	690	234	191	59	1174	−109		608		−294		0			751
Biscuit Brook[h]	2	W	1081	382	119	22	1604	−190	−100	403	55	−167	−44	67	98	926 (G)	1111
Upper Midwest and southern Ontario																	
Marcell WE-2[i]	3	W+D	−412	−97	−57	−30	−596	−183	−86	−233	−53	−156	−98	−48	−50		−342
Turkey[j]	2	B	1806	308	192	62	2367	−261	−97	342	46	−153	−40	36	76	1579 (G)	1880
Batchawana[j]	2	B	669	167	105	18	959	−259	−97	152	21	−254	−64	15	29	786 (G)	787
Clearwater 1[k]	2	B	730	357	−72	26	1041	−251	−96	700	57	−368	−97	−13	−13	980 (TI)	471
Clearwater 3[k]	2	B	652	302	6	23	983	−256	−98	566	46	−373	−97	53	50	992 (TI)	481
Middle 1[k]	2	B	1410	840	96	76	2422	−274	−93	1490	87	−366	−90	228	173	1419 (TI)	796
Nelson 1[k]	2	B	912	314	−26	28	1228	−218	−93	665	55	−332	−92	9	11	966 (TI)	668
Nelson 2[k]	2	B	1000	296	1	8	1305	−226	−97	745	62	−355	−99	−15	−19	960 (TI)	704
Nelson 3[k]	2	B	823	307	−30	27	1127	−219	−94	355	29	−340	−95	−12	−15	1178 (TI)	905
Nelson 4[k]	2	B	1290	550	48	53	1941	−224	−93	1185	98	−350	−97	1	1	1171 (TI)	881
Nelson 5[k]	2	B	1400	606	52	63	2121	−189	−81	1095	91	−315	−88	2	3	1456 (TI)	1150
Nelson 6[k]	2	B	1090	880	39	52	2061	−225	−96	715	59	−338	−94	56	71	1320 (TI)	1403
Filson Creek[l]	1.5	B	216	313	91	−15	605			−211	−60			−9	−7		
Lake 239 NW[m]	3	B	61	81	57	33	232	−156	−98	154	66	−139	−89	−28	−39	101 (G)	89
Lake 239 E[m]	3	B	130	85	78	37	330	−157	−98	−2	−1	−148	−94	−44	−62	303 (G)	367
Lake 239 NE[m]	3	B	34	47	51	18	150	−154	−98	−126	−53	−154	−100	−12	−17	−22 (G)	288
Dunnigan[n]	1	W+D	57	107	35	39	238			−99	−46	−149	−97			74 (G)	
August[n]	1	W+D	200	321	70	12	603			−128	−59	−143	−93			362 (G)	

Southeast																1058
White Oak Run[o]	3	W+D;B	−2	144	43	122	307	−130	−99	−669	−67	−228	−95	16	23	1074 (TEP)
White Oak Run[p]	2	W	−67	50	−10	52	25	−124	−99	−385	−69	−236	−99	−51	−54	573
Deep Run[p]	2	W	−78	43	−3	67	29	−124	−99	−353	−63	−237	−99	−48	−51	543
Coweeta #2[q]	12	B	91	188	360	74	713	−126	−99	−520	−86	−190	−99	31	22	1266
Coweeta #14[q]	3+	B	72	195	184	58	509	−150	−98	−591	−83	−221	−98	−16	−8	1187
Coweeta #18[q]	12	B	151	217	293	75	736	−139	−99	−567	−85	−209	−99	16	10	1357
Coweeta #27[q]	12	B	174	222	170	62	628	−176	−97	−382	−46	−289	−92	7	−3	1130
Coweeta #32[q]	3+	B	199	311	259	74	843	−176	−98	−686	−81	−266	−99	4	2	1615
Coweeta #34[q]	3+	B	245	290	323	79	937	−157	−98	−609	−84	−227	−99	22	12	1594?
Coweeta #36[q]	12	B	320	312	430	84	1146	−150	−97	−390	−53	−222	−96	105	60	1503
Brier Creek[r]	1	W;B	151	111	239	70	571	−134	−93	−327	−77	−162	−93	50	88	841 (G) 876
Rhode R. #10[s]	4	B	87	216	46	50	399	−233	−97	−207	−22	−397	−97	−130	−39	900
West																
Loch Vale[t]	2	W	267	66	78	20	431	−73	−85	−14	−7	−39	−26	0	0	411

[a] Deposition collection method: W = wet only; B = bulk deposition; W + D = wet + dry, including deposition of gaseous species of S and N.

[b] Method of alkalinity determination in parentheses: G = Gran titration; TI = total inflection point alkalinity (OME 1979); TEP = two-endpoint titration (APHA 1971).

[c] Calculated alkalinity = $Ca^{2+} + Mg^{2+} + Na^{+} + K^{+} + NH_4^{+} - SO_4^{2-} - NO_3^{-} - Cl^{-}$.

[d] Driscoll and Schafran (1987).

[e] Galloway et al. (1983). Retention of individual ions calculated from Figures 17.1 through 17.6.

[f] Likens et al. (1977), Table 17.11.

[g] Hemond and Eshleman (1984). Net export of individual ions estimated from Tables 17.2 and 17.3 and Figure 17.5.

[h] Chapter 8, Stoddard and Murdoch, this volume.

[i] Urban 1987, Table 6.6. Watershed is a bog with some upland drainage.

[j] Jeffries et al. 1919–86. Export for terrestrial basins only, calculated using columns 3 and 4 of Tables III and IV. Deposition and terrestrial export values were converted to areal basis using lake areas and watershed areas, respectively.

[k] Jeffries et al. 1985. Gross export from Tables 6.8 and 6.9. Net export from Table 6.18E, except for Cl^-, SO_4^{2-}, and alkalinity, which were calculated from Tables 6.11 and 6.12. Deposition of SO_2 from ambient air concentrations (from Jeffries et al., Table 6.19) are included in estimates of SO_4^{2-} and alkalinity deposition used in this table.

[l] Siegel and Pfannkuch (1984). Ion budgets were determined for a 17-month period; values are expressed on annual basis. Nitrate and ammonium budgets were not compiled.

[m] Bayley and Schindler (1985). All three watersheds had been burned prior to 3-year budget period presented.

[n] Modified from Table 10 of Wonson-Liukkonen (1987). Fluxes calculated on whole watershed basis (lake + terrestrial drainage basin); t_w for August Lake recalculated to achieve Cl^- balance (see Wonson-Liukkonen, p. 86).

[o] Cosby et al. (1984).

[p] Schaffer and Galloway (1982).

[q] Swank and Waide 1988, Table 4.11.

[r] Buell and Peters 1988, Table III. Deposition inputs represent the average between a high-elevation and a low-elevation wet deposition collector.

[s] Weller et al. 1986, Table 3. Alkalinity budgets (direct measurement) were not determined.

[t] Baron and Bricker (1987). Fluxes are average for 1984 and 1985, in Table 3. Export values were normalized to balance chloride budgets.

TABLE 17.8. Ion enrichment analysis for selected case study regions.

Lake type[c]	Subregion	N	Lake: deposition Cl ratio	Enrichment calculations[b] R_{Ca}	R_{Mg}	R_K	R_{Na}	R_{SO4}	R_{SO4}, %	R_{CB}	$ANCGEN_{Gran}$	$ANCGEN_{C_B\text{-}C_A}$
All drainage	ELS	344	1.68	76	26	3.8	27	−12	−13	133	173	184
			(1.25, 2.44)	(50, 108)	(10, 51)	(0.6, 6.3)	(17, 33)	(−66, 14)	(−47, 18)	(88, 199)	(118, 222)	(135, 239)
	ADR	81	1.61	75	12	2.0	22	−10	−8	115	141	156
			(1.27, 197)	(39, 116)	(2, 27)	(0.4, 4.4)	(17, 29)	(−41, 12)	(−27, 13)	(64, 175)	(93, 222)	(90, 235)
	ME	124	1.54	83	17	4.5	31	−7	−9	144	167	187
			(1.03, 2.72)	(58, 127)	(5, 34)	(2.5, 7.0)	(26, 38)	(−76, 27)	(−53, 55)	(97, 200)	(126, 222)	(143, 251)
	UMW	130	1.72	75	52	3.8	24	−4	−8	161	175	178
			(1.26, 2.16)	(50, 96)	(35, 74)	(0.2, 6.3)	(13, 32)	(−31, 8)	(−38, 13)	(105, 208)	(119, 209)	(136, 224)
Low DOC	ELS	144	1.65	65	13	3.0	23	−10	−8	101	158	157
			(1.23, 2.64)	(40, 90)	(4, 27)	(0.5, 5.7)	(14, 30)	(−70, 12)	(−50, 16)	(72, 148)	(99, 220)	(99, 214)
	ADR	55	1.53	71	11	1.8	21	−4	−3	107	124	124
			(1.21, 1.96)	(43, 116)	(4, 27)	(0.4, 3.6)	(17, 27)	(−21, 18)	(−15, 16)	(72, 167)	(89, 218)	(86, 224)
	ME	62	1.74	67	11	4.0	28	−21	−24	109	162	159
			(1.24, 3.18)	(52, 91)	(−9, 19)	(0.8, 6.1)	(23, 34)	(−89, 15)	(−53, 23)	(72, 148)	(114, 237)	(113, 219)
	UMW	23	1.92	40	30	5.7	13	−22	−21	89	182	164
			(1.07, 2.86)	(38, 73)	(24, 35)	(1.9, 5.9)	(8, 17)	(−83, 4)	(−52, 5)	(89, 108)	(94, 220)	(104, 178)
	SBRP	33	2.71	18	3	8.2	31	−181	−89	55	395	326
			(2.39, 3.97)	(2, 25)	(−9, 7)	(4.7, 13.0)	(25, 39)	(−251, −156)	(−92, −85)	(33, 75)	(289, 456)	(245, 393)
High DOC	ELS	200	1.67	87	42	4.1	29	−13	−16	173	179	208
			(1.26, 2.35)	(55, 119)	(19, 67)	(0.7, 7.5)	(21, 34)	(−53, 17)	(−41, 26)	(116, 217)	(131, 217)	(156, 252)
	ADR	26	1.71	91	13	3.6	27	−35	−27	131	163	189
			(1.44, 2.20)	(31, 117)	(0, 35)	(0.2, 5.2)	(15, 32)	(−69, −13)	(−39, −11)	(57, 182)	(115, 246)	(139, 272)
	ME	62	1.34	118	31	5.5	34	0	0	192	171	216
			(0.95, 2.42)	(79, 145)	(13, 50)	(2.9, 7.7)	(28, 40)	(−53, 32)	(−44, 72)	(140, 250)	(136, 220)	(173, 270)
	UMW	107	1.68	82	58	3.3	29	−4	−4	184	170	189
			(1.27, 2.14)	(58, 102)	(41, 80)	(0.2, 6.3)	(20, 34)	(−29, 9)	(−34, 13)	(129, 214)	(129, 200)	(143, 227)

Groundwater flow-through DOC >5	ELS	29	1.53	47	43	6.3	16	−19	−24	132	123	140
			(1.22, 2.11)	(30, 79)	(17, 64)	(3.6, 10.7)	(6, 31)	(−38, 0)	(−45, 0)	(66, 160)	(68, 180)	(85, 241)
	UMW	25	1.53	40	41	4.5	14	−19	−24	131	105	140
			(1.22, 1.95)	(30, 77)	(17, 61)	(3.3, 9.8)	(6, 23)	(−38, 0)	(−47, 0)	(66, 152)	(63, 172)	(85, 230)
All groundwater recharge	ELS	189	1.88	18	19	5.8	7	−38	−51	47	107	111
			(1.17, 3.13)	(−2, 35)	(5, 34)	(1.9, 11.6)	(3, 11)	(−95, −11)	(−75, −15)	(23, 92)	(59, 159)	(66, 157)
	UMW	159	1.58	23	23	9.1	7	−23	−37	62	95	106
			(1.09, 2.42)	(11, 35)	(11, 34)	(4.5, 13.5)	(4, 11)	(−76, −7)	(−59, −9)	(35, 94)	(57, 133)	(59, 138)
	FLA	18	5.08	−34	4	−7	6	−145	−77	−34	168	185
			(3.35, 7.07)	(−53, −4)	(3, 9)	(−10, −2)	(2, 11)	(−251, −90)	(−85, −67)	(−49, 7)	(107, 337)	(114, 326)
DOC <5	ELS	106	1.81	14	9	5.5	6	−30	−33	39	88	75
			(1.23, 3.52)	(−2, 29)	(−3, 23)	(−1.7, 11.5)	(2, 10)	(−91, −2)	(−59, −4)	(−7, 64)	(58, 133)	(51, 138)
	UMW	86	1.56	18	12	8.3	6	−16	−23	47	79	62
			(1.12, 2.55)	(10, 30)	(6, 26)	(4.6, 13.3)	(2, 9)	(−58, 9)	(−51, 14)	(33, 68)	(52, 133)	(47, 132)
	FLA	11	3.51	−31	4	−6	6	−91	−70	−30	109	114
			(2.83, 4.00)	(−43, −4)	(3, 7)	(−10, 0)	(2, 10)	(−104, −64)	(−77, −66)	(−48, 7)	(88, 168)	(86, 185)
DOC >5	ELS	83	2.00	33	32	6.9	10	−51	−56	79	108	125
			(1.11, 3.15)	(1, 42)	(17, 36)	(3.2, 12.2)	(4, 17)	(−97, −21)	(−85, −27)	(24, 101)	(78, 232)	(105, 195)
	UMW	73	1.67	33	33	9.2	10	−31	−56	88	108	115
			(0.94, 2.42)	(13, 45)	(19, 36)	(4.5, 13.6)	(4, 17)	(−80, −18)	(−78, −24)	(49, 101)	(58, 160)	(101, 171)

[a]The approach for ion enrichment analysis is described in Chapter 2 (Munson and Gherini, this volume) and Appendix A. Results are based on lakes in the National Surface Water Survey with ANC ≤200 μeq/L^{-1}, screened to remove systems with watershed disturbances.

[b]R_i = reaction of ion i, in μeq L^{-1} except where noted. Values shown are medians and quartiles (in parentheses).

[c]High DOC = >5 mg L^{-1} (417 μm). Groundwater flow-through lakes are seepage lakes have SiO_2 >1 mg L^{-1} or K^+ >15 μeq L^{-1} (in Florida); groundwater recharge lakes are the remaining seepage lakes.

TABLE 17.9. Comparison of evapoconcentration ratios (EC) for Adirondack lakes by several methods.[a]

Method		Mean	SD
#1	EC estimated from chloride ratios	1.48	1.39
#2	EC estimated from site-interpolated precipitation and regional runoff values from Knox and Nordenson (1955)	1.97	0.29
#3	EC estimated as in Method #2, but using elevation corrected precipitation estimate (developed by R. Bradley, University of Massachusetts)	1.81	0.21
#4	EC estimates from Direct/Delayed Response Project	1.50	0.37
#5	EC estimated as in Method #4, but with elevation corrected precipitation values	1.44	0.32

[a] Data from Charles, unpublished data for 65 clear drainage lakes.

(Chapter 12, Pollman and Canfield, this volume; Chapter 13, Cook and Jager, this volume; Rapp et al. 1985, Baker et al. 1988). Thus, we are confident that Cl^- ratios used in ion enrichment analysis are reasonably representative of evapoconcentration for most lakes.

Chloride ratios for low ANC drainage lakes in the Adirondack Mountains, Maine, and Upper Midwest are similar (medians = 1.6, 1.5, and 1.7, respectively; Table 17.8) and indicate a modest concentration effect. Evapoconcentration is somewhat greater in southeastern lakes and streams. The median chloride-based evapoconcentration estimated for several Southern Blue Ridge drainage lakes was 2.7 (Table 17.8). Evapoconcentration calculated from watershed budgets in the Southern Blue Ridge Province range from 1.3 (mean for several Coweeta watersheds, Swank and Waide 1988) to 2.8 (Chapter 10, Cosby et al., this volume).

Evapoconcentration plays a major role in the acid-base status of Florida lakes. The median Cl^- based evapoconcentration value for 18 Florida low ANC groundwater recharge lakes was 5.1; evapoconcentration values for seepage lakes generally increase from north to south (Chapter 12, Pollman and Canfield, this volume, Baker et al. 1988). High evapoconcentration values in Florida seepage lakes occur because a large fraction of precipitation input to seepage lakes is lost through evaporation from the lake surface. High evapoconcentration decreases the ANC of lakes with ANC ≤ 0 through a direct concentration effect. Conversely, for lakes with ANC > 0, evaporation tends to increase ANC. High evaporation rates also tend to increase hydrologic residence times, resulting in a greater neutralizing effect of in-lake ANC generation (Chapter 12, Pollman and Canfield, this volume).

Reactions of Major Ions

Base Cations

Variations in Base Cation Enrichment by Hydrologic Type

Ion enrichment calculations show that median base cation (C_B; defined here as $Ca^{2+} + Mg^{2+} + K^+ + Na^+$) enrichment is generally higher in drainage lakes than in groundwater recharge lakes (Figure 17.8). For the low ANC data set, median C_B enrichment was 133 μeq L^{-1} for drainage lakes and 47 μeq L^{-1} for groundwater recharge lakes. Groundwater flowthrough lakes are more similar to drainage lakes than to groundwater recharge lakes with respect to base cation enrichment. For the Upper Midwest, the only region with a significant number of groundwater flowthrough lakes, base cation enrichment for groundwater flowthrough lakes was similar to that for drainage lakes (medians were 131 μeq L^{-1} and 161 μeq L^{-1}, respectively, compared to a median C_B enrichment of 62 μeq L^{-1} for groundwater recharge lakes.

This comparison clearly shows that seepage hydrology alone (i.e., lack of channeled inflows) is not a good indicator of base cation enrichment or acid sensitivity. Seepage lakes that receive sufficient groundwater input (groundwater flowthrough systems) have high C_B enrichment and are not particularly sensitive to acidification, whereas seepage lakes that are hydraulically isolated from groundwater inputs (groundwater recharge systems) have low C_B enrichment and are potentially susceptible to acidification (Eilers et al. 1983, Anderson and Bowser 1986, Baker et al. 1986b, 1988a). The relationship between groundwater inputs and base cation concentration is illustrated in Figure 17.9.

Among drainage lakes, base cation enrichment is higher for high DOC systems than for low DOC systems (Table 17.8). For example, in Maine,

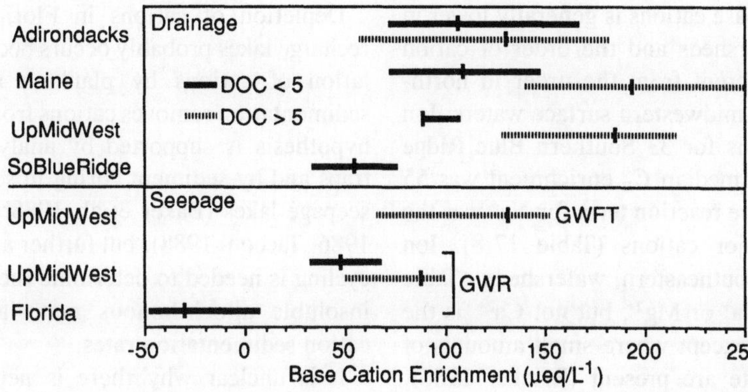

FIGURE 17.8. Bar chart showing net base cation enrichment ($Ca^{2+} + Mg^{2+} + K^{+} + Na^{+}$) for screened, low ANC ($\leq$ 200 μeq L^{-1}) lakes throughout the case study regions. Concentrations are net ion enrichment terms calculated by Equation 17-2. Drainage = drainage lakes; GWFT = groundwater flow-through lakes; GWR = groundwater recharge lakes. A DOC value of 5 mg L^{-1} corresponds to 417 μmol L^{-1}. Calculations and data sources are described in Appendix A.

median C_B enrichment was 109 μeq L^{-1} for low DOC drainage lakes, but 192 μeq L^{-1} for high DOC drainage lakes. Similarly, in the Upper Midwest, median C_B enrichment in high DOC drainage lakes (184 μeq L^{-1}) was more than twice as high as in low DOC drainage lakes (89 μeq L^{-1}). The observation that C_B enrichment is higher in high DOC lakes than in low DOC lakes probably reflects the higher WSA:LA ratios for the high DOC lakes (Table 17.1).

Interregional Differences Among Drainage Lakes

Calcium is the dominant cation in drainage lakes in the Upper Midwest, Maine, and the Adirondack Mountains. Calcium enrichment is similar among these three regions with median R_{Ca} values of 75, 83, and 75 μeq L^{-1}, respectively, for the low ANC lakes (Table 17.8). By contrast, Mg^{2+} enrichment is considerably higher in the Upper Midwest (median R_{Mg} = 52 μeq L^{-1}) than in Maine or the Adirondack Mountains (median R_{Mg} = 17 and 12 μeq L^{-1}, respectively), resulting in higher total C_B enrichment for the Upper Midwest lakes than for the northeastern lakes. The relative importance of Mg^{2+} in Upper Midwest lakes is reflected in lakewater composition: median Ca^{2+}:Mg^{2+} ratios are about 3:1 (eq/eq) throughout the Northeast, including the Catskills/Poconos streams (Chapter 8, Stoddard and Murdoch, this volume), but 1.5:1 in the Upper Midwest. The higher Mg^{2+} enrichment in the Upper Midwest compared to the Northeast apparently reflects geological differences between the two regions; Siegel and Pfannkuch (1984) suggested that high Mg^{2+} production in northern Minnesota is associated with readily weatherable ferromagnesium minerals.

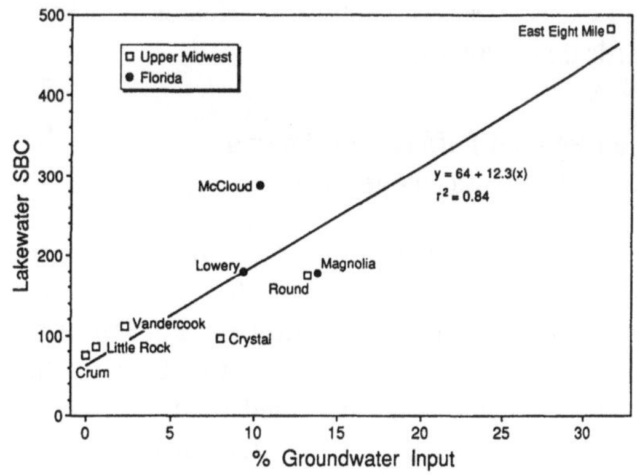

FIGURE 17.9. Groundwater input, as a percentage of total water input, versus SBC for seepage lakes in the Upper Midwest and Florida. Sources of data are: Round, East Eight Mile = S. Greb, Wisconsin Department of Natural Resources, pers. comm.; Vandercook = Lin et al. 1987; Crystal = Armstrong et al. 1987; McCloud = Baker et al. 1986b; Lowery and Magnolia = Baker et al. 1985c; Crum = unpublished data, Minnesota Pollution Control Agency, pers. comm.; Little Rock = Weir et al., in preparation.

Enrichment of base cations is generally lower in southeastern watersheds and the order of cation dominance is different from the order in northeastern and upper midwestern surface waters. Ion enrichment analysis for 33 Southern Blue Ridge lakes showed that median C_B enrichment was 55 μeq L^{-1} and that the reaction term for Na^+ was the largest of all major cations (Table 17.8). Ion budgets for 11 southeastern watersheds (Table 17.7) show that Na^+ or Mg^{2+}, but not Ca^{2+}, is the dominant cation, except where small amounts of calcite or dolomite are present (Bricker 1986). Relatively low cation export rates for watersheds in this region compared with temperate watersheds (Table 17.7) reflect the low weathering rates of underlying bedrock and the well-weathered status of southeastern soils (Bricker 1986, Shaffer and Galloway 1982, Cosby et al. 1985a; Chapter 11, Elwood et al., this volume).

Cation production in the Sierra Nevada appears to be dominated by feldspar weathering. In the Sierra Nevada lakes, Ca^{2+} and Na^+ are the dominant cations and the sum of $[Ca^{2+} + Na^+]$ nearly balances ANC in most systems (Chapter 15, Melack and Stoddard, this volume). In the few lakes where $Ca^{2+} + Na^+$ is not balanced by ANC, the imbalance is explained by weathering of minor deposits of easily weathered rocks containing Mg^{2+} (contributing Mg^{2+}) or gypsum (contributing SO_4^{2-} and excess Ca^{2+}) (Chapter 15, Melack and Stoddard, this volume).

Interregional Differences Among Groundwater Recharge Lakes

Base cation enrichment is very different in the two major groups of groundwater recharge lakes in Florida and the Upper Midwest. For low ANC groundwater recharge lakes in the Upper Midwest, base cations are enriched relative to atmospheric deposition (median C_B enrichment = 62 μeq L^{-1}); enrichment of Ca^{2+} and Mg^{2+} are nearly equal and much greater than Na^+ and K^+ enrichment. By contrast, Ca^{2+} and K^+ are depleted in low ANC Florida groundwater recharge lakes, whereas Mg^{2+} and Na^+ are nearly conservative and the net C_B reaction is negative (median R_{C_B} = −34 μeq L^{-1}) (Table 17.8). These results are similar to an analysis of Baker et al. (1988a) that applied somewhat different criteria for lake grouping.

Depletion of cations in Florida groundwater recharge lakes probably occurs because the assimilation of cations by plankton and subsequent sedimentation removes cations from solution. This hypothesis is supported by analysis of sediment traps and by sediment coring in several low ANC seepage lakes (Baker et al. 1988b, 1989, Bienert 1986, Tacconi 1988), but further analysis of cation cycling is needed to determine the contribution of insoluble allochthonous material to measured cation sedimentation rates.

It is unclear why there is net enrichment of base cations in Upper Midwest groundwater recharge lakes but net depletion of at least two major cations in Florida groundwater recharge lakes, but it may occur because (1) Florida lakes have more production and higher sedimentation rates or (2) there is greater groundwater input to the Upper Midwest lakes, partially offsetting in-lake retention processes.

Sulfate

Drainage Lakes

Median SO_4^{2-} enrichment values for low ANC drainage lakes in the Northeast, Maine, and the Upper Midwest are close to 0 (± 10%), indicating that SO_4^{2-} is typically at steady state (i.e., inputs ≃ outputs) in these systems (Table 17.8; Figure 17.10). The conclusion that SO_4^{2-} is at steady state in many temperate drainage systems is supported by a number of watershed studies (Table 17.7; Rochelle et al. 1987a) and by modeling analysis (Rochelle and Church 1987). In the Northeast, watershed SO_4^{2-} retention by adsorption is not expected on the basis of soil mineralogy; furthermore, in-lake retention is expected to be low because water *RT* values are typically < 1 year (Rochelle et al. 1987a). Although regional analysis of SO_4^{2-} retention in several groups of drainage lakes shows that inputs ≃ outputs, SO_4^{2-} is certainly retained to a substantial degree in some drainage lakes. Most notably, this has been documented in several lakes in the Experimental Lakes Area (Cook et al. 1986, Schindler et al. 1986). The major difference between these lakes and many case study region lakes that would account for their higher SO_4^{2-} retention is longer hydrologic residence times.

The lack of a clear difference in SO_4^{2-} retention between low DOC and high DOC lakes in temperate regions (Table 17.8) may reflect counterbalanc-

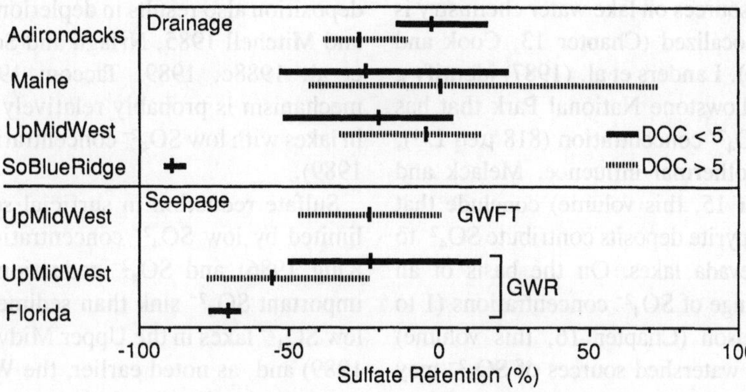

FIGURE 17.10. Bar chart showing net SO_4^{2-} enrichment for screened, low ANC (≤ 200 μeq L^{-1}) lakes throughout the case study regions. Concentrations are net enrichment terms calculated by equation 17.2. Drainage = drainage lakes; GWFT = groundwater flow-through lakes; GWR = groundwater recharge lakes. A DOC value of 5 mg L^{-1} corresponds to 417 μmol L^{-1}. Calculations and data sources are described in Appendix A.

ing tendencies. On one hand, one would expect greater SO_4^{2-} retention in high DOC lakes, since the watersheds of many high DOC lakes include wetlands, which are known to retain SO_4^{2-} (Bayley et al. 1986, Urban 1987, Hemond 1980). On the other hand, high DOC lakes tend to have lower *RT* values (see section on hydrologic characteristics) and should have less in-lake retention of SO_4^{2-} (Baker et al. 1986a).

Limited evidence suggests that SO_4^{2-} is at steady state in most case study region lakes in the West. Intensive studies at Emerald and Gem lakes in the Sierra Nevada (Chapter 15, Melack and Stoddard, this volume) indicate no major sinks of SO_4^{2-}. Sulfate to Cl^- ratios indicate that SO_4^{2-} is generally at steady state in most Rocky Mountain lakes (Chapter 14, Turk and Spahr, this volume). Despite the lack of quantitative data, the general characteristics of western lakes, including thin, poorly developed watershed soils, low SO_4^{2-} concentrations, and rapid flushing, suggest that SO_4^{2-} retention is unlikely.

In striking contrast to the steady state behavior seen in most temperate drainage systems, southeastern watersheds (western Virginia mountains and Southern Blue Ridge Province) typically retain a large fraction of SO_4^{2-} input (Johnson 1984, Johnson et al. 1986; Chapter 10, Cosby et al., this volume; Chapter 11, Elwood et al., this volume; Rochelle et al. 1987, Rochelle and Church 1987). Ion enrichment analysis of 12 Southern Blue Ridge lakes shows that median retention of SO_4^{2-} was 83% (Table 17.8). Cosby et al. (Chapter 10, this volume) concluded that 65% to 70% of the SO_4^{2-} input is retained in the watersheds of the western Virginia mountains. Ion budgets for a number of southeastern watersheds show that retention typically exceeds 50% of atmospheric input (Table 17.7). The difference in behavior of SO_4^{2-} in northeastern and upper midwestern temperate watersheds compared to southeastern watersheds is related to the extent of soil weathering. Soils in temperate glaciated regions (e.g., Spodosols) are relatively young and lack adsorption sites, whereas highly weathered soils (e.g., Ultisols) contain clay minerals and amorphous Al-oxides capable of anion adsorption (Rajan 1978, Hingstorn et al. 1967, Johnson and Henderson 1979, Johnson 1984). The SO_4^{2-} adsorption capacity of southeastern watersheds is limited. Breakthrough may occur on a time scale of decades to a century (Shaffer and Galloway 1982, Cosby et al. 1985b; Chapter 10, Cosby et al., this volume) and there is some evidence that breakthrough may already be occurring in some southeastern streams (Chapter 10, Cosby et al., this volume).

Deposits of sulfur-bearing minerals occur in several case study regions (see individual chapters), but watershed sources apparently have little influence on SO_4^{2-} concentrations in most low ANC systems in the case study regions. For the Northeast and the Upper Midwest, this conclusion is supported by ion enrichment analysis as well as by individual watershed budgets. In regions where sulfide minerals are present, the influence of bedrock sulfur sources on lake water chemistry is probably highly localized (Chapter 13, Cook and Jager, this volume). Landers et al. (1987) identify a lake (Fern) in Yellowstone National Park that has extremely high SO_4^{2-} concentration (818 μeq L^{-1}), as a result of geothermal influence. Melack and Stoddard (Chapter 15, this volume) conclude that both gypsum and pyrite deposits contribute SO_4^{2-} to several Sierra Nevada lakes. On the basis of an unusually wide range of SO_4^{2-} concentrations (1 to 44 μeq L^{-1}), Nelson (Chapter 16, this volume) hypothesizes that watershed sources of SO_4^{2-} may be important in some Cascade Mountain lakes. Several Cascade lakes have been identified that have elevated SO_4^{2-} and/or low ANC as a result of watershed sources (Welch et al. 1986, Loranger et al. 1986). However, with the exception of Fern Lake, none of the case study lakes with known watershed SO_4^{2-} sources is acidic. Although no acid mine drainage streams were sampled in the case study regions, they are numerous throughout much of the Appalachians and Interior Southeast. These, of course, have large watershed sources of SO_4^{2-} from oxidation of pyritic minerals (Herlihy et al. 1990).

Groundwater Recharge Lakes

The median SO_4^{2-} reaction for all low ANC groundwater recharge lakes was −51% (Q1 = −75%; Q3 = −15%). The high SO_4^{2-} retention values calculated by ion enrichment analysis are consistent with values obtained from whole lake SO_4^{2-} budgets for seepage lakes (Brezonik et al. 1987). Modeling studies (Baker et al. 1986a, Baker and Brezonik 1988, Kelly et al. 1987) have shown that net SO_4^{2-} retention is a function of mean depth and water residence time. Since groundwater recharge lakes typically have outflow-based hydrologic residence times longer than 5 years (Baker and Brezonik 1988), predicted SO_4^{2-} retention would be expected to exceed 50%, in agreement with the results of ion enrichment analysis.

Two mechanisms account for SO_4^{2-} losses in seepage lakes: (1) SO_4^{2-} reduction, and (2) seston deposition. The occurrence of SO_4^{2-} reduction in surficial sediments and anoxic hypolimnia of softwater lakes and its importance in alkalinity production have been widely documented (Kelly and Rudd 1984, Cook et al. 1986, Carignan 1985, Rudd et al. 1986a, Baker et al. 1985a,b, 1986b, 1989; see review by Brezonik et al. 1987). Seston deposition also results in depletion of SO_4^{2-} (David and Mitchell 1985, Nriagu and Soon 1985, Baker et al. 1988c, 1989, Tacconi 1988). The latter mechanism is probably relatively more important in lakes with low SO_4^{2-} concentrations (Baker et al. 1989).

Sulfate reduction in surficial sediments can be limited by low SO_4^{2-} concentrations (Lovely and Klug 1986) and SO_4^{2-} reduction may be a less important SO_4^{2-} sink than sedimentation in some low SO_4^{2-} lakes in the Upper Midwest (Baker et al. 1989) and, as noted earlier, the West. The importance of other limiting factors—such as iron supply for sequestering reduced sulfur, inhibition of SO_4^{2-} reduction by sediment acidification, and availability of energy (oxidizable forms of carbon) for SO_4^{2-} reducers—has been addressed for a few systems (Rudd et al. 1986b, Baker et al. 1989, Carignan 1985) but we do not know whether SO_4^{2-} reduction, on a regional basis, is limited by factors other than SO_4^{2-} concentration.

There is a clear regional difference in SO_4^{2-} retention between the two large groups of groundwater recharge lakes in the Upper Midwest and Florida (Figure 17.10). Ion enrichment calculations for low ANC groundwater recharge lakes show that median SO_4^{2-} depletion is higher on a concentration basis (median = −144 μeq L^{-1}) and as a fraction of input (median = −77%) in Florida than in the Upper Midwest (medians = −23 μeq L^{-1} and −37%, respectively). The reasons for this discrepancy are not clear. Compared to low ANC groundwater recharge lakes in the Upper Mid-west, Florida lakes are generally larger (median = 17.0 ha, versus 9.0 ha for the Upper Midwest) and shallower (median = 3.4 m, versus 6.1 m for the Upper Midwest), but there is considerable overlap in hydrologic characteristics. It is possible that the differences in calculated SO_4^{2-} retention reflect systematic errors in dry deposition estimates. However, sensitivity analysis showed that median SO_4^{2-} retention for the Upper Midwest low ANC groundwater recharge lakes varied from −27% (using wet-only deposition) to −42% (with a doubling of dry deposition), whereas the median SO_4^{2-} retention for Florida low ANC groundwater recharge lakes varied from −77% (with wet-only deposition) to −82% (with doubled dry deposition). We conclude that calculated retention values are not particularly sensitive to dry deposition estimates and that errors in dry deposi-

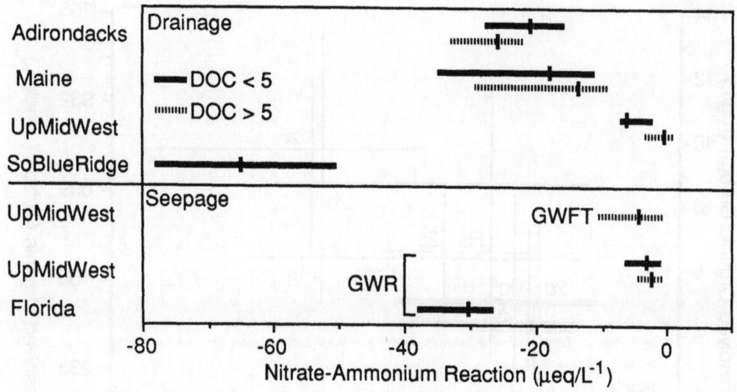

FIGURE 17.11. Bar chart showing the difference between net NO_3^- and net NH_4^+ enrichment for screened, low ANC ($\leq$200 µeq L^{-1}) lakes throughout the case study regions. Concentrations are net enrichment terms calculated by Equation 17-2. Drainage = drainage lakes; GWFT = groundwater flow-through lakes; GWR = groundwater recharge lakes. A DOC value of 5 mg L^{-1} corresponds to 147 µmol L^{-1}. Calculations and data sources are described in Appendix A.

tion estimates probably do not cause the observed regional differences in SO_4^{2-} retention.

Florida lakes may be more efficient in retaining SO_4^{2-} because they are warmer than Upper Midwest lakes and may therefore have higher rates of microbial activity. Enhanced microbial activity could increase the efficiency of SO_4^{2-} reduction directly, by increasing the activity of SO_4^{2-} reducers, or indirectly, by increasing the net efficiency of SO_4^{2-} reduction through continued maintenance of reducing conditions throughout the year. Available evidence is inadequate to evaluate the mechanisms that control the efficiency of SO_4^{2-} retention in these lakes.

Nitrate and Ammonium

Ion enrichment analysis shows that nearly all of the NO_3^- and NH_4^+ that enters lake-watershed systems in the case study regions is retained, regardless of lake type or region (Figure 17.11). In this regard, it should be noted that our ion enrichment analysis overestimates NO_3^- retention because the calculations are based on fall index samples collected at a time when NO_3^- concentrations were likely to be very low relative to winter and spring concentrations. However, the conclusion that NH_4^+ is efficiently retained on an annual basis is also supported by numerous watershed ion budgets (Table 17.7). Nitrate is also efficiently retained in most systems, but there are important exceptions (e.g., Loch Vale, Hubbard Brook, Darts Lake, and Emerald Lake). One factor that may account for low calculated NO_3^- retention for some intensively studied watersheds is a failure to account for dry deposition (HNO_3; NO_2), which can be comparable to wet deposition at some sites (Appendix B). Likens et al. (1977) reached this conclusion concerning their NO_3^- budget for Hubbard Brook, and this problem may in part account for the low calculated NO_3^- retention at Darts Lake. However, one would expect relatively little dry NO_3^- deposition at Loch Vale or Emerald Lake, yet NO_3^- retention was < 50% in both systems. Low NO_3^- retention appears to be associated with systems that have large spring pulses of NO_3^- from snowmelt and short flow paths (Galloway et al. 1987, Schafran and Driscoll 1987; Chapter 15, Melack and Stoddard, this volume). For this reason, we would expect low NO_3^- retention in most high-elevation western lakes and in some Adirondack Mountains lakes, but direct evidence is limited to input-output budgets for only a few systems. However, even minor changes in the efficiency of NO_3^- retention may have an important effect on acid-base chemistry.

DOC and Organic Acids

For some regions in North America, notably Florida, the Upper Midwest, the Adirondack Mountains, and Maine, median DOC concentrations are high (> 300 µmol L^{-1}; Table 17.3) and DOC

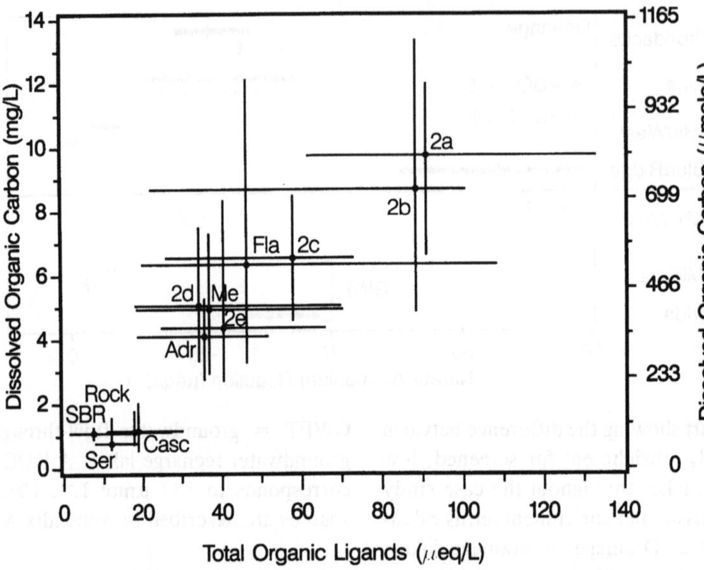

FIGURE 17.12. Median total organic ligands (A_t) as a function of median DOC for the case study lakes; lines represent quartile ranges for each variable. Abbreviations are: SBR = Southern Blue Ridge Province, Ser = Sierra Mountains, Rock = Rocky Mountains, Casc = Cascade Mountains, 2a = northern Minnesota, 2b = eastcentral Minnesota, 2c = northwest Wisconsin, 2d = northcentral Wisconsin-western Upper Peninsula of Michigan, 2e = eastern Upper Peninsula of Michigan, Fla = Florida (subregion 3b), Me = Maine, Adr = Adirondack Park.

represents a significant portion of the total dissolved solids. It consists largely of humic material (i.e., humic and fulvic acids), which is a mixture of complex high molecular weight organic polymers containing carboxyl and phenolic functional groups. Although the acid-base characteristics of DOC were not specifically determined in any of the case study regions, DOC is likely to exert control on pH and ANC in those regions where DOC is a large fraction of the total dissolved solids. In addition, DOC may also influence trace metal speciation, mineral weathering, cation leaching, and nutrient availability (Chapter 1, Munson and Gherini, this volume).

Streams in Virginia and the Southern Blue Ridge and lakes in the western United States have low median DOC concentrations and the 75th percentile is below 150 µmol L^{-1} (Table 17.6). Although median DOC values in the Adirondack Mountains and Maine are high (300 to 400 µmol L^{-1}), very few lakes have DOC concentrations > 1,000 µmol L^{-1}. In contrast, lakes in Florida and the Upper Midwest have median DOC concentrations > 500 µmol L^{-1} (Figure 17.5).

Concentrations of DOC and total organic ligands are related to the occurrence of wetlands in the case study regions. The Upper Midwest, Florida, Maine, and, to a lesser extent, the Adirondack Mountains, are characterized by numerous wetlands and correspondingly high DOC (Chapter 13, Cook and Jager, this volume; Chapter 12, Pollman and Canfield, this volume; Chapter 7, Kahl et al., this volume; Chapter 6, Driscoll et al., this volume). In the mountainous regions in the western United States, the Southern Blue Ridge and Virginia wetlands are uncommon and watersheds are often steeply sloped; consequently, DOC is generally low.

Total organic ligands are strongly related to DOC concentration, with an acid functional group content of 10 µeq A_t/mg DOC (Figure 17.12). The slope of this regional data set agrees quite closely with the value of 10 µeq A_t/mg DOC for charge density measured by Oliver et al. (1983) on samples that were stripped of all metals, and is in the range of charge densitites for total organic acids reported in the literature (summarized in Cook et al. 1987). Similarly, an analysis of median lakewater chemistry from the ELS-I showed that the median charge density was 11.7 μeq A^-/mg DOC (Driscoll et al. 1989). The variance within subregions increases dramatically as DOC increases (Figure 17.12), indicating the heterogeneous nature of DOC. Also, the concentration of A^- is related to pH, in addition to DOC (Oliver et al. 1983, Driscoll et al. 1989), so that the range in pH within a region causes variations in A^-.

The effect of organic acids on pH was investigated using the difference (δpH) between air-equilibrated pH and the pH expected if all ANC were derived from bicarbonate species (Chapter 2, Munson and Gherini, this volume; Chapter 13, Cook and Jager, this volume). In Florida, the Upper Midwest, Maine, and the Adirondack Mountains, δpH is low for most acidic lakes (Figure 17.13), suggesting that these lakes are acidic due to mineral acids, not organic acids. The role of organic acids in affecting acid-base chemistry in these four regions is greater for moderate pH values (pH 6.0 to 7.0) than for lower or higher values (Figure 17.13). Values of δpH are typically small for surface waters in western Virginia, the Southern Blue Ridge Province, and the western United States.

Total ANC Production

For the low ANC data set, net ANC production ($ANCGEN_{Gran}$) was highest in the drainage lakes (median = 173 μeq L^{-1}), intermediate in the groundwater flowthrough lakes (median = 123 μeq L^{-1}), and lowest in the groundwater recharge lakes (median = 107 μeq L^{-1}) (Figure 17.14). Ion enrichment analysis showed that ANC (both measured and calculated from reactions) was produced in virtually all the 579 low ANC lakes in the data set.* It is important to understand that there is net ANC production even in acidic (ANC ≤ 0) lakes: median $ANCGEN_{Gran}$ was 56 μeq L^{-1} in acidic drainage lakes and 36 μeq L^{-1} in acidic groundwater recharge lakes. Published watershed and lake budgets also indicate net ANC production in nearly all systems that have been studied, with the exception of a peat bog (Table 17.7). These results support the conclusion that case study lakes generally have ANC < 0 only when the capacity to neutralize acid inputs is exceeded by atmospheric inputs of H^+. This conclusion also is supported by detailed ion budgets and by empirical relationships (see Table 17.7 and the section entitled Current Status of Surface Water Chemistry).

Naturally acidic surface waters are not common in the case study regions, with the exception of very high DOC systems that have limited contact with mineral soils (omnotrophic systems). These are fairly common in the Florida case study region and, to a lesser extent, in the Upper Midwest. Natural acidification by organic acids is illustrated by a group of 10 lakes sampled during the ELS in the Okefenokee Swamp (not included in the RCS) that had DOC > 2,000 μmol L^{-1}, but SO_4^{2-} concentrations < 13 μeq L^{-1}. These lakes had ANC values of −63 to −209 μeq L^{-1} and pH values of 3.8 to 4.3 (Eilers et al. 1988b). Median measured ANC production ($ANCGEN_{Gran}$) was only about 14% lower than C_B-C_A ANC production ($ANCGEN_{C_B\text{-}C_A}$) for high DOC systems (Table 17.8), indicating that even for high DOC systems, base cation production and other processes are important in regulating ANC. Natural acidification can also result from weathering of sulfide minerals, but only one lake in the case study regions (Fern Lake, Wyoming) was clearly acidic as a result of sulfur mineralization.

Drainage Lakes

Ion enrichment analysis shows that base cation enrichment is the dominant source of ANC in most drainage lakes, accounting for a median of 93% of net ANC production ($ANCGEN_{Gran}$) for the low ANC data set (Figure 17.15). On the basis of median contributions, base cation enrichment accounted for 83% of $ANCGEN_{Gran}$ in the Adirondack Mountains, 94% of $ANCGEN_{Gran}$ in Maine, and 102% of $ANCGEN_{Gran}$ in the Upper Midwest. Watershed ion budgets (Table 17.6) and empirical relationships (e.g., plots of C_B versus ANC-SO_4^{2-}) confirm the importance of base cation enrichment for these regions (see regional chapters). In regions where ion enrichment analysis was not conducted (Catskills Mountains, Sierra Nevada, Cascade Mountains, and Rocky Mountains; see

*$ANCGEN_{Gran}$ was negative in two lakes, Swamp (2B3-076) and unnamed (2B2-055). However, in both cases the lake:deposition Cl^- ratios were < 1, indicating that the assumption of Cl^- conservancy may be invalid or that deposition inputs were not correct.

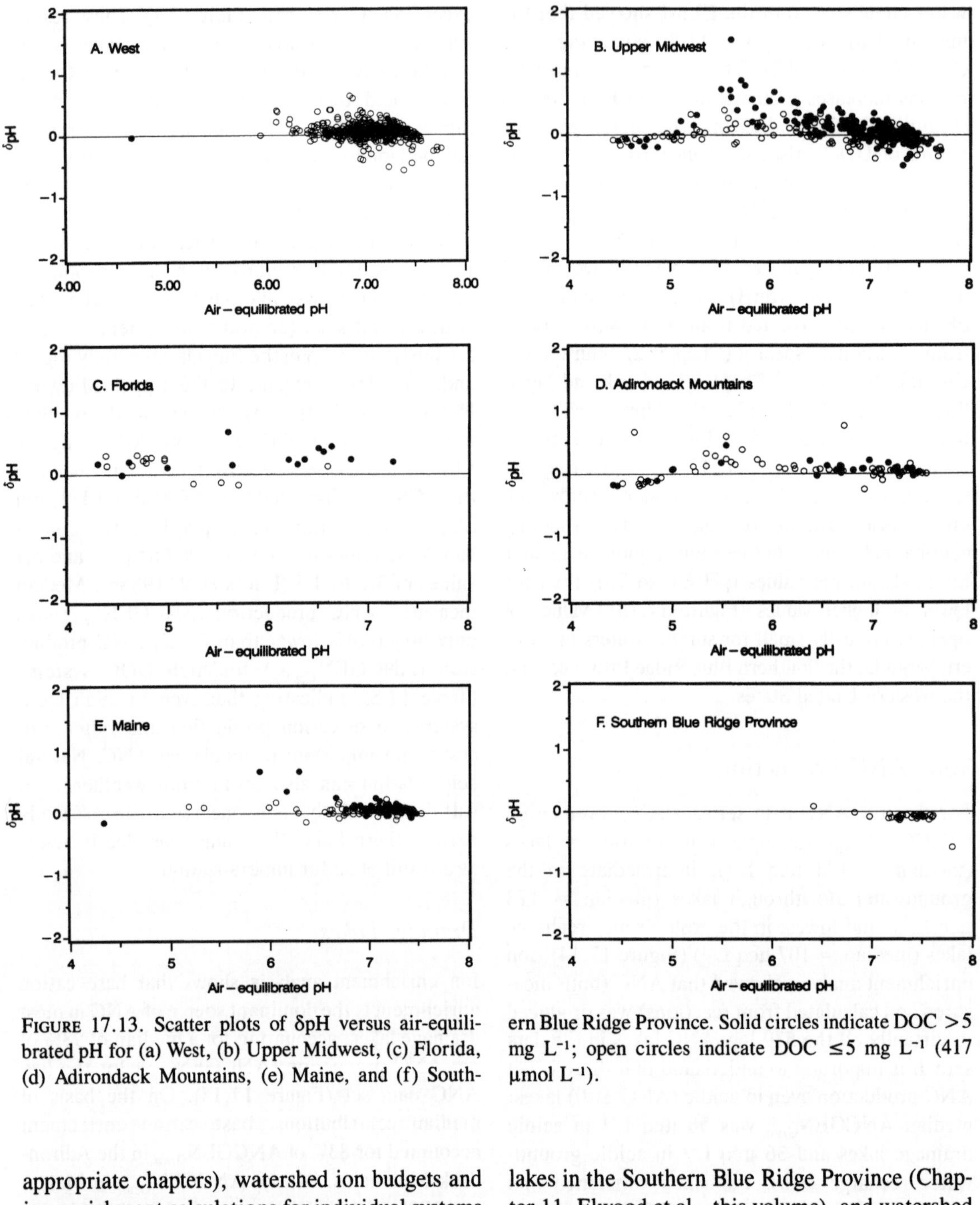

FIGURE 17.13. Scatter plots of δpH versus air-equilibrated pH for (a) West, (b) Upper Midwest, (c) Florida, (d) Adirondack Mountains, (e) Maine, and (f) Southern Blue Ridge Province. Solid circles indicate DOC >5 mg L^{-1}; open circles indicate DOC ≤5 mg L^{-1} (417 μmol L^{-1}).

appropriate chapters), watershed ion budgets and ion enrichment calculations for individual systems also confirm the importance of base cation production in ANC regulation.

In contrast, base cation production is a much less important component of net ANC production in the Southeast than in the Northeast or Upper Midwest (Figure 17.15). Base cation production accounted for < 25% of $ANCGEN_{Gran}$ in 12 lakes in the Southern Blue Ridge Province (Chapter 11, Elwood et al., this volume), and watershed ion budgets for this region show that C_B production is generally < 50% of net alkalinity production (Table 17.6).

Sulfate retention is unimportant with respect to net alkalinity production in most low residence time temperate drainage systems because SO_4^{2-} is typically at steady state in these systems (Shaffer et

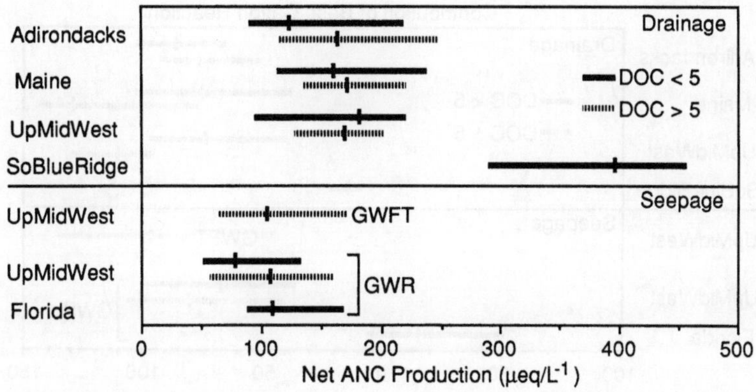

FIGURE 17.14. Bar chart showing net ANC production ($ANCGEN_{Gran}$) for screened, low ANC (≤200 µeq L^{-1}) lakes throughout the case study regions. Concentrations are net enrichment terms calculated by equation 17-2. Drainage = drainage lakes; GWFT = groundwater flow-through lakes; GWR = groundwater recharge lakes. A DOC value of 5 mg L^{-1} corresponds to 417 µmol L^{-1}. Calculations and data sources are described in Appendix A.

al. 1988). However, SO_4^{2-} adsorption is a very important source of ANC in the Southeast watersheds. Ion enrichment for Southern Blue Ridge Province drainage lakes showed that SO_4^{2-} retention accounted for 57% (median) of net ANC production ($ANCGEN_{Gran}$) and ion budgets for watersheds throughout the Southeast show that SO_4^{2-} adsorption accounts for over half the net watershed ANC production (Chapter 11, Elwood et al., this volume; Chapter 10, Cosby et al., this volume). As stated earlier, SO_4^{2-} production in watersheds is uncommon and there is little evidence to suggest that oxidation of metal sulfide minerals is an important mechanism of ANC depletion in many case study region lakes.

Most of the remaining ANC production in drainage systems is made up of the balance of NO_3^- retention, which produces ANC, and NH_4^+ retention, which consumes ANC. Because both processes are very efficient (typically > 90% efficiency), net ANC production resulting from consumption of these ions is closely related to their relative abundance in atmospheric deposition. For the east coast, $NO_3^-:NH_4^+$ ratios in wet deposition are typically 1.2 to 1.9 (Figure 17.1) and when dry inputs are considered, $NO_3^-:NH_4^+$ ratios exceed 2.0. Accordingly, ion enrichment analysis shows that the net difference between ANC production resulting from NO_3^- retention and ANC consumption resulting from NH_4^+ uptake accounts for 12% to 16% of $ANCGEN_{Gran}$ in drainage lakes in the Adirondacks, Maine, and the Southern Blue Ridge Province. In contrast, wet NO_3^- and NH_4^+ inputs in the Upper Midwest, including southern Ontario, are nearly balanced, and ion enrichment analysis shows that $NO_3^- - NH_4^+$ retention is < 1% of $ANCGEN_{Gran}$ (Figure 17.15). Again we caution that conclusions regarding NO_3^- and NH_4^+ retention and the influence of these processes on net ANC contribution are uncertain, because of the uncertainty in calculated retention efficiencies for these ions.

Aluminum solubility and speciation are important from a standpoint of toxicity to aquatic organisms, but Al production accounts for a small fraction of net ANC production in most case study region drainage systems in the East. Median Al enrichment was < 1% for all regions in which ion enrichment analysis was conducted.

Groundwater Flowthrough Lakes

Ion enrichment analysis shows that ANC regulation in groundwater flowthrough lakes (Upper Midwest only) is also dominated by base cation production, with C_B reactions accounting for 93% (median) of measured ANC production. Sulfate retention accounts for an additional 20% of measured ANC production; $NO_3^- - NH_4^+$ retention accounts for < 5% of $ANCGEN_{Gran}$. Thus, with the exception of a modest additional contribution

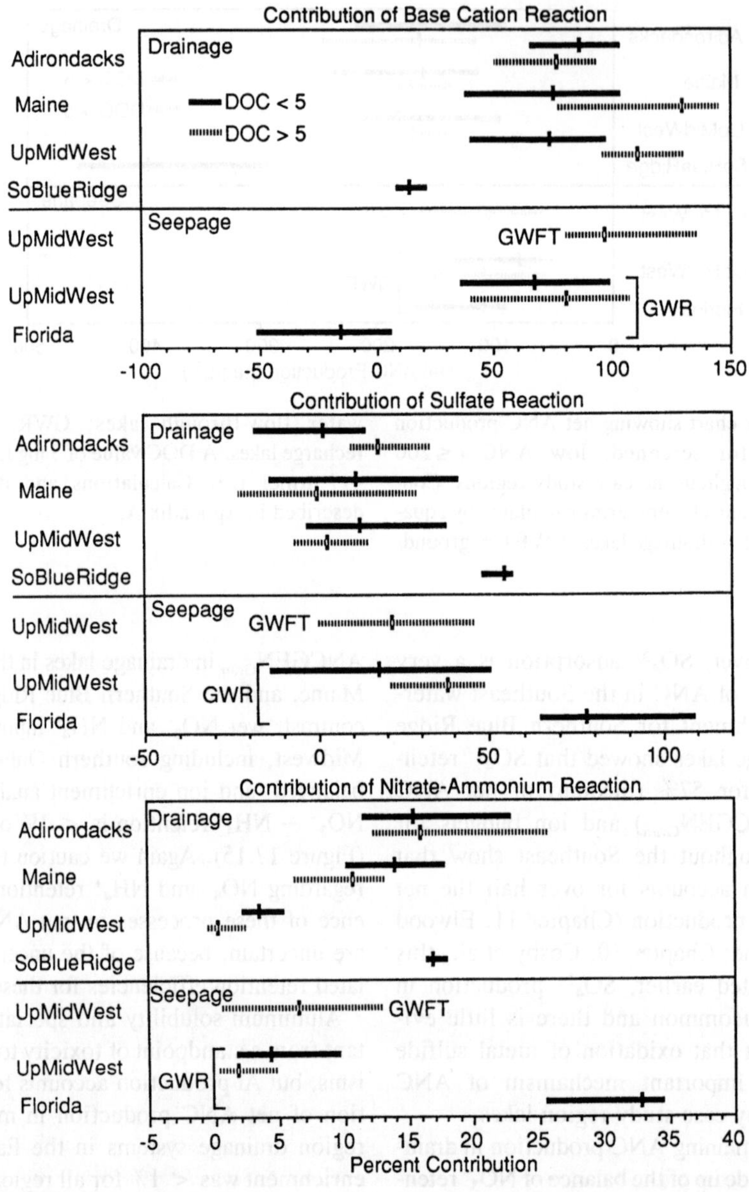

FIGURE 17.15. Contribution of ion enrichment reactions to net ANC production for screened, low ANC ($\leq$200 μeq L^{-1}) lakes throughout the case study regions. (a) Sum of base cations, (b) SO_4^{2-}, (c) $NO_3^- - NH_4^+$. Positive contributions imply that the net ion reaction contributes to ANC; negative contributions imply that the net reaction depletes ANC. Drainage = drainage lakes; GWFT = groundwater flow-through lakes; GWR = groundwater recharge lakes. A DOC value of 5 mg L^{-1} corresponds to 417 μmol L^{-1}. Calculations and data sources are described in Appendix A.

of ANC from SO_4^{2-} retention, ANC production processes in groundwater flowthrough lakes are similar to those in drainage lakes.

Groundwater Recharge Lakes

The magnitude and extent of ANC regulation in the two groups of groundwater recharge lakes (Florida and the Upper Midwest) are quite different from ANC regulation in drainage or groundwater flowthrough lakes. For the Upper Midwest group, the median C_B contribution to $ANCGEN_{Gran}$ is 73% and the median SO_4^{2-} contribution is 30%. Because NO_3^- and NH_4^+ inputs are nearly balanced and retention of both is > 90%, the median net contribution of these ions to $ANCGEN_{Gran}$ is only 3% (Table 17.8). Median $ANCGEN_{C_B-C_A}$ (105 μeq L^{-1}) is slightly greater than median $ANCGEN_{Gran}$ (95 μeq L^{-1}), suggesting that strong organic acid contributions have a small negative influence on ANC production. This is consistent with ion budgets for several Upper Midwest groundwater recharge lakes (Lin et al. 1987, Tacconi 1988; Chapter 13, Cook and Jager, this volume).

In Florida groundwater recharge lakes, SO_4^{2-} retention is the most important ANC regulation process, accounting for 81% of $ANCGEN_{Gran}$. As said earlier, there is net depletion of base cations in this group of lakes; C_B depletion accounts for a median loss of 12% of $ANCGEN_{Gran}$. Nitrate retention exceeds NH_4^+ retention (reflecting greater NO_3^- inputs), so NO_3^- − NH_4^+ retention accounts for another 32% of $ANCGEN_{Gran}$. Overall, median $ANCGEN_{C_B-C_A}$ was 185 μeq L^{-1}, slightly greater than $ANCGEN_{Gran}$ (168 μeq L^{-1}).

Limitations of Ion Enrichment Analysis

Because detailed watershed studies can be conducted for only a small number of systems, we have used ion enrichment analysis to derive inferences about the nature and extent of ANC regulation for whole populations of lakes. As noted previously (Chapter 2, Munson and Gherini, this volume), the validity of ion enrichment analysis depends upon several underlying assumptions. In this analysis, we attempted to minimize deviations from the assumption that Cl^- is conservative and is derived from atmospheric deposition by eliminating lakes with anomalously high Cl^- concentrations. The general agreement between lake and deposition Cl^- ratios and expected evapoconcentration suggests that the assumption that Cl^- is conservative for the screened data set is not seriously violated. A second assumption is that lakes are at hydrologic steady state. This assumption is probably very nearly true for lakes with long hydraulic residence times and for lakes with steady-state inputs and outputs. However, for many low residence time drainage lakes, particularly in the Northeast, this assumption may not be strictly correct because stream inputs show strong seasonal variation, notably a spring snowmelt pulse. This seasonal variation is potentially a serious problem, but results from ion enrichment calculations compare well with results from annual or multi-annual ion budgets compiled for several Northeast systems (Hubbard Brook, Bisquit Brook, Darts Lake, Panther Lake, Woods Lake, and Sagamore Lake) and with the regional scale input/output SO_4^{2-} budgets of Rochelle et al. (1987b). These comparisons suggest that the problem of using a fall index sample to represent steady-state conditions is not as serious as one might at first assume, at least for a synoptic evaluation of major biogeochemical processes.

Finally, there is considerable uncertainty in estimating atmospheric deposition. For the eastern regions and the Upper Midwest, errors in interpolated wet deposition values are probably < 20% (Chapter 3, Husar et al., this volume). There is much greater uncertainty in dry deposition estimates. It is impossible to conduct a true error analysis of dry deposition rates, because our dry deposition estimates are based upon either an inferential approach or surrogate collector measurements (Appendix B, Baker, this volume). The major uncertainty, then, is not the analytical error associated with the measurement itself, but the validity of the extrapolation from measured values to watershed dry deposition rates. Therefore, in lieu of formal error analysis, we conducted a sensitivity analysis to determine the extent to which calculated reaction terms are dependent upon dry deposition estimates. In this analysis, we calculated reaction terms for the entire data set using wet-only deposition (a theoretical lower limit estimate of atmospheric deposition rates) and using twice the best estimate of dry deposition. A

TABLE 17.10. Sensitivity analysis showing the effect of dry deposition estimates on ion enrichment calculations.

Lake type	Deposition	Lake: deposition Cl ratio	SO_4 µeq L^{-1}	SO_4 %	NO_3 µeq L^{-1}	NO_3 %	C_B µeq L^{-1}	C_B %	$ANCGEN_{calc}$ µeq L^{-1}	$ANCGEN_{C_B\text{-}C_A}$ µeq L^{-1}
Groundwater recharge	Wet only	2.24	−25	−43	−39	−99	72	132	110	114
	Best estimate	1.88	−38	−51	−45	−99	47	72	106	114
	Wet + 2x dry	1.62	−44	−55	−47	−99	34	45	105	105
Drainage	Wet only	1.98	6	10	−37	−99	161	365	171	181
	Best estimate	1.67	−10	11	−42	−99	133	212	171	182
	Wet + 2x dry	1.36	−18	−18	−43	−99	124	165	159	176

summary of this analysis (Table 17.10) illustrates several key points regarding major ions:

- Estimates of evapoconcentration based upon Cl^- are only moderately sensitive to dry deposition estimates. For example, the median Cl^- ratio for groundwater recharge lakes was 1.9; this increased to 2.2 for wet-only deposition and decreased to 1.6 when dry Cl^- deposition was doubled. The relative insensitivity of Cl^- ratios to dry deposition reflects the fact that dry deposition is a minor component (typically < 30%) of total Cl^- deposition.
- The SO_4^{2-} reaction term is moderately sensitive to dry deposition estimates. Nevertheless, all three deposition scenarios show SO_4^{2-} depletion for groundwater recharge lakes, with median depletion ranging from −43% (with wet-only deposition) to −55% (doubling dry deposition). Conclusions regarding SO_4^{2-} reactions in drainage lakes are also robust. The median reaction calculated with wet-only deposition is +10% (e.g., some SO_4^{2-} is produced in the watersheds), but the median reaction changes to −20% when dry deposition is doubled. In either case, the median reaction is not large, and the conclusion that Northeast and Upper Midwest drainage systems are more or less at steady state with respect to SO_4^{2-} is supported by ion enrichment analysis.
- Calculated NO_3^- retention values are close to 100% in all three deposition scenarios. The insensitivity to NO_3^- dry deposition occurs because deposition concentrations (by any scenario) are nearly two orders of magntitude higher than lake water concentrations, and for most regions, dry deposition of NO_3^- is relatively minor (dry deposition < 50% of wet deposition in all regions except Florida).
- Calculated C_B enrichment is highly dependent upon dry deposition rates. For groundwater recharge lakes, median C_B enrichment ranges from 72 µeq L^{-1} with wet-only deposition to 34 µeq L^{-1} with doubled dry deposition. Drainage lakes show similar sensitivity: the median reaction calculated with wet-only deposition is 161 µeq L^{-1}, compared with a median reaction of 124 µeq L^{-1} with doubled dry deposition.
- Median ANC reactions are insensitive to simultaneous increases or decreases of dry deposition rates for all ions. However, this is a conservative assessment, since our best dry deposition estimates may be high for some ions and low for others. For example, if actual SO_2 deposition were higher and actual C_B dry deposition were lower than our estimates, the effect on calculated ANC generation would be greater than if dry deposition rates of both ions were increased or decreased simulateously.

Overall, sensitivity analysis for dry deposition rates shows that most of our stated conclusions regarding biogeochemical processes are robust and we would not expect improved estimates of dry deposition to substantially change our conclusions.

Regional ion enrichment analyses were not conducted where key assumptions clearly were not met. Regional ion enrichment analysis for lakes in the West was precluded by an inability to accurately interpolate wet deposition rates to individual lakes. In the Catskill Mountains road salting and other local anthropogenic influences resulted in the elimination of most lakes during the screening process.

Regional Synopses

Adirondack Mountains (Chapter 6)

Atmospheric deposition in the Adirondack Mountains is highly acidic (pH < 4.5) and it is widely believed to be responsible for lake acidification in this region. Many Adirondack lakes are currently acidic: 18% of the ELS-I lakes had ANC ≤ 0 and 10% had pH values ≤ 5.0 (Linthurst et al. 1986a). Furthermore, the highest proportion of lakes in the ELS-I survey with extractable Al > 50 μg L^{-1} is in the Adirondack Mountains (18%; Linthurst et al. 1986a), and there is clear evidence that the deleterious effects of low pH in this region are exacerbated by Al toxicity (Chapter 4, Baker and Christiansen, this volume). There is also clear evidence of severe pH depressions during snowmelt in Adirondack lakes, with some lakes experiencing pH depressions > 1 unit. Eshleman (1988) has estimated that 35% of the lakes in the Adirondack Mountains could experience ANC depressions to ≤ 0 during snowmelt.

Adirondack lakes sampled in the ELS-I are typically rapidly flushed drainage systems. Processes that supply base cations are the dominant source of acid neutralization, and the extent of acid neutralization has been related to the length of the flowpath of precipitation through the watershed (Chen et al. 1984a). Available evidence suggests that SO_4^{2-} is generally at steady state in Adirondack lakes. Ion enrichment analysis shows that NO_3^- and NH_4^+ are efficiently retained, resulting in some net ANC production, since NO_3^- inputs are higher than NH_4^+ inputs. Numerous small (≤ 4 ha) seepage lakes in the Adirondack Mountains also were included in the Adirondack Lake Survey Corporation study (Kretser et al. 1989). Geochemical processes in these lakes have not been studied.

Maine (Chapter 7)

The median ANC of Maine ELS-I lakes was 138 μeq L^{-1}; less than 2% were acidic. However, two other surveys that focused on lakes potentially more sensitive to acidification than the population as a whole show that there are numerous acidic lakes in Maine. Thirteen percent of the 90 lakes in the High Elevation Lake Monitoring study were acidic and 30% of the 125 lakes in the Aquifer Lake Study were acidic (Chapter 7, Kahl et al., this volume).

Most (> 90%) of the Maine ELS-I lakes are drainage lakes and the dominant source of ANC is associated with cation production. Many Maine drainage lakes have high DOC. In these lakes, organic acids may play a major role in determining the acid-base status.

Catskill Mountain Streams (Chapter 8)

The Catskill Mountains receive the highest wet deposition loadings of any of the case study regions in the United States. The sensitivity of streams in this region to acidic deposition is of major interest, because many streams are sources of drinking water for the City of New York. Results of the National Stream Survey (NSS) and the U.S. Geological Survey/City of New York (USGS/NYC) study show that a small fraction of the streams are currently acidic during baseflow conditions; however, many more streams are acidic during snowmelt and storm events.

The NSS showed that 6% of the upstream ends of reaches were acidic (ANC ≤ 0) and 5% had pH values ≤ 5.0 during baseflow, whereas < 1% of the downstream ends had ANC ≤ 0 or pH ≤ 5.0 (Kaufman et al. 1988). The USGS/NYC survey of 51 streams, which had a different sampling approach (not population based), indicated that 8% of the streams had ANC ≤ 0 at baseflow and that this fraction nearly doubled (to 16%) during episodic events. Weathering and cation exchange provide most of the acid neutralization in these streams, and the extent of chronic acidification is largely controlled by the base cation supply rate.

Southeastern Canada (Chapter 9)

Wet SO_4^{2-} deposition throughout southeastern Canada ranges from < 100 eq ha^{-1} yr^{-1} to 730 eq ha^{-1} yr^{-1}, with higher deposition rates occurring in southern Ontario, Quebec, and southern New Brunswick, and lower deposition rates occurring to the north, west, and east. A large portion of this region is underlain by the noncarbonate bedrocks of the Canadian Shield and overlain by thin, coarsely textured soils. Consequently, base cation concentrations are generally low. As a result of low base cation production and acidic deposition, most lakes have ANC $\leq$ 100 $\mu eq\ L^{-1}$ and pH values betweeen 6.0 and 7.0. Fourteen thousand lakes are estimated to be acidic, defined in this case as having pH $\leq$ 5.0. Episodic ANC depressions associated with snowmelt are common. Most low ANC lakes in southeastern Canada are drainage lakes. The ANC supply to these lakes is dominated by base cation production. Sulfate adsorption is unimportant, but SO_4^{2-} reduction may be important, primarily in lakes with long hydraulic residence times. Organic anions may be important in areas with extensive wetlands, particularly Nova Scotia, Labrador, and northwestern Ontario, but there are almost no acidic lakes with high DOC in areas that receive low levels of acidic deposition.

Western Virginia Mountains (Chapter 10)

Sulfate inputs to the western Virginia mountains (ca. 500 eq $ha^{-1}\ yr^{-1}$) are among the highest for the case study regions, whereas H^+ inputs (ca. 400 eq $ha^{-1}\ yr^{-1}$) are comparable to those of much of the eastern United States. The dominant acid-sensitive resource in this region is headwater streams. In a near census of natural brook trout streams in the western Virginia mountains (Virginia Trout Stream Sensitivity Study), 93% of the streams were classified as sensitive (ANC $\leq$ 200 $\mu eq\ L^{-1}$), 49% were very sensitive (ANC $\leq$ 50 $\mu eq\ L^{-1}$), and 10% were acidic (ANC $\leq$ 0). Cation exchange and SO_4^{2-} adsorption are the dominant biogeochemical mechanisms of ANC regulation and stream ANC is related to bedrock geology. Average SO_4^{2-} retention was 68% for the VTSSS streams. Because SO_4^{2-} adsorption removes SO_4^{2-} and contributes ANC to stream water, these streams have not reached steady state with respect to SO_4^{2-} or ANC. Breakthrough of SO_4^{2-} adsorption capacity is projected to occur within a few decades at current loading rates. Following breakthrough, SO_4^{2-} concentrations should increase markedly with a concomitant decrease in ANC of 30 to 90 $\mu eq\ L^{-1}$ (Webb et al. 1990). Trend analysis for Deer Run indicates that SO_4^{2-} breakthrough may have started.

Southern Blue Ridge Province (Chapter 11)

The dominant acid sensitive resources in the Southern Blue Ridge Province are low-order streams and rapidly flushed reservoirs. None of the streams and lakes sampled during the NSS and ELS-I in the Southern Blue Ridge Province was chronically acidic, but low median ANC values (100 $\mu eq\ L^{-1}$ for upstream reach ends, 250 $\mu eq\ L^{-1}$ for lakes) show that these systems are potentially susceptible to acidification. Episodic acidification has been observed in several streams and is potentially threatening to trout populations. Eshleman (1988) estimated that 2.2% of Southern Blue Ridge Province streams should experience ANC depressions $\leq$ 0.

Anion retention, particularly SO_4^{2-} adsorption, is a more important source of ANC to these systems than base cation production, reflecting the well-weathered status of the soils and the base-poor bedrock (granite or sandstone). Because SO_4^{2-} adsorption is an important ANC regulation process, eventual saturation of the SO_4^{2-} adsorption capacity of the soils may be followed by a period of rapid acidification. The base cation composition of low ANC waters in this region is often dominated by Na^+ rather than Ca^{2+}. Most of these systems have low DOC, and organic anions are relatively unimportant in regulating ANC or pH.

Florida (Chapter 12)

A substantial fraction (22%) of the Florida ELS-I lakes are currently acidic. Most of the low ANC and acidic lakes in Florida are groundwater recharge lakes inherently sensitive to acidification because they receive very little groundwater input and therefore little ANC input from the terrestrial watershed. Internal processes – reduction of SO_4^{2-} and removal of NO_3^-, NH_4^+, Ca^{2+}, and K^+ by seston deposition – are more important than terrestrial weathering inputs in regulating ANC in these lakes. In reaching these conclusions in this book and in related papers (Baker et al. 1988b), we have excluded lakes with major anthropogenic activities in their watersheds in order to examine the chemistry of lakes that are undisturbed by human activities other than acidic deposition. However, the majority of Florida ELS-I lakes are impacted by agriculture and other human activities. Lakes with watershed distubances have higher ANC and are less sensitive to acidification, but it is unclear whether this is a result of anthropogenic disturbances (i.e., fertilizer inputs) or whether the edaphic characteristics of a watershed that encourage anthropogenic activities also render the lake inherently less susceptible to acidification.

Upper Midwest (Chapter 13)

In the Upper Midwest, there is an east-west gradient in atmospheric deposition: wet SO_4^{2-} loadings decrease from 400 eq ha^{-1} yr^{-1} in the eastern Upper Penninsula of Michigan to 200 eq ha^{-1} yr^{-1} in northeastern Minnesota. The easternmost subregion (2E) has the largest fraction of low ANC lakes (9.8% with ANC $\leq$ 0), yet no lakes in northern Minnesota (2A) have ANC $\leq$ 0. This east–west gradient in lakewater ANC is related not only to the gradient in atmospheric deposition but also to varying base cation concentrations. Nearly 90% of the lakes with ANC $\leq$ 50 μeq L^{-1} are groundwater recharge lakes and these systems are clearly the most sensitive resource. There is no clear evidence of severe episodic ANC depressions (i.e., ANC $\leq$ 0) in this region, but decreased ANC resulting from base cation dilution has been observed for a number of systems during snowmelt.

For the majority of lakes in the Upper Midwest, base cation production is the dominant mechanism of ANC production, but SO_4^{2-} reduction also contributes significant ANC to low ANC seepage lakes. Because NO_3^- and NH_4^+ deposition rates are nearly balanced, efficient retention of both ions by lake and watershed processes results in no net consumption or production of ANC.

West (Chapters 14, 15, and 16)

Of the 720 lakes sampled in the WLS, only one was acidic–Fern, which receives effluent from a geothermal spring. Sulfate concentrations in the West are generally low and ANC is nearly balanced by base cations. Chronic, severe acidification has not occurred in this region. Nevertheless, many of the western lakes have low ANC and low base cation concentrations that make them potentially susceptible to acidification (Landers et al. 1987).

The majority of lakes in the West are drainage lakes, but approximately 20% are seepage lakes. Because base cation production from the watersheds in the West is typically low, the chemical differences between seepage lakes and drainage lakes are less pronounced than in other regions. Base cation production is undoubtedly the dominant source of ANC for most of these lakes, as seen in the close correlations between ANC and base cations (Chapter 14, Turk and Spahr, this volume; Chapter 15, Melack and Stoddard, this volume; Chapter 16, Nelson, this volume). Sulfate adsorption or reduction is probably not important in this region, and retention of inorganic nitrogen species may be relatively inefficient because of rapid flushing during the spring snowmelt. Most western lakes are clear, and the influence of DOC on ANC or pH is minimal.

Episodic acidification resulting from snowmelt results in ANC depressions in a few sites that have been intensively studied. Acidic episodes have been documented, but are apparently uncommon. Nevertheless, because of the importance of snowmelt on the chemistry of these lakes, we would expect a close relationship between acidity of the snowpack and the magnitude of ANC and pH depression during snowmelt.

Summary of Major Biogeochemical Processes

Major biogeochemical processes that regulate surface water ANC are summarized in Table 17.11. Each process shown in Table 17.11 has been ranked by the RCS authors according to its importance in each region. The table thus represents a composite view of the relative importance of these processes in controlling ANC and pH throughout the case study regions. In summary:

- Hydrologic processes are extremely important in regulating ANC in all surface waters. Variation in flowpath through soils is a major determinant of pH and ANC in drainage lakes, largely because contact with soils is necessary for base cation production to occur. The extent of hydrologic isolation from the groundwater systems is a major factor controlling base cation and ANC concentrations in seepage lakes. Throughout most of the case study regions, evapoconcentration is relatively uniform, as indicated by lake:deposition Cl^- ratios around 1.5 in the Upper Midwest, Maine, and the Adirondacks. In the southeast, and particularly in Florida, evapoconcentration is much higher, with a median lake/deposition Cl^- ratio of 5.0 in Florida. In Florida lakes, high evapoconcentration ratios probably exacerbate

TABLE 17.11. Relative importance of processes influencing acid-base chemistry of low ANC lakes and streams in the case study regions.[a]

	ADR	ME	CAT	VIR	SBR	FLA	UMW	ROC	SIER	CASC
Acidic deposition	●	◑	●	●	●	◑	◑	☼	☼	☼
Watershed processes										
Mineral acid episodes	●	–	●	●	◑	☼	◑	☼	☼	☼
Evapoconcentration	◑	◑	◑	◑	◑	●	◑	☼	☼	☼
Surface water runoff	●	◑	●	●	●	☼	☼	●	●	●
Soil process										
Base cation mobility	●	●	●	●	◑	☼	◑	●	●	●
Sulfate adsorption	☼	☼	☼	◑	●	☼	☼	☼	☼	☼
Nitrogen retention	●	●	◑	◑	●	●	●	–	–	–
DOC production	◑	◑	☼	☼	☼	●	●	☼	☼	☼
Al dynamics	●	☼	●	●	☼	☼	☼	☼	☼	☼
Surface water processes										
CO_2 degassing	◑	◑	☼	●	●	☼	◑	◑	◑	◑
Sulfate reduction	☼	☼	☼	☼	☼	●	●	☼	☼	☼
N retention	☼	●	☼	☼	◑	●	●	●	●	●

[a] Processes are ranked by regional authors according to their overall importance within and among regions. ● = very important; ◑ = moderately important; ☼ = unimportant; – = insufficient information or not applicable; ADK = Adirondacks; ME = Maine; CAT = Catskills; VIR = Virginia; SBR = Southern Blue Ridge; SIER = Sierra Nevada; CASC = Cascade Mountains. Rankings refer to lakes only for all regions except CAT, VIR, and SBR, where they refer to lakes and streams.

the effects of acidic deposition by concentrating H^+. Episodic acidification associated with storms or snowmelt is important in many case study lakes, except for seepage lakes, and acidic episodes occur in most of the case study regions in the eastern United States and in southcentral Canada.

- Base cation production is the dominant source of ANC in all lakes and streams in the case study regions except in the streams in the western Virginia mountains and the Southern Blue Ridge Province, where SO_4^{2-} adsorption assumes greater importance, and in Florida groundwater recharge lakes where anion reduction processes are the dominant source of ANC. Primary weathering is the dominant source of base cations in the West; weathering and cation exchange contribute to cation production in the Upper Midwest and the Northeast; and cation exchange is the dominant source of cations in the Southeast. Calcium is the dominant cation in the Northeast and West; Ca^{2+} and Mg^{2+} are co-dominant in the Upper Midwest; and Na^+, Ca^{2+}, or Mg^{2+} is dominant in the Southeast. Immobilization of cations by in-lake processes (seston deposition) appears to cause a net sink of Ca^{2+} and K^+ in Florida groundwater recharge lakes.
- Sulfate is generally at steady state in the watersheds of drainage lakes throughout the Northeast and Upper Midwest. Sulfate adsorption is a very important process in well-weathered soils of the western Virginia mountains and Southern Blue Ridge Province that can account for losses of > 60% of input SO_4^{2-}. Sulfate adsorption is the dominant source of ANC in many low ANC lakes and streams in these regions. In-lake SO_4^{2-} retention is an important SO_4^{2-} sink in seepage lakes, accounting for nearly 40% of SO_4^{2-} input to Upper Midwest groundwater recharge lakes and nearly 80% of input to Florida groundwater recharge lakes. In-lake SO_4^{2-} sinks occur by SO_4^{2-} reduction in sediments, which is first order with respect to lakewater SO_4^{2-} concentration, and by seston deposition, which is probably important only in lakes with low SO_4^{2-} concentrations. In-lake SO_4^{2-} retention is the dominant source of ANC in Florida groundwater recharge lakes and the second most important source of ANC in the Upper Midwest groundwater recharge lakes.
- Inorganic nitrogen species (NO_3^- and NH_4^+) are efficiently retained in all low ANC systems for which ion enrichment analysis was conducted, but efficiency may be reduced (perhaps to under

50%) in rapidly flushed systems with large snowmelt inputs. Biological assimilation is undoubtedly the dominant mechanism for retention for both of these nutrient species, although some loss of NO_3^- may also occur by denitrification in wetlands or lake sediments. Atmospheric inputs of NO_3^- exceed NH_4^+ inputs in the eastern case study regions, so ANC production from NO_3^- retention is greater than ANC destruction by NH_4^+ retention. Nitrate inputs roughly balance NH_4^+ inputs in the Upper Midwest and ANC production from NO_3^- retention counterbalances ANC consumption resulting from NH_4^+ retention.

- Organic acids contribute to decreased pH in many high DOC lakes throughout the Upper Midwest, Maine, Florida, and eastern Canada; some lakes in these regions are naturally acidic from organic acids. For the low ANC lakes, Gran ANC production was about 14% lower than C_B-C_A ANC production, indicating that the effect of organic acids on ANC is rather modest overall. Most lakes with high DOC levels also have high base cation enrichment. Because of this, most high DOC systems in the case study regions have relatively high ANC compared with low DOC systems.

Acknowledgments. We would like to thank Donald Charles, Jack Cosby, Steven Gherini, Dean Jeffries, Ronald Munson, David Schindler, Hans Martin Seip, and John Stoddard for their critical reviews of this chapter. We also thank Susan Christie for editing and Joe Bernert, Mark Mitch, and Jeremy Smith for data analysis and statistical support.

18 Long-Term Temporal Trends in Surface Water Chemistry

Timothy J. Sullivan

ABSTRACT. This chapter (1) synthesizes the data presented in the foregoing 11 case study chapters regarding past changes in surface water chemistry in response to acidic deposition, (2) quantifies the acidification that has occurred, where possible, and (3) evaluates likely future responses of surface waters in the study regions to projected changes in acidic deposition. Based on spatial evaluation of surface water chemical ion ratios within the regions under investigation, waters that are currently acidic largely as a consequence of high SO_4^{2-} relative to base cation concentrations are most prevalent in the southwestern Adirondack Mountains, the Catskill Mountains, Virginia, Florida, Maine (small aquifer and high-elevation lakes), and northern Michigan and Wisconsin. The SO_4^{2-} concentrations in most of the low ANC ($\leq$ 100 $\mu eq\ L^{-1}$) waters in these regions can generally be ascribed to evapoconcentration of atmospheric inputs.

Empirical distributions of lakewater chemistry across a depositional gradient in the Adirondack and Maine regions suggest that concentrations of ANC, base cations, inorganic Al, and possibly organic acid anions have changed in response to acidic deposition in drainage lakes with current ANC $\leq$ 50 $\mu eq\ L^{-1}$.

Historical reconstructions of lakewater chemistry based on diatom remains in sediment are presented for 50 lakes in the eastern, upper midwestern, and western United States, with the greatest attention to the Adirondack region. Estimated ANC change from preindustrial times to the present for Adirondack lakes with ANC $\leq$ 50 $\mu eq\ L^{-1}$ ranged from a loss of 35 $\mu eq\ L^{-1}$ to a gain of 3 $\mu eq\ L^{-1}$, with a median value of -13 $\mu eq\ L^{-1}$. Assuming that all inorganic monomeric Al (Al_i) in these lakes is attributable to increased mineral acidity, biologically relevant change in chemistry is given by ($\Delta ANC - \Delta Al_i$), and shows a median value of -24 $\mu eq\ L^{-1}$, assuming an average Al valence of +2. Only two lakes in the Adirondack paleoecological data set had current ANC > 50 $\mu eq\ L^{-1}$, and both showed no indication of historical acidification.

Paleoecological data for lakes in Maine, the Upper Midwest, and Florida suggested that some lakes had decreased in pH since preindustrial times, but the data were not as consistent as data for the Adirondacks in suggesting widespread acidification.

Introduction

Numerous attempts have been made to quantify past changes in surface water chemistry in response to acidic deposition of sulfur and/or nitrogen compounds. These attempts have included application of basic geochemical principles (e.g., Henriksen 1979, 1980, 1982, Dickson 1980, Christophersen and Wright 1981, Thompson 1982b, Wright 1983), evaluation of historical water quality data (e.g., Schofield 1976c, 1982, Pfeiffer and Festa 1980, Lewis 1982, Smith and Alexander 1983, Kramer et al., 1986, Eilers et al. 1989; Chapter 8, Stoddard and Murdoch, this volume), and mechanistic model assessment (e.g., Cosby et al. 1985b, Lin and Schnoor 1986). However, a rigorous regional assessment of historical change in surface water chemistry attributable to acidic deposition has not been achieved in North America. Such a regional assessment was difficult because (1) a statistical regional sampling of current surface water chemistry had not been performed, (2) although current [SO_4^{2-}] (and in some cases [NO_3^-]) could be measured, data were insufficient to make defendable estimates of background, preindustrial concentrations, (3) paleoecological reconstructions of historical water chemistry had been completed for only a few lakes, (4) aqueous labile monomeric Al, an important component of chemical change in response to acidic deposition, had not been measured on a regional scale, (5) process-level predictive models had not been validated, and (6) high quality historical measurements of surface water chemistry were generally lacking.

Recent research has to a large degree rectified the first four limitations listed above, in particular the U.S. EPA's Eastern and Western Lake Surveys (Linthurst et al. 1986a, Landers et al. 1987), Phase II of the Eastern Lake Survey (ELS-II), the Paleoecological Investigation of Recent Lake Acidification (PIRLA) study (Charles et al. 1986, Charles and Whitehead, 1986), a recent continuation of the PIRLA research effort (PIRLA-II, Charles and Smol, in press), and numerous smaller scale regional investigations. Comparison of paleoecological assessments and dynamic model hindcasts are currently in progress in an attempt to further validate some of the model approaches (Sullivan et al. 1990). Substantial problems remain, however, with respect to the use of long-term historical records of surface water chemistry (Kramer et al. 1986, Asbury et al. 1989, Eilers et al. in 1989). Although these historical data are useful in conjunction with other techniques, the uncertainties are large relative to the expected changes, particularly with respect to the methyl orange titration endpoints used. Monitoring data are now becoming available with which to assess more recent trends that have occurred over the last decade (Driscoll et al., this volume). Thus, the assessment of historical changes in surface water chemistry in response to acidic deposition is still difficult, but for the first time in the United States, scientists are in a position to make valid regional assessments of past change. The objectives of this chapter are to synthesize the data presented in the previous 11 case study chapters regarding historical change in water quality, to quantify where possible the amount of change that has occurred, and to present some indication of the likely future responses of North American surface waters to acidic deposition.

The approach is based on the following suppositions:

- Surface waters differ markedly in their response to acidic deposition, largely because of differences in hydrologic routing, water residence times, in-lake processes, organic acid influence, soils, geology, and vegetation (see Chapter 1, Munson and Gherini, this volume). These differences result in considerable variation in lake and stream chemistry among and within regions. To minimize heterogeneity, past or future changes in water quality must be assessed on a subpopulational basis (i.e., using stratified data sets).
- A single method does not exist that is appropriate for regional assessment of past or future change attributable to acidic deposition. Only by the use of a variety of assessment tools can defendable population-level estimates be made.
- Many regions of interest contain large numbers of surface waters for which either acidic deposition effects cannot be evaluated because of other perturbations or watershed influences, or for which quantification of change would be meaningless biologically. This chapter, therefore, focuses on systems for which changes can and should be evaluated.

Five principal questions must be addressed for the surface water subpopulations of interest:

1. What is the geographical distribution of surface waters that are currently influenced by anthropogenic SO_4^{2-} (or SO_4^{2-} plus NO_3^-)?
2. What proportion of the current surface water SO_4^{2-} is anthropogenic in origin, that is, what were the preindustrial background SO_4^{2-} levels?
3. Once it has been established that certain aquatic systems have been altered, what are the chemical constituents that have changed and by how much?
4. What percentage of the currently acidic (ANC ≤ 0 or pH ≤ 5.2) systems have become acidic as a result of acidic deposition?
5. What degree of future change should be expected?

Each of the above questions is addressed for the case study regions (Chapters 6 to 16) for which there are sufficient data for making a regional assessment. The discussion focuses primarily on the regions that have been studied most intensively, such as the Adirondack Mountains. Treatment is minimized where uncertainties are large (e.g., seepage lakes of Florida and Upper Midwest). Such bias in treatment reflects data availability and should not be construed as indicating the relative importance of either the resources at risk or the extent of past damage.

FIGURE 18.1. Ratio of acid neutralizing capacity (ANC) to the sum of base cations (Ca^{2+} + Mg^{2+} + Na^{+} + K^{+}) for low ANC ($\leq$ 200 μeq L^{-1}) ELS lakes in Maine. The base cation sum was corrected for cations charge balanced by Cl^{-}.

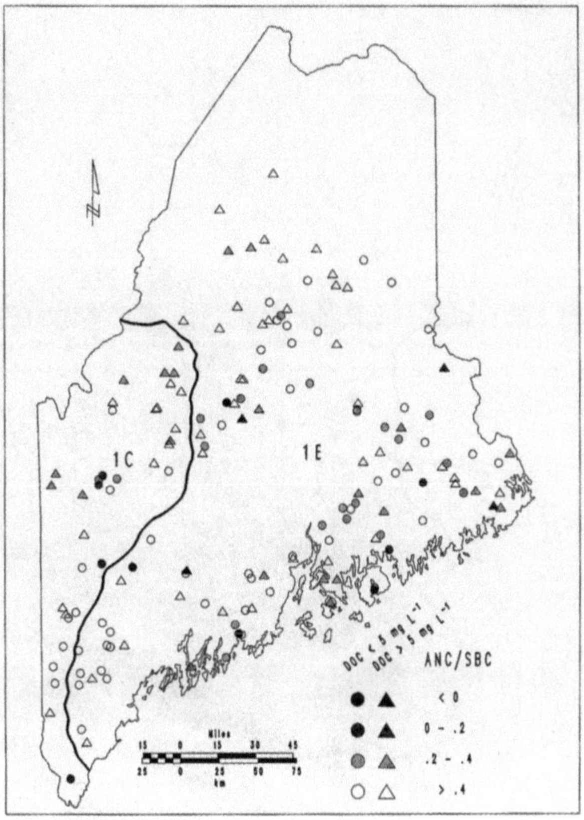

Geographical Distribution of Waters Likely to Have Acidified

Ion ratios provide a qualitative assessment of chemical change in surface waters potentially attributable to acidic deposition. Although such ratios must be interpreted with caution and have often been overinterpreted (Chapter 3, Husar et al., this volume), they constitute a useful regional assessment tool. The ANC/[C_B] ratio is frequently used to infer acidification or chemical change, as is the variation [C_B] − ANC. Interpretation of the ratio is based upon the assumption that pristine, clear (low DOC; $\leq$ 417 μmol L^{-1}) surface waters typically exhibit a ratio of ANC to base cations near 1:1 (Henriksen 1979). A ratio << 1 does not necessarily indicate that acidification (loss of ANC) has occurred, however, but only suggests that the water chemistry may have been altered. In many waters, a discrepancy from the 1:1 relationship may be ascribed solely to increased base cation release. This ratio may also reflect watershed sources of SO_4^{2-}, organic anions, or marine influence (Chapter 3, Husar et al., this volume). For example, a substantial number of ELS lakes in Maine exhibited low ANC/[C_B] (Figure 18.1), even after correction of the base cation term for cations that are charge-balanced by Cl^{-}. Many lakes in this region that exhibited a low ANC:[C_B] ratio are either high in DOC (> 417 μmol L^{-1}) or are located near the coast. Examination of the [SO_4^{*}]/[C_B*]† ratio for the same lakes (Figure 18.2) illustrates that nonmarine SO_4^{2-} does not account for the observed difference between ANC and [C_B]. Many lakes that exhibited a low ANC:[C_B]

†Placement of an asterisk as a superscript indicates that the chemical species has been corrected to reflect only the estimated nonmarine component. Marine contributions were estimated using surface water Cl^{-} concentrations and assuming that all Cl^{-} was of marine origin and that other ionic contributions from sea spray occurred in the same proportion as in sea water. Brackets denote aqueous concentrations, in units of μeq L^{-1}, unless otherwise noted.

FIGURE 18.2. Map of the SO_4^{2-}:SBC ion ratio for low ANC ($\leq$ 200 μeq L^{-1}) surface waters in: (a) Maine (ELS), (b) Maine (HELM and ALPS), (c) Adirondacks, (d) Catskill Mountains, (e) western Virginia, (f) Florida, (g) Upper Midwest, (h) Cascade Mountains, and (i) Sierra Nevada. SBC refers to the base cation sum (Ca^{2+} + Mg^{2+} + Na^{+} + K^{+}), which was corrected for cations charge balanced by Cl^{-}. SO_4^{2-} was sea-salt corrected for Maine and Florida lakes. SBC in (b) does not include K^{+}, which was not measured in the HELM and ALPS studies.

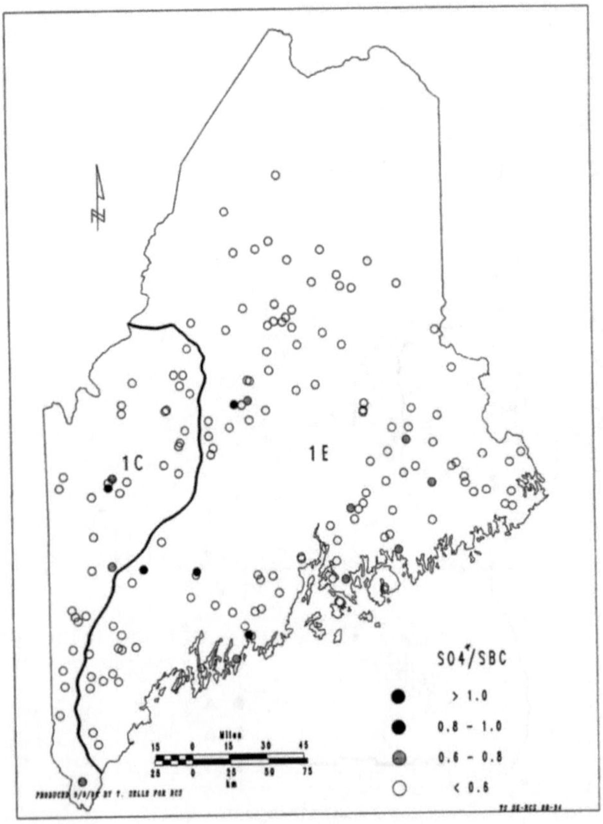

a

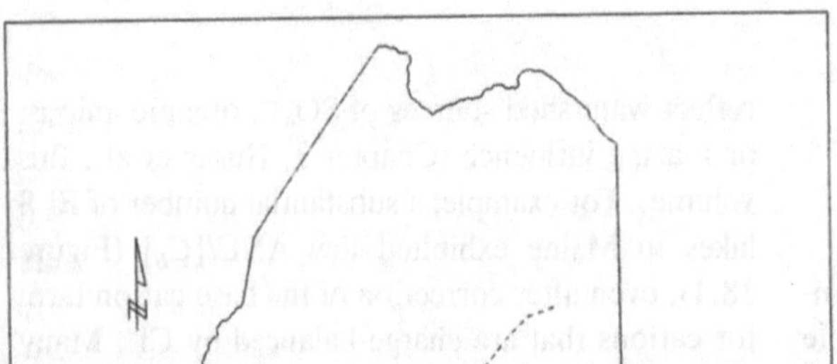

b

FIGURE 18.2. *Continued.*

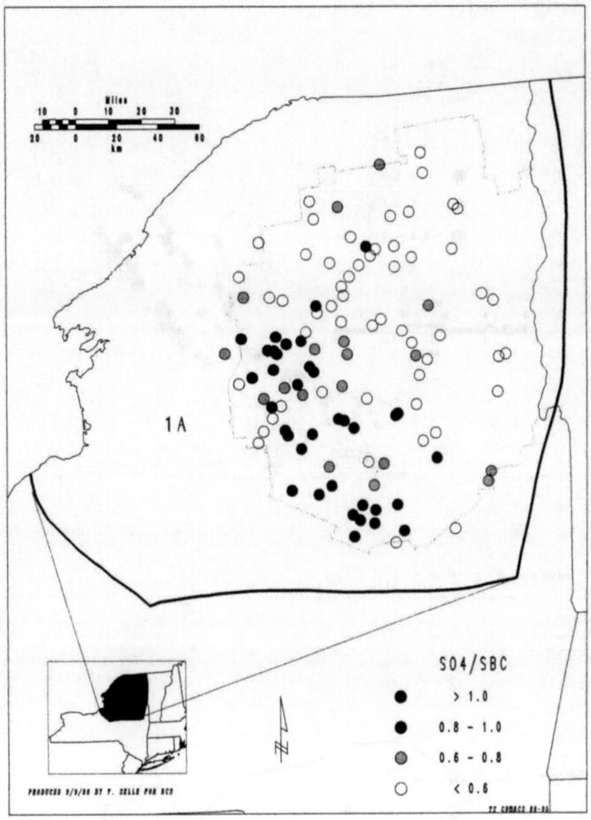

c

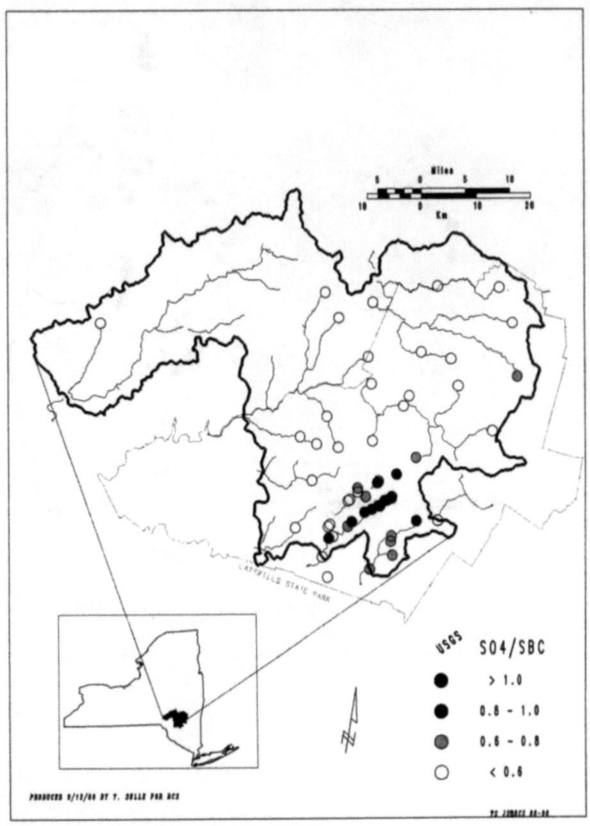

d

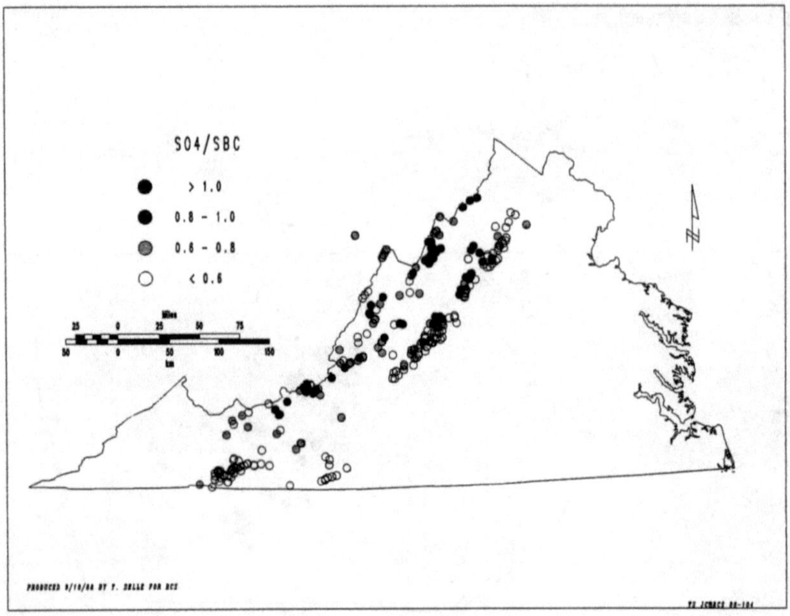

FIGURE 18.2. *Continued.*

e

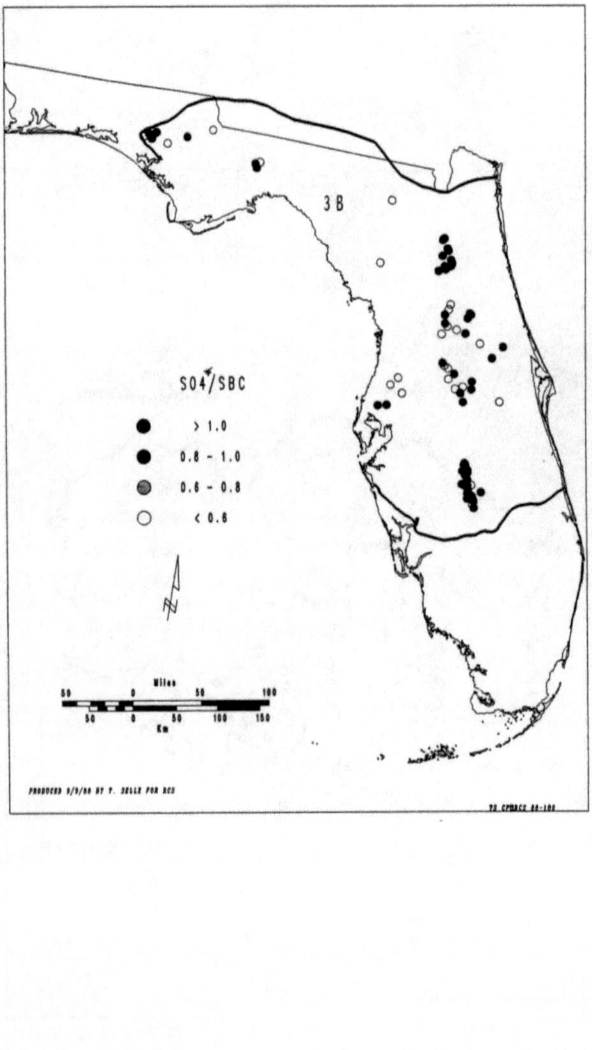

f

FIGURE 18.2. *Continued.*

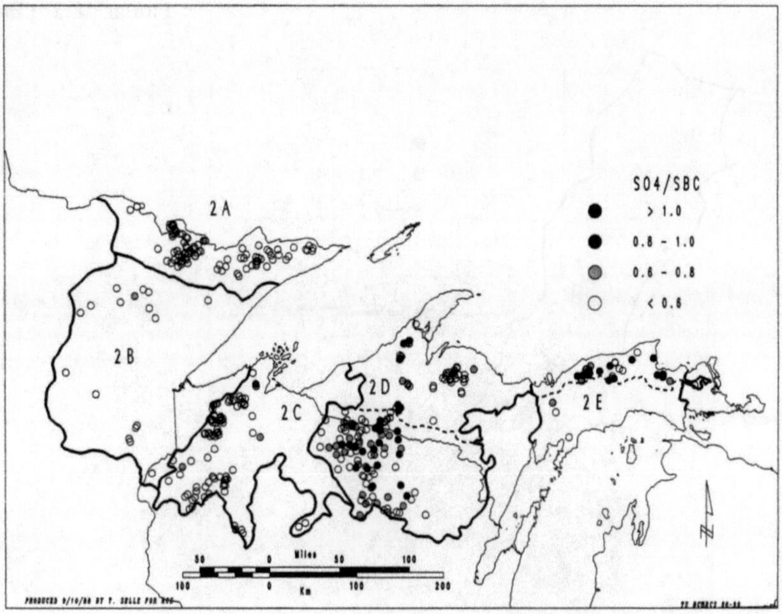

g

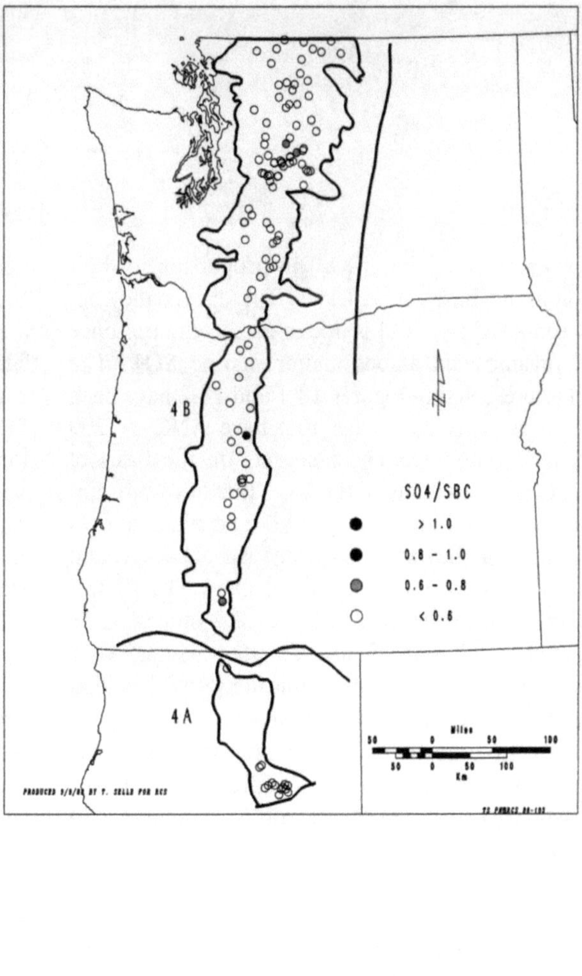

h

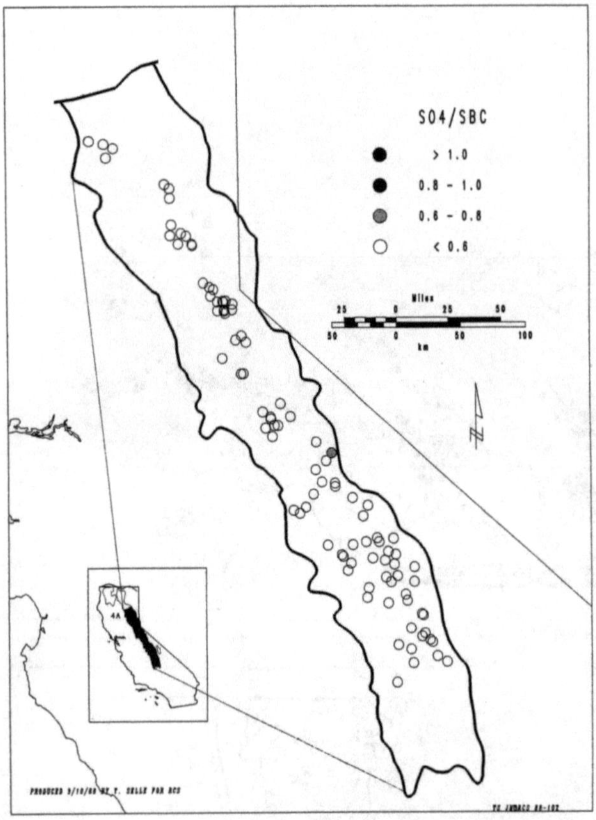

FIGURE 18.2. *Continued.*

ratio have fairly low $[SO_4^*]$ in proportion to their nonmarine base cation levels. The discrepancy between ANC and $[C_B]$ is due in part to the presence of organic acid anions and/or marine SO_4^{2-}. The lakes presented in Figures 18.1 and 18.2 have been screened to remove lakes that have ANC > 200 μeq L^{-1}. This creating of subsets on the basis of ANC removes many of the lakes that have substantial watershed sources of SO_4^{2-} and alkalinity, as discussed in the section on preindustrial background surface water $[SO_4^{2-}]$ in this chapter. The $[SO_4^*]/[C_B^*]$ ratio presented in Figure 18.2 indicates the spatial distribution of surface waters having ANC ≤ 200 μeq L^{-1}, where nonmarine $[SO_4^{2-}]$ is high relative to a system's inherent sensitivity, as reflected by base cation concentrations. Where this ratio exceeds 1.0, water is acidic (ignoring Al), solely because of SO_4^{2-}, regardless of organic acid concentrations. Lakes that are acidic because of high $[SO_4^{2-}]$ relative to $[C_B]$ are most prevalent in the southwestern Adirondacks, Florida, Maine (small aquifer and high-elevation lakes), and northern Michigan and Wisconsin (Figure 18.2).

Streams in Virginia also exhibit high $[SO_4^{2-}]:[C_B]$ (Figure 18.2e). Such conditions are absent from the Southern Blue Ridge Province and from all regions in the western United States (e.g., Figures 18.2h,i). In the Catskill Mountains, few samples exhibited $[SO_4^{2-}] > [C_B]$ (Figure 18.2d). Nitrate concentrations were elevated in many Catskill streams, however, in contrast with most other regions studied. Incorporation of $[NO_3^-]$ into the ratio $([SO_4^{2-} + NO_3^-]/[C_B])$ showed the prevalence of Catskill stream samples that were acidic because of $[SO_4^{2-} + NO_3^-]$ (Figure 18.3). However, most of the samples having ratios > 1 were taken from only a few stream systems in the southern part of the Catskills region. Adirondack lakes were the only other waters studied that exhibited elevated $[NO_3^-]$. Incorporation of NO_3^- into the ratio $([SO_4^{2-} + NO_3^-]/[C_B])$ did not change the pattern observed when only $[SO_4^{2-}]$ was included in the numerator, however, and therefore the revised ratios are not included in the map of Adirondack lakes (Figure 18.2c).

The contrast between Eastern Lake Survey (ELS-I) data from Maine (Figure 18.2a) and data

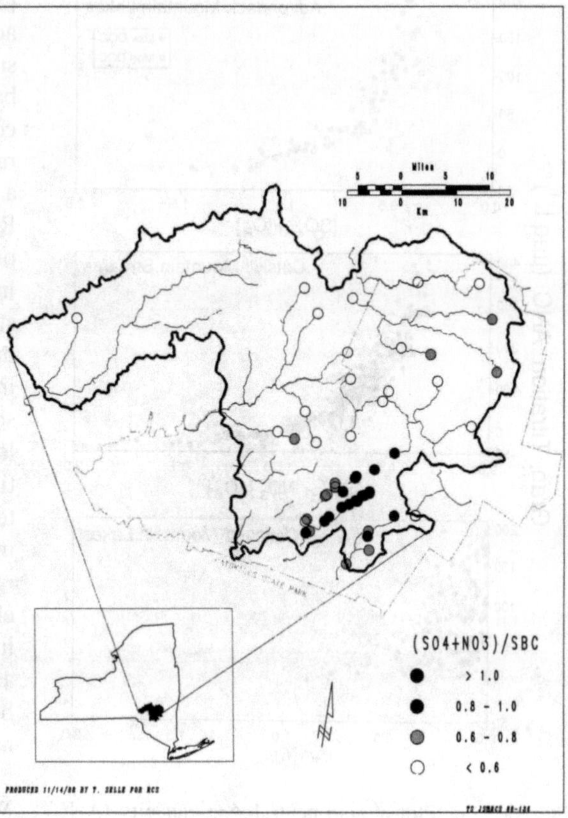

FIGURE 18.3. Map of the $[SO_4^{2-} + NO_3^-]$:SBC ion ratio for low ANC ($\leq$ 200 μeq L^{-1}) streams in the Catskill Mountains.

from the ALPS and HELM lake studies in Maine (Figure 18.2b) is particularly noteworthy. Whereas only two ELS lakes in Maine showed $[SO_4^*]/[C_B^*] > 1$, a large proportion of the ALPS and HELM lakes fall into this category (Figure 18.2b). These studies focused on the most poorly buffered lakes in Maine and those most likely to have been impacted (Chapter 7, Kahl et al., this volume), whereas ELS-I sampled the general population of lakes and did not represent the small (< 4 ha), sometimes very low pH, systems included by ALPS and HELM.

All ELS lakes sampled in the Southern Blue Ridge Province exhibited a $[SO_4^{2-}]/[C_B]$ ratio < 0.6 (data not shown). The very low $[SO_4^{2-}]$ relative to $[C_B]$ for these lakes is attributable to SO_4^{2-} adsorption in soils of this region (Galloway et al. 1983, Rochelle and Church 1987; Chapter 11, Elwood et al., this volume). The region contains a high percentage of systems having low concentrations of base cations, however, and receives relatively high levels of acidic deposition (Chapter 3, Husar et al., this volume). In contrast, the western regions generally exhibited low ratios because of low levels of acidic deposition. Base cation concentrations in Sierra Nevada and Cascade Mountain lakes were extremely low, and these are undoubtedly among the most poorly buffered aquatic systems in the world. Low levels of acidic deposition, however, yield lake $[SO_4^{2-}]$ that is typically 10% to 30% of concentrations found in the eastern United States.

Thus, based on the $[SO_4^{2-}]/[C_B]$ or $[SO_4^{2-} + NO_3^-]/[C_B]$ ratio, historical changes attributable to acidic deposition are most likely to have occurred in the Adirondack Mountains, Maine, Catskill Mountains, Michigan, Wisconsin, Florida, and Virginia. Similar conclusions are drawn, based on the ANC/$[C_B]$ ratio (figures presented in Chapters 6 to 16 for surface waters having ANC $\leq$ 400 μeq L^{-1}), although, in addition to surface water acidification, organic, marine, and watershed influences may cause low values of this ratio. The above list of regions is based only on the areas represented by the regional case studies and is not intended to be a complete listing of impacted areas in the United States.

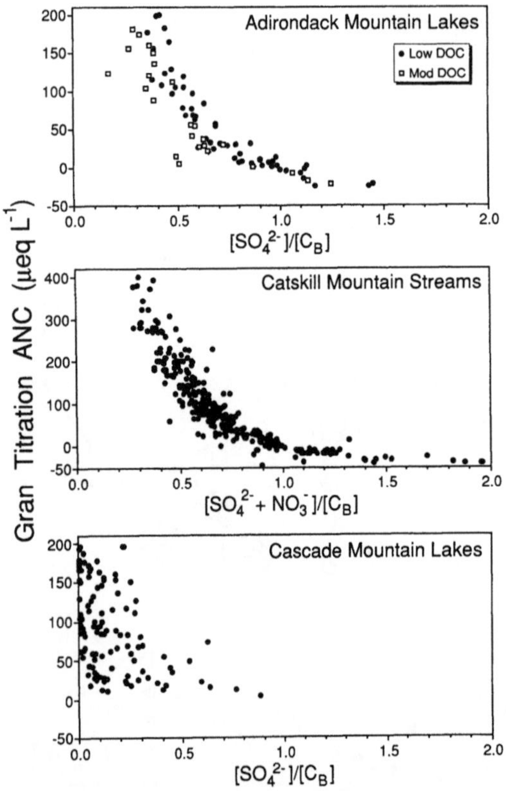

FIGURE 18.4. Plot of acid neutralizing capacity (ANC) versus the ion ratio SO_4^{2-}:SBC for low ANC ($\leq$ 200 μeq L^{-1}) surface waters in (a) Adirondacks, (b) Catskill Mountains, and (c) Cascade Mountains. Because $[NO_3^-]$ is relatively high at some Catskill stream sites, the ratio includes NO_3^- in (b) and is presented as $[SO_4^{2-} + NO_3^-]$:SBC. SBC is the base cation sum, $Ca^{2+} + Mg^{2+} + Na^+ + K^+$.

The $[SO_4^{2-}]/[C_B]$ ratio is useful not only for illustrating lakes or streams that are acidic because of high $[SO_4^{2-}]$ relative to inherent sensitivity, but also for illustrating the degree to which $[SO_4^{2-}]$ and $[C_B]$ influence surface water ANC, H^+, and inorganic Al (Al_i). For example, in the Adirondacks, ANC decreases with increasing $[SO_4^{2-}]/[C_B]$ in a very systematic fashion for low ANC lakes ($\leq$ 200 μeq L^{-1}) (Figure 18.4a). Organic acid anions do not appear to play a major role in this relationship. Negative ANC values are apparent only where $[SO_4^{2-}]/[C_B]$ exceeds approximately 0.8. Similar results are obtained in the Catskill streams when NO_3^- is also included in the ratio along with SO_4^{2-} (Figure 18.4b). In contrast, in western regions, Minnesota, and the Southern Blue Ridge (e.g., Figure 18.4c), $[SO_4^{2-}]/[C_B]$ is less than ~0.8 for all lakes, and no acidic systems (ANC $\leq$ 0) were observed. Many small acidic lakes in the ALPS and HELM data bases in Maine are acidic largely because of high concentrations of organic acid anions ($RCOO^-$), rather than SO_4^{2-} (Figure 18.5). For these analyses, a charge density of 7.5 μeq/mg was assumed for $RCOO^-$, based on an estimate of organic influences on Gran titration results for northeastern lakes included in the Eastern Lakes Survey–Phase II study (ELS-II) (Chapter 3, Husar et al., this volume). Organic anion influence was also pronounced in Florida and Minnesota (data not shown), where some lakes appeared to be acidic (Florida only) or low ANC because of high organic acid anions relative to base cations ($[RCOO^-]/[C_B]$). Similar patterns were observed when organic anions were estimated by the Oliver et al. (1983) method.

Acidic deposition impacts fish primarily via elevated surface water concentrations of H^+ and inorganic monomeric Al (Al_i) (Chapter 4, Baker and Christensen, this volume), and ion ratios can be used to illustrate the combined influence of mineral acid anions on these potentially toxic parameters (e.g., Driscoll and Newton 1985). Monomeric Al was not fractionated into labile (mainly inorganic) and nonlabile (mainly organically complexed) components in ELS-I. In ELS-II, however, a subset of 145 northeastern ELS-I lakes were resampled in autumn 1986, and methodological protocols included Al fractionation. Elevated $[Al_i]$ (> 5 μeq L^{-1}) was observed only in lakes having elevated $[SO_4^{*}]$ (> 75 μeq L^{-1}) (Figure 18.6a). Furthermore, an empirical relationship was developed between total monomeric Al (Al_m), DOC, and Al_i from the ELS-II data base, yielding:

$$Al_i = \underset{(2.057)}{10.906} + \underset{(0.017)}{0.944}\, Al_m - \underset{(0.402)}{2.549}\, DOC \quad (18\text{-}1)$$

$$r^2 = 0.96,\ n = 143$$

where Al is in μg L^{-1} and DOC is in mg L^{-1}, and standard errors of the parameter estimates are presented in parentheses. This relationship was used to estimate $[Al_i]$ for the full northeastern ELS-I data set, and the resulting estimates also illustrate that elevated $[Al_i]$ is found only in the presence of elevated $[SO_4^{*}]$ for northeastern lakes (Figure 18.6b). The importance of $[SO_4^{*}]$ as a cause of elevated concentrations of H^+ and Al_i is illustrated in Figure

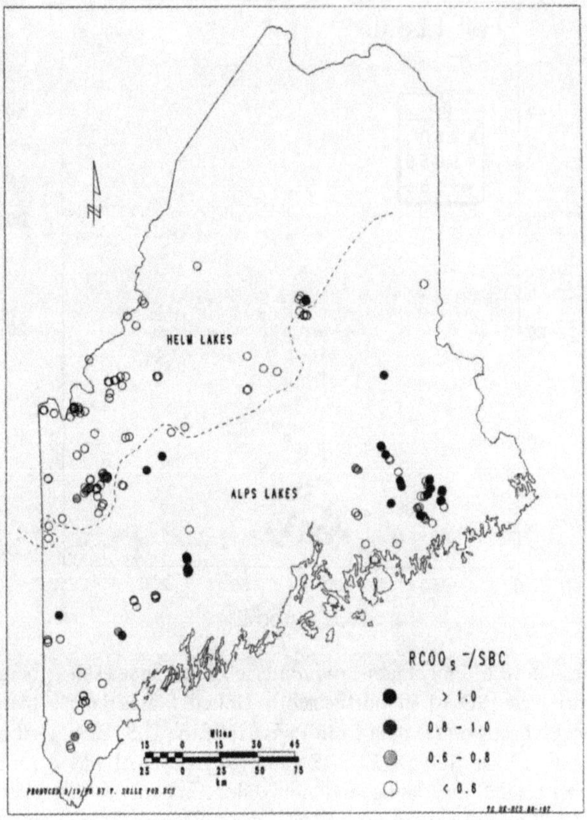

FIGURE 18.5. Map of the organic acid anion (RCOO$^-$):SBC ion ratio for low ANC ($\leq$ 200 μeq L^{-1}) lakes included in the HELM and ALPS studies in Maine. RCOO$^-$ was estimated from measurements of dissolved organic carbon (DOC), using a charge density of 7.5 μeq/mg C to approximate the quantitative influence of organic acid anions on Gran titration measurements of ANC (see Chapter 3, Husar et al., this volume).

18.7. Concentrations of [$H^+ + Al_i$] are generally < 5 μeq L^{-1} for lakes that have [SO_4^*]/[Ca* + Mg*] < 0.8. Many lakes that have [SO_4^*]/[Ca* + Mg*] > 0.8 exhibit elevated concentrations of [$H^+ + Al_i$], which may be deleterious to fish (Chapter 4, Baker and Christensen, this volume).

Preindustrial Background Surface Water Chemistry

Sulfate

Data that illustrate the relationship between surface water SO_4^{2-} concentration and other parameters such as ANC, H^+, or Al_i are of little value in assessing changes attributable to acidic deposition unless inferences can be made and defended regarding the source of SO_4^{2-} in surface waters. Sullivan et al. (1988a) presented data that illustrated an approximately linear relationship between median wet SO_4^{2-} deposition values interpolated to lake sites and median lake SO_4^{2-} concentration for lakes in the eastern, upper midwestern, and western United States. Data from the Southern Blue Ridge Province exhibited the only substantial deviation from the observed pattern, and this was attributed by Sullivan et al. (1988a) to SO_4^{2-} adsorption in southeastern soils (Galloway et al. 1983a, Rochelle and Church 1987a). Although watershed sources of SO_4^{2-} may influence concentrations in certain waters, the overall pattern of surface water [SO_4^{2-}] was in agreement with a hypothesis of atmospheric deposition as the primary source of lake [SO_4^{2-}]. Similar conclusions were reached by Neary and Dillon (1988) in an evaluation of Ontario data. Geologic sources of SO_4^{2-} to surface waters are most often evident in systems with ANC $>$ 100 to 200 μeq L^{-1}. For example, spline diagrams of anion and cation concentration variations across a gradient of ANC show extremely stable [SO_4^{2-}] in low ANC systems (Brakke et al. 1988; Chapter 15, Melack and Stoddard, this volume; Chapter 14, Turk and Spahr, this volume). Melack and Stoddard (this

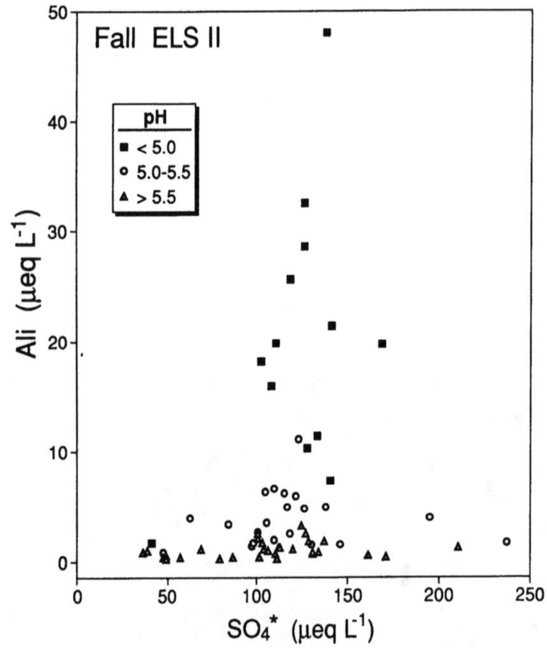

FIGURE 18.6. Plot of labile monomeric Al versus sea salt corrected [SO_4^{2-}] in northeastern United States lake water, based on (a) data from Phase II of the U.S. EPA Eastern Lake Survey (ELS-II), and (b) ELS-I. Al was fractionated into labile and nonlabile components in ELS-II using the methods of Driscoll (1984) and Røgeberg and Henriksen (1985). Al fractionation was estimated for ELS-I lakes from measured total monomeric Al and DOC, using equation 18-1. Data are presented for lakes having ANC ≤ 50 μeq L^{-1}. Equivalence of Al species was calculated using the chemical equilibrium model ALCHEMI (Schecher and Driscoll 1987).

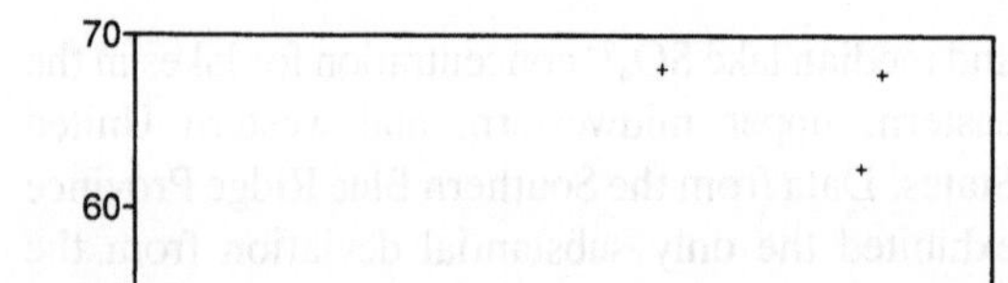

volume) attributed the observed coincident peaks in Ca^{2+} and SO_4^{2-} to small amounts of gypsum ($CaSO_4$) or oxidizable pyrite (FeS_2) and soluble calcite ($CaCO_3$) in the drainage basins of some high ANC Sierra lakes. Almost all Sierra lakes having ANC ≤ 100 μeq L^{-1} have [SO_4^{2-}] < 10 μeq L^{-1}. Similarly, almost all Cascade and Rocky Mountain low ANC systems have [SO_4^{2-}] < 30 μeq L^{-1} (Chapter 14, Turk and Spahr, this volume; Chapter 16, Nelson, this volume) (Figure 18.8).

Figure 18.8 shows that low ANC (≤ 100 μeq L^{-1}) Adirondack and Maine drainage lakes had [SO_4^{2-}] generally < 150 μeq L^{-1}. Most Adirondack lakes had concentrations of 100 to 150 μeq L^{-1}, whereas most Maine lakes had 50 to 100

◄

FIGURE 18.7. Plot of [$H^+ + Al_i$] versus the ion ratio [SO_4^{2-}]:[$Ca^{2+} + Mg^{2+}$] for northeastern ELS-I lakes. Al_i was estimated as in Figure 18.6b. All ions in the ratio were sea-salt corrected.

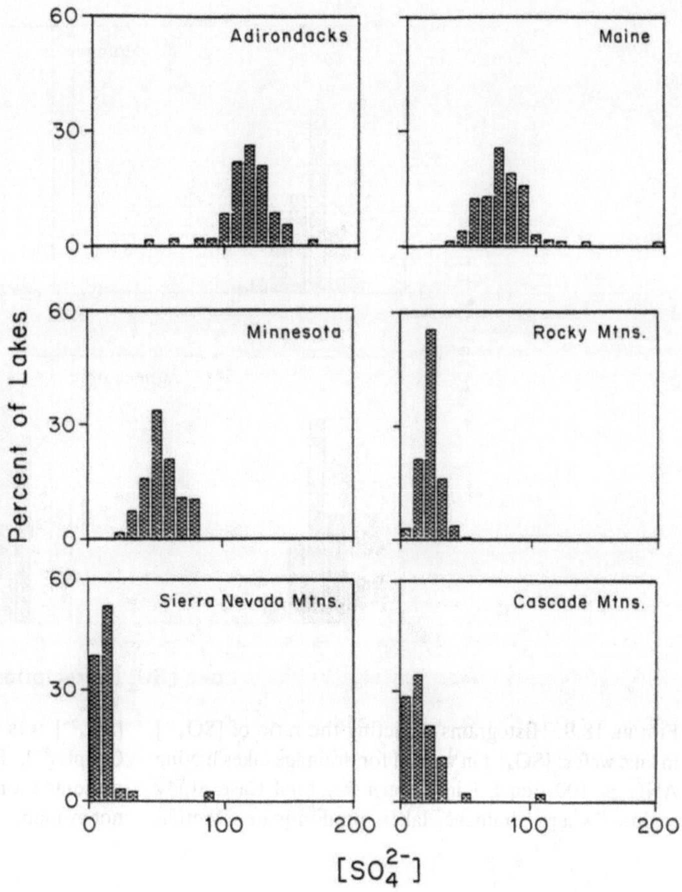

FIGURE 18.8. Histograms depicting $[SO_4^{2-}]$ in drainage lakes having ANC $\leq$ 100 µeq L^{-1} in Regional Case Studies regions where low ANC drainage lakes are common.

µeq L^{-1} of $[SO_4^{*}]$. When higher ANC (> 100 µeq L^{-1}) drainage lakes in the Adirondacks and Maine were included, 13 sampled lakes in ELS-I had $[SO_4^{*}]$ > 160 µeq L^{-1} (ranging up to 375 µeq L^{-1}). None of these 13 lakes had $[H^+ + Al_i]$ > 2 µeq L^{-1}, thus illustrating that watershed contributions of SO_4^{2-} are not typically associated with acidified or acid-sensitive systems in the northeastern United States.

After deleting stream systems impacted by acid mine drainage (e.g., Herlihy et al. 1990), regional distributions of $[SO_4^{2-}]$ for low ANC ($\leq$ 100 µeq L^{-1}) drainage lakes and streams generally show a high degree of homogeneity, despite small intraregional differences in such factors as wet deposition, local emission sources, land use, and canopy type, thus affecting levels of dry deposition (Figure 18.8). Such homogeneity would not occur if substantial watershed sources of SO_4^{2-} were scattered among these low ANC systems.

The spatial distribution of lakewater $[SO_4^{2-}]$ in Minnesota is influenced by bedrock sources of sulfur (Rapp et al. 1985; Chapter 13, Cook and Jager, this volume). For the low ANC drainage lake subset, however, the $[SO_4^{2-}]$ distribution is narrow, with almost all lakes having $[SO_4^{2-}]$ of 40 to 80 µeq L^{-1} (Figure 18.8). Similarly, Cook and Jager (this volume) found that median $[SO_4^{2-}]$ for the 58 upper midwestern ELS lakes underlain by granite and having ANC of $\leq$ 200 µeq L^{-1} was 60 µeq L^{-1}, with an interquartile range of 50 to 69 µeq L^{-1}.

Based on regional estimates of dry deposition and evapoconcentration, surface water $[SO_4^{2-}]$ would be expected to increase relative to wet deposition $[SO_4^{2-}]$ input by a factor of about 1.5 to 3.0. The distribution of values for the ratio of lakewater $[SO_4^{2-}]$ to precipitation $[SO_4^{2-}]$ is presented for drainage lakes having ANC $\leq$ 100 µeq L^{-1} in Maine, the Adirondacks, northeastern Minnesota,

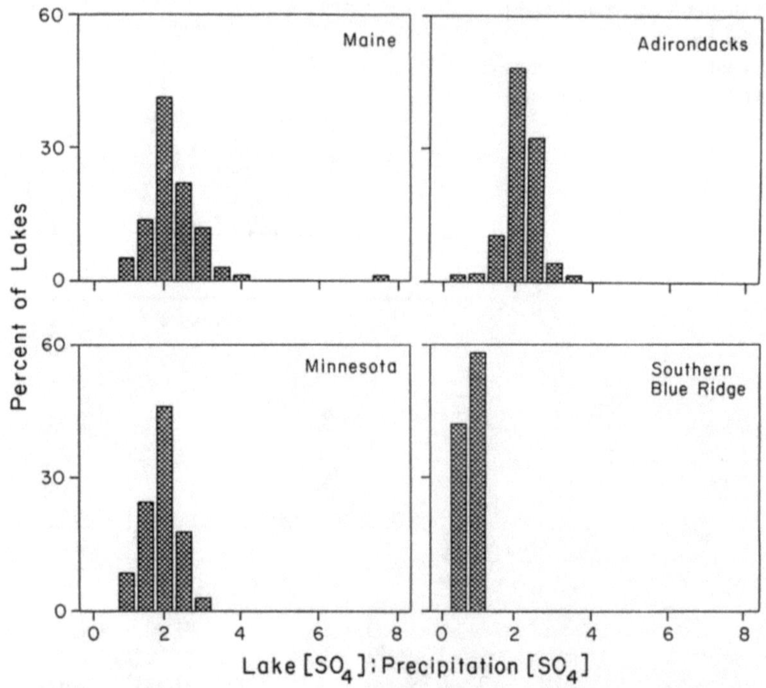

FIGURE 18.9. Histograms depicting the ratio of $[SO_4^{2-}]$ in lake water: $[SO_4^{2-}]$ in wetfall for drainage lakes having ANC $\leq$ 100 μeq L^{-1} in eastern Regional Case Study regions where drainage lakes predominate. Wetfall $[SO_4^{2-}]$ was interpolated to individual lake sites (see Chapter 3, Husar et al., this volume). Data are not presented for western regions because wetfall data were not available as interpolated concentrations at lake sites.

and the Southern Blue Ridge in Figure 18.9. As expected, values for the ratio are generally between 1.5 and 3.0, except in the Southern Blue Ridge, where sulfur retention in soils results in ratios generally < 1.0. Dry deposition estimates for SO_4^{2-} vary widely (OTA 1984), but estimates of 50% to 100% of wet deposition are typically used (Summers et al. 1986; Chapter 10, Cosby et al., this volume; Appendix B, Baker, this volume). Evapotranspiration estimates are also highly variable regionally. Evapoconcentration is probably low (< 1.5) for western mountainous regions (Chapters 14 to 16, Turk and Spahr, Melack and Stoddard, Nelson, this volume). Based on Cl^- ratios, Baker et al. (this volume) estimated median evapoconcentration in the Adirondacks, Maine, and the Upper Midwest to be very similar: 1.6, 1.5, and 1.7, respectively. These estimates generally agree well with regional runoff estimates of evapoconcentration (Chapters 13 and 17, Cook and Jager, Baker et al., this volume; Rapp et al. 1985, Baker et al. 1988b). In contrast, evapoconcentration in southeastern systems can be considerably higher. The median ratio estimated from Cl^- values for 11 Florida low ANC groundwater recharge lakes was 4.37 (Chapter 17, Baker et al., this volume), attributable to low water loading rates, high evaporation, and long water residence times (Chapter 12, Pollman and Canfield, this volume).

Thus, differences between $[SO_4^{2-}]$ in wetfall and surface water concentrations can easily be attributed to dry deposition and evapotranspiration effects for low ANC drainage lakes. The lake $[SO_4^{2-}]$ to precipitation $[SO_4^{2-}]$ ratio was approximately 2.0 to 2.5 in most low ANC drainage lakes in the Adirondacks, Maine, and Minnesota (Figure 18.9). Low ANC drainage lakes were not common in the Upper Midwest outside Minnesota, and $[SO_4^{2-}]$ in precipitation was not available as interpolated values to lakes and streams in other regions. There is no indication that watershed sources of SO_4^{2-} are significant to the populations of low ANC lakes. Similar conclusions were reached by Brakke et al. (1989) on the basis of observed SO_4^{2-} concentrations in surface waters in regions that do not receive acidic deposition. Brakke et al. (1989) estimated a background SO_4^{2-} level for lakes in the northeastern United States of 10 to 15 μeq L^{-1}, in general agreement with the Cosby et al. (Chapter 10, this volume) estimate of 22 μeq L^{-1} for Virginia streams. Given the caveat that the estimate applies only to low ANC drainage systems, a background [SO_4^{2-}] value of 10 to 20 μeq L^{-1} seems reasonable for the eastern regions under consideration here. The background concentrations are most likely to be lower in groundwater recharge lakes than in drainage lakes, due to the increased importance of SO_4^{2-} reduction resulting from long hydrologic residence times (Rudd et al. 1986b) and the decreased importance of dry deposition as a mechanism for sulfur input (Norton et al. 1988a).

Inorganic Monomeric Aluminum

Inorganic monomeric Al (Al_i) is often mobilized and transported from soils to surface waters in response to atmospheric deposition of mineral acids (Cronan and Schofield 1979, Dickson 1980, Driscoll and Schecher 1988). Some forms of Al_i are toxic to fish (Muniz and Leivestad 1980, Baker and Schofield 1982, Baker 1982, Rosseland and Skogheim 1984), whereas organically complexed Al is generally regarded as nontoxic to aquatic biota. Inorganic monomeric Al cannot be measured directly, however, but is estimated based on operationally defined labile (mainly inorganic) and nonlabile (mainly organic) fractions (Driscoll 1984). The procedure involves measurement of total monomeric Al (Al_m) by complexation with either 8-hydroxyquinoline (Barnes 1975) or pyrocatechol violet (Dougan and Wilson 1974, Røgeberg and Henriksen 1985), followed by colorimetric determination, or sometimes in the case of 8-hydroxyquinoline complexation, atomic absorption spectroscopy. Nonlabile monomeric Al (Al_o) is measured in a similar fashion using a sample aliquot that has passed through a cation exchange column. Based on measurements of Al_m in ELS-I and WLS-I, aqueous inorganic Al appears to be important primarily in Adirondack lakes (Linthurst et al. 1986). This contention is supported by results of Al fractionation for a subset of 155 northeastern ELS-I lakes in ELS-II. Inorganic monomeric Al was estimated in the National Stream Survey (NSS) (Kaufmann et al. 1988), but was found to be elevated only in mid-Atlantic subregions and Florida. Streamwater [Al_i] data from the NSS have not yet been thoroughly evaluated, however.

Data from both ELS-I and ELS-II clearly show that Al_i is associated with high concentrations of nonmarine SO_4^{2-} (> approx. 75 μeq L^{-1}) (Figure 18.6a,b) in lakes that have pH values < 5.5. Similar results were found for 1,005 lakes surveyed in Norway (Henriksen et al. 1988), where elevated Al_i was generally restricted to low pH lakes in high deposition areas of southernmost Norway.

Mobilization and transport of Al within the soil profile is thought to be an important component of the podzolization process (Driscoll and Schecher 1988). In pristine environments, Al appears to be retained largely within the lower soil horizons. Dissolved Al concentrations are low in most natural waters due to the low solubility of inorganic Al under circumneutral conditions (Stumm and Morgan 1970). It has been hypothesized, however, that mineral acids from atmospheric deposition have altered the natural process of podzolization by transporting inorganic Al from soils to surface waters (Cronan and Schofield 1979). In addition to the northeastern United States, elevated Al concentrations have been reported in low pH waters within regions receiving high inputs of acidic substances in Sweden, Norway, Belgium, The Netherlands, Germany, and Canada, as summarized by Driscoll and Schecher (1988).

Based on empirical observations and solubility considerations, we assume that Al_i concentrations are substantial only for low pH (< 5.5) waters with high concentrations of mineral acid anions, primarily SO_4^{2-} and/or NO_3^-. Although high [Cl^-] could contribute to elevated [Al_i], [Cl^-] is generally low (≤ 10 μeq L^{-1}) for northeastern lakes with elevated Al_i (> 5 μeq L^{-1}), and therefore does not appear to be an important consideration. Thus, the most appropriate estimate for preindustrial background [Al_i] for most northeastern lakes is that the concentrations were negligible (i.e., < 5 μeq L^{-1}).

Quantification of Changes in Surface Water Chemistry

Data presented in the preceding sections provide evidence suggesting that surface water chemistry has been substantially altered by SO_4^{2-} in some of the regions under consideration here, most notably the Adirondacks, Florida, Maine (particularly small aquifer and high-elevation lakes), northern

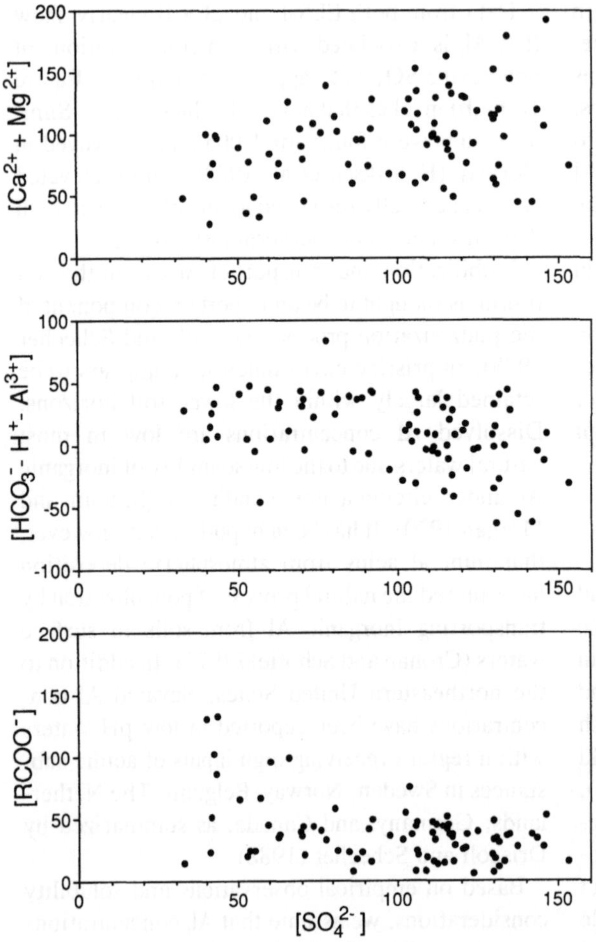

FIGURE 18.10. Plots of lakewater ionic components versus sea salt corrected $[SO_4^{2-}]$ for a combined data set, including lakes in the Adirondacks and Maine (ELS-I data) having ANC $\leq$ 50 μeq L^{-1}: (a) $[Ca^{2+} + Mg^{2+}]$ versus $[SO_4^{2-}]$, (b) $[HCO_3^- - H^+ - Al^{3+}]$ versus $[SO_4^{2-}]$, and (c) $[RCOO^-]$ versus $[SO_4^{2-}]$. $[Ca^{2+}]$, $[Mg^{2+}]$, and $[SO_4^{2-}]$ were sea salt corrected. $RCOO^-$ was estimated from DOC, assuming a charge density of 7.5 μeq/mg C (see Chapter 3, Husar et al., this volume).

Michigan and Wisconsin, western Virginia, and the Catskill Mountains. The data also illustrate that some surface waters in these regions are currently acidic solely because of high $[SO_4^{2-}]$, or $[SO_4^{2-} + NO_3^-]$ as in the Catskills, relative to base cation concentrations. The data further indicate that regional patterns of surface water $[SO_4^{2-}]$ are generally attributable to atmospheric deposition, and that for the systems of most interest with respect to acidic deposition effects (i.e., those having ANC < 100 μeq L^{-1}), background preindustrial $[SO_4^{2-}]$ was likely to be low, about 10 to 20 μeq L^{-1} for drainage systems in the eastern United States. Although watershed sources of SO_4^{2-} are substantial in some systems, they occur most often in insensitive high ANC systems or streams receiving acid mine drainage. The next tasks are to identify the chemical parameters likely to have changed as a consequence of acidic deposition and to quantify the amount of change that has occurred.

Identifying Constituents that Have Changed

It is generally assumed that watersheds and surface waters respond to increased concentrations of SO_4^{2-} and/or NO_3^- primarily by increasing Ca^{2+}, Mg^{2+}, and Al^{n+} export, and/or decreasing carbonate alkalinity (Henriksen 1979, 1980, 1982, Wright 1983; Chapter 1, Munson and Gherini, this volume; Chapter 3, Husar et al., this volume). A growing body of indirect evidence also suggests that organic acid anion concentrations may have decreased in response to acidic deposition (Almer et al. 1974, Krug and Frink 1983, Marmorek et al. 1988), although a general consensus has not been achieved. The parameters most likely to have changed thus include SO_4^{2-}, and in a few cases NO_3^-, HCO_3^-, Ca^{2+}, Mg^{2+}, Al^{n+}, H^+, and possibly $RCOO^-$. Husar et al. (Chapter 3, this volume) present an empirical model for estimating ANC change in response to acidic deposition, as:

$$\Delta ANC = (F_{BC} + F_{AL} + F_{org} - 1) \times \Delta SO_4^* \quad (18\text{-}2)$$

where the F terms are proportionality factors for change in $[Ca^{2+} + Mg^{2+}]$, $[Al^{n+}]$, and $[RCOO^-]$, respectively, as initially proposed by Henriksen (1982) for $[Ca^{2+} + Mg^{2+}]$. Thus, for base cations:

$$F_{BC} = \Delta[Ca^{2+} + Mg^{2+}]/\Delta SO_4^* \quad (18\text{-}3)$$

and similarly for the other proportionality factors. Equation 18-2 is based on a definition of ANC that includes Al^{n+} as a base cation (Chapter 2, Munson and Gherini, this volume) and corresponds most closely with Gran ANC determinations (Sullivan et al. 1989). Increased $[Al_i]$ alone therefore leads to increased ANC, although the Al_i may have deleterious biological consequences (Chapter 4, Baker and Christensen, this volume).

Evidence from Empirical Evaluations

As discussed previously, evaluation of changes in water chemistry in response to acidic deposition must involve quantification of changes in carbonate ANC ($HCO_3^- - H^+$), base cations (principally Ca^{2+} and Mg^{2+}), inorganic Al, and possibly organic acid anions. Empirical evidence for the northeastern United States suggests that these changes have indeed occurred. In order to examine lakes across a range of SO_4^{2-} deposition, data were combined from the Adirondack and Maine regions, which are generally similar with respect to geologic, edaphic, and vegetational characteristics, and consequently are considered to be in the same major land resource area (MLRA) (Austin 1972). Both the MLRA and the ecoregions as defined by Omernik (1987) classify the Adirondacks and much of Maine together. The southeastern portion of the Maine coastal zone is classified differently.

Figure 18.10b shows a significant ($p < 0.0001$) decreasing trend in $[HCO_3^- - H^+ - Al^{3+}]$, with increasing lake $[SO_4^*]$, for drainage lakes with ANC $\leq$ 50 $\mu eq\ L^{-1}$. This decrease in $[ANC - Al^{3+}]$ is most pronounced at $[SO_4^*] > 75\ \mu eq\ L^{-1}$. Similarly, a significant ($p < 0.0001$) increasing trend is observed for $[Ca^* + Mg^*]$ with increasing lake $[SO_4^*]$ (Figure 18.10a). A less clear, but nevertheless significant ($p < 0.01$) trend of decreasing $RCOO^-$ was also observed (Figure 18.10c). The observed trend of decreasing DOC

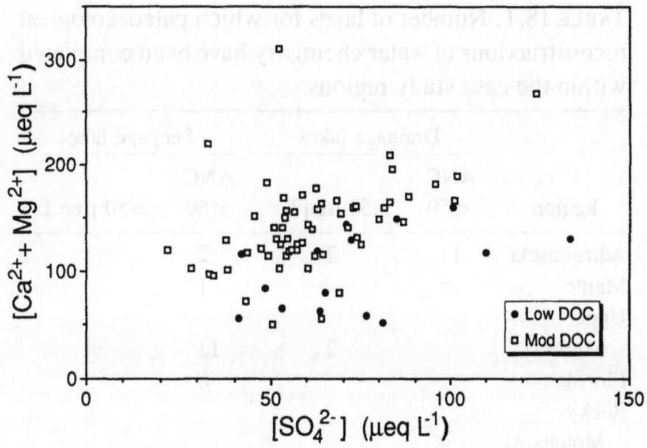

FIGURE 18.11. Plot of $[Ca^{2+} + Mg^{2+}]$ versus $[SO_4^{2-}]$ for upper midwestern drainage lakes having ANC $\leq$ 100 $\mu eq\ L^{-1}$. Data points are coded by dissolved organic carbon (DOC): $\leq$ 5 mg L^{-1} (417 $\mu mol\ L^{-1}$), •; > 5 mg L^{-1}, □.

and organic acid anions with increasing lake $[SO_4^*]$ may be driven, however, at least to some extent, by wetland processes. In catchments dominated by wetlands, sulfur retention via biological reduction reactions may remove SO_4^{2-} from solution during low flow periods (e.g., autumn sampling) (Kerekes et al. 1986a,b) and at the same time contribute high concentrations of DOC to drainage waters. Only drainage systems were included in the analyses depicted in Figure 18.10 because seepage lakes are expected to respond differently to SO_4^{2-} loading (Chapter 17, Baker et al., this volume). For systems with ANC > 50 $\mu eq\ L^{-1}$, only the increasing trend in base cations is significant.

Low ANC drainage lakes in the Upper Midwest also showed an increasing trend in $[Ca^{2+} + Mg^{2+}]$ with increasing $[SO_4^{2-}]$ (Figure 18.11), although the pattern was influenced by organic acid concentrations. High DOC lakes generally exhibited higher base cation concentrations within this ANC stratum. Few drainage lakes in the Upper Midwest had ANC $\leq$ 50, however, precluding empirical evaluation of this subset.

Groundwater flowthrough lakes in the Upper Midwest were defined by Baker et al. (Chapter 17, this volume) as those having SiO_2 > 1 mg L^{-1}. Classification of seepage lakes into groundwater flowthrough and recharge categories on the basis of $[SiO_2]$ results in a very clear distinction of upper midwestern subpopulations and marked differences in $[C_B]$ and ANC between these subpopulations.

TABLE 18.1. Number of lakes for which paleoecological reconstructions of water chemistry have been completed within the case study regions.

	Drainage lakes		Seepage lakes	
Region	ANC <50	>50 μeq L⁻¹	ANC ≤50	>50 μeq L⁻¹
Adirondacks	11	2	2	
Maine	8		1	
Upper Midwest		2	12	1
Florida			6	
Rocky Mountains	4			
Sierra Nevada	1			

Groundwater flow-through lakes in the Upper Midwest were generally high in pH and ANC, due largely to substantial groundwater inputs of base cations (Chapter 17, Baker et al., this volume). Only 6% of these lakes had ANC ≤ 50 μeq L^{-1} and none were acidic. They are therefore of relatively little interest with respect to past changes in acid-base status that might have occurred in response to acidic deposition. Groundwater recharge lakes in the Upper Midwest were more common (71% of the seepage lakes), and were more frequently low in pH and ANC. Five percent were acidic and 9% had pH < 5.5. Nearly 90% of upper midwestern lakes that had ANC ≤ 50 μeq L^{-1} were in this hydrological category (Chapter 17, Baker et al., this volume). Upper midwestern low ANC (≤ 100 μeq L^{-1}) groundwater recharge lakes exhibited a significant ($p \leq 0.001$) negative relationship between [HCO_3^- − H^+] and [SO_4^{2-}] ($r^2 = 0.07$, $n = 200$, 4 outliers deleted) and also a significant ($p \leq 0.0001$) negative relationship between total organic anions, estimated by the Oliver et al. (1983) method, and SO_4^{2-} ($r^2 = 0.15$, $n = 200$, 4 outliers deleted). No relationship was evident, however, between [Ca^{2+} + Mg^{2+}] and [SO_4^{2-}]. These data are consistent with a hypothesis of decreased ANC and organic anions in response to acidic deposition loading for this upper midwestern hydrologic lake type, and suggest that increased base cation release may not be as important in these low ANC groundwater recharge systems as it appears to be for northeastern and upper midwestern drainage lakes. Such an interpretation is consistent with current understanding of important processes that modify drainage and seepage lake chemistry (Chapter 1, Munson and Gherini, this volume).

Evidence from Paleoecological Studies

Historical reconstructions of water chemistry, based on diatom remains in sediment cores (Chapter 3, Husar et al., this volume) (Table 18.1), have been completed for a total of about 50 lakes within the eastern, upper midwestern, and western United States. By far the most complete paleoecological assessment data are available for the Adirondacks, where chemical reconstructions for pH and ANC have been performed for 15 lakes (Table 18.2; Chapter 6, Driscoll et al., this volume; Chapter 3, Husar et al., this volume). Several general observations can be made concerning these data. Estimated ANC change for the 11 low ANC (≤ 50 μeq L^{-1}) drainage lakes was variable, ranging from a loss of 35 μeq L^{-1} to a gain of 3 μeq L^{-1}, with a median value of −13 μeq L^{-1}. Assuming that all Al_i now present in these lakes is attributable to mineral acidity (e.g., preindustrial $Al_i = 0$), "biologically relevant" change in chemistry is given by (ΔANC − ΔAl_i), and showed a median value of −24 μeq L^{-1} for drainage systems, assuming an average Al valence of +2. Only two lakes in the paleoecological data set have current ANC > 50 μeq L^{-1}, and neither of these showed historical loss of ANC. Only two lakes were found to have been historically acidic ($ANC_o \leq 0$ and $pH_o \leq 5.2$) (see section on inorganic monomeric aluminum) and both were small seepage lakes. As expected, the two seepage lakes showed low [Al_i].

Diatom reconstructions of lakewater pH have been performed on cores from nine lakes in Maine (Charles and Norton 1986; Chapter 7, Kahl et al., this volume). These reconstructions suggested no, or slight, acidification since preindustrial times. Five of the seven Maine lakes reported by Charles and Norton (1986) are currently acidic and tend to have both high [SO_4^{2-}] and low [Ca^{2+} + Mg^{2+}] relative to ELS-I population estimates. These acidic lakes are primarily very small (4 ≤ 6 ha; 3 ≤ 4 ha), apparently naturally acidic (pre-1850 pH < 5.5) systems.

In the Upper Midwest region, diatom-inferred pH reconstructions have been completed for 15 lakes,

TABLE 18.2. Current lakewater chemistry and results of diatom-inferred changes in chemistry from pre-1900 to the present for lakes in the Adirondack region.

Lake	Area (ha)	Type[a]	Current water chemistry[b]						Diatom-inferred chemistry[c]			
			pH	ANC	SO_4^{2-}	DOC	Al_i	$(Ca^{2+} + Mg^{2+})$	pH_o	ANC_o	ΔpH	ΔANC
Deep	12	D	4.76	−27	130	192	11	58	5.4	5	−0.6	−14
Merriam	8	D	4.88	−13	89	458	9	48	5.4	−1	−0.3	−5
Bear	22	D	5.01	−5	142	58	2	119	5.9	8	−0.4	−13
Upper Wallface	6	D	5.0	−2	111	179	5	80	5.3	9	−0.4	−11
Big Moose	515	D	5.24	1	132	200	5	114	6.0	32	−1.0	−27
West Pond	12	D	5.31	3	95	408	2	91	5.6	7	0.0	+3
Queer	53	D	5.63	4	130	158	0	124	6.4	42	−1.0	−35
Darts	50	D	5.57	5	130	208	2	122	5.9	17	−0.4	−8
Woods	26	D	4.9	−10	ND	167	10	ND	5.2	24	−0.4	−11
Sagamore	66	D	6.16	37	147	492	1	195	6.8	74	−0.5	−22
Rondaxe	68	D	6.40	44	127	200	0	157	6.8	79	−0.3	−29
Windfall	2	D	7.39	71	113	275	7	184	6.8	65	0.0	+6
Clear	71	D	6.88	99	104	233	0	203	7.0	105	0.0	+5
Barnes	2	S	4.62	−20	55	408	1	34	5.2	−14	−0.5	−11
Little Echo	1	S	4.18	−63	68	1192	0	50	4.6	−51	0.0	−2

[a] D = drainage; S = seepage.
[b] ANC, SO_4^{2-}, and $(Ca^{2+} + Mg^{2+})$ in μeq L^{-1}, DOC and Al_m in μmol L^{-1}. Chemical data for Upper Wallface are from Charles (1982); data for Bear and Windfall are from ALSC; data for Woods are from Davis et al. (1988); data for all other lakes are from Linthurst et al. (1986). Al data are from ELS-II, Driscoll (personal communication), ELS-I (converted to Al_i using Equation 18-1), ALSC, and Schofield et al. (1985).
[c] Diatom-inferred chemistry is from Charles et al. (in press) and Charles (unpublished).

as summarized by Cook and Jager (Chapter 13, this volume). Four lakes showed a pH decline of 0.2 to 0.4 pH units during the past 50 to 100 years, all of which have current pH ≤ 5.7. Diatom-inferred pH increased in one lake by 0.2 pH units. No change was inferred for the other 10 lakes, including 4 lakes with current pH > 6.0. One lake, McNearney Lake, Michigan, was inferred to have been historically acidic, defined as pH ≤ 5.2. McNearney Lake is currently buffered against pH change by high concentrations of inorganic Al (Cook et al. 1990).

Paleoecological reconstructions of pH and ANC have been completed for six seepage lakes in northern Florida as part of the PIRLA study. Two of the lakes, Barco and Suggs, have become more acidic since about 1900 (Chapter 12, Pollman and Canfield, this volume). The inferred pH declines correspond with increased SO_4^{2-} emissions in the southeastern United States (Husar 1986) and with increases in Pb and PAH within the core profiles (Sweets et al. 1990; Chapter 12, Pollman and Canfield, this volume).

In the western United States, diatom-inferred pH reconstructions have been reported for four lakes in Rocky Mountain National Park, Colorado (Baron et al. 1986) and for Emerald Lake in Sequoia National Park, California (Chapter 15, Melack and Stoddard, this volume). No evidence of significant historical acidification has been found for western lakes (Charles et al. 1989).

The emerging pattern of historical change in lakewater chemistry based on diatom stratigraphies suggests:

1. Lakes that currently have pH > 6.0 (ANC > approx. 50 μeq L^{-1}) do not exhibit evidence of recent acidification. Additional data are needed, however, because few lakes high in pH and ANC have been evaluated in areas that receive high levels of acidic deposition. Empirical relationships between $[SO_4^{2-}]$ and ANC, however, also support this preliminary assessment that these systems may not have been acidified.
2. Inferred decreases in ANC have been small (≤ 35 μeq L^{-1}) for low ANC lakes (those having current ANC ≤ 50 μeq L^{-1}). The median inferred loss of ANC for low ANC Adirondack lakes was 13 μeq L^{-1}.
3. Some lakes were historically acidic, but these are most often small (< 6 ha) lakes that were

not well represented in the ELS (Johnson et al. 1989).

4. Some seepage lakes in Florida and the Upper Midwest have acidified to some extent, but more data are needed for further assessment of change. Interpretation of inferred acidification of Florida lakes is complicated by coincident changes in hydrology that may have contributed to the inferred ANC declines (Chapter 12, Pollman and Canfield, this volume).
5. Few data are available for western United States lakes, but they are consistent in suggesting no change in acid-base status.

Evidence from Comparison of Historical and Modern Measurements

There are a number of problems associated with the interpretation of historical surface water chemistry data. The most significant difficulty is the lack of documentation of historical sampling and analytical procedures (Kramer 1986). Alkalinity and pH measurements were generally made colorimetrically with indicator dyes prior to 1950. Methyl orange was often the indicator. It changes color across a pH gradient and historical records are unclear regarding the exact endpoint used. The best approach for assessment of these data has been to specify the most likely endpoints used by the original analysts and to "correct" historical measurements prior to their comparison with present-day electronic measurements (e.g., Kramer et al. 1986, Asbury et al. 1989, Eilers et al. 1989). Additional difficulties in interpretation include potential sample contamination from glass collection bottles (Kramer and Tessier 1982), climatological variability, land use changes, and statistical uncertainties.

Some rigorous evaluations of historical data have been undertaken, however, that included data quality assurance and correction of colorimetric measurements (e.g., Kramer et al. 1986, Asbury et al. 1989, Eilers et al. 1989). Estimated changes in alkalinity and pH in New York lakes were sensitive to the assumption made about the pH of the methyl orange endpoint used. Assuming an endpoint pH of 4.2, substantial acidification of New York lakes over the past 50 years was indicated (Kramer et al. 1986, Asbury et al. 1989). Alternatively, use of an endpoint near 4.0 suggested little change in acid-base status (Kramer et al. 1986). Eilers et al. (1989) selected a consistent set of data from 145 of the northern Wisconsin lakes that had been surveyed by Birge, Juday, and coworkers during the period 1925 to 1941. After correction of historical data to enable direct comparison with recent measurements by Eilers et al. (1983), the data suggested, in general, that conductivity, Ca^{2+}, pH, and alkalinity had increased during the intervening 50-year period. The increases were attributed to watershed development, and in two cases to major watershed fires. When only the 19 lakes that had not experienced development were considered, it was estimated that 7 had experienced a pH decline. Because of the uncertainties involved, the authors could not attribute the apparent decline in pH of some lakes to acidic deposition alone.

Evaluations of some historical chemical data are presented in several of the preceding case study chapters in this volume (e.g., Chapter 6, Driscoll et al.; Chapter 7, Kahl et al.; Chapter 8, Stoddard and Murdoch; Chapter 12, Pollman and Canfield). Although these data do not lead to regional conclusions, they may be useful in comparison with other approaches. Kahl et al. (Chapter 7, this volume) reported an unpublished resurvey by Haines and coworkers in 1985 of 50 Maine lakes originally studied by Fuller and Cooper (1946). Methods and colorimetric indicators were duplicated to the extent possible in the resurvey. The data did not suggest a decline in lake pH. Limited historical water quality data were presented by Pollman and Canfield (Chapter 12, this volume) for softwater lakes in the Trail Ridge Lake District of Florida. The most extensive data record in Florida was for McCloud Lake. It suggested a pH decline of about 0.3 units from 4.90 to 4.64 during the period 1968–69 to 1985. Regression analysis indicated an increase in $[H^+]$ of 1.25 μeq L^{-1} yr^{-1} in this lake since 1968. The authors could not conclude that this trend was related to acidic deposition. $[H^+]$ was clearly correlated with the $[SO_4^*]/[Cl^-]$ ratio, however, suggesting that increased deposition of nonmarine SO_4^{2-}, rather than increased evapoconcentration, had promoted higher lake concentrations of SO_4^{2-} and H^+.

Stoddard and Murdoch (Chapter 8, this volume) presented historical data for Schoharie Creek in the Catskill region from 1922 to the present. Although $[SO_4^{2-}]$ had to be estimated, the data were generally internally consistent, and showed marked increases in both [Ca^{2+} + Mg^{2+}] and ANC since the 1950s. A decrease in [SO_4^{2-}] was accompanied by a concomitant increase in [NO_3^-]. The Schoharie Creek basin has undergone substantial development since the period after World War II. These data suggest watershed impacts as a likely cause of the apparent changes in water chemistry, in agreement with the interpretation by Eilers et al. (1989) of estimated changes in water chemistry of many Wisconsin lakes.

Percentage of Waters that Are Acidic

Although it is difficult to quantify the magnitude of past changes in important chemical parameter concentrations, such as pH and ANC, it is somewhat easier and perhaps more important to attempt to quantify the number of surface waters that are currently very low in pH or ANC as a direct consequence of acidic deposition. It is these very low pH/ANC waters that are most important biologically, although specification of precise cutoff values of pH or ANC for biological effects is not possible (e.g., Schindler 1988b; Chapter 4, Baker and Christensen, this volume). The focus of this discussion is on waters inferred to have had preindustrial ANC ≤ 0 or pH ≤ 5.2. Sullivan et al. (1988a) analyzed lake chemistry data from eight areas of North America that receive low levels of wet SO_4^{2-} deposition (< 10 kg ha^{-1} yr^{-1}). With the exception of two lakes in northwestern Ontario and one lake in Yellowstone National Park (influenced by a hot spring), acidic lakes (ANC ≤ 0) were absent from the 2,130 lakes investigated. The authors cautioned, however, that small lakes, especially those < 4 ha and, to a lesser extent, those 4 to 10 ha, were underrepresented. This caution seems especially pertinent in view of the previously discussed results of diatom-inferred acid-base status of several small Adirondack and Maine lakes that suggested historical acidity, and the preponderance of organically acidic small lakes in Maine (Figure 18.5; Chapter 7, Kahl et al., this volume). For the lakes in the ELS-I frame population, however, which did not include lakes < 4 ha and underrepresented those from 4 to ~ 12 ha in area (Johnson et al. 1989), the data suggested that acidic northeastern drainage lakes sampled in ELS-I would generally not be acidic in the absence of nonmarine SO_4^{2-}. This conclusion agrees with results of paleoecological data available to date (see section on Paleoecology) and with ion ratios illustrating that most acidic lakes sampled by ELS-I are currently acidic because of high [SO_4^*] relative to base cation concentrations (see section entitled Geographical Distribution of Impacted Waters). This does not imply that naturally acidic lakes do not exist, however. Organically acidic small lakes in Maine (Chapter 7, Kahl et al., this volume), McNearney Lake, Michigan (Chapter 13, Cook and Jager, this volume), and the highly acidic systems in the Okefenokee Swamp of Georgia (Linthurst et al. 1986a) are examples of naturally acidic lakes. The considerable number of lakes and streams that currently have [SO_4^*]/[C_B^*] > 1 (Figures 18.2, 18.4, and 18.7) are acidic because of [SO_4^*], however, and it is likely that most of these have become acidic because of SO_4^{2-} deposition. In contrast, many of the ALPS and HELM lakes that now have [SO_4^*] > [C_B] (Figure 18.2b) may have been acidic previously because of high levels of organic anions (Figure 18.5).

Complicating Factors in the Evaluation of Acidification

There are two major difficulties in assessing regional and temporal trends in surface water chemistry in response to acidic deposition. The first concerns the extent to which factors other than acidic deposition, particularly those related to land use, have influenced surface water acid-base chemistry. Land use changes, such as forest regrowth or recovery from watershed disturbance, can cause acidification of surface waters. Second, the regional water chemistry data available for evaluating temporal trends are based primarily on single water samples collected during autumn (lakes) or spring (streams). Data that can be used to evaluate seasonal variability or episodic pH and ANC depressions associated with hydrological episodes are not available on a regional scale. The complications associated with land use changes and seasonal variability are discussed in the following sections.

Land Use

An earlier hypothesis proposed that recent lake-water acidification could be attributed to changes in land use (Rosenqvist 1978b, Krug and Frink 1983). Rosenqvist's (1978b) contention that increased acidic humus formation, in response to decreased upland agriculture, had caused the observed regional acidification in southernmost Norway was not substantiated by subsequent research. Drabløs and Sevaldrud (1980) examined data for dilute Norwegian lakes that had experienced pH decline and loss of fish populations between 1965 and 1980. They eliminated changes in forestry and drainage as sources of the acidification in four of five study areas because most of the affected lakes were above treeline. Dairy farming, sheep grazing, and changes in reindeer pasturing did not correlate with loss of fish populations.

Deforestation can acidify runoff, but changes in surface water acidity after clear-cutting are due primarily to nitric acid (Likens et al. 1970, Borman et al. 1968, Lawrence and Driscoll 1988). The chemical effects of deforestation last only a few years (Borman and Likens 1979), and cannot be of regional significance to chronic acid-base status because $[NO_3^-]$ is very low in most acidic surface waters in the United States (Linthurst et al. 1986a, Kaufmann et al. 1988; Chapter 6, Driscoll et al., this volume). Similarly, forest growth can increase acidity by the formation of organic acids (Richter 1984) and uptake of base cations (Nilsson et al. 1982), but does not play an important role in determining the acid-base status of most acidic lakes and streams in the Adirondacks, Upper Peninsula of Michigan, Catskills, and Virginia, because DOC concentrations are low in most of the acidic systems and base cation uptake cannot cause waters to become acidic without an excess of acid anions.

Activities such as road building and agriculture disturb natural vegetation and can expose soil or bedrock to increased leaching of chemical constituents. Such disturbances frequently result in increased surface water concentrations of base cations, ANC, and pH (Haines and Akielaszek 1983, Havas et al. 1984, Eilers et al. 1989). Recovery from disturbance generally leads to decreased pH and ANC as the system returns to predisturbance conditions. Thus, a short-term investigation of an ecosystem in the process of recovering from a watershed disturbance might erroneously conclude that acidification was occurring in response to changes in atmospheric deposition or some other cause external to the watershed. Analysis of paleolimnological data avoids such potential problems in interpretation, because preindustrial chemistry is inferred for a time period that predates significant anthropogenic activities, and recovery from disturbance is unlikely to proceed beyond predisturbance conditions. In addition, paleolimnological analyses often document evidence of forest fires and thus can rule out such natural disturbances as a cause of chemical change. Diatom-inferred pH reconstructions for some lakes do indicate disturbance effects and subsequent recovery, for example, at Woods Lake in the Adirondacks (Davis et al. 1988).

Diatom analyses of sediments from acidified lakes have shown that, in almost all cases, the preacidification floras of the lakes were very stable. The same diatom species occurred in approximately the same proportions at each site for hundreds of years, despite shifting land management patterns and climatic changes (Charles et al. 1989). At all acidified sites in the United Kingdom, the beginning of acidification of lakes was consistent with the expansion of industry in Britain from the late eighteenth to early nineteenth century. At no sites did the acidification predate sedimentary trace metal contamination (Battarbee et al. 1988). The onset of recent acidification inferred for Adirondack lakes in the PIRLA-I study began in all cases after 1900 (Charles et al., in press). Although watershed disturbance may have played a limited role, it was not a primary cause of recent acidification.

Thus, land use changes and watershed disturbances can have a large effect on surface water chemistry, but the regional acidification of surface waters in parts of Europe and North America cannot be attributed to land use practices. Where these changes have been substantial, it may be difficult, or impossible, to detect the effects of acidic deposition.

Seasonal Variation

A major limitation of this evaluation of past changes in water chemistry is that quantifications are based largely on values for index chemistry. For lakes, these are from autumn (e.g., ELS-I) or summer (e.g., ALSC, PIRLA-I) measurements. The chemistry of lakes and streams often varies seasonally and also annually in response to changes in precipitation, temperature, hydrology, and biological processes. Surface waters are generally most acidic (lowest in pH and ANC, highest in Al_i) during periods of high discharge, such as spring snowmelt or rainstorm events. Peak concentrations of [NO_3^-] and [Al_i] frequently occur as a pulse during the early phases of snowmelt. Seasonal pH/ANC depressions can be attributed to a number of factors, however. These include natural processes, such as dilution of base cations, nitrogen mineralization, and flushing of organic materials from the terrestrial to aquatic environment, and also anthropogenic factors related to acidic deposition. The SO_4^{2-} response to episodic increases in discharge is highly variable as a consequence of such factors as ionic fractionation within the snowpack, S-mineralization, dry deposition during the pre-event period, and wetland dynamics. The end result of these complex interactions is that drainage water [SO_4^{2-}] can decrease (Sullivan et al. 1986a, 1987), remain constant (Driscoll et al. 1985, Seip et al., 1985, Sullivan et al. 1986a), or increase (Galloway et al. 1980, LaZerte and Dillon 1984, Sullivan et al. 1987) during hydrological episodes.

It is not possible at this time to adequately describe the relationship between seasonal variability in surface water chemistry and chronic acidification due to atmospheric deposition. It is likely, however, that the quantifications of historic change in acid-base status presented here are somewhat conservative (i.e., changes based on spring chemistry would probably be larger). To some extent, paleolimnology integrates chemical seasonality in the resulting inferences of chemical change. Nevertheless, the algal microfossils are calibrated to summer index chemistry, and this should result in an overall underestimation of maximum high-discharge acidification.

Seasonal changes in hydrology can also have a large impact on the relative importance of mineral and organic acids. For example, LaZerte and Dillon (1984) conducted a detailed mass balance study of the seasonal importance of SO_4^{2-} and organic anions as sources of acidity in the Plastic Lake catchment in the Muskoka-Haliburton district of Ontario. Organic acidity was dominant during summer low-flow periods, but SO_4^{2-} was more important during periods of high discharge and on an annual basis. Other studies have also documented similar seasonal patterns (Kerekes et al. 1982, 1986a, 1986b, Visser 1984, Kessel-Taylor 1986, Cook et al. 1988).

Seasonal changes in Adirondack lakewater chemistry have been discussed by several investigators (Schofield 1982, Driscoll and Schafran 1984, Driscoll et al. 1987a,b; Chapter 6, Driscoll et al., this volume; Schaefer et al. 1990), and are summarized here. Acidic episodes associated with spring snowmelt were found in Panther Lake, and were associated with decreased pH, base cation concentrations, and ANC, and increased NO_3^- and Al_m concentrations (Schofield 1982). The Panther Lake watershed is dominated by thick glacial till deposits and exhibits high baseflow pH and ANC. However, the dominant hydrologic flowpath changes during periods of rapid discharge, from one of deep flow through glacial till to shallow flow through upper soil horizons. This occurs when groundwater capacity is exceeded and downward percolation is restricted (Chapter 6, Driscoll et al., this volume). Short-term changes in the pH of chronically acidic Woods Lake were far less variable, however. Although pH depressions were observed coincident with snowmelt, the changes were small because of the logarithmic nature of the pH scale and the strong pH buffering that resulted from high concentrations of aqueous Al.

Schaefer et al. (1990) examined changes in outlet chemistry for 11 Adirondack lakes in 1986 and 10 lakes in 1987 (also summarized by Driscoll et al., Chapter 6, this volume). The study lakes included a gradient in baseline ANC, which was consistently 10% to 15% lower than the annual maximum ANC that was recorded in late summer or fall. Snowmelt ANC depressions for all of the lakes revealed three general patterns. Lakes with intermediate baseline ANC exhibited the largest episodic changes. Lakes with the highest baseline ANC had the largest contribution of ΔC_B as a fraction of ΔANC (i.e., the episode was largely driven by dilution). And thirdly, the proportional contribution of NO_3^- to the ANC depression increased sharply with decreasing baseline ANC. Seasonal inputs of acidity in the low ANC lakes were neutralized via dissolution of Al because these watersheds are dominated by thin deposits of glacial till. The mobilization of Al minimized the extent of ANC and pH depressions (Chapter 6, Driscoll et al., this volume).

Summary of Historical Change Evaluation

Past change in surface water chemistry in response to atmospheric deposition has been most clearly demonstrated for the Adirondack region. Paleoecological studies in the Adirondacks suggest (1) that ANC and pH changes have generally been restricted to lakes with current ANC $\leq$ 50 µeq L^{-1} (pH $\leq$ 6.0), and (2) that ANC changes in this low ANC subset have been relatively small (median loss of 13 µeq L^{-1}). Inorganic Al (Al_i) has increased in many of these waters in response to acidic deposition and must also be considered in an evaluation of historical changes in water quality. A more biologically meaningful change is given by ΔANC $-$ ΔAl_i, which showed a median value of -24 for the Adirondack drainage lakes with ANC $\leq$ 50 µeq L^{-1} included in the paleoecological investigations (Table 18.2; assuming an average valence of +2 for Al_i). These data are supported by empirical observations of spatial patterns in [Ca^{2+} + Mg^{2+}] and [ANC $-$ Al_i] along a gradient of lake [SO_4^{2-}] in the Adirondacks and Maine (Figure 18.10). The empirical data implied a loss of ANC, an increase in [Al_i], and an increase in base cations in response to increased [SO_4^{2-}] in waters with ANC $\leq$ 50 µeq L^{-1}. For higher ANC systems, only the base cation trend was statistically significant. Although many small, highly sensitive lakes, such as those included in the HELM and ALPS studies in Maine (Chapter 7, Kahl et al., this volume), were probably always acidic, most acidic lakes in the northeastern United States > 4 ha are currently acidic because of high [SO_4^{2-}] relative to base cation concentrations. Paleoecological data and comparison with data from areas that do not receive substantial atmospheric loadings suggest that, except for the small lakes that were not included in the ELS, most currently acidic lakes in the Adirondacks and Maine were not acidic in preindustrial times but have become acidifed by atmospheric deposition of sulfur.

A substantial percentage of the streams in western Virginia (~10%) are currently acidic because of high [SO_4^{2-}] relative to base cations (Figure 18.2e; Chapter 10, Cosby et al., this volume). Although paleoecological data are not available for these stream systems, the most logical conclusion is that many of these systems were acidified by atmospheric deposition.

In the Upper Midwest and in Florida, data regarding likely historical acidification are most convincing for groundwater recharge lakes. Complications related to potential hydrological and climatological variations, scarcity of data on pristine seepage lakes, and limited process-level research on these seepage systems preclude rigorous evaluation of historical change in these lakes. The data are generally consistent, however, with a hypothesis of acidification by atmospheric deposition of some groundwater recharge lakes in Florida, northern Michigan, and northern Wisconsin.

Chronic acidification has not been demonstrated in the western United States or the Southern Blue Ridge, due largely to low levels of acidic deposition in the former and sulfur retention by soils in the latter. Although episodic acidification may occur in some western systems, quantitative data are scarce.

Future Considerations

A rigorous evaluation of possible future changes in surface water quality in response to constant or changing deposition loadings is beyond the scope of this chapter and book. Some generalizations can be made, however, about future considerations, based on the premise that the key to the future can be found in an evaluation of the past. Recognizing the limits of our understanding of the past changes in water quality discussed previously, the following discussion is a first attempt to approximate the levels of change that might be expected.

In the northeastern United States, atmospheric deposition of sulfur has been decreasing slightly (Chapter 3, Husar et al., this volume). Assuming a nearly constant loading in the future, changes in surface water chemistry in these predominantly drainage systems are likely to be determined primarily by changes in base cation supply, which are poorly understood. Sulfate appears to be generally conservative in these systems (see section on sulfate), and proportional changes in ANC relative to past increases in SO_4^{2-} have been small. Based on paleoecological data and on an inferred background [SO_4^{2-}] of 20 µeq L^{-1}, < 35% of the added SO_4^{2-} has resulted in an equivalent loss of ANC for any of the lakes studied, and a value of 13% appears more typical for the lakes with current ANC $\leq$ 50 µeq L^{-1}. Including Al_i in these estimates, the change in [ANC $-$ Al_i] increased to 24% of added SO_4^{2-} for the typical cored Adirondack lake with current ANC $\leq$ 50 µeq L^{-1}. Thus, small changes in SO_4^{2-} deposition would not be expected to result in appreciable changes in ANC or [ANC − Al_i]. The major difference between acidified and nonacidified waters in this region is the concentration of base cations. Quantification of future changes in base cation supply is difficult and uncertain.

Past changes in lakewater chemistry in response to atmospheric deposition in the Upper Midwest and Florida appear to have been largely confined to groundwater recharge lakes. Base cation supply is less important in these precipitation-dominated seepage lakes (Chapters 1 and 17, Munson and Gherini, Baker et al., this volume), and lakewater response to deposition is more closely related to the amount of deposition received, in-lake processes, and climatological and hydrological variations (e.g., Chapter 12, Pollman and Canfield, and Chapter 13, Cook and Jager, this volume). Since atmospheric deposition of sulfur has been decreasing in the Upper Midwest (Chapter 3, Husar et al., this volume), continued acidification seems unlikely. Groundwater recharge lakes also constitute the major resource of interest in Florida (Chapter 12, Pollman and Canfield, this volume). It is reasonable to assume that projected increases in acidic deposition in this region may result in further acidification of sensitive Florida lakes.

Evaluation of possible future changes in surface water chemistry in western Virginia and the Southern Blue Ridge is complicated by the phenomenon of sulfur retention in soils (Galloway et al. 1983a, Rochelle and Church 1987; Chapter 10, Cosby et al., this volume; Chapter 11, Elwood et al., this volume) in addition to the projections of future increases in acidic deposition. As the sulfur retention capacity becomes depleted, continued acidification of some systems in Virginia is likely (e.g., Cosby et al. 1985). Cosby et al. (Chapter 10, this volume) estimated that [SO_4^{2-}] in stream water at Deep Run, Virginia, is currently increasing at a rate of 2.4 $\mu eq\ L^{-1}\ yr^{-1}$, accompanied by an increase in [H^+] of 0.4 $\mu eq\ L^{-1}$ and in base cations of 2.0 $\mu eq\ L^{-1}$.

Surface waters in the western United States are clearly the most susceptible of the regions included in this study to acidic deposition effects (Chapter 14, Turk and Spahr, Chapter 15, Melack and Stoddard, Chapter 16, Nelson, this volume). In particular, high-elevation lakes in the Sierra Nevada and Cascade Mountain ranges are undoubtedly among the most dilute in the world (e.g., Eilers et al. in press) and therefore presumably highly susceptible to atmospheric inputs. That no chronic acidification has been demonstrated to date in western systems is attributable to the low levels of deposition received (Hidy and Young 1986). If the western United States received atmospheric deposition comparable to levels now observed in the eastern states, changes in water chemistry would undoubtedly be substantial. These systems can perhaps best be compared to those of southernmost Norway, where 70% of the lakes are now acidic (ANC ≤ 0) in response to atmospheric deposition similar to that received in the northeastern United States (Wright 1988, Henriksen et al. 1988).

Acknowledgments. I am grateful to all of the participants in the Regional Case Studies project for sharing their data, ideas, and expertise in numerous project meetings over the past two years. The scientific exchange was of great benefit to me, and the need to establish group consensus on "key issues" (a need forced upon us relentlessly by project coordinator, D.F. Charles, despite great pain, hardship, and duress) vastly improved our collective understanding of the acidification process. I thank J.M. Eilers, R.B. Cook, H.M. Seip, S.J. Christie, J.L. Stoddard, and D.F. Charles for thoughtful review comments; J. Bernert, D. Kugler, and M. DeHaan for data analyses; and R. Royce for kindly typing the manuscript. Thanks also to A.R. Selle for providing the maps. Special thanks to D.F. Charles for a job well done.

Appendix A
Ion Enrichment Analysis for the Regional Case Studies Project

Lawrence A. Baker

Introduction

In many studies, researchers have determined the mechanisms and extent of ANC regulation within individual lakes and watersheds by constructing ion budgets (see reviews by Schindler 1986, Brezonik et al. 1987; Munson and Gherini, Chapter 2, this volume). Few studies, however, have attempted a systematic examination of ANC regulation within large populations of lakes. In several chapters of this book, we have used ion enrichment analysis to evaluate the mechanisms and extent of alkalinity regulation. A major advantage of ion enrichment analysis is that data requirements are minimal: calculations are based entirely upon the chemistry of atmospheric deposition and lakewater chemistry. A potential drawback, though, is that a number of simplifying assumptions must be made that may not be valid for some lakes.

Theory

The first step in ion enrichment analysis is to calculate the chemical composition of lake water that would occur in the absence of any biogeochemical reactions in the lake/watershed system, that is, the composition that would occur if lake water were merely precipitation concentrated by evaporation, as in a "Teflon" lake. In the Regional Case Studies analysis, Cl^- ratios were used to calculate evapotranspiration. Assuming that Cl^- is conservative, the expected concentration of ion i (i_{exp}) would be

$$[i]_{exp} = [i]_d * [Cl^-]_L/[Cl^-]_D \qquad (A\text{-}1)$$

where: $[i]_d$ = volume-weighted atmospheric deposition of ion i (wet + dry deposition divided by annual precipitation volume)

$[Cl^-]_L$ = lakewater Cl^- concentration

$[Cl^-]_D$ = Cl^- concentration in atmospheric deposition

The extent of biogeochemical reaction is then calculated by comparing expected lakewater composition with measured composition. Enrichment or depletion can be calculated on the basis of concentration (μeq L^{-1}) and as a percentage of the expected concentration:

$$R_i\ (\mu eq\ L^{-1}) = ([i]_m - [i]_{exp}) \qquad (A\text{-}2a)$$

$$R_i(\%) = \frac{([i]_m - [i]_{exp})}{[i]_{exp}} \times 100 \qquad (A\text{-}2b)$$

where: R_i = reaction for ion i (μeq L^{-1})

$[i]_m$ = measured lakewater concentration of ion i

In our analysis, net ANC generation was calculated in two ways: (1) by the sum of reactions and (2) from Gran ANC data. Based on electroneutrality considerations, we assumed that reactions resulting in gains in cations increase ANC, whereas reactions that remove cations from solution consume alkalinity. Conversely, reactions that result in gains of anions decrease alkalinity, whereas reactions that consume ions generate alkalinity:

$$ANCGEN_{C_B\text{-}C_A} = R_{Ca} + R_{Mg} + R_K + R_{Na} + R_{NH4} + R_{Al} + R_{Mn} - R_{SO4} - R_{NO3} - F^- \qquad (A\text{-}3)$$

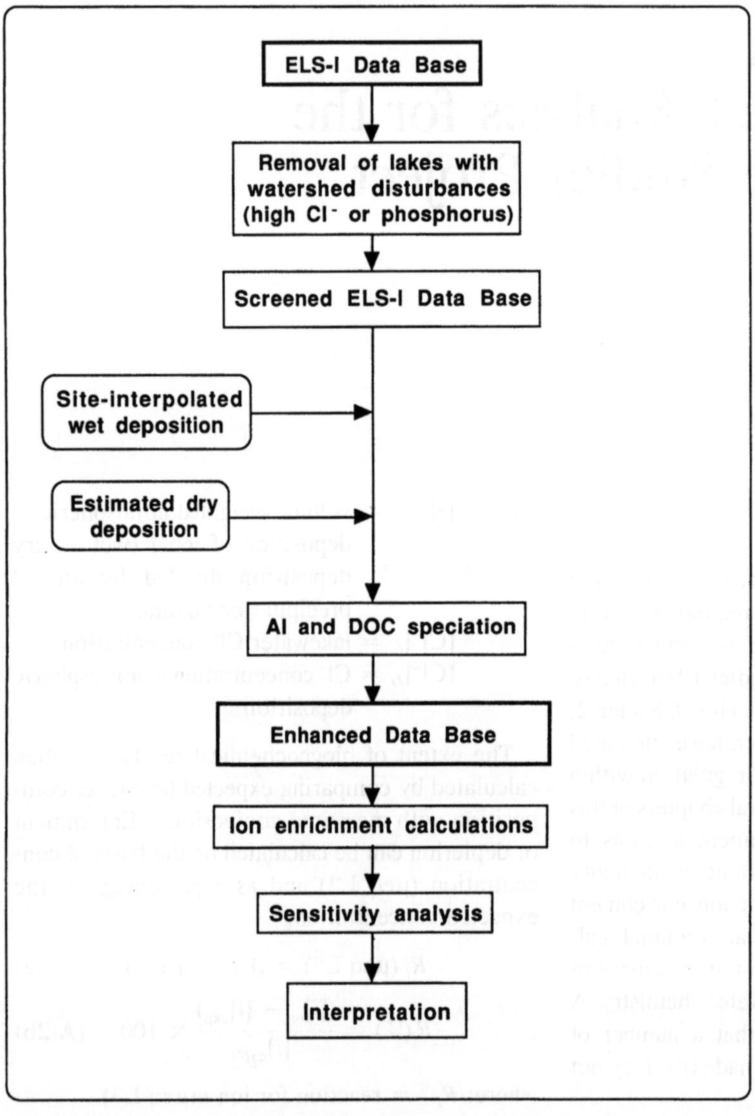

FIGURE A.1. Flow chart showing development of ion enrichment analysis.

In these calculations, the contribution of Al (R_{Al}) was computed using ALCHEMI (Schecher and Driscoll 1987); R_{Al} represents the sum of inorganic monomeric aluminum species:

$$R_{Al} = Al(OH)_4^- + Al(OH)_2^+ + Al(OH)^{2+} + AlF^{2+} + AlF_2^+ + Al^{3+} \quad \text{(A-4)}$$

We also calculated measured ANC generation ($ANCGEN_{Gran}$), which is simply the reaction term for Gran ANC (Equation A-2a). In calculating $ANCGEN_{Gran}$, we assumed that the Gran ANC of precipitation was $-[H^+]$; the Gran ANC of dry deposition was estimated by difference from fluxes of strong acid and base ions (Appendix B).

The validity of this approach depends upon several assumptions: (1) Cl^- is a conservative tracer, (2) all inputs to the lake/watershed system occur by atmospheric deposition, and (3) the lake/watershed system is at steady state. To ensure the validity of these assumptions, the ELS-I data were screened to remove lakes for which assumptions (1) and (2) were potentially invalid, as described later in this discussion. The steady-state assumption is best met for lakes with long residence times; exceptions to this assumption are discussed as they pertain to individual groups of lakes.

Implementation

Implementation of ion enrichment analysis required a number of steps: (1) calculation of atmospheric inputs, including dry deposition, (2) screening to remove lakes with watershed disturbances, (3) subsetting by hydrogeochemical classes, (4) calculation of Al speciation, (5) model calculations, (6) sensitivity analysis, and (7) interpretation. These steps are represented schematically in Figure A.1.

Atmospheric Inputs

Wet deposition chemistry and volume were determined by geometric interpolation from National Atmospheric Deposition Program (NADP) stations to individual study lakes using a 3-year mean (WY 1982–84) (Appendix C). We estimated dry deposition from region-specific dry:wet ratios, as described in Appendix B. Because there is considerable uncertainty in estimating dry deposition fluxes, sensitivity analysis was conducted to determine the response of ion enrichment calculations to changes in dry deposition estimates. Two cases were examined. First, all calculations were made using wet-only deposition (i.e., dry deposition fluxes = 0). This case represents the lower limit for ion inputs to the watersheds. In the high input case, we doubled the original dry deposition estimates. Results from this analysis are summarized in Chapter 17 (Baker et al., this volume).

Data Base Screening to Remove Anthropogenic Disturbances

Although the ELS-I data base was screened to remove lakes with known point sources, some watersheds contain anthropogenic disturbances that could alter lakewater chemistry (e.g., salted roads, lakeshore dwellings). Therefore, a screening procedure was developed to remove lakes with unusually high Cl^- or total phosphorus concentrations indicating watershed disturbances (e.g., road salt, septic tanks). For all subregions except Southern New England, Maine, and Florida, screening criteria were developed by examining cumulative frequency diagrams of Cl^- and total phosphorus and rejecting obvious outliers. Exclusion criteria for Cl^- ranged from 20 to 40 μeq L^{-1}, and exclusion criteria for total phosphorus ranged from 20 to 30 μg L^{-1} (Table A.1). Screening for Florida lakes was done by inspecting low altitude aerial photographs taken during the ELS-I. For Southern New England and Maine, where many lakes are near the coastline, the Cl^- rejection criterion was the upper 95% confidence limit from a predictive relationship between distance from the coast and lakewater [Cl^-] (T. Sullivan unpublished data).

TABLE A.1. Exclusion criteria for ion enrichment calculations.[a]

Subregion	Total phosphorus[b] (μg L^{-1})	Chloride[b] (μeq L^{-1})	% of total lakes in screened subset: All	% of total lakes in screened subset: Low ANC
Adirondacks	20	20	67	56
Maine[c]	20	**	81	58
Upper Midwest	30	20	73	55
Southern Blue Ridge	20	40	32	32
Florida[d]	**	**	21	19

[a] Data are from the ELS-I data base; total *P* variable was PTL11; Cl^- variable was CL16.
[b] ** = criterion not applicable.
[c] Maine = 1E and Maine portion of 1C. Exclusion criteria based on distance from coast (see Sullivan et al. 1987).
[d] Florida lakes were screened on the basis of detailed land use analysis. No Cl^- criterion was used because there are no road salt inputs.

Hydrologic Classification

To examine the roles of hydrology and DOC in alkalinity regulation, we segregated lakes into two DOC classes and three hydrologic classes. The low DOC class included lakes with DOC ≤ 5 mg L^{-1} (417 μmol L^{-1}) and the high DOC class included lakes with DOC > 5 mg L^{-1}. The three hydrologic classes were drainage, groundwater flow-through, and groundwater recharge. The drainage category was expanded from the original data base definition to include reservoirs and closed basins. Lakes classified as seepage in the ELS-I data base were subdivided into groundwater recharge (GWR) and groundwater flow-through (GWFT) systems, based on chemical characteristics. For all lakes except

TABLE A.2. Occurrence of low ANC ($\leq$ 200 μeq L^{-1}) lakes by subregion and lake type in screened data set for enrichment factor analysis.[a]

	Drainage				Groundwater flow-through				Groundwater recharge			
	Low DOC		High DOC		Low DOC		High DOC		Low DOC		High DOC	
	All	Low ANC	All	Low ANC	All	Low ANC	All	Low ANC	All	Low ANC	All	Low ANC
Adirondacks	65	55	31	26	0	0	0	0	6	6	2	1
Maine	76	62	100	62	1	1	3	2	3	3	2	2
Upper Midwest	42	23	159	107	15	4	43	25	91	86	77	73
Florida	4	4	5	4	0	0	4	2	11	11	7	7
S. Blue Ridge	40	33	0	0	0	0	0	0	0	0	0	[illegible]

[a] All = All ANC levels; low ANC = ANC $\leq$ 200 μeq L^{-1}; low DOC = $\leq$ 5 mg L^{-1} (417 μmol L^{-1}); high DOC = > [illegible] mg L^{-1}. Enrichment calculations were not performed for lake groups with subpopulations containing fewer than 10 lakes.

those in Florida, SiO_2 was used as the classification criterion ($SiO_2 \leq 1.0$ mg L^{-1} = GWR; $SiO_2 > 1.0$ mg L^{-1} = GWFT), but for Florida lakes, SiO_2 data were widely scattered and K^+ was used instead ($[K^+] < 15$ μeq L^{-1} = GWR; $[K^+] > 15$ μeq L^{-1} = GWFT). The two DOC classes and three hydrologic classes resulted in six hydrogeochemical groups (Table A.2). Although ion enrichment calculations were performed on all screened regional case study lakes, most of the results presented in the chapters are based upon lakes with ANC $\leq$ 200 μeq L^{-1}. Graphical representation of ion enrichment analysis is discussed by Munson and Gherini (see Figure 2.10, Chapter 2, this volume).

Appendix B
Regional Estimates of Atmospheric Dry Deposition

Lawrence A. Baker

Introduction

Ion enrichment analyses (see Chapters 2 and 17 and Appendix A) were conducted using total (wet + dry) atmospheric deposition. Wet deposition at each lake site was estimated by kriging, using wet deposition data from the National Atmospheric Deposition Program (NADP) (see Appendix C). Although we know that dry deposition inputs can be a significant portion of total atmospheric deposition (Galloway and Whelpdale 1983, Johannes et al. 1984, Lindberg et al. 1986, Barrie and Sirois 1986, Sirois and Barrie 1988), we have no single, reliable method for estimating dry deposition rates on a regional basis. We therefore used several approaches in an attempt to derive the best estimates of dry deposition for all major ions. These estimates were used to calculate region-specific dry:wet ratios, which were then used in conjunction with site-interpolated wet deposition fluxes to estimate total atmospheric deposition to each lake in the data set.

Four approaches were used: (1) measured fluxes to surrogate surfaces (NADP dry buckets), (2) inferred deposition fluxes (cross product of ambient air measurements and the appropriate deposition velocities, (3) regional-scale mass balance studies, and (4) the opinions of experts, who may have used one or more of these methods for their estimates. Expert opinions were solicited through questionnaires sent to approximately a dozen atmospheric chemists and watershed modelers; additional comments were solicited by letters or phone calls.

Compilation of NADP Dry Bucket Data

What Do Dry Bucket Measurements Represent?

The National Atmospheric Deposition Program (NADP) has compiled a substantial "dry bucket" data base, but these data have not been widely used, in part because of uncertainties in extrapolating from surrogate surface collectors to natural surfaces (see review by Hicks 1986). However, the effect of surface type diminishes as particle size increases, and the importance of gravitational settling increases relative to other deposition mechanisms (adsorption, impaction). Therefore, on a theoretical basis, one would expect a dry bucket sampler to be a reasonable surrogate for coarse particle deposition to natural surfaces but a poor surrogate for fine particle deposition. Thus, for ions that are primarily associated with large particles (e.g., Ca^{2+}, Mg^{2+}, K^{+}, Na^{+}, and Cl^{-}; see review of particle size distribution by Milford and Davidson 1985), dry bucket measurements should be a rough approximation of deposition rates to natural surfaces.

Several lines of evidence support this contention. The relative importance of gravitational settling has been studied by measuring deposition to upward-facing and downward-facing surfaces. The hypothesis in these experiments is that the higher deposition rates to upward-facing surfaces directly show the importance of gravitational settling. Coe and Lindberg (1987) reported that the total particle density was 10 times higher on upper surfaces of leaves than on their downward-facing surfaces. Dasch (1983) reported that deposition rates of most ions were higher on upward-facing surfaces (several types of media) than on downward-facing surfaces. For example, Ca^{2+} deposition to upper surfaces accounted for > 90% of total deposition (Dasch 1983). There are few direct intercomparisons of coarse particle deposition rates between surrogate surfaces and natural surfaces. Lindberg et al. (1986) showed that deposition of Ca^{2+} measured by a bulk collector was within 30% of measured wet + dry deposition in a Tennessee forest. Measured dry deposition rates in this study were based upon throughfall enrichment. Cadle et al. (1986) showed that, for most ions, dry deposition rates measured by an Aerochemetrics wet/dry collector were similar to measurements of dry deposition to snow surfaces.

Interpretation of dry bucket SO_4^{2-} and NO_3^- fluxes is more difficult, because both species occur as gases as well as aerosols in the atmosphere. Furthermore, SO_4^{2-} is generally associated with submicron aerosols, whereas NO_3^- aerosols typically have a bimodal size distribution (Wolff 1984). Removal of submicron particles from the atmosphere is controlled more by impaction than by gravitational settling, so extrapolation of natural fluxes from dry bucket measurements is more tenuous than for large particles. However, Davidson et al. (1985) have noted that although the mass median diameter (MMD) of atmospheric SO_4^{2-} aerosol is < 1 μm, the MMD of SO_4^{2-} particles on surfaces is often >> 1 μm; these investigators concluded that sedimentation of SO_4^{2-} aerosols may be a more important deposition mechanism than generally believed.

Clean dry buckets probably do not collect SO_2 (Dasch 1983) and SO_4^{2-} collected in dry buckets is generally believed to originate from atmospheric aerosols rather than from direct adsorption of SO_2 to the collector surface. There may be some adsorption of SO_2 in dry buckets that have accumulated alkaline materials during exposure (E. Edgerton, Environmental Sciences and Engineering, Inc., pers. comm.), but the relative contribution of SO_2 on exposed dry buckets is unknown and is probably small. Studies that have compared dry bucket SO_4^{2-} fluxes with fluxes to natural surfaces have yielded inconsistent results. Davidson et al. (1985) concluded that conventional dry buckets were more efficient collectors of aerosol SO_4^{2-} than Teflon plates or Petri dishes and that deposition to the latter two surfaces closely approximated modeled deposition to a grass surface. Sickles et al. (1983) concluded that dry bucket SO_4^{2-} fluxes overestimated fluxes to natural surfaces (leaves of *Ligustrum*). Cadle et al. (1986) concluded that dry SO_4^{2-} deposition in dry buckets was 28% less than direct dry deposition to snow because the snow also collected some SO_2, which the dry buckets did not. In another study of dry deposition to snow, Ibrahim et al. (1983) concluded that all four types of dryfall samplers that they examined underestimated fluxes of submicron particles and overestimated fluxes of supermicron particles; however, they presented no comparison of total SO_4^{2-} deposition. In conclusion, dry bucket SO_4^{2-} probably originates from atmospheric aerosols rather than SO_2, but dry bucket SO_4^{2-} deposition rates are only a very rough approximation of aerosol SO_4^{2-} deposition to natural surfaces.

The atmospheric origin of dry bucket NO_3^- is less certain. Nitric acid vapor (HNO_3) is highly reactive and some investigators believe that it is readily adsorbed to polyethylene and other surrogate surfaces (e.g., Johannes and Altwicker 1983, Lindberg and Lovett 1985). Others, however, have concluded that NO_3^- collected on surrogate surfaces is derived primarily from particulate NO_3^- (Dasch 1983, Huebert and Robert 1985).

Data Compilation

In this study, we used dry bucket measurements as the primary method of deriving regional-scale estimates of dry deposition for Ca^{2+}, Mg^{2+}, Na^+, K^+, Cl^-, and NH_4^+. Dry bucket SO_4^{2-} and NO_3^- fluxes were assumed to represent aerosol fluxes of these constituents for regions where no other estimates were available and also to provide a lower bound of total dry sulfur and nitrogen fluxes for regions where estimates of total dry sulfur and nitrogen were available from modeling studies or inferential approaches. NADP wet and dry bucket data were compiled for the Upper Midwest and the Northeast for the entire period of record. Sites with short periods of record (< 2 years) were not used because of the high variability in dry bucket deposition rates among sampling periods. Wet and dry bucket data for Florida were obtained from the Florida Acid Deposition Study (FADS; ES&E 1985).

For regions where NADP data were used, average annual wet deposition was calculated from NADP annual mean concentrations and rain gauge precipitation. Mean dry bucket fluxes were calculated for all sampling periods, regardless of season. The only screening step used in this analysis was the exclusion of dry bucket samples with > 200 mL water. A substantial fraction (roughly 30%) of the remaining samples had smaller volumes of water. This screening criterion was based on a balance between the desire to maintain a large sample size and the error that would be introduced by including wet deposition as dry deposition. On one hand, exclusion of all samples containing small amounts of water would have resulted in a much smaller number of valid dry samples, since roughly 30% of all dry bucket samples contained 1 to 200 mL of water. Given the extreme variability among samples, inclusion of samples with small amounts of water to increase the total number of samples was judged to be a reasonable way of increasing the total number of samples used to calculate the means. On the other hand, 200 mL corresponds to a precipitation volume of only 0.3 cm, which is < 1% of the annual precipitation volume in all regions. Thus, the addition of a small amount of wet deposition probably did not significantly alter our estimate of dry deposition.

A recent paper by Gatz et al. (1988) examined screening criteria for dry buckets and concluded that screening for wet samples or visible contamination (bird droppings, etc.) had either no effect or a minor effect on Ca^{2+}, Na^+, Cl^-, and SO_4^{2-}. Nevertheless, removal of contaminated samples significantly decreased loadings of K^+, NH_4^+, and PO_4^{3-}; removal of wet samples decreased loadings of H^+ and, to a lesser extent, NO_3^-. Our screening procedure did remove samples with significant precipitation, so we would not expect our data base to be biased with regard to H^+ and NO_3^-. However, our dry bucket deposition rates may be positively biased for K^+, NH_4^+, and PO_4^{3-}. This bias generally has had little effect on our inferences about alkalinity regulation in watersheds, because watershed reactions of K^+ and PO_4^{3-} are unimportant in this regard. Ammonium is an important ion in watershed alkalinity regulation, but most NH_4^+ inputs occur as wet inputs. A modest bias in dry bucket fluxes should not have a serious effect on ion enrichment calculations.

Dry bucket fluxes calculated for individual ions were surprisingly consistent among individual collector sites within a given region. Thus, mean deposition rates from all collectors within each region were averaged to calculate region-specific dry:wet ratios for each major ion (Table B.1). Dry:wet ratios generally follow the pattern:

$$K^+ > Ca^{2+} > Mg^{2+} > Na \geq Cl^- \approx SO_4^{2-} \geq NH_4^+ \geq NO_3^-$$

Dry bucket deposition exceeds wet deposition for K^+ in all regions and for Ca^{2+} in all regions except the Upper Midwest. Dry:wet ratios range from 0.2 to 0.8 for Mg^{2+}, Na^+, NH_4^+, SO_4^{2-}, and Cl^- and are ≤ 0.2 for NO_3^- throughout the four regions (Table B.1).

Inferential Estimates of Dry Deposition

For several regions, ambient air concentration data were available for gas and particulate sulfur and nitrogen species. These were used to calculate average annual dry deposition of individual species by means of the equation:

$$F_i = C_i \times v_d \qquad \text{(B-1)}$$

where: F_i = flux
C_i = ambient air concentration
v_d = deposition velocity

Appropriate deposition velocities were derived from the literature. For Florida, ambient air concentrations were available for SO_4^{2-} aerosol, SO_2, NO_2, and HNO_3 (ES&E 1985), and the following deposition velocities were used: SO_4^{2-} = 0.2 cm sec^{-1}; SO_2 = 0.5 cm sec^{-1}; NO_2 = 0.1 cm sec^{-1}; and HNO_3 = 1 cm sec^{-1} (see Voldner and Sirois 1986, Voldner et al. 1986, Lee and Schwartz 1981, Lindberg et al. 1986). For the Upper Midwest, limited data were available for SO_2; the annual mean deposition velocity was assumed to be 0.3 cm sec^{-1} based on the work of Voldner and Sirois (1986), which showed that deposition velocities are lower (ca. 0.1 cm sec^{-1}) during the snow-covered portion of the year than in the growing season (ca. 0.5 cm sec^{-1}).

TABLE B.1. Mean dry bucket:wet bucket ratios for Florida, the Upper Midwest, the Northeast, and the Southern Blue Ridge case study regions.

	Northeast[a]		S. Blue Ridge[b]		Florida[c]		Upper Midwest[d]	
Ion	Mean	SD	Mean	SD	Mean	SD	Mean	SD
K^+	2.21	0.89	1.64	0.63	2.05	0.81	1.64	0.18
Ca^{2+}	1.19	0.98	1.15	0.55	1.75	0.93	0.54	0.10
Mg^{2+}	0.75	0.38	0.67	0.33	0.63	0.17	0.65	0.14
Na^+	0.37	0.17	0.24	0.08	0.47	0.16	0.22	0.09
NH_4^+	0.34	0.28	0.38	0.10	0.39	0.19	0.19	0.10
SO_4^{2-}[e]	0.22	0.09	0.21	0.07	0.28	0.05	0.24	0.04
NO_3^-[e]	0.10	0.05	0.09	0.03	0.19	0.06	0.10	0.04
Cl^-	0.24	0.18	0.15	0.09	0.42	0.14	0.19	0.03

[a] Northeast data are based on NADP sites at Huntington, Bennett Bridge, and Knobit, NY; Bennington, VT; Hubbard Brook, NH; and Caribou, ME.
[b] Southern Blue Ridge dry:wet ratios are based on NADP stations at Coweeta, Walker Branch, Clemson, and Great Smoky Mt. National Park.
[c] Deposition estimates taken from Baker et al. (1988) using data from the FADS network (ES&E 1985).
[d] Midwest data are based on 8 NADP stations (2 in Minnesota, 3 in Wisconsin, and 3 in northern Michigan).
[e] Dry:wet ratios for sulfate and nitrate in this table are based upon wet bucket:dry bucket measurements at the NADP sites. Dry bucket fluxes of these ions are probably roughly equivalent to aerosol deposition rates for these species but include very little SO_2 and an unknown fraction of HNO_3 inputs (see text).

Summary of Regional Dry:Wet Ratios

Dry:wet ratios used in ion enrichment analyses are presented by region in Table B.2. The following discussion summarizes the basis for these ratios.

Northeast

Galloway and Whelpdale (1983) reported a dry:wet ratio of 1.3:1 for the northeastern United States based on a subcontinental mass balance, but more recent studies have reported lower dry:wet ratios. Sirois and Barrie (1988) reported dry:wet ratios of 0.59 for Long Point (just north of Lake Ontario) and 0.31 for Chalk River (southeastern Quebec). These ratios are similar to RELMAP model results (EPA 1987) showing that dry:wet ratios in this region are 0.4 to 1.0. Galloway et al. (1983) reported that dry deposition of SO_4^{2-} was 30% of total deposition for three Adirondack watersheds (dry:wet ratio = 0.42). For the same watersheds, Tetra Tech has used a dry:wet ratio of 0.7 for ILWAS model applications (R. Munson, Tetra Tech., Inc., pers. comm.). Considering all these, a dry:wet ratio of 0.5 for the Northeast appears reasonable.

For NO_3^-, the dry bucket to wet bucket ratio of 0.10 calculated for the Northeast (Table B.3) represents a lower limit for dry:wet ratios. Sirois and Barrie (1988) reported dry:wet ratios of total NO_3^- (excluding NO_2) of 0.23 and 0.28 for Long Point and Chalk River; this agrees relatively well with the value of 0.4 used in ILWAS model applications (R. Munson, Tetra Tech., Inc., pers. comm.). Little other information appears to be available, so we used a dry:wet ratio of 0.4 for ion enrichment calculations.

Dry:wet ratios for other aerosols were calculated from NADP wet:dry buckets at five sites in New York and New England. The NADP derived

TABLE B.2. Summary of dry:wet ratios used in ion enrichment analysis.[a]

Ion	Northeast	S. Blue Ridge	Florida	Midwest
Ca^{2+}	1.19	1.15	1.75	0.54
Mg^{2+}	1.75	0.67	0.63	0.65
K^+	2.21	1.64	2.05	1.64
Na^+	0.37	0.24	0.47	0.22
NH_4^+	0.34	0.38	0.39	0.19
Cl^-	0.24	0.15	0.42	0.19
SO_4^{2-}	0.50	1.0	0.70	0.37
NO_3^-	0.40	1.5	0.96	0.30

[a] Ratios are based on total (aerosol + gas) dry deposition. See regional discussions.

dry:wet ratios are compared with dry:wet ratios calculated by Johannes et al. (1984) during the ILWAS study (Table B.3). Dry:wet ratios from the two sources are in close agreement for Na^+, NH_4^+, Cl^-, NO_3^-, and SO_4^{2-}, but the NADP derived ratios are higher for Ca^{2+}, Mg^{2+}, and K^+. The discrepancy probably occurs because Johannes et al. had more rigorous siting requirements than NADP (A.J. Johannes, U. of Oklahoma, pers. comm.). The NADP derived ratios are also in general agreement with values used in the Direct/Delayed Response Project (DDRP) (B. Rochelle, NSI Technology Services, pers. comm.).

TABLE B.3. Comparison of dry:wet ratios from Johannes et al. (1984) with NADP derived dry:wet ratios for the Northeast.[a]

Ion	Johannes et al. Wet	Johannes et al. Dry	Johannes et al. Dry:Wet	NADP derived dry:wet ratios
Ca^{2+}	140	60	0.43	1.19
Mg^{2+}	46	15	0.33	0.75
Na^+	56	17	0.30	0.37
K^+	26	11	0.42	2.21
NH_4^+	246	28	0.11	0.34
Cl^-	55	13	0.24	0.24
SO_4^{2-}	729	95	0.13	0.22
NO_3^-	420	49	0.12	0.10

[a] Johannes et al. (1984) data collected by dry and wet buckets at Woods Lake, 1978–81. NADP dry:wet ratios are from Table B.1.

Southern Blue Ridge

Lindberg et al. (1986) reported that dry deposition accounted for 55% of total SO_4^{2-} input to the Walker Branch watershed in 1981–83. Vapor accounted for about 70% of the dry input (62 meq m^{-2} yr^{-1}) and particle inputs accounted for the remaining 30% (26 meq m^{-2} yr^{-1}). Particle fluxes from this study compare well with the long-term mean for dry bucket SO_4^{2-} flux at the Walker Branch NADP station (20 meq m^{-2} yr^{-1}). The dry (SO_2 + aerosol) to wet ratio calculated from the Lindberg et al study was 1.3. More recent studies by Lindberg and colleagues have resulted in total dry:wet ratios of 0.9 at a 300 m site (loblolly pine) in the Tennessee Valley and 1.3 for an 1,800 m site in the Smoky Mountain National Park (S. Lindberg, Oak Ridge National Lab, pers. comm.). Variations depend upon altitude and, apparently, upon year. Overall, a dry:wet ratio of 1.0 appears to be reasonable for SO_4^{2-} deposition in this region.

Dry NO_3^- deposition accounted for 62% of total NO_3^- deposition at the Walker Branch watershed (Lindberg et al. 1986). Most of the dry deposition (76%) resulted from gaseous deposition, with smaller contributions from coarse particles (24%) and even less from fine particles (< 1%). The dry:wet ratio calculated from these data is 1.7. Lindberg has derived dry:wet ratios for the Tennessee Valley site (1.4) and the Smoky Mountain site (5.2) for 1986–87. Since the ELS-I watersheds are lower in elevation than the Smoky Mountain site, a total dry:wet ratio of 1.5 appears to be reasonable.

Data from the NADP sites were used to compute average dry:wet ratios for other ions (Table B.1). The studies of Lindberg et al. and the NADP data were used to derive the total dry (gas + aerosol):wet ratios used in ion enrichment calculations for the Southern Blue Ridge Province (Table B.3).

Florida

Spatial patterns of ambient air concentrations of gases and aerosols in Florida are relatively well-known as a result of an extensive atmospheric monitoring program conducted by ES&E, Inc. between 1981–85 (ES&E 1984). This monitoring network included nine stations in the area circumscribed by the ELS-I lakes; sampling included wet deposition, dry bucket measurements, and ambient air concentration measurements. In the study of Baker et al. (1988), sampling stations were grouped into five subregions, each represented by one to four stations. Dry deposition of aerosol SO_4^{2-} was calculated from the geometric mean of ambient air measurements and a deposition velocity of 0.2 cm sec^{-1} (dry bucket SO_4^{2-} was measured, but was not used in our final calculations of dry:wet SO_4^{2-} ratios); deposition of SO_2 was calculated from the geometric mean of ambient air concentrations and a deposition velocity of 0.5 cm sec^{-1} (see foregoing discussion). Details of sampling and data analysis are presented in ES&E (1984) and Baker et al. (1988).

Table B.4a shows that the mean ratio of SO_2 deposition:wet deposition is 0.51 and that this ratio is relatively constant (SD = 0.08) throughout the state. The mean ratio of aerosol:wet deposition is 0.20 and this is also quite consistent among subregions (SD = 0.03). On the average, aerosol

TABLE B.4. Wet and dry deposition of sulfur and nitrogen in Florida.[a]

	All fluxes in eq ha^{-1} yr^{-1}				Ratios		
Region	Wet	Aerosol	Gas	Total	Aer.:Wet	Gas:Wet	Dry:Wet
				a. Sulfur			
A	357	68	149	574	0.19	0.42	0.61
B	287	62	125	474	0.22	0.44	0.65
C	328	60	156	544	0.18	0.48	0.66
D	381	60	223	664	0.16	0.59	0.74
E	211	48	131	390	0.23	0.62	0.85
Mean	312	60	157	529	0.20	0.51	0.70
SD	60	7	35	93	0.03	0.08	0.09
				b. Nitrogen			
A	156	16	113	285	0.10	0.72	0.83
B	132	31	81	244	0.24	0.61	0.85
C	148	33	129	310	0.22	0.87	1.10
D	153	39	139	331	0.26	0.91	1.16
E	117	15	87	219	0.13	0.74	0.87
Mean	141	27	110	278	0.19	0.77	0.96
SD	15	10	23	41	0.06	0.11	0.14

[a] Sampling stations A and B are in the Panhandle; C is in Gainesville; D is in Tampa, and E is in Lake Placid. See Baker et al. (1988); also Chapter 12, Pollman and Canfield, this volume, for discussion of regional estimates of atmospheric deposition in Florida.

deposition accounts for 38% of total dry sulfur inputs, and SO_2 deposition accounts for the remaining 62%. Total dry deposition of SO_4^{2-} (SO_2 + SO_4^{2-} aerosol) is 0.70 × wet (± 0.09), representing 41% of total atmospheric input.

Fluxes of NO_2 and HNO_3 were based on annual geometric means of ambient air concentrations measured in the FADS study using deposition velocities of 0.1 cm sec^{-1} and 1.0 cm sec^{-1} to estimate fluxes. We assumed that dry bucket NO_3^- fluxes represented coarse aerosol inputs and computed total dry nitrogen deposition as the sum of particulate and gas fluxes. Mean dry bucket NO_3^- was 0.19 × wet deposition (± 0.06) (Table B.4b), and gaseous nitrogen inputs (HNO_3 + NO_2) represent 0.77 × wet deposition (± 0.11) of wet input. Mean total dry nitrogen flux was 0.96 × wet deposition (± 0.14) of wet deposition for the region (Table B.4b).

Upper Midwest

Several studies of ambient air concentrations have been conducted in the Upper Midwest, but unfortunately no region-wide studies exist. Measurement techniques among studies are not necessarily similar, and the duration of measurements varies from < 1 year to several years (Eisenreich et al. 1978, Pratt and Krupa 1985, APIOS 1982–1985, Barrie and Sirois 1986, MPCA unpublished data).

Ambient air concentrations of SO_2 were available for several sites in Minnesota (MPCA 1988) and southern Ontario (APIOS 1982–1985). Annual geometric means from these studies were used together with a region-wide deposition velocity of 0.3 cm sec^{-1} (see previous discussion) to estimate deposition of SO_2. Aerosol SO_4^{2-} data were not available and NADP dry bucket data were used to estimate aerosol SO_4^{2-} fluxes. The regional mean ratio of aerosol (dry bucket) SO_4^{2-} deposition:wet deposition was 0.23 ± 0.08 for the region and the mean ratio of SO_2 deposition:wet deposition was 0.14 ± 0.04 (Table B.5). The average ratio of total dry SO_4^{2-} deposition to wet deposition is 0.37 ± 0.09; that is, 27% of total SO_4^{2-} deposition is dry deposition. These estimates compare reasonably well with estimates from other studies. Barrie and Sirois (1986) estimated deposition of dry sulfur to be 0.34 × wet at the Experimental Lakes Area (north of the Minnesota border) and 0.19 × wet SO_4^{2-} deposition at Algoma, Ontario (near the eastern tip of the Upper Peninsula of Michigan). Similarly, RELMAP results show that the ratio of dry to total SO_4^{2-} deposition is 0.2 to 0.25 (corre-

TABLE B.5. Wet and dry deposition of sulfur for the Upper Midwest.

	All fluxes in eq ha^{-1} yr^{-1}							
			Gas			Ratios		
Site	Wet	Aerosol	μg m^{-3}	Flux[a]	Total	Aer.:Wet	Gas:Wet	Dry:Wet
Marcell	22	1	1.1	3	26	0.05	0.15	0.20
Fernberg	18	6	1.1	3	27	0.32	0.17	0.49
Spooner	29	7	1.5	4	41	0.24	0.15	0.39
Trout L.	30	7	1.5	4	41	0.24	0.15	0.39
Chassell	25	7	1.5	4	36	0.29	0.18	0.47
Douglass L.	40	9	1.6	4	54	0.22	0.12	0.34
Wellston	50	12	1.6	5	66	0.24	0.10	0.33
Mean	30	7	1.4	4	42	0.23	0.14	0.37
SD	10	3	0.2	0.6	13	0.08	0.04	0.09

[a]V_{SO_2} = 0.1 for 6 months; annual mean = 0.3. Annual geometric means of ambient air SO_2 concentrations were used for calculating SO_2 fluxes. For most sites (except Marcell), SO_2 concentrations were extrapolated from other sites. The Minnesota Pollution Control Agency's Isabella site was used to estimate SO_2 at the NADP Fernberg site; the APIOS Turkey Lake site was used to estimate SO_2 at Chassell and Wellston, MI; the MPCA Sandstone, MN, site was used to estimate SO_2 at Spooner, WI; SO_2 values at the other sites were extrapolated from nearby Canadian sites in the APIOS network.

sponding to dry:wet ratios of 0.11 to 0.33) throughout the Upper Midwest (EPA 1987).

Dry NO_3^- is more difficult to estimate because there appear to be virtually no NO_2 and HNO_3 measurements for the region. Dry bucket NO_3^- deposition may be considered a lower limit, since dry buckets probably catch primarily coarse NO_3^-, plus some HNO_3. The average dry bucket:wet ratio for the region was 0.10 ± 0.04 (Table B.1). Sirois and Barrie (1988) estimated that the ratio of (HNO_3 + NO_3^-) deposition to wet deposition was 0.35 at the Experimental Lakes site and 0.19 at the Algoma site. These numbers agree well with the dry:wet ratio of 0.20 used in the RILWAS model applications in this region (R. Munson, Tetra Tech, Inc., pers. comm.). Based on these estimates, we used a dry:wet ratio of 0.30 for the Upper Midwest (Table B.2). Dry:wet ratios for species other than SO_4^{2-} and NO_3^- were based on NADP dry bucket:wet bucket data from Table B.1; all dry:wet ratios are summarized in Table B.2.

Potential Error

The accuracy of the dry:wet ratios presented in Table B.3 varies among ions and regions, depending upon the quantity of data (particularly ambient air concentrations) and the extent to which modeling studies have been conducted within the region. In general, estimates are probably best for Florida, which has the most extensive available ambient air concentration data base, flat topography (no orographic influences), little seasonal variation, and the least complex terrain (deposition is largely to lake surfaces). Estimates for the Midwest have the greatest uncertainty, because there has been little effort to model dry deposition inputs to watersheds in this region and because the data base for ambient air quality is sparse.

Among ions, the extent of error in total atmospheric deposition (ultimately the number of interest) depends upon the magnitude of the dry:wet ratio. If the true dry:wet ratio is low, than a sizable error in dry:wet ratios may have relatively little effect on the total deposition estimate. To illustrate this, consider two cases in which the error in dry:wet ratios is a factor of 2. If we estimate a dry:wet ratio of 0.2 and the true value is 0.1, the error in total deposition input will be only 9%. If, however, the estimated dry:wet ratio were 2.0 and the real value were 1.0, the error in total deposition input would be 50%.

Based on these considerations, the magnitude of error in total deposition probably follows the order:

Largest (50% or more error):	NO_3^-, K^+
Intermediate (25% to 50% error likely):	SO_4^{2-}, Ca^{2+}, Mg^{2+}
Smallest (error probably < 25%):	Na^+, Cl^-, NH_4^+

Dry:Wet Ratios in Enrichment Factor Calculations

Calculation of total deposition of ion i at a given site was based on site-interpolated wet deposition (see Appendix C) and the region-specific dry:wet ratio for the ion of interest:

$$D_i = [i]_w * P*(D/W)_i \qquad \text{(B-2)}$$

where: $[i]_w$ = concentration of ion i in precipitation, µeq L^{-1}
P = corresponding precipitation volume (NADP raingauge), m
$(D/W)_i$ = region-specific dry:wet ratio for ion i (from Table B.5)

Dry deposition of H^+ was calculated by difference from the fluxes for measured ions:

$$D_{H+} = D_{cl} + D_{so4} + D_{no3} - D_{ca} - D_{mg} - D_k - D_{na} - D_{nh4}$$

Dry deposition of H^+ was generally positive (in other words, net dry deposition was acidic) because of the influence of gas absorption (HNO_3, NO_x, SO_2). For Midwest sites, however, dry H^+ deposition was slightly negative (in other words, net dry deposition was slightly alkaline) because gaseous inputs were low and there was substantial dry deposition of alkaline dustfall.

Corrected precipitation concentrations (volume-weighted wet + dry deposition) were then calculated for each ion:

$$[i]_c = [i]_w + D_i/P \qquad \text{(B-3)}$$

Ion balance calculations for these corrected precipitation concentrations showed that anion:cation ratios were very close to 1.0.

Sensitivity Analysis

Recognizing that there is considerable uncertainty in dry deposition estimates, and hence in corrected precipitation concentrations, we conducted a sensitivity analysis to evaluate the potential for error in ion enrichment analysis. A lower limit of atmospheric inputs was assumed to be wet-only deposition (i.e., dry deposition of all ions = 0). The upper limit for dry deposition was assumed to be twice the best estimate. The results of our sensitivity analysis for ion enrichment analysis are presented in Chapter 17.

Acknowledgments. The following individuals commented on methodologies used to estimate dry deposition: Robbins Church, Jack Cosby, Bruce Hicks, Rudy Husar, A.J. Johannes, Steve Lindberg, Ron Munson, and Gregg Pratt.

Appendix C
Estimation of Precipitation Amount and Chemistry at Eastern Lake Survey–Phase I Lakes

Rudolf B. Husar

Introduction

The central theme of this book is understanding how atmospheric deposition affects the chemistry of surface waters. Several procedures are used to show the changes that occur in precipitation from the time it is deposited on a watershed until it reaches a waterbody. The most important procedure is the ion enrichment analysis described in Chapter 2 and Appendix B. A critical component of these analyses is the determination of the amount and chemical characteristics of precipitation falling on lakes and their watersheds. This appendix describes the method used to interpolate precipitation amount and chemical data to Eastern Lake Survey lakes, and evaluate the appropriateness and accuracy of these procedures.

Data Sources and Extrapolation Procedures

Two major data sources were used for the calculations, the National Atmospheric Deposition Program (NADP) (Figure C.1a) and the National Weather Service Monitoring Network (Figure C.1b). Precipitation amount data come from the records of the National Weather Service, which routinely measures precipitation amounts hourly at over 6,000 sites in the United States. The Oak Ridge National Laboratory Carbon Dioxide Information Analysis Center has compiled a historical climatology data base composed of measurements from 1,219 National Weather Service stations for the past century. The distribution of the National Weather Service sites used in this analysis is shown in Figure C.1b. In this appendix, we refer to this group of sites as the CLIMA network. In addition to this data set, which is limited to sites with consistently high quality measurements, we used data from a denser network of sites that are also a part of the national hourly cooperative precipitation network. This network contains the highest density of stations that routinely measure precipitation (Figure C.1c). This data set is maintained by the National Climatic Center, Asheville, North Carolina. We refer to this network as the North Eastern (NE) network.

The source of precipitation chemistry data for the deposition calculations is the NADP network. This network measures both precipitation chemistry and amount. The weekly precipitation chemistry data for the NADP network were obtained from the Acid Deposition System (ADS) at Pacific Northwest Laboratories (Watson and Olsen 1984). The ADS contains several precipitation chemistry data bases in a uniform data format.

The average distance between stations in the NADP, CLIMA, and NE networks is 200, 80, and 20 km, respectively. Hence, these three networks range over about one order of magnitude in spatial resolution. The NADP precipitation chemistry data were used without any checks on the quality of the chemistry data. However, other filtering criteria were imposed. Only stations that had been operated for at least one year during the period 1982–84 were used.

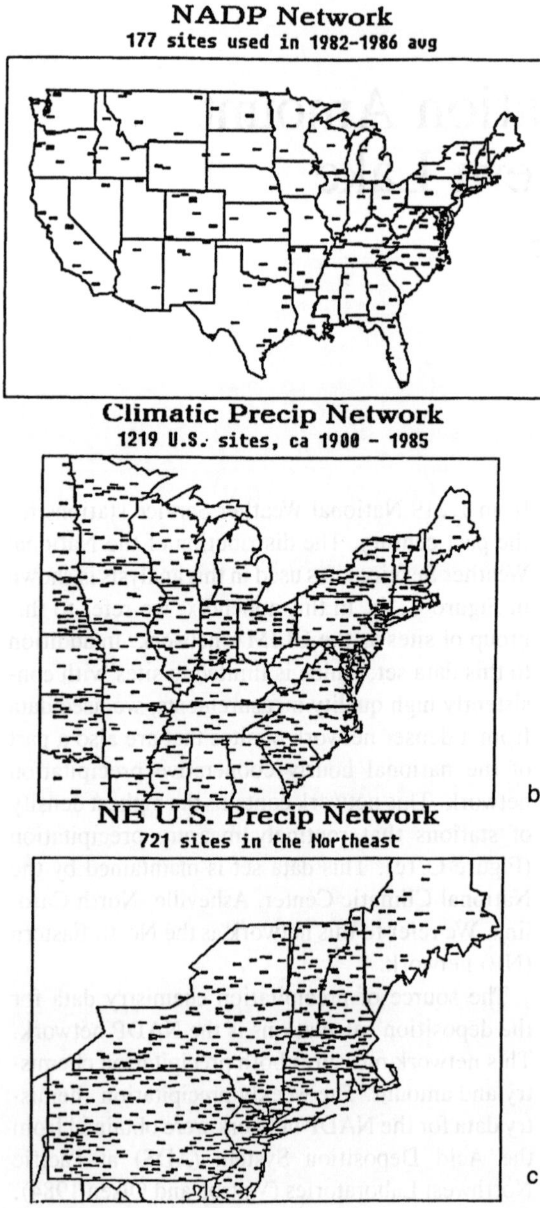

a
b
c

FIGURE C.1. Location of sites for three precipitation networks used in this study: (a) NADP, (b) NWS (CLIMA), and (c) NE.

Determination of the relationship between precipitation chemistry and lake chemistry is a key aspect of this book and is thoroughly evaluated in all the case study chapters. The chemistry of several hundred lakes has been characterized by the Eastern and Western Lake Surveys (Introduction, Charles, this volume) and can be used to evaluate these relationships. However, although precipitation is being monitored at several hundred sites in North America, corresponding precipitation data are not available at specific ELS and WLS sites. The nearest precipitation chemistry site may be 50 to 100 km away from a specific lake. Because of the significant spatial variability in precipitation amount and chemistry, the nearest precipitation chemistry site may not accurately represent deposition to the lake. Therefore, it is necessary to interpolate values to specific lake sites using all

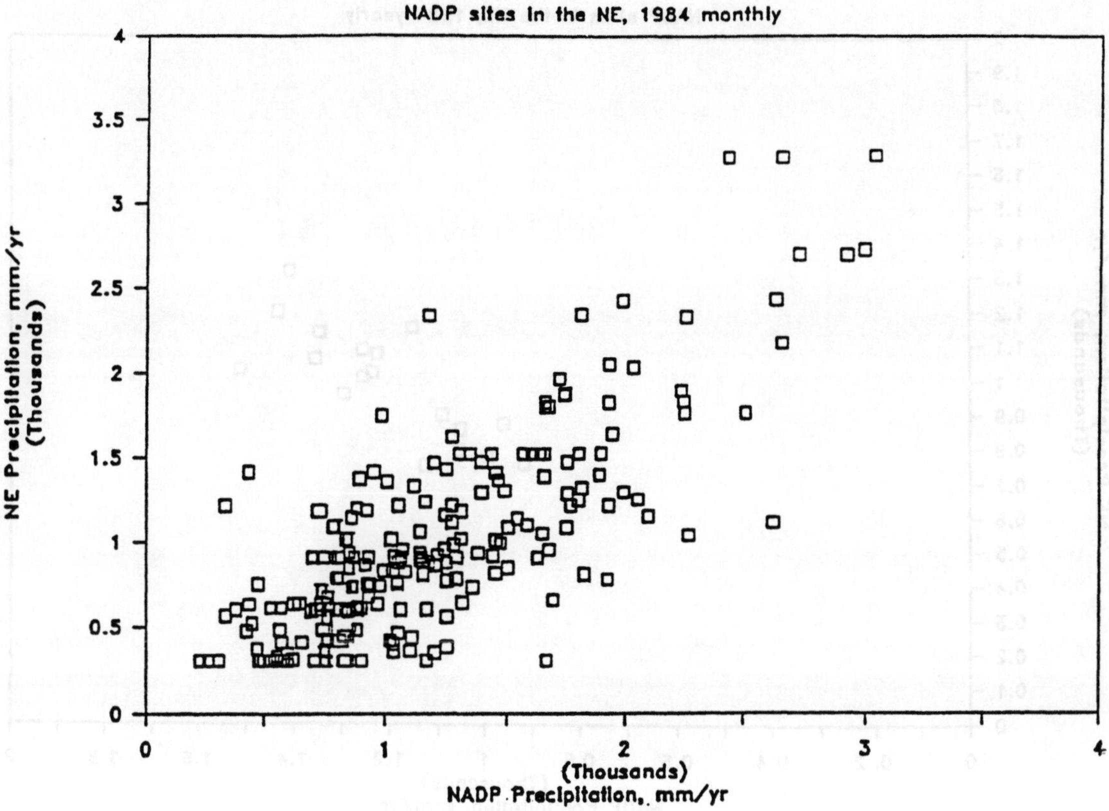

FIGURE C.2. Comparison of monthly precipitation interpolated from the NE network and measured at NADP sites.

appropriate nearby precipitation stations. The major considerations in making these interpolations are the sites and networks to be used, and the weighting procedure for calculating the interpolated values. In the ELS regions, the NADP data set is the only one with sufficient chemistry data; nevertheless, the NADP, CLIMA, and NE data sets can all be used to provide precipitation amount data. For calculations in this book, precipitation amount is calculated using the National Weather Service precipitation amount data because there are more available stations. Precipitation chemistry is interpolated using NADP data. An analysis of the various combinations of interpolated chemistry and amount data used to calculate loadings is presented later in this appendix.

The Eastern Lake Survey was conducted during the fall of 1984. Based on watershed characteristics and lake hydrologic retention time, it was concluded that 1984 lake chemistry was probably most affected by precipitation falling from 1982 to 1984. Therefore precipitation data for this time period were used for the interpolations. Interpolation to a particular site can be performed by calculating a weighted average using values from surrounding stations. In the simplest case, the weighting factors may be spatially constant, thus taking a simple functional form (e.g., $1/R^2$; e^{-R}). These methods are generally referred to as *deterministic*, because the weighting factors are invariant in space and the functional form of the distance function is obtained empirically from "past experience." The distance dependent weighting function can also be obtained from the actual observations using the statistical techniques known as kriging, which also automatically provides estimates for the uncertainty of the extrapolation. There is considerable debate at present about the relative merits of these extrapolation procedures (Vong et al. 1989). Interpolated data presented

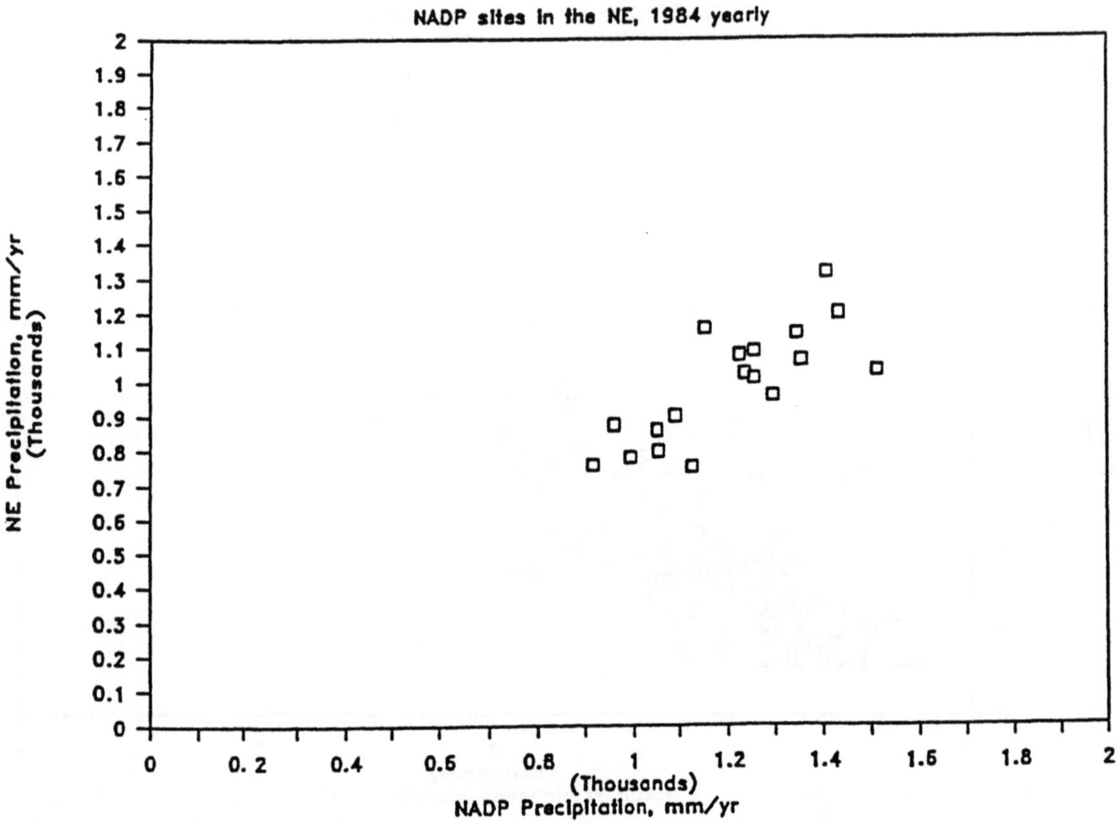

FIGURE C.3. Comparison of yearly precipitation interpolated from the NE network and measured at NADP sites.

in this book were calculated using a $1/R^2$ weighting factor.

The weighted average concentration for monthly and yearly values were calculated as follows:

$$C = \Sigma\, C_i P_i / \Sigma P_i$$

where C is the weighted average concentration; C_i is the reported sample concentration; and P_i is the rain gauge volume in mm.

Special attention was given to the records that contained missing data. We have used a simple rule-based expert system to handle the problem of missing data:

1. If the rain gauge datum was missing, then it was replaced by the sample volume datum.
2. If the sample volume was missing, then it was replaced by the rain gauge datum.
3. If both rain gauge and sample volumes were absent, then the entire record was ignored.

Following the application of these filters, there were 177 qualified stations in the NADP network.

We, and others, have observed that the seasonally weighted average concentrations of H^+, SO_4^{2-}, and NO_3^- exhibit a smoother pattern than the corresponding deposition maps. This is to be expected, since rainfall itself is highly variable in space. Hence, we use the smooth, weighted average concentration fields obtained from the precipitation chemistry networks (177 stations), which warrants a grid size of about 200 km.

In order to interpolate the wet deposition to regions that have no stations, we use a scheme that incorporates additional precipitation data. The precipitation data totals from 1,219 CLIMA stations warrant an effective grid resolution of about 80 km.

The specific algorithm used for the interpolation is the *inverse distance squared*. In this procedure, the surrounding stations at any target location are weighted by the inverse of the squared distance

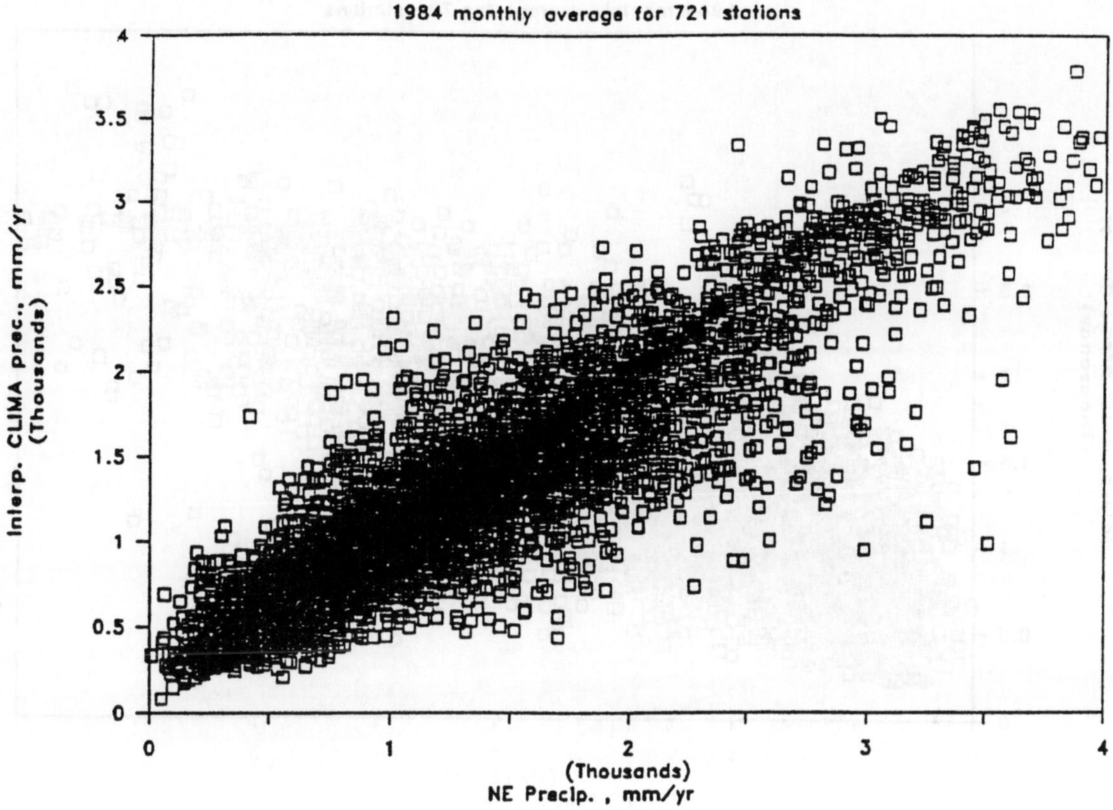

FIGURE C.4. Comparison of monthly precipitation interpolated from the CLIMA network and measured at NE sites.

between the "source" and "target" locations. For exploratory sensitivity analysis, we have also used $1/R$ and $1/R^3$ weighting schemes. Only the nearest eight stations were used, restricting the search radius to about 200 to 400 km.

Evaluation of Interpolation Results

Precipitation Interpolated from High to Low Station Density

Given a high station density network such as the NE precipitation network, it should be possible to estimate precipitation with reasonably high accuracy at any location within the domain of the network. This statement can be tested by estimating the precipitation amount at 18 NADP sites in the northeastern United States using values interpolated from the 721 stations in the NE network. Because the precipitation amount as well as the chemistry is monitored at the NADP sites, the measured and interpolated precipitation can be compared and the deviations may yield the interpolation uncertainty. This procedure of interpolating from high to low density also provides a test for the interpolation scheme and for the absolute compatibilities of the two networks.

It is worth noting that even if stations from the high and low density networks are nearby and instrumental deviations are absent, the interpolated and the measured values should not be identical. The interpolation takes into consideration the values at neighboring sites resulting from the $1/R^2$ weighting. Hence, the interpolated values are always less variable as a group than the measurements.

Figure C.2 presents a scatter diagram of measured and interpolated monthly average precipitation. A scattergram of measured and interpolated yearly average precipitation (Figure C.3) shows

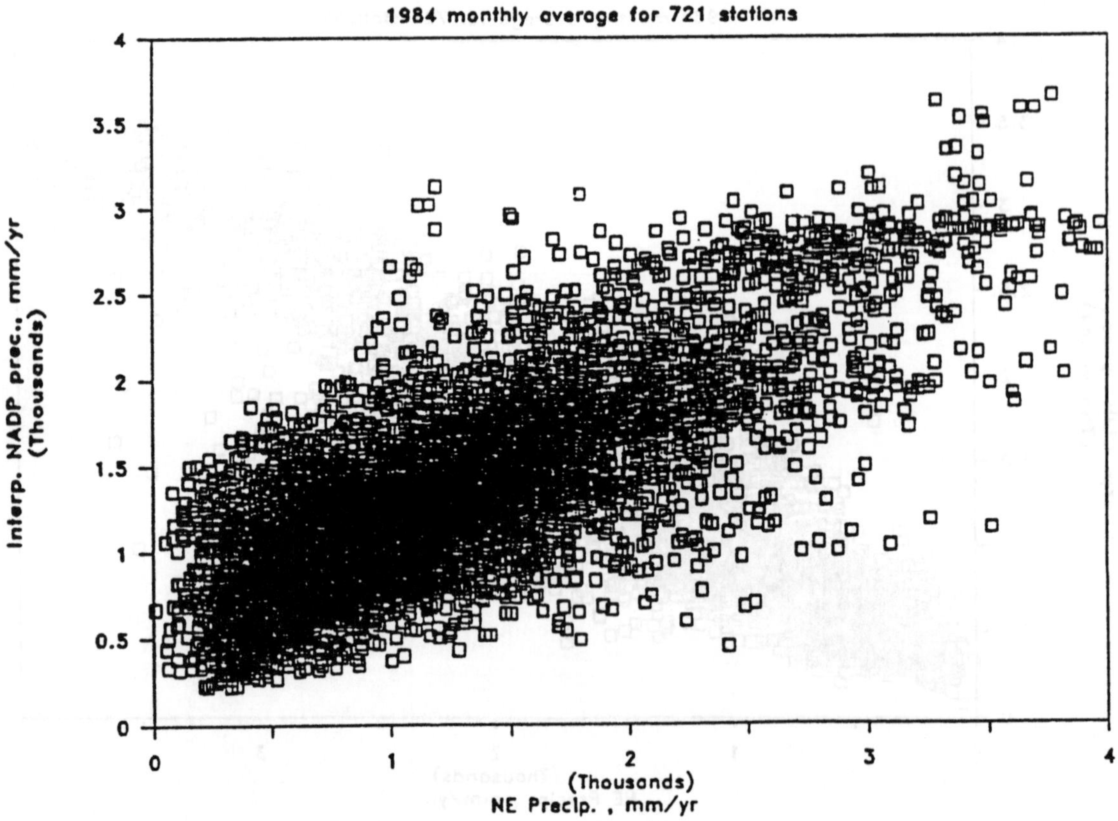

FIGURE C.5. Comparison of monthly precipitation interpolated from the NADP network and measured at NE sites.

much less scatter than the monthly values. In fact, for the 18 NADP sites, the ratio of interpolated and measured values is between 0.7 and 1.1.

The preceding data comparison shows that the interpolation of the precipitation amount on a monthly time scale is rather uncertain (± 25%) even if measured data are available from a high density network. However, interpolation of yearly average precipitation amounts can be made with much more certainty (± 10%), if high resolution data are available. Both monthly and yearly averages indicate that the measured precipitation at the NADP sites is generally higher by about 15% than the values interpolated from the NE network. The cause of this discrepancy in the absolute precipitation amount is not clear. It should be noted that the range of the yearly average precipitation amount at the 18 northeastern NADP sites is within a factor of 1.5. This is in sharp contrast with the yearly average precipitaiton range measured at the 721 NE site network (factor of 2 to 3). We may conclude that, given a monitoring network with a spatial resolution of 20 km, the yearly average precipitation can be estimated within about 10% at any location within the network.

Precipitation Interpolated from Low to High Station Density

When the measured data density is low (e.g., NADP) and the density of interpolated sites (e.g., ELS lakes) is high, then large interpolation uncertainties are expected. This arises from the data void between distant measuring sites. The uncertainty of interpolation is also influenced by the "texture" of precipitation between the measured values. We know that precipitation data have much texture at the 100 km scale. The high-density NE network constitutes a data set that is suitable for

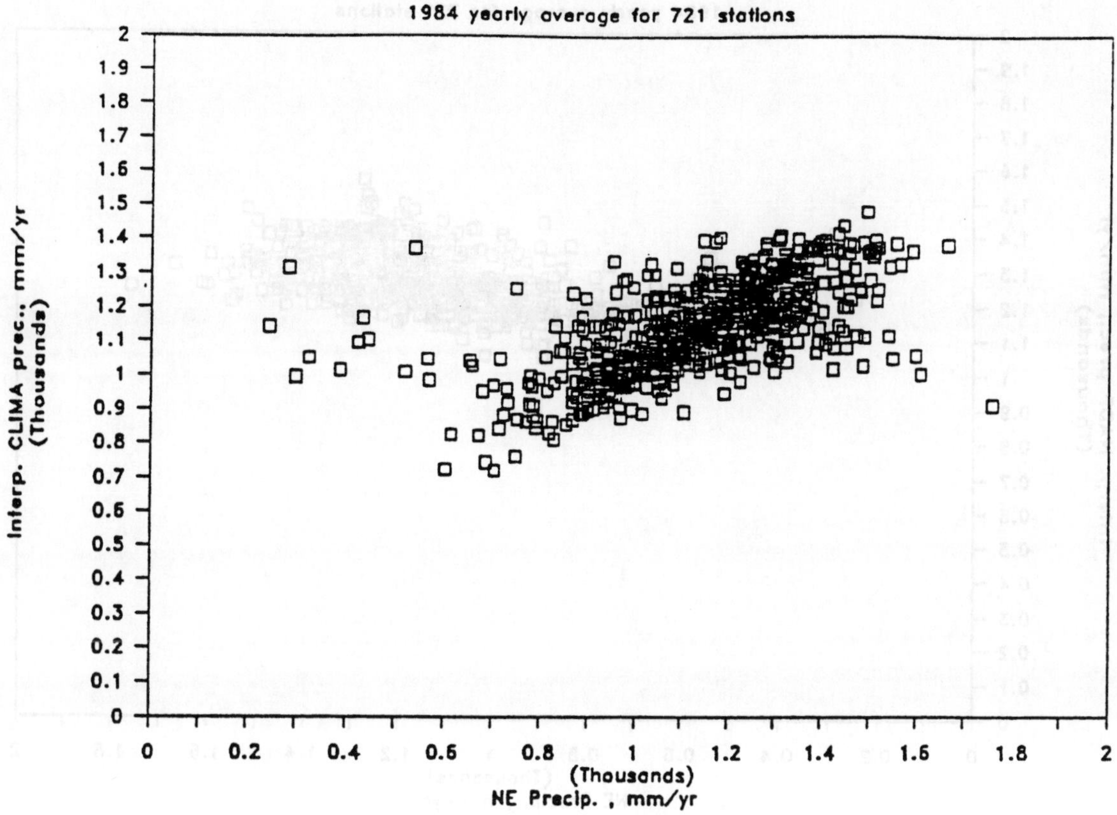

FIGURE C.6. Comparison of yearly precipitation interpolated from the CLIMA network and measured at NE sites.

testing the quality of interpolation from low to high density.

In another comparison, the monthly precipitation amounts from NADP and CLIMA sites were interpolated to 721 NE sites and compared to the measured values at each NE site. In a sense, the NE sites represent the numerous lakes for which interpolations are sought. The interpolated CLIMA (about 100 sites) monthly precipitation was compared to the measured data at the 721 NE sites (Figure C.4). The corresponding comparison of interpolated precipitation from NADP sites is shown in Figure C.5. The scatter plot shows significant deviations particularly at the values < 100 cm yr^{-1} measured precipitation. In this range, the NADP interpolations significantly exceed the measured values.

It is also useful to compare the yearly average precipitation at the NE sites. The scattergram comparing the interpolated CLIMA values to the measured values (Figure C.6) shows rather poor correlation. The comparison of the interpolated NADP values to the measured yearly average NE data (Figure C.7) is most revealing. The measured annual precipitation at NE sites ranges between 70 and 170 cm yr^{-1} (disregarding a few outliers in the 20- to 70-cm range). Remarkably, the NADP interpolated values are in the narrow range of 110 to 150 cm yr^{-1}. Hence, there appears to be no significant correlation between the two sets. The narrow range of yearly NADP precipitation values for 18 northeastern sites has been pointed out previously. The lack of variation of the interpolated NE precipitation appears to be the direct consequence of the rather uniform NADP annual precipitation in 1984.

Discussion and Conclusion

Precipitation amounts interpolated to ELS lakes using NADP site data appear to be about 15% higher than the precipitation amounts estimated

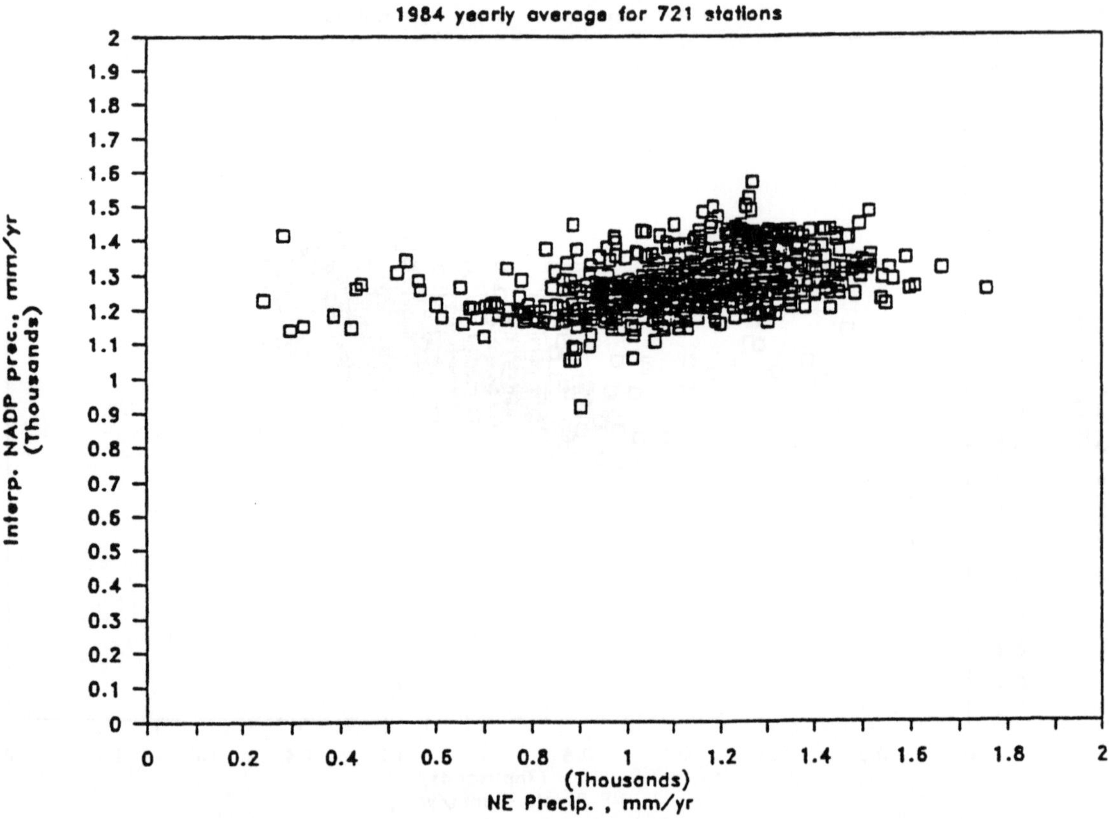

FIGURE C.7. Comparison of yearly precipitation interpolated from the NADP network and measured at NE sites.

using the high-density weather station network data. The interpolation of precipitation from a sparse network can yield only spatially smooth patterns with little spatial texture. In fact, the spatial scale in the texture of the interpolated values should be comparable to or larger than the characteristic distance between the monitoring sites. This observation is illustrated by plotting the time charts of interpolated precipitation at four lakes in the Adirondack region. The time charts of NADP interpolated precipitation (Figure C.8a) for the four lakes show almost identical patterns, because the lakes are < 200 km apart, and the driving data set does not have high spatial resolution.

The interpolated precipitation for the same lakes, using the NE data set, shows significant deviation among the four lakes (Figure C.8b), indicating that the NE network spatial scale is smaller than the distance among the lakes and that the monthly precipitation indeed varies in that spatial scale. This exploration clearly demonstrates the benefits of using high spatial resolution precipitation data.

FIGURE C.8. Comparison of interpolated precipitation estimates: (a) NADP interpolated precipitation and (b) NE interpolated precipitation (see text for explanation).

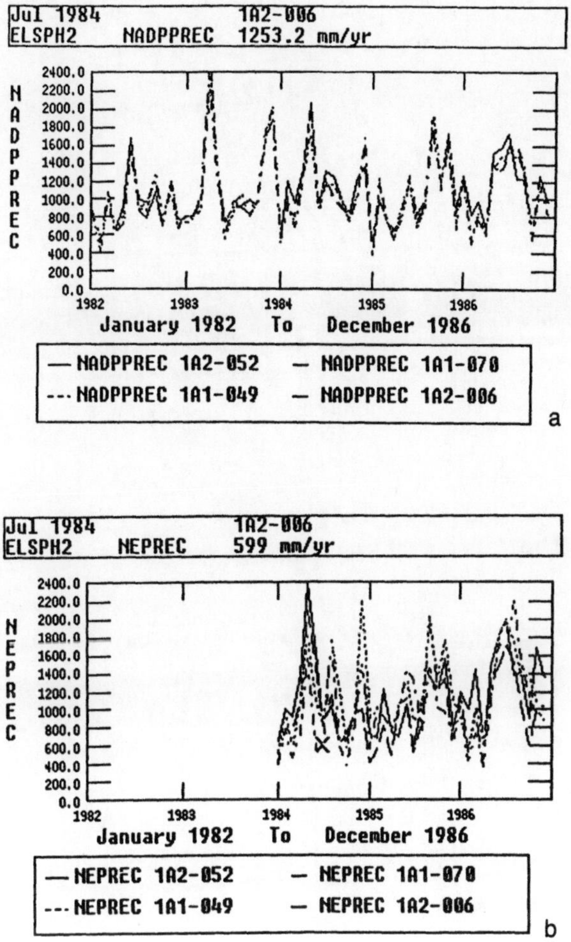

References

Aber, J.D., K.J. Nadelhoffer, P. Steudler, and J.M. Melillo. 1989. Nitrogen saturation in northern forest ecosystems. Bioscience 39:378–386.

Abrahamson, W.C. 1984. Species responses to fire on the Florida Lake Wales Ridge: A five year study. Am. J. Bot. 71:35–43.

Acidic Precipitation in Ontario Study. 1984. Annual statistics of concentration ambient air monitoring network, 1982. ARB-165-84-AQM.

Acidic Precipitation in Ontario Study. 1985. Annual statistics of concentration ambient air monitoring network, 1983. ARB-089-85-AQM.

Acidic Precipitation in Ontario Study. 1986. Annual statistics of concentration ambient air monitoring network, 1984. ARB-237-85-AQM.

Acidic Precipitation in Ontario Study. 1987. Annual statistics of concentration ambient air monitoring network, 1985. ARB-086-87-AQM.

Acidification Today and Tomorrow. 1982. Proceedings of the 1982 Stockholm Conference on the Acidification of the Environment, Stockholm, Sweden.

Allard M., and G. Moreau. 1986. Influence of acidification and aluminum on the density and biomass of lotic benthic invertebrates. Water Air Soil Pollut. 30: 673–679.

Almer, B., W. Dickson, C. Ekström, E. Hörnström, and U. Miller. 1974. Effects of acidification on Swedish lakes. Ambio 3:30–36.

Almer, B., W. Dickson, C. Ekström, and E. Hörnström. 1978. Sulfur pollution and the aquatic ecosystem. Pages 271–311 in J.O. Nriagu, ed. Sulfur in the Environment. Part II: Ecological Impacts. John Wiley and Sons, New York.

Altshuller, A.P., and R.A. Linthurst. 1984. The Acidic Deposition Phenomenon and Its Effects: Critical Assessment Review Papers. EPA/600/8-83/016BF. U.S. Environmental Protection Agency, Washington, D.C.

American Public Health Association (APHA). 1976. Standard Methods for the Examination of Water and Wastewater. 14th ed. American Public Health Association, Washington, D.C.

Amundsen, R.G., J. Harte, H. Michaels, and E. Pendall. 1988. The Role of Sediments in Controlling the Chemistry of Subalpine Lakes in the Sierra Nevada, California. Final Report. Contract A4-42-32, California Air Resources Board, Sacramento, CA.

Andersen, R., I.P. Muniz, and J. Skurdal. 1984. Effects of acidification on age class composition in arctic char (*Salvelinus alpinus* L.) and brown trout (*Salmo trutta* L.) in a coastal area, southwestern Norway. Instit. Freshw. Res. Drottningholm 61:5–15.

Anderson, M.P., and C.J. Bowser. 1986. The role of groundwater in delaying lake acidification. Water Resour. Res. 22:1101–1108.

Anderson, N.H. 1989. Department of Entomology, Oregon State University, Corvallis, OR (personal communication).

Anderson, R.S. 1971. Crustacean plankton of 146 alpine and subalpine lakes in western Canada. J. Fish. Res. Board Can. 28:311–321.

Anderson, R.S. 1974. Crustacean plankton communities of 340 lakes and ponds in and near the national parks of the Canadian Rocky Mountains. J. Fish. Res. Board Can. 31:855–869.

Andersson, B., and P. Andersson. 1984. The distribution of trout (*Salmo trutta* L.) in relation to pH–an inventory of small streams in Delsbo, central Sweden. Freshw. Res. Inst. Drottningholm 61:28–33.

Andersson, B.I., I. Alenas, and H. Hultberg. 1984. Liming of a small acidified river (River Anraseau) in southwestern Sweden, promoting successful reproduction of sea trout (*Salmo trutta* L.) Freshw. Res. Inst. Drottningholm 61:16–27.

Andersson, G., S. Fleischer, and W. Graneli. 1978. Influence of acidification on decomposition processes in lake sediment. Verh. Internat. Verein. Limnol. 20:802–807.

Andersson, I., O. Grahn, H. Hultberg, and L. Landner. 1974. Institute for Water and Air Research, STU Report 73-3651, Stockholm, Sweden (unpublished).

APIOS. 1984. Annual statistics of concentration cumulative ambient air monitoring network, 1982. ARB-165-84-AQM. Acidic Precipitation in Ontario Study.

APIOS. 1985. Annual statistics of concentration cumulative ambient air monitoring network, 1983. ARB-089-85-AQM. Acidic Precipitation in Ontario Study.

APIOS. 1986. Annual statistics of concentration cumulative ambient air monitoring network, 1984. ARB-237-86-AQM. Acidic Precipitation in Ontario Study.

APIOS. 1987. Annual statistics of concentration cumulative ambient air monitoring network, 1985. ARB-086-87-AQM. Acidic Precipitation in Ontario Study.

Appelberg, M. 1987. Some factors regulating the crayfish Astacus astacus L. in acid and neutralized waters. Ann. Soc. R. Zool. Belg. 117(Suppl. 1):167–179.

Appleby, P.G., and F. Oldfield. 1978. The calculation of lead-210 dates assuming a constant rate of supply of unsupported ^{210}Pb to the sediment. Catena 5:1–8.

April, R.A., and R.M. Newton. 1983. Mineralogy and chemistry of some Adirondack Spodosols. Soil Sci. 135:301–307.

April, R.A., and R.M. Newton. 1984. The geology and geochemistry of the ILWAS lake-watersheds. In The Integrated Lake Watershed Acidification Study 4:41–420.

April, R.A., and R.M. Newton. 1985. Influence of geology on lake acidification in the ILWAS watersheds. Water Air Soil Pollut. 26:373–386.

April, R.A., R.M. Newton, and L.T. Coles. 1986. Chemical weathering in two Adirondack watersheds: Past and present-day rates. Geol. Soc. Am. Bull. 97: 1232–1238.

Armstrong, D.E., J.P. Hurley, D.L. Swackhammer, and M.M. Shafer. 1987. Cycles of nutrient elements, hydrophobic organic compounds, and metals in Crystal Lake: Role of particle-mediated processes in regulation. Pages 491–578 in R.A. Hites and S.J. Eisenreich, eds. Sources and Fate of Aquatic Pollutants. Advances in Chemistry Series, No. 216. American Chemical Society, Washington, D.C.

Arno, S.F., and R.P. Hammerly. 1984. Timberline: Mountain and Arctic Forest Frontiers. The Mountaineers, Seattle, WA.

Arnold, D.E., R.W. Light, and E.A. Paul. 1985. Vulnerability of Selected Lakes and Streams in the Middle Atlantic Region to Acidification: A Regional Survey. FWS/OBS-80/40.19. U.S. Fish and Wildlife Service report. Kearneysville, WV.

Arrington, D.V., and R.C. Lindquist. 1987. Thickly mantled karst of the Interlachen, Florida, area. Pages 31–39 in B.F. Beck and W.L. Wilson, eds. Karst Hydrogeology: Engineering and Environmental Applications. A.A. Balkema, Rotterdam, The Netherlands.

Asbury, C.E., F.A. Vertucci, M.D. Mattson, and G.E. Likens. 1989. Acidification of Adirondack lakes. Environ. Sci. Technol. 23:362–368.

Austin, M.E. 1972 (Revised 1978). Land Resource Regions and Major Land Resource Areas of the United States. Rev. ed. Agricultural Handbook 292. U.S. Department of Agriculture, Soil Conservation Service, Washington, D.C.

Baas, J., F.D. Westerdahl, and R.L. Perrine. 1976. Nonpoint Source Water Quality Monitoring, Inyo National Forest, 1975. Contribution No. 156. Water Resources Center, University of California, Davis, CA.

Bache, B.W. 1983. The role of soil in determining surface water composition. Water Sci. and Technol. 15:33–45.

Backman, E.A. 1984. Pages 1–99 in 1000-year Record of Fire-Vegetation Interaction in the Northeastern U.S.M.S. Thesis. University of Massachusetts, Amherst, MA.

Bailey, R.A., G.B. Dalrymple, and M.A. Lanphere. 1988. Volcanism, structure, and geochronology of Long Valley Caldera, Mono County, California. J. Geophys. Res. 81:725–743.

Bailey, R.G. 1976. Ecoregions of the United States. One map (scale 1:7,500,000). U.S. Department of Agriculture, Forest Service, Ogden, UT.

Baker, J.P. 1982. Effects on fish of metals associated with acidification. Pages 165–176 in T.A. Haines and R.E. Johnson, eds. Acid Rain/Fisheries. American Fisheries Society, Bethesda, MD.

Baker, J.P. 1984a. The acidic deposition phenomenon and its effects on aquatic biology. Pages 5–74 to 5–129 in A.P. Altshuller and R.A. Linthurst, eds. The Acidic Deposition Phenomenon and Its Effects: Critical Assessment and Review Papers. EPA/600/8-83/016BF. U.S. Environmental Protection Agency, Washington, D.C.

Baker, J.P., and T.B. Harvey. 1984. Critique of Acid Lakes and Fish Population Status in the Adirondack Region of New York State. EPA/600/3-86/046. Final Report. U.S. Environmental Protection Agency, Washington, D.C.

Baker, J.P., and J.J. Magnuson. 1976. Limnological ressponses of Crystal Lake (Vilas County, Wisconsin) to intensive recreational use, 1924–1973. Trans. Wis. Acad. Sci. Arts Lett. 64:47–61.

Baker, J.P., and C.L. Schofield. 1982. Aluminum toxicity to fish in acidic waters. Water Air Soil Pollut. 18: 289–309.

Baker, J.P., T.B. Harvey, and J.P. Nicolette. 1984. Compilation of Available Data on the Status of Fish Populations in Regions of the Northeastern U.S. Susceptible to Acidic Deposition. Final Report. U.S. Environmental Protection Agency, Corvallis, OR.

Baker, J.P., C.S. Creager, W. Warren-Hicks, S.W. Christensen, and L. Godbout. 1988a. Identification of Critical Values for Effects of Acidification on Fish Populations. Final Report. Brookhaven National Laboratory and U.S. Environmental Protection Agency, Washington D.C.

Baker, L.A. 1984b. Mineral and Nutrient Cycles and Their Effect on the Proton Balance of a Softwater, Acidic Lake. Ph.D. Dissertation. University of Florida, Gainesville, FL.

Baker, L.A., and P.L. Brezonik. 1985. Sulfate mass balance for a small, acidic lake. Pages 297–306 in D.E. Caldwell, J.A. Brierley, and C.L. Brierley, eds. Planetary Ecology. Van Nostrand, Reinhold, New York.

Baker, L.A., and P.L. Brezonik. 1988. Dynamic model of internal alkalinity generation. Water Resour. Res. 24:65–74.

Baker, L.A., P.L. Brezonik, E.S. Edgerton, and R.W. Ogburn, III. 1985a. Sediment acid neutralization in softwater lakes. Water Air Soil Pollut. 25:215–230.

Baker, L.A., T.E. Perry, and P.L. Brezonik. 1985b. Neutralization of acid precipitation in softwater lakes. Lake Reserv. Manage. 1:356–360.

Baker, L.A., C.D. Pollman, and P.L. Brezonik. 1985c. Development of Interpretive Models for Lake Acidification in Florida. Florida Acid Deposition Phase IV Report. ESE 83-152-0602-0120. Environmental Sciences and Engineering, Gainesville, FL.

Baker, L.A., P.L. Brezonik, and C.D. Pollman. 1986a. Model of internal alkalinity generation: Sulfate retention component. Water Air Soil Pollut. 31:89–94.

Baker, L.A., P.L. Brezonik, and E.S. Edgerton. 1986b. Sources and sinks of ions in a softwater, acidic lake. Water Resour. Res. 24:65–74.

Baker, L.A. C.D. Pollman, and J.M. Eilers. 1988b. Alkalinity regulation in softwater Florida lakes. Water Resour. Res. 24:1069–1082.

Baker, L.A., J.E. Tacconi, and P.L. Brezonik. 1988c. Role of sedimentation in regulating major ion composition of softwater lakes. Verh. Int. Verein. Limnol. 23:346–350.

Baker, L.A., N.R. Urban, P.L. Brezonik, and L.A. Sherman. 1989. Sulfur cycling in an experimentally acidified seepage lake. Pages 79–100 in E. Saltzman and W. Cooper, eds. Biogenic Sulfur in the Environment. Advances in Chemistry Series, No. 393. American Chemical Society, Washington, D.C.

Baker, M.D., W.E. Inniss, C.I. Mayfield, and P.T.S. Wong. 1983. Effects of acidification, metals, and metalloids on sediment microorganisms. Water Res. 17:925–930.

Baldwin, E.M. 1981. Geology of Oregon, Third Ed. Kendall-Hunt Publ. Co., Dubuque, IA.

Baldwin, J.L. 1973. Climates of the United States. U.S. Department of Commerce, Washington, D.C.

Barie, L.A., and J.M. Hales. 1984. The spatial distributions of precipitation acidity and major ion wet deposition in North America during 1980. Tellus 36B: 333–355.

Barker, J.L., and E.E. Witt III. 1990. Effects of Acid Precipitation on the Water Quality of Streams on Laurel Hill, Somerset County, Pennsylvania, 1983–86. Water Resources Investigations Report No. 89-4113. U.S. Geological Survey, Harrisburg, PA.

Barnard, T.E., and J.J. Bisogni. 1985. Errors in Gran function analysis of titration data for dilute acidified water. Water Res. 19:393–399.

Barnes, R.B. 1975. The determination of specific forms of aluminum in natural water. Chem. Geol. 15:177–191.

Baron, J. 1983. Comparative water chemistry of four lakes in Rocky Mountain National Park. Water Resour. Bull. 19:897–902.

Baron, J., and O.P. Bricker. 1987. Hydrologic and chemical flux in Loch Vale watershed, Rocky Mountain National Park. Pages 141–156 in R.C. Averett and D.M. McKnight, eds. Chemical Quality of Water and the Hydrologic Cycle. Lewis Publishers, Chelsea, MI.

Baron, J., S.A. Norton, D.R. Beeson, and R. Herrmann. 1986. Sediment diatom and metal stratigraphy from Rocky Mountain lakes with special reference to atmospheric deposition. Can. J. Fish. Aquat. Sci. 43: 1350–1362.

Barrie, L.A. 1984. The spatial distributions of precipitation acidity and major ion wet deposition in North America during 1980. Tellus 36B:333–355.

Barrie, L.A., and A. Sirois. 1986. Wet and dry deposition of sulphates and nitrates in eastern Canada: 1979–1982. Water Air Soil Pollut. 30:303–310.

Barrie, L.A., H.A. Weibe, K. Anlaur, and P. Fellin. 1980. The Canadian Air and Precipitation Monitoring Network (APN). Pages 355–360 in M.M. Benarie, ed. Atmospheric Pollution, Proceedings of the 14th International Colloquium (Studies in Environmental Science). Paris.

Bateman, P.C. 1961. Granitic formations in the east-central Sierra Nevada near Bishop, California. Geol. Soc. Am. Bull. 72:1521–1538.

Battarbee, R.W. 1984. Diatom analysis and the acidification of lakes. Philos. Trans. R. Soc. Lond. Ser. B Biol. Sci. 305:451–477.

Battarbee, R.W., and D.F. Charles. 1987. The use of diatom assemblages in lake sediments as a means of assessing the timing, trends, and causes of lake acidification. Prog. Phys. Geog. 11:552–580.

Battarbee, R.W., J.P. Smol, and J. Meriläinen. 1986. Diatoms as indicators of pH: An historical review. Pages 5–14 in J.P. Smol, R.W. Battarbee, R.B. Davis, and J. Meriläinen, eds. Diatoms and Lake Acidity. Dr. W. Junk, Dordrecht, The Netherlands.

Battarbee R.W. et al. 1988. Lake Acidification in the United Kingdom 1800–1986. Evidence from Analysis of Lake Sediments. Department of the Environment, ENSIS Publishing, London.

Bayley, S.E., and D.W. Schindler. 1987. Sources of alkalinity in Precambrian Shield watersheds under natural conditions and after fire or acidification. Pages 531–548 in T.C. Hutchinson and K.M. Meema, eds. Effects of Atmospheric Pollutants on Forests, Wetlands, and Agricultural Ecosystems. Springer-Verlag, New York.

Bayley, S.E., R.S. Behr, and C.A. Kelly. 1986. Retention and release of S from a freshwater wetland. Water Air Soil Pollut. 30:101–114.

Beamish, R.J. 1976. Acidification of lakes in Canada by acid precipitation and the resulting effects on fishes. Water Air Soil Pollut. 6:501–514.

Beamish, R.J., and H.H. Harvey. 1972. Acidification of the La Cloche Mountain lakes, Ontario, and resulting fish mortalities. J. Fish. Res. Board Can. 29:1131–1143.

Beamish, R.J., W.L. Lockhart, J.C. Van Loon, and H.H. Harvey. 1975. Long-term acidification of a lake and resulting effects on fishes. Ambio 4:98–102.

Beaver, J.R., and T.L. Crisman. 1981. Acid precipitation and the response of ciliated protozoans in Florida lakes. Verh. Int. Verein. Limnol. 21:353–358.

Beaver, J.R., and T.L. Crisman. 1982. The trophic response of ciliated protozoans in freshwater lakes. Limnol. Oceanogr. 27:246–253.

Beaver, J.R., T.L. Crisman, and J.S. Bays. 1981. Thermal regimes of Florida lakes. Hydrobiologia 83:267–273.

Beggs, G.L., and J.M. Gunn. 1986. Response of lake trout (*Salvelinus namaycush*) and brook trout (*S. fontinalis*) to surface water acidification in Ontario. Water Air Soil Pollut. 30:711–717.

Beggs, G.L., J.M. Gunn, and C.H. Olver. 1985. The Sensitivity of Ontario Lake Trout (*Salvelinus namaycush*) and Lake Trout Lakes to Acidification. Ontario Fisheries Technical Report Series No. 17. Ontario Ministry of Natural Resources, Toronto, Ontario.

Beggs, G.L., J.M. Gunn, B.J. Shuter, and P.E. Ihssen. 1987a. The response of four Ontario sportfish species to surface water acidification. Unpublished manuscript.

Beggs, G.L., J.A. MacLean, T. Stewart, and F. Hicks. 1987b. A review of evidence for effects of acidic deposition on Ontario fisheries. Unpublished manuscript.

Behnke, R.J. 1976. Biology and Management of Threatened and Endangered Western Trouts. General Technical Report RM-28. USDA Forest Service, Rocky Mountain Forest and Range Experiment Station, Ft. Collins, CO.

Bell, H.L. 1971. Effect of low pH on the survival and emergence of aquatic insects. Water Res. 5:313–319.

Bennett, P.S. 1965. An Investigation of the Impact of Grazing on Ten Meadows in Sequoia and Kings Canyon National Parks. M.S. Thesis. San Jose State College, San Jose, CA.

Berntsen, C.M., J.F. Corliss, I.J. Fernandez, D.W. Johnson, H.C. Jones, III, W.H. Smith, and J.R. Lyons. 1984. Pages 1–48 in Acidic Deposition and Forests. Report of the SAF Task Force on the Effects of Acidic Deposition on Forest Ecosystems. Society of American Foresters, Bethesda, MD.

Berrill M., L. Hollett, A. Margosian, and J. Hudson. 1985. Variation in tolerance to low environmental pH by the crayfish *Orconectes rusticus O. propinquus*, and *Cambarus robustus*. Can. J. Zool. 63:2586–2589.

Berrill, M., L. Rowe, L. Hollett, and J. Hudson. 1987. Response of some aquatid benthic arthropods to low pH. Ann. Soc. R. Zool. Belg. 117(Suppl. 1):117–128.

Best, M.D., L.W. Creelman, S.K. Drousé, and D.J. Chaloud. 1986. National Surface Water Survey, Eastern Lake Survey–Phase I, Analytical Methods Manual. EPA/600/4-86/011. U.S. Environmental Protection Agency, Las Vegas, NV.

Bienert, R.W., Jr. 1986. Sediment Geochemistry of Northern Florida Lakes. Unpublished report. Department of Environmental Sciences, University of Florida, Gainesville, FL.

Blackwelder, E. 1931. Pleistocene glaciation in the Sierra Nevada and basin ranges. Bull. Geol. Soc. Am. 42:865–922.

Blaedel, W.J., and V.W. Meloche. 1963. Elementary Quantitative Analysis: Theory and Practice, Second edition. Harper and Row, New York.

Blanchard, C., H. Michaels, A. Bradman, and J. Harte. 1988. Episodic Acidification of a Low-alkalinity Pond in Colorado. University of California Energy and Resources Group, ERG-88-1. Berkeley, CA.

Bleiwas, A.S.H., P.M. Stokes, and M.M. Olaveson. 1984. Six years of plankton studies in the LaCLoche region of Ontario, with special reference to acidification. Verh. Internat. Verein. Limnol. 22:332–337.

Blick, D.J., and M.S. DeHaan. 1985. Unpublished analyses of Eastern and Western Lake Survey data. U.S. EPA Environmental Research Laboratory, Corvallis, OR.

Blick, D.J., Jr., J.J. Messer, D.H. Landers, and W.S. Overton. 1987. Statistical basis for the design and interpretation of the National Surface Water Survey, Phase I: Lakes and streams. Lake Reserv. Manage. 3:470–475.

Bloom, P.R. 1988. ALBIOS: Dissolution Kinetics of Minerals. Annual Report to Electric Power Research Institute, Palo Alto, CA (unpublished).

Bockheim, J.G., J.E. Leide, C.S. Li, K.K. Sorsa, K.J. Zuelsdorff, D.R. Knauer, and R.A. Goldstein. 1988. Acid precipitation and major and trace element biogeochemistry of terrestrial ecosystems in the Upper

Great Lakes Region, U.S.A. Pages 380-387 in R. Perry, R.M. Harrison, J.N. B. Bell, and J.N. Lester, eds. Acid Rain: Scientific and Technical Advances. Selper Ltd., London.

Bolt, G.H., and M.G.M. Bruggenwert. 1976. Soil Chemistry, A. Basic Elements. Elsevier, Amsterdam.

Booty, W.G., and J.R. Kramer. 1984. Sensitivity analysis of a watershed acidification model. Philos. Trans. R. Soc. Lond. Series B Biol. Sci. 305:441–449.

Booty, W.G., J.V. DePinto, and R.D. Scheffe. 1988. Drainage basin control of acid loadings to two Adirondack lakes. Water Resour. Res. 24:1024–1036.

Bormann, F.H., and G.E. Likens. 1979. Pattern and Process in a Forested Ecosystem. Springer-Verlag, New York.

Bormann, F.H., G.E. Likens, D.W. Fisher, and R.S. Pierce. 1968. Nutrient loss accelerated by clear-cutting of a forest ecosystem. Science 159:882–884.

Bormann, F.H., G.E. Likens, and J.M. Mellilo. 1977. Nitrogen budget for an aggrading northern hardwood forest ecosystem. Science 196:981–983.

Born, S.M., S.A. Smith, and D.A. Stephenson. 1979. Hydrogeology of glacial terrain lakes, with management and planning applications. J. Hydrol. 43:7–43.

Bortleson, G.C., G.T. Higgins, and G.W. Hill. 1974. Data on Selected Lakes in Washington, Part 2. USGS Water-Supply Bulletin 42. Washington State Dept. of Ecology in cooperation with U.S. Department of Interior, Washington, D.C.

Bortleson, G.C., G.T. Higgins, J.B. McConnell, and J.K. Innes. 1976. Data on Selected Lakes in Washington, Part 3. USGS Water-Supply Bulletin. Washington State Department of Ecology in cooperation with U.S. Department of Interior, Washington, D.C.

Boylen, C., M. Shick, D. Roberts, and R. Singer. 1983. Microbiological survey of Adirondack lakes with various pH values. Applic. Environ. Microbiol. 45:46–56.

Bradford, G.R., F.L. Bair, and V. Hunsaker. 1968. Trace and major element content of 170 high Sierra lakes in California. Limnol. Oceanog. 13:526–530.

Bradford, G.R., A.L. Page, and I.R. Straughan. 1981. Are Sierra lakes becoming acid? May-June: 6–7. Calif. Agric.

Brakke, D.F. 1984. Chemical Surveys of North Cascade Lakes. Institute for Watershed Studies, Western Washington University, Bellingham, WA.

Brakke, D.F., and T.J. Loranger. 1986. Acid neutralizing capacity of lakes in the north Cascades area of Washington State. Water Air Soil Pollut. 30:1045–1053.

Brakke, D.F., J.M. Eilers, and D.H. Landers. 1987. Hydrologic and chemical characteristics of darkwater, clearwater, and acidic lakes in the United States. Pages 281–290 in Acidification and Water Pathways. Proceedings of UNESCO/IHP Symposium, May 4–5. Norwegian National Committee for Hydrology, Bolkesjø, Norway.

Brakke, D.F., D.H. Landers, and J.M. Eilers. 1988. Chemical and physical characteristics of lakes in the northeastern United States. Environ. Sci. Technol. 22:155–163.

Brakke, D.F., A. Henriksen, and S.A. Norton. 1989. Estimated background concentrations of sulfate in dilute lakes. Water Resour. Bull. 25:247–253.

Brenner, M., M.W. Binford, and E.S. Deevey. 1990. Lakes. Pages 364–391 in R.L. Myers and J.J. Ewel, eds. Ecosystems of Florida. University of Central Florida Press, Orlando, FL.

Brewer, G.F. 1986. Sulfur, Heavy Metal, and Major Element Chemistry of Sediments from Four Eastern Maine Ponds. M.S. Thesis. Department of Geological Science, University of Maine, Orono, ME.

Brezonik, P.L., E.S. Edgerton, and C.D. Hendry, Jr. 1980. Acid precipitation and sulfate deposition in Florida. Science 208:1027–1029.

Brezonik, P.L., S.D. Preston, and T.L. Crisman. 1981. Limnological Studies on Lake Apopka and the Oklawaha Chain of Lakes 4. Water Quality in 1980. Report No. ENV-07-81-02. Department of Environmental Engineering Sciences, University of Florida, Gainesville, FL.

Brezonik, P.L., C.D. Hendry, Jr., E.S. Edgerton, R.L. Schulze, and T.L. Crisman. 1983. Acidity, Nutrients, and Minerals in Atmospheric Precipitation Over Florida: Deposition Patterns, Mechanisms, and Ecological Effects. EPA/600/3-84/004, NTIS No. P8 83 165837. U.S. Environmental Protection Agency, Corvallis, OR.

Brezonik, P.L., T.L. Crisman, and R.L. Schulze. 1984. Planktonic communities in Florida softwater lakes of varying pH. Can. J. Fish. Aquat. Sci. 41:46– 56.

Brezonik, P.L., L.A. Baker, J.R. Eaton, T.M. Frost, P. Garrison, T.K. Kratz, J.J. Magnuson, W.J. Rose, B.K. Shephard, W.A. Swenson, C.J. Watras, and K.E. Webster. 1986. Experimental acidification of Little Rock Lake, Wisconsin. Water Air Soil Pollut. 31: 115–121.

Brezonik, P.L., L.A. Baker, and T.E. Perry. 1987. Mechanisms of alkalinity generation in acid-sensitive softwater lakes. Pages 229–260 in R.A. Hites and S.J. Eisenreich, eds. Sources and Fates of Aquatic Pollutants. Advances in Chemistry Series, No. 216. American Chemical Society, Washington, D.C.

Bricker, O.P. 1986. Geochemical investigations of selected eastern United States watersheds affected by acid deposition. J. Geol. Soc. Lond. 143:621–626.

Bricker, O.P., and K.C. Rice. 1989. Acidic deposition to streams: A geology-based method predicts their sensitivity. Environ. Sci. Technol. 23:379–385.

Broberg, O., and G. Persson. 1984. External budgets for phosphorus, nitrogen, and dissolved organic carbon for the acidified Lake Gärdsjo/n. Arch. Hydrobiol. 99:160–175.

Bronsted, J.N. 1923. Rec. Trav. Chim. 30:777.

Brooks, H.K. 1982. Geologic Map of Florida. Institute of Food and Agricultural Sciences, University of Florida, Gainesville, FL.

Brown, C.J.D. 1971. Fishes of Montana. Big Sky Books, Montana State University, Bozeman.

Brown, D. 1983a. The relationship between surface water acidification and loss of fisheries. Pages 365–371 in Acid Precipitation–Origin and Effects. VDI Berichte Nr. 500 FDI Verglag GmbH, Dusseldorf, FRG.

Brown, D.J.A. 1983b. Effect of calcium and aluminum concentrations on the survival of brown trout (*Salmo trutta*) at low pH. Bull. Environ. Contam. Toxicol. 30:582–587.

Brown, E. 1985. The Forest Preserve of New York State: A Handbook for Conservationists. Adirondack Mountain Club, Glens Falls, NY.

Brown, J.C., and C.M. Skau. 1975. Chemical Composition of Snow in the East-central Sierra Nevada. Cooperative Report Series Publication AG-1:1–12.

Brugam, R.B. 1981. Chemistry of lake water and groundwater in areas of contrasting glacial drifts. Hydrobiologia 80:47–62.

Buddington, A.F. 1968. Adirondack anorthositic series. Pages 215–231 in Y.W. Isachsen, ed. Origin of Anorthosite and Related Rocks. State Museum and Science Service, Albany, NY.

Buell, G.R., and N.E. Peters. 1988. Atmospheric deposition effects on the chemistry of a stream in northeastern Georgia. Water Air Soil Pollut. 39:275–291.

Buikema, A.L., B.I. Chevone, F.M. Hawkridge, E.L. Dropp, E.E. Mason, J.M. McManus, W.W. Parks, G.L. Pellet, D.D. Salkovitz, and P. Toczydlowski. 1985. Virginia Acid Precipitation Network Summary Report: 1982–1984. Report to State Air Pollution Control Board, Richmond, VA.

Buikema, A.L., B.I. Chevone, P. Toczydlowski, L. Monroe, J.G. Webster, and G. McCreary. 1986. Virginia Acid Precipitation Network 1985 Rain Water Analyses. Report to State Air Pollution Control Board, Richmond, VA.

Buikema, A.L., B.I. Chevone, J.A. Melendez, and G. McCreary. 1987. Virginia Acid Precipitation Network 1986 Rain Water Analyses. Report to State Air Pollution Control Board, Richmond, VA.

Buikema, A.L., B.I. Chevone, and J. Doughton. 1988. Virginia Acid Precipitation Network 1987 Rain Water Analyses. Report to State Air Pollution Control Board, Richmond, VA.

Bukaveckas, P.A. 1988. Effects of calcite treatment on primary producers in acidified Adirondack lakes. I. Response of macrophyte communities. Lake Reserv. Manage. 4:107–113.

Bukaveckas, P.A. 1989. Effects of calcite treatment on primary producers in acidified Adirondack lakes. II. Short-term response by phytoplankton communities. Can. J. Fish. Aquat. Sci. 46:352–359.

Burke, M.T. 1977. The Flora and Vegetation of the Rae Lakes Basin, Southern Sierra Nevada: An Ecological Overview. M.A. Thesis. University of California, Davis.

Burns, D.A., J.N. Galloway, and G.R. Hendrey. 1981. Acidification of surface waters in two areas of the eastern United States. Water Air Soil Pollut. 16:277–285.

Burton, T.M. 1977. Personal communication. Professor of Oceanography, Florida State University, Tallahassee, FL. Cited in ESE 1987.

Burton, T.M., R.M. Stanford, and J.W. Allan. 1985. Acidification effects on stream biota and organic matter processing. Can. J. Fish. Aquat. Sci. 42:669–675.

Butler, J.N. 1982. Carbon Dioxide Equilibria and Their Application. Addison-Wesley, Reading, MA.

Butz, C. 1940. Geology of the Appalachian Valley in Virginia. Virginia Geologic Survey Bulletin 52.

Byers, G.E. 1984. The Effects of Acid Rain on the Movement of Ions in a Typic Quartzipsamment Soil under Natural Vegetation in Florida. Ph.D. Dissertation. University of Florida, Gainesville, FL.

Bytnerowicz, A., and D.M. Olszyk. 1988. Measurement of Atmospheric Dry Deposition at Emerald Lake in Sequoia National Park. Final Report. Contract A7-32-39. California Air Resources Board, Sacramento, CA.

Cadle, S.H., J.M. Dasch, and N.E. Grossnickle. 1984a. Retention and release of chemical species by a northern Michigan snowpack. Water Air Soil Pollut. 22: 303–319.

Cadle, S.H., J.M. Dasch, and N.E. Grossnickle. 1984b. Northern Michigan snowpack–a study of acid stability and release. Atmos. Environ. 18:807–816.

Cadle, S.H., J.M. Dasch, and R.V. Kopple. 1986. Wintertime wet and dry deposition in northern Michigan. Atmos. Environ. 20:1171–1178.

Cadle, S.H., J.M. Dasch, and R.V. Kopple. 1987. Composition of snowmelt and runoff in northern Michigan. Environ. Sci. Technol. 21:295–299.

Cahill, T.A., H.J. Annegard, D. Ewell, and P.J. Feeney. 1986. Particulate Monitoring for Acid Deposition Research at Sequoia National Park, California. Final Report. Contract A4-124-32, California Air Resources Board, Sacramento, CA.

Caldwell, D.W., and P.T. Davis. 1983. The timing of alpine glaciation of Mt. Katahdin, Maine. Pages 79–86

in Proceedings of NEIGC 75th annual meeting, October, Greenville, ME (unpublished).
Caldwell, R.E., and R.W. Johnson. 1982. General Soil Map of Florida. Soil Conservation Service, U.S. Department of Agriculture, Washington, D.C.
California Air Resources Board. 1988. The Fifth Annual Report to the Governor and Legislature on the Air Resources Board's Acid Deposition Research and Monitoring Program. California Air Resources Board, Sacramento, CA.
California Cooperative Snow Survey. 1986. Water Conditions in California. Bulletin 120-86. California Department of Water Resources, Sacramento, CA.
California Department of Water Resources. 1987. California snow survey measurement schedule. California Cooperative Snow Surveys 1-33, Sacramento, CA.
Campbell, P.G.C., and P.M. Stokes. 1985. Acidification and toxicity of metals to aquatic biota. Can. J. Fish. Aquat. Sci. 42:2034–2049.
Canada/U.S. 1983. Memorandum of Intent on Transboundary Air Pollution, Impact Assessment. Working Group I, Section 3 – Aquatic Effects.
Canfield, D.E., Jr. 1981. Chemical and Trophic State Characteristics of Florida Lakes in Relation to Regional Geology. Final Report. Cooperative Fish and Wildlife Research Unit, University of Florida, Gainesville, FL.
Canfield, D.E., Jr. 1983. Sensitivity of Florida lakes to acidic precipitation. Water Resour. Res. 19:833–839.
Canfield, D.E., Jr., H.L. Schramm, J.V. Shireman, and W.T. Haller. 1983. Sensitivity of Florida lakes to acid deposition. Pages 283–306 in A.E.S. Green and W. H. Smith, eds. Acid Deposition Causes and Effects: A State Assessment Model. Government Institutes, Rockville, MD.
Canfield, D.E., Jr., M.J. Maceina, F.G. Nordlie, and J. V. Shireman. 1985. Plasma osmotic and electrolyte concentrations of largemouth bass from some acidic Florida lakes. Trans. Am. Fish. Soc. 114:423–429.
Carignan, R. 1985. Quantitative importance of alkalinity flux from the sediments of acidic lakes. Nature 317:158–160.
Carignan, R. 1988. Seasonal dynamics of sulfate and hydrogen sulfide near the sediment-water interface of an oligotrophic acid lake. Verh. Int. Verein. Limnol. 23:106–115.
Carignan, R., and A. Tessier. 1985. Zinc deposition in acid lakes: The role of diffusion. Science 228:1524–1526.
Carignan, R., and A. Tessier. 1988. The co-diagenesis of sulfur and iron in acid lake sediments of southwestern Quebec. Geochim. Cosmochim. Acta 52:1179–1188.
Carlisle, V.W., R.E. Caldwell, F. Sodek, L.C. Hammond, F.G. Calhoun, M.A. Granger, and H.L. Breland. 1978. Characterization Data for Selected Florida Soils. Soil Science Research Report No. 78-1.
Castelle, A.J. 1986. Carbon Dioxide Dynamics and Base Cation Leaching from Acid Forest Soils in the White Oak Run Watershed, Shenandoah National Park, Virginia. M.S. Thesis. Department of Environmental Sciences, University of Virginia, Charlottesville.
Catling, P.M., B. Freedman, C. Stewart, J.J. Kerekes, and L.P. Lefkovitch. 1986. Aquatic plants of acid lakes in Kejimkujik National Park, Nova Scotia; floristic composition and relation to water chemistry. Can. J. Bot. 64:724–729.
Ceraso, J., C.B. Epstein, S.L. Clarke, and M. Oppenheimer. 1986. New York City's Water Supply: Acid Deposition, Inorganic Pollution, and the Catskill Reservoirs. Environmental Defense Fund, New York.
Chabot, B.F., and W.D. Billings. 1972. Origin and ecology of the Sierran alpine flora and vegetation. Ecol. Monogr. 42:163–199.
Chapman, C.A. 1970. The Geology of Acadia National Park. The Chatham Press. Old Greenwich, CT.
Chappell, W.R., R.R. Meglen, G.A. Swanson, L.A. Taylor, R.J. Sistko, R.B. McNelly, and R.W. Klusman. 1985. Pages 1–70 in Acidification Status of Colorado Lakes. Part I: Chemical Classification. University of Colorado, Denver, CO.
Chappell, W.R., R.R. Meglen, G.A. Swanson, L.A. Taylor, R.J. Sistko, R.B. McNelly, E.L. Hartman, M.L. Rottman, R.W. Klusman, and T.D. Margulies. 1986. Pages 1–96 in Acidification Status of Colorado Lakes. Part II: Chemical Classification. University of Colorado, Denver, CO.
Charles, D.F. 1984. Recent pH history of Big Moose Lake (Adirondack Mountains, New York, USA) inferred from sediment diatom assemblages. Verh. Int. Verein. Limnol. 22:559–566.
Charles, D.F. 1985. Relationships between surface sediment diatom assemblages and lakewater characteristics in Adirondack lakes. Ecology 66:994–1011.
Charles, D.F., and S.A. Norton. 1986. Paleolimnological evidence for trends in atmospheric deposition of acids and metals. Pages 335–431 in Acid Deposition: Long-term Trends. Committee on Monitoring and Assessment of Trends in Acid Deposition. National Academy Press, Washington, D.C.
Charles, D.F., and J.P. Smol. 1988. New methods for using diatoms and chrysophytes to infer past pH of low-alkalinity lakes. Limnol. Oceanogr. 33:1451–1462.
Charles, D.F., and J.P. Smol. (In press) The PIRLA-II project: Regional assessment of lake acidification trends. Verh. Int. Verein. Limnol.
Charles, D.F., and D.R. Whitehead. 1986. The PIRLA project: Paleoecological investigation of recent lake acidification. Hydrobiologia 143:13–20.

Charles, D.F., D.R. Whitehead, D.S. Anderson, R.W. Bienert, Jr., K.E. Camburn, R.B. Cook, T.L. Crisman, R.B. Davis, B.D. Fry, R.A. Hites, J.S. Kahl, J.C. Kingston, R.G. Kreis, Jr., M.J. Mitchell, S.A. Norton, L.A. Roll, J.P. Smol, P.R. Sweets, A.J. Uutala, J.R. White, M.C. Whiting, and R.J. Wise. 1986. The PIRLA project (Paleoecological Investigation of Recent Lake Acidification): Preliminary results for the Adirondacks, New England, northern Great Lakes states, and northern Florida. Water Air Soil Pollut. 30:355–365.

Charles, D.F., D.R. Whitehead, D.R. Engstrom, B.D. Fry, R.A. Hites, S.A. Norton, J.W. Owen, L.A. Roll, S.C. Schindler, J.P. Smol, A.J. Uutala, J.R. White, and R.J. Wise. 1987. Paleolimnological evidence for recent acidification of Big Moose Lake, Adirondack Mountains, NY (U.S.A.). Biogeochemistry 3:267–296.

Charles, D.F., R.W. Battarbee, I. Renberg, H. van Dam, and J.P. Smol. 1989. Paleoecological analysis of lake acidification trends in North America and Europe using diatoms and chrysophytes. Pages 207–276 in S.A. Norton, S.E. Lindberg, and A.L. Page, eds. Acid Precipitation, Volume 4. Soils, Aquatic Processes, and Lake Acidification. Advances in Environmental Sciences. Springer-Verlag, New York.

Charles, D.F., M.W. Binford, E.J. Furlong, R.A. Hites, M.J. Mitchell, S.A. Norton, F. Oldfield, M.J. Paterson, J.P. Smol, A.J. Uutala, J.R. White, D.R. Whitehead, and R.J. Wise. 1990a. Paleoecological investigation of recent lake acidification in the Adirondack Mountains, New York. J. Paleolimnol.

Charles, D.F., S.S. Dixit, B.F. Cumming, and J.P. Smol. 1990b. Variability in diatom and chrysophyte assemblages and inferred pH: Paleolimnological studies of Big Moose L., N.Y.J. Paleolimnol.

Charlson, R.J., and H. Rodhe. 1982. Factors controlling the acidity of natural rainwater. Nature 295:683–685.

Chen, C.W., R.J.M. Hudson, S.A. Gherini, J.D. Dean, R.A. Goldstein. 1983a. Acid rain model: Canopy module. J. Environ. Engin. 109:585–603.

Chen, C.W., S.A. Gherini, R.J.M. Hudson, and J.D. Dean. 1983b. The Integrated Lake-Watershed Acidification Study. Volume 1. Model Principles and Application Procedures. EA-3221. Electric Power Research Institute, Palo Alto, CA.

Chen, C.W., S.A. Gherini, N.E. Peters, P.S. Murdoch, R.M. Newton, and R.A. Goldstein. 1984a. Hydrologic analyses of acidic and alkaline lakes. Water Resour. Res. 20:1875–1882.

Chen, C.W., S.A. Gherini, J.D. Dean, R.J.M. Hudson, and R.A. Goldstein. 1984b. Development and calibration of the Integrated Lake-Watershed Acidification Study model. Pages 175–203 in J.L. Schnoor, ed. Modeling of Total Acid Precipitation Impacts. Butterworth Publishers, Stoneham, MA.

Chen, C.W., S.A. Gherini, and R.A. Goldstein. 1984c. Modeling the lake acidification process. Pages 5:1–5:43 in M.J. Wood, ed. Ecological Effects of Acid Precipitation. EA-79-6-LD. Electric Power Research Institute, Palo Alto, CA.

Chen, C.W., S.A. Gherini, R.K. Munson, L. Gomez, and C. Donkers. 1988. Sensitivity of Meander Lake to acid deposition. J. Environ. Engin. 114:1200–1216.

Christensen, S.W., J.E. Breck, and W. Van Winkle. 1988. Predicting acidification effects on fish populations using laboratory data and field information. Environ. Toxicol. Chem. 7:735–747.

Christophersen, N., and R.F. Wright. 1981. Sulfate flux and a model for sulfate concentrations in streamwater at Birkenes, a small forested catchment in southernmost Norway. Water Resour. Res. 17:377–389.

Chronic, H. 1986. Pages of Stone, Geology of Western National Parks and Monuments. The Mountaineers, Seattle, WA.

Church, M.R. 1983. Analysis of trends based on historic measurements of surface water quality. Pages 4–53 to 4–89 in A.P. Altshuller and R.A. Linthurst, eds. Acidic Deposition Phenomenon and Its Effects: Critical Assessment Review Papers. Volume II. Effects Sciences. EPA/600/8-83/016a. U.S. Environmental Protection Agency, Washington, D.C.

Church, M.R., and J.N. Galloway. 1984. Application of Henriksen's 'acidification indicator' and 'predictor nomograph' to two Adirondack lakes. Water Air Soil Pollut. 22:111–120.

Church, M.R., P.W. Shaffer, K.N. Eshleman, and B.P. Rochelle. 1990. Potential effects of sulphur deposition on stream chemistry in the Southern Blue Ridge Province. Water Air Soil Pollut. 50:39–48.

Claassen, H.C., M.M. Reddy, and A.F. White. 1983. Relationship between precipitation and vadose-zone chemistry in a high-altitude watershed in Colorado. CONF-8308106. Proceedings of the Fifth Annual Participants' Information Meeting. DOE Low-level Waste Manage Program. Denver, CO.

Clark, W.E., R.H. Musgrove, C.G. Menke, and J.W. Cagle, Jr. 1962. Interim Report on the Water Resources of Alachua, Bradford, Clay, and Union Counties. Circular No. 36. Florida Bureau of Geology Information, Tallahassee, FL.

Clark, W.E., R.H. Musgrove, C.G. Menke, and J.W. Cagle, Jr. 1964. Water Resources of Alachua, Bradford, Clay, and Union Counties. Report of Investigation No. 35. Florida Bureau of Geology Information, Tallahassee, FL.

Clayton, J.L. 1988. Some observations on the stoichiometry of feldspar hydrolysis in granitic soil. J. Environ. Qual. 17:153–157.

Cleveland, L., E. Little, S. Hamilton, D. Buckler, and

J. Hunn. 1986. Interactive toxicity of aluminum and acidity to early life stages of brook trout. Trans. Am. Fish. Soc. 115:610–620.

Clewell, A.F. 1981. Natural Setting and Vegetation of the Florida Panhandle: An Account of the Environments and Plant Communities of Northern Florida Wesst of the Suwannee River. Volume 1. Report to the U.S. Army Corps of Engineers, Mobile, AL.

Clow, D. 1987. Geologic Controls on the Neutralization of Acid Deposition and on the Chemical Evolution of Surface and Ground Waters in the Emerald Lake Watershed, Sequoia National Park, California. M.S. Thesis. California State University, Fresno, CA.

Coe, J.M., and S.E. Lindberg. 1987. The morphology and size distribution of atmospheric particles deposited on foliage and inert surfaces. J. Air Poll. Control Assoc. 37:237–243.

Colbeck, S.C. 1981. A simulation of the enrichment of atmospheric pollutants in snow cover runoff. Water Resour. Res. 17:1383–1388.

Cole, K. 1983. Late Pleistocene vegetation of Kings Canyon, Sierra Nevada, California. Quat. Res. 19: 117–129.

Collins, N.C., A.P. Zimmerman, and R. Knoechel. 1981. Comparisons of benthic infauna and epifauna in acidified and nonacidified Ontario lakes. Pages 35–48 in R. Singer, ed. Effects of Acidic Precipitation on Benthos. North American Benthological Society, Hamilton, NY.

Colquhoun, J.R., J. Symula, M. Pfeiffer, and J. Feuer. 1981. Preliminary Report of Stream Sampling for Acidification Studies – 1980. Technical Report 81-2. New York State Deparatment of Environmental Conservation, Albany, NY.

Colquhoun, J.R., J. Symula, J. Mellon, G. Aylesworth, and R.W. Karcher, Jr. 1983. Results (1981–1982) of Bioassays of Fish to Determine Sensitivity in Acidified Waters. Draft Analysis. New York State Department of Environmental Conservation, Division of Fish and Wildlife, Field Toxicant Research Unit, Rome, NY.

Colquhoun, J.R., W. Kretser, and M. Pfeiffer. 1984. Acidity Status Update of Lakes and Streams in New York State. WM P-83 (6/84). New York Department of Environmental Conservation, Albany, NY.

Confer, J.L., T. Kaaret, and G.E. Likens. 1983. Zooplankton diversity and biomass in recently acidified lakes. Can. J. Fish. Aquat. Sci. 40:36–42.

Conover, C.S., J.J. Geraghty, and G.G. Parker, Sr. 1984. Ground water. Pages 35–53 in E.A. Fernald and D.J. Patton, eds. Water Resources Atlas of Florida. Florida State University, Tallahassee, FL.

Conway, H.L., J. Tokos, and G.R. Hendrey. 1983. The Effect of Acidification on the Biomass and Production of Benthic Algae in Three Adirondack Mountain Lakes. Final Report to NYS-ERDA.

Cook, R.B. 1989. (In press) Controls on sulfur cycling in small lakes. In E.T. Degens, S. Kempe, A. Lein, and Y. Sorolein, eds. Interactions of Biogeochemical Cycles in Aqueous Ecosystems. University of Hamburg, FRG.

Cook, R.B., and D.W. Schindler. 1983. The biogeochemistry of sulfur in an experimentally acidified lake. Environ. Biogeochem. Ecol. Bull. (Stockholm) 35:115–127.

Cook, R.B., C.A. Kelly, D.W. Schindler, and M.A. Turner. 1986. Mechanisms of hydrogen ion neutralization in am experimentally acidified lake. Limnol. Oceanogr. 31:134–148.

Cook, R.B., C.A. Kelley, J.C. Kingston, and R.G. Kreis, Jr. 1987. Chemical limnology of soft water lakes in the Upper Midwest. Biogeochemistry 4:97– 117.

Cook, R.B., M.L. Jones, D.R. Marmorek, J.W. Elwood, J.L. Malanchuk, R.S. Turner, and J.P. Smol. 1988. The Effects of Acidic Deposition on Aquatic Resources in Canada: An Analysis of Past, Present, and Future Effects. Publication ORNL/TM- 10405. Oak Ridge National Laboratory, Environmental Sciences Division, Oak Ridge, TN.

Cook, R.B., R.G. Kreis, J.C. Kingston, K.E. Camburn, S.A. Norton, M.J. Mitchell, B.D. Fry, and L.C. Shane. 1990. Paleolimnology of McNearney Lake: An acidic lake in northern Michigan. J. Paleolimnol. 3: 13–34.

Cooper, E.L., and C.C. Wagner. 1973. The Effects of Acid Mine Drainage on Fish Populations. Fish and Food Organisms in Acid Mine Waters of Pennsylvania. EPA-R-3-73-032. U.S. Environmental Protection Agency, Washington, D.C.

Cooper, S.D., T.M. Jenkins, and C. Soiseth. 1988a. Integrated Watershed Study: An Investigation of Fish and Amphibian Populations in the Vicinity of the Emerald Lake Basin, Sequoia National Park. Final Report. Contract A4-122-32. California Air Resources Board, Sacramento, CA.

Cooper, S.D., K. Kratz, R.W. Holmes, and J.M. Melack. 1988b. An Integrated Watershed Study: An Investigation of the Biota in the Emerald Lake System and Stream Channel Experiments. Final Report. Contract A5-139-32. California Air Resources Board, Sacramento, CA.

Cosby, B.J., R.F. Wright, G.M. Hornberger, and J.N. Galloway. 1985a. Modeling the effects of acid deposition: Assessment of a lumped parameter model of soil water and streamwater chemistry. Water Resour. Res. 21:51–63.

Cosby, B.J., R.F. Wright, G.M. Hornberger, and J.N. Galloway. 1985b. Modeling the effects of acid deposition: Estimation of long-term water quality responses in a small forested catchment. Water Resour. Res. 21: 1591–1601.

Cosby, B.J., G.M. Hornberger, J.N. Galloway, and R.F. Wright. 1985c. Time scales of catchment acidification: A quantitative model for estimating freshwater acidification. Environ. Sci. Technol. 19: 1144–1149.

Cosby, B.J., G.M. Hornberger, R.F. Wright, and J.N. Galloway. 1986. Modelling the effects of acid deposition: Control of long-term sulfate dynamics by soil sulfate adsorption. Water Resour. Res. 22:1283–1291.

Cougan, K.A., D.W. Sutton, D.V. Peck, V. Miller, and J. E. Pollard. 1988. National Surface Water Survey: National Stream Survey (Middle Atlantic Phase I and Southeast Screening Surveys). Quality Assurance Report. EPA/600/4-88/018. U.S. Environmental Protection Agency, Las Vegas, NV.

Cowing, D.J., and M. Scott. 1975. Limnological Data Report for the Maine Department of Protection–USGS Cooperative Lakes Studies Report. Open File Report. U.S. Geological Survey, Augusta, ME.

Cowing, D.J., and M. Scott. 1976. Limnological Data Report for the Maine Department of Protection–USGS Cooperative Lakes Studies Report. Open File Report. U.S. Geological Survey, Augusta, ME.

Cowing, D.J., and M. Scott. 1977. Limnological Data Report for the Maine Department of Protection–USGS Cooperative Lakes Studies Report. Open File Report. U.S. Geological Survey, Augusta, ME.

Cozzarelli, I.M. 1986. A Two Water Model of Aluminum Mobilization and Immobilization in a Natural Soil System. M.S. Thesis. Department of Environmental Sciences, University of Virginia, Charlottesville, VA.

Craft, J.T. 1976. Pleistocene Local Glaciation in the Adirondack Mountains, New York. Ph.D. Dissertation. University of Western Ontario, London, Ontario, Canada.

Creager, C.S. 1985. Characterization of the Influence of Soil Particulates on Precipitation Chemistry at Five Sites in the Western United States. Report No. 0785-224-F. Kilkelly Environmental Associates, Raleigh, NC.

Crisman, T.L. 1984. Temperature and subtropical biotic responses to lake acidity: A preliminary assessment of the importance of direct and indirect pH effects. Pages 139–157 in M.E. Vittes and H. Kennedy, eds. Proceedings of Conference on Acid Rain: Impact on Florida and the Southeast. University of Central Florida, Orlando, FL.

Crisman, T.L., R.L. Schulze, L. Brezonik, and S. Bloom. 1980. Acid precipitation: The biotic response in Florida lakes. Pages 296–297 in D. Drabløs and A. Tollan, eds. Ecological Impact of Acid Precipitation. Proceedings of an International Conference, Sandefjord, Norway. SNSF Project, Oslo, Norway.

Crisman, T.L., R.W. Bienert, J.A. Foran, M.A. Gunn, P. Scheuerman, N. Gourlie, R.A. Garren, M.W. Binford, R.W. Ogburn, and P.L. Brezonik. 1982. Current and Projected Effects of Acid Precipitation on the Biota of a Softwater Florida Lake. NCSU Grant APP-007-02-1980. Annual Report to U.S. Environmental Protection Agency, Washington, D.C.

Crisman, T.L., C.L. Clarkson, A.E. Kellar, R.A. Garren, and R.W. Bienert. 1986. A preliminary assessment of the importance of littoral and benthic autotrophic communities in acidic lakes. Pages 17–27 in B.G. Isom, S.D. Dennis, and J.M. Bates, eds. Impact of Acid Rain and Deposition on Aquatic Biological Systems. ASTM STP 928. American Society for Testing and Materials, Philadelphia, PA.

Cronan, C.S. 1984. Vegetation and soil chemistry of the ILWAS watersheds. The Integrated Lake-Watershed Acidification Study, Vol. 4: Summary of Major Results. EPRI Final Report. EA-3221. Electric Power Research Institute, Palo Alto, CA.

Cronan, C.S. 1985. Biogeochemical influence of vegetation and soils in the ILWAS watersheds. Water Air Soil Pollut. 26:354–371.

Cronan, C.S., and W.A. Reiners. 1983. Canopy processing of acidic precipitation by conifers and hardwood forests in New England. Oecologia (Berlin) 5:216–223.

Cronan, C.S., and C.L. Schofield. 1979. Aluminum leaching response to acid precipitation: Effects on high-elevation watersheds in the Northeast. Science 204:304–306.

Cronan, C.S., W.A. Reiners, R.C. Reynolds, Jr., and G. E. Lang. 1978. Forest floor leaching: Contributions from mineral, organic, and carbonic acids in New Hampshire subalpine forests. Science 200:309– 311.

Cronan, C.S., W.J. Walker, and P.R. Bloom. 1986. Predicting aqueous aluminum concentrations in natural waters. Nature 324:140–143.

Cronan, C.S., J.C. Conlan, and S. Skibinski. 1987. Forest vegetation in relation to surface water chemistry in the North Branch of the Moose River, Adirondack Park, New York. Biogeochemistry 3:121–128.

Cronan, C.S., C.T. Driscoll, R.M. Newton, J.M. Kelly, C.L. Schofield, R.J. Bartlett, and R. April. 1990. A comparative analysis of aluminum biogeochemistry in a northeastern and a southeastern forested watershed. Water Resour. Res. 26:1413–1430.

Culp, T.R. 1980. Aquatic Biology of Schoharie Reservoir and Associated Water Bodies: Prattsville Pumped Storage Project. Report prepared by Ichthyological Associated, Inc., Ithaca, NY.

Cummins, C.P. 1986. Effects of aluminium and low pH on growth and development in Rana temporaria tadpoles. Oecologia 69:248–252.

Cunningham, R.L., and E.J. Ciolkosz. 1984. Pages 1–47 in Soils of the Northeastern United States. Bulletin

848. The Pennsylvania State University Agricultural Experiment Station, University Park, PA.

Curry, R.R. 1969. Holocene climatic and glacial history of the central Sierra Nevada, California. Geol. Soc. Am. Special Paper 123:1–47.

Cusimano, R.F., J.P. Baker, W.J. Warren-Hicks, V. Lesser, W.W. Taylor, M.C. Frabrizio, D.B. Hayes, and B.P. Baldigo. 1990. Fish Communities in Lakes in Subregion 2B (Upper Peninsula of Michigan) in Relation to Lake Acidity. EPA/600/3-89/021. U.S. EPA Environmental Research Laboratory, Corvallis, OR.

Darcy, H. 1856. Les Fontaines Publiques De La Ville De Dijon. V. Delmont, Paris.

Dasch, J.M. 1983. A comparison of surrogate surfaces for dry deposition collection. Pages 883–900 in H.R. Pruppacher, R.G. Semonin, and W.G. N. Slinn, eds. Precipitation, Scavenging, Dry Deposition, and Resuspension, Vol. 2. Elsevier Science Publishing Co., New York.

David, M.B., and M.J. Mitchell. 1985. Sulfur constituents and cycling in waters, seston, and sediments of an oligotrophic lake. Limnol. Oceanogr. 30:1196–1207.

Davidson, C.I., S.E. Lindberg, J.A. Schmidt, L.G. Cartwright, and L.R. Landis. 1985. Dry deposition of sulfate onto surrogate surfaces. J. Geophys. Res. 90: 2123–2130.

Davis, G.D. 1988. Fulfilling the Promise of the Adirondack Park. Vol. I. Biological Diversity: Saving All the Pieces. The Adirondack Council, Elizabethtown, NY.

Davis, G.F., J.J. Whipple, S.A. Gherini, C.W. Chen, R.A. Goldstein, R.W.H. Chan, and R.K. Munson. 1987. Big Moose Basin: Simulation of response to acidic deposition. Biogeochemistry 3:141–161.

Davis, O.K., R.S. Anderson, P.L. Fall, M.K. O'Rourke, and R.S. Thompson. 1985a. Palynological evidence for early Holocene aridity in the southern Sierra Nevada, California. Quat. Res. 24:322–332.

Davis, R.B. 1987. Paleolimnological diatom studies of acidification of lakes by acid rain: An application of quaternary science. Quat. Sci. Rev. 6:147–163.

Davis, R.B., and D.S. Anderson. 1985. Methods of pH calibration of sedimentary diatom remains for reconstructing history of pH in lakes. Hydrobiologia 120: 69–87.

Davis, R.B., M.O. Smith, J.H. Bailey, and S.A. Norton. 1978. Acidification of Maine (U.S.A.) lakes by acidic precipitation. Verh. Int. Verein. Limnol. 20:532–537.

Davis, R.B., S.A. Norton, C.T. Hess, and D.F. Brakke. 1983. Paleolimnological reconstruction of the effects of atmospheric deposition of acids and heavy metals on the chemistry and biology of lakes in New England and Norway. Hydrobiologia 103:113–123.

Davis, R.B., D.S. Anderson, and F. Berge. 1985b. Palaeolimnological evidence that lake acidification is accompanied by loss of organic matter. Nature 316: 436–438.

Davis, R.B., D.S. Anderson, D.F. Charles, and J.N. Galloway. 1988. Two-hundred-year pH history of Woods, Sagamore, and Panther lakes in the Adirondack Mountains, New York State. Pages 89–111 in W.J. Adams, G.A. Chapman, and W.G. Landis, eds. Aquatic Toxicology and Hazard Assessment: 10th Volume, ASTM STP971. American Society for Testing and Materials, Philadelphia, PA.

Deevey, E.S., Jr. 1989. Estimation of downward leakage from Florida lakes. Limnol. Oceanogr. 33:1308–1320.

Deffeyes, K.S. 1965. Carbonate equilibria: A graphic and algebraic approach. Limnol. Oceanogr. 10:412–426.

Del Prete, A., and J.N. Galloway. 1983. Temporal Trends in the pH of Woods, Sagamore, and Panther Lakes Determined by Analysis of Diatom Populations. The Integrated Lake-Watershed Acidification Study: Proceedings of the ILWAS Annual Review Conference. Electric Power Research Institute, Palo Alto, CA.

Dermott, R.M. 1985. Benthic fauna in a series of lakes displaying a gradient of pH. Hydrobiologia 128:31–38.

Detenbeck, N.E. 1987. Nutrient Cycling and the Growth of Benthic Algae in Experimentally Acidified Little Rock Lake, Wisconsin. Ph.D. Dissertation. University of Minnesota, Minneapolis, MN.

Dethier, D.P. 1979. Atmospheric contribution to stream water chemistry in the north Cascade Range, Washington. Water Resour. Res. 15:787–794.

Dickinson, J.C., Jr. 1948. An ecological reconnaissance of the biota of some ponds and ditches in northern Florida. Q.J. Fla. Acad. Sci. 11:1–28.

Dickson, R.R. 1960. Climates of the United States: Tennessee. Climatology of the U.S., No. 60-40. Weather Bureau, U.S. Department of Commerce, Washington, D.C.

Dickson, W. 1980. Properties of acidified water. Pages 75–83 in D. Drabløs and A. Tollan, eds. Ecological Impact of Acid Precipitation. Proceedings of an International Conference, Sandefjord, Norway. SNSF Project, Oslo, Norway.

Dietrich, R.V. 1970. Geology and Virginia. University of Virginia, Charlottesville, VA.

Dillon, D.T., D.S. Jeffries, and N.D. Yan. 1980. Some aspects of acidification in southern Ontario. Pages 212–213 in D. Drabls and A. Tollan, eds. Ecological Impact of Acid Precipitation. Proceedings of an International Conference, Sandefjord, Norway. SNSF Project, Oslo, Norway.

Dillon, P.J., N.D. Yan, and H.H., Harvey. 1984. Acidic deposition: Effects on aquatic ecosystems. Critical Rev. Environ. Cont. 13:167–194.

Dillon, P.J., R.A. Reid, and R. Girard. 1986. Changes in the chemistry of lakes near Sudbury, Ontario, following reduction of SO_2 emission. Water Air Soil Pollut. 31:59–65.

Dillon, P.J., R.A. Reid, and E. de Grosbois. 1987. The rate of acidification of aquatic ecosystems in Ontario, Canada. Nature 329:45–48.

DiNunzio, M.G. 1984. Adirondack Wildlife. Brodock Press, Inc., Utica, NY.

Dise, N.B. 1984. A Synoptic Survey of Headwater Streams in Shenandoah National Park, Virginia, to Evaluate Sensitivity to Acidification by Acid Deposition. M.S. Thesis. Department of Environmental Sciences, University of Virginia, Charlottesville, VA.

Dixit, S.S., A.S. Dixit, and R.D. Evans. 1987. Paleolimnological evidence of recent acidification in two Sudbury (Canada) lakes. Sci. Total Environ. 67:53–67.

Donkers, C., S. Gherini, C. Chen, and R. Munson. 1986. Hydrologic Characteristics of Crum Lake. Northern States Power Company, Minneapolis, MN (unpublished).

Dorn, R.I., B.D. Turrin, A.J.T. Jull, T.W. Linick, and D. J. Donahue. 1987. Radiocarbon and cation-ratio ages for rock varnish on Tioga and Tahoe marainal boulders of Pine Creek, eastern Sierra Nevada, California, and their paleoclimatic implications. Quat. Res. 28:38–49.

Dougan, W.K., and A.L. Wilson. 1974. The absorptiomnetric determination of aluminum in water. Analyst 99:413–430.

Dozier, J., J.M. Melack, and D. Marks. 1987. Snow Deposition, Melt, Runoff, and Chemistry in a Small Alpine Watershed, Emerald Lake Basin, Sequoia National Park. Final Report. Contract A3-106-32, California Air Resources Board, Sacramento, CA.

Dozier, J., J.M. Melack, R. Kattelmann, K. Elder, M. Williams, S. Peterson, and D. Marks. 1989. Snow, Snowmelt, Rain, Runoff, and Chemistry in a Sierra Nevada Watershed. Final Report. Contract A4-147-32, California Air Resources Board, Sacramento, CA.

Drabløs, D., and I.H. Sevaldrud. 1980. Lake acidification, fish damage, and utilization of outfields. A comparative survey of six highland areas, southeastern Norway. Pages 354–355 in D. Drabløs and A. Tollan, eds. Ecological Impact of Acid Precipitation. Proceedings of an International Conference, Sandefjord, Norway. SNSF Project, Oslo, Norway.

Draper, N.R., and H. Smith. 1981. Applied Regression Analysis. John Wiley and Sons, New York.

Drever, J.I. 1982. The Geochemistry of Natural Waters. Prentice-Hall, Englewood Cliff, NJ.

Drever, J.I., and D.R. Hurcomb. 1986. Neutralization of atmospheric acidity in an alpine drainage basin by chemical weathering in the north Cascade Mountains. Geology 14:221–224.

Driscoll, C.T. 1984. A procedure for the fractionation of aqueous aluminum in dilute acidic waters. Int. J. Environ. Anal. Chem. 16:267–283.

Driscoll, C.T., and J.J. Bisogni. 1984. Weak acid/base systems in dilute acidified lakes and streams of the Adirondack region of New York State. Pages 53–72 in J.L. Schnoor, ed. Modeling of Total Acidification Impacts. Butterworth Publishers, Boston, MA.

Driscoll, C.T., and R.M. Newton. 1985. Chemical characteristics of Adirondack lakes. Environ. Sci. Technol. 19:1018–1024.

Driscoll, C.T., and G.C. Schafran. 1984. Short-term changes in the base neutralizing capacity of an acid Adirondack lake, New York. Nature 310:308–310.

Driscoll, C.T., and W.D. Schecher. 1988. Aluminum and its role in biology. Pages 59–122 in H. Sigel and A. Sigel, eds. Metal Ions in Biological Systems. Marcel Dekker, New York.

Driscoll, C.T., J.P. Baker, J.J. Bisogni, and C.L. Schofield. 1980. Effect of aluminum speciation on fish in dilute acidified waters. Nature 284:161–164.

Driscoll, C.T., J.P. Baker, J.J. Bisogni, and C.L. Schofield. 1984. Aluminum speciation and equilibria in dilute acidic surface waters of the Adirondack region of New York State. Pages 55–74 in J.I. Teasly, ed. Geological Aspects of Acid Deposition. 1984 Acid Precipitation Series, Vol. 7. Butterworth Publishers, Boston, MA.

Driscoll, C.T., N. van Breeman, and J. Mulder. 1985. Aluminum chemistry in a forested spodosol. Soil Sci. Soc. Am. J. 49:437.

Driscoll, C.T., B.J. Wyskowski, C.C. Cosentini, and M. E. Smith. 1987a. Processes regulating temporal and longitudinal variations in the chemistry of a low-order woodland stream in the Adirondack Mountains region of New York. Biogeochemistry 3:225–241.

Driscoll, C.T., C.P. Yatsko, and F.J. Unangst. 1987b. Longitudinal and temporal trends in the water chemistry of the North Branch of the Moose River. Biogeochemistry 3:37–61.

Driscoll, C.T., B.J. Wyskowski, P. DeStaffan, and R.M. Newton. 1989a. Chemistry and transfer of aluminum in the Adirondack region of New York, USA. Pages 83–105 in T. Lewis, ed. The Environmental Chemistry and Toxicology of Aluminum. Lewis Publisher, New York.

Driscoll, C.T., W.A. Ayling, G.F. Fordham, and L.M. Oliver. 1989b. Chemical response of lakes treated with $CaCO_3$ to reacidification. Can. J. Fish. Aquat. Sci. 46:258–267.

Driscoll, C.T., G.E. Likens, L.O. Hedin, J.S. Eaton, and F.H. Bormann. 1989c. Changes in the chemistry of surface waters: 25-year results at the Hubbard Brook Experimental Forest. Environ. Sci. Technol. 23:137–143.

Driscoll, C.T., R.D. Fuller, and W.D. Schecher. 1989d. The role of organic acids in the acidification of surface waters in the eastern U.S. Water Air Soil Pollut. 43: 21–40.

Drousé, S.K. 1987. Evaluation of Quality Assurance and Quality Control Sample Data for the National Stream Survey Phase I–Pilot Survey. EPA/600/8-87/057.

Lockheed Engineering and Management Services Company, Inc., Las Vegas, NV.
Duever, L.C. 1981. The Florida scrub: Misunderstood habitat. ENFO 81:1-3.
Duncan, L.C., and W. Ausserer. 1984. Acid Precipitation in the Washington Cascades Phase II. Project A-119-WASH, U.S. Department of Interior Grant No. G-876, State of Washington Water Research Center, Washington State University, Pullman, WA.
Duncan, L.C., and K. Lillquist. 1986. The Chemical and Physical Characteristics of Three Cascade Mountain Lakes. Final Report, Contract #C0085152, Washington State Department of Ecology, Olympia, WA.
Duncan, L.C., and P. McGinty. 1988. Assessment of the composition of winter precipitation at Snoqualmie and Stevens Pass sites, winter 1986–87. Final Report, Contract #C0087021, Washington State Department of Ecology, Olympia, WA.
Duncan, L.C., E.B. Welch, and W. Ausserer. 1986. The composition of precipitation at Snoqualmie Pass and Stevens Pass in the central Cascades of Washington State. Water Air Soil Pollut. 30:217–229.
Edgerton, E.S., and P.L. Brezonik. 1981. Acid Rainfall in Florida: Results of a Monitoring Network for 1980. Department of Environmental Engineering Sciences, University of Florida, Gainesville, FL.
Edmunds, W.J., D.D. Rector, N.O. Wilson, and T.L. Arnold. 1986. Properties, Classification, and Upland Oak Site Quality for Residual Soils Derived from Shales, Phyllites, Siltstones, and Sandstones in Southwestern Virginia. Bulletin 86-5. July. Virginia Agricultural Experiment Station.
Effler, S.W., and E.M. Owens. 1985. Impact of lake acidification on stratification. J. Environ. Eng. ASCE 111:822-832.
Effler, S., G. Schafran, and C. Driscoll. 1985. Partitioning light attenuation in an acidic lake. Can J. Fish. Aquat. Sci. 42:1707-1711.
Eilers, J.M. 1984. Distribution and Inventory of Softwater Seepage Lakes in Wisconsin. ERL-D Report CR809484. U.S. Environmental Protection Agency, Duluth, MN.
Eilers, J.M., and J.A. Bernert. In review. An Evaluation of National Lake Survey Boundaries for the Study of Lake Acidification. U.S. Environmental Protection Agency, Environmental Research Laboratory, Corvallis, OR.
Eilers, J.M., G.E. Glass, K.E. Webster, and J.A. Rogalla. 1983. Hydrologic control of lake susceptibility to acidification. Can. J. Fish Aquat. Sci. 40:1896–1904.
Eilers, J.M., G.J. Lien, and R.G. Berg. 1984. Aquatic organisms in acidic environments: A literature review. Wisc. Nat. Resour. Tech. Bull. 150:1–18.
Eilers, J.M., P. Kanciruk, R.A. McCord, W.S. Overton, L. Hook, D.J. Blick, D.F. Brakke, P.E. Kellar, M.S. DeHaan, M.E. Silverstein, and D.H. Landers. 1987a. Characteristics of Lakes in the Western United States. Volume II, Data Compendium for Selected Physical and Chemical Variables. EPA/600/3- 86/054b, U.S. Environmental Protection Agency, Washington D.C.
Eilers, J.M., D.H. Landers, D.F. Brakke, R.A. Linthurst, and M.S. DeHaan. 1987b. Factors contributing to differences in acid neutralizing capacity among lakes in the western United States. Pages 403–418 in R.F. Dworsky, ed. Water Resources Related to Mining and Energy—Preparing for the Future. American Water Resources Association, Bethesda, MD.
Eilers, J.M., D.F. Brakke, D.H. Landers, and W.S. Overton. 1987c. Chemistry of lakes in designated wilderness areas in the western United States. Environ. Monit. Assess. 12:13–21.
Eilers, J.M., D.H. Landers, and D.F. Brakke. 1988a. Chemcal and physical characteristics of lakes in the southeastern United States. Environ. Sci. Technol. 22: 72–177.
Eilers, J.M., D.F. Brakke, and D.H. Landers. 1988b. Chemical and physical characteristics of lakes in the Upper Midwest, United States. Environ. Sci. Technol. 22:164–172.
Eilers, J.M., D.F. Brakke, D.H. Landers, and P.E. Kellar. 1988c. Characteristics of lakes in mountainous areas of the western United States. Verh. Int. Verein. Limnol. 23:144–151.
Eilers, J.M., G.E. Glass, A.K. Pollack, and J.A. Sorenson. 1989. Changes in conductivity, alkalinity, calcium, and pH during a fifty-year period in selected northern Wisconsin lakes. Can J. Fish. Aquat. Sci. 46:929–1944.
Eilers, J.M., T.J. Sullivan, and K.C. Hurley. (1990) The most dilute lake in the world? Hydrobiologia 199:1–6.
Eisenreich, S.J., G.J. Hollad, and S. Langevin. 1978. Precipitation Chemistry and Atmospheric Deposition of Trace Elements in Northeastern Minnesota. Prepared for the Minnesota Environmental Quality Council and Minnesota State Planning Agency, St. Paul, MN.
Electric Power Research Institute (EPRI). 1984. The Integrated Lake-Watershed Acidification Study, Vol. 4: Summary of Major Results. EA-3221. Palo Alto, CA.
Ellis, B.K., J.A. Stanford, G.R. Gregory, and L.F. Marnell. 1986. Pages 1–59 in Monitoring Water Quality of Selected Lakes in Glacier National Park, Montana. Open File Report. Flathead Lake Biological Station, Polson, MT.
Elwood, J.W., and P.J. Mulholland. 1989. Effects of acidic precipitation on stream ecosystems. Pages 85–135 in D.C. Adriano and A.H. Johnson, eds. Acid Precipitation: Volume 2. Biological and Ecological Effects. Springer-Verlag, New York.
Elwood, J.W., M.A. Bogle, H.L. Boston, C.W. Boylen, C.M. Brooks, R.B. Cook, C.C. Cosentini, C.T. Driscoll, P.J. Mulholland, M.P. Osgood, A.V. Palumbo,

A.D. Rosemond, C.L. Schofield, M.E. Smith, R.R. Turner, and B.J. Wyskowski. 1985. Ecological Effects of Acidification on Low-Order Woodland Streams, with Particular Emphasis on the Chemistry and Effects of Aluminum (ALSS). EPRI Project RP2326-1. Annual Progress Report to Electric Power Research Institute, Palo Alto, CA, Period Sept. 1984–Aug. 1985. Oak Ridge National Laboratory, Oak Ridge, TN.

Engblom, E., and P. Lingdell. 1984. The mapping of short-term acidification with the help of biological pH indicators. Instit. Freshw. Res. Drottningholm 61:60–68.

English, M.C., D.S. Jeffries, N.W. Foster, R.G. Semkin, and P.W. Hazlett. 1986. A preliminary assessment of the chemical and hydrological interaction of acidic snowmelt water with the terrestrial portion of a Canadian Shield catchment. Water Air Soil Pollut. 31: 27–34.

Engstrom, D.R., and H.E. Wright, Jr. 1985. Chemical stratigraphy of lake sediments as a record of environmental change. Pages 11–67 in E.Y. Howarth and J.W.G. Lund, eds. Lake Sediments and Environmental History. Leicester University Press, Leicester, MA.

Environment Canada. 1988. Acid Rain: A National Sensitivity Assessment. One map (scale 1:7,500,000). Environmental Fact Sheet 88-1. Inland Waters and Lands Direcorate, Ottawa, Ontario.

Environmental Science and Engineering, Inc. (ESE). 1983. Florida Acid Deposition Study. Phase II Report Addendum: Acid Deposition Monitoring Program Analysis. ESE No. 81-631-600. Gainesville, FL.

Environmental Science and Engineering, Inc. (ESE). 1984. Florida Acid Deposition Study. Phase III Report, Volume I. ESE No. 82-615-0600-2110. Gainesville, FL.

Environmental Science and Engineering, Inc. (ESE). 1985. Florida Acid Deposition Study. Phase IV Report, Volume I. ESE No. 83-152-0106/0207/0307. Gainesville, FL.

Environmental Science and Engineering, Inc. (ESE). 1987. Florida Acid Deposition Study. Five-Year Data Summary. ESE No. 85-186-0106-2110. Gainesville, FL.

EPA. 1987. Major Scientific Findings on Acid Deposition. Volume II. Emissions, Deposition Monitoring, Atmospheric Processes, and Forest Effects. EPA Staff Paper. U.S. Environmental Protection Agency, Office of Research and Development, Washington, D.C.

Epstein, C.B., and M. Oppenheimer. 1987. Empirical relation between sulfur dioxide emissions and acid deposition derived fron monthly data. Nature 323: 245–247.

Eshleman, K.N. 1988. Predicting regional episodic acidification of surface waters using empirical models. Water Resour. Res. 34:1118–1126.

Eshleman, K.N., and H.F. Hemond. 1985. The role of organic acids in the acid-base status of surface waters at Bickford watershed, Massachusetts. Water Resour. Res. 21:1503–1510.

Eshleman, K.N., and P.R. Kaufmann. 1988. Assessing the regional effects of sulfur deposition on surface water shemistry: The Southern Blue Ridge. Environ. Sci. Technol. 22:685–690.

Ethridge, F.G. 1977. Petrology, transport, and environment in isochronous Upper Devonian sandstone and siltstone units, New York. J. Sediment. Petrol. 47: 63–65.

Evans, R.A. 1989. Response of limnetic insect populations of two acidic, fishless lakes to liming and brook trout (*Salvelinus fontinalis*). Can. J. Fish. Aquat. Sci. 46:342–351.

Ewell, D.M., R.G. Flocchini, and L.O. Myrup. 1989. Aerosol transport in the southern Sierra Nevada. J. Appl. Met. 28:112–129.

Farag, A.M. 1988. Effects of Acid and Aluminum on Survival, Whole Body Ion Content, and Weights of Yellowstone Cutthroat Trout (*Salmo clarki bouvieri*). M.S. Thesis. University of Wyoming, Laramie, WY.

Farmer, G.J., T.R. Goff, D. Ashfield, and H.S. Samant. 1980. Some effects of the acidification of Atlantic salmon rivers in Nova Scotia. Can. Tech. Rep. Fish. Aquat. Sci. No. 972.

Fay, J.A., D. Golomb, and S. Kumar. 1986. Modeling of the 1900–1980 trend of precipitation acidity at Hubbard Brook, New Hampshire. Atmos. Environ. 20: 1825–1828.

Feely, H.W., D.C. Bogen, S.J. Nagourney, and C.C. Torquato. 1985. Rates of dry deposition determined using wet/dry collectors. J. Geophys. Res. 90:2161– 2165.

Fenneman, N.M. 1928. Physiographic divisions of the United States. Ann. Assoc. Am. Geog. 18:261–353.

Fenneman, N.M. 1938. Physiography of Eastern United States. McGraw-Hill, New York.

Fenneman, N.M. 1946. Physical Divisions of the United States. One map (scale 1:7,500,000). U.S. Geological Survey, Reston, VA.

Fernald, E.A. 1981. Atlas of Florida. Florida State University, Tallahassee, FL.

Fernald, E.A., and D.J. Patton. 1984. Water Resources Atlas of Florida. Florida State University, Tallahassee, FL.

Fernandez, I.J. 1983. Pages 1–87 in Field Study Elements to Assess the Sensitivity of Soils to Acidic Deposition Induced Alterations in Forest Productivity. Technical Bulletin No. 404. National Council of the Paper Industry for Air and Stream Improvement.

Fernandez, I.J. 1985a. Acid deposition and forest soils: Potential impacts and sensitivity. Pages 223–239 in D.D. Adams and W. Page, eds. Acid Deposition—Environmental, Economic, and Policy Issues. Plenum Publishing Corporation, New York.

Fernandez, I.J. 1985b. Potential effects of atmospheric deposition on forest soils. Pages 237–250 in Symposium on the Effects of Air Pollutants on Forest Ecosystems. The Acid Rain Foundation, St. Paul, MN.

Fernandez, I.J. 1986. Air pollution: Synthesis of the role of major air pollutants in determining forest health and productivity. Pages 217–239 in T.C. Hennessey, ed. The Role of Stress Management in Maintaining and Increasing Forest Stand Productivity. Martinus Nijhoff, New York.

Fernandez, I.J. 1987. Pages 1–20 in Vertical Trends in the Chemistry of Forest Soil Microcosms Following Experimental Acidification. Maine Agricultural Experiment Station Technical Bulletin No. 126. University of Maine, Orono, ME.

Fernandez, I.J., and M.M. Czapowskyj. 1985. Levels of trace metals in the forest floors of low elevation, commercial spruce-fir sites in Maine. Environ. Sci. Technol. 4:1–7.

Fernandez, I.J., and M.M. Czapowskyj. 1986. Pages 1–23 in Selected Relationships for Trace Metals in Maine Low Elevation Spruce-Fir Forest Floors. Maine Agricultural Experiment Station Technical Bulletin No. 119. University of Maine, Orono, ME.

Fernandez, I.J., and S.M. Goltz. 1988. An Integrated Study of Atmospheric Deposition and Nutrient Cycling in Commercial Northeast Spruce-Fir Ecosystems. U.S. Forest Service Co-op Agreement 23-209. University of Maine, Orono, ME.

Fernandez, I.J., and P.A. Kosian. 1986. Chemical response of soil leachate to alternative approaches to experimental acidification. Comm. Soil Sci. Plant Anal. 17:953–973.

Fernandez, I.J., and P.A. Kosian. 1987. Soil air CO_2 concentrations in a forested New England spruce-fir forest. Soil Sci. Soc. Am. J. 51:261–263.

Fernandez, I.J., and R.A. Struchtemeyer. 1985. Chemical characteristics of soils under spruce-fir forests in eastern Maine. Can. J. Soil Sci. 65:61–69.

Fernandez, I.J., L. Wortman, and S.A. Norton. 1986. Composition of Precipitation at the National Atmospheric Deposition Program/National Trends Network (NADP/NTN) Site in Greenville, Maine–November 1979 to December 1982. Maine Agricultural Experiment Station Technical Bulletin No. 118. University of Maine, Orono, ME.

Feth, J.H., S.M. Rogers, and G.E. Roberson. 1964a. Chemical Composition of Snow in the Northern Sierra Nevada and Other Areas. USGS Water Supply Paper 1535-J. U.S. Geological Survey, Reston, VA.

Feth, J.H., G.E. Roberson, and W.L. Polzer. 1964b. Sources of Mineral Constituents in Water from Granitic Rocks, Sierra Nevada, California and Nevada. USGS Water Supply Paper 1535-M.U.S. Geological Survey, Reston, VA.

Fiance, S.B. 1978. Effects of pH on the biology and distribution of *Ephemerella funeralis* (Ephemeroptera). Oikos 31:332–339.

Findlay, D.L., and S.E.M. Kasian. 1986. Phytoplankton community responses to acidification of Lake 223, Experimental Lakes Area, northwestern Ontario. Water Air Soil Pollut. 30:719–726.

Fiske, R.S., C.A. Hopson, and A.C. Waters. 1964. Geologic map and section of Mount Rainier National Park Washington. Miscellaneous Investigations Series Map I-432. U.S. Geological Survey, Department of the Interior, Reston, VA.

Flocchini, R.G., T.A. Cahill, D.J. Shadoan, S.J. Lange, R.A. Eldred, P.J. Feeney, G.W. Wolfe, D.C. Simmeroth, and J.K. Suder. 1976. Monitoring California's aerosols by size and elemental composition. Environ. Sci. Technol. 10:74–82.

Florida Department of Environmental Regulation (FDER). 1984. An Analysis of Acid Deposition Issues: The Impacts of Proposed National Acid Deposition Control Legislation on Florida. Florida Department of Environmental Regulation, Tallahassee, FL.

Florida Electric Power Coordinating Group, Inc. (FCG). 1986. Florida Acid Deposition Study. Final Report: A Synthesis of the Florida Acid Deposition Study. Florida Electric Power Coordinating Group, Inc., Tampa, FL.

Fowler, D.L. 1985. Fish Assemblage Characteristics of Acid-Sensitive Streams in the Southern Appalachian Mountains. M.S. Thesis. University of Georgia, Athens, GA.

Fowler, D.L., M.J. Van Den Avyle, and M. Hudy. (1987) Fish assemblage characteristics of acid-sensitive streams in the southern Appalachian mountains. Proceedings of the Annual Conference of the Southeastern Association of Fish and Wildlife Agencies 41: 126–135.

France, R.L. 1987. Reproductive impairment of the crayfish *Orconectes virilis* in response to acidification of Lake 223. Can J. Fish. Aquat. Sci. 44:97–106.

France, R.L., and P.M. Stokes. 1987. Life stage and population variation in resistance and tolerance of *Hyalella azteca* (Amphipoda) to low pH. Can. J. Fish. Aquat. Sci. 44:1102–1111.

Francis, A.J., H. Quinby, and G.R. Hendrey. 1984. Effect of lake pH on microbial decomposition of allochthonous litter. Pages 1–22 in G.R. Hendrey, ed. Early Biotic Responses to Advancing Lake Acidification. Butterworth Publishers, Boston, MA.

Franklin, J.F., and C.T. Dyrness. 1973. Natural vegetation of Oregon and Washington. General Technical Report PNW-8. Pacific Northwest Forest and Range Experiment Station, USDA Forest Service, Portland, OR.

Frenette, J.-J., Y. Richard, and G. Moreau. 1986. Fish responses to acidity in Quebec lakes: A review. Water Air Soil Pollut. 30:461–475.

Frey, D.G., ed. 1966. Limnology in North America. University of Wisconsin Press, Madison, WI.

Friberg, F.C. C. Otto, and B.S. Svensson. 1980. Effects of acidification on the dynamics of allochthonous leaf material and benthic invertebrate communities in running waters. Pages 304–305 in D. Drabløs and A. Tollan, eds. Ecological Impact of Acid Precipitation. Proceedings of an International Conference, Sandefjord, Norway. SNSF Project, Oslo, Norway.

Frost, T.M., and P.K. Montz. 1988. Early zooplankton responses to experimental acidification in Little Rock Lake, Wisconsin, U.S.A. Verh. Internat. Verein. Limnol. 23:2279–2285.

Frost, T.M., P.J. Garrison, J.A. Perry, M.E. Sierszen, W.A. Swenson, C.J. Watras, K.E. Webster, and J.G. Eaton. 1988. Early biotic responses to the experimental acidification of Little Rock Lake, Wisconsin. University of Wisconsin, Madison, WI (unpublished report).

Fuller, S.L., and G.P. Cooper. 1946. Fish Survey Report Series. Maine Department of Inland Fish and Game, Augusta, ME.

Fuller, S.W., M.B. David, and C.T. Driscoll. 1985. Sulfate adsorption relationships in some northern forest Spodosols. J. Soil Sci. Soc. Am. 49:1034–1040.

Funk, W.H., et al. 1985. Baseline Study of Reflection Lakes, Mount Rainier National Park. Report 66. State of Washington Water Research Center, Washington State University, Pullman, WA.

Furlong, E.T., L.R. Cessar, and R.A. Hites. 1987. Accumulation of polycyclic aromatic hydrocarbons in acid sensitive lakes. Geochim. Cosmochim. Acta 51:2965–2975.

Gallant, A.L., T.R. Whittier, D.P. Larsen, J.M. Omernik, and R.T. Hughes. 1989. Regionalization as a Tool for Managing Environmental Resources. EPA/600/3-89/060. U.S. Environmental Protection Agency, Environmental Research Laboratory, Corvallis, OR.

Galloway, J.N., and E.B. Cowling. 1978. The effects of precipitation on aquatic and terrestrial ecosystems – a proposed precipitation chemistry network. J. Air Pollut. Control Assoc. 28:229–235.

Galloway, J.N., and G.E. Likens. 1979. Atmospheric enhancement of metal deposition in Adirondack lake sediments. Limnol. Oceanogr. 24:427–433.

Galloway, J.N., and D.M. Whelpdale. 1983. An atmospheric sulfur budget for eastern North America. Atmos. Environ. 14:409–417.

Galloway, J.N., C.L. Schofield, G.R. Hendrey, N.E. Peters, and A.H. Johannes. 1980. Sources of acidity in three lakes acidified during snowmelt. Pages 264–265 in D. Drabløs and A. Tollan, eds. Ecological Impact of Acid Precipitation. Proceedings of an International Conference, Sandefjord, Norway. SNSF Project, Oslo, Norway.

Galloway, J.N., G.E. Likens, W.C. Keene, and J.M. Miller. 1982. The composition of precipitation in remote areas of the world. J. Geophys. Res. 87:8771–8787.

Galloway, J.N., S.A. Norton, and M.R. Church. 1983a. Freshwater acidification from atmospheric deposition of sulfuric acid: A conceptual model. Environ. Sci. Technol. 17:541–545.

Galloway, J.N., C.L. Schofield, N.E. Peters, G.R. Hendrey, and E.R. Altwicker. 1983b. Effect of atmospheric sulfur on the composition of three Adirondack lakes. Can. J. Fish. Aquat. Sci. 40:799- 806.

Galloway, J.N., G.E. Likens, and M.E. Hawley. 1984. Acid precipitation: Natural versus anthropogenic components. Science 226:829–831.

Galloway, J.N., G.R. Hendrey, C.L. Schofield, N.E. Peters, and A.H. Johannes. 1987. Processes and causes of lake acidification during spring snowmelt in the west-central Adirondack Mountains, New York. Can. J. Fish. Aquat. Sci. 44:1595–1602.

Garland, J.A. 1981. Enrichment of sulfate in maritime aerosols. Atmos. Environ. 15:787–791.

Garrels, R.M. 1967. Genesis of some groundwaters from igneous rocks. Pages 405–420 in P.H. Abelson, ed. Researches in Geochemistry. John Wiley and Sons, New York.

Garrels, R.M., and F.T. MacKenzie. 1967. Origin of the chemical compositions of some springs and lakes. Pages 222–242 in W. Stumm, ed. Advances in Chemistry Services 67. Equilibrium Concepts in Natural Water Systems. American Chemical Society, Washington, D.C.

Garren, R.A., C.L. Clarkson, and T.L. Crisman. 1989. Chemistry of the lakes of the Ocala National Forest, Florida. Pages 12–1 to 12–29 in D.F. Charles and D.R. Whitehead, eds. Paleoecological Investigation of Recent Lake Acidification (PIRLA): Interim Report. Research Project No. 2174-10. Indiana University Foundation, Bloomington, IN.

Garrison, P.J., S.R. Greb, D.R. Knauer, D.A. Wentz, J.T. Krohelski, J.G. Bockheim, S.A. Gherini, and C.W. Chen. 1987. Application of the ILWAS model to the northern Great Lakes states. Lake Reserv. Manage. 3: 356–364.

Gathright, T.M., II. 1976. Geology of the Shenandoah National Park, Virginia. Virginia Division of Mineral Resources Bulletin. Charlottesville, VA.

Gatz, D.F., W.R. Barnard, and G.J. Stensland. 1986. The role of alkaline materials in precipitation chemistry: A brief review of the issues. Water Air Soil Pollut. 30: 245–251.

Gatz, D.F., V.C. Bowersox, and J. Su. 1988. Screening criteria for NADP dry-bucket sample data. Presented at the 81st annual meeting of the Air Pollution Control Association, June 19–24, 1988, Dallas, TX.

Gebert, W.A., D.J. Graczyk, and W.R. Krug. 1987. Average annual runoff in the United States: 1951–1980. Hydrologic Investigations Atlas. U.S. Geological Survey, Reston, VA.

Geraghty, J.J., D.W. Miller, F. van der Leeden, and F.L. Troise. 1973. Water Atlas of the United States. Water Information Center, Inc., Port Washington, NY.

Gherini, S.A., L. Mok, R.J. Hudson, G.F. Davis, C.W. Chen, and R.A. Goldstein. 1985. The ILWAS model: Formulation and application. Water Air Soil Pollut. 26:425–459.

Gherini, S.A., R. Munson, P. Chan, G.F. Davis, C.W. Chen, and J. Whipple. 1986. Simulation of the Response of Meander Lake to Changes in Acid Deposition Loading. Tetra Tech, Inc., Lafayette, CA (unpublished).

Gherini, S., R. Munson, E. Altwicker, R. April, C. Chen, N. Clesceri, C. Cronan, C. Driscoll, A. Johannes, R. Newton, N. Peters, and C. Schofield. 1989. Regional Integrated Lake-Watershed Acidification Study (RILWAS): Summary of Major Findings. EPRI RP2174-1. Electric Power Research Institute, Palo Alto, CA.

Ghirelli, R.P., M.P. Lo, and L. Margler. 1977. Bacterial Water Quality in Wilderness Areas. Contribution 162. Water Resources Center, University of California, Davis, CA.

Gibson, J.H., J.N. Galloway, C.L. Schofield, W. McFee, R. Johnson, S. McCorley, N. Dise, and D. Herzog. 1983. Pages 1–137 in Rocky Mountain Acidification Study. FWS/OBS-80/40. U.S. Fish and Wildlife Service, Division of Biological Services Eastern Energy and Land Use Team, Washington, D.C.

Gilbert, D.A., T.H. Sagraves, M.M. Lang, R.K. Munson, and S.A. Gherini. 1989. Blue Lake Acidification Study. Pacific Gas and Electric Company, San Ramon, CA.

Glass, G.E., and O.L. Loucks. 1986. Implications of a gradient in acid and ion deposition across the northern lake states. Environ. Sci. Technol. 20:35–43.

Gloss, S.P., C.L. Schofield, R.L. Spateholts, and B.A. Plonski. 1989. Survival, growth, reproduction, and diet of brook trout (*Salvelinus fontinalis*) stocked into lakes after liming to mitigate acidity. Can. J. Fish. Aquat. Sci. 46:277–286.

Goldstein, R.A., S.A. Gherini, C.W. Chen, L. Mok, and R.J.M. Hudson. 1984. Integrated Lake-Watershed Acidification Study (ILWAS): A mechanistic ecosystem analysis. Trans. R. Soc. Lond. Ser. B 305:409–425.

Goldstein, R.A., C.W. Chen, and S.A. Gherini. 1985. Integrated lake-watershed acidification study: Summary. Water Air Soil Pollut. 26:327–337.

Goldstein, R.A., S.A. Gherini, and R.K. Munson. 1987a. Principles of lake-watershed acidification. Pages 273–280 in R. Perry, R.M. Harrison, J.N.B. Bell, and J.N. Lester, eds. Acid Rain: Scientific and Technical Advances. Selper Ltd., Ealing, London.

Goldstein, R.A., S.A. Gherini, C.T. Driscoll, R. April, C.L. Schofield, and C.W. Chen. 1987b. Lake-watershed acidification in the north branch of the Moose River: Introduction. Biogeochemistry 3:5–20.

Goodyear, C.P. 1980. Compensation in fish populations. Pages 253–280 in C.H. Hocutt and J.R. Stauffer, Jr., eds. Biological Monitoring of Fish. Lexington Books, D.C. Heath, Lexington, MA.

Gorham, E., P.M. Vitousek, and W.A. Reiners. 1979. The regulation of chemical budgets over the course of terrestrial ecosystem succession. Annu. Rev. Ecol. Syst. 10:53–84.

Gorham, E., W.E. Dean, and J.E. Sanger. 1983. The chemical composition of lakes in the northcentral United States. Limnol. Oceanogr. 28:287–301.

Gorham, E., F.B. Martin, and J.T. Litzau. 1984. Acid rain:ionic correlations in the eastern United States. Science 225:407–409.

Gorham, E., S.J. Eisenreich, J. Ford, and M.V. Santelmann. 1985. The chemistry of bog waters. Pages 339–363 in W. Stumm, ed. Chemical Processes in Lakes. Wiley-Interscience, New York.

Gosz, J.R., D.G. Brookins, and D.I. Moore. 1983. Using strontium isotope ratios to estimate inputs to ecosystems. Bioscience 33:23–30.

Graetz, D.A., C.D. Pollman, B. Roof, and E. Will. 1985. Effects of acidic treatments on soil chemistry and microbiology. Pages 4-5 to 4–123 in Florida Acid Deposition Study. Phase IV Report, Volume I. ESE No. 83-152-0106/0207/0307. Environmental Science and Engineering, Inc., Gainesville, FL.

Grahn, O. 1977. Macrophyte succession in Swedish lakes caused by deposition of airborne acid substances. Water Air Soil Pollut. 7:295–305.

Grahn, O., H. Hultberg, and L. Landner. 1974. Oligotrophication – a self-accelerating process in lakes subjected to excessive supply of acid substances. Ambio 3:93–94.

Gran, G. 1952. Determination of the equivalence point in potentiometric titrations. ANALYST: Int. Congr. Anal. Chem. 77:661–671.

Granat, L., H. Rodhe, and R.O. Hallberg. 1976. The global sulphur cycle. In B.H. Svensson and R. Soderlund, eds. Nitrogen, Phosphorus, and Sulphur – Global Cycles. SCOPE Report 7. Ecol. Bull. (Stockholm) 22:89–134.

Grande, M., I.P. Muniz, and S. Andersen. 1978. Relative tolerance of some salmonids to acid waters. Verh. Internat. Verein. Limnol. 20:2076–2084.

Grant, M.C., and W.M. Lewis. 1982. Chemical loading rates from precipitation in the Colorado Rockies. Tellus 34:74–88.

Gray, H.A., G.R. Cass, J.J. Huntzicker, E.K. Heyerdahl, and J.A. Rau. 1986. Characteristics of atmospheric organic and elemental carbon particle concentrations in Los Angeles. Environ. Sci. Technol. 20:580–589.

Greb, S., R. Munson, S.A. Gherini, L. Gomez, C.W. Chen, P. Garrison, and D.R. Knauer. 1987. RILWAS-Wisconsin: Simulation of the Effects of Acid Deposition on Crystal, Vandercook, and Little Rock Lakes. Electric Power Research Institute, Palo Alto, CA.

Green, D., J. Matuszek, and D. Wales. 1987. Extensive Survey Data Report. Ontario Fisheries Acidification Report Series No. 87-12. Ontario Ministry of Natural Resources, Toronto, Ontario.

Green, J.C. 1982. Geologic map of Minnesota, Two Harbors sheet. Scale 1:250,000. Minnesota Geological Survey, Minneapolis, MN.

Gschwandtner, G., K.C. Gschwandtner, and K. Eldridge. 1985. Historic Emissions of Sulfur and Nitrogen Oxides in the United States from 1900 to 1980. EPA/600/7-85/009a. U.S. Environmental Protection Agency, Research Triangle Park, NC.

Gschwend, P.M., and R.A. Hites. 1981. Fluxes of polycyclic aromatic hydrocarbons to marine and lacustrine sediments in the northeastern United States. Geochim. Cosmochim. Acta 45:2359–2367.

Gubala, C.P. 1988. The Cycling of Iron and Trace Metals in the Sediments of Acidic Lakes. Ph.D. Dissertation. School of Public and Environmental Affairs, Indiana University, Bloomington, IN.

Gunn, J.M. 1986. Behavior and ecology of salmonid fishes exposed to episodic pH depressions. Environ. Biol. Fish 12:241–252.

Gunn, J.M., and W. Keller. 1984. Spawning site water chemistry and lake trout (*Salvelinus namaycush*) sac fry survival during spring snowmelt. Can. J. Fish. Aquat. Sci. 41:319–329.

Gunn, J.M., M.J. McMurtry, J.N. Bowlby, J.M. Casselman, and V.A. Liimatainen. 1987. Survival and growth of stocked lake trout in relation to body size, stocking season, lake acidity, and biomass of competitors. Trans. Am. Fish. Soc. 116:618–627.

Haan, C.T. 1975. Evaluation of a Model for Simulating Monthly Water Yields from Small Watersheds. Southern Cooperative Series Bulletin No. 201. Kentucky Agricultural Experiment Station, Lexington, KY.

Hack, J.T. 1965. Geomorphology of the Shenandoah Valley and West Virginia and Origin of the Residual Ore Deposits. Professional Paper 484. U.S. Geological Survey, Reston, VA.

Hack, J.T. 1969. The area, its geology: Cenozoic development of the southern Appalachians. Pages 1–17 in P.C. Holt, ed. The Distributional History of the Biota of the Southern Appalachians, Part I. Invertebrates. Research Division Monograph 1, Virginia Polytechnic Institute, Blacksburg, VA.

Hack, J.T., and J.C. Goodlett. 1960. Geomorphology and Forest Ecology of a Mountain Region in the Central Appalachians. Professional Paper 347. U.S. Geological Survey, Reston, VA.

Haelen, T.R. 1984. Hydrological and Solution Chemistry Response of a Tributary Valley in the White Oak Run Catchment, Shenandoah National Park, Virginia. M.S. Thesis. Department of Environmental Sciences, University of Virginia, Charlottesville, VA.

Hagley, C.A., C.L. Mayer, and R. Hoenicke. 1988. National Surface Water Survey: National Stream Survey (Phase I, Southeast Screening, and Episodes Pilot). Field Operations Report. EPA/600/4-88/023. U.S. Environmental Protection Agency, Las Vegas, NV.

Haines, T.A. 1981. Acidic precipitation and its consequences for aquatic ecosystems: A review. Trans. Am. Fish. Soc. 110:669–707.

Haines, T.A. 1986. Fish population trends in response to surface water acidification. Pages 300–334 in National Research Council, Committee on Monitoring and Assessment. Acid Deposition: Long-term Trends. National Academy Press, Washington, D.C.

Haines, T.A., and J.J. Akielaszek. 1983. A Regional Survey of the Chemistry of Headwater Lakes and Streams in New England: Vulnerability to Acidification. Air Pollution and Acid Rain. Report No. 15. FWS/OBS-80. U.S. Fish and Wildlife Service, Department of the Interior, Washington, D.C. 141 pp.

Haines, T.A., and J.P. Baker. 1986. Evidence of fish population responses to acidification in the eastern United States. Water Air Soil Pollut. 31:605–629.

Haines, T.A., and J.S. Kahl. 1983. Pages 1–55 in Identification and Evaluation of Potential Sites for a Calibrated Watershed/Acidification Study in the Northeastern U.S. Project Report. U.S. Environmental Protection Agency, Corvallis, OR.

Haines, T.A., J.J. Akielaszek, S.A. Norton, and R.B. Davis. 1983. Errors in pH measurement with colorimetric indicators inn low alkalinity waters. Hydrobiologia 107:57–61.

Haines, T.A., J.S. Kahl, and C.H. Jagoe. 1984. Long-Term Chemical and Biological Monitoring of Maine Lakes. Project Report. U.S. Environmental Protection Agency, Corvallis, OR.

Haines, T.A., C.H. Jagoe, and S.J. Pauwels. 1985. Fish species distribution as a function of lake acidity in the northeastern United States. Pages 82–95 in P. Rago and R. Schreiber, eds. Acid Rain and Fisheries: A Debate of Issues. FWS/OBS-80/40.21. U.S. Fish and Wildlife Service, Washington, D.C.

Haines, T.A., S.J. Pauwels, and C.H. Jagoe. 1986. Predicting and Evaluating the Effects of Acidic Precipitation on Water Chemistry and Endemic Fish Populations in the Northeastern United States. FWS/ OBS-80/40.23. U.S. Fish and Wildlife Service, Washington, D.C.

Haines, T.A., S.A. Norton, J.S. Kahl, S.J. Pauwels, C.H. Jagoe, and C.W. Fay. 1990. Intensive Studies of Stream Fish Populations in Maine. EPA/600/3-90/043. U.S. EPA Environmental Research Laboratory, Corvallis, OR.

Hall, R.J., and F.P. Ide. 1987. Evidence of acidification effects on stream insect communities in central Ontario between 1937 and 1985. Can. J. Fish. Aquat. Sci. 44:1652–1657.

Hall, R.J., G.E. Likens, S.B. Fiance, and G.R. Hendrey. 1980. Experimental acidification of a stream in Hubbard Brook Experimental Forest, New Hampshire. Ecology 61:976–989.

Hall, T.J. 1973. A Limnological Study of Shadow Lake, a Subalpine Lake at Mount Rainier National Park, Washington. M.S. Thesis, Central Washington State College, Ellensburg, WA.

Halliwell, D.B. 1989. A Classification of Streams in Massachusetts. Ph.D. Dissertation. Department of Forestry and Wildlife Management, University of Massachusetts, Amherst, MA.

Hansen, D.A., and G.M. Hidy. 1982. Review of questions regarding rain acidity data. Atmos. Environ. 16:2107–2126.

Hansen, W.R. 1978. Climatology of the Front Range Urban Corridor and Vicinity, Colorado. U.S. Geological Survey Professional Paper 1019, U.S. Geological Survey, Reston, VA.

Harrell, F.E., Jr. 1983. The LOGIST procedure. Pages 181–202 in SUGI Supplementary Library User's Guide. SAS Institute, Inc., Cary, NC.

Harriman, R., and B. Morrison. 1982. Ecology of streams draining forested and nonforested catchments in an area of central Scotland subject to acid precipitation. Hydrobiologia 88:251–263.

Harris, D.V., and E.P. Kiver. 1985. The Geologic Story of the National Parks and Monuments, Fourth Edition. John Wiley and Sons, New York.

Harrison, E. 1987. A Simulation Model of Forest Succession in Shaver Hollow. Ph.D. Dissertation. Department of Environmental Sciences, University of Virginia, Charlottesville, VA.

Harte, J., G.P. Lockett, R.A. Schneider, H. Michaels, and C. Blanchard. 1985. Acid precipitation and surface water vulnerability on the western slope of the high Colorado Rockies. Water Air Soil Pollut. 25:313–320.

Hartmann, H., and C. Steinberg. 1987. *Mallomonadacea* (*Chrysophyceae*) scales: Early biotic paleoindicators of lake acidification. Hydrobiologia 143:87–91.

Hartshorne, R. 1939. The nature of geography, a critical survey of current thought in the light of the past. Ann. Assoc. Am. Geogr. 29:173–658.

Harvey, H.H. 1975. Fish populations in a large group of acid-stressed lakes. Verh. Int. Verein. Limnol. 19: 2406–2417.

Harvey, H.H. 1979. The acid deposition problem and emerging research needs in the toxicology of fishes. Pages 115–128 in Proceedings of the Fifth Annual Aquatic Toxicity Workshop. November 7–9, 1978, Hamilton, Ontario. Fish. Mar. Serv. Tech. Rep. 862.

Harvey, H.H., and C. Lee. 1982. Historical fisheries changes related to surface water pH changes in Canada. Pages 45–46 in R. Johnson, ed. Acid Rain/Fisheries. America Fisheries Society, Bethesda, MD.

Harvey, H.H., and J.M. McArdle. 1986. Composition of the benthos in relation to pH in the LaCloche lakes. Water Air Soil Pollut. 30:529–536.

Harvey, H.H., and D.M. Whelpdale. 1986. The effects on fishes of acidic deposition and snowmelt events. Water Air Soil Pollut. 30:579–586.

Harvey, H.H., R.C. Pierce, P.J. Dillon, J.R. Kramer, and D.M. Whelpdale. 1981. Acidification of the Canadian Aquatic Environment: Scientific Criterion for Assessment of the Effects of Acidic Deposition on Aquatic Ecosystems. Report No. 18475. National Research Council of Canada, Ottawa, Ontario.

Hasler, A.D. 1938. Fish biology and limnology of Crater Lake, Oregon. J. Wildl. Manage. 2:94–103.

Hasler, A.D. 1975. Coupling of Land and Water Systems. Springer-Verlag, Berlin, FRG.

Hatcher, R.D., Jr. 1988. Bedrock geology and regional geologic setting of Coweeta Hydrologic Laboratory in the eastern Blue Ridge. Pages 81–92 in W.T. Swank and D.A. Crossley, Jr., eds. Forest Hydrology and Ecology at Coweeta. Springer-Verlag, New York.

Havas, M., and T.C. Hutchinson. 1982. Aquatic invertebrates from the Smoking Hills, N.W.T.: Effect of pH and metals on mortality. Can. J. Fish. Aquat. Sci. 39: 890–903.

Havas, M., T.C. Hutchinson, and G.E. Likens. 1984. Red herrings in acid rain research. Environ. Sci. Technol. 18:176–186.

Havens, K.E., and J. DeCosta. 1986. The role of aluminium contamination in determining phytoplankton and zooplankton responses to acidification. Water Air Soil Pollut. 33:277–293.

Hayes, T.P., J.J.R. Kinney, and N.J.M. Wheeler. 1984. California Surface Wind Climatology. California Air Resources Board, Sacramento, CA.

Healy, H.G. 1974. Potentiometric Surface and Areas of Artesian Flow of the Floridan Aquifer in Florida. Florida Bureau of Geology Map Series No. 73. Tallahassee, FL.

Healy, H.G. 1975. Terraces and Shorelines of Florida. Florida Bureau of Geology Map Series No. 71. Tallahassee, FL.

Heath, R.C., and C.S. Conover. 1981. Hydrologic Almanac of Florida. Open File Report 81-1107. U.S. Geological Survey, Reston, VA.

Hedin, L.L., G.E. Likens, and F.H. Bormann. 1987. Decrease in precipitation acidity resulting from decreased sulfate concentration. Nature 325:244–246.

Heiskary, S.A., M.E. Hora, and J.D. Thorton. 1982. Acid precipitation impact assessment in Minnesota derived from current and historical data. Pages 147–175 in L.H. Keith, ed. Energy and Environmental Chemistry, Vol. 2. Ann Arbor Science Publishers, Ann Arbor, MI.

Heit, M., Y. Tan, C. Klusek, and J.C. Burke. 1981. Anthropogenic trace elements and polycyclic aromatic hydrocarbon levels in sediment cores from two lakes in the Adirondack acid lake region. Water Air Soil Pollut. 15:441–464.

Heit, M., C. Klusek, and J. Baron. 1984. Evidence of deposition of anthropogenic pollutants in remote Rocky Mountain lakes. Water Air Soil Pollut. 22:403–416.

Hemond, H.F. 1980. Biogeochemistry of Thoreau's Bog, Concord, Massachusetts. Ecol. Monog. 50:507–526.

Hemond, H.F., and K.N. Eshleman. 1984. Neutralization of acid deposition by nitrate retention at Bickford Watershed, Massachusetts. Water Resour. Res. 20: 1718–1724.

Hendrey, G.R. 1976. Effects of Low pH on the Growth of Periphytic Algae in Artificial Stream Channels. Report IR 2576. SNSF Project, Oslo, Norway.

Hendrey, G.R., and F.A. Vertucci. 1980. Benthic plant communities in acidic Lake Colden, New York: Sphagnum and the algal mat. Pages 314–315 in D. Drabløs and A. Tollan, eds. Ecological Impact of Acid Precipitation. Proceedings of an International Conference, Sandefjord, Norway. SNSF Project, Oslo, Norway.

Hendrey, G.W., and E.B. Welch. 1974. Phytoplankton productivity in Findley Lake. Hydrobiologia 45:45–63.

Hendrey, G.R., J.N. Galloway, S.A. Norton, C.L. Schofield, P.W. Shaffer, and D.A. Burns. 1980. Geological and Hydrochemical Sensitivity of the Eastern United States to Acid Precipitation. EPA/600/3-80/024. U.S. Environmental Protection Agency, Corvallis, OR.

Hendry, C.D., Jr. 1977. The Chemistry of Precipitation in Northcentral Florida. M.S. Thesis. University of Florida, Gainesville, FL.

Hendry, C.D., Jr. 1983. Spatial and Temporal Variations in Bulk, Wet, and Dry Atmospheric Deposition of Acidity and Minerals Across Florida. Ph.D. Dissertation. University of Florida, Gainesville, FL.

Hendry, C.D., Jr., and P.L. Brezonik. 1980. Chemistry of precipitation at Gainesville, Florida. Environ. Sci. Technol. 14:843–849.

Hendry, C.D., Jr., and P.L. Brezonik. 1984. Chemical composition of softwater Florida lakes and their sensitivity to acid precipitation. Water Resour. Bull. 20: 75–86.

Hendry, C.W., Jr., and C.R. Sproul. 1966. Geology and Ground Water Resources of Leon County, Florida. Bulletin No. 47. Florida Geological Survey, Tallahassee, FL.

Henriksen, A. 1979. A simple approach for identifying and measuring acidification of fresh water. Nature 278: 542–545.

Henriksen, A. 1980. Acidification of freshwaters – a large scale titration. Pages 68–74 in D. Drabløs and A. Tollan, eds. Ecological Impact of Acid Precipitation. Proceedings of an International Conference, Sandefjord, Norway. SNSF Project, Oslo, Norway.

Henriksen, A. 1982. Changes in Base Cation Concentrations Due to Freshwater Acidification. Acid Rain Research Report. Norwegian Institute for Water Research, Oslo, Norway.

Henriksen, A. 1984. Changes in base cation concentrations due to freshwater acidification. Verh. Int. Verein. Limnol. 22:692–698.

Henriksen, A. 1987. 1000 Lake Survey, Norway. Report. Norwegian Institute for Water Research, Oslo, Norway.

Henriksen, A., and D.F. Brakke. 1988a. Sulfate deposition to surface waters. Environ. Sci. Technol. 22:8–14.

Henriksen, A., and D.F. Brakke. 1988b. Increasing contributions of nitrogen to the acidity of surface water in Norway. Water Air Soil Pollut. 42:183–201.

Henriksen, A., O.K. Skogheim, and B.O. Rosseland. 1984. Episodic changes in pH and aluminium-speciation kill fish in a Norwegian salmon river. Vatten 40: 255–260.

Henriksen, A., L. Lien, T.S. Traaen, I.S. Sevaldrud, and D.F. Brakke. 1988. Lake acidification in Norway – present and predicted chemical status. Ambio 17:259–266.

Henrikson, L., H.G. Oscarson, and J.A. E. Stenson. 1980. Does the change of predator system contribute to the biotic development in acidified lakes? Pages 316–317 in D. Drabløs and A. Tollan, eds. Ecological Impact of Acid Precipitation. Proceedings of an International Conference, Sandefjord, Norway. SNSF Project, Oslo, Norway.

Herczeg, A.L., and R.H. Hesslein. 1984. Determination of hydrogen ion in softwater lakes using carbon dioxide equilibria. Geochim. Cosmochim. Acta 48: 837–845.

Herczeg, A.L., W.S. Broecker, R.F. Anderson, S.L. Schiff, and D.W. Schindler. 1985. A new method for monitoring temporal trends in the acidity of fresh waters. Nature 315:133–135.

Herlihy, A.T., P.R. Kaufmann, M.E. Mitch, and D.D. Brown. 1990. Regional estimates of acid mine drainage impact on streams in the mid-Atlantic and southeastern United States. Water Air Soil Pollut. 50:91–107.

Hermann, R., E.L. Morgan, and R.L. Green. 1976. Aluminum precipitation, Beech Flats and Walkers

Prong Creeks, Great Smoky Mountains National Park. Pages 715–718 in Proceedings of the 1st Conference on Scientific Research in the National Parks.

Hesthagen, T. 1985. Fish kill of Atlantic salmon (*Salmo salar*) and brown trout (*Salmo trutta*) in the acidified Vikedal River, western Norway, and its impact on juvenile and adult fish stocks. Pages 121–122 in *Abstracts of the International Symposium on Acidic Precipitation*, Muskoka, Ontario.

Heubert, B.J., and C.H. Robert. 1985. The dry deposition of nitric acid to grass. J. Geophys. Res. 90:2085–2090.

Hicks, B.B. 1986. Measuring dry deposition: A reassessment of the state of the art. Water Air Soil Pollut. 30:75–90.

Hicks, B.B. 1989. Overview of deposition processes. Pages 3-1 to 3-21 in J. Malanchuk and J. Nilson, eds. The Role of Nitrogen in the Acidification of Soils and Surface Waters. Miljørapport 1989:10. Nordic Council of Ministers, Copenhagen, Denmark.

Hicks, B.B., and D.R. Matt. 1988. Combining biology, chemistry, and meteorology in modeling and measuring dry deposition. J. Atmos. Chem. 6:117–131.

Hidy, G.M., and J.R. Young. 1986. Acid Deposition and the West, a Scientific Assessment. WEST Associates, Newbury Park, CA.

Hill, D.P., R.A. Bailey, and A.S. Ryall. 1988. Active tectonic and magmatic processes beneath Long Valley Caldera, eastern California: An overview. J. Geophys. Res. 90:11–111.

Hill, M. 1975. Geology of the Sierra Nevada. University of California Press, Berkeley, CA.

Hillman, D.C.J., J.F. Potter, and S.J. Simon. 1986. National Surface Water Survey, Eastern Lake Survey (Phase I -Synoptic Chemistry) Analytical Methods Manual. EPA/600/4-86/009. U.S. Environmental Protection Agency, Las Vegas, NV.

Hingston, F.J., R.J. Atkinson, A.M. Posner, and J.P. Quirk. 1967. Specific adsorption of anions. Nature 215:1459–1461.

Hirsch, R.M., and N.E. Peters. 1988. Short-term trends in sulfate deposition of selected bulk precipitation stations in New York. Atmos. Environ. 22:1175– 1178.

Hirsch, R.M., and J.R. Slack. 1984. A nonparametric trend test for seasonal data with serial dependence. Water Resour. Res. 20:727–732.

Hirsch, R.M., J.R. Slack, and R.A. Smith. 1982. Techniques of trend analysis for monthly water quality analysis. Water Resour. Res. 18:107–121.

Hites, R.A. 1981. Sources and fates of atmospheric polycyclic aromatic hydrocarbons. American Chemical Society Symposium Service 167:187–196.

Hobaek, A., and G.G. Raddum. 1980. Zooplankton Communities in Acidified Lakes in South Norway. Report IR 75/80. SNSF Project, Oslo, Norway.

Hodgson, J.R., S.A. Lyden, P.J. Anders, and C.P. Frigo. 1987. Biota of a clear-water acidic lake. Jack Pine Warbler 65:3–13.

Hoffman, M.R., J.L. Collett, Jr., and B.C. Daube, Jr. 1989. Characterization of Cloud Chemistry and Frequency of Canopy Exposure to Clouds in the Sierra Nevada Mountains. Final Report. Contract A6-185-32. California Air Resources Board, Sacramento, CA.

Hoffman, R.W., C.R. Goldman, S. Paulson, and G.R. Winters. 1981. Aquatic impacts of deicing salts in the central Sierra Nevada Mountains, California. Water Resour. Bull. 17:280–285.

Hoham, R.W., and W.W. Mohn. 1985. The optimum pH of four strains of acidophilic snow algae in the genus Chloromonas (*Chlorophyta*) and possible effects of acid precipitation. J. Phycol. 21:603–609.

Holdren, G.R., Jr., T.M. Brunelle, G. Matisoff, and M. Wahlen. 1984. Timing the increase in atmospheric sulphur deposition in the Adirondack Mountains. Nature 311:245–248.

Holland, H.D. 1978. The Chemistry of the Atmosphere and Oceans. John Wiley and Sons, New York.

Holmes, R.W. 1986. Calibration of Diatom-pH-alkalinity Methodology for the Interpretation of the Sedimentary Record in Emerald Lake Integrated Watershed Study. Final Report. Contract A4-118-32. California Air Resources Board, Sacramento, CA.

Holmes, R.W., M.C. Whiting, and J.L. Stoddard. 1989. Diatom-inferred pH and acid neutralizing capacity changes since 1825 A.D. in a dilute, high elevation, Sierra Nevada lake. Freshwater Biol. 21:295– 310.

Holt-Jensen, A. 1980. Geography, Its History and Concepts. English translation. Barnes and Noble, Totowa, NJ.

Holtze, K.E. 1984. Effects of pH and Ionic Strength on Aluminum Toxicity to Early Developmental Stages of Rainbow Trout (*Salmo gairdneri Richardson*). Ontario Ministry of the Environment, Quality Protection Section, Water Resources Branch, Rexdale, Ontario.

Holtze, K.E., and N.J. Hutchinson. 1989. Lethality of low pH and Al to early life stages of six fish species inhibiting Precambrian Shield waters in Ontario. Can. J. Fish. Aquat. Sci. 46:1188–1202.

Hooper, R.P., C.T. West, and N.E. Peters. 1989. Assessing the Response of Emerald Lake, an Alpine Watershed in Sequoia National Park, California, to Acidification During Snowmelt Using a Simple Hydrochemical Model. Investigation Report. U.S. Geological Survey Water Resources, Denver, CO.

Hopkins, P.S. 1987. Invertebrate Drift and the Effects of an Acid Pulse on Invertebrates in High Altitude Sierra Nevada Streams. M.A. Thesis. University of California, Santa Barbara, CA.

Hopkins, P.S., K.W. Kratz, and S.D. Cooper. 1989. Effects of an experimental acid pulse on invertebrates in a high altitude Sierra Nevada stream. Hydrobiologia 171:45–58.

Hornberger, G.M., K.J. Beven, B.J. Cosby, and D.E. Sappington. 1985. Shenandoah watershed study: Calibration of a topography-based, variable contributing area hydrological model of a small forested catchment. Water Resour. Res. 21:1841–1850.

Hornberger, G.M., B.J. Cosby, and J.N. Galloway. 1986. Modelling the effects of acid deposition: Uncertainty and spatial variability in estimation of long-term sulfate dynamics in a region. Water Resour. Res. 22:1293–1302.

Hornbrook, E.H.W., I.M. Kettles, and W.W. Shilts. 1986. Geochemistry of aquatic and terrestrial sediments, Precambrian Shield of southeastern Ontario. Water Air Soil Pollut. 31:969–979.

Hough, B.K. 1957. Basic Soils Engineering. Ronald Press, New York.

Huckabee, J.W., C.P. Goodyear, and R.D. Jones. 1975. Acid rock in the Great Smokies: Unanticipated impact on aquatic biota of road construction in regions of sulfide mineralization. Trans. Am. Fish. Soc. 104:677–684.

Hudelson, R., G. Boyer, and J. McMillan. 1980. High Mountain Lake and Stream Survey of the Bridger Wilderness Area: 1969–1975. D.J. Reports F-1-R-8 to F-1-R-12. Wyoming Game and Fish Department, Cheyenne, WY.

Huebert, B.J., R.B. Norton, M.J. Bollinger, D.D. Parrish, C. Hahn, Y.A. Bush, P.C. Murphy, F.C. Fehsenfeld, and D.L. Albritton. 1983. Gas phase and precipitation acidities in the Colorado Mountains. Pages 17–23 in R. Hermann and A.I. Johnson, eds. Acid Rain—a Water Resources Issue for the 80's. AWRA International Symposium, Denver, CO.

Huete, A.R., and J.G. McColl. 1984. Soil cation leaching by "acid rain" with varying nitrate-to-sulfate ratios. J. Environ. Qual. 13:366–371.

Hulsman, P.F., P.M. Fowles, and J.M. Gunn. 1983. Mortality of walleye eggs and rainbow trout yolk-sac larvae in low-pH waters of the LaCLoche Mountain Area, Ontario. Trans. Am. Fish. Soc. 112:680–683.

Hunn, J.B., L. Cleveland. and E.E. Little. 1987. Influence of pH and aluminum on developing brook trout in a low calcium water. Environ. Pollut. 43:63–73.

Hunsaker, C.T., J.L. Malanchuk, R.J. Olson, S.W. Christensen, and R.S. Turner. 1986. Adirondack Mountains headwater lake chemistry relationships with watershed characteristics. Water Air Soil Pollut. 31:79–88.

Hunt, C.B. 1967. Physiography of the United States. W.H. Freeman and Co., San Francisco, CA.

Hunt, C.B. 1974. Natural Regions of the United States and Canada. W.H. Freeman and Co., San Francisco, CA.

Hunt, C.B. 1986. Surficial Deposits of the United States. Van Nostrand Rheinhold, New York.

Hunter, M.L., Jr., J.J. Jones, K.E. Gibbs, J.R. Moring, and M. Brett. 1985. Interactions among Waterfowl, Fishes, Invertebrates, and Macrophytes in Four Maine Lakes of Different Acidity. FWS/OBS- 80/40.20. U.S. Fish and Wildlife Service, Washington, D.C.

Hunter, M.L., Jr., J.J. Jones, J.W. Witham, and T.M. Mingo. 1986a. Biomass and species richness of aquatic macrophytes in four Maine (U.S.A.) lakes of different acidity. Aquat. Bot. 24:91–95.

Hunter, M.L., Jr., J.J. Jones, K.E. Gibbs, and J.R. Moring. 1986b. Duckling responses to lake acidification: Do black ducks and fish compete? Oikos 47:26–32.

Huntington, G.L., and M. Akeson. 1987. Pedologic Investigations in Support of Acid Rain Studies, Sequoia National Park, California. Department of Land, Air, and Water Resources, University of California, Davis, CA.

Hurley, J.P., D.E. Armstrong, G.J. Kenoyer, and C.J. Bowser. 1985. Ground water as a silica source for diatom production in a precipitation-dominated lake. Science 227:1576–1578.

Husar, R.B. 1986. Emissions of sulfur dioxide and nitrogen oxides and trends for eastern North America. Pages 48–92 in National Research Council, Committee on Monitoring and Assessment of Trends in Acid Deposition. Acid Deposition: Long-term Trends. National Academy Press, Washington, D.C.

Husar, R.B. 1988a. Trends of Seasonal Haziness and Sulfur Emissions Over the Eastern U.S. Project #CR813357. Report to U.S. Environmental Protection Agency, Washington, D.C.

Husar, R.B. 1988b. Trends of Seasonal Haziness and Sulfur Emissions Over the Eastern U.S. Project #CR813357. Year 02 Report. U.S. Environmental Protection Agency, Washington, D.C.

Husar, R.B., and J.M. Holloway. 1982. Sulfur and nitrogen over North America. Pages 95–115 in Ecological Effects of Acid Deposition. Report PM 1636. National Board for Environmental Protection, Solna, Sweden.

Husar, R.B., and D.E. Patterson. 1984a. SO_2 Concentration Estimates for New York City, 1880–1980. Project Number DW 14930338-01-1. Report to U.S. Environmental Protection Agency, Washington, D.C.

Husar, R.B., and D.E. Patterson. 1984b. SO_2 Concentration Estimates for New York City, 1880–1980. Project Summary 68-02-3746. U.S. Environmental Protection Agency, Washington, D.C.

Hustedt, F. 1939. Systematische und okologische untersuchungen uber die diatomeen-flora von Java, Bali, und Sumatra nach dem Material der Deutschen Limnologischen Sunda-Expedition III. Die okologischen factor in und ihr einfluss auf die diatomeenflora. Archiv. füer Hydrobiol. Suppl. 16:274–394.

Hutchinson, G.E. 1957. A Treatise on Limnology. Volume I. Geography, Physics and Chemistry. John Wiley and Sons, New York.

Hutchinson, G.E. 1967. A Treatise on Limnology. Volume II. John Wiley and Sons, New York.

Hutchinson, N.J., and J.B. Sprague. 1986. Toxicity of trace metal mixtures to American flagfish (*Jordenella floridae*) in soft, acidic water and implications for cultural acidification. Can. J. Fish. Aquat. Sci. 43:647–655.

Hutchinson, N.J., J.R. Munro, K.E. Holtze, T.W. Pawson. 1985. Biological Effects of Acidification. X. Utility of Laboratory Toxicity Testing for Describing in Situ Responses to Salmonids to Acidification. Presented at International Symposium on Acidic Precipitation, September 1985, Muskoka, Ontario.

Hutchinson, N.J., K.E. Holtze, J.R. Munro, and T.W. Pawson. 1989. Modifying effects of life stage, ionic strength, and post-exposure mortality on lethality of H^+ and Al to lake trout and brook trout. Aquat. Toxicol. 15:1–26.

Hutchinson, T.C., and M. Havas. 1986. Recovery of previously acidified lakes near Coniston, Canada, following reductions in atmospheric sulphur and metal emissions. Water Air Soil Pollut. 28:319–333.

Ibrahim, M., L.A. Barrie, and F. Fanaki. 1983. An experimental and theoretical investigation of the dry deposition of particles to snow, pine trees, and artificial collectors. Atmos. Environ. 17:781–788.

Ingersoll, C.G. 1986. The Effects of pH, Aluminum, and Calcium on Survival and Growth of Brook Trout (*Salvelinus fontinalis*) Early Life Stages. Ph.D. Dissertation. University of Wyoming, Laramie, WY.

Jackson, M. 1985. Filamentous algae in Ontario softwater lakes. Pages 117–118 in Abstracts of the International Symposium on Acidic Precipitation, Muskoka, Ontario.

Jackson, S.T., and D.F. Charles. 1988. Aquatic macrophytes in Adirondack (New York) lakes: Patterns of species composition in relation to environment. Can. J. Bot. 66:1449–1460.

Jeffries, D.S. 1984. Atmospheric deposition of pollutants in the Sudbury area. Pages 117–154 in J.O. Nriagu, ed. Environmental Impacts of Smelters. Advances in Environmental Science and Technology, No. 15. John Wiley and Sons, New York.

Jeffries, D.S. 1986. Evaluation of the regional acidification of lakes in eastern Canada using ion ratios. Pages 17–37 in Proceedings of the ECE Workshop on Acidification of Rivers and Lakes, April 28–30, Grafenau, FRG.

Jeffries, D.S. 1990. Snowpack storage of pollutants, release during snowmelt, and impact on receiving surface waters. Pages 107–132 in S.A. Norton, S.E. Lindberg, and A.L. Page, eds. Acid Precipitation, Volume 4. Soils, Aquatic Processes, and Lake Acidification. Springer-Verlag, New York.

Jeffries, D.S., and R.G. Semkin. 1983. Changes in snowpack, stream, and lake chemistry during snowmelt in the Turkey Lakes watershed. VDI-Berichte 500:377–386.

Jeffries, D.S., C.M. Cox, and P.J. Dillon. 1979. Depression of pH in lakes and streams in central Ontario during snowmelt. J. Fish. Res. Board Can. 36:640–646.

Jeffries, D.S., W.A. Scheider, and W.R. Snyder. 1984. Geochemical interactions of watersheds with precipitation in area affected by smelter emissions near Sudbury, Ontario. Pages 195–241 in J.O. Nriagu, ed. Advances in Environmental Science and Technology. Volume 15. John Wiley and Sons, New York.

Jeffries, D.S., R.G. Semkin, R. Neureuther, and M. Seymour. 1986a. Influence of atmospheric deposition on lake mass balances in the Turkey Lakes watershed, central Ontario. Water Air Soil Pollut. 30:1033–1044.

Jeffries, D.S., D.L. Wales, J.R.M. Kelso, and R.A. Linthurst. 1986b. Regional chemical characteristics of lakes in North America: Part 1—Eastern Canada. Water Air Soil Pollut. 31:551–568.

Jeffries, D.S., R.G. Semkin, R. Neureuther, and M. Seymour. 1988. Ion mass budgets for lakes in the Turkey Lakes watershed, June 1981–May 1983. Can. J. Fish. Aquat. Sci. 45(Suppl.1):47–58.

Jenne, E.A., L.E. Eary, L.W. Vail, D.C. Girvin, A.M. Liebtrau, L.F. Hibler, T.B. Miley, and M.J. Monsour. 1989. An Evaluation and Analysis of Three Dynamic Watershed Acidification Codes (MAGIC, ETD, and ILWAS). PNL-6687/UC-11. Richland, WA.

Jensen, J.H. 1987. Valley poljes in Florida karst. Pages 47–51 in B.F. Beck and W.L. Wilson, eds. Karst Hydrogeology: Engineering and Environmental Applications. A.A. Balkema, Rotterdam, The Netherlands.

Jensen, K.W., and E. Snekvik. 1972. Low pH levels wipe out salmon and trout populations in southernmost Norway. Ambio 1:223–225.

Jensen, M.J., N.K. Kaushik, P.T.S. Wong, and J.B. Robinson. 1988. Effect of acid precipitation on microbial decomposition processes in sediment from streams in the Turkey Lakes Watershed. Can. J. Fish. Aquat. Sci. 45(Suppl. 1):159–169.

Jewell, M.E. 1939. An ecological study of fresh-water sponges of Wisconsin. II. The influence of calcium. Ecology 20:11–28.

Johannes, A.H., and E.R. Altwicker. 1983. Relationship between dry deposition as measured via collection with a dry bucket vs. ambient air concentrations.

Pages 903–912 in H.R. Pruppacher, R.G. Semonin, and W.G. N. Slinn, eds. Precipitation, Scavenging, Dry Deposition, and Resuspension, Vol. 2. Elsevier Science Publishing Co., New York.

Johannes A.H., J.N. Galloway, and D.E. Troutman. 1980. Snow pack storage and ion release. Pages 260–261 in D. Drabløs and A. Tollan, eds. Ecological Impact of Acid Precipitation. Proceedings of an International Conference, Sandefjord, Norway. SNSF Project, Oslo, Norway.

Johannes, A.H., E.R. Altwicker, and N.L. Clesceri. 1984. Atmospheric inputs to the ILWAS watersheds. In The Integrated Lake-Watershed Acidification Study, Vol. 4: Summary of Major Results. EPRI EA-3221. Electric Power Research Institute, Palo Alto, CA.

Johannes, A.H., E.R. Altwicker, and N.L. Clesceri. 1985. The Integrated Lake-Watershed Acidification Study: Atmospheric inputs. Water Air Soil Pollut. 26: 339–353.

Johannes, A.H., E.R. Altwicker, and M. Garrity. 1986. Use of existing network data to determine wet deposition. Trans. Air Pollut. Contr. Assoc. March: 381–392.

Johannessen, M., and A. Henriksen. 1978. Chemistry in snowmelt: Changes in concentrations during melting. Water Resour. Res. 14:615–619.

Johnson, A.H., D.F. Charles, and S.B. Anderson. 1987a. Lake acidification: The authors' reply. Environment 29:4–5.

Johnson, C.B., T.J. Sullivan, and D.J. Blick. 1989. Defining regional populations of lakes for the assessment of surface water quality. Water Resour. Bull. 25:565–572.

Jeffries, D.S., C.M. Cox, and P.J. Dillon. 1979. Depression of pH in lakes and streams in central Ontario during snowmelt. J. Fish. Res. Board Can. 36:640–646.

Jeffries, D.S., W.A. Scheider, and W.R. Snyder. 1984. Geochemical interactions of watersheds with precipitation in area affected by smelter emissions near Sudbury, Ontario. Pages 195–241 in J.O. Nriagu, ed. Advances in Environmental Science and Technology. Volume 15. John Wiley and Sons, New York.

Jeffries, D.S., R.G. Semkin, R. Neureuther, and M. Seymour. 1986a. Influence of atmospheric deposition on lake mass balances in the Turkey Lakes watershed, central Ontario. Water Air Soil Pollut. 30:1033–1044.

Jeffries, D.S., D.L. Wales, J.R.M. Kelso, and R.A. Linthurst. 1986b. Regional chemical characteristics of lakes in North America: Part 1–Eastern Canada. Water Air Soil Pollut. 31:551–568.

Jeffries, D.S., R.G. Semkin, R. Neureuther, and M. Seymour. 1988. Ion mass budgets for lakes in the Turkey Lakes watershed, June 1981–May 1983. Can. J. Fish. Aquat. Sci. 45(Suppl.1):47–58.

Jenne, E.A., L.E. Eary, L.W. Vail, D.C. Girvin, A.M. Liebtrau, L.F. Hibler, T.B. Miley, and M.J. Monsour. 1989. An Evaluation and Analysis of Three Dynamic Watershed Acidification Codes (MAGIC, ETD, and ILWAS). PNL-6687/UC-11. Richland, WA.

Jensen, J.H. 1987. Valley poljes in Florida karst. Pages 47–51 in B.F. Beck and W.L. Wilson, eds. Karst Hydrogeology: Engineering and Environmental Applications. A.A. Balkema, Rotterdam, The Netherlands.

Jensen, K.W., and E. Snekvik. 1972. Low pH levels wipe out salmon and trout populations in southernmost Norway. Ambio 1:223–225.

Jensen, M.J., N.K. Kaushik, P.T.S. Wong, and J.B. Robinson. 1988. Effect of acid precipitation on microbial decomposition processes in sediment from streams in the Turkey Lakes Watershed. Can. J. Fish. Aquat. Sci. 45(Suppl. 1):159–169.

Jewell, M.E. 1939. An ecological study of fresh-water sponges of Wisconsin. II. The influence of calcium. Ecology 20:11–28.

Johannes, A.H., and E.R. Altwicker. 1983. Relationship between dry deposition as measured via collection with a dry bucket vs. ambient air concentrations. Pages 903–912 in H.R. Pruppacher, R.G. Semonin, and W.G. N. Slinn, eds. Precipitation, Scavenging, Dry Deposition, and Resuspension, Vol. 2. Elsevier Science Publishing Co., New York.

Johannes A.H., J.N. Galloway, and D.E. Troutman. 1980. Snow pack storage and ion release. Pages 260–261 in D. Drabløs and A. Tollan, eds. Ecological Impact of Acid Precipitation. Proceedings of an International Conference, Sandefjord, Norway. SNSF Project, Oslo, Norway.

Johannes, A.H., E.R. Altwicker, and N.L. Clesceri. 1984. Atmospheric inputs to the ILWAS watersheds. In The Integrated Lake-Watershed Acidification Study, Vol. 4: Summary of Major Results. EPRI EA-3221. Electric Power Research Institute, Palo Alto, CA.

Johannes, A.H., E.R. Altwicker, and N.L. Clesceri. 1985. The Integrated Lake-Watershed Acidification Study: Atmospheric inputs. Water Air Soil Pollut. 26:339–353.

Johannes, A.H., E.R. Altwicker, and M. Garrity. 1986. Use of existing network data to determine wet deposition. Trans. Air Pollut. Contr. Assoc. March: 381–392.

Johannessen, M., and A. Henriksen. 1978. Chemistry in snowmelt: Changes in concentrations during melting. Water Resour. Res. 14:615–619.

Johnson, A.H., D.F. Charles, and S.B. Anderson. 1987a. Lake acidification: The authors' reply. Environment 29:4–5.

Johnson, C.B., T.J. Sullivan, and D.J. Blick. 1989. Defining regional populations of lakes for the assessment of surface water quality. Water Resour. Bull. 25:565–572.

Johnson, D.M., R.P. Petersen, D.R. Lycan, J.W. Sweet, M.E. Neuhaus, and A.L. Schaedel. 1985a. Atlas of

Oregon Lakes. Oregon State University Press, Corvallis, OR.

Johnson, D.W. 1984a. Sulfur cycling in forests. Biogeochemistry 1:29–43.

Johnson, D.W. 1985. Forest Nutrient Cycles as Affected by Climate, Species Composition, Stand Age, and Intensive Harvesting. Forestry Energy Agreement, International Energy Agency. Government of Canada, Canadian Forestry Service.

Johnson, D.W., and D.W. Cole. 1980. Anion mobility in soils: Relevance to nutrient transport from forest ecosystems. Environ. Internat. 3:79–90.

Johnson, D.W., and G.S. Henderson. 1979. Sulfate adsorption and sulfur fractions in a highly weathered soil under a mixed deciduous forest. Soil Sci. 128:34–40.

Johnson, D.W., and D.A. Webster. 1977. Avoidance of low pH in selection of spawning sites by brook trout (*Salvelinus fontinalis*). J. Fish. Res. Board Can. 34: 2215–2218.

Johnson, D.W., J. Turner, and J.M. Kelly. 1982. The effects of acid rain on forest nutrient status. Water Resour. Res. 18:449–461.

Johnson, D.W., D.D. Richter, G.M. Lovett, and S.E. Lindberg. 1985b. The effects of atmospheric deposition on potassium, calcium and magnesium cycling in two deciduous forests. Can. J. Forest Res. 15:773–782.

Johnson, D.W., D.D. Richter, H. Van Miegroet, D.W. Cole, and J.M. Kelly. 1986. Sulfur cycling in five forest ecosystems. Water Air Soil Pollut. 30:965–980.

Johnson, D.W., H.A. Simonin, J.R. Colquhoun, and F.M. Flock. 1987b. In situ toxicity tests of fishes in acid waters. Biogeochemistry 3:181–208.

Johnson, D.W., A.J. Friedland, H. Van Miegroet, R.B. Harrison, E. Miller, S.E. Lindberg, D.W. Cole, D.A. Schaefer, and D.E. Todd. 1988. Nutrient status of some contrasting high-elevation forests in the eastern and western United States. Proceedings of the United States-German Symposium. Burlington, VT.

Johnson, J.E. 1987. Protected Fishes of the United States and Canada. American Fisheries Society, Bethesda, MD.

Johnson, N.M. 1979. Acid rain: Neutralization within the Hubbard Brook ecosystem and regional implications. Science 204:497–499.

Johnson, N.M. 1984b. Acid rain neutralization by geologic materials. Pages 37–53 in O.P. Bricker, ed. Geological Aspects of Acid Deposition. Butterworth Publishers, Boston, MA.

Johnson, N.M., G.E. Likens, F.H. Bormann, D.W. Fisher, and R.S. Pierce. 1969. A working model for the variation in streamwater chemistry at Hubbard Brook Experimental Forest, New Hampshire. Water Resour. Res. 5:1353–1363.

Johnson, N.M., C.T. Driscoll, J.S. Eaton, G.E. Likens, and W.H. McDowell. 1981. "Acid rain," dissolved aluminum, and chemical weathering at the Hubbard Brook Experimental Forest, New Hampshire. Geochim. Cosmochim. Acta 45:1421–1437.

Johnston, J. 1977. Management needs of alpine lakes in Washington. Pages 39–47 in A Symposium of the Management of High Mountain Lakes in California's National Parks. California Trout, Inc., San Francisco, CA.

Jones, H.C., III, J.C. Noggle, R.C. Young, J.M. Kelley, H. Olem, R.J. Ruane, R.W. Pasch, G.J. Hufantis, and W.J. Parkhurst. 1983. Investigations of the Cause of Fishkills in Fish-Rearing Facilities in Raven Fork Watershed. TVA/ONR/WR-83/9. Tennessee Valley Authority, Office of Natural Resources, Division of Air and Water Resources, Knoxville, TN.

Jones, H.G., and W.J. Orville-Thomas. 1987. Seasonal snowcovers: Physics, chemistry, hydrology. NATO ASI Series C: Mathematical and Physical Science. Volume 211. D. Reidel, Dordrecht, The Netherlands.

Jones, H.G., W. Sochanska, J.-Y. Charette, J. Stein, J. Roberge, and A.P. Plamondon. 1986. Snowmelt in a boreal forest site: An integrated model of meltwater quality (SNOQUAL 1). Water Air Soil Pollut. 31: 431–440.

Jordan, C.L. 1984. Florida's weather and climate: Implications for water. Pages 18–35 in E.A. Fernald and D.J. Patton, eds. Water Resources Atlas for Florida. Florida State University, Tallahassee, FL.

Joslin, D., P.A. Mays, M.H. Wolfe, J.M. Kelly, R.W. Garber, and P.F. Brewer. 1987. The chemistry of tension lysimeter versus laterally flowing soil water in steeply sloping spruce and hardwood stands. J. Environ. Qual. 16:152–160.

Juday, C. 1943. The summer standing crop of plants and animals in four Wisconsin lakes. Trans. Wis. Acad. Sci. Arts Lett. 34:103–135.

Juday, C., et al. 1938. Mineral content of lake waters of northeastern Wisconsin. Trans. Wis. Acad. Sci. Arts Lett. 31:223–276.

Junge, C.E., and R.T. Werby. 1958. The concentration of chloride, sodium, potassium, calcium, and sulfate in rain water over the United States. J. Meteorol. 15: 417–425.

Kahl, J.S. 1982. Present and inferred historic atmospheric deposition of heavy metals at two eastern Maine sites. Northeast. Environ. Sci. 1:170–175.

Kahl, J.S. 1987. Preliminary chemical comparisons of Maine, U.S.A., small lakes with the results of the U.S. EPA Eastern Lake Survey. In Proceedings of an International Symposium on Acidification and Water Pathways, UNESCO II (unpublished).

Kahl, J.S., and M. Scott. 1988. Chemistry of Maine's high elevation lakes: Results from the HELM project. Lake Reserv. Manage. 4:33–40.

Kahl, J.S., J.L. Andersen, and S.A. Norton. 1985. Pages 1–123 in Water Resource Baseline Data and Assessment of Impacts from Acidic Precipitation, Acadia National Park, Maine. National Park Service Technical Report #16.

Kahl, J.S., S.A. Norton, T.A. Haines, C.W. Fay, and R. B. Davis. 1989. The influence of organic acidity on the chemistry of Maine surface waters. Water Air Soil Pollut. 46:221–233.

Kanciruk, P., J.M. Eilers, R.A. McCord, D.H. Landers, D.F. Brakke, and R.A. Linthurst. 1986. Characteristics of Lakes in the Eastern United States. Volume III. Data Compendium of Site Characteristics and Chemical Variables. EPA/600/4-86/007c. U.S. Environmental Protection Agency, Washington, D.C.

Kane, D.A., and C.F. Rabeni. 1987. Effects of aluminum and pH on the early life stages of smallmouth bass (*Micropterus dolemieui*). Water Res. 21:633– 639.

Kaplan, E., H.C. Thode, Jr., and A. Protas. 1981. Rocks, soils, and water quality. Relationships and implications for effects of acid precipitation on surface water in the northeastern United States. Environ. Sci. Technol. 15:539–544.

Karlsson-Norrgren, L., I. Bjorklund, O. Ljungberg, and P. Runn. 1986. Acid water and aluminum exposure: Experimentally induced gill lesions in brown trout, *Salmo trutta* L.J. Fish. Diseases 9:11–25.

Karnaugh, E. 1988. Structure, Abundance, and Distribution of the Pelagic Zooplankton in Crater Lake, Oregon. CPSU/OSU 88-3, Cooperative Agreement No. CA 9000-3-0003, Subagreement 12. National Park Service, Cooperative Park Studies Unit, College of Forestry, Oregon State University, Corvallis, OR.

Karner, F.R. 1968. Compositional variation in the Tunk Lake Granite Pluton, southeastern Maine. Geol. Soc. Am. Bull. 79:193–222.

Katz, B.G., O.P. Bricker, and M.M. Kennedy. 1985. Geochemical mass-balance relationships for selected ions in precipitation and stream water, Catoctin Mountains, Maryland. Am. J. Sci. 285:931–962.

Kaufmann, P.R., A.T. Herlihy, J.W. Elwood, M.E. Mitch, W.S. Overton, M.J. Sale, J.J. Messer, K.A. Cougan, D.V. Peck, K.H. Reckhow, A.J. Kinney, S.J. Christie, D.D. Brown, C.A. Hagley, and H.I. Jager. 1988. Chemical Characteristics of Streams in the Mid-Atlantic and Southeastern United States. Volume I: Population Descriptions and Physico-Chemical Relationships. EPA/600/3-88/021a. U.S. Environmental Protection Agency, Washington, D.C.

Keller, A.E. 1984. Fish Communities in Florida Lakes: Relationship to Physico-Chemical Parameters. M.S. Thesis. Environmental Engineering Sciences Department, University of Florida, Gainesville, FL.

Keller, W., and J.R. Pitblado. 1984. Crustacean plankton in northeastern Ontario lakes subjected to acidic deposition. Water Air Soil Pollut. 23:271–291.

Keller, W., and J.R. Pitblado. 1986. Water quality changes in Sudbury area lakes: A comparison of synoptic surveys in 1974–1976 and 1981–1983. Water Air Soil Pollut. 29:285–296.

Kelly, C.A. 1988. Toward improving comparisons of alkalinity generation in lake basins. Limnol. Oceanogr. 33:1635–1637.

Kelly, C.A., and J.W.M. Rudd. 1984. Epilimnetic sulfate reduction and its relationship to lake acidification. Biogeochemistry 1:63–77.

Kelly, C.A., J.W.M. Rudd, R.B. Cook, and D.W. Schindler. 1982. The potential importance of bacterial processes in regulating rate of lake acidification. Limnol. Oceanogr. 27:868–882.

Kelly, C.A., J.W.M. Rudd, A. Furutani, and D.W. Schindler. 1984. Effects of lake acidification on rates of organic matter decomposition in sediments. Limnol. Oceanogr. 29:687–694.

Kelly, C.A., J.W.M. Rudd, R.H. Hesslein, C.T. Driscoll, S.A. Gherini, and R.E. Hecky. 1987. Prediction of biological acid neutralization in acid-sensitive lakes. Biogeochemistry 3:129–140.

Kelly, G.A., J.S. Griffith, and R.D. Jones. 1980. Changes in the Distribution of Trout in Great Smoky Mountains National Park, 1900–1977. Technical Paper 102. U.S. Fish and Wildlife Service, Washington, D.C.

Kelly, J.M., and P.A. Mays. 1989. Root zone physical and chemical characteristics in southeastern spruce-fir stands. J. Soil Sci. Soc. Am. 53:1248–1255.

Kelly, T.J., and D.H. Stedman. 1980. Effect of urban sources on acid precipitation in the western United States. Science 210:1043.

Kelly, W.H. 1973. Trout Fishing in the Catskills. Conservationist April–May 1973:103–108.

Kelly, W.H., and M.C. Gann. 1979. Esopus Creek Fisheries Investigations 1975–1978. New York Department of Environmental Conservation, Division of Fish and Wildlife, New Paltz, NY.

Kelso, J.R.M., and D.S. Jeffries. 1988. Response of headwater lakes to varying atmospheric deposition in north-central Ontario, 1979–1985. Can. J. Fish. Aquat. Sci. 45:1905–1911.

Kelso, J.R.M., C.K. Minns, J.H. Lipsit, and D.S. Jeffries. 1986a. Headwater lake chemistry during the spring freshet in north-central Ontario. Water Air Soil Pollut. 29:245–259.

Kelso, J.R.M., C.K. Minns, J.E. Gray, and M.L. Jones. 1986b. Acidification of Surface Waters in Eastern Canada and Its Relationship to Aquatic Biota. No. 87. Canadian Special Publication of Fisheries and Aquatic Sciences, Ottawa, Ontario.

Kennedy, V.C., G.W. Zellweger, and R.J. Avanzino. 1979. Variation of rain chemistry during storms at two sites in northern California. Water Resour. Res. 15: 687–702.

Kenoyer, G.J. 1986. Groundwater/Lake Dynamics and Chemical Evolution in a Sandy Silicate Aquifer in Northern Wisconsin. Ph.D. Dissertation. University of Wisconsin, Madison, WI.

Kerekes, J.J., G. Howell, S. Beauchamp, and T. Pollock. 1982. Characterization of three lake basins sensitive to acid precipitation in central Nova Scotia (June 1979 to May 1980). Int. Rev. Gesamten Hydrobiol. 67: 679–694.

Kerekes, J.J., S. Beauchamp, R. Tordon, C. Tremblay, and T. Pollock. 1986a. Organic versus anthropogenic acidity in tributaries of the Kejimkujik watersheds in western Nova Scotia. Water Air Soil Pollut. 31:165–173.

Kerekes, J.J., S. Beauchamp, R. Tordon, and T. Pollock. 1986b. Sources of sulfate and acidity in wetlands and lakes in Nova Scotia. Water Air Soil Pollut. 31:207–214.

Kerfoot, H.B., and M.L. Faber. 1987. National Surface Water Survey, Western Lake Survey (Phase I–Synoptic Chemistry) Analytical Methods Manual. EPA/600/8-87/038. U.S. Environmental Protection Agency, Las Vegas, NV.

Kessel-Taylor, I. 1986. An Examination of Alternative Causes of Atlantic Salmon Decline and Surface Water Acidification in Southwest Nova Scotia. Working Paper No. 46. Land Directorate, Environment Canada, Ottawa, Ontario.

Kilgore, B.M. 1973. The ecological role of fire in Sierran conifer forests: Its application to national park management. Quat. Res. 3:496–513.

Kilgore, B.M., and D.R. Taylor. 1979. Fire history of a Sequoia-mixed conifer forest. Ecology 60:129–142.

Kilham, P. 1982. Acid precipitation: Its role in the alkalization of a lake in Michigan. Limnol. Oceanogr. 27:856–867.

Kimmel, W.G., D.J. Murphey, W.E. Sharpe, and D.R. DeWalle. 1985. Macroinvertebrate community structure and detritus processing rates in two southeastern Pennsylvania streams acidified by atmospheric deposition. Hydrobiologia 124:97–102.

King, B.P., R.B. Neumann, and J.B. Hadley. 1968. Geology of the Great Smoky Mountains National Park, Tennessee and North Carolina. Geology Survey Professional Paper 587. U.S. Government Printing Office, Washington, D.C.

Kingston, J.C., R.B. Cook, R.G. Kreis, K.E. Camburn, S.A. Norton, P.R. Sweets, M.W. Binford, M.J. Mitchell, S.C. Schindler, L. Shane, and G. King. (In press) Paleoecological investigation of recent lake acidification in the Northern Great Lakes states. J. Paleolimnol.

Kincaid, C.T., J.R. Morrey, S.B. Yabusaki, A.R. Felmy, and J.E. Rodgers. 1984. Geohydrochemical Models for Solute Migration: Volumes 1 and 2. EPRI EA-3417. Electric Power Research Institute, Palo Alto, CA.

Klikoff, L.G. 1965. Microenvironmental influence on vegetational pattern near timberline in the central Sierra Nevada. Ecol. Monogr. 35:187–211.

Kling, G.W., and M.C. Grant. 1984. Acid precipitation in the Colorado front range: An overview with time predictions for significant effects. Arct. Alp. Res. 16: 321–329.

Knapp, C.M., C.L. Mayer, D.V. Peck, J.R. Baker, and G.J. Filbin. 1987. Field Operations Report, National Surface Water Survey, National Stream Survey, Pilot Survey. EPA/600/8-87/019. Lockheed Engineering and Management Services Company, Inc., Las Vegas, NV.

Knapp, C.M., D.G. Heimbuch, H.S. Greening, and G.J. Filbin. 1988. Maryland Synoptic Stream Chemistry Survey. Ad-88-2. International Science and Technology, Inc., Maryland Department of Natural Resources, Power Plant Research Program, Annapolis, MD.

Knauer, D.R., J.G. Bockheim, A.J. Prey, J. Flickinger, J.E. Leide, J.A. Morton, L.C. Sheng, E.A. Jepsen, J.M. Esser, D.A. Wentz, W.A. Rose, J.T. Krohelski, P.J. Garrison, S.A. Gherini, and C.W. Chen. 1989. Wisconsin RILWAS Report. Prepared for Electric Power Research Institute, Palo Alto, CA.

Knox, C.E., and T.J. Nordenson. 1955. Average Annual Runoff and Precipitation in the New England-New York Area. Hydrologic Investigations Atlas HA-7. U.S. Geological Survey. U.S. Government Printing Office, Washington, D.C.

Knudson, D.A. 1986. Estimated Monthly Emissions of Sulfur Dioxide and Oxides of Nitrogen for the 48 Contiguous States, 1975–1984. ANL/EES-TN-318, Vol. 1. U.S. Department of Energy, Washington, D.C.

Kramer, J.R. 1985. Correction of colorimetric alkalinity, acidity and pH data and comparison of changes over time. Pages 43–50 in Proceedings of the American Water Works Association Conference on Acid Rain. June 23, 1985, Washington, D.C.

Kramer, J.R. 1986. Appendix D: Historical correction factors for alkalinity and acid status of surfaced waters. Pages 471–481 in Acid Deposition: Long-Term Trends. National Academy of Sciences, Washington, D.C.

Kramer, J.R., and S.S. Davies. 1988. Estimation of noncarbonato protolytes for selected lakes in the Eastern Lake Survey. Environ. Sci. Technol. 22:182– 185.

Kramer, J.R., and A. Tessier. 1982. Acidification of aquatic systems: A critique of chemical approaches. Environ. Sci. Technol. 16:606–615.

Kramer, J.R., A.W. Andren, R.A. Smith, A.H. Johnson, R.B. Alexander, and G. Oehlert. 1986. Streams and lakes. Pages 231–299 in National Research Council, Committee on Monitoring and Assessment of Trends in Acid Deposition. Acid Deposition: Long-term Trends. National Academy Press, Washington, D.C.

Kratz, T.K., R.B. Cook, C.J. Bowser, and P.L. Brezonik. 1987. Winter and spring pH depressions in northern Wisconsin lakes caused by increases in pCO_2. Can. J. Fish. Aquat. Sci. 44:1082–1088.

Kreis, R.G., Jr., J.C. Kingston, K.E. Camburn, R.B. Cook, S.A. Norton, G. Blake, R.A. Hites, L.A. Roll, S.C. Schindler, L.C. Shane, and G.A. King. (1989) Paleolimnology of McNearney Lake, Michigan: An acidic lake in the northern Great Lakes region. Section 11 in D.F. Charles and D.R. Whitehead, eds. Paleoecological Investigation of Recent Lake Acidification (PIRLA): 1983–1985. EN-6526, Project 2174-10. Interim Report. Electric Power Research Institute, Palo Alto, CA.

Kretser, W.A., and J.R. Colquhoun. 1984. Treatment of New York's Adirondack lakes by liming. Fisheries 9: 36–41.

Kretser, W.A., J. Gallagher, and J. Nicolette. 1989. Adirondack Lakes Study 1984–1987: An Evaluation of Fish Communities and Water Chemistry. Adirondack Lakes Survey Corporation, Ray Brook, NY.

Krug, E.C. 1987. Acid deposition/watershed interactions: The importance of humic substances and biological processes. Amer. Assoc. Adv. Sci., Annual Meeting, February 14–18, 1987, Chicago, IL.

Krug, E.C. 1989. Assessment of the Theory and Hypothesis of the Acidification of Watersheds. Illinois State Water Survey, Champaign, IL. SWS Contract Report 457. U.S. Department of Energy, Washington, D.C.

Krug, E.C., and C.R. Frink. 1983. Acid rain on acid soil: A new perspective. Science 221:520–525.

Kudish, M. 1985. Forest history of Frost Valley. Adirondack Magazine April 1985:16–18.

Kwiatkowski, R., and J. Roff. 1976. Effects of acidity on the phytoplankton and primary productivity of selected northern Ontario lakes. Can. J. Bot. 54: 2546–2561.

Lacroix, G.L. 1985. Survival of eggs and alevins of Atlantic salmon (*Salmo salar*) in relation to the chemistry of interstitial water in redds in some acidic streams of Atlantic Canada. Can. J. Fish. Aquat. Sci. 42:292–299.

Lacroix, G.L., and D.R. Townsend. 1987. Responses of juvenile Atlantic salmon (*Salmo salar*) to episodic increases in acidity of Nova Scotia rivers. Can. J. Fish. Aquat. Sci. 44:1475–1484.

Laessle, A.M. 1942. The Plant Communities of the Welaka Area, with Special Reference to Correlations between the Soils and Vegetational Succession. Volume IV, No.1, Biological Science Series, University of Florida, Gainesville, FL.

Laessle, A.M. 1958. The origin and successional relationship of sandhill vegetation and sand pine scrub. Ecol. Mono. 28:361–387.

Laidler, K.J. 1965. Chemical Kinetics. Ed. 2. McGraw-Hill, New York.

Laird, L.B., H.E. Taylor, and V.C. Kennedy. 1986. Snow chemistry of the Cascade-Sierra Nevada Mountains. Environ. Sci. Technol. 20:275–290.

LaMarche, V.C., Jr. 1973. Holocene climatic variations inferred from treeline fluctuations in the White Mountains, California. Quat. Res. 3:632–660.

Lamberti, G.A., and V.H. Resh. 1983. Stream periphyton and insect herbivores: An experimental study of grazing by a caddisfly population. Ecology 64:1124–1135.

Lamia, J.A. 1987. The Limnological and Biological Characteristics of Cue Lake: An Acidic Lake in North Florida. M.S. Thesis. University of Florida, Gainesville, FL.

Landers, D.H., J.M. Eilers, D.F. Brakke, W.S. Overton, P.E. Kellar, M.E. Silverstein, R.D. Schonbrod, R.E. Crowe, R.A. Linthurst, J.M. Omernik, S.A. Teague, and E.P. Meier. 1987. Characteristics of Lakes in the Western United States. Volume I. Population Descriptions and Physico-Chemical Relationships. EPA/600/3-86/054a. U.S. Environmental Protection Agency, Washington, D.C.

Landers, D.H., J.M. Eilers, D.F. Brakke, and P.E. Kellar. 1988a. Characteristics of acidic lakes in the eastern United States. Int. Verein. Theor. Ange. Limnol. Verh. 23:152–162.

Landers, D.H., W.S. Overton, R.A. Linthurst, and D.F. Brakke. 1988b. Eastern Lake Survey: Regional estimates of lake chemistry. Environ. Sci. Technol. 22: 128–135.

Langdon, R.W. 1983. Fisheries Status in Relation to Acidity in Selected Vermont Lakes. Agency of Environmental Conservation, Department of Water Resources and Environmental Engineering, State of Vermont, Montpelier, VT.

Langdon, R.W. 1984. Fisheries Status in Relation to Acidity in Selected Vermont Lakes 1983. Department of Water Resources and Environmental Engineering, Water Quality Division, State of Vermont, Montpelier, VT.

Langdon, R.W. 1985. Fisheries Status in Relation to Acidity in Selected Vermont Streams. Draft Final Report to NCSU Acid Deposition Program. Department of Water Resources and Environmental Engineering, Water Quality Division, State of Vermont, Montpelier, VT.

Larmie, F.M. 1955. A Survey of Thirty-one Cascade Mountain lakes of Oregon with Reference to the Game

Fish Populations. M.S. thesis. Oregon State College, Corvallis, OR.

Larson, D.W. 1972. Comparative linmology and phytoplankton ecology of four oligotrophic lakes in Oregon, U.S.A., with emphasis on lake typology. Northwest Sci. 46:149–163.

Larson, G.L. 1973. A limnology study of a high mountain lake in Mount Rainier National Park, Washington State, U.S.A. Arch. Hydrolbiol. 72:10–48.

Larson, G.L. 1988. Crater Lake Limnological Studies, 1987 Annual Report. Cooperative Agreement No. CA 9000-3-0003, Subagreement 12, Cooperative Park Studies Unit, College of Forestry, Oregon State University, Corvallis, OR.

Larson, G.L., and S.E. Moore. 1985. Encroachment of exotic rainbow trout into stream populations of native brook trout in the southern Appalachian Mountains. Trans. Am. Fish. Soc. 114:195–203.

Lasier, P.J. 1986. The Use of Benthic Macroinvertebrates as Indicators of Acid Sensitivity in Headwater Streams of the Southern Blue Ridge Province. M.S. Thesis. University of Georgia, Athens, GA.

Lawrence, G.B., and C.T. Driscoll. 1988. Aluminum chemistry downstream of a whole-tree-harvested watershed. Environ. Sci. Technol. 22:1293–1299.

Lawrence, G.B., R.D. Fuller, and C.T. Driscoll. 1988. Release of aluminum following whole-tree harvesting at the Hubbard Brook Experimental Forest, New Hampshire. J. Environ. Qual. 16:383–390.

Lawson, D.R., and J.G. Wendt. 1982. Acid Deposition in California. SAE Technical Paper Series. Warrendale, PA.

LaZerte, B.D., and P.J. Dillon. 1984. Relative importance of anthropogenic versus natural sources of acidity in lakes and streams of central Ontario. Can. J. Fish. Aquat. Sci. 41:1664–1677.

Leach, S.D. 1982. Estimated Water Use in Florida, 1980. Florida Bureau of Geology Map Series No. 103. Tallahassee, FL.

Lee, S., and J.L. Schnoor. 1988. Reactions that modify chemistry in lakes of the National Surface Water Survey. Environ. Sci. Technol. 22:190–195.

Lee Y., and S.E. Schwartz. 1981. Evaluation of rate of uptake of nitrogen dioxide by atmospheric and surface liquid water. J. Geophys. Res. 86:11,971–11,983.

Lefohn, A.S., and G.O. Klock. 1985. The possible importance of forest soil processes in defining surface water pH depression. J. Air Pollut. Control Assoc. 35: 632–637.

LeGrand, H.E. 1958. Chemical character of water in the igneous and metamorphic rocks of North Carolina. Econ. Geol. 54:178–189.

Leivestad, H., and I.P. Muniz. 1976. Fish kill at low pH in a Norwegian river. Nature 259:391–392.

Leivestad, H., G. Hendrey, I.P. Muniz, and E. Snekvik. 1976. Effects of acid precipitation on freshwater organisms. Pages 87–111 in F.H. Braekke, ed. Impact of Acid Precipitation on Forest and Freshwater Ecosystems in Norway. SNSF Project FR 6/76. Aas, Norway.

Lennon, R.E. 1961. The Trout Fishery in Shenandoah National Park. Special Scientific Report–Fisheries No. 295. U.S. Bureau of Sport Fisheries and Wildlife.

Lennon, R.E. 1962. An annotated list of the fishes of Great Smoky Mountains National Park. J. Tennessee Acad. Sci. 37:5–7.

Leonard, R.L., C.R. Goldman, and G.E. Likens. 1981. Some measurements of pH and chemistry of precipitation at Davis and Lake Tahoe, California. Water Air Soil Pollut. 15:153–167.

Lerman, A., and G.J. Brunskill. 1971. Migration of major constituents from lake sediments into lake water and its bearing on lake water composition. Limnol. Oceanogr. 16:880–890.

Leslie, P.H. 1945. On the use of matrices in certain population mathematics. Biometrika 33:183–212.

Leuven, R.S.E.W., S.E. Wendelaar Bonga, F.G.F. Oyen, and W. Hagemeijer. 1987. Effects of acid stress on the distribution and reproductive success of freshwater fish in Dutch soft waters. Ann. Soc. Roy. Zool. Belg. 117:231–242.

Lewis, W.M., Jr. 1982. Changes in pH and buffering capacity of lakes in the Colorado Rockies. Limnol. Oceanogr. 27:167–172.

Lewis, W.M., Jr., and M.C. Grant. 1979. Change in the output of ions from a watershed as a result of the acidification of precipitation. Ecology 60:1093–1097.

Lewis, W.M., Jr., and M.C. Grant. 1980. Acid precipitation in the western United States. Science 207: 176–177.

Lewis, W.M., Jr., M.C. Grant, and J.F. Saunders, III. 1984. Chemical patterns of atmospheric deposition in the state of Colorado. Water Resour. Res. 20:1691–1704.

Likens, G.E. 1985. An Ecosystems Approach to Aquatic Ecology: Mirror Lake and Its Environment. Springer-Verlag, New York.

Likens, G.E., F.H. Bormann, N.M. Johnson, D.W. Fisher, and R.S. Pierce. 1970. Effects of forest cutting and herbicide treatment on nutrient budgets in the Hubbard Brook watershed-ecosystem. Ecol. Monogr. 40:23–47.

Likens, G.E., F.H. Bormann, R.S. Pierce, J.S. Eaton, and N.M. Johnson. 1977. Biogeochemistry of a Forested Ecosystem. Input-output budgets, pages 65–86. Springer-Verlag, New York.

Liljestrand, H.M., and J.J. Morgan. 1981. Spatial variation of acid precipitation in southern California. Environ. Sci. Technol. 15:333–338.

Lin, J.C., and J.L. Schnoor. 1986. Acid precipitation model for seepage lakes. J. Environ. Engin. 112:677–695.

Lin, J.C., J.L. Schnoor, and G.E. Glass. 1987. Ion budgets in a seepage lake. Pages 209–228 in R.A. Hites and S.J. Eisenreich, eds. Sources and Fates of Aquatic Pollutants. Advances in Chemistry Series, No. 216. American Chemical Society, Washington, D.C.

Lindberg, S.E., and D.W. Johnson. 1989. 1987 Annual Group Leaders' Reports of the Integrated Forest Study. ORNL/TM-11052. Environmental Sciences Division, Oak Ridge National Laboratory, Oak Ridge, TN.

Lindberg, S.E., and G.M. Lovett. 1985. Field measurements of particle deposition rates to foliage and inert surfaces in a forest canopy. Environ. Sci. Technol. 19:238–244.

Lindberg, S.E., G.M. Lovett, D.D. Richter, and D.W. Johnson. 1986. Atmospheric deposition and canopy interactions of major ions in a forest. Science 231: 141–145.

Lindberg, S.E., D. Silsbee, D.A. Schaefer, J.G. Owens, and W. Petty. 1988. A comparison of atmospheric exposure conditions at high- and low-elevation forests in the Southern Appalachian Mountains. Pages 321–344 in M.H. Unsworth and D. Fowler, eds. Acid Deposition at High Elevation Sites. Kluwer Academic Publishers, New York.

Lindstrom, T., W. Dickson, and G. Andersson. 1984. Reclaiming acid high mountain lakes by liming: A progress report. Freshw. Res. Inst. Drottningholm 61: 128–137.

Lins, H.F. 1986. Recent patterns of sulfate variability in pristine streams. Atmos. Environ. 20:367–375.

Linsey, G.A., D.W. Schindler, and M.P. Stainton. 1987. Atmospheric deposition of nutrients and major ions at the Experimental Lakes Area, northwestern Ontario, in 1970–1982. Can. J. Fish. Aquat. Sci. 44 (Suppl. 1):206–214.

Linsley, R.K., M.A. Kohler, and J.H.L. Paulhus. 1975. Hydrology for Engineers. Second edition. McGraw-Hill, New York.

Linthurst, R.A., D.H. Landers, J.M. Eilers, D.F. Brakke, W.S. Overton, E.P. Meier, and R.E. Crowe. 1986a. Characteristics of Lakes in the Eastern United States. Volume I. Population Descriptions and Physico-Chemical Relationships. EPA/600/4- 86/007a. U.S. Environmental Protection Agency, Washington, D.C.

Linthurst, R.A., D.H. Landers, J.M. Eilers, P.E. Kellar, D.F. Brakke, W.S. Overton, R.E. Crowe, E.P. Meier, P. Kanciruk, and D.S. Jeffries. 1986b. Regional chemical characteristics of lakes in North America: Part II—eastern United States. Water Air Soil Pollut. 31:577–591.

Litaor, M.I. 1987. The influence of eolian dust on the genesis of alpine soils in the Front Range, Colorado. Soil Sci. Soc. Am. J. 51:142–147.

Loar, J.M. 1985. Application of Habitat Evaluation Models in Southern Appalachian Trout Streams. ORNL/TM-9323. Oak Ridge National Laboratory, Oak Ridge, TN.

Lodge, J.P., J.B. Pate, W. Baskergill, G.S. Swanson, K.C. Hill, E. Lorange, and A.L. Lazrus. 1968. Chemistry of United States Precipitation: Final Report on the National Precipitation Sampling Network. National Center for Atmospheric Research, Boulder, CO.

Loftis, J.C., and C.H. Taylor. 1989. Detecting acid precipitation impacts on lake water quality. Environ. Manage. 13:529–539.

Logan, R.M., J.D. Derby, and L.C. Duncan. 1982. Acid precipitation and lake susceptibility in the central Washington Cascades. Environ. Sci. Technol. 16:771–775.

Longcore, J.R., and D. McAuley. 1988. Aquatic and Terrestrial Resource Investigations of Acid Precipitation and Associated Metals. Final Report for Co-op Agreement 14-16-009-1538. U.S. Fish and Wildlife Service, Washington, D.C.

Loranger, T.J. 1986. Temporal variability of water and snowpack chemistry in the north Cascade Mountains. Thesis. Western Washington University, Bellingham, WA.

Loranger, T.J., D.F. Brakke, M.B. Bonoff, and B.F. Gall. 1986. Temporal variability of lake waters in the north Cascade Mountains (Washington, U.S.A.). Water Air Soil Pollut. 31:123–129.

Love, J.D., and A.C. Christiansen. 1985. Geologic Map of Wyoming. U.S. Geological Survey, Reston, VA.

Lovett, G.M., and J.D. Kingsman. 1990. Atmospheric pollutant deposition to high-elevation ecosystems. Atmos. Environ. (in press).

Lovett, G.M., W.A. Reiners, and R.K. Olsen. 1982. Cloud droplet deposition in sub-alpine balsam fir forests: Hydrological and chemical inputs. Science 218:1303–1304.

Lovley, D.R., and M.J. Klug. 1986. Model for the distribution of sulfate reduction and methanogenesis in freshwater sediments. Geochim. Cosmochim. Acta 50:11–18.

Lund, L.J., and A.D. Brown. 1989. Integrated Soil Processes Studies at Emerald Lake Watershed. Final Report. Contract A5-204-32. California Air Resources Board, Sacramento, CA.

Lund, L.J., A.D. Brown, M.A. Lueking, S.C. Nodvin, A.L. Page, and G. Sposito. 1987. Soil Processes at Emerald Lake Watershed. Final Report. Contract A3-105-32. California Air Resources Board, Sacramento, CA.

Lynch, D.D., and N.B. Dise. 1985. Sensitivity of Stream Basins in Shenandoah National Park to Acid Deposition. Water Resources Investigations Report 85-4115. U.S. Geological Survey, Washington, D.C., and Department of Environmental Sciences, University of Virginia, Richmond, VA.

Lynch, J.A., E.S. Corbett, and G.B. Rishel. 1985. Atmospheric Deposition–Spatial and Temporal Trends in Pennsylvania, 1984. Institute for Research on Land and Water Resources, No. LW8405A. Pennsylvania State University, University Park, PA.

Lynch, J.A., C.M. Hanna, and E.S. Corbett. 1986. Predicting pH, alkalinity and total acidity in stream water during episodic events. Water Resour. Res. 22:905–912.

McAvoy, D.C., and C.T. Driscoll. 1989. The chemical response following base application to Little Simon Pond. J. Water Pollut. Control Fed. 61:1552–1563.

MacDonald, G.A. 1966. Geology of the Cascade Range and Modoc Plateau. In E.H. Bailey, ed. Geology of Northern California. USGS Bulletin 190, California Division of Mines and Geology, San Francisco, CA.

MacIsaac, H.J., T.C. Hutchinson, and W. Keller. 1987. Analysis of planktonic rotifer assemblages from Sudbury, Ontario, area lakes of varying chemical composition. Can. J. Fish. Aquat. Sci. 44:1692–1701.

MacKay, R.J., and K.E. Kersey. 1985. A preliminary study of aquatic insect communities and leaf decomposition in acid streams near Dorset, Ontario. Hydrobiologia 122:3–11.

Mackie, G.L. 1987. Effects of acidifying environments on freshwater mollusks in southern Ontario, Canada. Amer. Malacol. Bull. 5:31–39.

Mairs, D.F. 1966. A total alkalinity atlas for Maine lake waters. Limnol. Oceanogr. 11:68–72.

Mairs, D.F. 1967. Surface chloride distribution in Maine lakes. Water Resour. Res. 3:1090–1092.

Majors, J., and D.W. Taylor. 1977. Alpine. Pages 601–675 in M.G. Barbour and J. Major, eds. Terrestrial Vegetation of California. John Wiley and Sons, New York.

Malanchuk, J.L., and R.S. Turner. 1987. Effects on aquatic systems. Pages 8–81 in Interim Assessment: The Causes and Effects of Acidic Deposition. Volume IV. Effects of Acidic Deposition. The National Acid Precipitaion Assessment Program, Washington, D.C.

Malander, M.W. 1983. A Stable Isotopic Study of Waters within and Surrounding the Boundary Waters Canoe Area Wilderness, Minnesota. M.S. Thesis. Northern Illinois University, De Kalb, IL.

Malley, D.F., D.L. Findlay, and P.S. Chang. 1982. Ecological effects of acid precipitation on zoo-plankton. Pages 297–327 in F. D'Itri, ed. Acid Precipitation: Effects on Ecological Systems. Ann Arbor Science, Ann Arbor, MI.

Marmorek, D.R. 1983. The Effects of Lake Acidification on Zooplankton Community Structure and Phytoplankton-Zooplankton Interactions: An Experimental Approach. M.S. Thesis. University of British Columbia, Vancouver, British Columbia.

Marmorek, D.R. 1984. Changes in the temporal behavior and size structure of plankton systems in acid lakes. Pages 23–41 in G.R. Hendrey, ed. Early Biotic Responses to Advancing Lake Acidification. Butterworth Publishers, Boston, MA.

Marmorek, D.R., and D.P. Bernard. 1987. The Use of Zooplankton to Detect Lake Acidification and Recovery. Final Report. U.S. Environmental Protection Agency, Corvallis, OR.

Marmorek, D.R., G.B. Cunningham, M.L. Jones, and P. Bunnel. 1984. Snowmelt Effects Related to Acidic Precipitation: A Structured Review of Existing Knowledge and Current Research Activities. Final Report. Environment Canada, Environment and Social Systems Analysts, Inc., Vancouver, British Columbia.

Marmorek, D.R., K.W. Thornton, J.P. Baker, D.P. Bernard, M.L. Jones, and B.S. Reuber. 1986. Acidic Episodes in Surface Waters: The State of Science. Final Report. U.S. Environmental Protection Agency, Corvallis, OR.

Marmorek, D.R., D.P. Bernard, M.L. Jones, L.P. Rattie, and T.J. Sullivan. 1988. The effects of mineral acid deposition on concentrations of dissolved organic acids in surface waters. ERL-COR-500AP. U.S. Environmental Protection Agency, Corvallis, OR.

Martin, C.W. 1979. Precipitation and streamwater chemistry in an undisturbed forested watershed in New Hampshire. Ecology 60:36–42.

Martin, L. 1965. The Physical Geography of Wisconsin, Ed. 3. University of Wisconsin Press, Madison, WI.

Mast, M.A. 1989. A Field and Laboratory Study of Chemical Weathering with Special Reference to Acid Deposition. Thesis. Department of Geology and Geophysics, University of Wyoming, Laramie, WY.

Mathews, R.C., Jr. 1978. Ecological Survey of Abrams Creek in the Great Smoky Mountains National Park. NPS-SER Research/Resources Management Report No. 28. U.S. Department of Interior, National Park Service, Washington, D.C.

Mathews, R.C., Jr., and E.L. Morgan. 1982. Toxicity of Anakeesta formation leachates to shovel-nosed salamander, Great Smoky Mountains National Park. J. Environ. Qual. 11:102–106.

Matthes, F.E. 1930. Geologic History of the Yosemite Valley (Calif.) (with a chapter on the granitic rocks of the Yosemite region by F.C. Calkins). Professional Paper 160. U.S. Geological Survey, Reston, VA.

Matuszek, J.E., and G.L. Beggs. 1988. Fish species richness in relation to lake area, pH, and other abiotic factors in Ontario lakes. Can. J. Fish. Aquat. Sci. 45:1931–1941.

Matuszek, J.E., J. Goodier, and D.L. Wales. 1988. The Occurrence of Cyprinids and Other Small Fish Species in Relation to pH in Ontario Lakes. Ontario Fisheries Acidification Report Series No. 88-14, Ontario Ministry of Natural Resources, Toronto, Ontario.

Mayewski, P.A., G.H. Denton, and T.J. Hughes. 1981. Late Wisconsin ice sheets of North America. In G.H. Denton and T.J. Hughes, eds. The Last Great Ice Sheets. John Wiley and Sons, New York.

McBean, G.A., and S. Nikleva. 1986. Composition of snow in Pacific coastal mountains. Atmos. Environ. 20:1161–1164.

McCleneghan, K., J.L. Nelson, J.T. King, and S.J. Baumgartner. 1985. Statewide Survey of Aquatic Ecosystem Chemistry. Laboratory Report No. 85-5. California Dept. of Fish and Game, Rancho Cordova, CA.

McCleneghan, K., R.H. Imai, J.T. King, and S.J. Boggs. 1987. Statewide Survey of Aquatic Ecosystem Chemistry, 1986. Laboratory Report No. 86-5. California Department of Fish and Game, Rancho Cordova, CA.

McColl, J.G. 1980. A Survey of Acid Precipitation in Northern California. Final Report. Contract A8-136-31. California Air Resources Board, Sacramento, CA.

McColl, J.G. 1981a. Increasing hydrogen ion activity of water in two reservoirs supplying the San Francisco Bay area, California. Water Resour. Res. 17:1510–1516.

McColl, J.G. 1981b. Acid rain in northern California. Fremontia 8:3–5.

McColl, J.G., and D.S. Bush. 1978. Precipitation and throughfall chemistry in the San Francisco Bay area. J. Environ. Qual. 7:352–357.

McConnell, J.B., G.C. Bortleson, and J.K. Innes. 1976. Data on selected lakes in Washington, part 4. USGS Water-Supply Bulletin 42. Washington State Department of Ecology, Olympia, WA, in cooperation with U.S. Department of Interior, Washington, D.C.

McCormick, J.H., K.M. Jensen, and L.E. Anderson. 1989. Chronic effects of low pH and elevated aluminum on survival, maturation, spawning, and embryolarval development of the fathead minnow in soft water. Water Air Soil Pollut. 43:293–307.

McDonald, D.G. 1983. The effects of H^+ upon the gills of freshwater fish. Can. J. Zool. 61:691–703.

McFee, W.W. 1980. Sensitivity of Soil Regions to Acid Precipitation. EPA/600/3-80/013. U.S. Environmental Protection Agency, Corvallis, OR.

McIntire, C.D., and M.K. Debacon. 1988. Taxonomy and ecology of the phytoplankton of Crater Lake. Final Report, Cooperative Agreement No. CA-9000-3-0003, Subagreement No. 13, National Park Service, Pacific Northwest Region, Seattle, WA.

McKee, B. 1972. Cascadia, The Geologic Evolution of the Pacific Northwest. McGraw-Hill, New York.

McKinley, V.L., and J.R. Vestal. 1982. Effects of acid on litter decomposition in an Arctic lake. Appl. Environ. Microbiol. 43:1188–1195.

McMartin, B. 1985. Citizen's Guide to Adirondack Forestry. Adirondack Park Agency, State of New York, Ray Brook, NY.

McVeety, B.T., and R.A. Hites. 1988. Atmospheric deposition of polycyclic aromatic hydrocarbons to Siskiwit Lake, Isle Royale: A mass balance approach. Atmos. Environ. 22:511–536.

Means, D.B. 1987. Impacts on Diversity of the 1985 Land and Resource Management Plan for National Forests in Florida. Coastal Plains Institute, Tallahassee, FL.

Means, D.B., and G. Grow. 1985. The endangered longleaf pine community. ENFO, September, 1985. Florida Conservation Foundation, Winter Park, FL.

Meehean, O.L. 1941. Fish populations of five Florida lakes. Trans. Am. Fish. Soc. 71:184–194.

Meinert, D.L., and F.A. Miller, III. 1981. A Review of Water Quality Data in Acid Sensitive Watersheds within the Tennessee Valley. Division of Water Resources, Office of Natural Resources, Tennessee Valley Authority, Chattanooga, TN.

Melack, J.M., and F.V. Setaro. 1986. Survey of Sensitivity of Southern California Lakes to Acid Deposition. Final Report. Contract A3-107-32. California Air Resources Board, Sacramento, CA.

Melack, J.M., J.L. Stoddard, and D.R. Dawson. 1983a. Acid precipitation and buffer capacity of lakes in the Sierra Nevada, California. Pages 465–471 in J.A. Johnson and R.A. Clarke, eds. International Symposium on Hydrometeorology. American Water Resources Association, Bethesda, MD.

Melack, J.M., J.L. Stoddard, D.R. Dawson, and C.A. Ochs. 1983b. Atmospheric Deposition and the Chemical and Biological Status of Lakes in the Sierra Nevada, California. Technical Completion Report. NTIS PB83-255521. California Water Resources Center, University of California, Davis, CA.

Melack, J.M., J.L. Stoddard, and C.A. Ochs. 1985. Major ion chemistry and sensitivity to acid precipitation of Sierra Nevada lakes. Water Resour. Res. 21: 27–32.

Melack, J.M., S.D. Cooper, and R.W. Holmes. 1987. Chemical and Biological Survey of Lakes and Streams in the Emerald Lake Watershed, Sequoia National Park. Final Report. Contract A3-096-32. California Air Resources Board, Sacramento, CA.

Melack, J.M., S.D. Cooper, and T.M. Jenkins. 1989. Chemical and Biological Characteristics of Emerald Lake and the Streams in Its Watershed, and Responses of the Lake and Streams to Acidic Deposition. Final Report. Contract A6-184-32. California Air Resources Board, Sacramento, CA.

Melillo, J.M., J.D. Aber, P.A. Steudler, and J.P. Schimel. 1983. Denitrification potentials in a successional sequency of northern hardwood forest stands. Environ. Biogeochem. Ecol. Bull. (Stockholm) 35: 217–228.

Meriläinen, J. 1967. The diatom flora and the hydrogen-ion concentration of the water. Ann. Bot. Fenn. 4:51–58.

Messer, J.J., and K.N. Eshleman. 1987. The Feasibility of Quantifying the Regional Extent, Magnitude, Duration, and Frequency of Episodes: Results of the National Surface Water Survey Pilot Studies. Presentation at NAPAP Aquatic Effects Task Group VI Peer Review, May 17–23, New Orleans, LA.

Messer, J.J., C.W. Ariss, R. Baker, S.K. Drousé, K.N. Eshleman, P.R. Kaufmann, R.A. Linthurst, J.M. Omernik, W.S. Overton, M.J. Sale, R.D. Schonbrod, S.M. Stambaugh, and J.R. Tuschall, Jr. 1986. National Surface Water Survey: National Stream Survey Phase I–Pilot Survey. EPA/600/4-86/026. U.S. Environmental Protection Agency, Washington, D.C.

Messer, J.J., C.W. Ariss, J.R. Baker, S.K. Drousé, K.N. Eshleman, A.J. Kinney, W.S. Overton, M.J. Sale, and R.D. Schonbrod. 1988. Stream chemistry in the Southern Blue Ridge: Feasibility of a regional synoptic sampling approach. Water Resour. Bull. 24: 821–829.

Michaels, P.J., P.J. Stenger, D.E. Sappington, and P.C. Knappenberger. 1988. Origin and Destination of Pollutant-Bearing Airstreams Impacting and Exiting the Commonwealth of Virginia. University of Virginia, State Climatology Office of the Environmental Sciences, Charlottesville, VA.

Mierle, G., K. Clark, and R.L. France. 1986. The impact of acidification on aquatic biota in North America: A comparison of field and laboratory results. Water Air Soil Pollut. 31:593–604.

Milford, J.B., and C.I. Davidson. 1985. The sizes of particulate trace elements in the atmosphere–a review. J. Air Poll. Control Assoc. 35:1249–1260.

Miller, C.D. 1988. Holocene eruptions at the Inyo volcanic chain, California: Implications for possible eruptions in Long Valley caldera. Geology 13:14–17.

Miller, D.H. 1955. Snow Cover and Climate in the Sierra Nevada, California. No. 11. University of California, Publishing in Geography, Los Angeles, CA.

Miller, D.W. 1987. Weathering and Cation Exchange in Soils of Shenandoah National Park, Virginia. M.S. Thesis. Department of Environmental Sciences, University of Virginia, Charlottesville, VA.

Miller, P. 1980. Effects of artificial acidification on the growth of periphyton. Can. J. Fish. Aquat. Sci. 37: 355–363.

Miller, P.R., M.H. McCutchan, and H.P. Milligan. 1972. Oxidant air pollution in the Central Valley, Sierra Nevada Foothills, and Mineral King Valley of California. Atmos. Environ. 6:623–633.

Mills, K.H., and D.W. Schindler. 1986. Biological indicators of lake acidification. Water Air Soil Pollut. 30:779–789.

Mills, K.H., S.M. Chalanchuk, L.C. Mohr, and I.J. Davies. 1987. Responses of fish populations in Lake 223 to 8 years of experimental acidification. Can. J. Fish. Aquat. Sci. 44:114–125.

Minns, C.K., J.R.M. Kelso, and M.G. Johnson. 1986. Large-scale risk assessment of acid rain impacts on fisheries: Models and lessons. Can. J. Fish. Aquat. Sci. 43:900–921.

Mitchell, M.J., M.B. David, and A.J. Uutala. 1985. Sulfur distribution in lake sediment profiles as an index of historical depositional patterns. Hydrobiologia 121:121–127.

Mitchell, M.J., S.C. Schindler, J.S. Owen, and S.A. Norton. 1988. Comparison of sulfur concentrations within lake sediment profiles. Hydrobiologia 157: 219–229.

Mohn, L.O., and P.E. Bugas, Jr. 1979. Virginia Trout Stream and Environmental Inventory. Virginia Commission of Game and Inland Fisheries, Richmond, VA.

Mohn, L.O., J.W. Kaufman, P.M. Bugas, and P.P. Smith. 1988. Effects of acidification on bottom fauna and fish populations in St. Mary's River, Augusta County, Virginia. Abstract of oral presentation at the Northeast American Fisheries Society Meeting, March, 1988. Northeast American Fisheries Society (Abstract).

Mohnen, V.A. 1988. Mountain Cloud Chemistry Project–Wet, Dry, and Cloud Water Deposition. U.S. Environmental Protection Agency, AREAL, Research Triangle Park, NC.

Molliter, A.V., and D.J. Raynal. 1982. Acid precipitation and ionic movements in Adirondack forest soil. J. Soil Sci. Am. 46:137–141.

Monk, C.D. 1960. A preliminary study on the relationship between the vegetation of a mesic hammock community and a sandhill community. J. Fla. Acad. Sci. 23:1–12.

Monk, C.D. 1968. Successional and environmental relationships of the forest vegetation of northcentral Florida. Am. Midl. Nat. 79:441–457.

Moore, J.G., and C. Wahrhaftig. 1984. Geology of Emerald Lake Basin in Relation to Acid Precipitation. Open File Report 84-400, pages 36–39. U.S. Geological Survey, Reston, VA.

Moore, R.B. 1937. Forest type map write-up by watersheds, Shenandoah National Park. U.S. Department of the Interior, National Park Service, Shenandoah National Park, Luray, VA.

Moore, S.E., B. Ridley, and G.L. Larson. 1983. Standing crops of brook trout concurrent with removal of rainbow trout from selected streams in the Great

Smoky Mountains National Park. North Am. J. Fish. Manage. 3:72–80.

Moore, S.E., G.L. Larson, and B. Ridley. 1986. Population control of exotic rainbow trout in streams of a natural area park. Environ. Manage. 10:215–219.

Morel, F.M.M. 1983. Principles of Aquatic Chemistry. Wiley-Interscience, New York. Morris, F.A., D.V. Peck, M.B. Bonoff, K.J. Cabble, and S.L. Pierett. 1986. National Surface Water Survey, Eastern Lake Survey (Phase I–Synoptic Chemistry) Field Operations Report. EPA/600/4-86/010. U.S. Environmental Protection Agency, Las Vegas, NV.

Mount, D.R. 1987. Physiological and Toxicological Effects of Long-Term Exposure to Acid, Aluminum and Low Calcium on Adult Brook Trout (*Salvelinus fontinalis*) and Rainbow Trout (*Salmo gairdneri*). Ph.D. Dissertation. University of Wyoming, Laramie, WY.

Mount, D.R., and M.D. Marcus (eds.) 1988. Physiological, Toxicological, and Population Responses of Brook Trout to Acidification. Research Report. Electric Power Research Institute, Palo Alto, CA.

Mount, D.R., J.E. Breck, S.W. Christensen, W.A. Gern, M.D. Marcus, C.G. Ingersoll, D.D. Gulley, D.G. McDonald, B.R. Parkhurst, W. Van Winkle, C.M. Wood, and H.L. Bergman. 1989. Physiologic, Toxicologic, and Population Responses of Brook Trout to Acidification. EPRI EN-6238. Electric Power Research Institute, Palo Alto, CA.

Moyse, D.W., and I.J. Fernandez. 1987. Trace metals in the forest floor at Saddleback Mt., Maine. Water Air Soil Pollut. 34:385–397.

MPCA. 1988. Unpublished data supplied by G. Pratt, Minnesota Pollution Control Agency, St. Paul, MN.

Mudrey, M.G., Jr., B.A. Brown, and J.K. Greenburg. 1982. Bedrock geologic map of Wisconsin. University of Wisconsin-Extention and Wisconsin Geological and Natural History Survey, Madison, WI.

Mulholland, P.J., J.D. Newbold, J.W. Elwood, and C.L. Hom. 1983. The effect of grazing intensity on phosphorus spiralling in autotrophic streams. Oecologia 66:199–206.

Mulholland, P.J., J.W. Elwood, and A.V. Palumbo. 1986. Effect of stream acidification on periphyton composition, chlorophyll, and productivity. Can. J. Fish. Aquat. Sci. 43:1846–1858.

Müller, P. 1980. Effects of artificial acidification on the growth of periphyton. Can. J. Fish. Aquat. Sci. 37: 355–363.

Munger, J.W. 1982. Chemistry of atmospheric precipitation in the north-central United States: Influence of sulfate, nitrate, ammonia, and calcareous soil particulates. Environment 16:1633–1645.

Munger, J.W., and S.J. Eisenreich. 1983. Continental scale variations in precipitation chemistry. Environ. Sci. Technol. 17:32–42.

Muniz, I.P., and H. Leivestad. 1980. Acidification effects on freshwater fish. Pages 84–92 in D. Drabløs and A. Tollan, eds. Ecological Impact of Acid Precipitation. Proceedings of an International Conference, Sandefjord, Norway. SNSF Project, Oslo, Norway.

Muniz, I.P., H.M. Seip, and I.H. Sevaldrud. 1984. Relationship between fish populations and pH for lakes in southernmost Norway. Water Air Soil Pollut. 23:97–113.

Muniz, I.P., R. Andersen, and T.J. Sullivan. 1987. Physiological response of brown trout (*Salmo trutta*) spawners and post-spawners exposed to acidic, aluminum-rich stream water. Water Air Soil Pollut. 36: 371–379.

Munson, R.K., and S.A. Gherini. 1986. Simulation of Pancake-Hall Creek. Electric Power Research Institute, Palo Alto, CA.

Munson, R.K., S.A. Gherini, M.M. Lang, L.E. Gomez, C.W. Chen, R.A. Goldstein, and D.R. Knauer. 1987. ILWAS model applications: Response of various surface waters to deposition acidity. Pages 301–308 in R. Perry, R.M. Harrison, J.N.B. Bell, and J.N. Lester, eds. Acid Rain: Scientific and Technical Advances. Selper Ltd., Ealing, London.

Munz, P.A., and D.D. Keck. 1959. A California Flora. University of California Press, Berkeley, CA.

Murdoch, P.S. 1988. Chemical Budgets and Stream-Chemistry Dynamics at Biscuit Brook, Catskill Mountains, New York, 1984–85. Water Resources Investigations Report 88-4035. U.S. Geological Survey, Reston, VA.

Murdoch, P.S., and C.R. Barnes. (In press) Stream acidification in the Catskill Mountains of New York. Water Resources Investigations Report. U.S. Geological Survey, Reston, VA.

Murdoch, P.S., N.E. Peters, and R.M. Newton. 1987. Hydrologic Analysis of Two Headwater Lake Basins of Differring Lake pH in the West-Central Adirondack Mountains of New York. Water Resources Investigations Report 84-4313. U.S. Geological Survey, Reston, VA.

Musgrove, R.H., J.B. Foster, and L.G. Toler. 1965. Water Resources of the Econfina Creek Basin Area in Northwestern Florida. Report of Investigations No. 41. Florida Geological Survey, Tallahassee, FL.

Myers, R.L. 1985. Fire and the dynamic relationship between Florida sandhill and sand pine scrub vegetation. Bull. Torrey Bot. Club 112:241–252.

Myers, R.L., and D.L. White. 1987. Landscape history and changes in sandhill vegetation in northcentral and southcentral Florida. Bull. Torrey Bot. Club 114:21–32.

Nalewajko, C., and M.A. O'Mahony. 1988. Effects of acid pH shock on phosphate concentrations and microbial phosphate uptake in an acidifying and circumneutral lake. Can. J. Fish. Aquat. Sci. 45:254–260.

National Atmospheric Deposition Program (NADP). 1986. NADP/NTN Annual Data Summary: Precipitation Chemistry in the United States, 1984. NADP/NTN Coordinator's Office, Natural Resource Ecology Laboratory, Colorado State University, Fort Collins, CO.

National Atmospheric Deposition Program (NADP). 1987a. NADP/NTN Annual Data Summary: Precipitation Chemistry in the United States, 1985. Vol. IX, No. 1. NADP/NTN Coordinator's Office, Natural Resource Ecology Laboratory, Colorado State University, Fort Collins, CO.

National Atmospheric Deposition Program (NADP). 1987b. NADP/NTN Annual Data Summary: Precipitation Chemistry in the United States, 1986. NADP/NTN Coordinator's Office, Natural Resource Ecology Laboratory, Colorado State University, Fort Collins, CO.

National Atmospheric Deposition Program (NADP). 1988. NADP/NTN Annual Data Summary: Precipitation Chemistry in the United States 1987. NADP/NTN Coordinator's Office, Natural Resource Ecology Laboratory, Colorado State University, Fort Collins, CO.

National Atmospheric Deposition Program (NADP). 1989. NADP/NTN Annual Data Summary: Precipitation Chemistry in the United States 1988. NADP/NTN Coordinator's Office, Natural Resource Ecology Laboratory, Colorado State University, Fort Collins, CO.

National Oceanic and Atmospheric Administration (NOAA). 1956. Climatological Data Annual Summary 1955. U.S. Department of Commerce, Washington, D.C.

National Oceanic and Atmospheric Administration (NOAA). 1957. Climatological Data Annual Summary 1956. U.S. Department of Commerce, Washington, D.C.

National Oceanic and Atmospheric Administration (NOAA). 1982. Monthly normals of temperature, precipitation, and heating and cooling degree days 1951–1980, Virginia-District of Columbia. NOAA National Climatic Center, Asheville, NC.

National Oceanic and Atmospheric Administration (NOAA). 1986. Climatological Data Annual Summary 1985. U.S. Department of Commerce, Washington, D.C.

National Oceanic and Atmospheric Administration (NOAA). Climatic Summaries. Asheville, NC.

National Research Council (NRC). 1983. Acid Deposition. Atmospheric Processes in Eastern North American. National Academy Press, Washington, D.C.

National Research Council (NRC). 1984. Acid Deposition: Processes of Lake Acidification. National Academy Press, Washington, D.C.

National Research Council (NRC). 1986. Acid Deposition: Long-Term Trends. National Academy Press, Washington, D.C.

National Research Council Canada. 1981. Acidification in the Canadian Aquatic Environment: Scientific Criteria for Assessing the Effects of Acidic Deposition on Aquatic Ecosystems. No. 18475. Environmental Secretariat, Ottawa, Canada.

Neal, C. 1988. Determination of dissolved CO_2 in upland streamwater. J. Hydrol. 99:127–142.

Neary, B.P., and P.J. Dillon. 1988. Effects of sulphur deposition on lake-water chemistry in Ontario, Canada. Nature 333:340–343.

Nelson, P.O. 1985. Acid Rain Buffering Potential in Oregon Cascade Lakes: Secondary Mineral Solubility Control of Solution Ionic Composition. Report No. WRRI-100. Water Resources Research Institute, Oregon State University, Corvallis, OR.

Nelson, P.O., and R.P. Baumgartner. 1988. Major Ion and Dissolved Aluminium Chemistry of Selected Lakes in Mt. Rainier National Park. Final Report. Cooperative Agreement No. CA 9000 3-0003, Subagreement No. 15, CPSU, College of Forestry, Oregon State University, Corvallis, OR.

Nelson, P.O., and G.K. Delwiche. 1983. Sensitivity of Oregon's Cascade Lakes to Acid Precipitation. Report No. WRRI-85. Water Resources Research Institute, Oregon State University, Corvallis, OR.

Nelson, W.C. 1987. Survival and Growth of Fingerling Trout Planted in High Lakes of Colorado. Technical Publication No. 36. Colorado Division of Wildlife, Denver, CO.

Nero, R.W., and D.W. Schindler. 1983. Decline of *Mysis relicta* during the acidification of Lake 223. Can. J. Fish. Aquat. Sci. 40:1905–1911.

Neves, R.J., and G.B. Pardue. 1983. Abundance and production of fishes in a small Appalachian stream. Trans. Am. Fish. Soc. 112:21–26.

Neville, C.M. 1985. Physiological response of juvenile rainbow trout, *Salmo gairdneri*, to acid and aluminum—prediction of field responses from laboratory data. Can. J. Fish. Aquat. Sci. 42:2004–2019.

Newell, A.D. 1987. Predicting spring lake chemistry from fall samples. Pages 353–363 in R. Perry, R.M. Harrison, J.N.B. Bell, and J.N. Lester, eds. Acid Rain: Scientific and Technical Advances. Selper Ltd., Ealing, London.

Newman, L. 1975. Acidity in rainwater: Has an explanation been presented? Science 185:957–958.

Newton, R.M., and R.A. April. 1982. The relationship between surface water chemistry and geology in the North Branch of the Moose River. Biogeochemistry 3:21–35.

Newton, R.M., J. Weintraub, and R. April. 1987. The relationship between surface water chemistry and geology in the north branch of the Moose River. Biogeochemistry 3:21-35.

New York State Department of Conservation. 1930. A Biological Survey of the Champlain Watershed, Supplement to Nineteenth Annual Report, 1929. J.B. Lyon Company, Albany, NY.

New York State Department of Conservation. 1931. A Biological Survey of the St. Lawrence Watershed, Supplement to Twentieth Annual Report, 1930. J.B. Lyon Company, Albany, NY.

New York State Department of Conservation. 1932. A Biological Survey of the Oswegatchie and Black River Systems, Supplement to Twenty-first Annual Report, 1931. J.B. Lyon Company, Albany, NY.

New York State Department of Conservation. 1933. A Biological Survey of the Upper Hudwon Watershed, Supplement to Twenty-second Annual Report, 1932. J.B. Lyon Company, Albany, NY.

New York State Department of Conservation. 1934. A Biological Survey of the Racquette Watershed, Supplement to Twenty-third Annual Report, 1933. J.B. Lyon Company, Albany, NY.

New York State Department of Conservation. 1935. A Biological Survey of the Mohawk-Hudson Watershed, Supplement to Twenty-fourth Annual Report, 1934. J.B. Lyon Company, Albany, NY.

New York State Department of Conservation. 1936. A Biological Survey of the Delaware and Susquehanna Watersheds, Supplement to Twenty-fifth Annual Report, 1935. J.B. Lyon Co., Albany, NY.

New York State Department of Conservation. 1937. A Biological Survey of the Lower Hudson Watershed, Supplement to Twenty-sixth Annual Report, 1936. J.B. Lyon Co., Albany, NY.

Nichols, D.S., and R.E. McRoberts. 1986. Relations between lake acidification and sulfate deposition in northern Minnesota, Wisconsin, and Michigan. Water Air Soil Pollut. 31:197-206.

Nikolaidis, N. 1987. Modeling the Direct Versus Delayed Response of Surface Waters to Acid Deposition in the Northeastern United States. Ph.D. Dissertation. University of Iowa, Ames, IA.

Nilssen, J.P. 1980. Acidification of a small watershed in southern Norway and some characteristics of acidic aquatic environments. Int. Rev. Gesamten Hydrobiol. 65:177-207.

Nilsson, S.I., H.G. Miller, and J.D. Miller. 1982. Forest growth as a possible cause of soil and water acidification: An examination of the concepts. Oikos 39:40-49.

Nishida, A.I., and J.L. Schnoor. 1989. Steady State Model to Determine Lake Resources at Risk to Acid Deposition in the Sierra Nevada, California. Final Report. Contract A7-32-36. California Air Resources Board, Sacramento, CA.

Nodvin, S.C. 1987. Processes affecting the chemistry of waters passing through a high elevation Sierra Nevada watershed. Pages 101-106 in Acidification and Water Pathways. Proceedings of UNESCO/IHP International Symposium, Volume I. Norwegian National Commission for Hydrology, Oslo, Norway.

Nodvin, S.C., and J.S. Kahl. 1987. Chemical characteristics of headwater streams in central Maine. Annual Meeting of the Ecological Society of America, August, Minneapolis, MN (unpublished abstract).

Nodvin, S.C., C.T. Driscoll, and G.E. Likens. 1986a. The effect of pH on sulfate adsorption by a forest soil. Soil Sci. 142:69-75.

Nodvin, S.C., L.B. Weeks, E.P.E. Thomas, and L.J. Lund. 1986b. Alkalization of a high-elevation Sierra Nevada stream. Water Resour. Res. 22:1077-1082.

Nodvin, S.C., C.T. Driscoll, and G.E. Likens. 1988. Soil processes and sulfate loss at the Hubbard Brook Experimental Forest. Biogeochemistry 5:185-199.

Norris, R.M., and R.W. Webb. 1976. Geology of California. John Wiley and Sons, New York.

Norton, S.A. 1986. A review of the chemical record in lake sediment of energy-related air pollution and its effects on lakes. Water Air Soil Pollut. 30:331-346.

Norton, S.A., and J.S. Kahl. 1982. Reduction of terrestrial alkalinity production: A major mechanism of surface water acidification. Symposium on Acidic Precipitation and Atmospheric Deposition in Western North America, June, Bellingham, WA (unpublished abstract).

Norton, S.A., and J.S. Kahl. 1986. A comparison of lake sediments and ombrotrophic peat deposits as long-term monitors of atmospheric pollution. Pages 40-57 in T. Boyle, ed. New Approaches to Monitoring Aquatic Ecosystems. ASTM STP 940. American Society for Testing and Materials, Philadelphia, PA.

Norton, S.A., R.B. Davis, and D.F. Brakke. 1981. Responses of Northern New England Lakes to Atmospheric Inputs of Acids and Heavy Metals. Completion Report Project A-048-ME. Land and Water Resources Center, University of Maine, Orono, ME.

Norton, S.A., J.J. Akielaszek, T.A. Haines, K.J. Stromborg, and J.R. Longcore. 1982. Pages 1-13 in Bedrock Geologic Control of Sensitivity of Aquatic Ecosystems in the United States to Acidic Deposition. National Atmospheric Deposition Program, Fort Collins, CO.

Norton, S.A., J.S. Kahl, D.F. Brakke, G.F. Brewer, T.A. Haines, and S.C. Nodvin. 1988a. Regional patterns and local variability of dry and occult deposition strongly influence sulfate concentrations in Maine lakes. Sci. Total Environ. 72:183-196.

Norton, S.A., D.F. Brakke, and J.S. Kahl. 1988b. Major Influences on the Chemistry of Lake Waters in Maine. Volume III. Maine Geological Survey. C.T. Jackson.

Norton, S.A., M.J. Mitchell, J.S. Kahl, and G.F. Brewer. 1988c. In-lake alkalinity generation by sulfate reduction – a paleolimnological assessment. Water Air Soil Pollut. 39:33–45.

Norton, S.A., D.F. Brakke, and A. Henriksen. 1989. Red herring lakes and streams in the acid-rain literature. Sci. Total Environ. 83:113–125.

Noss, R.F. 1988. The longleaf pine landscape of the southeast: Almost gone and almost forgotten. Endangered Species Update 5:1–5.

Nriagu, J.O. 1984. Environmental Impacts of Smelters. Advances in Environmental Science and Technology, No. 15. John Wiley and Sons, New York.

Nriagu, J.O., and R.D. Coker. 1983. Sulfur in sediments chronicles past changes in lake acidification. Nature 303:692–694.

Nriagu, J.O., and Y.K. Soon. 1985. Distribution and isotopic composition of sulfur in lake sediments of northern Ontario. Geochim. Cosmochim. Acta 49: 823–834.

Odum, H.W. 1951. The promise of regionalism. Chapter 15 in M. Jensen, ed. Regionalism in America. University of Wisconsin Press, Madison, WI.

Office of Technology Assessment. 1984. Acid Rain and Transported Air Pollutants: Implications for Public Policy. U.S. Government Printing Office, Washington, D.C.

Ogburn, R.W., III. 1984. Phosphorus Dynamics in an Acidic Softwater Florida Lake. Ph.D. Dissertation. University of Florida, Gainesville, FL.

Okland, J. 1980. Environment and snails (Gastropoda): Studies of 1,000 lakes in Norway. Pages 322–323 in D. Drabløs and A. Tollan, eds. Ecological Impact of Acid recipitation. Proceedings of an International Conference, Sandefjord, Norway. SNSF Project, Oslo, Norway.

Okland, J., and K.A. Okland. 1986. The effects of acid deposition on benthic animals in lakes and streams. Experientia 42:471–486.

Oksanen, J., E. Läära, P. Huttunen, and J. Meriläinen. 1988. Estimation of pH optima and tolerances of diatoms in lake sediments by the methods of weighted averaging, least squares and maximum likelihood, and their use for the prediction of lake acidity. J. Paleolimnol. 1:39–49.

Olem, H. 1985. Acidification Trends in Surface Waters of the Southern Appalachians. TVA/ONRED/AWR-85/13. Tennessee Valley Authority, Office of Natural Resources and Economic Development, Division of Air and Water Resources, Chattanooga, TN.

Olem, H. 1986. Episodic Changes in Stream Water Quality in Five Watersheds in the Southern Blue Ridge Province. Interim Report. Interagency Agreement No. DW64930283-01, TV-61968A. Tennessee Valley Authority, Office of Natural Resources and Economic Development, Division of Air and Water Resources, Water Quality Branch, Chattanooga, TN.

Oliver, B.G., E.M. Thurman, and R.L. Malcolm. 1983. The contribution of humic substances to the acidity of colored natural waters. Geochim. Cosmochim. Acta 47:2031–2035.

Olsen, A.R., and C.R. Watson. 1984. Acid Deposition Annual Data Summaries: 1980, 1981, 1982. EPA/600/7-84/097. U.S. Environmental Protection Agency, Washington, D.C.

Olson, J.J., D.W. Johnson, and D.S. Shriner. 1982. Regional Assessment of Potential Sensitivity of Soils in the Eastern United States to Acid Precipitation. ORNL/TM-8374. Oak Ridge National Laboratory, Oak Ridge, TN.

Omernik, J.M. 1985. Total Alkalinity of Surface Waters: A Map of the Appalachian Region. U.S. Environmental Protection Agency, Corvallis, OR.

Omernik, J.M. 1987. Ecoregions of the conterminous United States. Ann. Assoc. Am. Geog. 77:118–125.

Omernik, J.M., and A.L. Gallant. 1986. Ecoregions of the Pacific Northwest. EPA/600/3-86/033, U.S. EPA Environmental Research Laboratory, Corvallis, OR.

Omernik, J.M., and A.L. Gallant. 1988. Ecoregions of the Upper Midwest States. EPA/600/3-88/037. U.S. Environmental Protection Agency, Corvallis, OR.

Omernik, J.M., and G.E. Griffith. 1986. Total alkalinity of surface waters: A map of the Upper Midwest region of the United States. Environ. Manage. 10: 829–839.

Omernik, J.M., and C.F. Powers. 1982. Total Alkalinity of Surface Waters – A National Map. EPA/600/D-82/333, U.S. EPA Environmental Research Laboratory, Corvallis, OR.

Omernik, J.M., and C.F. Powers. 1983. Total alkalinity of surface waters – a national map. Ann. Assoc. Am. Geog. 73:133–136.

Oppenheimer, M., C.B. Epstein, and R.E. Yuhnke. 1985. Acid deposition, smelter emissions, and the linearity issue in the western United States. Science 229:859–862.

Ormerod, S.J., and R.W. Edwards. 1987. The ordination and classification of macroinvertebrate assemblages in the catchment of the River Wye in relation to environmental factors. Freshwater Biol. 17:533–546.

Ormerod, S.J., N.S. Weatherley, P.V. Varallo, and P.G. Whitehead. 1988. Preliminary empirical models of the historical and future impact of acidification on the ecology of the Welsh streams. Freshwater Biol. 20:127–140.

Osberg, P.H., A.H. Hussey, and G.M. Boone. 1985.

Geologic Map of Maine. Maine Geological Survey, Augusta, ME.

OTA. 1984. Acid Rain and Transported Air Pollutants: Implications for Public Policy. Report OTA-0-204. Office of Technology Assessment, U.S. Congress, Washington, D.C.

Overpeck, J.T. 1985. A pollen study of a late Quaternary peat bog, south-central Adirondack Mountains, New York. Geol. Soc. Am. Bull. 96:145–154.

Overrein, L.N., H.M. Seip, and A. Tollan. 1980. Acid Precipitation – Effects on Forest and Fish. Final Report of the SNSF Project, 1972–1980. Oslo, Norway.

Overton, W.S., P. Kanciruk, L.A. Hook, J.M. Eilers, D.H. Landers, D.F. Brakke, D.J. Blick, Jr., R.A. Linthurst, M.S. DeHaan, and J.M. Omernik. 1986. Characteristics of Lakes in the Eastern United States. Volume II. Lakes Sampled and Descriptive Statistics for Physical and Chemical Variables. EPA/600/4-86/007b. U.S. Environmental Protection Agency, Washington, D.C.

Palmer, S.L. 1984. Surface water. Pages 54–68 in E.A. Fernald and D.J. Patton, eds. Water Resources Atlas of Florida. Florida State University, Tallahassee, FL.

Palumbo, A.V., P.J. Mulholland, and J.W. Elwood. 1987. Microbial communities on leaf material protected from macroinvertebrate grazing in acidic and circumneutral streams. Can. J. Fish. Aquat. Sci. 44:1064–1070.

Parent, L., M. Allard, D. Planas, and G. Moreau. 1986. The effects of short-term and continuous experimental acidification on biomass and productivity of running water periphytic algae. Pages 28-41 in B.G. Isom, D. Denis, and J.M. Bates, eds. Impact of Acid Rain and Deposition on Aquatic Systems. ASTM STP 928. American Society for Testing and Materials, Philadelphia, PA.

Parker, G.G. 1981. Factors Controlling the Quality and Variability of Summertime through Fall in Two Virginia Piedmont Forests. M.S. Thesis. Department of Environmental Sciences, University of Virginia, Charlottesville, VA.

Parker, G.G., A.G. Hely, W.B. Keighton, F.H. Olmstead, and others. 1964. Water Resources of the Delaware River Basin. Professional Paper 381. U.S. Geological Survey, Reston, VA.

Parkhurst, B., and M. Marcus. 1985. Lake Acidification and Fisheries: Summary of Discussions and Conclusions from a Recent Workshop. Electric Power Research Institute, Palo Alto, CA.

Parsons, D.J. 1981. The historical role of fire in the foothill communities of Sequoia National Park. Madrono 28:111–120.

Parsons, D.J. 1983. Wilderness protection: An example from the southern Sierra Nevada, USA. Environ. Conserv. 10:23–30.

Parsons, D.J., and T.J. Stohlgren. 1987. Impacts of Visitor Use on Backcountry Campsites in Sequoia and Kings Canyon National Parks, California. Technical Report 25. Cooperative National Park Resources Studies Unit, University of California, Davis, CA.

Pastor, J., and J.G. Bockheim. 1984. Distribution and cycling of nutrients in an aspen-mixed-hardwood-spodosol ecosystem in northern Wisconsin. Ecology 64:339–353.

Patalas, K. 1964. The crustacean plankton communities of 52 lakes of different altitudinal zones of northern Colorado. Int. Verein. Limnol. Verh. 15:719–726.

Patrick, R., V.P. Binett, and S.G. Halterman. 1981. Acid lakes from natural and anthropogenic causes. Science 211:446–448.

Patterson, D.E., R.B. Husar, W.E. Wilson, and L.F. Smith. 1981. Monte Carlo simulation of daily regional sulfur distribution: Comparison with SURE sulfate data and visual range observations during August 1977. J. Appl. Meteorol. 20:404–420.

Pauwels, S.J., and T.A. Haines. 1986. Fish species distribution in relation to water chemistry in selected Maine lakes. Water Air Soil Pollut. 30:477–488.

Payer, R.D. 1985. Indexing Minnesota Fish Lakes Relative to Potential Susceptivity to Acidic Deposition. Special Publication No. 138. Minnesota Dept. of Natural Resources, Division of Fish and Wildlife, St. Paul, MN.

Pederson, G.L., E.B. Welch, and A.H. Litt. 1976. Plankton secondary production and biomass: Their relation to lake trophic state. Hydrobiologia 50:129–144.

Perdue, M. 1985. Acidic functional groups of humic substances. Pages 493–526 in G.R. Aiken, D.M. McKnight, R.L. Wershaw, and P. McCarthy, eds. Humus Substances in Soil, Sediment, and Waters. Wiley-Interscience, New York.

Perdue, E.M., K.C. Beck, and J.H. Reuter. 1976. Organic complexes of iron and aluminum in natural waters. Nature 260:418–420.

Peroni, P.A., and W.G. Abrahamson. 1989. Post-Settlement Vegetation Loss on the Southern Lake Wales Ridge, Florida. Department of Biology, Bucknell University, Lewisburg, PA.

Perry, J.A., N.H. Troelstrup, M. Newsom, and B. Shelley. 1987. Whole-ecosystem manipulation experiments: The search for generality. Water Sci. Technol. 19:55–71.

Perry, T.E. 1987. Sediment-Water Interactions in Softwater Lakes in the Upper Midwest. M.S. Thesis. University of Minnesota, Minneapolis, MN.

Perry, T.E., C.D. Pollman, and P.L. Brezonik. 1986. Buffering capacity of softwater lake sediments in Florida. Pages 67–83 in B.G. Isom, S.D. Dennis, and J.M. Bates, eds. Impact of Acid Rain and Deposition on Aquatic Biological Systems. ASTM STP 928. American Society for Testing and Materials, Philadelphia, PA.

Peters, N.E., and C.T. Driscoll. 1987. Hydrogeologic controls on surface water chemistry in the Adirondack region of New York State. Biogeochemistry 3:163–180.

Peters, N.E., and P.S. Murdoch. 1985. Hydrogeologic comparison of an acidic lake basin with a neutral lake basin in the west central Adirondack Mountains, New York. Water Air Soil Pollut. 26:387–402.

Peters, N.E., R.A. Schroeder, and D.E. Troutman. 1982. Temporal Trends in the Acidity of Precipitation and Surface Waters of New York. Water Supply Paper 2188. U.S. Geological Survey, Reston, VA.

Petersen, L. 1986. Effects of acid deposition on soil and sensitivity of the soil to acidification. Experientia 42: 340–344.

Peterson, R.H., P.G. Daye, and J.L. Metcalfe. 1980. Inhibition of Atlantic salmon (*Salmo salar*) hatching at low pH. Can. J. Fish. Aquat. Sci. 44:1432–1442.

Pezzetta, J.M., and I.K. Iskandar. 1975. Sediment characteristics in the vicinity of the Pulliam Power Plant, Green Bay, Wisconsin. Environ. Geol. 1:155–165.

Pfeiffer, M.H., and P.J. Festa. 1980. Acidity Status of Lakes in the Adirondack Region of New York in Relation to Fish Resources. New York State Department of Environmental Conservation, Albany, NY.

Phillips, K.N., and A.S. Van Denburgh. 1968. Hydrology of Crater, East, and Davis Lakes, Oregon. USGS Water Supply Paper 1859-E.U.S. Geological Survey, U.S. Department of Interior, Washington, D.C.

Pirkle, E.C., W.A. Pirkle, and W.H. Yoho. 1977. The Highland Heavy Mineral Sand Deposit on Trail Ridge in Northern Peninsular Florida. Report of Investigation No. 84. Florida Geological Survey, Tallahassee, FL.

Platt, W.J., and M. Schwartz. 1990. Temperate hardwood forests. Pages 194–229 in R.L. Myers and J. Ewel, eds. Ecosystems of Florida. University of Central Florida Press, Orlando, FL.

Pollman, C.D. 1986. The Florida Acid Deposition Study, Final Report: A Synthesis of the Florida Acid Deposition Study. Ecological Effects. Florida Electric Power Coordinating Group, Inc., Tampa FL.

Pollman, C.D., and R.E. Dickinson. (In preparation) Regionalization of the IAG model and its application to forecasting lake acidification in Florida and the Upper Midwest.

Pollman, C.D., and C.D. Hendry. 1983. The Florida Acid Deposition Study, Phase II: Distribution and analysis of lake and soil buffering characteristics in Florida. Pages 222–240 in A.E. S. Green and W.H. Smith, eds. Acid Deposition Causes and Effects: A State Assessment Model. Government Institutes, Rockville, MD.

Pollman, C.D., and P.R. Sweets. (In review) Model hindcasting of pre-cultural acidity in softwater seepage lakes in the Adirondacks, Upper Midwest, and Florida.

Pollman, C.D., T.L. Crisman, P.L. Brezonik, and P. Sacco. 1980. Limnological Studies on Lake Apopka and the Oklawaha Chain of Lakes, 3. Water Quality in 1979. Report No. 07-80-02. Department of Environmental Engineering and Science, University of Florida, Gainesville, FL.

Pollman, C.D., P.L. Brezonik, and L.A. Baker. 1985. Florida Acid Deposition Study, Phase IV Report. Nutrient Loadings and Trophic Conditions in Softwater Lakes and Their Relationships to pH. ESE No. 83-152-0106/0207/0307. Environmental Science and Engineering, Inc., Gainesville, FL.

Potkin, M., and L. Munn. 1988. Subalpine and Alpine Plant Communities in the Bridger Wilderness, Wind River Range, Wyoming. Report to U.S. Forest Service. Contract No. 53-8555-3-000015. University of Wyoming, Laramie, WY.

Powers, C.F., and D.L. Rambo. 1981. The occurrence of acid precipitation on the west coast of the United States. Environ. Monit. Assess. 1:93–105.

Pratt, G.C., and S.V. Krupa. 1985. Aerosol chemistry in Minnesota and Wisconsin and its relation to rainfall chemistry. Atmos. Environ. 19:961–971.

Price, E.E., and M.C. Swift. 1985. Inter- and intraspecific variability in the response of zooplankton to acid stress. Can. J. Fish. Aquat. Sci. 42:1749–1754.

Prugh, B.J., Jr., F.J. Easton, and D.D. Lynch. 1986. Water Resources Data, Virginia, Water Year 1986. Data Report VA-86-1. U.S. Geological Survey, Reston, VA.

Puri, H.S., and R.O. Vernon. 1964. Summary of the Geology of Florida and a Guidebook to the Classic Exposures. Special Publication No. 5. Florida Bureau of Geology, Tallahassee, FL.

Pyle, C. 1988. The type and extent of anthropogenic vegetation disturbance in the Great Smoky Mountains before National Park Service acquisition. Castanea 53:225–235.

Quay, P.D., W.S. Broecker, R.H. Hesslein, and D.W. Schindller. 1980. Vertical diffusion rates determined by tritium tracer experiments in the thermocline and hypolimnion of two lakes. Limnol. Oceanogr. 25:201–218.

Raddum, G.G., and A. Fjellheim. 1984. Acidification and early warning organisms in freshwater in western Norway. Verh. Int. Verein. Limnol. 22:1973–1980.

Raddum, G., A. Hobaek, E. Lomsland, and T. Johnsen. 1980. Phytoplankton and zooplankton in acidified lakes in southern Norway. Pages 332–333 in D. Drabløs and A. Tollan, eds. Ecological Impact of Acid Precipitation. Proceedings of an International Conference, Sandefjord, Norway. SNSF Project, Oslo, Norway.

Rago, P.J., and J.G. Wiener. 1986. Does pH affect fish species richness when lake area is considered? Trans. Am. Fish. Soc. 115:438–447.

Rahel, F.J. 1986. Biogeographic influences on fish species composition of northern Wisconsin lakes with applications for lake acidification studies. Can. J. Fish. Aquat. Sci. 43:124–134.

Rahel, F.J., and J.J. Magnuson. 1983. Low pH and the absence of fish species in naturally acidic Wisconsin lakes: Inferences for cultural acidification. Can. J. Fish. Aquat. Sci. 40:3–9.

Raisz, E. 1957. Landforms of the United States. Sixth revised edition. One map (scale 1:10,000,000).

Rajan, S.S.S. 1978. Sulfate adsorbed on hydrous alumina, ligands displaced, and changes in surface charge. Soil Sci. Soc. Am. J. 42:39–44.

Rapp, G., J.D. Allert, B.W. Liukkonen, J.A. Ilse, G.E. Glass, and O.L. Loucks. 1985. Acidic deposition and watershed characteristics in relation to lake chemistry in northeastern Minnesota. Environ. Internat. 11: 425–440.

Rapp, G., B.W. Liukkonen, J.D. Allert, J.A. Sorensen, G.E. Glass, and O.L. Loucks. 1987. Geologic and atmospheric input factors affecting watershed chemistry in Upper Michigan. Environ. Geol. Water Sci. 9:155–171.

Rascher, C.M., C.T. Driscoll, and N.E. Peters. 1987. Concentration and flux of solutes from snow and forest floor during snowmelt in the west central Adirondack region of New York. Biogeochemistry 3:209–224.

Rask, M., and J. Raitaniemi. 1988. The growth of perch, *Perca fluviatilis* L., in recently acidified lakes of southern Finland—a comparison with unaffected waters. Arch. Hydrobiol. 112:387–397.

Rasmussen, L.A., and W.V. Tangborn. 1976. Hydrology of the north Cascades region, Washington: 1. Runoff, precipitation, and storage characteristics. Water Resour. Res. 12:187–202.

Ravera, O. 1986. Effects of experimental acidification on freshwater environments. Experientia 42:507–516.

Raynal, D.J., B.T. Fitzgerald, and R.D. Masters. 1987. Characterization of Atmospheric Deposition at Huntington Forest, Adirondack Mountains, New York, 1978–1985. College of Environmental Science and Forestry, State University of New York, Syracuse, NY.

Reckhow, K.H. 1987. Robust Bayes models of fish response to lake acidification. Pages 61–72 in M.B. Beck, ed. Systems Analysis in Water Quality Management. Pergamon Press, New York.

Reckhow, K.H. 1988. A comparison of robust Bayes and classical estimators for regional lake models of fish response to acidification. Water Resour. Res. 24: 1061–1068.

Reckhow, K.H., R.W. Black, T.B. Stockton, Jr., J.D. Vogt, and J.G. Wood. 1987. Empirical models of fish response to lake acidification. Can. J. Fish. Aquat. Sci. 45:1432–1442.

Reddy, M.M., T.D. Liebermann, J.C. Jelinski, and N. Caine. 1985. Variation in pH during summer storms near the Continental Divide in central Colorado, U.S.A. Arct. Alp. Res. 17:79–88.

Redmond, K.T. In press. Crater Lake climate and lake level variability. In The Clarity of Crater Lake: An Ecosystem Study. AAAS Symposium, Oregon State University, Corvallis, OR.

Reeder, S.W., B. Hitchon, and A.A. Levinson. 1972. Hydrogeochemistry of the surface waters of the McKenzie River drainage basin, Canada. I. Factors controlling inorganic composition. Geochim. Cosmochim. Acta 36:825–865.

Reilly, J.F. 1989. A Chemical Mass Balance of Crater Lake, Oregon. Master's Project Report. Civil Engineering Department, Oregon State University, Corvallis, OR.

Reilly, T.A., and R. Zasoski. 1989. A Survey of the Soils of the Sierra Nevada for Sensitivity to Acid Deposition. Final Report. Contract A7-32-37. California Air Resources Board, Sacramento, CA.

Reimers, N., J.A. Maciolek, and E.P. Pister. 1955. Limnological study of the lakes in Convict Creek Basin, Mono County, California. U.S. Fish and Wildlife Service, Fishery Bulletin 56:437–503.

Reuss, J.O., and D.W. Johnson. 1985. Effect of soil processes on the acidification of water by acid deposition. J. Environ. Qual. 14:26–31.

Reuss, J.O., and D.W. Johnson. 1986. Acid Deposition and the Acidification of Soils and Waters. Ecological Studies Volume 59. Springer-Verlag, New York.

Reuss, J.O., N. Christopherson, and H.M. Seip. 1986. A critique of models for freshwater and soil acidification. Water Air Soil Pollut. 30:909–930.

Reuss, J.O., B.J. Cosby, and R.F. Wright. 1988. Acid deposition: Chemical processes governing soil and water acidification. Nature.

Reynolds, R.C., Jr., and N.M. Johnson. 1972. Chemical weathering in the temperate glacial environment of the northern Cascade Mountains. Geochim. Cosmochim. Acta 36:537–554.

Rich, J.L. 1934. Glacial Geology of the Catskills. New York State Museum Bulletin 299.

Richter, D.D. 1984. Comment on acid precipitation in historical perspective and effects of acid precipitation. Environ. Sci. Technol. 18:632–634.

RMCC. 1986. Assessment of the State of Knowledge on the Long-Range Transport of Air Pollutants and Acid

Deposition. Part 3. Aquatic Effects. Federal-Provincial Research and Monitoring Coordinating Committee, Ottawa, Ontario.

Robarge, W., and I.J. Fernandez. 1986. Quality Assurance Methods Manual for Laboratory Analytical Techniques. U.S. Environmental Protection Agency and U.S. Forest Service Forest Response Program, Corvallis, OR.

Roberts, D.A., and C.W. Boylen. 1988. Patterns of epipelic algal distribution in an acidic Adirondack lake. J. Phycol. 24:146–152.

Roberts, D.A., and C.W. Boylen. 1989. Effects of liming on the epipelic algae community of Woods Lake, New York. Can. J. Fish. Aquat. Sci. 46:287- 294.

Roberts, D.A., R. Singer, and C.W. Boylen. 1985. The submerged macrophyte communities of Adirondack lakes (New York, USA) of varying degrees of acidity. Aquat. Bot. 21:219–235.

Robins, C.R., R.M. Bailey, C.E. Bond, J.R. Booker, E.A. Lachner, R.N. Lea, and W.B. Scott. 1980. A List of Common and Scientific Names of Fishes from the United States and Canada. Special Publication No. 12. American Fisheries Society, Bethesda, MD.

Rochelle, B.P., and M.R. Church. 1987. Regional patterns of sulfur retention in watersheds of the eastern U.S. Water Air Soil Pollut. 36:61–73.

Rochelle, B.P., M.R. Church, and M.B. David. 1987a. Sulfur retention at intensively studied sites in the U.S. and Canada. Water Air Soil Pollut. 33:73–83.

Rochelle, B.P., M.R. Church, D.H. Landers, J.M. Eilers, and J.J. Messer. 1987b. Sulfur retention in watersheds: Relationship of acid deposition and surface water chemistry. Unpublished manuscript.

Rodhe, H., P. Crutzen, and A. Vanderpool. 1981. Formation of sulfuric and nitric acid in the atmosphere during long range transport. Tellus 33:132–141.

Rogalla, J.A., P.L. Brezonik, and G.E. Glass. 1986. Empirical models for lake acidification in the Upper Great Lakes region. Water Air Soil Pollut. 31:95–100.

Røgeberg, E.J.S., and A. Henriksen. 1985. An automated method for fractionation and determination of aluminum species in freshwaters. Vatten 41:48–53.

Rooke, J.B., and G.L. Mackie. 1984. Mollusca of six low-alkalinity lakes in Ontario. Can. J. Fish. Aquat. Sci. 41:777–782.

Rosemond, A.D. 1987. The Effects of Acidification on the Macroinvertebrate Community of Streams. M.S. Thesis. University of North Carolina, Chapel Hill, NC.

Rosemond, A.D., J.W. Elwood, and S.R. Reice. Submitted. The effects of acidification on the macroinvertebrate communities of streams in the Great Smoky Mountains National Park. Can. J. Fish. Aquat. Sci.

Rosenau, J.C., G.L. Faulkner, C.W. Hendry, Jr., and R.W. Hull. 1977. Springs of Florida. Bulletin No. 31 (revised). Florida Bureau of Geology, Tallahassee, FL.

Rosenqvist, I.T. 1978a. Alternative sources for acidification of river water in Norway. Sci. Total Environ. 10:39–49.

Rosenqvist, I.T. 1978b. Acid precipitation and other possible sources for acidification of rivers and lakes. Sci. Total Environ. 10:271–272.

Ross, C.P., D.A. Andrews, and I.J. Witking. 1955. Geologic Map of Montana. U.S. Geological Survey, Reston, VA.

Rosseland, B.O. 1986. Ecological effects of acidification on tertiary consumers. Water Air Soil Pollut. 30:451–460.

Rosseland, B.O., and O.K. Skogheim. 1984. A comparative study of salmonid fish species in acid aluminum-rich water. Rep. Inst. Freshwater Res. Drottningholm. 61:186–194.

Rosseland, B.O., I. Savaldrud, D. Svalastog, and I.P. Muniz. 1980. Studies on freshwater fish populations —effects of acidification on reproduction, population structure, growth, and food selection. Pages 336–337 in D. Drabløs and A. Tollan, eds. Ecological Impacts of Acid Precipitation. Proceedings of an International Conference, Sandefjord, Norway. SNSF Project, Oslo, Norway.

Roth, P., C. Blanchard, J. Harte, H. Michaels, and M.T. El-Ashry. 1985. The American West's Acid Rain Test. World Resources Institute, Washington, D.C.

Rourke, R.W., J.A. Ferwerda, and K.J. LaFlamme. 1978. The Soils of Maine. Publication 676. Report No. 203. Maine LSA Experiment Station, Orono, ME.

Rudd, J.W.M., C.A. Kelly, V. St. Louis, R.H. Hesslein, A. Furutani, and M.H. Holoka. 1986a. Microbial consumption of nitric and sulfuric acids in north temperate lakes. Limnol. Oceanogr. 31:1267–1280.

Rudd, J.W.M., C.A. Kelly, and A. Furutani. 1986b. The role of sulfate reduction in long-term accumulation of organic and inorganic sulfur in lake sediments. Limnol. Oceanogr. 31:1281–1291.

Rudd, J.W.M., C.A. Kelly, D.W. Schindler, and M.S. Turner. 1988. Disruption of the nitrogen cycle in acidified lakes. Science 240:1515–1517.

Ruggaber, G.J. 1988. Development and Application of a Thermal Stratification Model to Adirondack Lakes. M.S. Thesis. Syracuse University, Syracuse, NY.

Rundel, P.W., D.J. Parsons, and D.T. Gordon. 1977. Montane and subalpine vegetation of the Sierra Nevada and Cascade ranges. Pages 559–599 in M.G. Barbour and J. Major, eds. Terrestrial Vegetation of California. John Wiley and Sons, New York.

Rundel, P.W., T.V. St. John, and W.L. Berry. 1989. Vegetation Process Studies. Final Report. Contract A4-121-32. California Air Resources Board, Sacramento, CA.

Rustad, L.E. 1988. The Biogeochemistry of Al in a Red Spruce Forest Floor in Maine. Ph.D. Dissertation. University of Maine, Orono, ME.

Rustad, S., N. Christophersen, and H.M. Seip. 1986. Model for streamwater chemistry of a tributary to Harp Lake, Ontario. Can. J. Fish. Aquat. Sci. 43: 625–633.

Ruttner, F. 1963. Fundamentals of Limnology. University of Toronto Press, Toronto, Ontario.

Ryan, P.M., and H.H. Harvey. 1977. Growth of rock bass, Ambioplites rupestris, in relation to the morphoedaphic index as an indicator of an environmental stress. J. Fish Res. Board Can. 34:2079–2088.

Ryan, P.M., and H.H. Harvey. 1980. Growth responses of yellow perch, *Perca flavescens* (Mitchill), to lake acidification in the La Cloche Mountain lakes of Ontario. Environ. Biol. Fish 5:97–108.

Ryan, P.F., J.N. Galloway, B.J. Cosby, G.M. Hornberger, and J.R. Webb. 1989. Changes in the chemical composition of streamwater in two catchments in the Shenandoah National Park, Virginia, in response to atmospheric deposition of sulfur. Water Resour. Res. 25:2091–2099.

Sadler, K. 1983. A model relating the reuslts of low pH bioassay experiments to the fishery status of Norwegian lakes. Freshwater Ecol. 13:453–463.

Sadler, K., and S. Lynam. 1986. Some effects of low pH and calcium on the growth and tissue mineral content of yearling brown trout, *Salmo trutta*. J. Fish Biol. 29:313–324.

Saila, S.B., X. Chen, K. Erzini, and B. Martin. 1987. Compensatory Mechanisms in Fish Populations: Literature Reviews. Volume 1. Critical Evaluation of Case Histories of Fish Populations Experiencing Chronic Exploitation or Impact. EA-5200. Research Project 1633-6. Electric Power Research Institute, Palo Alto, CA.

Sale, M.J., P.R. Kaufmann, H.I. Jager, J.M. Coe, K.A. Cougan, A.J. Kinney, M.E. Mitch, and W.S. Overton. 1988. Chemical Characteristics of Streams in the Mid-Atlantic and Southeastern United States. Volume II. Streams Sampled, Descriptive Statistics, and Compendium of Physical and Chemical Data. EPA/600/3-88/021b. U.S. Environmental Protection Agency, Washington, D.C.

Schaefer, D.A., C.T. Driscoll, R.S. Van Dreason, and C. P. Yatsko. 1990. The episodic acidification of Adirondack Lakes during snowmelt. Water Resour. Res. 26:1639–1647.

Schafran, G.C., and C.T. Driscoll. 1987. Comparison of terrestrial and hypolimnetic sediment generation of acid neutralizing capacity for an aciaic Adirondack lake. Environ. Sci. Technol. 21:988–993.

Schecher, W.D., and C.T. Driscoll. 1987. An evaluation of uncertainty associated with aluminum equilibrium calculations. Water Resour. Res. 23:525–535.

Scheider, W.A., J. Moss, and P.J. Dillon. 1978. Measurement and uses of hydraulic and nutrient budgets. Pages 77–83 in Proceedings of National Conference on Lake Restoration, August 22–24, 1978, Minneapolis, MN. U.S. EPA 440/5-79/001.

Schiff, S.L., and R.F. Anderson. 1986. Alkalinity production in epilimnetic sediments: Acidic and nonacidic lakes. Water Air Soil Pollut. 30:941–948.

Schiff, S., and R.F. Anderson. 1987. Limnocorral studies of chemical and biological acid neutralization in two freshwater lakes. Can. J. Fish. Aquat. Sci. 44(Suppl. 1):173–187.

Schindler, D.W. 1971. Light, temperature, and oxygen regimes of selected lakes in the Experimental Lakes Area, northwestern Ontario. J. Fish. Res. Board Can. 28:157–169.

Schindler, D.W. 1986a. The significance of in-lake production of alkalinity. Water Air Soil Pollut. 30:931–946.

Schindler, D.W. 1986b. Recovery of Canadian lakes from acidification. Pages 11–22 in Effects of Air Pollution on Terrestrial and Aquatic Ecosystems (Cost 612), Working Group II: Effects of Air Pollution on Aquatic Ecosystems. Proceedings of the Workshop on Reversibility of Acidification, Royal Norwegian Council for Scientific and Industrial Research and the Commission of the European Communities, June 9– 11, Grimstad, Norway.

Schindler, D.W. 1987. Detecting ecosystem responses to anthropogenic stress. Can. J. Fish. Aquat. Sci. 446: 6–25.

Schindler, D.W. 1988a. Confusion over the origin of alkalinity in lakes. Limnol. Oceanogr. 33:1637–1640.

Schindler, D.W. 1988b. The effects of acid rain on freshwater ecosystems. Science 239:149–157.

Schindler, D.W., G.J. Brunskill, S. Emerson, W.S. Broecker, and T.-H. Peng. 1972. Atmospheric carbon dioxide: Its role in maintaining phytoplankton standing crops. Science 177:1192–1194.

Schindler, D.W., R.W. Newbury, K.G. Beaty, and P. Campbell. 1976. Natural water and chemical budgets for a small Precambrian lake basin in central Canada. J. Fish. Res. Board Can. 33:2526–2543.

Schindler, D.W., R. Wegemann, R.B. Cook, T.J. Ruszczynski, and J. Prokopowich. 1980. Experimental acidification of Lake 223, Experimental Lakes Area: Background data and the first three years of acidification. Can J. Fish. Aquat. Sci. 37:342–354.

Schindler, D.W., K.H. Mills, D.F. Malley, D.L. Findlay, J.A. Shearer, I.J. Davies, M.A. Turner, G.A. Linsey, and D.R. Cruikshank. 1985a. Long-term ecosystem stress: The effects of years of experimental acidification on a small lake. Science 228:1395–1401.

Schindler, D.W., M.A. Turner, and R.H. Hesslein. 1985b. Acidification and alkalinization of lakes by experimental addition of nitrogen compounds. Biogeochemistry 1:117–133.

Schindler, D.W., M.A. Turner, M.P. Stainton, and G.A. Linsey. 1986. Natural sources of acid neutralizing capacity in low-alkalinity lakes of the Precambrian Shield. Science 232:844–847.

Schmidt, W., and M.W. Clark. 1980. Geology of Bay County, Florida. Bulletin No. 57. Florida Geological Survey, Tallahassee, FL.

Schneider, R. 1984. Acidic precipitation and surface water vulnerability on the western slope of the high Colorado Rockies. Pages 28–34 in Acid Rain in the Rocky Mountain West. Colorado Department of Health, Denver, CO.

Schnoor, J.L., and N.P. Nikolaidis. 1989. Assessment of Episodic Freshwater Acidification in the Sierra Nevada, California. Final Report. Contract A7-32-36. California Air Resources Board, Sacramento, CA.

Schnoor, J.L., and W. Stumm. 1984. Acidification of aquatic and terrestrial systems. Pages 311–338 in W. Stumm, ed. Chemical Processes in Lakes. John Wiley and Sons, New York.

Schnoor, J.L., and W. Stumm. 1986. The role of chemical weathering in the neutralization of acidic deposition. Schweiz. Z. Hydrol. 48:171–195.

Schnoor, J.L., W.D. Palmer, Jr., and G.E. Glass. 1984. Modeling impacts of acid precipitation for northeastern Minnesota. Pages 155–173 in J.L. Schnoor, ed. Modeling of Total Acid Precipitation Impacts. Butterworth Publishers, Boston, MA.

Schnoor, J.L., N.P. Nikolaidis, and G.E. Glass. 1986. Lake resources at risk to acidic deposition in the Upper Midwest. J. Water Pollut. Control Fed. 58: 139–148.

Schofield, C.L. 1965. Water quality in relation to survival of brook trout. *Salvelinus fontinalis* (Mitchell). Trans. Am. Fish. Soc. 94:277–235.

Schofield, C.L. 1972. The ecological significance of air pollution induced changes in water quality in dilute lake districts in the Northeast. Trans. Northeast. Fish Wildlife Conf. Pages 98–112.

Schofield, C.L. 1976a. Acidification of Adirondack Lakes by Atmospheric Precipitation: Extent and Magnitude of the Problem. Final report, Project F-28-R. New York State Department of Environmental Conservation, Albany, NY.

Schofield, C.L. 1976b. Acidification of Adirondack Lakes by Atmospheric Precipitation: Long-term and Seasonal Trends. Final report, Project F-28-R. New York State Department of Environmental Conservation, Albany, NY.

Schofield, C.L. 1976c. Dynamics and Management of Adirondack Fish Populations. Project Report April 1, 1975–March 31, 1976. No. F-28-R-4. Department of Environmental Conservation, Albany, NY.

Schofield, C.L. 1976d. Acid precipitation: Effects on fish. Ambio 5:228–230.

Schofield, C.L. 1980. Processes limiting fish populations in acidified lakes. Pages 345–355 in D.S. Shriner, C.R. Richmond, and S.E. Lindberg, eds. Atmospheric Sulfur Deposition: Environmental Impact and Health Effects. Ann Arbor Science Publishers, Inc., Ann Arbor, MI.

Schofield, C.L. 1982. Historical fisheries changes in the United States related to decreases in surface water pH. Pages 57–67 in Johnson, R.E., ed. Acid Rain/Fisheries. American Fisheries Society, Bethesda, MD.

Schofield, C.L. 1984. Surface water chemistry in the ILWAS basins. The Integrated Lake-Watershed Acidification Study, Vol. 4: Summary of Major Results. EA-3221. Electric Power Research Institute, Palo Alto, CA.

Schofield, C.L., and C.T. Driscoll. 1987. Fish species distribution in relation to water quality gradients in the North Branch of the Moose River Basin. Biogeochemistry 3:63–85.

Schofield, C.L., and J.R. Trojnar. 1980. Aluminum toxicity to fish in acidified waters. Pages 347–366 in T. Y. Toribara, M.W. Miller, and P.E. Morrow, eds. Polluted Rain. Plenum Press, New York.

Schofield, C.L., D.A. Webster, C.A. Guthrie, and W.A. Flick. 1981. Management of Acidified Waters. Final Report C-164480. New York State Department of Environmental Conservation, Albany, NY.

Schofield, C.L., J.N. Galloway, and G.R. Hendrey. 1985. Surface water chemistry in the ILWAS basins. Water Air Soil Pollut. 26:403–424.

Schofield, C.L., S.P. Gloss, and D. Josephson. 1986. Extensive Evaluation of Lake Liming, Restocking Strategies, and Fish Population Response in Acidic Lakes Following Neutralization by Liming. Interim Progress Report NEC-86/18. Air Pollution and Acid Rain. U.S. Fish and Wildlife Service, Washington, D.C.

Schofield, C.L., S.P. Gloss, B. Plonski, and R. Spateholts. 1989. Production and growth efficiency of brook trout (*Salvelinus fontinalis*) in two Adirondack Mountain (New York) lakes following liming. Can. J. Fish. Aquat. Sci. 46:333–341.

Schulze, R.L. 1980. The Biotic Response to Acid Precipitation in Florida Lakes. M.S. Thesis. University of Florida, Gainesville, FL.

Schuurkes, J.A.A.R. 1986. Atmospheric ammonium sulfate deposition and its role in the acidification and nitrogen enrichment of poorly buffered aquatic systems. Experimentia 42:351–357.

Schweinforth, R.L., R.C. Young, J.R. Wright, Jr., R.W. Garber, J.E. Persall, and T.A. McDonough. 1989. Effects of pH and Aluminum on Early Life Stages of Smallmouth Bass and Rainbow Trout. TV-68154A. Tennessee Valley Authority, Nashville, TN.

Scott, T.M. 1983. The Hawthorn Formation of Northeastern Florida, Part I: The Geology of the Hawthorn Formation of Northeastern Florida. Report of Investigation No. 94. Florida Bureau of Geology, Tallahassee, FL.

Scott, T.M., and P.L. McGill. 1981. The Hawthorn Formation of Central Florida. Report of Investigation No. 91. Florida Bureau of Geology, Tallahassee, FL.

Scuderi, L.A. 1987a. Glacier variations in the Sierra Nevada, California, as related to a 1200-year tree-ring chronology. Quat. Res. 27:220–231.

Scuderi, L.A. 1987b. Late-Holocene upper timberline variation in the southern Sierra Nevada. Nature 325: 242–244.

Seip, H.M. 1980. Acidification of freshwater–sources and mechanisms. Pages 358–365 in D. Drabløs and A. Tollan, eds. Ecological Impact of Acid Precipitation. Proceedings of an International Conference, Sandefjord, Norway. SNSF Project, Oslo, Norway.

Seip, H., R. Seip, P. Dillon, and E. de Grosbois. 1985. Model of sulphate concentrations in a small stream in the Harp Lake catchment, Ontario. Can. J. Fish. Aquat. Sci. 42:927–937.

Semkin, R.G., and D.S. Jeffries. 1986. Storage and release of major ionic contaminants from the snowpack in the Turkey Lakes watershed. Water Air Soil Pollut. 31:215–221.

Senectutus. 1943. Fifty-four Years on the Beaverkill. Angler's Club Bull. 22:3–13.

Servos, M.R., and G.L. Mackie. 1986. The effect of short-term acidification during spring snowmelt on selected Mollusca in south-central Ontario. Can. J. Zool. 64:1690–1695.

Servos, M.R., J.B. Rooke, and G.L. Mackie. 1985. Reproduction of selected mollusca in some low alkalinity lakes in south-central Ontario. Can. J. Zool. 63:511–515.

Sevaldrud, I.H., and O.K. Skogheim. 1986. Changes in fish populations in southernmost Norway during the last decade. Water Air Soil Pollut. 30:381–386.

Shafer, M.D., R.E. Dickinson, J.P. Heaney, and W.C. Huber. 1986. Gazeteer of Florida Lakes. Publication No. 96. Florida Water Resources Research Center, University of Florida, Gainesville, FL.

Shaffer, P.W., and J.N. Galloway. 1982. Acid precipitation: The impact on two headwater streams in Shenandoah National Park, Virginia. Pages 43–53 in R. Herrmann and A.I. Johnson, eds. International Symposium on Hydrometeorology. American Water Resources Association, Denver, CO.

Shaffer, P.W., and M.R. Church. 1989. Terrestrial and in-lake contributions to alkalinity budgets of drainage lakes: An assessment of regional differences. Can. J. Fish. Aquat. Sci. 46:509–515.

Shaffer, P.W., R.P. Hooper, K.N. Eshleman, and M.R. Church. 1988. Watershed vs in-lake alkalinity generation: A comparison of rates using input-output studies. Water Air Soil Pollut. 39:263–273.

Shair, F.H. 1985. Atmospheric tracer experiments aimed at characterizing the transport and dispersion of airborn pollutants in Sequoia and Kings Canyon National Parks. Air Pollution Research Meeting. Sequoia National Park. National Park Service, California Air Resources Board, Sacramento, CA.

Shanks, R.E. 1954. Climates of the Great Smoky Mountains. Ecology 35:354–361.

Shannon, E.E. 1970. Eutrophication Factors in North-central Florida Lakes. Ph.D. Dissertation. University of Florida, Gainesville, FL.

Shannon, E.E., and P.L. Brezonik. 1972. Limnological characteristics of Florida lakes. Limnol. Oceanogr. 17:97–110.

Sharp, R.P., and J.H. Birman. 1963. Additions to classical sequence of Pleistocene glaciations, Sierra Nevada, California. Geol. Soc. Am. Bull. 74:1079– 1086.

Sharpe, W.E., W.G. Kimmel, E.S. Young, Jr., and D.R. DeWalle. 1983. In situ bioassays of fish mortality in two Pennsylvania streams acidified by atmospheric deposition. Northeast. Environ. Sci. 2:171–178.

Sharpe, W.E., D.R. DeWalle, R.T. Leibfried, R.S. Dinicola, W.G. Kimmel, and L.S. Sherwin. 1984. Causes of acidification of four streams on Laurel Hill in southwestern Pennsylvania. J. Environ. Qual. 13: 619–631.

Sharpe, W.E., V.G. Leibfried, W.G. Kimmel, and D.R. DeWalte. 1987. The relationship of water quality and fish occurrence to soils and geology in an area of high hydrogen and sulfate ion deposition. Water Resour. Bull. 23:37–46.

Shearer, J.A., and E.R. DeBruyn. 1986. Phytoplankton productivity responses to direct addition of sulfuric and nitric acids to the waters of a double-basin lake. Water Air Soil Pollut. 30:695–702.

Shearer, J.A., E.J. Fee, E.R. DeBruyn, and D.R. DeClercq. 1987. Phytoplankton primary production and light attenuation responses to the experimental acidification of a small Canadian shield lake. Can. J. Fish. Aquat. Sci. 44:83–90.

Sheath, R.G., M. Havas, J.A. Hellebust, and T.C. Hutchinson. 1982. Effects of long-term natural acidification on the algal communities of tundra ponds at the Smoking Hills, N.W.T., Canada. Can. J. Bot..60:58–72.

Sheldon, A.L. 1985. Perlid stoneflies (*Plecoptera*) in an Appalachian drainage: A multivariate approach to mapping stream communities. Am. Midl. Nat. 113:334–342.

Shepard, J.P., M.J. Mitchell, T. Scott, Y. Zang, and D.J. Raynal. 1989. Measurements of wet and dry deposition in a northern hardwood forest. Water Air Soil Pollut. 48:225–238.

Sheridan, J.R. 1971. Annual Project Report, Fisheries Management Program–Shenandoah National Park. U.S. Bureau of Sport Fisheries and Wildlife.

Sherman, L.A. 1988. Sediment Porewater Diagenesis in a Low Alkalinity Lake. M.S. Thesis. University of Minnesota, Minneapolis, MN.

Shilts, W.W. 1981. Sensitivity of Bedrock to Acid Precipitation: Modification by Glacial Processes. Paper 81-14. Geological Survey of Canada, Ottawa, Ontario.

Shirahata, H., R.W. ELias, C.C. Patterson, and M. Koide. 1980. Chronological variations in concentrations and isotopic compositions of anthropogenic atmospheric lead in sediments of a remote subalpine pond. Geochim. Cosmochim. Acta 44:149–162.

Shireman, J.V., D.E. Colle, and D.F. Durant. 1981. Efficiency of rotenone sampling with large and small blocknets in vegetated and open-water habitats. Trans. Am. Fish. Soc. 110:77–80.

Sickles, J.E., and W.D. Bach. 1983. Comparison of several techniques for determining dry deposition flux. Pages 979–989 in H.R. Pruppacher, R.G. Semonin, and W.G.N. Slinn, eds. Precipitation, Scavenging, Dry Deposition, and Resuspension, Vol. 2. Elsevier Science Publishing Co., New York.

Sickman, J.O., and J.M. Melack. 1989. Characterization of Year-round Sensitivity of California's Montane Lakes to Acidic Deposition. Final Report. Contract A5-203-43. California Air Resources Board, Sacramento, CA.

Siddens, L.K., W.K. Seim, L.R. Curtis, and G.A. Chapman. 1986. Comparison of continuous and episodic exposure to acidic, aluminum-contaminated waters of brook trout (*Salvelinus fontinalis*). Can. J. Fish. Aquat. Sci. 43:2036–2040.

Siegel, D.I. 1981. The effect of snowmelt on the water quality of Filson Creek and Omaday Lake, northeastern Minnesota. Water Resour. Res. 17:238–242.

Siegel, D.I., and H.O. Pfannkuch. 1984. Silicate dissolution influence on Filson Creek chemistry, northeastern Minnesota. Geol. Soc. Am. Bull. 95:1446–1453.

Siegfried, C.A., J.W. Sutherland, S.D. Quinn, and J.A. Bloomfield. 1984. Lake acidification and the biology of Adirondack lakes: I. rotifer communities. Verh. Int. Verein. Limnol. 22:549–558.

Siegfried, C.A., J.W. Sutherland, and J.A. Bloomfield. 1988. Analysis of plankton community structure in Adirondack lakes in relation to acidification. Pages 445–450 in R. Perry, R.M. Harrison, J.N.B. Bell, and J.N. Lester, eds. Acid Rain: Scientific and Technical Advances. Selper Limited, London.

Siegfried, C.A., J.A. Bloomfield, and J.W. Sutherland. 1989. Plankton rotifer community structure in Adirondack, New York, U.S.A. lakes in relation to acidity, trophic status and related water quality characteristics. Hydrobiologia 175:33–48.

Sierszen, M.S. 1988. Zooplankton Feeding Ecology and the Experimental Acidification of Little Rock Lake. Ph.D. Dissertation. University of Wisconsin, Madison, WI.

Sievering, H. 1987. Midwest/western/eastern U.S. precipitation and aerosol sulfate: Differences attributable to natural source inputs. Atmos. Environ. 21: 2525–2530.

Sievering, H., and N. Ton. 1985. A study of sulfate aerosol dry loading collection efficiency by passive samplers. Air Pollut. Control Assoc. J. 33:1072–1074.

Silsbee, D.G., and G.L. Larson. 1981. Physical, Chemical, and Bacteriological Characteristics of Streams in the Great Smoky Mountains National Park. Research/Resources Management Report No. 47. Department of the Interior, National Park Service, Gatlinburg, TN.

Silsbee, D.G., and G.L. Larson. 1982. Water quality of streams in the Great Smoky Mountains National Park. Hydrobiologia 89:97–115.

Silsbee, D.G., and G.L. Larson. 1983. A comparison of streams in logged and unlogged areas of the Great Smoky Mountains National Park. Hydrobiologia 102:99–111.

Silver, P.A. 1988. Distribution of scaled chrysophytes in 17 Adirondack (New York) lakes with special reference to pH. Can. J. Bot. 66:1391–1403.

Silverman, G., and D.C. Erman. 1979. Alpine lakes in Kings Canyon National Park, California: Baseline conditions and possible effects of visitor use. J. Environ. Manage. 8:73–87.

Simmons, C.E., and R.C. Heath. 1982. Water Quality Characteristics of Streams in Forested and Rural Areas of North Carolina. U.S. Geological Survey Water Supply Paper 2185. U.S. Government Printing Office, Washington, D.C.

Simpkins, W.W. 1978. Interrelationships Between Drift Composition and Water Chemistry in Forest County, Wisconsin. Unpublished report. Wisconsin Dept. of Natural Resources, Rhinelander, WI.

Simpkins, W.W., M.C. McCartney, and D.M. Mickelson. 1978. Glacial Geology of Forest County, Wisconsin. Open File report. Wisconsin Geological and Natural History Survey, Madison, WI.

Sims, P.K., and G.B. Morey. 1972. Geology of Minnesota: A Centennial Volume. Minnesota Geological Survey, St. Paul, MN.

Sinclair, W.C., and J.W. Stewart. 1985. Sinkhole Type, Development, and Distribution in Florida. Map Series No. 110. Florida Bureau of Geology, Tallahassee, FL.

Singer, R. 1982. Effects of acidic precipitation on benthos. Pages 329–363 in F. D'Itri, ed. Acid Precipitation. Effects on Ecological Systems. Ann Arbor Science, Ann Arbor, MI.

Singer, R., and C.W. Boylen. 1984. Biological Field Survey of Northeastern Acidified Lakes. Final Report. EPA/NCSU Acid Deposition Program, Raleigh, NC. Freshwater Institute and Department of Biology, Rensselaer Polytechnic Institute, Troy, NY.

Singer, R., and D. Smith. 1987. The Use of Benthic Macroinvertebrates as Indicators of Acidification. Report. U.S. Environmental Protection Agency, Corvallis, OR.

Singer, R., G.L. Evans, and N.C. Pratt. 1984. Phytoplankton limitation by phosphorus and zooplankton grazing in an acidic Adirondack lake. J. Freshwater Ecol. 2:423–434.

Sirois, A., and L.A. Barrie. 1988. An estimate of the importance of dry deposition as a pathway of acidic substances from the atmosphere to the biosphere in eastern Canada. Tellus 40B:59–80.

Sisson, T.W., and J.G. Moore. 1984. Geology of the Giant Forest–Lodgepole Area, Sequoia National Park, California. Open File Report No. 84-254. U.S. Geological Survey, Reston, VA.

Skartveit, A. 1980. Observed relationships between ionic composition of precipitation and snowmelt. Pages 242–243 in D. Drabløs and A. Tollan, eds. Ecological Impact of Acid Precipitation. Proceedings of an International Conference, Sandefjord, Norway. SNSF Project, Oslo, Norway.

Skogheim, O.K., B.O. Rosseland, and I.H. Sevaldrud. 1984. Deaths of spawners of Atlantic salmon (*Salmo salar* L.) in River Ogna, southwestern Norway, caused by acidic aluminum-rich water. Freshwater Res. Inst. Drottningholm 61:16–27.

Smith, D.L., J.K. Underwood, J.G. Ogden, III, and B.C. Sabean. 1986. Fish species distribution and water chemistry in Nova Scotia lakes. Water Air Soil Pollut. 30:489–496.

Smith, M.E., B.J. Wyskowski, C.M. Brooks, C.T. Driscoll, and C.C. Cosentini. 1990. Relationships between acidity and benthic invertebrates of low-order woodland streams in the Adirondack Mountains, New York. Can. J. Fish. Aquat. Sci. 47:1318–1329.

Smith, R.A., and R.B. Alexander. 1983. Evidence for acid precipitation induced trends in stream chemistry at hydrologic benchmark stations. Pages 1–12 in U.S. Geological Survey Circular 910. U.S. Government Printing Office, Washington, D.C.

Smith, R.A., and R.B. Alexander. 1986. Correlations between stream sulphate and regional SO_2 emissions. Nature 322:722–724.

Smith, R.A., R.B. Alexander, and M.G. Wolman. 1987. Water-quality trends in the nation's rivers. Science 235:1607–1615.

Smol, J.P. 1984. *Chrysophyceae*. Proceedings of a Workshop on Paleolimnological Studies of the History and Effects of Acidic Precipitation. #CR-811631-01-0: 298-307. U.S. Environmental Protection Agency, Washington, D.C.

Smol, J.P. 1986. Chrysophycean microfossils as indicators of lakewater pH. Pages 275–287 in J.P. Smol, R.W. Battarbee, R.B. Davis, and J. Meriläinen, eds. Diatoms and Lake Acidity. Dr. W. Junk, Dordrecht, The Netherlands.

Smol, J.P., D.F. Charles, and D.R. Whitehead. 1984a. Mallomonadacean microfossils provide evidence of recent lake acidification. Nature 307:628–630.

Smol, J.P., D.F. Charles, and D.R. Whitehead. 1984b. Mallomonadacean (*Chrysophyceae*) assemblages and their relationships with limnological characteristics in 38 Adirondack (New York) lakes. Can. J. Bot. 62: 911–923.

Smol, J.P., R.W. Battarbee, R.B. Davis, and J. Meriläinen. 1986. Diatoms and Lake Acidity. Dr. W. Junk, Dordrecht, The Netherlands.

Sonzogni, W.C., P.C. Uttormark, and G.F. Lee. 1976. A phosphorus residence time model and application. Water Res. 10:429–435.

Spahr, N.E., and J.T. Turk. 1985. Estimation of Evaporation from Ned Wilson Lake, Flat Tops Wilderness Area, Colorado. Water Resources Investigations Report 85-4244. U.S. Geological Survey, Denver, CO.

Springer, M.E. 1984. Soils in the spruce-fir region of the Great Smoky Mountains. Pages 201–210 in P.S. White, ed. The Southern Appalachian Spruce-Fir Ecosystem: Its Biology and Threats. Research/ Resources Management Report SER-71. U.S. Department of Interior, National Park Service, Southeast Region, Atlanta, GA.

Sprules, W.G. 1972. Effects of size-selective predation and food competition in high altitude zooplankton communities. Ecology 53:375–386.

Sprules, W.G. 1975. Midsummer crustacean zooplankton communities in acid-stressed lakes. J. Fish. Res. Board Can. 32:389–395.

St. John, T.V., and P.W. Rundel. 1976. The role of fire as a mineralizing agent in a Sierran coniferous forest. Oecologia 25:35–45.

Stark, W.H., and E. McCoy. 1938. Distribution of bacteria in certain lakes of northern Wisconsin. Zentbl. Bakt. Parasitkde, 2 Abt. 98:201–209.

State of Maine Water Quality Assessment. 1988. Maine Department of Environmental Protection, Augusta, ME.

Staubitz, W.W., and P.J. Zarriello. 1989. Hydrology of two headwater lakes in the Adirondack Mountains of New York. Can. J. Fish. Aquat. Sci. 46:268–276.

Stauffer, R.E. 1985. Use of solute tracers released by weathering to estimate groundwater inflow to seepage lakes. Environ. Sci. Technol. 19:405–411.

Stauffer, R.E. A chemical typology of low alkalinity lakes in Florida and southeast Georgia. Unpublished manuscript.

Stensland, G.J.D.M. Whelpdale, and G. Oehert. 1986. Chapter 5: Precipitation chemistry. Pages 128–199 in Acid Deposition: Long-Term Trends. National Academy Press, Washington, D.C.

Stephens, L.A. 1969. A Comparison of Climatic Events at Four Elevations in the Great Smoky Mountains National Park. Ph.D. Dissertation. University of Tennessee, Knoxville, TN.

Stephenson, M., and G.L. Mackie. 1986. Lake acidification as a limiting factor in the distribution of the freshwater amphipod Hyalella azteca. Can. J. Fish. Aquat. Sci. 43:288–292.

Stephenson, N.L. 1988. Climatic Control of Vegetation Distribution: The Role of the Water Balance with Examples from North America and Sequoia National Park, California. Ph.D. Thesis. Cornell University, Ithaca, NY.

Stevenson, A.C., H.J.B. Birks, R.J. Flower, and R.W. Battarbee. 1989. Diatom-based pH reconstruction of lake acidification using canonical correspondence analysis. Ambio 13:228–233.

Stevenson, F.J. 1982. Humus Chemistry: Genesis, Composition, Reactions. Wiley-Interscience, New York.

Stoddard, J.L. 1986. Nutritional Status, Microcrustacean Communities and Susceptibility to Acid Precipitation of High Elevation Lakes in the Sierra Nevada, California. Ph.D. Thesis. University of California, Santa Barbara, CA.

Stoddard, J.L. 1987a. Alkalinity dynamics in an unacidified alpine lake, Sierra Nevada, California. Limnol. Oceanog. 32:825–839.

Stoddard, J.L. 1987b. Microcrustacean communities of high elevation lakes in the Sierra Nevada, California. J. Plankt. Res. 9:631–650.

Stoddard, J.L. 1988. Are Sierran lakes different? A reply to the comments of Kelly and Schindler. Limnol. Oceanog. 33:1641–1646.

Stoddard, J.L. (In review) Trends in Catskill streamwater quality: Evidence from historical data.

Stohlgren, T., and D.J. Parsons. 1987. Variation of wet deposition chemistry in Sequoia National Park, California. Atmos. Environ. 21:1369–1375.

Stokes, P.M. 1986. Ecological effects of acidification on primary producers in aquatic systems. Water Air Soil Pollut. 30:421–438.

Stoneburner, D.L. 1977. Preliminary observations of the aquatic insects of the Smoky Mountains: Altitudinal zonation in the spring. Hydrobiologia 56:137– 143.

Strayer, R., C.-J. Lin, and M. Alexander. 1981. Effects of simulated acid rain on nitrification and nitrogen mineralization in forest soils. J. Environ. Qual. 10:547–551.

Stuart, S. 1984. Hydrology and aquatic chemistry of monitor lake watersheds in the Wind River Mountains. Pages 130–173 in Air Quality and Acid Deposition Potential in the Bridger and Fitzpatrick Wildernesses. U.S. Forest Service, Ogden, UT.

Stumm, W., and G. Furrer. 1987. The dissolution of oxides and aluminum silicates; examples of surface-coordination-kinetics. Pages 197–219 in W. Stumm, ed. Aquatic Surface Chemistry: Chemical Processes at the Particle-Water Interface. Wiley-Interscience, New York.

Stumm, W., and J.J. Morgan. 1970. Aquatic Chemistry. Wiley-Interscience, New York.

Stumm, W., and J.J. Morgan. 1981. Aquatic Chemistry. 2nd Ed. Wiley-Interscience, New York.

Stumm, W., L. Sigg, and J.L. Schnoor. 1987. Aquatic chemistry of acid deposition. Environ. Sci. Technol. 21:8–13.

Sullivan, T.J., N. Christophersen, I.P. Muniz, H.M. Seip, and P.D. Sullivan. 1986. Aqueous aluminum chemistry response to episodic increases in discharge. Nature 323:324–327.

Sullivan, T.J., N. Christophersen, R.P. Hooper, H.M. Seip. I.P. Muniz, P.D. Sullivan, and R.D. Vogt. 1987. Episodic variation in streamwater chemistry at Birkenes, southernmost Norway: Evidence for the importance of water flow paths. Pages 269–279 in Proceedings of UNESCO-IHP Symposium, Acidification and Water Pathways, Vol. I. Bolkesj, Norway.

Sullivan, T.J., J.M. Eilers, M.R. Church, D.J. Blick, K.N. Eshleman, D.H. Landers, and M.S. DeHaan. 1988a. Atmospheric wet sulphate deposition and lakewater chemistry. Nature 331:607–609.

Sullivan, T.J., C.T. Driscoll, J.M. Eilers, and D.H. Landers. 1988b. Evaluation of the role of sea salt inputs in the long-term acidification of coastal New England lakes. Environ. Sci. Technol. 22:185–190.

Sullivan, T.J., C.T. Driscoll, S.A. Gherini, R.K. Munson, R.B. Cook, D.F. Charles, and C.P. Yatsko. 1989. The influence of organic acid anions and aqueous aluminum on measurements of acid neutralizing capacity in surface waters. Nature 338:408–410.

Sullivan, T.J., B.J. Cosby, J.A. Bernert, E.A. Jenne, J.M. Eilers, D.F. Charles, and A.R. Selle. 1990a. Comparison of paleolimnological and MAGIC model hindcasts of lakewater acidification in the Adirondack Mountains. Poster presentation, National Acid Precipitation Assessment Program Conference on Acidic Deposition: State of Science and Technology, February 1990, Hilton Head, SC.

Sullivan, T.J., D.L. Kugler, M.J. Small, C.B. Johnson, D.H. Landers, B.J. Rosenbaum, W.S. Overton, W.A. Kretser, and J. Gallagher. 1990b. Variation in Adirondack, New York, lakewater chemistry as a function of surface area. Water Resour. Bull. 26:1–10.

Summers, P.W., V.C. Bowersox, and G.J. Stensland. 1986. The geographical distribution and temporal variation of acidic deposition in eastern North America. Water Air Soil Pollut. 31:523–535.

Sutcliffe, D.W., and T.R. Carrick. 1973. Studies on mountain streams in the English Lake District. Freshwater Biol. 3:437–462.

Sutherland, J.W., ed. 1989. Field Surveys of the Biota and Selected Water Chemistry Parameters in 50 Adirondack Lakes. New York Department of Environmental Conservation, Albany, NY.

Swank, W.T., and D.A. Crossley, Jr. 1988. Introduction and site characterization. Pages 3–16 in W.T. Swank and D.A. Crossley, Jr. eds. Ecological Studies, Volume 66: Forest Hydrology and Ecology at Coweeta. Springer-Verlag, New York.

Swank, W.T., and J.E. Douglass. 1977. Nutrient budgets for undisturbed and manipulated hardwood forest ecosystems in the mountains of North Carolina. Pages 343–362 in D.L. Correll, ed. Watershed Research in Eastern North America, Volume I. Chesapeake Bay Center for Environmental Studies, Smithsonian Institute, Edgewater, MD.

Swank, W.T., and J.B. Waide. 1988. Characterization of baseline precipitation and stream chemistry and nutrient budgets for control watersheds. Pages 57–79 in W.T. Swank and D.A. Crossley, Jr., eds. Ecological Studies, Volume 66: Forest Hydrology and Ecology at Coweeta. Springer-Verlag, New York.

Sweets, P.R., C.D. Pollman, R. Bienert, T.L. Crisman, and M.W. Binford. 1987. Paleolimnology and Watershed Modelling: Examining the Hypotheses for Current Lake Acidity in North Florida. Presented at the ISEM/ESA sponsored symposium, "Recent Lake Acidification Trends in North America: Paleoecological Evidence." AIBS annual meeting, Columbus, OH.

Sweets, P.R., R.W. Bienert, Jr., T.L. Crisman, and M.W. Binford. (1990) Paleoecological investigations of recent lake acidification in northern Florida. J. Paleolimnol. 3:.

Swift, L.W., Jr., G.B. Cunningham, and J.E. Douglass. 1988. Climatology and hydrology. Pages 35–55 in W.T. Swank and D.A. Crossley, Jr., eds. Ecological Studies, Volume 66: Forest Hydrology and Ecology at Coweeta. Springer-Verlag, New York.

Tacconi, J.E. 1988. Ion Budgets and Cycling Processes in an Acid-Sensitive Seepage Lake. M.S. Thesis. University of Minnesota, Minneapolis, MN.

Talbot, R.W., and A.W. Elzerman. 1985. Acidification of southern Appalachian lakes. Environ. Sci. Technol. 19:552–557.

Taylor, T.P., and D.C. Erman. 1979. The response of benthic plants to past levels of human use in high mountain lakes in Kings Canyon National Park, California, U.S.A.J. Environ. Manage. 9:271–278.

Taylor, T.P., and D.C. Erman. 1980. The littoral bottom fauna of high elevation lakes in Kings Canyon National Park. Calif. Fish Game 66:112–119.

Tebo, M. 1985. The Southeastern Piney Woods: Describers, Destroyers, Survivors. M.S. Thesis. Florida State University, Tallahassee, FL.

Tenbus, F.J. 1987. Hydrological Characteristics of the Alluvial Infill at the White Oak Run Catchment, Shenandoah National Park. M.S. Thesis. Department of Environmental Sciences, University of Virginia, Charlottesville, VA.

Tennessee Valley Authority (TVA). 1971. Tennessee Valley Streams: Their Fish, Bottom Fauna, and Aquatic Habitat. Upper Little Tennessee River Drainage Basin. Division of Forestry, Fisheries, and Wildlife Development, Fisheries and Waterfowl Resources Branch.

Ter Braak, C.J.F. 1986. Canonical correspondence analysis: A new eigenvector technique for multivariate direct gradient analysis. Ecology 67:1167–1179.

Tessier, A.J., and R.J. Horwitz. 1988a. Analysis and Interpretation of Zooplankton Samples Collected during Phase II of the National Lake Survey. Report 88-18. Academy of Natural Sciences, Philadelphia, PA.

Tessier, A.J., and R.J. Horwitz. 1988b. Zooplankton Community Patterns and Relationships to Environmental Parameters in High Elevation Lakes in Maine. Report 88-25. Academy of Natural Sciences, Philadelphia, PA.

Thie, J., and G. Ironside, eds. 1976. Ecological (Biophysical) Land Classification in Canada. Proceedings of the first meeting, Canada Committee on Ecological (Biophysical) Land Classification. May 25–28, 1976, Petewawa, Ontario. Ecological Land Classification Series No. 1. Environment Canada, Ottawa, Ontario.

Thompson, M.E. 1982a. Exchange of marine sodium for calcium during chemical weathering in the Isle Aux Morts River Basin, Newfoundland. Geochim. Cosmochim. Acta 46:361–365.

Thompson, M.E. 1982b. The cation denudation rate as a quantitative index of sensitivity of eastern Canadian rivers to acidic atmospheric precipitation. Water Air Soil Pollut. 18:215–226.

Thompson, M.E. 1986. The cation denudation rate model—its continued validity. Water Air Soil Pollut. 31:17–26.

Thompson, M.E., and M.B. Hutton. 1985. Sulfate in lakes of eastern Canada: Calculated yields compared with measured wet and dry deposition. Water Air Soil Pollut. 24:77.

Thornbury, W.D. 1965. Regional Geomorphology of the United States. John Wiley and Sons, New York.

Thornthwaite, C.W. 1948. An approach toward a rational classification of climate. Geog. Rev. 38:55–94.

Thornton, J.D., and S.J. Eisenreich. 1982. Impact of land use on the acid and trace element composition of precipitation in the northcentral U.S. Atmos. Environ. 16:1945–1955.

Tonn, W.M., and J.J. Magnuson. 1982. Patterns in the species composition and richness of fish assemblages in northern Wisconsin lakes. Ecology 63:1149–1166.

Tonnessen, K.A. 1983. The Potential Effects of Acid Deposition on Aquatic Ecosystems of the Sierra Nevada, California. Ph.D. Thesis. University of California, Berkeley, CA.

Tonnessen, K.A. 1984. Potential for aquatic ecosystem acidification in the Sierra Nevada, California. Pages 147–149 in G.R. Hendrey, ed. Early Biotic Response to Advancing Lake Acidification. Butterworth Publishers, Boston, MA.

Tonnessen, K.A., and J. Harte. 1980. The potential for acid precipitation damage to aquatic ecosystems of the Sierra Nevada, California (U.S.A.) Pages 338–339 in D. Drabløs and A. Tollan, eds. Ecological Impact of Acid Precipitation. Proceedings of an International Conference, Sandefjord, Norway. SNSF Project, Oslo Norway.

Tonnessen, K.A., and J. Harte. 1982. Acid rain and ecological damage: Implications of Sierra Nevada lake studies. Public Affairs Report 23:1–19.

Tornes, L.A. 1979. Soil Survey of Ulster County, New York. Soil Conservation Service, U.S. Department of Agriculture, Washington, D.C.

Traaen, T.S. 1980. Effects of acidity on decomposition of organic matter in aquatic ecosystems. Pages 340–341 in D. Drabls and A. Tollan, eds. Ecological Impact of Acid Precipitation. Proceedings of an International Conference, Sandefjord, Norway. SNSF Project, Oslo, Norway.

Trewartha, G.T. 1954. Introduction to Climate. McGraw-Hill, New York.

Trippel, E.A., and H.H. Harvey. 1987. Reproductive responses of five white sucker (*Catostomus commersoni*) populations in relation to lake acidity. Can. J. Fish. Aquat. Sci. 44:1018–1023.

Trojnar, J.R. 1977. Egg and larval survival of white suckers (*Catastomus commersoni*) at low pH. J. Fish. Res. Board Canada 34:262–266.

Turk, J.T. 1983. An Evaluation of Trends in the Acidity of Precipitation and the Related Acidification of Surface Water in North America. Water Supply Paper 2249. U.S. Geological Survey, Reston, VA.

Turk, J.T., and D.B. Adams. 1983. Sensitivity to acidification of lakes in the Flat Tops Wilderness Area, Colorado. Water Resour. Res. 19:346–350.

Turk, J.T., and D.C. Campbell. 1987. Estimates of acidification of lakes in the Mt. Zirkel Wilderness Area, Colorado. Water Resour. Res. 23:1757–1761.

Turk, J.T., and N.E. Spahr. 1989. Chemistry of Rocky Mountain lakes. Pages 181–208 in D.C. Adriano and M. Havas, eds. Acid Precipitation, Volume 1: Case Studies. Springer-Verlag, New York.

Turner, M.A., M.B. Jackson, D.L. Findlay, R.W. Graham, E.R. DeBruyn, and E.M. Vandermeer. 1987. Early responses of periphyton to experimental lake acidification. Can. J. Fish. Aquat. Sci. 44:135–149.

Turner, R.S., R.J. Olsen, and C.C. Brandt. 1986. Pages 1–63 in Areas Having Soil Characteristics that May Indicate Sensitivity to Acidic Deposition under Alternate Forest Damage Hypotheses. Publication No. 2720. Oak Ridge National Laboratory Environmental Science Division, Oak Ridge, TN.

Turney, G.L., N.P. Dion, and S.S. Sumioka. 1986. Water Quality of Selected Lakes in Mount Rainier National Park, Washington, with Respect to Lake Acidification. USGS Water-Resources Investigations Report 85-4254. U.S. Dept. of Interior, Tacoma, WA.

Tweto, O. 1979. Geologic Map of Colorado. Scale 1:500,000. U.S. Geological Survey, Reston, VA.

United Nations Educational, Scientific, and Cultural Organization. 1978. World Water Balance and Water Resources of the Earth. UNESCO Press, Paris.

United States–Canada Memorandum of Intent on Transboundary Air Pollution. 1982. Atmospheric Sciences and Analysis Work Group 2. Report No. 2F-M. Washington, D.C.–Ottawa, Ontario.

United States Environmental Data Service. 1987. Climatological data, New York. Vol. 99, No. 13. New York, NY.

Urban, N.R. 1987. The Nature and Origins of Acidity in Bogs. Ph.D. Dissertation. University of Minnesota, Minneapolis, MN.

U.S. Department of Agriculture (USDA). 1979. General soil map–Virginia. U.S. Department of Agriculture, Soil Conservation Service, Lanham, MD.

U.S. Department of Agriculture (USDA). 1981. Land Resource Regions and Major Land Resource Areas of the United States. Soil Conservation Service, Washington, D.C.

U.S. Department of Commerce. 1968. Pages 1–80 in Climatic Atlas of the United States. Environmental Sciences Services Administration, Washington, D.C.

U.S. Environmental Protection Agency (EPA). 1983. Methods for Chemical Analysis of Water and Wastes. EPA/600/4-79/020, revised March 1983. U.S. Environmental Protection Agency, Corvallis, OR.

U.S. Fish and Wildlife Service. 1982. Effects of Acid Precipitation on Aquatic Resources: Results of Modeling Workshops. FWS/OBS-80/40.123. Biological Services Program, U.S. Fish and Wildlife Service, Washington, D.C.

U.S. Geological Survey. 1970. National Atlas of the United States. Washington, D.C.

Valentini, J.L., and S.A. Gherini. 1986. The Integrated Lake-Watershed Acidification Study: Database Documentation. EPRI EA-3221, Volume 5. Electric Power Research Institute, Palo Alto, CA.

van Belle, G., and J.P. Hughes. 1984. Nonparametric testss for trend in water quality. Water Resour. Res. 20:127–136.

van Breemen, N., J. Mulder, and C.T. Driscoll. 1983. Acidification and alkalinization of soils. Plant Soil 75:383–408.

Van Coillie, R., C. Thellen, P.G.C. Campbell, and Y. Vigneault. 1983. Effets Toxiques de l'Aluminium Chez les Salmonides en Relation avec des Conditions

Physico-chimiques Acides. Rapp. Tech. Can. Sci. Halieut. et Aquat. No. 1237. Ministere des Peches et des Oceans, Quebec, Canada.

Van Diver B. 1985. Roadside Geology of New York. Mountain Press, Missoula, MT.

van Frankenhuyze, K., G.H. Geen, and C. Koivisto. 1985. Direct and indirect effects of low pH on the transformation of detrital energy by the shredding caddisfly, *Clistoronia magnifica* (Banks) (Limnephilidae). Can. J. Zool. 63:2298–2304.

Vankat, J.L. 1977. Fire and man in Sequoia National Park. Ann. Assoc. Am. Geog. 67:17–27.

Vankat, J.L., and J. Major. 1978. Vegetation changes in Sequoia National Park, California. J. Biogeog. 5:377–402.

Van Winkle, W., S.W. Christensen, and J.E. Breck. 1986. Linking laboratory and field responses of fish populations to acidification. Water Air Soil Pollut. 30:639–648.

Velbel, M.A. 1985a. Hydrogeochemical constraints on mass balances in forested watersheds of the southern Appalachians. Pages 231–247 in J.I. Drever, ed. The Chemistry of Weathering. NATO Advanced Research Workshop, July 1984, Rodez, France. D. Reidel, The Netherlands.

Velbel, M.A. 1985b. Geochemical mass balances and weathering rates in forested watersheds of the Southern Blue Ridge. Am. J. Sci. 285:904–930.

Veno, P.A. 1976. Successional relationships of five Florida plant communities. Ecology 57:498–508.

Vernon, R.O. 1942. Geology of Holmes and Washington Counties, Florida. Geological Bulletin No. 21. Florida Geological Survey, Tallahassee, FL.

Verry, E.S. 1983. Precipitation chemistry at the Marcell Experimental Forest in northcentral Minnesota. Water Resour. Res. 19:454–462.

Verry, E.S., and A.R. Harris. 1988. A description of low- and high-acid deposition. Water Resour. Res. 24:481–492.

Virginia Department of Conservation and Economic Development. 1969. James River Basin, Comprehensive Water Resources Plan. Planning Bulletin 213, Richmond, VA.

Visser, S.A. 1984. Seasonal changes in the concentration and colour of humic substances in some aquatic environments. Freshwater Biol. 14:79–87.

Vitousek, P.M. 1977. The regulation of element concentrations in mountain streams in the northeastern United States. Ecol. Monogr. 47:65–87.

Vitousek, P.M., and W.A. Reiners. 1975. Ecosystem succession and nutrient retention: A hypothesis. Bioscience 25:376-381.

Voldner, E.C., and A. Sirois. 1986. Monthly mean spatial variations dry deposition velocities of oxides of sulfur and nitrogen. Water Air Soil Pollut. 30:179–186.

Voldner, E.C., L.A. Barrie, and A. Sirois. 1986. A literature review of dry deposition of oxides of sulphur and nitrogen with emphasis of long-range transport modelling in North America. Atmos. Environ. 20:2101–2123.

Volk, R.G., and G.E. Byers. 1983. Effects of acid precipitation on a typic quartzipsamment soil in Florida. Pages 265–282 in A.E.S. Green and W.H. Smith, eds. Acid Deposition Causes and Effects: A State Assessment Model. Government Institutes, Rockville, MD.

Volk, R.G., and C.D. Pollman. 1985. Florida Acid Deposition Study, Phase III Report. Sulfate Adsorption. ESE No. 82-615-0600-2110. Environmental Science and Engineering, Inc., Gainesville, FL.

Vong, R.J., et al. 1988. Changes in rainwater acidity associated with closure of a copper smelter. J. Geophys. Res. 93:7169–7179.

Vong, R.J., S. Cline, G. Reams, J. Bernert, D.F. Charles, J. Gibson, T. Haas, H. Moore, R. Husar, A. Olsen, J. Simpson, and S. Seilkop. 1989. Regional Analysis of Wet Deposition for Effects Research. EPA/600/S3-89/030. U.S. Environmental Protection Agency, Environmental Research Laboratory, Corvallis, OR.

Vreeland, J.L. 1983. The Role of Shallow Groundwater in the Hydrogeochemical Response of the White Oak Run Catchment, Shenandoah National Park. M.S. Thesis. Department of Environmental Sciences, University of Virginia, Charlottesville, VA.

Wahrhaftig, C. 1984. Geomorphology and Glacial Geology of Wolverton and Crescent Meadow Areas and Vicinity, Sequoia National Park, California. Open File Report 84-400. U.S. Geological Survey, Reston, VA.

Waide, J.B., and W.T. Swank. 1987. Patterns and Trends in Precipitation and Stream Chemistry at the Coweeta Hydrologic Laboratory. Aquatic Effects Task Group VI Peer Review. Volume II. May 17–23, New Orleans, LA.

Waldman, J.M., J.W. Munger, D.J. Jacob, R.C. Flagan, J.J. Morgan, and M.R. Hoffmann. 1982. Chemical composition of acid fog. Science 218:677–680.

Wales, D.L., and V.A. Liimatainen. 1987. Preliminary Assessment of the Current Impact and Potential Risk of Acidic Deposition on Walleye Populations in Ontario. Ontario Fisheries Acidification Report Series No. 87-11. Ministry of Natural Resources, Toronto, Ontario.

Walker, G.W., and P.B. King. 1969. Geologic Map of Oregon. Miscellaneous Geologic Investigators Map I-595. U.S. Geological Survey, Department of Interior, Reston, VA.

Walker, W.J., C.S. Cronan, and P.R. Bloom. 1990. Aluminum solubility in organic soil horizons from northern and southern forested watersheds. Soil Sci. Soc, Am. J. 54:369–374.

Watras, C.J., and T.M. Frost. 1989. Little Rock Lake (Wisconsin): Perspectives on an experimental ecosystem approach to seepage lake acidification. Arch. Environ. Contam. Toxicol. 18:157–165.

Watson, C.R., and A.R. Olsen. 1984. Acid Deposition System (ADS) for Statistical Reporting. EPA/600/8-84/023. U.S. Environmental Protection Agency, Research Triangle Park, NC.

Watt, W.D., C. Scott, and W.H. White. 1983. Evidence of acidification of some Nova Scotian rivers and its impact on Atlantic salmon, *Salmo salar.* Can. J. Fish. Aquat. Sci. 40:462–473.

Way, J.H. 1972. A More Detailed Discussion of the Depositional Environmental Analyses–Middle and Upper Devonian Sedimentary Rocks, Catskill Mountain Area, New York. Ph.D. Dissertation. Rensselaer Polytechnic Institute, Troy, NY.

Weatherley, N.S., and S.J. Ormerod. 1987. The impact of acidification on macroinvertebrate assemblages in Welsh streams: Towards an empirical model. Environ. Pollut. 46:223–240.

Weathers, K.C., G.E. Likens, F.H. Bormann, J.S. Eaton, W.B. Bowden, J.L. Anderson, D.A. Cass, J.N. Galloway, W.C. Keene, K.D. Kimball, P. Huth, and D. Smiley. 1986. A regional acidic cloud/fog water event in the eastern United States. Nature 319: 657–658.

Webb, J.R. 1987a. Virginia Trout Stream Sensitivity Study, April 1987. Synoptic Survey, Study Handbook. Department of Environmental Sciences, University of Virginia, Charlottesville, VA.

Webb, J.R. 1987b. Retention of Atmospheric Sulfate by Catchments in Shenandoah National Park, Virginia. M.S. Thesis (submitted). Department of Environmental Sciences, University of Virginia, Charlottesville, VA.

Webb, J.R. 1988. Sulfate Retention by Catchments in Shenandoah National Park. M.S. Thesis. Department of Environmental Sciences, University of Virginia, Charlottesville, VA.

Webb, J.R., B.J. Cosby, J.N. Galloway, and G.M. Hornberger. 1989. Acidification of native brook trout streams in Virginia. Water Resour. Res. 25:1367–1377.

Weiner, G.S., C.B. Schrek, and H.W. Li. 1986. Effects of low pH on reproduction of rainbow trout. Trans. Am. Fish. Soc. 115:75–82.

Weiner, R.W., J.M. McLelland, Y.W. Isachsen, and L.M. Hall. 1984. Stratigraphy and Structural Geology of the Adirondack Mountains, New York: Review and Synthesis. Special Paper 194. Geological Society of America, Boulder, CO.

Weinstein, D.A. 1982. The Long-Term Nutrient Retention Properties of Forest Ecosystems: S Simulation Investigation. Ph.D. Dissertation. University of Tennessee, Knoxville, TN.

Weintraub, J. 1986. An Assessment of the Susceptibility of Two Alpine Watersheds to Surface Water Acidification: Sierra Nevada, California. M.S. Thesis. Department of Geology, Indiana University, Bloomington, IN.

Welch, E.B., and W.H. Chamberlain. 1981. Initial Detection of Acid Lakes in Washington State. Final Report for the National Park Service, U.S. Department of Interior, Pacific Northwest Region, Seattle, WA.

Welch, E.B., W.H. Chamberlain, and D.E. Spyridakis. 1984. Chemical content of snow and effect of melting on Cascade Mountain lakes. Northwest Sci. 58:85–93.

Welch, E.B., D.E. Spyridakis, and T. Smayda. 1986. Temporal chemical variability in acid sensitive high elevation lakes. Water Air Soil Pollut. 31:35–44.

Welch, E.B., D.E. Spyridakis, and T. Smayda. 1987. Acidification Potential in High Mountain Lakes in Washington: Part 4. Final Report. State of Washington, Department of Ecology, Olympia, WA.

Wentz, D.A., and W.J. Rose. 1989. Interrelationships among hydrologic budget components of a northern Wisconsin seepage lake and implications for acid deposition modeling. Arch. Environ. Contam. Toxicol. 18:147–155.

Wentz, D.A. J.T. Krohelski, W.J. Rose, J.G. Bockheim, P.J. Garrison, D.R. Knauer, and R.A. Goldstein. 1987. Hydrologic and chemical budgets of Wisconsin seepage lakes receiving acid deposition. Pages 309–316 in R. Perry, R.M. Harrison, J.N.B. Bell, and J.N. Lester, eds. Acid Rain: Scientific and Technical Advances. Selper Ltd., Ealing, London.

Wetzel, R.G. 1975 (second edition 1983). Limnology. W.B. Saunders Company, Philadelphia, PA.

White, W.H. 1970. The Geomorphology of Florida Peninsula. Bulletin No. 51. Florida Bureau of Geology, Tallahassee, FL.

Whitehead, D.R., D.F. Charles, S.E. Reed, S.T. Jackson, and M.C. Sheehan. 1986. Late-glacial and Holocene acidity changes in Adirondack (NY) lakes. Pages 251–274 in J.P. Smol, R.W. Battarbee, R.B. Davis, and J. Meriläinen, eds. Diatoms and Lake Acidity. Developments in Hydrobiology 29. Dr. W. Junk, Dordrecht, The Netherlands.

Whiting, M.C., D.R. Whitehead, R.W. Holmes, and S.A. Norton. 1989. Paleolimnological reconstruction of recent acidity changes in four Sierra Nevada lakes. J. Paleolimnol. 2:285–304.

Whitney, S. 1979. The Sierra Nevada. Sierra Club Books, San Francisco, CA.

Whittaker, R.H. 1956. Vegetation of the Great Smoky Mountains. Ecol Monogr. 26:1–80.

Whitworth, W.E., and R.J. Strange. 1983. Growth and production of sympatric brook and rainbow trout in an

Appalachian stream. Trans. Am. Fish. Soc. 112:469–475.

Wiener, J.G. 1983. Comparative Analyses of Fish Populations in Naturally Acidic and Circumneutral Lakes in Northern Wisconsin. FWS/OBS-80/40.16. U.S. Fish and Wildlife Service Report, Kearneysville, WV.

Wiener, J.G., and J.M. Eilers. 1987. Chemical and biological status of lakes and streams in the Upper Midwest: Assessment of acidic deposition effects. Lake Reserv. Manage. 3:365–378.

Wiener, J.G., P.J. Rago, and J.M. Eilers. 1984. Species composition of fish communities in northern Wisconsin lakes: Relation to pH. Pages 133–146 in G.R. Hendrey, ed. Early Biotic Response to Advancing Lake Acidification. Butterworth Publishers, Boston, MA.

Wiken, E. 1986. Terrestrial Ecozones of Canada. Ecological Land Classification Series No. 19. Environment Canada, Ottawa, Ontario.

Wiklander, L. 1964. Cation and anion exchange phenomena. Pages 163–205 in F.E. Bear, ed. Chemistry of the Soil. Van Nostrand Reinhold, New York.

Wile, I., and G.E. Miller. 1983. The Macrophyte Flora of 46 Acidified and Acid-Sensitive Soft Water Lakes in Ontario. Ontario Ministry of Environment, Rexdale, Ontario, Canada.

Wile, I., G.E. Miller, G.G. Hitchin, and N.D. Yan. 1985. Species composition and biomass of the macrophyte vegetation of one acidified and two acid-sensitive lakes in Ontario. Can. Field Nat. 99:308–312.

Williams, E.H. 1976. Distributional Patterns of High Altitude Zooplankton. Ph.D. Thesis. Princeton, University, Princeton, NJ.

Williams, W.T. 1979. Long distance transport of air pollution into Sequoia, Kings Canyon, Yosemite, and Lassen National Parks. Pages 165–169 in Proceedings, Second National Conference for Scientific Research in the National Parks. U.S. National Park Service, Washington, D.C., and American Institute of Biological Science, Washington, D.C.

Williams, W.T., M. Brady, and S.C. Willison. 1977. Air pollution damage to the forests of the Sierra Nevada mountains of California. J. Air Pollut. Control Assoc. 27:230–234.

Winger, P.V., P.J. Lasier, M. Hudy, D.L. Fowler, and M. J. Van Den Avyle. 1987. Sensitivity of high-elevation streams in the Southern Blue Ridge Province to acidic deposition. Water Resour. Bull. 23: 379–386.

Winter, T.C. 1977. Classification hydrologic settings of lakes in the northcentral United States. Water Resour. Res. 13:753–767.

Wismar, et al. 1982. Chemical changes of lakes within the Mount St. Helens blast zone. Science 216:172–178.

Wolcott, E.E. 1961. Lakes of Western Washington. USGS Water Supply Bulletin, No. 4. U.S. Geological Survey, U.S. Department of Interior, Reston, VA.

Wolff, E.W., and D.A. Peel. 1985. The record of global pollution in polar snow and ice. Nature 313: 535–540.

Wolff, G.T. 1984. On the nature of nitrate in coarse continental aerosols. Atmos. Environ. 18:977–981.

Wonson-Liukkonen, B. 1987. Geochemistry of Two Small Lakes in Northeastern Minnesota in Relation to Atmospheric Inputs. M.S. Thesis. University of Minnesota, Duluth, MN.

Wood, L.W. 1978. Limnology of Remote Lakes in the Adirondack Region of New York State with Emphasis on Acidification Problems. New York State Department of Health, Albany, NY.

Wood, S.H. 1977. Distribution, correlation, and radiocarbon dating of late Holocene tephra, Mono and Inyo craters, eastern California. Geol. Soc. Am. Bull. 88:89–95.

Woodward, D.F., A.M. Farag, M.E. Mueller, E.E. Little, and F.A. Vertucci. 1989. Sensitivity of endemic Snake River cutthroat trout to acidity and elevated aluminum. Trans. Am. Fish. Soc. 118:630– 643.

Wright, H.E., Jr. 1972. Physiography of Minnesota. Pages 561–578 in P.K. Sims and G.B. Morey, eds. Geology of Minnesota: A Centennial Volume. Minnesota Geological Survey, St. Paul, MN.

Wright, R.F. 1976. The impact of forest fire on the nutrient influxes to small lakes in northeastern Minnesota. Ecology 57:649–663.

Wright, R.F. 1983. Predicting Acidification of North American Lakes, Acid Rain Research Report 4/1983. Report No. 0-81036. Norwegian Institute for Water Research, Oslo, Norway.

Wright, R.F. 1988. Acidification of lakes in the eastern United States and southern Norway: A comparison. Environ. Sci. Technol. 22:178–182.

Wright, R.F., and A. Henriksen. 1978. Chemistry of small Norwegian lakes, with special reference to acid precipitation. Limnol. Oceanog. 23:487–498.

Wright, R.F., N. Conroy, W.T. Dickson, R. Harriman, A. Henriksen, and C.L. Schofield. 1980. Acidified lake districts of the world: A comparison of water chemistry of lakes in southern Norway, southern Sweden, southwestern Scotland, the Adirondack Mountains of New York, and southeastern Ontario. Pages 377–379 in D. Drabløs and A. Tollan, eds. Ecological Impact of Acid Precipitation. Proceedings of an International Conference, Sandefjord, Norway. SNSF Project, Oslo, Norway.

Wright, R.F., B.J. Cosby, G.M. Hornberger, and J.N. Galloway. 1986a. Commparison of paleolimnological with MAGIC model reconstructions of water acidification. Water Air Soil Pollut. 30:367–380.

Wright, R.F., E. Gjessing, N. Christophersen, E. Lotse, H.M. Seip, A. Semb, B. Sletaune, R. Storhaug, and K. Wedum. 1986. Project rain: Changing acid deposition to whole catchments. The first year of treatment. Water Air Soil Pollut. 30:47–64.

Yan, N.D. 1979. Phytoplankton of an acidified, heavy metal-contaminated lake near Sudbury, Ontario, 1973–1977. Water Air Soil Pollut. 11:43–55.

Yan, N.D. 1983. Effects of changes in pH on transparency and thermal regimes of Lohi Lake, near Sudbury, Ontario. Can. J. Fish. Aquat. Sci. 40:621- 626.

Yan, N.D. 1986. Empirical prediction of crustacean biomass in nutrient poor Canadian Shield lakes. Can. J. Fish. Aquat. Sci. 43:788–796.

Yan, N.D., and P.M. Stokes. 1978. Phytoplankton of an acidic lake, and its responses to experimental alterations of pH. Environ. Conserv. 5:93–100.

Yan, N.D., and R. Strus. 1980. Crustacean zooplankton communities of acidic, metal-contaminated lakes near Sudbury, Ontario. Can. J. Fish. Aquat. Sci. 37:2282–2293.

Yan, N.D., G.E. Miller, I. Wile, and G.G. Hutchinson. 1985a. Richness of aquatic macrophyte floras of soft water lakes of differing pH and trace metal content in Ontario, Canada. Aquat. Bot. 23:27–41.

Yan, N.D., R.W. Nero, W. Keller, and D.C. Lasenby. 1985b. Are Chaoborus larvae more abundant in acidified than in non-acidified lakes in central Canada? Holarctic Ecol. 8:93–99.

Yobbi, D.K., and G.C. Chappell. 1979. Summary of the Hydrology of the Upper Etonia Creek Basin. Technical Publication No. SJ79-5. St. Johns River Water Management District, Palatka, FL.

Young, J.R., E.C. Ellis, and G.M. Hidy. 1988. Deposition of air-borne acidifiers in the western environment. J. Environ. Qual. 17:1–26.

Zar, J.H. 1984. Biostatistical Analysis. Second edition. Prentice-Hall, Englewood Cliffs, NJ.

Zardus, M., T. Blank, and D. Schulz. 1977. Status of Fishes in 137 Lakes in Sequoia and Kings Canyon National Parks, California. National Park Service, Three Rivers, CA.

Zischke, J.A., J.W. Arthur, K.J. Nordlie, P.O. Hermanutz, D.A. Standen, and T.P. Henry. 1983. Acidification effects on macroinvertebrates and fathead minnows (*Pimephales promelas*) in outdoor experimental channels. Water Res. 17:47–63.

Glossary

Abbreviations and Acronyms

ADS: Acid Deposition System
AERP: Aquatic Effects Research Program
ALCHEMI: A solution equilibrium model developed to perform solution phase acid-base calculations
ALP: Aquifer Lakes Project
ALSC: Adirondack Lake Survey Corporation
ALSS: Aluminum in Streams Study
ALTM: Adirondack Long-Term Monitoring study
AMD: Acid mine drainage
ANC: Acid neutralizing capacity
ANC_{Gran}: ANC determined by Gran titration
ANCGEN: Net ANC production; $ANCGEN_{CB\text{-}CA}$ = sum of ionic reactions; $ANCGEN_{Gran}$ = ANCGEN based on the Gran ANC reaction
ANCOVA: Analysis of covariance
ANDEF: Anion deficit
ANOVA: Analysis of variance
APHA: American Public Health Association
ARC/INFO: A Geographic Information System software package used to store, manipulate, and display spatially distributed data. Copyright of Environmental Systems Research Institute (ESRI).

BNC: Base neutralizing capacity
BP: Before present

CADMP: California Acid Deposition Monitoring Program
CAPITA model: A regional Monte Carlo model, developed by the Center for Air Pollution Impact and Trend Analysis
CCA: Canonical correspondence analysis
CEC: Cation exchange capacity
CFD: Cumulative frequency distribution
CL: Confidence limit

DDRP: Direct/Delayed Response Project
DIC: Dissolved inorganic carbon
DOC: Dissolved organic carbon

ELA: Experimental Lakes Area, Ontario, Canada
ELS-I: Eastern Lake Survey–Phase I
ELS-II: Eastern Lake Survey–Phase II
EMS: Eastern Maine Streams Project
ENAMAP: Eastern North America Model of Air Pollution
EPA: U.S. Environmental Protection Agency
EPRI: Electric Power Research Institute
ERL-C: U.S. EPA Environmental Research Laboratory, Corvallis, Oregon
ESE: Environmental Science and Engineering
ET: Evapotranspiration
ET/MB: Evapotranspiration/mass balance
ETD: Enhanced Trickle Down model

FADS: Florida Acid Deposition Study
FCG: Florida Electric Power Coordinating Group
FDER: Florida Department of Environmental Regulation
FIN: Fish Information Network data base

GIS: Geographic Information System

HBEF: Hubbard Brook Experimental Forest
HELM: High Elevation Lake Monitoring Project

IAG: In-lake Alkalinity Generation model
ID: Identification number
ILWAS: Integrated Lake Watershed Acidification Study

LAF: Lake Acidification and Fisheries project
LAI: Leaf area index
LAMP: Lake Acidification Mitigation Project

MAGIC: Model of Acidification of Groundwater in Catchments
MIBK: Methyl-isobutyl-ketone
MIBK-HQ: Methyl-isobutyl-ketone-hydroxy-quinoline

NADP: National Atmospheric Deposition Program
NADP/NTN: National Atmospheric Deposition Program/National Trends Network
NAPAP: National Acid Precipitation Assessment Program
NBS: National Bureau of Standards
NLS: National Lake Survey
NOAA: National Oceanographic and Atmospheric Administration
NPS: National Park Service
NRC: National Research Council
NSS-I: National Stream Survey–Phase I
NSS-MAR: National Stream Survey–Mid-Atlantic Region subset
NSS-VR: National Stream Survey–Virginia Ridge subset
NSWS: National Surface Water Survey
NTN: National Trends Network
NTU: Nephelometric turbidity unit
NY DEC: New York State Department of Environmental Conservation (sometimes NYSDEC)
NYCDEP: New York City Department of Environmental Protection

ORNL: Oak Ridge National Laboratory
OTA: Office of Technology Assessment, U.S. Congress

P/E: Precipitation to evaporation ratio
PCA: Principal component analysis
PCU: Platinum cobalt unit
PCV: Pyrocatechol violet
PHCALC: A solution equilibrium model developed to perform solution phase acid-base calculations
PIRLA: Paleoecological Investigation of Recent Lake Acidification study
ppm: Parts per million
PSC: Public Service Commission

QA: Quality assurance
QC: Quality control

RCS: Regional Case Studies
RILWAS: Regional Integrated Lake Watershed Acidification Study
RRP: Relative reproductive potential, output from the FISHEGGS model
RT: Residence time

SAS: Statistical Analysis System
SBC: Sum of base cations; see also C_B
SBR: Southern Blue Ridge
SBRP: Southern Blue Ridge Province
SD: Standard deviation
SE: Standard error
SNP: Shenandoah National Park (Synoptic Survey, Phase I)
SWSNP: Southwest Shenandoah National Park (Synoptic Survey, Phase II)

TMWS: Tunk Mountain Watershed Study
TVA: Tennessee Valley Authority

UAPSP: Utilities Acid Precipitation Sampling Program
UCL: 95% upper confidence limit
USDA: U.S. Department of Agriculture
USFS: U.S. Forest Service
USFWS: U.S. Fish and Wildlife Service
USGS: U.S. Geological Survey

VTSSS: Virginia Trout Stream Sensitivity Survey

WA: Watershed area
WDNR: Wisconsin Department of Natural Resources
WLS: Western Lake Survey
WMP: Watershed Manipulation Project
WSA:LA: Watershed area to lake area ratio
WY: Water year

Symbols

A_s: Strong organic anions
A_t: Total analytic concentration of organic anions
A_w: Weak organic anions
Al: Aluminum
Al_{ext}: MIBK-extractable aluminum (total monomeric); sometimes Al_{MIBK}
Al_{im}: Inorganic monomeric aluminum; sometimes Al_i
Al_m: Total monomeric aluminum; sometimes Al_T
Al_{MIBK}: Methyl-isobutyl-ketone aluminum (total monomeric); see Al_{ext}
Al_{om}: Aluminum associated with organic ligands; organic monomeric aluminum; sometimes Al_o
Al_T: Total monomeric aluminum; sometimes Al_m
Alk: Carbonate alkalinity
An^-: Nonprotonated organic anions

C_A: Sum of strong acid anions
C_B: Base cations or sum of base cations; may be defined differently in different chapters (e.g., $Ca^{2+} + Mg^{2+}$); generally used in chemical equations; see SBC
Ca^{2+}: Calcium ion
$CaCO_3$: Calcium carbonate or calcite
$CaSO_4$: Gypsum
Cl^-: Chloride ion
cm: Centimeter
CO_2: Carbon dioxide
CO_3^{2-}: Carbonate
C_T: Total dissolved inorganic carbon

F^-: Fluoride ion
Fe: Iron
$Fe(OH)_3$: Ferric hydroxide
FeS_2: Pyrite (ferrous sulfide)
ft: Feet
F(x): Cumulative frequency distribution

g: Gram
G(x): Cumulative areal distribution

H^+: Hydrogen ion
H_2CO_3: Carbonic acid
H_2O: Water
H_2SO_4: Sulfuric acid
H_4SiO_4: Silicic acid
ha: Hectare
HCl: Hydrochloric acid
HCO_3^-: Bicarbonate
HNO_3: Nitric acid

K^+: Potassium ion
K_a or K_A: Acid dissociation constant
kg: Kilogram
km^2: Square kilometers

m: Meter
MAX: Maximum value of variable *X*
mg: Milligram
mg L^{-1}: Milligrams per liter, unit of concentration
Mg^{2+}: Magnesium ion
mi: Mile
mi^2: Square miles
MIN: Minimum value of variable *X*
mL: Milliliter
mm: Millimeter
msl: Minimum survival level

n: Number of samples
N: Number of lakes in a target population
Na^+: Sodium ion
NaCl: Sodium chloride
NH_4^+: Ammonium
NO_3^-: Nitrate
n.s.: Not significant

O_2: Oxygen
OH^-: Hydroxide

p: Significance level (sometimes *P*)
PAH: Polycyclic aromatic hydrocarbons
pCO_2: Partial pressure of CO_2; carbonic acid
pH: The negative logarithm of hydrogen ion activity
pH-ae: Air-equilibrated pH (also pH_{ae})
pK_a: Negative logarithm (base 10) of acid dissociation constant

r: Pearson's correlation coefficient
R_T: Total organic acid anion
r^2: Coefficient of determination
R^2: Coefficient of multiple determination
$RCOO^-$: Organic anion

SiO_2: Silicon dioxide; silica
SO_4^{2-}: Sulfate

t_w: Hydraulic (or hydrologic) residence time (also τ_w)
Tg yr^{-1}: Total grams per year

yr: Year

°C: Degrees Celsius
Δ: Delta; denotes an increment of change; sometimes δ
ΔANC: Change in acid neutralizing capacity
ΔpH: Difference between closed-system pH and air-equilibrated pH (sometimes δpH)
ΔSBC: Change in sum of base cations; sometimes ΔC_B

μeq L^{-1}: Microequivalents per liter; unit of concentration
μg L^{-1}: Micrograms per liter; unit of concentration
μmol L^{-1}: Micromoles per liter, unit of concentration
μS cm^{-1}: Microsiemens per centimeter; unit of electrical conductance

ΣC_A: Sum of strong acid anions
ΣC_B: Sum of base cations

τ_w: Hydraulic (or hydrologic) residence time (also t_w)

ELS Subregions

1A: Adirondack Mountains
1B: The Poconos/Catskill Mountains
1C: Central New England
1D: Southern New England
1E: Maine
2A: Northeastern Minnesota subregion of ELS-I
2B: Westcentral Minnesota subregion of ELS-I
2C: Northwestern Wisconsin subregion of ELS-I
2D: Northcentral Wisconsin subregion of ELS-I
2E: Northeast Michigan subregion of ELS-I
3A: Southern Blue Ridge
3B: Florida

WLS Subregions

4A: California
4B: Pacific Northwest
4C: Northern Rocky Mountains
4D: Central Rocky Mountains
4E: Southern Rocky Mountains

NSS Subregions

1D: Poconos/Catskill Mountains
2As: Southern Blue Ridge
2Bn: Valley and Ridge
2Cn: Northern Appalachians
2D: Ozark/Ouachita Mountains
2X: Southern Appalachians
3A: Piedmont
3B: Mid-Atlantic Coastal Plain
3C: Florida

Definitions

absence: when used in the context of the presence or absence of a fish species or other organism, refers to a species not caught during sampling (generally using a standardized sampling regime for all waters sampled).
abundance: the number of organisms per unit area or volume.
accuracy: the closeness of a measured value to the true value of an analyte.
acid anion: negatively charged ion that does not react with hydrogen ion in the pH range of most natural waters.
acid-base chemistry: the reaction of acids (proton donors) with bases (proton acceptors). For this book, this means the result of reactions of natural and anthropogenic acids and bases described in terms of pH and acid neutralizing capacity of the system.
acid cation: hydrogen ion or metal ion that can hydrolyze water to produce hydrogen ions (e.g., ionic forms of aluminum, manganese, and iron).
acid mine drainage: runoff with high concentrations of metals and sulfate and high levels of acidity resulting from the oxidation of sulfide minerals that have been exposed to air and water by mining activities.
acid neutralizing capacity (ANC): the equivalent capacity of a water sample to neutralize acids, determined by titration with a strong acid. The components of ANC include weak bases (car-

bonate species, dissociated organic acids, alumino-hydroxides, borates, and silicates) and strong bases (primarily OH^-). In the NSWS, as in most other recent studies of acid-base chemistry, ANC was measured by the Gran titration procedure, often termed Gran ANC.

acidic deposition: rain, snow, or dry fallout containing high concentrations of sulfuric acid, nitric acid, or hydrochloric acid, usually produced by atmospheric transformation of the by-products of fossil fuel combustion (power plants, smelters, automobiles, etc.); precipitation with $pH < 5.0$ is generally considered to be unnaturally acidic.

acidic lake or stream: for this book, a lake or stream with acid neutralizing capacity ≤ 0.

acidification: the temporary or permanent decrease of acid neutralizing capacity in water or base saturation in soil caused by natural or anthropogenic processes.

acidophilic: describes organisms that thrive in an acidic environment ($pH < 7.0$).

air-equilibrated sample: a water sample that has been brought to equilibrium with the atmosphere or with standard air (300 ppm CO_2) before analysis; used with some pH measurements.

algorithm: a set of numerical steps or routines followed to obtain a numerical output from a numerical input.

aliasing: occurrence of an apparent shift in frequency of a periodic phenomenon, arising as a result of the choice of discrete space or time sampling points to represent a continuous process. The choice may introduce a spurious periodic solution or mask a real periodic phenomenon.

alkalinity: generally the equivalent sum of HCO_3^- + CO^2 + OH^- − H^+ (i.e., buffering conferred by the carbonate system). Acid neutralizing capacity, sometimes used interchangeably with the term total alkalinity, includes alkalinity plus additional buffering from dissociated organic acids and other compounds (see Chapter 1).

alkalinity map class: in the NSWS, a geographic area defined by the expected alkalinity of a majority of surface waters (not necessarily reflecting measured alkalinity); used as a stratification factor in the lake selection design. Three alkalinity map classes were used: < 100, 100 < 200, and 200 ≤ 400 $\mu eq\ L^{-1}$.

analyte: a chemical species measured in a water sample.

analytical model: mathematical model in which the solution of the governing equations is obtained by mathematical analysis rather than numerical manipulation.

ANC deficit: the difference in concentration between the sum of nonmarine base cations and ANC.

anion: a negatively charged ion.

anion-cation balance: a method of assessing whether all anions have been accounted for and measured accurately; in an electrically neutral solution, such as water, the total charge of positive ions (cations) equals the total charge of negative ions (anions).

anion deficit: the concentration in $\mu eq\ L^{-1}$ of measured cations minus measured anions; usually a result of unmeasured organic anions or analytical uncertainty and often used as a surrogate for organic anion concentration.

background sulfate: estimated pre-industrial (pre-1850) concentration of sulfate in surface waters or precipitation.

base cation: an alkali or alkaline earth metal cation; usually calcium, magnesium, sodium, and potassium. See **protolyte**.

base cation supply: (1) the pool of base cations (Ca^{2+}, Mg^{2+}, K^+, Na^+) in the soil available for exchange with acid cations. The base cation pool is determined by the cation exchange capacity of the soil and the percentage of exchange sites occupied by base cations; (2) the rate at which base cations can be supplied to buffer incoming acid cations; this rate is determined by the relative rate of mineral weathering, the availability of base cations on exchange sites, and the rate of mobile anion leaching.

benthic: refers to bottom zones or bottom-dwelling organisms in a lake or stream.

benthic invertebrates: organisms lacking a spinal column that live on the bottom of a lake or stream; by one definition, benthic macroinvertebrates are organisms large enough to be retained by a 0.595-mm sieve.

bioassay: measurement of the response of an organism or group of organisms upon exposure to in situ environmental conditions or environmental conditions simulated in the laboratory; also sometimes referred to as toxicity test.

biological response variables: quantifiable measurements of biological response.

biomass: the total quantity of organic matter in units of weight or mass.

buffering capacity: the property of a solution that permits the relative concentrations of hydrogen and hydroxyl ions to be maintained by neutralizing added acids or bases, within limits.

calcium bicarbonate system: see **carbonate system**.

calculated conductance: the sum of the theoretical specific conductances of all measured ions in a sample.

calibration: process of checking, adjusting, or standardizing operating characteristics of instruments and model appurtenances on a physical model or coefficients in a mathematical model with empirical data of known quality. The process of evaluating the scale readings of an instrument with a known standard in terms of the physical quantity to be measured.

carbonate system: a lake in which the major part of acid neutralizing capacity is composed of carbonate; organic or other weak anions contribute less than 10% to the total anion charge.

cation: a positively charged ion.

cation exchange: a reversible process occurring in soil and sediments in which acidic cations (e.g., hydrogen ions) are adsorbed and base cations are released; the interchange between a cation in solution and another cation on the surface of any surface-active material such as clay or organic matter.

cation exchange capacity: the sum total of exchangeable cations that a soil can adsorb.

cation retention: the physical, biological, and geochemical processes by which cations in watersheds are held, retained, or prevented from reaching receiving surface waters.

chelator: a class of compounds, organic and inorganic, that can bind metal ions and change their biological availability.

circumneutral: close to neutrality with respect to pH (pH = 7.0).

cirque: a steep-walled, usually semicircular basin excavated by the head of a glacier.

clearwater lake: for this book, a lake with DOC concentration $\leq 417\ \mu mol\ L^{-1}$ ($\leq 5\ mg\ L^{-1}$).

closed lake: a lake having a surface water inlet but no surface water outlet.

closed system: method of measurement in which a water sample is collected and analyzed for pH and dissolved inorganic carbon without exposure to the atmosphere.

cluster analysis: a multivariate classification technique for identifying similar (or dissimilar) groups of observations.

colored lake: for this book, a lake with DOC concentration $> 417\ \mu mol\ L^{-1}$ ($> 5\ mg\ L^{-1}$).

conceptual model: simplified or symbolic representation of prototype or system behavior and responses.

conductance (or conductivity): the ability of a substance (e.g., aqueous solution) to carry an electrical current. For natural waters, conductance is closely related to the total concentration of dissolved ions. Conductance is usually measured under very specific conditions.

confidence limits: a statistical expression, based on a specified probability, that estimates the upper and/or lower value (limit) or the interval expected to contain the true population mean.

cumulative frequency distribution: a function, $F(x)$, such that for any reference value X, $F(x)$ is the estimated proportion of lakes in the population having a value $x \leq X$.

degassing: the loss of a gas from water; the loss of dissolved CO_2 from water can result in changes in pH and DIC.

denitrification: biologically mediated conversion of nitrate to gaseous forms of nitrogen (N_2, NO, N_2O); denitrification occurs during decomposition of organic matter.

dissolved inorganic carbon (DIC): a measure of the dissolved carbon dioxide, carbonic acid, bicarbonate, and carbonate anions. The latter two often constitute the major part of acid neutralizing capacity.

dissolved organic carbon (DOC): the organic fraction of carbon in a water sample that is dissolved or unfilterable, often operationally defined as the component that can pass through a filter of 0.45 µm pore size.

dose-response relationship: the association between the environmental concentration to which an organism (or population) is exposed and the expected response or biological effect.

drainage lake: a lake with a permanent surface water inlet and outlet, or a lake with an outlet but no inlet.

electroneutrality: having no net electric charge.

episodic acidification: an episode in a lake or stream in which acidification to acid neutralizing capacity ≤ 0 occurs.

equivalent: unit of ionic charge; the quantity of a substance that either gains or loses one mole of protons or electrons.

eutrophication: a process of accelerated aquatic primary production in response to nutrient enrichment, which ultimately can result in oxygen depletion and changes in biological community structure and function.

evapotranspiration: loss of water from the soil by evaporation and by transpiration from the plants growing in the soil.

extractable aluminum: an operationally defined aluminum fraction determined using MIBK extraction. It is an estimate of total monomeric aluminum that includes both labile (mainly inorganic) and nonlabile (mainly organically complexed) components.

frame: a structural representation of a population providing a sampling capability.

geomorphic unit: an area with similar physiographic or geologic characteristics; see **physiography**.

gibbsite: a mineral, $Al(OH)_3$; a principal constituent of many bauxites (aluminum ores).

glacial till: glacially deposited debris of all sizes mixed indiscriminately together.

Gran analysis: a mathematical procedure used to determine the equivalence points of a titration curve for acid neutralizing capacity.

groundwater: water in a saturated zone within soil or rock.

groundwater flowthrough lake: a seepage lake that receives significant groundwater input; for this book, sometimes defined as a lake with H_4SiO_4 concentrations ≥ 60 μmol L^{-1}, or SiO_2 concentrations ≥ 1.0 mg L^{-1}, or, in Florida, K^+ concentrations ≥ 15 μeq L^{-1}.

groundwater recharge lake: a seepage lake that receives little groundwater input; sometimes defined as a lake with H_4SiO_2 concentrations ≤ 60 μmol L^{-1}, or SiO_2 concentrations < 1.0 mg L^{-1}, or, in Florida, K^+ concentrations < 15 μeq L^{-1}.

hectare (ha): a measure of surface area equal to 10,000 m^2 or 2.47 acres.

high DOC lake or stream: for this book, a lake with a dissolved organic carbon concentration ≥ 417 μmol L^{-1}, which is equivalent to 5 mg L^{-1}.

hydraulic (or hydrologic) residence time: the average length of time water resides in a lake; the time required to exchange the total volume of water in a lake.

hydrologic flowpaths: the routes taken by precipitation as it moves from the surface of the land, into the soil, and through underlying geologic materials within a watershed.

hydrologic type: in the NSWS, lake hydrologic types were determined from USGS topographic maps: **seepage**–has no inlets or outlets; **drainage**–has outlets; **closed**–has inlets, but no outlets; **reservoir**–human-made impoundment. See Chapter 1 for further descriptions.

hypolimnion: the bottom layer of cold water below the thermocline in a stratified lake.

in situ: referring to measurements collected within the water column at a lake or stream.

inclusion probability: the chance of a lake being selected for sampling.

index sample: in the NSWS, a sample or group of samples collected from a certain place at a lake or stream reach at a particular time of the year and used to represent chemical conditions in that water body. In the NLS, the index sample was a single sample collected from the center of each lake at a depth of 1.5 m during the fall turnover period. For the NSS, the index sample was the average of 2 or 3 samples collected during the spring baseflow period at each end of a stream reach.

initial DIC: a measurement of dissolved inorganic carbon made on an aliquot immediately before it is titrated for acid neutralizing capacity.

inorganic aluminum: the sum of free aluminum ions (Al^{3+}) and dissolved aluminum bound to inorganic ligands; operationally defined by labile monomeric aluminum.

inverse cumulative frequency distribution: a function, $1 - F(x)$, such that for any reference value X, $1 - F(x)$ is the estimated number of lakes or streams in the population having a value $x \geq X$.

ionic strength: a measure of the inter-ionic effect resulting from the electrical attraction and repulsion between various ions. In very dilute solutions, ions behave independently of each other and the ionic strength can be calculated from the measured concentrations of anions and cations present in the solution. Units are in moles L^{-1}.

isothermal: in the NLS, refers to a lake with a temperature difference of < 4°C between the reading at 1.5 m below the surface and 1.5 m above the lake bottom (synonymous with nonstratified or mixed).

labile monomeric aluminum: operationally defined as aluminum that can be retained on a cation exchange column and measured by one of the two extraction procedures used to measure monomeric aluminum. Labile monomeric aluminum is assumed to represent inorganic monomeric aluminum (Al_i or Al_{im}).

long-term acidification: a partial or complete permanent loss of acid neutralizing capacity from a lake or stream. Chronic acidification.

low DOC lake or stream: for this book, a lake or stream with a dissolved organic carbon (DOC) concentration ≤ 417 μmol L^{-1}, which is equivalent to 5 mg L^{-1}.

macrophyte: macroscopic forms of aquatic vegetation, including macroalgae, aquatic mosses and ferns, and angiosperms.

matrix: the physical and chemical makeup of a sample being analyzed.

mechanistic model: mathematical model in which actual ecosystem processes are explicitly included.

median (M): the value of the middle item in a set of data arranged in rank order; the value of x such that $F(x)$ or $G(x) = 0.5$; the 50th percentile.

mineral acids: inorganic acids (H_2SO_4, HNO_3, HCl).

mineral weathering: dissolution of rocks and minerals by chemical and physical processes.

mobile anions: anions that flow in solutions through watershed soils, wetlands, streams, or lakes without being adsorbed or retained through physical, biological, or geochemical processes.

model: an abstraction or representation of a prototype or system, generally on a smaller scale.

monomeric aluminum: aluminum that occurs as a free ion (Al^{3+}), in simple inorganic complexes (e.g., $Al(OH)_n^{3-n}$, AlF_n^{3-n}), or in simple organic complexes, but not in polymeric forms; operationally, extractable aluminum measured by the pyrocatechol violet method or the methylisobutyl-ketone method (also referred to as the MIBK or "oxine" method) is assumed to represent total monomeric aluminum (Al_m or Al_{ext}). Monomeric aluminum can be divided into labile and nonlabile components using cation exchange columns.

Monte Carlo method: technique of stochastic sampling or selection of random numbers to generate synthetic data.

morphometry: lake shape, size, depth; physical description of lake basin.

nitrification: oxidation of ammonium to nitrite or nitrate by microorganisms. A by-product of this reaction is H^+.

node: the points identifying either an upstream or downstream end of a reach, as used in the NSS-I; the point above or below the confluence of two streams.

nonlabile monomeric aluminum: operationally defined as aluminum that passes through a cation exchange column and is measured by one of the two extraction procedures used to measure monomeric aluminum; assumed to represent organic monomeric aluminum (Al_o or Al_{om}).

nonlinear model: mathematical model using one or more nonlinear equations.

nonstratified: see **isothermal**.

nontarget lake or stream: a lake or stream that either was not the focus of the NSWS objectives or could not be sampled within the constraints of the NSWS.

nomograph: a graph reducing a mathematical formula to curves so that the resulting value can be read on the chart coordinates for any value assigned to the variable involved.

NTU: nephelometric turbidity unit; a measure of light scattered by a solution of suspended materials.

numerical model: mathematical model in which the governing equations are not solved analytically, but approximately, using discrete numerical values to represent the variables involved and using arithmetic operations.

nutrient cycling: the movement or transfer of chemicals required for biological maintenance or growth among components of the ecosystem by physical, chemical, or biological processes.

oligotrophic lake: a lake with low primary productivity.

open system: a measurement of pH or dissolved inorganic carbon obtained from a sample that was exposed to the atmosphere during collection or measurement.

organic acids: acids possessing a carboxyl (−COOH) group or phenolic (C−OH) group; includes fulvic and humic acids.

organic anion: a dissociated organic molecule with a negative net ionic charge.

organic aluminum: aluminum bound to organic matter; operationally defined as that fraction of aluminum determined by colorimetry after a sample is passed through a strong cation exchange column.

orographic: associated with or induced by the presence of mountains.

outliers: observations not typical of the population from which the sample is drawn.

percent ion balance difference: a quality assurance procedure used to see that the sum of the anion equivalents equals the sum of the cation equivalents. See **anion-cation balance**.

percent relative standard deviation (% RSD): the standard deviation divided by the mean, multiplied by 100.

periphyton: plants that live attached to or closely associated with surfaces (e.g., on the bottom sediments or macrophytes).

pH: the negative logarithm of the hydrogen ion activity. The pH scale extends from 1 (most acidic) to 14 (most alkaline); a difference of one pH unit indicates a tenfold change in hydrogen ion activity.

physiography: the elevation, slope, and shape of the landform, including its spatial relationships with other physical features.

plankton: plant or animal species that spend part or all of their lives carried passively by water currents.

platinum cobalt unit (PCU): measure of the color of a water sample defined by a potassium hexachloroplatinate and cobalt chloride standard color series.

population estimate: a statistical estimate that applies to a specific population of target lakes or streams, not only to the sample.

precision: a measure of the capacity of a method to provide reproducible measurements of a particular analyte.

probability sample: a sample in which each unit has a known probability of being selected.

protolyte: that portion of a molecule that reacts with either H^+ or OH^- in solution.

prototype: the full-sized structure, system process, or phenomenon being modeled.

quality assurance (QA): a program designed to provide assurance that a product (e.g., data base) meets a defined standard of quality with a stated level of confidence.

quality control (QC): steps taken during sample collection and analysis to ensure that the data quality meets the minimum standards established by the quality assurance plan.

quartile: any of the three values that divide the population of a frequency distribution into four equal classes, each representing 25%; used to express range of variation.

reference value (X_c): a concentration of interest for a given chemical variable.

relative bias: the expected difference between a measured value and the true value, expressed as a percentage of the true value.

representativeness: a measure of data quality; the degree to which sample data accurately and precisely reflect the characteristics of a population.

root mean square: a summary statistic of the relative or absolute standard deviation; a pooled standard deviation of the % relative standard deviation (% RSD), calculated by the formula: $[x^2_{\%RSD} + SD^2_{\%RSD}\,(n-1/n)]^{1/2}$.

Secchi disk: a black-and-white or all-white disk, 20 cm in diameter, used to measure water transparency.

Secchi disk transparency: a measure of lakewater clarity determined by lowering a Secchi disk into the lake, recording the depth at which it disappears from view, raising it until it comes into view, and recording that depth; the mean value of these depths is then calculated.

seepage lake: a lake with no permanent surface water inlets or outlets.

sensitivity: (1) the tendency of a system to experience effects as a result of changes in acid-base chemistry, particularly biological effects; (2) a

measure of the degree to which a model's output responds to small changes in an input value.
simulation: description of a prototype or system response to different conditions or inputs using a model rather than actually observing the response to the conditions or inputs.
soft water: water with low concentrations of calcium and magnesium salts.
special interest lake: in the NLS, a lake selected nonrandomly that is not part of the probability sample; selection was based on quality and amount of data available or on the lake's involvement in other research programs.
species richness: the number of species occurring in a given aquatic ecosystem, generally estimated by the number of species caught using a standard sampling regime.
specific conductance: the conductivity between two plates with an area of 1 cm^2 across a distance of 1 cm at 25°C.
spring baseflow index period: in the NSS-I, a period of the year when streams are expected to exhibit chemical characteristics most closely linked to acidic deposition. The time period between snowmelt and leafout (March 15 to May 15 in the NSS-I) when stream reaches were visited, coinciding with expected periods of highest geochemical interest (i.e., low seasonal pH and sensitive life stages of biota).
standard deviation: the square root of the variance of a given statistic, calculated by the equation $[\Sigma x^2 - (\Sigma x)^2]^{\frac{1}{2}}/(n-1)^{\frac{1}{2}}$.
standard error: standard deviation divided by the square root of the sample size.
steady state: the condition that occurs when a property (e.g., mass, volume, concentration) of a system does not change appreciably with time. This condition requires that sources and sinks of the property are in balance (e.g., inputs equal outputs; production equals consumption).
stochastic: an occurrence that is probabilistic in nature.
stratification factors: factors used to define NSWS lake and stream strata prior to lake selection; the factors used in the NLS were region, subregion, and alkalinity map class.
stratified design: a statistical design in which the population is divided into strata, and a sample is selected from each stratum.
stratified lake: in the NLS, a lake with a temperature difference > 4°C between the water layers at 1.5 m below the surface and 1.5 m above the lake bottom. If the temperature difference is also > 4°C between the water layers at 1.5 m below the surface and 60% of site depth, then the lake is strongly stratified; if not, it is weakly stratified.
stratum: in the NSWS, a subpopulation of lakes or streams within an alkalinity map class within a subregion and within a region, as defined by the stratified design.
stream reach: a stream segment of relatively uniform morphology. In the NSS, segments of the stream network were represented as blue lines on 1:250,000-scale USGS maps. Each reach was defined as the length of stream between two blue-line confluences.
strong acids: acids with a high tendency to donate protons or to completely dissociate in natural waters (e.g., H_2SO_4, HNO_3, HCl^-, and some organic acids).
strong bases: bases with a high tendency to accept protons or to completely dissociate in natural waters (e.g., NaOH).
subpopulation: any defined subset of the target population.
subregions: in the NSWS, areas within each region that are similar in water quality, physiography, vegetation, climate, and soil; used as a stratification factor in NSWS design.
sulfate adsorption: the process by which sulfate is chemically exchanged (for OH^-) or adsorbed onto positively charged sites on the soil matrix; under some conditions this process is reversible and the sulfate may be desorbed.
sulfate reduction: (1) the conversion of sulfate to sulfide during the decomposition of organic matter under anaerobic conditions (dissimilatory sulfate reduction); (2) the formation of organic compounds containing reduced sulfur compounds (assimilatory sulfate reduction).
sulfate retention: the physical, biological, and geochemical processes by which sulfate in watersheds is held, retained, or prevented from reaching receiving surface waters.
sum of base cations (SBC or C_B): usually refers to the equivalent sum of Ca^{2+}, Mg^{2+}, Na^+, and K^+, but may be defined differently in some chapters (e.g., $Ca^{2+} + Mg^{2+} - Cl^-$).
surface water: streams and lakes.
surficial geology: characteristics of the earth's

surface, especially consisting of unconsolidated residual, alluvial, or glacial deposits lying on the bedrock.

synoptic: relating to or displaying conditions as they exist simultaneously over a broad area.

systematic error: a consistent error introduced in the measuring process. Such an error commonly results in biased estimations.

systematic random sampling: a sampling technique in which the units in the population are ordered. A first sampling unit is randomly drawn from the first K units, and every kth unit afterward is included in the sample (K being equal to N divided by the sample size).

target population: in the NSWS, the lake or stream population of interest from which lakes or streams were selected for sampling. This population was defined by the sampling protocol.

thermocline: the area of most rapid temperature change with depth in a stratified lake.

total aluminum: in the NSWS, the total aluminum, measured by atomic absorption spectroscopy, in an unfiltered, acidified sample.

total monomeric aluminum: the simple unpolymerized form of aluminum present in inorganic or organic complexes (Al_T or Al_m); an operationally defined aluminum fraction determined by colorimetry (PCV technique).

transparency: the clarity of unfiltered water, measured with a Secchi disk.

true color: the color of water that has been filtered or centrifuged to remove particles that may impart an apparent color; true color ranges from clear blue to blackish-brown.

turnover: a period of water circulation in lakes, during which few or no differences in temperature are observed with depth.

upper confidence limit (95%): a value that, in association with a statistic, has a 95% chance of being above the true value of the population parameter of interest.

watershed: the geographic area from which surface water drains into a particular lake or point along a stream.

watershed disturbance: for this book, a disturbance of the natural environment in a watershed, including roads, houses, logging, mining, and livestock.

weak acids: acids that have a low proton-donating tendency or that tend to dissociate only partially in natural waters (e.g., H_2CO_3, H_4SiO_4, and most organic acids).

weak bases: bases with a low proton-accepting tendency or that tend to dissociate only partially in natural waters (e.g., HCO_3^-, $Al(OH)_4^-$).

Index

B

I

MIX
Papier aus verantwortungsvollen Quellen
Paper from responsible sources
FSC® C105338

If you have any concerns about our products,
you can contact us on
ProductSafety@springernature.com

In case Publisher is established outside the EU,
the EU authorized representative is:
Springer Nature Customer Service Center GmbH
Europaplatz 3, 69115 Heidelberg, Germany

Printed by Libri Plureos GmbH
in Hamburg, Germany